# Geometric Formulas

### Rectangle

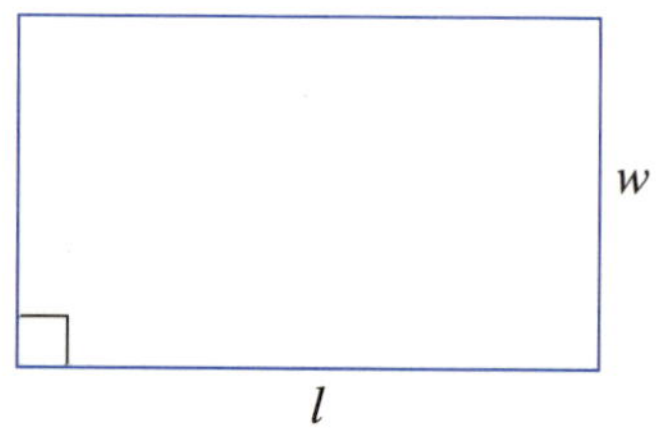


Perimeter: $P = 2l + 2w$

Area: $A = lw$

### Square

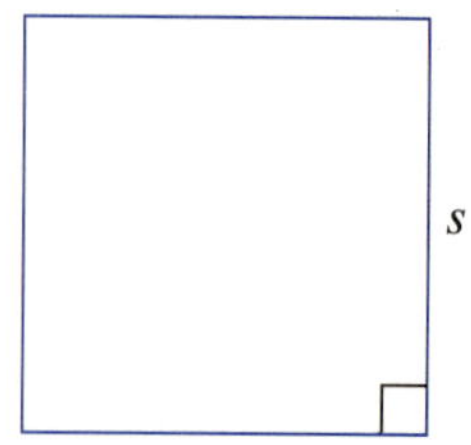


Perimeter: $P = 4s$

Area: $A = s^2$

### Triangle

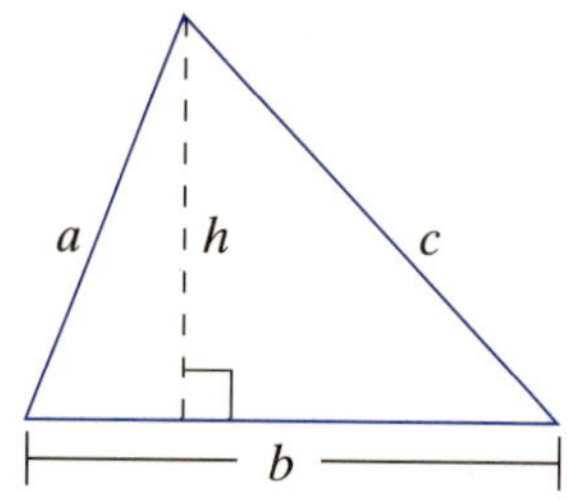


Perimeter: $P = a + b + c$

Area: $A = \frac{1}{2}bh$

### Sum of Angles of Triangle

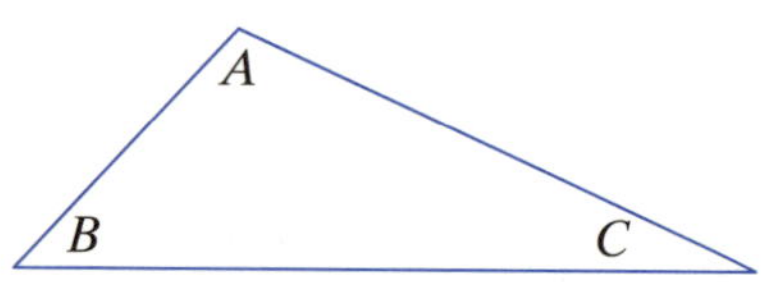


$A + B + C = 180°$

The sum of the measures of the three angles is 180°.

### Right Triangle

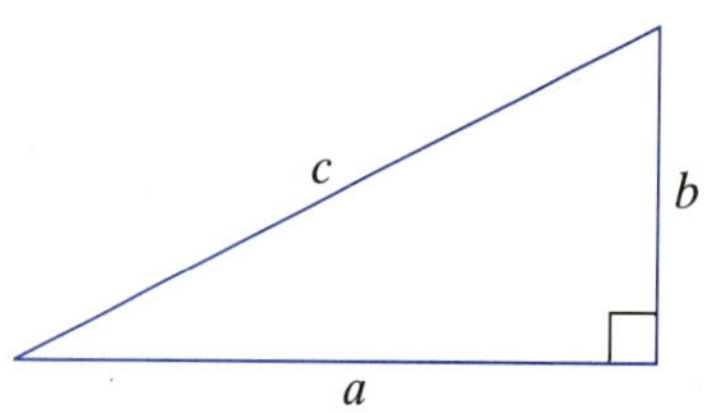


Perimeter: $P = a + b + c$

Area: $A = \frac{1}{2}ab$

One 90° (right) angle

### Pythagorean Theorem (for Right Triangles)

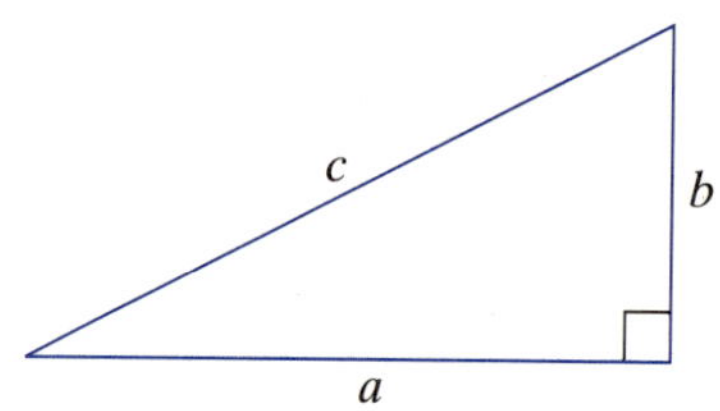


$a^2 + b^2 = c^2$

### Isosceles Triangle

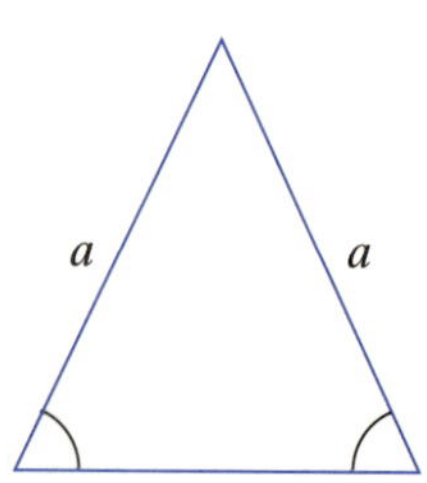


Triangle has:
two equal sides and
two equal angles.

### Equilateral Triangle

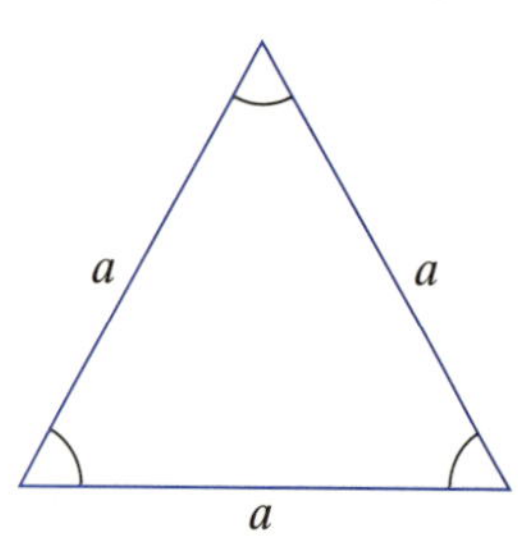


Triangle has:
three equal sides and
three equal angles.
Measure of each angle is 60°.

### Trapezoid

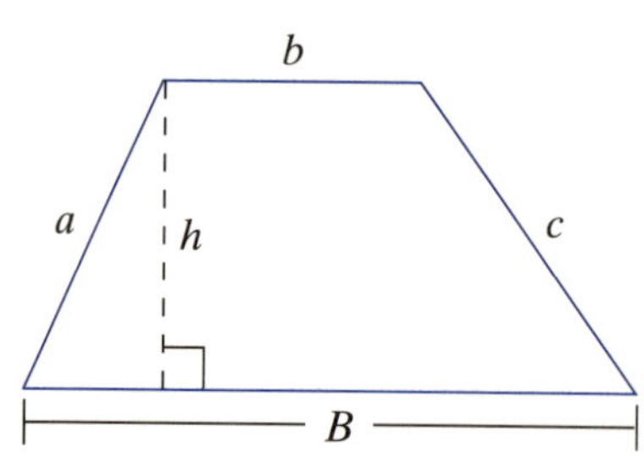


Perimeter: $P = a + b + c + B$

Area: $A = \frac{1}{2}h(B + b)$

## Parallelogram

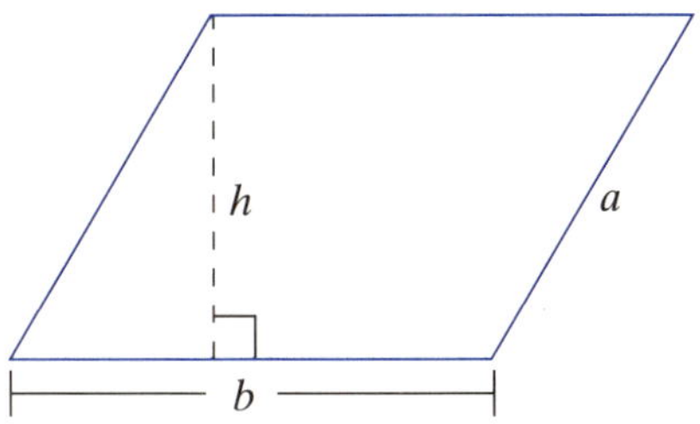


Perimeter: $P = 2a + 2b$
Area: $A = bh$

## Circle

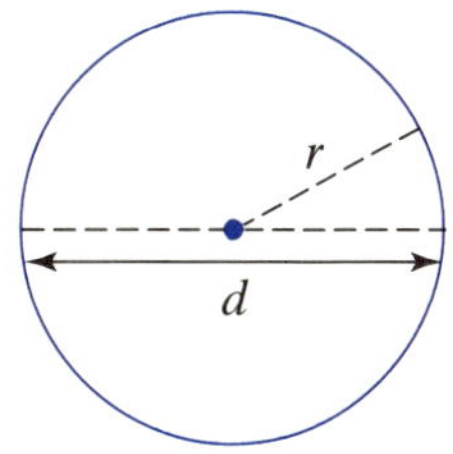


Circumference: $C = \pi d$
$C = 2\pi r$
Area: $A = \pi r^2$

## Rectangular Solid

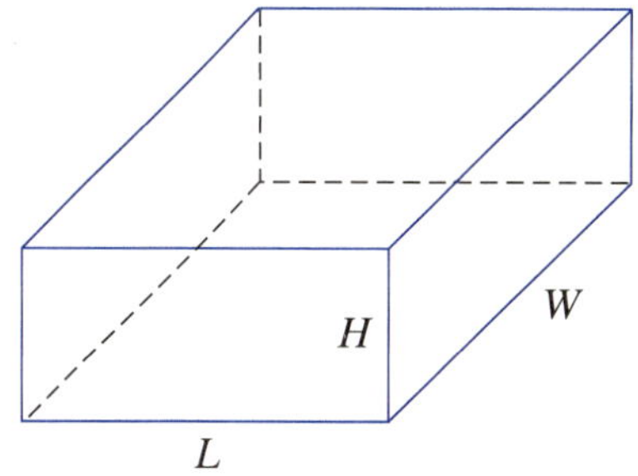


Volume: $V = LWH$
Surface Area: $A = 2HW + 2LW + 2LH$

## Cube

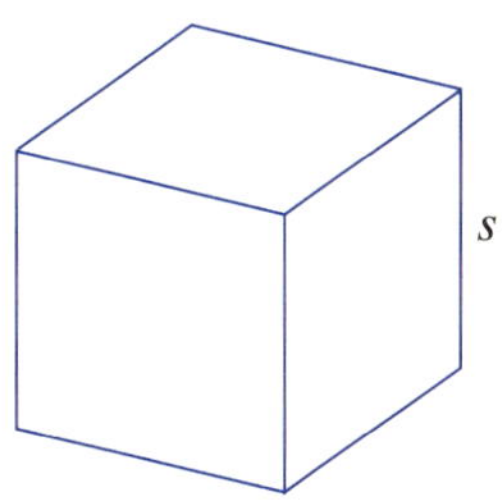


Volume: $V = s^3$
Surface Area: $A = 6s^2$

## Cone

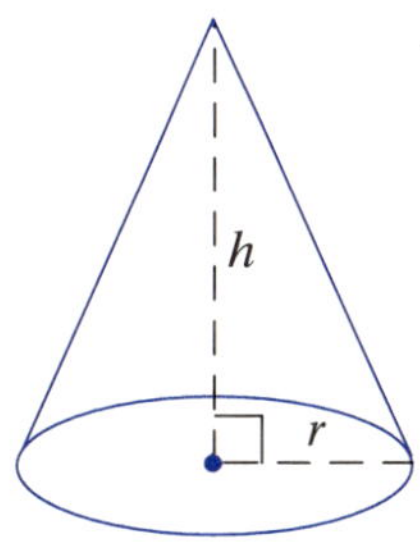


Volume: $V = \frac{1}{3}\pi r^2 h$
Lateral Surface Area: $A = \pi r\sqrt{r^2 + h^2}$

## Right Circular Cylinder

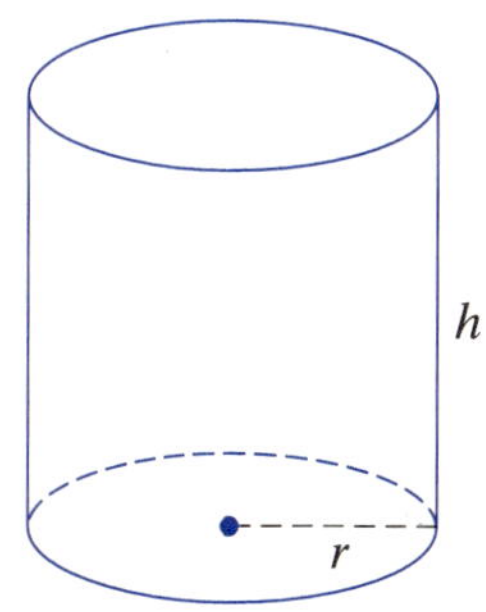


Volume: $V = \pi r^2 h$
Surface Area:
$A = 2\pi rh + 2\pi r^2$

## Other Formulas

Distance: $d = rt$ ($r$ = rate, $t$ = time)

Temperature: $F = \frac{9}{5}C + 32 \qquad C = \frac{5}{9}(F - 32)$

Simple Interest: $I = Prt$
($P$ = principal, $r$ = annual interest rate, $t$ = time in years)

Compound Interest: $A = P\left(1 + \frac{r}{n}\right)^{nt}$
($P$ = principal, $r$ = annual interest rate, $t$ = time in years, $n$ = number of compoundings per year)

Companion Website
http://www.prenhall.com/martin-gay

# INTERMEDIATE ALGEBRA

# INTERMEDIATE ALGEBRA

Third Edition

K. ELAYN MARTIN-GAY
*University of New Orleans*

Prentice Hall

Upper Saddle River, New Jersey 07458

**Library of Congress Cataloging-in-Publication Data**

Martin-Gay, K. Elayn
Intermediate algebra/K. Elayn Martin-Gay.—3rd ed.
p. cm.
Includes index.
ISBN 0-13-016631-6 (alk. paper)
1. Algebra. I. Title.
QA152.2.M368 2001 00-035954
512.9—dc21 CIP

**Executive Acquisition Editor:** Karin E. Wagner
**Editor in Chief:** Christine Hoag
**Project Manager:** Ann Marie Jones
**Assistant Vice President of Production and Manufacturing:** David W. Riccardi
**Executive Managing Editor:** Kathleen Schiaparelli
**Senior Managing Editor:** Linda Mihatov Behrens
**Project Management:** Elm Street Publishing Services, Inc.
**Manufacturing Buyer:** Alan Fischer
**Manufacturing Manager:** Trudy Pisciotti
**Senior Marketing Manager:** Eilish Collins Main
**Marketing Assistant:** Dan Auld
**Director of Marketing:** John Tweeddale
**Development Editors:** Tony Palermino/Emily Keaton
**Associate Editor, Mathematics/Statistics Media:** Audra J. Walsh
**Editorial Assistant/Supplements Editor:** Kate Marks
**Art Director:** Maureen Eide
**Assistant to the Art Director:** John Christiana
**Interior Designer:** Donna Wickes
**Cover Designer:** Joseph Sengotta
**Art Editor:** Grace Hazeldine
**Art Manager:** Gus Vibal
**Director of Creative Services:** Paul Belfanti
**Photo Researcher:** Kathy Ringrose
**Photo Editor:** Beth Boyd
**Cover Photo:** Alex Demyan/Alex Demyan Photographs
**Art Studio:** Academy Artworks
**Compositor:** Preparé Inc., Italy

Prentice Hall



Printed in the United States of America
10 9 8 7 6

ISBN 0-13-016631-6

Prentice-Hall International (UK) Limited, *London*
Prentice-Hall of Australia Pty. Limited, *Sydney*
Prentice-Hall Canada, Inc., *Toronto*
Prentice-Hall Hispanoamericana, S.A., *Mexico*
Prentice-Hall of India Private Limited, *New Delhi*
Prentice-Hall of Japan, Inc., *Tokyo*
Pearson Education Asia, Pte. Ltd.
Editora Prentice-Hall do Brasil, Ltda., *Rio de Janeiro*

*To my husband, Clayton,*
*and our sons,*
*Eric and Bryan*

# CONTENTS


Preface xi

## 1 REAL NUMBERS AND ALGEBRAIC EXPRESSIONS 2

1.1 Tips for Success in Mathematics 4
1.2 Algebraic Expressions and Sets of Numbers 8
1.3 Operations on Real Numbers 19
1.4 Properties of Real Numbers 30
*Chapter 1 Project: Analyzing Newspaper Circulation 39*
*Chapter 1 Vocabulary Check 40*
*Chapter 1 Highlights 41*
*Chapter 1 Review 43*
*Chapter 1 Test 44*

## 2 EQUATIONS, INEQUALITIES, AND PROBLEM SOLVING 46

2.1 Linear Equations in One Variable 48
2.2 An Introduction to Problem Solving 56
2.3 Formulas and Problem Solving 67
2.4 Linear Inequalities and Problem Solving 75
2.5 Compound Inequalities 86
2.6 Absolute Value Equations 93
2.7 Absolute Value Inequalities 98
*Chapter 2 Project: Analyzing Municipal Budgets 104*
*Chapter 2 Vocabulary Check 106*
*Chapter 2 Highlights 106*
*Chapter 2 Review 111*
*Chapter 2 Test 113*
*Chapter 2 Cumulative Review 114*

## 3 GRAPHS AND FUNCTIONS 116

3.1 Graphing Equations 118
3.2 Introduction to Functions 128
3.3 Graphing Linear Functions 142
3.4 The Slope of a Line 149
3.5 Equations of Lines 162
3.6 Graphing Linear Inequalities 172
*Chapter 3 Project: Modeling Real Data 177*
*Chapter 3 Vocabulary Checks 178*
*Chapter 3 Highlights 179*
*Chapter 3 Review 184*
*Chapter 3 Test 187*
*Chapter 3 Cumulative Review 188*


## 4 SYSTEMS OF EQUATIONS 190

4.1 Solving Systems of Linear Equations in Two Variables 192
4.2 Solving Systems of Linear Equations in Three Variables 204
4.3 Systems of Linear Equations and Problem Solving 209
4.4 Solving Systems of Equations by Matrices 219
4.5 Solving Systems of Equations by Determinants 225
*Chapter 4 Project: Locating Lightning Strikes 232*
*Chapter 4 Vocabulary Check 233*
*Chapter 4 Highlights 234*
*Chapter 4 Review 239*
*Chapter 4 Test 240*
*Chapter 4 Cumulative Review 241*


## 5 EXPONENTS, POLYNOMIALS, AND POLYNOMIAL FUNCTIONS 242

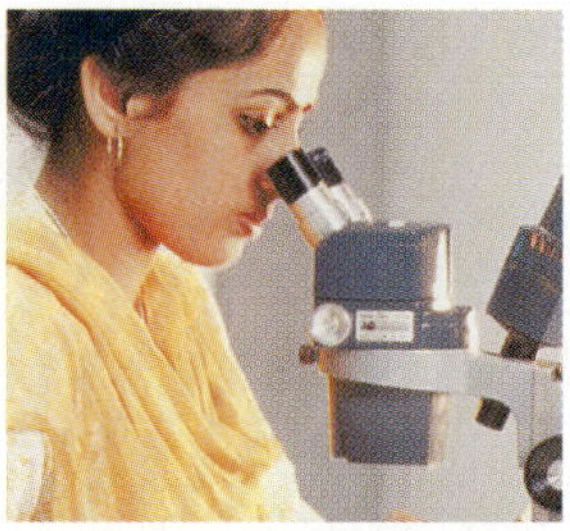


5.1 Exponents and Scientific Notation 244
5.2 More Work with Exponents and Scientific Notation 252
5.3 Polynomials and Polynomial Functions 258
5.4 Multiplying Polynomials 268
5.5 The Greatest Common Factor and Factoring by Grouping 276
5.6 Factoring Trinomials 282
5.7 Factoring by Special Products and Factoring Strategies 289
5.8 Solving Equations by Factoring and Problem Solving 297
5.9 An Introduction to Graphing Polynomial Functions 309
*Chapter 5 Project: Investigating Earth's Water 317*
*Chapter 5 Vocabulary Check 318*
*Chapter 5 Highlights 319*
*Chapter 5 Review 322*
*Chapter 5 Test 325*
*Chapter 5 Cumulative Review 326*

## 6 RATIONAL EXPRESSIONS 328

6.1 Rational Functions and Multiplying and Dividing Rational Expressions 330
6.2 Adding and Subtracting Rational Expressions 341
6.3 Simplifying Complex Fractions 348
6.4 Dividing Polynomials 354
6.5 Synthetic Division and the Remainder Theorem 360
6.6 Solving Equations Containing Rational Expressions 364
6.7 Rational Equations and Problem Solving 372
6.8 Variation and Problem Solving 382
*Chapter 6 Project: Modeling Electricity Production 390*
*Chapter 6 Vocabulary Check 392*
*Chapter 6 Highlights 393*
*Chapter 6 Review 397*
*Chapter 6 Test 400*
*Chapter 6 Cumulative Review 402*


## 7 RATIONAL EXPONENTS, RADICALS, AND COMPLEX NUMBERS 404

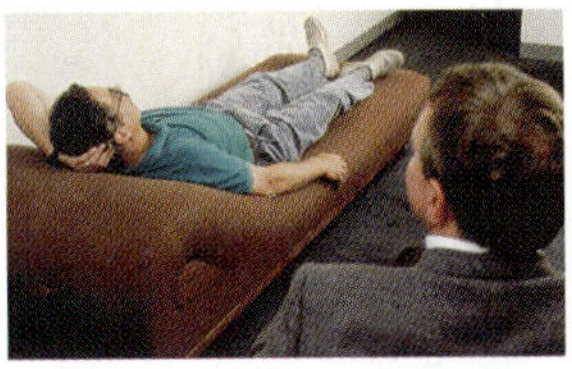


7.1 Radicals and Radical Functions 406
7.2 Rational Exponents 413
7.3 Simplifying Radical Expressions 420
7.4 Adding, Subtracting, and Multiplying Radical Expressions 426
7.5 Rationalizing Denominators and Numerators of Radical Expressions 430
7.6 Radical Equations and Problem Solving 436
7.7 Complex Numbers 445
*Chapter 7 Project: Calculating the Length and Period of a Pendulum 452*
*Chapter 7 Vocabulary Checks 453*
*Chapter 7 Highlights 454*
*Chapter 7 Review 457*
*Chapter 7 Test 459*
*Chapter 6 Cumulative Review 460*


## 8 QUADRATIC EQUATIONS AND FUNCTIONS 462

8.1 Solving Quadratic Equations by Completing the Square 464
8.2 Solving Quadratic Equations by the Quadratic Formula 474
8.3 Solving Equations by Using Quadratic Methods 483
8.4 Nonlinear Inequalities in One Variable 492
8.5 Quadratic Functions and Their Graphs 499
8.6 Further Graphing of Quadratic Functions 507
*Chapter 8 Project: Fitting a Quadratic Model to Data 515*
*Chapter 8 Vocabulary Check 516*
*Chapter 8 Highlights 516*
*Chapter 8 Review 519*
*Chapter 8 Test 521*
*Chapter 8 Cumulative Review 522*

## 9 CONIC SECTIONS 524

9.1 The Parabola and the Circle 526
9.2 The Ellipse and the Hyperbola 535
9.3 Solving Nonlinear Systems of Equations 544
9.4 Nonlinear Inequalities and Systems of Inequalities 550
*Chapter 9 Project: Modeling Conic Sections 554*
*Chapter 9 Vocabulary Checks 555*
*Chapter 9 Highlights 556*
*Chapter 9 Review 558*
*Chapter 9 Test 560*
*Chapter 9 Cumulative Review 561*

## 10 EXPONENTIAL AND LOGARITHMIC FUNCTIONS 564

10.1 The Algebra of Functions; Composite Functions 566
10.2 Inverse Functions 571
10.3 Exponential Functions 582
10.4 Logarithmic Functions 590
10.5 Properties of Logarithms 597
10.6 Common Logarithms, Natural Logarithms, and Change of Base 603
10.7 Exponential and Logarithmic Equations and Applications 608
*Chapter 10 Project: Modeling Temperature 615*
*Chapter 10 Vocabulary Checks 617*
*Chapter 10 Highlights 617*
*Chapter 10 Review 620*
*Chapter 10 Test 624*
*Chapter 10 Cumulative Review 625*

## 11 SEQUENCES, SERIES, AND THE BINOMIAL THEOREM 628

11.1 Sequences 630
11.2 Arithmetic and Geometric Sequences 634
11.3 Series 641
11.4 Partial Sums of Arithmetic and Geometric Sequences 646
11.5 The Binomial Theorem 653
*Chapter 11 Project: Modeling College Tuition 658*
*Chapter 11 Vocabulary Checks 659*
*Chapter 11 Highlights 659*
*Chapter 11 Review 661*
*Chapter 11 Test 663*
*Chapter 11 Cumulative Review 664*

## APPENDICES 665

A. Review of Angles, Lines, and Special Triangles 665
B. Review of Geometric Figures 673
C. Review of Volume and Surface Area 677
D. An Introduction to Using a Graphing Utility 683

**ANSWERS TO SELECTED EXERCISES A-1**
**INDEX I-1**
**PHOTO CREDITS P-1**

# PREFACE

## ABOUT THIS BOOK

*Intermediate Algebra, Third Edition* was written to provide a **solid foundation in algebra** as well as to develop students' problem-solving skills. Specific care has been taken to ensure that students have the most **up-to-date and relevant** text preparation for their next mathematics course, as well as to help students succeed in nonmathematical courses that require a grasp of algebraic fundamentals. I have tried to achieve this by writing a user-friendly text that is keyed to objectives and contains many worked-out examples. The basic concepts of graphs and functions are introduced early, and problem solving techniques, real-life and real-data applications, data interpretation, appropriate use of technology, mental mathematics, number sense, critical thinking, decision-making, and geometric concepts are emphasized and integrated throughout the book.

The many factors that contributed to the success of the first two editions have been retained. In preparing this edition, I considered the comments and suggestions of colleagues throughout the country, students, and many users of the prior editions. The AMATYC Crossroads in Mathematics: Standards for Introductory College Mathematics before Calculus and the MAA and NCTM standards (plus Addenda), together with advances in technology, also influenced the writing of this text.

*Intermediate Algebra, Third Edition* is **part of a series of texts** that can include *Basic College Mathematics*, *Prealgebra, Third Edition*, and *Beginning Algebra, Third Edition*. Also available are *Beginning and Intermediate Algebra, Second Edition*, a combined algebra text and *Intermediate Algebra: A Graphing Approach, Second Edition*. Throughout the series, pedagogical features are designed to develop student proficiency in algebra and problem solving, and to prepare students for future courses.

## KEY PEDAGOGICAL FEATURES IN THE THIRD EDITION

**Readability and Connections** I have tried to make the writing style as clear as possible while still retaining the mathematical integrity of the content. When a new topic is presented, an effort has been made to **relate the new ideas to those that students**

**may already know**. Constant reinforcement and connections within problem solving strategies, data interpretation, geometry, patterns, graphs, and situations from everyday life can help students gradually master both new and old information.

**Problem Solving Process** This is formally introduced in Chapter 2 with a **new four-step process that is integrated throughout the text**. The four steps are Understand, Translate, Solve, and Interpret. The repeated use of these steps throughout the text in a variety of examples shows their wide applicability. Reinforcing the steps can increase students' confidence in beginning problems.

**Applications and Connections** Every effort was made to include as many accessible, interesting and relevant real-life applications as possible throughout the text in both worked-out examples and exercise sets. The applications **strengthen students' understanding of mathematics in the real world** and help to motivate students. They show connections to a wide range of fields including agriculture, allied health, art, astronomy, automotive ownership, aviation, biology, business, chemistry, communication, computer technology, construction, consumer affairs, demographics, earth science, education, entertainment, environmental issues, finance and economics, food service, geography, government, history, hobbies, labor and career issues, life science, medicine, music, nutrition, physics, political science, population, recreation, sports, technology, transportation, travel, weather, and important related mathematical areas such as geometry and statistics. (See the Index of Applications on page xxi.) Many of the applications are based on **recent and interesting real-life data**. For instance, see Section 4.3, exercise 44 , Section 5.3 exercise 85, or Section 6.4, exercise 58 for a variety of ways real data is used. Sources for data include newspapers, magazines, government publications, publicly held companies, special interest groups, research organizations, and reference books. Opportunities for obtaining your own real data with and without using the internet are also included.

**Helpful Hints** Helpful Hints, formerly Reminders, contain practical advice on applying mathematical concepts. These are found throughout the text and **strategically placed** where students are most likely to need immediate reinforcement. They are highlighted in a box for quick reference and, as appropriate, an indicator line is used to precisely identify the particular part of a problem or concept being discussed. For instance, see pages 88 and 293.

**Visual Reinforcement of Concepts** The text contains numerous graphics, models, and illustrations to visually clarify and reinforce concepts. These include **new and updated** bar graphs and circle graphs in two and three dimensions, line graphs, calculator screens, application illustrations, photographs, and geometric figures. There are now **approximately 1,000 figures**.

**Real World Chapter Openers** The new two-page chapter opener focuses on how math is used in a specific career, provides links to the World Wide Web, and references a "Spotlight on Decision Making" feature within the chapter for further exploration of the **career and the relevance of algebra**. For example, look at the opener for Chapter 3. The opening pages also contain a list of section titles, and an introduction to the mathematics to be studied together with mathematical connections to previous chapters in the text.

**Student Resource Icons** At the beginning of each section, videotape, tutorial software CD Rom, Student Solutions Manual, and Study Guide icons are displayed. These icons help remind students that these learning aids are available should they choose to use them to review concepts and skills at their own pace. These items have **direct correlation to the text** and emphasize the text's methods of solution.

**Chapter Highlights** Found at the end of each chapter, the Chapter Highlights contain key definitions, concepts, *and* examples to **help students understand and retain** what they have learned.

**Chapter Project** This feature occurs at the end of each chapter, often serving as a chapter wrap-up. For **individual or group completion**, the multi-part Chapter Project, usually hands-on or data based, allows students to problem solve, make interpretations, and to think and write about algebra.

In addition, a reference to alternative or additional Real World Activities is given. This **internet option** invites students to find and retrieve real data for use in solving problems. Visit the Real World Activities Website by going to http://www.prenhall.com/martin-gay.

**Functional Use of Color and New Design** Elements of this text are highlighted with color or design to make it easier for students to read and study. Special care has been taken to use color within solutions to examples or in the art to **help clarify, distinguish, or connect concepts**. For example, look at page 261 in Section 5.3.

## EXERCISE SETS

Each text section ends with an exercise set, usually divided into two parts. Both parts contain graded exercises. The **first part is carefully keyed** to at least one worked example in the text. Once a student has gained confidence in a skill, the **second part contains exercises not keyed to examples**. Exercises and examples marked with a video icon () have been worked out step-by-step by the author in the videos that accompany this text.

Throughout the text exercises there is an emphasis on **data and graphical interpretation** via tables, charts, and graphs. The ability to interpret data and read and create a variety of types of graphs is developed gradually so students become comfortable with it. Similarly, throughout the text there is integration of **geometric concepts**, such as perimeter and area. Exercises and examples marked with a geometry icon (△) have been identified for convenience.

Each exercise set contains one or more of the following features.

**Spotlight on Decision Making** These unique **new, specially designed applications** help students develop their decision-making and problem solving abilities, skills useful in mathematics and in life. Appropriately placed before an exercise set begins, students have an opportunity to immediately practice and reinforce basic algebraic concepts found in the accompanying section in relevant, accessible contexts. There is an emphasis on workplace or job-related career situations (such as the decisions of a Meteorologist in Section 3.1, a phychologist in Section 7.6, or a Webmaster in Section 10.4) as well as decision-making in general (such as choosing a long-distance telephone plan in Section 4.3 or choosing an online service in Section 5.4 or deciding between two job offers in Section 11.1).

**Mental Mathematics** These problems are found at the beginning of many exercise sets. They are mental warm-ups that **reinforce concepts** found in the accompanying section and increase students' confidence before they tackle an exercise set. By relying on their own mental skills, students increase not only their confidence in themselves, but also their number sense and estimation ability.

**Writing Exercises** These exercises now found in almost every exercise set are marked with the icon (). They require students to **assimilate information** and provide a written response to explain concepts or justify their thinking. Guidelines recommended by the American Mathematical Association of Two Year Colleges (AMATYC) and other professional groups recommend incorporating writing in mathematics courses to reinforce concepts. Writing opportunities also occur within features such as Spotlight on Decision Making and Chapter Projects.

**Data and Graphical Interpretation** Throughout the text there is an emphasis on data interpretation in exercises via tables, bar charts, line graphs, or circle graphs. The ability to interpret data and read and create a variety of graphs is **developed gradually** so students become comfortable with it.

**Calculator Explorations and Exercises** These optional explorations offer guided instruction, through examples and exercises, on the proper use of **scientific and graphing calculators or computer graphing utilities as tools in the mathematical problem-solving process**. Placed appropriately throughout the text, these explorations reinforce concepts or motivate discovery learning.

Additional exercises building on the skills developed in the Explorations may be found in exercise sets throughout the text, and are marked with the icon for scientific calculator use or with the icon for graphing calculator use.

**Review Exercises** These exercises occur in each exercise set (except for those in Chapter 1). These problems are **keyed to earlier sections** and review concepts learned earlier in the text that are needed in the next section or in the next chapter. These exercises show the **links between earlier topics and later material**.

**A Look Ahead** These exercises occur at the end of some exercise sets. This section contains examples and problems similar to those found in a subsequent algebra course. "A Look Ahead" is presented as **a natural extension of the material** and contains an example followed by advanced exercises.

In addition to the approximately 5,000 exercises within sections, exercises may also be found in the Vocabulary Checks, Chapter Reviews, Chapter Tests, and Cumulative Reviews.

**Vocabulary Checks** Vocabulary checks, **new to this edition**, provide an opportunity for students to become more familiar with the use of mathematical terms as they strengthen their verbal skills.

**Chapter Review and Chapter Test** The end of each chapter contains a review of topics introduced in the chapter. The review problems are keyed to sections. The chapter test is not keyed to sections.

**Cumulative Review** Each chapter after the first contains a **cumulative review of all chapters beginning with the first** up through the chapter at hand. Each problem contained in the cumulative review is actually an earlier worked example in the text that is referenced in the back of the book along with the answer. Students who need to see a complete worked-out solution, with explanation, can do so by turning to the appropriate example in the text.

## KEY CONTENT FEATURES IN THE THIRD EDITION

**Overview** This new edition retains many of the factors that have contributed to its success. Even so, **every section of the text was carefully re-examined**. Throughout the new edition you will find numerous new applications, examples, and many real-life applications and exercises. For example, look at Sections 1.1, 3.4, or 8.2. Some sections have internal re-organization to better clarify and enhance the presentation.

**Increased Integration of Geometry Concepts** In addition to the traditional topics in beginning algebra courses, this text contains a strong emphasis on problem solving, and geometric concepts are integrated throughout. The geometry concepts presented are those most important to a students' understanding of algebra, and I have included **many applications and exercises** devoted to this topic. These are marked with the icon △. Also, geometric figures, a review of angles, lines, and special triangles, are covered in the appendices. The inside front cover provides a quick reference of geometric formulas.

**Real Numbers and Algebraic Expressions** Chapter 1 now begins with Tips for Success in Mathematics (Section 1.1). Chapter 1 has been streamlined and refreshed for **greater efficiency and relevance**. New applications and real data enhance the chapter.

**Early and Intuitive Introduction to Graphs and Functions** As bar and line graphs are gradually introduced in Chapters 1 and 2, an emphasis is placed on the notion of paired data. This leads naturally to the concepts of ordered pair and the rectangular coordinate system introduced in Chapter 3. This edition offers more real data and conceptual type applications and further strengthens the introduction to slope.

Once students are comfortable with graphing equations, functions are introduced in Chapter 3. The concept of function is illustrated in numerous ways to ensure student understanding: by listing ordered pairs of data, showing rectangular coordinate system graphs, visually representing set correspondences, and including numerous real-data and conceptual examples. **The importance of a function is continuously reinforced** by not treating it as a single, stand-alone topic but by constantly integrating functions in appropriate sections of this text.

**Increased Attention to Problem Solving** Building on the strengths of the prior editions, a special emphasis and strong commitment is given to contemporary, accessible, and practical applications of algebra. **Real data** was drawn from a variety of sources including internet sources, magazines, newspapers, government publications, and reference books. **Unique Spotlight on Decision Making exercises and a new four-step problem solving process are incorporated throughout** to focus on helping to build students problem-solving skills.

**Increased Opportunities for Using Technology** Optional explorations for a calculator or graphing calculator (or graphing utility such as Texas Instruments Interactive), are integrated appropriately **throughout the text** in Calculator Explorations features and in exercises marked with a calculator icon. The Martin-Gay **Companion Website** includes links to internet sites to allow opportunities for finding data and using it for problem solving such as with the accompanying on-line Real World Activities. The Website also includes links to search potential mathematically related careers branching from the chapter openers. Instructors may also choose from a variety of **distance learning or on-line delivery options** including Blackboard or Web CT.

**New Examples** Detailed step-by-step examples were added, deleted, replaced, or updated as needed. Many of these reflect real life. **Examples are used in two ways.** Often there are numbered, formal examples, and occasionally an example or application is used to introduce a topic or informally discuss the topic.

**New Exercises** A significant amount of time was spent on the exercise sets. New exercises and examples **help address a wide range of student learning styles and abilities.** The text now includes the following types of exercises: spotlight on decision making exercises, mental math, computational exercises, real-life applications, writing exercises, multi-part exercises, review exercises, a look ahead exercises, optional calculator or graphing calculator exercises, data analysis from tables and graphs, vocabulary checks, and projects for individual or group assignment. Also available are new on-line Real World Activities accessed via this textbook's companion website, and a selection of group activities in a worksheet ready, easy to use format, found in the Instructor's Resource Manual with Tests.

**Enhanced Supplements Package** The new Third Edition is supported by a wealth of supplements designed for **added effectiveness and efficiency**. New items include the MathPro 4.0 Explorer tutorial software together with a unique video clip feature, a new computerized testing system TestGenEQ, and an expanded and improved Martin-Gay companion website. Some highlights in print materials include the addition of teaching tips in the Annotated Instructor's Edition, and an expanded Instructor's Resource Manual with Tests including additional exercises and short group activities in a ready-to use-format. Please see the list of supplements for descriptions.

## OPTIONS FOR ON-LINE AND DISTANCE LEARNING

For maximum convenience, Prentice Hall offers on-line interactivity and delivery options for a variety of distance learning needs. Instructors may access or adopt these in conjunction with this text, *Intermediate Algebra*.

### *Companion Website*

Visit *http://www.prenhall.com/martin-gay*
The companion Website includes basic distance learning access to provide links to the text's Real World Activities, career-related sites referenced in the chapter opening pages and a selection of on-line self quizzes. E-mail is available. For quick reference, the inside front cover of this text also lists the companion Website URL.

### *WebCT*

WebCT includes distance learning access to content found in the Martin-Gay Companion Website plus more. WebCT provides tools to create, manage, and use on-line course materials. Save time and take advantage of items such as on-line help, communication tools, and access to instructor and student manuals. Your college may already have WebCT's software installed on their server or you may choose to download it. Contact your local Prentice Hall sales representative for details.

### *Blackboard*

Visit *http://www.prenhall.com/demo*
For distance learning access to content and features from the Martin-Gay Companion Website plus more, Blackboard provides simple templates and tools to create, manage, and use on-line course materials. Save time and take advantage of items such as on-line help, course management tools, communication tools, and access to instructor and student manuals. No technical experience required. Contact your local Prentice Hall sales representative for details.

For a *complete* computer-based internet course . . .
Prentice Hall Interactive Math
Visit *http://www.prenhall.com/interactive_math*

Prentice Hall Interactive Math is an exciting, proven choice to help students succeed in math. Created for a computer-based course, it provides the effective teaching philosophy of K. Elayn Martin-Gay in an Internet-based course format. Interactive Math, Intermediate Algebra, takes advantage of state-of-the-art technology to provide highly flexible and user-friendly course management tools and an engaging, highly interactive student learning program that easily accommodates the variety of learning styles and broad spectrum of students presented by the typical intermediate algebra class. Personalized learning includes reading, writing, watching video clips, and exploring concepts through interactive questions and activities. Contact your local Prentice Hall sales representative for details.

## SUPPLEMENTS FOR THE INSTRUCTOR

### *Printed Supplements*

Annotated Instructor's Edition (ISBN 0-13-016632-4)

- Answers to exercises on the same text page or in Graphing Answer Section
- Graphing Answer Section contains answers to exercises requiring graphical solutions, chapter projects, and Spotlight on Decision Making exercises.
- Teaching Tips throughout the text placed at key points in the margin where students historically need extra help. These tips provide ideas on how to help students through these concepts, as well as ideas for expanding upon a certain concept, or ideas for classroom activities

Instructor's Solutions Manual (ISBN 0-13-017331-2)

- Detailed step-by-step solutions to even-numbered section exercises
- Solutions to every Spotlight on Decision Making exercise
- Solutions to every Calculator Exploration exercise
- Solutions to every Chapter Test and Chapter Review exercise
- Solution methods reflect those emphasized in the textbook

Instructor's Resource Manual with Tests (ISBN 0-13-017332-0)

- Notes to the Instructor that includes an introduction to Interactive Learning, Interpreting Graphs and Data, Alternative Assessment, Using Technology and Helping Students Succeed
- Eight Chapter Tests per chapter (5 free response, 3 multiple choice)
- Two Cumulative Review Tests (one free response, one multiple choice) following every two chapters
- Eight Final Exams (4 free response, 4 multiple choice)
- Twenty additional exercises per section for added test exercises or worksheets, if needed
- Group Activities by Bettie A. Truitt, Ph.D. (on average of two per chapter; providing short group activities in a convenient ready-to-use handout format)
- Answers to all items

### *Media Supplements*

TestGen EQ CD-Rom (Windows/Macintosh) (ISBN 0-13-018575-2)

- Algorithmically driven, text specific testing program
- Networkable for administering tests and capturing grades on-line
- Edit or add your own questions to create a nearly unlimited number of tests and worksheets
- Use the new "Function Plotter" to create graphs
- Tests can be easily exported to HTML so they can be posted to the Web for student practice

Computerized Tutorial Software Course Management Tools
MathPro 4.0 Explorer Network CD-Rom (ISBN 0-13-018577-9)

- Enables instructors to create either customized or algorithinically generated practice tests from any section of a chapter, or a test of random items

- Includes an e-mail function for network users, enabling instructors to send a message to a specific student or to an entire group
- Network based reports and summaries for a class or student and for cumulative or selected scores are available

Companion Website: *http://www.prenhall.com/martin-gay*

- Create a customized online syllabus with Syllabus Manager
- Assign Internet-based Real World Activities, wherein students find and retrieve real data for use in guided problem solving
- Assign quizzes or monitor student self quizzes by having students e-mail results, such as true/false reading quizzes or vocabulary check quizzes
- Destination links provide additional opportunities to explore related sites

## SUPPLEMENTS FOR THE STUDENT

### *Printed Supplements*

Student Solutions Manual (ISBN 0-13-017329-0)

- Detailed step-by-step solutions to odd-numbered section exercises
- Solutions to every (odd and even) Mental Math exercise
- Solutions to odd-numbered Calculator Exploration exercises
- Solutions to every (odd and even) exercise found in the Chapter Reviews and Chapter Tests
- Solution methods reflect those emphasized in the textbook
- Ask your bookstore about ordering

Student Study Guide (ISBN 0-13-017320-7)

- Additional step-by-step worked out examples and exercises
- Practice tests and final examination
- Includes Study Skills and Note-taking suggestions
- Includes Hints and Warnings section
- Solutions to all exercises, tests, and final examination
- Solution methods reflect those emphasized in the text
- Ask your bookstore about ordering

How to Study Mathematics

- Have your instructor contact the local Prentice Hall sales representative

Math on the Internet: A Student's Guide

- Have your instructor contact the local Prentice Hall sales representative

Prentice Hall/New York Times, Theme of the Times Newspaper Supplement

- Have your instructor contact the local Prentice Hall sales representative

### *Media Supplements*

Computerized Tutorial Software
MathPro 4.0 Explorer Network CD-Rom (ISBN 0-13-018577-9)
MathPro 4.0 Explorer Student CD-Rom (ISBN 0-13-018578-7)

- Keyed to each section of the text for text-specific tutorial exercises and instruction
- Warm-up exercises and graded Practice Problems
- Video clips, providing a problem (similar to the one being attempted) being explained and worked out on the board

- Explorations, allowing explorations of concepts associated with objectives in more detail
- Algorithmically generated exercises, and includes bookmark, on-line help, glossary, and summary of scores for the exercises tried
- Interactive feedback
- Have your instructor contact the local Prentice Hall sales representative—also available for home use

Videotape Series (ISBN 0-13-018583-3)

- Written and presented by textbook author K. Elayn Martin-Gay
- Keyed to each section of the text
- Presentation and step-by-step solutions to exercises from each section of the text.
- Examples or exercises taken directly from the text are marked with a video icon .
- Key concepts are explained

Companion Website: *www.prenhall.com/martin-gay*

- Offers Warm-ups, Real World Activities, True/False Reading Quizzes, Chapter Quizzes, and Vocabulary Check Quizzes
- Includes a link to the Real World Activities referenced in each chapter of this text
- Option to e-mail results to your instructor
- Destination links provide additional opportunities to explore other related sites, such as those mentioned in this text's chapter opening pages

## ACKNOWLEDGMENTS

First, as usual, I would like to thank my husband, Clayton, for his constant encouragement. I would also like to thank my children, Eric and Bryan, for continuing to eat my burnt meals. Thankfully, they have started to cook a little themselves.

I would also like to thank my extended family for their invaluable help and wonderful sense of humor. Their contributions are too numerous to list. They are Rod and Karen Pasch; Peter, Michael, Christopher, Matthew, and Jessica Callac; Stuart, Earline, Melissa, Mandy, and Bailey Martin; Mark, Sabrina, and Madison Martin; Leo, Barbara, Aaron, and Andrea Miller; and Jewett Gay.

A special thank you to all the users of the first and second editions of this text and for their suggestions for improvements that were incorporated into the third edition. I would also like to thank the following reviewers for their input and suggestions:

Ken Araujo, *Florence-Darlington Technical College*
Ron Beeler, *Southern Illinois University–Edwardsville*
Debra Bryant, *Tennessee Technological University*
William Culbertson, *Northwest State Community College*
Lisa Finley, *Craven Community College*
Teresa Hasenauer, *Indian River Community College*
Robert Hervey, *Hillsborough Community College*
Cheryl Hobneck, *Illinois Valley Community College*
Helene Humphrey, *San Joaquin Delta College*
Stanley Max, *University of Southern Maine*

Cheryl McAllister, *Southeast Missouri State University*
Carol McVey, *Florence-Darlington Technical College*
Marcia Mollé, *Metropolitan Community College*
Magdala Ray, *Palm Beach Community College*
Robert Rider, *Ohio University–Zanesville*
Heidi Staebler, *Texas A & M University–Commerce*
Peggy Tibbs, *Arkansas Tech University*
Douglas Wood, *New Hampshire Community Technical College*

There were many people who helped me develop this text and I will attempt to thank some of them here. Cheryl Cantwell was invaluable for contributing to the overall accuracy of this text. Emily Keaton was also invaluable for her many suggestions and contributions during the development and writing of this text. I thank Tony Palermino for his many suggestions during the editing process and a special thank you to Ann Marie Jones, my project manager. She kept us all organized and on task. Ingrid Mount at Elm Street Publishing Services provided guidance throughout the production process. I very much appreciated the writers, formatters, and accuracy checkers of the supplements. I thank Emily Keaton for her fine job on the Student Study Guide, and Bettie A. Truitt for the selection of group activities. I thank Terri Bittner, Carrie Green, Cindy Trimble, Jeff Rector, and Teri Lovelace at Laurel Technical Services for all their work on some of the supplements, and providing a thorough accuracy check. Lastly, a special thank you to my editor Karin Wagner for her support and assistance throughout the development and production of this text and to all the staff at Prentice Hall: Chris Hoag, Linda Behrens, Alan Fischer, Maureen Eide, Grace Hazeldine, Gus Vibal, Audra Walsh, Kate Marks, Eilish Main, Daniel Auld, Elise Schneider, Stephanie Szolusha, John Tweedale, Paul Corey, and Tim Bozik.

*K. Elayn Martin-Gay*

## ABOUT THE AUTHOR

K. Elayn Martin-Gay has taught mathematics at the University of New Orleans for more than 20 years. Her numerous teaching awards include the local University Alumni Association's Award for Excellence in Teaching, and Outstanding Developmental Educator at University of New Orleans, presented by the Louisiana Association of Developmental Educators.

Prior to writing textbooks, K. Elayn Martin-Gay developed an acclaimed series of lecture videos to support developmental mathematics students in their quest for success. These highly successful videos originally served as the foundation material for her texts. Today the tapes specifically support each book in the Martin-Gay series.

Elayn is the author of nine published textbooks as well as multimedia interactive mathematics, all specializing in developmental mathematics courses such as basic mathematics, prealgebra, beginning and intermediate algebra. She has provided author participation across the broadest range of materials: textbook, videos, tutorial software, and Interactive Math courseware. All the components are designed to work together. This offers an opportunity of various combinations for an integrated, consistent teaching and learning package.

# APPLICATIONS INDEX

Agriculture
combine machine rental, 663
milk cows/sheep on farms, 203
Architecture
arch height/width for bridge, 535, 561
bridge design, 553
circular pool with fountain, 535
Astronomy
astronomical unit, 74
distance between Jupiter and the sun, 251
eccentricity, 542, 543
elliptical path of planet, 543
probe speed, 74
pulsars, 252
surface area of Moon, 443
Automobiles
car rentals, 85, 111, 148, 217
driving time, 74
minivan rental, 185, 186
new car model, 652
radius of curvature, 459
rate of speed, 210–11, 239
safe-driver discounts, 16
skid marks and speed of, 443

Business
annual revenue, 359
average cost, 369
boat manufacture, 177
bookshelves manufacture, 28
break-even point, 66, 111, 113, 214–15, 217, 218, 239, 240
camera manufacturing costs, 148
cellular telephone subscriptions, 418, 585
cellular telephone users, 589
cigarette consumption, 82–83
commission calculation, 81–82
company profit, 170
computer-related employees, 113
cost and revenue equations, 202
cost for pressing compact discs, 330
cost/revenue/profit functions, 217, 218, 267, 571
dairy machines, 380
delivery service operation, 444
demand equation, 425, 474
earnings from small business, 662
fax charges, 652
fax machine pre-tax price, 64
federal hourly minimum wage, 126
floor space for displays, 17
holiday rush rental fee, 652
home health aides, 160
hotel rooms, 240
Internet retail revenues, 589
manufacturing costs, 307, 397, 513
manufacturing time, 148
market equilibrium, 549
maximum profit, 514
medical assistants employed, 171
municipal budgets, 104–5
net income, 482
net revenue, 420
newspaper circulation analysis, 39–40
operating income, 98
original computer price, 60
paging subscribers, 186
paralegals, 160
production cost, 389
research/development expenditures, 134–36
retail sales, 251
sales prediction, 170
straight-line depreciation, 127
supply and demand, 202
support stockings order, 175
systems analysts employed, 171
telephone company operators decrease, 64
television ad campaign, 652
total cost in producing tapes, 267
total profit function, 498
total revenues, 251, 341
tour packages, 27
Web site usage analysis, 595
work problems, 380, 389

Chemistry
acid solutions, 218, 219
alcohol solutions, 239
liters of solutions, 218
mixing solutions, 212–13
phenobarbital solution, 217
plant food solution, 65
Construction
board length, 358
cable placement, 443
ceiling tiles, 72
floor tiles, 72, 112
framing home, 302–3
guy wire for TV tower, 443
pipe connections, 458
roof slope, 160
spotlight mounting, 443
wire attached to telephone pole, 443
wire placement, 440
work problems, 381

Distance. *See* Time and distance

Education
ACT assessment, 219
beginning salaries for candidates for bachelor's degrees, 85
college enrollment, 633
college tuition modeling, 658
final course average, 92
final exam scores, 84, 92
learning curve, 614
pre-tax text book price, 64
quiz scores, 62, 188
SAT scores, 66
study/work time, 177
temporary classrooms funding, 83
time management, 8
tuition/fees, 148, 161
Electricity
cable length, 306
combined resistance, 399
current in circuit, 280
current/resistance, 389
modeling production, 391–92
parts list resistors, 337
Electronics
CD-player reliability, 601
frequency of oscillations, 419
resistance formula, 379
Entertainment
Disney World one-day pass price, 153–54
movie/television reviewing, 177
play tickets sold, 217
theater ticket prices, 513
Environment
deforestation in India, 65
greenhouse gases, 589
HDPE bottle recycling, 65
insecticide and weevils killed, 652
iodine isotope decay rate, 506
methane emissions, 514
mosquitoes killed by insecticide, 663
newsprint recycling, 66
pollutant removal, 340
radioactive waste decay, 645, 662
radioactive material estimate, 586–87, 588, 589, 641

Finance and economics
annual average income with associate's degree, 160
annual income of woman with bachelor's degree, 160
average annual earnings for high school graduate, 111
bottled water/filtered water comparison, 199
car savings plan, 112
ceiling paint, 388
checking accounts balances, 452
college tuition modeling, 658
compound interest, 72, 73, 112, 113, 470–71, 586, 589, 614, 622, 624–24
compound interest on inheritance savings, 326
continuous compound interest, 608, 623
credit union loans, 64
donation amount, 648
doubling investment, 611–12
doubling savings, 662
drafting supplies, 217
final loan payment, 606
interest rate calculation, 239, 473
Internet service providers, 274
investment risk over time, 29
job offers, 632, 645, 652, 662
laundry costs, 112
long-distance telephone plans, 216
losses on bets, 652
lottery prize division, 340
monthly salary, 640, 662
office supplies, 217
online tax returns, 65
rent increases, 158
room and board, 652
Roth IRA contributions, 651
salary calculation, 636
salary/commission arrangement, 217
sales salary/products sold, 241
savings accounts, 69–70, 281
telephone calling plans, 85
total amount deposited, 652
types of coins, 239
U.S. federal government income, 203
vacation allowance, 633
women's business earnings, 652
yearly earnings for high school graduates, 188
Food
broccoli price, 114
buttermilk, 217
candy mix, 239
chicken consumption, 515
chicken/pork consumption, 92
fruit drinks, 240
party mix, 217
poultry consumption, 141
red meat consumption, 202
Tootsie Roll Pops, 296
wedding cake, 307
whole milk/skim milk consumption, 85

Geometry
angles
measures of, 65, 215–16, 218, 668, 671, 672
measures of complementary, 66
measures of supplementary, 66
area
of circle, 28
of parallelogram, 9

of rectangle, 38
circle
area of, 140, 390, 391
circumference of, 390, 391
radius of, 457, 473
unknown fractional part of, 29, 43
circular column
weight of, 386–87
Clarke belt length, 73
cone
rationalizing numerator of radical expression for, 436
surface area of, 425
volume of, 390, 681
volume and surface area of, 680
cube
surface area of, 281
volume of, 141, 257, 288
volume and surface area of, 680
cylinder
surface area of, 386, 681
volume of, 73, 74, 276, 390
cylindrical solid
volume of, 324
cylindrical tank
dimensions of, 65
golden ratio, 482
hemisphere
volume of, 681
heptagon
perimeter of, 398
hexagon
length of each side of, 358
irregular shapes
perimeter of, 430
isosceles triangle
lengths of legs/sides of, 239, 241, 473, 480, 520
values in, 218
parallel lines cut by transversal, 218
parallelogram
area of, 257
height of, 358, 364
length of base, 340
perimeter
of floor plan, 63
of geometric figure, 141
of rectangle, 28, 63
of square, 63, 65
of triangle, 65
pyramid
volume of, 680
quadrilateral
length of sides of, 217
measures of angles in, 219
perimeter of, 346
rectangle
area of, 323, 340
dimensions of, 65, 107, 111, 112, 239, 307, 324, 480, 481, 514, 549
length of, 399
length and width of, 307
perimeter of, 429, 482, 549
width of, 307
rectangular solid
capacity of, 681
volume of, 681
volume and surface area of, 679, 680
width of, 364
right triangle
length of hypotenuse, 671
length of unknown leg of, 348, 440
lengths of legs of, 307
shaded regions
area of, 296
sphere
circumference of, 73
rational denominator of radical expression for, 436
surface area of, 400, 681
volume of, 73, 680
volume and surface area of, 679, 680
square
area of, 324, 326, 346
dimensions of, 473
perimeter of, 346, 549
square-based pyramid
volume of, 681
volume and surface area of, 680
trapezoid
perimeter of, 549
triangle
area of, 340
base and height of, 482
height of, 74
length of sides of, 61-62, 64
length of unknown sides/legs of, 442–43, 459, 521
measures of angles of, 65, 668
missing lengths in, 670, 672
perimeter of, 549
unknown angle of, 451

Medicine
AIDS cases reported, 267
arthritis reports, 187
basal metabolic rate (BMR), 419–20
births in United States, 171
body-mass index, 381
flu epidemic victims, 613
fluoride doses for children, 71
health care spending, 267
health maintenance organizations, 267
heartworm preventive, 141
infectious disease, 633
mumps eradication, 168
people infected from virus, 661
vitamin A intake for child, 480
Miscellaneous
amount of money, 63
amphitheater seats, 661
auditorium seats, 640
bounce of racquetball, 662
bouncing rubber ball, 640
cardboard dimensions, 491
carpet rolls stacked, 647
China's and Taiwan's armed forces, 257
conveyor belt use, 381
dog food consumption, 491
dog pen, 113
elevation differences, 29
federally owned land, 63, 64
fencing for land, 63
Fibonacci sequence, 633
first class mailing, 85
force exerted by tractors, 444
frames purchases, 217
households owning black-and-white televisions/microwaves, 224
ice cream container, 112
living arrangements for children, 177–78
machinery rental cost, 148
movie theater seats, 663
newspaper story angle, 63
oil can display, 662
paint for walls, 73
possibilities of selecting letter, 161
probability of event, 74
rabbit's dietary needs, 218
revolutions of rotating flywheel, 652
rotation in game of pool, 652
seagull flight, 161
shelf space for encyclopedias, 17
shopping mall parking garage, 85
slope for sewer pipe, 161
speaker wire, 307
spinner game, 29
square inches converted to square meters, 257
surfers in human pyramid, 645
theater seats, 633
thickness of paper stack, 662
time to fill pond, 491
time to fill tank, 491
tin can manufacture, 280
tourists in 2020, 18
tree planting, 644
triangular planting, 663
TV tube measurement, 473
washer cross-section, 296
work problems, 127, 375–77, 379, 380, 381, 399
work times, 487–88, 491, 520, 521, 652, 663
zip codes, 66

Physics
density of object, 257
Doppler effect, 381
length and period of pendulum, 452–53
pendulum swings, 640, 645, 662, 663
speed of wave over stretched string, 419
weight attached to spring, 383
Politics
gubernatorial salaries, 64
Population
baby gorillas born, 643–44
bacterial culture/colony, 633, 639, 640
bison, 588
Canada, 623
Cape Cod mosquitoes, 589
cranes born, 662
density for U.S., 257
Egypt, 623
endangered fish, 645
fungus culture, 645
India, 613
Indonesia, 623
Italy, 614
Japan, 623
lemmings, 611
mallard ducks, 623
Mexico City, 589
mice, 625
opossums killed, 645
otters born, 645
over age 65, 39
and pollutants produced, 388
prairie dog, 625
prison inmates, 370, 484-83, 514
rats, 588
Russia, 613
Senegal, 613
small town, 663
sparrow, 633
wolves, 613
wood duck, 625
world, 589
yeast colony/culture, 661, 662
Psychology
test subject screening, 441

Real estate
building appreciation, 170
median price of existing home, 171
property value, 640
sales prediction, 165–66

Science
atmospheric pressure, 390, 614
deepest hole in ocean floor, 72
density of water in Lake Superior, 257
earthquake magnitude, 604–5, 608
Earth's water, 317–18
escape velocity, 411, 413
femur/tibia bone measurements, 141
Kola Superdeep Borehole, 72
radiation problems, 623
speed of sound, 353
volcano height above sea level, 29
volcano surface area, 425
wind turbine generator power, 114
Sports
bicyclist's speed, 380
cyclist's workout rates, 380
exercise program, 641
Grand Prix auto race, 491–92
gymnastics competition, 112
jogger's speed, 490
jogging/biking speed, 490
NBA top-scoring player, 218
rate of speed, 217
running distance, 380
speed skating, 84
WNBA top scoring player, 218

Technology
addition by computer, 257
CD-ROM lifespan, 64
compact disc storage units, 192
homes with computers, 64
homes with online services, 64
space shuttle costs, 73
straight-line depreciation, 170
Temperature
Boyle's law, 385, 400
Celsius/Fahrenheit equivalencies, 28
Celsius/Fahrenheit interconversions, 92, 112
center of the sun, 252
Charles's law, 388
daily low, 482
for glass being liquid, 85
high/low, 72
modeling, 615–16
stibnite melting point, 85
of water around lava flow, 37
Time and distance
aircraft speed/distance, 17
average speed, 72, 112
bouncing rubber ball, 663
current of stream, 399
cycling time, 70
distance across lake, 534–35
distance across pond, 458
distance ball covers, 652
distance from lightning flash, 17
distance object falls, 633
distance olive falls, 661
distance saved, 479–80, 481
distance traveled by pendulum, 650
finding speed, 488–89
freely falling objects, 473, 482, 652
height and person's visibility, 444
height of object, 261, 266, 267, 281, 307, 324, 326, 401
height of projectile, 267, 281
horsepower transmitted, 390
humming bird wing beats, 43
interval time for projectile, 498
jet/car travel time, 381
jet record speed, 17
light source, 374–75, 389
maximum height, 512, 514, 520, 521
original speed/speed during rainfall, 491
original speed/speed returning home, 491
parachutist's free fall, 663
plane speed in still air, 380
rate ball thrown to travel given distance, 73
rocket launches, 380
rocket return time, 301
round-trip drive, 72
rowing in still water/current speed, 217
rowing time, 381
ships travel time/miles, 519
spacecraft speed, 74
speed of boat in still water, 380
speed of car, 388
speed of car in still air, 381
speed of current, 377–78, 380
speed of jet, 399
speed of plane, 379
speed of train, 380, 400
speed of truck, 379
speed of wind, 400
tank emptying, 399
velocity of dropped object, 444
walker's speed, 400
walking distance saved, 521
wind speed and sail, 390
Transportation
aircraft operation cost, 17
aircraft seats, 64
airport arrivals/departures, 57, 64
flatland and mountain rate for truck, 380
luggage on plane, 84
passengers aboard airline flights, 141, 251
radioaltimeter reading, 378
road slope, 160
slope of descent, 160
slope of jet climb, 160
speed of plane in still air/speed of wind, 217
subway system passengers, 252
subway volume, 257
urban planning, 223

Weather
hurricane tracking, 125
locating lightning strikes, 232–33
rainfall in inches, 219
tornado-related deaths, 64
Weight
of ball, 388
boxes on elevator, 84
of boy, 614
column, 386–87, 389
of individual above Earth, 389
puppy's gain in, 632
support by rectangular beam, 389, 390
of whales, 65

# Highlights of *Intermediate Algebra,* Third Edition

*Intermediate Algebra, Third Edition* has been written and designed to help you succeed in this course. Specific care has been taken to ensure you have the most up-to-date and relevant text features to provide you with a solid foundation in algebra, as many accessible real-world applications as possible, and to prepare you for future courses.

## Get Motivated!

The Number One Job in the United States

A Webmaster is responsible for creating and managing Web sites on the World Wide Web. According to the 1999 *Jobs Rated Almanac,* Webmaster is rated as the number one job in the United States based on factors such as income, stress, physical demands, potential growth, job security, and work environment.

Webmasters have technical, business, and visual design skills or understanding. Some Webmasters may focus on the computer programming aspects of developing Web sites, including graphics and Web site security issues. Others may focus on Web site content, such as company or organization history, product information, or other marketing topics. Webmasters should be comfortable with people management, project management, and strategic planning. Webmasters use math and problem-solving skills in tasks such as creating budgets, making hardware or software purchasing decisions, and compiling and analyzing Web site usage statistics.

*For more information about a career as a Webmaster or other Web professional, visit the World Organization of Webmasters Web site by first going to www.prenhall.com/martin-gay.*

*In the Spotlight on Decision Making feature on page 595, you will have the opportunity to make a decision about modeling Web site usage statistics as a Webmaster.*

EXPONENTIAL AND LOGARITHMIC FUNCTIONS

10


10.1 The Algebra of Functions; Composite Functions

10.2 Inverse Functions

10.3 Exponential Functions

10.4 Logarithmic Functions

10.5 Properties of Logarithms

10.6 Common Logarithms, Natural Logarithms, and Change of Base

10.7 Exponential and Logarithmic Equations and Applications


In this chapter, we discuss two closely related functions: exponential and logarithmic functions. These functions are vital to applications in economics, finance, engineering, the sciences, education, and other fields. Models of tumor growth and learning curves are two examples of the uses of exponential and logarithmic functions.

565

***pages 564, 565***

### REAL-WORLD CHAPTER OPENERS

New Real-World Chapter Openers focus on how algebraic concepts relate to the world around you so that you see the relevance and practical applications of algebra in daily life.

They also provide links to the World Wide Web and reference a *Spotlight on Decision Making* feature within the chapter for further exploration.

### SPOTLIGHT ON DECISION-MAKING

These unique new applications encourage you to develop your decision-making and problem solving abilities, and develop life skills, primarily using workplace or career-related situations.

**SPOTLIGHT ON DECISION MAKING**

Suppose you are the Webmaster for a small but growing company. One of your duties is to ensure that your company's newly established Web site can adequately handle the number of visitors to it. You decide to find a mathematical model for recent Web site usage statistics to help predict future numbers of visitors. Ultimately, you would like to use this model to predict when your Web site's server capacity must be expanded.

The first step in finding a model for the usage statistics is to decide what type of mathematical model to use: linear, quadratic, exponential, or logarithmic. The graph shows the number of visitors to your company's Web site in each of the first five months since it was established. Use the graph to decide which type of mathematical model to use. Explain your reasoning.

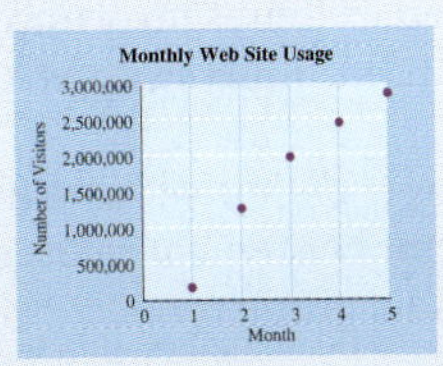


| Month | 1 | 2 | 3 | 4 | 5 |
|---|---|---|---|---|---|
| Visitors | 166,511 | 1,320,978 | 1,996,298 | 2,475,445 | 2,847,100 |

***page 595***

# Become a Confident Problem-Solver!

A goal of this text is to help you develop problem-solving abilities.

**Example 2** The three busiest airports in the United States are in the cities of Chicago, Atlanta, and Dallas/Ft. Worth. The airport in Atlanta has 7.7 million more arrivals and departures than the Dallas/Ft. Worth airport. The Chicago airport has 9.8 million more arrivals and departures than the Dallas/Ft Worth airport. Write the sum of the arrivals and departures from these three cities as a simplified algebraic expression. Let $x$ be the number of arrivals and departures from the Dallas/Ft. Worth airport.

*Solution* If $x$ = millions of arrivals and departures at the Dallas/Ft. Worth airport, then
$x + 7.7$ = millions of arrivals and departures at the Atlanta airport and
$x + 9.8$ = millions of arrivals and departures at the Chicago airport
Since we want their sum, we have

| | | | | | |
|---|---|---|---|---|---|
| In words: | arrivals and departures at Dallas/Ft. Worth | + | arrivals and departures at Atlanta | + | arrivals and departures at Chicago. |
| Translate: | $x$ | + | $(x + 7.7)$ | + | $(x + 9.8)$ |

Then $x + (x + 7.7) + (x + 9.8) = x + x + 7.7 + x + 9.8$
$= 3x + 17.5$ Combine like terms.

In Exercise 25, we will find the actual number of arrivals and departures at these airports.

2 Our main purpose for studying algebra is to solve problems. The following problem-solving strategy will be used throughout this text and may also be used to solve real-life problems that occur outside the mathematics classroom.

**page 57**

## GENERAL STRATEGY FOR PROBLEM-SOLVING

Save time by having a plan. This text's organization can help you. Note the outlined problem-solving steps, *Understand, Translate, Solve,* and *Interpret.*

Problem-solving is introduced early and emphasized and integrated throughout the book. The author provides patient explanations and illustrates how to apply the problem-solving procedure to the in-text examples.

## GEOMETRY

Geometric concepts are integrated throughout the text. Examples and exercises involving geometric concepts are identified with a triangle icon.

The inside front cover of this text contains *Geometric Formulas* for convenient reference, as well as appendices on geometry.

**40.** In 1945, Arthur C. Clarke, a scientist and science-fiction writer, predicted that an artificial satellite placed at a height of 22,248 miles directly above the equator would orbit the globe at the same speed with which the Earth was rotating. This belt along the equator is known as the Clarke belt. Use the formula for circumference of a circle and find the "length" of the Clarke belt. (*Hint:* Recall that the radius of the Earth is approximately 4000 miles. Round to the nearest whole mile.)

**page 73**

# Get Involved!

Real-world applications in this textbook will help to reinforce your problem-solving skills, and show you how algebra is connected to a wide range of fields like consumer affairs, sports, and business. You will be asked to evaluate and interpret real data in graphs, tables, and in context to solve applications. See also the Index of Applications located on page xxi for a quick way to locate those in your areas of interest.

**90.** In 1996, the number of U.S. paging subscribers (in millions) was 42. The number of subscribers in 1999 (in millions) was 58. Let $y$ be the number of subscribers (in millions) in the year $x$, where $x = 0$ represents 1996. (*Source:* Strategis Group for Personal Communications Asso.)

**a.** Write a linear equation that models the number of U.S. paging subscribers (in millions) in terms of the year $x$. [*Hint:* Write 2 ordered pairs of the form (years past 1996, number of subscribers).]

**b.** Use this equation to predict the number of U.S. paging subscribers in the year 2007. (Round to the nearest million.)

**page 186**

## INTERESTING, RELEVANT, AND PRACTICAL REAL-WORLD APPLICATIONS

Accessible applications reinforce concepts needed for success in this course, and relate those concepts to everyday life.

**22.** In 2006, the number of people employed as database administrators, computer support specialists, and all other computer scientists is expected to be 461,000 in the United States. This represents a 118% increase over the number of people employed in these occupations in 1996. Find the number of database administrators, computer support specialists, and all other computer scientists employed in 1996. (*Source:* U.S. Bureau of Labor Statistics)

**page 113**

## INTERESTING REAL DATA

Real-world applications include those based on real data.

Example 2 The three busiest airports in the United States are in the cities of Chicago, Atlanta, and Dallas/Ft. Worth. The airport in Atlanta has 7.7 million more arrivals and departures than the Dallas/Ft. Worth airport. The Chicago airport has 9.8 million more arrivals and departures than the Dallas/Ft Worth airport. Write the sum of the arrivals and departures from these three cities as a simplified algebraic expression. Let $x$ be the number of arrivals and departures from the Dallas/Ft. Worth airport.

of arrivals and departures at the Dallas/Ft. Worth airport, then
of arrivals and departures at the Atlanta airport and
of arrivals and departures at the Chicago airport
sum, we have

**page 57**

5 For additional Chapter Projects, visit the Real World Activities Website by going to http://www.prenhall.com/martin-gay.

CHAPTER PROJECT

### *Investigating Earth's Water*

Earth is covered by water. In fact, oceans cover nearly three-fourths of the surface of Earth. However, oceans aren't the only source of Earth's water. The melting of one of the other main sources of Earth's water, icecaps and glaciers, is expected to contribute to a global rise in ocean level due to global warming over the next 100 years. In this project, you will have the opportunity to investigate where Earth's water exists and how the ocean level will change. This project may be completed by working in groups or individually.

1. Refer to Table 1. Which accounts for more of Earth's water: groundwater or icecaps and glaciers?
2. Find the total volume of water that exists on planet Earth. Add this figure to the table.
3. Using the total you computed in Question 2, complete the percent column of the table. Discuss your findings.

Widespread industrialization during the nineteenth and twentieth centuries has led to an increase in the presence of carbon dioxide, methane, nitrous oxide, and chlorofluorocarbons in Earth's atmosphere. These so-called greenhouse gases are believed by some scientists to be responsible for an increase in the average global temperature of 0.2°C to 0.3°C in the last half of the twentieth century. If this global warming trend continues, one of its consequences may be a global increase in the level of the oceans. An overall increase in ocean level will be due to changes in icecaps and glaciers, as well as thermal expansion. Higher global temperatures lead to warming in the top layers of the ocean, causing the water to expand and elevate the sea level.

TABLE 1. WHERE EARTH'S WATER EXISTS

| | *Water Volume (cubic kilometers)* | *Percent* |
|---|---|---|
| Atmosphere | $1.3 \times 10^4$ | |
| Average in stream channels | $1.0 \times 10^3$ | |
| Freshwater lakes | $1.2 \times 10^5$ | |
| Groundwater | $8.3 \times 10^6$ | |
| Icecaps and glaciers | $2.9 \times 10^7$ | |
| Oceans | $1.32 \times 10^9$ | |
| Saline lakes and inland seas | $1.0 \times 10^5$ | |
| Water in soil above groundwater | $6.7 \times 10^5$ | |
| Total | | |

(*Source:* Data from B.J. Skinner, *Earth Resources*, 2nd Ed., Prentice Hall, 1976)

**page 317**

## CHAPTER PROJECT

New *Chapter Projects* for individuals or groups provide chapter wrap-up and extend the chapter's concepts in a multi-part application.

In addition, references to *Real World Activities* are given. This **internet option** invites you to find and retrieve real data for use in solving problems.

# Be Confident!

Several features of this text can be helpful in building your confidence and mathematical competence. As you study, also notice the connections the author makes to relate new material to ideas that you may already know.

1.1 TIPS FOR SUCCESS IN MATHEMATICS

CD-ROM SSM SSG Video

▶ OBJECTIVES

1. Getting ready for this course.
2. General tips for success.
3. How to use this text.
4. Get help as soon as you need it.
5. How to prepare for and take an exam.
6. Tips on time management.

page 4

## ◀ TIPS FOR SUCCESS

New coverage of study skills in Section 1.1 reinforces this important component to success in this course.

## MENTAL MATH ▶

*Mental Math* warm-up exercises reinforce concepts found in the accompanying section and can increase your confidence before beginning an exercise set.

MENTAL MATH

*State the slope and the y-intercept of each line with the given equation.*

**1.** $y = -4x + 12$ **2.** $y = \frac{2}{3}x - \frac{7}{2}$ **3.** $y = 5x$

**4.** $y = -x$ **5.** $y = \frac{1}{2}x + 6$ **6.** $y = -\frac{2}{3}x + 5$

*Decide whether the lines are parallel, perpendicular, or neither.*

**7.** $y = 12x + 6$, $y = 12x - 2$ **8.** $y = -5x + 8$, $y = -5x - 8$ **9.** $y = -9x + 3$, $y = \frac{3}{2}x - 7$ **10.** $y = 2x - 12$, $y = \frac{1}{2}x - 6$

page 169

HELPFUL HINT

Be very careful when simplifying expressions such as $-5^2$ and $(-5)^2$.

$$-5^2 = -(5 \cdot 5) = -25 \quad \text{and} \quad (-5)^2 = (-5)(-5) = 25$$

Without parentheses, the base to square is 5, not $-5$.

page 24

## ◀ HELPFUL HINTS

Found throughout the text, these contain practical advice on applying mathematical concepts. They are strategically placed where you are most likely to need immediate reinforcement.

## VOCABULARY CHECKS ▶

New *Vocabulary Checks* allow you to write your answers to questions about chapter content and strengthen verbal skills.

CHAPTER 8 VOCABULARY CHECK

*Fill in each blank with one of the words or phrases listed below.*

quadratic formula quadratic discriminant $\pm\sqrt{b}$
completing the square quadratic inequality

$(h, k)$ $(0, k)$ $(h, 0)$ $\frac{-b}{2a}$

1. The ________ helps us know find the number and type of solutions of a quadratic equation.
2. If $a^2 = b$, then $a =$ ______.
3. The graph of $f(x) = ax^2 + bx + c$ where $a$ is not 0 is a parabola whose vertex has $x$-value of ____.
4. A(n) ______________ is an inequality that can be written so that one side is a quadratic expression and the other side is 0.
5. The process of writing a quadratic equation so that one side is a perfect square trinomial is called ______________.
6. The graph of $f(x) = x^2 + k$ has vertex ______.
7. The graph of $f(x) = (x - h)^2$ has vertex ______.
8. The graph of $f(x) = (x - h)^2 + k$ has vertex ______.
9. The formula $x = \frac{-b \pm \sqrt{b^2 - 4ac}}{2a}$ is called the ______________.
10. A ________ equation is one that can be written in the form $ax^2 + bx + c = 0$ where $a, b$, and $c$ are real numbers and $a$ is not 0.

page 516

# Visualize It!

The Third Edition increases emphasis on visualization. Graphing is introduced early and intuitively. Knowing how to read and use graphs is a valuable skill in the workplace as well as in this and other courses.

## VISUALIZATION OF TOPICS

Many illustrations, models, photographs, tables, charts, and graphs provide visual reinforcement of concepts and opportunities for data interpretation.

*The following graph is called a circle graph or a pie chart. The circle represents a whole, or in this case, 100%. This particular graph shows the kind of loans that customers get from credit unions. Use this graph to answer Exercises 21–24.*

**Credit Union Loans**

*Source:* National Credit Union Administration.

**21.** What percent of credit union loans are for credit cards and other unsecured loans?

**22.** What types of loans make up most credit union loans?

**23.** If the University of New Orleans Credit Union processed 300 loans last year, how many of these might we expect to be automobile loans?

**24.** If Homestead's Credit Union processed 537 loans last year, how many of these do you expect to be either home mortgages or other real estate? (Round to the nearest whole.)

**page 64**

**95.** The following bar graph shows the top five countries with the projected number of tourists visiting in 2020.

**Tourists in 2020**

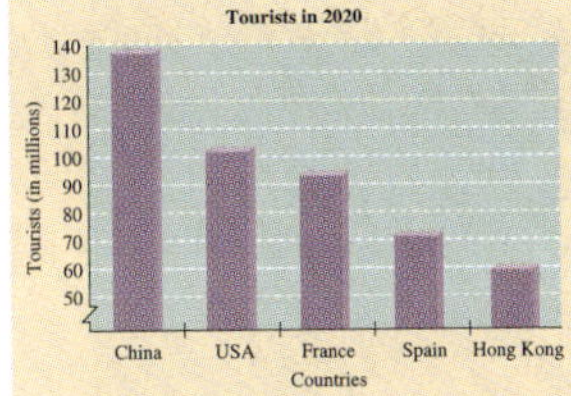


*Source:* Dan Smith, *The State of the World Atlas.*

Use the height of each bar to estimate the millions of tourists for each country.

**page 18**

*For income tax purposes, Jason Verges, owner of Copy Services, uses a method called* **straight-line depreciation** *to show the loss in value of a copy machine he recently purchased. Jason assumes that he can use the machine for 7 years. The following graph shows the value of the machine over the years. Use this graph to answer the following questions.*

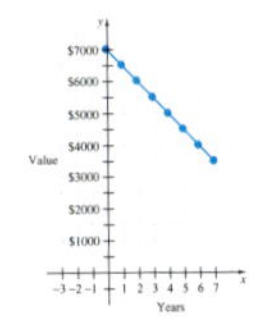

**63.** What was the purchase price of the copy machine?

**64.** What is the depreciated value of the machine in 7 years?

**65.** What loss in value occurred during the first year?

**66.** What loss in value occurred during the second year?

**67.** Why do you think that this method of depreciating is called straight-line depreciation?

**68.** Why is the line tilted downward?

**page 127**

## SCIENTIFIC AND GRAPHING CALCULATOR EXPLORATIONS

Enhanced *Explorations* contain examples and exercises to **reinforce** concepts, help **interpret** graphs, or **motivate** discovery learning. Scientific and graphing utility exercises can also be found in exercise sets.

**53.** The Cassini spacecraft mission to Saturn was launched October 15, 1997. It will take more than six and a half years to reach Saturn, arriving in July 2004. During its mission, Cassini will travel a total distance of 2 billion miles in 80.5 months. Find the average speed of the spacecraft in miles per hour. (*Hint:* Convert 80.5 months to hours using 1 month = 30 days and then use the formula $d = rt$.) (*Source:* NASA Jet Propulsion Laboratory)

**page 74**

### GRAPHING CALCULATOR EXPLORATIONS

It is possible to use a graphing calculator to sketch the graph of more than one equation on the same set of axes. For example, graph the functions $f(x) = x^2$ and $g(x) = x^2 + 4$ on the same set of axes.

To graph on the same set of axes, press the Y= key and enter the equations on the first two lines.

$$Y_1 = x^2$$

$$Y_2 = x^2 + 4$$

Then press the GRAPH key as usual. The screen should look like this.

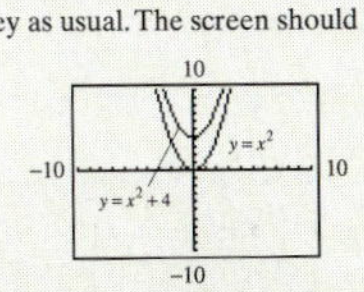


Notice that the graph of $y$ or $g(x) = x^2 + 4$ is the graph of $y = x^2$ moved 4 units upward.

*Graph each pair of functions on the same set of axes. Describe the similarities and differences in their graphs.*

**1.** $f(x) = |x|$
$g(x) = |x| + 1$

**2.** $f(x) = x^2$
$h(x) = x^2 - 5$

**3.** $f(x) = x$
$H(x) = x - 6$

**4.** $f(x) = |x|$
$G(x) = |x| + 3$

**5.** $f(x) = -x^2$
$F(x) = -x^2 + 7$

**6.** $f(x) = x$
$F(x) = x + 2$

**page 136**

# Discover the Best Supplemental Resource for You!

## *Integrated Learning Program*

All of the components of the Martin-Gay supplemental resources fit together to help you learn and understand algebra. Use these student resources based on your personal learning style to enhance what you learn from your instructor and textbook.

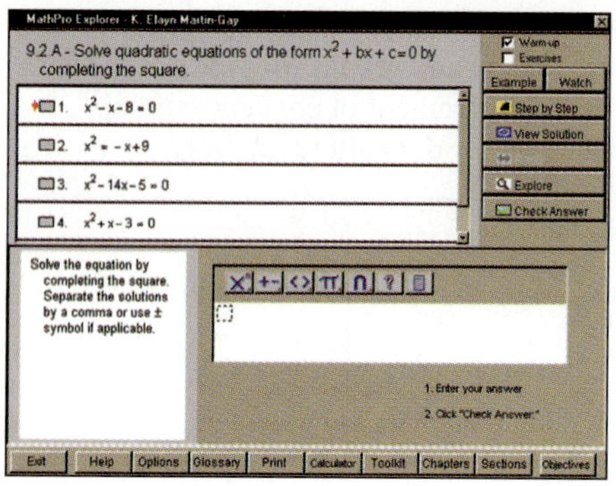


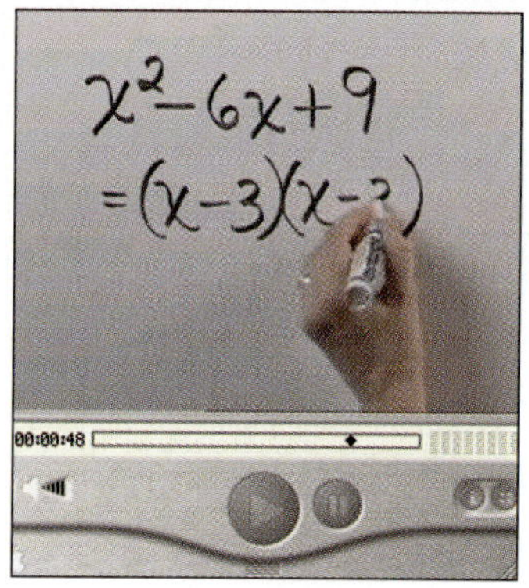


### MATHPRO EXPLORER 4.0

This **interactive** tutorial software is developed around the content and concepts of *Beginning Algebra.*
It provides:

- virtually unlimited practice problems with immediate feedback
- video clips
- step-by-step solutions
- exploratory activities
- on-line help
- summary of progress

*Available on CD-ROM*

### LECTURE VIDEO SERIES BY K. ELAYN MARTIN-GAY

Hosted by the award-winning teacher and author of *Intermediate Algebra,* these videos cover each objective in every chapter section as a supplementary review. Both problems taken directly from the text and problems similar to those found in the text are included. Look for the video icon next to examples and exercises in this text.

### WWW.PRENHALL.COM/MARTIN-GAY COMPANION WEBSITE!

The website offers warm-ups, real world activities, reading quizzes, vocabulary quizzes, and chapter quizzes, that you can send e-mail to your instructor, and links to explore related sites such as those noted in this text's chapter-opening pages.

### ALSO AVAILABLE

**The New York Times/ Themes of the Times**
Newspaper-format supplement

# INTERMEDIATE ALGEBRA

## Outdoor Opportunities

*Geology* comes from the Greek words *geo* meaning "the earth" and *logy* meaning "science" or "study." Geologists and other geoscientists study the natural resources, hazards, history, environments, and habitats of our planet. According to the National Science Foundation, over 125,000 geoscientists work in the United States, and as we continue to use up natural resources, the demand for geoscientists will increase.

People who like to work outdoors, travel, and solve puzzles make good geologists. Many spend a lot of their time "in the field" collecting data or rock specimens to be studied later in laboratories. Some geologists travel to far-flung locations to locate increasingly hard-to-find natural resources like oil, natural gas, or fresh water. Geologists must be able to take measurements, read charts and tables, use and prepare maps, and interpret data. They use these skills in activities like piecing together the geological history of a region, predicting the next eruption of a volcano, and reducing the impact of earthquakes on human life.

*For more information about careers in geology and the geosciences, visit the American Geological Institute Website by first going to www.prenhall.com/martin-gay.*

*In the Spotlight on Decision Making feature on page 37, you will have the opportunity to make color-coding decisions for a map of lava-heated ocean waters as a geologist.*

# 1 REAL NUMBERS AND ALGEBRAIC EXPRESSIONS

**1.1** Tips for Success in Mathematics

**1.2** Algebraic Expressions and Sets of Numbers

**1.3** Operations on Real Numbers

**1.4** Properties of Real Numbers

In arithmetic, we add, subtract, multiply, divide, raise to powers, and take roots of numbers. In algebra, we add, subtract, multiply, divide, raise to powers, and take roots of variables. Letters, such as $x$, that represent numbers are called **variables**. Understanding these algebraic expressions depends on your understanding of arithmetic expressions. This chapter reviews the arithmetic operations on real numbers and the corresponding algebraic expressions. Having done so, we will be prepared to explore how widely and diversely useful these algebraic expressions are for problem solving.

## 1.1 TIPS FOR SUCCESS IN MATHEMATICS

CD-ROM SSM SSG Video

**OBJECTIVES**

1. Getting ready for this course.
2. General tips for success.
3. How to use this text.
4. Get help as soon as you need it.
5. How to prepare for and take an exam.
6. Tips on time management.

Before reading this section, remember that your instructor is your best source for information. Please see your instructor for any additional help or information.

1 Now that you have decided to take this course, remember that a positive attitude will make all the difference in the world. Your belief that you can succeed is just as important as your commitment to this course. Make sure that you are ready for this course by having the time and positive attitude that it takes to succeed.

Next make sure that you have scheduled your math course at a time that will give you the best chance for success. For example, if you are also working, you may want to check with your employer to make sure that your work hours will not conflict with your course schedule.

Now you are ready for your first class period. Double-check your schedule and allow yourself extra time to arrive in case of traffic or in case you have trouble locating your classroom. Make sure that you bring at least your textbook, paper, and a writing instrument with you. Are you required to have a lab manual, graph paper, calculator, or some other supply besides this text? If so, bring this material with you also.

2 Below are some general tips that will increase your chance for success in a mathematics class. Many of these tips will also help you in other courses you may be registered for.

*Exchange names and phone numbers* with at least one other person in class. This contact person can be a great help in case you miss the class assignment or want to discuss math concepts or exercises that you find difficult.

*Choose to attend all class periods*. If possible, sit near the front of the classroom. This way, you will see and hear the presentation better. It may also be easier for you to participate in classroom activities.

*Do your homework*. You've probably heard the phrase "practice makes perfect" in relation to music and sports. It also applies to mathematics. You will find that the more time you spend solving mathematics problems, the easier the process becomes. Be sure to schedule enough time to complete your assignments before the next class period.

*Check your work*. Review the steps you made while working a problem. Learn to check your answers in the original problems. You may also compare your answers to the answers to selected exercises listed in the back of the book. If you have made a mistake, figure out what went wrong. Then correct your mistake. If you can't find your mistake, don't erase your work or throw it away. Bring your work to your instructor, a tutor in a math lab, or a classmate. Someone can help you find where you had trouble only if they have your work to look at.

*Learn from your mistakes.* Everyone, even your instructor, makes mistakes. (That definitely includes me—Elayn Martin-Gay. You usually don't see my mistakes because many other people double-check my work in this text. If I make a mistake on a videotape, it is edited out so that you are not confused by it.) Use your mistakes to learn and to become a better math student. The key is finding and understanding your mistakes. Was your mistake a careless mistake or did you make it because you can't read your own "math" writing? If so, try to work more slowly or write more neatly and make a conscious effort to carefully check your work. Did you make a mistake because you don't understand a concept? Take the time to review the concept or ask questions to better understand the concept.

*Know how to get help if you need it.* It's OK to ask for help. In fact, it's a good idea to ask for help whenever there is something that you don't understand. Make sure you know when your instructor has office hours and how to find his or her office. Find out if math tutoring services are available on your campus. Check out the hours, location, and requirements of the tutoring service. Know whether videotapes or software are available and how to access these resources.

*Organize your class materials*, including homework assignments, graded quizzes and tests, and notes from your class or lab. All of these items will make valuable references throughout your course and as you study for upcoming tests and your final exam. Make sure that you can locate any of these materials when you need them.

*Read your textbook before class.* Reading a mathematics textbook is unlike entertainment reading such as reading a newspaper. Your pace will be much slower. It is helpful to have a pencil and paper with you when you read. Try to work out examples on your own as you encounter them in your text. You may also write down any questions that you want to ask in class. I know that when you read a mathematics textbook, sometimes some of the information in a section will still be unclear. But once you hear a lecture or watch a video on that section, you will understand it much more easily than if you had not read your text.

*Don't be afraid to ask questions.* From experience, I can tell you that you are not the only person in class with questions. Other students are normally grateful that someone has spoken up.

*Hand in assignments on time.* This way you can be sure that you will not lose points needlessly for being late. Show every step of a problem and be neat and organized. Also be sure that you understand which problems are assigned for homework. You can always double-check this assignment with another student in your class.

**3** There are many helpful resources that are available to you in this text. It is important that you become familiar with and use these resources. This should increase your chances for success in this course. For example:

- If you need help in a particular section, check at the beginning of the section to see what videotapes or software is available. These resources are usually available to you in a tutorial lab, resource center, or library.
- Many of the exercises in this text are referenced by an example(s). Use this referencing in case you have trouble completing an assignment from the exercise set.
- Make sure that you understand the meaning of the icons that are beside many exercises. The video icon tells you that the corresponding exercise may be viewed on the videotape that corresponds to that section. The pencil icon tells you that this exercise is a writing exercise in which you should answer in complete

sentences. The calculator icons are placed by exercises that can be worked more efficiently with the use of a scientific calculator or a graphing calculator.

- There are many opportunities at the end of each chapter to help you understand the concepts of the chapter.

**Vocabulary Check** provides a vocabulary self-check to make sure that you know the vocabulary in that chapter.

**Highlights** contain chapter summaries with examples.

**Chapter Review** contains additional exercises that are keyed to sections of the chapter.

**Chapter Test** is a sample test to help you prepare for an exam.

**Cumulative Review** is a review consisting of material from the beginning of the book to the end of the particular chapter.

4 If you have trouble completing assignments or understanding the mathematics, get help as soon as you need it! This tip is presented as an objective on its own because it is *so* important. In mathematics, usually the material presented in one section builds on your understanding of the previous section. What does this mean? It means that if you don't understand the concepts covered during a class period, there is a good chance that you will not understand the concepts covered during the next class period. If this happens to you, get help as soon as you can.

Where can you get help? Many suggestions have been made in this section on where to get help, and now it is up to you to do it. Try your instructor, a tutoring center, or math lab, or you may want to form a study group with fellow classmates. If you do decide to see your instructor or go to a tutoring center, make sure that you have a neat notebook and be ready with your questions.

5 Make sure that you allow yourself plenty of time to prepare for a test. If you think that you are a little "math anxious," it may be that you are not preparing for a test in a way that will insure success. The way that you prepare for a test in mathematics is important. To prepare for a test,

1. Review your previous homework assignments.
2. Review any notes from class and section level quizzes you may have taken. (If this is a final exam, also review chapter tests you have taken.)
3. Review concepts and definitions by reading the Highlights at the end of each chapter.
4. Practice working exercises by completing the Chapter Review found at the end of each chapter. (If this is a final exam, work a Cumulative Review. There is one found at the end of each chapter (except Chapter 1). Choose the review found at the end of the latest chapter that you have covered in your course.) **Don't stop here!**
5. It is important that you place yourself in conditions similar to test conditions to see how you will perform. In other words, once you feel that you know the material, get out a few blank sheets of paper and take a sample test. There is a Chapter Test available at the end of each chapter, or you can work selected problems from the Chapter Review, or your instructor may provide you with a review sheet. During this sample test, do not use your notes or your textbook. Then check your sample test. If you are not satisfied with the results, study the areas that you are weak in and try again.

6. On the day of the test, allow yourself plenty of time to arrive to where you will be taking your exam.

When taking your test,

1. Read the directions on the test carefully.
2. Read each problem carefully as you take your test. Make sure that you answer the question asked.
3. Watch your time and pace yourself so that you may attempt each problem on your test.
4. If you have time, check your work and answers.
5. Do not turn your test in early. If you have extra time, spend it double-checking your work.

6 As a college student, you know the demands that classes, homework, work, and family place on your time. Some days you probably wonder how you'll ever get everything done. One key to managing your time is developing a schedule. Here are some hints for making a schedule:

1. Make a list of all of your weekly commitments for the term. Include classes, work, regular meetings, extracurricular activities, etc. You may also find it helpful to list such things as doing laundry, regular workouts, grocery shopping, etc.
2. Next, estimate the time needed for each item on the list. Also make a note of how often you will need to do each item. Don't forget to include time estimates for reading, studying, and homework you do outside of your classes. You may want to ask your instructor for help estimating the time needed for this item.
3. In the exercise set below, you are asked to block out a typical week on the schedule grid given. Start with items with fixed time slots, like classes and work.
4. Next, include the items on your list with flexible time slots. Think carefully about how best to schedule some items such as study time.
5. Don't fill up every time slot on the schedule. Remember that you need to allow time for eating, sleeping, and relaxing! You should also allow a little extra time in case things take longer than planned.
6. If you find that your weekly schedule is too full for you to handle, you may need to make some changes in your workload, class load, or in other areas of your life. You may want to talk to your advisor, manager or supervisor at work, or someone in your college's academic counseling center for help with such decisions.

## *Exercise Set 1.1*

**1.** What is your instructor's name?

**2.** What are your instructor's office location and office hours?

**3.** What is the best way to contact your instructor?

**4.** What does this icon  mean?

**5.** What does this icon  mean?

**6.** Do you have the name and contact information of at least one other student in class?

**7.** Will your instructor allow you to use a calculator in this class?

**8.** Are videotapes and/or tutorial software available to you?
**9.** Is there a tutoring service available? If so, what are its hours?
**10.** Have you attempted this course before? If so, write down ways that you may improve your chances of success during this attempt.
**11.** List some steps that you may take in case you begin having trouble understanding the material or completing an assignment.
**12.** Read or reread objective 6 and fill out the schedule grid below.

| | Monday | Tuesday | Wednesday | Thursday | Friday | Saturday | Sunday |
|---|---|---|---|---|---|---|---|
| 7:00 a.m. | | | | | | | |
| 8:00 a.m. | | | | | | | |
| 9:00 a.m. | | | | | | | |
| 10:00 a.m. | | | | | | | |
| 11:00 a.m. | | | | | | | |
| 12:00 p.m. | | | | | | | |
| 1:00 p.m. | | | | | | | |
| 2:00 p.m. | | | | | | | |
| 3:00 p.m. | | | | | | | |
| 4:00 p.m. | | | | | | | |
| 5:00 p.m. | | | | | | | |
| 6:00 p.m. | | | | | | | |
| 7:00 p.m. | | | | | | | |
| 8:00 p.m. | | | | | | | |
| 9:00 p.m. | | | | | | | |

## 1.2 ALGEBRAIC EXPRESSIONS AND SETS OF NUMBERS

CD-ROM SSM SSG Video

**OBJECTIVES**

1. Identify and evaluate algebraic expressions.
2. Identify natural numbers, whole numbers, integers, and rational and irrational real numbers.
3. Find the absolute value of a number.
4. Find the opposite of a number.
5. Write phrases as algebraic expressions.

1 Recall that letters that represent numbers are called **variables.** An **algebraic expression** is formed by numbers and variables connected by the operations of addition, subtraction, multiplication, division, raising to powers, and/or taking roots. For example,

$$2x + 3, \quad \frac{x+5}{6} - \frac{z^2}{y^2}, \quad \text{and} \quad \sqrt{y} - 1.6$$

are algebraic expressions or, more simply, expressions.

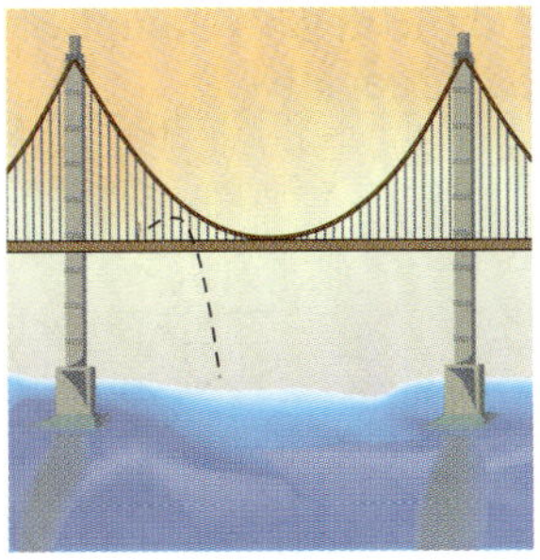

Algebraic expressions occur often during problem solving. For example, suppose that a television commercial for a watch is being filmed on the Golden Gate Bridge. A portion of this commercial consists of dropping a watch from the bridge. In order to determine the best camera angles and also whether the watch will survive the fall, it is important to know the speed of the watch at 1-second intervals. The algebraic expression

$$32t$$

gives the speed of the watch in feet per second for time $t$.

To find the speed of the watch at 1 second, for example, we replace the variable $t$ with 1 and perform the indicated multiplication. This process is called **evaluating** an expression, and the result is called the **value** of the expression for the given replacement value.

**HELPFUL HINT**
Recall that $32t$ means $32 \cdot t$.

When $t = 1$ second, $32t = 32 \cdot 1 = 32$ feet per second.
When $t = 2$ seconds, $32t = 32 \cdot 2 = 64$ feet per second.
When $t = 3$ seconds, $32t = 32 \cdot 3 = 96$ feet per second.

**Example 1** FINDING THE AREA OF A TILE

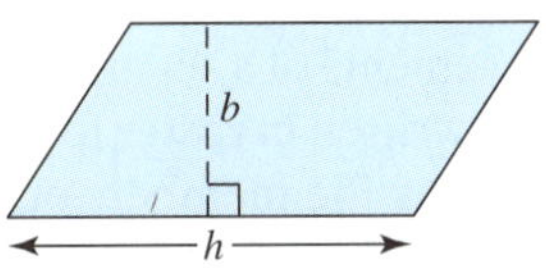


The research department of a flooring company is considering a new flooring design that contains parallelograms. The area of a parallelogram with base $b$ and height $h$ is $bh$. Find the area of a parallelogram with base 10 centimeters and height 8.2 centimeters.

*Solution* We replace $b$ with 10 and $h$ with 8.2 in the algebraic expression $bh$.

$$bh = 10 \cdot 8.2 = 82$$

The area is 82 square centimeters.

**Example 2** Evaluate: $3x - y$ when $x = 15$ and $y = 4$.

*Solution* We replace $x$ with 15 and $y$ with 4 in the expression.

$$3x - y = 3 \cdot 15 - 4 = 45 - 4 = 41$$

When evaluating an expression to solve a problem, we often need to think about the kind of number that is appropriate for the solution. For example, if we are asked to determine the maximum number of parking spaces for a parking lot to be constructed, an answer of $98\frac{1}{10}$ is not appropriate because $\frac{1}{10}$ of a parking space is not realistic.

2 Let's review some common sets of numbers and their graphs on a number line. To construct a number line, we draw a line and label a point 0 with which we associate the number 0. This point is called the **origin**. Choose a point to the right of 0 and label it 1. The distance from 0 to 1 is called the **unit distance** and can be used to locate more

points. The **positive numbers** lie to the right of the origin, and the **negative numbers** lie to the left of the origin. The number 0 is neither positive nor negative.

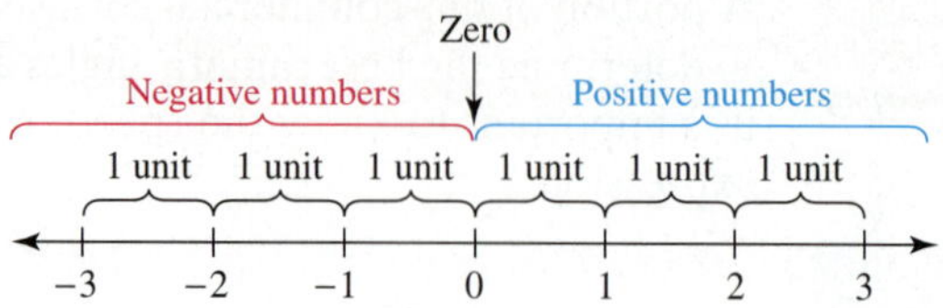


A number is **graphed** on a number line by shading the point on the number line that corresponds to the number. Some common sets of numbers and their graphs include:

***Natural numbers:*** $\{1, 2, 3, \dots\}$

***Whole numbers:*** $\{0, 1, 2, 3, \dots\}$

***Integers:*** $\{\dots, -3, -2, -1, 0, 1, 2, 3, \dots\}$

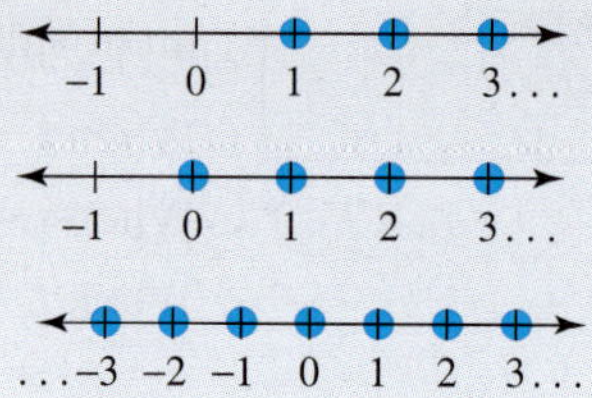


Each listing of three dots above, ..., is called an **ellipsis** and means to continue in the same pattern.

The members of a set are called its **elements**. When the elements of a set are listed, such as those displayed in the previous paragraph, the set is written in **roster** form. A set can also be written in **set builder notation**, which describes the members of a set but does not list them. The following set is written in set builder notation.

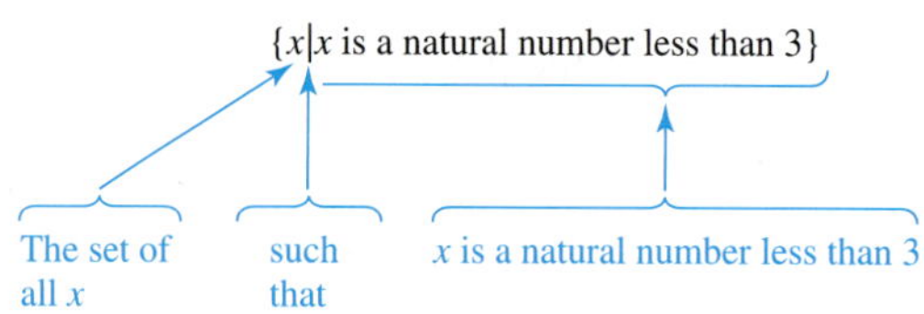


This same set written in roster form is $\{1, 2\}$.

A set that contains *no* elements is called the **empty set**, symbolized by { }, or the **null set**, symbolized by $\varnothing$.

$$\{x \mid x \text{ is a month with 32 days}\} \text{ is } \varnothing \text{ or } \{\ \ \}$$

because no month has 32 days. The set has no elements.

**HELPFUL HINT**

Use { } or $\varnothing$ to write the empty set. $\{\varnothing\}$ is **not** the empty set because it has one element: $\varnothing$.

**Example 3** List the elements in each set.

**a.** $\{x \mid x \text{ is a whole number between 1 and 6}\}$
**b.** $\{x \mid x \text{ is a natural number greater than 100}\}$

*Solution* **a.** $\{2, 3, 4, 5\}$ **b.** $\{101, 102, 103, \dots\}$

The symbol $\in$ is used to denote that an element is in a particular set. The symbol $\in$ is read as "is an element of." For example, the true statement

$$3 \text{ is an element of } \{1, 2, 3, 4, 5\}$$

can be written in symbols as

$$3 \in \{1, 2, 3, 4, 5\}$$

The symbol $\notin$ is read as "is not an element of." In symbols, we write the true statement "$p$ is not an element of $\{a, 5, g, j, q\}$" as

$$p \notin \{a, 5, g, j, q\}$$

**Example 4** Determine whether each statement is true or false.

**a.** $3 \in \{x \mid x \text{ is a natural number}\}$ **b.** $7 \notin \{1, 2, 3\}$

*Solution* **a.** True, since 3 is a natural number and therefore an element of the set.
**b.** True, since 7 is not an element of the set $\{1, 2, 3\}$.

We can use set builder notation to describe three other common sets of numbers.

***Real numbers:*** $\{x \mid x$ corresponds to a point on the number line$\}$

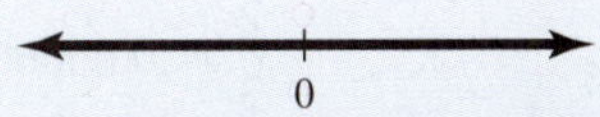


***Rational numbers:*** $\left\{\frac{a}{b} \,\middle|\, a \text{ and } b \text{ are integers and } b \neq 0\right\}$

***Irrational numbers:*** $\{x \mid x$ is a real number and $x$ is not a rational number$\}$

**HELPFUL HINT**
Notice from the definition that all real numbers are either rational or irrational.

Also notice that every integer is also a rational number since each integer can be written as the quotient of itself and 1:

$$3 = \frac{3}{1}, \quad 0 = \frac{0}{1}, \quad -8 = \frac{-8}{1}$$

Not every rational number, however, is an integer. The rational number $\frac{2}{3}$, for example, is not an integer. Some square roots are rational numbers and some are irrational numbers. For example, $\sqrt{2}$, $\sqrt{3}$ and $\sqrt{7}$ are irrational numbers while $\sqrt{25}$ is a rational number because $\sqrt{25} = 5 = \frac{5}{1}$. The number $\pi$ is an irrational number. To help you make the distinction between rational and irrational numbers, here are a few examples of each.

| ***Rational Numbers*** | | ***Irrational Numbers*** |
|---|---|---|
| ***Numbers*** | ***Equivalent Quotient of Integers,*** $\frac{a}{b}$ | |
| $-\frac{2}{3}$ | $\frac{-2}{3}$ or $\frac{2}{-3}$ | $\sqrt{5}$ |
| $\sqrt{36}$ | $\frac{6}{1}$ | $\frac{\sqrt{6}}{7}$ |
| 5 | $\frac{5}{1}$ | $-\sqrt{13}$ |
| 0 | $\frac{0}{1}$ | $\pi$ |
| 1.2 | $\frac{12}{10}$ | $\frac{2}{\sqrt{3}}$ |
| $3\frac{7}{8}$ | $\frac{31}{8}$ | |

Every rational number can be written as a decimal that either repeats or terminates. For example,

$$\frac{1}{2} = 0.5 \qquad \frac{5}{4} = 1.25$$

$$\frac{2}{3} = 0.6666666\ldots = 0.\overline{6} \qquad \frac{1}{11} = 0.090909\ldots = 0.\overline{09}$$

An irrational number written as a decimal neither terminates nor repeats. When we perform calculations with irrational numbers, we often use decimal approximations that have been rounded. For example, consider the following irrational numbers along with a 4-decimal-place approximations of each:

$$\pi \approx 3.1416 \qquad \sqrt{2} \approx 1.4142$$

Earlier we mentioned that every integer is also a rational number. In other words, all the elements of the set of integers are also elements of the set of rational numbers. When this happens, we say that the set of integers, set $Z$, is a subset of the set of rational numbers, set $I$. In symbols,

$$Z \underbrace{\subseteq}_{\text{is a subset of}} I$$

The natural numbers, whole numbers, integers, rational numbers, and irrational numbers are each a subset of the set of real numbers. The relationships among these sets of numbers are shown in the following diagram.

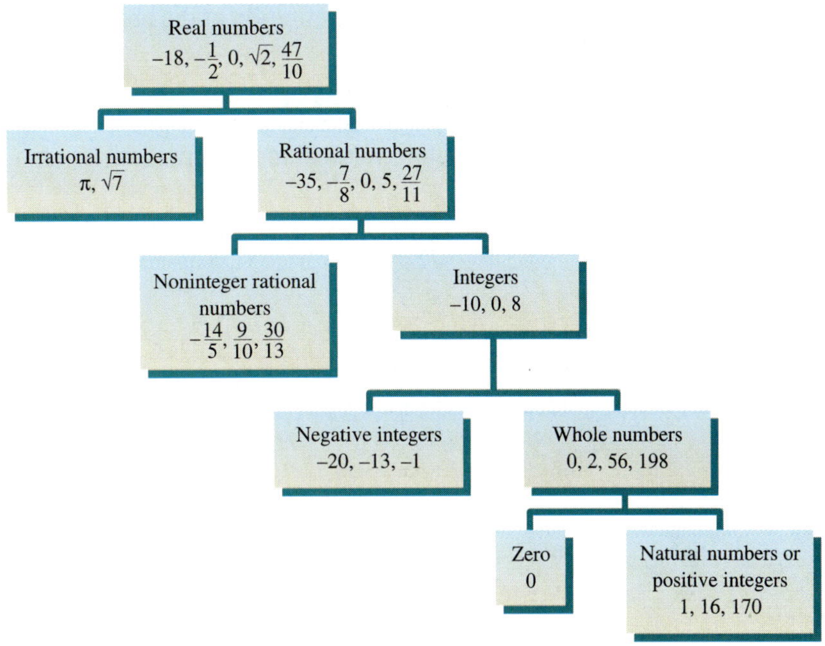


**Example 5** Determine whether the following statements are true or false.

**a.** 3 is a real number. **b.** $\frac{1}{5}$ is an irrational number.

**c.** Every rational number is an integer. **d.** $\{1, 5\} \subseteq \{2, 3, 4, 5\}$

*Solution* **a.** True. Every whole number is a real number.

**b.** False. The number $\frac{1}{5}$ is a rational number, since it is in the form $\frac{a}{b}$ with $a$ and $b$ integers and $b \neq 0$.

**c.** False. The number $\frac{2}{3}$, for example, is a rational number, but it is not an integer.

**d.** False since the element 1 in the first set is not an element of the second set.

3 The number line can also be used to visualize distance, which leads to the concept of absolute value. The **absolute value** of a real number $a$, written as $|a|$, is the distance between $a$ and 0 on the number line. Since distance is always positive or zero, $|a|$ is always positive or zero.

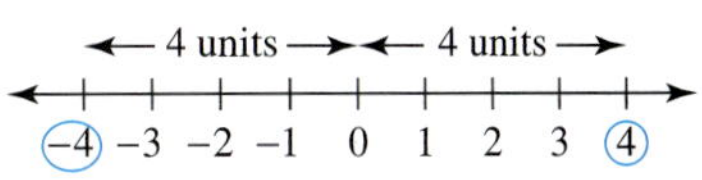


Using the number line, we see that

$$|4| = 4 \qquad \text{and also} \qquad |-4| = 4.$$

Why? Because both 4 and −4 are a distance of 4 units from 0.

An equivalent definition of the absolute value of a real number $a$ is given next.

**ABSOLUTE VALUE**

The absolute value of $a$, written as $|a|$, is

$$|a| = \begin{cases} a \text{ if } a \text{ is 0 or a positive number} \\ -a \text{ if } a \text{ is a negative number} \end{cases}$$

**Example 6** Find each absolute value.

**a.** $|3|$ **b.** $|-5|$ **c.** $-|2|$ **d.** $-|-8|$ **e.** $|0|$

*Solution*

**a.** $|3| = 3$ since 3 is located 3 units from 0 on the number line.

**b.** $|-5| = 5$ since $-5$ is 5 units from 0 on the number line.

**c.** $-|2| = -2$. The negative sign outside the absolute value bars means to take the opposite of the absolute value of 2.

**d.** $-|-8| = -8$. Since $|-8|$ is 8, $-|-8| = -8$.

**e.** $|0| = 0$ since 0 is located 0 units from 0 on the number line.

4 The number line can also help us visualize opposites. Two numbers that are the same distance from 0 on the number line but are on opposite sides of 0 are called **opposites**.

See the definition illustrated on the number lines below.

The opposite of 6 is $-6$

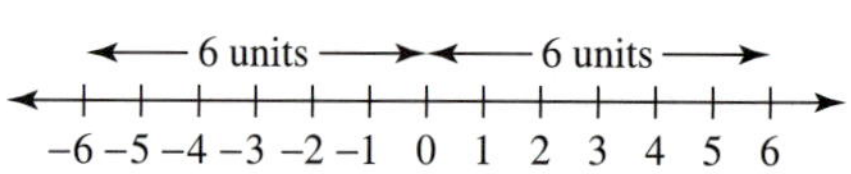


The opposite of $\frac{2}{3}$ is $-\frac{2}{3}$.

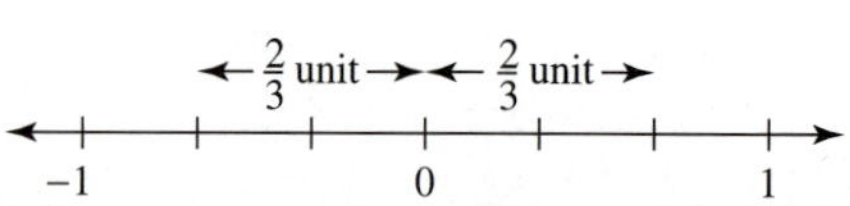


The opposite of $-4$ is 4.

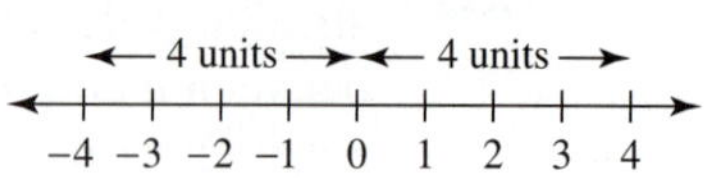


**OPPOSITE**

The opposite of a number $a$ is the number $-a$.

Above we state that the opposite of a number $a$ is $-a$. This means that the opposite of $-4$ is $-(-4)$. But from the number line above, the opposite of $-4$ is 4. This means that $-(-4) = 4$, and in general, we have the following property.

**HELPFUL HINT**
The opposite of 0 is 0.

**DOUBLE NEGATIVE PROPERTY**

For every real number $a$, $-(-a) = a$.

**Example 7** Write the opposite of each.

**a.** 8 **b.** $\frac{1}{5}$ **c.** $-9.6$

*Solution* **a.** The opposite of 8 is $-8$.

**b.** The opposite of $\frac{1}{5}$ is $-\frac{1}{5}$.

**c.** The opposite of $-9.6$ is $-(-9.6) = 9.6$.

5 Often, solving problems involves translating a phrase to an algebraic expression. The following is a list of key words and phrases and their translations.

| *Addition* | *Subtraction* | *Multiplication* | *Division* |
|---|---|---|---|
| sum | difference of | product | quotient |
| plus | minus | times | divide |
| added to | subtracted from | multiply | into |
| more than | less than | twice | ratio |
| increased by | decreased by | of | |
| total | less | | |

**Example 8** Translate each phrase to an algebraic expression. Use the variable $x$ to represent each unknown number.

**a.** Eight times a number

**b.** Three more than eight times a number

**c.** The quotient of a number and $-7$

**d.** One and six-tenths subtracted from twice a number

*Solution* **a.** $8 \cdot x$ or $8x$

**b.** $8x + 3$

**c.** $x \div -7$ or $\frac{x}{-7}$

**d.** $2x - 1.6$

## SPOTLIGHT ON DECISION MAKING

Suppose you work for an auto insurance company. Your company has just announced a new partnership with an automobile club that will allow club members to receive a 5% discount on their auto insurance. In addition, any auto club members who are safe drivers (rated 1 or 2 on the safety scale) will receive an additional 10% safe-driver discount. Your supervisor has asked you to compile a master mailing list of all current insurance clients who already belong to the auto club so they may be notified of their eligibility for the 5% discount. You must also identify the subset of auto club members who should receive a separate mailing about the additional 10% discount.

Using the given list of current insurance clients, decide who should be included in the master mailing list for notification of the 5% discount. Which of these clients should also be sent information about the 10% safe-driver discount?

CURRENT CLIENT DATABASE

| *Client Name* | *Age (years)* | *Safety Rating (1-5)* | *Airbags (0 = no, 1 = yes)* | *Annual Mileage (miles)* | *Auto Club (0 = no, 1 = yes)* |
|---|---|---|---|---|---|
| Alvarez, Wendy | 29 | 1 | 1 | 5000 | 1 |
| Brown, Keisha | 19 | 2 | 0 | 5000 | 0 |
| Cardoni, Anthony | 43 | 1 | 0 | 10,000 | 0 |
| Darden, Clay | 35 | 3 | 1 | 7500 | 1 |
| Evans, Gabriella | 26 | 2 | 1 | 5000 | 1 |
| Fonteneau, Monique | 38 | 4 | 1 | 7500 | 0 |
| Greenberg, Ira | 49 | 1 | 0 | 5000 | 1 |
| Hakkinen, Mika | 31 | 1 | 1 | 15,000 | 0 |
| Issacson, Maude | 55 | 2 | 0 | 2000 | 1 |
| Jones, Harold | 47 | 1 | 0 | 7500 | 1 |
| Khalosef, Avi | 52 | 3 | 1 | 5000 | 1 |
| Lee, Feng | 33 | 2 | 1 | 10,000 | 0 |
| Martinez, Ricardo | 25 | 4 | 0 | 5000 | 1 |
| Nunn, Destiny | 21 | 4 | 1 | 5000 | 0 |

## Exercise Set 1.2

*Find the value of each algebraic expression at the given replacement values. See Examples 1 and 2.*

**1.** $5x$ when $x = 7$

**2.** $3y$ when $y = 45$

**3.** $9.8z$ when $z = 3.1$

**4.** $7.1a$ when $a = 1.5$

**5.** $ab$ when $a = \frac{1}{2}$ and $b = \frac{3}{4}$

**6.** $yz$ when $y = \frac{2}{3}$ and $z = \frac{1}{5}$

**7.** $3x + y$ when $x = 6$ and $y = 4$

**8.** $2a - b$ when $a = 12$ and $b = 7$

**9.** The aircraft B737-400 flies an average speed of 400 miles per hour.

The expression

$$400t$$

gives the distance traveled by the aircraft in $t$ hours. Find the distance traveled by the B737-400 in 5 hours.

**10.** The algebraic expression $1.5x$ gives the total length of shelf space needed in inches for $x$ encyclopedias. Find the length of shelf space needed for a set of 30 encyclopedias.

△ **11.** Employees at Wal-Mart constantly reorganize and reshelve merchandise. In doing so, they calculate floor space needed for displays. The algebraic expression $l \cdot w$ gives the floor space needed in square units for a display that measures length $l$ units and width $w$ units. Calculate the floor space needed for a display whose length is 5.1 feet and whose width is 4 feet.

**12.** The algebraic expression $\frac{x}{5}$ can be used to calculate the distance in miles that you are from a flash of lightning, where $x$ is the number of seconds between the time you see a flash of lightning and the time you hear the thunder. Calculate the distance that you are from the flash of lightning if you hear the thunder 2 seconds after you see the lightning.

**13.** The B747-400 aircraft costs $7098 dollars per hour to operate. The algebraic expression

$$7098t$$

gives the total cost to operate the aircraft for $t$ hours. Find the total cost to operate the B747-400 for 5.2 hours.

**14.** Flying the SR-71A jet, Capt. Elden W. Joersz, USAF, set a record speed of 2193.16 miles per hour. At this speed, the algebraic expression $2193.16t$ gives the total distance flown in $t$ hours. Find the distance flown by the SR-71A in 1.7 hours.

*List the elements in each set. See Example 3.*

**15.** $\{x \mid x$ is a natural number less than $6\}$

**16.** $\{x \mid x$ is a natural number greater than $6\}$

**17.** $\{x \mid x$ is a natural number between 10 and $17\}$

**18.** $\{x \mid x$ is an odd natural number$\}$

**19.** $\{x \mid x$ is a whole number that is not a natural number$\}$

**20.** $\{x \mid x$ is a natural number less than $1\}$

**21.** $\{x \mid x$ is an even whole number less than $9\}$

**22.** $\{x \mid x$ is an odd whole number less than $9\}$

*Graph each set on a number line.*

**23.** $\{0, 2, 4, 6\}$

**24.** $\{-1, -2, -3\}$

**25.** $\left\{\frac{1}{2}, \frac{2}{3}\right\}$

**26.** $\{1, 3, 5, 7\}$

**27.** $\{-2, -6, -10\}$

**28.** $\left\{\frac{1}{4}, \frac{1}{3}\right\}$

**29.** In your own words, explain why the empty set is a subset of every set.

**30.** In your own words, explain why every set is a subset of itself.

*List the elements of the set* $\left\{3, 0, \sqrt{7}, \sqrt{36}, \frac{2}{5}, -134\right\}$ *that are also elements of the given set. See Example 4.*

**31.** Whole numbers

**32.** Integers

**33.** Natural numbers

**34.** Rational numbers

**35.** Irrational numbers

**36.** Real numbers

*Place* $\in$ *or* $\notin$ *in the space provided to make each statement true. See Example 4.*

**37.** $-11 \quad \{x \mid x$ is an integer$\}$

**38.** $-6 \quad \{2, 4, 6, \ldots\}$

**39.** $0 \quad \{x \mid x$ is a positive integer$\}$

**40.** $12 \quad \{1, 2, 3, \ldots\}$

**41.** $12 \quad \{1, 3, 5, \ldots\}$

**42.** $\frac{1}{2} \quad \{x \mid x$ is an irrational number$\}$

**43.** $0 \quad \{1, 2, 3, \ldots\}$

**44.** $0 \quad \{x \mid x$ is a natural number$\}$

*Determine whether each statement is true or false. See Examples 4 and 5. Use the following sets of numbers.*

$N$ = set of natural numbers
$Z$ = set of integers
$I$ = set of irrational numbers
$Q$ = set of rational numbers
$\mathbb{R}$ = set of real numbers

**45.** $Z \subseteq \mathbb{R}$ **46.** $\mathbb{R} \subseteq N$

**47.** $-1 \in Z$ **48.** $\frac{1}{2} \in Q$

**49.** $0 \in N$ **50.** $Z \subseteq Q$

**51.** $\sqrt{5} \notin I$ **52.** $\pi \notin \mathbb{R}$

**53.** $N \subseteq Z$ **54.** $I \subseteq N$

**55.** $\mathbb{R} \subseteq Q$ **56.** $N \subseteq Q$

**57.** In your own words, explain why every natural number is also a rational number but not every rational number is a natural number.

**58.** In your own words, explain why every irrational number is a real number but not every real number is an irrational number.

*Find each absolute value. See Example 6.*

**59.** $-|2|$ **60.** $|8|$

**61.** $|-4|$ **62.** $|-6|$

**63.** $|0|$ **64.** $|-1|$

**65.** $-|-3|$ **66.** $-|-11|$

**67.** Explain why $-(-2)$ and $-|-2|$ simplify to different numbers.

**68.** The boxed definition of absolute value states that $|a| = -a$ if $a$ is a negative number. Explain why $|a|$ is always nonnegative, even though $|a| = -a$ for negative values of $a$.

*Write the opposite of each number. See Example 7.*

**69.** $-6.2$ **70.** $-7.8$

**71.** $\frac{4}{7}$ **72.** $\frac{9}{5}$

**73.** $-\frac{2}{3}$ **74.** $-\frac{14}{3}$

**75.** 0 **76.** 10.3

*Write each phrase as an algebraic expression. Use the variable x to represent each unknown number. See Example 8.*

**77.** Twice a number. **78.** Six times a number.

**79.** Five more than twice a number.

**80.** One more than six times a number.

**81.** Ten less than a number.

**82.** A number minus seven.

**83.** The sum of a number and two.

**84.** The difference of twenty-five and a number.

**85.** A number divided by eleven.

**86.** The quotient of twice a number and thirteen.

**87.** Twelve added to three times a number.

**88.** Four subtracted from a number.

**89.** Seventeen subtracted from a number.

**90.** Four subtracted from three times a number.

**91.** Twice the sum of a number and three.

**92.** The quotient of four and the sum of a number and one.

**93.** The quotient of five and the difference of four and a number.

**94.** Eight times the difference of a number and 9.

**95.** The following bar graph shows the top five countries with the projected number of tourists visiting in 2020.

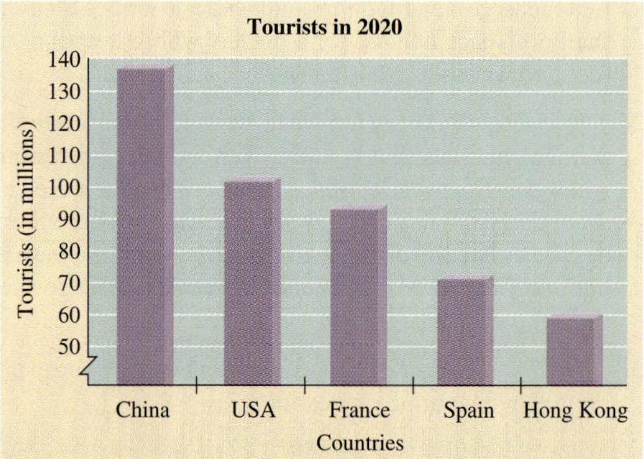


*Source:* Dan Smith, *The State of the World Atlas.*

Use the height of each bar to estimate the millions of tourists for each country.

| | |
|---|---|
| China | |
| USA | |
| France | |
| Spain | |
| Hong Kong | |

## 1.3 OPERATIONS ON REAL NUMBERS

CD-ROM SSM SSG Video

**OBJECTIVES**

1. Add and subtract real numbers.
2. Multiply and divide real numbers.
3. Simplify expressions containing exponents.
4. Find roots of numbers.
5. Use the order of operations.
6. Evaluate algebraic expressions.

1 When solving problems, we often have to add real numbers. For example, if the New Orleans Saints lose 5 yards in one play, then lose another 7 yards in the next play, their total loss may be described by $-5 + (-7)$.

The addition of two real numbers may be summarized by the following.

### ADDING REAL NUMBERS

1. To add two numbers with the *same sign*, add their absolute values and attach their common sign.
2. To add two numbers with *different signs*, subtract the smaller absolute value from the larger absolute value and attach the sign of the number with the larger absolute value.

For example, to add $-5 + (-7)$, first add their absolute values.

$$|-5| = 5, |-7| = 7, \quad \text{and} \quad 5 + 7 = 12$$

Next, attach their common negative sign.

$$-5 + (-7) = -12$$

(This represents a total loss of 12 yards for the New Orleans Saints in the example above.)

To find $(-4) + 3$, first subtract their absolute values.

$$|-4| = 4, |3| = 3, \quad \text{and} \quad 4 - 3 = 1$$

Next, attach the sign of the number with the larger absolute value.

$$(-4) + 3 = -1$$

**Example 1** Add.

**a.** $-3 + (-11)$ **b.** $3 + (-7)$ **c.** $-10 + 15$

**d.** $-8.3 + (-1.9)$ **e.** $-\frac{2}{3} + \frac{3}{7}$

*Solution* **a.** $-3 + (-11) = -(3 + 11) = -14$ **b.** $3 + (-7) = -4$

**c.** $-10 + 15 = 5$ **d.** $-8.3 + (-1.9) = -10.2$

**e.** $-\frac{2}{3} + \frac{3}{7} = -\frac{14}{21} + \frac{9}{21} = -\frac{5}{21}$

Subtraction of two real numbers may be defined in terms of addition.

**SUBTRACTING REAL NUMBERS**

If $a$ and $b$ are real numbers,

$$a - b = a + (-b)$$

In other words, to subtract a real number, we add its opposite.

**Example 2** Subtract.

**a.** $2 - 8$ **b.** $-8 - (-1)$ **c.** $-11 - 5$ **d.** $10.7 - (-9.8)$

**e.** $\frac{2}{3} - \frac{1}{2}$ **f.** $1 - 0.06$ **g.** Subtract 7 from 4.

*Solution*

Add the opposite

**a.** $2 - 8 = 2 + (-8) = -6$

Add the opposite

**b.** $-8 - (-1) = -8 + (1) = -7$

**c.** $-11 - 5 = -11 + (-5) = -16$ **d.** $10.7 - (-9.8) = 10.7 + 9.8 = 20.5$

**e.** $\frac{2}{3} - \frac{1}{2} = \frac{2 \cdot 2}{3 \cdot 2} - \frac{1 \cdot 3}{2 \cdot 3} = \frac{4}{6} + \left(-\frac{3}{6}\right) = \frac{1}{6}$

**f.** $1 - 0.06 = 1 + (-0.06) = 0.94$ **g.** $4 - 7 = 4 + (-7) = -3$

To add or subtract three or more real numbers, add or subtract from left to right.

**Example 3** Simplify the following expressions.

**a.** $11 + 2 - 7$ **b.** $-5 - 4 + 2$

*Solution* **a.** $11 + 2 - 7 = 13 - 7 = 6$ **b.** $-5 - 4 + 2 = -9 + 2 = -7$

**2** In order to discover sign patterns when you multiply real numbers, recall that multiplication by a positive integer is the same as repeated addition. For example,

$$3(2) = 2 + 2 + 2 = 6$$
$$3(-2) = (-2) + (-2) + (-2) = -6$$

Notice here that $3(-2) = -6$. This illustrates that the product of two numbers with different signs is negative. We summarize sign patterns for multiplying any two real numbers as follows.

**MULTIPLYING TWO REAL NUMBERS**

The product of two numbers with the same sign is positive.
The product of two numbers with different signs is negative.

Also recall that the product of zero and any real number is zero.

$$0 \cdot a = 0$$

**Example 4** Multiply.

**a.** $(-8)(-1)$ **b.** $(-2)\frac{1}{6}$ **c.** $3(-3)$ **d.** $0(11)$

**e.** $\left(\frac{1}{5}\right)\left(-\frac{10}{11}\right)$ **f.** $(7)(1)(-2)(-3)$ **g.** $8(-2)(0)$

*Solution*

**a.** Since the signs of the two numbers are the same, the product is positive. Thus $(-8)(-1) = +8$, or 8.

**b.** Since the signs of the two numbers are different or unlike, the product is negative. Thus $(-2)\frac{1}{6} = -\frac{2}{6} = -\frac{1}{3}$.

**c.** $3(-3) = -9$

**d.** $0(11) = 0$

**e.** $\left(\frac{1}{5}\right)\left(-\frac{10}{11}\right) = -\frac{10}{55} = -\frac{2}{11}$

**f.** To multiply three or more real numbers, multiply from left to right.

$$\begin{aligned}(7)(1)(-2)(-3) &= 7(-2)(-3)\\ &= -14(-3)\\ &= 42\end{aligned}$$

**g.** Since zero is a factor, the product is zero.

$$(8)(-2)(0) = 0$$

**HELPFUL HINT**

The following sign patterns may be helpful when we are multiplying.

1. An odd number of negative factors gives a negative product.
2. An even number of negative factors gives a positive product.

Recall that $\frac{8}{4} = 2$ because $2 \cdot 4 = 8$. Likewise, $\frac{8}{-4} = -2$ because $(-2)(-4) = 8$. Also, $\frac{-8}{4} = -2$ because $(-2)4 = -8$, and $\frac{-8}{-4} = 2$ because $2(-4) = -8$. From these examples, we can see that the sign patterns for division are the same as for multiplication.

**DIVIDING TWO REAL NUMBERS**

The quotient of two numbers with the same sign is positive.
The quotient of two numbers with different signs is negative.

Also recall that division by a nonzero real number $b$ is the same as multiplication by $\frac{1}{b}$. In other words,

$$\frac{a}{b} = a \cdot \frac{1}{b}$$

This means that to simplify $\frac{a}{b}$, we can divide by $b$ or multiply by $\frac{1}{b}$. The nonzero numbers $b$ and $\frac{1}{b}$ are called **reciprocals**. Notice that $b$ *must* be a nonzero number. We do not define division by 0. For example, $5 \div 0$, or $\frac{5}{0}$, is undefined. To see why, recall that if $5 \div 0 = n$, a number, then $n \cdot 0 = 5$. This is not possible since $n \cdot 0 = 0$ for any number $n$, and never 5. Thus far we have learned that we cannot divide 5 or any other nonzero number by 0.

Can we divide 0 by 0? By the same reasoning, if $0 \div 0 = n$, a number, then $n \cdot 0 = 0$. This is true for any number $n$ so that the quotient $0 \div 0$ would not be a single number. To avoid this, we say that

Division by 0 is undefined.

**Example 5** Divide.

**a.** $\frac{20}{-4}$ **b.** $\frac{-9}{-3}$ **c.** $-\frac{3}{8} \div 3$ **d.** $\frac{-40}{10}$ **e.** $\frac{-1}{10} \div \frac{-2}{5}$ **f.** $\frac{8}{0}$

*Solution* **a.** Since the signs are different or unlike, the quotient is negative and $\frac{20}{-4} = -5$.

**b.** Since the signs are the same, the quotient is positive and $\frac{-9}{-3} = 3$.

**c.** $-\frac{3}{8} \div 3 = -\frac{3}{8} \cdot \frac{1}{3} = -\frac{1}{8}$

**d.** $\frac{-40}{10} = -4$

**e.** $\frac{-1}{10} \div \frac{-2}{5} = -\frac{1}{10} \cdot -\frac{5}{2} = \frac{1}{4}$

**f.** $\frac{8}{0}$ is undefined.

With sign rules for division, we can understand why the positioning of the negative sign in a fraction does not change the value of the fraction. For example,

$$\frac{-12}{3} = -4, \quad \frac{12}{-3} = -4, \quad \text{and} \quad -\frac{12}{3} = -4$$

Since all the fractions equal −4, we can say that

$$\frac{-12}{3} = \frac{12}{-3} = -\frac{12}{3}$$

In general, the following holds true.

If $a$ and $b$ are real numbers and $b \neq 0$, then $\frac{a}{-b} = \frac{-a}{b} = -\frac{a}{b}$.

3 Recall that when two numbers are multiplied, they are called **factors**. For example, in $3 \cdot 5 = 15$, the 3 and 5 are called factors.

A natural number *exponent* is a shorthand notation for repeated multiplication of the same factor. This repeated factor is called the **base**, and the number of times it is used as a factor is indicated by the **exponent**. For example,

$$\underset{\text{base}}{4}^{\overset{\text{exponent}}{3}} = \underbrace{4 \cdot 4 \cdot 4}_{4 \text{ is a factor } 3 \text{ times}} = 64$$

Also,

$$\underset{\text{base}}{2}^{\overset{\text{exponent}}{5}} = \underbrace{2 \cdot 2 \cdot 2 \cdot 2 \cdot 2}_{2 \text{ is a factor } 5 \text{ times}} = 32$$

**EXPONENTS**

If $a$ is a real number and $n$ is a natural number, then the ***n*th power of *a***, or ***a* raised to the *n*th power**, written as $a^n$, is the product of $n$ factors, each of which is $a$.

$$\underset{\text{base}}{a}^{\overset{\text{exponent}}{n}} = \underbrace{a \cdot a \cdot a \cdot a \cdot \ldots \cdot a}_{a \text{ is a factor } n \text{ times}}$$

It is not necessary to write an exponent of 1. For example, 3 is assumed to be $3^1$.

**Example 6** Simplify each expression.

**a.** $3^2$ **b.** $\left(\frac{1}{2}\right)^4$ **c.** $-5^2$

**d.** $(-5)^2$ **e.** $-5^3$ **f.** $(-5)^3$

*Solution*

**a.** $3^2 = 3 \cdot 3 = 9$

**b.** $\left(\frac{1}{2}\right)^4 = \left(\frac{1}{2}\right)\left(\frac{1}{2}\right)\left(\frac{1}{2}\right)\left(\frac{1}{2}\right) = \frac{1}{16}$ **c.** $-5^2 = -(5 \cdot 5) = -25$

**d.** $(-5)^2 = (-5)(-5) = 25$ **e.** $-5^3 = -(5 \cdot 5 \cdot 5) = -125$

**f.** $(-5)^3 = (-5)(-5)(-5) = -125$

**HELPFUL HINT**

Be very careful when simplifying expressions such as $-5^2$ and $(-5)^2$.

$$-5^2 = -(5 \cdot 5) = -25 \quad \text{and} \quad (-5)^2 = (-5)(-5) = 25$$

Without parentheses, the base to square is 5, not −5.

4 The opposite of squaring a number is taking the **square root** of a number. For example, since the square of 4, or $4^2$, is 16, we say that a square root of 16 is 4. The notation $\sqrt{a}$ is used to denote the **positive**, or **principal, square root** of a nonnegative number $a$. We then have in symbols that

$$\sqrt{16} = 4$$

**Example 7** Find the square roots.

**a.** $\sqrt{9}$ **b.** $\sqrt{25}$ **c.** $\sqrt{\frac{1}{4}}$

*Solution*

**a.** $\sqrt{9} = 3$ since 3 is positive and $3^2 = 9$.

**b.** $\sqrt{25} = 5$ since $5^2 = 25$.

**c.** $\sqrt{\frac{1}{4}} = \frac{1}{2}$ since $\left(\frac{1}{2}\right)^2 = \frac{1}{4}$.

We can find roots other than square roots. Since 2 cubed, written as $2^3$, is 8, we say that the **cube root** of 8 is 2. This is written as

$$\sqrt[3]{8} = 2.$$

Also, since $3^4 = 81$ and 3 is positive,

$$\sqrt[4]{81} = 3.$$

**Example 8** Find the roots.

**a.** $\sqrt[3]{27}$ **b.** $\sqrt[5]{1}$ **c.** $\sqrt[4]{16}$

*Solution*

**a.** $\sqrt[3]{27} = 3$ since $3^3 = 27$.

**b.** $\sqrt[5]{1} = 1$ since $1^5 = 1$.

**c.** $\sqrt[4]{16} = 2$ since 2 is positive and $2^4 = 16$.

Of course, as mentioned in Section 1.2, not all roots simplify to rational numbers. We study radicals further in Chapter 7.

5 Expressions containing more than one operation are written to follow a particular agreed-upon **order of operations.** For example, when we write $3 + 2 \cdot 10$, we mean to multiply first, and then add.

**ORDER OF OPERATIONS**

Simplify expressions using the order that follows. If grouping symbols such as parentheses are present, simplify expressions within those first, starting with the innermost set. If fraction bars are present, simplify the numerator and denominator separately.

1. Raise to powers or take roots in order from left to right.
2. Multiply or divide in order from left to right.
3. Add or subtract in order from left to right.

**Example 9** Simplify.

**a.** $3 + 2 \cdot 10$ **b.** $2(1 - 4)^2$ **c.** $\dfrac{|-2|^3 + 1}{-7 - \sqrt{4}}$ **d.** $\dfrac{(6 + 2) - (-4)}{2 - (-3)}$

*Solution* **a.** First multiply; then add.

$$3 + 2 \cdot 10 = 3 + 20 = 23$$

**b.** $2(1 - 4)^2 = 2(-3)^2$ Simplify inside grouping symbols first.

$= 2(9)$ Write $(-3)^2$ as 9.

$= 18$ Multiply.

**c.** Simplify the numerator and the denominator separately; then divide.

$\dfrac{|-2|^3 + 1}{-7 - \sqrt{4}} = \dfrac{2^3 + 1}{-7 - 2}$ Write $|-2|$ as 2 and $\sqrt{4}$ as 2.

$= \dfrac{8 + 1}{-9}$ Write $2^3$ as 8.

$= \dfrac{9}{-9} = -1$ Simplify the numerator, then divide.

**d.** $\dfrac{(6 + 2) - (-4)}{2 - (-3)} = \dfrac{8 - (-4)}{2 - (-3)}$ Simplify inside grouping symbols first.

$= \dfrac{8 + 4}{2 + 3}$ Write subtractions as equivalent additions.

$= \dfrac{12}{5}$ Add in both the numerator and denominator.

6 Recall from Section 1.2 that an algebraic expression is formed by numbers and variables connected by the operations of addition, subtraction, multiplication, division, raising to powers, and/or taking roots. Also, if numbers are substituted for the variables in an algebraic expression and the operations performed, the result is called **the value of the expression** for the given replacement values. This entire process is called **evaluating an expression**.

**Example 10** Evaluate each algebraic expression when $x = 2$, $y = -1$, and $z = -3$.

**a.** $z - y$ **b.** $z^2$ **c.** $\frac{2x + y}{z}$

*Solution* **a.** $z - y = -3 - (-1) = -3 + 1 = -2$

**b.** $z^2 = (-3)^2 = 9$

**c.** $\frac{2x + y}{z} = \frac{2(2) + (-1)}{-3} = \frac{4 + (-1)}{-3} = \frac{3}{-3} = -1$

Sometimes variables such as $x_1$ and $x_2$ will be used in this book. The small 1 and 2 are called **subscripts.** The variable $x_1$ can be read as "$x$ sub 1," and the variable $x_2$ can be read as "$x$ sub 2." The important thing to remember is that they are two different variables. For example, if $x_1 = -5$ and $x_2 = 7$, then

$$x_1 - x_2 = -5 - 7 = -12.$$

**Example 11** The algebraic expression $\frac{5(x - 32)}{9}$ represents the equivalent temperature in degrees Celsius when $x$ is the temperature in degrees Fahrenheit. Complete the following table by evaluating this expression at the given values of $x$.

| | | | | |
|---|---|---|---|---|
| Degrees Fahrenheit | $x$ | $-4$ | 10 | 32 |
| Degrees Celsius | $\frac{5(x - 32)}{9}$ | | | |

*Solution* To complete the table, evaluate $\frac{5(x - 32)}{9}$ at each given replacement value.

When $x = -4$,

$$\frac{5(x - 32)}{9} = \frac{5(-4 - 32)}{9} = \frac{5(-36)}{9} = -20$$

When $x = 10$,

$$\frac{5(x - 32)}{9} = \frac{5(10 - 32)}{9} = \frac{5(-22)}{9} = -\frac{110}{9}$$

When $x = 32$,

$$\frac{5(x - 32)}{9} = \frac{5(32 - 32)}{9} = \frac{5 \cdot 0}{9} = 0$$

The completed table is

| | | | | |
|---|---|---|---|---|
| Degrees Fahrenheit | $x$ | $-4$ | 10 | 32 |
| Degrees Celsius | $\frac{5(x - 32)}{9}$ | $-20$ | $-\frac{110}{9}$ | 0 |

Thus, $-4$°F is equivalent to $-20$°C, 10°F is equivalent to $-\frac{110}{9}$°C, and 32°F is equivalent to 0°C.

## SPOTLIGHT ON DECISION MAKING

Suppose you are a travel agent. A tour company is offering bonuses to travel agents booking clients on selected tour packages for a limited time. However, prior to participating in this bonus program, you must select only one type of tour package for which you will receive the bonus. Information about the selected tour packages is shown in the table.

Based on client inquiries during the past week, you estimate that you could probably interest 30 clients in a cruise to Alaska, 60 in a Bermuda package, 50 in a trip to Cancun, and 40 in a Hawaii package. However, you also estimate that in each case, only half of the clients would book the trip if it cost over $1000 per person.

Which one of the tour packages would you choose for participating in the tour company's bonus program? Why?

**SELECTED TOUR PACKAGES**

| *Destination* | *Cost per person* | *Bonus per person booked* |
|---|---|---|
| Alaska cruise | $2029 | $100 |
| Bermuda | $ 699 | $ 25 |
| Cancun, Mexico | $1349 | $ 75 |
| Hawaii | $ 840 | $ 50 |

## *Exercise Set 1.3*

*Find each sum or difference. See Examples 1 through 3.*

**1.** $-3 + 8$

**2.** $-5 + (-9)$

**3.** $-14 + (-10)$

**4.** $12 + (-7)$

**5.** $-4.3 - 6.7$

**6.** $-8.2 - (-6.6)$

**7.** $13 - 17$

**8.** $15 - (-1)$

**9.** $\frac{11}{15} - \left(-\frac{3}{5}\right)$

**10.** $\frac{7}{10} - \frac{4}{5}$

**11.** $19 - 10 - 11$

**12.** $-13 - 4 + 9$

*Find each product or quotient. See Examples 4 and 5.*

**13.** $(-5)(12)$

**14.** $6(-3)$

**15.** $(-8)(-10)$

**16.** $7(0)$

**17.** $\frac{-12}{-4}$

**18.** $\frac{60}{-6}$

**19.** $\frac{0}{-2}$

**20.** $\frac{-2}{0}$

**21.** $(-4)(-2)(-1)$

**22.** $5(-3)(-2)$

**23.** $\frac{-6}{7} \div 2$

**24.** $\frac{-9}{13} \div (-3)$

**25.** $\left(-\frac{2}{7}\right)\left(-\frac{1}{6}\right)$

**26.** $\frac{5}{9}\left(-\frac{3}{5}\right)$

*Evaluate. See Example 6.*

**27.** $-7^2$

**28.** $(-7)^2$

**29.** $(-6)^2$

**30.** $-6^2$

**31.** $(-2)^3$

**32.** $-2^3$

**33.** Explain why $-3^2$ and $(-3)^2$ simplify to different numbers.

**34.** Explain why $-3^3$ and $(-3)^3$ simplify to the same number.

*Find the following roots. See Examples 7 and 8.*

**35.** $\sqrt{49}$

**36.** $\sqrt{81}$

**37.** $\sqrt{\frac{1}{9}}$

**38.** $\sqrt{\frac{1}{25}}$

**39.** $\sqrt[3]{64}$

**40.** $\sqrt[5]{32}$

**41.** $\sqrt[4]{81}$

**42.** $\sqrt[3]{1}$

*Simplify each expression. Round Exercises 61 and 62 to the nearest ten thousandth. See Example 9.*

**43.** $3(5-7)^4$

**44.** $7(3-8)^2$

**45.** $-3^2+2^3$

**46.** $-5^2-2^4$

**47.** $\dfrac{3-(-12)}{-5}$

**48.** $\dfrac{-4-(-8)}{-4}$

**49.** $|3.6-7.2|+|3.6+7.2|$

**50.** $|8.6-1.9|-|2.1+5.3|$

**51.** $\dfrac{(3-\sqrt{9})-(-5-1.3)}{-3}$

**52.** $\dfrac{-\sqrt{16}-(6-2.4)}{-2}$

**53.** $\dfrac{|3-9|-|-5|}{-3}$

**54.** $\dfrac{|-14|-|2-7|}{-15}$

**55.** $(-3)^2+2^3$

**56.** $(-15)^2-2^4$

**57.** $\dfrac{3(-2+1)}{5}-\dfrac{-7(2-4)}{1-(-2)}$

**58.** $\dfrac{-1-2}{2(-3)+10}-\dfrac{2(-5)}{-1(8)+1}$

**59.** $\dfrac{\frac{-3}{10}}{\frac{42}{50}}$

**60.** $\dfrac{\frac{-5}{21}}{\frac{-6}{42}}$

**61.** $\dfrac{-1.682-17.895}{(-7.102)(-4.691)}$

**62.** $\dfrac{(-5.161)(3.222)}{7.955-19.676}$

*Find the value of each expression when $x=-2$, $y=-5$, and $z=3$. See Example 10.*

**63.** $x^2+z^2$

**64.** $y^2-z^2$

**65.** $-5(-x+3y)$

**66.** $-7(-y-4z)$

**67.** $\dfrac{3z-y}{2x-z}$

**68.** $\dfrac{5x-z}{-2y+z}$

*Find the value of the expression when $x_1=2$, $x_2=4$, $y_1=-3$, $y_2=2$. See Example 10.*

**69.** $\dfrac{y_2-y_1}{x_2-x_1}$

**70.** $\sqrt{(x_2-x_1)^2+(y_2-y_1)^2}$

*See Example 11.*

△ **71.** The algebraic expression $8+2y$ represents the perimeter of a rectangle with width 4 and length $y$.

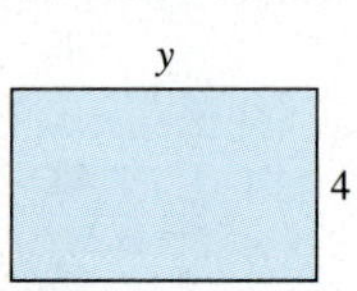


**a.** Complete the table that follows by evaluating this expression at the given values of $y$.

| Length | $y$ | 5 | 7 | 10 | 100 |
|---|---|---|---|---|---|
| Perimeter | $8+2y$ | | | | |

**b.** Use the results of the table in **a** to answer the following question. As the width of a rectangle remains the same and the length increases, does the perimeter increase or decrease? Explain how you arrived at your answer.

△ **72.** The algebraic expression $\pi r^2$ represents the area of a circle with radius $r$.

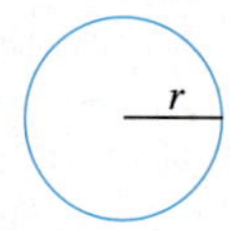


**a.** Complete the table below by evaluating this expression at given values of $r$. (Use 3.14 for $\pi$.)

| Radius | $r$ | 2 | 3 | 7 | 10 |
|---|---|---|---|---|---|
| Area | $\pi r^2$ | | | | |

**b.** As the radius of a circle increases, does its area increase or decrease? Explain your answer.

**73.** The algebraic expression $\dfrac{100x+5000}{x}$ represents the cost per bookshelf (in dollars) of producing $x$ bookshelves.

**a.** Complete the table below.

| Number of Bookshelves | $x$ | 10 | 100 | 1000 |
|---|---|---|---|---|
| Cost per Bookshelf | $\dfrac{100x+5000}{x}$ | | | |

**b.** As the number of bookshelves manufactured increases, does the cost per bookshelf increase or decrease? Why do you think that this is so?

**74.** If $c$ is degrees Celsius, the algebraic expression $1.8c+32$ represents the equivalent temperature in degrees Fahrenheit.

**a.** Complete the table below.

| Degrees Celsius | $c$ | −10 | 0 | 50 |
|---|---|---|---|---|
| Degrees Fahrenheit | $1.8c+32$ | | | |

**b.** As degrees Celsius increase, do degrees Fahrenheit increase or decrease?

*Each circle below represents a whole, or 1. Determine the unknown fractional part of each circle.*

**75.**

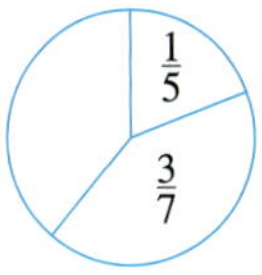


**76.**

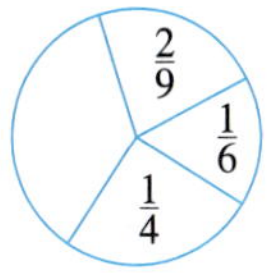


**77.** Most of Mount Kea, a volcano on Hawaii, lies below sea level. If this volcano begins at 5998 meters below sea level and then rises 10,203 meters, find the height of the volcano above sea level.

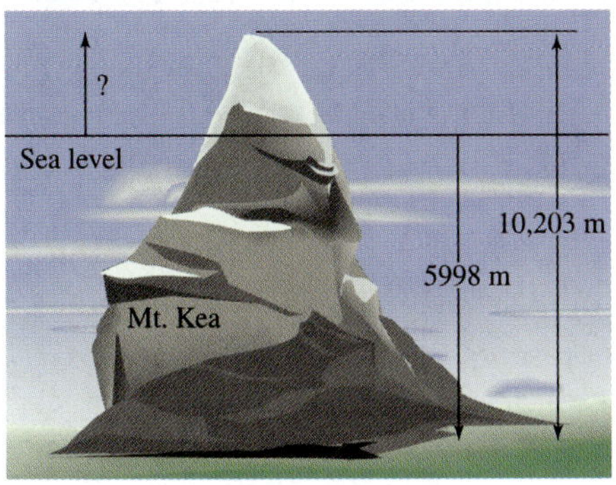


**78.** The highest point on land on Earth is the top of Mt. Everest in the Himalayas, at an elevation of 29,028 feet above sea level. The lowest point on land is the Dead Sea, between Israel and Jordan, at 1319 feet below sea level. Find the difference in elevations.

*A fair game is one in which each team or player has the same chance of winning. Suppose that a game consists of three players taking turns spinning a spinner. If the spinner lands on yellow, player 1 gets a point. If the spinner lands on red, player 2 gets a point, and if the spinner lands on blue, player 3 gets a point. After 12 spins, the player with the most points wins.*

**a.**

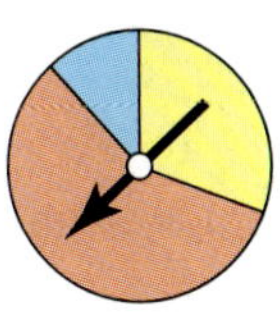

**b.**

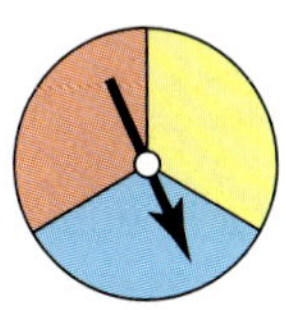

**c.**

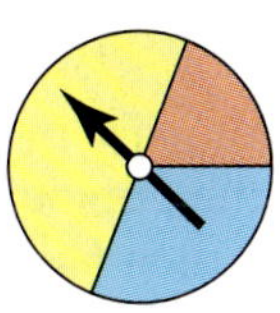

**d.**

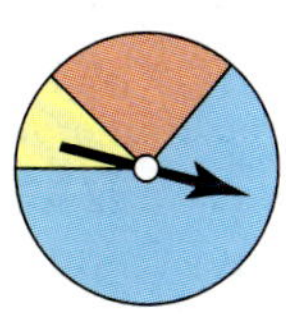

**79.** Which spinner would lead to a fair game?

**80.** If you are player 2 and want to win the game, which spinner would you choose?

**81.** If you are player 1 and want to lose the game, which spinner would you choose?

**82.** Is it possible for the game to end in a three-way tie? If so, list the possible ending scores.

**83.** Is it possible for the game to end in a two-way tie? If so, list the possible ending scores.

*Use a calculator to approximate each square root. Round to four decimal places.*

**84.** $\sqrt{10}$

**85.** $\sqrt{273}$

**86.** $\sqrt{7.9}$

**87.** $\sqrt{19.6}$

*Investment firms often advertise their gains and losses in the form of bar graphs such as the one that follows. This graph shows investment risk over time for the S&P 500 Index by showing average annual compound returns for 1 year, 5 years, 15 years, and 25 years. For example, after one year, the annual compound return in percent for an investor is anywhere from a gain of 181.5% to a loss of 64%. Use this graph to answer the questions below.*

**88.** A person investing in the S&P 500 Index may expect at most an average annual gain of what percent after 15 years?

**89.** A person investing in the S&P 500 Index may expect to lose at most an average per year of what percent after 5 years?

**90.** Find the difference in percent of the highest average annual return and the lowest average annual return after 15 years.

**91.** Find the difference in percent of the highest average annual return and the lowest average annual return after 25 years.

**92.** Do you think that the type of investment shown in the figure is recommended for short-term investments or long-term investments? Explain your answer.

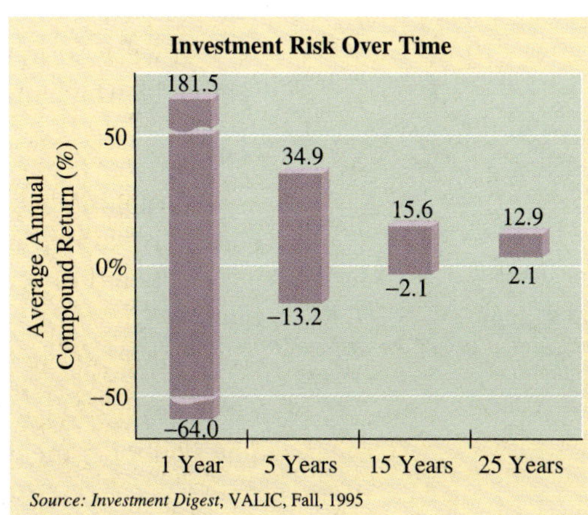

## 1.4 PROPERTIES OF REAL NUMBERS

CD-ROM SSM SSG Video

▶ **OBJECTIVES**

1. Use operation and order symbols to write mathematical sentences.
2. Identify identity numbers and inverses.
3. Identify and use the commutative, associative, and distributive properties.
4. Write algebraic expressions.
5. Simplify algebraic expressions.

**1** In Section 1.2, we used the symbol = to mean "is equal to." All of the following key words and phrases also imply equality.

**EQUALITY**

| | | | |
|---|---|---|---|
| equals | is/was | represents | is the same as |
| gives | yields | amounts to | is equal to |

**Example 1** Write each sentence using mathematical symbols.

**a.** The sum of $x$ and 5 is 20.
**b.** Two times the sum of 3 and $y$ amounts to 4.
**c.** Subtract 8 from $x$, and the difference is the same as the product of 2 and $x$.
**d.** The quotient of $z$ and 9 is 3 times the difference of $z$ and 5.

*Solution* **a.** The sum of $x$ and 5 can be written as "$x + 5$," and the word "is" means "is equal to" in this sentence, so we write $x + 5 = 20$.

**b.** $2(3 + y) = 4$

**c.** $x - 8 = 2x$

**d.** $\frac{z}{9} = 3(z - 5)$

If we want to write in symbols that two numbers are not equal, we can use the symbol ≠, which means "**is not equal to.**" For example,

$$3 \neq 2$$

Graphing two numbers on a number line gives us a way to compare two numbers. For two real numbers $a$ and $b$, we say ***a* is less than *b*** if on the number line $a$ lies to the left of $b$. Also, if $b$ is to the right of $a$ on the number line, then ***b* is greater than *a*.**

The symbol > means "**is greater than.**" Since $b$ is greater than $a$, we write

$$b > a$$

**Example 2** Insert $<$, $>$, or $=$ between each pair of numbers to form a true statement.

**a.** $-1 \quad -2$ **b.** $\frac{12}{4} \quad 3$ **c.** $-5 \quad 0$ **d.** $-3.5 \quad -3.05$

*Solution* **a.** $-1 > -2$ since $-1$ lies to the right of $-2$ on the number line.

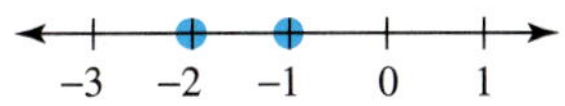


**b.** $\frac{12}{4} = 3$.

**c.** $-5 < 0$ since $-5$ lies to the left of 0 on the number line.

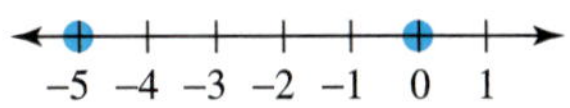


**d.** $-3.5 < -3.05$ since $-3.5$ lies to the left of $-3.05$ on the number line.

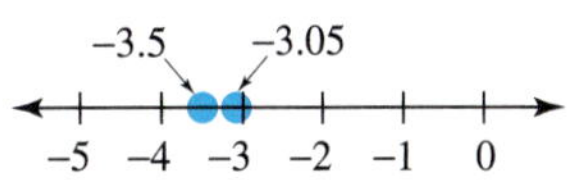


**HELPFUL HINT**

When inserting the $>$ or $<$ symbol, think of the symbols as arrowheads that "point" toward the smaller number when the statement is true.

In addition to $<$ and $>$, there are the inequality symbols $\leq$ and $\geq$. The symbol

$\leq$ means **"is less than or equal to"**

and the symbol

$\geq$ means **"is greater than or equal to"**

For example, the following are true statements.

| | | |
|---|---|---|
| $10 \leq 10$ | since | $10 = 10$ |
| $-8 \leq 13$ | since | $-8 < 13$ |
| $-5 \geq -5$ | since | $-5 = -5$ |
| $-7 \geq -9$ | since | $-7 > -9$ |

**Example 3** Write each sentence using mathematical symbols.

**a.** The sum of 5 and $y$ is greater than or equal to 7.
**b.** 11 is not equal to $z$.
**c.** 20 is less than the difference of 5 and twice $x$.

*Solution* **a.** $5 + y \geq 7$ **b.** $11 \neq z$ **c.** $20 < 5 - 2x$

2 Of all the real numbers, two of them stand out as extraordinary: 0 and 1. Zero is the only number that when *added* to any real number, the result is the same real number. Zero is thus called the **additive identity**. Also, one is the only number that when *multiplied* by any real number, the result is the same real number. One is thus called the **multiplicative identity.**

| | *Addition* | *Multiplication* |
|---|---|---|
| Identity Properties | The additive identity is 0.<br>$a + 0 = 0 + a = a$ | The multiplicative identity is 1.<br>$a \cdot 1 = 1 \cdot a = a$ |

In section 1.2, we learned that $a$ and $-a$ are opposites.

Another name for opposites is **additive inverse.** For example, the additive inverse of 3 is −3. Notice that the sum of a number and its opposite is always 0.

In section 1.3, we learned that, for a nonzero number, $b$ and $\frac{1}{b}$ are reciprocals. Another name for reciprocal is **multiplicative inverse.** For example, the multiplicative inverse of $-\frac{2}{3}$ is $-\frac{3}{2}$. Notice that the product of a number and its reciprocal is always 1.

| | | |
|---|---|---|
| Inverse Properties | For each number $a$, there is a unique number $-a$ called the **additive inverse** or **opposite** of $a$ such that<br>$a + (-a) = (-a) + a = 0$ | For each nonzero $a$, there is a unique number $\frac{1}{a}$ called the **multiplicative inverse** or **reciprocal** of $a$ such that<br>$a \cdot \frac{1}{a} = \frac{1}{a} \cdot a = 1$ |

**Example 4** Write the additive inverse, or opposite, of each.

**a.** 8 **b.** $\frac{1}{5}$ **c.** −9.6

*Solution* **a.** The opposite of 8 is −8. **b.** The opposite of $\frac{1}{5}$ is $-\frac{1}{5}$.

**c.** The opposite of −9.6 is $-(-9.6) = 9.6$.

**Example 5** Write the multiplicative inverse, or reciprocal, of each.

**a.** 11 **b.** −9 **c.** $\frac{7}{4}$

*Solution* **a.** The reciprocal of 11 is $\frac{1}{11}$.

**b.** The reciprocal of $-9$ is $-\frac{1}{9}$.

**c.** The reciprocal of $\frac{7}{4}$ is $\frac{4}{7}$ because $\frac{7}{4} \cdot \frac{4}{7} = 1$.

3 In addition to these special real numbers, all real numbers have certain properties that allow us to write equivalent expressions—that is, expressions that have the same value. These properties will be especially useful in Chapter 2 when we solve equations.

The **commutative properties** state that the order in which two real numbers are added or multiplied does not affect their sum or product.

**COMMUTATIVE PROPERTIES**

For real numbers $a$ and $b$,

***Addition*** $a + b = b + a$

***Multiplication*** $a \cdot b = b \cdot a$

The **associative properties** state that regrouping numbers that are added or multiplied does not affect their sum or product.

**ASSOCIATIVE PROPERTIES**

For real numbers $a$, $b$, and $c$,

***Addition*** $(a + b) + c = a + (b + c)$

***Multiplication*** $(a \cdot b) \cdot c = a \cdot (b \cdot c)$

**Example 6** Use the commutative property of addition to write an expression equivalent to $7x + 5$.

*Solution* $7x + 5 = 5 + 7x$.

**Example 7** Use the associative property of multiplication to write an expression equivalent to $4 \cdot (9y)$. Then simplify this equivalent expression.

*Solution* $4 \cdot (9y) = (4 \cdot 9)y = 36y$.

The **distributive property** states that multiplication distributes over addition.

**DISTRIBUTIVE PROPERTY**

For real numbers $a$, $b$, and $c$,

$$a(b + c) = ab + ac$$

**Example 8** Use the distributive property to multiply.

**a.** $3(2x + y)$ **b.** $-(3x - 1)$

*Solution* **a.** $3(2x + y) = 3 \cdot 2x + 3 \cdot y$ Apply the distributive property.

$= 6x + 3y$ Apply the associative property of multiplication.

**b.** Recall that $-(3x - 1)$ means $-1(3x - 1)$.

$$-1(3x - 1) = -1(3x) + (-1)(-1)$$

$$= -3x + 1$$

**4** As mentioned earlier, an important step in problem solving is to be able to write algebraic expressions from word phrases. Sometimes this involves a direct translation, but often an indicated operation is not directly stated but rather implied.

**Example 9** Write each as an algebraic expression.

**a.** A vending machine contains $x$ quarters. Write an expression for the *value* of the quarters.
**b.** The number of grams of fat in $x$ pieces of bread if each piece of bread contains 2 grams of fat.
**c.** The cost of $x$ desks if each desk costs \$156.
**d.** Sales tax on a purchase of $x$ dollars if the tax rate is 9%.

*Solution* Each of these examples implies finding a product.

**a.** The value of the quarters is found by multiplying the value of a quarter (0.25 dollar) by the number of quarters.

| | | | | |
|---|---|---|---|---|
| In words: | Value of a quarter | · | Number of quarters | |
| Translate: | 0.25 | · | $x$, | or $0.25x$ |

**b.**

| | | | | |
|---|---|---|---|---|
| In words: | Number of grams of fat in one piece of bread | · | Number of pieces of bread | |
| Translate: | 2 | · | $x$, | or $2x$ |

**c.**

| | | | | |
|---|---|---|---|---|
| In words: | Cost of a desk | · | Number of desks | |
| Translate: | 156 | · | $x$, | or $156x$ |

**d.** In words: Sales tax rate · purchase price

Translate: 0.09 · $x$, or $0.09x$

(Here, we wrote 9% as a decimal, 0.09.)

Two or more unknown numbers in a problem may sometimes be related. If so, try letting a variable represent one unknown number and then represent the other unknown number or numbers as expressions containing the same variable.

## Example 10

Write each as an algebraic expression.

**a.** Two numbers have a sum of 20. If one number is $x$, represent the other number as an expression in $x$.

**b.** The older sister is 8 years older than her younger sister. If the age of the younger sister is $x$, represent the age of the older sister as an expression in $x$.

△ **c.** Two angles are complementary if the sum of their measures is 90°. If the measure of one angle is $x$ degrees, represent the measure of the other angle as an expression in $x$.

**d.** If $x$ is the first of two consecutive integers, represent the second integer as an expression in $x$.

*Solution*

**a.** If two numbers have a sum of 20 and one number is $x$, the other number is "the rest of 20."

In words: Twenty minus $x$

Translate: 20 − $x$

**b.** The older sister's age is

In words: Eight years added to younger sister's age

Translate: 8 + $x$

**c.** In words: Ninety minus $x$

Translate: 90 − $x$

**d.** The next consecutive integer is always one more than the previous integer.

In words: The first integer plus one

Translate: $x$ + 1

5 Often, an expression may be **simplified** by removing grouping symbols and combining any like terms. The **terms** of an expression are the addends of the expression. For example, in the expression $3x^2 + 4x$, the terms are $3x^2$ and $4x$.

| ***Expression*** | ***Terms*** |
|---|---|
| $-2x + y$ | $-2x, y$ |
| $3x^2 - \frac{y}{5} + 7$ | $3x^2, -\frac{y}{5}, 7$ |

Terms with the same variable(s) raised to the same power are called **like terms.** We can add or subtract like terms by using the distributive property. This process is called **combining like terms.**

**Example 11** Use the distributive property to simplify each expression.

**a.** $3x - 5x + 4$ **b.** $7yz + yz$ **c.** $4z + 6.1$

*Solution* **a.** $3x - 5x + 4 = (3 - 5)x + 4$ Apply the distributive property.

$= -2x + 4$

**b.** $7yz + yz = (7 + 1)yz = 8yz$

**c.** $4z + 6.1$ cannot be simplified further since $4z$ and $6.1$ are not like terms.

Let's continue to use properties of real numbers to simplify expressions. Recall that the distributive property can also be used to multiply. For example,

$$-2(x + 3) = -2(x) + (-2)(3) = -2x - 6$$

The associative and commutative properties may sometimes be needed to rearrange and group like terms when we simplify expressions.

$$\begin{aligned} -7x^2 + 5 + 3x^2 - 2 &= -7x^2 + 3x^2 + 5 - 2 \\ &= (-7 + 3)x^2 + (5 - 2) \\ &= -4x^2 + 3 \end{aligned}$$

**Example 12** Simplify each expression.

**a.** $3xy - 2xy + 5 - 7 + xy$ **b.** $7x^2 + 3 - 5(x^2 - 4)$

**c.** $(2.1x - 5.6) - (-x - 5.3)$

*Solution* **a.** $3xy - 2xy + 5 - 7 + xy = 3xy - 2xy + xy + 5 - 7$ Apply the commutative property.

$= (3 - 2 + 1)xy + (5 - 7)$ Apply the distributive property.

$= 2xy - 2$ Simplify.

**b.** $7x^2 + 3 - 5(x^2 - 4) = 7x^2 + 3 - 5x^2 + 20$ Apply the distributive property.

$= 2x^2 + 23$ Simplify.

**c.** Think of $-(-x - 5.3)$ as $-1(-x - 5.3)$ and use the distributive property.

$$(2.1x - 5.6) - 1(-x - 5.3) = 2.1x - 5.6 + 1x + 5.3$$

$= 3.1x - 0.3$ Combine like terms.

## SPOTLIGHT ON DECISION MAKING

Suppose you are a geologist studying Hawaiian volcanoes. When lava from a volcano flows into the ocean, it heats the water around it. You are color-coding a map of ocean water temperatures around a lava flow. If the water temperature is less than or equal to 22°C, the map will be colored dark blue. If the water temperature is greater than 29°C, the map will be colored tan. Otherwise, the map will be colored green. Decide what color on the map should be used for the following temperatures.

**a.** 25°C

**b.** 18°C

**c.** 29°C

**d.** 22°C

**e.** 31°C

## *Exercise Set 1.4*

*Write each statement using mathematical symbols. See Example 1.*

**1.** The product of 4 and $c$ is 7.

**2.** The sum of 10 and $x$ is $-12$.

**3.** 3 times the sum of $x$ and 1 amounts to 7.

**4.** 9 times the difference of 4 and $m$ amounts to 1.

**5.** The quotient of $n$ and 5 is 4 times $n$.

**6.** The quotient of 8 and $y$ is 3 more than $y$.

**7.** The difference of $z$ and 2 is the same as the product of $z$ and 2.

**8.** Five added to twice $q$ is the same as 4 more than $q$.

*Insert $<$, $>$, or $=$ in the space provided to form a true statement. See Example 2.*

**9.** $0 \quad -2$

**10.** $-5 \quad 0$

**11.** $\frac{12}{3} \quad \frac{8}{2}$

**12.** $\frac{20}{5} \quad \frac{20}{4}$

**13.** $-7.9 \quad -7.09$

**14.** $-13.07 \quad -13.7$

*Write each sentence using mathematical symbols. See Example 3.*

**15.** The product of 7 and $x$ is less than or equal to $-21$.

**16.** 10 subtracted from $x$ is greater than 0.

**17.** The sum of $-2$ and $x$ is not equal to 10.

**18.** The quotient of $y$ and 3 is less than or equal to $y$.

**19.** Twice the difference of $x$ and 6 is greater than the reciprocal of 11.

**20.** Four times the sum of 5 and $x$ is not equal to the opposite of 15.

**21.** 7 subtracted from $y$ is 6.

**22.** The sum of $z$ and $w$ is 12.

**23.** Twice the difference of $x$ and 6 is $-27$

**24.** 5 times the sum of 6 and $y$ is $-35$.

*Write the opposite (additive inverse) of each number. Then write the reciprocal (multiplicative inverse) of each number if one exists. See Examples 4 and 5.*

**25.** 5

**26.** 9

**27.** $-8$

**28.** $-4$

**29.** $-\frac{1}{4}$

**30.** $\frac{1}{9}$

**31.** 0

**32.** $\frac{0}{6}$

**33.** $\frac{7}{8}$

**34.** $-\frac{23}{5}$

**35.** Name the only real number that has no reciprocal, and explain why this is so.

**36.** Name the only real number that is its own opposite, and explain why this is so.

*Use a commutative property to write an equivalent expression. See Example 6.*

**37.** $7x + y$

**38.** $3a + 2b$

**39.** $z \cdot w$

**40.** $r \cdot s$

**41.** $\frac{1}{3} \cdot \frac{x}{5}$

**42.** $\frac{x}{2} \cdot \frac{9}{10}$

**43.** Is subtraction commutative? Explain why or why not.

**44.** Is division commutative? Explain why or why not.

*Use an associative property to write an equivalent expression. See Example 7.*

**45.** $5 \cdot (7x)$

**46.** $3 \cdot (10z)$

**47.** $(x + 1.2) + y$

**48.** $5q + (2r + s)$

**49.** $(14z) \cdot y$

**50.** $(9.2x) \cdot y$

**51.** Evaluate $12 - (5 - 3)$ and $(12 - 5) - 3$. Use these two expressions and discuss whether subtraction is associative.

**52.** Evaluate $24 \div (6 \div 3)$ and $(24 \div 6) \div 3$. Use these two expressions and discuss whether division is associative.

*Use the distributive property to find the product. See Example 8.*

**53.** $3(x + 5)$

**54.** $7(y + 2)$

**55.** $-(2a + b)$

**56.** $-(c + 7d)$

**57.** $2(6x + 5y + 2z)$

**58.** $5(3a + b + 9c)$

*Write each sentence using mathematical symbols.*

**59.** 6 subtracted from twice $y$ is the reciprocal of 8.

**60.** 7 subtracted from the product of 5 and $n$ is the opposite of $n$.

**61.** The sum of $n$ and 5, divided by 2, is greater than twice $n$.

**62.** The product of 8 and $x$, divided by 5, is less than 3 more than $x$.

*Complete the statement to illustrate the given property.*

**63.** $3x + 6 =$ ________ Commutative property of addition

**64.** $8 + 0 =$ ___ Additive identity property

**65.** $\frac{2}{3} + \left(-\frac{2}{3}\right) =$ ___ Additive inverse property

**66.** $4(x + 3) =$ ________ Distributive property

**67.** $7 \cdot 1 =$ ___ Multiplicative identity property

**68.** $0 \cdot (-5.4) =$ ___ Multiplication property of zero

**69.** $10(2y) =$ ________ Associative property

**70.** $9y + (x + 3z) =$ ________ Associative property

**71.** To demonstrate the distributive property geometrically, represent the area of the larger rectangle in two ways: First as length $a$ times width $b + c$, and second as the sum of the areas of the smaller rectangles.

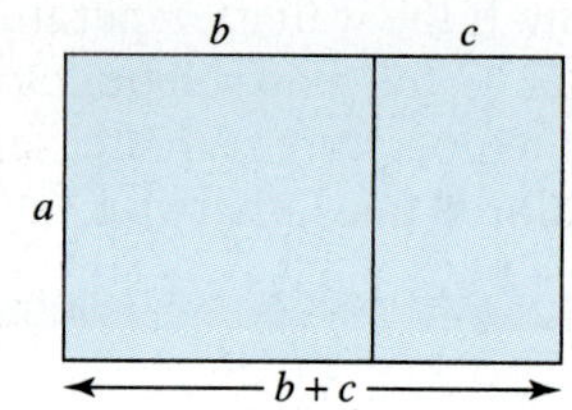


*Write each of the following as an algebraic expression. See Examples 9 and 10.*

**72.** Write an expression for the amount of money (in dollars) in $n$ nickels.

**73.** Write an expression for the amount of money (in dollars) in $d$ dimes.

**74.** Two numbers have a sum of 25. If one number is $x$, represent the other number as an expression in $x$.

**75.** Two numbers have a sum of 112. If one number is $x$, represent the other number as an expression in $x$.

**76.** Two angles are supplementary if the sum of their measures is 180°. If the measure of one angle is $x$ degrees, represent the measure of the other angle as an expression in $x$.

**77.** If the measure of an angle is $5x$ degrees, represent the measure of its complement as an expression in $x$.

**78.** The cost of $x$ compact discs if each compact disc costs \$6.49.

**79.** The cost of $y$ books if each book costs \$35.61.

**80.** If $x$ is an odd integer, represent the next odd integer as an expression in $x$.

**81.** If $2x$ is an even integer, represent the next even integer as an expression in $x$.

*Simplify each expression. See Examples 11 and 12.*

**82.** $-9 + 4x + 18 - 10x$

**83.** $5y - 14 + 7y - 20y$

**84.** $5k - (3k - 10)$

**85.** $-11c - (4 - 2c)$

**86.** $(3x + 4) - (6x - 1)$

**87.** $(8 - 5y) - (4 + 3y)$

**88.** $3(x - 2) + x + 15$

**89.** $-4(y + 3) - 7y + 1$

**90.** $-(n + 5) + (5n - 3)$

**91.** $-(8 - t) + (2t - 6)$

**92.** $4(6n - 3) - 3(8n + 4)$

**93.** $5(2z - 6) + 10(3 - z)$

**94.** $3x - 2(x - 5) + x$

**95.** $7n + 3(2n - 6) - 2$

**96.** $-1.2(5.7x - 3.6) + 8.75x$

**97.** $5.8(-9.6 - 31.2y) - 18.65$

**98.** $8.1z + 7.3(z + 5.2) - 6.85$

**99.** $6.5y - 4.4(1.8x - 3.3) + 10.95$

**100.** △ Do figures with the same surface area always have the same volume? To see, take two $8\frac{1}{2}$-by-11-inch sheets of paper and construct two cylinders using the following figures as a guide. Working with a partner, measure the height and the radius of each resulting cylinder and use the expression $\pi r^2 h$ to approximate each volume to the nearest tenth of a cubic inch. Explain your results.

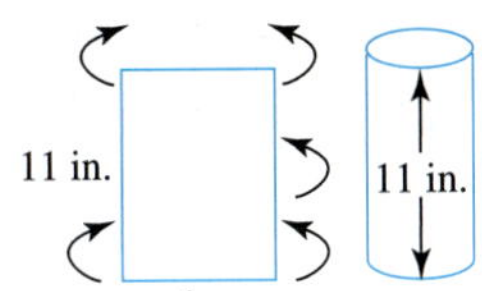


Cylinder 1

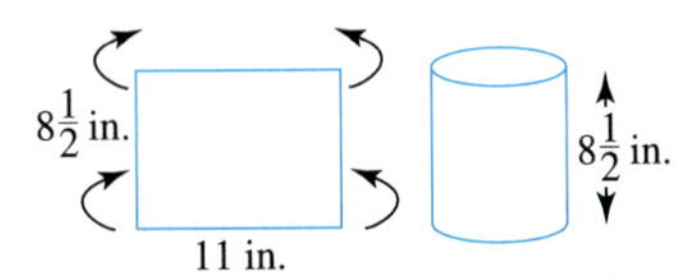


Cylinder 2

△ **101.** Use the same idea as in Exercise 100, work with a partner, and discover whether two rectangles with the same perimeter always have the same area. Explain your results.

*The following graph is called a broken-line graph, or simply a line graph. This particular graph shows past, present, and future predicted population over 65. Just as with a bar graph, to find the population over 65 for a particular year, read the height of the corresponding point. To read the height, follow the point horizontally to the left until you reach the vertical axis.*

**102.** Estimate the population over 65 in the year 1940.

**103.** Estimate the predicted population over 65 in the year 2030.

**104.** Estimate the predicted population over 65 in the year 2010.

**105.** Estimate the population over 65 in the year 1993.

**106.** Is the population over 65 increasing as time passes or decreasing? Explain how you arrived at your answer.

**107.** The percent of Americans over 65 approximately tripled from 1900 to 1993. If this percent in 1900 was 4.1 %, estimate the percent of Americans over 65 in the year 1993.

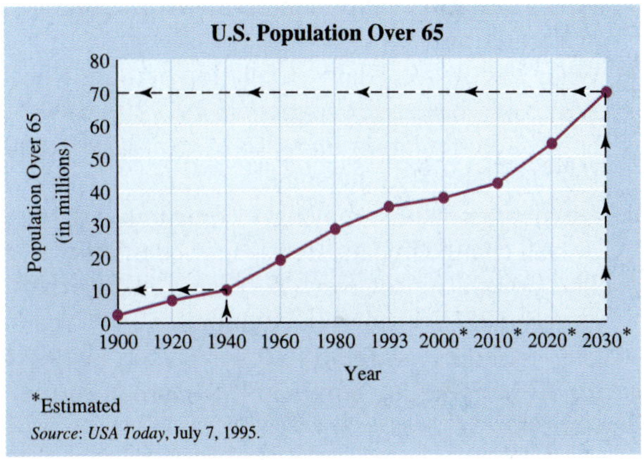


For additional Chapter Projects, visit the Real World Activities Website by going to http://www.prenhall.com/martin-gay.

# 1 CHAPTER PROJECT

## *Analyzing Newspaper Circulation*

The number of daily newspapers in business in the United States has declined steadily recently, continuing a trend that started in the mid-1970s. In 1900, there were roughly 2300 daily newspapers in operation. However, by 1998, that number had dropped to only 1509 newspapers in existence. Average overall daily newspaper circulation also continues to decline, from 60.2 million in 1992 to 56.7 million in 1997.

The table below gives data about daily newspaper circulation for New York City–based newspapers for the years 1997 and 1998. In this project, you will have the opportunity to analyze this data. This project may be completed by working in groups or individually.

| *New York–based Newspaper* | *1997 Daily Circulation* | *1998 Daily Circulation* | *Daily Edition Newsstand Price* |
|---|---|---|---|
| *Wall Street Journal* | 1,774,880 | 1,740,450 | \$0.75 |
| *New York Times* | 1,074,741 | 1,066,658 | \$1.00 |
| *New York Daily News* | 721,256 | 723,143 | \$0.50 |
| *New York Newsday* | 568,914 | 572,444 | \$0.50 |
| *New York Post* | 436,226 | 437,467 | \$0.35 |

1. Find the change in circulation from 1997 to 1998 for each newspaper. Did any newspaper gain circulation? If so, which one(s) and by how much?
2. Construct a bar graph showing the change in circulation from 1997 to 1998 for each newspaper. Which newspaper experienced the largest change in circulation?
3. What was the total daily circulation of these New York City–based newspapers in 1997? In 1998? Did total circulation increase or decrease from 1997 to 1998? By how much?
4. Discuss factors that may have contributed to the overall change in daily newspaper circulation.
5. The population of New York City is approximately 7,381,000. Find the number of New York City–based newspapers sold per person in 1998 for New York City. (*Source:* U.S. Bureau of the Census)
6. The population of the New York City metropolitan area is approximately 16,332,000. Find the number of New York City–based newspapers sold per person in 1998 for the New York City metropolitan area. (*Source:* The United Nations)
7. Which of the figures found in Questions 5 and 6 do you think is more meaningful? Why? Why might neither of these figures be capable of describing the full circulation situation?
8. Assuming that each copy was sold from a newsstand in the New York City metropolitan area, use the daily edition newspaper prices given in the table to find the total amount spent each day on these New York City–based newspapers in 1998. Find the total amount spent annually on these daily newspapers in 1998. (Remember: Daily editions are published only on weekdays.)
9. How accurate do you think the figures you found in Question 8 are? Explain your reasoning.

## CHAPTER 1 VOCABULARY CHECK

*Fill in each blank with one of the words or phrases listed below.*

| | | | |
|---|---|---|---|
| distributive | absolute value | inequality | algebraic expression |
| real | opposite | commutative | exponent |
| reciprocals | associative | whole | variable |

1. A(n) ________________ is formed by numbers and variables connected by the operations of addition, subtraction, multiplication, division, raising to powers, and/or taking roots.
2. The ________ of a number $a$ is $-a$.
3. $3(x - 6) = 3x - 18$ by the ___________ property.
4. The _____________ of a number is the distance between that number and 0 on the number line.
5. A(n) _________ is a shorthand notation for repeated multiplication of the same factor.
6. A letter that represents a number is called a ________.
7. The symbols < and > are called __________ symbols.
8. If $a$ is not 0, then $a$ and $1/a$ are called __________.
9. $A + B = B + A$ by the ___________ property.
10. $(A + B) + C = A + (B + C)$ by the ___________ property.
11. The numbers $0, 1, 2, 3, \ldots$ are called _______ numbers.
12. If a number corresponds to a point on the number line, we know that number is a _____ number.

# CHAPTER 1 HIGHLIGHTS

| DEFINITIONS AND CONCEPTS | EXAMPLES |
|---|---|
| **Section 1.2 Algebraic Expressions and Sets of Numbers** | |
| Letters that represent numbers are called **variables.** | Examples of variables are<br>$x, a, m, y$ |
| An **algebraic expression** is formed by numbers and variables connected by the operations of addition, subtraction, multiplication, division, raising to powers, and/or taking roots. | Examples of algebraic expressions are<br>$7y, -3, \frac{x^2 - 9}{-2} + 14x, \sqrt{3} + \sqrt{m}$ |
| To **evaluate** an algebraic expression containing variables, substitute the given numbers for the variables and simplify. The result is called the **value** of the expression. | Evaluate $2.7x$ if $x = 3$.<br>$2.7x = 2.7(3)$<br>$= 8.1$ |
| **Natural numbers:** $\{1, 2, 3, \ldots\}$<br>**Whole numbers:** $\{0, 1, 2, 3, \ldots\}$<br>**Integers:** $\{\ldots, -3, -2, -1, 0, 1, 2, 3, \ldots\}$<br>Each listing of three dots above is called an **ellipsis.**<br>The members of a set are called its **elements.**<br>**Set builder notation** describes the elements of a set but does not list them.<br>**Real numbers:** $\{x \mid x$ corresponds to a point on the number line.$\}$<br>**Rational numbers:** $\{\frac{a}{b} \mid a$ and $b$ are intergers and $b \neq 0\}$.<br>**Irrational numbers:** $\{x \mid x$ is a real number and $x$ is not a rational number$\}$.<br>If all the elements of set $A$ are also in set $B$, we say that set $A$ is a **subset** of set $B$, and we write $A \subseteq B$. | Given the set $\{-9.6, -5, -\sqrt{2}, 0, \frac{2}{5}, 101\}$ list the elements that belong to the set of<br>Natural numbers 101<br>Whole numbers 0, 101<br>Integers −5, 0, 101<br>Real numbers $-9.6, -5, -\sqrt{2}, 0, \frac{2}{5}, 101$<br>Rational numbers $-9.6, -5, 0, \frac{2}{5}, 101$<br>Irrational numbers $-\sqrt{2}$<br>List the elements in the set $\{x \mid x$ is an integer between $-2$ and $5\}$.<br>$\{-1, 0, 1, 2, 3, 4\}$<br>$\{1, 2, 4\} \subseteq \{1, 2, 3, 4\}$. |
| Absolute value:<br>$\lvert a\rvert = \begin{cases} a \text{ if } a \text{ is 0 or a positive number} \\ -a \text{ if } a \text{ is a negative number} \end{cases}$ | $\lvert 3\rvert = 3, \lvert 0\rvert = 0, \lvert -7.2\rvert = 7.2$ |
| The opposite of a number $a$ is the number $-a$. | The opposite of 5 is −5. The opposite of −11 is 11. |
| **Section 1.3 Operations on Real Numbers** | |
| Adding real numbers:<br>1. To add two numbers with the same sign, add their absolute values and attach their common sign.<br>2. To add two numbers with different signs, subtract the smaller absolute value from the larger absolute value and attach the sign of the number with the larger absolute value. | $\frac{2}{7} + \frac{1}{7} = \frac{3}{7}$<br>$-5 + (-2.6) = -7.6$<br>$-18 + 6 = -12$<br>$20.8 + (-10.2) = 10.6$ |
| Subtracting real numbers:<br>$a - b = a + (-b)$ | $18 - 21 = 18 + (-21) = -3$ |
| Multiplying and dividing real numbers:<br>The product or quotient of two numbers with the same sign is positive.<br>The product or quotient of two numbers with different signs is negative. | $(-8)(-4) = 32 \qquad \frac{-8}{-4} = 2$<br>$8 \cdot 4 = 32 \qquad \frac{8}{4} = 2$<br>$-17 \cdot 2 = -34 \qquad \frac{-14}{2} = -7$<br>$4(-1.6) = -6.4 \qquad \frac{22}{-2} = -11$ |

*(continued)*

| DEFINITIONS AND CONCEPTS | EXAMPLES |
|---|---|
| **Section 1.3 Operations on Real Numbers, continued** | |
| A natural number **exponent** is a shorthand notation for repeated multiplication of the same factor. | $3^4 = 3 \cdot 3 \cdot 3 \cdot 3 = 81$ |
| The notation $\sqrt{a}$ is used to denote the **positive**, or **principal, square root** of a nonnegative number $a$. $\sqrt{a} = b$ if $b^2 = a$ and $b$ is positive. | $\sqrt{49} = 7$ |
| Also, $\sqrt[3]{a} = b$ if $b^3 = a$ | $\sqrt[3]{64} = 4$ |
| $\sqrt[4]{a} = b$ if $b^4 = a$ and $b$ is positive | $\sqrt[4]{16} = 2$ |
| ***Order of Operations*** Simplify expressions using the order that follows. If grouping symbols such as parentheses are present, simplify expressions within those first, starting with the innermost set. If fraction bars are present, simplify the numerator and denominator separately. 1. Raise to powers or take roots in order from left to right. 2. Multiply or divide in order from left to right. 3. Add or subtract in order from left to right. | Simplifly $\frac{42 - 2(3^2 - \sqrt{16})}{-8}$. $\frac{42 - 2(3^2 - \sqrt{16})}{-8} = \frac{42 - 2(9 - 4)}{-8} = \frac{42 - 2(5)}{-8} = \frac{42 - 10}{-8} = \frac{32}{-8} = -4$ |
| **Section 1.4 Properties of Real Numbers** | |
| Symbols: $=$ is equal to | $-5 = -5$ |
| $\neq$ is not equal to | $-5 \neq -3$ |
| $>$ is greater than | $1.7 > 1.2$ |
| $<$ is less than | $-1.7 < -1.2$ |
| $\geq$ is greater than or equal to | $\frac{5}{3} \geq \frac{5}{3}$ |
| $\leq$ is less than or equal to | $-\frac{1}{2} \leq \frac{1}{2}$ |
| Identity: $a + 0 = a \quad 0 + a = a$ | $3 + 0 = 3 \quad 0 + 3 = 3$ |
| $a \cdot 1 = a \quad 1 \cdot a = a$ | $-1.8 \cdot 1 = -1.8 \quad 1 \cdot -1.8 = -1.8$ |
| Inverse: $a + (-a) = 0 \quad -a + a = 0$ | $7 + (-7) = 0 \quad -7 + 7 = 0$ |
| $a \cdot \frac{1}{a} = 1 \quad \frac{1}{a} \cdot a = 1, a \neq 0$ | $5 \cdot \frac{1}{5} = 1 \quad \frac{1}{5} \cdot 5 = 1$ |
| Commutative: $a + b = b + a$ | $x + 7 = 7 + x$ |
| $a \cdot b = b \cdot a$ | $9 \cdot y = y \cdot 9$ |
| Associative: $(a + b) + c = a + (b + c)$ | $(3 + 1) + 10 = 3 + (1 + 10)$ |
| $(a \cdot b) \cdot c = a \cdot (b \cdot c)$ | $(3 \cdot 1) \cdot 10 = 3(1 \cdot 10)$ |
| Distributive: $a(b + c) = ab + ac$ | $6(x + 5) = 6 \cdot x + 6 \cdot 5 = 6x + 30$ |

# CHAPTER 1 REVIEW

***(1.2)*** *Find the value of each algebraic expression at the given replacement values.*

**1.** $7x$ when $x = 3$

**2.** $st$ when $s = 1.6$ and $t = 5$

**3.** The humming bird has an average wing speed of 90 beats per second. The expression $90t$ gives the number of wing beats in $t$ seconds. Calculate the number of wing beats in *1 hour* for the hummingbird.

*List the elements in each set.*

**4.** $\{x \mid x$ is an odd integer between $-2$ and $4\}$

**5.** $\{x \mid x$ is an even integer between $-3$ and $7\}$

**6.** $\{x \mid x$ is a negative whole number$\}$

**7.** $\{x \mid x$ is a natural number that is not a rational number

**8.** $\{x \mid x$ is a whole number greater than $5\}$

**9.** $\{x \mid x$ is an integer less than $3\}$

*Determine whether each statement is true or false if* $A = \{6, 10, 12\}$, $B = \{5, 9, 11\}$, $C = \{\ldots, -3, -2, -1, 0, 1, 2, 3, \ldots\}$, $D = \{2, 4, 6, \ldots, 16\}$, $E = \{x \mid x$ *is a rational number*$\}$, $F = \{\ \}$, $G = \{x \mid x$ is an irrational number$\}$, and $H = \{x \mid x$ is a real number$\}$.

**10.** $10 \in D$

**11.** $B \in 9$

**12.** $\sqrt{169} \notin G$

**13.** $0 \notin F$

**14.** $\pi \in E$

**15.** $\pi \in H$

**16.** $\sqrt{4} \in G$

**17.** $-9 \in E$

**18.** $A \subseteq D$

**19.** $C \not\subseteq B$

**20.** $C \not\subseteq E$

**21.** $F \subseteq H$

**22.** $B \subseteq B$

**23.** $D \subseteq C$

**24.** $C \subseteq H$

**25.** $G \subseteq H$

**26.** $\{5\} \in B$

**27.** $\{5\} \subseteq B$

*List the elements of the set* $\left\{5, -\frac{2}{3}, \frac{8}{2}, \sqrt{9}, 0.3, \sqrt{7}, 1\frac{5}{8}, -1, \pi\right\}$ *that are also elements of each given set.*

**28.** Whole numbers

**29.** Natural numbers

**30.** Rational numbers

**31.** Irrational numbers

**32.** Real numbers

**33.** Integers

*Find the opposite.*

**34.** $-\frac{3}{4}$

**35.** $0.6$

**36.** $0$

**37.** $1$

*Find the reciprocal.*

**38.** $-\frac{3}{4}$

**39.** $0.6$

**40.** $0$

**41.** $1$

***(1.3)*** *Simplify.*

**42.** $-7 + 3$

**43.** $-10 + (-25)$

**44.** $5(-0.4)$

**45.** $(-3.1)(-0.1)$

**46.** $-7 - (-15)$

**47.** $9 - (-4.3)$

**48.** $(-6)(-4)(0)(-3)$

**49.** $(-12)(0)(-1)(-5)$

**50.** $(-24) \div 0$

**51.** $0 \div (-45)$

**52.** $(-36) \div (-9)$

**53.** $60 \div (-12)$

**54.** $\left(-\frac{4}{5}\right) - \left(-\frac{2}{3}\right)$

**55.** $\left(\frac{5}{4}\right) - \left(-2\frac{3}{4}\right)$

**56.** Determine the unknown fractional part.

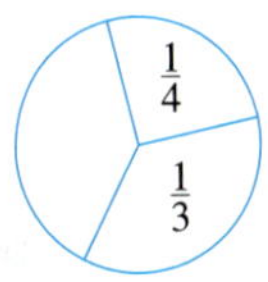


*Simplify.*

**57.** $-5 + 7 - 3 - (-10)$

**58.** $8 - (-3) + (-4) + 6$

**59.** $3(4 - 5)^4$

**60.** $6(7 - 10)^2$

**61.** $\left(-\frac{8}{15}\right) \cdot \left(-\frac{2}{3}\right)^2$

**62.** $\left(-\frac{3}{4}\right)^2 \cdot \left(-\frac{10}{21}\right)$

**63.** $\dfrac{-\frac{6}{15}}{\frac{8}{25}}$

**64.** $\dfrac{\frac{4}{9}}{-\frac{8}{45}}$

**65.** $-\frac{3}{8} + 3(2) \div 6$

**66.** $5(-2) - (-3) - \frac{1}{6} + \frac{2}{3}$

**67.** $|2^3 - 3^2| - |5 - 7|$

**68.** $|5^2 - 2^2| + |9 \div (-3)|$

**69.** $(2^3 - 3^2) - (5 - 7)$

**70.** $(5^2 - 2^4) + [9 \div (-3)]$

**71.** $\dfrac{(8 - 10)^3 - (-4)^2}{2 + 8(2) \div 4}$

**72.** $\dfrac{(2 + 4)^2 + (-1)^5}{12 \div 2 \cdot 3 - 3}$

**73.** $\dfrac{(4 - 9) + 4 - 9}{10 - 12 \div 4 \cdot 8}$

**74.** $\dfrac{3 - 7 - (7 - 3)}{15 + 30 \div 6 \cdot 2}$

**75.** $\dfrac{\sqrt{25}}{4 + 3 \cdot 7}$

**76.** $\dfrac{\sqrt{64}}{24 - 8 \cdot 2}$

*Find the value of each expression when $x = 0$, $y = 3$, and $z = -2$.*

**77.** $x^2 - y^2 + z^2$

**78.** $\dfrac{5x + z}{2y}$

**79.** $\dfrac{-7y - 3z}{-3}$

**80.** $(x - y + z)^2$

△ **81.** The algebraic expression $2\pi r$ represents the circumference of (distance around) a circle of radius $r$.

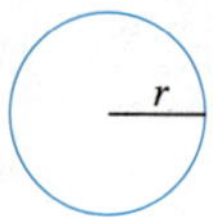


**a.** Complete the table below by evaluating the expression at given values of $r$. (Use 3.14 for $\pi$)

| Radius | $r$ | 1 | 10 | 100 |
|---|---|---|---|---|
| Circumference | $2\pi r$ | | | |

**b.** As the radius of a circle increases, does the circumference of the circle increase or decrease?

***(1.4)*** *Simplify each expression.*

**82.** $5xy - 7xy + 3 - 2 + xy$

**83.** $4x + 10x - 19x + 10 - 19$

**84.** $6x^2 + 2 - 4(x^2 + 1)$

**85.** $-7(2x^2 - 1) - x^2 - 1$

**86.** $(3.2x - 1.5) - (4.3x - 1.2)$

**87.** $(7.6x + 4.7) - (1.9x + 3.6)$

*Write each statement using mathematical symbols.*

**88.** Twelve is the product of $x$ and negative 4.

**89.** The sum of $n$ and twice $n$ is negative fifteen.

**90.** Four times the sum of $y$ and three is −1.

**91.** The difference of $t$ and five, multiplied by six is four.

**92.** Seven subtracted from $z$ is six.

**93.** Ten less than the product of $x$ and nine is five.

**94.** The difference of $x$ and 5 is at least 12.

**95.** The opposite of four is less than the product of $y$ and seven.

**96.** Two-thirds is not equal to twice the sum of $n$ and one-fourth.

**97.** The sum of $t$ and six is not more than negative twelve.

*Name the property illustrated.*

**98.** $(M + 5) + P = M + (5 + P)$

**99.** $5(3x - 4) = 15x - 20$

**100.** $(-4) + 4 = 0$

**101.** $(3 + x) + 7 = 7 + (3 + x)$

**102.** $(XY)Z = (YZ)X$

**103.** $\left(-\dfrac{3}{5}\right) \cdot \left(-\dfrac{5}{3}\right) = 1$

**104.** $T \cdot 0 = 0$

**105.** $(ab)c = a(bc)$

**106.** $A + 0 = A$

**107.** $8 \cdot 1 = 8$

*Complete the equation using the given property.*

**108.** $5x - 15z =$ __________ Distributive property

**109.** $(7 + y) + (3 + x) =$ __________ Commutative property

**110.** $0 =$ ______________________ Additive inverse property

**111.** $1 =$ __________ Multiplicative inverse property

**112.** $[(3.4)(0.7)]5 =$ __________ Associative property

**113.** $7 =$ ______ Additive identity property

*Insert $<$, $>$, or $=$ to make each statement true.*

**114.** $-9 \quad -12$

**115.** $0 \quad -6$

**116.** $-3 \quad -1$

**117.** $7 \quad |-7|$

**118.** $-5 \quad -(-5)$

**119.** $-(-2) \quad -2$

## CHAPTER 1 TEST

*Determine whether each statement is true or false.*

**1.** $-2.3 > -2.33$

**2.** $-6^2 = (-6)^2$

**3.** $-5 - 8 = -(5 - 8)$

**4.** $(-2)(-3)(0) = \dfrac{-4}{0}$

**5.** All natural numbers are integers.

**6.** All rational numbers are integers.

*Simplify.*

**7.** $5 - 12 \div 3(2)$

**8.** $|4 - 6|^3 - (1 - 6^2)$

**9.** $(4 - 9)^3 - |-4 - 6|^2$

**10.** $[3|4 - 5|^5 - (-9)] \div (-6)$

**11.** $\dfrac{6(7 - 9)^3 + (-2)}{(-2)(-5)(-5)}$

*Evaluate each expression when $q = 4$, $r = -2$, and $t = 1$.*

**12.** $q^2 - r^2$

**13.** $\dfrac{5t - 3q}{3r - 1}$

**14.** The algebraic expression $5.75x$ represents the total cost for $x$ adults to attend the theater.

**a.** Complete the table that follows.

**b.** As the number of adults increases does the total cost increase or decrease?

| Adults | $x$ | 1 | 3 | 10 | 20 |
|---|---|---|---|---|---|
| Total Cost | $5.75x$ | | | | |

*Write each statement using mathematical symbols..*

**15.** Twice the absolute value of the sum of $x$ and five is 30.

**16.** The square of the difference of six and $y$, divided by seven, is less than $-2$.

**17.** The product of nine and $z$, divided by the absolute value of $-12$, is not equal to 10.

**18.** Three times the quotient of $n$ and five is the opposite of $n$.

**19.** Twenty is equal to 6 subtracted from twice $x$.

**20.** Negative two is equal to $x$ divided by the sum of $x$ and five.

*Name each property illustrated.*

**21.** $6(x - 4) = 6x - 24$

**22.** $(4 + x) + z = 4 + (x + z)$

**23.** $(-7) + 7 = 0$

**24.** $(-18)(0) = 0$

**25.** Write an expression for the total amount of money (in dollars) in $n$ nickels and $d$ dimes.

*Simplify each expression.*

**26.** $4y^2 + 10 - 2(y^2 + 10)$

**27.** $(8.3x - 2.9) - (9.6x - 4.8)$

## Keeping Us Informed

The field of journalism includes newspaper, radio, and television reporters, as well as photographers, graphic artists, copy editors, and other communications specialists. In the United States, the news media are responsible for keeping us informed about current national and international events.

Excellent communication skills are the top requirement for journalists. However, journalists also must be informed about a wide variety of subjects so they can report effectively on their news assignments. A liberal-arts education, including courses in economics, statistics, and sciences, can be a huge asset for journalists. Journalists must be able to do research, conduct interviews, study documents, and interpret numerical data and statistics.

*For more information about careers in journalism, visit the American Society of Newspaper Editors Website by first going to www.prenhall.com/martin-gay.*

*In the Spotlight on Decision Making feature on page 63, you will have the opportunity to use percents to make a decision about how to approach a story as a reporter.*

# EQUATIONS, INEQUALITIES, AND PROBLEM SOLVING

# 2


2.1 Linear Equations in One Variable

2.2 An Introduction to Problem Solving

2.3 Formulas and Problem Solving

2.4 Linear Inequalities and Problem Solving

2.5 Compound Inequalities

2.6 Absolute Value Equations

2.7 Absolute Value Inequalities


Mathematics is a tool for solving problems in such diverse fields as transportation, engineering, economics, medicine, business, and biology. We solve problems using mathematics by modeling real-world phenomena with mathematical equations or inequalities. Our ability to solve problems using mathematics, then, depends in part on our ability to solve equations and inequalities. In this chapter, we solve linear equations and inequalities in one variable and graph their solutions on number lines.

## 2.1 LINEAR EQUATIONS IN ONE VARIABLE

CD-ROM SSM SSG Video

**OBJECTIVES**

1. Solve linear equations using properties of equality.
2. Solve linear equations that can be simplified by combining like terms.
3. Solve linear equations containing fractions.
4. Recognize when an equation is an identity and when it has no solution.

**1** Linear equations model many real-life problems. For example, we can use a linear equation to calculate the increase in digital camera sales.

With the help of your computer, digital cameras allow you to see your pictures and make copies immediately, send them in e-mail or use them on a Web page. Current sales and projected sales of these cameras are shown in the graph below.

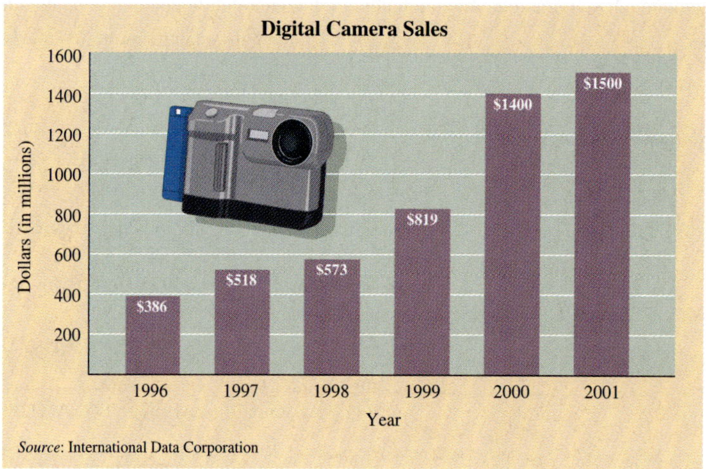


To find the increase in sales from 1999 to 2000, for example, we can use the equation below.

| | | | | | |
|---|---|---|---|---|---|
| In words: | Increase in sales | is | sales in 2000 | minus | sales in 1999 |
| Translate: | $x$ | $=$ | 1400 | $-$ | 819 |

Since our variable $x$ (increase in sales) is by itself on one side of the equation, we can find the value of $x$ by simplifying the right side.

$$x = 581$$

The increase in sales of digital cameras from 1999 to 2000 is $581 million.

The **equation**, $x = 1400 - 819$, like every other equation, is a statement that two expressions are equal. Oftentimes, the unknown variable is not by itself on one side of the equation. In these cases, we will use properties of equality to write equivalent equations so that a solution many be found. This is called **solving the equation**. In this section, we concentrate on solving equations such as this one, called **linear equations** in one variable. Linear equations are also called **first-degree equations** since the exponent on the variable is 1.

***Linear Equations in One Variable***

$$3x = -15 \qquad 7 - y = 3y \qquad 4n - 9n + 6 = 0 \qquad z = -2$$

### LINEAR EQUATIONS IN ONE VARIABLE

A linear equation in one variable is an equation that can be written in the form

$$ax + b = c$$

where $a$, $b$, and $c$ are real numbers and $a \neq 0$.

When a variable in an equation is replaced by a number and the resulting equation is true, then that number is called a **solution** of the equation. For example, 1 is a solution of the equation $3x + 4 = 7$, since $3(1) + 4 = 7$ is a true statement. But 2 is not a solution of this equation, since $3(2) + 4 = 7$ is not a true statement. The **solution set** of an equation is the set of solutions of the equation. For example, the solution set of $3x + 4 = 7$ is $\{1\}$.

To **solve an equation** is to find the solution set of an equation. Equations with the same solution set are called **equivalent equations**. For example,

$$3x + 4 = 7 \qquad 3x = 3 \qquad x = 1$$

are equivalent equations because they all have the same solution set, namely $\{1\}$. To solve an equation in $x$, we start with the given equation and write a series of simpler equivalent equations until we obtain an equation of the form

$$\boldsymbol{x = \textbf{number}}$$

Two important properties are used to write equivalent equations.

### THE ADDITION AND MULTIPLICATION PROPERTIES OF EQUALITY

If $a$, $b$, and $c$, are real numbers, then

$a = b$ and $a + c = b + c$ are equivalent equations.

Also, $a = b$ and $ac = bc$ are equivalent equations as long as $c \neq 0$.

The **addition property of equality** guarantees that the same number may be added to (or subtracted from) both sides of an equation, and the result is an equivalent equation. The **multiplication property of equality** guarantees that both sides of an equation may be multiplied by (or divided by) the same nonzero number, and the result is an equivalent equation.

For example, to solve $2x + 5 = 9$, use the addition and multiplication properties of equality to isolate $x$—that is, to write an equivalent equation of the form

$$x = \text{number}$$

**Example 1** Solve for $x$: $2x + 5 = 9$.

*Solution* First, use the addition property of equality and subtract 5 from both sides. We do this so that our only variable term, $2x$, is by itself on one side of the equation.

$$2x + 5 = 9$$

$$2x + 5 - 5 = 9 - 5 \qquad \text{Subtract 5 from both sides.}$$

$$2x = 4 \qquad \text{Simplify.}$$

Now that the variable term is isolated, we can finish solving for $x$ by using the multiplication property of equality and dividing both sides by 2.

$$\frac{2x}{2} = \frac{4}{2} \quad \text{Divide both sides by 2.}$$

$$x = 2 \quad \text{Simplify.}$$

*Check* To see that 2 is the solution, replace $x$ in the original equation with 2.

$$\begin{aligned} 2x + 5 &= 9 && \text{Original equation.} \\ 2(2) + 5 &\stackrel{?}{=} 9 && \text{Let } x = 2. \\ 4 + 5 &\stackrel{?}{=} 9 \\ 9 &= 9 && \text{True.} \end{aligned}$$

Since we arrive at a true statement, 2 is the solution or the solution set is $\{2\}$.

**Example 2** Solve: $0.6 = 2 - 3.5c$.

*Solution* We use both the addition property and the multiplication property of equality.

$$\begin{aligned} 0.6 &= 2 - 3.5c \\ 0.6 - 2 &= 2 - 3.5c - 2 && \text{Subtract 2 from both sides.} \\ -1.4 &= -3.5c && \text{Simplify. The variable term is now isolated.} \\ \frac{-1.4}{-3.5} &= \frac{-3.5c}{-3.5} && \text{Divide both sides by } -3.5. \\ 0.4 &= c && \text{Simplify } \frac{-1.4}{-3.5}. \end{aligned}$$

**HELPFUL HINT**
Don't forget that

$0.4 = c$ and $c = 0.4$ are equivalent equations.

We may solve an equation so that the variable is alone on either side of the equation.

*Check*

$$\begin{aligned} 0.6 &= 2 - 3.5c \\ 0.6 &\stackrel{?}{=} 2 - 3.5(0.4) && \text{Replace } c \text{ with } 0.4. \\ 0.6 &\stackrel{?}{=} 2 - 1.4 && \text{Multiply.} \\ 0.6 &= 0.6 && \text{True.} \end{aligned}$$

The solution is 0.4.

**2** Often, an equation can be simplified by removing any grouping symbols and combining any like terms.

**Example 3** Solve: $-6x - 1 + 5x = 3$.

*Solution* First, the left side of this equation can be simplified by combining like terms $-6x$ and $5x$. Then use the addition property of equality and add 1 to both sides of the equation.

$$\begin{aligned} -6x - 1 + 5x &= 3 \\ -x - 1 &= 3 && \text{Combine like terms.} \\ -x - 1 + 1 &= 3 + 1 && \text{Add 1 to both sides of the equation.} \\ -x &= 4 && \text{Simplify.} \end{aligned}$$

Notice that this equation is not solved for $x$ since we have $-x$ or $-1x$, not $x$. To solve for $x$, divide both sides by $-1$.

$$\frac{-x}{-1} = \frac{4}{-1} \quad \text{Divide both sides by } -1.$$
$$x = -4 \quad \text{Simplify.}$$

Check to see that the solution is $-4$.

If an equation contains parentheses, use the distributive property to remove them.

**Example 4** Solve: $2(x - 3) = 5x - 9$.

*Solution* First, use the distributive property.

$$2(x - 3) = 5x - 9$$
$$2x - 6 = 5x - 9 \quad \text{Use the distributive property.}$$

Next, get variable terms on the same side of the equation by subtracting $5x$ from both sides.

$$2x - 6 - 5x = 5x - 9 - 5x \quad \text{Subtract } 5x \text{ from both sides.}$$
$$-3x - 6 = -9 \quad \text{Simplify.}$$
$$-3x - 6 + 6 = -9 + 6 \quad \text{Add 6 to both sides.}$$
$$-3x = -3 \quad \text{Simplify.}$$
$$\frac{-3x}{-3} = \frac{-3}{-3} \quad \text{Divide both sides by } -3.$$
$$x = 1$$

Let $x = 1$ in the original equation to see that 1 is the solution.

**3** If an equation contains fractions, we first clear the equation of fractions by multiplying both sides of the equation by the *least common denominator* (LCD) of all fractions in the equation.

**Example 5** Solve for $y$: $\frac{y}{3} - \frac{y}{4} = \frac{1}{6}$.

*Solution* First, clear the equation of fractions by multiplying both sides of the equation by 12, the LCD of denominators 3, 4, and 6.

$$\frac{y}{3} - \frac{y}{4} = \frac{1}{6}$$
$$12\left(\frac{y}{3} - \frac{y}{4}\right) = 12\left(\frac{1}{6}\right) \quad \text{Multiply both sides by the LCD 12.}$$
$$12\left(\frac{y}{3}\right) - 12\left(\frac{y}{4}\right) = 2 \quad \text{Apply the distributive property.}$$
$$4y - 3y = 2 \quad \text{Simplify.}$$
$$y = 2 \quad \text{Simplify.}$$

**Check:** To check, let $y = 2$ in the original equation.

$$\frac{y}{3} - \frac{y}{4} = \frac{1}{6} \qquad \text{Original equation.}$$

$$\frac{2}{3} - \frac{2}{4} \stackrel{?}{=} \frac{1}{6} \qquad \text{Let } y = 2.$$

$$\frac{8}{12} - \frac{6}{12} \stackrel{?}{=} \frac{1}{6} \qquad \text{Write fractions with the LCD.}$$

$$\frac{2}{12} \stackrel{?}{=} \frac{1}{6} \qquad \text{Subtract.}$$

$$\frac{1}{6} = \frac{1}{6} \qquad \text{Simplify.}$$

This is a true statement, so the solution is 2.

As a general guideline, the following steps may be used to solve a linear equation in one variable.

### SOLVING A LINEAR EQUATION IN ONE VARIABLE

**Step 1:** Clear the equation of fractions by multiplying both sides of the equation by the least common denominator (LCD) of all denominators in the equation.
**Step 2:** Use the distributive property to remove grouping symbols such as parentheses.
**Step 3:** Combine like terms on each side of the equation.
**Step 4:** Use the addition property of equality to rewrite the equation as an equivalent equation with variable terms on one side and numbers on the other side.
**Step 5:** Use the multiplication property of equality to isolate the variable.
**Step 6:** Check the proposed solution in the original equation.

**Example 6** Solve for $x$: $\frac{x+5}{2} + \frac{1}{2} = 2x - \frac{x-3}{8}$.

*Solution* Multiply both sides of the equation by 8, the LCD of 2 and 8.

$$8\left(\frac{x+5}{2} + \frac{1}{2}\right) = 8\left(2x - \frac{x-3}{8}\right) \qquad \text{Multiply both sides by 8.}$$

$$8\left(\frac{x+5}{2}\right) + 8 \cdot \frac{1}{2} = 8 \cdot 2x - 8\left(\frac{x-3}{8}\right) \qquad \text{Apply the distributive property.}$$

$$4(x+5) + 4 = 16x - (x-3) \qquad \text{Simplify.}$$

$$4x + 20 + 4 = 16x - x + 3 \qquad \text{Use the distributive property to remove parentheses.}$$

$$4x + 24 = 15x + 3 \qquad \text{Combine like terms.}$$

$$-11x + 24 = 3 \qquad \text{Subtract } 15x \text{ from both sides.}$$

$$-11x = -21 \qquad \text{Subtract 24 from both sides.}$$

$$\frac{-11x}{-11} = \frac{-21}{-11} \qquad \text{Divide both sides by } -11.$$

$$x = \frac{21}{11} \qquad \text{Simplify.}$$

To check, verify that replacing $x$ with $\frac{21}{11}$ makes the original equation true. The solution is $\frac{21}{11}$.

If an equation contains decimals, you may want to first clear the equation of decimals.

**Example 7** Solve: $0.3x + 0.1 = 0.27x - 0.02$.

*Solution* To clear this equation of decimals, we multiply both sides of the equation by 100. Recall that multiplying a number by 100 moves its decimal point two places to the right.

$$100(0.3x + 0.1) = 100(0.27x - 0.02)$$

$$100(0.3x) + 100(0.1) = 100(0.27x) - 100(0.02) \quad \text{Use the distributive property.}$$

$$30x + 10 = 27x - 2 \quad \text{Multiply.}$$

$$30x - 27x = -2 - 10 \quad \text{Subtract 27x and 10 from both sides.}$$

$$3x = -12 \quad \text{Simplify.}$$

$$\frac{3x}{3} = \frac{-12}{3} \quad \text{Divide both sides by 3.}$$

$$x = -4 \quad \text{Simplify.}$$

Check to see that the solution is $-4$.

4 So far, each linear equation that we have solved has had a single solution. A linear equation in one variable that has exactly one solution is called a **conditional equation**. We will now look at two other types of equations: contradictions and identities.

An equation in one variable that has no solution is called a **contradiction**, and an equation in one variable that has every number (for which the equation is defined) as a solution is called an **identity**. The next examples show how to recognize contradictions and identities.

**Example 8** Solve for $x$: $3x + 5 = 3(x + 2)$.

*Solution* First, use the distributive property and remove parentheses.

$$3x + 5 = 3(x + 2)$$

$$3x + 5 = 3x + 6 \quad \text{Apply the distributive property.}$$

$$3x + 5 - 3x = 3x + 6 - 3x \quad \text{Subtract } 3x \text{ from both sides.}$$

$$5 = 6$$

The equation $5 = 6$ is a false statement no matter what value the variable $x$ might have. Thus, the original equation has no solution. Its solution set is written either as { } or $\varnothing$. This equation is a contradiction.

**Example 9** Solve for $x$: $6x - 4 = 2 + 6(x - 1)$.

*Solution* First, use the distributive property and remove parentheses.

$$6x - 4 = 2 + 6(x - 1)$$

$$6x - 4 = 2 + 6x - 6 \quad \text{Apply the distributive property.}$$

$$6x - 4 = 6x - 4 \quad \text{Combine like terms.}$$

At this point we might notice that both sides of the equation are the same, so replacing $x$ by any real number gives a true statement. Thus the solution set of this equation is the set of real numbers, and the equation is an identity. Continuing to "solve" $6x - 4 = 6x - 4$, we eventually arrive at the same conclusion.

$$6x - 4 + 4 = 6x - 4 + 4 \quad \text{Add 4 to both sides.}$$

$$6x = 6x \quad \text{Simplify.}$$

$$6x - 6x = 6x - 6x \quad \text{Subtract } 6x \text{ from both sides.}$$

$$0 = 0 \quad \text{Simplify.}$$

Since $0 = 0$ is a true statement for every value of $x$, all real numbers are solutions. The solution set is the set of all real numbers or, $\mathbb{R}$, $\{x \mid x \text{ is a real number}\}$, and the equation is called an identity.

**HELPFUL HINT**

For linear equations, *any* false statement such as $5 = 6$, $0 = 1$, or $-2 = 2$ informs us that the original equation has no solution. Also, *any* true statement such as $0 = 0$, $2 = 2$, or $-5 = -5$ informs us that the original equation is an identity.

## MENTAL MATH

*Simplify each expression by combining like terms.*

**1.** $3x + 5x + 6 + 15$
**2.** $8y + 3y + 7 + 11$
**3.** $5n + n + 3 - 10$
**4.** $m + 2m + 4 - 8$
**5.** $8x - 12x + 5 - 6$
**6.** $4x - 10x + 13 - 16$

## Exercise Set 2.1

*Solve for the variable. See Examples 1 and 2.*

**1.** $-3x = 36$
**2.** $8x = -40$
**3.** $x + 2.8 = 1.9$
**4.** $y - 8.6 = -6.3$
**5.** $5x - 4 = 26$
**6.** $2y - 3 = 11$
**7.** $-4 = 3x + 11$
**8.** $-9 = 5x + 11$
**9.** $-4.1 - 7z = 3.6$
**10.** $10.3 - 6x = -2.3$
**11.** $5y + 12 = 2y - 3$
**12.** $4x + 14 = 6x + 8$

*Solve for the variable. See Examples 3 and 4.*

**13.** $8x - 5x + 3 = x - 7 + 10$
**14.** $6 + 3x + x = -x + 2 - 26$
**15.** $5x + 12 = 2(2x + 7)$
**16.** $2(x + 3) = x + 5$
**17.** $3(x - 6) = 5x$
**18.** $6x = 4(5 + x)$
**19.** $-2(5y - 1) - y = -4(y - 3)$
**20.** $-3(2w - 7) - 10 = 9 - 2(5w + 4)$
**21.** **a.** Simplify the expression $4(x + 1) + 1$.
**b.** Solve the equation $4(x + 1) + 1 = -7$.
**c.** Explain the difference between solving an equation for a variable and simplifying an expression.
**22.** Explain why the multiplication property of equality does not include multiplying both sides of an equation by 0. (*Hint:* Write down a false statement and then multiply both sides by 0. Is the result true or false? What does this mean?)

*Solve for the variable. See Examples 5 through 7.*

**23.** $\frac{x}{2} + \frac{2}{3} = \frac{3}{4}$

**24.** $\frac{x}{2} + \frac{x}{3} = \frac{5}{2}$

**25.** $\frac{3t}{4} - \frac{t}{2} = 1$

**26.** $\frac{4r}{5} - 7 = \frac{r}{10}$

**27.** $\frac{n-3}{4} + \frac{n+5}{7} = \frac{5}{14}$

**28.** $\frac{2+h}{9} + \frac{h-1}{3} = \frac{1}{3}$

**29.** $0.6x - 10 = 1.4x - 14$

**30.** $0.3x + 2.4 = 0.1x + 4$

*Solve the following. See Examples 8 and 9.*

**31.** $4(n + 3) = 2(6 + 2n)$

**32.** $6(4n + 4) = 8(3 + 3n)$

**33.** $3(x - 1) + 5 = 3x + 7$

**34.** $5x - (x + 4) = 5 + 4(x - 2)$

**35.** In your own words, explain why the equation $x + 7 = x + 6$ has no solution while the solution set of the equation $x + 7 = x + 7$ contains all real numbers.

**36.** In your own words, explain why the equation $x = -x$ has one solution, namely 0, while the solution set of the equation $x = x$ is all real numbers.

*Solve the following.*

**37.** $-9x = -72$

**38.** $-7x = 56$

**39.** $x - 1.7 = -7.6$

**40.** $y - 9.3 = -12.6$

**41.** $6x + 9 = 51$

**42.** $4x + 11 = 47$

**43.** $-5x + 1.5 = -19.5$

**44.** $-3x - 4.7 = 11.8$

**45.** $x - 10 = -6x + 4$

**46.** $4x - 7 = 2x - 7$

**47.** $3x - 4 - 5x = x + 4 + x$

**48.** $13x - 15x + 8 = 4x + 2 - 24$

**49.** $5(y + 4) = 4(y + 5)$

**50.** $6(y - 4) = 3(y - 8)$

**51.** $0.6x - 10 = 1.4x - 14$

**52.** $0.3x + 2.4 = 0.1x + 4$

**53.** $6x - 2(x - 3) = 4(x + 1) + 4$

**54.** $10x - 2(x + 4) = 8(x - 2) + 6$

**55.** $\frac{3}{8} + \frac{b}{3} = \frac{5}{12}$

**56.** $\frac{a}{2} + \frac{7}{4} = 5$

**57.** $z + 3(2 + 4z) = 6(z + 1) + 5z$

**58.** $4(m - 6) - m = 8(m - 3) - 5m$

**59.** $\frac{3t+1}{8} = \frac{5+2t}{7} + 2$

**60.** $4 - \frac{2z+7}{9} = \frac{7-z}{12}$

**61.** $\frac{m-4}{3} - \frac{3m-1}{5} = 1$

**62.** $\frac{n+1}{8} - \frac{2-n}{3} = \frac{5}{6}$

**63.** $5(x - 2) + 2x = 7(x + 4) - 38$

**64.** $3x + 2(x + 4) = 5(x + 1) + 3$

**65.** $y + 0.2 = 0.6(y + 3)$

**66.** $-(w + 0.2) = 0.3(4 - w)$

**67.** $2y + 5(y - 4) = 4y - 2(y - 10)$

**68.** $9c - 3(6 - 5c) = c - 2(3c + 9)$

**69.** $2(x - 8) + x = 3(x - 6) + 2$

**70.** $4(x + 5) = 3(x - 4) + x$

**71.** $\frac{3x-1}{9} + x = \frac{3x+1}{3} + 4$

**72.** $\frac{2z+7}{8} - 2 = z + \frac{z-1}{2}$

**73.** $1.5(4 - x) = 1.3(2 - x)$

**74.** $2.4(2x + 3) = -0.1(2x + 3)$

**75.** $-2(b - 4) - (3b - 1) = 5b + 3$

**76.** $4(t - 3) - 3(t - 2) = 2t + 8$

**77.** $\frac{1}{4}(a + 2) = \frac{1}{6}(5 - a)$

**78.** $\frac{1}{3}(8 + 2c) = \frac{1}{5}(3c - 5)$

*Find the value of K such that the equations are equivalent.*

**79.** $3.2x + 4 = 5.4x - 7$
$3.2x = 5.4x + K$

**80.** $-7.6y - 10 = -1.1y + 12$
$-7.6y = -1.1y + K$

**81.** $\frac{x}{6} + 4 = \frac{x}{3}$
$x + K = 2x$

**82.** $\frac{5x}{4} + \frac{1}{2} = \frac{x}{2}$
$5x + K = 2x$

*Solve and check.*

**83.** $2.569x = -12.48534$

**84.** $-9.112y = -47.537304$

**85.** $2.86z - 8.1258 = -3.75$

**86.** $1.25x - 20.175 = -8.15$

**87.** Recall from Section 1.3 that a game is fair if each team or player has an equal chance of winning. Cut or tear a sheet of paper into 10 pieces, numbering each piece from 1 to 10. Place the pieces into a bag. Draw 2 pieces from the bag, record their sum, and then return them to the bag. If the sum is 10 or less, player 1 gets a point. If their sum is more than 10, player 2 gets a point. Is this a fair game? Try it and see.

## REVIEW EXERCISES

*Translate each phrase into an expression. Use the variable x to represent each unknown number. See Section 1.2.*

**88.** the quotient of 8 and a number

**89.** the sum of 8 and a number

**90.** the product of 8 and a number

**91.** the difference of 8 and a number

**92.** 2 more than three times a number

**93.** 5 subtracted from twice a number

### A Look Ahead

**Example**

Solve for $x$: $5x(x-1)+14=x(4x-3)+x^2$.

*Solution*

$$\begin{aligned}5x^2-5x+14&=4x^2-3x+x^2\\5x^2-5x+14&=5x^2-3x\\-5x+14&=-3x\\14&=2x\\7&=x\end{aligned}$$

*Solve the following. See example.*

**94.** $x(x-6)+7=x(x+1)$

**95.** $7x^2+2x-3=6x(x+4)+x^2$

**96.** $3x(x+5)-12=3x^2+10x+3$

**97.** $x(x+1)+16=x(x+5)$

## 2.2 AN INTRODUCTION TO PROBLEM SOLVING

CD-ROM SSM SSG Video

**OBJECTIVES**

1. Write algebraic expressions that can be simplified.
2. Apply the steps for problem solving.

**1** In order to prepare for problem solving, we practice writing algebraic expressions that can be simplified.

Our first example involves consecutive integers and perimeter. Recall that *consecutive integers* are integers that follow one another in order. Study the examples of consecutive, even, and odd integers and their representations.

***Consecutive Integers:***

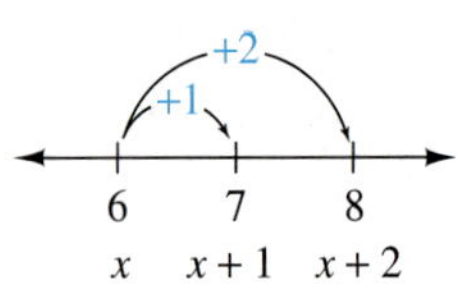


***Consecutive Even Integers:***

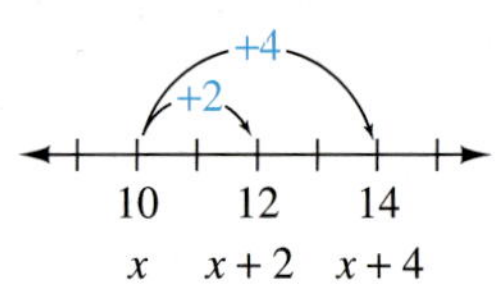


***Consecutive Odd Integers:***

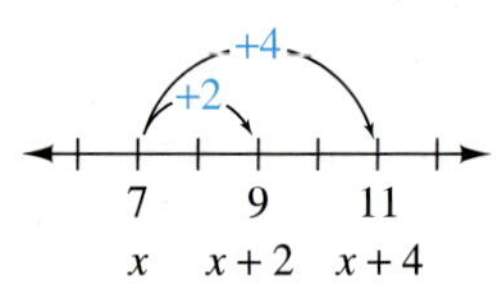


**Example 1** Write the following as algebraic expressions. Then simplify.

**a.** The sum of two consecutive integers, if $x$ is the first consecutive integer.

△**b.** The perimeter of the triangle with sides of length $x$, $5x$, and $6x-3$.

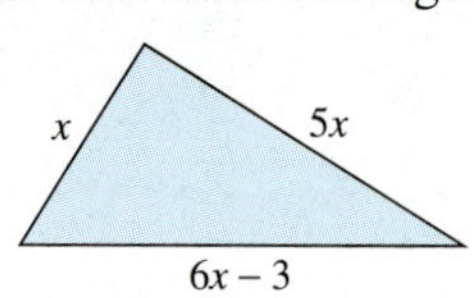


*Solution* **a.** Recall that if $x$ is the first integer, then the next consecutive integer is 1 more, or $x+1$.

| In words: | first integer | plus | next consecutive integer |
|---|---|---|---|
| Translate: | $x$ | $+$ | $(x+1)$ |

Then $x+(x+1)=x+x+1$

$=2x+1$ Simplify by combining like terms.

**b.** The perimeter of a triangle is the sum of the lengths of the sides.

In words: side + side + side

Translate: $x \quad + \quad 5x \quad + (6x - 3)$

Then $x + 5x + (6x - 3) = x + 5x + 6x - 3$

$= 12x - 3$ Simplify.

**Example 2** The three busiest airports in the United States are in the cities of Chicago, Atlanta, and Dallas/Ft. Worth. The airport in Atlanta has 7.7 million more arrivals and departures than the Dallas/Ft. Worth airport. The Chicago airport has 9.8 million more arrivals and departures than the Dallas/Ft Worth airport. Write the sum of the arrivals and departures from these three cities as a simplified algebraic expression. Let $x$ be the number of arrivals and departures from the Dallas/Ft. Worth airport.

*Solution* If $x$ = millions of arrivals and departures at the Dallas/Ft. Worth airport, then
$x + 7.7$ = millions of arrivals and departures at the Atlanta airport and
$x + 9.8$ = millions of arrivals and departures at the Chicago airport
Since we want their sum, we have

| | | | | | |
|---|---|---|---|---|---|
| In words: | arrivals and departures at Dallas/Ft. Worth | + | arrivals and departures at Atlanta | + | arrivals and departures at Chicago. |
| Translate: | $x$ | + | $(x + 7.7)$ | + | $(x + 9.8)$ |

Then $x + (x + 7.7) + (x + 9.8) = x + x + 7.7 + x + 9.8$

$= 3x + 17.5$ Combine like terms.

In Exercise 25, we will find the actual number of arrivals and departures at these airports.

2 Our main purpose for studying algebra is to solve problems. The following problem-solving strategy will be used throughout this text and may also be used to solve real-life problems that occur outside the mathematics classroom.

**HELPFUL HINT**

You may want to begin this section by studying key words and phrases and their translations in Sections 1.2 Objective 5 and 1.4 Objective 4.

**GENERAL STRATEGY FOR PROBLEM SOLVING**

1. UNDERSTAND the problem. During this step, become comfortable with the problem. Some ways of doing this are:

   Read and reread the problem.

   Choose a variable to represent the unknown.

   Construct a drawing.

   Propose a solution and check. Pay careful attention to how you check your proposed solution. This will help when writing an equation to model the problem.
2. TRANSLATE the problem into an equation.
3. SOLVE the equation.
4. INTERPRET the results: *Check* the proposed solution in the stated problem and *state* your conclusion.

Let's review this strategy by solving a problem involving unknown numbers.

**Example 3** **FINDING UNKNOWN NUMBERS**

Find two numbers such that the second number is 3 more than twice the first number and the sum of the two numbers is 72.

*Solution*

1. UNDERSTAND the problem. First let's read and reread the problem and then propose a solution. For example, if the first number is 25, then the second number is 3 more than twice 25, or 53. The sum of 25 and 53 is 78, not the required sum, but we have gained some valuable information about the problem. First, we know that the first number is less than 25 since our guess led to a sum greater than the required sum. Also, we have gained some information as to how to model the problem with an equation.

**HELPFUL HINT**

The purpose of proposing a solution is not to guess correctly but to gain confidence and to help understand the problem and how to model it.

Next let's assign a variable and use this variable to represent any other unknown quantities. If we let

the first number $= x$, then

the second number $= \underbrace{2x}_{\text{twice the second number}} + \underbrace{3}_{\text{3 more than}}$

2. TRANSLATE the problem into an equation. To do so, we use the fact that the sum of the numbers is 72. First let's write this relationship in words and then translate to an equation.

| In words: | first number | added to | second number | is | 72 |
|---|---|---|---|---|---|
| | ↓ | ↓ | ↓ | ↓ | ↓ |
| Translate: | $x$ | $+$ | $(2x + 3)$ | $=$ | 72 |

3. SOLVE the equation.

$$x + (2x + 3) = 72$$

$$x + 2x + 3 = 72 \quad \text{Remove parentheses.}$$

$$3x + 3 = 72 \quad \text{Combine like terms.}$$

$$3x = 69 \quad \text{Subtract 3 from both sides.}$$

$$x = 23 \quad \text{Divide both sides by 3.}$$

4. INTERPRET. Here, *we check* our work and *state* the solution. Recall that if the first number $x = 23$, then the second number $2x + 3 = 2 \cdot 23 + 3 = 49$.

**Check:** Is the second number 3 more than twice the first number? Yes, since 3 more than twice 23 is $46 + 3$, or 49. Also, their sum, $23 + 49 = 72$, is the required sum.

**State:** The two numbers are 23 and 49.

Many of today's rates and statistics are given as percents. Interest rates, tax rates, nutrition labeling, and percent of households in a given category are just a few examples. Before we practice solving problems containing percents, let's take a moment and review the meaning of percent and how to find a percent of a number.

The word *percent* means "per hundred," and the symbol % is used to denote percent. This means that 23% is 23 per hundred, or $\frac{23}{100}$. Also,

$$41\% = \frac{41}{100} = 0.41$$

To find a percent of a number, we multiply.

## Example 4 Find 16% of 25.

*Solution* To find 16% of 25, we find the product of 16% (written as a decimal) and 25.

$$16\% \cdot 25 = 0.16 \cdot 25$$
$$= 4$$

Thus, 16% of 25 is 4.

Next, we solve a problem containing percent.

## Example 5 FINDING THE ORIGINAL PRICE OF A COMPUTER

Suppose that The Digital Store just announced an 8% decrease in the price of their Compaq Presario computers. If one particular computer model sells for \$2162 after the decrease, find the original price of this computer.

*Solution*

1. UNDERSTAND. Read and reread the problem. Recall that a percent decrease means a percent of the original price. Let's guess that the original price of the computer is \$2500. The amount of decrease is then 8% of \$2500, or $(0.08)(\$2500) = \$200$. This means that the new price of the computer is the original price minus the decrease, or $\$2500 - \$200 = \$2300$. Our guess is incorrect, but we now have an idea of how to model this problem. In our model, we will let $x =$ the original price of the computer.

2. TRANSLATE.

| | | | | | |
|---|---|---|---|---|---|
| In words: | original price of computer | minus | 8% of original price | is | new price |
| | ↓ | ↓ | ↓ | ↓ | ↓ |
| Translate: | $x$ | $-$ | $0.08x$ | $=$ | 2162 |

3. SOLVE the equation.

$$x - 0.08x = 2162$$

$$0.92x = 2162 \qquad \text{Combine like terms.}$$

$$x = \frac{2162}{0.92} = 2350 \qquad \text{Divide both sides by 0.92.}$$

4. INTERPRET.

**Check:** If the original price of the computer was \$2350, the new price is

$$\$2350 - (0.08)(\$2350) = \$2350 - \$188$$

$$= \$2162 \qquad \text{The given new price.}$$

**State:** The original price of the computer was \$2350.

## △ Example 6 FINDING THE LENGTHS OF A TRIANGLE'S SIDES

A pennant in the shape of an isosceles triangle is to be constructed for the Slidell High School Athletic Club and sold at a fund-raiser. The company manufacturing the pennant charges according to perimeter, and the athletic club has determined that a perimeter of 149 centimeters should make a nice profit. If each equal side of the triangle is twice the length of the third side, increased by 12 centimeters, find the lengths of the sides of the triangular pennant.

*Solution*

1. UNDERSTAND. Read and reread the problem. Recall that the perimeter of a triangle is the distance around. Let's guess that the third side of the triangular pennant is 20 centimeters. This means that each equal side is twice 20 centimeters, increased by 12 centimeters, or $2(20) + 12 = 52$ centimeters.

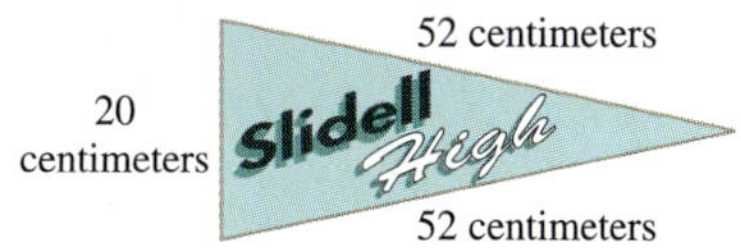


This gives a perimeter of $20 + 52 + 52 = 124$ centimeters. Our guess is incorrect, but we now have a better understanding of how to model this problem.
Now we let the third side of the triangle $= x$

| the first side = | twice | the third side | increased by 12 |
|---|---|---|---|
| | ↓ | ↓ | ↓ |
| = | 2 | $x$ | + 12, |

or $2x + 12$
the second side $= 2x + 12$

2. TRANSLATE.

| | | | | | | | |
|---|---|---|---|---|---|---|---|
| In words: | first side | + | second side | + | third side | = | 149 |
| | ↓ | | ↓ | | ↓ | | ↓ |
| Translate: | $(2x + 12)$ | + | $(2x + 12)$ | + | $x$ | = | 149 |

3. SOLVE the equation.

$$(2x + 12) + (2x + 12) + x = 149$$
$$2x + 12 + 2x + 12 + x = 149 \quad \text{Remove parentheses.}$$
$$5x + 24 = 149 \quad \text{Combine like terms.}$$
$$5x = 125 \quad \text{Subtract 24 from both sides.}$$
$$x = 25 \quad \text{Divide both sides by 5.}$$

4. INTERPRET. If the third side is 25 centimeters, then the first side is $2(25) + 12 = 62$ centimeters and the second side is 62 centimeters also.

**Check:** The first and second sides are each twice 25 centimeters increased by 12 centimeters or 62 centimeters. Also, the perimeter is $25 + 62 + 62 = 149$ centimeters, the required perimeter.

**State:** The lengths of the sides of the triangle are 25 centimeters, 62 centimeters, and 62 centimeters.

**Example 7** Kelsey Ohleger was helping her friend Benji Burnstine study for an algebra exam. Kelsey told Benji that her two latest art history quiz scores are two consecutive even integers whose sum is 174. Help Benji find the scores.

*Solution*

1. UNDERSTAND. Read and reread the problem. Since we are looking for consecutive even integers, let

$$x = \text{the first integer. Then}$$
$$x + 2 = \text{the next consecutive even integer.}$$

2. TRANSLATE.

| | | | | | |
|---|---|---|---|---|---|
| In words: | first integer | + | next even integer | = | 174 |
| | ↓ | | ↓ | | ↓ |
| Translate: | $x$ | + | $(x + 2)$ | = | 174 |

3. SOLVE.

$$x + (x + 2) = 174$$
$$2x + 2 = 174 \quad \text{Combine like terms.}$$
$$2x = 172 \quad \text{Subtract 2 from both sides.}$$
$$x = 86 \quad \text{Divide both sides by 2.}$$

4. INTERPRET. If $x = 86$, then $x + 2 = 86 + 2$ or 88.

**Check:** The numbers 86 and 88 are two consecutive even integers. Their sum is 174, the required sum.

**State:** Kelsey's art history quiz scores are 86 and 88.

## SPOTLIGHT ON DECISION MAKING

Suppose you are a reporter for the *Marston Gazette*, a daily newspaper serving the medium-sized industrial city of Marston. Your editor has assigned you to a feature story about food pantries, soup kitchens, and other efforts to alleviate hunger among the city's homeless, poor, and working poor.

While researching your story assignment, you find that, according to the U.S. Department of Agriculture, 10.2% of all households in the United States do not have access to enough food to meet their basic needs. A survey conducted by Marston Social Services reveals that approximately 4950 of the estimated 39,400 Marston households must cope with hunger.

Armed with these basic facts, you now need to decide what angle to take with your story. Which of the following approaches would you choose? Why? What other information would you want to consider?

**a.** Marston lags behind nation in fight against hunger.
**b.** Marston mirrors national hunger picture.
**c.** Marston makes progress against hunger.

## Exercise Set 2.2

*Write the following as algebraic expressions. Then simplify. See Examples 1 and 2.*

△ **1.** The perimeter of the square with side length $y$.

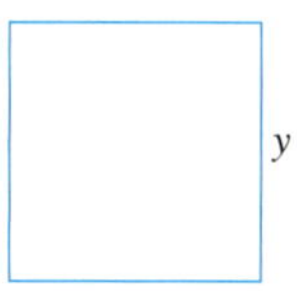


△ **2.** The perimeter of the rectangle with length $x$ and width $x - 5$.

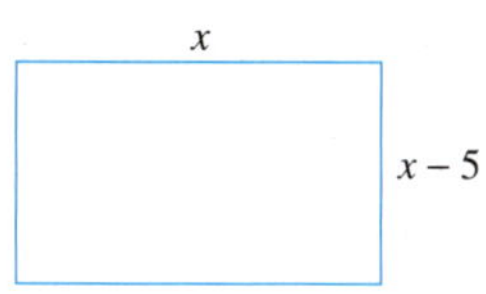


**3.** The sum of three consecutive integers if the first integer is $z$.

**4.** The sum of three consecutive odd integers if the first integer is $x$.

**5.** The total amount of money (in cents) in $x$ nickels and $(x + 3)$ dimes. (*Hint:* the value of a nickel is 5 cents and the value of a dime is 10 cents.)

**6.** The total amount of money (in cents) in $y$ quarters and $(2y - 1)$ nickels. (Use the hint for Exercise 5.)

△ **7.** A piece of land along Bayou Liberty is to be fenced and subdivided as shown so that each rectangle has the same dimensions. Express the total amount of fencing needed as an algebraic expression in $x$.

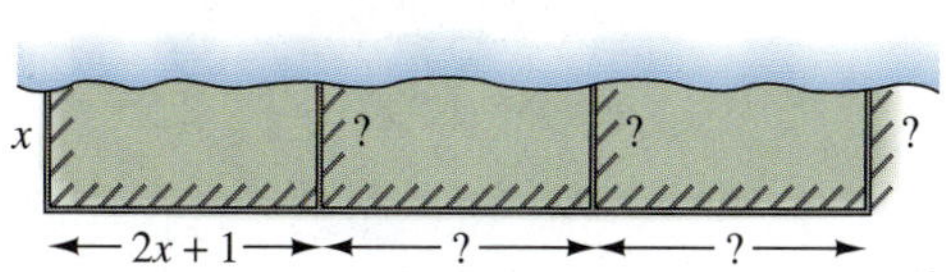


△ **8.** Write the perimeter of the floor plan shown as an algebraic expression in $x$.

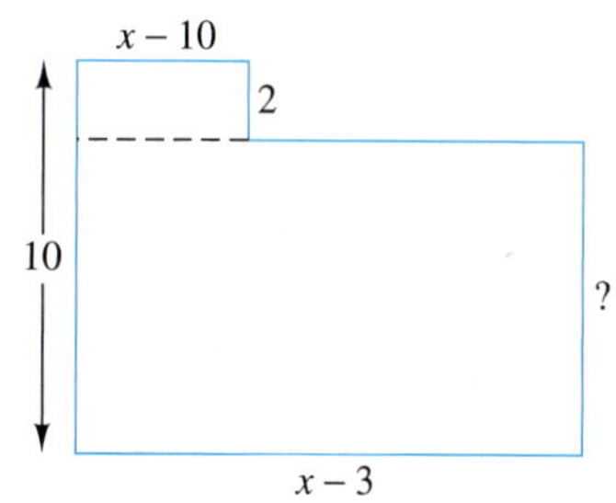


*Solve. See Example 3.*

**9.** Four times the difference of a number and 2 is the same as 6 times the number, increased by 2. Find the number.

**10.** Twice the sum of a number and 3 is the same as 1 subtracted from the number. Find the number.

**11.** One number is 5 times another number. If the sum of the two numbers is 270, find the numbers.

**12.** One number is 6 less than another number. If the sum of the two numbers is 150, find the numbers.

*Solve. See Example 4.*

**13.** Find 30% of 260.

**14.** Find 70% of 180.

**15.** Find 12% of 16.

**16.** Find 22% of 12.

**17.** The United States consists of 2271 million acres of land. Approximately 29% of this land is federally owned. Find the number of acres that are federally owned. (*Source:* U.S. General Services Administration)

**18.** The state of Nevada contains the most federally owned acres of land in the United States. If 90% of the state's 70 million acres of land is federally owned, find the number of federally owned acres. (*Source:* U.S. General Services Administration)

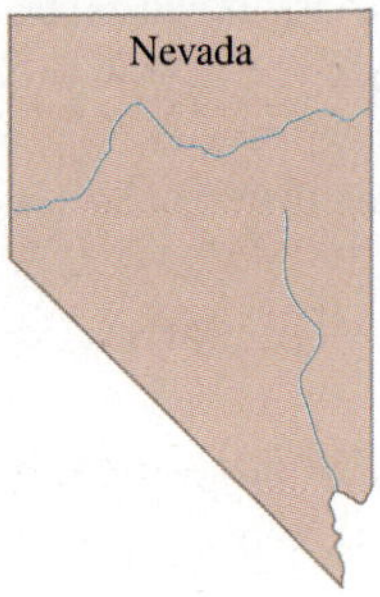


**19.** Recently, 47% of homes in the United States contained computers. If Charlotte, North Carolina, contains 110,000 homes, how many of these homes would you expect to have computers? (*Source:* Telecommunication Research survey)

**20.** Recently, 26% of homes in the United States contained online services. If Abilene, Texas, contains 40,000 homes, how many of these homes would you expect to have online services? (*Source:* TelecommunicationResearch survey)

*The following graph is called a circle graph or a pie chart. The circle represents a whole, or in this case, 100%. This particular graph shows the kind of loans that customers get from credit unions. Use this graph to answer Exercises 21–24.*

**Credit Union Loans**

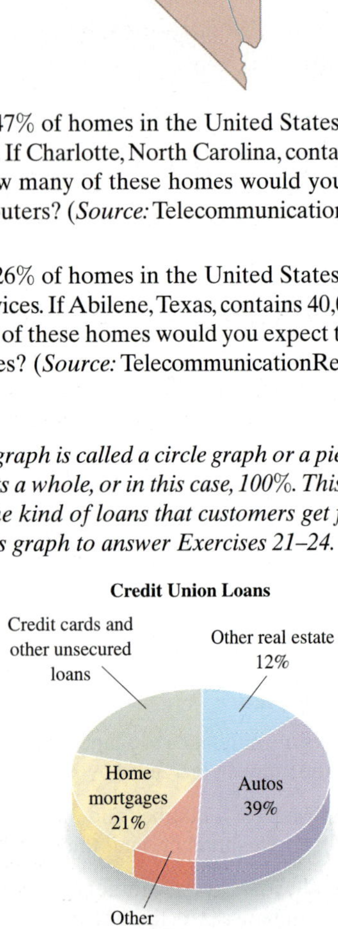


*Source:* National Credit Union Administration.

**21.** What percent of credit union loans are for credit cards and other unsecured loans?

**22.** What types of loans make up most credit union loans?

**23.** If the University of New Orleans Credit Union processed 300 loans last year, how many of these might we expect to be automobile loans?

**24.** If Homestead's Credit Union processed 537 loans last year, how many of these do you expect to be either home mortgages or other real estate? (Round to the nearest whole.)

*Solve. See Examples 5 through 7.*

**25.** The airports in Chicago, Atlanta, and Dallas/Ft. Worth have a total of 199 million annual arrivals and departures. Use this information and Example 2 in this section to find the number from each individual airport.

**26.** The perimeter of the triangle in Example 1b is 483 feet. Find the length of each side.

**27.** The B767-300ER aircraft has 104 more seats than the B737-200 aircraft. If their total number of seats is 328, find the number of seats for each aircraft. (*Source:* Air Transport Association of America)

**28.** The governor of Connecticut makes $29,000 less per year than the governor of Delaware. If the total of these salaries is $185,000, find the salary of each governor. (*Source: 2000 World Almanac*)

**29.** A new fax machine was recently purchased for an office in Hopedale for $464.40 including tax. If the tax rate in Hopedale is 8%, find the price of the fax machine before taxes.

**30.** A premedical student at a local university was complaining that she had just paid $86.11 for her human anatomy book, including tax. Find the price of the book before taxes if the tax rate at this university is 9%.

**31.** According to government statistics, the number of telephone company operators in the United States is expected to decrease to 26,000 by the year 2006. This represents a decrease of 47% from the number of telephone operators in 1996. (*Source:* U.S. Bureau of Labor Statistics)

**a.** Find the number of telephone company operators in 1996. Round to the nearest whole number.

**b.** In your own words, explain why you think that the need for telephone company operators is decreasing.

**32.** The number of deaths by tornadoes from the 1940s to the 1980s has decreased by 70.86%. There were 521 deaths from tornadoes in the 1980s. (*Source:* National Weather Service)

**a.** Find the number of deaths by tornadoes in the 1940s. Round to the nearest whole number.

**b.** In your own words, explain why you think that the number of deaths by tornadoes has decreased so much since the 1940s.

**33.** Manufacturers claim that a CD-ROM disc will last 20 years. Recently, statements made by the U.S. National Archives and Records Administration suggest that 20 years decreased by 75% is a more realistic lifespan because the aluminum substratum on which the data is recorded can be affected by oxidation. Find the lifespan of a CD-ROM according to the U.S. National Archives and Records Administration.

**34.** In one year, 2.7% of India's forest was lost to deforestation. This percent represents 10,000 square kilometers of forest. Find the total square kilometers of forest in India before this decrease. (Round to the nearest whole square kilometer.)

**35.** Americans used computers to electronically file 21.1 million federal income tax returns in 1999. This represented a 24% increase over the previous year. How many federal income tax returns were filed electronically in 1998? Round to the nearest million. (*Source: Associated Press*, April, 1999)

**36.** HDPE (high-density polyethylene) plastics are used to make milk and water jugs, as well as bottles for juices and laundry detergents. When these types of bottles are recycled, they can be used to make bags, recycling bins, motor oil bottles, and agricultural pipe. HDPE bottle recycling increased 7% to 704 million pounds from 1996 to 1997. How many pounds of HDPE bottles were recycled in 1996? Round to the nearest million. (*Source:* American Plastics Council)

△ **37.** Two frames are needed with the same outside perimeter: one frame in the shape of a square and one in the shape of an equilateral triangle. Each side of the triangle is 6 centimeters longer than each side of the square. Find the dimensions of each frame.

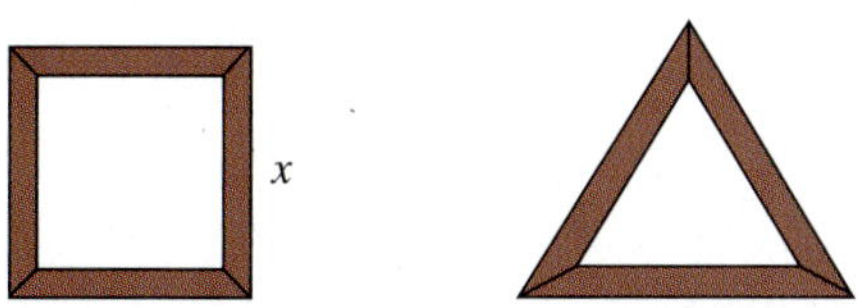


△ **38.** The length of a rectangular sign is 2 feet less than three times its width. Find the dimensions if the perimeter is 28 feet.

△ **39.** In a blueprint of a rectangular room, the length is to be 2 centimeters greater than twice its width. Find the dimensions if the perimeter is to be 40 centimeters.

**40.** A plant food solution contains 5 cups of water for every 1 cup of concentrate. If the solution contains 78 cups of these two ingredients, find the number of cups of concentrate in the solution.

△ **41.** The external tank of a NASA space shuttle contains the propellants used for the first 8.5 minutes after launch. Its height is 5 times the sum of its width and 1. If the sum of the height and width is 55.4 meters, find the dimensions of this tank.

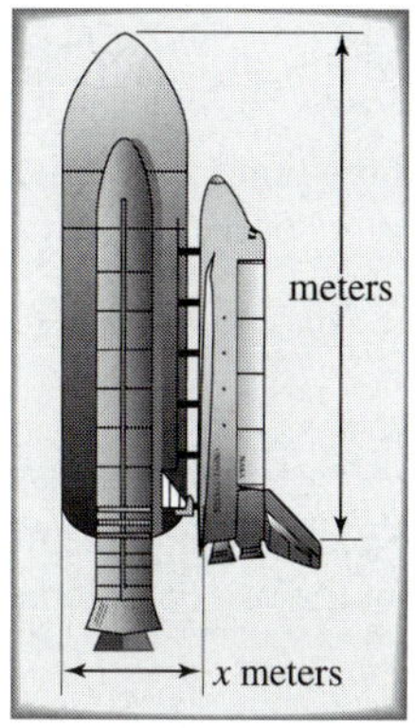


**42.** The blue whale is the largest of the whales. Its average weight is 3 times the difference of the average weight of a humpback whale and 5 tons. If the total of the average weights is 117 tons, find the average weight of each type of whale.

*Recall that the sum of the angle measures of a triangle is 180°.*

△ **43.** Find the measures of the angles of a triangle if the measure of one angle is twice the measure of a second angle and the third angle measures 3 times the second angle decreased by 12.

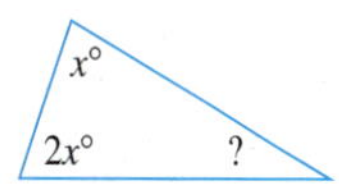


△ **44.** Find the angles of a triangle whose two base angles are equal and whose third angle is 10° less than three times a base angle.




*Recall that two angles are complements of each other if their sum is 90°. Two angles are supplements of each other if their sum is 180°. Find the measure of each angle.*

△ **45.**

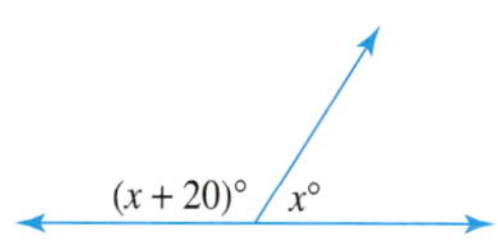


△ **46.**

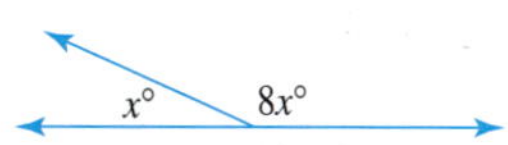


△ **47.**

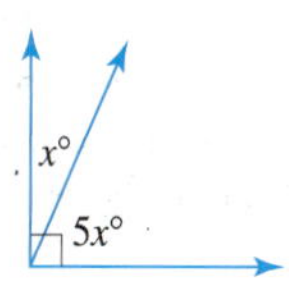

△ **48.**

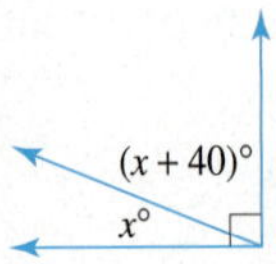


△ **49.** One angle is three times its supplement increased by 20°. Find the measures of the two supplementary angles.

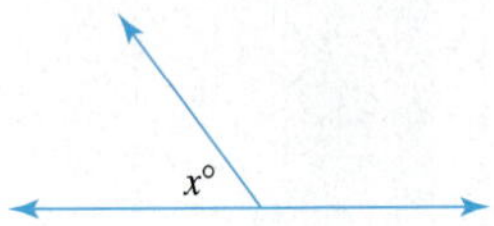


△ **50.** One angle is twice its complement increased by 30°. Find the measures of the two complementary angles.

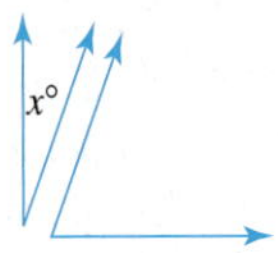


*Solve. Exercises 51–56 involve consecutive integers.*

**51.** The sum of three consecutive integers is 228. Find the integers.

**52.** The sum of three consecutive odd integers is 327. Find the integers.

**53.** The zip codes of three Nevada locations—Fallon, Fernley, and Gardnerville Ranchos—are three consecutive even integers. If twice the first integer added to the third is 268,222, find each zip code.

**54.** During a recent year, the average SAT scores in math for the states of Alabama, Louisiana, and Michigan were 3 consecutive integers. If the sum of the first integer, second integer, and three times the third integer is 2637, find each score.

**55.** Determine whether there are three consecutive integers such that their sum is three times the second integer.

**56.** Determine whether there are two consecutive odd integers such that 7 times the first exceeds 5 times the second by 54.

*To break even in a manufacturing business, income or revenue R must equal the cost of production C. Use this information for Exercises 57 through 62.*

**57.** The cost $C$ to produce $x$ number of skateboards is $C = 100 + 20x$. The skateboards are sold wholesale for \$24 each, so revenue $R$ is given by $R = 24x$. Find how many skateboards the manufacturer needs to produce and sell to break even. (*Hint:* Set the cost expression equal to the revenue expression and solve for $x$.)

**58.** The revenue $R$ from selling $x$ number of computer boards is given by $R = 60x$, and the cost $C$ of producing them is given by $C = 50x + 5000$. Find how many boards must be sold to break even. Find how much money is needed to produce the break-even number of boards.

**59.** The cost $C$ of producing $x$ number of paperback books is given by $C = 4.50x + 2400$. Income $R$ from these books is given by $R = 7.50x$. Find how many books should be produced and sold to break even.

**60.** Find the break-even quantity for a company that makes $x$ number of computer monitors at a cost $C$ given by $C = 875 + 70x$ and receives revenue $R$ given by $R = 105x$.

**61.** In your own words, explain what happens if a company makes and sells fewer products than the break-even point.

**62.** In your own words, explain what happens if more products than the break-even point are made and sold.

△ **63.** Newsprint is either discarded or recycled. Americans recycle about 27% of all newsprint, but an amount of newsprint equivalent to 30 million trees is discarded every year. About how many trees' worth of newsprint is *recycled* in the United States each year? (*Source:* The Earth Works Group)

△ **64.** Find an angle such that its supplement is equal to twice its complement increased by 50°.

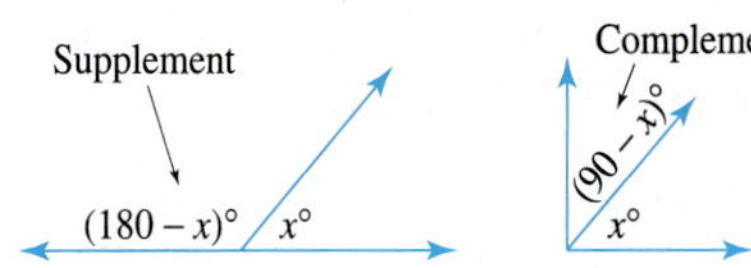


## REVIEW EXERCISES

*Find the value of the following expressions for the given values. See Section 1.3.*

**65.** $2a + b - c$; $a = 5, b = -1$, and $c = 3$

**66.** $-3a + 2c - b$; $a = -2$; $b = 6$, and $c = -7$

**67.** $4ab - 3bc$; $a = -5, b = -8$, and $c = 2$

**68.** $ab + 6bc$; $a = 0, b = -1$ and $c = 9$

**69.** $n^2 - m^2$; $n = -3$ and $m = -8$

**70.** $2n^2 + 3m^2$; $n = -2$ and $m = 7$

*Use a calculator to find the value of the following expressions for the given values. See Section 1.3.*

**71.** $P + Prt$; $P = 3000, r = 0.0325, t = 2$

**72.** $\frac{1}{3}lwh$; $l = 37.8, w = 5.6, h = 7.9$

**73.** $\frac{1}{3}Bh$; $B = 53.04, h = 6.89$

**74.** $\left(1 + \frac{r}{n}\right)^{nt}$; $r = 0.06, n = 3, t = 1$

## 2.3 FORMULAS AND PROBLEM SOLVING

CD-ROM SSM SSG Video

▶ **OBJECTIVES**

1. Solve a formula for a specified variable.
2. Use formulas to solve problems.

**1** Solving problems that we encounter in the real world sometimes requires us to express relationships among measured quantities. A **formula** is an equation that describes a known relationship among quantities such as time, area, and gravity. Some examples of formulas are

| ***Formula*** | ***Meaning*** |
|---|---|
| $I = PRT$ | Interest = principal · rate · time |
| $A = lw$ | Area of a rectangle = length · width |
| $d = rt$ | Distance = rate · time |
| $C = 2\pi r$ | Circumference of a circle = $2 \cdot \pi \cdot$ radius |
| $V = lwh$ | Volume of a rectangular solid = length · width · height |

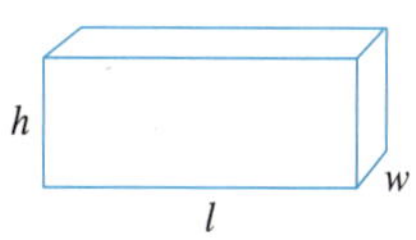


Other formulas are listed in the front cover of this text. Notice that the formula for the volume of a rectangular solid $V = lwh$ is solved for $V$ since $V$ is by itself on one side of the equation with no $V$'s on the other side of the equation. Suppose that the volume of a rectangular solid is known as well as its width and its length, and we wish to find its height. One way to find its height is to begin by solving the formula $V = lwh$ for $h$.

**△ Example 1** Solve $V = lwh$ for $h$.

*Solution* To solve $V = lwh$ for $h$, isolate $h$ on one side of the equation. To do so, divide both sides of the equation by $lw$.

$$V = lwh$$

$$\frac{V}{lw} = \frac{lwh}{lw} \quad \text{Divide both sides by } lw.$$

$$\frac{V}{lw} = h \quad \text{Simplify.}$$

Then to find the height of a rectangular solid, divide the volume by the product of its length and its width. ■

The following steps may be used to solve formulas and equations in general for a specified variable.

**SOLVING EQUATIONS FOR A SPECIFIED VARIABLE**

**Step 1:** Clear the equation of fractions by multiplying each side of the equation by the least common denominator.
**Step 2:** Use the distributive property to remove grouping symbols such as parentheses.
**Step 3:** Combine like terms on each side of the equation.
**Step 4:** Use the addition property of equality to rewrite the equation as an equivalent equation with terms containing the specified variable on one side and all other terms on the other side.
**Step 5:** Use the distributive property and the multiplication property of equality to isolate the specified variable.

**Example 2** Solve $3y - 2x = 7$ for $y$.

*Solution* This is a linear equation in two variables. Often an equation such as this is solved for $y$ in order to reveal some properties about the graph of this equation, which we will learn more about in Chapter 3. Since there are no fractions or grouping symbols, we begin with Step 4 and isolate the term containing the specified variable $y$ by adding $2x$ to both sides of the equation.

$$3y - 2x = 7$$

$$3y - 2x + 2x = 7 + 2x \quad \text{Add } 2x \text{ to both sides.}$$

$$3y = 7 + 2x$$

To solve for $y$, divide both sides by 3.

$$\frac{3y}{3} = \frac{7 + 2x}{3} \quad \text{Divide both sides by 3.}$$

$$y = \frac{2x + 7}{3} \quad \text{or} \quad y = \frac{2x}{3} + \frac{7}{3}$$

**Example 3** Solve $A = \frac{1}{2}(B + b)h$ for $b$.

*Solution* Since this formula for finding the area of a trapezoid contains fractions, we begin by multiplying both sides of the equation by the LCD 2.

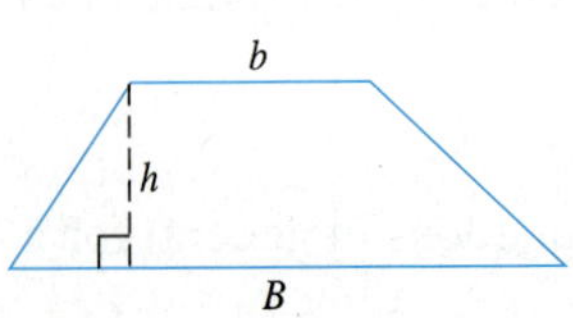


$$A = \frac{1}{2}(B + b)h$$

$$2 \cdot A = 2 \cdot \frac{1}{2}(B + b)h \quad \text{Multiply both sides by 2.}$$

$$2A = (B + b)h \quad \text{Simplify.}$$

Next, use the distributive property and remove parentheses.

$$2A = (B + b)h$$

$$2A = Bh + bh \quad \text{Apply the distributive property.}$$

$$2A - Bh = bh \quad \text{Isolate the term containing } b \text{ by subtracting } Bh \text{ from both sides.}$$

$$\frac{2A - Bh}{h} = \frac{bh}{h} \quad \text{Divide both sides by } h.$$

$$\frac{2A - Bh}{h} = b, \quad \text{or} \quad b = \frac{2A - Bh}{h}$$

**HELPFUL HINT**

Remember that we may isolate the specified variable on either side of the equation.

2 In this section, we also solve problems that can be modeled by known formulas. We use the same problem-solving steps that were introduced in the previous section.

Formulas are very useful in problem solving. For example, the compound interest formula

$$A = P\left(1 + \frac{r}{n}\right)^{nt}$$

is used by banks to compute the amount $A$ in an account that pays compound interest. The variable $P$ represents the principal or amount invested in the account, $r$ is the annual rate of interest, $t$ is the time in years, and $n$ is the number of times compounded per year.

## Example 4 FINDING THE AMOUNT IN A SAVINGS ACCOUNT

Marial Callier just received an inheritance of \$10,000 and plans to place all the money in a savings account that pays 5% compounded quarterly to help her son go to college in 3 years. How much money will be in the account in 3 years?

*Solution*

1. UNDERSTAND. Read and reread the problem. The appropriate formula needed to solve this problem is the compound interest formula

$$A = P\left(1 + \frac{r}{n}\right)^{nt}$$

Make sure that you understand the meaning of all the variables in this formula.

$A$ = amount in the account after $t$ years
$P$ = principal or amount invested
$t$ = time in years
$r$ = annual rate of interest
$n$ = number of times compounded per year

2. TRANSLATE. Use the compound interest formula and let $P = \$10{,}000$, $r = 5\% = 0.05$, $t = 3$ years, and $n = 4$ since the account is compounded quarterly, or 4 times a year.

Formula: $A = P\left(1 + \frac{r}{n}\right)^{nt}$

Substitute: $A = 10{,}000\left(1 + \frac{0.05}{4}\right)^{4\cdot 3}$

3. SOLVE. We simplify the right side of the equation.

$$A = 10{,}000\left(1 + \frac{0.05}{4}\right)^{4\cdot 3}$$

$$A = 10{,}000(1.0125)^{12}$$ Simplify $1 + \frac{0.05}{4}$ and write $4 \cdot 3$ as 12.

$$A \approx 10{,}000(1.160754518)$$ Approximate $(1.0125)^{12}$.

$$A \approx 11{,}607.55$$ Multiply and round to two decimal places.

4. INTERPRET.

**Check:** Repeat your calculations to make sure that no error was made. Notice that \$11,607.55 is a reasonable amount to have in the account after 3 years.

**State:** In 3 years, the account will contain \$11,607.55.

## Example 5 FINDING CYCLING TIME

The fastest average speed by a cyclist across the continental United States is 15.4 mph, by Pete Penseyres. If he traveled a total distance of about 3107.5 miles at this speed, find his time cycling. Write the time in days, hours, and minutes. *(Source: The Guinness Book of World Records*, 1998)

*Solution*

1. UNDERSTAND. Read and reread the problem. The appropriate formula needed is the distance formula

$d = rt$ where

$d$ = distance traveled $r$ = rate and $t$ = time

2. TRANSLATE. Use the distance formula and let $d = 3107.5$ miles and $r = 15.4$ mph.

**Formula:** $d = rt$

**Substitute:** $3107.5 = 15.4t$

3. SOLVE.

$$\frac{3107.5}{15.4} = \frac{15.4t}{15.4}$$ Divide both sides by 15.4.

$$201.79 \approx t$$

The time is approximately 201.79 hours. Since there are 24 hours in a day, we divide 201.79 by 24 and find that the time is approximately 8.41 days. Now, let's convert the decimal part of 8.41 days back to hours. To do this, multiply 0.41 by 24 and the result is 9.84 hours. Next, we convert the decimal part of 9.84 hours to minutes by multiplying by 60 since there an 60 minutes in an hour. We have $0.84 \cdot 60 \approx 50$ minutes rounded to the nearest whole. The time is then approximately

8 days, 9 hours, 50 minutes.

4. INTERPRET.

**Check:** Repeat your calculations to make sure that an error was not made.

**State:** Pete Penseyres's cycling time was approximately 8 days, 9 hours, 50 minutes.

## SPOTLIGHT ON DECISION MAKING

Suppose you are a dentist. Although fluoride can play an important part in a treatment plan to prevent tooth decay, you know that large doses of fluoride can be lethal. The formula $F = 10qpr$ can be used to calculate the number of milligrams (mg) of fluoride $F$ ingested by a patient who receives $q$ milliliters of a fluoride solution with $p$ percent concentration and molecular weight ratio $r$. The molecular weight ratios for common fluoride compounds are given in Table 1. Table 2 shows the maximum safe doses of fluoride for children, along with certainly lethal doses.

Decide whether or not a fluoride treatment of 5 milliliters of an 8 percent $SnF_2$ solution is safe for a 50-pound child.

MOLECULAR WEIGHT RATIOS

| *Fluoride Compound* | *Ratio* |
|---|---|
| NaF | $\frac{1}{2.2}$ |
| $Na_2FPO_3$ | $\frac{1}{7.6}$ |
| $SnF_2$ | $\frac{1}{4.1}$ |

FLUORIDE DOSES FOR CHILDREN

| *Weight (pounds)* | *Maximum Safe Dose (mg)* | *Certainly Lethal Dose (mg)* |
|---|---|---|
| 20 | 73 | 291 |
| 30 | 109 | 436 |
| 40 | 146 | 582 |
| 50 | 182 | 727 |
| 60 | 218 | 873 |
| 70 | 255 | 1018 |
| 80 | 291 | 1163 |
| 90 | 327 | 1309 |
| 100 | 363 | 1454 |

(*Source*: Based on data from S.B. Heifetz and H.S Horowitz. "The Amounts of Fluoride in Current Fluoride Therapies: Safety Considerations for Children," *ASDC J. Dent. Child.*, July–Aug. 1984.)

## MENTAL MATH

*Solve each equation for the specified variable. See Examples 1 through 3.*

**1.** $2x + y = 5$; for $y$

**2.** $7x - y = 3$; for $y$

**3.** $a - 5b = 8$; for $a$

**4.** $7r + s = 10$; for $s$

**5.** $5j + k - h = 6$; for $k$

**6.** $w - 4y + z = 0$; for $z$

## Exercise Set 2.3

*Solve each equation for the specified variable. See Examples 1–3.*

**1.** $D = rt$; for $t$

**2.** $W = gh$; for $g$

**3.** $I = PRT$; for $R$

**4.** $V = lwh$; for $l$

**5.** $9x - 4y = 16$; for $y$

**6.** $2x + 3y = 17$; for $y$

△ **7.** $P = 2L + 2W$; for $W$

**8.** $A = 3M - 2N$; for $N$

**9.** $J = AC - 3$; for $A$

**10.** $y = mx + b$; for $x$

**11.** $W = gh - 3gt^2$; for $g$

**12.** $A = Prt + P$; for $P$

**13.** $T = C(2 + AB)$; for $B$

**14.** $A = 5H(b + B)$; for $b$

△ **15.** $C = 2\pi r$; for $r$

△ **16.** $S = 2\pi r^2 + 2\pi rh$; for $h$

**17.** $E = I(r + R)$; for $r$

**18.** $A = P(1 + rt)$; for $t$

**19.** $s = \frac{n}{2}(a + L)$; for $L$

**20.** $\frac{3}{4}(b - 2c) = a$; for $b$

**21.** $N = 3st^4 - 5sv$; for $v$

**22.** $L = a + (n - 1)d$; for $d$

△ **23.** $S = 2LW + 2LH + 2WH$; for $H$

**24.** $T = 3vs - 4ws + 5vw$; for $v$

*In this exercise set, round all dollar amounts to two decimal places. Solve. See Example 4.*

**25.** Complete the table and find the balance $A$ if \$3500 is invested at an annual percentage rate of 3% for 10 years and compounded $n$ times a year.

| $n$ | 1 | 2 | 4 | 12 | 365 |
|---|---|---|---|---|---|
| $A$ | | | | | |

**26.** Complete the table and find the balance $A$ if \$5000 is invested at an annual percentage rate of 6% for 15 years and compounded $n$ times a year.

| $n$ | 1 | 2 | 4 | 12 | 365 |
|---|---|---|---|---|---|
| $A$ | | | | | |

**27.** If you are investing money in a savings account paying a rate of $r$, which account should you choose—an account compounded 4 times a year or 12 times a year? Explain your choice.

**28.** To borrow money at a rate of $r$, which bank should you choose—one compounding 4 times a year or 12 times a year? Explain your choice.

**29.** A principal of \$6000 is invested in an account paying an annual percentage rate of 4%. Find the amount in the account after 5 years if the account is compounded

**a.** semiannually
**b.** quarterly
**c.** monthly

**30.** A principal of \$25,000 is invested in an account paying an annual percentage rate of 5%. Find the amount in the account after 2 years if the account is compounded

**a.** semiannually
**b.** quarterly
**c.** monthly

*Solve. See Examples 4 and 5.*

**31.** The day's high temperature in Phoenix, Arizona, was recorded as 104°F. Write 104°F as degrees Celsius.

**32.** The annual low temperature in Nome, Alaska, was recorded as −15°C. Write −15°C as degrees Fahrenheit.

**33.** Omaha, Nebraska, is about 90 miles from Lincoln, Nebraska. Irania must go to the law library in Lincoln to get a document for the law firm she works for. Find how long it takes her to drive round-trip if she averages 50 mph.

**34.** It took the Selby family $5\frac{1}{2}$ hours round-trip to drive from their house to their beach house 154 miles away. Find their average speed.

△ **35.** A package of floor tiles contains 24 one-foot-square tiles. Find how many packages should be bought to cover a square ballroom floor whose side measures 64 feet.

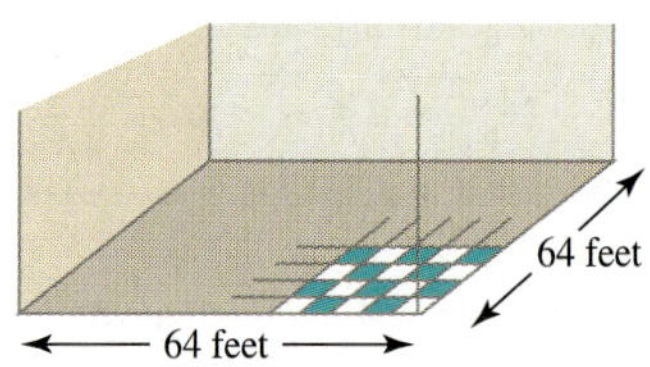


△ **36.** One-foot-square ceiling tiles are sold in packages of 50. Find how many packages must be bought for a rectangular ceiling 18 feet by 12 feet.

△ **37.** The deepest hole in the ocean floor is beneath the Pacific Ocean and is called Hole 504B. It is located off the coast of Ecuador. Scientists are drilling it to learn more about the Earth's history. Currently, the hole is in the shape of a cylinder whose volume is approximately 3800 cubic feet and whose length is 1.3 miles. Find the radius of the hole to the nearest hundredth of a foot. (*Hint:* Make sure the same units of measurement are used.)

**38.** The deepest man-made hole is called the Kola Superdeep Borehole. It is approximately 8 miles deep and is located near a small Russian town in the Arctic Circle. If it takes 7.5 hours to remove the drill from the bottom of the hole, find the rate that the drill can be retrieved in feet per second. Round to the nearest tenth. (*Hint:* Write 8 miles as feet, 7.5 hours as seconds, then use the formula $d = rt$.)

**39.** On April 1, 1985, *Sports Illustrated* published an April Fool's story by writer George Plimpton. He wrote that the New York Mets had discovered a man who could throw a 168-miles-per-hour fast ball. If the distance from the pitcher's mound to the plate is 60.5 feet, how long would it take for a ball thrown at that rate to travel that distance? (*Hint:* Write the rate 168 miles per hour in feet per second.

$$168 \text{ miles per hour} = \frac{168 \text{ miles}}{1 \text{ hour}}$$
$$= \frac{\_\_ \text{ feet}}{\_\_ \text{ seconds}}$$
$$= \frac{\_\_ \text{ feet}}{1 \text{ second}}$$
$$= \_\_ \text{ feet per second.}$$

Then use the formula $d = r \cdot t$.)

**40.** In 1945, Arthur C. Clarke, a scientist and science-fiction writer, predicted that an artificial satellite placed at a height of 22,248 miles directly above the equator would orbit the globe at the same speed with which the Earth was rotating. This belt along the equator is known as the Clarke belt. Use the formula for circumference of a circle and find the "length" of the Clarke belt. (*Hint:* Recall that the radius of the Earth is approximately 4000 miles. Round to the nearest whole mile.)

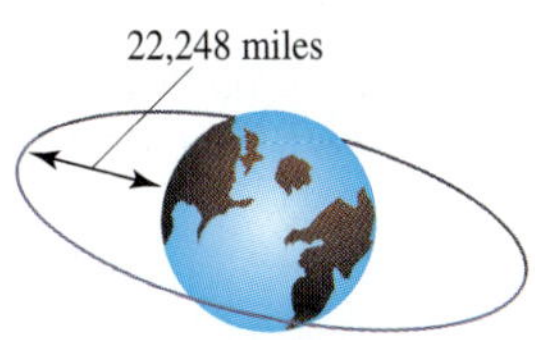


**41.** Eartha is the world's largest globe. It is located at the headquarters of DeLorme, a mapmaking company in Yarmouth, Maine. Eartha is 41.125 feet in diameter. Find its exact circumference (distance around) and then approximate its circumference using 3.14 for $\pi$. (*Source:* DeLorme)

**42.** Eartha is in the shape of a sphere. Its radius is about 20.6 feet. Approximate its volume to the nearest cubic foot. (*Source:* DeLorme)

**43.** How much do you think it costs each American to build a space shuttle? Write down your estimate. The space shuttle *Endeavour* was completed in 1992 and cost approximately \$1.7 billion. If the population of the United States in 1992 was 250 million, find the cost per person to build the *Endeavour*. How close was your estimate?

**44.** An orbit such as Clarke's belt in Exercise 40 is called a geostationary orbit. In your own words, why do you think that communications satellites are placed in geostationary orbits?

**45.** Find *how much interest* \$10,000 earns in 2 years in a certificate of deposit paying 8.5% interest compounded quarterly.

**46.** Bryan, Eric, Mandy, and Melissa would like to go to Disneyland in 3 years. Their total cost should be \$4500. If each invests \$1000 in a savings account paying 5.5% interest, compounded semiannually, will they have enough in 3 years?

**47.** A gallon of latex paint can cover 500 square feet. Find how many gallon containers of paint should be bought to paint two coats on each wall of a rectangular room whose dimensions are 14 feet by 16 feet (assume 8-foot ceilings).

**48.** A gallon of enamel paint can cover 300 square feet. Find how many gallon containers of paint should be bought to paint three coats on a wall measuring 21 feet by 8 feet.

**49.** A portion of the external tank of the Space Shuttle *Endeavour* is a liquid hydrogen tank. If the ends of the tank are hemispheres, find the volume of the tank. To do so, answer parts a through c.

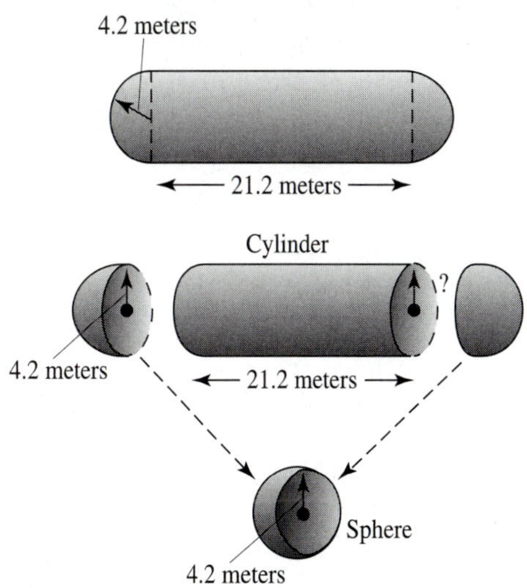


**a.** Find the volume of the cylinder shown. Round to 2 decimal places.

**b.** Find the volume of the sphere shown. Round to 2 decimal places.

**c.** Add the results of parts a and b. This sum is the approximate volume of the tank.

**50.** The space probe *Pioneer 10* traveled from Mars to Jupiter, a distance of 619 million miles, in 21 months. Find the average speed of the probe in miles per hour. (*Hint:* Convert 21 months to hours [using 1 month = 30 days] and then use the formula $d = rt$.)

**51.** Find how long it takes Mark to drive 135 miles on I-10 if he merges onto I-10 at 10 A.M. and drives nonstop with his cruise control set on 60 mph.

△ **52.** If the area of a triangular kite is 18 square feet and its base is 4 feet, find the height of the kite.

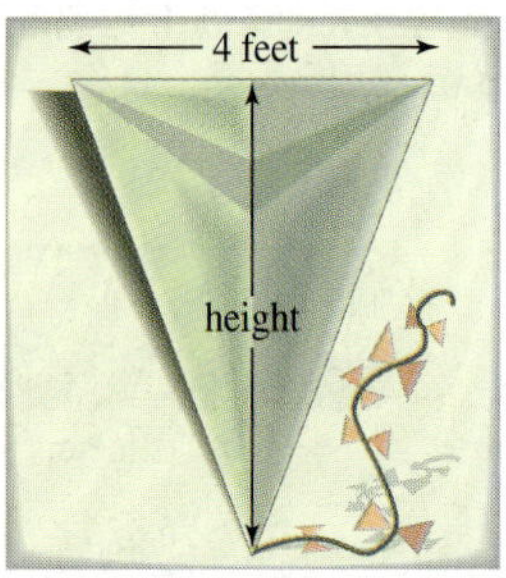


**53.** The Cassini spacecraft mission to Saturn was launched October 15, 1997. It will take more than six and a half years to reach Saturn, arriving in July 2004. During its mission, Cassini will travel a total distance of 2 billion miles in 80.5 months. Find the average speed of the spacecraft in miles per hour. (*Hint:* Convert 80.5 months to hours using 1 month = 30 days and then use the formula $d = rt$.) (*Source:* NASA Jet Propulsion Laboratory)

△ **54.** The Space Shuttle *Endeavour* has a cargo bay that is in the shape of a cylinder whose length is 18.3 meters and whose diameter is 4.6 meters. Find its volume.

**55.** Solar system distances are so great that units other than miles or kilometers are often used. For example, the astronomical unit (AU) is the average distance between the Earth and the Sun, or 92,900,000 miles. Use this information to convert each planet's distance in miles from the Sun to astronomical units. Round to three decimal places.

| | ***Miles from the Sun*** | ***AU from the Sun*** |
|---|---|---|
| Mercury | 36 million | |
| Venus | 67.2 million | |
| Earth | 92.9 million | |
| Mars | 141.5 million | |
| Jupiter | 483.3 million | |
| Saturn | 886.1 million | |
| Uranus | 1783 million | |
| Neptune | 2793 million | |
| Pluto | 3670 million | |

*The measure of the chance or likelihood of an event occurring is its* ***probability.*** *A formula basic to the study of probability is the formula for the probability of an event when all the outcomes are equally likely. This formula is*

$$\text{Probability of an event} = \frac{\text{number of ways that the event can occur}}{\text{number of possible outcomes}}$$

*For example, to find the probability that a single spin on the spinner will result in red, notice first that the spinner is divided into 8 parts, so there are 8 possible outcomes. Next, notice that there is only one sector of the spinner colored red, so the number of ways that the spinner can land on red is 1. Then this probability denoted by P(red) is*

$$P(red) = \frac{1}{8}$$

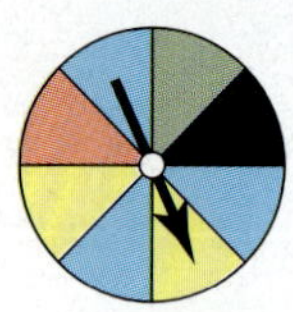

*Find each probability in simplest form.*

**56.** P(green)

**57.** P(yellow)

**58.** P(black)

**59.** P(blue)

**60.** P(green or blue)

**61.** P(black or yellow)

62. P(red, green, or black)

63. P(yellow, blue, or black)

64. P(white)

65. P(red, yellow, green, blue, or black)

66. From the previous probability formula, what do you think is always the probability of an event that is impossible occuring?

67. What do you think is always the probability of an event that is sure to occur?

### REVIEW EXERCISES

*Determine which numbers in the set* $\{-3, -2, -1, 0, 1, 2, 3\}$ *are solutions of each inequality.*

**68.** $x < 0$

**69.** $x > 1$

**70.** $x + 5 \le 6$

**71.** $x - 3 \ge -7$

**72.** In your own words, explain what real numbers are solutions of $x < 0$.

**73.** In your own words, explain what real numbers are solutions of $x > 1$.

## 2.4 LINEAR INEQUALITIES AND PROBLEM SOLVING

CD-ROM SSM SSG Video

▶ **OBJECTIVES**

1. Use interval notation.
2. Solve linear inequalities using the addition property of inequality.
3. Solve linear inequalities using the multiplication property of inequality.
4. Solve problems that can be modeled by linear inequalities.

**1** Relationships among measureable quantities are not always described by equations. For example, suppose that a salesperson earns a base of \$600 per month plus a commission of 20% of sales. Find the minimum amount of sales needed to receive a total income of *at least* \$1500 per month. Here, the phrase "at least" implies that an income of \$1500 *or more* is acceptable. In symbols, we can write

$$\text{income} \ge 1500$$

This is an example of an inequality, and we will solve this problem in Example 8.

A **linear inequality** is similar to a linear equation except that the equality symbol is replaced with an inequality symbol, such as $<, >, \le$, or $\ge$.

***Linear Inequalities in One Variable***

| $3x + 5 \ge 4$ | $2y < 0$ | $3(x - 4) > 5x$ | $\frac{x}{3} \le 5$ |
|---|---|---|---|
| ↑ is greater than or equal to | ↑ is less than | ↑ is greater than | ↑ is less than or equal to |

### LINEAR INEQUALITY IN ONE VARIABLE

A linear inequality in one variable is an inequality that can be written in the form

$$ax + b < c$$

where $a$, $b$, and $c$ are real numbers and $a \ne 0$.

In this section, when we make definitions, state properties, or list steps about an inequality containing the symbol $<$, we mean that the definition, property, or steps apply to inequalities containing the symbols $>$, $\leq$ and $\geq$ also.

A **solution** of an inequality is a value of the variable that makes the inequality a true statement. The **solution set** of an inequality is the set of all solutions. Notice that the solution set of the inequality $x > 2$, for example, contains all numbers greater than 2. Its graph is an interval on the number line since an infinite number of values satisfy the variable. If we use open/closed-circle notation, the graph of $\{x | x > 2\}$ looks like the following.

In this text, a convenient notation, called **interval notation**, will be used to write solution sets of inequalities. To help us understand this notation, a different graphing notation will be used. Instead of an open circle, we use a parenthesis; instead of a closed circle, we use a bracket. With this new notation, the graph of $\{x | x > 2\}$ now looks like

and can be represented in interval notation as $(2, \infty)$. The symbol $\infty$ is read "infinity" and indicates that the interval includes *all* numbers greater than 2. The left parenthesis indicates that 2 *is not* included in the interval. Using a left bracket, [, would indicate that 2 *is* included in the interval. The following table shows three equivalent ways to describe an interval: in set notation, as a graph, and in interval notation.

| *Set Notation* | *Graph* | *Interval Notation* |
|---|---|---|
| $\{x \| x < a\}$ |  $a$ | $(-\infty, a)$ |
| $\{x \| x > a\}$ |  $a$ | $(a, \infty)$ |
| $\{x \| x \leq a\}$ |  $a$ | $(-\infty, a]$ |
| $\{x \| x \geq a\}$ |  $a$ | $[a, \infty)$ |
| $\{x \| a < x < b\}$ |  $a$ $b$ | $(a, b)$ |
| $\{x \| a \leq x \leq b\}$ |  $a$ $b$ | $[a, b]$ |
| $\{x \| a < x \leq b\}$ |  $a$ $b$ | $(a, b]$ |
| $\{x \| a \leq x < b\}$ |  $a$ $b$ | $[a, b)$ |

**HELPFUL HINT**
Notice that a parenthesis is always used to enclose $\infty$ and $-\infty$.

**Example 1** Graph each set on a number line and then write in interval notation.

**a.** $\{x|x \geq 2\}$ **b.** $\{x|x < -1\}$ **c.** $\{x|0.5 < x \leq 3\}$

*Solution*

**a.** 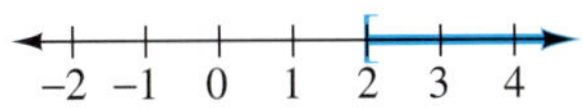

$[2, \infty)$

**b.** 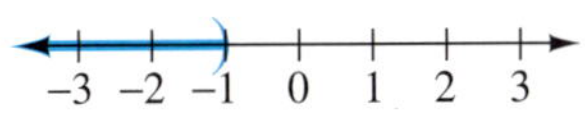

$(-\infty, -1)$

**c.** 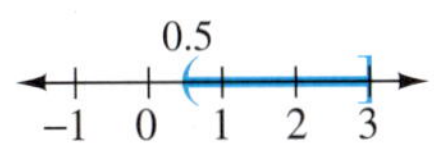

$(0.5, 3]$

2 Interval notation can be used to write solutions of linear inequalities. To solve a linear inequality, we use a process similar to the one used to solve a linear equation. We use properties of inequalities to write equivalent inequalities until the variable is isolated.

**ADDITION PROPERTY OF INEQUALITY**

If $a$, $b$, and $c$ are real numbers, then

$$a < b \quad \text{and} \quad a + c < b + c$$

are equivalent inequalities.

In other words, we may add the same real number to both sides of an inequality and the resulting inequality will have the same solution set. This property also allows us to subtract the same real number from both sides.

**Example 2** Solve $x - 2 < 5$. Graph the solution set.

*Solution*

$$x - 2 < 5$$

$$x - 2 + 2 < 5 + 2 \quad \text{Add 2 to both sides.}$$

$$x < 7 \quad \text{Simplify.}$$

The solution set is $\{x|x < 7\}$, which in interval notation is $(-\infty, 7)$. The graph of the solution set is

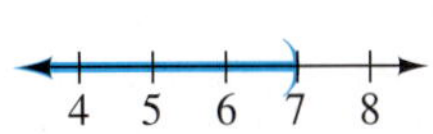

**HELPFUL HINT**

In Example 2, the solution set is $\{x | x < 7\}$. This means that *all* numbers less than 7 are solutions. For example, 6.9, 0, $-\pi$, 1, and $-56.7$ are solutions, just to name a few. To see this, replace $x$ in $x - 2 < 5$ with each of these numbers and see that the result is a true inequality.

**Example 3** Solve $3x + 4 \geq 2x - 6$. Graph the solution set.

*Solution*

$$3x + 4 \geq 2x - 6$$

$$3x + 4 - 2x \geq 2x - 6 - 2x \quad \text{Subtract } 2x \text{ from both sides.}$$

$$x + 4 \geq -6 \quad \text{Combine like terms.}$$

$$x + 4 - 4 \geq -6 - 4 \quad \text{Subtract 4 from both sides.}$$

$$x \geq -10 \quad \text{Simplify.}$$

The solution set is $\{x | x \geq -10\}$, which in interval notation is $[-10, \infty)$. The graph of the solution set is

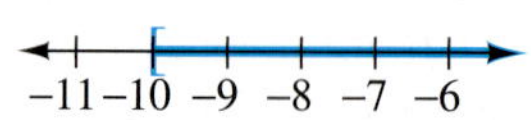


3 Next, we introduce and use the multiplication property of inequality to solve linear inequalities. To understand this property, let's start with the true statement $-3 < 7$ and multiply both sides by 2.

$$-3 < 7$$

$$-3(2) < 7(2) \quad \text{Multiply by 2.}$$

$$-6 < 14 \quad \text{True.}$$

The statement remains true.

Notice what happens if both sides of $-3 < 7$ are multiplied by $-2$.

$$-3 < 7$$

$$-3(-2) < 7(-2) \quad \text{Multiply by } -2.$$

$$6 < -14 \quad \text{False.}$$

The inequality $6 < -14$ is a false statement. However, **if the direction of the inequality sign is reversed**, the result is

$$6 > -14 \quad \text{True.}$$

These examples suggest the following property.

**MULTIPLICATION PROPERTY OF INEQUALITY**

If $a$, $b$, and $c$ are real numbers and $c$ is **positive**, then
$a < b$ and $ac < bc$ are equivalent inequalities.
If $a$, $b$, and $c$ are real numbers and $c$ is **negative**, then
$a < b$ and $ac > bc$ are equivalent inequalities.

In other words, we may multiply both sides of an inequality by the same positive real number and the result is an equivalent inequality.

We may also multiply both sides of an inequality by the same **negative number** and **reverse the direction of the inequality symbol**, and the result is an equivalent inequality. The multiplication property holds for division also, since division is defined in terms of multiplication.

> **HELPFUL HINT**
> Whenever both sides of an inequality are multiplied or divided by a negative number, the direction of the inequality symbol **must be** reversed to form an equivalent inequality.

## Example 4

Solve and graph the solution set.

**a.** $\frac{1}{4}x \leq \frac{3}{8}$ **b.** $-2.3x < 6.9$

*Solution* **a.**

$$\frac{1}{4}x \leq \frac{3}{8}$$

> **HELPFUL HINT**
> The inequality symbol is the same since we are multiplying by a *positive* number.

$$4 \cdot \frac{1}{4}x \leq 4 \cdot \frac{3}{8} \quad \text{Multiply both sides by 4.}$$

$$x \leq \frac{3}{2} \quad \text{Simplify.}$$

The solution set is $\left\{x \mid x \leq \frac{3}{2}\right\}$, which in interval notation is $\left(-\infty, \frac{3}{2}\right]$. The graph of the solution set is

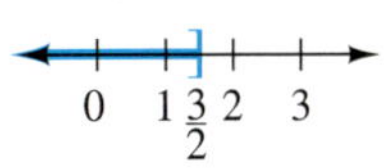


**b.**

$$-2.3x < 6.9$$

> **HELPFUL HINT**
> The inequality symbol is *reversed* since we divided by a *negative* number.

$$\frac{-2.3x}{-2.3} > \frac{6.9}{-2.3} \quad \text{Divide both sides by } -2.3 \text{ and reverse the inequality symbol.}$$

$$x > -3 \quad \text{Simplify.}$$

The solution set is $\{x \mid x > -3\}$, which is $(-3, \infty)$ in interval notation. The graph of the solution set is

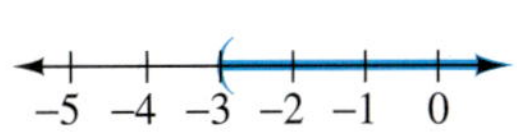


To solve linear inequalities in general, we follow steps similar to those for solving linear equations.

### SOLVING A LINEAR INEQUALITY IN ONE VARIABLE

**Step 1:** Clear the equation of fractions by multiplying both sides of the inequality by the least common denominator (LCD) of all fractions in the inequality.
**Step 2:** Use the distributive property to remove grouping symbols such as parentheses.
**Step 3:** Combine like terms on each side of the inequality.
**Step 4:** Use the addition property of inequality to write the inequality as an equivalent inequality with variable terms on one side and numbers on the other side.
**Step 5:** Use the multiplication property of inequality to isolate the variable.

**Example 5** Solve $-(x-3)+2 \le 3(2x-5)+x$.

*Solution*

$$-(x-3)+2 \le 3(2x-5)+x$$

$$-x+3+2 \le 6x-15+x \quad \text{Apply the distributive property.}$$

$$5-x \le 7x-15 \quad \text{Combine like terms.}$$

$$5-x+x \le 7x-15+x \quad \text{Add } x \text{ to both sides.}$$

$$5 \le 8x-15 \quad \text{Combine like terms.}$$

$$5+15 \le 8x-15+15 \quad \text{Add 15 to both sides.}$$

$$20 \le 8x \quad \text{Combine like terms.}$$

$$\frac{20}{8} \le \frac{8x}{8} \quad \text{Divide both sides by 8.}$$

$$\frac{5}{2} \le x, \quad \text{or } x \ge \frac{5}{2} \quad \text{Simplify.}$$

**HELPFUL HINT**

Don't forget that $\frac{5}{2} \le x$ means the same as $x \ge \frac{5}{2}$.

The solution set written in interval notation is $\left[\frac{5}{2}, \infty\right)$, and its graph is

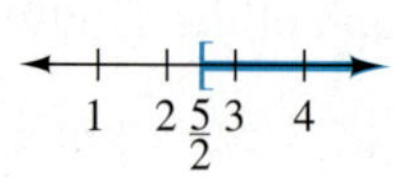


**Example 6** Solve $\frac{2}{5}(x-6) \ge x-1$

*Solution*

$$\frac{2}{5}(x-6) \geq x-1$$

$$5\left[\frac{2}{5}(x-6)\right] \geq 5(x-1) \quad \text{Multiply both sides by 5 to eliminate fractions.}$$

$$2x - 12 \geq 5x - 5 \quad \text{Apply the distributive property.}$$

$$-3x - 12 \geq -5 \quad \text{Subtract } 5x \text{ from both sides.}$$

$$-3x \geq 7 \quad \text{Add 12 to both sides.}$$

$$\frac{-3x}{-3} \leq \frac{7}{-3} \quad \text{Divide both sides by } -3 \text{ and reverse the inequality symbol.}$$

$$x \leq -\frac{7}{3} \quad \text{Simplify.}$$

The solution set written in interval notation is $\left(-\infty, -\frac{7}{3}\right]$, and its graph is

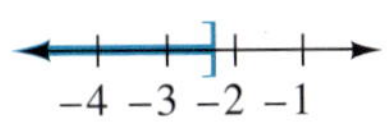


**Example 7** Solve $2(x+3) > 2x + 1$.

*Solution*

$$2(x+3) > 2x + 1$$

$$2x + 6 > 2x + 1 \quad \text{Distribute on the left side.}$$

$$2x + 6 - 2x > 2x + 1 - 2x \quad \text{Subtract } 2x \text{ from both sides.}$$

$$6 > 1 \quad \text{Simplify.}$$

$6 > 1$ is a true statement for all values of $x$, so this inequality and the original inequality are true for all numbers. The solution set is $\{x \mid x \text{ is a real number}\}$, or $(-\infty, \infty)$ in interval notation, and its graph is

4 Application problems containing words such as "at least," "at most," "between," "no more than," and "no less than" usually indicate that an inequality be solved instead of an equation. In solving applications involving linear inequalities, we use the same procedure as when we solved applications involving linear equations.

**Example 8** **CALCULATING INCOME WITH COMMISSION**

A salesperson earns \$600 per month plus a commission of 20% of sales. Find the minimum amount of sales needed to receive a total income of at least \$1500 per month.

*Solution* 1. UNDERSTAND. Read and reread the problem. Let $x =$ amount of sales

2. TRANSLATE. As stated in the beginning of this section, we want the income to be greater than or equal to \$1500. To write an inequality, notice that the sales person's income consists of \$600 plus a commission (20% of sales).

| | | | | | | |
|---|---|---|---|---|---|---|
| In words: | 600 | + | commission (20% of sales) | $\geq$ | 1500 |
| | ↓ | | ↓ | | ↓ |
| Translate: | 600 | + | $0.20x$ | $\geq$ | 1500 |

3. SOLVE the inequality for $x$.

$$600 + 0.20x \geq 1500$$
$$600 + 0.20x - 600 \geq 1500 - 600$$
$$0.20x \geq 900$$
$$x \geq 4500$$

4. INTERPRET.

**Check:** The income for sales of \$4500 is

$$600 + 0.20(4500), \text{ or } 1500.$$

Thus, if sales are greater than or equal to \$4500, income is greater than or equal to \$1500.

**State:** The minimum amount of sales needed for the salesperson to earn at least \$1500 per month is \$4500 per month.

## Example 9 FINDING THE ANNUAL CONSUMPTION

In the United States, the annual consumption of cigarettes is declining. The consumption $c$ in billions of cigarettes per year since the year 1985 can be approximated by the formula

$$c = -14.25t + 598.69$$

where $t$ is the number of years after 1985. Use this formula to predict the years that the consumption of cigarettes will be less than 200 billion per year.

*Solution*

1. UNDERSTAND. Read and reread the problem. To become familiar with the given formula, let's find the cigarette consumption after 20 years, which would be the year 1985 + 20, or 2005. To do so, we substitute 20 for $t$ in the given formula.

$$c = -14.25(20) + 598.69 = 313.69$$

Thus, in 2005, we predict cigarette consumption to be about 313.69 billion.

Variables have already been assigned in the given formula. For review, they are

$c$ = the annual consumption of cigarettes in the United States in billions of cigarettes

$t$ = the number of years after 1985

2. TRANSLATE. We are looking for the years that the consumption of cigarettes $c$ is less than 200. Since we are finding years $t$, we substitute the expression in the formula given for $c$, or

$$-14.25t + 598.69 < 200$$

3. SOLVE the inequality.

$$-14.25t + 598.69 < 200 \quad \text{Subtract 598.69 from both sides.}$$
$$-14.25t < -398.69 \quad \text{Divide both sides by } -14.25 \text{ and round the result.}$$
$$t > 27.98$$

4. INTERPRET.

**Check:** We substitute a number greater than 27.98 and see that $c$ is less than 200.

**State:** The annual consumption of cigarettes will be less than 200 billion for the years more than 27.98 years after 1985, or after approximately $28 + 1985 = 2013$.

## SPOTLIGHT ON DECISION MAKING

Suppose you are the superintendent of Copley Public Schools. You are aware that the general population of Copley is increasing and that enrollment at the schools is steadily rising. The high school can house a maximum of 1200 students. Once this maximum has been exceeded, temporary classrooms must be erected to handle the overflow.

You have been studying the changes in population and conclude that the equation $y = 30x + 1025$ models the high school enrollment $x$ years from now. As you prepare a long-term planning report, you must decide whether temporary classrooms will be needed in the next 10 years. If so, when is the latest that funding for temporary classrooms could be added to the annual budget?

## MENTAL MATH

*Solve each inequality mentally and write it in set notation.*

**1.** $x - 2 < 4$
**2.** $x - 1 > 6$
**3.** $x + 5 \geq 15$
**4.** $x + 1 \leq 8$
**5.** $3x > 12$
**6.** $5x < 20$
**7.** $\frac{x}{2} \leq 1$
**8.** $\frac{x}{4} \geq 2$

## Exercise Set 2.4

*Graph the solution set of each inequality and write it in interval notation. See Example 1.*

**1.** $\{x | x < -3\}$
**2.** $\{x | x \geq -7\}$
**3.** $\{x | x \geq 0.3\}$
**4.** $\{x | x < -0.2\}$
**5.** $\{x | 5 < x\}$
**6.** $\{x | -7 \geq x\}$
**7.** $\{x | -2 < x < 5\}$
**8.** $\{x | -5 \leq x \leq -1\}$
**9.** $\{x | 5 > x > -1\}$
**10.** $\{x | -3 \geq x \geq -7\}$
**11.** When graphing the solution set of an inequality, explain how you know whether to use a parenthesis or a bracket.

**12.** Explain what is wrong with the interval notation $(-6, -\infty)$

*Determine which number(s) listed below are solutions to each inequality. A number may be used more than once.*

$$0, 5, -5, \tfrac{3}{4}, -\tfrac{3}{4}, \pi, 100, -1{,}000{,}000$$

**13.** $x > 5$
**14.** $x < 5$
**15.** $y \leq -1$
**16.** $y \geq -\frac{1}{2}$

*Solve. Graph the solution set and write it in interval notation. See Examples 2 through 4.*

**17.** $x - 7 \geq -9$

**18.** $x + 2 \leq -1$

**19.** $7x < 6x + 1$

**20.** $11x < 10x + 5$

**21.** $8x - 7 \leq 7x - 5$

**22.** $7x - 1 \geq 6x - 1$

**23.** $2 + 4x > 5x + 6$

**24.** $7 + 8x > 9x + 12$

**25.** $\frac{3}{4}x \geq 2$

**26.** $\frac{5}{6}x \geq -8$

**27.** $5x < -23.5$

**28.** $4x > -11.2$

**29.** $-3x \geq 9$

**30.** $-4x \geq 15$

**31.** $-x < -4$

**32.** $-x > -2$

*Solve. Write the solution set using interval notation. See Examples 5 through 7.*

**33.** $-2x + 7 \geq 9$

**34.** $8 - 5x \leq 23$

**35.** $15 + 2x \geq 4x - 7$

**36.** $10 + x < 6x - 10$

**37.** $3(x - 5) < 2(2x - 1)$

**38.** $5(x + 4) \leq 4(2x + 3)$

**39.** $\frac{1}{2} + \frac{2}{3} \geq \frac{x}{6}$

**40.** $\frac{3}{4} - \frac{2}{3} > -\frac{x}{6}$

**41.** $4(x - 1) \geq 4x - 8$

**42.** $3x + 1 < 3(x - 2)$

**43.** $7x < 7(x - 2)$

**44.** $8(x + 3) \leq 7(x + 5) + x$

**45.** $4(2x + 1) > 4$

**46.** $6(2 - x) \geq 12$

**47.** $\frac{x + 7}{5} > 1$

**48.** $\frac{2x - 4}{3} \leq 2$

**49.** $\frac{-5x + 11}{2} \leq 7$

**50.** $\frac{4x - 8}{7} < 0$

**51.** $8x - 16.4 \leq 10x + 2.8$

**52.** $18x - 25.6 < 10x + 60.8$

**53.** $2(x - 3) > 70$

**54.** $3(5x + 6) \geq -12$

**55.** Explain how solving a linear inequality is similar to solving a linear equation.

**56.** Explain how solving a linear inequality is different from solving a linear equation.

*Solve. Write the solution set using interval notation.*

**57.** $-5x + 4 \leq -4(x - 1)$

**58.** $-6x + 2 < -3(x + 4)$

**59.** $\frac{1}{4}(x - 7) \geq x + 2$

**60.** $\frac{3}{5}(x + 1) \leq x + 1$

**61.** $\frac{2}{3}(x + 2) < \frac{1}{5}(2x + 7)$

**62.** $\frac{1}{6}(3x + 10) > \frac{5}{12}(x - 1)$

**63.** $4(x - 6) + 2x - 4 \geq 3(x - 7) + 10x$

**64.** $7(2x + 3) + 4x \leq 7 + 5(3x - 4)$

**65.** $\frac{5x + 1}{7} - \frac{2x - 6}{4} \geq -4$

**66.** $\frac{1 - 2x}{3} + \frac{3x + 7}{7} > 1$

**67.** $\frac{-x + 2}{2} - \frac{1 - 5x}{8} < -1$

**68.** $\frac{3 - 4x}{6} - \frac{1 - 2x}{12} \leq -2$

**69.** $0.8x + 0.6x \geq 4.2$

**70.** $0.7x - x > 0.45$

**71.** $\frac{x + 5}{5} - \frac{3 + x}{8} \geq -\frac{3}{10}$

**72.** $\frac{x - 4}{2} - \frac{x - 2}{3} > \frac{5}{6}$

**73.** $\frac{x + 3}{12} + \frac{x - 5}{15} < \frac{2}{3}$

**74.** $\frac{3x + 2}{18} - \frac{1 + 2x}{6} \leq -\frac{1}{2}$

*Solve. See Examples 8 and 9*

**75.** Shureka has scores of 72, 67, 82, and 79 on her algebra tests. Use an inequality to find the minimum score she can make on the final exam to pass the course with an average of 60 or higher, given that the final exam counts as two tests.

**76.** In a Winter Olympics speed-skating event, Hans scored times of 3.52, 4.04, and 3.87 minutes on his first three trials. Use an inequality to find the maximum time he can score on his last trial so that his average time is under 4.0 minutes.

**77.** A small plane's maximum takeoff weight is 2000 pounds. Six passengers weigh an average of 160 pounds each. Use an inequality to find the maximum weight of luggage and cargo the plane can carry.

**78.** A clerk must use the elevator to move boxes of paper. The elevator's weight limit is 1500 pounds. If each box of paper weighs 66 pounds and the clerk weighs 147 pounds, use an inequality to find the maximum number of boxes she can move on the elevator at one time.

**79.** To mail an envelope first class, the U.S. Post Office charges 33 cents for the first ounce and 22 cents per ounce for each additional ounce. Use an inequality to find the maximum number of whole ounces that can be mailed for \$4.00

**80.** A shopping mall parking garage charges \$2 for the first half hour and \$1.20 for each additional half hour or a portion of a half hour. Use an inequality to find how long you can park if you have \$8.00 in cash.

**81.** Northeast Telephone Company offers two billing plans for local calls. Plan 1 charges \$25 per month for unlimited calls, and plan 2 charges \$13 per month plus 6 cents per call. Use an inequality to find the number of monthly calls for which plan 1 is more economical than plan 2.

**82.** A car rental company offers two subcompact rental plans. Plan A charges \$32 per day for unlimited mileage, and plan B charges \$24 per day plus 15 cents per mile. Use an inequality to find the number of daily miles for which plan A is more economical than plan B.

**83.** At room temperature, glass used in windows actually has some properties of a liquid. It has a very slow, viscous flow. (Viscosity is the property of a fluid that resists internal flow. For example, lemonade flows more easily than fudge syrup. Fudge syrup has a higher viscosity than lemonade.) Glass does not become a true liquid until temperatures are greater than or equal to 500°C. Find the Fahrenheit temperatures for which glass is a liquid. (Use the formula $F = \frac{9}{5}C + 32$.)

**84.** Stibnite is a silvery white mineral with a metallic luster. It is one of the few minerals that melts easily in match flame or at temperatures of approximately 977°F or greater. Find the Celsius temperatures for which stibnite melts. (Use the formula $C = \frac{5}{9}[F - 32]$.)

**85.** Although beginning salaries vary greatly according to your field of study, the equation $s = 2806.6t + 32{,}558$ can be used to approximate and to predict average beginning salaries for candidates for bachelor's degrees. The variable $s$ is the starting salary and $t$ is the number of years after 1995.

**a.** Approximate when beginning salaries for candidates will be greater than \$50,000.

**b.** Determine the year you plan to graduate from college. Use this year to find the corresponding value of $t$ and approximate your beginning salary.

**86.** Use the formula in Example 9 to estimate the years that the consumption of cigarettes will be less than 50 billion per year.

*The average consumption per person per year of whole milk $w$ in gallons can be approximated by the equation*

$$w = -0.18t + 8.72$$

*where $t$ is the number of years after 1994. The average consumption of skim milk $s$ per person per year can be approximated by the equation*

$$s = 0.26t + 5.86$$

*where $t$ is the number of years after 1994. The consumption of whole milk is shown on the graph in blue and the consumption of skim milk is shown on the graph in red. Use this information to answer Exercises 87–95.*

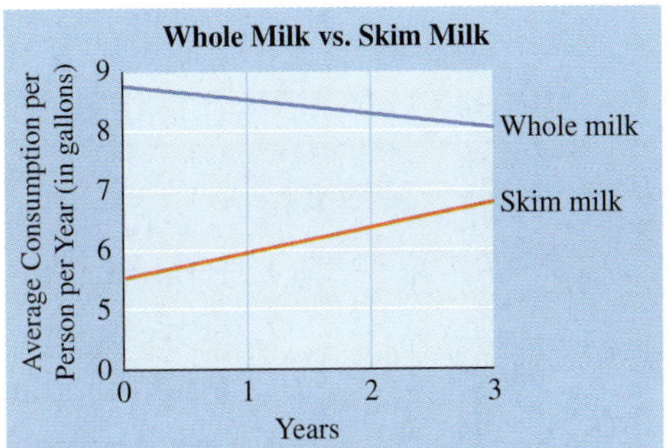


*Source:* Based on data from Economic Research Service, U.S. Department of Agriculture, *Agricultural Outlook*, June/July 1998.

**87.** Is the consumption of whole milk increasing or decreasing over time? Explain how you arrived at your answer.

**88.** Is the consumption of skim milk increasing or decreasing over time? Explain how you arrived at your answer.

**89.** Predict the consumption of whole milk in the year 2005 (*Hint:* Find the value of $t$ that corresponds to the year 2005.)

**90.** Predict the consumption of skim milk in the year 2005 (*Hint:* Find the value of $t$ that corresponds to the year 2005.)

**91.** Determine when the consumption of whole milk will be less than 7 gallons per person per year.

**92.** Determine when the consumption of skim milk will be greater than 8 gallons per person per year.

**93.** For 1994 through 1997 the consumption of whole milk was greater than the consumption of skim milk. Explain how this can be determined from the graph.

**94.** How will the two lines in the graph appear when the consumption of whole milk is the same as the consumption of skim milk?

**95.** The consumption of whole milk will be the same as the consumption of skim milk when $w = s$. Find when this will occur, by substituting the given equivalent expression for $w$ and the given equivalent expression for $s$ and solving for $t$. Round the value of $t$ to the nearest whole and estimate the year when this will occur.

## REVIEW EXERCISES

*List or describe the integers that make both inequalities true.*

**96.** $x < 5$ and $x > 1$

**97.** $x \geq 0$ and $x \leq 7$

**98.** $x \geq -2$ and $x \geq 2$

**99.** $x < 6$ and $x < -5$

*Graph each set on a number line and write it in interval notation. See Section 2.4.*

**100.** $\{x | 0 \leq x \leq 5\}$

**101.** $\{x | -7 < x \leq 1\}$

**102.** $\left\{x \middle| -\frac{1}{2} < x < \frac{3}{2}\right\}$

**103.** $\{x | -2.5 \leq x < 5.3\}$

## 2.5 COMPOUND INEQUALITIES

CD-ROM SSM SSG Video

**OBJECTIVES**

1. Find the intersection of two sets.
2. Solve compound inequalities containing **and**.
3. Find the union of two sets.
4. Solve compound inequalities containing **or**.

Two inequalities joined by the words **and** or **or** are called **compound inequalities.**

***Compound Inequalities***

$$x + 3 < 8 \text{ and } x > 2$$

$$\frac{2x}{3} \geq 5 \text{ or } -x + 10 < 7$$

1 The solution set of a compound inequality formed by the word **and** is the **intersection** of the solution sets of the two inequalities.

### INTERSECTION OF TWO SETS

The intersection of two sets, $A$ and $B$, is the set of all elements common to both sets. $A$ intersect $B$ is denoted by

$$A \cap B$$

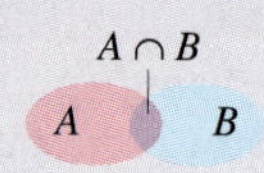


**Example 1** Find the intersection: $\{2, 4, 6, 8\} \cap \{3, 4, 5, 6\}$

*Solution* The numbers 4 and 6 are in both sets. The intersection is $\{4, 6\}$.

2 A value is a solution of a compound inequality formed by the word **and** if it is a solution of *both* inequalities. For example, the solution set of the compound inequality $x \leq 5$ and $x \geq 3$ contains all values of $x$ that make the inequality $x \leq 5$ a true statement **and** the inequality $x \geq 3$ a true statement. The first graph shown below is the

graph of $x \le 5$, the second graph is the graph of $x \ge 3$, and the third graph shows the intersection of the two graphs. The third graph is the graph of $x \le 5$ **and** $x \ge 3$.

$\{x \mid x \le 5\}$
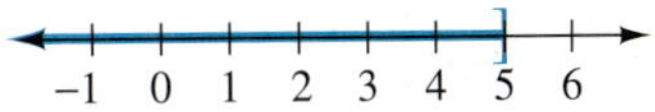
$(-\infty, 5]$

$\{x \mid x \ge 3\}$
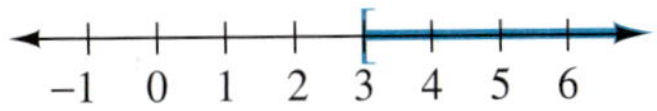
$[3, \infty)$

$\{x \mid x \le 5 \text{ and } x \ge 3\}$
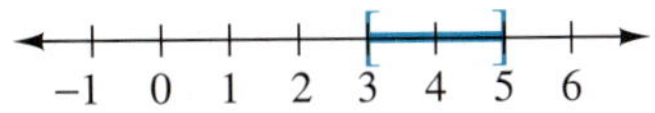
$[3, 5]$

The compound inequality $x \le 5$ and $x \ge 3$ can be written in a more compact form as $3 \le x \le 5$. The solution set $\{x \mid 3 \le x \le 5\}$ includes all numbers that are less than or equal to 5 and at the same time greater than or equal to 3. In interval notation, the solution set is $[3, 5]$.

**Example 2** Solve $x - 7 < 2$ and $2x + 1 < 9$.

*Solution* First we solve each inequality separately.

$$\begin{aligned} x - 7 &< 2 \quad \text{and} \quad & 2x + 1 &< 9 \\ x &< 9 \quad \text{and} \quad & 2x &< 8 \\ x &< 9 \quad \text{and} \quad & x &< 4 \end{aligned}$$

Now we can graph the two intervals on two number lines and find their intersection. Their intersection is shown on the third number line.

$\{x \mid x < 9\}$
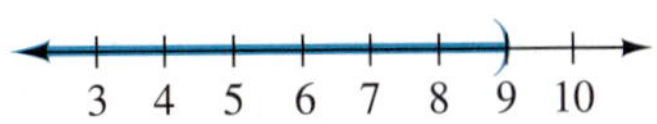
$(-\infty, 9)$

$\{x \mid x < 4\}$
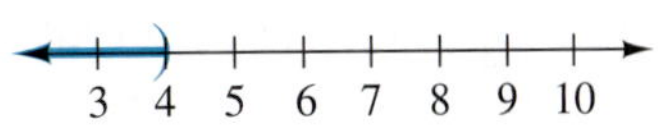
$(-\infty, 4)$

$\{x \mid x < 9 \text{ and } x < 4\}$
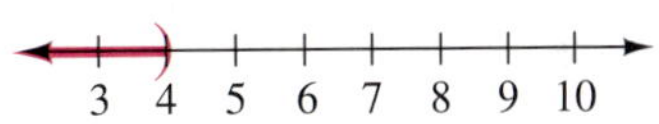
$(-\infty, 4)$

$= \{x \mid x < 4\}$

The solution set is $(-\infty, 4)$.

**Example 3** Solve $2x \ge 0$ and $4x - 1 \le -9$.

*Solution* First we solve each inequality separately.

$$\begin{aligned} 2x &\ge 0 \quad \text{and} \quad & 4x - 1 &\le -9 \\ x &\ge 0 \quad \text{and} \quad & 4x &\le -8 \\ x &\ge 0 \quad \text{and} \quad & x &\le -2 \end{aligned}$$

Now we can graph the two intervals and find their intersection.

$\{x \mid x \ge 0\}$
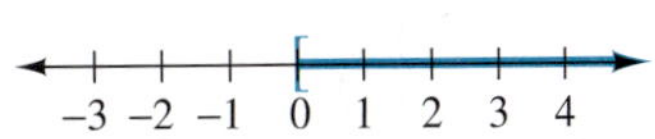
$[0, \infty)$

$\{x \mid x \le -2\}$
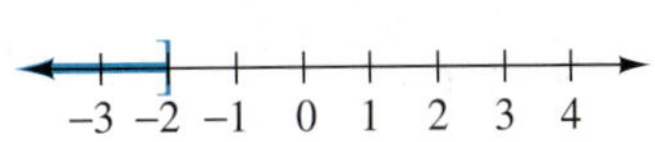
$(-\infty, -2]$

$\{x \mid x \ge 0 \text{ and } x \le -2\} = \varnothing$
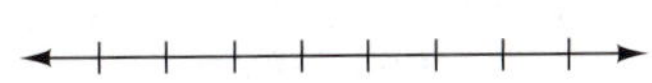
$\varnothing$

There is no number that is greater than or equal to 0 *and* less than or equal to −2. The solution set is ∅.

**HELPFUL HINT**
Example 3 shows that some compound inequalities have no solution. Also, some have all real numbers as solutions.

To solve a compound inequality written in a compact form, such as $2 < 4 - x < 7$, we get $x$ alone in the "middle part." Since a compound inequality is really two inequalities in one statement, we must perform the same operations on all three parts of the inequality.

**Example 4** Solve $2 < 4 - x < 7$.

*Solution* To get $x$ alone, we first subtract 4 from all three parts.

$$2 < 4 - x < 7$$

$$2 - 4 < 4 - x - 4 < 7 - 4 \quad \text{Subtract 4 from all three parts.}$$

$$-2 < -x < 3 \quad \text{Simplify.}$$

$$\frac{-2}{-1} > \frac{-x}{-1} > \frac{3}{-1} \quad \text{Divide all three parts by } -1 \text{ and reverse the inequality symbols.}$$

$$2 > x > -3$$

**HELPFUL HINT**
Don't forget to reverse both inequality symbols.

This is equivalent to $-3 < x < 2$.
The solution set in interval notation is $(-3, 2)$, and its graph is shown.

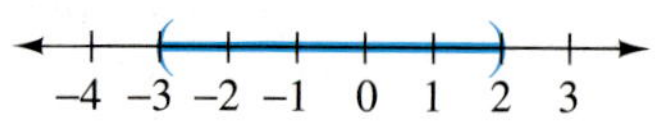


**Example 5** Solve $-1 \le \frac{2x}{3} + 5 \le 2$.

*Solution* First, clear the inequality of fractions by multiplying all three parts by the LCD of 3.

$$-1 \le \frac{2x}{3} + 5 \le 2$$

$$3(-1) \le 3\left(\frac{2x}{3} + 5\right) \le 3(2) \quad \text{Multiply all three parts by the LCD of 3.}$$

$$-3 \le 2x + 15 \le 6 \quad \text{Use the distributive property and multiply.}$$

$$-3 - 15 \le 2x + 15 - 15 \le 6 - 15 \quad \text{Subtract 15 from all three parts.}$$

$$-18 \le 2x \le -9 \quad \text{Simplify.}$$

$$\frac{-18}{2} \le \frac{2x}{2} \le \frac{-9}{2} \quad \text{Divide all three parts by 2.}$$

$$-9 \le x \le -\frac{9}{2} \quad \text{Simplify.}$$

The graph of the solution is shown.

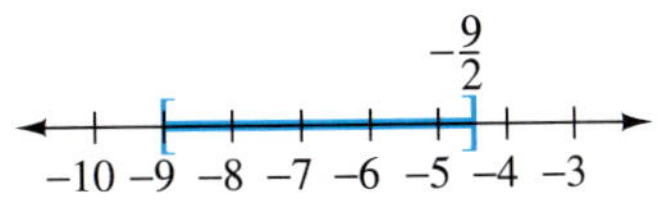


The solution set in interval notation is $\left[-9, -\frac{9}{2}\right]$.

3 The solution set of a compound inequality formed by the word **or** is the **union** of the solution sets of the two inequalities.

**UNION OF TWO SETS**

The union of two sets, $A$ and $B$, is the set of elements that belong to *either* of the sets. $A$ union $B$ is denoted by

$$A \cup B$$

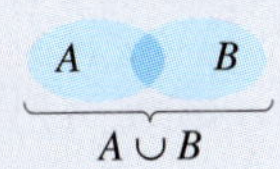


**Example 6** Find the union: $\{2, 4, 6, 8\} \cup \{3, 4, 5, 6\}$

*Solution* The numbers that are in either set or both sets are $\{2, 3, 4, 5, 6, 8\}$. This set is the union.

4 A value is a solution of a compound inequality formed by the word **or** if it is a solution of **either** inequality. For example, the solution set of the compound inequality $x \le 1$ **or** $x \ge 3$ contains all numbers that make the inequality $x \le 1$ a true statement **or** the inequality $x \ge 3$ a true statement.

$\{x \mid x \le 1\}$

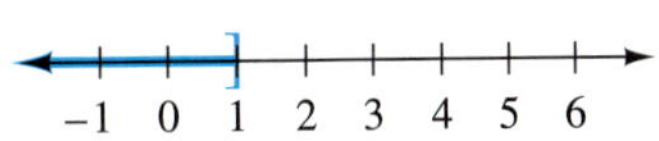


$(-\infty, 1]$

$\{x \mid x \ge 3\}$

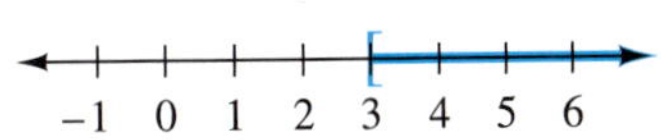


$[3, \infty)$

$\{x \mid x \le 1 \text{ or } x \ge 3\}$

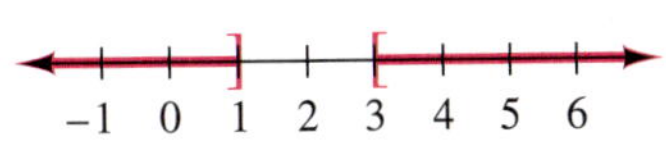


$(-\infty, 1] \cup [3, \infty)$

In interval notation, the set $\{x \mid x \le 1 \text{ or } x \ge 3\}$ is written as $(-\infty, 1] \cup [3, \infty)$.

**Example 7** Solve $5x - 3 \le 10$ or $x + 1 \ge 5$.

*Solution* First we solve each inequality separately.

$$5x - 3 \le 10 \quad \text{or} \quad x + 1 \ge 5$$
$$5x \le 13 \quad \text{or} \quad x \ge 4$$
$$x \le \frac{13}{5} \quad \text{or} \quad x \ge 4$$

Now we can graph each interval and find their union.

$\left\{x \middle| x \leq \frac{13}{5}\right\}$

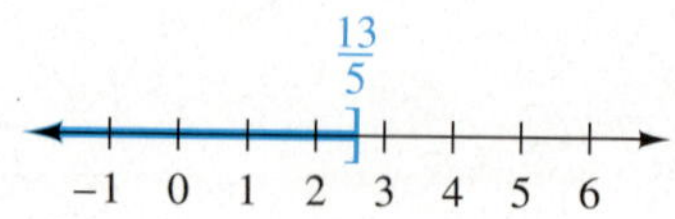


$\left(-\infty, \frac{13}{5}\right]$

$\{x | x \geq 4\}$

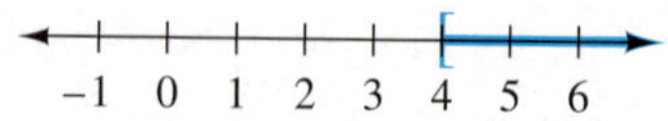


$[4, \infty)$

$\left\{x \middle| x \leq \frac{13}{5} \text{ or } x \geq 4\right\}$

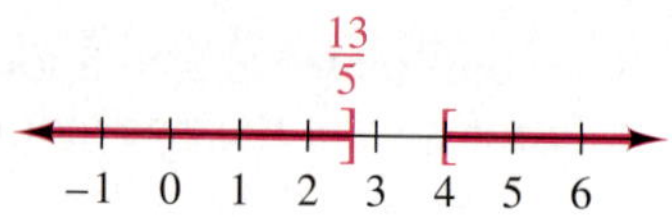


$\left(-\infty, \frac{13}{5}\right] \cup [4, \infty)$

The solution set is $\left(-\infty, \frac{13}{5}\right] \cup [4, \infty)$.

## Example 8

Solve: $-2x - 5 < -3$ or $6x < 0$.

*Solution* First we solve each inequality separately.

$$\begin{aligned} -2x - 5 &< -3 \quad \text{or} \quad 6x < 0 \\ -2x &< 2 \quad \text{or} \quad x < 0 \\ x &> -1 \quad \text{or} \quad x < 0 \end{aligned}$$

Now we can graph each interval and find their union.

$\{x | x > -1\}$

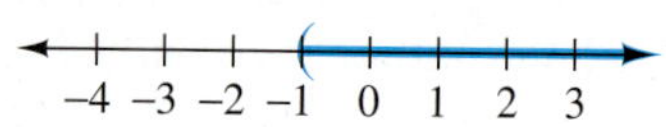


$(-1, \infty)$

$\{x | x < 0\}$

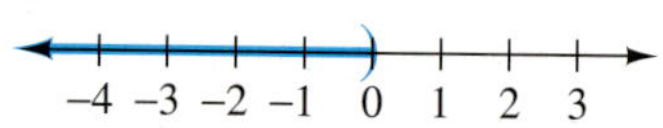


$(-\infty, 0)$

$\{x | x > -1 \text{ or } x < 0\}$
= all real numbers

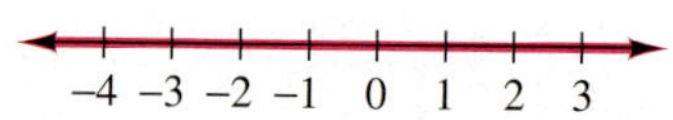


$(-\infty, \infty)$

The solution set is $(-\infty, \infty)$.

## Exercise Set 2.5

*If $A = \{x | x$ is an even integer$\}$, $B = \{x | x$ is an odd integer$\}$, $C = \{2, 3, 4, 5\}$, and $D = \{4, 5, 6, 7\}$, list the elements of each set. See Examples 1 and 6.*

**1.** $C \cup D$
**2.** $C \cap D$
**3.** $A \cap D$
**4.** $A \cup D$
**5.** $A \cup B$
**6.** $A \cap B$
**7.** $B \cap D$
**8.** $B \cup D$
**9.** $B \cup C$
**10.** $B \cap C$
**11.** $A \cap C$
**12.** $A \cup C$

*Solve each compound inequality. Graph the solution set and write it in interval notation. See Examples 2 and 3.*

**13.** $x < 5$ and $x > -2$
**14.** $x \leq 7$ and $x \leq 1$
**15.** $x + 1 \geq 7$ and $3x - 1 \geq 5$
**16.** $-2x < -8$ and $x - 5 < 5$
**17.** $4x + 2 \leq -10$ and $2x \leq 0$
**18.** $x + 4 > 0$ and $4x > 0$

*Solve each compound inequality. Graph the solution set and write it in interval notation. See Examples 4 and 5.*

**19.** $5 < x - 6 < 11$

**20.** $-2 \le x + 3 \le 0$

**21.** $-2 \le 3x - 5 \le 7$

**22.** $1 < 4 + 2x < 7$

**23.** $1 \le \frac{2}{3}x + 3 \le 4$

**24.** $-2 < \frac{1}{2}x - 5 < 1$

**25.** $-5 \le \frac{x + 1}{4} \le -2$

**26.** $-4 \le \frac{2x + 5}{3} \le 1$

*Solve each compound inequality. Graph the solution set and write it in interval notation. See Examples 7 and 8.*

**27.** $x < -1$ or $x > 0$

**28.** $x \le 1$ or $x \le -3$

**29.** $-2x \le -4$ or $5x - 20 \ge 5$

**30.** $x + 4 < 0$ or $6x > -12$

**31.** $3(x - 1) < 12$ or $x + 7 > 10$

**32.** $5(x - 1) \ge -5$ or $5 - x \le 11$

**33.** Explain how solving an and-compound inequality is similar to finding the intersection of two sets.

**34.** Explain how solving an or-compound inequality is similar to finding the union of two sets.

*Solve each compound inequality. Graph the solution set and write it in interval notation.*

**35.** $x < 2$ and $x > -1$

**36.** $x < 5$ and $x < 1$

**37.** $x < 2$ or $x > -1$

**38.** $x < 5$ or $x < 1$

**39.** $x \ge -5$ and $x \ge -1$

**40.** $x \le 0$ or $x \ge -3$

**41.** $x \ge -5$ or $x \ge -1$

**42.** $x \le 0$ and $x \ge -3$

**43.** $0 \le 2x - 3 \le 9$

**44.** $3 < 5x + 1 < 11$

**45.** $\frac{1}{2} < x - \frac{3}{4} < 2$

**46.** $\frac{2}{3} < x + \frac{1}{2} < 4$

**47.** $x + 3 \ge 3$ and $x + 3 \le 2$

**48.** $2x - 1 \ge 3$ and $-x > 2$

**49.** $3x \ge 5$ or $-x - 6 < 1$

**50.** $\frac{3}{8}x + 1 \le 0$ or $-2x < -4$

**51.** $0 < \frac{5 - 2x}{3} < 5$

**52.** $-2 < \frac{-2x - 1}{3} < 2$

**53.** $-6 < 3(x - 2) \le 8$

**54.** $-5 < 2(x + 4) < 8$

**55.** $-x + 5 > 6$ and $1 + 2x \le -5$

**56.** $5x \le 0$ and $-x + 5 < 8$

**57.** $3x + 2 \leq 5$ or $7x > 29$

**58.** $-x < 7$ or $3x + 1 < -20$

**59.** $5 - x > 7$ and $2x + 3 \geq 13$

**60.** $-2x < -6$ or $1 - x > -2$

**61.** $-\frac{1}{2} \leq \frac{4x - 1}{6} < \frac{5}{6}$

**62.** $-\frac{1}{2} \leq \frac{3x - 1}{10} < \frac{1}{2}$

**63.** $\frac{1}{15} < \frac{8 - 3x}{15} < \frac{4}{5}$

**64.** $-\frac{1}{4} < \frac{6 - x}{12} < -\frac{1}{6}$

**65.** $0.3 < 0.2x - 0.9 < 1.5$

**66.** $-0.7 \leq 0.4x + 0.8 < 0.5$

*The formula for converting Fahrenheit temperatures to Celsius temperatures is $C = \frac{5}{9}(F - 32)$. Use this formula for Exercises 67 and 68.*

**67.** During a recent year, the temperatures in Chicago ranged from $-29°$ to $35°C$. Use a compound inequality to convert these temperatures to Fahrenheit temperatures.

**68.** In Oslo, the average temperature ranges from $-10°$ to $18°$ Celsius. Use a compound inequality to convert these temperatures to the Fahrenheit scale.

*Solve.*

**69.** Christian D'Angelo has scores of 68, 65, 75, and 78 on his algebra tests. Use a compound inequality to find the scores he can make on his final exam to receive a C in the course. The final exam counts as two tests, and a C is received if the final course average is from 70 to 79.

**70.** Wendy Wood has scores of 80, 90, 82, and 75 on her chemistry tests. Use a compound inequality to find the range of scores she can make on her final exam to receive a B in the course. The final exam counts as two tests, and a B is received if the final course average is from 80 to 89.

*Use the graph to answer Exercises 71 and 72.*

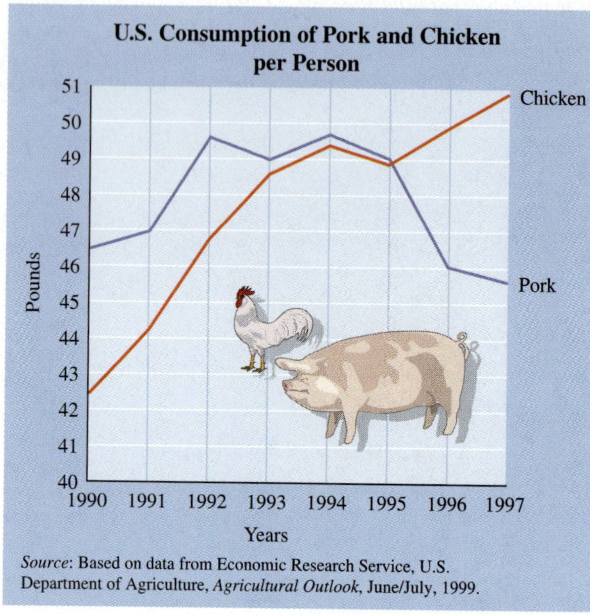


**71.** For what years was the consumption of pork greater than 48 pounds per person *and* the consumption of chicken greater than 48 pounds per person?

**72.** For what years was the consumption of pork less than 48 pounds per person *or* the consumption of chicken greater than 49 pounds per person?

## REVIEW EXERCISES

*Evaluate the following. See Sections 1.2 and 1.3.*

**73.** $|-7| - |19|$

**74.** $|-7 - 19|$

**75.** $-(-6) - |-10|$

**76.** $|-4| - (-4) + |-20|$

*Find by inspection all values for x that make each equation true.*

**77.** $|x| = 7$

**78.** $|x| = 5$

**79.** $|x| = 0$

**80.** $|x| = -2$

## *A Look Ahead*

**Example**
Solve $x - 6 < 3x < 2x + 5$.

*Solution:*
Notice that this inequality contains a variable not only in the middle, but also on the left and the right. When this occurs, we solve by rewriting the inequality using the word ***and***.

$$x - 6 < 3x \quad \text{and} \quad 3x < 2x + 5$$
$$-6 < 2x \quad \text{and} \quad x < 5$$
$$-3 < x$$
$$x > -3 \quad \text{and} \quad x < 5$$

$x > -3$

$x < 5$

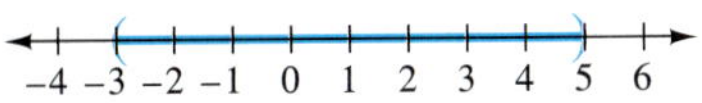


$-3 < x < 5$, or $(-3, 5)$

*Solve each compound inequality for x. See the example.*

**81.** $2x - 3 < 3x + 1 < 4x - 5$

**82.** $x + 3 < 2x + 1 < 4x + 6$

**83.** $-3(x - 2) \le 3 - 2x \le 10 - 3x$

**84.** $7x - 1 \le 7 + 5x \le 3(1 + 2x)$

**85.** $5x - 8 < 2(2 + x) < -2(1 + 2x)$

**86.** $1 + 2x < 3(2 + x) < 1 + 4x$

## 2.6 ABSOLUTE VALUE EQUATIONS

CD-ROM SSM SSG Video

▶ **OBJECTIVE**

1. Solve absolute value equations.

**1** In Chapter 1, we defined the absolute value of a number as its distance from 0 on a number line.

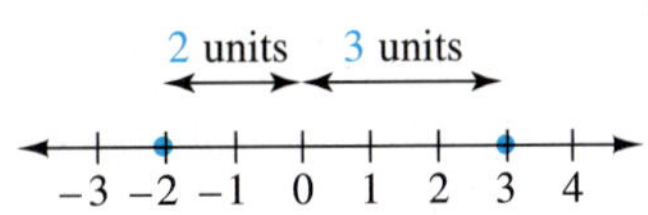


$$|-2| = 2 \text{ and } |3| = 3$$

In this section, we concentrate on solving equations containing the absolute value of a variable or a variable expression. Examples of absolute value equations are

$$|x| = 3 \qquad -5 = |2y + 7| \qquad |z - 6.7| = |3z + 1.2|$$

Since distance and absolute value are so closely related, absolute value equations and inequalities (see Section 2.7) are extremely useful in solving distance-type problems, such as calculating the possible error in a measurement. For the absolute value equation $|x| = 3$, its solution set will contain all numbers whose distance from 0 is 3 units. Two numbers are 3 units away from 0 on the number line: 3 and −3.

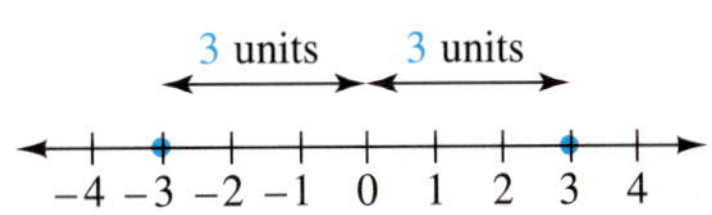


Thus, the solution set of the equation $|x| = 3$ is $\{3, -3\}$. This suggests the following:

**SOLVING EQUATIONS OF THE FORM $|x| = a$**

If $a$ is a positive number, then $|x| = a$ is equivalent to $x = a$ or $x = -a$.

**Example 1** Solve $|p| = 2$.

*Solution* Since 2 is positive, $|p| = 2$ is equivalent to $p = 2$ or $p = -2$.
To check, let $p = 2$ and then $p = -2$ in the original equation.

| | | | |
|---|---|---|---|
| $\lvert p\rvert = 2$ | Original equation. | $\lvert p\rvert = 2$ | Original equation. |
| $\lvert 2\rvert = 2$ | Let $p = 2$. | $\lvert -2\rvert = 2$ | Let $p = -2$. |
| $2 = 2$ | True. | $2 = 2$ | True. |

The solutions are 2 and $-2$ or the solution set is $\{2, -2\}$.

If the expression inside the absolute value bars is more complicated than a single variable $x$, we can still apply the absolute value property.

**Example 2** Solve $|5w + 3| = 7$.

*Solution* Here the expression inside the absolute value bars is $5w + 3$. If we think of the expression $5w + 3$ as $x$ in the absolute value property, we see that $|x| = 7$ is equivalent to

$$x = 7 \quad \text{or} \quad x = -7$$

Then substitute $5w + 3$ for $x$, and we have

$$5w + 3 = 7 \quad \text{or} \quad 5w + 3 = -7$$

Solve these two equations for $w$.

$$\begin{aligned} 5w + 3 &= 7 \quad \text{or} \quad 5w + 3 = -7 \\ 5w &= 4 \quad \text{or} \quad 5w = -10 \\ w &= \frac{4}{5} \quad \text{or} \quad w = -2 \end{aligned}$$

*Check* To check, let $w = -2$ and then $w = \frac{4}{5}$ in the original equation.

Let $w = -2$

$$\begin{aligned} |5(-2) + 3| &= 7 \\ |-10 + 3| &= 7 \\ |-7| &= 7 \\ 7 &= 7 \quad \text{True.} \end{aligned}$$

Let $w = \frac{4}{5}$

$$\begin{aligned} \left|5\left(\frac{4}{5}\right) + 3\right| &= 7 \\ |4 + 3| &= 7 \\ |7| &= 7 \\ 7 &= 7 \quad \text{True.} \end{aligned}$$

Both solutions check, and the solutions are $-2$ and $\frac{4}{5}$.

**Example 3** Solve $\left|\frac{x}{2} - 1\right| = 11$.

*Solution* $\left|\frac{x}{2} - 1\right| = 11$ is equivalent to

$$\frac{x}{2} - 1 = 11 \quad \text{or} \quad \frac{x}{2} - 1 = -11$$

$$2\left(\frac{x}{2} - 1\right) = 2(11) \quad \text{or} \quad 2\left(\frac{x}{2} - 1\right) = 2(-11) \qquad \text{Clear fractions.}$$

$$x - 2 = 22 \quad \text{or} \quad x - 2 = -22 \qquad \text{Apply the distributive property.}$$

$$x = 24 \quad \text{or} \quad x = -20$$

The solutions are 24 and $-20$.

To apply the absolute value rule, first make sure that the absolute value expression is isolated.

**HELPFUL HINT**
If the equation has a single absolute value expression containing variables, isolate the absolute value expression first.

**Example 4** Solve $|2x| + 5 = 7$.

*Solution* We want the absolute value expression alone on one side of the equation, so begin by subtracting 5 from both sides. Then apply the absolute value property.

$$|2x| + 5 = 7$$

$$|2x| = 2 \qquad \text{Subtract 5 from both sides.}$$

$$2x = 2 \quad \text{or} \quad 2x = -2$$

$$x = 1 \quad \text{or} \quad x = -1$$

The solutions are $-1$ and 1.

**Example 5** Solve $|y| = 0$.

*Solution* We are looking for all numbers whose distance from 0 is zero units. The only number is 0. The solution is 0.

The next two examples illustrate a special case for absolute value equations. This special case occurs when an isolated absolute value is equal to a negative number.

**Example 6** Solve $2|x| + 25 = 23$.

*Solution* First, isolate the absolute value.

$$2|x| + 25 = 23$$

$$2|x| = -2 \qquad \text{Subtract 25 from both sides.}$$

$$|x| = -1 \qquad \text{Divide both sides by 2.}$$

The absolute value of a number is never negative, so this equation has no solution. The solution set is $\{\ \}$ or $\varnothing$.

**Example 7** Solve $\left|\frac{3x+1}{2}\right| = -2$.

*Solution* Again, the absolute value of any expression is never negative, so no solution exists. The solution set is { } or ∅.

Given two absolute value expressions, we might ask, when are the absolute values of two expressions equal? To see the answer, notice that

$$|2| = |2|, \quad |-2| = |-2|, \quad |-2| = |2|, \quad \text{and} \quad |2| = |-2|$$

same, same, opposites, opposites

Two absolute value expressions are equal when the expressions inside the absolute value bars are equal to or are opposites of each other.

**Example 8** Solve $|3x + 2| = |5x - 8|$.

*Solution* This equation is true if the expressions inside the absolute value bars are equal to or are opposites of each other.

$$3x + 2 = 5x - 8 \quad \text{or} \quad 3x + 2 = -(5x - 8)$$

Next, solve each equation.

$$\begin{aligned} 3x + 2 &= 5x - 8 && \text{or} & 3x + 2 &= -5x + 8 \\ -2x + 2 &= -8 && \text{or} & 8x + 2 &= 8 \\ -2x &= -10 && \text{or} & 8x &= 6 \\ x &= 5 && \text{or} & x &= \frac{3}{4} \end{aligned}$$

The solutions are $\frac{3}{4}$ and 5.

**Example 9** Solve $|x - 3| = |5 - x|$.

*Solution*

$$\begin{aligned} x - 3 &= 5 - x && \text{or} & x - 3 &= -(5 - x) \\ 2x - 3 &= 5 && \text{or} & x - 3 &= -5 + x \\ 2x &= 8 && \text{or} & x - 3 - x &= -5 + x - x \\ x &= 4 && \text{or} & -3 &= -5 \quad \text{False.} \end{aligned}$$

Recall from Section 2.1 that when an equation simplifies to a false statement, the equation has no solution. Thus, the only solution for the original absolute value equation is 4.

The following box summarizes the methods shown for solving absolute value equations.

**ABSOLUTE VALUE EQUATIONS**

| | |
|---|---|
| $\lvert x\rvert = a$ | If $a$ is positive, then solve $x = a$ or $x = -a$. <br> If $a$ is 0, solve $x = 0$. <br> If $a$ is negative, the equation $\lvert x\rvert = a$ has no solution. |
| $\lvert x\rvert = \lvert y\rvert$ | Solve $x = y$ or $x = -y$. |

## MENTAL MATH

*Simplify each expression.*

1. $\lvert -7\rvert$
2. $\lvert -8\rvert$
3. $-\lvert 5\rvert$
4. $-\lvert 10\rvert$
5. $-\lvert -6\rvert$
6. $-\lvert -3\rvert$
7. $\lvert -3\rvert + \lvert -2\rvert + \lvert -7\rvert$
8. $\lvert -1\rvert + \lvert -6\rvert + \lvert -8\rvert$

## *Exercise Set 2.6*

*Solve each absolute value equation. See Examples 1 through 7.*

1. $\lvert x\rvert = 7$
2. $\lvert y\rvert = 15$
3. $\lvert 3x\rvert = 12.6$
4. $\lvert 6n\rvert = 12.6$
5. $\lvert 2x - 5\rvert = 9$
6. $\lvert 6 + 2n\rvert = 4$
7. $\left\lvert \frac{x}{2} - 3\right\rvert = 1$
8. $\left\lvert \frac{n}{3} + 2\right\rvert = 4$
9. $\lvert z\rvert + 4 = 9$
10. $\lvert x\rvert + 1 = 3$
11. $\lvert 3x\rvert + 5 = 14$
12. $\lvert 2x\rvert - 6 = 4$
13. $\lvert 2x\rvert = 0$
14. $\lvert 7z\rvert = 0$
15. $\lvert 4n + 1\rvert + 10 = 4$
16. $\lvert 3z - 2\rvert + 8 = 1$
17. $\lvert 5x - 1\rvert = 0$
18. $\lvert 3y + 2\rvert = 0$
19. Write an absolute value equation representing all numbers $x$ whose distance from 0 is 5 units.
20. Write an absolute value equation representing all numbers $x$ whose distance from 0 is 2 units.

*Solve. See Examples 8 and 9.*

21. $\lvert 5x - 7\rvert = \lvert 3x + 11\rvert$
22. $\lvert 9y + 1\rvert = \lvert 6y + 4\rvert$
23. $\lvert z + 8\rvert = \lvert z - 3\rvert$
24. $\lvert 2x - 5\rvert = \lvert 2x + 5\rvert$
25. Describe how solving an absolute value equation such as $\lvert 2x - 1\rvert = 3$ is similar to solving an absolute value equation such as $\lvert 2x - 1\rvert = \lvert x - 5\rvert$.
26. Describe how solving an absolute value equation such as $\lvert 2x - 1\rvert = 3$ is different from solving an absolute value equation such as $\lvert 2x - 1\rvert = \lvert x - 5\rvert$.

*Solve each absolute value equation.*

27. $\lvert x\rvert = 4$
28. $\lvert x\rvert = 1$
29. $\lvert y\rvert = 0$
30. $\lvert y\rvert = 8$
31. $\lvert z\rvert = -2$
32. $\lvert y\rvert = -9$
33. $\lvert 7 - 3x\rvert = 7$
34. $\lvert 4m + 5\rvert = 5$
35. $\lvert 6x\rvert - 1 = 11$
36. $\lvert 7z\rvert + 1 = 22$
37. $\lvert 4p\rvert = -8$
38. $\lvert 5m\rvert = -10$
39. $\lvert x - 3\rvert + 3 = 7$
40. $\lvert x + 4\rvert - 4 = 1$
41. $\left\lvert \frac{z}{4} + 5\right\rvert = -7$
42. $\left\lvert \frac{c}{5} - 1\right\rvert = -2$
43. $\lvert 9v - 3\rvert = -8$
44. $\lvert 1 - 3b\rvert = -7$
45. $\lvert 8n + 1\rvert = 0$
46. $\lvert 5x - 2\rvert = 0$
47. $\lvert 1 + 6c\rvert - 7 = -3$
48. $\lvert 2 + 3m\rvert - 9 = -7$
49. $\lvert 5x + 1\rvert = 11$
50. $\lvert 8 - 6c\rvert = 1$
51. $\lvert 4x - 2\rvert = \lvert -10\rvert$
52. $\lvert 3x + 5\rvert = \lvert -4\rvert$

**53.** $|5x + 1| = |4x - 7|$

**54.** $|3 + 6n| = |4n + 11|$

**55.** $|6 + 2x| = -|-7|$

**56.** $|4 - 5y| = -|-3|$

**57.** $|2x - 6| = |10 - 2x|$

**58.** $|4n + 5| = |4n + 3|$

**59.** $\left|\frac{2x - 5}{3}\right| = 7$

**60.** $\left|\frac{1 + 3n}{4}\right| = 4$

**61.** $2 + |5n| = 17$

**62.** $8 + |4m| = 24$

**63.** $\left|\frac{2x - 1}{3}\right| = |-5|$

**64.** $\left|\frac{5x + 2}{2}\right| = |-6|$

**65.** $|2y - 3| = |9 - 4y|$

**66.** $|5z - 1| = |7 - z|$

**67.** $\left|\frac{3n + 2}{8}\right| = |-1|$

**68.** $\left|\frac{2r - 6}{5}\right| = |-2|$

**69.** $|x + 4| = |7 - x|$

**70.** $|8 - y| = |y + 2|$

**71.** $\left|\frac{8c - 7}{3}\right| = -|-5|$

**72.** $\left|\frac{5d + 1}{6}\right| = -|-9|$

**73.** Explain why some absolute value equations have two solutions.

**74.** Explain why some absolute value equations have one solution.

### REVIEW EXERCISES

*The circle graph shows the sources of Walt Disney Company's operating income for 1999. Use this graph to answer Exercises 75–77. See Section 2.2.*

**Walt Disney Company Operating Income 1999**

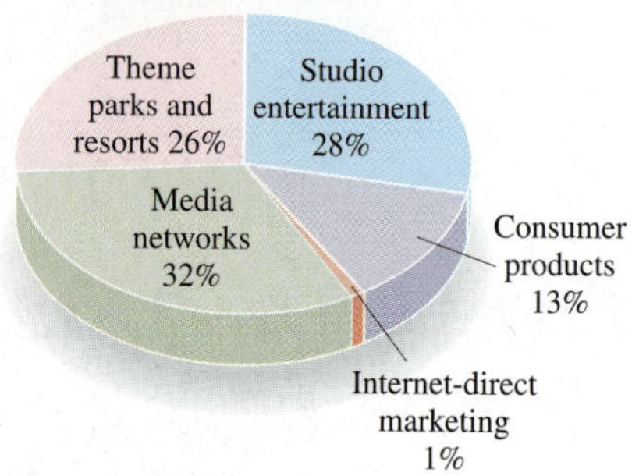


*Source:* Walt Disney Company.

**75.** What percent of Disney's operating income came from the consumer products?

**76.** A circle contains 360°. Find the number of degrees found in the 26% sector for theme parks and resorts.

**77.** If Disney's operating income for all of 1999 was \$3.4 billion, find the amount of income expected from the media networks segment.

*List five integer solutions of each inequality.*

**78.** $|x| \le 3$

**79.** $|x| \ge -2$

**80.** $|y| > -10$

**81.** $|y| < 0$

## 2.7 ABSOLUTE VALUE INEQUALITIES

CD-ROM SSM SSG Video

**OBJECTIVES**

1. Solve absolute value inequalities of the form $|x| < a$.
2. Solve absolute value inequalities of the form $|x| > a$.

**1** The solution set of an absolute value inequality such as $|x| < 2$ contains all numbers whose distance from 0 is less than 2 units, as shown below.

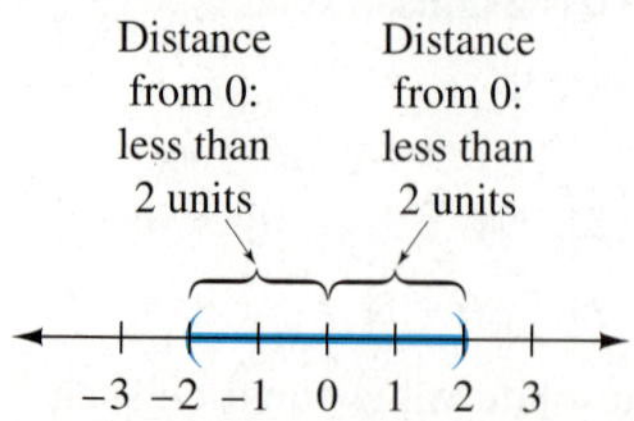


The solution set is $\{x|-2 < x < 2\}$, or $(-2, 2)$ in interval notation.

**Example 1** Solve $|x| \le 3$.

*Solution* The solution set of this inequality contains all numbers whose distance from 0 is less than or equal to 3. Thus 3, $-3$, and all numbers between 3 and $-3$ are in the solution set.

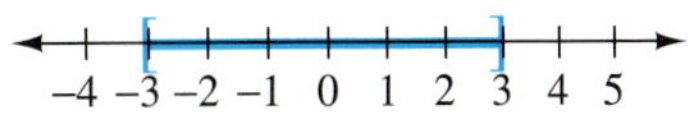


The solution set is $[-3, 3]$.

In general, we have the following.

**SOLVING ABSOLUTE VALUE INEQUALITIES OF THE FORM $|x| < a$**

If $a$ is a positive number, then $|x| < a$ is equivalent to $-a < x < a$.

This property also holds true for the inequality symbol $\le$.

**Example 2** Solve for $m$: $|m - 6| < 2$.

*Solution* Replace $x$ with $m - 6$ and $a$ with 2 in the preceding property, and we see that

$$|m - 6| < 2 \quad \text{is equivalent to} \quad -2 < m - 6 < 2$$

Solve this compound inequality for $m$ by adding 6 to all three parts.

$$-2 < m - 6 < 2$$

$$-2 + 6 < m - 6 + 6 < 2 + 6 \quad \text{Add 6 to all three parts.}$$

$$4 < m < 8 \quad \text{Simplify.}$$

The solution set is $(4, 8)$, and its graph is shown.

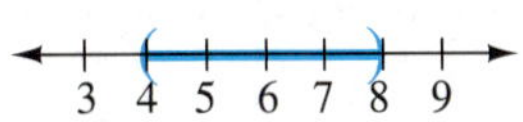


**HELPFUL HINT**
Before using an absolute value inequality property, isolate the absolute value expression on one side of the inequality.

**Example 3** Solve for $x$: $|5x + 1| + 1 \le 10$.

*Solution* First, isolate the absolute value expression by subtracting 1 from both sides.

$$|5x + 1| + 1 \le 10$$

$$|5x + 1| \le 10 - 1 \quad \text{Subtract 1 from both sides.}$$

$$|5x + 1| \le 9 \quad \text{Simplify.}$$

Since 9 is positive, we apply the absolute value property for $|x| \le a$.

$$-9 \le 5x + 1 \le 9$$

$$-9 - 1 \le 5x + 1 - 1 \le 9 - 1 \quad \text{Subtract 1 from all three parts.}$$

$$-10 \le 5x \le 8 \quad \text{Simplify.}$$

$$-2 \le x \le \frac{8}{5} \quad \text{Divide all three parts by 5.}$$

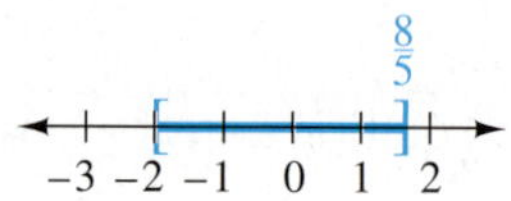


The solution set is $\left[-2, \frac{8}{5}\right]$, and the graph is shown above.

**Example 4** Solve for $x$: $\left|2x - \frac{1}{10}\right| < -13$.

*Solution* The absolute value of a number is always nonnegative and can never be less than −13. Thus this absolute value inequality has no solution. The solution set is { } or ∅.

**2** Let us now solve an absolute value inequality of the form $|x| > a$, such as $|x| \ge 3$. The solution set contains all numbers whose distance from 0 is 3 or more units. Thus the graph of the solution set contains 3 and all points to the right of 3 on the number line or −3 and all points to the left of −3 on the number line.

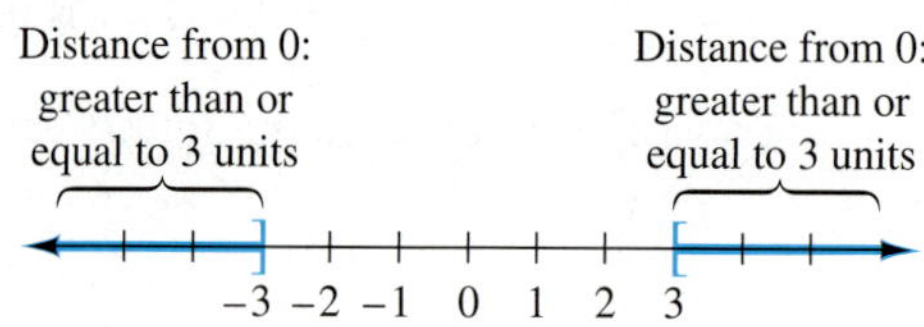


This solution set is written as $\{x \mid x \le -3 \text{ or } x \ge 3\}$. In interval notation, the solution is $(-\infty, -3] \cup [3, \infty)$, since "or" means "union." In general, we have the following.

**SOLVING ABSOLUTE VALUE INEQUALITIES OF THE FORM $|x| > a$**

If $a$ is a positive number, then $|x| > a$ is equivalent to $x < -a$ or $x > a$.

This property also holds true for the inequality symbol ≥.

**Example 5** Solve for $y$: $|y - 3| > 7$.

*Solution* Since 7 is positive, we apply the property for $|x| > a$.

$$|y - 3| > 7 \text{ is equivalent to } y - 3 < -7 \text{ or } y - 3 > 7$$

Next, solve the compound inequality.

$$y - 3 < -7 \quad \text{or} \quad y - 3 > 7$$

$$y - 3 + 3 < -7 + 3 \quad \text{or} \quad y - 3 + 3 > 7 + 3 \quad \text{Add 3 to both sides.}$$

$$y < -4 \quad \text{or} \quad y > 10 \quad \text{Simplify.}$$

The solution set is $(-\infty, -4) \cup (10, \infty)$, and its graph is shown.

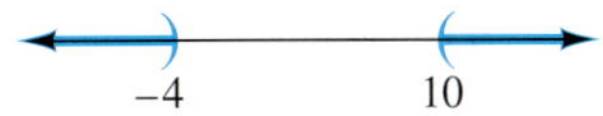


Examples 6 and 8 illustrate special cases of absolute value inequalities. These special cases occur when an isolated absolute value expression is less than, less than or equal to, greater than, or greater than or equal to a negative number or 0.

**Example 6** Solve $|2x + 9| + 5 > 3$.

*Solution* First isolate the absolute value expression by subtracting 5 from both sides.

$$|2x + 9| + 5 > 3$$

$$|2x + 9| + 5 - 5 > 3 - 5 \quad \text{Subtract 5 from both sides.}$$

$$|2x + 9| > -2 \quad \text{Simplify.}$$

The absolute value of any number is always nonnegative and thus is always greater than $-2$. This inequality and the original inequality are true for all values of $x$. The solution set is $\{x \mid x \text{ is a real number}\}$ or $(-\infty, \infty)$ and its graph is shown.

**Example 7** Solve $\left|\frac{x}{3} - 1\right| - 7 \geq -5$.

*Solution* First, isolate the absolute value expression by adding 7 to both sides.

$$\left|\frac{x}{3} - 1\right| - 7 \geq -5$$

$$\left|\frac{x}{3} - 1\right| - 7 + 7 \geq -5 + 7 \quad \text{Add 7 to both sides.}$$

$$\left|\frac{x}{3} - 1\right| \geq 2 \quad \text{Simplify.}$$

Next, write the absolute value inequality as an equivalent compound inequality and solve.

$$\frac{x}{3} - 1 \leq -2 \quad \text{or} \quad \frac{x}{3} - 1 \geq 2$$

$$3\left(\frac{x}{3} - 1\right) \leq 3(-2) \quad \text{or} \quad 3\left(\frac{x}{3} - 1\right) \geq 3(2) \quad \text{Clear the inequalities of fractions.}$$

$$x - 3 \leq -6 \quad \text{or} \quad x - 3 \geq 6 \quad \text{Apply the distributive property.}$$

$$x \leq -3 \quad \text{or} \quad x \geq 9 \quad \text{Add 3 to both sides.}$$

The solution set is $(-\infty, -3] \cup [9, \infty)$, and its graph is shown.

**Example 8** Solve for $x$: $\left|\frac{2(x+1)}{3}\right| \le 0$

*Solution* Recall that "≤" means "less than or equal to." The absolute value of any expression will never be less than 0, but it may be equal to 0. Thus, to solve $\left|\frac{2(x+1)}{3}\right| \le 0$ we solve $\left|\frac{2(x+1)}{3}\right| = 0$

$$\frac{2(x+1)}{3} = 0$$

$$3\left[\frac{2(x+1)}{3}\right] = 3(0) \quad \text{Clear the equation of fractions.}$$

$$2x + 2 = 0 \quad \text{Apply the distributive property.}$$

$$2x = -2 \quad \text{Subtract 2 from both sides.}$$

$$x = -1 \quad \text{Divide both sides by 2.}$$

The solution set is $\{-1\}$.

The following box summarizes the types of absolute value equations and inequalities.

**SOLVING ABSOLUTE VALUE EQUATIONS AND INEQUALITIES WITH $a > 0$**

| ***Algebraic Solution*** | ***Solution Graph*** |
|---|---|
| $\lvert x\rvert = a$ is equivalent to $x = a$ or $x = -a$. |  |
| $\lvert x\rvert < a$ is equivalent to $-a < x < a$. |  |
| $\lvert x\rvert > a$ is equivalent to $x < -a$ or $x > a$. |  |

## Exercise Set 2.7

*Solve each inequality. Then graph the solution set. See Examples 1 through 4.*

1. $|x| \le 4$
2. $|x| < 6$
3. $|x - 3| < 2$
4. $|y| \le 5$
5. $|x + 3| < 2$
6. $|x + 4| < 6$
7. $|2x + 7| \le 13$
8. $|5x - 3| \le 18$
9. $|x| + 7 \le 12$
10. $|x| + 6 \le 7$
11. $|3x - 1| < -5$
12. $|8x - 3| < -2$
13. $|x - 6| - 7 \le -1$
14. $|z + 2| - 7 < -3$

*Solve each inequality. Graph the solution set. See Examples 5 through 7.*

15. $|x| > 3$
16. $|y| \ge 4$
17. $|x + 10| \ge 14$
18. $|x - 9| \ge 2$
19. $|x| + 2 > 6$
20. $|x| - 1 > 3$
21. $|5x| > -4$
22. $|4x - 11| > -1$
23. $|6x - 8| + 3 > 7$
24. $|10 + 3x| + 1 > 2$

*Solve each inequality. Graph the solution set. See Example 8.*

25. $|x| \le 0$
26. $|x| \ge 0$

**27.** $|8x + 3| > 0$

**28.** $|5x - 6| < 0$

**29.** Write an absolute value inequality representing all numbers $x$ whose distance from 0 is less than 7 units.

**30.** Write an absolute value inequality representing all numbers $x$ whose distance from 0 is greater than 4 units.

**31.** Write $-5 \le x \le 5$ as an equivalent inequality containing an absolute value.

**32.** Write $x > 1$ or $x < -1$ as an equivalent inequality containing an absolute value.

*Solve each inequality. Graph the solution set.*

**33.** $|x| \le 2$

**34.** $|z| < 6$

**35.** $|y| > 1$

**36.** $|x| \ge 10$

**37.** $|x - 3| < 8$

**38.** $|-3 + x| \le 10$

**39.** $|0.6x - 3| > 0.6$

**40.** $|1 + 0.3x| \ge 0.1$

**41.** $5 + |x| \le 2$

**42.** $8 + |x| < 1$

**43.** $|x| > -4$

**44.** $|x| \le -7$

**45.** $|2x - 7| \le 11$

**46.** $|5x + 2| < 8$

**47.** $|x + 5| + 2 \ge 8$

**48.** $|-1 + x| - 6 > 2$

**49.** $|x| > 0$

**50.** $|x| < 0$

**51.** $9 + |x| > 7$

**52.** $5 + |x| \ge 4$

**53.** $6 + |4x - 1| \le 9$

**54.** $-3 + |5x - 2| \le 4$

**55.** $\left|\frac{2}{3}x + 1\right| > 1$

**56.** $|5x - 1| \ge 2$

**57.** $|5x + 3| < -6$

**58.** $|4 + 9x| \ge -6$

**59.** $|8x + 3| \ge 0$

**60.** $|5x - 6| \le 0$

**61.** $|1 + 3x| + 4 < 5$

**62.** $|7x - 3| - 1 \le 10$

**63.** $|x| - 3 \ge -3$

**64.** $|x| + 6 < 6$

**65.** $|8x| - 10 > -2$

**66.** $|6x| - 13 \ge -7$

**67.** $\left|\frac{x + 6}{3}\right| > 2$

**68.** $\left|\frac{7 + x}{2}\right| \ge 4$

**69.** $|2(3 + x)| > 6$

**70.** $|5(x - 3)| \ge 10$

**71.** $\left|\frac{5(x + 2)}{3}\right| < 7$

**72.** $\left|\frac{6(3 + x)}{5}\right| \le 4$

**73.** $-15 + |2x - 7| \le -6$

**74.** $-9 + |3 + 4x| < -4$

**75.** $\left|2x + \frac{3}{4}\right| - 7 \le -2$

**76.** $\left|\frac{3}{5} + 4x\right| - 6 < -1$

*Solve each equation or inequality for x.*

**77.** $|2x - 3| < 7$

**78.** $|2x - 3| > 7$

**79.** $|2x - 3| = 7$

**80.** $|5 - 6x| = 29$

**81.** $|x - 5| \ge 12$

**82.** $|x + 4| \ge 20$

**83.** $|9 + 4x| = 0$

**84.** $|9 + 4x| \geq 0$

**85.** $|2x + 1| + 4 < 7$

**86.** $8 + |5x - 3| \geq 11$

**87.** $|3x - 5| + 4 = 5$

**88.** $|8x| = -5$

**89.** $|x + 11| = -1$

**90.** $|4x - 4| = -3$

**91.** $\left|\frac{2x - 1}{3}\right| = 6$

**92.** $\left|\frac{6 - x}{4}\right| = 5$

**93.** $\left|\frac{3x - 5}{6}\right| > 5$

**94.** $\left|\frac{4x - 7}{5}\right| < 2$

**95.** Describe how solving $|x - 3| = 5$ is different from solving $|x - 3| < 5$.

**96.** Describe how solving $|x + 4| = 0$ is similar to solving $|x + 4| \leq 0$.

*The expression $|x_T - x|$ is defined to be the absolute error in x, where $x_T$ is the true value of a quantity and x is the measured value or value as stored in a computer.*

**97.** If the true value of a quantity is 3.5 and the absolute error must be less than 0.05, find the acceptable measured values.

**98.** If the true value of a quantity is 0.2 and the approximate value stored in a computer is $\frac{51}{256}$, find the absolute error.

## REVIEW EXERCISES

*Recall the formula:*

$$\text{Probability of an event} = \frac{\text{number of ways that the event can occur}}{\text{number of possible outcomes}}$$

*Find the probability of rolling each number on a single toss of a die. (Recall that a die is a cube with each of its six sides containing 1, 2, 3, 4, 5, and 6 black dots, respectively.) See Section 2.3.*

**99.** $P$(rolling a 2)

**100.** $P$(rolling a 5)

**101.** $P$(rolling a 7)

**102.** $P$(rolling a 0)

**103.** $P$(rolling a 1 or 3)

**104.** $P$(rolling a 1, 2, 3, 4, 5, or 6)

*Consider the equation $3x - 4y = 12$. For each value of x or y given, find the corresponding value of the other variable that makes the statement true. See Section 2.3.*

**105.** If $x = 2$, find $y$

**106.** If $y = -1$, find $x$.

**107.** If $y = -3$, find $x$

**108.** If $x = 4$, find $y$

For additional Chapter Projects, visit the Real World Activities Website by going to http://www.prenhall.com/martin-gay.

# 2 CHAPTER PROJECT

## *Analyzing Municipal Budgets*

Nearly all cities, towns, and villages operate with an annual budget. Budget items might include expenses for fire and police protection as well as for street maintenance and parks. No matter how big or small the budget, city officials need to know if municipal spending is over or under budget. In this project, you will have the opportunity to analyze a municipal budget and make budgetary recommendations. This project may be completed by working in groups or individually.

Suppose that each year your town creates a municipal budget. The next year's annual municipal budget is submitted for approval by the town's citizens at the annual town meeting. This year's budget was printed in the town newspaper earlier in the year.

You have joined a group of citizens who are concerned about your town's budgeting and spending processes. Your group plans to analyze this year's budget along with what was actually spent by the town this year. You hope to present your findings at the annual town meeting and make some budgetary recommendations for next year's budget. The municipal budget contains many different areas of spending. To help focus your group's analysis, you have decided to research spending habits only for categories in which the actual expenses differ from the budgeted amount by more than 12% of the budgeted amount.

1. For each category in the budget, write a specific absolute value inequality that describes the condition that must be met before your group will research spending habits for that category. In each case, let the variable $x$ represent the actual expense for a budget category.
2. For each category in the budget, write an equivalent compound inequality for the condition described in Question 1. Again, let the variable $x$ represent the actual expense for a budget category.
3. Below is a listing of the actual expenditures made this year for each budget category. Use the inequalities from either Question 1 or Question 2 to complete the Budget Worksheet given at the end of this project. (The first category has been filled in.) From the Budget Worksheet, decide which categories must be researched.

| | *Department/Program* | *Actual Expenditure* |
|---|---|---|
| I. | **Board of Health** | |
| | Immunization Programs | $14,800 |
| | Inspections | $41,900 |
| II. | **Fire Department** | |
| | Equipment | $375,000 |
| | Salaries | $268,500 |
| III. | **Libraries** | |
| | Book/Periodical Purchases | $107,300 |
| | Equipment | $29,000 |
| | Salaries | $118,400 |
| IV. | **Parks and Recreation** | |
| | Maintenance | $82,500 |
| | Playground Equipment | $45,000 |
| | Salaries | $118,000 |
| | Summer Programs | $96,200 |
| V. | **Police Department** | |
| | Equipment | $328,000 |
| | Salaries | $405,000 |
| VI. | **Public Works** | |
| | Recycling | $48,100 |
| | Sewage | $92,500 |
| | Snow Removal & Road Salt | $268,300 |
| | Street Maintenance | $284,000 |
| | Water Treatment | $94,100 |
| | **TOTAL** | $2,816,600 |

## *THE TOWN CRIER*

### Annual Budget Set at Town Meeting

ANYTOWN, USA (MG)—This year's annual budget is as follows:

| | | **Amount Budgeted** |
|---|---|---|
| **BOARD OF HEALTH** | | |
| | *Immunization Programs* | $15,000 |
| | *Inspections* | $50,000 |
| **FIRE DEPARTMENT** | | |
| | *Equipment* | $450,000 |
| | *Salaries* | $275,000 |
| **LIBRARIES** | | |
| | *Book/Periodical Purchases* | $90,000 |
| | *Equipment* | $30,000 |
| | *Salaries* | $120,000 |
| **PARKS AND RECREATION** | | |
| | *Maintenance* | $70,000 |
| | *Playground Equipment* | $50,000 |
| | *Salaries* | $140,000 |
| | *Summer Programs* | $80,000 |
| **POLICE DEPARTMENT** | | |
| | *Equipment* | $300,000 |
| | *Salaries* | $400,000 |
| **PUBLIC WORKS** | | |
| | *Recycling* | $50,000 |
| | *Sewage* | $100,000 |
| | *Snow Removal & Road Salt* | $200,000 |
| | *Street Maintenance* | $250,000 |
| | *Water Treatment* | $100,000 |
| **TOTAL** | | **$2,770,000** |

4. Can you think of possible reasons why spending in the categories that must be researched were over or under budget?
5. Based on this year's municipal budget and actual expenses, what recommendations would you make for next year's budget? Explain your reasoning.
6. (Optional) Research the annual budget used by your own town or your college or university. Conduct a similar analysis of the budget with respect to actual expenses. What can you conclude?

### BUDGET WORKSHEET

| **Budget category** | **Budgeted amount** | **Minimum allowed** | **Actual expense** | **Maximum allowed** | **Within budget?** | **Amt over/ under budget** |
|---|---|---|---|---|---|---|
| Immunization Programs | $15,000 | $13,200 | $14,800 | $16,800 | Yes | Under $200 |
| | | | | | | |
| | | | | | | |
| | | | | | | |
| | | | | | | |

## CHAPTER 2 VOCABULARY CHECK

*Fill in each blank with one of the words or phrases listed below.*

| | | | | | |
|---|---|---|---|---|---|
| contradiction | formula | consecutive integers | linear equation in one variable | identity | solution |
| absolute value | linear inequality in one variable | | compound inequality | intersection | union |

1. The statement "$x < 5$ or $x > 7$" *is called a*(*n*) ____________________.
2. An equation in one variable that has no solution is called a(n) ______________.
3. The ____________ of two sets is the set of all elements common to both sets.
4. The _______ of two sets is the set of all elements that belong to either of the sets.
5. An equation in one variable that has every number (for which the equation is defined) as a solution is called a(n) ________.
6. The equation $d = rt$ is also called a(n) ________.
7. A number's distance from 0 is called its ______________.
8. When a variable in an equation is replaced by a number and the resulting equation is true, then that number is called a(n) ________ of the equation.
9. The integers 17, 18, 19 are examples of __________________.
10. The statement $5x - 0.2 < 7$ is an example of a(n) ____________________________.
11. The statement $5x - 0.2 = 7$ is an example of a(n) ____________________________.

## CHAPTER 2 HIGHLIGHTS

| DEFINITIONS AND CONCEPTS | EXAMPLES |
|---|---|
| **Section 2.1 Linear Equations in One Variable** | |
| An **equation** is a statement that two expressions are equal. | Equations:<br>$5 = 5$ $\quad 7x + 2 = -14 \quad 3(x-1)^2 = 9x^2 - 6$ |
| A **linear equation in one variable** is an equation that can be written in the form $ax + b = c$, where $a$, $b$, and $c$ are real numbers and $a$ is not 0. | Linear equations:<br>$7x + 2 = -14 \quad x = -3$<br>$5(2y - 7) = -2(8y - 1)$ |
| A **solution** of an equation is a value for the variable that makes the equation a true statement. | Check to see that $-1$ is a solution of<br>$3(x - 1) = 4x - 2$.<br>$3(-1 - 1) = 4(-1) - 2$<br>$3(-2) = -4 - 2$<br>$-6 = -6$ True.<br>Thus, $-1$ is a solution. |
| **Equivalent equations** have the same solution. | $x - 12 = 14$ and $x = 26$ are equivalent equations. |
| The **addition property of equality** guarantees that the same number may be added to (or subtracted from) both sides of an equation, and the result is an equivalent equation.<br>The **multiplication property of equality** guarantees that both sides of an equation may be multiplied by (or divided by) the same nonzero number, and the result is an equivalent equation. | Solve for $x$: $-3x - 2 = 10$.<br>$-3x - 2 + 2 = 10 + 2$ Add 2 to both sides.<br>$-3x = 12$<br>$\frac{-3x}{-3} = \frac{12}{-3}$ Divide both sides by $-3$.<br>$x = -4$ |
| To solve linear equations in one variable: | Solve for $x$:<br>$x - \frac{x-2}{6} = \frac{x-7}{3} + \frac{2}{3}$ |

*(continued)*

| DEFINITIONS AND CONCEPTS | EXAMPLES |
|---|---|
| **Section 2.1 Linear Equations in One Variable** | |
| 1. Clear the equation of fractions. | 1. $6\left(x - \frac{x-2}{6}\right) = 6\left(\frac{x-7}{3} + \frac{2}{3}\right)$ Multiply both sides by 6.<br>$6x - (x-2) = 2(x-7) + 2(2)$ |
| 2. Remove grouping symbols such as parentheses. | 2. $6x - x + 2 = 2x - 14 + 4$ Remove grouping symbols. |
| 3. Simplify by combining like terms. | 3. $5x + 2 = 2x - 10$ |
| 4. Write variable terms on one side and numbers on the other side using the addition property of equality. | 4. $5x + 2 - 2 = 2x - 10 - 2$ Subtract 2.<br>$5x = 2x - 12$<br>$5x - 2x = 2x - 12 - 2x$ Subtract $2x$.<br>$3x = -12$ |
| 5. Isolate the variable using the multiplication property of equality. | 5. $\frac{3x}{3} = \frac{-12}{3}$ Divide by 3.<br>$x = -4$ |
| 6. Check the proposed solution in the original equation. | 6. $-4 - \frac{-4-2}{6} \stackrel{?}{=} \frac{-4-7}{3} + \frac{2}{3}$ Replace $x$ with $-4$ in the original equation.<br>$-4 - \frac{-6}{6} \stackrel{?}{=} \frac{-11}{3} + \frac{2}{3}$<br>$-4 - (-1) \stackrel{?}{=} \frac{-9}{3}$<br>$-3 = -3$ True. |
| **Section 2.2 An Introduction to Problem Solving** | |
| ***Problem-Solving Strategy*** | Colorado is shaped like a rectangle whose length is about 1.3 times its width. If the perimeter of Colorado is 2070 kilometers, find its dimensions. |
| 1. UNDERSTAND the problem. | 1. Read and reread the problem. Guess a solution and check your guess.<br>Let $x$ = width of Colorado in kilometers. Then<br>$1.3x$ = length of Colorado in kilometers<br>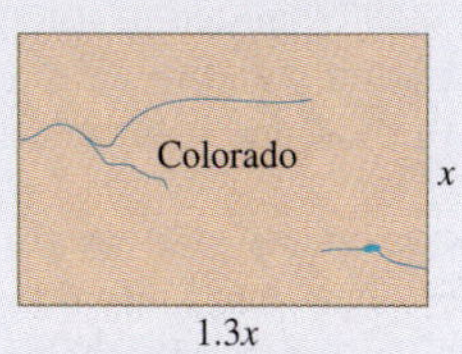 |
| 2. TRANSLATE the problem. | 2. In words:<br>twice the length + twice the width = perimeter<br>Translate: $2(1.3x) + 2x = 2070$ |

*(continued)*

| DEFINITIONS AND CONCEPTS | EXAMPLES |
|---|---|
| **Section 2.2 An Introduction to Problem Solving** | |
| **3.** SOLVE the equation. | **3.** $2.6x + 2x = 2070$<br>$4.6x = 2070$<br>$x = 450$ |
| **4.** INTERPRET the results. | **4.** If $x = 450$ kilometers, then $1.3x = 1.3(450) = 585$ kilometers. *Check*: The perimeter of a rectangle whose width is 450 kilometers and length is 585 kilometers is $2(450) + 2(585) = 2070$ kilometers, the required perimeter. *State*: The dimensions of Colorado are 450 kilometers by 585 kilometers |
| **Section 2.3 Formulas and Problem Solving** | |
| An equation that describes a known relationship among quantities is called a **formula.** | ***Formulas:***<br>$A = \pi r^2$ (area of a circle)<br>$I = PRT$ (interest = principal · rate · time) |
| **To solve a formula for a specified variable**, use the steps for solving an equation. Treat the specified variable as the only variable of the equation. | Solve $A = 2HW + 2LW + 2LH$ for $H$<br>$A - 2LW = 2HW + 2LH$ Subtract $2LW$.<br>$A - 2LW = H(2W + 2L)$ Factor out $H$.<br>$\frac{A - 2LW}{2W + 2L} = \frac{H(2W + 2L)}{2W + 2L}$ Divide by $2W + 2L$.<br>$\frac{A - 2LW}{2W + 2L} = H$ Simplify. |
| **Section 2.4 Linear Inequalities and Problem Solving** | |
| A **linear inequality in one variable** is an inequality that can be written in the form $ax + b < c$, where $a$, $b$, and $c$ are real numbers and $a \neq 0$. (The inequality symbols $\leq, >$, and $\geq$ also apply here.) | Linear inequalities:<br>$5x - 2 \leq -7$ $3y > 1$ $\frac{z}{7} < -9(z - 3)$ |
| The **addition property of inequality** guarantees that the same number may be added to (or subtracted from) both sides of an inequality, and the resulting inequality will have the same solution set. | $x - 9 \leq -16$<br>$x - 9 + 9 \leq -16 + 9$ Add 9.<br>$x \leq -7$ |
| The **multiplication property of inequality** guarantees that both sides of an inequality may be multiplied by (or divided by) the same **positive** number, and the resulting inequality will have the same solution set. We may also multiply (or divide) both sides of an inequality by the same **negative** number and **reverse the direction of the inequality symbol**, and the result is an inequality with the same solution set. | $6x < -66$<br>$\frac{6x}{6} < \frac{-66}{6}$ Divide by 6. Do not reverse direction of inequality symbol.<br>$x < -11$<br>$-6x < -66$<br>$\frac{-6x}{-6} > \frac{-66}{-6}$ Divide by −6. Reverse direction of inequality symbol.<br>$x > 11$ |

*(continued)*

**DEFINITIONS AND CONCEPTS** | **EXAMPLES**

## Section 2.4 Linear Inequalities and Problem Solving

To solve a linear inequality in one variable:

Solve for $x$:

$$\frac{3}{7}(x-4) \geq x+2$$

1. Clear the equation of fractions.

$$\text{1.}\quad 7\left[\frac{3}{7}(x-4)\right] \geq 7(x+2) \qquad \text{Multiply by 7.}$$

$$3(x-4) \geq 7(x+2)$$

2. Remove grouping symbols such as parentheses.
3. Simplify by combining like terms.

$$\text{2.}\quad 3x - 12 \geq 7x + 14 \qquad \text{Apply the distributive property.}$$

4. Write variable terms on one side and numbers on the other side using the addition property of inequality.

$$\text{4.}\quad -4x - 12 \geq 14 \qquad \text{Subtract } 7x.$$

$$-4x \geq 26 \qquad \text{Add 12.}$$

$$\frac{-4x}{-4} \leq \frac{26}{-4} \qquad \text{Divide by } -4. \text{ Reverse direction of inequality symbol.}$$

5. Isolate the variable using the multiplication property of inequality.

$$x \leq -\frac{13}{2}$$

## Section 2.5 Compound Inequalities

Two inequalities joined by the words **and** or **or** are called **compound inequalities.**

Compound inequalities:

$$x - 7 \leq 4 \quad \text{and} \quad x \geq -21$$

$$2x + 7 > x - 3 \quad \text{or} \quad 5x + 2 > -3$$

The solution set of a compound inequality formed by the word **and** is the **intersection** $\cap$ of the solution sets of the two inequalities.

Solve for $x$:

$x < 5$ and $x < 3$

$\{x|x < 5\}$  $(-\infty, 5)$

$\{x|x < 3\}$ 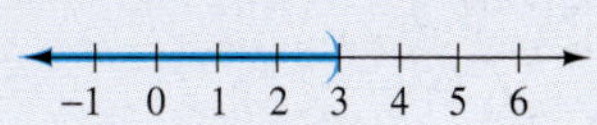 $(-\infty, 3)$

$\{x|x < 3$ and $x < 5\}$ 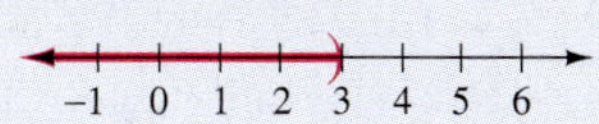 $(-\infty, 3)$

The solution set of a compound inequality formed by the word **or** is the **union**, $\cup$, of the solution sets of the two inequalities.

Solve for $x$:

$$x - 2 \geq -3 \quad \text{or} \quad 2x \leq -4$$

$$x \geq -1 \quad \text{or} \quad x \leq -2$$

$\{x|x \geq -1\}$  $[-1, \infty)$

$\{x|x \leq -2\}$ 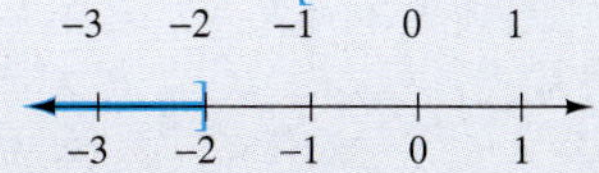 $(-\infty, -2]$

$\{x|x \leq -2$ or $x \geq -1\}$ 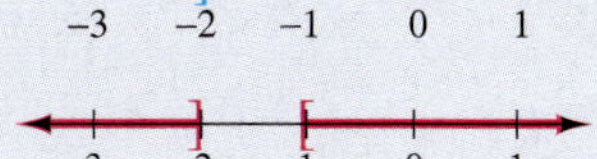 $(-\infty, -2] \cup [-1, \infty)$

*(continued)*

| DEFINITIONS AND CONCEPTS | EXAMPLES |
|---|---|

## Section 2.6 Absolute Value Equations

If $a$ is a positive number, then $|x| = a$ is equivalent to $x = a$ or $x = -a$.

Solve for $y$:

$$|5y - 1| - 7 = 4$$

$$\begin{aligned} |5y - 1| &= 11 && \text{Add 7.} \\ 5y - 1 = 11 \quad &\text{or} \quad 5y - 1 = -11 \\ 5y = 12 \quad &\text{or} \quad 5y = -10 && \text{Add 1.} \\ y = \frac{12}{5} \quad &\text{or} \quad y = -2 && \text{Divide by 5.} \end{aligned}$$

The solutions are $-2$ and $\frac{12}{5}$.

If $a$ is negative, then $|x| = a$ has no solution.

Solve for $x$:

$$\left|\frac{x}{2} - 7\right| = -1$$

The solution set is { } or $\varnothing$.

If an absolute value equation is of the form $|x| = |y|$, solve $x = y$ or $x = -y$.

Solve for $x$:

$$\begin{aligned} &|x - 7| = |2x + 1| \\ &x - 7 = 2x + 1 \quad \text{or} \quad x - 7 = -(2x + 1) \\ &x = 2x + 8 \qquad\qquad x - 7 = -2x - 1 \\ &-x = 8 \qquad\qquad\quad x = -2x + 6 \\ &x = -8 \quad \text{or} \quad 3x = 6 \\ &\qquad\qquad\qquad\quad x = 2 \end{aligned}$$

The solutions are $-8$ and 2.

## Section 2.7 Absolute Value Inequalities

If $a$ is a positive number, then $|x| < a$ is equivalent to $-a < x < a$.

Solve for $y$:

$$\begin{aligned} |y - 5| &\le 3 \\ -3 \le y - 5 &\le 3 \\ -3 + 5 \le y - 5 + 5 &\le 3 + 5 && \text{Add 5.} \\ 2 \le y &\le 8 \end{aligned}$$

The solution set is $[2, 8]$.

If $a$ is a positive number, then $|x| > a$ is equivalent to $x < -a$ or $x > a$.

Solve for $x$:

$$\left|\frac{x}{2} - 3\right| > 7$$

$$\begin{aligned} \frac{x}{2} - 3 < -7 \quad &\text{or} \quad \frac{x}{2} - 3 > 7 \\ x - 6 < -14 \quad &\text{or} \quad x - 6 > 14 && \text{Multiply by 2.} \\ x < -8 \quad &\text{or} \quad x > 20 && \text{Add 6.} \end{aligned}$$

The solution set is $(-\infty, -8) \cup (20, \infty)$.

## CHAPTER 2 REVIEW

**(2.1)** *Solve each linear equation.*

**1.** $4(x - 5) = 2x - 14$

**2.** $x + 7 = -2(x + 8)$

**3.** $3(2y - 1) = -8(6 + y)$

**4.** $-(z + 12) = 5(2z - 1)$

**5.** $n - (8 + 4n) = 2(3n - 4)$

**6.** $4(9v + 2) = 6(1 + 6v) - 10$

**7.** $0.3(x - 2) = 1.2$

**8.** $1.5 = 0.2(c - 0.3)$

**9.** $-4(2 - 3h) = 2(3h - 4) + 6h$

**10.** $6(m - 1) + 3(2 - m) = 0$

**11.** $6 - 3(2g + 4) - 4g = 5(1 - 2g)$

**12.** $20 - 5(p + 1) + 3p = -(2p - 15)$

**13.** $\frac{x}{3} - 4 = x - 2$

**14.** $\frac{9}{4}y = \frac{2}{3}y$

**15.** $\frac{3n}{8} - 1 = 3 + \frac{n}{6}$

**16.** $\frac{z}{6} + 1 = \frac{z}{2} + 2$

**17.** $\frac{y}{4} - \frac{y}{2} = -8$

**18.** $\frac{2x}{3} - \frac{8}{3} = x$

**19.** $\frac{b - 2}{3} = \frac{b + 2}{5}$

**20.** $\frac{2t - 1}{3} = \frac{3t + 2}{15}$

**21.** $\frac{2(t + 1)}{3} = \frac{2(t - 1)}{3}$

**22.** $\frac{3a - 3}{6} = \frac{4a + 1}{15} + 2$

**23.** $\frac{x - 2}{5} + \frac{x + 2}{2} = \frac{x + 4}{3}$

**24.** $\frac{2z - 3}{4} - \frac{4 - z}{2} = \frac{z + 1}{3}$

**(2.2)** *Solve.*

**25.** Twice the difference of a number and 3 is the same as 1 added to three times the number. Find the number.

**26.** One number is 5 more than another number. If the sum of the numbers is 285, find the numbers.

**27.** Find 40% of 130.

**28.** Find 1.5% of 8.

**29.** In 1998, the average annual earnings for a worker with an associate's degree was \$29,872. This represents a 30.47% increase over the average annual earnings for a high school graduate in 1998. Find the average annual earnings for a high school graduate in 1998. Round to the nearest whole dollar. (*Source:* U.S. Bureau of the Census)

**30.** Find four consecutive integers such that twice the first subtracted from the sum of the other three integers is sixteen.

**31.** Determine whether there are two consecutive odd integers such that 5 times the first exceeds 3 times the second by 54.

△ **32.** The length of a rectangular playing field is 5 meters less than twice its width. If 230 meters of fencing goes around the field, find the dimensions of the field.

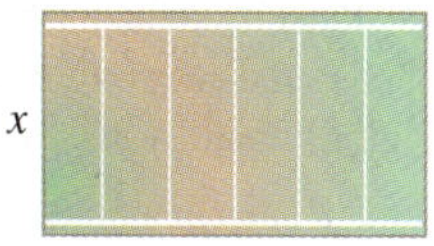


**33.** A car rental company charges \$29.95 per day for a compact car plus 15 cents per mile for every mile over 100 miles driven per day. If Mr. Woo's bill for 2 days' use is \$83.60 before taxes, find how many miles to the nearest whole mile he drove.

**34.** The cost $C$ of producing $x$ number of scientific calculators is given by $C = 4.50x + 3000$, and the revenue $R$ from selling them is given by $R = 16.50x$. Find the number of calculators that must be sold to break even. (To break even, revenue = cost.)

**35.** An entrepreneur can sell her musically vibrating plants for \$40 each, while her cost $C$ to produce $x$ number of plants is given by $C = 20x + 100$. Find her break-even point. Find her revenue if she sells exactly that number of plants.

*Solve each equation for the specified variable.*

△ **36.** $V = lwh; w$

△ **37.** $C = 2\pi r; r$

**38.** $5x - 4y = -12; y$

**39.** $5x - 4y = -12; x$

**40.** $y - y_1 = m(x - x_1); m$

**41.** $y - y_1 = m(x - x_1); x$

**42.** $E = I(R + r); r$

**43.** $S = vt + gt^2; g$

**44.** $T = gr + gvt; g$

**45.** $I = Prt + P; P$

△ **46.** $A = \frac{h}{2}(B + b); B$

△ **47.** $V = \frac{1}{3}\pi r^2 h; h$

**48.** $R = \frac{r_1 + r_2}{2}; r_1$

**49.** $\frac{V_1}{T_1} = \frac{V_2}{T_2}; T_2$

*Solve.*

**50.** A principal of \$3000 is invested in an account paying an annual percentage rate of 3%. Find the amount in the account after 7 years if the amount is compounded

**a.** semiannually

**b.** weekly

(Approximate to the nearest cent.)

**51.** The high temperature in Slidell, Louisiana, one day was 90° Fahrenheit. Convert this temperature to degrees Celsius.

△ **52.** Angie Applegate has a photograph in which the length is 2 inches longer than the width. If she increases each dimension by 4 inches, the area is increased by 88 square inches. Find the original dimensions.

△ **53.** One-square-foot floor tiles come 24 to a package. Find how many packages are needed to cover a rectangular floor 18 feet by 21 feet.

△ **54.** Determine which container holds more ice cream, an 8 inch by 5 inch by 3 inch box or a cylinder with radius of 3 inches and height of 6 inches.

**55.** Erasmos Gonzales left Los Angeles at 11 A.M. and drove nonstop to San Diego, 130 miles away. If she arrived at 1:15 P.M., find her average speed, rounded to the nearest mile per hour.

**(2.4)** *Solve each linear inequality.*

**56.** $3(x - 5) > -(x + 3)$

**57.** $-2(x + 7) \geq 3(x + 2)$

**58.** $4x - (5 + 2x) < 3x - 1$

**59.** $3(x - 8) < 7x + 2(5 - x)$

**60.** $24 \geq 6x - 2(3x - 5) + 2x$

**61.** $48 + x \geq 5(2x + 4) - 2x$

**62.** $\frac{x}{3} + \frac{1}{2} > \frac{2}{3}$

**63.** $x + \frac{3}{4} < -\frac{x}{2} + \frac{9}{4}$

**64.** $\frac{x - 5}{2} \leq \frac{3}{8}(2x + 6)$

**65.** $\frac{3(x - 2)}{5} > \frac{-5(x - 2)}{3}$

*Solve.*

**66.** George Boros can pay his housekeeper \$25 per week to do his laundry, or he can have the laundromat do it at a cost of 90 cents per pound for the first 10 pounds and 80 cents for each additional pound. Use an inequality to find the weight at which it is more economical to use the housekeeper than the laundromat.

**67.** Ceramic firing temperatures usually range from 500° to 1000° Fahrenheit. Use a compound inequality to convert this range to the Celsius scale. Round to the nearest degree.

**68.** In the Olympic gymnastics competition, Nana must average a score of 9.65 to win the silver medal. Seven of the eight judges have reported scores of 9.5, 9.7, 9.9, 9.7, 9.7, 9.6, and 9.5. Use an inequality to find the minimum score that the last judge can give so that Nana wins the silver medal.

**69.** Carol would like to pay cash for a car when she graduates from college and estimates that she can afford a car that costs between \$4000 and \$8000. She has saved \$500 so far and plans to earn the rest of the money by working the next two summers. If Carol plans to save the same amount each summer, use a compound inequality to find the range of money she must save each summer to buy the car.

**(2.5)** *Solve each inequality.*

**70.** $1 \leq 4x - 7 \leq 3$

**71.** $-2 \leq 8 + 5x < -1$

**72.** $-3 < 4(2x - 1) < 12$

**73.** $-6 < x - (3 - 4x) < -3$

**74.** $\frac{1}{6} < \frac{4x - 3}{3} \leq \frac{4}{5}$

**75.** $0 \leq \frac{2(3x + 4)}{5} \leq 3$

**76.** $x \leq 2$ and $x > -5$

**77.** $x \leq 2$ or $x > -5$

**78.** $3x - 5 > 6$ or $-x < -5$

**79.** $-2x \leq 6$ and $-2x + 3 < -7$

**(2.6)** *Solve each absolute value equation.*

**80.** $|x - 7| = 9$

**81.** $|8 - x| = 3$

**82.** $|2x + 9| = 9$

**83.** $|-3x + 4| = 7$

**84.** $|3x - 2| + 6 = 10$

**85.** $5 + |6x + 1| = 5$

**86.** $-5 = |4x - 3|$

**87.** $|5 - 6x| + 8 = 3$

**88.** $|7x| - 26 = -5$

**89.** $-8 = |x - 3| - 10$

**90.** $\left|\frac{3x - 7}{4}\right| = 2$

**91.** $\left|\frac{9 - 2x}{5}\right| = -3$

**92.** $|6x + 1| = |15 + 4x|$

**93.** $|x - 3| = |7 + 2x|$

*(2.7) Solve each absolute value inequality. Graph the solution set and write in interval notation.*

**94.** $|5x - 1| < 9$

**95.** $|6 + 4x| \geq 10$

**96.** $|3x| - 8 > 1$

**97.** $9 + |5x| < 24$

**98.** $|6x - 5| \leq -1$

**99.** $|6x - 5| \geq -1$

**100.** $\left|3x + \frac{2}{5}\right| \geq 4$

**101.** $\left|\frac{4x - 3}{5}\right| < 1$

**102.** $\left|\frac{x}{3} + 6\right| - 8 > -5$

**103.** $\left|\frac{4(x - 1)}{7}\right| + 10 < 2$

## CHAPTER 2 TEST

*Solve each equation.*

**1.** $8x + 14 = 5x + 44$

**2.** $3(x + 2) = 11 - 2(2 - x)$

**3.** $3(y - 4) + y = 2(6 + 2y)$

**4.** $7n - 6 + n = 2(4n - 3)$

**5.** $\frac{z}{2} + \frac{z}{3} = 10$

**6.** $\frac{7w}{4} + 5 = \frac{3w}{10} + 1$

**7.** $|6x - 5| = 1$

**8.** $|8 - 2t| = -6$

*Solve each equation for the specified variable.*

**9.** $3x - 4y = 8; y$

**10.** $4(2n - 3m) - 3(5n - 7m) = 0; n$

**11.** $S = gt^2 + gvt; g$

**12.** $F = \frac{9}{5}C + 32; C$

*Solve each inequality.*

**13.** $3(2x - 7) - 4x > -(x + 6)$

**14.** $8 - \frac{x}{2} \leq 7$

**15.** $-3 < 2(x - 3) \leq 4$

**16.** $|3x + 1| > 5$

**17.** $x \geq 5$ and $x \geq 4$

**18.** $x \geq 5$ or $x \geq 4$

**19.** $-x > 1$ and $3x + 3 \geq x - 3$

**20.** $6x + 1 > 5x + 4$ or $1 - x > -4$

**21.** Find 12% of 80.

*Solve.*

**22.** In 2006, the number of people employed as database administrators, computer support specialists, and all other computer scientists is expected to be 461,000 in the United States. This represents a 118% increase over the number of people employed in these occupations in 1996. Find the number of database administrators, computer support specialists, and all other computer scientists employed in 1996. (*Source:* U.S. Bureau of Labor Statistics)

△ **23.** A circular dog pen has a circumference of 78.5 feet. Approximate $\pi$ by 3.14 and estimate how many hunting dogs could be safely kept in the pen if each dog needs at least 60 square feet of room.

**24.** The company that makes Photoray sunglasses figures that the cost $C$ to make $x$ number of sunglasses weekly is given by $C = 3910 + 2.8x$, and the weekly revenue $R$ is given by $R = 7.4x$. Use an inequality to find the number of sunglasses that must be made and sold to make a profit. (Revenue must exceed cost in order to make a profit.)

**25.** Find the amount of money in an account after 10 years if a principal of $2500 is invested at 3.5% interest compounded quarterly. (Round to the nearest cent.)

**26.** The average fresh market price for broccoli in 1998 was \$1.10 per pound. A renewed demand for broccoli had caused the price to jump 27% to this level from 1995 market prices. What was the average fresh market price for broccoli in 1995? Round to the nearest cent. *(Source: Associated Press*, April, 1999)

**27.** The formula for the power generated by a wind turbine generator is $P = 0.5rAV^3$, where $P$ is the power generated in watts, $r$ is the density of air (1.225 kilograms per cubic meter at sea level), $A$ is the area swept by the wind turbine rotor in square meters, and $V$ is the wind speed in meters per second. Find the power generated by a wind turbine generator, at sea level, whose rotor sweeps an area of 3.14 square meters, when the wind is blowing at a speed of 9 meters per second. Round to the nearest whole watt. (Note: A wind speed of 9 meters per second is roughly equivalent to 20 mph.)

## CHAPTER 2 CUMULATIVE REVIEW

**1.** List the elements in each set.

**a.** $\{x \mid x \text{ is a whole number between 1 and 6}\}$

**b.** $\{x \mid x \text{ is a natural number greater than 100}\}$

**2.** Find each absolute value.

**a.** $|3|$ **b.** $|-5|$

**c.** $-|2|$ **d.** $-|-8|$

**e.** $|0|$

**3.** Add.

**a.** $-3 + (-11)$ **b.** $3 + (-7)$

**c.** $-10 + 15$ **d.** $-8.3 + (-1.9)$

**e.** $-\frac{2}{3} + \frac{3}{7}$

**4.** Find the square roots.

**a.** $\sqrt{9}$ **b.** $\sqrt{25}$

**c.** $\sqrt{\frac{1}{4}}$

**5.** Evaluate each algebraic expression when $x = 2$, $y = -1$, and $z = -3$.

**a.** $z - y$ **b.** $z^2$

**c.** $\frac{2x + y}{z}$

**6.** Write each sentence using mathematical symbols.

**a.** The sum of $x$ and 5 is 20.

**b.** Two times the sum of 3 and $y$ amounts to 4.

**c.** Subtract 8 from $x$, and the difference is the same as the product of 2 and $x$.

**d.** The quotient of $z$ and 9 is 3 times the difference of $z$ and 5.

**7.** Use the commutative property of addition to write an expression equivalent to $7x + 5$.

*Solve for x.*

**8.** $2x + 5 = 9$

**9.** $6x - 4 = 2 + 6(x - 1)$

**10.** Write the following as algebraic expressions. Then simplify.

**a.** The sum of two consecutive integers, if $x$ is the first consecutive integer.

**b.** The perimeter of the triangle with sides of length $x$, $5x$, and $6x - 3$.

**11.** Find two numbers such that the second number is 3 more than twice the first number and the sum of the two numbers is 72.

**12.** Solve $3y - 2x = 7$ for $y$.

**13.** Solve $A = \frac{1}{2}(B + b)h$ for $b$.

**14.** Graph each set on a number line and then write in interval notation.

**a.** $\{x \mid x \geq 2\}$ **b.** $\{x \mid x < -1\}$

**c.** $\{x \mid 0.5 < x \leq 3\}$

*Solve.*

**15.** $-(x - 3) + 2 \leq 3(2x - 5) + x$

**16.** $2(x + 3) > 2x + 1$

**17.** Find the intersection: $\{2, 4, 6, 8\} \cap \{3, 4, 5, 6\}$

**18.** Solve: $x - 7 < 2$ and $2x + 1 < 9$

**19.** Find the union: $\{2, 4, 6, 8\} \cup \{3, 4, 5, 6\}$

**20.** Solve: $-2x - 5 < -3$ or $6x < 0$

*Solve:*

**21.** $|p| = 2$

**22.** $\left|\frac{x}{2} - 1\right| = 11$

**23.** $|x - 3| = |5 - x|$

**24.** $|x| \leq 3$

**25.** $|2x + 9| + 5 > 3$

## Helping Prepare for Possible Dangerous Situations

Weather affects many aspects of our daily lives. An afternoon thunderstorm can put the damper on picnic plans or yard work. A cold snap can drive up our heating bills or affect the price of orange juice. Heavy rains can cause dangerous driving conditions or catastrophic flooding. High winds can damage roofs or bring down power lines. Accurately predicting the weather can allow people to prepare for possible damage or avoid dangerous weather situations.

Meteorology is the study of the atmosphere, including the science of weather forecasting. Meteorologists must be able to collect and interpret data, read maps and graphs, plot coordinates, make mathematical computations, use mathematical and physical models, and understand basic statistics. They should also have a solid background in chemistry, physics, earth science, and geography, as well as good computer and communication skills. Meteorologists often work as part of a team.

*For more information about careers in meteorology and the atmospheric sciences, visit the National Weather Association Website by first going to www.prenhall.com/martin-gay.*

*In the Spotlight on Decision Making feature on page 125, you will have the opportunity to make a decision about keeping track of a hurricane as a meteorologist.*

# GRAPHS AND FUNCTIONS

# 3

**3.1** Graphing Equations

**3.2** Introduction to Functions

**3.3** Graphing Linear Functions

**3.4** The Slope of a Line

**3.5** Equations of Lines

**3.6** Graphing Linear Inequalities

The linear equations and inequalities we explored in Chapter 2 are statements about a single variable. This chapter examines statements about two variables: linear equations and inequalities in two variables. We focus particularly on graphs of those equations and inequalities which lead to the notion of relation and to the notion of function, perhaps the single most important and useful concept in all of mathematics.

## 3.1 GRAPHING EQUATIONS

### OBJECTIVES

1. Plot ordered pairs.
2. Determine whether an ordered pair of numbers is a solution to an equation in two variables.
3. Graph linear equations.
4. Graph nonlinear equations.

**1** When two varying quantities are measured simultaneously and repeatedly, we can record the pairs of measurements and try to detect a pattern. If we recognize a pattern, we may be able to express the pattern as a two-variable equation. For example, suppose that the two quantities, products sold $x$ and monthly salary $y$, are measured simultaneously and repeatedly for an employee and recorded in the following table.

| | | | | | | | |
|---|---|---|---|---|---|---|---|
| *PRODUCTS SOLD* | $x$ | 0 | 100 | 200 | 300 | 400 | 1000 |
| *MONTHLY SALARY* | $y$ | 1500 | 1510 | 1520 | 1530 | 1540 | 1600 |

After studying the measurements, you may notice that there is a pattern in the quantities. Notice that the monthly salary $y$ is always equal to $1500 + \frac{1}{10}x$. We can use this information to write the equation $y = 1500 + \frac{1}{10}x$. This equation is called a **linear equation in two variables**. Before we discuss such equations further, let's first review the rectangular coordinate system.

In order to visualize the data in the table, the data pairs can be listed as ordered pairs of numbers. These ordered pairs of numbers can then be plotted on a **rectangular coordinate system**, which is also called a **Cartesian coordinate system** after its inventor, Rene Descartes (1596–1650).

The Cartesian coordinate system consists of two number lines that intersect at right angles at their 0 coordinates. We position these axes on paper such that one number line is horizontal and the other number line is then vertical. The horizontal number line is called the ***x*-axis** (or the axis of the **abscissa**), and the vertical number line is called the ***y*-axis** (or the axis of the **ordinate**). The point of intersection of these axes is named the **origin.**

Notice in the left figure below that the axes divide the plane into four regions. These regions are called **quadrants.** The top-right region is quadrant I. Quadrants II, III, and IV are numbered counterclockwise from the first quadrant as shown. The $x$-axis and the $y$-axis are not in any quadrant.

Each point in the plane can be located, or **plotted**, or graphed by describing its position in terms of distances along each axis from the origin. An **ordered pair**, represented by the notation $(x, y)$, records these distances.

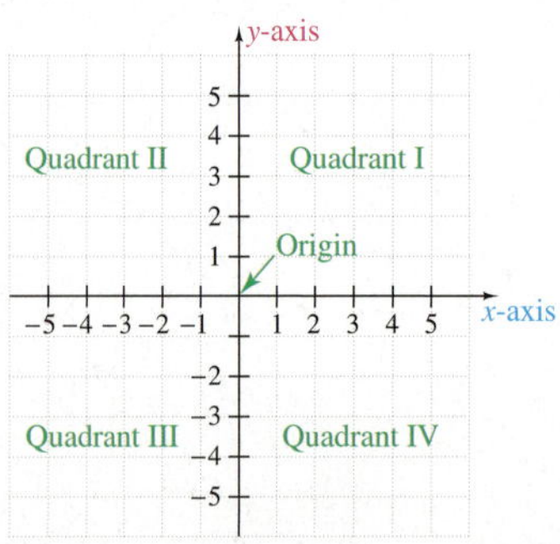


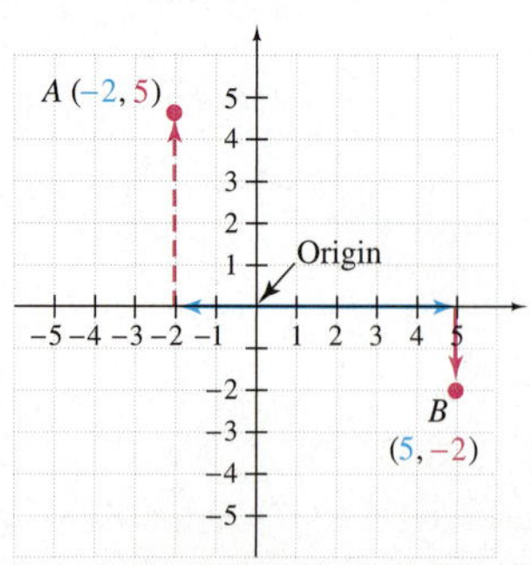

For example, the location of point $A$ in the figure on the right on page 118 is described as 2 units to the left of the origin along the $x$-axis and 5 units upward parallel to the $y$-axis. Thus, we identify point $A$ with the ordered pair $(-2, 5)$. Notice that the order of these numbers is critical. The $x$-value $-2$ is called the ***x*-coordinate** and is associated with the $x$-axis. The $y$-value 5 is called the ***y*-coordinate** and is associated with the $y$-axis. Compare the location of point $A$ with the location of point $B$, which corresponds to the ordered pair $(5, -2)$.

Keep in mind that **each ordered pair corresponds to exactly one point in the real plane and that each point in the plane corresponds to exactly one ordered pair.** Thus, we may refer to the ordered pair $(x, y)$ as the point $(x, y)$.

**Example 1** Plot each ordered pair on a Cartesian coordinate system and name the quadrant in which the point is located.

**a.** $(2, -1)$ **b.** $(0, 5)$ **c.** $(-3, 5)$ **d.** $(-2, 0)$ **e.** $\left(-\frac{1}{2}, -4\right)$ **f.** $(1.5, 1.5)$

*Solution* The six points are graphed as shown.

**a.** $(2, -1)$ lies in quadrant IV.

**b.** $(0, 5)$ is not in any quadrant.

**c.** $(-3, 5)$ lies in quadrant II.

**d.** $(-2, 0)$ is not in any quadrant.

**e.** $\left(-\frac{1}{2}, -4\right)$ is in quadrant III.

**f.** $(1.5, 1.5)$ is in quadrant I.

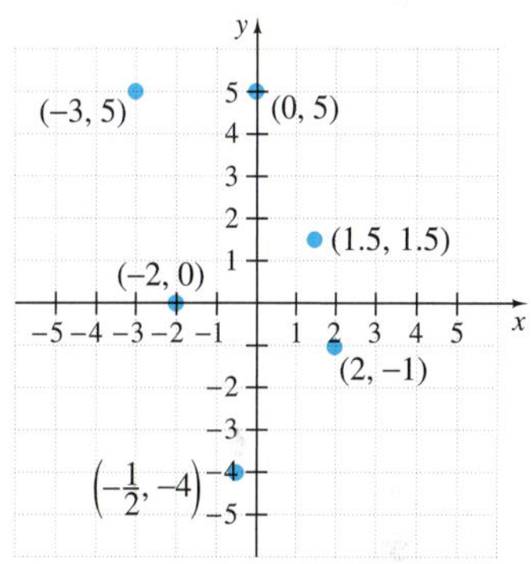


Notice that the $y$-coordinate of any point on the $x$-axis is 0. For example, the point with coordinates $(-2, 0)$ lies on the $x$-axis. Also, the $x$-coordinate of any point on the $y$-axis is 0. For example, the point with coordinates $(0, 5)$ lies on the $y$-axis. These points that lie on the axes do not lie in any quadrants.

2 **Solutions** of equations in two variables consist of two numbers that form a true statement when substituted into the equation. A convenient notation for writing these numbers is as ordered pairs. For example, we say that the ordered pair $(100, 1510)$ is a solution of the equation $y = 1500 + \frac{1}{10}x$ because when $x$ is replaced with 100 and $y$ is replaced with 1510, a true statement results.

$$y = 1500 + \frac{1}{10}x$$

$$1510 \stackrel{?}{=} 1500 + \frac{1}{10}(100) \qquad \text{Let } x = 100 \text{ and } y = 1510.$$

$$1510 \stackrel{?}{=} 1500 + 10$$

$$1510 = 1510 \qquad \text{True.}$$

**Example 2** Determine whether (0, −12), (1, 9), and (2, −6) are solutions of the equation $3x - y = 12$.

*Solution* To check each ordered pair, replace $x$ with the $x$-coordinate and $y$ with the $y$-coordinate and see whether a true statement results.

| Let $x = 0$ and $y = -12$. | Let $x = 1$ and $y = 9$. | Let $x = 2$ and $y = -6$. |
|---|---|---|
| $3x - y = 12$ | $3x - y = 12$ | $3x - y = 12$ |
| $3(0) - (-12) \stackrel{?}{=} 12$ | $3(1) - 9 \stackrel{?}{=} 12$ | $3(2) - (-6) \stackrel{?}{=} 12$ |
| $0 + 12 \stackrel{?}{=} 12$ | $3 - 9 \stackrel{?}{=} 12$ | $6 + 6 \stackrel{?}{=} 12$ |
| $12 = 12$ True. | $-6 = 12$ False. | $12 = 12$ True. |

Thus, (1, 9) is not a solution but both (0, −12) and (2, −6) are solutions.

3 In fact, the equation $3x - y = 12$ has an infinite number of ordered pair solutions. Since it is impossible to list all solutions, we visualize them by graphing them.

A few more ordered pairs that satisfy $3x - y = 12$ are (4, 0), (3, −3), (5, 3), and (1, −9). These ordered pair solutions along with the ordered pair solutions from Example 2 are plotted on the following graph. The graph of $3x - y = 12$ is the single line containing these points. Every ordered pair solution of the equation corresponds to a point on this line, and every point on this line corresponds to an ordered pair solution.

| $x$ | $y$ | $3x - y = 12$ |
|---|---|---|
| 5 | 3 | $3 \cdot 5 - 3 = 12$ |
| 4 | 0 | $3 \cdot 4 - 0 = 12$ |
| 3 | −3 | $3 \cdot 3 - (-3) = 12$ |
| 2 | −6 | $3 \cdot 2 - (-6) = 12$ |
| 1 | −9 | $3 \cdot 1 - (-9) = 12$ |
| 0 | −12 | $3 \cdot 0 - (-12) = 12$ |

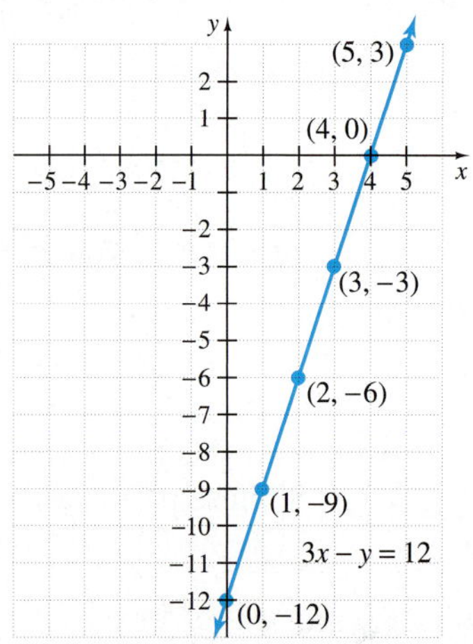


The equation $3x - y = 12$ is called a linear equation in two variables, and **the graph of every linear equation in two variables is a line.**

## LINEAR EQUATION IN TWO VARIABLES

A linear equation in two variables is an equation that can be written in the form

$$Ax + By = C$$

where $A$ and $B$ are not both 0. This form is called **standard form.**

As we mentioned earlier, the equation $y = 1500 + \frac{1}{10}x$ is also a linear equation in two variables. This means that the equation $y = 1500 + \frac{1}{10}x$ can be written in standard form $Ax + By = C$ and that its graph is a line. Its solutions in ordered pair form are shown in the next example along with a portion of its graph. Recall that $x$ is products sold and $y$ is monthly salary. Since we assume that the smallest amount of product sold is none, or 0, then $x$ must be greater than or equal to 0. Therefore, only the part of the graph that lies in Quadrant I is shown. Notice that the graph gives a visual picture of the correspondence between products sold and salary.

**HELPFUL HINT**
A line contains an infinite number of points and each ordered pair is a solution of its corresponding equation.

**Example 3** Use the graph of $y = 1500 + \frac{1}{10}x$ to answer the following questions.

**a.** If the salesperson has \$800 of products sold for a particular month, what is the salary for that month?

**b.** If the salesperson wants to make more than \$1600 per month, what must be the total amount of products sold?

*Solution* **a.** Since $x$ is products sold, find 800 along the $x$-axis and move vertically up until you reach a point on the line. From this point on the line, move horizontally to the left until you reach the $y$-axis. Its value on the $y$-axis is 1580, which means if \$800 worth of products is sold, the salary for the month is \$1580.

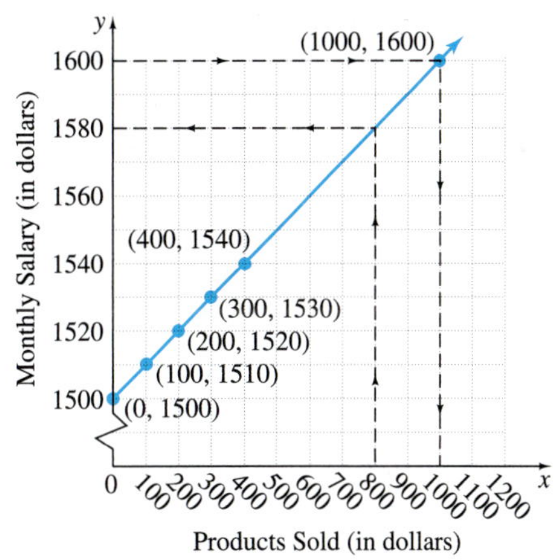


**b.** Since $y$ is monthly salary, find 1600 along the $y$-axis and move horizontally to the right until you reach a point on the line. Either read the corresponding $x$-value from the labeled ordered pair, or move vertically downward until you reach the $x$-axis. The corresponding $x$-value is 1000. This means that \$1000 worth of products sold gives a salary of \$1600 for the month. For the salary to be greater than \$1600, products sold must be greater that \$1000.

Recall from geometry that a line is determined by two points. This means that to graph a linear equation in two variables, just two solutions are needed. We will find a third solution, just to check our work. To find ordered pair solutions of linear equations in two variables, we can choose an $x$-value and find its corresponding $y$-value, or we can choose a $y$-value and find its corresponding $x$-value. The number 0 is often a convenient value to choose for $x$ and also for $y$.

**Example 4** Graph the equation $y = -2x + 3$.

*Solution* Find three ordered pair solutions, and plot the ordered pairs. The line through the plotted points is the graph. Since the equation is solved for $y$, let's choose three $x$-values. Let's let $x$ be 0, 2, and then $-1$ to find our three ordered pair solutions.

| Let $x = 0$ | Let $x = 2$ | Let $x = -1$ |
|---|---|---|
| $y = -2x + 3$ | $y = -2x + 3$ | $y = -2x + 3$ |
| $y = -2 \cdot 0 + 3$ | $y = -2 \cdot 2 + 3$ | $y = -2(-1) + 3$ |
| $y = 3$ Simplify. | $y = -1$ Simplify. | $y = 5$ Simplify. |

The three ordered pairs $(0, 3)$, $(2, -1)$ and $(-1, 5)$ are listed in the table and the graph is shown.

| $x$ | $y$ |
|---|---|
| 0 | 3 |
| 2 | $-1$ |
| $-1$ | 5 |

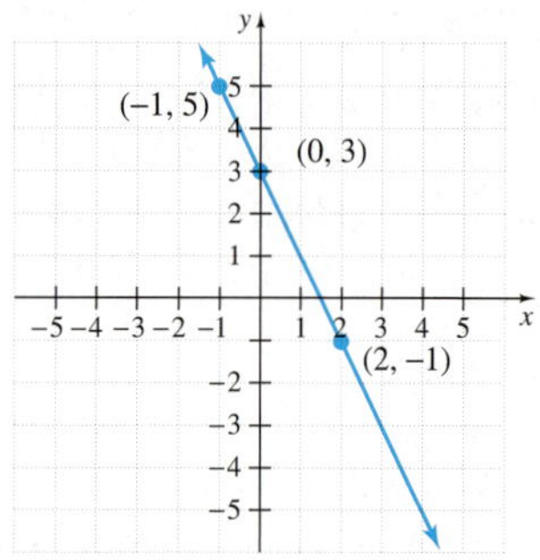


Notice that the graph crosses the $y$-axis at the point $(0, 3)$. This point is called the ***y*-intercept**. (You may sometimes see just the number 3 called the $y$-intercept.) This graph also crosses the $x$-axis at the point $(\frac{3}{2}, 0)$. This point is called the ***x*-intercept**. (You may also see just the number $\frac{3}{2}$ called the $x$-intercept.)

**Example 5** Graph the linear equation $y = \frac{1}{3}x$.

*Solution* To graph, we find ordered pair solutions, graph the solutions, and draw a line through the solutions. We will choose $x$-values and substitute in the equation. To avoid fractions, we choose $x$-values that are multiples of 3.

> **HELPFUL HINT**
> Notice that by using multiples of 3, we avoid fractions.

$$y = \frac{1}{3}x$$

↓

If $x = 6$, then $y = \frac{1}{3}(6)$, or 2.

↓

If $x = 0$, then $y = \frac{1}{3}(0)$, or 0.

↘

If $x = -3$, then $y = \frac{1}{3}(-3)$, or $-1$.

| $x$ | $y$ |
|---|---|
| 6 | 2 |
| 0 | 0 |
| $-3$ | $-1$ |

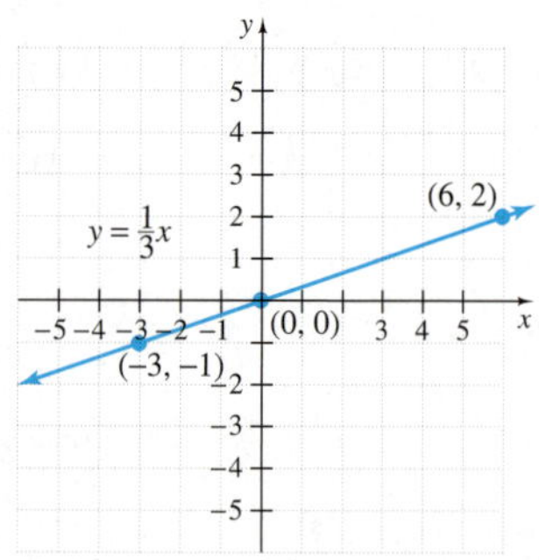


> **HELPFUL HINT**
> Since the equation $y = \frac{1}{3}x$ is solved for $y$, we choose $x$-values for finding points. This way, we simply need to evaluate an expression to find the $x$-value, as shown.

This graph crosses the $x$-axis at $(0, 0)$ and the $y$-axis at $(0, 0)$. This means that the $x$-intercept is $(0, 0)$ and that the $y$-intercept is $(0, 0)$.

**4** Not all equations in two variables are linear equations, and not all graphs of equations in two variables are lines.

## Example 6

Graph $y = x^2$.

*Solution* This equation is not linear, and its graph is not a line. We begin by finding ordered pair solutions. Because this graph is solved for $y$, we choose $x$-values and find corresponding $y$-values.

If $x = -3$, then $y = (-3)^2$, or 9.
If $x = -2$, then $y = (-2)^2$, or 4.
If $x = -1$, then $y = (-1)^2$, or 1.
If $x = 0$, then $y = 0^2$, or 0.
If $x = 1$, then $y = 1^2$, or 1.
If $x = 2$, then $y = 2^2$, or 4.
If $x = 3$, then $y = 3^2$, or 9.

| $x$ | $y$ |
|---|---|
| −3 | 9 |
| −2 | 4 |
| −1 | 1 |
| 0 | 0 |
| 1 | 1 |
| 2 | 4 |
| 3 | 9 |

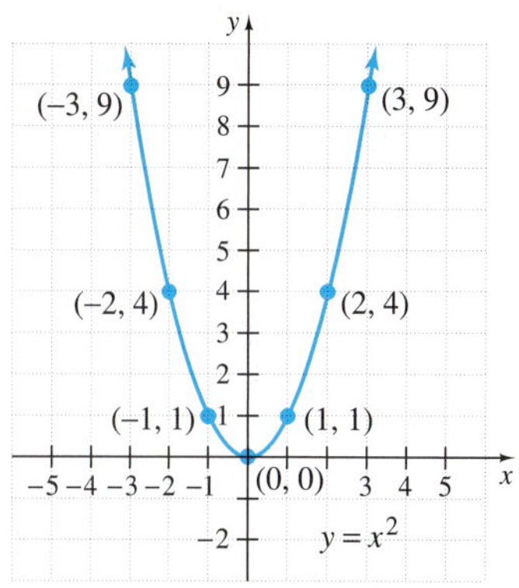


Study the table a moment and look for patterns. Notice that the ordered pair solution $(0, 0)$ contains the smallest $y$-value because any other $x$-value squared will give a positive result. This means that the point $(0, 0)$ will be the lowest point on the graph. Also notice that all other $y$-values correspond to two different $x$-values. For example, $3^2 = 9$ and also $(-3)^2 = 9$. This means that the graph will be a mirror image of itself across the $y$-axis. Connect the plotted points with a smooth curve to sketch its graph.

This curve is given a special name, a parabola. We will study more about parabolas in later chapters.

## Example 7

Graph the equation $y = |x|$.

*Solution* This is not a linear equation, and its graph is not a line. Because we do not know the shape of this graph, we find many ordered pair solutions. We will choose $x$-values and substitute to find corresponding $y$-values.

If $x = -3$, then $y = |-3|$, or 3.
If $x = -2$, then $y = |-2|$, or 2.
If $x = -1$, then $y = |-1|$, or 1.
If $x = 0$, then $y = |0|$, or 0.
If $x = 1$, then $y = |1|$, or 1.
If $x = 2$, then $y = |2|$, or 2.
If $x = 3$, then $y = |3|$, or 3.

| $x$ | $y$ |
|---|---|
| −3 | 3 |
| −2 | 2 |
| −1 | 1 |
| 0 | 0 |
| 1 | 1 |
| 2 | 2 |
| 3 | 3 |

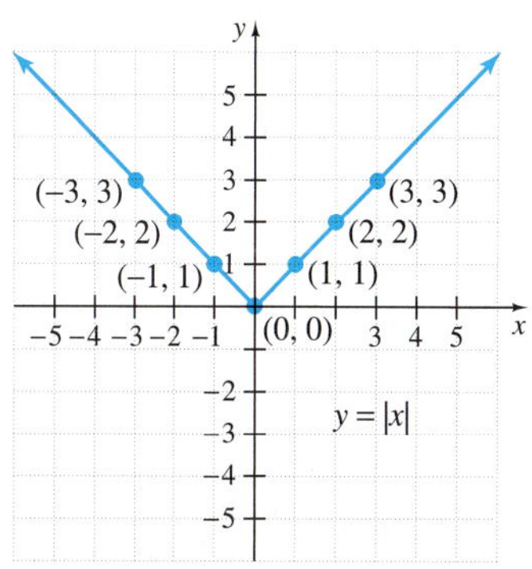


Again, study the table of values for a moment and notice any patterns.

From the plotted ordered pairs, we see that the graph of this absolute value equation is V-shaped.

## GRAPHING CALCULATOR EXPLORATIONS

In this section, we begin a study of graphing calculators and graphing software packages for computers. These graphers use the same point plotting technique that we introduced in this section. The advantage of this graphing technology is, of course, that graphing calculators and computers can find and plot ordered pair solutions much faster than we can. Note, however, that the features described in these boxes may not be available on all graphing calculators.

The rectangular screen where a portion of the rectangular coordinate system is displayed is called a **window**. We call it a **standard window** for graphing when both the $x$- and $y$-axes display coordinates between $-10$ and 10. This information is often displayed in the window menu on a graphing calculator as

Xmin = −10
Xmax = 10
Xscl = 1 The scale on the $x$-axis is one unit per tick mark.
Ymin = −10
Ymax = 10
Yscl = 1 The scale on the $y$-axis is one unit per tick mark.

To use a graphing calculator to graph the equation $y = -5x + 4$, press the [Y =] key and enter the keystrokes

[(−)] [5] [$x$] [+] [4].

↑

(Check your owner's manual to make sure the "negative" key is pressed here and not the "subtraction" key.)

The top row should now read $Y_1 = -5x + 4$. Next press the [GRAPH] key, and the display should look like this:

10

$y = -5x + 4$

−10 10

−10

*Use a standard window and graph the following equations. (Unless otherwise stated, we will use a standard window when graphing.)*

**1.** $y = -3.2x + 7.9$

**2.** $y = -x + 5.85$

**3.** $y = \frac{1}{4}x - \frac{2}{3}$

**4.** $y = \frac{2}{3}x - \frac{1}{5}$

**5.** $y = |x - 3| + 2$

**6.** $y = |x + 1| - 1$

**7.** $y = x^2 + 3$

**8.** $y = (x + 3)^2$

## SPOTLIGHT ON DECISION MAKING

Suppose you are a meteorologist. You are tracking Hurricane Felix. Hurricane position information is issued every 6 hours. The table lists Felix's most recent positions, given in latitude (vertical scale on the hurricane tracking chart) and longitude (horizontal scale on the hurricane tracking chart). Plot the position of the hurricane on the tracking chart and decide whether Felix is a threat to the United States. If so, what part?

NATIONAL HURRICANE CENTER
RECONNAISSANCE REPORT

*Felix Coordinates*

| *Date* | *Time* | *Latitude* | *Longitude* |
|---|---|---|---|
| 9/7 | 4 PM | 23.2 | 88.0 |
| 9/7 | 10 PM | 23.7 | 88.6 |
| 9/8 | 4 AM | 23.4 | 89.1 |
| 9/8 | 10 AM | 22.8 | 89.9 |
| 9/8 | 4 PM | 22.2 | 91.2 |
| 9/8 | 10 PM | 21.7 | 91.8 |
| 9/9 | 4 AM | 21.5 | 92.4 |
| 9/9 | 10 AM | 21.2 | 93.1 |

Hurricane Tracking Chart

## MENTAL MATH

*Determine the coordinates of each point on the graph.*

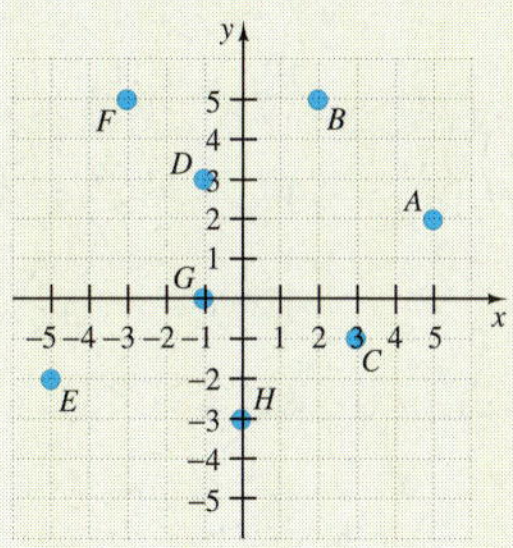


1. Point *A*
2. Point *B*
3. Point *C*
4. Point *D*
5. Point *E*
6. Point *F*
7. Point *G*
8. Point *H*

## Exercise Set 3.1

*Plot each point and name the quadrant or axis in which the point lies. See Example 1.*

1. $(3, 2)$
2. $(2, -1)$
3. $(-5, 3)$
4. $(-3, -1)$
5. $\left(5\frac{1}{2}, -4\right)$
6. $\left(-2, 6\frac{1}{3}\right)$
7. $(0, 3.5)$
8. $(-2, 4)$
9. $(-2, -4)$
10. $(-4.2, 0)$

*Given that x is a positive number and that y is a positive number, determine the quadrant or axis in which each point lies.*

11. $(x, -y)$
12. $(-x, y)$
13. $(x, 0)$
14. $(0, -y)$
15. $(-x, -y)$
16. $(0, 0)$

*Determine whether each ordered pair is a solution of the given equation. See Example 2.*

**17.** $y = 3x - 5$; $(0, 5)$, $(-1, -8)$

**18.** $y = -2x + 7$; $(1, 5)$, $(-2, 3)$

**19.** $-6x + 5y = -6$; $(1, 0)$, $\left(2, \frac{6}{5}\right)$

**20.** $5x - 3y = 9$; $(0, 3)$, $\left(\frac{12}{5}, -1\right)$

**21.** $y = 2x^2$; $(1, 2)$, $(3, 18)$

**22.** $y = 2|x|$; $(-1, 2)$, $(0, 2)$

**23.** $y = x^3$; $(2, 8)$, $(3, 9)$

**24.** $y = x^4$; $(-1, 1)$, $(2, 16)$

**25.** $y = \sqrt{x} + 2$; $(1, 3)$, $(4, 4)$

**26.** $y = \sqrt[3]{x} - 4$; $(1, -3)$, $(8, 6)$

*Determine whether each equation is linear or not. Then graph the equation. See Examples 3 through 7.*

**27.** $x + y = 3$

**28.** $y - x = 8$

**29.** $y = 4x$

**30.** $y = 6x$

**31.** $y = 4x - 2$

**32.** $y = 6x - 5$

**33.** $y = |x| + 3$

**34.** $y = |x| + 2$

**35.** $2x - y = 5$

**36.** $4x - y = 7$

**37.** $y = 2x^2$

**38.** $y = 3x^2$

**39.** $y = x^2 - 3$

**40.** $y = x^2 + 3$

**41.** $y = -2x$

**42.** $y = -3x$

**43.** $y = -2x + 3$

**44.** $y = -3x + 2$

**45.** $y = |x + 2|$

**46.** $y = |x - 1|$

**47.** $y = x^3$

Hint: Let $x = -3, -2, -1, 0, 1, 2$.

**48.** $y = x^3 - 2$

Hint: Let $x = -3, -2, -1, 0, 1, 2$.

**49.** $y = -|x|$

**50.** $y = -x^2$

**51.** $y = \frac{1}{3}x - 1$

**52.** $y = \frac{1}{2}x - 3$

**53.** $y = -\frac{3}{2}x + 1$

**54.** $y = -\frac{2}{3}x + 1$

**55.** Graph $y = x^2 - 4x + 7$. Let $x = 0, 1, 2, 3, 4$ to generate ordered pair solutions.

**56.** Graph $y = x^2 + 2x + 3$. Let $x = -3, -2, -1, 0, 1$ to generate ordered pair solutions.

△ **57.** The perimeter $y$ of a rectangle whose width is a constant 3 inches and whose length is $x$ inches is given by the equation

$$y = 2x + 6$$

**a.** Draw a graph of this equation.

**b.** Read from the graph the perimeter $y$ of a rectangle whose length $x$ is 4 inches.

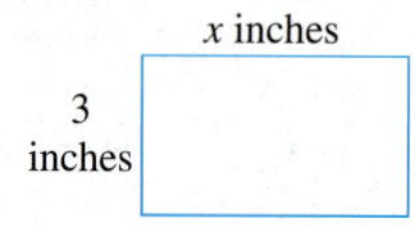


**58.** The distance $y$ traveled in a train moving at a constant speed of 50 miles per hour is given by the equation

$$y = 50x$$

where $x$ is the time in hours traveled.

**a.** Draw a graph of this equation.

**b.** Read from the graph the distance $y$ traveled after 6 hours.

*This graph shows hourly minimum wages and the years it increased. Use this graph for Exercises 59 through 62.*

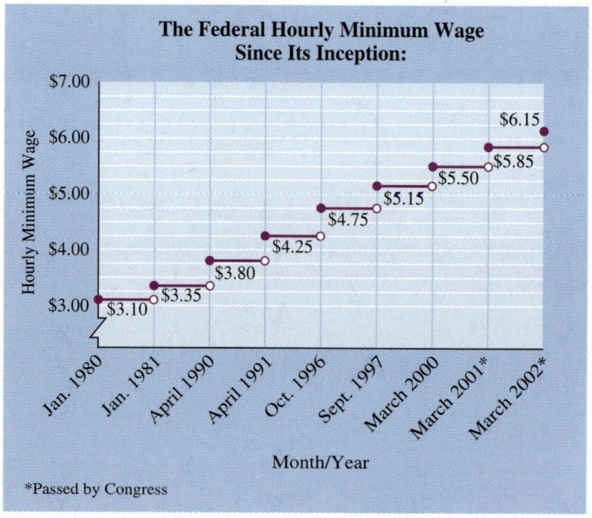


**59.** What was the first year that the minimum hourly wage rose above $4.00?

**60.** What was the first year that the minimum hourly wage rose above $5.00?

**61.** Why do you think that this graph is shaped the way it is?

**62.** The federal hourly minimum wage started in 1938 at $0.25. How much will it have increased by in 2002?

*For income tax purposes, Jason Verges, owner of Copy Services, uses a method called* **straight-line depreciation** *to show the loss in value of a copy machine he recently purchased. Jason assumes that he can use the machine for 7 years. The following graph shows the value of the machine over the years. Use this graph to answer the following questions.*

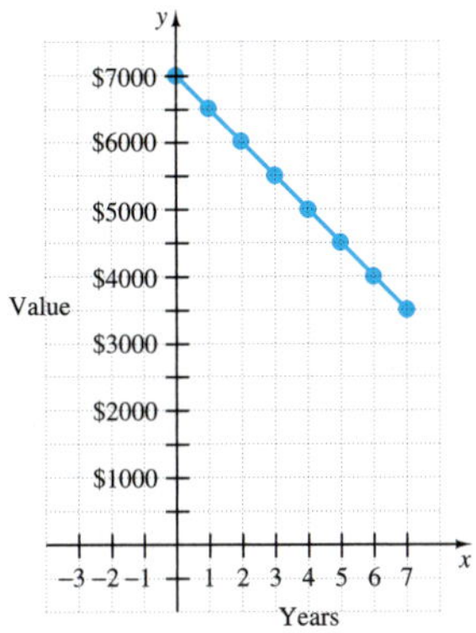


**63.** What was the purchase price of the copy machine?

**64.** What is the depreciated value of the machine in 7 years?

**65.** What loss in value occurred during the first year?

**66.** What loss in value occurred during the second year?

**67.** Why do you think that this method of depreciating is called straight-line depreciation?

**68.** Why is the line tilted downward?

*For exercises 69 through 72, match each description with the graph that best illustrates it.*

**69.** Moe worked 40 hours per week until the fall semester started. He quit and didn't work again until he worked 60 hours a week during the Christmas break.

**70.** Kawana worked 40 hours a week for her father during the summer. She slowly cut back her hours to not working at all during the fall semester. During the Christmas break, she started working again and increased her hours to 60 hours per week.

**71.** Wendy worked from July through February, never quitting. She worked between 10 and 30 hours per week.

**72.** Bartholomew worked from July through February, never quitting. He worked between 10 and 30 hours per week except during Christmas. At that time, he worked 40 hours per week.

**A.**

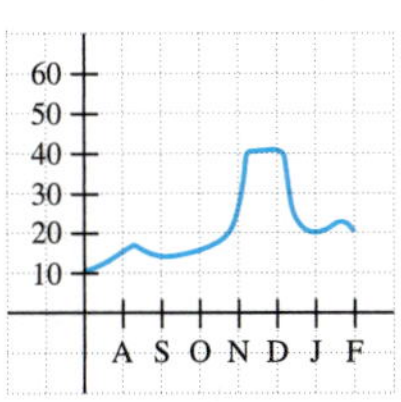


**B.**

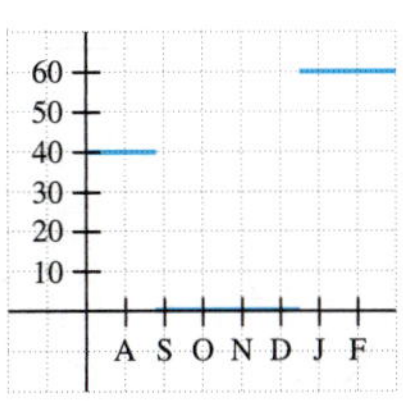


**C.**

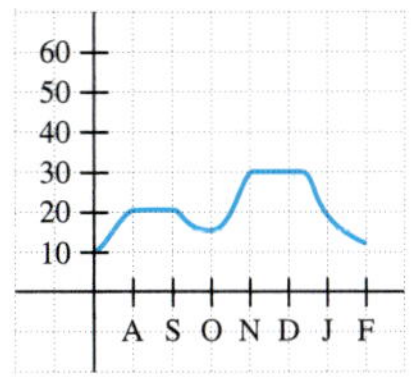


**D.**

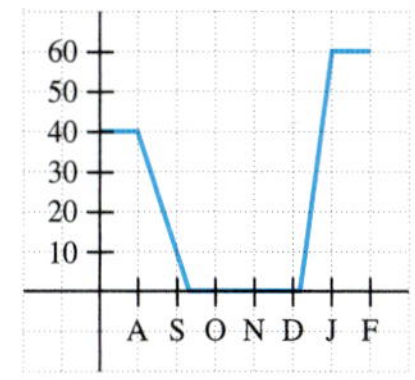


**73.** On the same set of axes, graph $y = 2x$, $y = 2x - 5$, and $y = 2x + 5$. What patterns do you see in these graphs?

**74.** On the same set of axes, graph $y = 2x$, $y = x$, and $y = -2x$. Describe the differences and similarities in these graphs.

**75.** Explain why we generally use three points to graph a line, when only two points are needed.

*Write each statement as an equation in two variables. Then graph each equation.*

**76.** The $y$-value is 5 more than three times the $x$-value.

**77.** The $y$-value is $-3$ decreased by twice the $x$-value.

**78.** The $y$-value is 2 more than the square of the $x$-value.

**79.** The $y$-value is 5 decreased by the square of the $x$-value.

*Use a graphing calculator to verify the graphs of the following exercises.*

**80.** Exercise 39

**81.** Exercise 40

**82.** Exercise 47

**83.** Exercise 48

## REVIEW EXERCISES

*Solve the following equations. See Section 2.1.*

**84.** $3(x - 2) + 5x = 6x - 16$

**85.** $5 + 7(x + 1) = 12 + 10x$

**86.** $3x + \frac{2}{5} = \frac{1}{10}$

**87.** $\frac{1}{6} + 2x = \frac{2}{3}$

*Solve the following inequalities. See Section 2.4.*

**88.** $3x \le -15$

**89.** $-3x > 18$

**90.** $2x - 5 > 4x + 3$

**91.** $9x + 8 \le 6x - 4$

## 3.2 INTRODUCTION TO FUNCTIONS

▶ **OBJECTIVES**

1. Define relation, domain, and range.
2. Identify functions.
3. Use the vertical line test for functions.
4. Find the domain and range of a function.
5. Use function notation.

**1** Equations in two variables, such as $y = 2x + 1$, describe **relations** between $x$-values and $y$-values. For example, if $x = 1$, then this equation describes how to find the $y$-value related to $x = 1$. In words, the equation $y = 2x + 1$ says that twice the $x$-value increased by 1 gives the corresponding $y$-value. The $x$-value of 1 corresponds to the $y$-value of $2(1) + 1 = 3$ for this equation, and we have the ordered pair $(1, 3)$.

There are other ways of describing relations or correspondences between two numbers or, in general, a first set (sometimes called the set of *inputs*) and a second set (sometimes called the set of *outputs*). For example.

| ***First Set: Input*** | ***Correspondence*** | ***Second Set: Output*** |
|---|---|---|
| People in a certain city | Each person's age | The set of nonnegative integers |

A few examples of ordered pairs from this relation might be (Ana, 4); (Bob, 36); (Trey, 21); and so on.

Below are just a few other ways of describing relations between two sets and the ordered pairs that they generate.

***First Set: Input*** ***Second Set: Output***

***Correspondence***

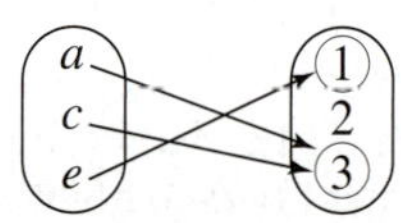


***Ordered Pairs***
$(a, 3), (c, 3), (e, 1)$

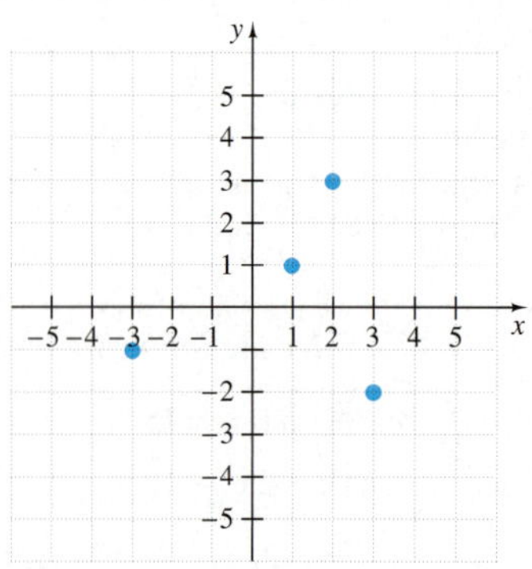


***Ordered Pairs***
$(-3, -1), (1, 1), (2, 3), (3, -2)$

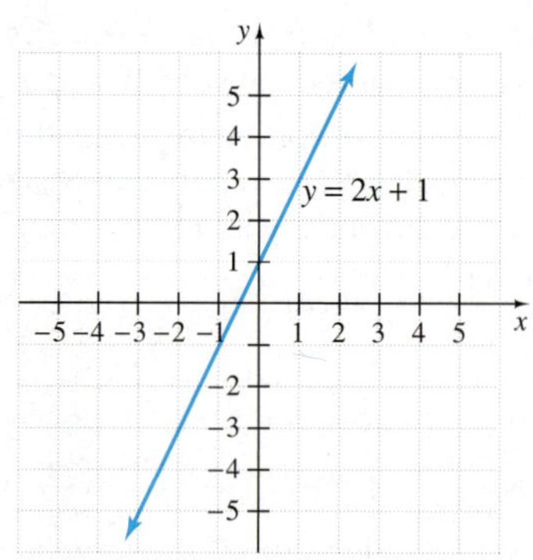


***Some Ordered Pairs***
$(1, 3), (0, 1)$ and so on

## RELATION, DOMAIN, AND RANGE

A **relation** is a set of ordered pairs.
The **domain** of the relation is the set of all first components of the ordered pairs.
The **range** of the relation is the set of all second components of the ordered pairs.

For example, the domain for our relation in the middle of the previous page is $\{a, c, e\}$ and the range is $\{1, 3\}$. Notice that the range does not include the element 2 of the second set. This is because no element of the first set is assigned to this element. If a relation is defined in terms of $x$- and $y$-values, we will agree that the domain corresponds to $x$-values and that the range corresponds to $y$-values.

**Example 1** Determine the domain and range of each relation.

**a.** $\{(2, 3), (2, 4), (0, -1), (3, -1)\}$

**b.**

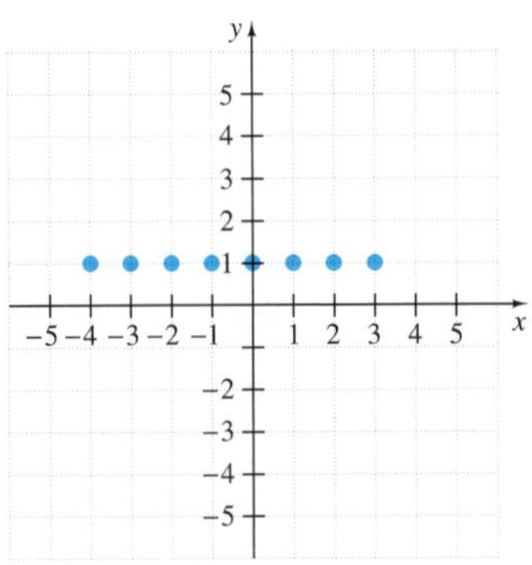


**c.**

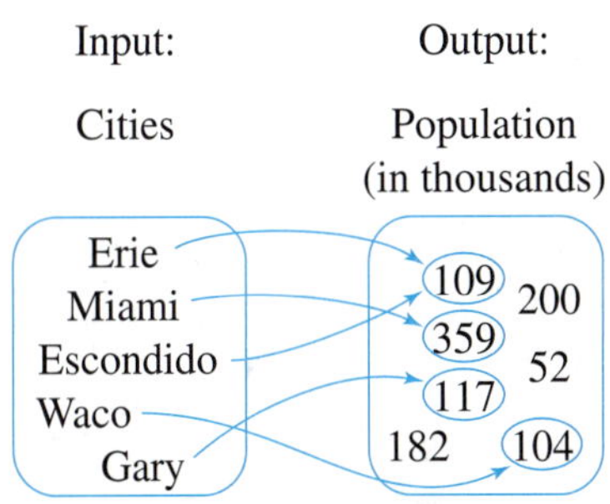


*Solution* **a.** The domain is the set of all first coordinates of the ordered pairs, $\{2, 0, 3\}$. The range is the set of all second coordinates, $\{3, 4, -1\}$.

**b.** Ordered pairs are not listed here, but are given in graph form. The relation is $\{(-4, 1), (-3, 1), (-2, 1), (-1, 1), (0, 1), (1, 1), (2, 1), (3, 1)\}$. The domain is $\{-4, -3, -2, -1, 0, 1, 2, 3\}$.
The range is $\{1\}$.

**c.** The domain is the first set, {Erie, Escondido, Gary, Miami, Waco}.
The range is the numbers in the second set that correspond to elements in the first set $\{104, 109, 117, 359\}$.

2 Now we consider a special kind of relation called a function.

## FUNCTION

A **function** is a relation in which each first component in the ordered pairs corresponds to *exactly* one second component.

**HELPFUL HINT**
A function is a special type of relation, so all functions are relations, but not all relations are functions.

**Example 2** Which of the following relations are also functions?

**a.** $\{(-2, 5), (2, 7), (-3, 5), (9, 9)\}$

**b.**

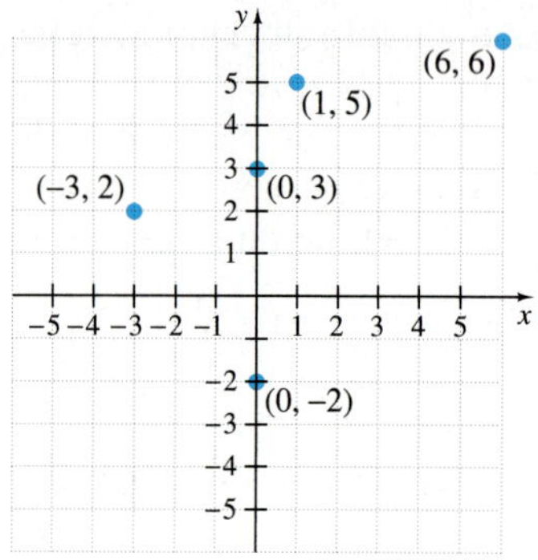


**c.**

| Input | Correspondence | Output |
|---|---|---|
| People in a certain city | Each person's age | The set of nonnegative integers |

*Solution* **a.** Although the ordered pairs $(-2, 5)$ and $(-3, 5)$ have the same $y$-value, each $x$-value is assigned to only one $y$-value, so this set of ordered pairs is a function.

**b.** The $x$-value 0 is assigned to two $y$-values, $-2$ and 3, in this graph so this relation does not define a function.

**c.** This relation is a function because although two different people may have the same age, each person has only one age. This means that each element in the first set is assigned to only one element in the second set.

We will call an equation such as $y = 2x + 1$ a **relation** since this equation defines a set of ordered pair solutions.

**Example 3** Is the relation $y = 2x + 1$ also a function?

*Solution* The relation $y = 2x + 1$ is a function if each $x$-value corresponds to just one $y$-value. For each $x$-value substituted in the equation $y = 2x + 1$, the multiplication and addition performed on each gives a single result, so only one $y$-value will be associated with each $x$-value. Thus, $y = 2x + 1$ is a function.

**Example 4** Is the relation $x = y^2$ also a function?

*Solution* In $x = y^2$, if $y = 3$, then $x = 9$. Also, if $y = -3$, then $x = 9$. In other words, the $x$-value 9 corresponds to two $y$-values, 3 and $-3$. Thus, $x = y^2$ is not a function.

**3** As we have seen so far, not all relations are functions. Consider the graphs of $y = 2x + 1$ and $x = y^2$ shown next. On the graph of $y = 2x + 1$, notice that each

$x$-value corresponds to only one $y$-value. Recall from Example 3 that $y = 2x + 1$ is a function.

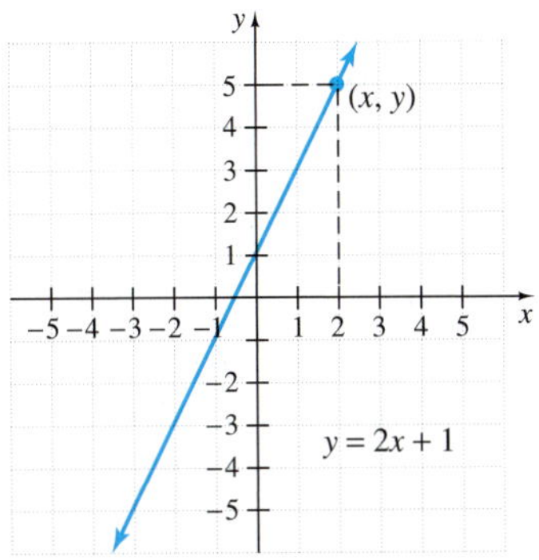


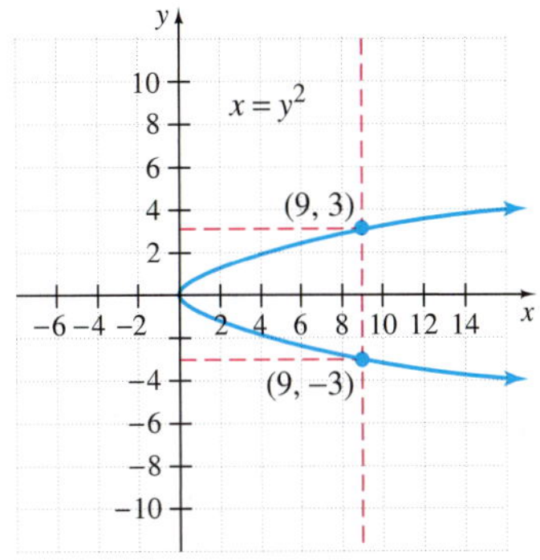


On the graph of $x = y^2$ the $x$-value 9, for example, corresponds to two $y$-values, 3 and $-3$, as shown by the vertical line. Recall from Example 4 that $x = y^2$ is not a function.

Graphs can be used to help determine whether a relation is also a function by the following vertical line test.

### VERTICAL LINE TEST

If no vertical line can be drawn so that it intersects a graph more than once, the graph is the graph of a function.

**Example 5** Which of the following graphs are graphs of functions?

**a.**

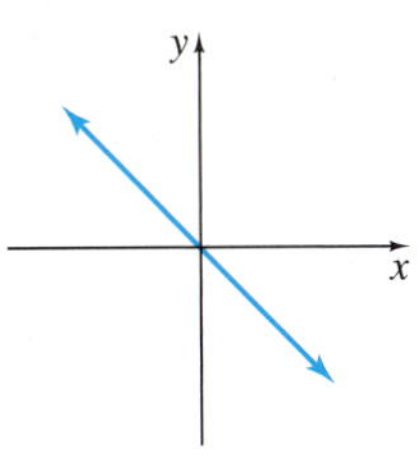

**b.**

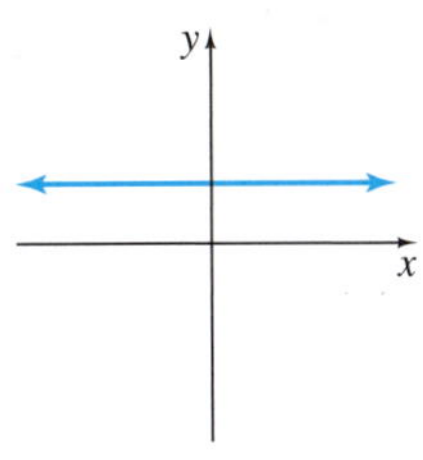

**c.**

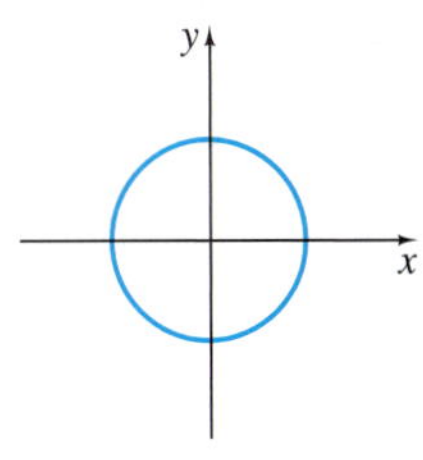

**d.**

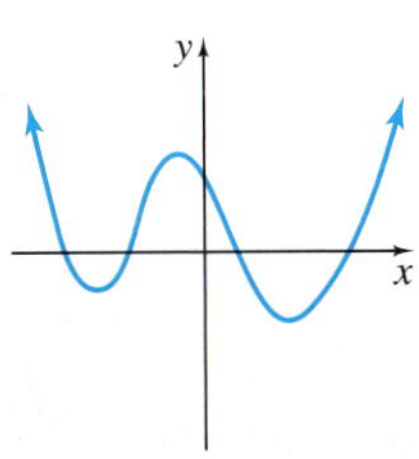

**e.**

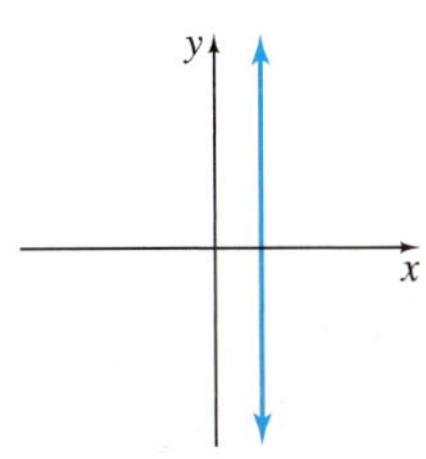

*Solution* **a.** This graph is the graph of a function since no vertical line will intersect this graph more than once.

**b.** This graph is also the graph of a function.

**c.** This graph is not the graph of a function. Note that vertical lines can be drawn that intersect the graph in two points.

**d.** This graph is the graph of a function.

**e.** This graph is not the graph of a function. A vertical line can be drawn that intersects this line at every point.

Recall that the graph of a linear equation in two variables is a line, and a line that is not vertical will pass the vertical line test. Thus, **all linear equations are functions except those whose graph is a vertical line.**

4 Next, we practice finding the domain and range of a relation from its graph.

**Example 6** Find the domain and range of each relation. Determine whether the relation is also a function.

**a.**

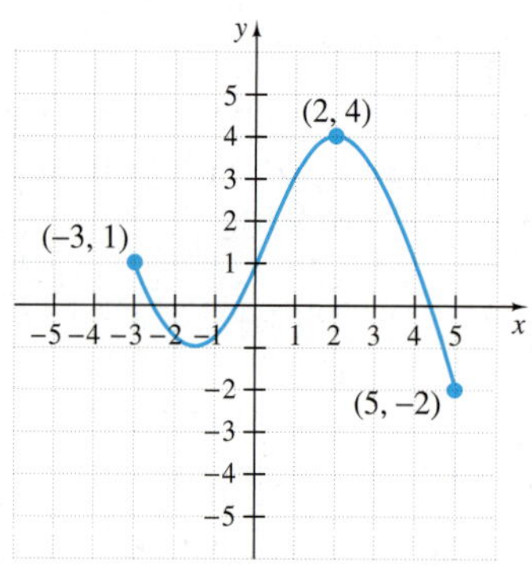


**b.**

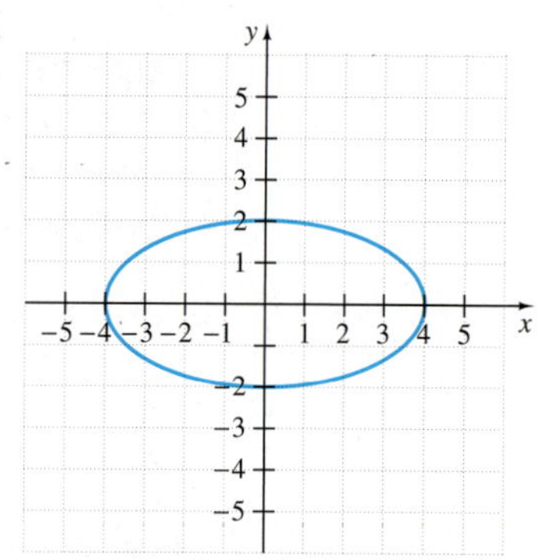


**c.**

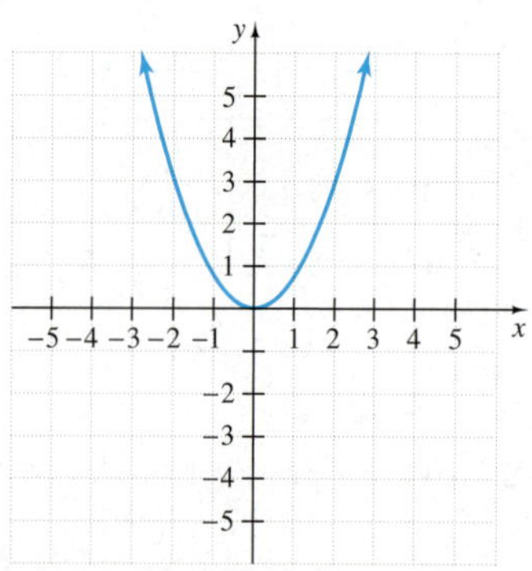


**d.**

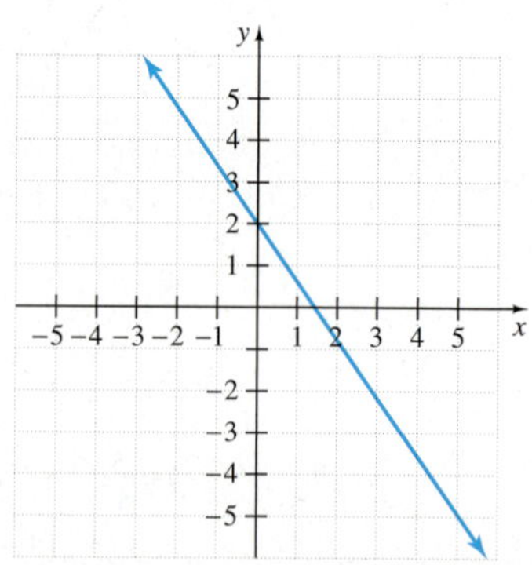


**HELPFUL HINT**

In Example 6, Part **a**, notice that the graph contains the endpoints $(-3, 1)$ and $(5, -2)$ whereas the graphs in Parts **c** and **d** contain arrows that indicate that they continue forever.

*Solution* By the vertical line test, graphs **a**, **c**, and **d** are graphs of functions. The domain is the set of values of $x$ and the range is the set of values of $y$. We read these values from each graph.

**a.**

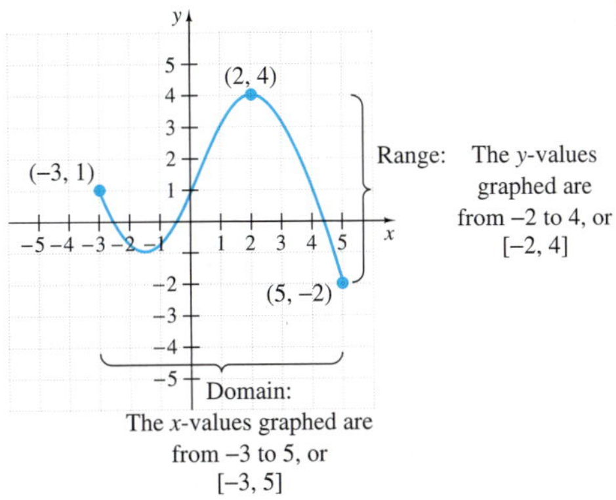


**b.**

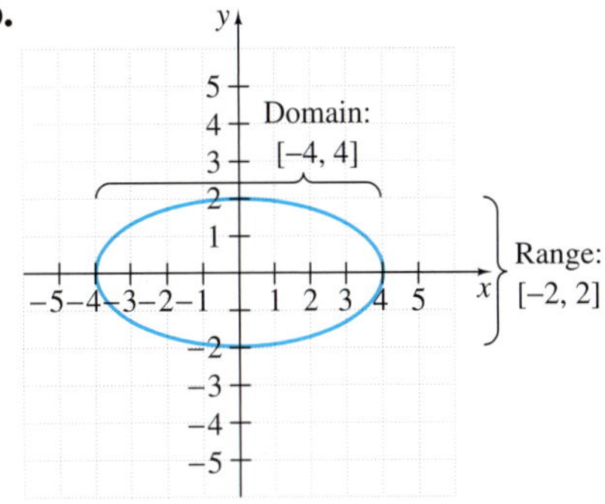


**c.**

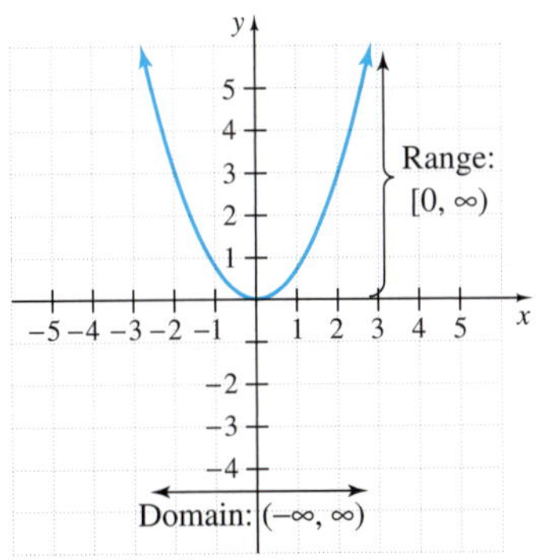


**d.**

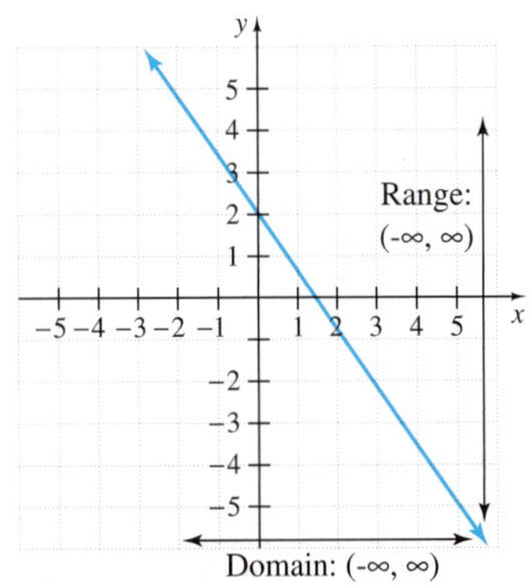


5 Many times letters such as $f$, $g$, and $h$ are used to name functions. To denote that $y$ is a function of $x$, we can write

$$y = f(x)$$

This means that **$y$ is a function of $x$** or that $y$ *depends on* $x$. For this reason, $y$ is called the **dependent variable** and $x$ the **independent variable**. The notation $f(x)$ is read "$f$ of $x$" and is called **function notation**.

For example, to use function notation with the function $y = 4x + 3$, we write $f(x) = 4x + 3$. The notation $f(1)$ means to replace $x$ with 1 and find the resulting $y$ or function value. Since

$$f(x) = 4x + 3$$

then

$$f(1) = 4(1) + 3 = 7$$

This means that when $x = 1$, $y$ or $f(x) = 7$. The corresponding ordered pair is $(1, 7)$. Here, the input is 1 and the output is $f(1)$ or 7. Now let's find $f(2)$, $f(0)$, and $f(-1)$.

$$\begin{aligned} f(x) &= 4x + 3 \\ f(2) &= 4(2) + 3 \\ &= 8 + 3 \\ &= 11 \end{aligned} \qquad \begin{aligned} f(x) &= 4x + 3 \\ f(0) &= 4(0) + 3 \\ &= 0 + 3 \\ &= 3 \end{aligned} \qquad \begin{aligned} f(x) &= 4x + 3 \\ f(-1) &= 4(-1) + 3 \\ &= -4 + 3 \\ &= -1 \end{aligned}$$

***Ordered Pairs:***

$(2, 11)$ $\qquad (0, 3)$ $\qquad (-1, -1)$

**HELPFUL HINT**

Note that $f(x)$ is a special symbol in mathematics used to denote a function. The symbol $f(x)$ is read "$f$ of $x$." It does *not* mean $f \cdot x$ ($f$ times $x$).

**Example 7** If $f(x) = 7x^2 - 3x + 1$ and $g(x) = 3x - 2$, find the following.

**a.** $f(1)$ **b.** $g(1)$ **c.** $f(-2)$ **d.** $g(0)$

*Solution* **a.** Substitute 1 for $x$ in $f(x) = 7x^2 - 3x + 1$ and simplify.

$$f(x) = 7x^2 - 3x + 1$$
$$f(1) = 7(1)^2 - 3(1) + 1 = 5$$

**b.**
$$g(x) = 3x - 2$$
$$g(1) = 3(1) - 2 = 1$$

**c.**
$$f(x) = 7x^2 - 3x + 1$$
$$f(-2) = 7(-2)^2 - 3(-2) + 1 = 35$$

**d.**
$$g(x) = 3x - 2$$
$$g(0) = 3(0) - 2 = -2$$

If it helps, think of a function, $f$, as a machine that has been programmed with a certain correspondence or rule. An input value (a member of the domain) is then fed into the machine, the machine does the correspondence or rule and the result is the output (a member of the range).

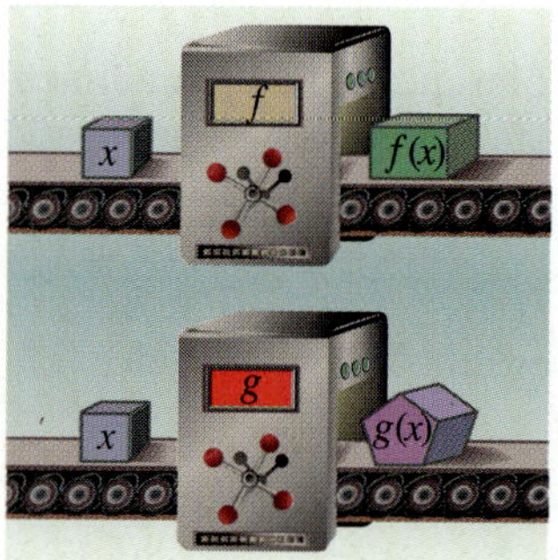


Many types of real-world paired data form functions. The broken-line graph on the next page shows the research and development spending by the Pharmaceutical Manufacturers Association.

**Example 8** The following graph shows the research and development expenditures by the Pharmaceutical Manufacturers Association as a function of time.

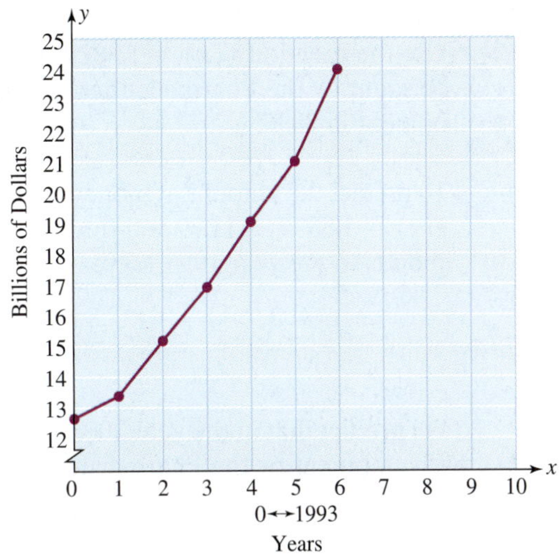


*Source*: Pharmaceutical Manufacturers Association
*The notation 0 ↔ 1993 means that year 0 corresponds to the year 1993, 1 corresponds to 1994, and so on.

**a.** Approximate the money spent on research and development in 1997.
**b.** In 1958, research and development expenditures were \$200 million. Find the increase in expenditures from 1958 to 1999.

*Solution* **a.** On the graph, since 0 corresponds to the year 1993, then 4 $(0 + 4)$ corresponds to the year 1997 $(1993 + 4)$. In 1997, approximately \$19 billion was spent.
**b.** In 1999, approximately \$24 billion, or \$24,000 million, was spent. The increase in spending from 1958 to 1999 is $\$24{,}000 - \$200 = \$23{,}800$ million.

Notice that the graph in Example 8 is the graph of a function since each year there is only one total amount of money spent by the Pharmaceutical Manufacturers Association on research and development. Also notice that the graph resembles the graph of a line. Often, businesses depend on equations that "closely fit" data-defined functions like this one in order to model the data and predict future trends. For example, by a method called **least squares**, the function $f(x) = 1.882x + 11.79$ approximates the data shown. Its graph and the actual data function are shown next.

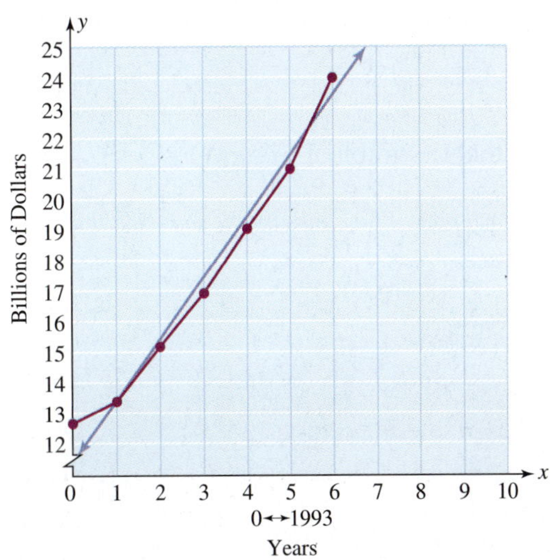

**Example 9** Use the function $f(x) = 1.882x + 11.79$ to predict the amount of money that will be spent by the Pharmaceutical Manufacturers Association on research and development in 2006.

*Solution* To predict the amount of money that will be spent in the year 2006 we use $f(x) = 1.882x + 11.79$ and find $f(13)$. (Notice that year 0 on the graph corresponds to the year 1993, so year 13 corresponds to the year 2006.)

$$\begin{aligned} f(x) &= 1.882x + 11.79 \\ f(13) &= 1.882(13) + 11.79 \\ &= 36.256 \end{aligned}$$

We predict that in the year 2006, \$36.256 billion dollars will be spent on research and development by the Pharmaceutical Manufacturers Association.

## GRAPHING CALCULATOR EXPLORATIONS

It is possible to use a graphing calculator to sketch the graph of more than one equation on the same set of axes. For example, graph the functions $f(x) = x^2$ and $g(x) = x^2 + 4$ on the same set of axes.

To graph on the same set of axes, press the [Y=] key and enter the equations on the first two lines.

$$Y_1 = x^2$$

$$Y_2 = x^2 + 4$$

Then press the [GRAPH] key as usual. The screen should look like this.

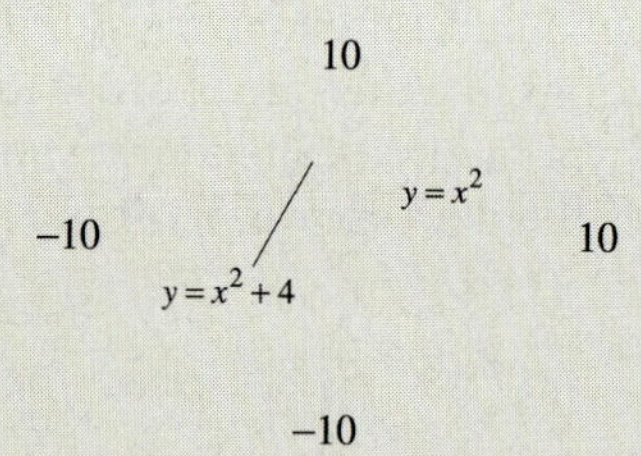


Notice that the graph of $y$ or $g(x) = x^2 + 4$ is the graph of $y = x^2$ moved 4 units upward.

*Graph each pair of functions on the same set of axes. Describe the similarities and differences in their graphs.*

**1.** $f(x) = |x|$
$g(x) = |x| + 1$

**2.** $f(x) = x^2$
$h(x) = x^2 - 5$

**3.** $f(x) = x$
$H(x) = x - 6$

**4.** $f(x) = |x|$
$G(x) = |x| + 3$

**5.** $f(x) = -x^2$
$F(x) = -x^2 + 7$

**6.** $f(x) = x$
$F(x) = x + 2$

## Exercise Set 3.2

*Find the domain and the range of each relation. Also determine whether the relation is a function. See Examples 1 and 2.*

1. $\{(-1, 7), (0, 6), (-2, 2), (5, 6)\}$
2. $\{(4, 9), (-4, 9), (2, 3), (10, -5)\}$
3. $\{(-2, 4), (6, 4), (-2, -3), (-7, -8)\}$
4. $\{(6, 6), (5, 6), (5, -2), (7, 6)\}$
5. $\{(1, 1), (1, 2), (1, 3), (1, 4)\}$
6. $\{(1, 1), (2, 1), (3, 1), (4, 1)\}$
7. $\left\{\left(\frac{3}{2}, \frac{1}{2}\right), \left(1\frac{1}{2}, -7\right), \left(0, \frac{4}{5}\right)\right\}$
8. $\{(\pi, 0), (0, \pi), (-2, 4), (4, -2)\}$
9. $\{(-3, -3), (0, 0), (3, 3)\}$
10. $\left\{\left(\frac{1}{2}, \frac{1}{4}\right), \left(0, \frac{7}{8}\right), (0.5, \pi)\right\}$

11.

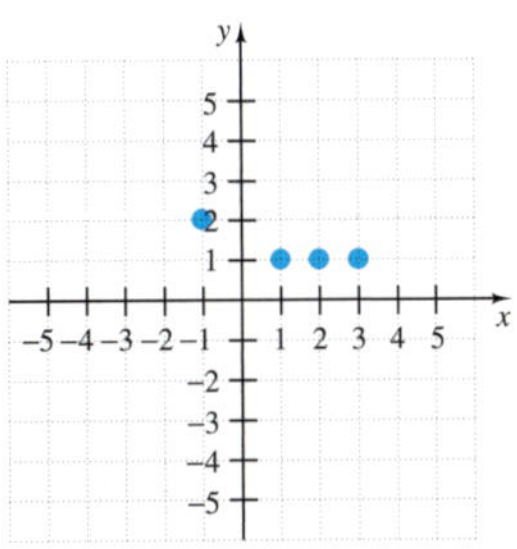

12.

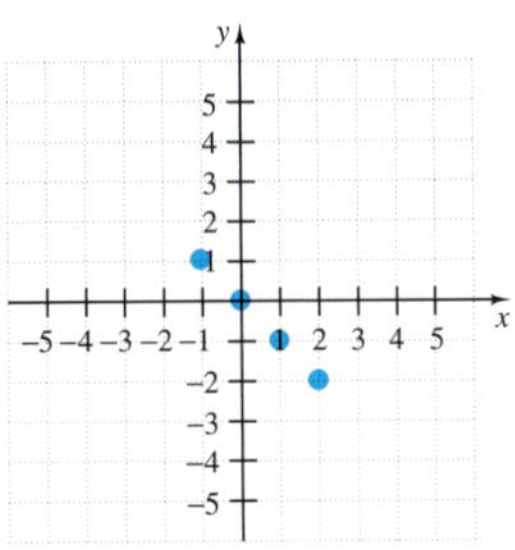

13. Input: State  Output: Number of Congressional Representatives

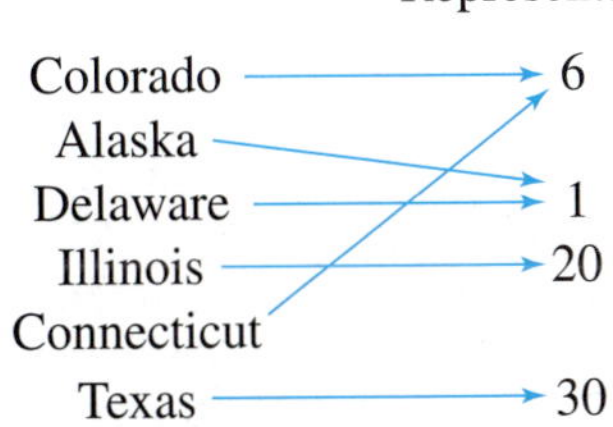


14. Input: Animal  Output: Average Life Span (in years)

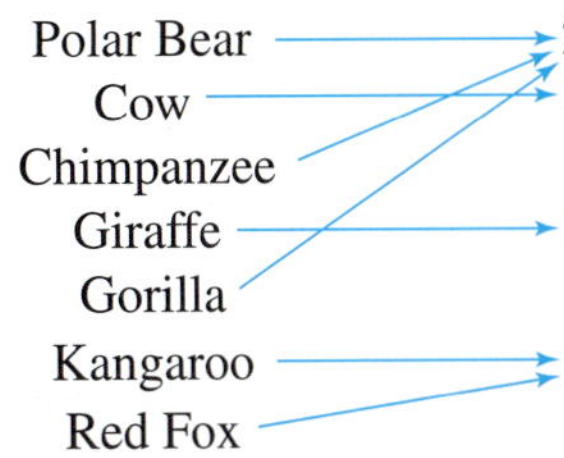


15. Input: Degrees Fahrenheit  Output: Degrees Celsius

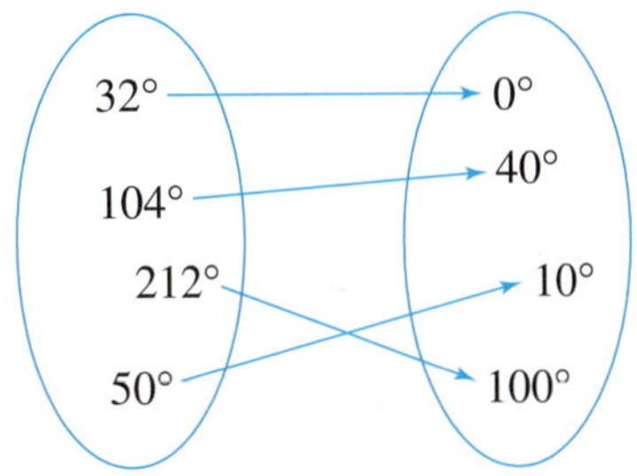


16. Input: Words  Output: Number of Letters

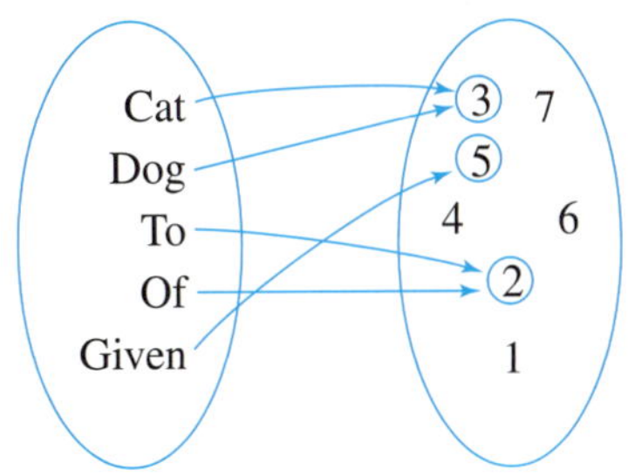


17. Input:  Output:

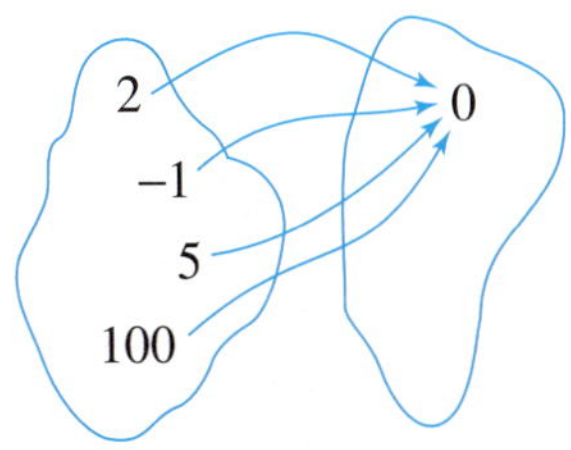

**18.** Input: Output:

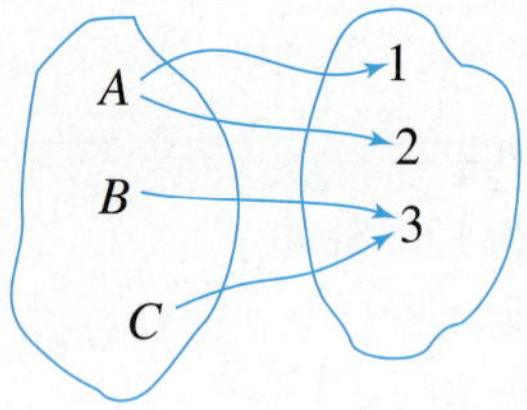


*In Exercises 19 and 20, determine whether the relation is a function.*

**19.**

| First set: Input | Correspondence | Second set: Output |
|---|---|---|
| Class of algebra students | Grade average | Set of nonnegative numbers |

**20.**

| First set: Input | Correspondence | Second set: Output |
|---|---|---|
| People in New Orleans (population 500,000) | Birthdate | Days of the year |

**21.** Describe a function whose domain is the set of people in your hometown.

**22.** Describe a function whose domain is the set of people in your algebra class.

*Use the vertical line test to determine whether each graph is the graph of a function. See Example 5.*

**23.**

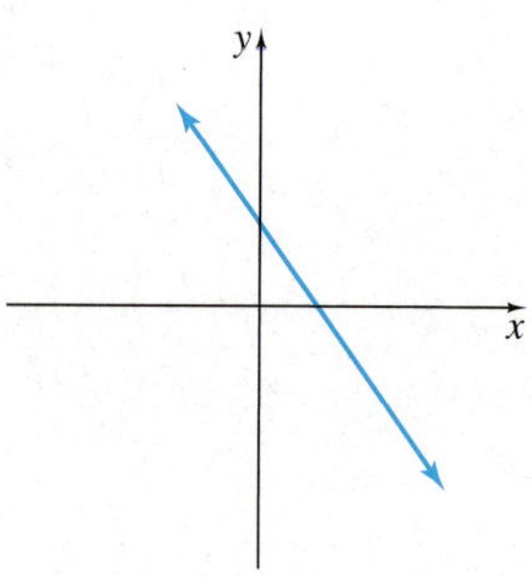


**24.**

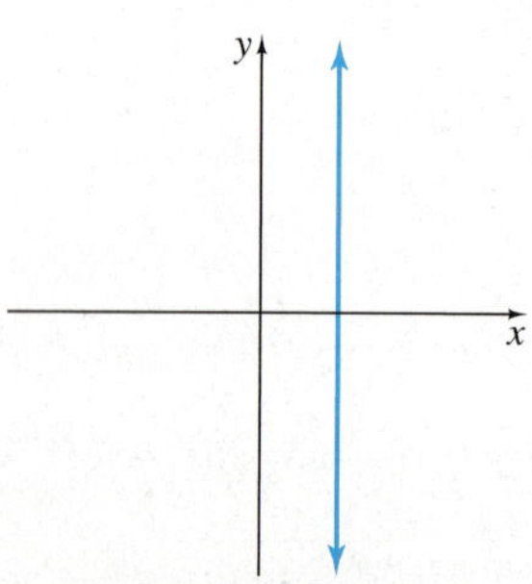


**25.**

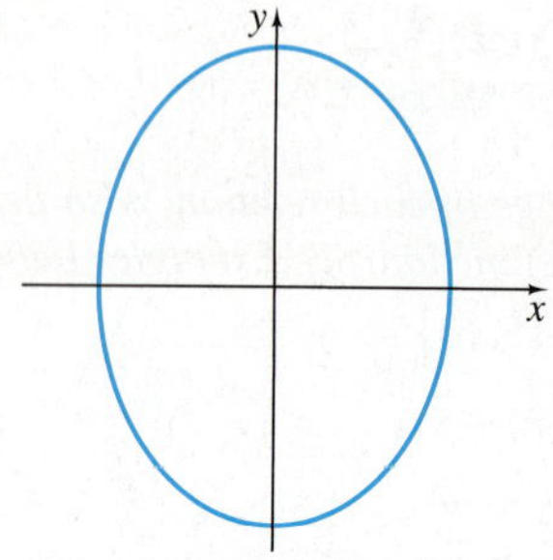


**26.**

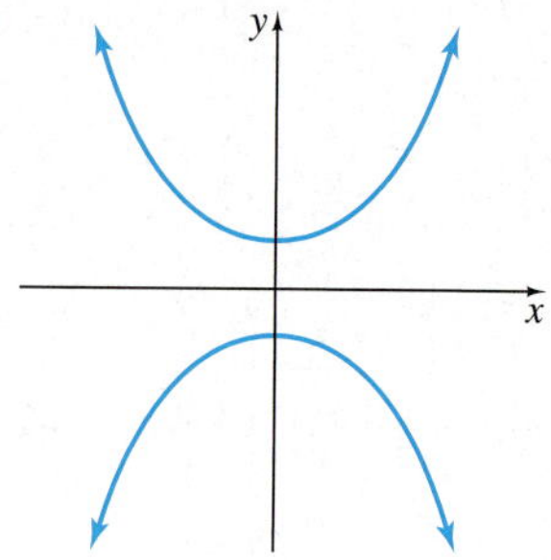


**27.**

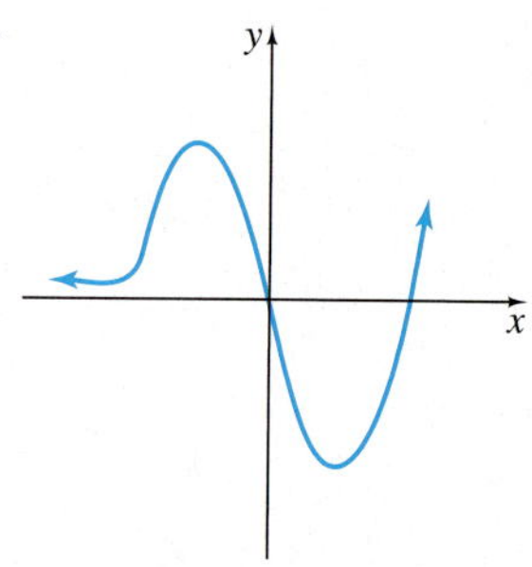


**28.**

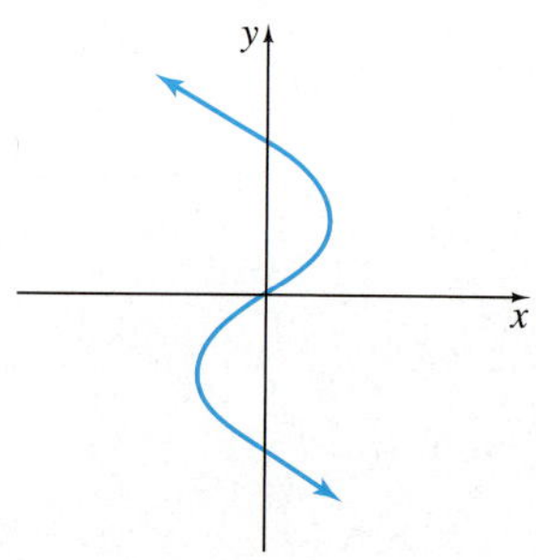


*Find the domain and the range of each relation. Use the vertical line test to determine whether each graph is the graph of a function. See Example 6.*

**29.**

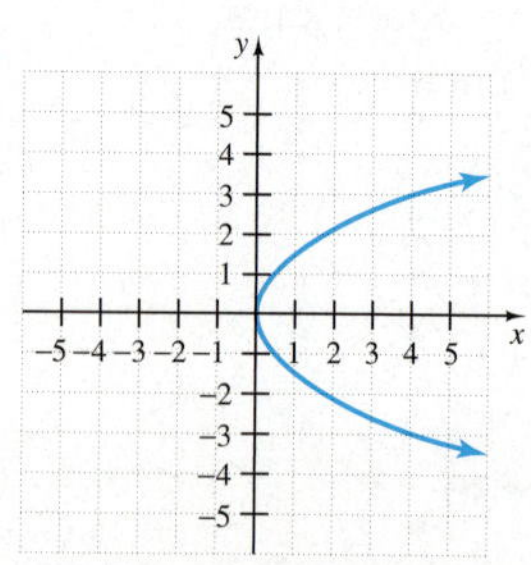

**30.**

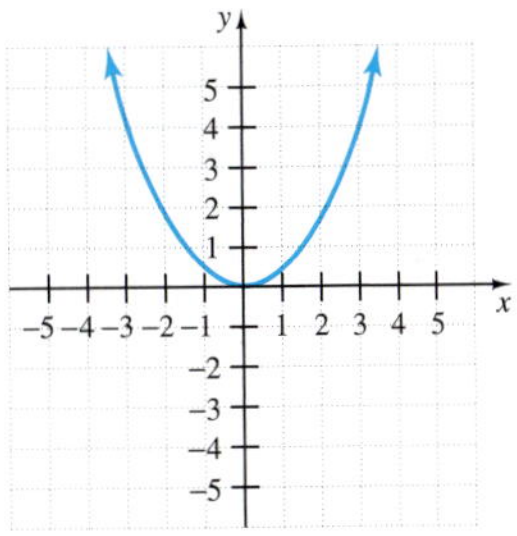

**31.**

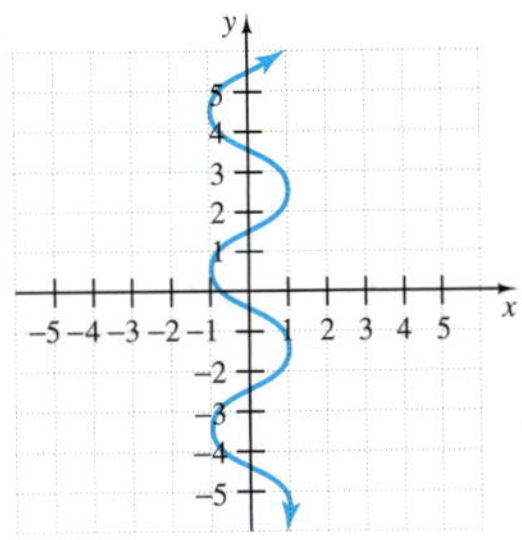

**32.**

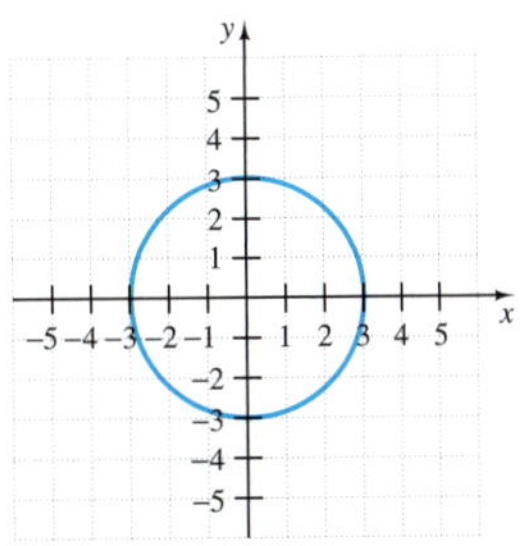

**33.**

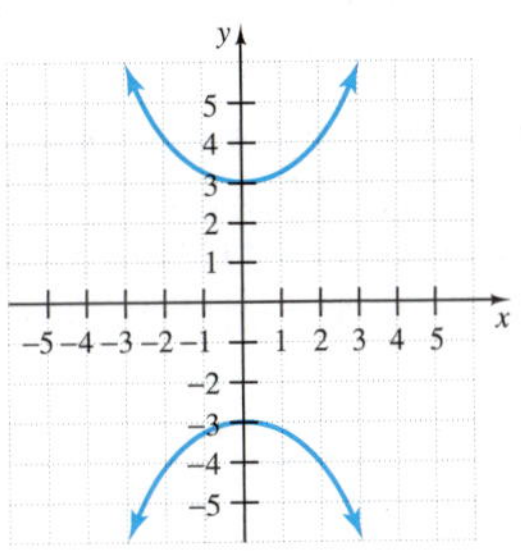

**34.**

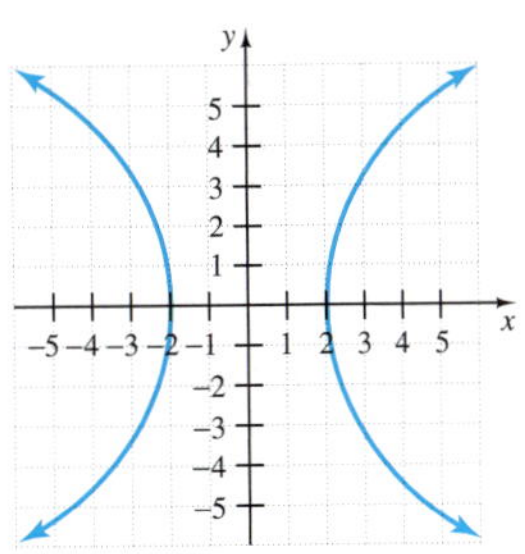

**35.**

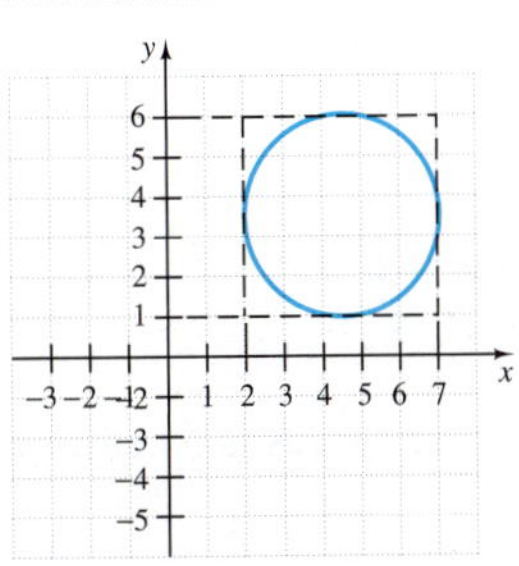

**36.**

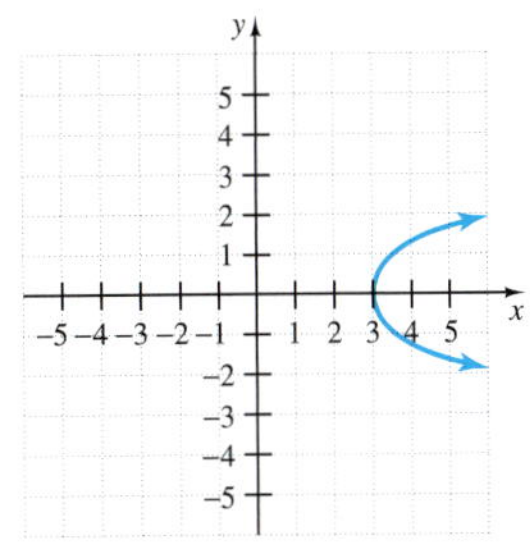

**37.**

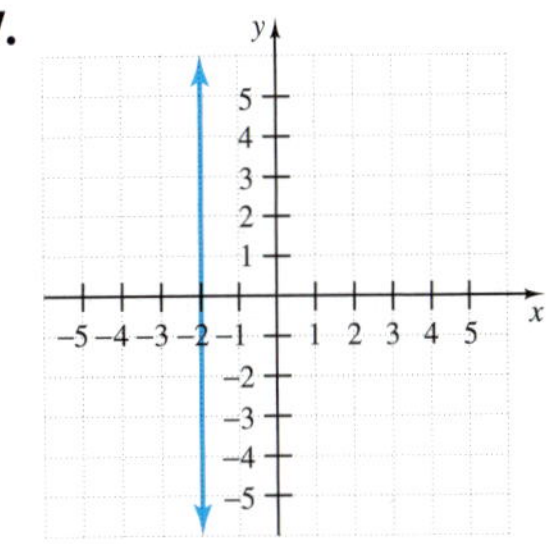

**38.**

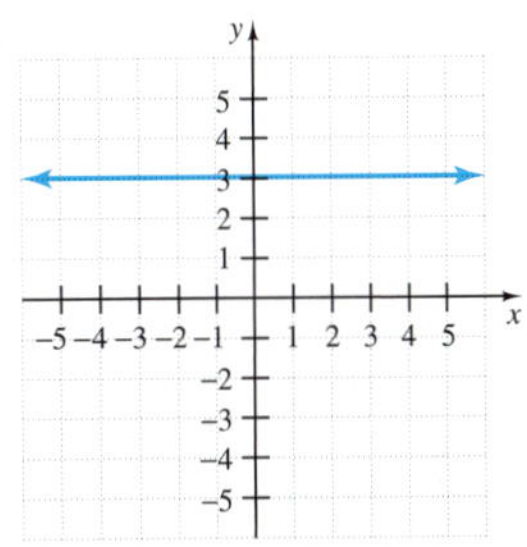

**39.**

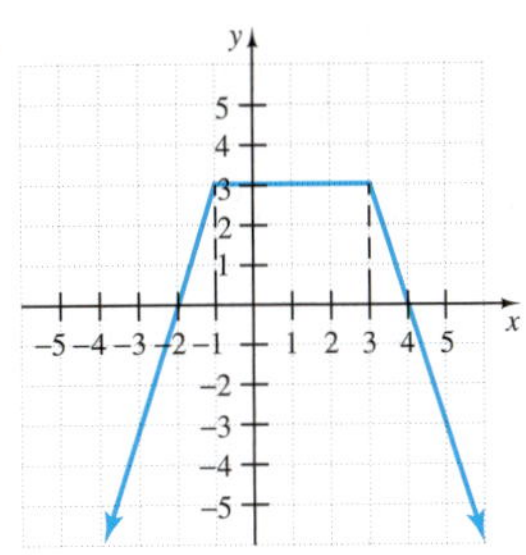

**40.**

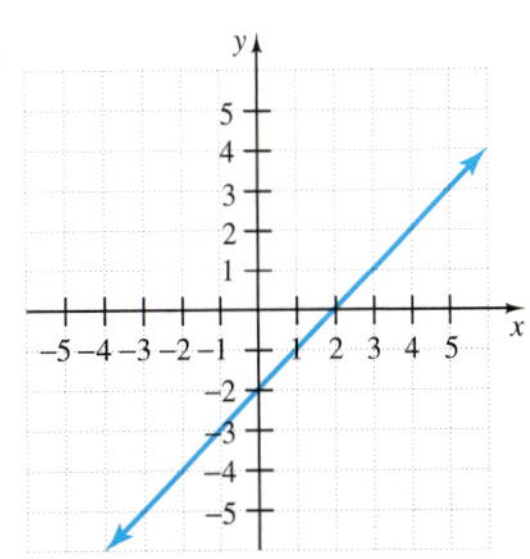

**41.** In your own words define **(a)** function; **(b)** domain; **(c)** range.

**42.** Explain the vertical line test and how it is used.

*Decide whether each is a function. See Examples 3 and 4.*

**43.** $y = x + 1$

**44.** $y = x - 1$

**45.** $x = 2y^2$

**46.** $y = x^2$

**47.** $y - x = 7$

**48.** $2x - 3y = 9$

**49.** $y = \frac{1}{x}$

**50.** $y = \frac{1}{x - 3}$

**51.** $y = 5x - 12$

**52.** $y = \frac{1}{2}x + 4$

**53.** $x = y^2$

**54.** $x = |y|$

*If $f(x) = 3x + 3$, $g(x) = 4x^2 - 6x + 3$, and $h(x) = 5x^2 - 7$, find the following. See Example 7.*

**55.** $f(4)$

**56.** $f(-1)$

**57.** $h(-3)$

**58.** $h(0)$

**59.** $g(2)$

**60.** $g(1)$

**61.** $g(0)$

**62.** $h(-2)$

*Given the following functions, find the indicated values. See Example 7.*

**63.** $f(x) = \frac{1}{2}x$; **a.** $f(0)$ **b.** $f(2)$ **c.** $f(-2)$

**64.** $g(x) = -\frac{1}{3}x$; **a.** $g(0)$ **b.** $g(-1)$ **c.** $g(3)$

**65.** $g(x) = 2x^2 + 4$; **a.** $g(-11)$ **b.** $g(-1)$ **c.** $g\left(\frac{1}{2}\right)$

**66.** $h(x) = -x^2$; **a.** $h(-5)$ **b.** $h\left(-\frac{1}{3}\right)$ **c.** $h\left(\frac{1}{3}\right)$

**67.** $f(x) = -5$; **a.** $f(2)$ **b.** $f(0)$ **c.** $f(606)$

**68.** $h(x) = 7$; **a.** $h(7)$ **b.** $h(542)$ **c.** $h\left(-\frac{3}{4}\right)$

**69.** $f(x) = 1.3x^2 - 2.6x + 5.1$ **a.** $f(2)$ **b.** $f(-2)$ **c.** $f(3.1)$

**70.** $g(x) = 2.7x^2 + 6.8x - 10.2$ **a.** $g(1)$ **b.** $g(-5)$ **c.** $g(7.2)$

*Use the graph of the function below to answer Exercises 71 through 78.*

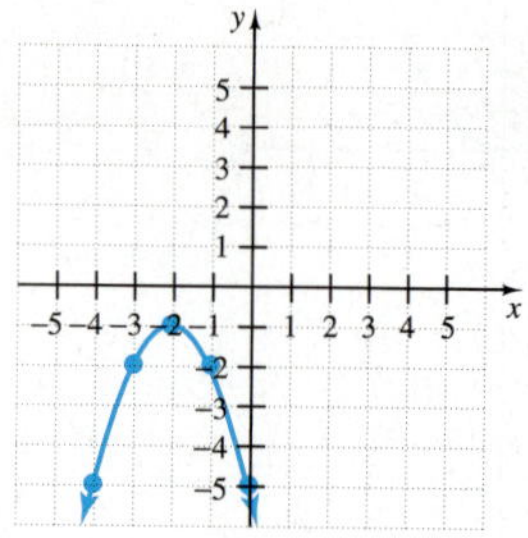


**71.** If $f(1) = -10$ write the corresponding ordered pair.

**72.** If $f(-5) = -10$, write the corresponding ordered pair.

**73.** Find $f(-1)$.

**74.** Find $f(-2)$.

**75.** Find all values of $x$ such that $f(x) = -5$.

**76.** Find all values of $x$ such that $f(x) = -2$.

**77.** What is the greatest number of $x$-intercepts that a function may have? Explain your answer.

**78.** What is the greatest number of $y$-intercepts that a function may have? Explain your answer.

*Use the graph in Example 8 to answer the following.*

**79. a.** Use the graph to approximate the money spent on research and development in 1994.

**b.** Recall that the function $f(x) = 1.882x + 11.79$ approximates the graph of Example 8. Use this equation to approximate the money spent on research and development in 1994. [*Hint:* Find $f(1)$.]

**80. a.** Use the graph to approximate the money spent on research and development in 1998.

**b.** Use the function $f(x) = 1.882x + 11.79$ to approximate the money spent on research and development in 1998. [*Hint:* Find $f(5)$.]

**81.** Use the function $f(x) = 1.882x + 11.79$ to predict the money that will be spent on research and development in 2005.

**82.** Use the function $f(x) = 1.882x + 11.79$ to predict the money that will be spent on research and development in 2010.

**83.** Since $y = x + 7$ describes a function, rewrite the equation using function notation.

**84.** In your own words, explain how to find the domain of a function given its graph.

*The function $A(r) = \pi r^2$ may be used to find the area of a circle if we are given its radius.*

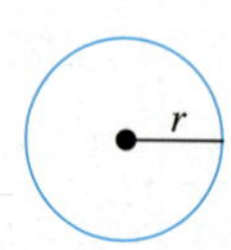


**85.** Find the area of a circle whose radius is 5 centimeters. (Do not approximate $\pi$.)

△ **86.** Find the area of a circular garden whose radius is 8 feet. (Do not approximate $\pi$.)

*The function $V(x) = x^3$ may be used to find the volume of a cube if we are given the length x of a side.*

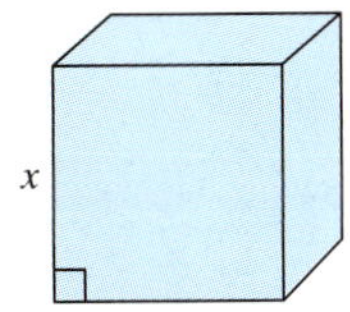


**87.** Find the volume of a cube whose side is 14 inches.

**88.** Find the volume of a die whose side is 1.7 centimeters.

*Forensic scientists use the following functions to find the height of a woman if they are given the height of her femur bone f or her tibia bone t in centimeters.*

$$H(f) = 2.59f + 47.24$$
$$H(t) = 2.72t + 61.28$$

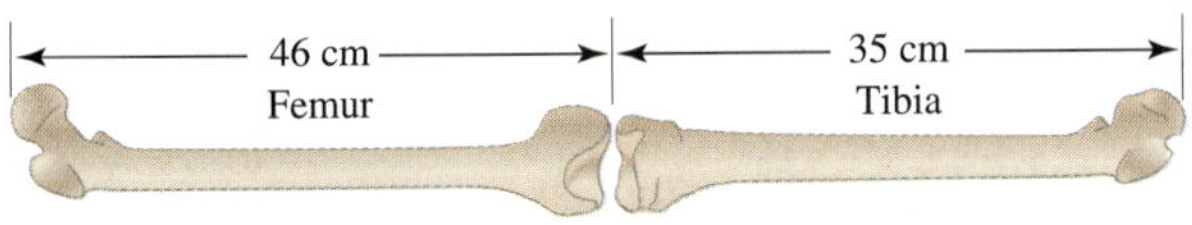


**89.** Find the height of a woman whose femur measures 46 centimeters.

**90.** Find the height of a woman whose tibia measures 35 centimeters.

*The dosage in milligrams D of Ivermectin, a heartworm preventive, for a dog who weighs x pounds is given by*

$$D(x) = \frac{136}{25}x;$$

**91.** Find the proper dosage for a dog that weighs 30 pounds.

**92.** Find the proper dosage for a dog that weighs 50 pounds.

**93.** The per capita consumption (in pounds) of all poultry in the United States is given by the function $C(x) = 1.7x + 88$, where $x$ is the number of years since 1995. (*Source:* Based on actual and estimated data from the Economic Research Service, U.S. Department of Agriculture, 1995–1999)

**a.** Find and interpret $C(2)$.

**b.** Predict the per capita consumption of all poultry in the United States in 2006.

**94.** The number of passengers (in millions) aboard airline flights in the United States is given by the function $P(x) = 25.4x + 448.4$, where $x$ is the number of years since 1991. (*Source:* Based on data from the Air Transport Association of America, 1991–1997)

**a.** Find and interpret $P(4)$.

**b.** Predict the number of airline passengers in 2005.

## REVIEW EXERCISES

*Complete the given table and use the table to graph the linear equation. See Section 3.1.*

**95.** $x - y = -5$

| | | | |
|---|---|---|---|
| $x$ | 0 | | 1 |
| $y$ | | 0 | |

**96.** $2x + 3y = 10$

| | | | |
|---|---|---|---|
| $x$ | 0 | | |
| $y$ | | 0 | 2 |

**97.** $7x + 4y = 8$

| | | | |
|---|---|---|---|
| $x$ | 0 | | |
| $y$ | | 0 | −1 |

**98.** $5y - x = -15$

| | | | |
|---|---|---|---|
| $x$ | 0 | | −2 |
| $y$ | | 0 | |

**99.** $y = 6x$

| | | | |
|---|---|---|---|
| $x$ | 0 | | −1 |
| $y$ | | 0 | |

**100.** $y = -2x$

| | | | |
|---|---|---|---|
| $x$ | 0 | | −2 |
| $y$ | | 0 | |

△ **101.** Is it possible to find the perimeter of the following geometric figure? If so, find the perimeter.

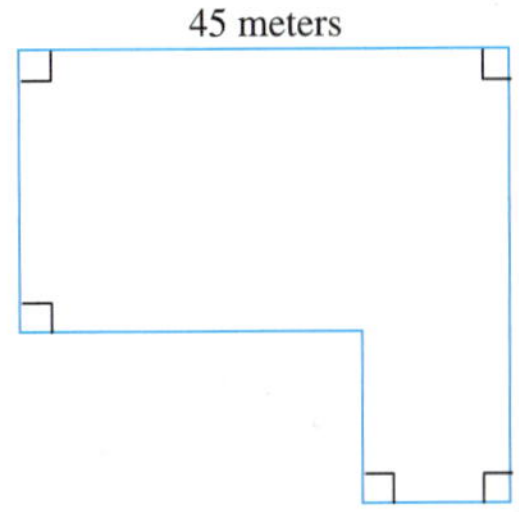


## A Look Ahead

**Example**

If $f(x) = x^2 + 2x + 1$, find the following.

**a.** $f(\pi)$ **b.** $f(c)$

*Solution*

**a.** $f(x) = x^2 + 2x + 1$
$f(\pi) = \pi^2 + 2\pi + 1$

**b.** $f(x) = x^2 + 2x + 1$
$f(c) = (c)^2 + 2(c) + 1$
$= c^2 + 2c + 1$

*Given the following functions, find the indicated values. See the previous example.*

**102.** $f(x) = 2x + 7;$
**a.** $f(2)$ **b.** $f(a)$

**103.** $g(x) = -3x + 12;$
**a.** $g(s)$ **b.** $g(r)$

**104.** $h(x) = x^2 + 7;$
**a.** $h(3)$ **b.** $h(a)$

**105.** $f(x) = x^2 - 12;$
**a.** $f(12)$ **b.** $f(a)$

## 3.3 GRAPHING LINEAR FUNCTIONS

**OBJECTIVES**

1. Graph linear functions.
2. Graph linear functions by finding intercepts.
3. Graph vertical and horizontal lines.

**1** In this section, we identify and graph linear functions. By the vertical line test, we know that all linear equations except those whose graphs are vertical lines are functions. For example, we know from Section 3.1 that $y = 2x$ is a linear equation in two variables. Its graph is shown.

| $x$ | $y = 2x$ |
|---|---|
| 1 | 2 |
| 0 | 0 |
| −1 | −2 |

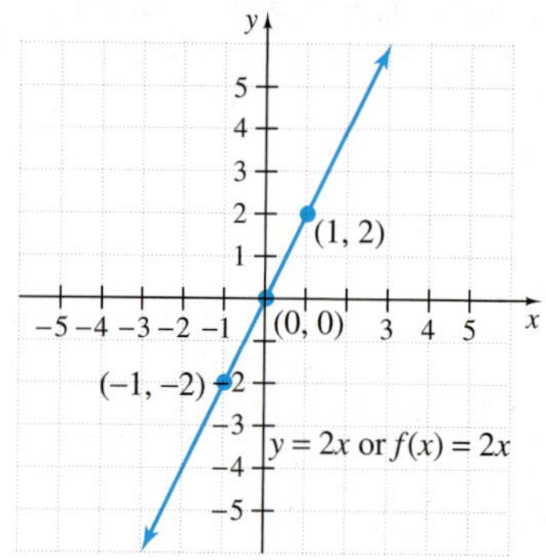


Because this graph passes the vertical line test, we know that $y = 2x$ is a function. If we want to emphasize that this equation describes a function, we may write $y = 2x$ as $f(x) = 2x$.

**Example 1** Graph $g(x) = 2x + 1$. Compare this graph with the graph of $f(x) = 2x$.

*Solution* To graph $g(x) = 2x + 1$, find three ordered pair solutions.

add 1

| $x$ | $f(x) = 2x$ | $g(x) = 2x + 1$ |
|---|---|---|
| 0 | 0 | 1 |
| −1 | −2 | −1 |
| 1 | 2 | 3 |

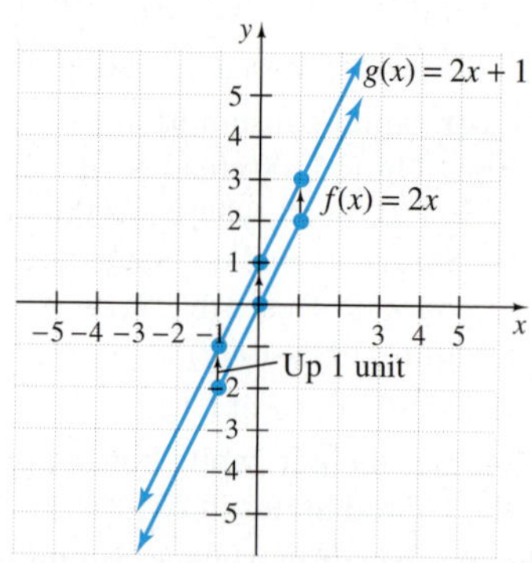


Notice that $y$-values for the graph of $g(x) = 2x + 1$ are obtained by adding 1 to each $y$-value of each corresponding point of the graph of $f(x) = 2x$. The graph of $g(x) = 2x + 1$ is the same as the graph of $f(x) = 2x$ shifted upward 1 unit.

In general, a **linear function** is a function that can be written in the form $f(x) = mx + b$. For example, $g(x) = 2x + 1$ is in this form, with $m = 2$ and $b = 1$.

**Example 2** Graph the linear functions $f(x) = -3x$ and $g(x) = -3x - 6$ on the same set of axes.

*Solution* To graph $f(x)$ and $g(x)$, find ordered pair solutions.

subtract 6

| $x$ | $f(x) = -3x$ | $g(x) = -3x - 6$ |
|---|---|---|
| 0 | 0 | −6 |
| 1 | −3 | −9 |
| −1 | 3 | −3 |
| −2 | 6 | 0 |

subtract 6

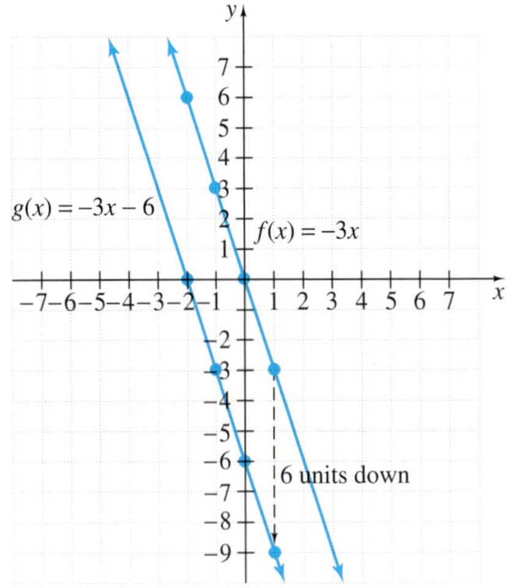


Each $y$-value for the graph of $g(x) = -3x - 6$ is obtained by subtracting 6 from the $y$-value of the corresponding point of the graph of $f(x) = -3x$. The graph of $g(x) = -3x - 6$ is the same as the graph of $f(x) = -3x$ shifted down 6 units.

2 Notice that the $y$-intercept of the graph of $g(x) = -3x - 6$ in the preceding figure is $(0, -6)$. In general, if *a linear function is written in the form* $f(x) = mx + b$ *or* $y = mx + b$, *the y-intercept is* $(0, b)$. This is because if $x$ is 0, then $f(x) = mx + b$ becomes $f(0) = m \cdot 0 + b = b$, and we have the ordered pair solution $(0, b)$. We will study this form more in the next section.

**Example 3** Find the $y$-intercept of the graph of each equation.

**a.** $f(x) = \frac{1}{2}x + \frac{3}{7}$ **b.** $y = -2.5x - 3.2$

*Solution* **a.** The $y$-intercept of $f(x) = \frac{1}{2}x + \frac{3}{7}$ is $\left(0, \frac{3}{7}\right)$.

**b.** The $y$-intercept of $y = -2.5x - 3.2$ is $(0, -3.2)$.

In general, to find the $y$-intercept of the graph of an equation not in the form $y = mx + b$, let $x = 0$ since any point on the $y$-axis has an $x$-coordinate of 0. To find the $x$-intercept of a line, let $y = 0$ or $f(x) = 0$ since any point on the $x$-axis has a $y$-coordinate of 0.

**FINDING $x$- AND $y$-INTERCEPTS**

To find an $x$-intercept, let $y = 0$ or $f(x) = 0$ and solve for $x$.
To find a $y$-intercept, let $x = 0$ and solve for $y$.

Intercepts are usually easy to find and plot since one coordinate is 0.

**Example 4** Graph $x - 3y = 6$ by plotting intercepts.

*Solution* Let $y = 0$ to find the $x$-intercept and $x = 0$ to find the $y$-intercept.

| If $y = 0$ then | If $x = 0$ then |
|---|---|
| $x - 3(0) = 6$ | $0 - 3y = 6$ |
| $x - 0 = 6$ | $-3y = 6$ |
| $x = 6$ | $y = -2$ |

The $x$-intercept is $(6, 0)$ and the $y$-intercept is $(0, -2)$. We find a third ordered pair solution to check our work. If we let $y = -1$, then $x = 3$. Plot the points $(6, 0)$, $(0, -2)$, and $(3, -1)$. The graph of $x - 3y = 6$ is the line drawn through these points, as shown.

| $x$ | $y$ |
|---|---|
| 6 | 0 |
| 0 | −2 |
| 3 | −1 |

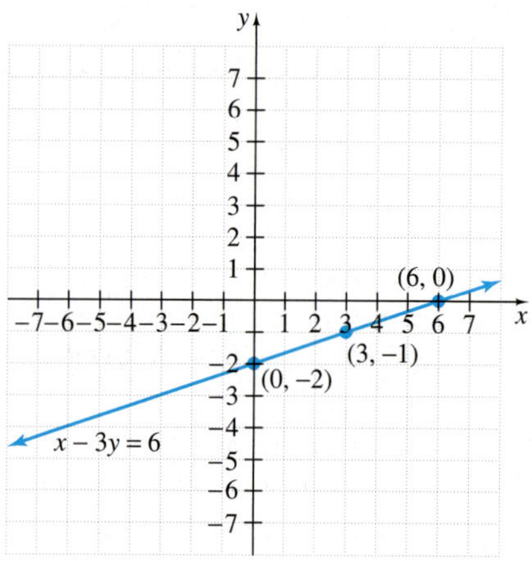


Notice that the equation $x - 3y = 6$ describes a linear function—"linear" because its graph is a line and "function" because the graph passes the vertical line test.

If we want to emphasize that the equation $x - 3y = 6$ from Example 4 describes a function, first solve the equation for $y$.

$$x - 3y = 6$$

$$-3y = -x + 6 \quad \text{Subtract } x \text{ from both sides.}$$

$$\frac{-3y}{-3} = \frac{-x}{-3} + \frac{6}{-3} \quad \text{Divide both sides by } -3.$$

$$y = \frac{1}{3}x - 2 \quad \text{Simplify.}$$

Next, let $y = f(x)$.

$$f(x) = \frac{1}{3}x - 2$$

**HELPFUL HINT**

Any linear equation that describes a function can be written using function notation. To do so, solve the equation for $y$ and then replace $y$ with $f(x)$, as we did above.

**Example 5** Graph $x = -2y$ by plotting intercepts.

*Solution* Let $y = 0$ to find the $x$-intercept and $x = 0$ to find the $y$-intercept.

| | | | |
|---|---|---|---|
| If $y = 0$ | then | If $x = 0$ | then |
| $x = -2(0)$ | or | $0 = -2y$ | or |
| $x = 0$ | | $0 = y$ | |
| $(0, 0)$ | | $(0, 0)$ | |

*Ordered pairs*

Both the $x$-intercept and $y$-intercept are $(0, 0)$. This happens when the graph passes through the origin. Since two points are needed to determine a line, we must find at least one more ordered pair that satisfies $x = -2y$. Let $y = -1$ to find a second ordered pair solution and let $y = 1$ as a check point.

| | | | |
|---|---|---|---|
| If $y = -1$ | then | If $y = 1$ | then |
| $x = -2(-1)$ | or | $x = -2(1)$ | or |
| $x = 2$ | | $x = -2$ | |

The ordered pairs are $(0, 0)$, $(2, -1)$, and $(-2, 1)$. Plot these points to graph $x = -2y$.

| $x$ | $y$ |
|---|---|
| 0 | 0 |
| 2 | −1 |
| −2 | 1 |

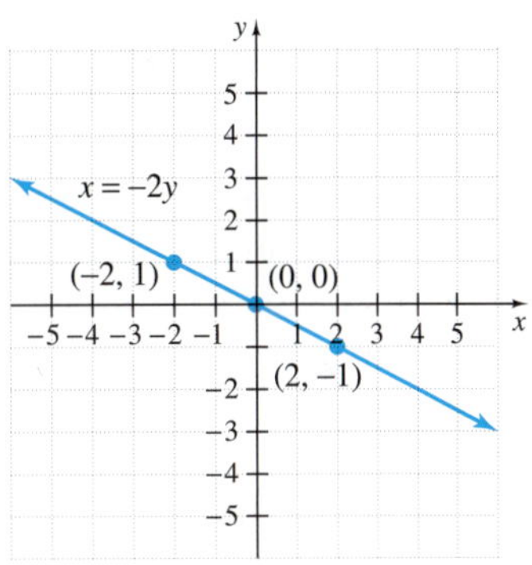


**3** The equations $x = c$ and $y = c$, where $c$ is a real number constant, are both linear equations in two variables. Why? Because $x = c$ can be written as $x + 0y = c$ and $y = c$ can be written as $0x + y = c$. We graph these two special linear equations below.

**Example 6** Graph $x = 2$.

*Solution* The equation $x = 2$ can be written as $x + 0y = 2$. For any $y$-value chosen, notice that $x$ is 2. No other value for $x$ satisfies $x + 0y = 2$. Any ordered pair whose $x$-coordinate is 2 is a solution to $x + 0y = 2$ because 2 added to 0 times any value of $y$ is $2 + 0$, or 2. We will use the ordered pairs (2, 3), (2, 0) and (2, −3) to graph $x = 2$.

| $x$ | $y$ |
|---|---|
| 2 | 3 |
| 2 | 0 |
| 2 | −3 |

$x$-intercept → (2, 0)

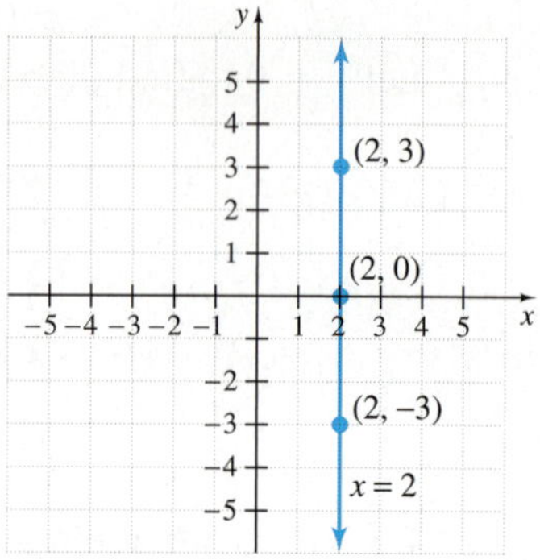


The graph is a vertical line with $x$-intercept (2, 0). Notice that this graph is not the graph of a function, and it has no $y$-intercept because $x$ is never 0.

**Example 7** Graph $y = -3$.

*Solution* The equation $y = -3$ can be written as $0x + y = -3$. For any $x$-value chosen, $y$ is −3. If we choose 4, 0, and −2 as $x$-values, the ordered pair solutions are (4, −3), (0, −3), and (−2, −3). We will use these ordered pairs to graph $y = -3$.

| $x$ | $y$ |
|---|---|
| 4 | −3 |
| 0 | −3 |
| −2 | −3 |

← $y$-intercept (0, −3)

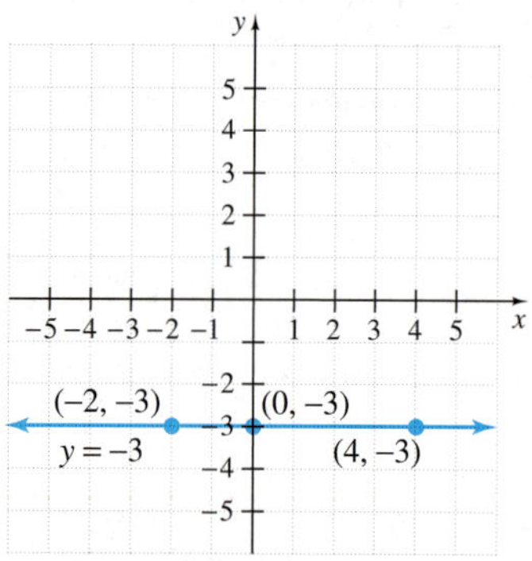


The graph is a horizontal line with $y$-intercept (0, −3) and no $x$-intercept. Notice that this graph is the graph of a function.

From Examples 6 and 7, we have the following generalization.

## GRAPHING VERTICAL AND HORIZONTAL LINES

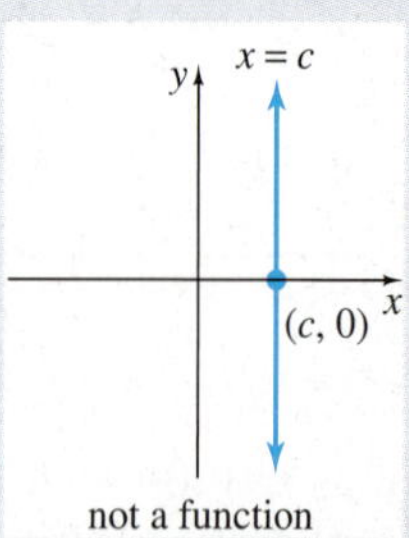


The graph of $x = c$, where $c$ is a real number, is a vertical line with $x$-intercept $(c, 0)$.
The graph of $y = c$, where $c$ is a real number, is a horizontal line with $y$-intercept $(0, c)$.

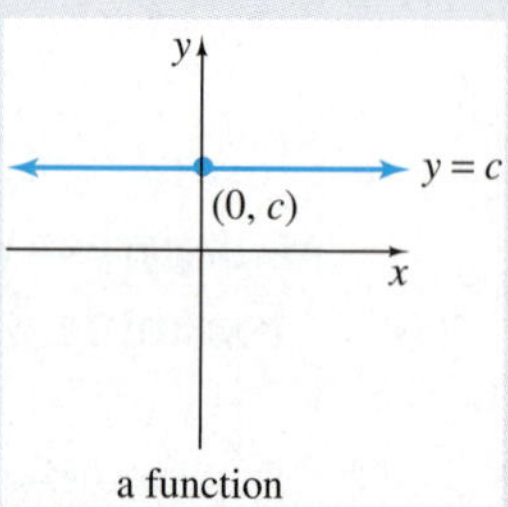

## GRAPHING CALCULATOR EXPLORATIONS

You may have noticed by now that to use the [Y=] key on a graphing calculator to graph an equation, the equation must be solved for $y$.

*Graph each function by first solving the function for y.*

1. $x = 3.5y$
2. $-2.7y = x$
3. $5.78x + 2.31y = 10.98$
4. $-7.22x + 3.89y = 12.57$
5. $y - |x| = 3.78$
6. $3y - 5x^2 = 6x - 4$
7. $y - 5.6x^2 = 7.7x + 1.5$
8. $y + 2.6|x| = -3.2$

## *Exercise Set 3.3*

*Graph each linear function. See Examples 1 and 2.*

1. $f(x) = -2x$
2. $f(x) = 2x$
3. $f(x) = -2x + 3$
4. $f(x) = 2x + 6$
5. $f(x) = \frac{1}{2}x$
6. $f(x) = \frac{1}{3}x$
7. $f(x) = \frac{1}{2}x - 4$
8. $f(x) = \frac{1}{3}x - 2$

*The graph of $f(x) = 5x$ follows. Use this graph to match each linear function with its graph. See Examples 1 through 3.*

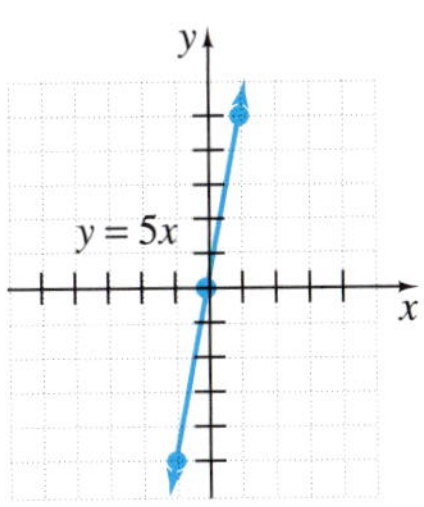


A

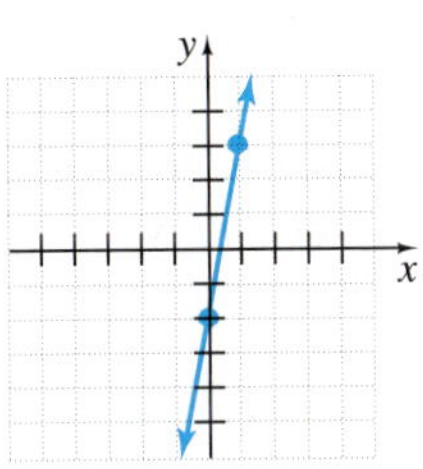


B

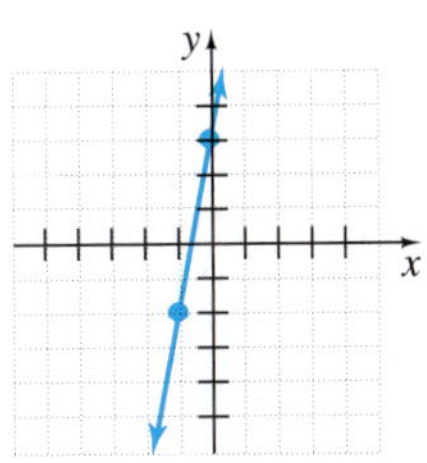


C

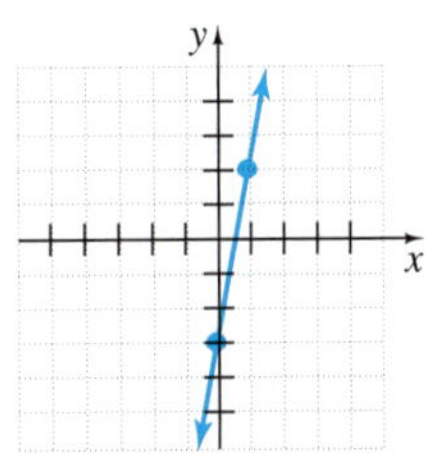


D

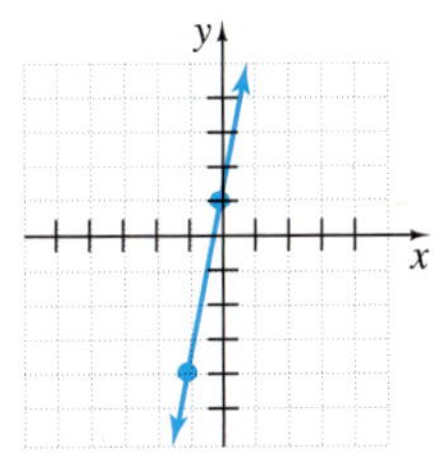


9. $f(x) = 5x - 3$
10. $f(x) = 5x - 2$
11. $f(x) = 5x + 1$
12. $f(x) = 5x + 3$

*Graph each linear function by finding x-and y-intercepts. See Examples 4 and 5.*

13. $x - y = 3$
14. $x - y = -4$
15. $x = 5y$
16. $2x = y$
17. $-x + 2y = 6$
18. $x - 2y = -8$
19. $2x - 4y = 8$
20. $2x + 3y = 6$
21. In your own words, explain how to find $x$- and $y$-intercepts.
22. Explain why it is a good idea to use three points to graph a linear equation.

*Graph each linear equation. See Examples 6 and 7.*

23. $x = -1$
24. $y = 5$
25. $y = 0$
26. $x = 0$
27. $y + 7 = 0$
28. $x - 3 = 0$

*Match each equation with its graph.*

**A**

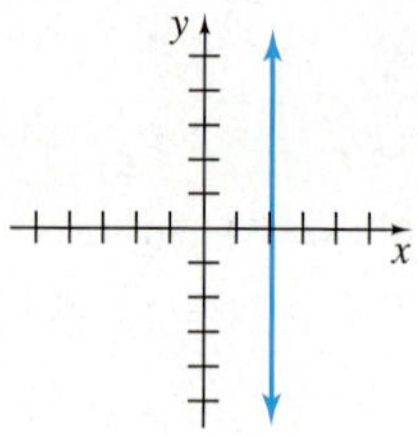


**B**

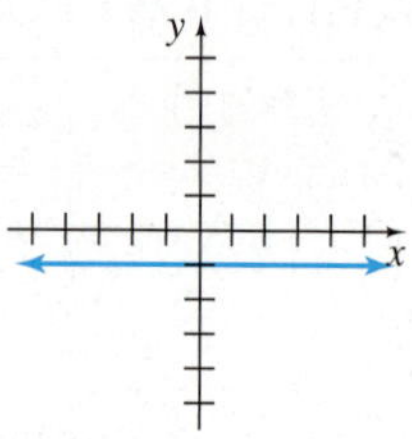


**C**

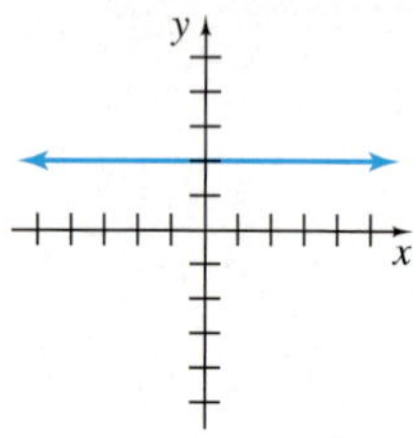


**D**

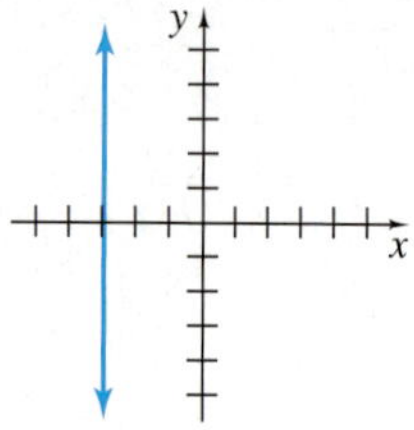


**29.** $y = 2$ **30.** $x = -3$

**31.** $x - 2 = 0$ **32.** $y + 1 = 0$

**33.** Discuss whether a vertical line ever has a $y$-intercept.

**34.** Discuss whether a horizontal line ever has an $x$-intercept.

*Graph each linear equation.*

**35.** $x + 2y = 8$ **36.** $x - 3y = 3$

**37.** $f(x) = \frac{3}{4}x + 2$ **38.** $f(x) = \frac{4}{3}x + 2$

**39.** $x = -3$ **40.** $f(x) = 3$

**41.** $3x + 5y = 7$ **42.** $3x - 2y = 5$

**43.** $f(x) = x$ **44.** $f(x) = -x$

**45.** $x + 8y = 8$ **46.** $x - 3y = 9$

**47.** $5 = 6x - y$ **48.** $4 = x - 3y$

**49.** $-x + 10y = 11$ **50.** $-x + 9 = -y$

**51.** $y = 1$ **52.** $x = 1$

**53.** $f(x) = \frac{1}{2}x$ **54.** $f(x) = -2x$

**55.** $x + 3 = 0$ **56.** $y - 6 = 0$

**57.** $f(x) = 4x - \frac{1}{3}$ **58.** $f(x) = -3x + \frac{3}{4}$

**59.** $2x + 3y = 6$ **60.** $4x + y = 5$

*Solve.*

**61.** Broyhill Furniture found that it takes 2 hours to manufacture each table for one of its special dining room sets. Each chair takes 3 hours to manufacture. A total of 1500 hours is available to produce tables and chairs of this style. The linear equation that models this situation is $2x + 3y = 1500$, where $x$ represents the number of tables produced and $y$ the number of chairs produced.

**a.** Complete the ordered pair solution (0, ) of this equation. Describe the manufacturing situation this solution corresponds to.

**b.** Complete the ordered pair solution ( , 0) for this equation. Describe the manufacturing situation this solution corresponds to.

**c.** If 50 tables are produced, find the greatest number of chairs the company can make.

**62.** While manufacturing two different camera models, Kodak found that the basic model costs \$55 to produce, whereas the deluxe model costs \$75. The weekly budget for these two models is limited to \$33,000 in production costs. The linear equation that models this situation is $55x + 75y = 33{,}000$, where $x$ represents the number of basic models and $y$ the number of deluxe models.

**a.** Complete the ordered pair solution (0, ) of this equation. Describe the manufacturing situation this solution corresponds to.

**b.** Complete the ordered pair solution ( ,0) of this equation. Describe the manufacturing situation this solution corresponds to.

**c.** If 350 deluxe models are produced, find the greatest number of basic models that can be made in one week.

**63.** The cost of renting a car for a day is given by the linear function $C(x) = 0.2x + 24$, where $C(x)$ is in dollars and $x$ is the number of miles driven.

**a.** Find the cost of driving the car 200 miles.

**b.** Graph $C(x) = 0.2x + 24$.

**c.** How can you tell from the graph of $C(x)$ that as the number of miles driven increases, the total cost increases also?

**64.** The cost of renting a piece of machinery is given by the linear function $C(x) = 4x + 10$, where $C(x)$ is in dollars and $x$ is given in hours.

**a.** Find the cost of renting the piece of machinery for 8 hours.

**b.** Graph $C(x) = 4x + 10$.

**c.** How can you tell from the graph of $C(x)$ that as the number of hours increases, the total cost increases also?

**65.** The yearly cost of tuition and required fees for attending a public two-year college full time can be estimated by the linear function $f(x) = 72.9x + 785.2$, where $x$ is the number of years after 1990 and $f(x)$ is the total cost. (*Source:* U.S. National Center for Education Statistics)

**a.** Use this function to approximate the yearly cost of attending a two-year college in the year 2010. [*Hint:* Find $f(20)$.]

**b.** Use the given function to predict in what year the yearly cost of tuition and required fees will exceed \$2000. [*Hint:* Let $f(x) = 2000$ and solve for $x$.]

**c.** Use this function to approximate the yearly cost of attending a two-year college in the present year. If you attend a two-year college, is this amount greater than or less than the amount that is currently charged by the college that you attend?

**66.** The yearly cost of tuition and required fees for attending a public four-year college can be estimated by the linear function $f(x) = 186.1x + 2030$, where $x$ is the number of years after 1990 and $f(x)$ is the total cost in dollars. (*Source:* U.S. National Center for Education Statistics)

**a.** Use this function to approximate the yearly cost of attending a four-year college in the year 2010. [*Hint:* Find $f(20)$.]

**b.** Use the given function to predict in what year the yearly cost of tuition and required fees will exceed \$5000. [*Hint:* Let $f(x) = 5000$ and solve for $x$.]

**c.** Use this function to approximate the yearly cost of attending a four-year college in the present year. If you attend a four-year college, is this amount greater than or less than the amount that is currently charged by the college that you attend?

*Use a graphing calculator to verify the results of each exercise.*

**67.** Exercise 9 **68.** Exercise 10

**69.** Exercise 17 **70.** Exercise 18

## REVIEW EXERCISES

*Solve the following. See Sections 2.6 and 2.7.*

**71.** $|x - 3| = 6$ **72.** $|x + 2| < 4$

**73.** $|2x + 5| > 3$ **74.** $|5x| = 10$

**75.** $|3x - 4| \leq 2$ **76.** $|7x - 2| \geq 5$

*Simplify.*

**77.** $\frac{-6 - 3}{2 - 8}$ **78.** $\frac{4 - 5}{-1 - 0}$

**79.** $\frac{-8 - (-2)}{-3 - (-2)}$ **80.** $\frac{12 - 3}{10 - 9}$

**81.** $\frac{0 - 6}{5 - 0}$ **82.** $\frac{2 - 2}{3 - 5}$

# 3.4 THE SLOPE OF A LINE

CD-ROM SSM SSG Video

**OBJECTIVES**

1. Find the slope of a line given two points on the line.
2. Find the slope of a line given the equation of a line.
3. Interpret the slope–intercept form in an application.
4. Find the slopes of horizontal and vertical lines.
5. Compare the slopes of parallel and perpendicular lines.

**1** You may have noticed by now that different lines often tilt differently. It is very important in many fields to be able to measure and compare the tilt, or **slope**, of lines. For example, a wheelchair ramp with a slope of $\frac{1}{12}$ means that the ramp rises 1 foot for every 12 horizontal feet. A road with a slope or grade of 11% (or $\frac{11}{100}$) means that the road rises 11 feet for every 100 horizontal feet.

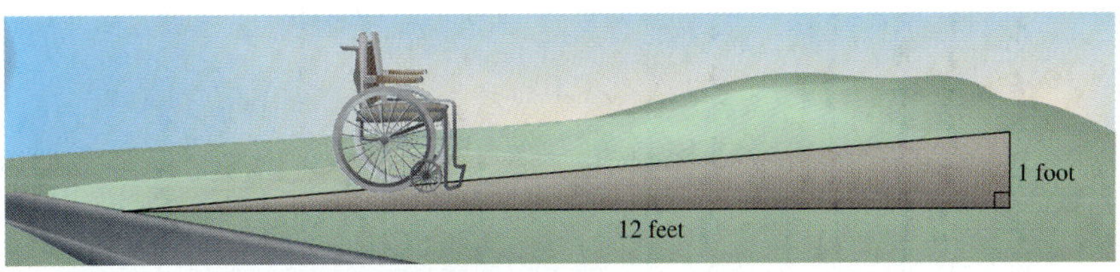


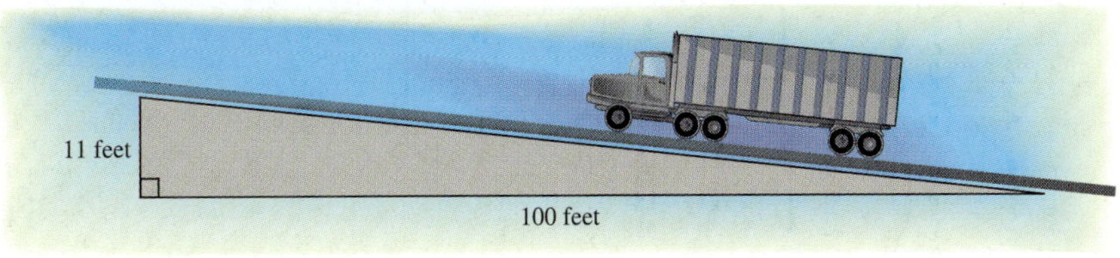

We measure the slope of a line as a ratio of **vertical change** to **horizontal change**. Slope is usually designated by the letter $m$.

Suppose that we want to measure the slope of the following line.

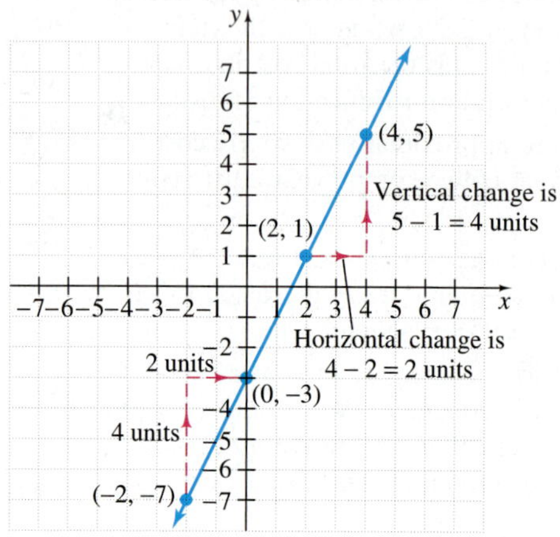


The vertical change between both pairs of points on the line is 4 units per horizontal change of 2 units. Then

$$\text{slope } m = \frac{\text{change in } y \text{ (vertical change)}}{\text{change in } x \text{ (horizontal change)}} = \frac{4}{2} = 2$$

Notice that slope is a rate of change between points. A slope of 2 or $\frac{2}{1}$ means that between pairs of points on the line, the rate of change is a vertical change of 2 units per horizontal change of 1 unit.

Consider the line below, which passes through the points $(x_1, y_1)$ and $(x_2, y_2)$. (The notation $x_1$ is read "$x$-sub-one.") The vertical change, or *rise*, between these points is the difference in the $y$-coordinates: $y_2 - y_1$. The horizontal change, or *run*, between the points is the difference of the $x$-coordinates: $x_2 - x_1$.

## SLOPE OF A LINE

Given a line passing through points $(x_1, y_1)$ and $(x_2, y_2)$ the **slope** $m$ of the line is

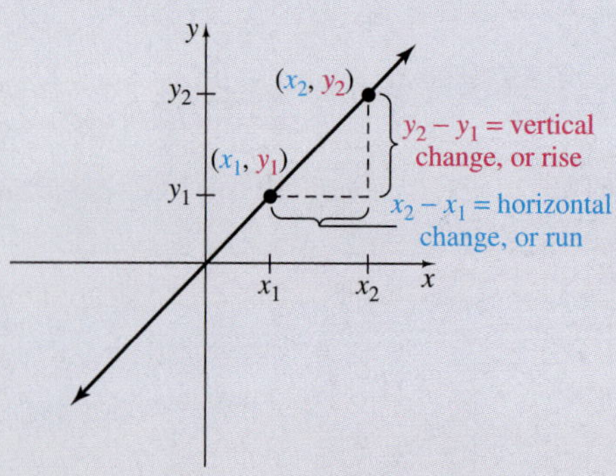


$$m = \frac{\text{rise}}{\text{run}} = \frac{y_2 - y_1}{x_2 - x_1},$$

as long as $x_2 \neq x_1$.

**Example 1** Find the slope of the line containing the points (0, 3) and (2, 5). Graph the line.

*Solution* We use the slope formula. It does not matter which point we call $(x_1, y_1)$ and which point we call $(x_2, y_2)$. We'll let $(x_1, y_1) = (0, 3)$ and $(x_2, y_2) = (2, 5)$.

$$m = \frac{y_2 - y_1}{x_2 - x_1}$$

$$= \frac{5 - 3}{2 - 0} = \frac{2}{2} = 1$$

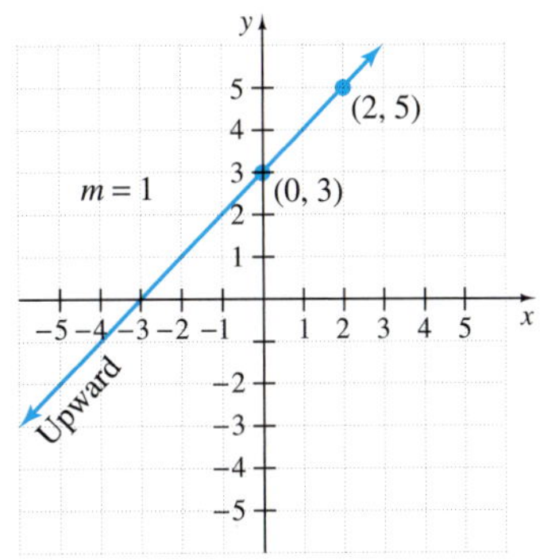


Notice in this example that the slope is positive and that the graph of the line containing (0, 3) and (2, 5) moves upward, or increases, as we go from left to right.

**HELPFUL HINT**
When we are trying to find the slope of a line through two given points, it makes no difference which given point is called $(x_1, y_1)$ and which is called $(x_2, y_2)$. Once an $x$-coordinate is called $x_1$, however, make sure its corresponding $y$-coordinate is called $y_1$.

**Example 2** Find the slope of the line containing the points (5, −4) and (−3, 3). Graph the line.

*Solution* We use the slope formula, and let $(x_1, y_1) = (5, -4)$ and $(x_2, y_2) = (-3, 3)$.

$$m = \frac{y_2 - y_1}{x_2 - x_1}$$

$$= \frac{3 - (-4)}{-3 - 5} = \frac{7}{-8} = -\frac{7}{8}$$

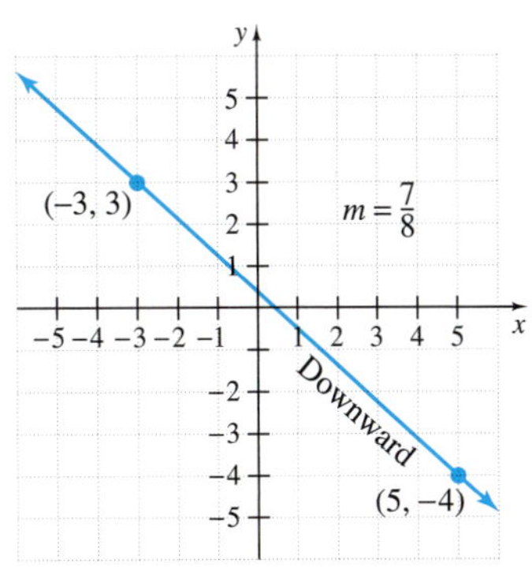


Notice in this example that the slope is negative and that the graph of the line through (5, −4) and (−3, 3) moves downward, or decreases, as we go from left to right.

2 As we have seen, the slope of a line is defined by two points on the line. Thus, if we know the equation of a line, we can find its slope.

**Example 3** Find the slope of the line whose equation is $f(x) = \frac{2}{3}x + 4$.

*Solution* Two points are needed on the line defined by $f(x) = \frac{2}{3}x + 4$ or $y = \frac{2}{3}x + 4$ to find its slope. We will use intercepts as our two points.

| If $x = 0$, then | If $y = 0$, then | |
|---|---|---|
| $y = \frac{2}{3} \cdot 0 + 4$ | $0 = \frac{2}{3}x + 4$ | |
| $y = 4$ | $-4 = \frac{2}{3}x$ | Subtract 4. |
| | $\frac{3}{2}(-4) = \frac{3}{2} \cdot \frac{2}{3}x$ | Multiply by $\frac{3}{2}$. |
| | $-6 = x$ | |

Use the points $(0, 4)$ and $(-6, 0)$ to find the slope. Let $(x_1, y_1)$ be $(0, 4)$ and $(x_2, y_2)$ be $(-6, 0)$. Then

$$m = \frac{y_2 - y_1}{x_2 - x_1} = \frac{0 - 4}{-6 - 0} = \frac{-4}{-6} = \frac{2}{3}$$

Analyzing the results of Example 3, you may notice a striking pattern:

The slope of $y = \frac{2}{3}x + 4$ is $\frac{2}{3}$, the same as the coefficient of $x$.

Also, the $y$-intercept is $(0, 4)$, as expected.

When a linear equation is written in the form $f(x) = mx + b$ or $y = mx + b$, $m$ is the slope of the line and $(0, b)$ is its $y$-intercept. The form $y = mx + b$ is appropriately called the **slope–intercept form**.

### SLOPE–INTERCEPT FORM

When a linear equation in two variables is written in slope–intercept form,

slope ↓ $y$-intercept is $(0, b)$ ↓

$$y = mx + b$$

then $m$ is the slope of the line and $(0, b)$ is the $y$-intercept of the line.

**Example 4** Find the slope and the $y$-intercept of the line $3x - 4y = 4$.

*Solution* We write the equation in slope–intercept form by solving for $y$.

| | |
|---|---|
| $3x - 4y = 4$ | |
| $-4y = -3x + 4$ | Subtract $3x$ from both sides. |
| $\frac{-4y}{-4} = \frac{-3x}{-4} + \frac{4}{-4}$ | Divide both sides by $-4$. |
| $y = \frac{3}{4}x - 1$ | Simplify. |

The coefficient of $x$, $\frac{3}{4}$, is the slope, and the $y$-intercept is $(0, -1)$.

3 Below is the graph of one-day ticket prices at Disney World for the years shown.

Notice that the graph resembles the graph of a line. Recall that businesses often depend on equations that "closely fit" graphs like this one to model the data and predict future trends. By the **least squares** method, the linear function $f(x) = 1.505x + 32.56$ approximates the data shown, where $x$ is the number of years since 1990 and $y$ is the ticket price for that year.

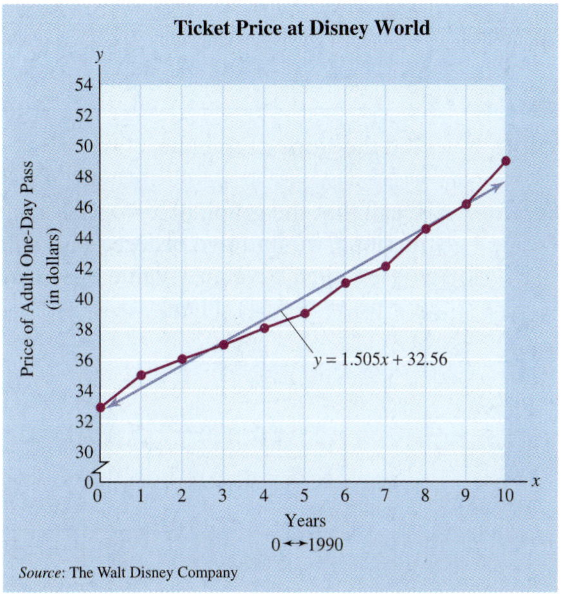


**HELPFUL HINT**
The notation $0 \leftrightarrow 1990$ means that the number 0 corresponds to the year 1990, 1 corresponds to the year 1991, and so on.

## Example 5 PREDICTING FUTURE PRICES

The adult one-day pass price $f(x)$ for Disney World is given by

$$f(x) = 1.505x + 32.56$$

where $x$ is the number of years since 1990

**a.** Use this equation to predict the ticket price for the year 2004.
**b.** What does the slope of this equation mean?
**c.** What does the $y$-intercept of this equation mean?

*Solution* **a.** To predict the price of a pass in 2004, we need to find $f(14)$. (Since year 1990 corresponds to $x = 0$, year 2004 corresponds to $x = 14$.)

$$\begin{aligned} f(x) &= 1.505x + 32.56 \\ f(14) &= 1.505(14) + 32.56 \qquad \text{Let } x = 14. \\ &= 53.63 \end{aligned}$$

We predict that in the year 2004 the price of an adult one-day pass to Disney World will be about \$53.63.

**b.** The slope of $f(x) = 1.505x + 32.56$ is 1.505. We can think of this number as $\frac{\text{rise}}{\text{run}}$ or $\frac{1.505}{1}$. This means that the ticket price increases on the average by \$1.505 every 1 year.

**c.** The $y$-intercept of $y = 1.505 + 32.56$ is $(0, 32.56)$.

year price

This means that at year $x = 0$ or 1990, the ticket price was about \$32.56.

**4** Next we find the slopes of two special types of lines: vertical lines and horizontal lines.

## Example 6

Find the slope of the line $x = -5$.

*Solution* Recall that the graph of $x = -5$ is a vertical line with $x$-intercept $(-5, 0)$. To find the slope, we find two ordered pair solutions of $x = -5$. Of course, solutions of $x = -5$ must have an $x$-value of $-5$. We will let $(x_1, y_1) = (-5, 0)$ and $(x_2, y_2) = (-5, 4)$. Then

$$m = \frac{y_2 - y_1}{x_2 - x_1} = \frac{4 - 0}{-5 - (-5)} = \frac{4}{0}$$

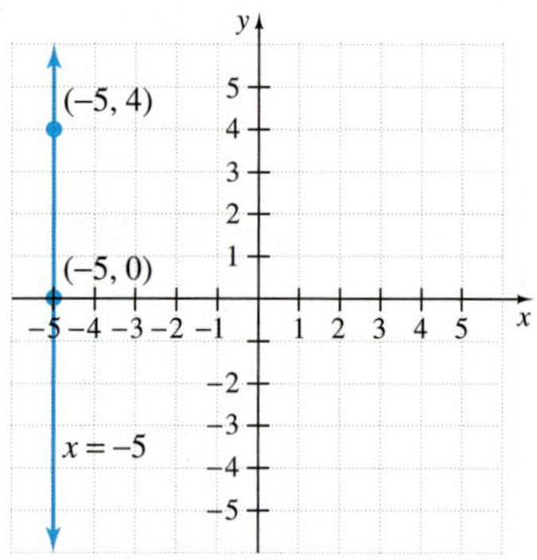


Since $\frac{4}{0}$ is undefined, we say that the slope of the vertical line $x = -5$ is undefined.

## Example 7

Find the slope of the line $y = 2$.

*Solution* Recall that the graph of $y = 2$ is a horizontal line with $y$-intercept $(0, 2)$. To find the slope, we find two points on the line, such as $(0, 2)$ and $(1, 2)$, and use these points to find the slope.

$$m = \frac{2 - 2}{1 - 0} = \frac{0}{1} = 0$$

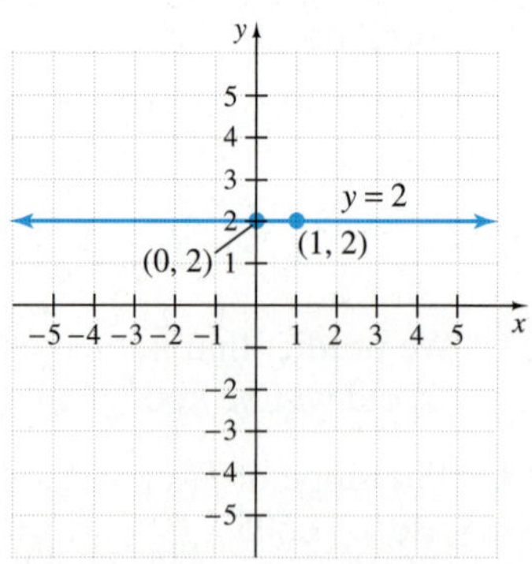


The slope of the horizontal line $y = 2$ is 0.

From the previous two examples, we have the following generalization.

The slope of any vertical line is undefined.
The slope of any horizontal line is 0.

**HELPFUL HINT**
Slope of 0 and undefined slope are not the same. Vertical lines have undefined slope, whereas horizontal lines have slope of 0.

The following four graphs summarize the overall appearance of lines with positive, negative, zero, or undefined slopes.

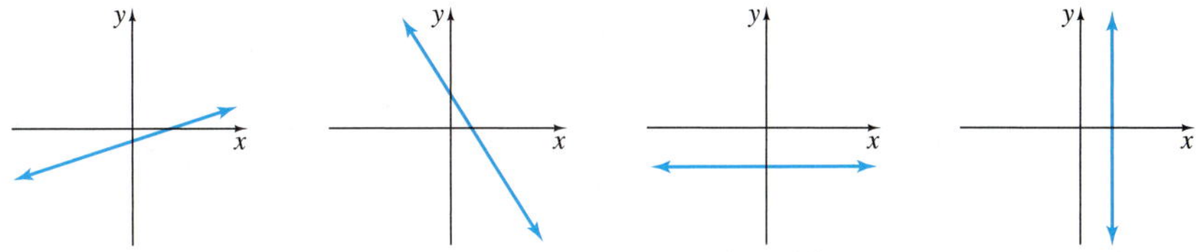


The appearance of a line can give us further information about its slope.

The graphs of $y = \frac{1}{2}x + 1$ and $y = 5x + 1$ are shown to the right. Recall that the graph of $y = \frac{1}{2}x + 1$ has a slope of $\frac{1}{2}$ and that the graph of $y = 5x + 1$ has a slope of 5.

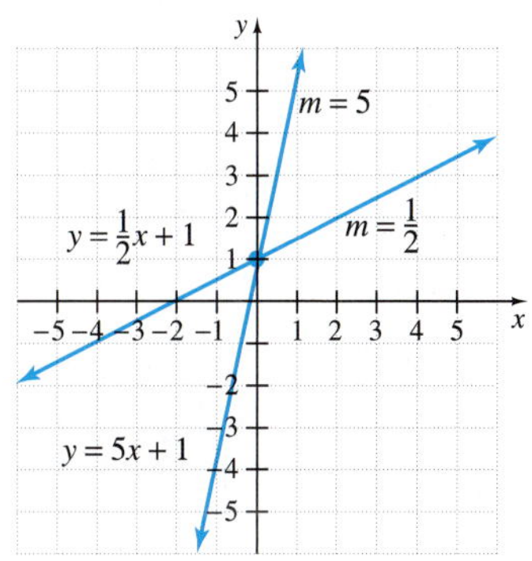


Notice that the line with the slope of 5 is steeper than the line with the slope of $\frac{1}{2}$. This is true in general for positive slopes.

For a line with positive slope $m$, as $m$ increases, the line becomes steeper.

5 Slopes of lines can help us determine whether lines are parallel. Parallel lines are distinct lines with the same steepness, so it follows that they have the same slope.

**PARALLEL LINES**

Two nonvertical lines are parallel if they have the same slope and different $y$-intercepts.

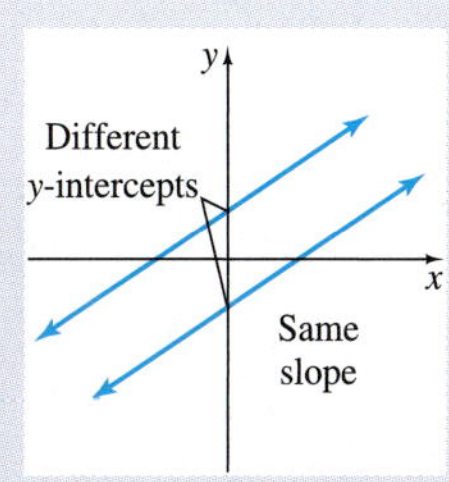

How do the slopes of perpendicular lines compare? (Two lines intersecting at right angles are called **perpendicular lines.**) Suppose that a line has a slope of $\frac{a}{b}$. If the line is rotated 90°, the rise and run are now switched, except that the run is now negative. This means that the new slope is $-\frac{b}{a}$. Notice that

$$\left(\frac{a}{b}\right)\cdot\left(-\frac{b}{a}\right) = -1$$

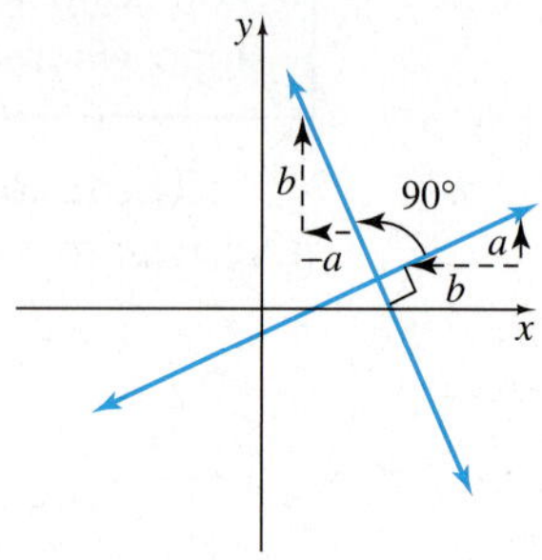


This is how we tell whether two lines are perpendicular.

**PERPENDICULAR LINES**

Two nonvertical lines are perpendicular if the product of their slopes is −1.

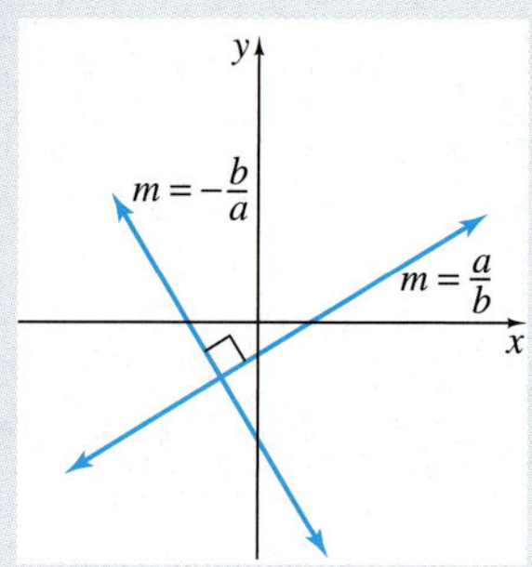


In other words, two nonvertical lines are perpendicular if the slope of one is the negative reciprocal of the slope of the other.

**Example 8** Are the following pairs of lines parallel, perpendicular, or neither?

**a.** $3x + 7y = 4$
$6x + 14y = 7$

**b.** $-x + 3y = 2$
$2x + 6y = 5$

*Solution* Find the slope of each line by solving each equation for $y$.

**a.** $3x + 7y = 4$

$$7y = -3x + 4$$
$$\frac{7y}{7} = \frac{-3x}{7} + \frac{4}{7}$$
$$y = -\frac{3}{7}x + \frac{4}{7}$$

↑ slope ↖ y-intercept $\left(0, \frac{4}{7}\right)$

$6x + 14y = 7$

$$14y = -6x + 7$$
$$\frac{14y}{14} = \frac{-6x}{14} + \frac{7}{14}$$
$$y = -\frac{3}{7}x + \frac{1}{2}$$

↑ slope ↑ y-intercept $\left(0, \frac{1}{2}\right)$

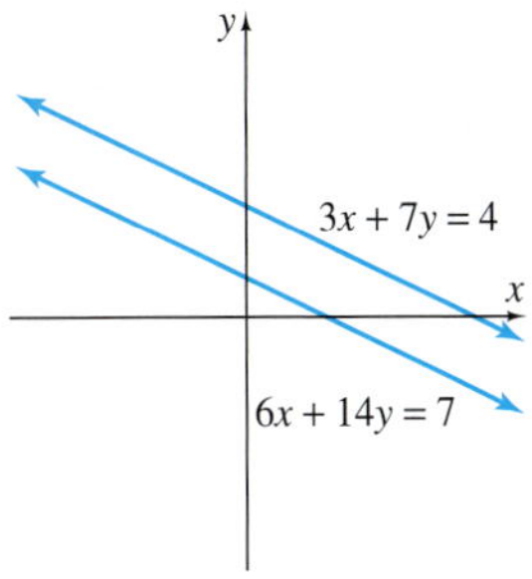


The slopes of both lines are $-\frac{3}{7}$.

The $y$-intercepts are different.
Therefore, the lines are parallel.

**b.** $-x + 3y = 2$ $\qquad$ $2x + 6y = 5$

$$3y = x + 2 \qquad\qquad 6y = -2x + 5$$

$$\frac{3y}{3} = \frac{x}{3} + \frac{2}{3} \qquad\qquad \frac{6y}{6} = \frac{-2x}{6} + \frac{5}{6}$$

$$y = \frac{1}{3}x + \frac{2}{3} \qquad\qquad y = -\frac{1}{3}x + \frac{5}{6}$$

slope: $\frac{1}{3}$; $y$-intercept: $\left(0, \frac{2}{3}\right)$ $\qquad$ slope: $-\frac{1}{3}$; $y$-intercept: $\left(0, \frac{5}{6}\right)$

The slopes are not the same and their product is not $-1$. $\left[\left(\frac{1}{3}\right)\cdot\left(-\frac{1}{3}\right) = -\frac{1}{9}\right]$
Therefore, the lines are neither parallel nor perpendicular.

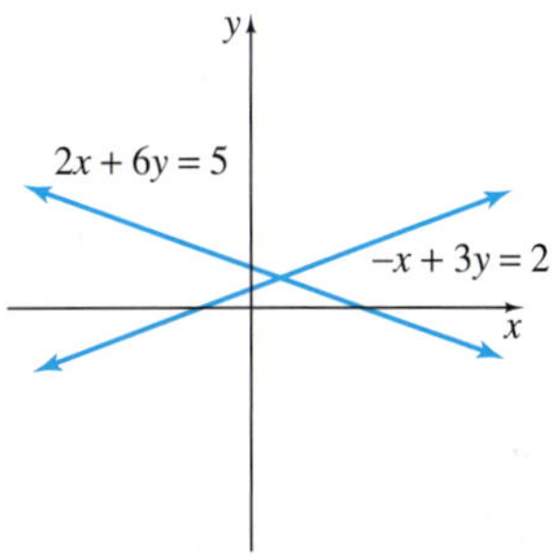


## GRAPHING CALCULATOR EXPLORATIONS

Many graphing calculators have a TRACE feature. This feature allows you to trace along a graph and see the corresponding $x$- and $y$-coordinates appear on the screen. Use this feature for the following exercises.

Graph each function and then use the TRACE feature to complete each ordered pair solution. (Many times the tracer will not show an exact $x$- or $y$-value asked for. In each case, trace as closely as you can to the given $x$- or $y$-coordinate and approximate the other, unknown coordinate to one decimal place.)

**1.** $y = 2.3x + 6.7$
$x = 5.1, y = ?$

**2.** $y = -4.8x + 2.9$
$x = -1.8, y = ?$

**3.** $y = -5.9x - 1.6$
$x = ?, y = 7.2$

**4.** $y = 0.4x - 8.6$
$x = ?, y = -4.4$

**5.** $y = x^2 + 5.2x - 3.3$
$x = 2.3, y = ?$
$x = ?, y = 36$
(There will be two answers here.)

**6.** $y = 5x^2 - 6.2x - 8.3$
$x = 3.2, y = ?$
$x = ?, y = 12$
(There will be two answers here.)

## SPOTLIGHT ON DECISION MAKING

Suppose you are the manager of an apartment complex. You have just notified residents of a rent increase. Some residents think that the increase may be unjustified and out of line with recent increases. A group of concerned residents asks you to hold an open meeting to answer questions about the increase. You are preparing a set of overheads to use during the meeting to show the history of rent increases at the apartment complex. Which overhead would you use and why?

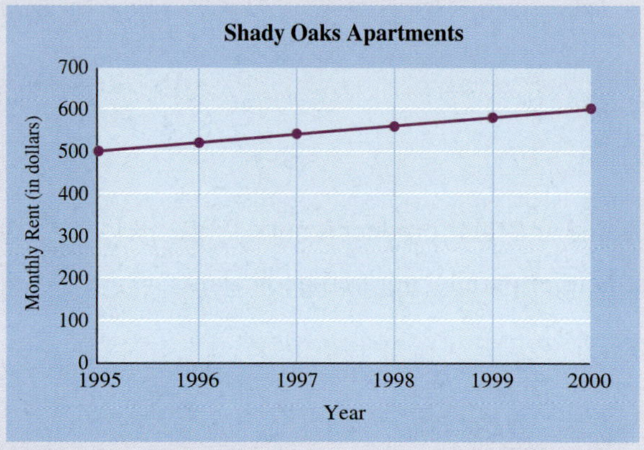


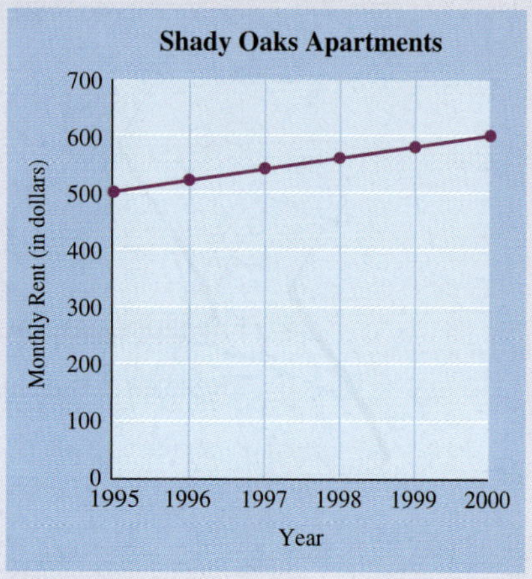


## MENTAL MATH

*Decide whether a line with the given slope slants upward, downward, horizontally, or vertically from left to right.*

**1.** $m = \frac{7}{6}$

**2.** $m = -3$

**3.** $m = 0$

**4.** $m$ is undefined

## Exercise Set 3.4

*Find the slope of the line that goes through the given points. See Examples 1 and 2.*

**1.** $(3, 2), (8, 11)$

**2.** $(1, 6), (7, 11)$

**3.** $(3, 1), (1, 8)$

**4.** $(2, 9), (6, 4)$

**5.** $(-2, 8), (4, 3)$

**6.** $(3, 7), (-2, 11)$

**7.** $(-2, -6), (4, -4)$

**8.** $(-3, -4), (-1, 6)$

**9.** $(-3, -1), (-12, 11)$

**10.** $(3, -1), (-6, 5)$

**11.** $(-2, 5), (3, 5)$

**12.** $(4, 2), (4, 0)$

**13.** $(-1, 1), (-1, -5)$

**14.** $(-2, -5), (3, -5)$

**15.** $(0, 6), (-3, 0)$

**16.** $(5, 2), (0, 5)$

**17.** $(-1, 2), (-3, 4)$

**18.** $(3, -2), (-1, -6)$

*Two lines are graphed on each set of axes. Decide whether $l_1$ or $l_2$ has the greater slope.*

**19.**

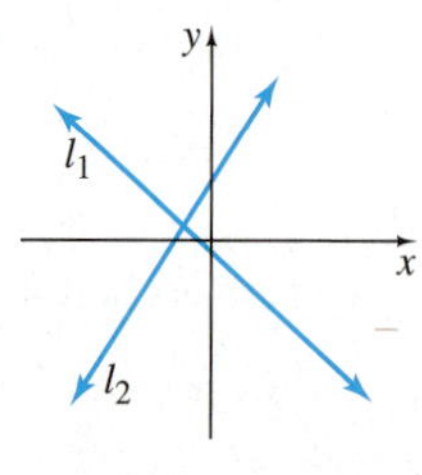


**20.**

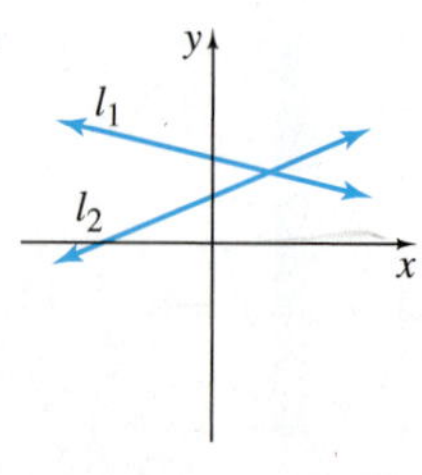

**21.**

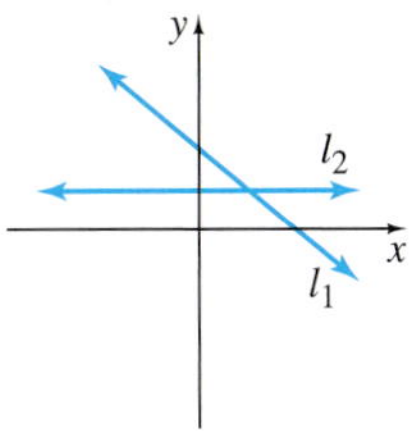


**22.**

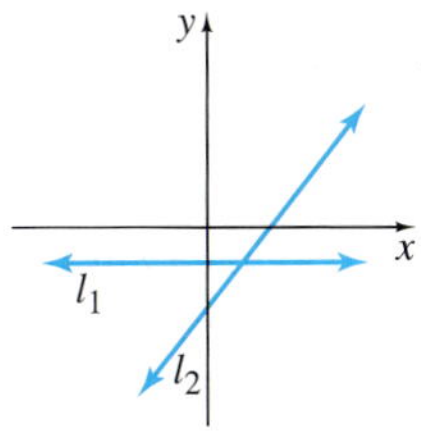


**23.**

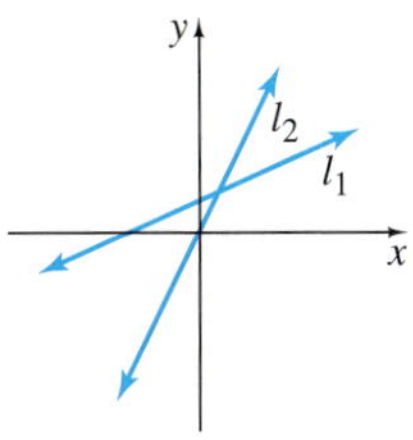


**24.**

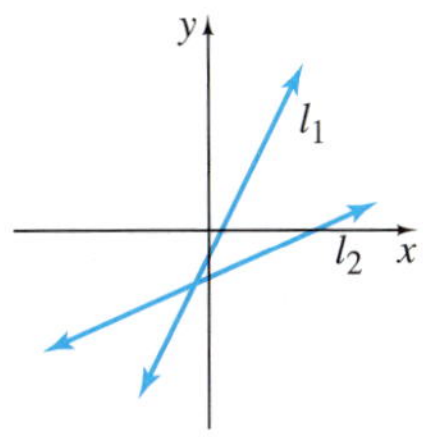


**25.** Each line below has negative slope.

**a.** Find the slope of each line.

**b.** Use the result of Part **a** to fill in the blank. For lines with negative slopes, the steeper line has the ____ (greater/lesser) slope.

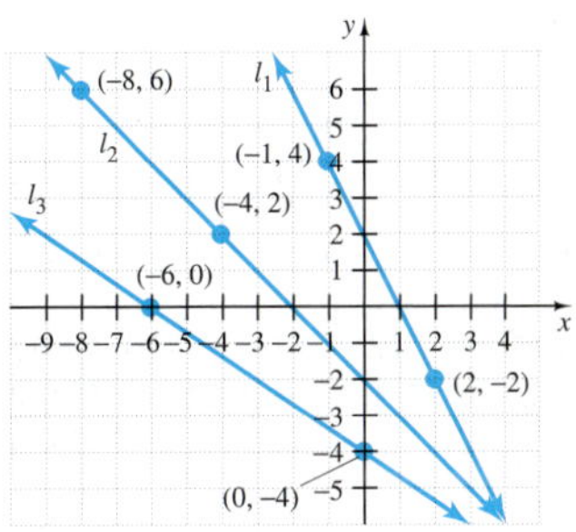


*Find the slope and the y-intercept of each line. See Examples 3 and 4.*

**26.** $f(x) = 5x - 2$

**27.** $f(x) = -2x + 6$

**28.** $2x + y = 7$

**29.** $-5x + y = 10$

**30.** $2x - 3y = 10$

**31.** $-3x - 4y = 6$

**32.** $f(x) = \frac{1}{2}x$

**33.** $f(x) = -\frac{1}{4}x$

*Match each graph with its equation.*

**A**

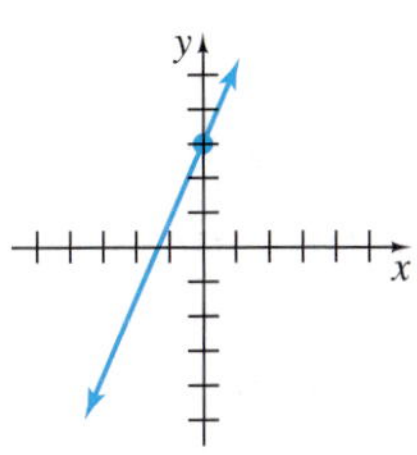


**B**

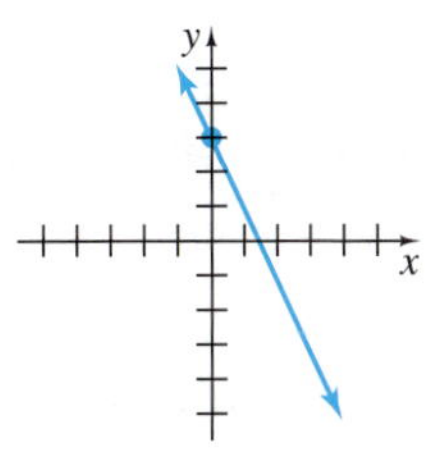


**C**

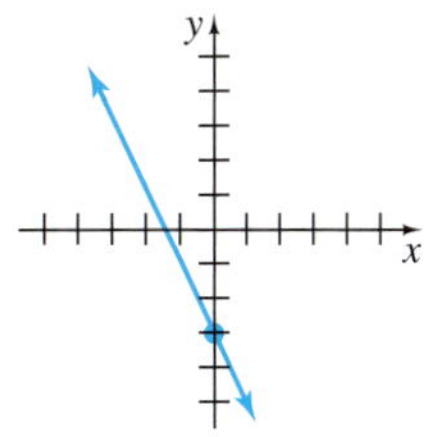


**D**

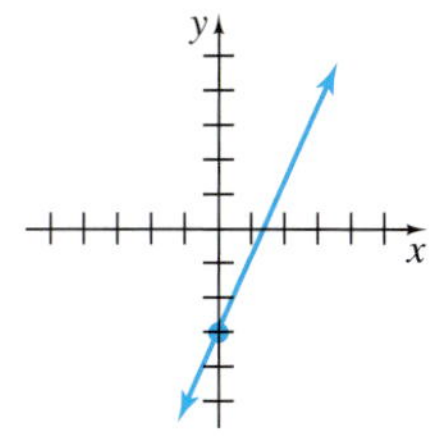


**34.** $f(x) = 2x + 3$

**35.** $f(x) = 2x - 3$

**36.** $f(x) = -2x + 3$

**37.** $f(x) = -2x - 3$

*Find the slope of each line. See Examples 6 and 7.*

**38.** $x = 1$

**39.** $y = -2$

**40.** $y = -3$

**41.** $x = 4$

**42.** $x + 2 = 0$

**43.** $y - 7 = 0$

**44.** Explain how merely looking at a line can tell us whether its slope is negative, positive, undefined, or zero.

**45.** Explain why the graph of $y = b$ is a horizontal line.

*Find the slope and the y-intercept of each line.*

**46.** $f(x) = -x + 5$

**47.** $f(x) = x + 2$

**48.** $-6x + 5y = 30$

**49.** $4x - 7y = 28$

**50.** $3x + 9 = y$

**51.** $2y - 7 = x$

**52.** $y = 4$

**53.** $x = 7$

**54.** $f(x) = 7x$

**55.** $f(x) = \frac{1}{7}x$

**56.** $6 + y = 0$

**57.** $x - 7 = 0$

**58.** $2 - x = 3$

**59.** $2y + 4 = -7$

*Determine whether the lines are parallel, perpendicular, or neither. See Example 8.*

**60.** $f(x) = -3x + 6$
$g(x) = 3x + 5$

**61.** $f(x) = 5x - 6$
$g(x) = 5x + 2$

**62.** $-4x + 2y = 5$
$2x - y = 7$

**63.** $2x - y = -10$
$2x + 4y = 2$

**64.** $-2x + 3y = 1$
$3x + 2y = 12$

**65.** $x + 4y = 7$
$2x - 5y = 0$

**66.** Explain whether two lines, both with positive slopes, can be perpendicular.

**67.** Explain why it is reasonable that nonvertical parallel lines have the same slope.

*Determine the slope of each line.*

**68.**

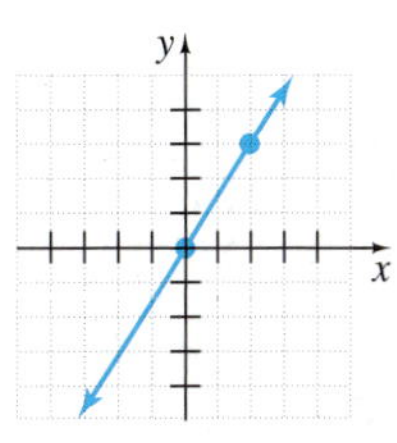


**69.**

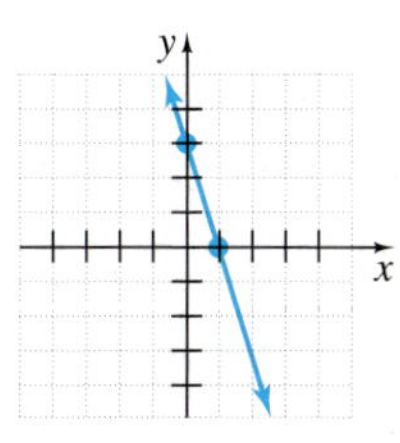

**70.**

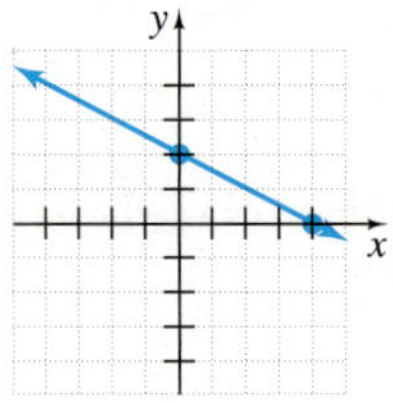


**71.**

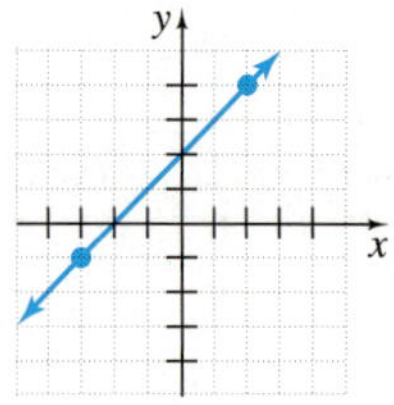


*Find each slope.*

**72.** Find the pitch, or slope, of the roof shown.

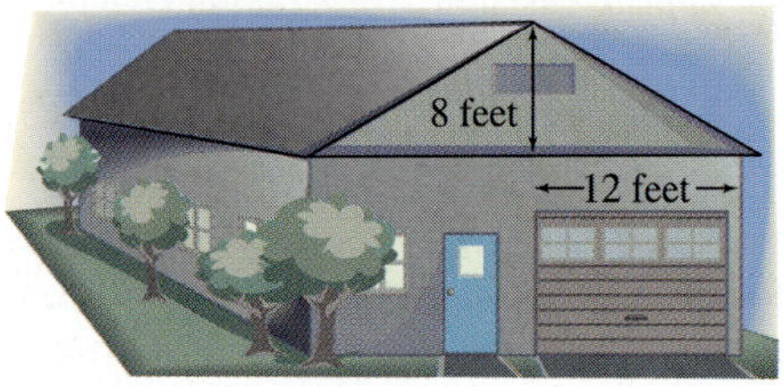


**73.** Upon takeoff, a Delta Airlines jet climbs to 3 miles as it passes over 25 miles of land below it. Find the slope of its climb.

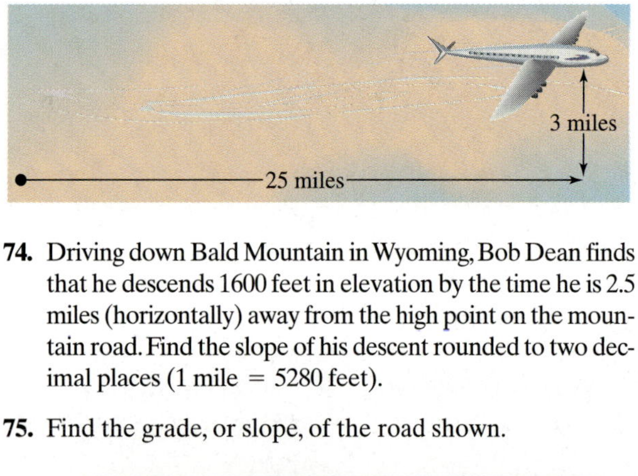


**74.** Driving down Bald Mountain in Wyoming, Bob Dean finds that he descends 1600 feet in elevation by the time he is 2.5 miles (horizontally) away from the high point on the mountain road. Find the slope of his descent rounded to two decimal places (1 mile = 5280 feet).

**75.** Find the grade, or slope, of the road shown.

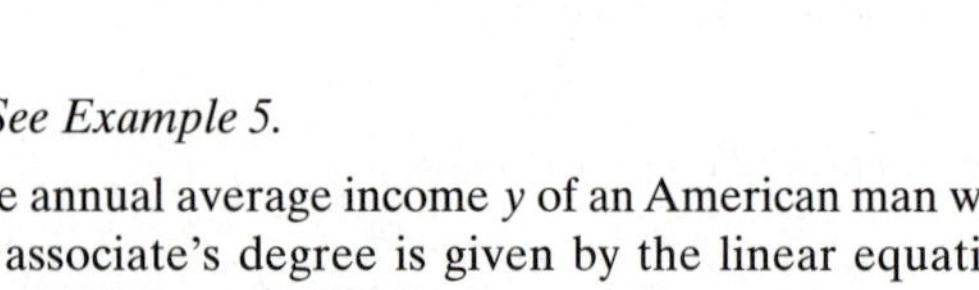


*Solve. See Example 5.*

**76.** The annual average income $y$ of an American man with an associate's degree is given by the linear equation $y = 1431.5x + 31{,}775.2$, where $x$ is the number of years after 1992. (*Source:* Based on data from the U.S. Bureau of the Census, 1992–1996)

**a.** Find the average income of an American man with an associate's degree in 1996.

**b.** Find and interpret the slope of the equation.

**c.** Find and interpret the $y$-intercept of the equation.

**77.** The annual income of an American woman with a bachelor's degree is given by the linear equation $y = 1054.7x + 23{,}285.9$, where $x$ is the number of years after 1991. (*Source:* Based on data from the U.S. Bureau of the Census, 1991–1996)

**a.** Find the average income of an American woman with a bachelor's degree in 1996.

**b.** Find and interpret the slope of the equation.

**c.** Find and interpret the $y$-intercept of the equation.

**78.** One of the top ten occupations in terms of job growth in the next few years is expected to be home health aide. The number of people $y$ in thousands employed as home health aides in the United States can be estimated by the linear equation

$$378x - 10y = -4950,$$

where $x$ is the number of years after 1996. (*Source:* Based on projections from the U.S. Bureau of Labor Statistics, 1996–2006)

**a.** Find the slope and $y$-intercept of the linear equation.

**b.** What does the slope mean in this context?

**c.** What does the $y$-intercept mean in this context?

**79.** One of the faster growing occupations over the next few years is expected to be paralegal. The number of people $y$ in thousands employed as paralegals in the United States can be estimated by the linear equation $-76x + 10y = 1130$, where $x$ is the number of years after 1996. (*Source:* Based on projections from the U.S. Bureau of Labor Statistics, 1996–2006)

**a.** Find the slope and $y$-intercept of the linear equation.

**b.** What does the slope mean in this context?

**c.** What does the $y$-intercept mean in this context?

**80.** In an earlier section, it was given that the yearly cost of tuition and required fees for attending a public four-year college full-time can be estimated by the linear function

$$f(x) = 186.1x + 2030$$

where $x$ is the number of years after 1990 and $f(x)$ is the total cost. (*Source:* U.S. National Center for Education Statistics)

**a.** Find and interpret the slope of this equation.

**b.** Find and interpret the $y$-intercept of this equation.

**81.** If an earlier section, it was given that the yearly cost of tuition and required fees for attending a public two-year college full-time can be estimated by the linear function

$$f(x) = 72.9x + 785.2$$

where $x$ is the number of years after 1990 and $f(x)$ is the total cost. (*Source:* U.S. National Center for Education Statistics)

**a.** Find and interpret the slope of this equation.
**b.** Find and interpret the $y$-intercept of this equation.

*Solve.*

△ **82.** Find the slope of a line parallel to the line $f(x) = -\frac{7}{2}x - 6$.

△ **83.** Find the slope of a line parallel to the line $f(x) = x$.

△ **84.** Find the slope of a line perpendicular to the line $f(x) = -\frac{7}{2}x - 6$.

△ **85.** Find the slope of a line perpendicular to the line $f(x) = x$.

△ **86.** Find the slope of a line parallel to the line $5x - 2y = 6$.

△ **87.** Find the slope of a line parallel to the line $-3x + 4y = 10$.

△ **88.** Find the slope of a line perpendicular to the line $5x - 2y = 6$.

**89.** The following graph shows the altitude of a seagull in flight over a time period of 30 seconds.

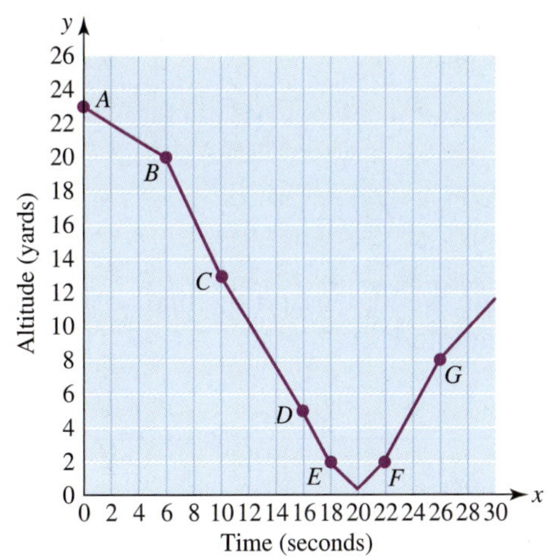


**a.** Find the coordinates of point $B$.
**b.** Find the coordinates of point $C$.
**c.** Find the rate of change of altitude between points $B$ and $C$. (Recall that the rate of change between points is the slope between points. This rate of change will be in yards per second.)
**d.** Find the rate of change of altitude (in yards per second) between points $F$ and $G$.

**90.** Professional plumbers suggest that a sewer pipe should be sloped 0.25 inch for every foot. Find the recommended slope for a sewer pipe. (*Source: Rules of Thumb* by Tom Parker, 1983, Houghton Mifflin Company)

**91.** Support the result of Exercise 62 by graphing the pair of equations on a graphing calculator.

**92.** Support the result of Exercise 63 by graphing the pair of equations on a graphing calculator. (*Hint:* Use the window showing $[-15, 15]$ on the $x$-axis and $[-10, 10]$ on the $y$-axis.)

**93. a.** On a single screen, graph $y = \frac{1}{2}x + 1$, $y = x + 1$ and $y = 2x + 1$. Notice the change in slope for each graph.
**b.** On a single screen, graph $y = -\frac{1}{2}x + 1$, $y = -x + 1$ and $y = -2x + 1$. Notice the change in slope for each graph.
**c.** Determine whether the following statement is true or false for slope $m$ of a given line. As $|m|$ becomes greater, the line becomes steeper.

## REVIEW EXERCISES

*Recall the formula*

$$\textit{Probability of an event} = \frac{\textit{number of ways that the event can occur}}{\textit{number of possible outcomes}}$$

*Suppose these cards are shuffled and one card is turned up. Find the possibility of selecting each letter.*

**94.** P(R)
**95.** P(B)
**96.** P(E)
**97.** P(I or T)
**98.** P(selecting a letter of the alphabet)
**99.** P(vowel)

*Simplify and solve for y. See Section 2.3.*

**100.** $y - 2 = 5(x + 6)$
**101.** $y - 0 = -3[x - (-10)]$
**102.** $y - (-1) = 2(x - 0)$
**103.** $y - 9 = -8[x - (-4)]$

## 3.5 EQUATIONS OF LINES

CD-ROM SSM SSG Video

▶ **OBJECTIVES**

1. Use the slope–intercept form to write the equation of a line.
2. Graph a line using its slope and $y$-intercept.
3. Use the point–slope form to write the equation of a line.
4. Write equations of vertical and horizontal lines.
5. Find equations of parallel and perpendicular lines.

**1** In the last section, we learned that the slope–intercept form of a linear equation is $y = mx + b$. When an equation is written in this form, the slope of the line is the same as the coefficient $m$ of $x$. Also, the $y$-intercept of the line is the same as the constant term $b$. For example, the slope of the line defined by $y = 2x + 3$ is, 2, and its $y$-intercept is 3.

We may also use the slope–intercept form to write the equation of a line given its slope and $y$-intercept.

**Example 1** Write an equation of the line with $y$-intercept $(0, -3)$ and slope of $\frac{1}{4}$.

*Solution* We are given the slope and the $y$-intercept. Let $m = \frac{1}{4}$ and $b = -3$, and write the equation in slope–intercept form, $y = mx + b$.

$$y = mx + b$$

$$y = \frac{1}{4}x + (-3) \qquad \text{Let } m = \frac{1}{4} \text{ and } b = -3.$$

$$y = \frac{1}{4}x - 3 \qquad \text{Simplify.}$$

**2** Given the slope and $y$-intercept of a line, we may graph the line as well as write its equation. Let's graph the line from Example 1.

**Example 2** Graph $y = \frac{1}{4}x - 3$.

*Solution* Recall that the slope of the graph of $y = \frac{1}{4}x - 3$ is $\frac{1}{4}$ and the $y$-intercept is $(0, -3)$. To graph the line, we first plot the $y$-intercept $(0, -3)$. To find another point on the line, we recall that slope is $\frac{\text{rise}}{\text{run}} = \frac{1}{4}$. Another point may then be plotted by starting at $(0, -3)$, rising 1 unit up, and then running 4 units to the right. We are now at the point $(4, -2)$. The graph is the line through these two points.

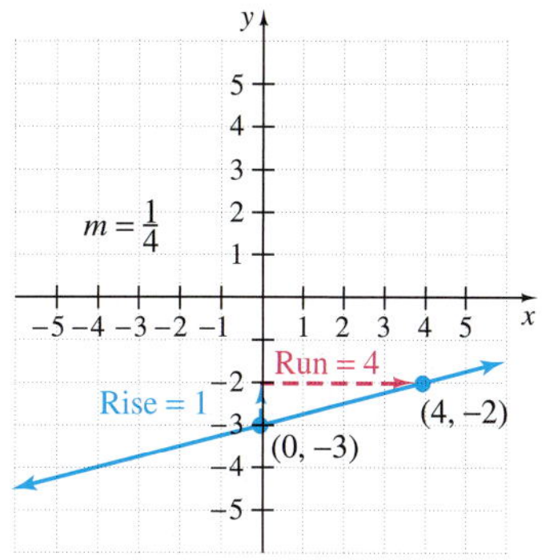


## Example 3 Graph $2x + 3y = 12$.

*Solution* First, we solve the equation for $y$ to write it in slope–intercept form. In slope–intercept form, the equation is $y = -\frac{2}{3}x + 4$. Next we plot the $y$-intercept $(0, 4)$. To find another point on the line, we use the slope $-\frac{2}{3}$, which can be written as $\frac{\text{rise}}{\text{run}} = \frac{-2}{3}$. We start at $(0, 4)$ and move down 2 units since the numerator of the slope is $-2$; then we move 3 units to the right since the denominator of the slope is 3. We arrive at the point $(3, 2)$. The line through these points is the graph.

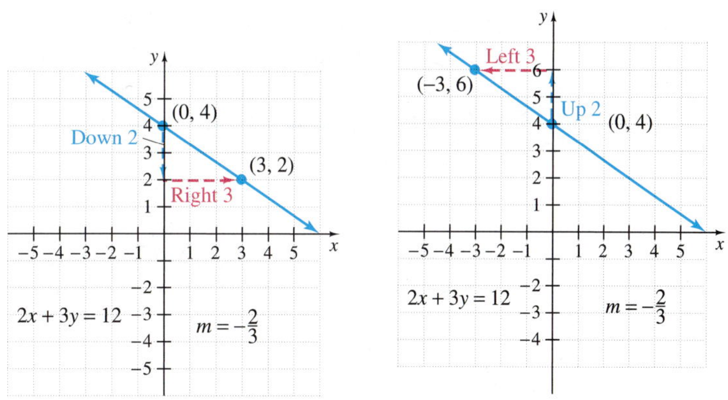


The slope $\frac{-2}{3}$ can also be written as $\frac{2}{-3}$, so to find another point in Example 3 we could start at $(0, 4)$ and move up 2 units and then 3 units to the left. We would stop at the point $(-3, 6)$. The line through $(-3, 6)$ and $(0, 4)$ is the same line as shown previously through $(3, 2)$ and $(0, 4)$.

**3** When the slope of a line and a point on the line are known, the equation of the line can also be found. To do this, use the slope formula to write the slope of a line that passes through points $(x_1, y_1)$ and $(x, y)$. We have

$$m = \frac{y - y_1}{x - x_1}$$

Multiply both sides of this equation by $x - x_1$ to obtain

$$y - y_1 = m(x - x_1)$$

This form is called the **point–slope form** of the equation of a line.

### POINT–SLOPE FORM OF THE EQUATION OF A LINE

The point–slope form of the equation of a line is $y - y_1 = m(x - x_1)$, where $m$ is the slope of the line and $(x_1, y_1)$ is a point on the line.

**Example 4** Find an equation of the line with slope −3 containing the point (1, −5). Write the equation in slope–intercept form $y = mx + b$.

*Solution* Because we know the slope and a point of the line, we use the point–slope form with $m = -3$ and $(x_1, y_1) = (1, -5)$.

$$\begin{aligned} y - y_1 &= m(x - x_1) && \text{Point–slope form.} \\ y - (-5) &= -3(x - 1) && \text{Let } m = -3 \text{ and } (x_1, y_1) = (1, -5). \\ y + 5 &= -3x + 3 && \text{Apply the distributive property.} \\ y &= -3x - 2 && \text{Write in slope–intercept form.} \end{aligned}$$

In slope–intercept form, the equation is $y = -3x - 2$.

**Example 5** Find an equation of the line through points (4, 0) and (−4, −5). Write the equation using function notation.

*Solution* First, find the slope of the line.

$$m = \frac{-5 - 0}{-4 - 4} = \frac{-5}{-8} = \frac{5}{8}$$

Next, make use of the point–slope form. Replace $(x_1, y_1)$ by either (4, 0) or (−4, −5) in the point–slope equation. We will choose the point (4, 0). The line through (4, 0) with slope $\frac{5}{8}$ is

$$\begin{aligned} y - y_1 &= m(x - x_1) && \text{Point–slope form.} \\ y - 0 &= \frac{5}{8}(x - 4) && \text{Let } m = \frac{5}{8} \text{ and } (x_1, y_1) = (4, 0). \\ 8y &= 5(x - 4) && \text{Multiply both sides by 8.} \\ 8y &= 5x - 20 && \text{Apply the distributive property.} \end{aligned}$$

To write the equation using function notation, we solve for $y$.

$$\begin{aligned} 8y &= 5x - 20 && \\ y &= \frac{5}{8}x - \frac{20}{8} && \text{Divide both sides by 8.} \\ f(x) &= \frac{5}{8}x - \frac{5}{2} && \text{Write using function notation.} \end{aligned}$$

The point–slope form of an equation is very useful for solving real-world problems.

## Example 6 PREDICTING SALES

Southern Star Realty is an established real estate company that has enjoyed constant growth in sales since 1990. In 1992 the company sold 200 houses, and in 1997 the company sold 275 houses. Use these figures to predict the number of houses this company will sell in the year 2006.

*Solution*

1. UNDERSTAND. Read and reread the problem. Then let

   $x$ = the number of years after 1990 and

   $y$ = the number of houses sold in the year corresponding to $x$.

   The information provided then gives the ordered pairs $(2, 200)$ and $(7, 275)$. To better visualize the sales of Southern Star Realty, we graph the linear equation that passes through the points $(2, 200)$ and $(7, 275)$.

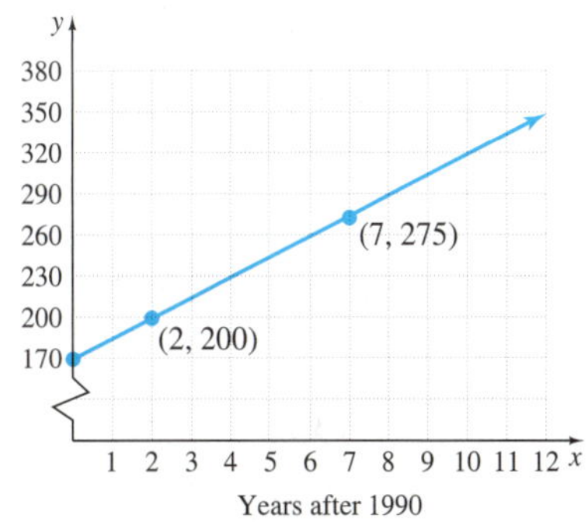


2. TRANSLATE. We write a linear equation that passes through the points $(2, 200)$ and $(7, 275)$. To do so, we first find the slope of the line.

$$m = \frac{275 - 200}{7 - 2} = \frac{75}{5} = 15$$

   Then, using the point–slope form to write the equation, we have

$$y - y_1 = m(x - x_1)$$

$$y - 200 = 15(x - 2) \quad \text{Let } m = 15 \text{ and } (x_1, y_1) = (2, 200).$$

$$y - 200 = 15x - 30 \quad \text{Multiply.}$$

$$y = 15x + 170 \quad \text{Add 200 to both sides.}$$

3. SOLVE. To predict the number of houses sold in the year 2006, we use $y = 15x + 170$ and complete the ordered pair $(16, \quad)$, since $2006 - 1990 = 16$.

$$y = 15(16) + 170 \quad \text{Let } x = 16.$$

$$y = 410$$

4. INTERPRET.

   **Check:** Verify that the point $(16, 410)$ is a point on the line graphed in step 1.

   **State:** Southern Star Realty should expect to sell 410 houses in the year 2006.

**4** A few special types of linear equations are linear equations whose graphs are vertical and horizontal lines.

**Example 7** Find the equation of the horizontal line containing the point (2, 3).

*Solution* Recall that a horizontal line has an equation of the form $y = b$. Since the line contains the point (2, 3), the equation is $y = 3$.

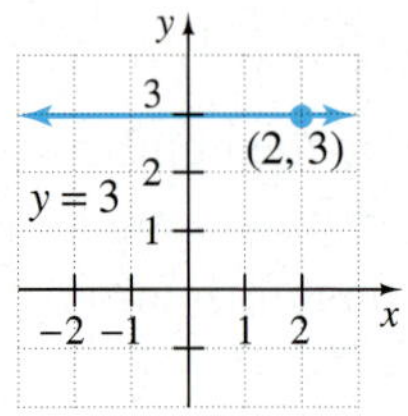


**Example 8** Find the equation of the line containing the point (2, 3) with undefined slope.

*Solution* Since the line has undefined slope, the line must be vertical. A vertical line has an equation of the form $x = c$, and since the line contains the point (2, 3), the equation is $x = 2$.

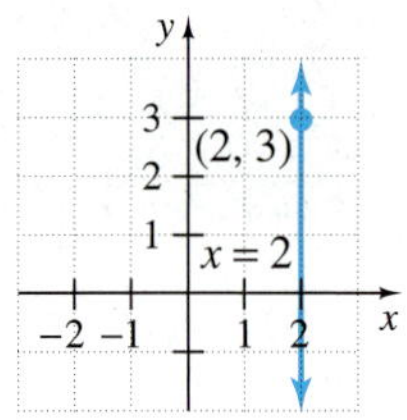


**5** Next, we find equations of parallel and perpendicular lines.

△ **Example 9** Find an equation of the line containing the point (4, 4) and parallel to the line $2x + 3y = -6$. Write the equation in standard form.

*Solution* Because the line we want to find is *parallel* to the line $2x + 3y = -6$, the two lines must have equal slopes. Find the slope of $2x + 3y = -6$ by writing it in the form $y = mx + b$.

$$2x + 3y = -6$$

$$3y = -2x - 6 \qquad \text{Subtract } 2x \text{ from both sides.}$$

$$y = \frac{-2x}{3} - \frac{6}{3} \qquad \text{Divide by 3.}$$

$$y = -\frac{2}{3}x - 2 \qquad \text{Write in slope-intercept form.}$$

The slope of this line is $-\frac{2}{3}$. Thus, a line parallel to this line will also have a slope of $-\frac{2}{3}$. The equation we are asked to find describes a line containing the point (4, 4) with a slope of $-\frac{2}{3}$. We use the point–slope form.

$$y - y_1 = m(x - x_1)$$

$$y - 4 = -\frac{2}{3}(x - 4) \qquad \text{Let } m = -\frac{2}{3},\ x_1 = 4, \text{ and } y_1 = 4.$$

$$3(y - 4) = -2(x - 4) \qquad \text{Multiply both sides by 3.}$$

$$3y - 12 = -2x + 8 \qquad \text{Apply the distributive property.}$$

$$2x + 3y = 20 \qquad \text{Write in standard form.}$$

**HELPFUL HINT**

Multiply both sides of the equation $2x + 3y = 20$ by $-1$, and it becomes $-2x - 3y = -20$. Both equations are in standard form, and their graphs are the same line.

**Example 10** Write a function that describes the line containing the point (4, 4) and is perpendicular to the line $2x + 3y = -6$.

*Solution* In the previous example, we found that the slope of the line $2x + 3y = -6$ is $-\frac{2}{3}$. A line perpendicular to this line will have a slope that is the negative reciprocal of $-\frac{2}{3}$, or $\frac{3}{2}$. From the point–slope equation, we have

$$y - y_1 = m(x - x_1)$$

$$y - 4 = \frac{3}{2}(x - 4) \qquad \text{Let } x_1 = 4,\ y_1 = 4 \text{ and } m = \frac{3}{2}.$$

$$2(y - 4) = 3(x - 4) \qquad \text{Multiply both sides by 2.}$$

$$2y - 8 = 3x - 12 \qquad \text{Apply the distributive property.}$$

$$2y = 3x - 4 \qquad \text{Add 8 to both sides.}$$

$$y = \frac{3}{2}x - 2 \qquad \text{Divide both sides by 2.}$$

$$f(x) = \frac{3}{2}x - 2 \qquad \text{Write using function notation.}$$

## FORMS OF LINEAR EQUATIONS

| | |
|---|---|
| $Ax + By = C$ | **Standard form** of a linear equation<br>$A$ and $B$ are not both 0. |
| $y = mx + b$ | **Slope–intercept form** of a linear equation<br>The slope is $m$, and the $y$-intercept is $(0, b)$. |
| $y - y_1 = m(x - x_1)$ | **Point–slope form** of a linear equation<br>The slope is $m$, and $(x_1, y_1)$ is a point on the line. |
| $y = c$ | **Horizontal line**<br>The slope is 0, and the $y$-intercept is $(0, c)$. |
| $x = c$ | **Vertical line**<br>The slope is undefined and the $x$-intercept is $(c, 0)$. |

## PARALLEL AND PERPENDICULAR LINES

Nonvertical parallel lines have the same slope. The product of the slopes of two nonvertical perpendicular lines is $-1$.

## SPOTLIGHT ON DECISION MAKING

Suppose you are a public health official. In 1993, the International Task Force for Disease Eradication (ITFDE) identified mumps as one of six infectious diseases that could probably be eradicated worldwide with current technology. The ITFDE defined "eradication" as reducing the incidence of a disease to zero. Does the graph of reported mumps cases in the United States support the possibility of U.S. mumps eradication? Explain.

Suppose U.S. officials would like to see mumps eradicated by 2010. If this goal does not currently seem possible, your department will increase eradication efforts with the launch of a new public awareness campaign. Will the new public awareness campaign be necessary? (*Hint:* Use the data for the years 1996 and 1997 to help you decide.)

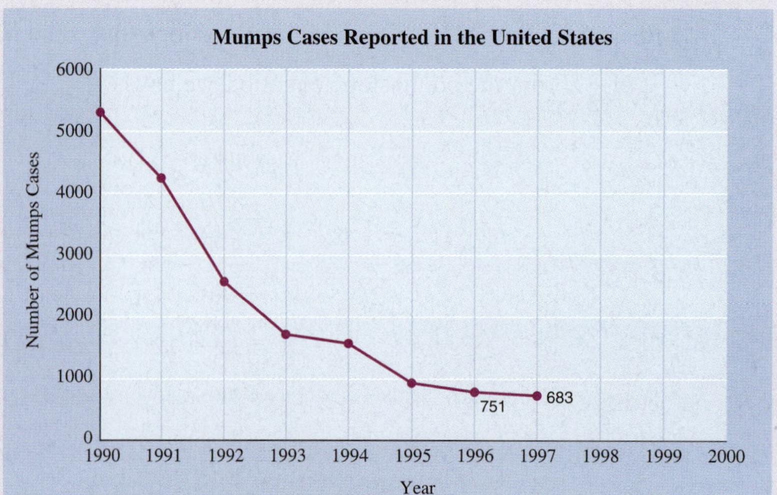


*Source*: :MMWE Summary of Notifiable Diseases, United States, 1997

## MENTAL MATH

*State the slope and the y-intercept of each line with the given equation.*

1. $y = -4x + 12$
2. $y = \frac{2}{3}x - \frac{7}{2}$
3. $y = 5x$
4. $y = -x$
5. $y = \frac{1}{2}x + 6$
6. $y = -\frac{2}{3}x + 5$

*Decide whether the lines are parallel, perpendicular, or neither.*

7. $y = 12x + 6$; $y = 12x - 2$
8. $y = -5x + 8$; $y = -5x - 8$
9. $y = -9x + 3$; $y = \frac{3}{2}x - 7$
10. $y = 2x - 12$; $y = \frac{1}{2}x - 6$

## Exercise Set 3.5

*Use the slope–intercept form of the linear equation to write the equation of each line with the given slope and y-intercept. See Example 1.*

1. Slope −1; y-intercept (0, 1)
2. Slope $\frac{1}{2}$; y-intercept (0, −6)
3. Slope 2; y-intercept $\left(0, \frac{3}{4}\right)$
4. Slope −3; y-intercept $\left(0, -\frac{1}{5}\right)$
5. Slope $\frac{2}{7}$; y-intercept (0, 0)
6. Slope $-\frac{4}{5}$; y-intercept (0, 0)

*Graph each linear equation. See Examples 2 and 3.*

7. $y = 5x$
8. $y = 2x + 12$
9. $x + y = 7$
10. $3x + y = 9$
11. $-3x + 2y = 3$
12. $-2x + 5y = -16$

*Find an equation of the line with the given slope and containing the given point. Write the equation in slope–intercept form. See Example 4.*

13. Slope 3; through (1, 2)
14. Slope 4; through (5, 1)
15. Slope −2; through (1, −3)
16. Slope −4; through (2, −4)
17. Slope $\frac{1}{2}$; through (−6, 2)
18. Slope $\frac{2}{3}$; through (−9, 4)
19. Slope $-\frac{9}{10}$; through (−3, 0)
20. Slope $-\frac{1}{5}$; through (4, −6)

*Find an equation of each line graphed. Write the equation in standard form.*

21.

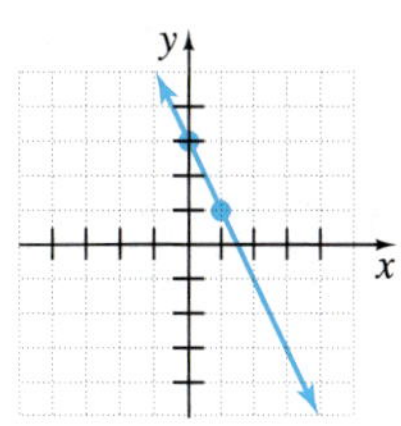


22.

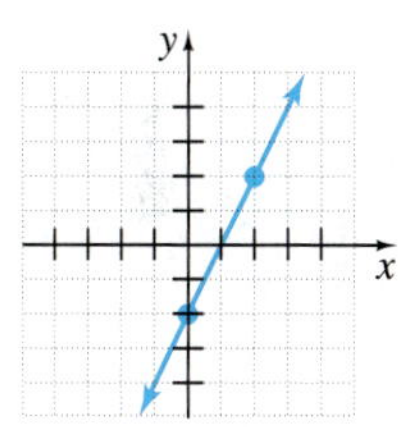


23.

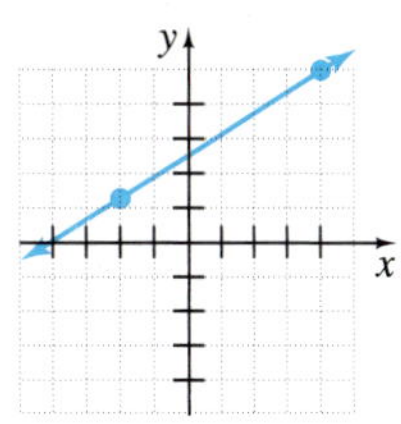


24.

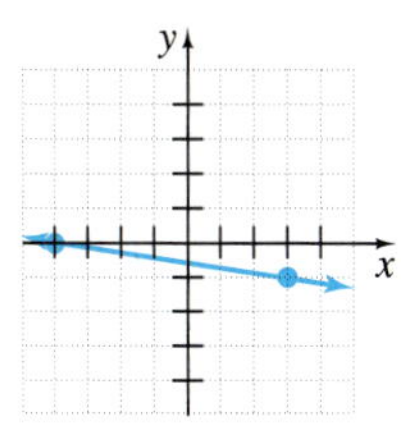


*Find an equation of the line passing through the given points. Use function notation to write the equation. See Example 5.*

25. (2, 0), (4, 6)
26. (3, 0), (7, 8)
27. (−2, 5), (−6, 13)
28. (7, −4), (2, 6)
29. (−2, −4), (−4, −3)
30. (−9, −2), (−3, 10)
31. (−3, −8), (−6, −9)
32. (8, −3), (4, −8)
33. Describe how to check to see if the graph of $2x - 4y = 7$ passes through the points (1.4, −1.05) and (0, −1.75). Then follow your directions and check these points.

*Use the graph of the following function $f(x)$ to find each value.*

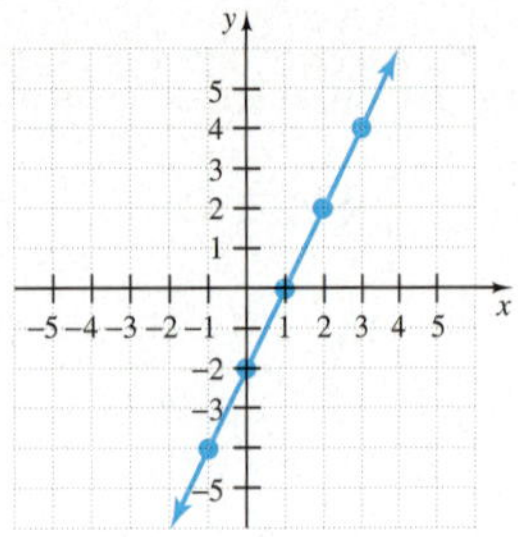


34. $f(1)$
35. $f(0)$
36. $f(-1)$
37. $f(2)$
38. Find $x$ such that $f(x) = 4$.
39. Find $x$ such that $f(x) = -6$.

*Write an equation of each line. See Examples 7 and 8.*

40. Vertical; through $(2, 6)$
41. Slope 0; through $(-2, -4)$
42. Horizontal; through $(-3, 1)$
43. Vertical; through $(4, 7)$
44. Undefined slope; through $(0, 5)$
45. Horizontal; through $(0, 5)$
46. Answer the following true or false. A vertical line is always perpendicular to a horizontal line.

*Find an equation of each line. Write the equation using function notation. See Examples 9 and 10.*

47. Through $(3, 8)$; parallel to $f(x) = 4x - 2$
48. Through $(1, 5)$; parallel to $f(x) = 3x - 4$
49. Through $(2, -5)$; perpendicular to $3y = x - 6$
50. Through $(-4, 8)$; perpendicular to $2x - 3y = 1$
51. Through $(-2, -3)$; parallel to $3x + 2y = 5$
52. Through $(-2, -3)$; perpendicular to $3x + 2y = 5$

*Find the equation of each line. Write the equation in standard form unless indicated otherwise.*

53. Slope 2; through $(-2, 3)$
54. Slope 3; through $(-4, 2)$
55. Through $(1, 6)$ and $(5, 2)$; use function notation.
56. Through $(2, 9)$ and $(8, 6)$
57. With slope $-\frac{1}{2}$; $y$-intercept 11
58. With slope $-4$; $y$-intercept $\frac{2}{9}$; use function notation.
59. Through $(-7, -4)$ and $(0, -6)$
60. Through $(2, -8)$ and $(-4, -3)$
61. Slope $-\frac{4}{3}$; through $(-5, 0)$
62. Slope $-\frac{3}{5}$; through $(4, -1)$
63. Vertical line; through $(-2, -10)$
64. Horizontal line; through $(1, 0)$
65. Through $(6, -2)$; parallel to the line $2x + 4y = 9$
66. Through $(8, -3)$; parallel to the line $6x + 2y = 5$
67. Slope 0; through $(-9, 12)$
68. Undefined slope; through $(10, -8)$
69. Through $(6, 1)$; parallel to the line $8x - y = 9$
70. Through $(3, 5)$; perpendicular to the line $2x - y = 8$
71. Through $(5, -6)$; perpendicular to $y = 9$
72. Through $(-3, -5)$; parallel to $y = 9$
73. Through $(2, -8)$ and $(-6, -5)$; use function notation.
74. Through $(-4, -2)$ and $(-6, 5)$; use function notation.
75. Del Monte Fruit Company recently released a new applesauce. By the end of its first year, profits on this product amounted to \$30,000. The anticipated profit for the end of the fourth year is \$66,000. The ratio of change in time to change in profit is constant. Let $x$ be years and $P$ be profit.
    **a.** Write a linear function $P(x)$ that expresses profit as a function of time.
    **b.** Use this function to predict the company's profit at the end of the seventh year.
    **c.** Predict when the profit should reach \$126,000.
76. The value of a computer bought in 1996 depreciates, or decreases, as time passes. Two years after the computer was bought, it was worth \$2600; 4 years after it was bought, it was worth \$1000.
    **a.** If this relationship between number of years past 1996 and value of computer is linear, write an equation describing this relationship. [Use ordered pairs of the form (years past 1996, value of computer).]
    **b.** Use this equation to estimate the value of the computer in the year 2001.
77. The Pool Fun Company has learned that, by pricing a newly released Fun Noodle at \$3, sales will reach 10,000 Fun Noodles per day during the summer. Raising the price to \$5 will cause the sales to fall to 8000 Fun Noodles per day.
    **a.** Assume that the relationship between sales price and number of Fun Noodles sold is linear and write an equation describing this relationship.
    **b.** Predict the daily sales of Fun Noodles if the price is \$3.50.
78. The value of a building bought in 1980 appreciates, or increases, as time passes. Seven years after the building was bought, it was worth \$165,000; 12 years after it was bought, it was worth \$180,000.
    **a.** If this relationship between number of years past 1980 and value of building is linear, write an equation describing this relationship. [Use ordered pairs of the form (years past 1980, value of building).]
    **b.** Use this equation to estimate the value of the building in the year 2000.

**79.** In 1994, the median price of an existing home in the United States was $109,900. In 1998, the median price of an existing home was $128,400. Let $y$ be the median price of an existing home in the year $x$, where $x = 0$ represents 1994. (*Source:* National Association of REALTORS®)

**a.** Write a linear equation that models the median existing home price in terms of the year $x$. [*Hint:* The line must pass through the points (0, 109,900) and (4, 128,400)]

**b.** Use this equation to predict the median existing home price for the year 2008.

**80.** The number of births (in thousands) in the United States in 1997 was 3895. The number of births (in thousands) in the United States in 1991 was 4111. Let $y$ be the number of births (in thousands) in the year $x$, where $x = 0$ represents 1991. (*Source:* National Center for Health Statistics)

**a.** Write a linear equation that models the number of births (in thousands) in terms of the year $x$. (See hint for Exercise 79**a.**)

**b.** Use this equation to predict the number of births in the United States for the year 2010.

**81.** The number of people employed in the United States as medical assistants was 225 thousand in 1996. By the year 2006, this number is expected to rise to 391 thousand. Let $y$ be the number of medical assistants (in thousands) employed in the United States in the year $x$, where $x = 0$ represents 1996. (*Source:* Bureau of Labor Statistics)

**a.** Write a linear equation that models the number of people (in thousands) employed as medical assistants in the year $x$. (See hint for Exercise 79**a.**)

**b.** Use this equation to estimate the number of people who will be employed as medical assistants in the year 2004.

**82.** The number of people employed in the United States as systems analysts was 506 thousand in 1996. By the year 2006, this number is expected to rise to 1025 thousand. Let $y$ be the number of systems analysts (in thousands) employed in the United States in the year $x$, where $x = 0$ represents 1996. (*Source:* Bureau of Labor Statistics)

**a.** Write a linear equation that models the number of people (in thousands) employed as systems analysts in the year $x$. (See hint for Exercise 79**a.**)

**b.** Use this equation to estimate the number of people who will be employed as systems analysts in the year 2002.

*Use a graphing calculator with a TRACE feature to see the results of each exercise.*

**83.** Exercise 55; graph the function and verify that it passes through (1, 6) and (5, 2).

**84.** Exercise 56; graph the equation and verify that it passes through (2, 9) and (8, 6).

**85.** Exercise 61; graph the equation. See that it has a negative slope and passes through (−5, 0).

**86.** Exercise 62; graph the equation. See that it has a negative slope and passes through (4, −1).

## REVIEW EXERCISES

*Solve and graph the solution. See Section 2.4.*

**87.** $2x - 7 \leq 21$

**88.** $-3x + 1 > 0$

**89.** $5(x - 2) \geq 3(x - 1)$

**90.** $-2(x + 1) \leq -x + 10$

**91.** $\frac{x}{2} + \frac{1}{4} < \frac{1}{8}$

**92.** $\frac{x}{5} - \frac{3}{10} \geq \frac{x}{2} - 1$

## *A Look Ahead*

**Example**

Find an equation of the perpendicular bisector of the line segment whose endpoints are (2, 6) and (0, −2).

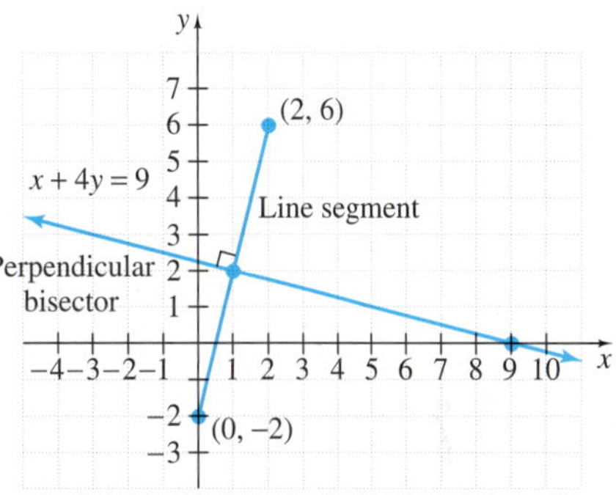


*Solution*

A perpendicular bisector is a line that contains the midpoint of the given segment and is perpendicular to the segment.

**Step 1:** The midpoint of the segment with endpoints (2, 6) and (0, −2) is (1, 2).

**Step 2:** The slope of the segment containing points (2, 6) and (0, −2) is 4.

**Step 3:** A line perpendicular to this line segment will have slope of $-\frac{1}{4}$.

**Step 4:** The equation of the line through the midpoint (1, 2) with a slope of $-\frac{1}{4}$ will be the equation of the perpendicular bisector. This equation in standard form is $x + 4y = 9$.

*Find an equation of the perpendicular bisector of the line segment whose endpoints are given. See the previous example.*

△ **93.** (3, −1); (−5, 1)

△ **94.** (−6, −3); (−8, −1)

△ **95.** (−2, 6); (−22, −4)

△ **96.** (5, 8); (7, 2)

△ **97.** (2, 3); (−4, 7)

△ **98.** (−6, 8); (−4, −2)

## 3.6 GRAPHING LINEAR INEQUALITIES

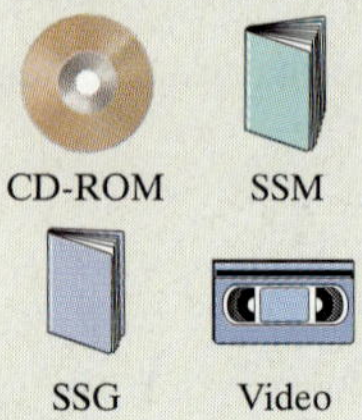


▶ **OBJECTIVES**

1. Graph linear inequalities.
2. Graph the intersection or union of two linear inequalities.

**1** Recall that the graph of a linear equation in two variables is the graph of all ordered pairs that satisfy the equation, and we determined that the graph is a line. Here we graph **linear inequalities** in two variables; that is, we graph all the ordered pairs that satisfy the inequality.

If the equal sign in a linear equation in two variables is replaced with an inequality symbol, the result is a linear inequality in two variables.

***Examples of Linear Inequalities in Two Variables***

$$3x + 5y \geq 6 \qquad 2x - 4y < -3$$

$$4x > 2 \qquad y \leq 5$$

To graph the linear inequality $x + y < 3$, for example, we first graph the related **boundary** equation $x + y = 3$. The resulting boundary line contains all ordered pairs the sum of whose coordinates is 3. This line separates the plane into two **half-planes**. All points "above" the boundary line $x + y = 3$ have coordinates that satisfy the inequality $x + y > 3$, and all points "below" the line have coordinates that satisfy the inequality $x + y < 3$.

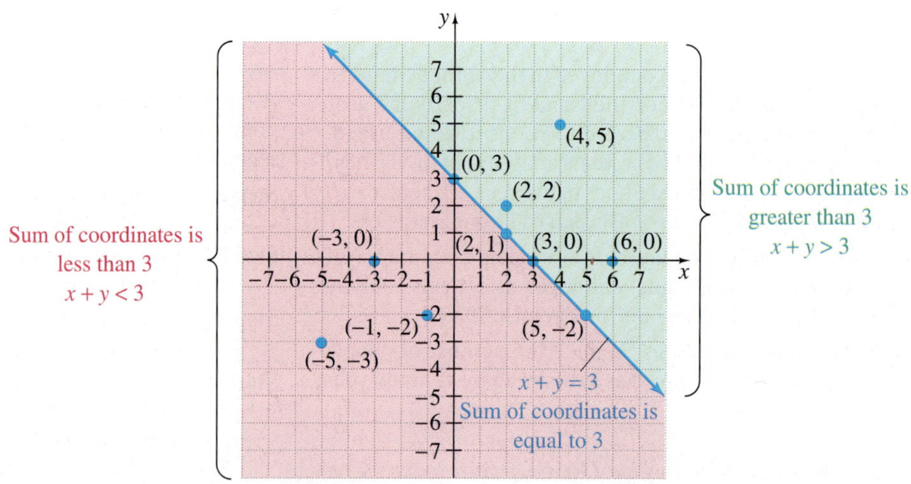


The graph, or **solution region**, for $x + y < 3$, then, is the half-plane below the boundary line and is shown shaded on page 173. The boundary line is shown dashed since it is not a part of the solution region. These ordered pairs on this line satisfy $x + y = 3$ and not $x + y < 3$.

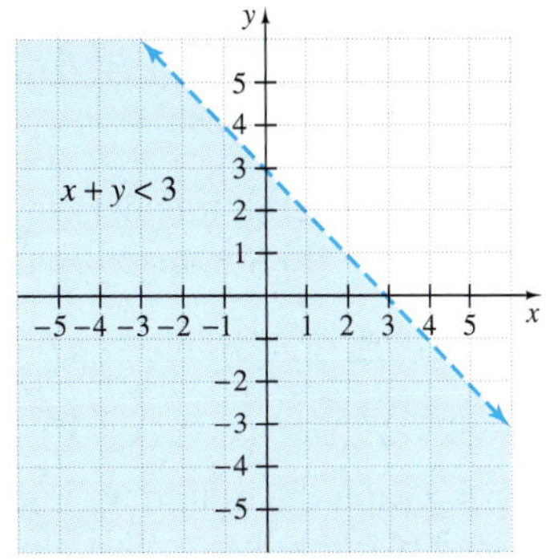


The following steps may be used to graph linear inequalities in two variables.

### GRAPHING A LINEAR INEQUALITY IN TWO VARIABLES

**Step 1:** Graph the boundary line found by replacing the inequality sign with an equal sign. If the inequality sign is $<$ or $>$, graph a dashed line indicating that points on the line are not solutions of the inequality. If the inequality sign is $\le$ or $\ge$, graph a solid line indicating that points on the line are solutions of the inequality.

**Step 2:** Choose a **test point not on the boundary line** and substitute the coordinates of this test point into the **original inequality.**

**Step 3:** If a true statement is obtained in Step 2, shade the half-plane that contains the test point. If a false statement is obtained, shade the half-plane that does not contain the test point.

**Example 1** Graph $2x - y < 6$.

*Solution* First, the boundary line for this inequality is the graph of $2x - y = 6$. Graph a dashed boundary line because the inequality symbol is $<$. Next, choose a test point on either side of the boundary line. The point $(0, 0)$ is not on the boundary line, so we use this point. Replacing $x$ with 0 and $y$ with 0 in the *original inequality* $2x - y < 6$ leads to the following:

$$2x - y < 6$$
$$2(0) - 0 < 6 \quad \text{Let } x = 0 \text{ and } y = 0.$$
$$0 < 6 \quad \text{True.}$$

Because $(0, 0)$ satisfies the inequality, so does every point on the same side of the boundary line as $(0, 0)$. Shade the half-plane that contains $(0, 0)$. The half-plane graph of the inequality is shown.

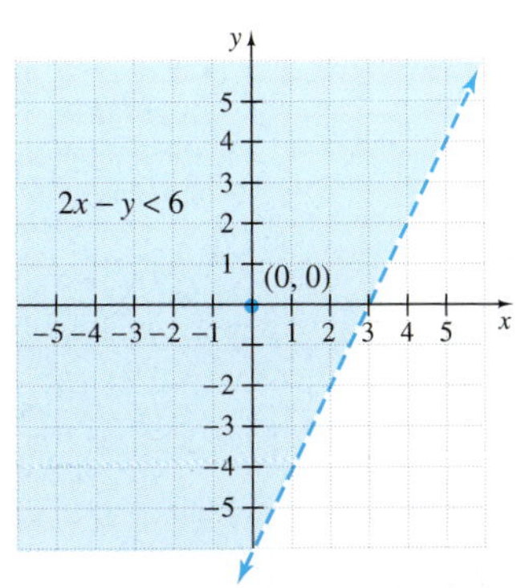

Every point in the shaded half-plane satisfies the original inequality. Notice that the inequality $2x - y < 6$ does not describe a function since its graph does not pass the vertical line test.

In general, linear inequalities of the form $Ax + By \leq C$, when $A$ and $B$ are not both 0, do not describe functions.

**Example 2** Graph $3x \geq y$.

*Solution* First, graph the boundary line $3x = y$. Graph a solid boundary line because the inequality symbol is $\geq$. Test a point not on the boundary line to determine which half-plane contains points that satisfy the inequality. We choose $(0, 1)$ as our test point.

$$3x \geq y$$
$$3(0) \geq 1 \quad \text{Let } x = 0 \text{ and } y = 1.$$
$$0 \geq 1 \quad \text{False.}$$

This point does not satisfy the inequality, so the correct half-plane is on the opposite side of the boundary line from $(0, 1)$. The graph of $3x \geq y$ is the boundary line together with the shaded region shown.

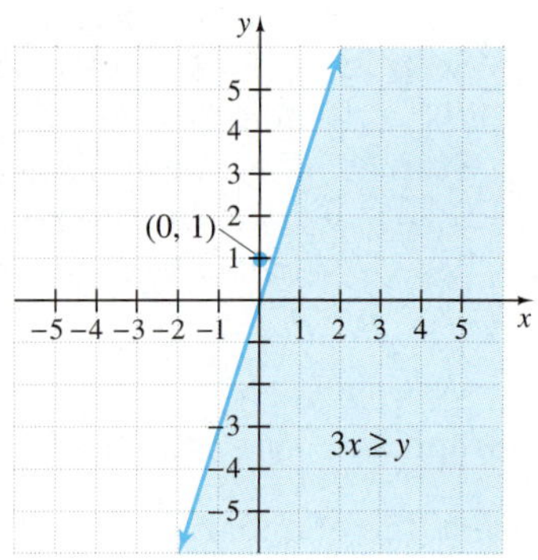


**2** The intersection and the union of linear inequalities can also be graphed, as shown in the next two examples.

**Example 3** Graph the intersection of $x \geq 1$ and $y \geq 2x - 1$.

*Solution* Graph each inequality. The intersection of the two graphs is all points common to both regions, as shown by the heaviest shading in the third graph.

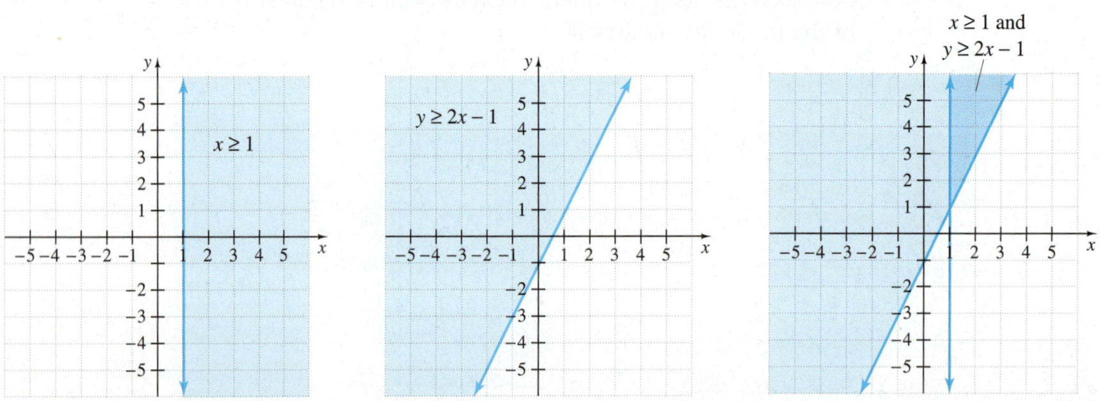

**Example 4** Graph the union of $x + \frac{1}{2}y \geq -4$ or $y \leq -2$.

*Solution* Graph each inequality. The union of the two inequalities is both shaded regions, including the solid boundary lines shown in the third graph.

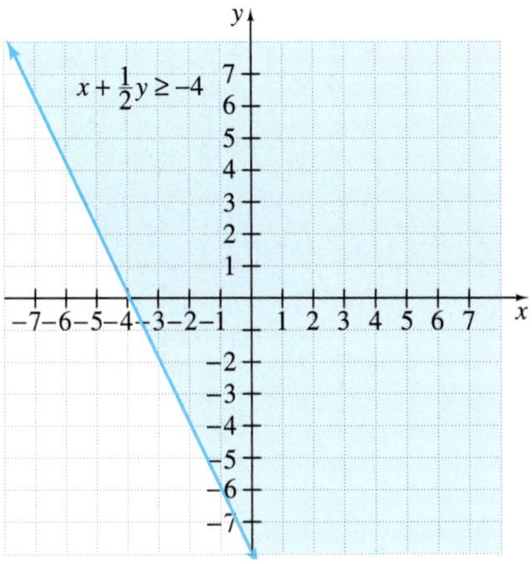


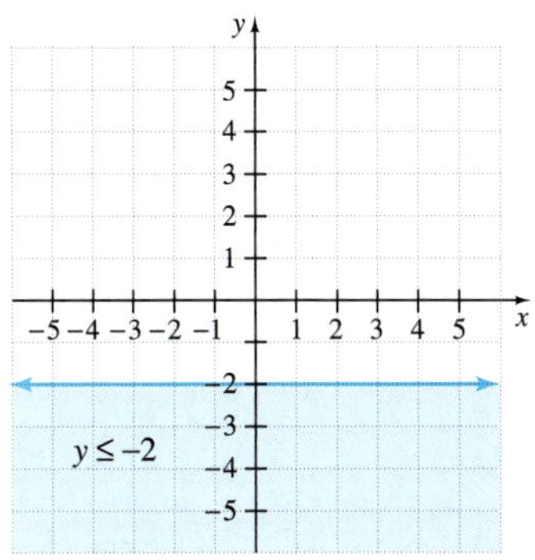


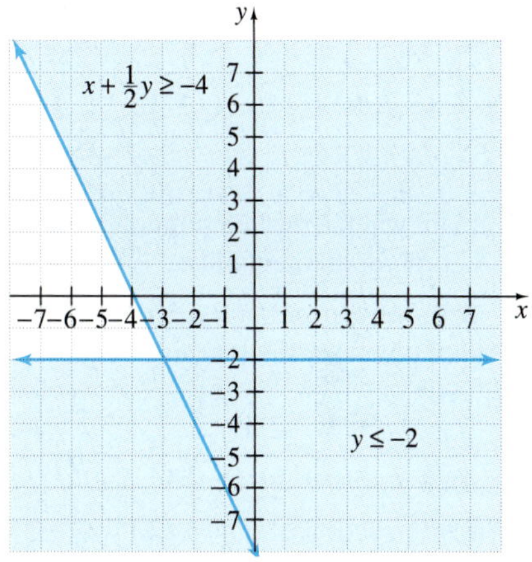


## SPOTLIGHT ON DECISION MAKING

Suppose you are a customer service representative for a mail-order medical supply company that sells support stockings. A customer, whose weight is 160 pounds and whose height is 5 feet 9 inches, places an order for support stockings and has asked your assistance in selecting the correct size. What size would you recommend that this customer order? Explain.

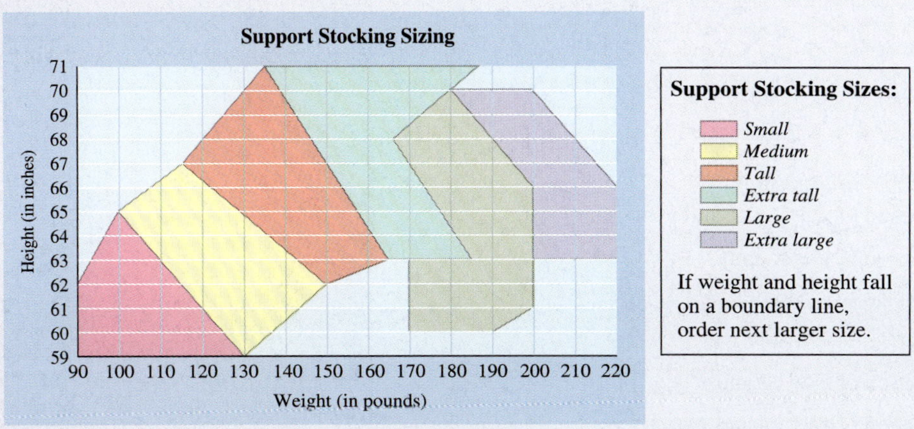

# Exercise Set 3.6

*Graph each inequality. See Examples 1 and 2.*

**1.** $x < 2$

**2.** $x > -3$

**3.** $x - y \geq 7$

**4.** $3x + y \leq 1$

**5.** $3x + y > 6$

**6.** $2x + y > 2$

**7.** $y \leq -2x$

**8.** $y \leq 3x$

**9.** $2x + 4y \geq 8$

**10.** $2x + 6y \leq 12$

**11.** $5x + 3y > -15$

**12.** $2x + 5y < -20$

**13.** Explain when a dashed boundary line should be used in the graph of an inequality.

**14.** Explain why, after the boundary line is sketched, we test a point on either side of this boundary in the original inequality.

*Graph each union or intersection. See Examples 3 and 4.*

**15.** The intersection of $x \geq 3$ and $y \leq -2$

**16.** The union of $x \geq 3$ or $y \leq -2$

**17.** The union of $x \leq -2$ or $y \geq 4$

**18.** The intersection of $x \leq -2$ and $y \geq 4$

**19.** The intersection of $x - y < 3$ and $x > 4$

**20.** The intersection of $2x > y$ and $y > x + 2$

**21.** The union of $x + y \leq 3$ or $x - y \geq 5$

**22.** The union of $x - y \leq 3$ or $x + y > -1$

*Graph each inequality.*

**23.** $y \geq -2$

**24.** $y \leq 4$

**25.** $x - 6y < 12$

**26.** $x - 4y < 8$

**27.** $x > 5$

**28.** $y \geq -2$

**29.** $-2x + y \leq 4$

**30.** $-3x + y \leq 9$

**31.** $x - 3y < 0$

**32.** $x + 2y > 0$

**33.** $3x - 2y \leq 12$

**34.** $2x - 3y \leq 9$

**35.** The union of $x - y \geq 2$ or $y < 5$

**36.** The union of $x - y < 3$ or $x > 4$

**37.** The intersection of $x + y \leq 1$ and $y \leq -1$

**38.** The intersection of $y \geq x$ and $2x - 4y \geq 6$

**39.** The union of $2x + y > 4$ or $x \geq 1$

**40.** The union of $3x + y < 9$ or $y \leq 2$

**41.** The intersection of $x \geq -2$ and $x \leq 1$

**42.** The intersection of $x \geq -4$ and $x \leq 3$

**43.** The union of $x + y \leq 0$ or $3x - 6y \geq 12$

**44.** The intersection of $x + y \leq 0$ and $3x - 6y \geq 12$

**45.** The intersection of $2x - y > 3$ and $x \geq 0$

**46.** The union of $2x - y > 3$ or $x \geq 0$

*Match each inequality with its graph.*

**A**

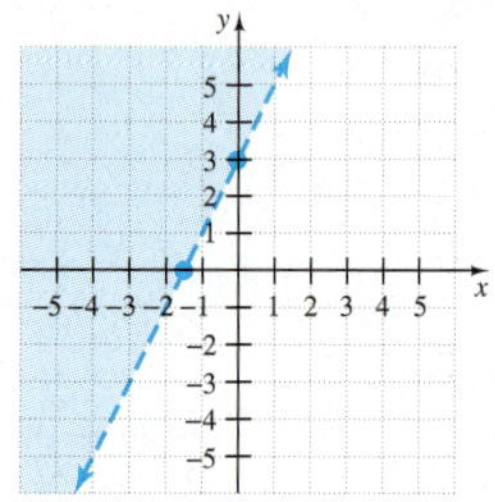

**B**

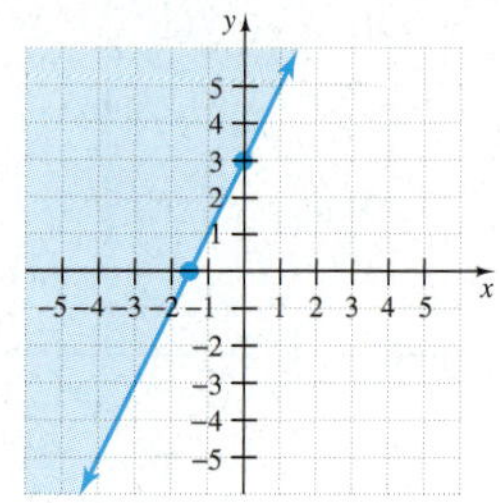

**C**

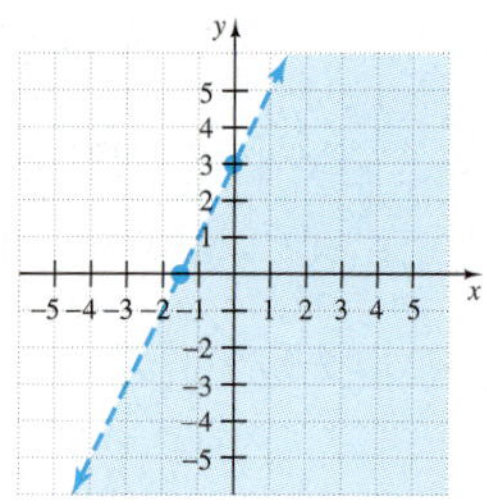

**D**

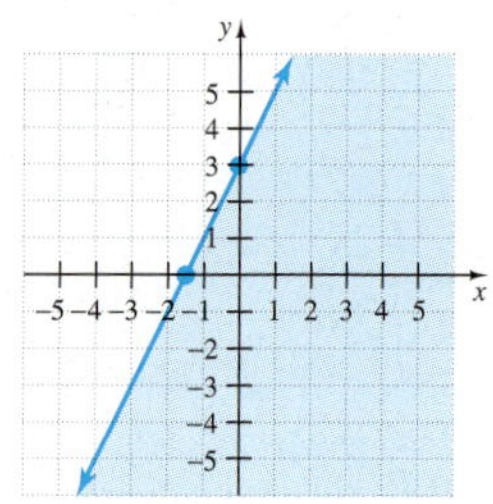

**47.** $y \leq 2x + 3$

**48.** $y < 2x + 3$

**49.** $y > 2x + 3$

**50.** $y \geq 2x + 3$

*Write the inequality whose graph is given.*

**51.**

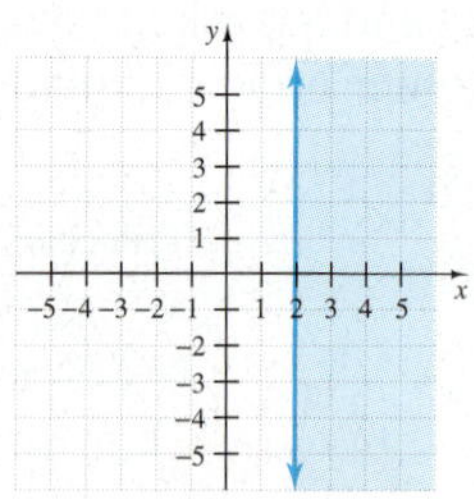

**52.**

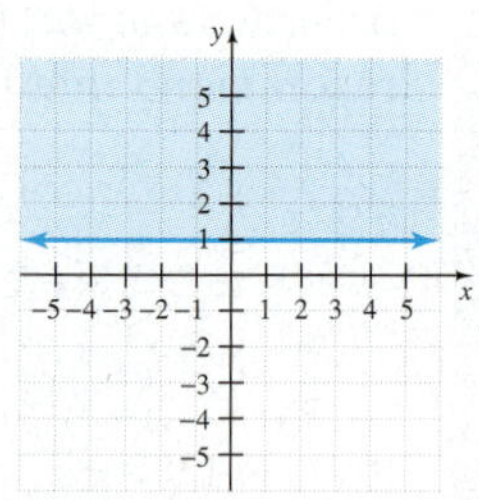

**53.**

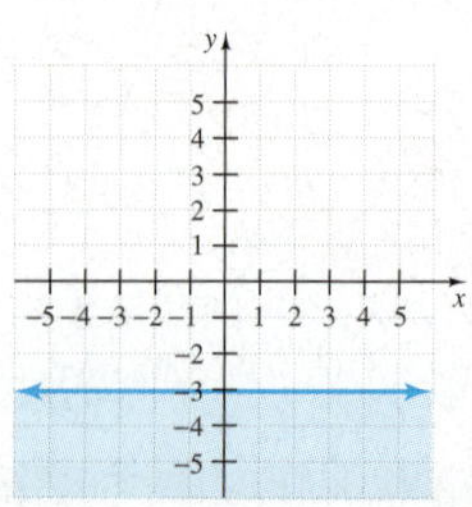

**54.**

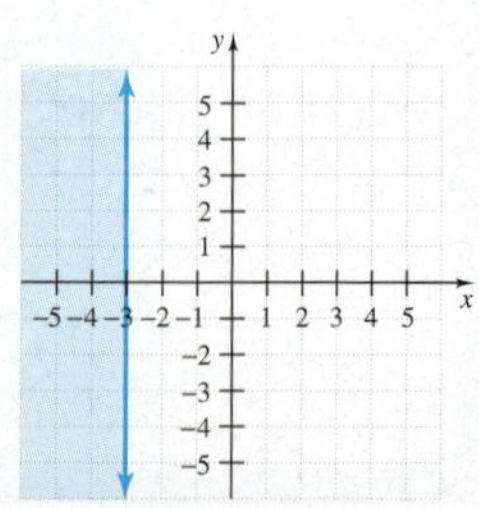

**55.**

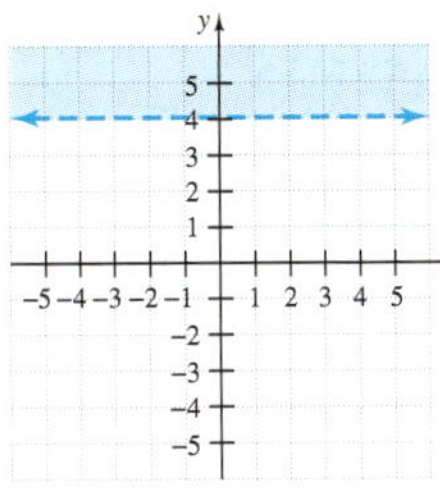

**56.**

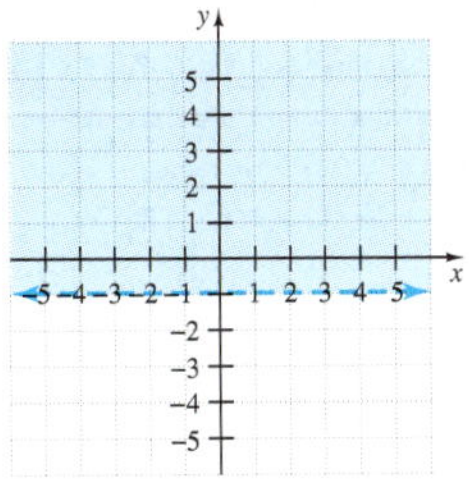

**57.**

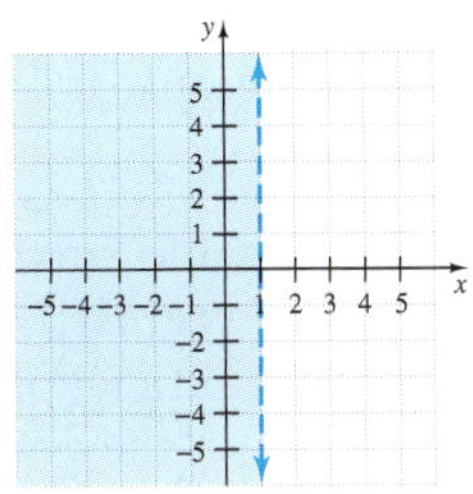

**58.**

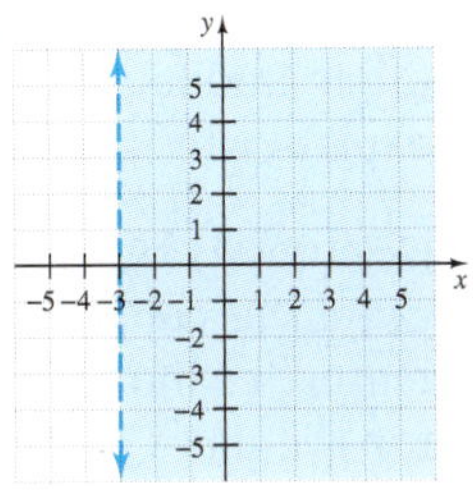

*Solve.*

**59.** Rheem Abo-Zahrah decides that she will study at most 20 hours every week and that she must work at least 10 hours every week. Let $x$ represent the hours studying and $y$ represent the hours working. Write two inequalities that model this situation and graph their intersection.

**60.** The movie and TV critic for the *New York Times* spends between 2 and 6 hours daily reviewing movies and fewer than 5 hours reviewing TV shows. Let $x$ represent the hours watching movies and $y$ represent the time spent watching TV. Write two inequalities that model this situation and graph their intersection.

**61.** Chris-Craft manufactures boats out of Fiberglas and wood. Fiberglas hulls require 2 hours work, whereas wood hulls require 4 hours work. Employees work at most 40 hours a week. The following inequalities model these restrictions, where $x$ represents the number of Fiberglas hulls produced and $y$ represents the number of wood hulls produced.

$$\begin{cases} x \geq 0 \\ y \geq 0 \\ 2x + 4y \leq 40 \end{cases}$$

*Graph the intersection of these inequalities.*

## REVIEW EXERCISES

*Evaluate each expression. See Sections 1.3 and 1.4.*

**62.** $2^3$

**63.** $3^2$

**64.** $-5^2$

**65.** $(-5)^2$

**66.** $(-2)^4$

**67.** $-2^4$

**68.** $\left(\frac{3}{5}\right)^3$

**69.** $\left(\frac{2}{7}\right)^2$

*Find the domain and the range of each relation. Determine whether the relation is also a function. See Section 3.2.*

**70.**

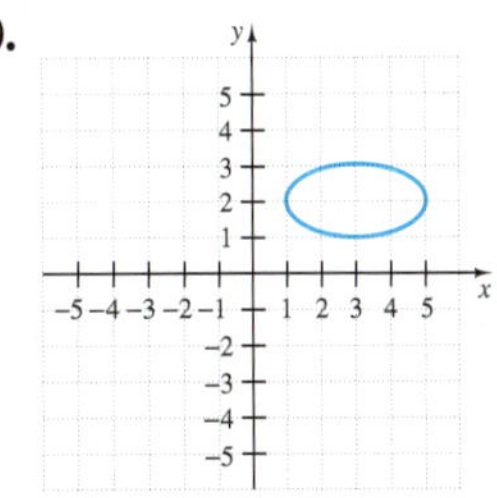

**71.**

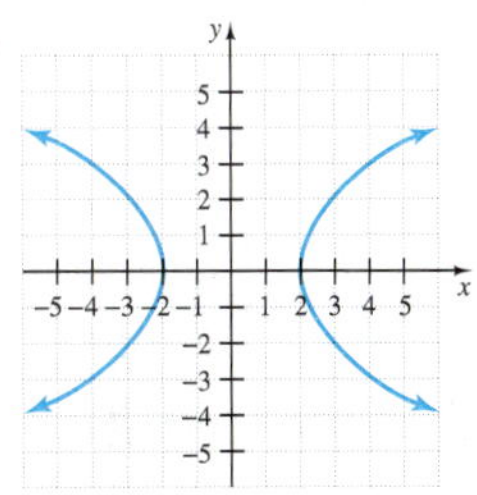

For additional Chapter Projects, visit the Real World Activities Website by going to http://www.prenhall.com/martin-gay.

# 3 CHAPTER PROJECT

## Modeling Real Data

The number of children who live with only one parent has been steadily increasing in the United States since the 1960s. According to the U.S. Bureau of the Census, the percent of children living with one parent varies widely by race/ethnic background. The trend since the 1960s also varies widely, with the data for white families being the most linear. The following table shows the percent of white children (under age 18) living with *both* parents during selected years from 1970 to 1998. In this project, you will have the opportunity to use the data in the table to find a linear function $f(x)$ that represents the data, reflecting the change in living arrangements for children. This project may be completed by working in groups or individually.

PERCENT OF U.S. CHILDREN (WHITE) WHO LIVE WITH BOTH PARENTS

| *Year* | *1970* | *1980* | *1990* | *1995* | *1996* | *1997* | *1998* |
|---|---|---|---|---|---|---|---|
| $x$ | 0 | 10 | 20 | 25 | 26 | 27 | 28 |
| *PERCENT, y* | 90 | 83 | 79 | 76 | 75 | 75 | 74 |

*Source*: U.S. Bureau of the Census

1. Plot the data given in the table as ordered pairs.
2. Use a straight edge to draw on your graph what appears to be the line that "best fits" the data you plotted.
3. Estimate the coordinates of two points that fall on your best-fitting line. Use these points to find a linear function $f(x)$ for the line.
4. What is the slope of your line? Interpret its meaning. Does it make sense in the context of this situation?
5. Find the value of $f(50)$. Write a sentence interpreting its meaning in context.
6. Compare your linear function with that of another student or group. Are they different? If so, explain why.

(Optional) Enter the data from the table into a graphing calculator. Use the linear regression feature of the calculator to find a linear function for the data. Compare this function to the one you found in Question 3. How are they alike or different? Find the value of $f(50)$ using the model you found with the graphing calculator. Compare it to the value of $f(50)$ you found in Question 5.

## CHAPTER 3 VOCABULARY CHECK

*Fill in each blank with one of the words or phrases listed below.*

| | | | | | |
|---|---|---|---|---|---|
| relation | line | function | standard | slope | domain |
| slope–intercept | $x$ | $y$ | range | parallel | linear function |
| point–slope | perpendicular | linear inequality | | | |

1. A ________ is a set of ordered pairs.
2. The graph of every linear equation in two variables is a _____.
3. The statement $-x + 2y > 0$ is called a ______________ in two variables.
4. _________ form of linear equation in two variables is $Ax + By = C$.
5. The _______ of a relation is the set of all second components of the ordered pairs of the relation.
6. ________ lines have the same slope and different $y$-intercepts.
7. ______________ form of a linear equation in two variables is $y = mx + b$.
8. A _________ is a relation in which each first component in the ordered pairs corresponds to exactly one second component.
9. In the equation $y = 4x - 2$, the coefficient of $x$ is the ______ of its corresponding graph.
10. Two lines are ______________ if the product of their slopes is $-1$.
11. To find the $x$-intercept of a linear equation, let ___ = 0 and solve for the other variable.
12. The ________ of a relation is the set of all first components of the ordered pairs of the relation.
13. A _____________ is a function that can be written in the form $f(x) = mx + b$.
14. To find the $y$-intercept of a linear equation, let _ = 0 and solve for the other variable.
15. The equation $y - 8 = -5(x + 1)$ is written in ___________ form.

# CHAPTER 3 HIGHLIGHTS

| DEFINITIONS AND CONCEPTS | EXAMPLES |
|---|---|

## Section 3.1 Graphing Equations

The **rectangular coordinate system,** or **Cartesian coordinate system,** consists of a vertical and a horizontal number line intersecting at their 0 coordinate. The vertical number line is called the **y-axis**, and the horizontal number line is called the **x-axis**. The point of intersection of the axes is called the **origin.** The axes divide the plane into four regions called **quadrants.**

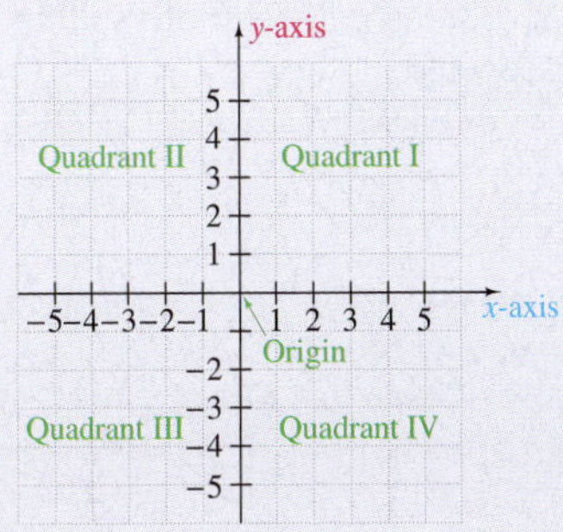


To **plot** or **graph** an ordered pair means to find its corresponding point on a rectangular coordinate system.

To plot or graph the ordered pair $(-2, 5)$, start at the origin. Move 2 units to the left along the $x$-axis, then 5 units upward parallel to the $y$-axis.

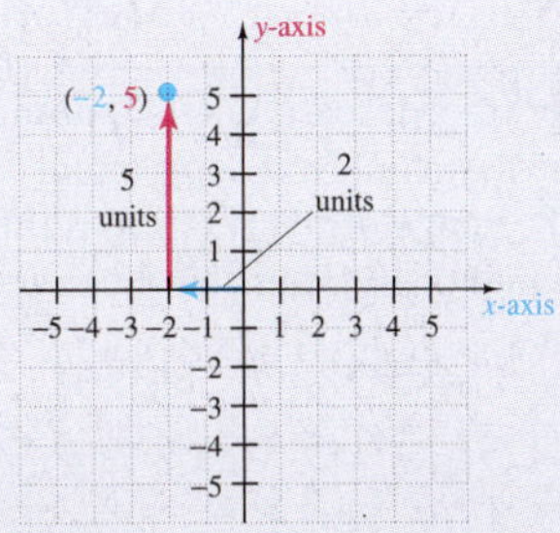


An ordered pair is a **solution** of an equation in two variables if replacing the variables by the corresponding coordinates results in a true statement.

Determine whether $(-2, 3)$ is a solution of $3x + 2y = 0$

$$3(-2) + 2(3) = 0$$

$$-6 + 6 = 0$$

$$0 = 0 \quad \text{True.}$$

$(-2, 3)$ is a solution.

A **linear equation in two variables** is an equation that can be written in the form $Ax + By = C$, where $A$, $B$, and $C$ are real numbers and $A$ and $B$ are not both 0. The form $Ax + By = C$ is called **standard form.**

***Linear Equations in Two Variables***

$$y = -2x + 5, \quad x = 7$$

$$y - 3 = 0, \quad 6x - 4y = 10$$

$6x - 4y = 10$ is in standard form.

The graph of a linear equation in two variables is a line. To graph a linear equation in two variables, find three ordered pair solutions. Plot the solution points, and draw the line connecting the points.

Graph $3x + y = -6$.

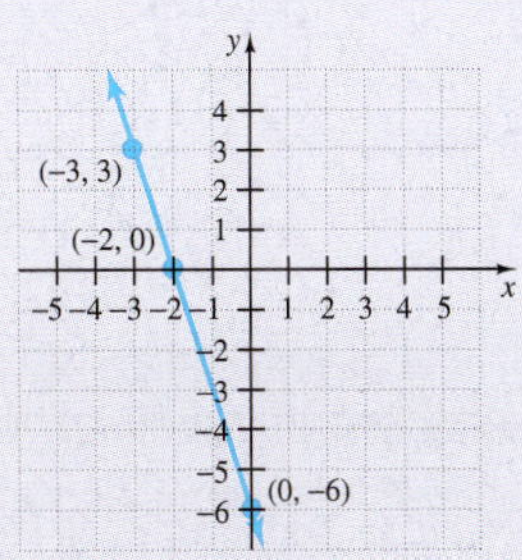


| $x$ | $y$ |
|---|---|
| 0 | −6 |
| −2 | 0 |
| −3 | 3 |

*(continued)*

**DEFINITIONS AND CONCEPTS** | **EXAMPLES**

## Section 3.1 Graphing Equations

To graph an equation that is not linear, find a sufficient number of ordered pair solutions so that a pattern may be discovered.

Graph $y = x^3 + 2$.

| $x$ | $y$ |
|---|---|
| −2 | −6 |
| −1 | 1 |
| 0 | 2 |
| 1 | 3 |
| 2 | 10 |

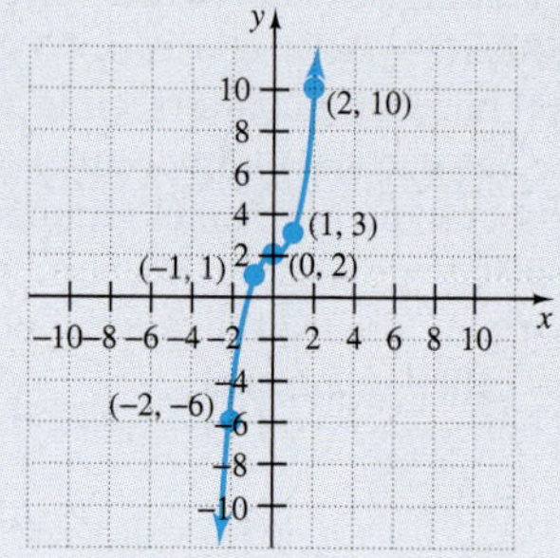


## Section 3.2 Introduction to Functions

A **relation** is a set of ordered pairs. The **domain** of the relation is the set of all first components of the ordered pairs. The **range** of the relation is the set of all second components of the ordered pairs.

***Relation***

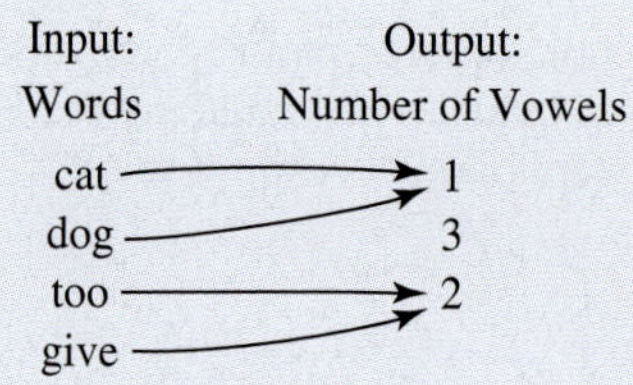


Domain: {cat, dog, too, give}
Range: {1, 2}

A **function** is a relation in which each element of the first set corresponds to exactly one element of the second set.

The previous relation is a function. Each word contains exactly one number of vowels.

***Vertical Line Test***

If no vertical line can be drawn so that it intersects a graph more than once, the graph is the graph of a function.

Find the domain and the range of the relation. Also determine whether the relation is a function.

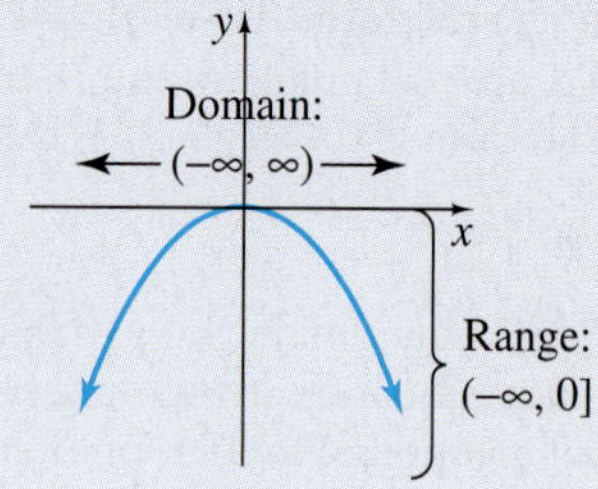


By the vertical line test, this graph is the graph of a function.

The symbol $f(x)$ means **function of $x$** and is called **function notation.**

If $f(x) = 2x^2 - 5$, find $f(-3)$.

$$f(-3) = 2(-3)^2 - 5 = 2(9) - 5 = 13$$

*(continued)*

| DEFINITIONS AND CONCEPTS | EXAMPLES |
|---|---|

## Section 3.3 Graphing Linear Functions

A **linear function** is a function that can be written in the form $f(x) = mx + b$.

***Linear Functions***

$f(x) = -3, g(x) = 5x, h(x) = -\frac{1}{3}x - 7$

To graph a linear function, find three ordered pair solutions. Graph the solutions and draw a line through the plotted points.

Graph $f(x) = -2x$.

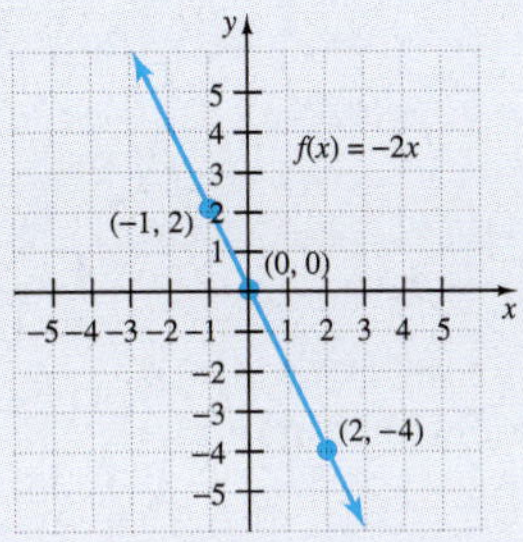


| $x$ | $y$ or $f(x)$ |
|---|---|
| −1 | 2 |
| 0 | 0 |
| 2 | −4 |

For any function $f(x)$, the graph of $y = f(x) + K$ is the same as the graph of $y = f(x)$ shifted $K$ units up if $K$ is positive and $|K|$ units down if $K$ is negative.

Graph $g(x) = -2x + 3$.
This is the same as the graph of $f(x) = -2x$ shifted 3 units up.

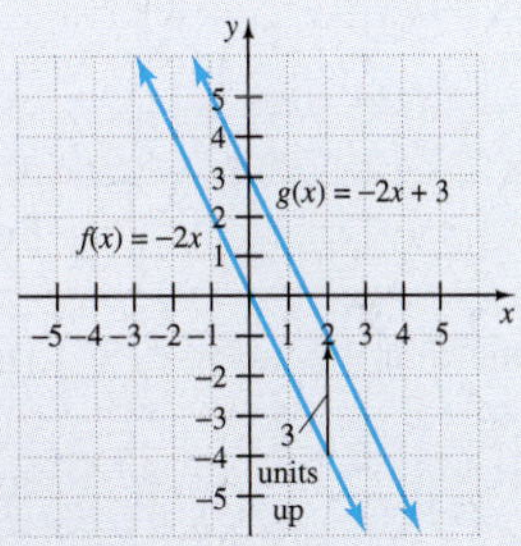


The $x$-coordinate of a point where a graph crosses the $x$-axis is called an ***x*-intercept**. The $y$-coordinate of a point where a graph crosses the $y$-axis is called a ***y*-intercept.**

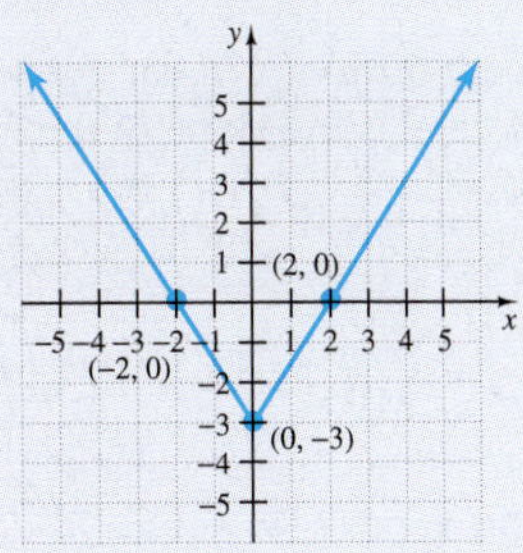


The $x$-intercepts of the graph are $(-2, 0)$ and $(2, 0)$.
The $y$-intercept is $(0, -3)$.

To find an $x$-intercept, let $y = 0$ or $f(x) = 0$ and solve for $x$.
To find a $y$-intercept, let $x = 0$ and solve for $y$.

Graph $5x - y = -5$ by finding intercepts.

If $x = 0$, then

$$5x - y = -5$$
$$5 \cdot 0 - y = -5$$
$$-y = -5$$
$$y = 5$$
$$(0, 5)$$

If $y = 0$, then

$$5x - y = -5$$
$$5x - 0 = -5$$
$$5x = -5$$
$$x = -1$$
$$(-1, 0)$$

*(continued)*

**DEFINITIONS AND CONCEPTS** | **EXAMPLES**

## Section 3.3 Graphing Linear Functions

Ordered pairs are $(0, 5)$ and $(-1, 0)$.

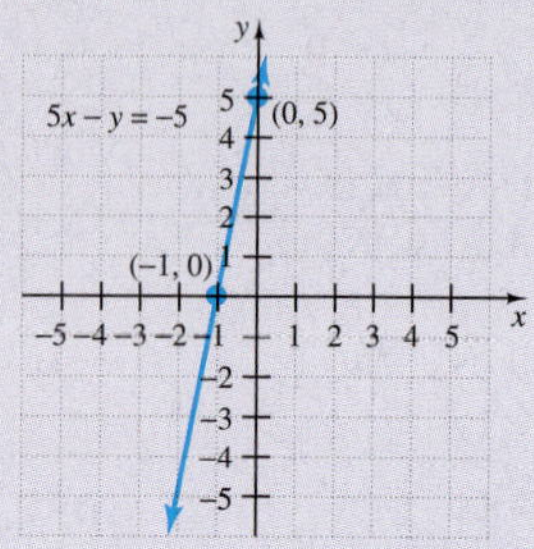


The graph of $x = c$ is a vertical line with $x$-intercept $(c, 0)$.

The graph of $y = c$ is a horizontal line with $y$-intercept $(0, c)$.

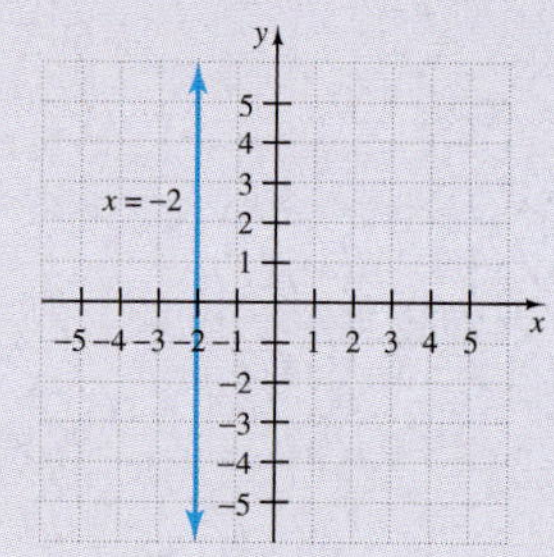


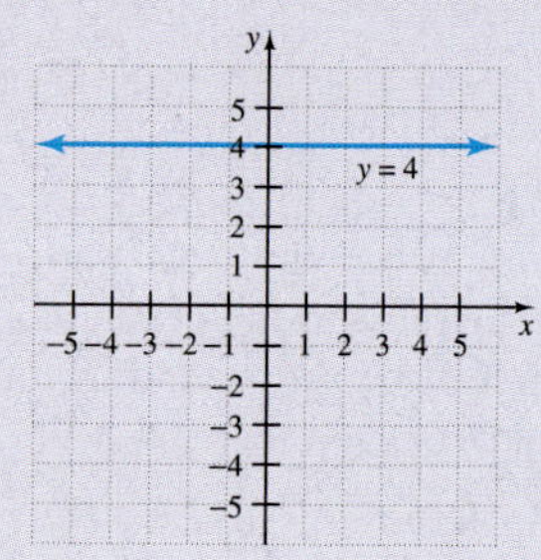


## Section 3.4 The Slope of a Line

The **slope** $m$ of the line through $(x_1, y_1)$ and $(x_2, y_2)$ is given by

$$m = \frac{y_2 - y_1}{x_2 - x_1} \text{ as long } x_2 \neq x_1$$

Find the slope of the line through $(-1, 7)$ and $(-2, -3)$.

$$m = \frac{y_2 - y_1}{x_2 - x_1} = \frac{-3 - 7}{-2 - (-1)} = \frac{-10}{-1} = 10$$

The **slope–intercept form** of a linear equation is $y = mx + b$, where $m$ is the slope of the line and $b$ is the $y$-intercept.

Find the slope and $y$-intercept of $-3x + 2y = -8$.

$$2y = 3x - 8$$

$$\frac{2y}{2} = \frac{3x}{2} - \frac{8}{2}$$

$$y = \frac{3}{2}x - 4$$

The slope the line is $\frac{3}{2}$, and the $y$-intercept is $(0, -4)$.

Nonvertical parallel lines have the same slope.

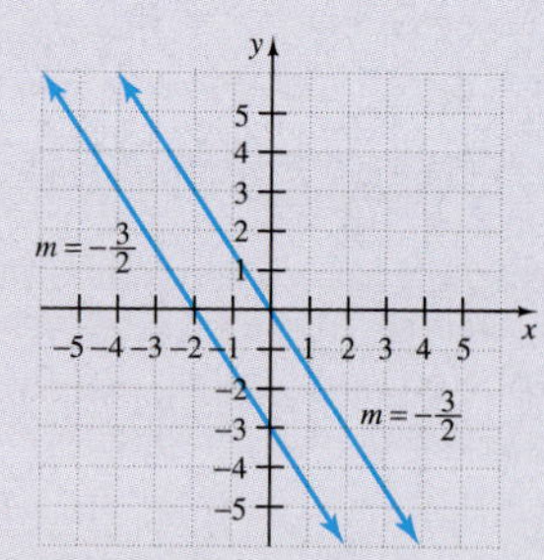


*(continued)*

| DEFINITIONS AND CONCEPTS | EXAMPLES |
|---|---|

## Section 3.4 The Slope of a Line

If the product of the slopes of two lines is $-1$, then the lines are perpendicular.

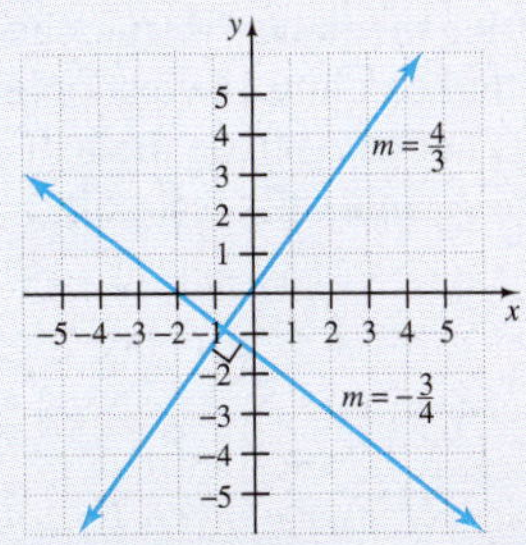


The slope of a horizontal line is 0.
The slope of a vertical line is undefined.

The slope of $y = -2$ is 0.
The slope of $x = 5$ is undefined.

## Section 3.5 Equations of Lines

We can use the slope–intercept form to write an equation of a line given its slope and $y$-intercept.

Write an equation of the line with $y$-intercept $(0, -1)$ and slope $\frac{2}{3}$.

$$y = mx + b$$

$$y = \frac{2}{3}x - 1$$

The point–slope form of the equation of a line is $y - y_1 = m(x - x_1)$, where $m$ is the slope of the line and $(x_1, y_1)$ is a point on the line.

Find an equation of the line with slope 2 containing the point $(1, -4)$. Write the equation in standard form: $Ax + By = C$.

$$y - y_1 = m(x - x_1)$$

$$y - (-4) = 2(x - 1)$$

$$y + 4 = 2x - 2$$

$$-2x + y = -6 \quad \text{Standard form.}$$

## Section 3.6 Graphing Linear Inequalities

If the equal sign in a linear equation in two variables is replaced with an inequality symbol, the result is a **linear inequality in two variables.**

***Linear Inequalities in Two Variables***

$$x \le -5 \qquad y \ge 2$$

$$3x - 2y > 7 \qquad x < -5$$

***To graph a linear inequality***

1. Graph the boundary line by graphing the related equation. Draw the line solid if the inequality symbol is $\le$ or $\ge$. Draw the line dashed if the inequality symbol is $<$ or $>$.
2. Choose a test point not on the line. Substitute its coordinates into the original inequality.

Graph $2x - 4y > 4$.

1. Graph $2x - 4y = 4$. Draw a dashed line because the inequality symbol is $>$.
2. Check the test point $(0, 0)$ in the inequality $2x - 4y > 4$.

$$2 \cdot 0 - 4 \cdot 0 > 4 \quad \text{Let } x = 0 \text{ and } y = 0.$$

$$0 > 4 \quad \text{False.}$$

*(continued)*

| DEFINITIONS AND CONCEPTS | EXAMPLES |
| --- | --- |
| **Section 3.6 Graphing Linear Inequalities** | |
| **3.** If the resulting inequality is true, shade the **half-plane** that contains the test point. If the inequality is not true, shade the half-plane that does not contain the test point. | **3.** The inequality is false, so we shade the half-plane that does not contain $(0, 0)$. |

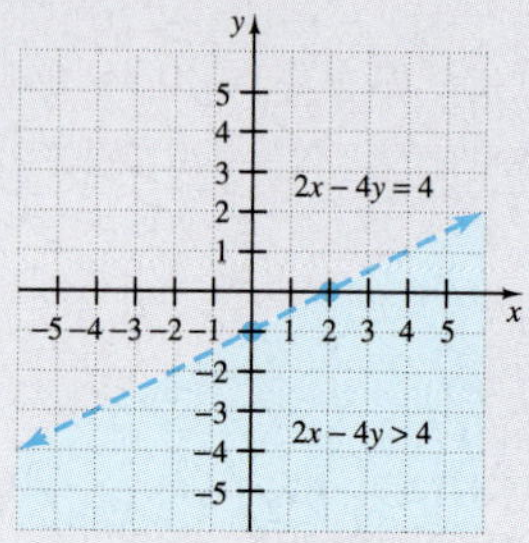


## CHAPTER 3 REVIEW

*(3.1) Plot the points and name the quadrant or axis in which each point lies.*

**1.** $A(2, -1), B(-2, 1), C(0, 3), D(-3, -5)$

**2.** $A(-3, 4), B(4, -3), C(-2, 0), D(-4, 1)$

*Determine whether each ordered pair is a solution to the given equation.*

**3.** $7x - 8y = 56$; $(0, 56), (8, 0)$

**4.** $-2x + 5y = 10$; $(-5, 0), (1, 1)$

**5.** $x = 13$; $(13, 5), (13, 13)$

**6.** $y = 2$; $(7, 2), (2, 7)$

*Determine whether each equation is linear or not. Then graph the equation already written below.*

**7.** $y = 3x$ **8.** $y = 5x$

**9.** $3x - y = 4$ **10.** $x - 3y = 2$

**11.** $y = |x| + 4$ **12.** $y = x^2 + 4$

**13.** $y = -\frac{1}{2}x + 2$ **14.** $y = -x + 5$

**15.** $y = 2x - 1$ **16.** $y = \frac{1}{3}x + 1$

**17.** $y = -1.36x$ **18.** $y = 2.1x + 5.9$

*(3.2) Find the domain and range of each relation. Also determine whether the relation is a function.*

**19.** $\left\{\left(-\frac{1}{2}, \frac{3}{4}\right), (6, 0.75), (0, -12), (25, 25)\right\}$

**20.** $\left\{\left(\frac{3}{4}, -\frac{1}{2}\right), (0.75, 6), (-12, 0), (25, 25)\right\}$

**21.**

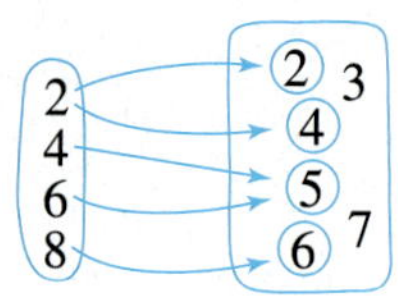


**22.**

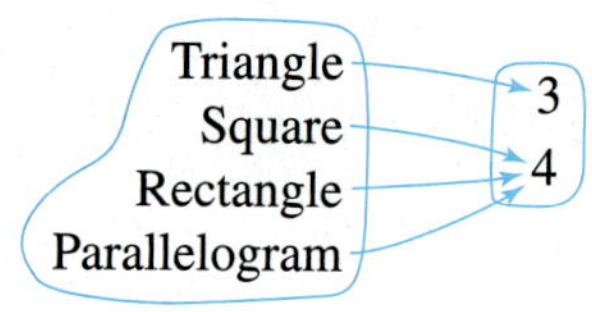


**23.**

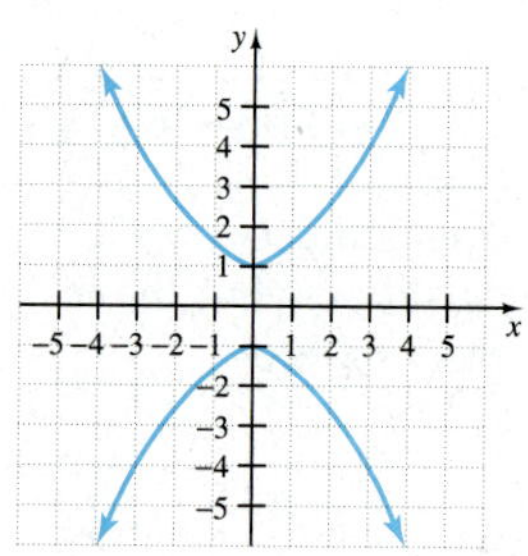

**24.**

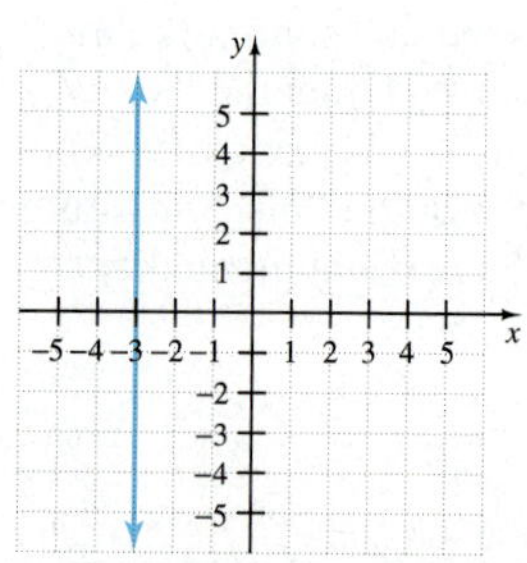

**25.**

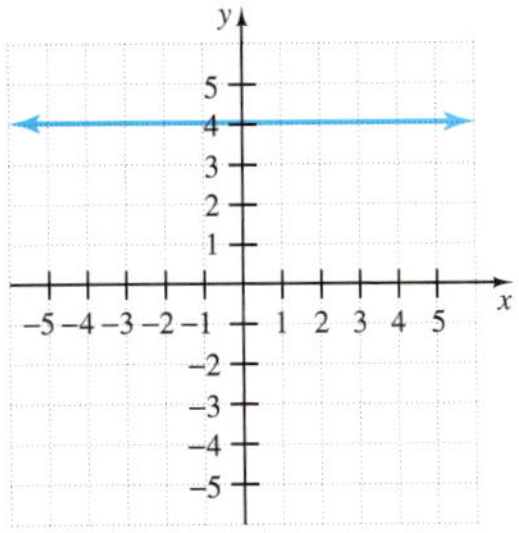

**26.**

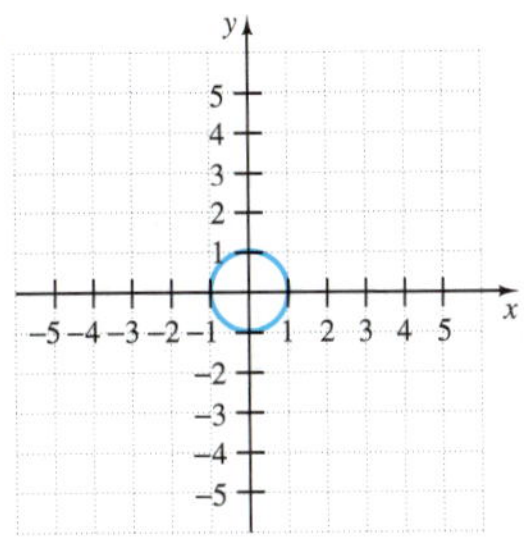

*If $f(x) = x - 5$, $g(x) = -3x$, and $h(x) = 2x^2 - 6x + 1$, find the following.*

**27.** $f(2)$

**28.** $g(0)$

**29.** $g(-6)$

**30.** $h(-1)$

**31.** $h(1)$

**32.** $f(5)$

*The function $J(x) = 2.54x$ may be used to calculate the weight of an object on Jupiter $J$ given its weight on Earth $x$.*

**33.** If a person weighs 150 pounds on Earth, find the equivalent weight on Jupiter.

**34.** A 2000-pound probe on Earth weighs how many pounds on Jupiter?

*Use the graph of the function below to answer Exercises 35 through 38.*

**35.** Find $f(-1)$.

**36.** Find $f(1)$.

**37.** Find all values of $x$ such that $f(x) = 1$.

**38.** Find all values of $x$ such that $f(x) = -1$.

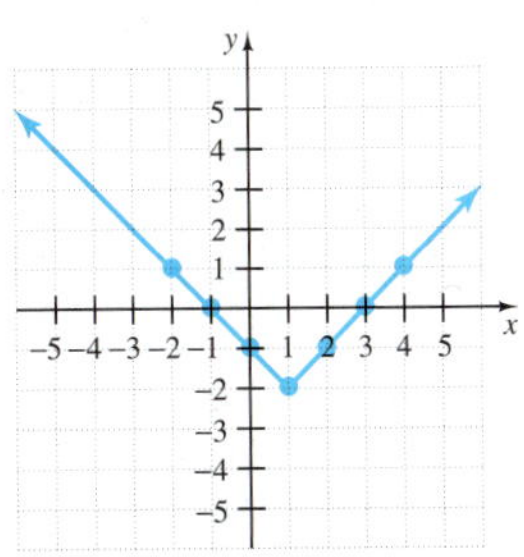

***(3.3)*** *Graph each linear function.*

**39.** $f(x) = x$

**40.** $f(x) = -\frac{1}{3}x$

**41.** $g(x) = 4x - 1$

*The graph of $f(x) = 3x$ is sketched below. Use this graph to match each linear function with its graph.*

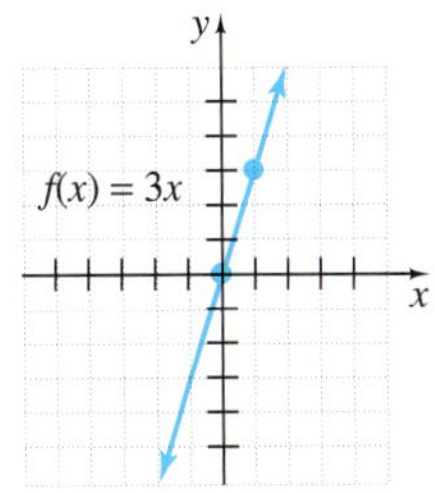


**A**

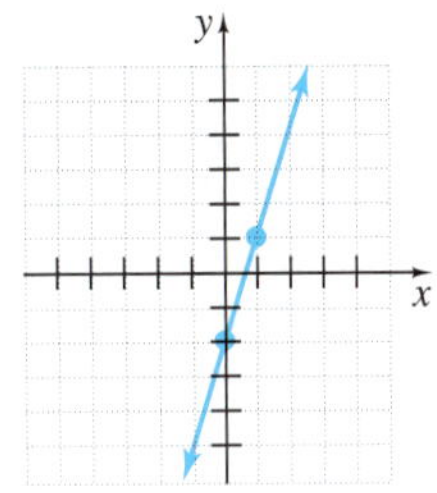

**B**

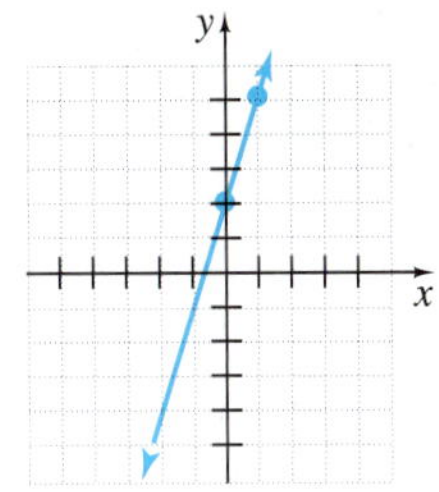

**C**

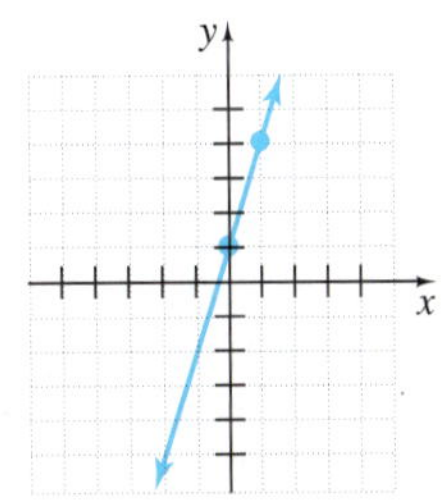

**D**

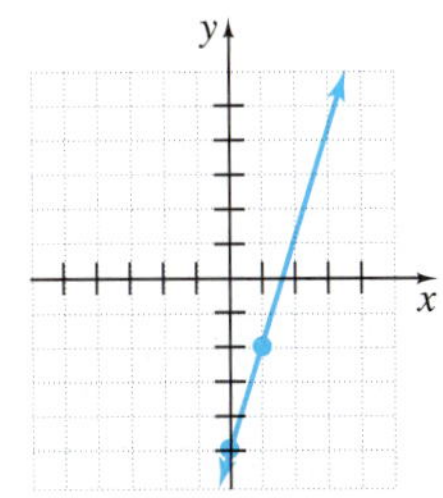

**42.** $f(x) = 3x + 1$

**43.** $f(x) = 3x - 2$

**44.** $f(x) = 3x + 2$

**45.** $f(x) = 3x - 5$

*Graph each linear equation by finding intercepts if possible.*

**46.** $4x + 5y = 20$

**47.** $3x - 2y = -9$

**48.** $4x - y = 3$

**49.** $2x + 6y = 9$

**50.** $y = 5$

**51.** $x = -2$

*Graph each linear equation.*

**52.** $x - 2 = 0$

**53.** $y + 3 = 0$

**54.** The cost $C$, in dollars, of renting a minivan for a day is given by the linear function $C(x) = 0.3x + 42$, where $x$ is number of miles driven.

**a.** Find the cost of renting the minivan for a day and driving it 150 miles.

**b.** Graph $C(x) = 0.3x + 42$.

***(3.4)*** *Find the slope of the line through each pair of points.*

**55.** $(2, 8)$ and $(6, -4)$

**56.** $(-3, 9)$ and $(5, 13)$

**57.** $(-7, -4)$ and $(-3, 6)$

**58.** $(7, -2)$ and $(-5, 7)$

*Find the slope and y-intercept of each line.*

**59.** $6x - 15y = 20$

**60.** $4x + 14y = 21$

*Find the slope of each line.*

**61.** $y - 3 = 0$

**62.** $x = -5$

*Two lines are graphed on each set of axes. Decide whether $l_1$ or $l_2$ has the greater slope.*

**63.**

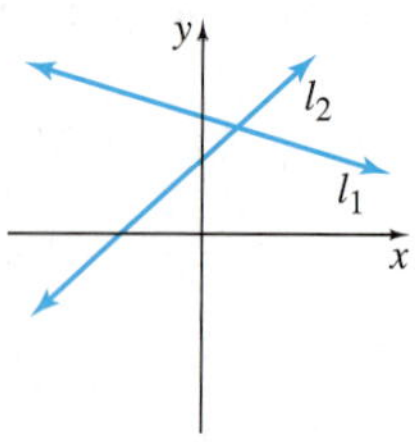


**64.**

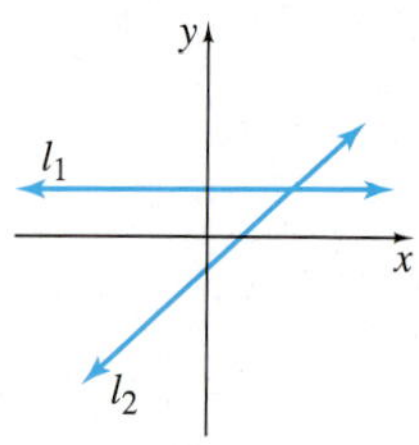


**65.**

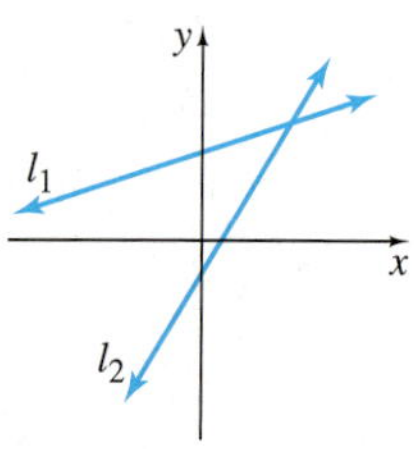


**66.**

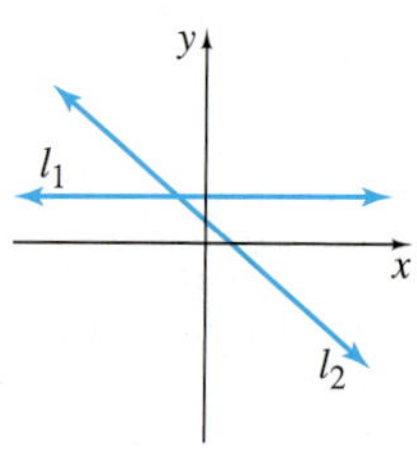


**67.** Recall from Exercise 54, that the cost $C$, in dollars, of renting a minivan for a day is given by the linear equation $y = 0.3x + 42$, where $x$ is number of miles driven.

**a.** Find and interpret the slope of this equation.

**b.** Find and interpret the $y$-intercept of this equation.

*Decide whether the lines are parallel, perpendicular, or neither.*

△ **68.** $f(x) = -2x + 6$
$g(x) = 2x - 1$

△ **69.** $-x + 3y = 2$
$6x - 18y = 3$

***(3.5)*** *Graph each linear equation using the slope and y-intercept.*

**70.** $y = -x + 1$

**71.** $y = 4x - 3$

**72.** $3x - y = 6$

**73.** $y = -5x$

*Find an equation of the line satisfying the conditions given.*

**74.** Horizontal; through $(3, -1)$

**75.** Vertical; through $(-2, -4)$

△ **76.** Parallel to the line $x = 6$; through $(-4, -3)$

**77.** Slope 0; through $(2, 5)$

*Find the standard form equation of each line satisfying the conditions given.*

**78.** Through $(-3, 5)$; slope 3

**79.** Slope 2; through $(5, -2)$

**80.** Through $(-6, -1)$ and $(-4, -2)$

**81.** Through $(-5, 3)$ and $(-4, -8)$

△ **82.** Through $(-2, 3)$; perpendicular to $x = 4$

△ **83.** Through $(-2, -5)$; parallel to $y = 8$

*Find the equation of each line satisfying the given conditions. Write each equation using function notation.*

**84.** Slope $-\frac{2}{3}$; $y$-intercept $(0, 4)$

**85.** Slope $-1$; $y$-intercept $(0, -2)$

△ **86.** Through $(2, -6)$; parallel to $6x + 3y = 5$

△ **87.** Through $(-4, -2)$; parallel to $3x + 2y = 8$

△ **88.** Through $(-6, -1)$; perpendicular to $4x + 3y = 5$

△ **89.** Through $(-4, 5)$; perpendicular to $2x - 3y = 6$

**90.** In 1996, the number of U.S. paging subscribers (in millions) was 42. The number of subscribers in 1999 (in millions) was 58. Let $y$ be the number of subscribers (in millions) in the year $x$, where $x = 0$ represents 1996. (*Source:* Strategis Group for Personal Communications Asso.)

**a.** Write a linear equation that models the number of U.S. paging subscribers (in millions) in terms of the year $x$. [*Hint:* Write 2 ordered pairs of the form (years past 1996, number of subscribers).]

**b.** Use this equation to predict the number of U.S. paging subscribers in the year 2007. (Round to the nearest million.)

**91.** In 1998, the number of people (in millions) reporting arthritis was 43. The number of people (in millions) predicted to be reporting arthritis in 2020 is 60. Let $y$ be the number of people (in millions) reporting arthritis in the year $x$, where $x = 0$ represents 1998. (*Source:* Arthritis Foundation)

**a.** Write a linear equation that models the number of people (in millions) reporting arthritis in terms of the year $x$ (See the hint for Exercise 90.)

**b.** Use this equation to predict the number of people reporting arthritis in 2010. (Round to the nearest million.)

*(3.6) Graph each linear inequality.*

**92.** $3x + y > 4$

**93.** $\frac{1}{2}x - y < 2$

**94.** $5x - 2y \le 9$

**95.** $3y \ge x$

**96.** $y < 1$

**97.** $x > -2$

**98.** Graph the union of $y > 2x + 3$ or $x \le -3$.

**99.** Graph the intersection of $2x < 3y + 8$ and $y \ge -2$.

## CHAPTER 3 TEST

**1.** Plot the points, and name the quadrant in which each is located: $A(6, -2)$, $B(4, 0)$, $C(-1, 6)$.

**2.** Complete the ordered pair solution $(-6, \quad)$ of the equation $2y - 3x = 12$.

*Graph each line.*

**3.** $2x - 3y = -6$

**4.** $4x + 6y = 7$

**5.** $f(x) = \dfrac{2}{3}x$

**6.** $y = -3$

**7.** Find the slope of the line that passes through $(5, -8)$ and $(-7, 10)$.

**8.** Find the slope and the $y$-intercept of the line $3x + 12y = 8$.

*Graph each nonlinear function. Suggested x-values have been given for ordered pair solutions.*

**9.** $f(x) = (x - 1)^2$
Let $x = -2, -1, 0, 1, 2, 3, 4$

**10.** $g(x) = |x| + 2$
Let $x = -3, -2, -1, 0, 1, 2, 3$

*Find an equation of each line satisfying the conditions given. Write Exercises 11–15 in standard form. Write Exercises 16–18 using function notation.*

**11.** Horizontal; through $(2, -8)$

**12.** Vertical; through $(-4, -3)$

△ **13.** Perpendicular to $x = 5$; through $(3, -2)$

**14.** Through $(4, -1)$; slope $-3$

**15.** Through $(0, -2)$; slope 5

**16.** Through $(4, -2)$ and $(6, -3)$

△ **17.** Through $(-1, 2)$; perpendicular to $3x - y = 4$

△ **18.** Parallel to $2y + x = 3$; through $(3, -2)$

△ **19.** Line $L_1$ has the equation $2x - 5y = 8$. Line $L_2$ passes through the points $(1, 4)$ and $(-1, -1)$. Determine whether these lines are parallel lines, perpendicular lines, or neither.

*Graph each inequality.*

**20.** $x \le -4$

**21.** $y > -2$

**22.** $2x - y > 5$

**23.** The intersection of $2x + 4y < 6$ and $y \le -4$

*Find the domain and range of each relation. Also determine whether the relation is a function.*

**24.**

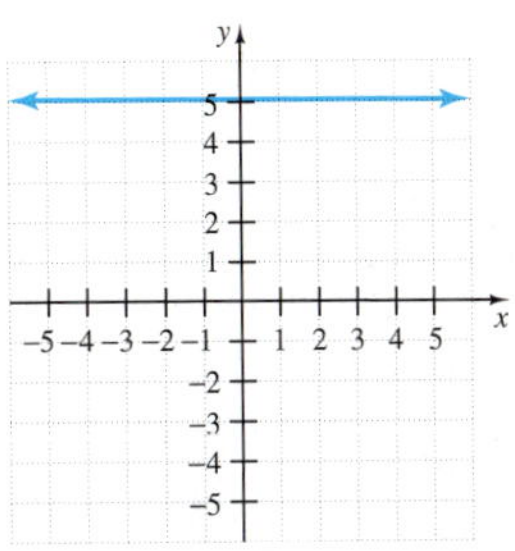

**25.**

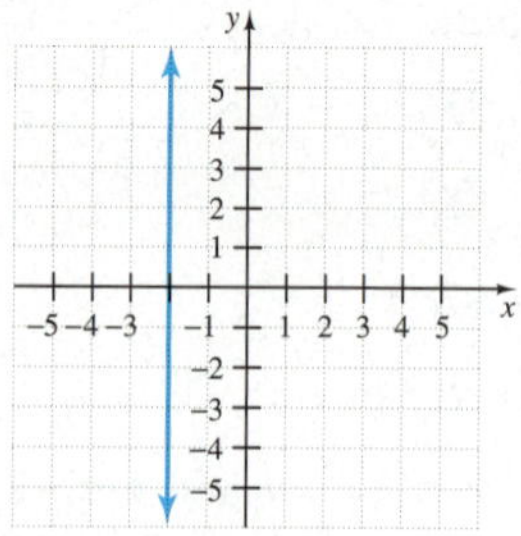

**26.**

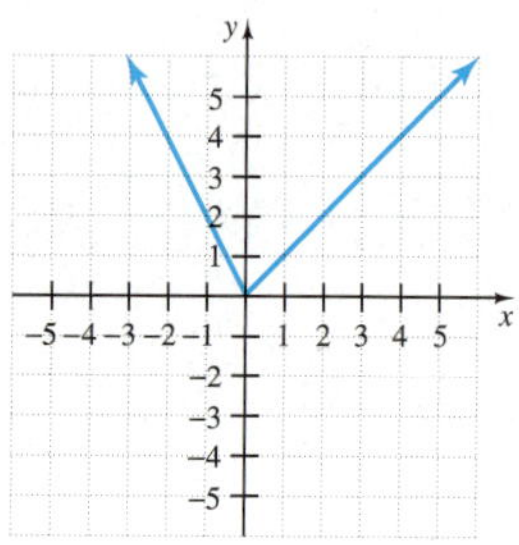

**27.**

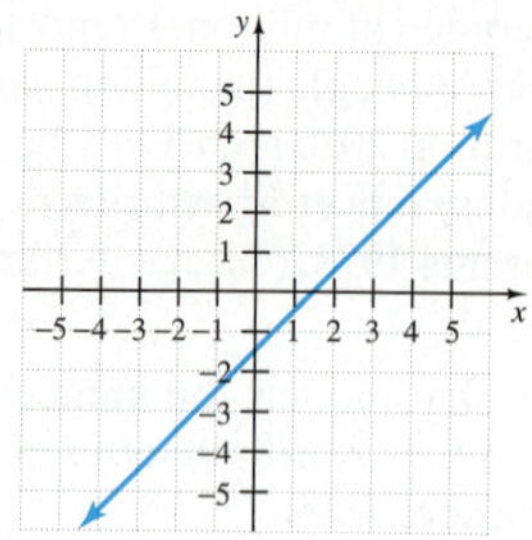

**28.** The average yearly earnings for high school graduates age 18 and older is given by the linear function

$$f(x) = 732x + 21{,}428$$

where $x$ is the number of years since 1996 that a person graduated. (*Source:* U.S. Census Bureau)

**a.** Find the average earnings in 1998 for high school graduates.

**b.** Predict the average earnings for high school graduates in the year 2005.

**c.** Predict the first year that the average earnings for high school graduates will be greater than $30,000.

**d.** Find and interpret the slope of this equation.

**e.** Find and interpret the $y$-intercept of this equation.

## CHAPTER 3 CUMULATIVE REVIEW

**1.** Evaluate: $3x - y$ when $x = 15$ and $y = 4$.

**2.** Determine whether the following statements are true or false.

**a.** 3 is a real number.

**b.** $\frac{1}{5}$ is an irrational number.

**c.** Every rational number is an integer.

**d.** $\{1, 5\} \subseteq \{2, 3, 4, 5\}$

**3.** Subtract.

**a.** $2 - 8$ **b.** $-8 - (-1)$

**c.** $-11 - 5$ **d.** $10.7 - (-9.8)$

**e.** $\frac{2}{3} - \frac{1}{2}$ **f.** $1 - 0.06$

**g.** Subtract 7 from 4.

**4.** Simplify each expression.

**a.** $3^2$ **b.** $\left(\frac{1}{2}\right)^4$

**c.** $-5^2$ **d.** $(-5)^2$

**e.** $-5^3$ **f.** $(-5)^3$

**5.** Insert $<, >$, or $=$ between each pair of numbers to form a true statement.

**a.** $-1 \quad -2$ **b.** $\frac{12}{4} \quad 3$

**c.** $-5 \quad 0$ **d.** $-3.5 \quad -3.05$

**6.** Write the multiplicative inverse, or reciprocal, of each.

**a.** 11 **b.** $-9$

**c.** $\frac{7}{4}$

**7.** Solve: $0.6 = 2 - 3.5c$

**8.** Solve for $x$: $3x + 5 = 3(x + 2)$

**9.** Find 16% of 25.

**10.** Kelsey Ohleger was helping her friend Benji Burnstine study for an algebra exam. Kelsey told Benji that her two latest art history quiz scores are two consecutive even integers whose sum is 174. Help Benji find the scores.

**11.** Solve $V = lwh$ for $h$.

**12.** Solve: $x - 2 < 5$. Graph the solution set.

**13.** Solve: $\frac{2}{5}(x - 6) \geq x - 1$

**14.** Solve: $2x \geq 0$ and $4x - 1 \leq -9$

**15.** Solve: $5x - 3 \leq 10$ or $x + 1 \geq 5$

**16.** Solve: $|5w + 3| = 7$

**17.** Solve: $|3x + 2| = |5x - 8|$

**18.** Solve for $x$: $|5x + 1| + 1 \leq 10$

**19.** Solve for $y$: $|y - 3| > 7$

**20.** Determine whether $(0, -12)$, $(1, 9)$, and $(2, -6)$ are solutions of the equation $3x - y = 12$.

**21.** Is the relation $y = 2x + 1$ also a function?

**22.** Find the $y$-intercept of the graph of each equation.

**a.** $f(x) = \frac{1}{2}x + \frac{3}{7}$

**b.** $y = -2.5x - 3.2$

**23.** Find the slope of the line whose equation is $f(x) = \frac{2}{3}x + 4$.

**24.** Write an equation of the line with $y$-intercept $(0, -3)$ and slope of $\frac{1}{4}$.

**25.** Graph: $2x - y < 6$.

## Planning for Growth

Development affects our lives and our environment in many ways. According to the Sierra Club Foundation, increased highway congestion lowered the average vehicle speed on the Washington D.C. beltway from 47 mph in 1981 to 23 mph in 1991. The American Farmland Trust reports that 50 acres of farmland are lost to development each hour. According to the Institute of Transportation Studies, at least half of the parking spaces in most American shopping malls are vacant at least 40% of the time.

Urban or regional planners tackle problems like these. They build or improve communities and cities and their environments. Planners analyze issues like population growth, housing needs, land use, urban or suburban sprawl, parks and recreation space, public transportation, and highways. They decide if current resources are adequate and, if not, propose ways to improve them. If their plans are approved by the community, planners then help with plan implementation. Planners must be problem solvers, understand how governments work, listen and communicate well, work well as part of a team, understand the principles of city design, and analyze and interpret data, as well as be able to identify trends in data.

*For more information about careers in urban and regional planning, visit the American Planning Association Website by first going to www.prenhall.com/martin-gay.*

*In the Spotlight on Decision Making feature on page 223, you will have the opportunity to make a decision about city bus routes as an urban planner.*

# SYSTEMS OF EQUATIONS

# 4


4.1 Solving Systems of Linear Equations in Two Variables

4.2 Solving Systems of Linear Equations in Three Variables

4.3 Systems of Linear Equations and Problem Solving

4.4 Solving Systems of Equations by Matrices

4.5 Solving Systems of Equations by Determinants


In this chapter, two or more equations in two or more variables are solved simultaneously. Such a collection of equations is called a **system of equations**. Systems of equations are good mathematical models for many real-world problems because these problems may involve several related patterns.

## 4.1 SOLVING SYSTEMS OF LINEAR EQUATIONS IN TWO VARIABLES

**OBJECTIVES**

1. Solve a system by graphing.
2. Solve a system by substitution.
3. Solve a system by elimination.

1 An important problem that often occurs in the fields of business and economics concerns the concepts of revenue and cost. For example, suppose that a small manufacturing company begins to manufacture and sell compact disc storage units. The revenue of a company is the company's income from selling these units, and the cost is the amount of money that a company spends to manufacture these units. The following coordinate system shows the graphs of revenue and cost for the storage units.

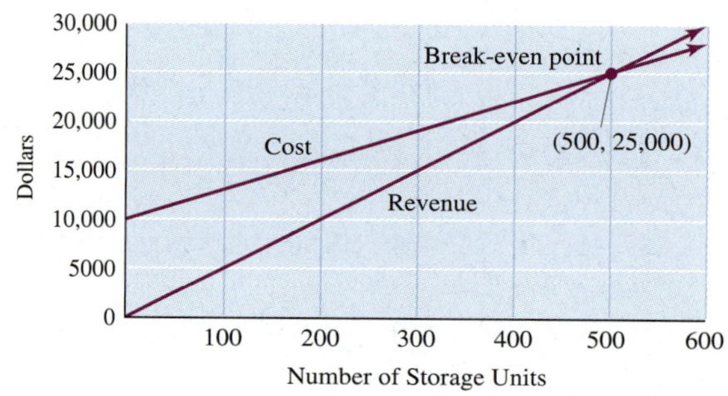


These lines intersect at the point (500, 25,000). This means that when 500 storage units are manufactured and sold, both cost and revenue are $25,000. In business, this point of intersection is called the **break-even point.** Notice that for $x$-values (units sold) less than 500, the cost graph is above the revenue graph, meaning that cost of manufacturing is greater than revenue, and so the company is losing money. For $x$-values (units sold) greater than 500, the revenue graph is above the cost graph, meaning that revenue is greater than cost, and so the company is making money.

Recall from Chapter 3 that each line is a graph of some linear equation in two variables. Both equations together form a **system of equations.** The common point of intersection is called the **solution of the system.** Some examples of systems of linear equations in two variables are

***Systems of Linear Equations in Two Variables***

$$\begin{cases} x - 2y = -7 \\ 3x + y = 0 \end{cases} \qquad \begin{cases} x = 5 \\ x + \dfrac{y}{2} = 9 \end{cases} \qquad \begin{cases} x - 3 = 2y + 6 \\ y = 1 \end{cases}$$

Recall that a solution of an equation in two variables is an ordered pair $(x, y)$ that makes the equation true. A **solution of a system** of two equations in two variables is an ordered pair $(x, y)$ that makes both equations true.

**Example 1** Determine whether the given ordered pair is a solution of the system.

**a.** $\begin{cases} -x + y = 2 \\ 2x - y = -3 \end{cases}$ $(-1, 1)$

**b.** $\begin{cases} 5x + 3y = -1 \\ x - y = 1 \end{cases}$ $(-2, 3)$

*Solution* **a.** We replace $x$ with $-1$ and $y$ with 1 in each equation.

$-x + y = 2$ First equation
$-(-1) + (1) \stackrel{?}{=} 2$ Let $x = -1$ and $y = 1$.
$1 + 1 \stackrel{?}{=} 2$
$2 = 2$ True.

$2x - y = -3$ Second equation
$2(-1) - (1) \stackrel{?}{=} -3$ Let $x = -1$ and $y = 1$.
$-2 - 1 \stackrel{?}{=} -3$
$-3 = -3$ True.

Since $(-1, 1)$ makes *both* equations true, it is a solution.

**b.** We replace $x$ with $-2$ and $y$ with 3 in each equation.

$5x + 3y = -1$ First equation
$5(-2) + 3(3) \stackrel{?}{=} -1$ Let $x = -2$ and $y = 3$.
$-10 + 9 \stackrel{?}{=} -1$
$-1 = -1$ True.

$x - y = 1$ Second equation
$(-2) - (3) \stackrel{?}{=} 1$ Let $x = -2$ and $y = 3$.
$-5 = 1$ False.

Since the ordered pair $(-2, 3)$ does not make *both* equations true, it is not a solution of the system.

We can estimate the solutions of a system by graphing each equation on the same coordinate system and estimating the coordinates of any point of intersection.

**Example 2** Solve each system by graphing. If the system has just one solution, estimate the solution.

**a.** $\begin{cases} x + y = 2 \\ 3x - y = -2 \end{cases}$

**b.** $\begin{cases} x - 2y = 4 \\ x = 2y \end{cases}$

**c.** $\begin{cases} 2x + 4y = 10 \\ x + 2y = 5 \end{cases}$

*Solution* Since the graph of a linear equation in two variables is a line, graphing two such equations yields two lines in a plane.

**a.** $\begin{cases} x + y = 2 \\ 3x - y = -2 \end{cases}$

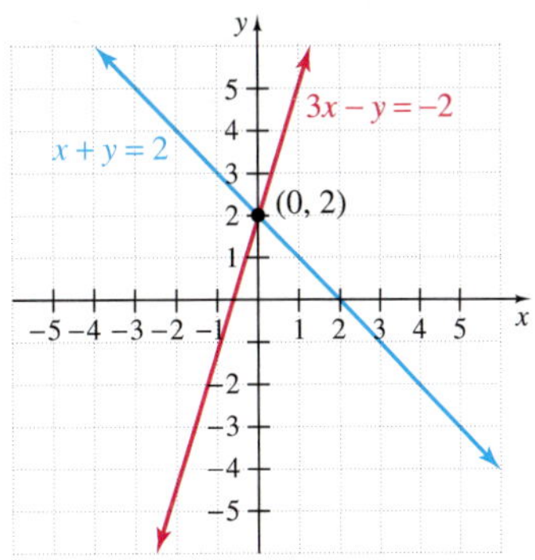


These lines intersect at one point as shown. The coordinates of the point of intersection appear to be $(0, 2)$. Check this estimated solution by replacing $x$ with 0 and $y$ with 2 in **both** equations.

$x + y = 2$ First equation.
$0 + 2 \stackrel{?}{=} 2$ Let $x = 0$ and $y = 2$.
$2 = 2$ True.

$3x - y = -2$ Second equation.
$3 \cdot 0 - 2 \stackrel{?}{=} -2$ Let $x = 0$ and $y = 2$.
$-2 = -2$ True.

The ordered pair $(0, 2)$ does satisfy both equations. We conclude therefore that $(0, 2)$ is the solution of the system. A system that has at least one solution, such as this one, is said to be **consistent.**

**b.** $\begin{cases} x - 2y = 4 \\ x = 2y \end{cases}$

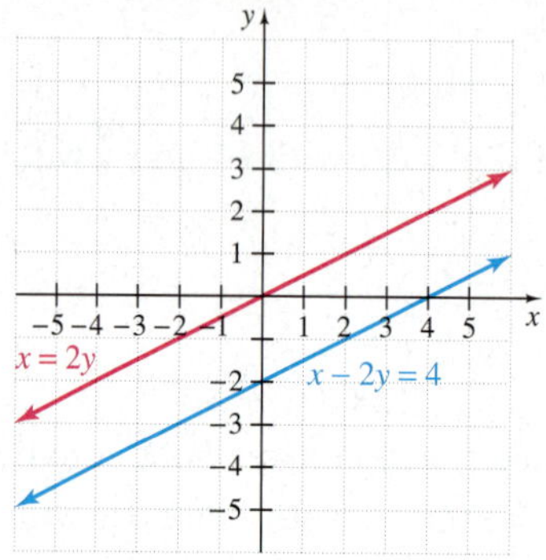


The lines appear to be parallel. To be sure, write each equation in point–slope form, $y = mx + b$.

| | | | |
|---|---|---|---|
| $x - 2y = 4$ | First equation. | $x = 2y$ | Second equation. |
| $-2y = -x + 4$ | Subtract $x$ from both sides. | $\frac{1}{2}x = y$ | Divide both sides by 2. |
| $y = \frac{1}{2}x - 2$ | Divide both sides by −2. | $y = \frac{1}{2}x$ | |

The graphs of these equations have the same slope, $\frac{1}{2}$, but different $y$-intercepts, so we have confirmed that these lines are parallel. Therefore, the system has no solution since the equations have no common solution (there are no intersection points). A system that has no solution is said to be **inconsistent.**

**c.** $\begin{cases} 2x + 4y = 10 \\ x + 2y = 5 \end{cases}$

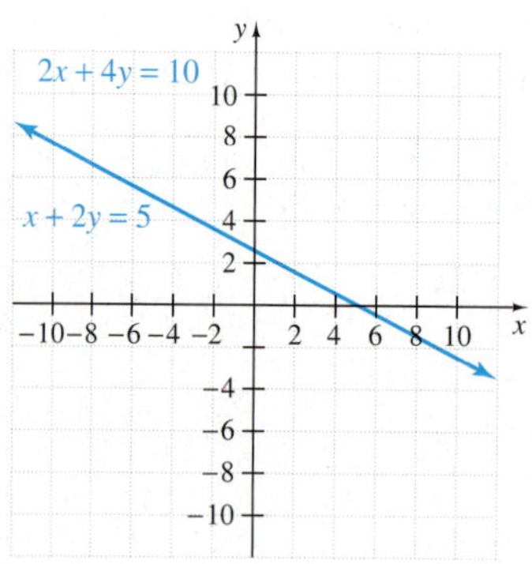


The graph of each equation is the same line. To confirm this, notice that if both sides of the second equation are multiplied by 2, the result is the first equation. This means that the equations have identical solutions. Any ordered pair solution of one equation satisfies the other equation also. Thus, these equations are said to be **dependent equations.** The solution set of the system is $\{(x, y) | x + 2y = 5\}$ or, equivalently, $\{(x, y) | 2x + 4y = 10\}$ since the lines describe identical ordered pairs. Written this way, the solution set is read "the set of all ordered pairs $(x, y)$, such that $2x + 4y = 10$." There are therefore an infinite number of solutions to this system.

We can summarize the information discovered in Example 2 as follows.

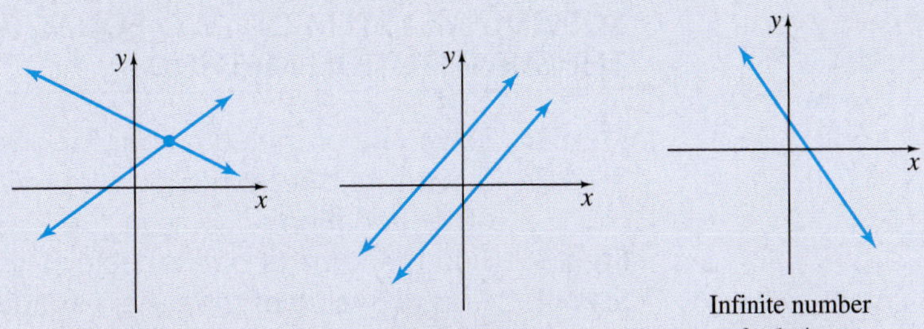


2 Graphing the equations of a system by hand is often a good method of finding approximate solutions of a system, but it is not a reliable method of finding exact solutions of a system. We turn instead to two algebraic methods of solving systems. We use the first method, the **substitution method,** to solve the system

$$\begin{cases} 2x + 4y = -6 & \text{First equation} \\ x = 2y - 5 & \text{Second equation} \end{cases}$$

**Example 3** Use the substitution method to solve the system.

$$\begin{cases} 2x + 4y = -6 & \text{First equation} \\ x = 2y - 5 & \text{Second equation} \end{cases}$$

*Solution* In the second equation, we are told that $x$ is equal to $2y - 5$. Since they are equal, we can *substitute* $2y - 5$ for $x$ in the first equation. This will give us an equation in one variable, which we can solve for $y$.

$$\begin{aligned} 2x + 4y &= -6 && \text{First equation} \\ 2(2y - 5) + 4y &= -6 && \text{Substitute } 2y - 5 \text{ for } x. \\ 4y - 10 + 4y &= -6 \\ 8y &= 4 \\ y &= \frac{4}{8} = \frac{1}{2} && \text{Solve for } y. \end{aligned}$$

The $y$-coordinate of the solution is $\frac{1}{2}$. To find the $x$-coordinate, we replace $y$ with $\frac{1}{2}$ in the second equation,

$$\begin{aligned} x &= 2y - 5. \\ x &= 2y - 5 \\ x &= 2\left(\frac{1}{2}\right) - 5 = 1 - 5 = -4 \end{aligned}$$

The ordered pair solution is $\left(-4, \frac{1}{2}\right)$. Check to see that $\left(-4, \frac{1}{2}\right)$ satisfies both equations of the system.

The steps below summarize the substitution method.

**SOLVING A SYSTEM OF TWO EQUATIONS USING THE SUBSTITUTION METHOD**

**Step 1:** Solve one of the equations for one of its variables.
**Step 2:** Substitute the expression for the variable found in Step 1 into the other equation.
**Step 3:** Find the value of one variable by solving the equation from Step 2.
**Step 4:** Find the value of the other variable by substituting the value found in Step 3 into the equation from Step 1.
**Step 5:** Check the ordered pair solution in *both* original equations.

**Example 4** Use the substitution method to solve the system.

$$\begin{cases} -\dfrac{x}{6} + \dfrac{y}{2} = \dfrac{1}{2} \\ \dfrac{x}{3} - \dfrac{y}{6} = -\dfrac{3}{4} \end{cases}$$

*Solution* First we multiply each equation by its least common denominator to clear the system of fractions. We multiply the first equation by 6 and the second equation by 12.

$$\begin{cases} 6\left(-\dfrac{x}{6} + \dfrac{y}{2}\right) = 6\left(\dfrac{1}{2}\right) \\ 12\left(\dfrac{x}{3} - \dfrac{y}{6}\right) = 12\left(-\dfrac{3}{4}\right) \end{cases} \quad \text{simplifies to} \quad \begin{cases} -x + 3y = 3 & \text{First equation} \\ 4x - 2y = -9 & \text{Second equation} \end{cases}$$

**HELPFUL HINT**
To avoid tedious fractions, solve for a variable whose coefficient is 1 or −1, if possible.

To use the substitution method, we now solve the first equation for $x$.

$$-x + 3y = 3 \quad \text{First equation}$$
$$3y - 3 = x \quad \text{Solve for } x.$$

Next we replace $x$ with $3y - 3$ in the second equation.

$$4x - 2y = -9 \quad \text{Second equation}$$

$$4(3y - 3) - 2y = -9$$
$$12y - 12 - 2y = -9$$
$$10y = 3$$
$$y = \frac{3}{10} \quad \text{Solve for } y.$$

To find the corresponding $x$-coordinate, we replace $y$ with $\dfrac{3}{10}$ in the equation $x = 3y - 3$. Then

$$x = 3\left(\frac{3}{10}\right) - 3 = \frac{9}{10} - 3 = \frac{9}{10} - \frac{30}{10} = -\frac{21}{10}$$

The ordered pair solution is $\left(-\frac{21}{10}, \frac{3}{10}\right)$. Check to see that this solution satisfies both original equations.

**HELPFUL HINT**

If a system of equations contains equations with fractions, first clear the equations of fractions.

3 The **elimination method,** or **addition method,** is a second algebraic technique for solving systems of equations. For this method, we rely on a version of the addition property of equality, which states that "equals added to equals are equal."

If $A = B$ and $C = D$ then $A + C = B + D$.

**Example 5** Use the elimination method to solve the system.

$$\begin{cases} x - 5y = -12 & \text{First equation} \\ -x + y = 4 & \text{Second equation} \end{cases}$$

*Solution* Since the left side of each equation is equal to the right side, we add equal quantities by adding the left sides of the equations and the right sides of the equations. This sum gives us an equation in one variable, $y$, which we can solve for $y$.

$$\begin{aligned} x - 5y &= -12 && \text{First equation} \\ -x + y &= 4 && \text{Second equation} \\ \hline -4y &= -8 && \text{Add.} \\ y &= 2 && \text{Solve for } y. \end{aligned}$$

The $y$-coordinate of the solution is 2. To find the corresponding $x$-coordinate, we replace $y$ with 2 in either original equation of the system. Let's use the second equation.

$$\begin{aligned} -x + y &= 4 && \text{Second equation} \\ -x + 2 &= 4 && \text{Let } y = 2. \\ -x &= 2 \\ x &= -2 \end{aligned}$$

The ordered pair solution is $(-2, 2)$. Check to see that $(-2, 2)$ satisfies both equations of the system.

The steps below summarize the elimination method.

**SOLVING A SYSTEM OF TWO LINEAR EQUATIONS USING THE ELIMINATION METHOD**

**Step 1:** Rewrite each equation in standard form, $Ax + By = C$.
**Step 2:** If necessary, multiply one or both equations by some nonzero number so that the coefficient of one variable in one equation is the opposite of its coefficient in the other equation.
**Step 3:** Add the equations.
**Step 4:** Find the value of one variable by solving the equation from Step 3.
**Step 5:** Find the value of the second variable by substituting the value found in Step 4 into either original equation.
**Step 6:** Check the proposed ordered pair solution in *both* original equations.

**Example 6** Use the elimination method to solve the system.

$$\begin{cases} 3x - 2y = 10 \\ 4x - 3y = 15 \end{cases}$$

*Solution* If we add the two equations, the sum will still be an equation in two variables. Notice, however, that we can eliminate $y$ when the equations are added if we multiply both sides of the first equation by 3 and both sides of the second equation by −2. Then

$$\begin{cases} 3(3x - 2y) = 3(10) \\ -2(4x - 3y) = -2(15) \end{cases} \quad \text{simplifies to} \quad \begin{cases} 9x - 6y = 30 \\ -8x + 6y = -30 \end{cases}$$

Next we add the left sides and add the right sides.

$$\begin{array}{r} 9x - 6y = 30 \\ -8x + 6y = -30 \\ \hline x \qquad = 0 \end{array}$$

To find $y$, we let $x = 0$ in either equation of the system.

$$\begin{aligned} 3x - 2y &= 10 && \text{First equation} \\ 3(0) - 2y &= 10 && \text{Let } x = 0. \\ -2y &= 10 \\ y &= -5 \end{aligned}$$

The ordered pair solution is $(0, -5)$. Check to see that $(0, -5)$ satisfies both equations.

**Example 7** Use the elimination method to solve the system.

$$\begin{cases} 3x + \dfrac{y}{2} = 2 \\ 6x + y = 5 \end{cases}$$

*Solution* If we multiply both sides of the first equation by −2, the coefficients of $x$ in the two equations will be opposites. Then

$$\begin{cases} -2\left(3x + \dfrac{y}{2}\right) = -2(2) \\ 6x + y \quad = 5 \end{cases} \quad \text{simplifies to} \quad \begin{cases} -6x - y = -4 \\ 6x + y = 5 \end{cases}$$

Now we can add the left sides and add the right sides.

$$\begin{array}{r} -6x - y = -4 \\ 6x + y = 5 \\ \hline 0 = 1 \end{array} \quad \text{False.}$$

The resulting equation, $0 = 1$, is false for all values of $y$ or $x$. Thus, the system has no solution. The solution set is $\{\ \}$ or $\varnothing$. This system is inconsistent, and the graphs of the equations are parallel lines.

**Example 8** Use the elimination method to solve the system.

$$\begin{cases} -5x - 3y = 9 \\ 10x + 6y = -18 \end{cases}$$

*Solution* To eliminate $x$ when the equations are added, we multiply both sides of the first equation by 2. Then

$$\begin{cases} 2(-5x - 3y) = 2(9) \\ 10x + 6y = -18 \end{cases} \quad \text{simplifies to} \quad \begin{cases} -10x - 6y = 18 \\ 10x + 6y = -18 \end{cases}$$

Next we add the equations.

$$\begin{array}{r} -10x - 6y = 18 \\ 10x + 6y = -18 \\ \hline 0 = 0 \end{array}$$

The resulting equation, $0 = 0$, is true for all possible values of $y$ or $x$. Notice in the original system that if both sides of the first equation are multiplied by $-2$, the result is the second equation. This means that the two equations are equivalent. They have the same solution set and there are an infinite number of solutions. Thus, the equations of this system are dependent, and the solution set of the system is

$$\{(x, y) \mid -5x - 3y = 9\} \quad \text{or, equivalently,} \quad \{(x, y) \mid 10x + 6y = -18\}.$$

## SPOTLIGHT ON DECISION MAKING

Suppose you have just signed a 12-month lease for an apartment. After moving in, you find that the tap water tastes terrible. You have two options: (a) buy bottled water or (b) buy a reusable water filter pitcher and filters to filter the tap water. You estimate that you use 10 gallons of drinking water each week. Buying bottled water costs \$0.50 per gallon. Using a water filter pitcher involves the following costs:

- A reusable water filter pitcher costs \$50.
- A water filter costs \$10 each. Each water filter lasts for 40 gallons, or 4 weeks in your situation. The cost of a water filter is \$2.50 per week.
- Tap water costs \$0.005 per gallon

An equation for the cumulative cost $y$ of the bottled water option after $x$ weeks is

$y$ = (cost per gallon of bottled water)(number of gallons used per week)$x$

An equation for the cumulative cost $y$ of filtered water option after $x$ weeks is

$y$ = (cost of reusable pitcher) + (cost of water filter per week)$x$ + (cost per gallon of tap water)(number of gallons used per week)$x$

Which option would you choose? Why?

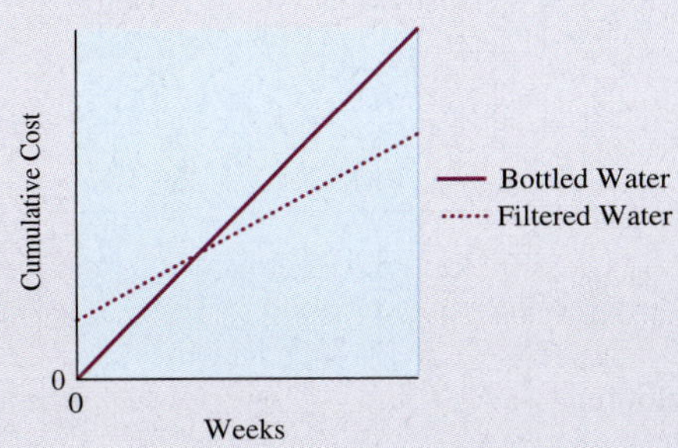

## GRAPHING CALCULATOR EXPLORATIONS

A graphing calculator may be used to approximate solutions of systems of equations by graphing each equation on the same set of axes and approximating any points of intersection. For example, approximate the solution of the system

$$\begin{cases} y = -2.6x + 5.6 \\ y = 4.3x - 4.9 \end{cases}$$

First use a standard window and graph both equations on a single screen.

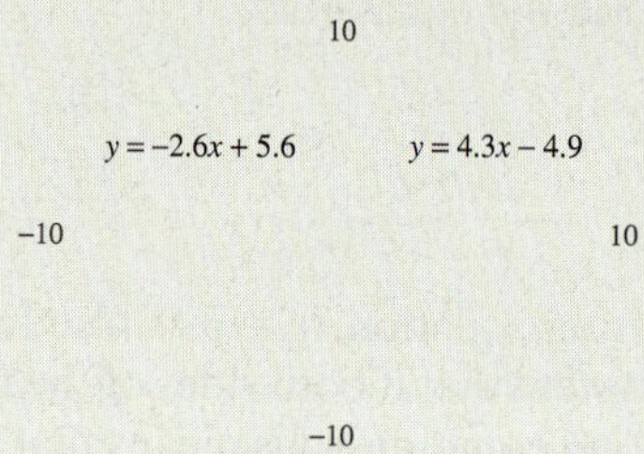


The two lines intersect. To approximate the point of intersection, trace to the point of intersection and use an Intersect feature of the graphing calculator or a Zoom In feature.

Using either method, we find that the approximate point of intersection is (1.52, 1.64).

*Solve each system of equations. Approximate the solutions to two decimal places.*

**1.** $y = -1.65x + 3.65$
$y = 4.56x - 9.44$

**2.** $y = 7.61x + 3.48$
$y = -1.26x - 6.43$

**3.** $2.33x - 4.72y = 10.61$
$5.86x + 6.22y = -8.89$

**4.** $-7.89x - 5.68y = 3.26$
$-3.65x + 4.98y = 11.77$

## MENTAL MATH

*Match each graph with the solution of the corresponding system.*

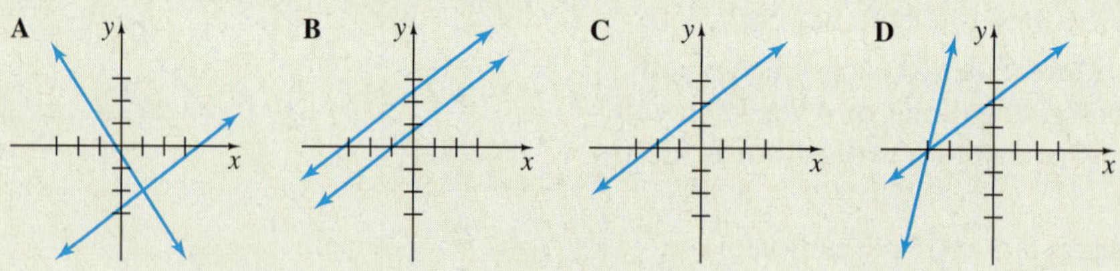


**1.** no solution

**2.** Infinite number of solutions

**3.** (1, −2)

**4.** (−3, 0)

## Exercise Set 4.1

*Determine whether each given ordered pair is a solution of each system. See Example 1.*

1. $\begin{cases} x - y = 3 \\ 2x - 4y = 8 \end{cases}$ $(2, -1)$

2. $\begin{cases} x - y = -4 \\ 2x + 10y = 4 \end{cases}$ $(-3, 1)$

3. $\begin{cases} 2x - 3y = -9 \\ 4x + 2y = -2 \end{cases}$ $(3, 5)$

4. $\begin{cases} 2x - 5y = -2 \\ 3x + 4y = 4 \end{cases}$ $(4, 2)$

5. $\begin{cases} y = -5x \\ x = -2 \end{cases}$ $(-2, 10)$

6. $\begin{cases} y = 6 \\ x = -2y \end{cases}$ $(-12, 6)$

*Solve each system by graphing. See Example 2.*

7. $\begin{cases} x + y = 1 \\ x - 2y = 4 \end{cases}$

8. $\begin{cases} 2x - y = 8 \\ x + 3y = 11 \end{cases}$

9. $\begin{cases} 2y - 4 = 0 \\ x + 2y = 5 \end{cases}$

10. $\begin{cases} 4x - y = 6 \\ x - y = 0 \end{cases}$

11. $\begin{cases} 3x - y = 4 \\ 6x - 2y = 4 \end{cases}$

12. $\begin{cases} -x + 3y = 6 \\ 3x - 9y = 9 \end{cases}$

13. Can a system consisting of two linear equations have exactly two solutions? Explain why or why not.

14. Suppose the graph of the equations in a system of two equations in two variables consists of a circle and a line. Discuss the possible number of solutions for this system.

*Solve each system of equations by the substitution method. See Examples 3 and 4.*

15. $\begin{cases} x + y = 10 \\ y = 4x \end{cases}$

16. $\begin{cases} 5x + 2y = -17 \\ x = 3y \end{cases}$

17. $\begin{cases} 4x - y = 9 \\ 2x + 3y = -27 \end{cases}$

18. $\begin{cases} 3x - y = 6 \\ -4x + 2y = -8 \end{cases}$

19. $\begin{cases} \frac{1}{2}x + \frac{3}{4}y = -\frac{1}{4} \\ \frac{3}{4}x - \frac{1}{4}y = 1 \end{cases}$

20. $\begin{cases} \frac{2}{5}x + \frac{1}{5}y = -1 \\ x + \frac{2}{5}y = -\frac{8}{5} \end{cases}$

21. $\begin{cases} \frac{x}{3} + y = \frac{4}{3} \\ -x + 2y = 11 \end{cases}$

22. $\begin{cases} \frac{x}{8} - \frac{y}{2} = 1 \\ \frac{x}{3} - y = 2 \end{cases}$

*Solve each system of equations by the elimination method. See Examples 5–8.*

23. $\begin{cases} 2x - 4y = 0 \\ x + 2y = 5 \end{cases}$

24. $\begin{cases} 2x - 3y = 0 \\ 2x + 6y = 3 \end{cases}$

25. $\begin{cases} 5x + 2y = 1 \\ x - 3y = 7 \end{cases}$

26. $\begin{cases} 6x - y = -5 \\ 4x - 2y = 6 \end{cases}$

27. $\begin{cases} 5x - 2y = 27 \\ -3x + 5y = 18 \end{cases}$

28. $\begin{cases} 3x + 4y = 2 \\ 2x + 5y = -1 \end{cases}$

29. $\begin{cases} 3x - 5y = 11 \\ 2x - 6y = 2 \end{cases}$

30. $\begin{cases} 6x - 3y = -3 \\ 4x + 5y = -9 \end{cases}$

31. $\begin{cases} x - 2y = 4 \\ 2x - 4y = 4 \end{cases}$

32. $\begin{cases} -x + 3y = 6 \\ 3x - 9y = 9 \end{cases}$

33. $\begin{cases} 3x + y = 1 \\ 2y = 2 - 6x \end{cases}$

34. $\begin{cases} y = 2x - 5 \\ 8x - 4y = 20 \end{cases}$

35. Write a system of two linear equations in $x$ and $y$ that has the ordered pair solution $(2, 5)$.

36. Which method would you use to solve the system

$$\begin{cases} 5x - 2y = 6 \\ 2x + 3y = 5 \end{cases}$$

Explain your choice.

*Solve each system of equations.*

37. $\begin{cases} 2x + 5y = 8 \\ 6x + y = 10 \end{cases}$

38. $\begin{cases} x - 4y = -5 \\ -3x - 8y = 0 \end{cases}$

39. $\begin{cases} x + y = 1 \\ x - 2y = 4 \end{cases}$

40. $\begin{cases} 2x - y = 8 \\ x + 3y = 11 \end{cases}$

41. $\begin{cases} \frac{1}{3}x + y = \frac{4}{3} \\ -\frac{1}{4}x - \frac{1}{2}y = -\frac{1}{4} \end{cases}$

42. $\begin{cases} \frac{3}{4}x - \frac{1}{2}y = -\frac{1}{2} \\ x + y = -\frac{3}{2} \end{cases}$

43. $\begin{cases} 2x + 6y = 8 \\ 3x + 9y = 12 \end{cases}$

44. $\begin{cases} x = 3y - 1 \\ 2x - 6y = -2 \end{cases}$

45. $\begin{cases} 4x + 2y = 5 \\ 2x + y = -1 \end{cases}$

46. $\begin{cases} 3x + 6y = 15 \\ 2x + 4y = 3 \end{cases}$

47. $\begin{cases} 10y - 2x = 1 \\ 5y = 4 - 6x \end{cases}$

48. $\begin{cases} 3x + 4y = 0 \\ 7x = 3y \end{cases}$

49. $\begin{cases} \frac{3}{4}x + \frac{5}{2}y = 11 \\ \frac{1}{16}x - \frac{3}{4}y = -1 \end{cases}$

50. $\begin{cases} \frac{2}{3}x + \frac{1}{4}y = -\frac{3}{2} \\ \frac{1}{2}x - \frac{1}{4}y = -2 \end{cases}$

**51.** $\begin{cases} x = 3y + 2 \\ 5x - 15y = 10 \end{cases}$

**52.** $\begin{cases} y = \frac{1}{7}x + 3 \\ x - 7y = -21 \end{cases}$

**53.** $\begin{cases} 2x - y = -1 \\ y = -2x \end{cases}$

**54.** $\begin{cases} x = \frac{1}{5}y \\ x - y = -4 \end{cases}$

**55.** $\begin{cases} 2x = 6 \\ y = 5 - x \end{cases}$

**56.** $\begin{cases} x = 3y + 4 \\ -y = 5 \end{cases}$

**57.** $\begin{cases} \frac{x+5}{2} = \frac{6-4y}{3} \\ \frac{3x}{5} = \frac{21-7y}{10} \end{cases}$

**58.** $\begin{cases} \frac{y}{5} = \frac{8-x}{2} \\ x = \frac{2y-8}{3} \end{cases}$

**59.** $\begin{cases} 4x - 7y = 7 \\ 12x - 21y = 24 \end{cases}$

**60.** $\begin{cases} 2x - 5y = 12 \\ -4x + 10y = 20 \end{cases}$

**61.** $\begin{cases} \frac{2}{3}x - \frac{3}{4}y = -1 \\ -\frac{1}{6}x + \frac{3}{8}y = 1 \end{cases}$

**62.** $\begin{cases} \frac{1}{2}x - \frac{1}{3}y = -3 \\ \frac{1}{8}x + \frac{1}{6}y = 0 \end{cases}$

**63.** $\begin{cases} 0.7x - 0.2y = -1.6 \\ 0.2x - y = -1.4 \end{cases}$

**64.** $\begin{cases} -0.7x + 0.6y = 1.3 \\ 0.5x - 0.3y = -0.8 \end{cases}$

**65.** $\begin{cases} 4x - 1.5y = 10.2 \\ 2x + 7.8y = -25.68 \end{cases}$

**66.** $\begin{cases} x - 3y = -5.3 \\ 6.3x + 6y = 3.96 \end{cases}$

*The concept of supply and demand is used often in business. In general, as the unit price of a commodity increases, the demand for that commodity decreases. Also, as a commodity's unit price increases, the manufacturer normally increases the supply. The point where supply is equal to demand is called the equilibrium point. The following graph shows the graph of a demand equation and a supply equation for ties. The x-axis represents number of ties in thousands, and the y-axis represents the cost of a tie. Use this graph for Exercises 67–70.*

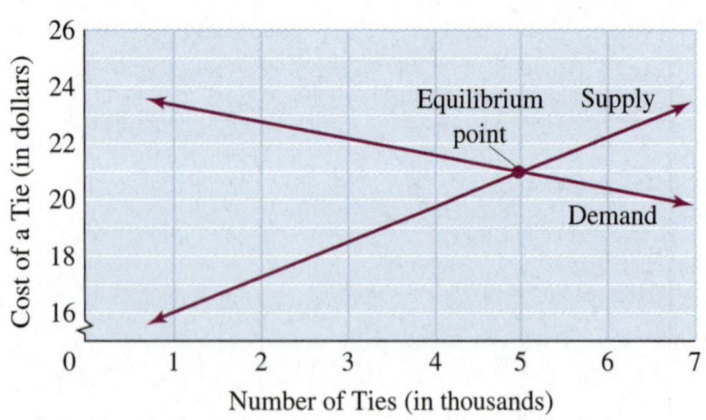


**67.** Find the number of ties and the price per tie when supply equals demand.

**68.** When $x$ is between 3 and 4, is supply greater than demand or is demand greater than supply?

**69.** When $x$ is greater than 7, is supply greater than demand or is demand greater than supply?

**70.** For what $x$-values are the $y$-values corresponding to the supply equation greater than the $y$-values corresponding to the demand equation?

*The revenue equation for a certain brand of toothpaste is $y = 2.5x$, where x is the number of tubes of toothpaste sold and y is the total income for selling x tubes. The cost equation is $y = 0.9x + 3000$, where x is the number of tubes of toothpaste manufactured and y is the cost of producing x tubes. The following set of axes shows the graph of the cost and revenue equations. Use this graph for Exercises 71–76.*

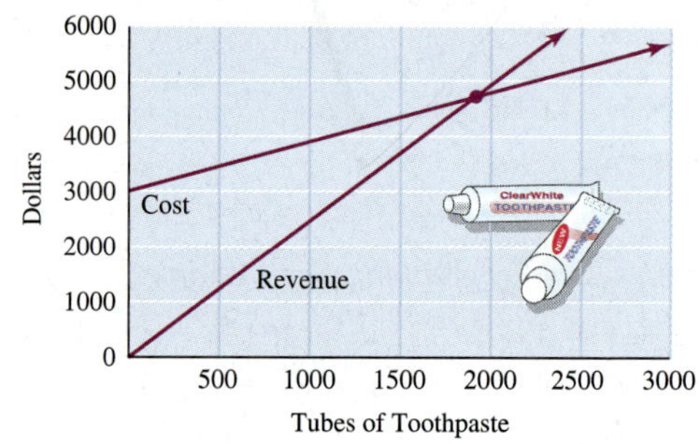


**71.** Find the coordinates of the point of intersection by solving the system

$$\begin{cases} y = 2.5x \\ y = 0.9x + 3000 \end{cases}$$

**72.** Explain the meaning of the ordered pair point of intersection.

**73.** If the company sells 2000 tubes of toothpaste, does the company make money or lose money?

**74.** If the company sells 1000 tubes of toothpaste, does the company make money or lose money?

**75.** For what $x$-values will the company make a profit? (*Hint:* For what $x$-values is the revenue graph "higher" than the cost graph?)

**76.** For what $x$-values will the company lose money? (*Hint:* For what $x$-values is the revenue graph "lower" than the cost graph?)

**77.** The amount $y$ of red meat consumed per person in the United States (in pounds) in the year $x$ can be modeled by the linear equation $y = -0.6x + 121.2$. The amount $y$ of all poultry consumed per person in the United States (in pounds) in the year $x$ can be modeled by the linear equation $y = 1.7x + 88$. In both models, $x = 0$ represents the year 1995. (*Source:* Based on data and forecasts from the Economic Research Service, U.S. Department of Agriculture, 1995–1999)

**a.** What does the slope of each equation tell you about the patterns of red meat and poultry consumption in the United States?

**b.** Solve this system of equations. (Round your final results to the nearest whole numbers.)

**c.** Explain the meaning of your answer to Part **b**.

**78.** The amount of U.S. federal government income $y$ (in billions of dollars) for the fiscal year $x$, from 1996 through 1999 ($x = 0$ represents 1996), can be modeled by the linear equation $y = 126.4x + 1455.4$. The amount of U.S. federal government expenditures (in billions of dollars) for the same period can be modeled by the linear equation $y = 48.6x + 1556.6$. Did expenses ever equal income during this period? If so, in what year? (*Source:* Financial Management Service, U.S. Dept. of the Treasury)

**79.** The number of milk cows $y$ (in thousands) on farms in the United States for the year $x$, from 1980 through 1995 ($x = 0$ represents 1980), can be modeled by the linear equation $107x + y = 11{,}096$. The number of sheep (in thousands) on farms in the United States for the same period can be modeled by the linear equation $y = -399x + 15{,}149$. In which year were there the same number of milk cows as sheep? (*Source:* National Agricultural Statistics Service)

## REVIEW EXERCISES

*Determine whether the given replacement values make each equation true or false. See Section 1.3.*

**80.** $3x - 4y + 2z = 5$; $x = 1$, $y = 2$, and $z = 5$

**81.** $x + 2y - z = 7$; $x = 2$, $y = -3$, and $z = 3$

**82.** $-x - 5y + 3z = 15$; $x = 0$, $y = -1$, and $z = 5$

**83.** $-4x + y - 8z = 4$; $x = 1$, $y = 0$, and $z = -1$

*Add the equations. See Section 4.1.*

**84.** $\begin{aligned} 3x + 2y - 5z &= 10 \\ -3x + 4y + z &= 15 \end{aligned}$

**85.** $\begin{aligned} x + 4y - 5z &= 20 \\ 2x - 4y - 2z &= -17 \end{aligned}$

**86.** $\begin{aligned} 10x + 5y + 6z &= 14 \\ -9x + 5y - 6z &= -12 \end{aligned}$

**87.** $\begin{aligned} -9x - 8y - z &= 31 \\ 9x + 4y - z &= 12 \end{aligned}$

## A Look Ahead

**Example**

Solve the system $\begin{cases} -\dfrac{4}{x} - \dfrac{4}{y} = -11 \\ \dfrac{1}{x} + \dfrac{1}{y} = 1 \end{cases}$

*Solution*

First, make the following substitution. Let $a = \dfrac{1}{x}$ and $b = \dfrac{1}{y}$ in both equations. Then

$$\begin{cases} -4\left(\dfrac{1}{x}\right) - 4\left(\dfrac{1}{y}\right) = -11 \\ \dfrac{1}{x} + \dfrac{1}{y} = 1 \end{cases}$$

is equivalent to $\begin{cases} -4a - 4b = -11 \\ a + b = 1 \end{cases}$

We solve by the elimination method. Multiplying both sides of the second equation by 4 and adding the left sides and the right sides of the equations,

$$\begin{cases} -4a - 4b = -11 \\ 4(a + b) = 4(1) \end{cases}$$

simplifies to $\begin{cases} -4a - 4b = -11 \\ 4a + 4b = 4 \end{cases}$

$$0 = -7 \quad \text{false}$$

The equation $0 = -7$ is false. Therefore, this system in $a$ and $b$ has no solution and hence the original system in $x$ and $y$ has no solution.

*Solve each system. See the preceding example.*

**88.** $\begin{cases} \dfrac{1}{x} + y = 12 \\ \dfrac{3}{x} - y = 4 \end{cases}$

**89.** $\begin{cases} x + \dfrac{2}{y} = 7 \\ 3x + \dfrac{3}{y} = 6 \end{cases}$

**90.** $\begin{cases} \dfrac{1}{x} + \dfrac{1}{y} = 5 \\ \dfrac{1}{x} - \dfrac{1}{y} = 1 \end{cases}$

**91.** $\begin{cases} \dfrac{2}{x} + \dfrac{3}{y} = 5 \\ \dfrac{5}{x} - \dfrac{3}{y} = 2 \end{cases}$

**92.** $\begin{cases} \dfrac{2}{x} + \dfrac{3}{y} = -1 \\ \dfrac{3}{x} - \dfrac{2}{y} = 18 \end{cases}$

**93.** $\begin{cases} \dfrac{3}{x} - \dfrac{2}{y} = -18 \\ \dfrac{2}{x} + \dfrac{3}{y} = 1 \end{cases}$

**94.** $\begin{cases} \dfrac{2}{x} - \dfrac{4}{y} = 5 \\ \dfrac{1}{x} - \dfrac{2}{y} = \dfrac{3}{2} \end{cases}$

**95.** $\begin{cases} \dfrac{5}{x} + \dfrac{7}{y} = 1 \\ -\dfrac{10}{x} - \dfrac{14}{y} = 0 \end{cases}$

## 4.2 SOLVING SYSTEMS OF LINEAR EQUATIONS IN THREE VARIABLES

CD-ROM SSM SSG Video

▶ **OBJECTIVE**

**1.** Solve a system of three linear equations in three variables.

In this section, the algebraic methods of solving systems of two linear equations in two variables are extended to systems of three linear equations in three variables. We call the equation $3x - y + z = -15$, for example, a **linear equation in three variables** since there are three variables and each variable is raised only to the power 1. A solution of this equation is an **ordered triple (*x*, *y*, *z*)** that makes the equation a true statement. For example, the ordered triple $(2, 0, -21)$ is a solution of $3x - y + z = -15$ since replacing $x$ with 2, $y$ with 0, and $z$ with $-21$ yields the true statement $3(2) - 0 + (-21) = -15$. The graph of this equation is a plane in three-dimensional space, just as the graph of a linear equation in two variables is a line in two-dimensional space.

Although we will not discuss the techniques for graphing equations in three variables, visualizing the possible patterns of intersecting planes gives us insight into the possible patterns of solutions of a system of three three-variable linear equations. There are four possible patterns.

1. Three planes have a single point in common. This point represents the single solution of the system. This system is **consistent**.

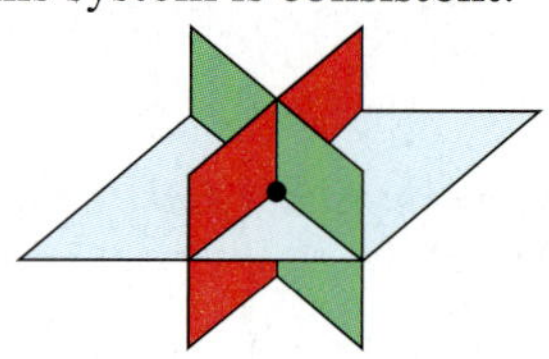

2. Three planes intersect at no point common to all three. This system has no solution. A few ways that this can occur are shown. This system is **inconsistent**.

3. Three planes intersect at all the points of a single line. The system has infinitely many solutions. This system is **consistent**.

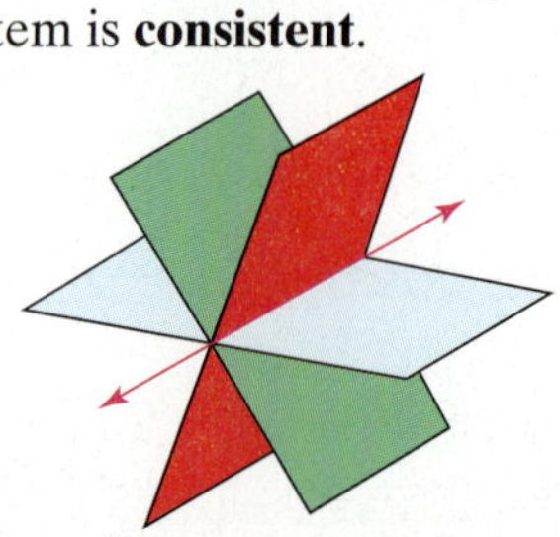

4. Three planes coincide at all points on the plane. The system is consistent, and the equations are **dependent**.

1 To use the elimination method to solve a system in three variables, we eliminate a variable and obtain a system in two variables.

**Example 1** Solve the system.

$$\begin{cases} 3x - y + z = -15 & \text{Equation (1)} \\ x + 2y - z = 1 & \text{Equation (2)} \\ 2x + 3y - 2z = 0 & \text{Equation (3)} \end{cases}$$

*Solution* Add equations (1) and (2) to eliminate $z$.

$$\begin{aligned} 3x - y + z &= -15 \\ x + 2y - z &= 1 \\ \hline 4x + y \quad &= -14 \qquad \text{Equation (4)} \end{aligned}$$

**HELPFUL HINT**
Don't forget to add two other equations besides equations (1) and (2) *and* to eliminate the same variable.

Next, add two *other* equations and *eliminate z again*. To do so, multiply both sides of equation (1) by 2 and add this resulting equation to equation (3). Then

$$\begin{cases} 2(3x - y + z) = 2(-15) \\ 2x + 3y - 2z = 0 \end{cases} \quad \text{simplifies to} \quad \begin{cases} 6x - 2y + 2z = -30 \\ 2x + 3y - 2z = 0 \end{cases}$$

$$8x + y = -30 \qquad \text{Equation (5)}$$

Now solve equations (4) and (5) for $x$ and $y$. To solve by elimination, multiply both sides of equation (4) by $-1$ and add this resulting equation to equation (5). Then

$$\begin{cases} -1(4x + y) = -1(-14) \\ 8x + y = -30 \end{cases} \quad \text{simplifies to} \quad \begin{cases} -4x - y = 14 \\ 8x + y = -30 \end{cases}$$

$$\begin{aligned} 4x &= -16 & \text{Add the equations.} \\ x &= -4 & \text{Solve for } x. \end{aligned}$$

Replace $x$ with $-4$ in equation (4) or (5).

$$\begin{aligned} 4x + y &= -14 & \text{Equation (4)} \\ 4(-4) + y &= -14 & \text{Let } x = -4. \\ y &= 2 & \text{Solve for } y. \end{aligned}$$

Finally, replace $x$ with $-4$ and $y$ with 2 in equation (1), (2), or (3).

$$\begin{aligned} x + 2y - z &= 1 & \text{Equation (2)} \\ -4 + 2(2) - z &= 1 & \text{Let } x = -4 \text{ and } y = 2. \\ -4 + 4 - z &= 1 \\ -z &= 1 \\ z &= -1 \end{aligned}$$

The solution is $(-4, 2, -1)$. To check, let $x = -4$, $y = 2$, and $z = -1$ in all three original equations of the system.

| ***Equation (1)*** | ***Equation (2)*** | ***Equation (3)*** |
|---|---|---|
| $3x - y + z = -15$ | $x + 2y - z = 1$ | $2x + 3y - 2z = 0$ |
| $3(-4) - 2 + (-1) = -15$ | $-4 + 2(2) - (-1) = 1$ | $2(-4) + 3(2) - 2(-1) = 0$ |
| $-12 - 2 - 1 = -15$ | $-4 + 4 + 1 = 1$ | $-8 + 6 + 2 = 0$ |
| $-15 = -15$ | $1 = 1$ | $0 = 0$ |
| True. | True. | True. |

All three statements are true, so the solution is $(-4, 2, -1)$.

**Example 2** Solve the system.

$$\begin{cases} 2x - 4y + 8z = 2 & (1) \\ -x - 3y + z = 11 & (2) \\ x - 2y + 4z = 0 & (3) \end{cases}$$

*Solution* Add equations (2) and (3) to eliminate $x$, and the new equation is

$$-5y + 5z = 11 \quad (4)$$

To eliminate $x$ again, multiply both sides of equation (2) by 2, and add the resulting equation to equation (1). Then

$$\begin{cases} 2x - 4y + 8z = 2 \\ 2(-x - 3y + z) = 2(11) \end{cases} \quad \text{simplifies to} \quad \begin{cases} 2x - 4y + 8z = 2 \\ -2x - 6y + 2z = 22 \end{cases}$$
$$-10y + 10z = 24 \quad (5)$$

Next, solve for $y$ and $z$ using equations (4) and (5). Multiply both sides of equation (4) by $-2$, and add the resulting equation to equation (5).

$$\begin{cases} -2(-5y + 5z) = -2(11) \\ -10y + 10z = 24 \end{cases} \quad \text{simplifies to} \quad \begin{cases} 10y - 10z = -22 \\ -10y + 10z = 24 \end{cases}$$
$$0 = 2 \quad \text{False.}$$

Since the statement is false, this system is inconsistent and has no solution. The solution set is the empty set { } or $\varnothing$.

The elimination method is summarized next.

### SOLVING A SYSTEM OF THREE LINEAR EQUATIONS BY THE ELIMINATION METHOD

**Step 1:** Write each equation in standard form $Ax + By + Cz = D$.
**Step 2:** Choose a pair of equations and use the equations to eliminate a variable.
**Step 3:** Choose any other pair of equations and eliminate the **same variable** as in Step 2.
**Step 4:** Two equations in two variables should be obtained from Step 2 and Step 3. Use methods from Section 4.1 to solve this system for both variables.
**Step 5:** To solve for the third variable, substitute the values of the variables found in Step 4 into any of the original equations containing the third variable.

**Example 3** Solve the system.

$$\begin{cases} 2x + 4y \qquad\quad = 1 & (1) \\ 4x \qquad\quad - 4z = -1 & (2) \\ \qquad\quad y - 4z = -3 & (3) \end{cases}$$

*Solution* Notice that equation (2) has no term containing the variable $y$. Let us eliminate $y$ using equations (1) and (3). Multiply both sides of equation (3) by $-4$, and add the resulting equation to equation (1). Then

$$\begin{cases} 2x + 4y \qquad\quad = 1 \\ \quad -4(y - 4z) = -4(-3) \end{cases} \quad \text{simplifies to} \quad \begin{cases} 2x + 4y \qquad\quad = 1 \\ \quad -4y + 16z = 12 \\ \hline 2x \qquad + 16z = 13 \end{cases} \quad (4)$$

Next, solve for $z$ using equations (4) and (2). Multiply both sides of equation (4) by $-2$ and add the resulting equation to equation (2).

$$\begin{cases} -2(2x + 16z) = -2(13) \\ \quad 4x - \;\; 4z = -1 \end{cases} \quad \text{simplifies to} \quad \begin{cases} -4x - 32z = -26 \\ \;\; 4x - \;\; 4z = -1 \\ \hline \qquad -36z = -27 \end{cases}$$

$$z = \frac{3}{4}$$

Replace $z$ with $\frac{3}{4}$ in equation (3) and solve for $y$.

$$y - 4\left(\frac{3}{4}\right) = -3 \qquad \text{Let } z = \frac{3}{4} \text{ in equation (3).}$$

$$y - 3 = -3$$

$$y = 0$$

Replace $y$ with 0 in equation (1) and solve for $x$.

$$2x + 4(0) = 1$$

$$2x = 1$$

$$x = \frac{1}{2}$$

The solution is $\left(\frac{1}{2}, 0, \frac{3}{4}\right)$. Check to see that this solution satisfies all three equations of the system.

**Example 4** Solve the system.

$$\begin{cases} x - 5y - 2z = 6 & (1) \\ -2x + 10y + 4z = -12 & (2) \\ \frac{1}{2}x - \frac{5}{2}y - z = 3 & (3) \end{cases}$$

*Solution* Multiply both sides of equation (3) by 2 to eliminate fractions, and multiply both sides of equation (2) by $-\frac{1}{2}$ so that the coefficient of $x$ is 1. The resulting system is then

$$\begin{cases} x - 5y - 2z = 6 & \text{(1)} \\ x - 5y - 2z = 6 & \text{Multiply (2) by } -\frac{1}{2}. \\ x - 5y - 2z = 6 & \text{Multiply (3) by 2.} \end{cases}$$

All three equations are identical, and therefore equations (1), (2), and (3) are all equivalent. There are infinitely many solutions of this system. The equations are dependent. The solution set can be written as $\{(x, y, z) \mid x - 5y - 2z = 6\}$.

## Exercise Set 4.2

*Solve each system. See Examples 1 and 3.*

**1.** $\begin{cases} x + y = 3 \\ 2y = 10 \\ 3x + 2y - 3z = 1 \end{cases}$

**2.** $\begin{cases} 5x = 5 \\ 2x + y = 4 \\ 3x + y - 4z = -15 \end{cases}$

**3.** $\begin{cases} 2x + 2y + z = 1 \\ -x + y + 2z = 3 \\ x + 2y + 4z = 0 \end{cases}$

**4.** $\begin{cases} 2x - 3y + z = 5 \\ x + y + z = 0 \\ 4x + 2y + 4z = 4 \end{cases}$

*Solve each system. See Examples 2 and 4.*

**5.** $\begin{cases} x - 2y + z = -5 \\ -3x + 6y - 3z = 15 \\ 2x - 4y + 2z = -10 \end{cases}$

**6.** $\begin{cases} 3x + y - 2z = 2 \\ -6x - 2y + 4z = -2 \\ 9x + 3y - 6z = 6 \end{cases}$

**7.** $\begin{cases} 4x - y + 2z = 5 \\ 2y + z = 4 \\ 4x + y + 3z = 10 \end{cases}$

**8.** $\begin{cases} 5y - 7z = 14 \\ 2x + y + 4z = 10 \\ 2x + 6y - 3z = 30 \end{cases}$

**9.** Write a system of linear equations in three variables that has $(-1, 2, -4)$ as a solution. (There are many possibilities.)

**10.** Write a system of three linear equations in three variables that has $(2, 1, 5)$ as a solution. (There are many possibilities.)

*Solve each system.*

**11.** $\begin{cases} x + 5z = 0 \\ 5x + y = 0 \\ y - 3z = 0 \end{cases}$

**12.** $\begin{cases} x - 5y = 0 \\ x - z = 0 \\ -x + 5z = 0 \end{cases}$

**13.** $\begin{cases} 6x - 5z = 17 \\ 5x - y + 3z = -1 \\ 2x + y = -41 \end{cases}$

**14.** $\begin{cases} x + 2y = 6 \\ 7x + 3y + z = -33 \\ x - z = 16 \end{cases}$

**15.** $\begin{cases} x + y + z = 8 \\ 2x - y - z = 10 \\ x - 2y - 3z = 22 \end{cases}$

**16.** $\begin{cases} 5x + y + 3z = 1 \\ x - y + 3z = -7 \\ -x + y = 1 \end{cases}$

**17.** $\begin{cases} x + 2y - z = 5 \\ 6x + y + z = 7 \\ 2x + 4y - 2z = 5 \end{cases}$

**18.** $\begin{cases} 4x - y + 3z = 10 \\ x + y - z = 5 \\ 8x - 2y + 6z = 10 \end{cases}$

**19.** $\begin{cases} 2x - 3y + z = 2 \\ x - 5y + 5z = 3 \\ 3x + y - 3z = 5 \end{cases}$

**20.** $\begin{cases} 4x + y - z = 8 \\ x - y + 2z = 3 \\ 3x - y + z = 6 \end{cases}$

**21.** $\begin{cases} -2x - 4y + 6z = -8 \\ x + 2y - 3z = 4 \\ 4x + 8y - 12z = 16 \end{cases}$

**22.** $\begin{cases} -6x + 12y + 3z = -6 \\ 2x - 4y - z = 2 \\ -x + 2y + \frac{z}{2} = -1 \end{cases}$

**23.** $\begin{cases} 2x + 2y - 3z = 1 \\ y + 2z = -14 \\ 3x - 2y = -1 \end{cases}$

**24.** $\begin{cases} 7x + 4y = 10 \\ x - 4y + 2z = 6 \\ y - 2z = -1 \end{cases}$

**25.** $\begin{cases} \frac{3}{4}x - \frac{1}{3}y + \frac{1}{2}z = 9 \\ \frac{1}{6}x + \frac{1}{3}y - \frac{1}{2}z = 2 \\ \frac{1}{2}x - y + \frac{1}{2}z = 2 \end{cases}$

**26.** $\begin{cases} \frac{1}{3}x - \frac{1}{4}y + z = -9 \\ \frac{1}{2}x - \frac{1}{3}y - \frac{1}{4}z = -6 \\ x - \frac{1}{2}y - z = -8 \end{cases}$

**27.** The fraction $\frac{1}{24}$ can be written as the following sum:

$$\frac{1}{24} = \frac{x}{8} + \frac{y}{4} + \frac{z}{3}$$

where the numbers $x$, $y$, and $z$ are solutions of

$$\begin{cases} x + y + z = 1 \\ 2x - y + z = 0 \\ -x + 2y + 2z = -1 \end{cases}$$

Solve the system and see that the sum of the fractions is $\frac{1}{24}$.

**28.** The fraction $\frac{1}{18}$ can be written as the following sum.

$$\frac{1}{18} = \frac{x}{2} + \frac{y}{3} + \frac{z}{9}$$

where the numbers $x$, $y$, and $z$ are solutions of

$$\begin{cases} x + 3y + z = -3 \\ -x + y + 2z = -14 \\ 3x + 2y - z = 12 \end{cases}$$

Solve the system and see that the sum of the fractions is $\frac{1}{18}$.

### REVIEW EXERCISES

*Solve. See Section 2.2.*

**29.** The sum of two numbers is 45 and one number is twice the other. Find the numbers.

**30.** The difference between two numbers is 5. Twice the smaller number added to five times the larger number is 53. Find the numbers.

*Solve. See Section 2.1.*

**31.** $2(x - 1) - 3x = x - 12$

**32.** $7(2x - 1) + 4 = 11(3x - 2)$

**33.** $-y - 5(y + 5) = 3y - 10$

**34.** $z - 3(z + 7) = 6(2z + 1)$

### *A Look Ahead*

*Solve each system.*

**35.** $\begin{cases} x + y - w = 0 \\ y + 2z + w = 3 \\ x - z = 1 \\ 2x - y - w = -1 \end{cases}$

**36.** $\begin{cases} 5x + 4y = 29 \\ y + z - w = -2 \\ 5x + z = 23 \\ y - z + w = 4 \end{cases}$

**37.** $\begin{cases} x + y + z + w = 5 \\ 2x + y + z + w = 6 \\ x + y + z = 2 \\ x + y = 0 \end{cases}$

**38.** $\begin{cases} 2x - z = -1 \\ y + z + w = 9 \\ y - 2w = -6 \\ x + y = 3 \end{cases}$

## 4.3 SYSTEMS OF LINEAR EQUATIONS AND PROBLEM SOLVING

CD-ROM SSM SSG Video

**▶ OBJECTIVES**

1. Solve problems that can be modeled by a system of two linear equations.
2. Solve problems with cost and revenue functions.
3. Solve problems that can be modeled by a system of three linear equations.

**1** Thus far, we have solved problems by writing one-variable equations and solving for the variable. Some of these problems can be solved, perhaps more easily, by writing a system of equations, as illustrated in this section. We begin with a problem about numbers.

### Example 1 FINDING UNKNOWN NUMBERS

A first number is 4 less than a second number. Four times the first number is 6 more than twice the second. Find the numbers.

*Solution* 1. UNDERSTAND. Read and reread the problem and guess a solution. If one number is 10 and this is 4 less than a second number, the second number is 14. Four times the first number is 4(10), or 40. This is not equal to 6 more than twice the second number, which is $2(14) + 6$ or 34. Although we guessed incorrectly, we now have a better understanding of the problem.

Since we are looking for two numbers, we will let

$$x = \text{first number}$$
$$y = \text{second number}$$

2. TRANSLATE. Since we have assigned two variables to this problem, we will translate the given facts into two equations. For the first statement we have

| In words: | the first number | is | 4 less than the second number |
|---|---|---|---|
| | ↓ | ↓ | ↓ |
| Translate: | $x$ | $=$ | $y - 4$ |

Next we translate the second statement into an equation.

| In words: | four times the first number | is | 6 more than twice the second number |
|---|---|---|---|
| | ↓ | ↓ | ↓ |
| Translate: | $4x$ | $=$ | $2y + 6$ |

3. SOLVE. Here we solve the system

$$\begin{cases} x = y - 4 \\ 4x = 2y + 6 \end{cases}$$

Since the first equation expresses $x$ in terms of $y$, we will use substitution. We substitute $y - 4$ for $x$ in the second equation and solve for $y$.

$$4x = 2y + 6 \quad \text{Second equation}$$
$$4(y - 4) = 2y + 6 \quad \text{Let } x = y - 4.$$
$$4y - 16 = 2y + 6$$
$$2y = 22$$
$$y = 11$$

Now we replace $y$ with 11 in the equation $x = y - 4$ and solve for $x$. Then $x = y - 4$ becomes $x = 11 - 4 = 7$. The ordered pair solution of the system is (7, 11).

4. INTERPRET. Since the solution of the system is (7, 11), then the first number we are looking for is 7 and the second number is 11.

**Check:** Notice that 7 *is* 4 less than 11, and 4 times 7 *is* 6 more than twice 11. The proposed numbers, 7 and 11, are correct.

**State:** The numbers are 7 and 11.

## Example 2 FINDING THE RATE OF SPEED

Two cars leave Indianapolis, one traveling east and the other west. After 3 hours they are 297 miles apart. If one car is traveling 5 mph faster than the other, what is the speed of each?

*Solution*

1. UNDERSTAND. Read and reread the problem. Let's guess a solution and use the formula $d = rt$ to check. Suppose that one car is traveling at a rate of 55 miles per hour. This means that the other car is traveling at a rate of 50 miles per hour since we are told that one car is traveling 5 mph faster than the other. To find the distance apart after 3 hours, we will first find the distance traveled by each car. One car's distance is rate · time $= 55(3) = 165$ miles. The other car's distance is rate · time $= 50(3) = 150$ miles. Since one car is traveling east and the other west, their distance apart is the sum of their distances, or 165 miles + 150 miles = 315 miles. Although this distance apart is not the required distance of 297 miles, we now have a better understanding of the problem.

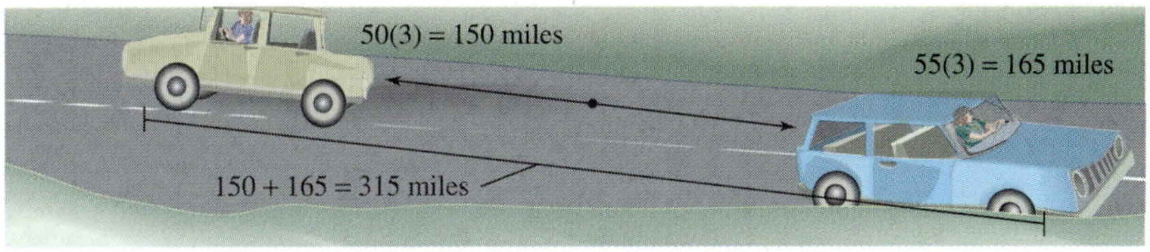


Let's model the problem with a system of equations. We will let

$x$ = speed of one car

$y$ = speed of the other car

We summarize the information on the following chart. Both cars have traveled 3 hours. Since distance = rate · time, their distances are $3x$ and $3y$ miles, respectively.

| | *Rate* · | *Time* = | *Distance* |
|---|---|---|---|
| ***ONE CAR*** | $x$ | 3 | $3x$ |
| ***OTHER CAR*** | $y$ | 3 | $3y$ |

2. TRANSLATE. We can now translate the stated conditions into two equations.

| | | | | | |
|---|---|---|---|---|---|
| In words: | one car's distance | added to | the other car's distance | is | 297 |
| | ↓ | ↓ | ↓ | ↓ | ↓ |
| Translate: | $3x$ | $+$ | $3y$ | $=$ | 297 |

| | | | |
|---|---|---|---|
| In words: | one car's speed | is | 5 mph faster than the other |
| | ↓ | ↓ | ↓ |
| Translate: | $x$ | $=$ | $y + 5$ |

3. SOLVE. Here we solve the system

$$\begin{cases} 3x + 3y = 297 \\ x = y + 5 \end{cases}$$

Again, the substitution method is appropriate. We replace $x$ with $y + 5$ in the first equation and solve for $y$.

$$3x + 3y = 297 \quad \text{First equation}$$

$$3(y + 5) + 3y = 297 \quad \text{Let } x = y + 5.$$

$$3y + 15 + 3y = 297$$

$$6y = 282$$

$$y = 47$$

To find $x$, we replace $y$ with 47 in the equation $x = y + 5$. Then $x = 47 + 5 = 52$. The ordered pair solution of the system is (52, 47).

4. INTERPRET. The solution (52, 47) means that the cars are traveling at 52 mph and 47 mph, respectively.

**HELPFUL HINT**
Don't forget to attach units, if appropriate.

**Check:** Notice that one car is traveling 5 mph faster than the other. Also, if one car travels 52 mph for 3 hours, the distance is $3(52) = 156$ miles. The other car traveling for 3 hours at 47 mph travels a distance of $3(47) = 141$ miles. The sum of the distances $156 + 141$ is 297 miles, the required distance.

**State:** The cars are traveling at 52 mph and 47 mph.

## Example 3 MIXING SOLUTIONS

Lynn Pike, a pharmacist, needs 70 liters of a 50% alcohol solution. She has available a 30% alcohol solution and an 80% alcohol solution. How many liters of each solution should she mix to obtain 70 liters of a 50% alcohol solution?

*Solution*

1. UNDERSTAND. Read and reread the problem. Next, guess the solution. Suppose that we need 20 liters of the 30% solution. Then we need $70 - 20 = 50$ liters of the 80% solution. To see if this gives us 70 liters of a 50% alcohol solution, let's find the amount of pure alcohol in each solution.

| number of liters | × | alcohol strength | = | amount of pure alcohol |
|---|---|---|---|---|
| ↓ | | ↓ | | ↓ |
| 20 liters | × | 0.30 | = | 6 liters |
| 50 liters | × | 0.80 | = | 40 liters |
| 70 liters | × | 0.50 | = | 35 liters |

Since 6 liters + 40 liters = 46 liters and not 35 liters, our guess is incorrect, but we have gained some insight as to how to model and check this problem. We will let

$x$ = amount of 30% solution, in liters
$y$ = amount of 80% solution, in liters

and use a table to organize the given data.

| | *Number of Liters* | *Alcohol Strength* | *Amount of Pure Alcohol* |
|---|---|---|---|
| ***30% SOLUTION*** | $x$ | 30% | $0.30x$ |
| ***80% SOLUTION*** | $y$ | 80% | $0.80y$ |
| ***50% SOLUTION NEEDED*** | 70 | 50% | $(0.50)(70)$ |

2. TRANSLATE. We translate the stated conditions into two equations.

| | | | | |
|---|---|---|---|---|
| In words: | amount of 30% solution | + | amount of 80% solution | = 70 |
| | ↓ | | ↓ | ↓ |
| Translate: | $x$ | + | $y$ | = 70 |
| In words: | amount of pure alcohol in 30% solution | + | amount of pure alcohol in 80% solution | = amount of pure alcohol in 50% solution |
| | ↓ | | ↓ | ↓ |
| Translate: | $0.30x$ | + | $0.80y$ | = $(0.50)(70)$ |

3. SOLVE. Here we solve the system

$$\begin{cases} x + y = 70 \\ 0.30x + 0.80y = (0.50)(70) \end{cases}$$

To solve this system, we use the elimination method. We multiply both sides of the first equation by −3 and both sides of the second equation by 10. Then

$$\begin{cases} -3(x + y) = -3(70) \\ 10(0.30x + 0.80y) = 10(0.50)(70) \end{cases} \quad \text{simplifies to} \quad \begin{cases} -3x - 3y = -210 \\ 3x + 8y = 350 \end{cases}$$

$$5y = 140$$

$$y = 28$$

Now we replace $y$ with 28 in the equation $x + y = 70$ and find that $x + 28 = 70$, or $x = 42$.

The ordered pair solution of the system is (42, 28).

4. INTERPRET.

**Check:** Check the solution in the same way that we checked our guess.

**State:** The pharmacist needs to mix 42 liters of 30% solution and 28 liters of 80% solution to obtain 70 liters of 50% solution.

2 Recall that businesses are often computing cost and revenue functions or equations to predict sales, to determine whether prices need to be adjusted, and to see whether the company is making or losing money. Recall also that the value at which revenue equals cost is called the break-even point. When revenue is less than cost, the company is losing money; when revenue is greater than cost, the company is making money.

## Example 4 FINDING A BREAK-EVEN POINT

A manufacturing company recently purchased \$3000 worth of new equipment to offer new personalized stationery to its customers. The cost of producing a package of personalized stationery is \$3.00, and it is sold for \$5.50. Find the number of packages that must be sold for the company to break even.

*Solution*

1. UNDERSTAND. Read and reread the problem. Notice that the cost to the company will include a one-time cost of \$3000 for the equipment and then \$3.00 per package produced. The revenue will be \$5.50 per package sold.
   To model this problem, we will let

$$\begin{aligned} x &= \text{number of packages of personalized stationery} \\ C(x) &= \text{total cost for producing } x \text{ packages of stationery} \\ R(x) &= \text{total revenue for selling } x \text{ packages of stationery} \end{aligned}$$

2. TRANSLATE. The revenue equation is

| | | | | | |
|---|---|---|---|---|---|
| In words: | revenue for selling $x$ packages of stationery | = | price per package | · | number of packages |
| | ↓ | | ↓ | | ↓ |
| Translate: | $R(x)$ | = | 5.5 | · | $x$ |

The cost equation is

| | | | | | | | |
|---|---|---|---|---|---|---|---|
| In words: | cost for producing $x$ packages of stationery | = | cost per of package | · | number of packages | + | cost for equipment |
| | ↓ | | ↓ | | ↓ | | ↓ |
| Translate: | $C(x)$ | = | 3 | · | $x$ | + | 3000 |

Since the break-even point is when $R(x) = C(x)$, we solve the equation

$$5.5x = 3x + 3000$$

3. SOLVE.

$$5.5x = 3x + 3000$$

$$2.5x = 3000 \qquad \text{Subtract } 3x \text{ from both sides.}$$

$$x = 1200 \qquad \text{Divide both sides by 2.5.}$$

4. INTERPRET.

**Check:** To see whether the break-even point occurs when 1200 packages are produced and sold, see if revenue equals cost when $x = 1200$. When $x = 1200$,
$R(x) = 5.5x = 5.5(1200) = 6600$ and
$C(x) = 3x + 3000 = 3(1200) + 3000 = 6600$.
Since $R(1200) = C(1200) = 6600$, the break-even point is 1200.

**State:** The company must sell 1200 packages of stationery to break even. The graph of this system is shown.

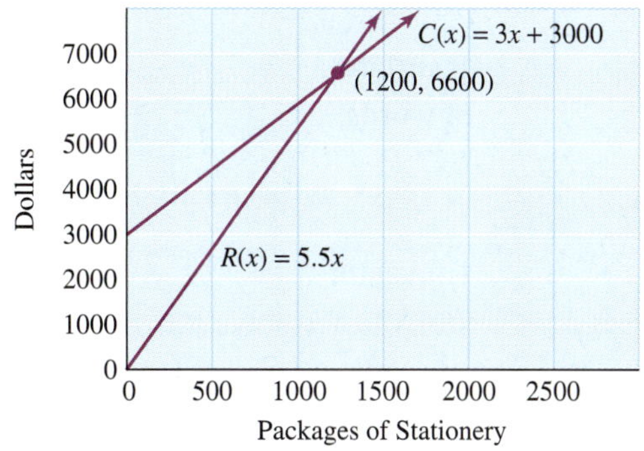


**3** To introduce problem solving by writing a system of three linear equations in three variables, we solve a problem about triangles.

## Example 5 FINDING ANGLE MEASURES

The measure of the largest angle of a triangle is 80° more than the measure of the smallest angle, and the measure of the remaining angle is 10° more than the measure of the smallest angle. Find the measure of each angle.

*Solution*

1. UNDERSTAND. Read and reread the problem. Recall that the sum of the measures of the angles of a triangle is 180°. Then guess a solution. If the smallest angle measures 20°, the measure of the largest angle is 80° more, or $20° + 80° = 100°$. The measure of the remaining angle is 10° more than the measure of the smallest angle, or $20° + 10° = 30°$. The sum of these three angles is $20° + 100° + 30° = 150°$, not the required 180°. We now know that the measure of the smallest angle is greater than 20°.

   To model this problem we will let

   $x$ = degree measure of the smallest angle

   $y$ = degree measure of the largest angle

   $z$ = degree measure of the remaining angle

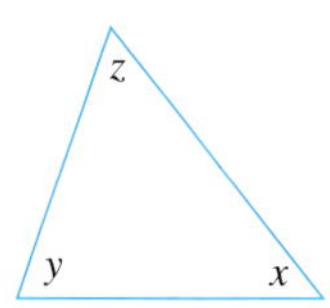


2. TRANSLATE. We translate the given information into three equations.

| | | | |
|---|---|---|---|
| In words: | the sum of the measures | = | 180 |
| | ↓ | | ↓ |
| Translate: | $x + y + z$ | = | 180 |

| In words: | the largest angle | is | 80 more than the smallest angle |
|---|---|---|---|
| | ↓ | ↓ | ↓ |
| Translate: | $y$ | $=$ | $x + 80$ |
| In words: | the remaining angle | is | 10 more than the smallest angle |
| | ↓ | ↓ | ↓ |
| Translate: | $z$ | $=$ | $x + 10$ |

3. SOLVE. We solve the system

$$\begin{cases} x + y + z = 180 \\ y = x + 80 \\ z = x + 10 \end{cases}$$

Since $y$ and $z$ are both expressed in terms of $x$, we will solve using the subsitution method. We substitute $y = x + 80$ and $z = x + 10$ in the first equation. Then

$$x + y + z = 180 \quad \text{First equation}$$

$$x + (x + 80) + (x + 10) = 180 \quad \text{Let } y = x + 80 \text{ and } z = x + 10.$$

$$3x + 90 = 180$$

$$3x = 90$$

$$x = 30$$

Then $y = x + 80 = 30 + 80 = 110$, and $z = x + 10 = 30 + 10 = 40$. The ordered triple solution is (30, 110, 40).

4. INTERPRET.

**Check:** Notice that $30° + 40° + 110° = 180°$. Also, the measure of the largest angle, 110°, is 80° more than the measure of the smallest angle, 30°. The measure of the remaining angle, 40°, is 10° more than the measure of the smallest angle, 30°.

**State:** The angles measure 30°, 110°, and 40°.

## SPOTLIGHT ON DECISION MAKING

Suppose you are choosing a long-distance telephone plan. You have narrowed your choices to the One Rate® 7¢ Plan offered by AT&T and the Qwest Communicator plan offered by Qwest Communications International, Inc. Under the AT&T One Rate® 7¢ Plan, calls made any time cost \$0.07 per minute and users are charged a \$5.95 monthly fee. Under the Qwest Communicator plan, calls made any time cost \$0.05 per minute and users are charged a \$9.95 monthly fee. Which long-distance plan would you choose? Why? What other factors would you want to consider? Would you change your choice if you knew that your long-distance calls averaged 4 hours per month? Explain your reasoning.

## Exercise Set 4.3

*Solve. See Examples 1–3 and 5.*

**1.** One number is two more than a second number. Twice the first is 4 less than 3 times the second. Find the numbers.

**2.** Three times one number minus a second is 8, and the sum of the numbers is 12. Find the numbers.

**3.** A Delta 727 traveled 560 mph with the wind and 480 mph against the wind. Find the speed of the plane in still air and the speed of the wind.

**4.** Terry Watkins can row about 10.6 kilometers in 1 hour downstream and 6.8 kilometers upstream in 1 hour. Find how fast he can row in still water, and find the speed of the current.

**5.** Find how many quarts of 4% butterfat milk and 1% butterfat milk should be mixed to yield 60 quarts of 2% butterfat milk.

**6.** A pharmacist needs 500 milliliters of a 20% phenobarbital solution but has only 5% and 25% phenobarbital solutions available. Find how many milliliters of each he should mix to get the desired solution.

**7.** Karen Karlin bought some large frames for $15 each and some small frames for $8 each at a closeout sale. If she bought 22 frames for $239, find how many of each type she bought.

**8.** Hilton University Drama Club sold 311 tickets for a play. Student tickets cost 50 cents each; nonstudent tickets cost $1.50. If total receipts were $385.50, find how many tickets of each type were sold.

**9.** One number is two less than a second number. Twice the first is 4 more than 3 times the second. Find the numbers.

**10.** Twice one number plus a second number is 42, and the one number minus the second number is −6. Find the numbers.

**11.** An office supply store in San Diego sells seven tablets and 4 pens for $6.40. Also, two tablets and 19 pens cost $5.40. Find the price of each.

**12.** A Candy Barrel shop manager mixes M&M's worth $2.00 per pound with trail mix worth $1.50 per pound. Find how many pounds of each she should use to get 50 pounds of a party mix worth $1.80 per pound.

**13.** An airplane takes 3 hours to travel a distance of 2160 miles with the wind. The return trip takes 4 hours against the wind. Find the speed of the plane in still air and the speed of the wind.

**14.** Two cyclists start at the same point and travel in opposite directions. One travels 4 mph faster than the other. In 4 hours they are 112 miles apart. Find how fast each is traveling.

△ **15.** The perimeter of a quadrilateral (four-sided polygon) is 29 inches. The longest side is twice as long as the shortest side. The other two sides are equally long and are 2 inches longer than the shortest side. Find the lengths of all four sides.

△ **16.** The perimeter of a triangle is 93 centimeters. If two sides are equally long and the third side is 9 centimeters longer than the others, find the lengths of the three sides.

**17.** The sum of three numbers is 40. One number is five more than a second and twice the third. Find the numbers.

**18.** The sum of the digits of a three-digit number is 15. The tens-place digit is twice the hundreds-place digit, and the ones-place digit is 1 less than the hundreds-place digit. Find the three-digit number.

**19.** Jack Reinholt, a car salesman, has a choice of two pay arrangements: a weekly salary of $200 plus 5% commission on sales, or a straight 15% commission. Find the amount of sales for which Jack's earnings are the same regardless of the pay arrangement.

**20.** Hertz car rental agency charges $25 daily plus 10¢ per mile. Budget charges $20 daily plus 25¢ per mile. Find the daily mileage for which the Budget charge for the day is twice that of the Hertz charge for the day.

**21.** Carroll Blakemore, a drafting student, bought 3 templates and a pencil one day for $6.45. Another day he bought 2 pads of paper and 4 pencils for $7.50. If the price of a pad of paper is three times the price of a pencil, find the price of each type of item.

*Given the cost function $C(x)$ and the revenue function $R(x)$, find the number of units $x$ that must be sold to break even. See Example 4.*

**22.** $C(x) = 30x + 10{,}000$, $R(x) = 46x$

**23.** $C(x) = 12x + 15{,}000$, $R(x) = 32x$

**24.** $C(x) = 1.2x + 1500$, $R(x) = 1.7x$

**25.** $C(x) = 0.8x + 900$, $R(x) = 2x$

**26.** $C(x) = 75x + 160{,}000$, $R(x) = 200x$

**27.** $C(x) = 105x + 70{,}000$, $R(x) = 245x$

**28.** The planning department of Abstract Office Supplies has been asked to determine whether the company should introduce a new computer desk next year. The department estimates that $6000 of new equipment will need to be purchased and that the cost of manufacturing each desk will be $200. The department also estimates that the revenue from each desk will be $450.

**a.** Determine the revenue function $R(x)$ from the sale of $x$ desks.

**b.** Determine the cost function $C(x)$ for manufacturing $x$ desks.

**c.** Find the break-even point.

**29.** Baskets, Inc., is planning to introduce a new woven basket. The company estimates that \$500 worth of new equipment will be needed to manufacture this new type of basket and that it will cost \$15 per basket to manufacture. The company also estimates that the revenue from each basket will be \$31.

**a.** Determine the revenue function $R(x)$ from the sale of $x$ baskets.

**b.** Determine the cost function $C(x)$ for manufacturing $x$ baskets.

**c.** Find the break-even point.

*Solve.*

△ **30.** Line $l$ and line $m$ are parallel lines cut by transversal $t$. Find the values of $x$ and $y$.

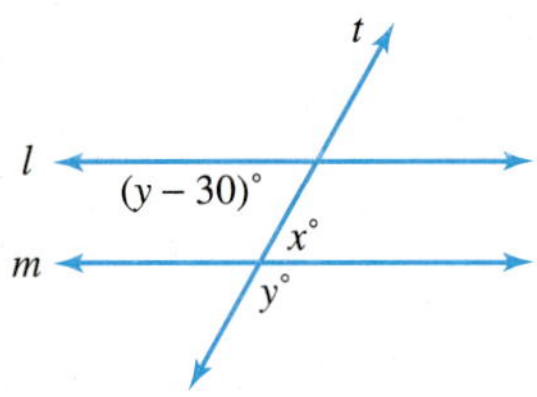


△ **31.** Find the values of $x$ and $y$ in the following isosceles triangle.

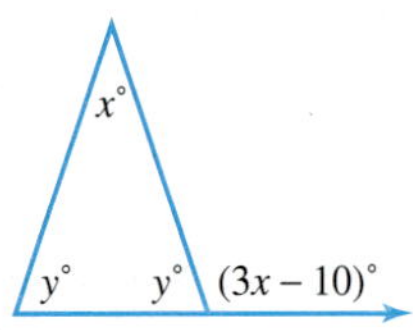


**32.** Two trains leave Tulsa, one traveling north and the other south. After 4 hours, they are 376 miles apart. If one train is traveling 10 mph faster than the other, what is the speed of each?

**33.** One solution contains 20% acid and a second solution contains 50% acid. How many ounces of each solution should be mixed in order to have 60 ounces of a 30% acid solution?

*Solve. See Example 5.*

**34.** Rabbits in a lab are to be kept on a strict daily diet to include 30 grams of protein, 16 grams of fat, and 24 grams of carbohydrates. The scientist has only three food mixes available with the following grams of nutrients per unit.

| | *Protein* | *Fat* | *Carbohydrate* |
|---|---|---|---|
| ***Mix A*** | 4 | 6 | 3 |
| ***Mix B*** | 6 | 1 | 2 |
| ***Mix C*** | 4 | 1 | 12 |

Find how many units of each mix are needed daily to meet each rabbit's dietary needs.

**35.** Gary Gundersen mixes different solutions with concentrations of 25%, 40%, and 50% to get 200 liters of a 32% solution. If he uses twice as much of the 25% solution as of the 40% solution, find how many liters of each kind he uses.

**36.** In 1999 the WNBA's top scorer was Cynthia Cooper of the Houston Comets. She scored a total of 686 points during the regular season. The number of two-point field goals Cooper made was 20 less than three times the number of three-point field goals she made. She also made 50 more free throws (each worth one point) than two-point field goals. Find how many free throws, two-point field goals, and three-point field goals Cynthia Cooper made during the 1999 season. (*Source:* Women's National Basketball Association)

**37.** During the 2000 NBA All-Star Game, the top-scoring player was Allen Iverson of the Philadelphia 76ers. Iverson, playing for the Eastern Conference All-Star Team, scored a total of 26 points during the All-Star Game. The number of free throws (each worth 1 point) he made was 2 more than the number of three-point field goals he made. Iverson also made 4 more two-point field goals than free throws. How many free throws, two-point field goals, and three-point field goals did Allen Iverson make during the 2000 NBA All-Star Game? (*Source:* National Basketball Association)

△ **38.** The measure of the largest angle of a triangle is 90° more than the measure of the smallest angle, and the measure of the remaining angle is 30° more than the measure of the smallest angle. Find the measure of each angle.

**39.** Suppose you mix an amount of 25% acid solution with an amount of 60% acid solution. You then calculate the acid strength of the resulting acid mixture. For which of the following results should you suspect an error in your calculation? Why?

**a.** 14% **b.** 32% **c.** 55%

**40.** Find the values of $a$, $b$, and $c$ such that the equation $y = ax^2 + bx + c$ has ordered pair solutions $(1, 6)$, $(-1, -2)$, and $(0, -1)$. To do so, substitute each ordered pair solution into the equation. Each time, the result is an equation in three unknowns: $a$, $b$, and $c$. Then solve the resulting system of three linear equations in three unknowns, $a$, $b$, and $c$.

**41.** Find the values of $a$, $b$, and $c$ such that the equation $y = ax^2 + bx + c$ has ordered pair solutions $(1, 2)$, $(2, 3)$ and $(-1, 6)$. (*Hint:* See Exercise 40.)

△ **42.** Find the values of $x$, $y$, and $z$ in the following triangle.

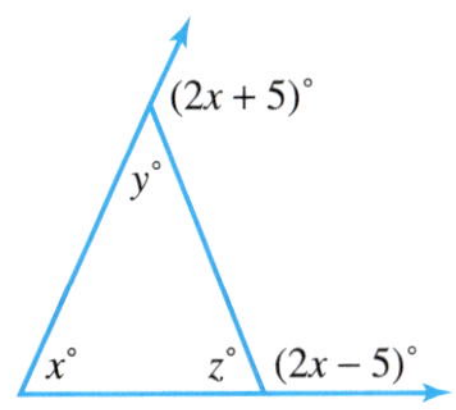


△ **43.** The sum of the measures of the angles of a quadrilateral is 360°. Find the values of $x$, $y$, and $z$ in the following quadrilateral.

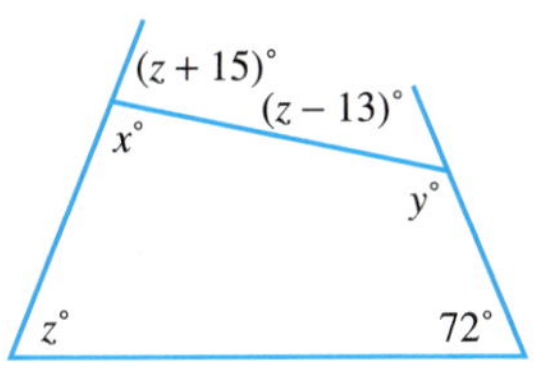


**44.** Data $(x, y)$ for the total number $y$ (in thousands) of college-bound students who took the ACT assessment in the year $x$ are $(5, 945)$, $(6, 925)$, and $(9, 1019)$, where $x = 5$ represents 1995 and $x = 9$ represents 1999. Find the values of $a$, $b$, and $c$ such that the equation $y = ax^2 + bx + c$ models this data. According to your model, how many students will take the ACT in 2005? (*Source:* ACT, Inc.)

**45.** Monthly normal rainfall data $(x, y)$ for Portland, Oregon, are $(4, 2.47)$, $(7, 0.6)$, $(8, 1.1)$, where $x$ represents time in months (with $x = 1$ representing January) and $y$ represents rainfall in inches. Find the values of $a$, $b$, and $c$ rounded to 2 decimal places such that the equation $y = ax^2 + bx + c$ models this data. According to your model, how much rain should Portland expect during September? (*Source:* National Climatic Data Center)

## REVIEW EXERCISES

*Multiply both sides of equation (1) by 2, and add the resulting equation to equation (2). See Section 4.2.*

**46.** $3x - y + z = 2$ (1)
$-x + 2y + 3z = 6$ (2)

**47.** $2x + y + 3z = 7$ (1)
$-4x + y + 2z = 4$ (2)

*Multiply both sides of equation (1) by −3, and add the resulting equation to equation (2). See Section 4.2.*

**48.** $x + 2y - z = 0$ (1)
$3x + y - z = 2$ (2)

**49.** $2x - 3y + 2z = 5$ (1)
$x - 9y + z = -1$ (2)

*Given the spinner below, find the probability of the spinner landing on the indicated color in one spin. See Section 2.3 Exercise set.*

**50.** $P$(red)

**51.** $P$(green)

**52.** $P$(white)

**53.** $P$(red or blue)

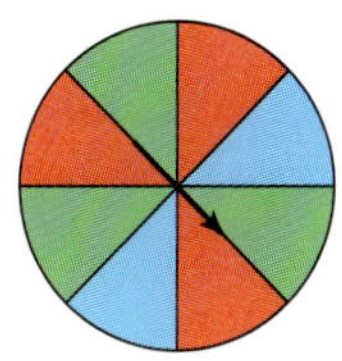

## 4.4 SOLVING SYSTEMS OF EQUATIONS BY MATRICES

CD-ROM SSM SSG Video

▶ **OBJECTIVES**

1. Use matrices to solve a system of two equations.
2. Use matrices to solve a system of three equations.

By now, you may have noticed that the solution of a system of equations depends on the coefficients of the equations in the system and not on the variables. In this section, we introduce solving a system of equations by a **matrix.**

**1** A matrix (plural: **matrices**) is a rectangular array of numbers. The following are examples of matrices.

$$\begin{bmatrix} 1 & 0 \\ 0 & 1 \end{bmatrix} \qquad \begin{bmatrix} 2 & 1 & 3 & -1 \\ 0 & -1 & 4 & 5 \\ -6 & 2 & 1 & 0 \end{bmatrix} \qquad \begin{bmatrix} a & b & c \\ d & e & f \end{bmatrix}$$

The numbers aligned horizontally in a matrix are in the same **row**. The numbers aligned vertically are in the same **column**.

$$\begin{matrix} \text{row 1} \rightarrow \\ \text{row 2} \rightarrow \end{matrix} \begin{bmatrix} 2 & 1 & 0 \\ -1 & 6 & 2 \end{bmatrix}$$

column 1, column 2, column 3

This matrix has 2 rows and 3 columns. It is called a 2 × 3 (read "two by three") matrix.

To see the relationship between systems of equations and matrices, study the example below.

***System of Equations***

$$\begin{cases} 2x - 3y = 6 & \text{Equation 1} \\ x + y = 0 & \text{Equation 2} \end{cases}$$

***Corresponding Matrix***

$$\left[\begin{array}{cc:c} 2 & -3 & 6 \\ 1 & 1 & 0 \end{array}\right] \begin{matrix} \text{Row 1} \\ \text{Row 2} \end{matrix}$$

Notice that the rows of the matrix correspond to the equations in the system. The coefficients of each variable are placed to the left of a vertical dashed line. The constants are placed to the right. Each of these numbers in the matrix is called an **element**.

The method of solving systems by matrices is to write this matrix as an equivalent matrix from which we easily identify the solution. Two matrices are equivalent if they represent systems that have the same solution set. The following **row operations** can be performed on matrices, and the result is an equivalent matrix.

### ELEMENTARY ROW OPERATIONS

1. Any two rows in a matrix may be interchanged.
2. The elements of any row may be multiplied (or divided) by the same nonzero number.
3. The elements of any row may be multiplied (or divided) by a nonzero number and added to their corresponding elements in any other row.

**HELPFUL HINT**

Notice that these *row* operations are the same operations that we can perform on *equations* in a system.

**Example 1** Use matrices to solve the system.

$$\begin{cases} x + 3y = 5 \\ 2x - y = -4 \end{cases}$$

*Solution* The corresponding matrix is $\left[\begin{array}{cc|c} 1 & 3 & 5 \\ 2 & -1 & -4 \end{array}\right]$. We use elementary row operations to write an equivalent matrix that looks like $\left[\begin{array}{cc|c} 1 & a & b \\ 0 & 1 & c \end{array}\right]$.

For the matrix given, the element in the first row, first column is already 1, as desired. Next we write an equivalent matrix with a 0 below the 1. To do this, we multiply row 1 by −2 and add to row 2. *We will change only row 2.*

$$\left[\begin{array}{cc|c} 1 & 3 & 5 \\ \underset{\text{row 1 element}}{-2(1)} + \underset{\text{row 2 element}}{2} & \underset{\text{row 1 element}}{-2(3)} + \underset{\text{row 2 element}}{(-1)} & \underset{\text{row 1 element}}{-2(5)} + \underset{\text{row 2 element}}{(-4)} \end{array}\right] \quad \text{simplifies to} \quad \left[\begin{array}{cc|c} 1 & 3 & 5 \\ 0 & -7 & -14 \end{array}\right]$$

Now we change the −7 to a 1 by use of an elementary row operation. We divide row 2 by −7, then

$$\left[\begin{array}{cc|c} 1 & 3 & 5 \\ \frac{0}{-7} & \frac{-7}{-7} & \frac{-14}{-7} \end{array}\right] \quad \text{simplifies to} \quad \left[\begin{array}{cc|c} 1 & 3 & 5 \\ 0 & 1 & 2 \end{array}\right]$$

This last matrix corresponds to the system

$$\begin{cases} x + 3y = 5 \\ \quad\quad y = 2 \end{cases}$$

To find $x$, we let $y = 2$ in the first equation, $x + 3y = 5$.

$$\begin{aligned} x + 3y &= 5 && \text{First equation} \\ x + 3(2) &= 5 && \text{Let } y = 2. \\ x &= -1 \end{aligned}$$

The ordered pair solution is $(-1, 2)$. Check to see that this ordered pair satisfies both equations.

**Example 2** Use matrices to solve the system.

$$\begin{cases} 2x - y = 3 \\ 4x - 2y = 5 \end{cases}$$

*Solution* The corresponding matrix is $\left[\begin{array}{cc|c} 2 & -1 & 3 \\ 4 & -2 & 5 \end{array}\right]$. To get 1 in the row 1, column 1 position, we divide the elements of row 1 by 2.

$$\left[\begin{array}{cc|c} \frac{2}{2} & -\frac{1}{2} & \frac{3}{2} \\ 4 & -2 & 5 \end{array}\right] \quad \text{simplifies to} \quad \left[\begin{array}{cc|c} 1 & -\frac{1}{2} & \frac{3}{2} \\ 4 & -2 & 5 \end{array}\right]$$

To get 0 under the 1, we multiply the elements of row 1 by −4 and add the new elements to the elements of row 2.

$$\left[\begin{array}{cc|c} 1 & -\frac{1}{2} & \frac{3}{2} \\ -4(1)+4 & -4\left(-\frac{1}{2}\right)-2 & -4\left(\frac{3}{2}\right)+5 \end{array}\right] \quad \text{simplifies to} \quad \left[\begin{array}{cc|c} 1 & -\frac{1}{2} & \frac{3}{2} \\ 0 & 0 & -1 \end{array}\right]$$

The corresponding system is $\begin{cases} x - \frac{1}{2}y = \frac{3}{2} \\ 0 = -1 \end{cases}$. The equation $0 = -1$ is false for all $y$ or $x$ values; hence the system is inconsistent and has no solution.

**2** To solve a system of three equations in three variables using matrices, we will write the corresponding matrix in the form

$$\left[\begin{array}{ccc|c} 1 & a & b & d \\ 0 & 1 & c & e \\ 0 & 0 & 1 & f \end{array}\right]$$

**Example 3** Use matrices to solve the system.

$$\begin{cases} x + 2y + z = 2 \\ -2x - y + 2z = 5 \\ x + 3y - 2z = -8 \end{cases}$$

*Solution* The corresponding matrix is $\left[\begin{array}{ccc|c} 1 & 2 & 1 & 2 \\ -2 & -1 & 2 & 5 \\ 1 & 3 & -2 & -8 \end{array}\right]$. Our goal is to write an equivalent matrix with 1's along the diagonal (see the numbers in red) and 0's below the 1's. The element in row 1, column 1 is already 1. Next we get 0's for each element in the rest of column 1. To do this, first we multiply the elements of row 1 by 2 and add the new elements to row 2. Also, we multiply the elements of row 1 by −1 and add the new elements to the elements of row 3. We *do not change row 1.* Then

$$\left[\begin{array}{ccc|c} 1 & 2 & 1 & 2 \\ 2(1)-2 & 2(2)-1 & 2(1)+2 & 2(2)+5 \\ -1(1)+1 & -1(2)+3 & -1(1)-2 & -1(2)-8 \end{array}\right] \quad \text{simplifies to} \quad \left[\begin{array}{ccc|c} 1 & 2 & 1 & 2 \\ 0 & 3 & 4 & 9 \\ 0 & 1 & -3 & -10 \end{array}\right]$$

We continue down the diagonal and use elementary row operations to get 1 where the element 3 is now. To do this, we interchange rows 2 and 3.

$$\left[\begin{array}{ccc|c} 1 & 2 & 1 & 2 \\ 0 & 3 & 4 & 9 \\ 0 & 1 & -3 & -10 \end{array}\right] \quad \text{is equivalent to} \quad \left[\begin{array}{ccc|c} 1 & 2 & 1 & 2 \\ 0 & 1 & -3 & -10 \\ 0 & 3 & 4 & 9 \end{array}\right]$$

Next we want the new row 3, column 2 element to be 0. We multiply the elements of row 2 by −3 and add the result to the elements of row 3.

$$\left[\begin{array}{ccc|c} 1 & 2 & 1 & 2 \\ 0 & 1 & -3 & -10 \\ -3(0)+0 & -3(1)+3 & -3(-3)+4 & -3(-10)+9 \end{array}\right] \quad \text{simplifies to} \quad \left[\begin{array}{ccc|c} 1 & 2 & 1 & 2 \\ 0 & 1 & -3 & -10 \\ 0 & 0 & 13 & 39 \end{array}\right]$$

Finally, we divide the elements of row 3 by 13 so that the final diagonal element is 1.

$$\left[\begin{array}{ccc|c} 1 & 2 & 1 & 2 \\ 0 & 1 & -3 & -10 \\ \frac{0}{13} & \frac{0}{13} & \frac{13}{13} & \frac{39}{13} \end{array}\right] \quad \text{simplifies to} \quad \left[\begin{array}{ccc|c} 1 & 2 & 1 & 2 \\ 0 & 1 & -3 & -10 \\ 0 & 0 & 1 & 3 \end{array}\right]$$

This matrix corresponds to the system

$$\begin{cases} x + 2y + z = 2 \\ y - 3z = -10 \\ z = 3 \end{cases}$$

We identify the $z$-coordinate of the solution as 3. Next we replace $z$ with 3 in the second equation and solve for $y$.

$$\begin{aligned} y - 3z &= -10 && \text{Second equation} \\ y - 3(3) &= -10 && \text{Let } z = 3. \\ y &= -1 \end{aligned}$$

To find $x$, we let $z = 3$ and $y = -1$ in the first equation.

$$\begin{aligned} x + 2y + z &= 2 && \text{First equation} \\ x + 2(-1) + 3 &= 2 && \text{Let } z = 3 \text{ and } y = -1. \\ x &= 1 \end{aligned}$$

The ordered triple solution is $(1, -1, 3)$. Check to see that it satisfies all three equations in the original system.

## SPOTLIGHT ON DECISION MAKING

Suppose you are an urban planner working for the public transportation authority of a large city. Currently, more commuters travel into the downtown area during the morning rush hour than travel out of it. However, a strong economy is creating new jobs in the suburbs. As suburban job growth continues, more city dwellers will travel out of the downtown area during the morning rush hour. You have been assigned to study the situation, focusing on how the trend will impact existing bus routes and schedules and how soon.

After a detailed study of public transportation utilization, you find that the number of commuters into downtown each morning can be described by the equation $y = 40{,}000 + 200x$, where $x$ is the number of months from now. The number of commuters out of downtown each morning can be described by the equation $y = 20{,}000 + 1000x$, where $x$ is the number of months from now. In the short term, as the numbers of commuters in each direction increase, additional buses can be put on the existing routes to handle the load. But once the number of commuters leaving downtown exceeds the number of commuters traveling into downtown, the bus routes must be totally revamped.

It usually takes the public transportation authority $1\frac{1}{2}$ years to plan major changes to bus routes. If changes are needed more quickly than that, a consulting firm can be hired to speed up the process. Decide whether a consulting firm will be needed to help revamp the bus routes.

# Exercise Set 4.4

*Solve each system of linear equations using matrices. See Example 1.*

**1.** $\begin{cases} x + y = 1 \\ x - 2y = 4 \end{cases}$

**2.** $\begin{cases} 2x - y = 8 \\ x + 3y = 11 \end{cases}$

**3.** $\begin{cases} x + 3y = 2 \\ x + 2y = 0 \end{cases}$

**4.** $\begin{cases} 4x - y = 5 \\ 3x - 3y = 0 \end{cases}$

*Solve each system of linear equations using matrices. See Example 2.*

**5.** $\begin{cases} x - 2y = 4 \\ 2x - 4y = 4 \end{cases}$

**6.** $\begin{cases} -x + 3y = 6 \\ 3x - 9y = 9 \end{cases}$

**7.** $\begin{cases} 3x - 3y = 9 \\ 2x - 2y = 6 \end{cases}$

**8.** $\begin{cases} 9x - 3y = 6 \\ -18x + 6y = -12 \end{cases}$

*Solve each system of linear equations using matrices. See Example 3.*

**9.** $\begin{cases} x + y = 3 \\ 2y = 10 \\ 3x + 2y - 4z = 12 \end{cases}$

**10.** $\begin{cases} 5x = 5 \\ 2x + y = 4 \\ 3x + y - 5z = -15 \end{cases}$

**11.** $\begin{cases} 2y - z = -7 \\ x + 4y + z = -4 \\ 5x - y + 2z = 13 \end{cases}$

**12.** $\begin{cases} 4y + 3z = -2 \\ 5x - 4y = 1 \\ -5x + 4y + z = -3 \end{cases}$

*Solve each system of linear equations using matrices.*

**13.** $\begin{cases} x - 4 = 0 \\ x + y = 1 \end{cases}$

**14.** $\begin{cases} 3y = 6 \\ x + y = 7 \end{cases}$

**15.** $\begin{cases} x + y + z = 2 \\ 2x - z = 5 \\ 3y + z = 2 \end{cases}$

**16.** $\begin{cases} x + 2y + z = 5 \\ x - y - z = 3 \\ y + z = 2 \end{cases}$

**17.** $\begin{cases} 5x - 2y = 27 \\ -3x + 5y = 18 \end{cases}$

**18.** $\begin{cases} 4x - y = 9 \\ 2x + 3y = -27 \end{cases}$

**19.** $\begin{cases} 4x - 7y = 7 \\ 12x - 21y = 24 \end{cases}$

**20.** $\begin{cases} 2x - 5y = 12 \\ -4x + 10y = 20 \end{cases}$

**21.** $\begin{cases} 4x - y + 2z = 5 \\ 2y + z = 4 \\ 4x + y + 3z = 10 \end{cases}$

**22.** $\begin{cases} 5y - 7z = 14 \\ 2x + y + 4z = 10 \\ 2x + 6y - 3z = 30 \end{cases}$

**23.** $\begin{cases} 4x + y + z = 3 \\ -x + y - 2z = -11 \\ x + 2y + 2z = -1 \end{cases}$

**24.** $\begin{cases} x + y + z = 9 \\ 3x - y + z = -1 \\ -2x + 2y - 3z = -2 \end{cases}$

**25.** Consider the system

$$\begin{cases} 2x - 3y = 8 \\ x + 5y = -3 \end{cases}$$

What is wrong with its corresponding matrix shown below?

$$\left[\begin{array}{cc|c} 2 & 3 & 8 \\ 0 & 5 & 3 \end{array}\right]$$

**26.** The percent $y$ of U.S. households that owned a black-and-white television set between the years 1980 and 1993 can be modeled by the linear equation $2.3x + y = 52$, where $x$ represents the number of years after 1980. Similarly, the percent $y$ of U.S. households that owned a microwave oven during this same period can be modeled by the linear equation $-5.4x + y = 14$. (*Source:* Based on data from the Energy Information Administration, U.S. Department of Energy)

**a.** The data used to form these two models was incomplete. It is impossible to tell from the data the year in which the percent of households owning black-and-white television sets was the same as the percent of households owning microwave ovens. Use matrix methods to estimate the year in which this occurred.

**b.** Did more households own black-and-white television sets or microwave ovens in 1980? In 1993? What trends do these models show? Does this seem to make sense? Why or why not?

**c.** According to the models, when will the percent of households owning black-and-white television sets reach 0%?

## REVIEW EXERCISES

*Determine whether each graph is the graph of a function. See Section 3.2.*

**27.**

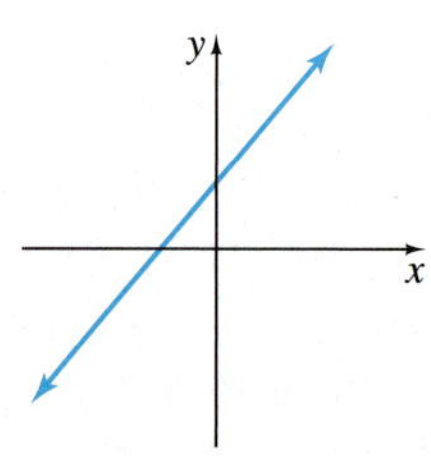


**28.**

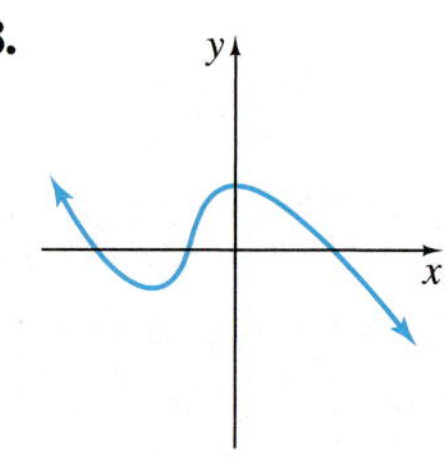


**29.**

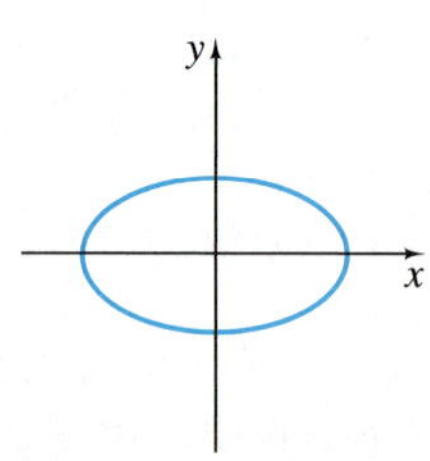


**30.**

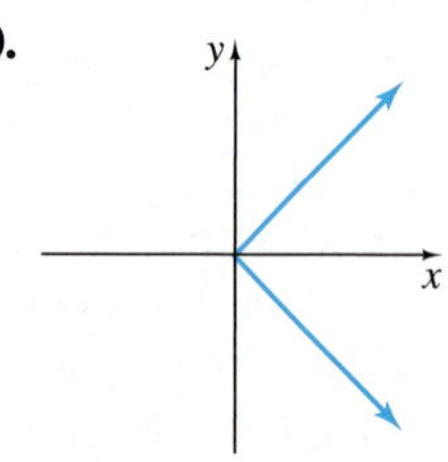


*Evaluate. See Section 1.3.*

**31.** $(-1)(-5) - (6)(3)$

**32.** $(2)(-8) - (-4)(1)$

**33.** $(4)(-10) - (2)(-2)$

**34.** $(-7)(3) - (-2)(-6)$

**35.** $(-3)(-3) - (-1)(-9)$

**36.** $(5)(6) - (10)(10)$

## 4.5 SOLVING SYSTEMS OF EQUATIONS BY DETERMINANTS

CD-ROM SSM SSG Video

**OBJECTIVES**

1. Define and evaluate a $2 \times 2$ determinant.
2. Use Cramer's rule to solve a system of two linear equations in two variables.
3. Define and evaluate a $3 \times 3$ determinant.
4. Use Cramer's rule to solve a linear system of three equations in three variables.

1 We have solved systems of two linear equations in two variables in four different ways: graphically, by substitution, by elimination, and by matrices. Now we analyze another method called **Cramer's rule**.

Recall that a matrix is a rectangular array of numbers. If a matrix has the same number of rows and columns, it is called a **square matrix**. Examples of square matrices are

$$\begin{bmatrix} 1 & 6 \\ 5 & 2 \end{bmatrix} \qquad \begin{bmatrix} 2 & 4 & 1 \\ 0 & 5 & 2 \\ 3 & 6 & 9 \end{bmatrix}$$

A **determinant** is a real number associated with a square matrix. The determinant of a square matrix is denoted by placing vertical bars about the array of numbers. Thus,

The determinant of the square matrix $\begin{bmatrix} 1 & 6 \\ 5 & 2 \end{bmatrix}$ is $\begin{vmatrix} 1 & 6 \\ 5 & 2 \end{vmatrix}$.

The determinant of the square matrix $\begin{bmatrix} 2 & 4 & 1 \\ 0 & 5 & 2 \\ 3 & 6 & 9 \end{bmatrix}$ is $\begin{vmatrix} 2 & 4 & 1 \\ 0 & 5 & 2 \\ 3 & 6 & 9 \end{vmatrix}$.

We define the determinant of a $2 \times 2$ matrix first. (Recall that $2 \times 2$ is read "two by two." It means that the matrix has 2 rows and 2 columns.)

### DETERMINANT OF A 2 × 2 MATRIX

$$\begin{vmatrix} a & b \\ c & d \end{vmatrix} = ad - bc$$

**Example 1** Evaluate each determinant.

**a.** $\begin{vmatrix} -1 & 2 \\ 3 & -4 \end{vmatrix}$ **b.** $\begin{vmatrix} 2 & 0 \\ 7 & -5 \end{vmatrix}$

*Solution* First we identify the values of $a$, $b$, $c$, and $d$. Then we perform the evaluation.

**a.** Here $a = -1$, $b = 2$, $c = 3$, and $d = -4$.

$$\begin{vmatrix} -1 & 2 \\ 3 & -4 \end{vmatrix} = ad - bc = (-1)(-4) - (2)(3) = -2$$

**b.** In this example, $a = 2, b = 0, c = 7$, and $d = -5$.

$$\begin{vmatrix} 2 & 0 \\ 7 & -5 \end{vmatrix} = ad - bc = 2(-5) - (0)(7) = -10$$

2 To develop Cramer's rule, we solve the system $\begin{cases} ax + by = h \\ cx + dy = k \end{cases}$ using elimination. First, we eliminate $y$ by multiplying both sides of the first equation by $d$ and both sides of the second equation by $-b$ so that the coefficients of $y$ are opposites. The result is that

$$\begin{cases} d(ax + by) = d \cdot h \\ -b(cx + dy) = -b \cdot k \end{cases} \quad \text{simplifies to} \quad \begin{cases} adx + bdy = hd \\ -bcx - bdy = -kb \end{cases}$$

We now add the two equations and solve for $x$.

$$\begin{aligned} adx + bdy &= hd \\ -bcx - bdy &= -kb \\ \hline adx - bcx &= hd - kb && \text{Add the equations.} \\ (ad - bc)x &= hd - kb \\ x &= \frac{hd - kb}{ad - bc} && \text{Solve for } x. \end{aligned}$$

When we replace $x$ with $\dfrac{hd - kb}{ad - bc}$ in the equation $ax + by = h$ and solve for $y$, we find that $y = \dfrac{ak - ch}{ad - bc}$.

Notice that the numerator of the value of $x$ is the determinant of

$$\begin{vmatrix} h & b \\ k & d \end{vmatrix} = hd - kb$$

Also, the numerator of the value of $y$ is the determinant of

$$\begin{vmatrix} a & h \\ c & k \end{vmatrix} = ak - hc$$

Finally, the denominators of the values of $x$ and $y$ are the same and are the determinant of

$$\begin{vmatrix} a & b \\ c & d \end{vmatrix} = ad - bc$$

This means that the values of $x$ and $y$ can be written in determinant notation.

$$x = \frac{\begin{vmatrix} h & b \\ k & d \end{vmatrix}}{\begin{vmatrix} a & b \\ c & d \end{vmatrix}} \quad \text{and} \quad y = \frac{\begin{vmatrix} a & h \\ c & k \end{vmatrix}}{\begin{vmatrix} a & b \\ c & d \end{vmatrix}}$$

For convenience, we label the determinants $D$, $D_x$, and $D_y$.

$x$-coefficients (first column of $D$ and $D_y$); $y$-coefficients (second column of $D$ and $D_x$)

$$\begin{vmatrix} a & b \\ c & d \end{vmatrix} = D \qquad \begin{vmatrix} h & b \\ k & d \end{vmatrix} = D_x \qquad \begin{vmatrix} a & h \\ c & k \end{vmatrix} = D_y$$

$x$-column replaced by constants (in $D_x$) — $y$-column replaced by constants (in $D_y$)

These determinant formulas for the coordinates of the solution of a system are known as **Cramer's rule**.

### CRAMER'S RULE FOR TWO LINEAR EQUATIONS IN TWO VARIABLES

The solution of the system $\begin{cases} ax + by = h \\ cx + dy = k \end{cases}$ is given by

$$x = \frac{\begin{vmatrix} h & b \\ k & d \end{vmatrix}}{\begin{vmatrix} a & b \\ c & d \end{vmatrix}} = \frac{D_x}{D} \qquad y = \frac{\begin{vmatrix} a & h \\ c & k \end{vmatrix}}{\begin{vmatrix} a & b \\ c & d \end{vmatrix}} = \frac{D_y}{D}$$

as long as $D = ad - bc$ is not 0.

When $D = 0$, the system is either inconsistent or the equations are dependent. When this happens, we need to use another method to see which is the case.

**Example 2** Use Cramer's rule to solve each system.

**a.** $\begin{cases} 3x + 4y = -7 \\ x - 2y = -9 \end{cases}$ **b.** $\begin{cases} 5x + y = 5 \\ -7x - 2y = -7 \end{cases}$

*Solution* **a.** First we find $D$, $D_x$, and $D_y$.

$$\begin{matrix} a & b & h \\ \downarrow & \downarrow & \downarrow \end{matrix}$$
$$\begin{cases} 3x + 4y = -7 \\ x - 2y = -9 \end{cases}$$
$$\begin{matrix} \uparrow & \uparrow & \uparrow \\ c & d & k \end{matrix}$$

$$D = \begin{vmatrix} a & b \\ c & d \end{vmatrix} = \begin{vmatrix} 3 & 4 \\ 1 & -2 \end{vmatrix} = 3(-2) - 4(1) = -10$$

$$D_x = \begin{vmatrix} h & b \\ k & d \end{vmatrix} = \begin{vmatrix} -7 & 4 \\ -9 & -2 \end{vmatrix} = (-7)(-2) - 4(-9) = 50$$

$$D_y = \begin{vmatrix} a & h \\ c & k \end{vmatrix} = \begin{vmatrix} 3 & -7 \\ 1 & -9 \end{vmatrix} = 3(-9) - (-7)(1) = -20$$

Then $x = \frac{D_x}{D} = \frac{50}{-10} = -5$ and $y = \frac{D_y}{D} = \frac{-20}{-10} = 2$. The ordered pair solution is $(-5, 2)$.

As always, check the solution in both original equations.

**b.** Find $D$, $D_x$, and $D_y$ for $\begin{cases} 5x + y = 5 \\ -7x - 2y = -7 \end{cases}$.

$$D = \begin{vmatrix} 5 & 1 \\ -7 & -2 \end{vmatrix} = 5(-2) - (1)(-7) = -3$$

$$D_x = \begin{vmatrix} 5 & 1 \\ -7 & -2 \end{vmatrix} = 5(-2) - (1)(-7) = -3$$

$$D_y = \begin{vmatrix} 5 & 5 \\ -7 & -7 \end{vmatrix} = 5(-7) - 5(-7) = 0$$

Then $x = \dfrac{D_x}{D} = \dfrac{-3}{-3} = 1$ and $y = \dfrac{D_y}{D} = \dfrac{0}{-3} = 0$.

The ordered pair solution is $(1, 0)$. Check this solution in both original equations.

3 A $3 \times 3$ determinant can be used to solve a system of three equations in three variables. The determinant of a $3 \times 3$ matrix, however, is considerably more complex than a $2 \times 2$ one.

## DETERMINANT OF A 3 × 3 MATRIX

$$\begin{vmatrix} a_1 & b_1 & c_1 \\ a_2 & b_2 & c_2 \\ a_3 & b_3 & c_3 \end{vmatrix} = a_1 \cdot \begin{vmatrix} b_2 & c_2 \\ b_3 & c_3 \end{vmatrix} - a_2 \cdot \begin{vmatrix} b_1 & c_1 \\ b_3 & c_3 \end{vmatrix} + a_3 \cdot \begin{vmatrix} b_1 & c_1 \\ b_2 & c_2 \end{vmatrix}$$

Notice that the determinant of a $3 \times 3$ matrix is related to the determinants of three $2 \times 2$ matrices. Each determinant of these $2 \times 2$ matrices is called a **minor**, and every element of a $3 \times 3$ matrix has a minor associated with it. For example, the minor of $c_2$ is the determinant of the $2 \times 2$ matrix found by deleting the row and column containing $c_2$.

$$\begin{matrix} a_1 & b_1 & \not{c_1} \\ \not{a_2} & \not{b_2} & \not{c_2} \\ a_3 & b_3 & \not{c_3} \end{matrix} \qquad \text{The minor of } c_2 \text{ is} \qquad \begin{vmatrix} a_1 & b_1 \\ a_3 & b_3 \end{vmatrix}$$

Also, the minor of element $a_1$ is the determinant of the $2 \times 2$ matrix that has no row or column containing $a_1$.

$$\begin{matrix} \not{a_1} & \not{b_1} & \not{c_1} \\ \not{a_2} & b_2 & c_2 \\ \not{a_3} & b_3 & c_3 \end{matrix} \qquad \text{The minor of } a_1 \text{ is} \qquad \begin{vmatrix} b_2 & c_2 \\ b_3 & c_3 \end{vmatrix}$$

So the determinant of a $3 \times 3$ matrix can be written as

$$a_1 \cdot (\text{minor of } a_1) - a_2 \cdot (\text{minor of } a_2) + a_3 \cdot (\text{minor of } a_3)$$

Finding the determinant by using minors of elements in the first column is called **expanding** by the minors of the first column. *The value of a determinant can be found by expanding by the minors of any row or column.* The following **array of signs** is helpful in determining whether to add or subtract the product of an element and its minor.

$$\begin{matrix} + & - & + \\ - & + & - \\ + & - & + \end{matrix}$$

If an element is in a position marked +, we add. If marked −, we subtract.

**Example 3** Evaluate by expanding by the minors of the given row or column.

$$\begin{vmatrix} 0 & 5 & 1 \\ 1 & 3 & -1 \\ -2 & 2 & 4 \end{vmatrix}$$

**a.** First column

**b.** Second row

*Solution* **a.** The elements of the first column are 0, 1, and −2. The first column of the array of signs is +, −, +.

$$\begin{vmatrix} 0 & 5 & 1 \\ 1 & 3 & -1 \\ -2 & 2 & 4 \end{vmatrix} = 0 \cdot \begin{vmatrix} 3 & -1 \\ 2 & 4 \end{vmatrix} - 1 \cdot \begin{vmatrix} 5 & 1 \\ 2 & 4 \end{vmatrix} + (-2) \cdot \begin{vmatrix} 5 & 1 \\ 3 & -1 \end{vmatrix}$$
$$= 0(12 - (-2)) - 1(20 - 2) + (-2)(-5 - 3)$$
$$= 0 - 18 + 16 = -2$$

**b.** The elements of the second row are 1, 3, and −1. This time, the signs begin with − and again alternate.

$$\begin{vmatrix} 0 & 5 & 1 \\ 1 & 3 & -1 \\ -2 & 2 & 4 \end{vmatrix} = -1 \cdot \begin{vmatrix} 5 & 1 \\ 2 & 4 \end{vmatrix} + 3 \cdot \begin{vmatrix} 0 & 1 \\ -2 & 4 \end{vmatrix} - (-1) \cdot \begin{vmatrix} 0 & 5 \\ -2 & 2 \end{vmatrix}$$
$$= -1(20 - 2) + 3(0 - (-2)) - (-1)(0 - (-10))$$
$$= -18 + 6 + 10 = -2$$

Notice that the determinant of the $3 \times 3$ matrix is the same regardless of the row or column you select to expand by.

4 A system of three equations in three variables may be solved with Cramer's rule also. Using the elimination process to solve a system with unknown constants as coefficients leads to the following.

### CRAMER'S RULE FOR THREE EQUATIONS IN THREE VARIABLES

The solution of the system $\begin{cases} a_1x + b_1y + c_1z = k_1 \\ a_2x + b_2y + c_2z = k_2 \\ a_3x + b_3y + c_3z = k_3 \end{cases}$ is given by

$$x = \frac{D_x}{D} \qquad y = \frac{D_y}{D} \qquad \text{and} \qquad z = \frac{D_z}{D}$$

*(continued)*

where

$$D = \begin{vmatrix} a_1 & b_1 & c_1 \\ a_2 & b_2 & c_2 \\ a_3 & b_3 & c_3 \end{vmatrix} \qquad D_x = \begin{vmatrix} k_1 & b_1 & c_1 \\ k_2 & b_2 & c_2 \\ k_3 & b_3 & c_3 \end{vmatrix}$$

$$D_y = \begin{vmatrix} a_1 & k_1 & c_1 \\ a_2 & k_2 & c_2 \\ a_3 & k_3 & c_3 \end{vmatrix} \qquad D_z = \begin{vmatrix} a_1 & b_1 & k_1 \\ a_2 & b_2 & k_2 \\ a_3 & b_3 & k_3 \end{vmatrix}$$

as long as $D$ is not 0.

**Example 4** Use Cramer's rule to solve the system.

$$\begin{cases} x - 2y + z = 4 \\ 3x + y - 2z = 3 \\ 5x + 5y + 3z = -8 \end{cases}$$

*Solution* First we find $D, D_x, D_y,$ and $D_z$. Beginning with $D$, we expand by the minors of the first column.

$$D = \begin{vmatrix} 1 & -2 & 1 \\ 3 & 1 & -2 \\ 5 & 5 & 3 \end{vmatrix} = 1 \cdot \begin{vmatrix} 1 & -2 \\ 5 & 3 \end{vmatrix} - 3 \cdot \begin{vmatrix} -2 & 1 \\ 5 & 3 \end{vmatrix} + 5 \cdot \begin{vmatrix} -2 & 1 \\ 1 & -2 \end{vmatrix}$$

$$= 1(3 - (-10)) - 3(-6 - 5) + 5(4 - 1)$$

$$= 13 + 33 + 15 = 61$$

$$D_x = \begin{vmatrix} 4 & -2 & 1 \\ 3 & 1 & -2 \\ -8 & 5 & 3 \end{vmatrix} = 4 \cdot \begin{vmatrix} 1 & -2 \\ 5 & 3 \end{vmatrix} - 3 \cdot \begin{vmatrix} -2 & 1 \\ 5 & 3 \end{vmatrix} + (-8) \cdot \begin{vmatrix} -2 & 1 \\ 1 & -2 \end{vmatrix}$$

$$= 4(3 - (-10)) - 3(-6 - 5) + (-8)(4 - 1)$$

$$= 52 + 33 - 24 = 61$$

$$D_y = \begin{vmatrix} 1 & 4 & 1 \\ 3 & 3 & -2 \\ 5 & -8 & 3 \end{vmatrix} = 1 \cdot \begin{vmatrix} 3 & -2 \\ -8 & 3 \end{vmatrix} - 3 \cdot \begin{vmatrix} 4 & 1 \\ -8 & 3 \end{vmatrix} + 5 \cdot \begin{vmatrix} 4 & 1 \\ 3 & -2 \end{vmatrix}$$

$$= 1(9 - 16) - 3(12 - (-8)) + 5(-8 - 3)$$

$$= -7 - 60 - 55 = -122$$

$$D_z = \begin{vmatrix} 1 & -2 & 4 \\ 3 & 1 & 3 \\ 5 & 5 & -8 \end{vmatrix} = 1 \cdot \begin{vmatrix} 1 & 3 \\ 5 & -8 \end{vmatrix} - 3 \cdot \begin{vmatrix} -2 & 4 \\ 5 & -8 \end{vmatrix} + 5 \cdot \begin{vmatrix} -2 & 4 \\ 1 & 3 \end{vmatrix}$$

$$= 1(-8 - 15) - 3(16 - 20) + 5(-6 - 4)$$

$$= -23 + 12 - 50 = -61$$

From these determinants, we calculate the solution.

$$x = \frac{D_x}{D} = \frac{61}{61} = 1 \qquad y = \frac{D_y}{D} = \frac{-122}{61} = -2 \qquad z = \frac{D_z}{D} = \frac{-61}{61} = -1$$

The ordered triple solution is $(1, -2, -1)$. Check this solution by verifying that it satisfies each equation of the system.

## Exercise Set 4.5

*Evaluate. See Example 1.*

1. $\begin{vmatrix} 3 & 5 \\ -1 & 7 \end{vmatrix}$

2. $\begin{vmatrix} -5 & 1 \\ 0 & -4 \end{vmatrix}$

3. $\begin{vmatrix} 9 & -2 \\ 4 & -3 \end{vmatrix}$

4. $\begin{vmatrix} 4 & 0 \\ 9 & 8 \end{vmatrix}$

5. $\begin{vmatrix} -2 & 9 \\ 4 & -18 \end{vmatrix}$

6. $\begin{vmatrix} -40 & 8 \\ 70 & -14 \end{vmatrix}$

*Use Cramer's rule, if possible, to solve each system of linear equations. See Example 2.*

7. $\begin{cases} 2y - 4 = 0 \\ x + 2y = 5 \end{cases}$

8. $\begin{cases} 4x - y = 5 \\ 3x - 3 = 0 \end{cases}$

9. $\begin{cases} 3x + y = 1 \\ 2y = 2 - 6x \end{cases}$

10. $\begin{cases} y = 2x - 5 \\ 8x - 4y = 20 \end{cases}$

11. $\begin{cases} 5x - 2y = 27 \\ -3x + 5y = 18 \end{cases}$

12. $\begin{cases} 4x - y = 9 \\ 2x + 3y = -27 \end{cases}$

*Evaluate. See Example 3.*

13. $\begin{vmatrix} 2 & 1 & 0 \\ 0 & 5 & -3 \\ 4 & 0 & 2 \end{vmatrix}$

14. $\begin{vmatrix} -6 & 4 & 2 \\ 1 & 0 & 5 \\ 0 & 3 & 1 \end{vmatrix}$

15. $\begin{vmatrix} 4 & -6 & 0 \\ -2 & 3 & 0 \\ 4 & -6 & 1 \end{vmatrix}$

16. $\begin{vmatrix} 5 & 2 & 1 \\ 3 & -6 & 0 \\ -2 & 8 & 0 \end{vmatrix}$

17. $\begin{vmatrix} 3 & 6 & -3 \\ -1 & -2 & 3 \\ 4 & -1 & 6 \end{vmatrix}$

18. $\begin{vmatrix} 2 & -2 & 1 \\ 4 & 1 & 3 \\ 3 & 1 & 2 \end{vmatrix}$

*Use Cramer's rule, if possible, to solve each system of linear equations. See Example 4.*

19. $\begin{cases} 3x + z = -1 \\ -x - 3y + z = 7 \\ 3y + z = 5 \end{cases}$

20. $\begin{cases} 4y - 3z = -2 \\ 8x - 4y = 4 \\ -8x + 4y + z = -2 \end{cases}$

21. $\begin{cases} x + y + z = 8 \\ 2x - y - z = 10 \\ x - 2y + 3z = 22 \end{cases}$

22. $\begin{cases} 5x + y + 3z = 1 \\ x - y - 3z = -7 \\ -x + y = 1 \end{cases}$

*Evaluate.*

23. $\begin{vmatrix} 10 & -1 \\ -4 & 2 \end{vmatrix}$

24. $\begin{vmatrix} -6 & 2 \\ 5 & -1 \end{vmatrix}$

25. $\begin{vmatrix} 1 & 0 & 4 \\ 1 & -1 & 2 \\ 3 & 2 & 1 \end{vmatrix}$

26. $\begin{vmatrix} 0 & 1 & 2 \\ 3 & -1 & 2 \\ 3 & 2 & -2 \end{vmatrix}$

27. $\begin{vmatrix} \frac{3}{4} & \frac{5}{2} \\ -\frac{1}{6} & \frac{7}{3} \end{vmatrix}$

28. $\begin{vmatrix} \frac{5}{7} & \frac{1}{3} \\ \frac{6}{7} & \frac{2}{3} \end{vmatrix}$

29. $\begin{vmatrix} 4 & -2 & 2 \\ 6 & -1 & 3 \\ 2 & 1 & 1 \end{vmatrix}$

30. $\begin{vmatrix} 1 & 5 & 0 \\ 7 & 9 & -4 \\ 3 & 2 & -2 \end{vmatrix}$

31. $\begin{vmatrix} -2 & 5 & 4 \\ 5 & -1 & 3 \\ 4 & 1 & 2 \end{vmatrix}$

32. $\begin{vmatrix} 5 & -2 & 4 \\ -1 & 5 & 3 \\ 1 & 4 & 2 \end{vmatrix}$

33. If all the elements in a single row of a determinant are zero, to what does the determinant evaluate? Explain your answer.

34. If all the elements in a single column of a determinant are 0, to what does the determinant evaluate? Explain your answer.

*Find the value of x such that each is a true statement.*

35. $\begin{vmatrix} 1 & x \\ 2 & 7 \end{vmatrix} = -3$

36. $\begin{vmatrix} 6 & 1 \\ -2 & x \end{vmatrix} = 26$

*Use Cramer's rule, if possible, to solve each system of linear equations.*

37. $\begin{cases} 2x - 5y = 4 \\ x + 2y = -7 \end{cases}$

38. $\begin{cases} 3x - y = 2 \\ -5x + 2y = 0 \end{cases}$

39. $\begin{cases} 4x + 2y = 5 \\ 2x + y = -1 \end{cases}$

40. $\begin{cases} 3x + 6y = 15 \\ 2x + 4y = 3 \end{cases}$

41. $\begin{cases} 2x + 2y + z = 1 \\ -x + y + 2z = 3 \\ x + 2y + 4z = 0 \end{cases}$

42. $\begin{cases} 2x - 3y + z = 5 \\ x + y + z = 0 \\ 4x + 2y + 4z = 4 \end{cases}$

43. $\begin{cases} \frac{2}{3}x - \frac{3}{4}y = -1 \\ -\frac{1}{6}x + \frac{3}{4}y = \frac{5}{2} \end{cases}$

44. $\begin{cases} \frac{1}{2}x - \frac{1}{3}y = -3 \\ \frac{1}{8}x + \frac{1}{6}y = 0 \end{cases}$

45. $\begin{cases} 0.7x - 0.2y = -1.6 \\ 0.2x - y = -1.4 \end{cases}$

46. $\begin{cases} -0.7x + 0.6y = 1.3 \\ 0.5x - 0.3y = -0.8 \end{cases}$

47. $\begin{cases} -2x + 4y - 2z = 6 \\ x - 2y + z = -3 \\ 3x - 6y + 3z = -9 \end{cases}$

48. $\begin{cases} -x - y + 3z = 2 \\ 4x + 4y - 12z = -8 \\ -3x - 3y + 9z = 6 \end{cases}$

**49.** $\begin{cases} x - 2y + z = -5 \\ 3y + 2z = 4 \\ 3x - y = -2 \end{cases}$

**50.** $\begin{cases} 4x + 5y = 10 \\ 3y + 2z = -6 \\ x + y + z = 3 \end{cases}$

**51.** Suppose you are interested in finding the determinant of a $4 \times 4$ matrix. Study the pattern shown in the array of signs for a $3 \times 3$ matrix. Use the pattern to expand the array of signs for use with a $4 \times 4$ matrix.

**52.** Why would expanding by minors of the second row be a good choice for the determinant $\begin{vmatrix} 3 & 4 & -2 \\ 5 & 0 & 0 \\ 6 & -3 & 7 \end{vmatrix}$?

## REVIEW EXERCISES

*Simplify each expression. See Section 1.4.*

**53.** $5x - 6 + x - 12$

**54.** $4y + 3 - 15y - 1$

**55.** $2(3x - 6) + 3(x - 1)$

**56.** $-3(2y - 7) - 1(11 + 12y)$

*Graph each function. See Section 3.3.*

**57.** $f(x) = 5x - 6$

**58.** $g(x) = -x + 1$

**59.** $h(x) = 3$

**60.** $f(x) = -3$

## *A Look Ahead*

**Example**

Evaluate the determinant.

$$\begin{vmatrix} 2 & 0 & -1 & 3 \\ 0 & 5 & -2 & -1 \\ 3 & 1 & 0 & 1 \\ 4 & 2 & -2 & 0 \end{vmatrix}$$

*Solution*

To evaluate a $4 \times 4$ determinant, select any row or column and expand by the minors. The array of signs for a $4 \times 4$ determinant is the same as for a $3 \times 3$ determinant except expanded. We expand using the fourth row.

$$\rightarrow \begin{vmatrix} 2 & 0 & -1 & 3 \\ 0 & 5 & -2 & -1 \\ 3 & 1 & 0 & 1 \\ 4 & 2 & -2 & 0 \end{vmatrix}$$

$$= -4 \cdot \begin{vmatrix} 0 & -1 & 3 \\ 5 & -2 & -1 \\ 1 & 0 & 1 \end{vmatrix} + 2 \cdot \begin{vmatrix} 2 & -1 & 3 \\ 0 & -2 & -1 \\ 3 & 0 & 1 \end{vmatrix}$$

$$-(-2) \cdot \begin{vmatrix} 2 & 0 & 3 \\ 0 & 5 & -1 \\ 3 & 1 & 1 \end{vmatrix} + 0 \cdot \begin{vmatrix} 2 & 0 & -1 \\ 0 & 5 & -2 \\ 3 & 1 & 0 \end{vmatrix}$$

Now find the value of each $3 \times 3$ determinant. The value of the $4 \times 4$ determinant is

$$-4(12) + 2(17) + 2(-33) + 0 = -80$$

*Find the value of each determinant. See the preceding example.*

**61.** $\begin{vmatrix} 5 & 0 & 0 & 0 \\ 0 & 4 & 2 & -1 \\ 1 & 3 & -2 & 0 \\ 0 & -3 & 1 & 2 \end{vmatrix}$

**62.** $\begin{vmatrix} 1 & 7 & 0 & -1 \\ 1 & 3 & -2 & 0 \\ 1 & 0 & -1 & 2 \\ 0 & -6 & 2 & 4 \end{vmatrix}$

**63.** $\begin{vmatrix} 4 & 0 & 2 & 5 \\ 0 & 3 & -1 & 1 \\ 0 & 0 & 2 & 0 \\ 0 & 0 & 0 & 1 \end{vmatrix}$

**64.** $\begin{vmatrix} 2 & 0 & -1 & 4 \\ 6 & 0 & 4 & 1 \\ 2 & 4 & 3 & -1 \\ 4 & 0 & 5 & -4 \end{vmatrix}$

4

For additional Chapter Projects, visit the Real World Activities Website by going to http://www.prenhall.com/martin-gay.

# CHAPTER PROJECT

## *Locating Lightning Strikes*

Lightning, most often produced during thunderstorms, is a rapid discharge of high-current electricity into the atmosphere. Around the world, lightning occurs at a rate of approximately 100 flashes per second. Because of lightning's potentially destructive nature, meteorologists track lightning activity by recording and plotting the positions of lightning strikes. In this project, you will have the opportunity to pinpoint the location of a lightning strike. This project may be completed by working in groups or individually.

Weather recording stations use a directional antenna to detect and measure the electromagnetic field emitted by a lightning bolt. The antenna can determine the angle between a fixed point and the position of the lightning strike but cannot determine the distance to the lightning strike. However, the angle measured by the antenna can be used to find the slope of the line connecting the positions of the weather station and the lightning strike. From there, the equation of the line connecting these points may be found.

If two such lines may be found—that is, if another weather station's antenna detects the same lightning flash—the coordinates of the lightning strike's position may be pinpointed.

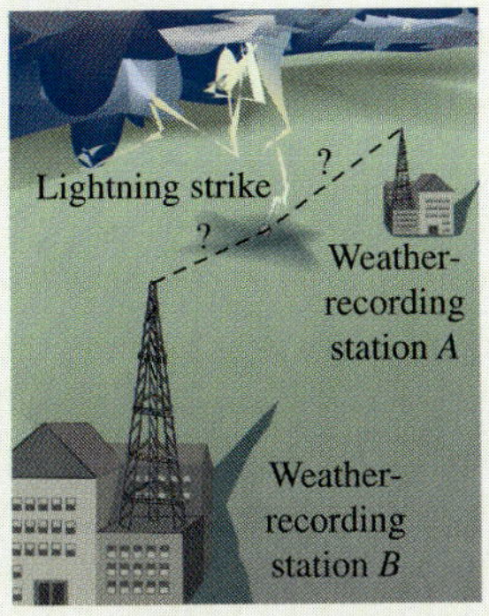

1. A weather recording station $A$ is located at the coordinates $(35, 28)$. A second weather recording station $B$ is located at the coordinates $(52, 12)$. Plot the positions of the two weather recording stations.
2. A lightning strike is detected by both stations. Station $A$ uses a measured angle to find the slope of the line from the station to the lightning strike as $m = -1.732$. Station $B$ computes a slope of $m = 0.577$ from the angle it measured. Use this information to find the equations of the lines connecting each station to the position of the lightning strike.
3. Solve the resulting system of equations in each of the following ways (or work with other students in your group so that each student solves the system in one of the following ways).
   (a) Using a graph. Graph the two equations on your plot of the positions of the two weather recording stations. Estimate the coordinates of their point of intersection.
   (b) Using either the method of substitution or of elimination (whichever you prefer)
   (c) Using matrices
   (d) Using Cramer's rule
   (e) (Optional) Using a graphing calculator to graph the lines and an intersect feature to estimate the coordinates of their point of intersection
4. Compare the results from each method. What are the coordinates of the lightning strike? Which method do you prefer? Why?

## CHAPTER 4 VOCABULARY CHECK

*Fill in each blank with one of the words or phrases listed below.*

| | | | |
|---|---|---|---|
| matrix | determinant | consistent | system of equations |
| solution | inconsistent | square | |

1. Two or more linear equations in two variables form a ________________.
2. A ________ of a system of two equations in two variables is an ordered pair that makes both equations true.
3. A(n) __________ system of equations has at least one solution.
4. If a matrix has the same number of rows and columns, it is called a ________ matrix.
5. A real number associated with a square matrix is called its ___________.
6. A(n) ___________ system of equations has no solution.
7. A _______ is a rectangular array of numbers.

# CHAPTER 4 HIGHLIGHTS

**DEFINITIONS AND CONCEPTS** | **EXAMPLES**

## Section 4.1 Solving Systems of Linear Equations in Two Variables

A **system of linear equations** consists of two or more linear equations.

A **solution** of a system of two equations in two variables is an ordered pair $(x, y)$ that makes both equations true.

Geometrically, a solution of a system in two variables is a point common to the graphs of the equations.

A system of equations with at least one solution is a **consistent system**. A system that has no solution is an **inconsistent system**.

If the graphs of two linear equations are identical, the equations are **dependent**.

If their graphs are different, the equations are **independent**.

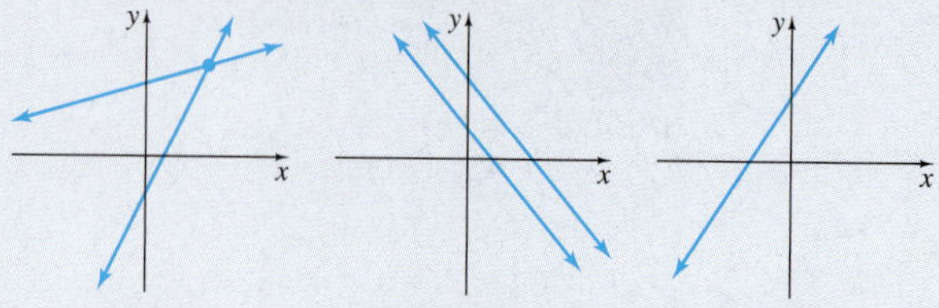


One solution: consistent and independent | No solution: inconsistent and independent | Infinite number of solutions; consistent and dependent

To solve a system of linear equations by the **substitution method:**

**Step 1:** Solve one equation for a variable.

**Step 2:** Substitute the expression for the variable into the other equation.

**Step 3:** Solve the equation from *Step 2* to find the value of one variable.

**Step 4:** Substitute the value from *Step 3* in either original equation to find the value of the other variable.

**Step 5:** Check the solution in both equations.

Solve by substitution:

$$\begin{cases} y = x + 2 \\ 3x - 2y = -5 \end{cases}$$

Substitute $x + 2$ for $y$ in the second equation.

$$\begin{aligned} 3x - 2y &= -5 \\ 3x - 2(x + 2) &= -5 \\ 3x - 2x - 4 &= -5 \\ x - 4 &= -5 && \text{Simplify.} \\ x &= -1 && \text{Add 4.} \end{aligned}$$

To find $y$, let $x = -1$ in $y = x + 2$, so $y = -1 + 2 = 1$. The solution $(-1, 1)$ checks.

To solve a system of linear equations by the **elimination method:**

**Step 1:** Rewrite each equation in standard form $Ax + By = C$.

**Step 2:** Multiply one or both equations by a nonzero number so that the coefficients of a variable are opposites.

**Step 3:** Add the equations.

**Step 4:** Find the value of one variable by solving the resulting equation.

**Step 5:** Substitute the value from *Step 4* into either original equation to find the value of the other variable.

**Step 6:** Check the solution in both equations.

Solve by elimination:

$$\begin{cases} x - 3y = -3 \\ -2x + y = 6 \end{cases}$$

Multiply both sides of the first equation by 2.

$$\begin{aligned} 2x - 6y &= -6 \\ -2x + y &= 6 \\ \hline -5y &= 0 && \text{Add.} \\ y &= 0 && \text{Divide by } -5. \end{aligned}$$

To find $x$, let $y = 0$ in an original equation.

$$\begin{aligned} x - 3y &= -3 \\ x - 3 \cdot 0 &= -3 \\ x &= -3 \end{aligned}$$

The solution $(-3, 0)$ checks.

| DEFINITIONS AND CONCEPTS | EXAMPLES |
|---|---|

## Section 4.2 Solving Systems of Linear Equations in Three Variables

A **solution** of an equation in three variables $x$, $y$, and $z$ is an **ordered triple** $(x, y, z)$ that makes the equation a true statement.

Verify that $(-2, 1, 3)$ is a solution of $2x + 3y - 2z = -7$.

Replace $x$ with $-2$, $y$ with 1, and $z$ with 3.

$$2(-2) + 3(1) - 2(3) = -7$$
$$-4 + 3 - 6 = -7$$
$$-7 = -7 \quad \text{True.}$$

$(-2, 1, 3)$ is a solution.

To solve a system of three linear equations by the elimination method:

**Step 1:** Write each equation in standard form, $Ax + By + Cz = D$.

**Step 2:** Choose a pair of equations and use the equations to eliminate a variable.

**Step 3:** Choose any other pair of equations and eliminate the same variable.

**Step 4:** Solve the system of two equations in two variables from Steps 1 and 2.

**Step 5:** Solve for the third variable by substituting the values of the variables from Step 4 into any of the original equations.

Solve

$$\begin{cases} 2x + y - z = 0 & (1) \\ x - y - 2z = -6 & (2) \\ -3x - 2y + 3z = -22 & (3) \end{cases}$$

1. Each equation is written in standard form.
2. $$\begin{array}{l} 2x + y - z = 0 \ (1) \\ x - y - 2z = -6 \ (2) \\ \hline 3x \quad - 3z = -6 \ (4) \end{array} \quad \text{Add.}$$
3. Eliminate $y$ from equations (1) and (3) also.
$$\begin{array}{l} 4x + 2y - 2z = 0 \\ -3x - 2y + 3z = -22 \ (3) \\ \hline x \quad + z = -22 \ (5) \end{array} \quad \begin{array}{l}\text{Multiply equation (1) by 2.} \\ \\ \text{Add.}\end{array}$$
4. Solve
$$\begin{cases} 3x - 3z = -6 & (4) \\ x + z = -22 & (5) \end{cases}$$
$$\begin{array}{l} x - z = -2 \\ x + z = -22 \ (5) \\ \hline 2x \quad = -24 \\ x = -12 \end{array} \quad \text{Divide equation (4) by 3.}$$
To find $z$, use equation (5).
$$x + z = -22$$
$$-12 + z = -22$$
$$z = -10$$
5. To find $y$, use equation (1).
$$2x + y - z = 0$$
$$2(-12) + y - (-10) = 0$$
$$-24 + y + 10 = 0$$
$$y = 14$$

The solution is $(-12, 14, -10)$.

| DEFINITIONS AND CONCEPTS | EXAMPLES |
|---|---|
| **Section 4.3 Systems of Linear Equations and Problem Solving** | |
| **1.** UNDERSTAND the problem. | Two numbers have a sum of 11. Twice one number is 3 less than 3 times the other. Find the numbers.<br>**1.** Read and reread.<br>$x$ = one number<br>$y$ = other number |
| **2.** TRANSLATE. | **2.** In words: sum of numbers / is / 11<br>Translate: $x + y$ = 11<br>In words: twice one number / is / 3 less than 3 times the other number<br>Translate: $2x$ = $3y - 3$ |
| **3.** SOLVE. | **3.** Solve the system: $\begin{cases} x + y = 11 \\ 2x = 3y - 3 \end{cases}$<br>In the first equation $x = 11 - y$. Substitute into the other equation.<br>$2x = 3y - 3$<br>$2(11 - y) = 3y - 3$<br>$22 - 2y = 3y - 3$<br>$-5y = -25$<br>$y = 5$<br>Replace $y$ with 5 in the equation $x = 11 - y$. Then $x = 11 - 5 = 6$. The solution is (6, 5). |
| **4.** INTERPRET. | **4.** *Check:* See that $6 + 5 = 11$ is the required sum and that twice 6 is 3 times 5 less 3. *State:* The numbers are 6 and 5. |
| **Section 4.4 Solving Systems of Equations by Matrices** | |
| A **matrix** is a rectangular array of numbers. | $\begin{bmatrix} -7 & 0 & 3 \\ 1 & 2 & 4 \end{bmatrix}$ $\begin{bmatrix} a & b & c \\ d & e & f \\ g & h & i \end{bmatrix}$ |
| The **corresponding matrix of the system** is obtained by writing a matrix composed of the coefficients of the variables and the constants of the system. | The corresponding matrix of the system<br>$\begin{cases} x - y = 1 \\ 2x + y = 11 \end{cases}$ is $\left[\begin{array}{cc\|c} 1 & -1 & 1 \\ 2 & 1 & 11 \end{array}\right]$ |

*(continued)*

| DEFINITIONS AND CONCEPTS | EXAMPLES |
|---|---|

## Section 4.4 Solving Systems of Equations by Matrices

The following **row operations** can be performed on matrices, and the result is an equivalent matrix.

***Elementary row operations***

1. Interchange any two rows.
2. Multiply (or divide) the elements of one row by the same nonzero number.
3. Multiply (or divide) the elements of one row by the same nonzero number and add to its corresponding elements in any other row.

Use matrices to solve: $\begin{cases} x - y = 1 \\ 2x + y = 11 \end{cases}$

The corresponding matrix is

$$\left[\begin{array}{cc|c} 1 & -1 & 1 \\ 2 & 1 & 11 \end{array}\right]$$

Use row operations to write an equivalent matrix with 1's along the diagonal and 0's below each 1 in the diagonal. Multiply row 1 by −2 and add to row 2. Change row 2 only.

$$\left[\begin{array}{cc|c} 1 & -1 & 1 \\ -2(1) + 2 & -2(-1) + 1 & -2(1) + 11 \end{array}\right]$$

$$\text{simplifies to} \quad \left[\begin{array}{cc|c} 1 & -1 & 1 \\ 0 & 3 & 9 \end{array}\right]$$

Divide row 2 by 3.

$$\left[\begin{array}{cc|c} 1 & -1 & 1 \\ \frac{0}{3} & \frac{3}{3} & \frac{9}{3} \end{array}\right] \quad \text{simplifies to} \quad \left[\begin{array}{cc|c} 1 & -1 & 1 \\ 0 & 1 & 3 \end{array}\right]$$

This matrix corresponds to the system

$$\begin{cases} x - y = 1 \\ \quad\;\; y = 3 \end{cases}$$

Let $y = 3$ in the first equation.

$$x - 3 = 1$$
$$x = 4$$

The ordered pair solution is (4, 3).

## Section 4.5 Solving Systems of Equations by Determinants

A **square matrix** is a matrix with the same number of rows and columns.

$$\begin{bmatrix} -2 & 1 \\ 6 & 8 \end{bmatrix} \quad \begin{bmatrix} 4 & -1 & 6 \\ 0 & 2 & 5 \\ 1 & 1 & 2 \end{bmatrix}$$

A **determinant** is a real number associated with a square matrix. To denote the determinant, place vertical bars about the array of numbers.

The determinant of $\begin{bmatrix} -2 & 1 \\ 6 & 8 \end{bmatrix}$ is $\begin{vmatrix} -2 & 1 \\ 6 & 8 \end{vmatrix}$.

The determinant of a $2 \times 2$ matrix is

$$\begin{vmatrix} a & b \\ c & d \end{vmatrix} = ad - bc$$

$$\begin{vmatrix} -2 & 1 \\ 6 & 8 \end{vmatrix} = -2 \cdot 8 - 1 \cdot 6 = -22$$

*(continued)*

| DEFINITIONS AND CONCEPTS | EXAMPLES |
|---|---|

## Section 4.5 Solving Systems of Equations by Determinants

***Cramer's Rule for Two Linear Equations in Two Variables***

The solution of the system $\begin{cases} ax + by = h \\ cx + dy = k \end{cases}$ is given by

$$x = \frac{\begin{vmatrix} h & b \\ k & d \end{vmatrix}}{\begin{vmatrix} a & b \\ c & d \end{vmatrix}} = \frac{D_x}{D} \qquad y = \frac{\begin{vmatrix} a & h \\ c & k \end{vmatrix}}{\begin{vmatrix} a & b \\ c & d \end{vmatrix}} = \frac{D_y}{D}$$

as long as $D = ad - bc$ is not 0.

Use Cramer's rule to solve

$$\begin{cases} 3x + 2y = 8 \\ 2x - \phantom{2}y = -11 \end{cases}$$

$$D = \begin{vmatrix} 3 & 2 \\ 2 & -1 \end{vmatrix} = 3(-1) - 2(2) = -7$$

$$D_x = \begin{vmatrix} 8 & 2 \\ -11 & -1 \end{vmatrix} = 8(-1) - 2(-11) = 14$$

$$D_y = \begin{vmatrix} 3 & 8 \\ 2 & -11 \end{vmatrix} = 3(-11) - 8(2) = -49$$

$$x = \frac{D_x}{D} = \frac{14}{-7} = -2 \qquad y = \frac{D_y}{D} = \frac{-49}{-7} = 7$$

The ordered pair solution is $(-2, 7)$.

***Determinant of a 3 × 3 Matrix***

$$\begin{vmatrix} a_1 & b_1 & c_1 \\ a_2 & b_2 & c_2 \\ a_3 & b_3 & c_3 \end{vmatrix} = a_1 \cdot \begin{vmatrix} b_2 & c_2 \\ b_3 & c_3 \end{vmatrix} - a_2 \cdot \begin{vmatrix} b_1 & c_1 \\ b_3 & c_3 \end{vmatrix} + a_3 \cdot \begin{vmatrix} b_1 & c_1 \\ b_2 & c_2 \end{vmatrix}$$

Each $2 \times 2$ matrix above is called a **minor**.

$$\begin{vmatrix} 0 & 2 & -1 \\ 5 & 3 & 0 \\ 2 & -2 & 4 \end{vmatrix} = 0\begin{vmatrix} 3 & 0 \\ -2 & 4 \end{vmatrix} - 2\begin{vmatrix} 5 & 0 \\ 2 & 4 \end{vmatrix} + (-1)\begin{vmatrix} 5 & 3 \\ 2 & -2 \end{vmatrix}$$

$$= 0(12 - 0) - 2(20 - 0) - 1(-10 - 6)$$

$$= 0 - 40 + 16 = -24$$

***Cramer's Rule for Three Equations in Three Variables***

The solution of the system $\begin{cases} a_1x + b_1y + c_1z = k_1 \\ a_2x + b_2y + c_2z = k_2 \\ a_3x + b_3y + c_3z = k_3 \end{cases}$ is given by

$$x = \frac{D_x}{D}, \qquad y = \frac{D_y}{D}, \qquad \text{and} \qquad z = \frac{D_z}{D}$$

where

$$D = \begin{vmatrix} a_1 & b_1 & c_1 \\ a_2 & b_2 & c_2 \\ a_3 & b_3 & c_3 \end{vmatrix} \qquad D_x = \begin{vmatrix} k_1 & b_1 & c_1 \\ k_2 & b_2 & c_2 \\ k_3 & b_3 & c_3 \end{vmatrix}$$

$$D_y = \begin{vmatrix} a_1 & k_1 & c_1 \\ a_2 & k_2 & c_2 \\ a_3 & k_3 & c_3 \end{vmatrix} \qquad D_z = \begin{vmatrix} a_1 & b_1 & k_1 \\ a_2 & b_2 & k_2 \\ a_3 & b_3 & k_3 \end{vmatrix}$$

as long as $D$ is not 0.

Use Cramer's rule to solve

$$\begin{cases} \phantom{x + {}} 3y + 2z = 8 \\ x + y + \phantom{2}z = 3 \\ 2x - y + \phantom{2}z = 2 \end{cases}$$

$$D = \begin{vmatrix} 0 & 3 & 2 \\ 1 & 1 & 1 \\ 2 & -1 & 1 \end{vmatrix} = -3$$

$$D_x = \begin{vmatrix} 8 & 3 & 2 \\ 3 & 1 & 1 \\ 2 & -1 & 1 \end{vmatrix} = 3$$

$$D_y = \begin{vmatrix} 0 & 8 & 2 \\ 1 & 3 & 1 \\ 2 & 2 & 1 \end{vmatrix} = 0$$

$$D_z = \begin{vmatrix} 0 & 3 & 8 \\ 1 & 1 & 3 \\ 2 & -1 & 2 \end{vmatrix} = -12$$

$$x = \frac{D_x}{D} = \frac{3}{-3} = -1 \qquad y = \frac{D_y}{D} = \frac{0}{-3} = 0$$

$$z = \frac{D_z}{D} = \frac{-12}{-3} = 4$$

The ordered triple solution is $(-1, 0, 4)$.

# CHAPTER 4 REVIEW

*(4.1) Solve each system of equations in two variables by each of three methods: (1) graphing, (2) substitution, and (3) elimination.*

**1.** $\begin{cases} 3x + 10y = 1 \\ x + 2y = -1 \end{cases}$

**2.** $\begin{cases} y = \frac{1}{2}x + \frac{2}{3} \\ 4x + 6y = 4 \end{cases}$

**3.** $\begin{cases} 2x - 4y = 22 \\ 5x - 10y = 16 \end{cases}$

**4.** $\begin{cases} 3x - 6y = 12 \\ 2y = x - 4 \end{cases}$

**5.** $\begin{cases} \frac{1}{2}x - \frac{3}{4}y = -\frac{1}{2} \\ \frac{1}{8}x + \frac{3}{4}y = \frac{19}{8} \end{cases}$

**6.** The revenue equation for a certain style of backpack is

$$y = 32x$$

where $x$ is the number of backpacks sold and $y$ is the income in dollars for selling $x$ backpacks. The cost equation for these units is

$$y = 15x + 25{,}500$$

where $x$ is the number of backpacks manufactured and $y$ is the cost in dollars for manufacturing $x$ backpacks. Find the number of units to be sold for the company to break even.

*(4.2) Solve each system of equations in three variables.*

**7.** $\begin{cases} x + z = 4 \\ 2x - y = 4 \\ x + y - z = 0 \end{cases}$

**8.** $\begin{cases} 2x + 5y = 4 \\ x - 5y + z = -1 \\ 4x - z = 11 \end{cases}$

**9.** $\begin{cases} 4y + 2z = 5 \\ 2x + 8y = 5 \\ 6x + 4z = 1 \end{cases}$

**10.** $\begin{cases} 5x + 7y = 9 \\ 14y - z = 28 \\ 4x + 2z = -4 \end{cases}$

**11.** $\begin{cases} 3x - 2y + 2z = 5 \\ -x + 6y + z = 4 \\ 3x + 14y + 7z = 20 \end{cases}$

**12.** $\begin{cases} x + 2y + 3z = 11 \\ y + 2z = 3 \\ 2x + 2z = 10 \end{cases}$

**13.** $\begin{cases} 7x - 3y + 2z = 0 \\ 4x - 4y - z = 2 \\ 5x + 2y + 3z = 1 \end{cases}$

**14.** $\begin{cases} x - 3y - 5z = -5 \\ 4x - 2y + 3z = 13 \\ 5x + 3y + 4z = 22 \end{cases}$

*(4.3) Use systems of equations to solve the following applications.*

**15.** The sum of three numbers is 98. The sum of the first and second is two more than the third number, and the second is four times the first. Find the numbers.

**16.** One number is 3 times a second number, and twice the sum of the numbers is 168. Find the numbers.

**17.** Two cars leave Chicago, one traveling east and the other west. After 4 hours they are 492 miles apart. If one car is traveling 7 mph faster than the other, find the speed of each.

△ **18.** The foundation for a rectangular Hardware Warehouse has a length three times the width and is 296 feet around. Find the dimensions of the building.

**19.** James Callahan has available a 10% alcohol solution and a 60% alcohol solution. Find how many liters of each solution he should mix to make 50 liters of a 40% alcohol solution.

**20.** An employee at a See's Candy Store needs a special mixture of candy. She has creme-filled chocolates that sell for \$3.00 per pound, chocolate-covered nuts that sell for \$2.70 per pound, and chocolate-covered raisins that sell for \$2.25 per pound. She wants to have twice as many raisins as nuts in the mixture. Find how many pounds of each she should use to make 45 pounds worth \$2.80 per pound.

**21.** Chris Kringler has \$2.77 in his coin jar—all in pennies, nickels, and dimes. If he has 53 coins in all and four more nickels than dimes, find how many of each type of coin he has.

**22.** If \$10,000 and \$4000 are invested such that \$1250 is earned in one year, and if the rate of interest on the larger investment is 2% more than that of the smaller investment, find the rates of interest.

△ **23.** The perimeter of an isosceles (two sides equal) triangle is 73 centimeters. If two sides are of equal length and the third side is 7 centimeters longer than the others, find the lengths of the three sides.

**24.** The sum of three numbers is 295. One number is five more than a second and twice the third. Find the numbers.

*(4.4) Use matrices to solve each system.*

**25.** $\begin{cases} 3x + 10y = 1 \\ x + 2y = -1 \end{cases}$

**26.** $\begin{cases} 3x - 6y = 12 \\ 2y = x - 4 \end{cases}$

**27.** $\begin{cases} 3x - 2y = -8 \\ 6x + 5y = 11 \end{cases}$

**28.** $\begin{cases} 6x - 6y = -5 \\ 10x - 2y = 1 \end{cases}$

**29.** $\begin{cases} 3x - 6y = 0 \\ 2x + 4y = 5 \end{cases}$

**30.** $\begin{cases} 5x - 3y = 10 \\ -2x + y = -1 \end{cases}$

**31.** $\begin{cases} 0.2x - 0.3y = -0.7 \\ 0.5x + 0.3y = 1.4 \end{cases}$

**32.** $\begin{cases} 3x + 2y = 8 \\ 3x - y = 5 \end{cases}$

**33.** $\begin{cases} x + z = 4 \\ 2x - y = 0 \\ x + y - z = 0 \end{cases}$

**34.** $\begin{cases} 2x + 5y = 4 \\ x - 5y + z = -1 \\ 4x - z = 11 \end{cases}$

**35.** $\begin{cases} 3x - y = 11 \\ x + 2z = 13 \\ y - z = -7 \end{cases}$

**36.** $\begin{cases} 5x + 7y + 3z = 9 \\ 14y - z = 28 \\ 4x + 2z = -4 \end{cases}$

**37.** $\begin{cases} 7x - 3y + 2z = 0 \\ 4x - 4y - z = 2 \\ 5x + 2y + 3z = 1 \end{cases}$

**38.** $\begin{cases} x + 2y + 3z = 14 \\ y + 2z = 3 \\ 2x - 2z = 10 \end{cases}$

*(4.5) Evaluate.*

**39.** $\begin{vmatrix} -1 & 3 \\ 5 & 2 \end{vmatrix}$

**40.** $\begin{vmatrix} 3 & -1 \\ 2 & 5 \end{vmatrix}$

**41.** $\begin{vmatrix} 2 & -1 & -3 \\ 1 & 2 & 0 \\ 3 & -2 & 2 \end{vmatrix}$

**42.** $\begin{vmatrix} -2 & 3 & 1 \\ 4 & 4 & 0 \\ 1 & -2 & 3 \end{vmatrix}$

*Use Cramer's rule, if possible, to solve each system of equations.*

**43.** $\begin{cases} 3x - 2y = -8 \\ 6x + 5y = 11 \end{cases}$

**44.** $\begin{cases} 6x - 6y = -5 \\ 10x - 2y = 1 \end{cases}$

**45.** $\begin{cases} 3x + 10y = 1 \\ x + 2y = -1 \end{cases}$

**46.** $\begin{cases} y = \frac{1}{2}x + \frac{2}{3} \\ 4x + 6y = 4 \end{cases}$

**47.** $\begin{cases} 2x - 4y = 22 \\ 5x - 10y = 16 \end{cases}$

**48.** $\begin{cases} 3x - 6y = 12 \\ 2y = x - 4 \end{cases}$

**49.** $\begin{cases} x + z = 4 \\ 2x - y = 0 \\ x + y - z = 0 \end{cases}$

**50.** $\begin{cases} 2x + 5y = 4 \\ x - 5y + z = -1 \\ 4x - z = 11 \end{cases}$

**51.** $\begin{cases} x + 3y - z = 5 \\ 2x - y - 2z = 3 \\ x + 2y + 3z = 4 \end{cases}$

**52.** $\begin{cases} 2x - z = 1 \\ 3x - y + 2z = 3 \\ x + y + 3z = -2 \end{cases}$

**53.** $\begin{cases} x + 2y + 3z = 14 \\ y + 2z = 3 \\ 2x - 2z = 10 \end{cases}$

**54.** $\begin{cases} 5x + 7y = 9 \\ 14y - z = 28 \\ 4x + 2z = -4 \end{cases}$

## CHAPTER 4 TEST

*Evaluate each determinant.*

**1.** $\begin{vmatrix} 4 & -7 \\ 2 & 5 \end{vmatrix}$

**2.** $\begin{vmatrix} 4 & 0 & 2 \\ 1 & -3 & 5 \\ 0 & -1 & 2 \end{vmatrix}$

*Solve each system of equations graphically and then solve by the elimination method or the substitution method.*

**3.** $\begin{cases} 2x - y = -1 \\ 5x + 4y = 17 \end{cases}$

**4.** $\begin{cases} 7x - 14y = 5 \\ x = 2y \end{cases}$

*Solve each system.*

**5.** $\begin{cases} 4x - 7y = 29 \\ 2x + 5y = -11 \end{cases}$

**6.** $\begin{cases} 15x + 6y = 15 \\ 10x + 4y = 10 \end{cases}$

**7.** $\begin{cases} 2x - 3y = 4 \\ 3y + 2z = 2 \\ x - z = -5 \end{cases}$

**8.** $\begin{cases} 3x - 2y - z = -1 \\ 2x - 2y = 4 \\ 2x - 2z = -12 \end{cases}$

**9.** $\begin{cases} \frac{x}{2} + \frac{y}{4} = -\frac{3}{4} \\ x + \frac{3}{4}y = -4 \end{cases}$

*Use Cramer's rule, if possible, to solve each system.*

**10.** $\begin{cases} 3x - y = 7 \\ 2x + 5y = -1 \end{cases}$

**11.** $\begin{cases} 4x - 3y = -6 \\ -2x + y = 0 \end{cases}$

**12.** $\begin{cases} x + y + z = 4 \\ 2x + 5y = 1 \\ x - y - 2z = 0 \end{cases}$

**13.** $\begin{cases} 3x + 2y + 3z = 3 \\ x - z = 9 \\ 4y + z = -4 \end{cases}$

*Use matrices to solve each system.*

**14.** $\begin{cases} x - y = -2 \\ 3x - 3y = -6 \end{cases}$

**15.** $\begin{cases} x + 2y = -1 \\ 2x + 5y = -5 \end{cases}$

**16.** $\begin{cases} x - y - z = 0 \\ 3x - y - 5z = -2 \\ 2x + 3y = -5 \end{cases}$

**17.** $\begin{cases} 2x - y + 3z = 4 \\ 3x - 3z = -2 \\ -5x + y = 0 \end{cases}$

**18.** Frame Masters, Inc., recently purchased \$5500 worth of new equipment in order to offer a new style of eyeglass frame. The marketing department of Frame Masters estimates that the cost of producing this new frame is \$18 and that the frame will be sold to stores for \$38. Find the number of frames that must be sold in order to break even.

**19.** A motel in New Orleans charges \$90 per day for double occupancy and \$80 per day for single occupancy. If 80 rooms are occupied for a total of \$6930, how many rooms of each kind are there?

**20.** The research department of a company that manufactures children's fruit drinks is experimenting with a new flavor. A 17.5% fructose solution is needed, but only 10% and 20% solutions are available. How many gallons of a 10% fructose solution should be mixed with a 20% fructose solution in order to obtain 20 gallons of a 17.5% fructose solution?

**21.** A company that manufactures boxes recently purchased \$2000 worth of new equipment to offer gift boxes to its customers. The cost of producing a package of gift boxes is \$1.50 and it is sold for \$4.00. Find the number of packages that must be sold for the company to break even.

## CHAPTER 4 CUMULATIVE REVIEW

1. Determine whether each statement is true or false.
   a. $3 \in \{x \mid x \text{ is a natural number}\}$
   b. $7 \notin \{1, 2, 3\}$
2. Simplify the following expressions.
   a. $11 + 2 - 7$
   b. $-5 - 4 + 2$
3. Write the additive inverse, or opposite, of each.
   a. $8$
   b. $\frac{1}{5}$
   c. $-9.6$
4. Use the distributive property to multiply.
   a. $3(2x + y)$
   b. $-(3x - 1)$
5. Use the distributive property to simplify each expression.
   a. $3x - 5x + 4$
   b. $7yz + yz$
   c. $4z + 6.1$

*Solve.*

6. $-6x - 1 + 5x = 3$
7. $0.3x + 0.1 = 0.27x - 0.02$
8. A pennant in the shape of an isosceles triangle is to be constructed for the Slidell High School Athletic Club and sold at a fund-raiser. The company manufacturing the pennant charges according to perimeter, and the athletic club has determined that a perimeter of 149 centimeters should make a nice profit. If each equal side of the triangle is twice the length of the third side, increased by 12 centimeters, find the lengths of the sides of the triangular pennant.
9. Solve: $3x + 4 \geq 2x - 6$. Graph the solution set.
10. Solve: $2 < 4 - x < 7$
11. Solve: $|2x| + 5 = 7$
12. Solve for $m$: $|m - 6| < 2$
13. Plot each ordered pair on a Cartesian coordinate system and name the quadrant in which the point is located.
    a. $(2, -1)$
    b. $(0, 5)$
    c. $(-3, 5)$
    d. $(-2, 0)$
    e. $\left(-\frac{1}{2}, -4\right)$
    f. $(1.5, 1.5)$
14. Use the graph of $y = 1500 + \frac{1}{10}x$ to answer the following questions:
    a. If the salesperson has \$800 of products sold for a particular month, what is the salary for that month?
    b. If the salesperson wants to make more than \$1600 per month, what must be the total amount of products sold?
15. If $f(x) = 7x^2 - 3x + 1$ and $g(x) = 3x - 2$, find the following.
    a. $f(1)$
    b. $g(1)$
    c. $f(-2)$
    d. $g(0)$
16. Graph $g(x) = 2x + 1$. Compare this graph with the graph of $f(x) = 2x$.
17. Find the slope and the $y$-intercept of the line $3x - 4y = 4$.
18. Are the following pairs of lines parallel, perpendicular, or neither?
    a. $3x + 7y = 4$
       $6x + 14y = 7$
    b. $-x + 3y = 2$
       $2x + 6y = 5$
19. Find an equation of the line through points $(4, 0)$ and $(-4, -5)$. Write the equation using function notation.
20. Graph $3x \geq y$.
21. Determine whether the given ordered pair is a solution of the system.
    a. $\begin{cases} -x + y = 2 \\ 2x - y = -3 \end{cases}$ $(-1, 1)$
    b. $\begin{cases} 5x + 3y = -1 \\ x - y = 1 \end{cases}$ $(-2, 3)$
22. Solve the system.
    $$\begin{cases} 3x - y + z = -15 \\ x + 2y - z = 1 \\ 2x + 3y - 2z = 0 \end{cases}$$
23. Use matrices to solve the system.
    $$\begin{cases} x + 3y = 5 \\ 2x - y = -4 \end{cases}$$
24. Evaluate each determinant.
    a. $\begin{vmatrix} -1 & 2 \\ 3 & -4 \end{vmatrix}$
    b. $\begin{vmatrix} 2 & 0 \\ 7 & -5 \end{vmatrix}$

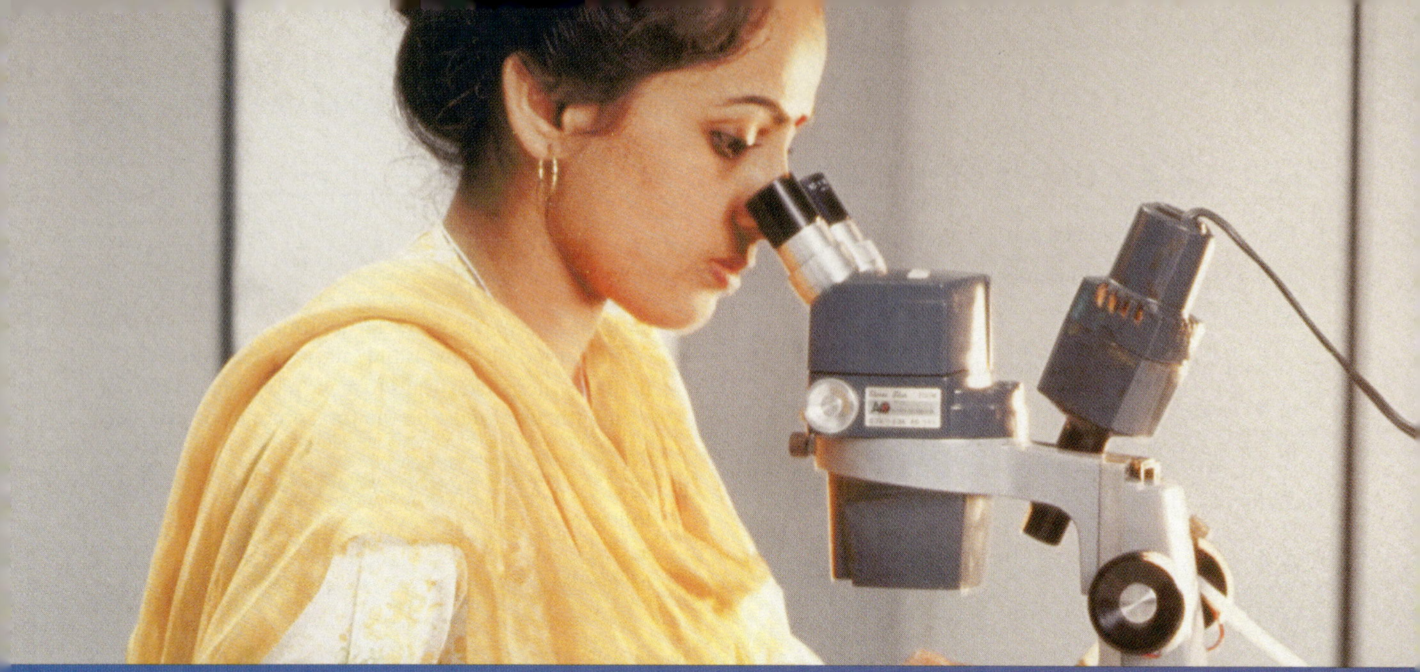

## Too Small to be Seen with the Naked Eye

A microbe is a tiny living organism that is too small to be seen with the naked eye. Microbes are everywhere, both indoors and outdoors. Some scientists estimate that a single gram of ordinary soil may contain up to one billion microbes belonging to over 10,000 different species of microbes. In addition, a single milliliter of coastal ocean water may contain up to one million microbes. We even find microbes in our food: Bread, chocolate, yogurt, and cheese are produced with the help of certain microbes.

A microbiologist is a person who studies microbes. Microbiologists work in both the public and private sectors, identifying harmful microbes in food and water, developing vaccines or treatments for disease, increasing crop yields, protecting the environment, or conducting research. They need an understanding of the sciences, especially biology, chemistry, and physics, as well as mathematics. Microbiologists use mathematics in tasks such as writing grant proposals to obtain project funding, summarizing research results with percents or statistics, modeling bacterial growth, and taking measurements of microbes.

*For more information about a career in microbiology, visit the American Society for Microbiology Website by first going to www.prenhall.com/martin-gay.*

*In the Spotlight on Decision Making feature on page 250, you will have the opportunity to make a decision about microscope magnification as a microbiologist.*

# EXPONENTS, POLYNOMIALS, AND POLYNOMIAL FUNCTIONS

# 5


5.1 Exponents and Scientific Notation

5.2 More Work with Exponents and Scientific Notation

5.3 Polynomials and Polynomial Functions

5.4 Multiplying Polynomials

5.5 The Greatest Common Factor and Factoring by Grouping

5.6 Factoring Trinomials

5.7 Factoring by Special Products and Factoring Strategies

5.8 Solving Equations by Factoring and Problem Solving

5.9 An Introduction to Graphing Polynomial Functions


Linear equations are important for solving problems. They are not sufficient, however, to solve all problems. Many real-world phenomena are modeled by polynomials. We begin this chapter by reviewing exponents. We will then study operations on polynomials and how polynomials can be used in problem solving. We conclude with a study of graphs of polynomial functions.

## 5.1 EXPONENTS AND SCIENTIFIC NOTATION

CD-ROM SSM SSG Video

**OBJECTIVES**

1. Use the product rule for exponents.
2. Evaluate expressions raised to the 0 power.
3. Use the quotient rule for exponents.
4. Evaluate expressions raised to the negative $n$th power.
5. Convert between scientific notation and standard notation.

1 Recall that exponents may be used to write repeated factors in a more compact form. As we have seen in the previous chapters, exponents can be used when the repeated factor is a number or a variable. For example,

$5^3$ (3 is the exponent, 5 is the base) means $\underbrace{5 \cdot 5 \cdot 5}$ — 3 factors; each factor 5

$y^6$ (6 is the exponent, $y$ is the base) means $\underbrace{y \cdot y \cdot y \cdot y \cdot y \cdot y}$ — 6 factors; each factor is $y$

Expressions such as $5^3$ and $y^6$ that contain exponents are called **exponential expressions.**

Exponential expressions can be multiplied, divided, added, subtracted, and themselves raised to powers. In this section, we review operations on exponential expressions.

We review multiplication first. To multiply $x^2$ by $x^3$, use the definition of $a^n$.

$$x^2 \cdot x^3 = \underbrace{(x \cdot x)(x \cdot x \cdot x)}_{x \text{ is a factor 5 times}} = x^5$$

Notice that the result is exactly the same if we add the exponents.

$$x^2 \cdot x^3 = x^{2+3} = x^5$$

This suggests the following.

**PRODUCT RULE FOR EXPONENTS**

If $m$ and $n$ are positive integers and $a$ is a real number, then

$$a^m \cdot a^n = a^{m+n}$$

In other words, the *product* of exponential expressions with a common base is the common base raised to a power equal to the *sum* of the exponents of the factors.

**Example 1** Use the product rule to simplify.

**a.** $2^2 \cdot 2^5$ **b.** $x^7x^3$ **c.** $y \cdot y^2 \cdot y^4$

*Solution*

**a.** $2^2 \cdot 2^5 = 2^{2+5} = 2^7$

**b.** $x^7x^3 = x^{7+3} = x^{10}$

**c.** $y \cdot y^2 \cdot y^4 = (y^1 \cdot y^2) \cdot y^4$

$= y^3 \cdot y^4$

$= y^7$

**Example 2** Use the product rule to multiply.

**a.** $(3x^6)(5x)$ **b.** $(-2x^3p^2)(4xp^{10})$

*Solution* Here, we use properties of multiplication to group together like bases.

**a.** $(3x^6)(5x) = 3(5)x^6x^1 = 15x^7$

**b.** $(-2x^3p^2)(4xp^{10}) = -2(4)x^3x^1p^2p^{10} = -8x^4p^{12}$

**2** The definition of $a^n$ does not include the possibility that $n$ might be 0. But if it did, then, by the product rule,

$$\underbrace{a^0 \cdot a^n} = a^{0+n} = a^n = \underbrace{1 \cdot a^n}.$$

From this, we reasonably define that $a^0 = 1$, as long as $a$ does not equal 0.

**ZERO EXPONENT**

If $a$ does not equal 0, then $a^0 = 1$.

**Example 3** Evaluate the following.

**a.** $7^0$ **b.** $-7^0$ **c.** $(2x + 5)^0$ **d.** $2x^0$

*Solution*

**a.** $7^0 = 1$

**b.** Without parentheses, only 7 is raised to the 0 power.

$$-7^0 = -(7^0) = -(1) = -1$$

**c.** $(2x + 5)^0 = 1$

**d.** $2x^0 = 2(1) = 2$

**3** To find quotients of exponential expressions, we again begin with the definition of $a^n$ to simplify $\frac{x^9}{x^2}$. For example,

$$\frac{x^9}{x^2} = \frac{x \cdot x \cdot x \cdot x \cdot x \cdot x \cdot x \cdot x \cdot x}{x \cdot x} = x^7$$

(Assume for the next two sections that denominators containing variables are not 0.)

Notice that the result is exactly the same if we subtract the exponents.

$$\frac{x^9}{x^2} = x^{9-2} = x^7$$

This suggests the following.

### QUOTIENT RULE FOR EXPONENTS

If $a$ is a nonzero real number and $n$ and $m$ are integers, then

$$\frac{a^m}{a^n} = a^{m-n}$$

In other words, the *quotient* of exponential expressions with a common base is the common base raised to a power equal to the *difference* of the exponents.

**Example 4** Use the quotient rule to simplify.

**a.** $\frac{x^7}{x^4}$ **b.** $\frac{5^8}{5^2}$ **c.** $\frac{20x^6}{4x^5}$ **d.** $\frac{12y^{10}z^7}{14y^8z^7}$

*Solution* **a.** $\frac{x^7}{x^4} = x^{7-4} = x^3$

**b.** $\frac{5^8}{5^2} = 5^{8-2} = 5^6$

**c.** $\frac{20x^6}{4x^5} = 5x^{6-5} = 5x^1$, or $5x$

**d.** $\frac{12y^{10}z^7}{14y^8z^7} = \frac{6}{7}y^{10-8} \cdot z^{7-7} = \frac{6}{7}y^2z^0 = \frac{6}{7}y^2$, or $\frac{6y^2}{7}$

4 When the exponent of the denominator is larger than the exponent of the numerator, applying the quotient rule yields a negative exponent. For example,

$$\frac{x^3}{x^5} = x^{3-5} = x^{-2}$$

Using the definition of $a^n$, though, gives us

$$\frac{x^3}{x^5} = \frac{x \cdot x \cdot x}{x \cdot x \cdot x \cdot x \cdot x} = \frac{1}{x^2}$$

From this, we reasonably define $x^{-2} = \frac{1}{x^2}$ or, in general, $a^{-n} = \frac{1}{a^n}$.

### NEGATIVE EXPONENTS

If $a$ is a real number other than 0 and $n$ is a positive integer, then

$$a^{-n} = \frac{1}{a^n}$$

**Example 5** Use only positive exponents to write the following. Simplify if possible.

**a.** $5^{-2}$ **b.** $2x^{-3}$ **c.** $(3x)^{-1}$ **d.** $\frac{m^5}{m^{15}}$

**e.** $\frac{3^3}{3^6}$ **f.** $2^{-1} + 3^{-2}$ **g.** $\frac{1}{t^{-5}}$

*Solution*

**a.** $5^{-2} = \frac{1}{5^2} = \frac{1}{25}$

**b.** $2x^{-3} = 2 \cdot \frac{1}{x^3} = \frac{2}{x^3}$ Without parentheses, only $x$ is raised to the $-3$ power.

**c.** $(3x)^{-1} = \frac{1}{(3x)^1} = \frac{1}{3x}$ With parentheses, both 3 and $x$ are raised to the $-1$ power.

**d.** $\frac{m^5}{m^{15}} = m^{5-15} = m^{-10} = \frac{1}{m^{10}}$

**e.** $\frac{3^3}{3^6} = 3^{3-6} = 3^{-3} = \frac{1}{3^3} = \frac{1}{27}$

**f.** $2^{-1} + 3^{-2} = \frac{1}{2^1} + \frac{1}{3^2} = \frac{1}{2} + \frac{1}{9} = \frac{9}{18} + \frac{2}{18} = \frac{11}{18}$

**g.** $\frac{1}{t^{-5}} = \frac{1}{\frac{1}{t^5}} = 1 \div \frac{1}{t^5} = 1 \cdot \frac{t^5}{1} = t^5$

**HELPFUL HINT**

Notice that when a factor containing an exponent is moved from the numerator to the denominator or from the denominator to the numerator, the sign of its exponent changes.

$$x^{-3} = \frac{1}{x^3}, \quad 5^{-2} = \frac{1}{5^2} = \frac{1}{25}$$

$$\frac{1}{y^{-4}} = y^4, \quad \frac{1}{2^{-3}} = 2^3 = 8$$

**Example 6** Simplify each expression. Use positive exponents to write the answers.

**a.** $\frac{x^{-9}}{x^2}$ **b.** $\frac{p^4}{p^{-3}}$ **c.** $\frac{2^{-3}}{2^{-1}}$ **d.** $\frac{2x^{-7}y^2}{10xy^{-5}}$ **e.** $\frac{(3x^{-3})(x^2)}{x^6}$

*Solution* **a.** $\dfrac{x^{-9}}{x^2} = x^{-9-2} = x^{-11} = \dfrac{1}{x^{11}}$

**b.** $\dfrac{p^4}{p^{-3}} = p^{4-(-3)} = p^7$

**c.** $\dfrac{2^{-3}}{2^{-1}} = 2^{-3-(-1)} = 2^{-2} = \dfrac{1}{2^2} = \dfrac{1}{4}$

**d.** $\dfrac{2x^{-7}y^2}{10xy^{-5}} = \dfrac{x^{-7-1} \cdot y^{2-(-5)}}{5} = \dfrac{x^{-8}y^7}{5} = \dfrac{y^7}{5x^8}$

**e.** Simplify the numerator first.

$$\frac{(3x^{-3})(x^2)}{x^6} = \frac{3x^{-3+2}}{x^6} = \frac{3x^{-1}}{x^6} = 3x^{-1-6} = 3x^{-7} = \frac{3}{x^7}$$

**Example 7** Simplify. Assume that $a$ and $t$ are nonzero integers and that $x$ is not 0.

**a.** $x^{2a} \cdot x^3$ **b.** $\dfrac{x^{2t-1}}{x^{t-5}}$

*Solution* **a.** $x^{2a} \cdot x^3 = x^{2a+3}$ Use the product rule.

**b.** $\dfrac{x^{2t-1}}{x^{t-5}} = x^{(2t-1)-(t-5)}$ Use the quotient rule.

$= x^{2t-1-t+5} = x^{t+4}$

5 Very large and very small numbers occur frequently in nature. For example, the distance between the Earth and the Sun is approximately 150,000,000 kilometers. A helium atom has a diameter of 0.000 000 022 centimeters. It can be tedious to write these very large and very small numbers in standard notation like this. **Scientific notation** is a convenient shorthand notation for writing very large and very small numbers.

Helium Atom

0.000 000 022 centimeters

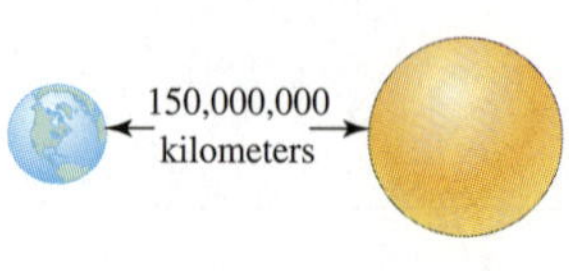


## SCIENTIFIC NOTATION

A positive number is written in **scientific notation** if it is written as the product of a number $a$, where $1 \le a < 10$ and an integer power $r$ of 10:

$$a \times 10^r$$

The following are examples of numbers written in scientific notation.

diameter of helium atom $\rightarrow 2.2 \times 10^{-8}$ cm; $1.5 \times 10^8$ Km $\leftarrow$ approximate distance between Earth and Sun

### WRITING A NUMBER IN SCIENTIFIC NOTATION

**Step 1:** Move the decimal point in the original number until the new number has a value between 1 and 10.

**Step 2:** Count the number of decimal places the decimal point was moved in Step 1. If the decimal point was moved to the left, the count is positive. If the decimal point was moved to the right, the count is negative.

**Step 3:** Write the product of the new number in Step 1 and 10 raised to an exponent equal to the count found in Step 2.

**Example 8** Write each number in scientific notation.

**a.** 730,000 **b.** 0.00000104

*Solution* **a. Step. 1:** Move the decimal point until the number is between 1 and 10.

730,000.

**Step 2:** The decimal point is moved 5 places to the left, so the count is positive 5.

**Step 3:** $730{,}000 = 7.3 \times 10^5$.

**b. Step 1:** Move the decimal point until the number is between 1 and 10.

0.00000104

**Step 2:** The decimal point is moved 6 places to the right, so the count is −6.

**Step 3:** $0.00000104 = 1.04 \times 10^{-6}$.

To write a scientific notation number in standard form, we reverse the preceding steps.

### WRITING A SCIENTIFIC NOTATION NUMBER IN STANDARD NOTATION

Move the decimal point in the number the same number of places as the exponent on 10. If the exponent is positive, move the decimal point to the right. If the exponent is negative, move the decimal point to the left.

**Example 9** Write each number in standard notation.

**a.** $7.7 \times 10^8$ **b.** $1.025 \times 10^{-3}$

*Solution* **a.** $7.7 \times 10^8 = 770{,}000{,}000$ Since the exponent is positive, move the decimal point 8 places to the right. Add zeros as needed.

**b.** $1.025 \times 10^{-3} = 0.001025$ Since the exponent is negative, move the decimal point 3 places to the left. Add zeros as needed.

## SCIENTIFIC CALCULATOR EXPLORATIONS

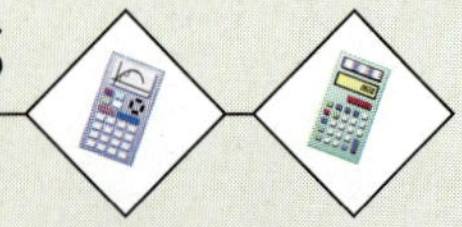

Multiply 5,000,000 by 700,000 on your calculator. The display should read [3.5  12] or [3.5 E 12], which is the product written in scientific notation. Both these notations mean $3.5 \times 10^{12}$.

To enter a number written in scientific notation on a calculator, find the key marked [EE]. (On some calculators, this key may be marked [EXP].)

To enter $7.26 \times 10^{13}$, press the keys

[7.26] [EE] [13]

The display will read [7.26  13] or [7.26 E 13].

*Use your calculator to perform each operation indicated.*

1. Multiply $3 \times 10^{11}$ and $2 \times 10^{32}$.
2. Divide $6 \times 10^{14}$ by $3 \times 10^{9}$.
3. Multiply $5.2 \times 10^{23}$ and $7.3 \times 10^{4}$.
4. Divide $4.38 \times 10^{41}$ by $3 \times 10^{17}$.

## SPOTLIGHT ON DECISION MAKING

Suppose you are a microbiologist. You know that when an image is viewed through a microscope, its magnification is the number of times the image is enlarged. For example, if a 4-millimeter-long object is viewed at five times magnification (denoted $5\times$ magnification), it appears as an object that is $5 \times 4 = 20$ millimeters long.

Suppose you are studying the *Ebola Zaire* virus, which has an average length of $9.2 \times 10^{-5}$ centimeters. You would like to view an Ebola virus with a microscope so that it appears to be 4 centimeters long. Decide what magnification setting (rounded to the nearest thousand) you will need to use on the microscope.

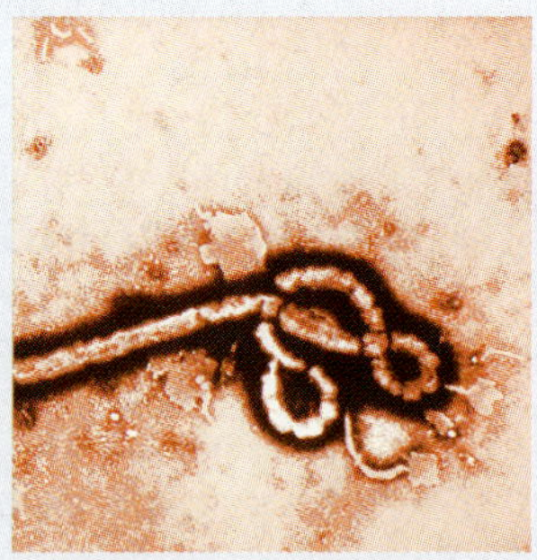

## MENTAL MATH

*Use positive exponents to state each expression.*

1. $5x^{-1}y^{-2}$
2. $7xy^{-4}$
3. $a^2b^{-1}c^{-5}$
4. $a^{-4}b^2c^{-6}$
5. $\dfrac{y^{-2}}{x^{-4}}$
6. $\dfrac{x^{-7}}{z^{-3}}$

## *Exercise Set 5.1*

*Use the product rule to simplify each expression. See Examples 1 and 2.*

**1.** $4^2 \cdot 4^3$
**2.** $3^3 \cdot 3^5$
**3.** $x^5 \cdot x^3$
**4.** $a^2 \cdot a^9$
**5.** $-7x^3 \cdot 20x^9$
**6.** $-3y \cdot -9y^4$
**7.** $(4xy)(-5x)$
**8.** $(7xy)(7aby)$
**9.** $(-4x^3p^2)(4y^3x^3)$
**10.** $(-6a^2b^3)(-3ab^3)$

*Evaluate the following. See Example 3.*

**11.** $-8^0$
**12.** $(-9)^0$
**13.** $(4x + 5)^0$
**14.** $8x^0 + 1$
**15.** $(5x)^0 + 5x^0$
**16.** $4y^0 - (4y)^0$
**17.** Explain why $(-5)^0$ simplifies to 1 but $-5^0$ simplifies to $-1$.
**18.** Explain why both $4x^0 - 3y^0$ and $(4x - 3y)^0$ simplify to 1.

*Find each quotient. See Example 4.*

**19.** $\frac{a^5}{a^2}$
**20.** $\frac{x^9}{x^4}$
**21.** $\frac{x^9y^6}{x^8y^6}$
**22.** $\frac{a^{12}b^2}{a^9b}$
**23.** $-\frac{26z^{11}}{2z^7}$
**24.** $\frac{16x^5}{8x}$
**25.** $\frac{-36a^5b^7c^{10}}{6ab^3c^4}$
**26.** $\frac{49a^3bc^{14}}{-7abc^8}$

*Simplify each expression. Write answers with positive exponents. See Examples 5 and 6.*

**27.** $4^{-2}$
**28.** $2^{-3}$
**29.** $\frac{x^7}{x^{15}}$
**30.** $\frac{z}{z^3}$
**31.** $5a^{-4}$
**32.** $10b^{-1}$
**33.** $\frac{x^{-2}}{x^5}$
**34.** $\frac{y^{-6}}{y^{-9}}$
**35.** $\frac{8r^4}{2r^{-4}}$
**36.** $\frac{3s^3}{15s^{-3}}$
**37.** $\frac{x^{-9}x^4}{x^{-5}}$
**38.** $\frac{y^{-7}y}{y^8}$

*Simplify the following. Write answers with positive exponents.*

**39.** $4^{-1} + 3^{-2}$
**40.** $1^{-3} - 4^{-2}$
**41.** $4x^0 + 5$
**42.** $-5x^0$
**43.** $x^7 \cdot x^8$
**44.** $y^6 \cdot y$
**45.** $2x^3 \cdot 5x^7$
**46.** $-3z^4 \cdot 10z^7$
**47.** $\frac{z^{12}}{z^{15}}$
**48.** $\frac{x^{11}}{x^{20}}$
**49.** $\frac{y^{-3}}{y^{-7}}$
**50.** $\frac{z^{-12}}{z^{10}}$
**51.** $3x^{-1}$
**52.** $(4x)^{-1}$
**53.** $3^0 - 3t^0$
**54.** $4^0 + 4x^0$
**55.** $\frac{r^4}{r^{-4}}$
**56.** $\frac{x^{-5}}{x^3}$
**57.** $\frac{x^{-7}y^{-2}}{x^2y^2}$
**58.** $\frac{a^{-5}b^7}{a^{-2}b^{-3}}$
**59.** $\frac{2a^{-6}b^2}{18ab^{-5}}$
**60.** $\frac{18ab^{-6}}{3a^{-3}b^6}$
**61.** $\frac{(24x^8)(x)}{20x^{-7}}$
**62.** $\frac{(30z^2)(z^5)}{55z^{-4}}$

*Write each number in scientific notation. See Example 8.*

**63.** 31,250,000
**64.** 678,000
**65.** 0.016
**66.** 0.007613
**67.** 67,413
**68.** 36,800,000
**69.** 0.0125
**70.** 0.00084
**71.** 0.000053
**72.** 98,700,000,000

*Write each number in standard notation, without exponents. See Example 9.*

**73.** $3.6 \times 10^{-9}$
**74.** $2.7 \times 10^{-5}$
**75.** $9.3 \times 10^7$
**76.** $6.378 \times 10^8$
**77.** $1.278 \times 10^6$
**78.** $7.6 \times 10^4$
**79.** $7.35 \times 10^{12}$
**80.** $1.66 \times 10^{-5}$
**81.** $4.03 \times 10^{-7}$
**82.** $8.007 \times 10^8$
**83.** Explain how to convert a number from standard notation to scientific notation.
**84.** Explain how to convert a number from scientific notation to standard notation.
**85.** Which numbers have values that are less than 1?
**a.** $3.5 \times 10^{-5}$ **b.** $3.5 \times 10^5$
**c.** $-3.5 \times 10^5$ **d.** $-3.5 \times 10^{-5}$
**86.** Which numbers are equal to 36,000? Of these, which is written in scientific notation?
**a.** $36 \times 10^3$ **b.** $360 \times 10^2$
**c.** $0.36 \times 10^5$ **d.** $3.6 \times 10^4$

*Write each number in scientific notation.*

**87.** The approximate distance between Jupiter and the sun is 778,300,000 kilometers. (*Source:* National Space Data Center)
**88.** Total revenues for Wal-Mart in fiscal year 2000 were \$166,808,000,000. (*Source:* Wal-Mart Stores, Inc.)
**89.** In February 2000, domestic airline flights carried a total of 43,141,000 passengers. (*Source:* Air Transport Association of America)
**90.** In 1998, the American toy industry had retail sales of \$27,200,000. (*Source:* Toy Manufacturers of America, Inc.)

**91.** In 1997, the New York City subway system carried a total of 1,130,000,000 passengers. (*Source:* New York City Transit Authority)

**92.** The center of the sun is about 27,000,000°F.

**93.** A pulsar is a rotating neutron star that gives off sharp, regular pulses of radio waves. For one particular pulsar, the rate of pulses is every 0.001 second.

**94.** To convert from cubic inches to cubic meters, multiply by 0.0000164.

*Simplify. Assume that variables in the exponent represent nonzero integers and that x, y, and z are not 0. See Example 7.*

**95.** $x^5 \cdot x^{7a}$

**96.** $y^{2p} \cdot y^{9p}$

**97.** $\dfrac{x^{3t-1}}{x^t}$

**98.** $\dfrac{y^{4p-2}}{y^{3p}}$

**99.** $x^{4a} \cdot x^7$

**100.** $x^{9y} \cdot x^{-7y}$

**101.** $\dfrac{z^{6x}}{z^7}$

**102.** $\dfrac{y^6}{y^{4z}}$

**103.** $\dfrac{x^{3t} \cdot x^{4t-1}}{x^t}$

**104.** $\dfrac{z^{5x} \cdot z^{x-7}}{z^x}$

**105.** $x^{9+b} \cdot x^{3a-b}$

**106.** $z^{2a-b} \cdot z^{5a-b}$

*Without calculating, determine which number is larger.*

**107.** $7^{11}$ or $7^{13}$

**108.** $5^{10}$ or $5^9$

**109.** $7^{-11}$ or $7^{-13}$

**110.** $5^{-10}$ or $5^{-9}$

### REVIEW EXERCISES

*Evaluate. See Section 1.3.*

**111.** $(5 \cdot 2)^2$

**112.** $5^2 \cdot 2^2$

**113.** $\left(\dfrac{3}{4}\right)^3$

**114.** $\dfrac{3^3}{4^3}$

**115.** $(2^3)^2$

**116.** $(2^2)^3$

**117.** $(2^{-1})^4$

**118.** $(2^4)^{-1}$

## 5.2 MORE WORK WITH EXPONENTS AND SCIENTIFIC NOTATION

CD-ROM SSM SSG Video

**OBJECTIVES**

1. Use the power rules for exponents.
2. Use exponent rules and definitions to simplify exponential expressions.
3. Compute, using scientific notation.

**1** The volume of the cube shown whose side measures $x^2$ units is $(x^2)^3$ cubic units. To simplify an expression such as $(x^2)^3$, we use the definition of $a^n$. Then

$$(x^2)^3 = \underbrace{(x^2)(x^2)(x^2)}_{x^2 \text{ is a factor 3 times}} = x^{2+2+2} = x^6$$

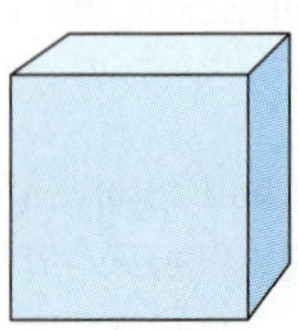
$x^2$ units

Notice that the result is exactly the same if the exponents are multiplied.

$$(x^2)^3 = x^{2 \cdot 3} = x^6$$

This suggests that the power of an exponential expression raised to a power is the product of the exponents. Two additional rules for exponents are given in the following box.

### THE POWER RULE AND POWER OF A PRODUCT OR QUOTIENT RULES FOR EXPONENTS

If $a$ and $b$ are real numbers and $m$ and $n$ are integers, then

$$(a^m)^n = a^{m \cdot n} \quad \text{Power rule}$$
$$(ab)^m = a^m b^m \quad \text{Power of a product}$$
$$\left(\frac{a}{b}\right)^n = \frac{a^n}{b^n}\ (b \neq 0) \quad \text{Power of a quotient}$$

**Example 1** Use the power rule to simplify the following expressions. Use positive exponents to write all results.

**a.** $(x^5)^7$ **b.** $(2^2)^3$ **c.** $(5^{-1})^2$ **d.** $(y^{-3})^{-4}$

*Solution*

**a.** $(x^5)^7 = x^{5 \cdot 7} = x^{35}$

**b.** $(2^2)^3 = 2^{2 \cdot 3} = 2^6 = 64$

**c.** $(5^{-1})^2 = 5^{-1 \cdot 2} = 5^{-2} = \dfrac{1}{5^2} = \dfrac{1}{25}$

**d.** $(y^{-3})^{-4} = y^{-3(-4)} = y^{12}$

**Example 2** Use the power rules to simplify the following. Use positive exponents to write all results.

**a.** $(5x^2)^3$ **b.** $\left(\frac{2}{3}\right)^3$ **c.** $\left(\frac{3p^4}{q^5}\right)^2$ **d.** $\left(\frac{2^{-3}}{y}\right)^{-2}$ **e.** $(x^{-5}y^2z^{-1})^7$

*Solution*

**a.** $(5x^2)^3 = 5^3 \cdot (x^2)^3 = 5^3 \cdot x^{2 \cdot 3} = 125x^6$

**b.** $\left(\dfrac{2}{3}\right)^3 = \dfrac{2^3}{3^3} = \dfrac{8}{27}$

**c.** $\left(\dfrac{3p^4}{q^5}\right)^2 = \dfrac{(3p^4)^2}{(q^5)^2} = \dfrac{3^2 \cdot (p^4)^2}{(q^5)^2} = \dfrac{9p^8}{q^{10}}$

**d.** $\left(\dfrac{2^{-3}}{y}\right)^{-2} = \dfrac{(2^{-3})^{-2}}{y^{-2}}$

$= \dfrac{2^6}{y^{-2}} = 64y^2$ Use the negative exponent rule.

**e.** $(x^{-5}y^2z^{-1})^7 = (x^{-5})^7 \cdot (y^2)^7 \cdot (z^{-1})^7$

$= x^{-35}y^{14}z^{-7} = \dfrac{y^{14}}{x^{35}z^7}$

**2** In the next few examples, we practice the use of several of the rules and definitions for exponents. The following is a summary of these rules and definitions.

**SUMMARY OF RULES FOR EXPONENTS**

If $a$ and $b$ are real numbers and $m$ and $n$ are integers, then

| | | |
|---|---|---|
| ***Product rule*** | $a^m \cdot a^n = a^{m+n}$ | |
| ***Zero exponent*** | $a^0 = 1$ | $(a \neq 0)$ |
| ***Negative exponent*** | $a^{-n} = \dfrac{1}{a^n}$ | $(a \neq 0)$ |
| ***Quotient rule*** | $\dfrac{a^m}{a^n} = a^{m-n}$ | $(a \neq 0)$ |
| ***Power rule*** | $(a^m)^n = a^{m \cdot n}$ | |
| ***Power of a product*** | $(ab)^m = a^m \cdot b^m$ | |
| ***Power of a quotient*** | $\left(\dfrac{a}{b}\right)^m = \dfrac{a^m}{b^m}$ | $(b \neq 0)$ |

**Example 3** Simplify each expression. Use positive exponents to write the answers.

**a.** $(2x^0y^{-3})^{-2}$ **b.** $\left(\dfrac{x^{-5}}{x^{-2}}\right)^{-3}$ **c.** $\left(\dfrac{2}{7}\right)^{-2}$ **d.** $\dfrac{5^{-2}x^{-3}y^{11}}{x^2y^{-5}}$

*Solution* **a.** $(2x^0y^{-3})^{-2} = 2^{-2}(x^0)^{-2}(y^{-3})^{-2}$

$= 2^{-2}x^0y^6$

$= \dfrac{1(y^6)}{2^2}$ Write $x^0$ as 1.

$= \dfrac{y^6}{4}$

**b.** $\left(\dfrac{x^{-5}}{x^{-2}}\right)^{-3} = \dfrac{(x^{-5})^{-3}}{(x^{-2})^{-3}} = \dfrac{x^{15}}{x^6} = x^{15-6} = x^9$

**c.** $\left(\dfrac{2}{7}\right)^{-2} = \dfrac{2^{-2}}{7^{-2}} = \dfrac{7^2}{2^2} = \dfrac{49}{4}$

**d.** $\dfrac{5^{-2}x^{-3}y^{11}}{x^2y^{-5}} = (5^{-2})\left(\dfrac{x^{-3}}{x^2}\right)\left(\dfrac{y^{11}}{y^{-5}}\right) = 5^{-2}x^{-3-2}y^{11-(-5)} = 5^{-2}x^{-5}y^{16}$

$= \dfrac{y^{16}}{5^2x^5} = \dfrac{y^{16}}{25x^5}$

**Example 4** Simplify each expression. Use positive exponents to write the answers.

**a.** $\left(\dfrac{3x^2y}{y^{-9}z}\right)^{-2}$ **b.** $\left(\dfrac{3a^2}{2x^{-1}}\right)^3\left(\dfrac{x^{-3}}{4a^{-2}}\right)^{-1}$

*Solution* There is often more than one way to simplify exponential expressions. Here, we will simplify inside the parentheses if possible before we apply the power rules for exponents.

**a.** $\left(\dfrac{3x^2y}{y^{-9}z}\right)^{-2} = \left(\dfrac{3x^2y^{10}}{z}\right)^{-2} = \dfrac{3^{-2}x^{-4}y^{-20}}{z^{-2}} = \dfrac{z^2}{3^2x^4y^{20}} = \dfrac{z^2}{9x^4y^{20}}$

**b.** $\left(\dfrac{3a^2}{2x^{-1}}\right)^3\left(\dfrac{x^{-3}}{4a^{-2}}\right)^{-1} = \dfrac{27a^6}{8x^{-3}} \cdot \dfrac{x^3}{4^{-1}a^2}$

$= \dfrac{27 \cdot 4 \cdot a^6x^3x^3}{8 \cdot a^2} = \dfrac{27a^4x^6}{2}$

**Example 5** Simplify each expression. Assume that $a$ and $b$ are integers and that $x$ and $y$ are not 0.

**a.** $x^{-b}(2x^b)^2$ **b.** $\dfrac{(y^{3a})^2}{y^{a-6}}$

*Solution* **a.** $x^{-b}(2x^b)^2 = x^{-b}2^2x^{2b} = 4x^{-b+2b} = 4x^b$

**b.** $\dfrac{(y^{3a})^2}{y^{a-6}} = \dfrac{y^{2(3a)}}{y^{a-6}} = \dfrac{y^{6a}}{y^{a-6}} = y^{6a-(a-6)} = y^{6a-a+6} = y^{5a+6}$

3 To perform operations on numbers written in scientific notation, we use properties of exponents.

**Example 6** Perform the indicated operations. Write each result in scientific notation.

**a.** $(8.1 \times 10^5)(5 \times 10^{-7})$ **b.** $\dfrac{1.2 \times 10^4}{3 \times 10^{-2}}$

*Solution* **a.** $(8.1 \times 10^5)(5 \times 10^{-7}) = 8.1 \times 5 \times 10^5 \times 10^{-7}$

$= 40.5 \times 10^{-2}$

$= (4.05 \times 10^1) \times 10^{-2}$

$= 4.05 \times 10^{-1}$

**b.** $\dfrac{1.2 \times 10^4}{3 \times 10^{-2}} = \left(\dfrac{1.2}{3}\right)\left(\dfrac{10^4}{10^{-2}}\right) = 0.4 \times 10^{4-(-2)}$

$= 0.4 \times 10^6 = (4 \times 10^{-1}) \times 10^6 = 4 \times 10^5$

**Example 7** Use scientific notation to simplify $\dfrac{2000 \times 0.000021}{700}$.

*Solution*

$$\frac{2000 \times 0.000021}{700} = \frac{(2 \times 10^3)(2.1 \times 10^{-5})}{7 \times 10^2} = \frac{2(2.1)}{7} \cdot \frac{10^3 \cdot 10^{-5}}{10^2}$$
$$= 0.6 \times 10^{-4}$$
$$= (6 \times 10^{-1}) \times 10^{-4}$$
$$= 6 \times 10^{-5}$$

## MENTAL MATH

*Simplify. See Example 1.*

1. $(x^4)^5$
2. $(5^6)^2$
3. $x^4 \cdot x^5$
4. $x^7 \cdot x^8$
5. $(y^6)^7$
6. $(x^3)^4$
7. $(z^4)^5$
8. $(z^3)^7$
9. $(z^{-6})^{-3}$
10. $(y^{-4})^{-2}$

# Exercise Set 5.2

*Simplify. Use positive exponents to write each answer. See Examples 1 and 2.*

1. $(3^{-1})^2$
2. $(2^{-2})^2$
3. $(x^4)^{-9}$
4. $(y^7)^{-3}$
5. $(y)^{-5}$
6. $(z^{-1})^{10}$
7. $(3x^2y^3)^2$
8. $(4x^3yz)^2$
9. $\left(\dfrac{2x^5}{y^{-3}}\right)^4$
10. $\left(\dfrac{3a^{-4}}{b^7}\right)^3$
11. $(a^2bc^{-3})^{-6}$
12. $(6x^{-6}y^7z^0)^{-2}$
13. $\left(\dfrac{x^7y^{-3}}{z^{-4}}\right)^{-5}$
14. $\left(\dfrac{a^{-2}b^{-5}}{c^{-11}}\right)^{-6}$

*Simplify. Use positive exponents to write each answer. See Examples 3 and 4.*

15. $\left(\dfrac{a^{-4}}{a^{-5}}\right)^{-2}$
16. $\left(\dfrac{x^{-9}}{x^{-4}}\right)^{-3}$
17. $\left(\dfrac{2a^{-2}b^5}{4a^2b^7}\right)^{-2}$
18. $\left(\dfrac{5x^7y^4}{10x^3y^{-2}}\right)^{-3}$
19. $\dfrac{4^{-1}x^2yz}{x^{-2}yz^3}$
20. $\dfrac{8^{-2}x^{-3}y^{11}}{x^2y^{-5}}$
21. Is there a number $a$ such that $a^{-1} = a^1$? If so, give the value of $a$.
22. Is there a number $a$ such that $a^{-2}$ is a negative number? If so, give the value of $a$.

*Simplify. Use positive exponents to write each answer.*

23. $(5^{-1})^3$
24. $(8^2)^{-1}$
25. $(x^7)^{-9}$
26. $(y^{-4})^5$
27. $\left(\dfrac{7}{8}\right)^3$
28. $\left(\dfrac{4}{3}\right)^2$
29. $(4x^2)^2$
30. $(-8x^3)^2$
31. $(-2^{-2}y)^3$
32. $(-4^{-6}y^{-6})^{-4}$
33. $\left(\dfrac{4^{-4}}{y^3x}\right)^{-2}$
34. $\left(\dfrac{7^{-3}}{ab^2}\right)^{-2}$
35. $\left(\dfrac{6p^6}{p^{12}}\right)^2$
36. $\left(\dfrac{4p^6}{p^9}\right)^3$
37. $(-8y^3xa^{-2})^{-3}$
38. $(-xy^0x^2a^3)^{-3}$
39. $\left(\dfrac{x^{-2}y^{-2}}{a^{-3}}\right)^{-7}$
40. $\left(\dfrac{x^{-1}y^{-2}}{5^{-3}}\right)^{-5}$
41. $\left(\dfrac{3x^5}{6x^4}\right)^4$
42. $\left(\dfrac{8^{-3}}{y^2}\right)^{-2}$
43. $\left(\dfrac{1}{4}\right)^{-3}$
44. $\left(\dfrac{1}{8}\right)^{-2}$
45. $\dfrac{(y^3)^{-4}}{y^3}$
46. $\dfrac{2(y^3)^{-3}}{y^{-3}}$
47. $\dfrac{8p^7}{4p^9}$
48. $\left(\dfrac{2x^4}{x^2}\right)^3$

**49.** $(4x^6y^5)^{-2}(6x^4y^3)$

**50.** $(5xy)^3(z^{-2})^{-3}$

**51.** $x^6(x^6bc)^{-6}$

**52.** $2(y^2b)^{-4}$

**53.** $\dfrac{2^{-3}x^2y^{-5}}{5^{-2}x^7y^{-1}}$

**54.** $\dfrac{7^{-1}a^{-3}b^5}{a^2b^{-2}}$

**55.** $\left(\dfrac{2x^2}{y^4}\right)^3 \cdot \left(\dfrac{2x^5}{y}\right)^{-2}$

**56.** $\left(\dfrac{3z^{-2}}{y}\right)^2 \cdot \left(\dfrac{9y^{-4}}{z^{-3}}\right)^{-1}$

*Perform indicated operations. Write each result in scientific notation. See Examples 6 and 7.*

**57.** $(5 \times 10^{11})(2.9 \times 10^{-3})$

**58.** $(3.6 \times 10^{-12})(6 \times 10^9)$

**59.** $(2 \times 10^5)^3$

**60.** $(3 \times 10^{-7})^3$

**61.** $\dfrac{3.6 \times 10^{-4}}{9 \times 10^2}$

**62.** $\dfrac{1.2 \times 10^9}{2 \times 10^{-5}}$

**63.** $\dfrac{0.0069}{0.023}$

**64.** $\dfrac{0.00048}{0.0016}$

**65.** $\dfrac{18{,}200 \times 100}{91{,}000}$

**66.** $\dfrac{0.0003 \times 0.0024}{0.0006 \times 20}$

**67.** $\dfrac{6000 \times 0.006}{0.009 \times 400}$

**68.** $\dfrac{0.00016 \times 300}{0.064 \times 100}$

**69.** $\dfrac{0.00064 \times 2000}{16{,}000}$

**70.** $\dfrac{0.00072 \times 0.003}{0.00024}$

**71.** $\dfrac{66{,}000 \times 0.001}{0.002 \times 0.003}$

**72.** $\dfrac{0.0007 \times 11{,}000}{0.001 \times 0.0001}$

**73.** $\dfrac{8.25 \times 10^{15}}{(2.5 \times 10^{-2})(2.2 \times 10^{-5})}$

**74.** $\dfrac{(2.6 \times 10^{-3})(4.8 \times 10^{-4})}{1.3 \times 10^{-12}}$

*Solve.*

**75.** A computer can add two numbers in about $10^{-8}$ second. Express in scientific notation how long it would take this computer to do this task 200,000 times.

**76.** The density $D$ of an object is equivalent to the quotient of its mass $M$ and volume $V$. Thus, $D = \dfrac{M}{V}$. Express in scientific notation the density of an object whose mass is 500,000 pounds and whose volume is 250 cubic feet.

**77.** The density of ordinary water is $3.12 \times 10^{-2}$ tons per cubic foot. The volume of water in the largest of the Great Lakes, Lake Superior, is $4.269 \times 10^{14}$ cubic feet. Use the formula $D = \dfrac{M}{V}$ (see Exercise 76) to find the mass (in tons) of the water in Lake Superior. Express your answer in scientific notation. (*Source:* National Ocean Service)

**78.** The estimated population of the United States in 1999 was $2.73 \times 10^8$. The land area of the United States is $3.536 \times 10^6$ square miles. Find the population density (number of people per square mile) for the United States in 1999. Round to the nearest whole number. (*Source:* U.S. Bureau of the Census)

△ **79.** Each side of the cube shown is $\dfrac{2x^{-2}}{y}$ meters. Find its volume.

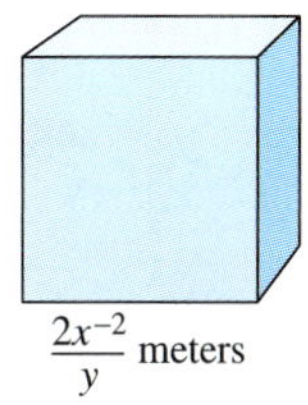


△ **80.** The lot shown is in the shape of a parallelogram with base $\dfrac{3x^{-1}}{y^{-3}}$ feet and height $5x^{-7}$ feet. Find its area.

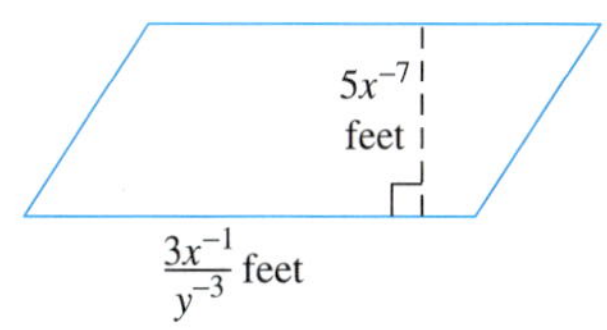


**81.** To convert from square inches to square meters, multiply by $6.452 \times 10^{-4}$. The area of the following square is $4 \times 10^{-2}$ square inches. Convert this area to square meters.

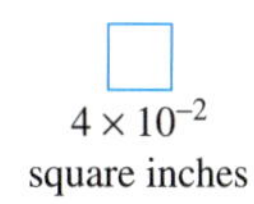


**82.** To convert from cubic inches to cubic meters, multiply by $1.64 \times 10^{-5}$. A grain of salt is in the shape of a cube. If the average size of a grain of salt is $3.8 \times 10^{-6}$ cubic inches, convert this volume to cubic meters.

**83.** Explain whether $0.4 \times 10^{-5}$ is written in scientific notation.

**84.** The subway system with the largest passenger volume in the world in 1997 was the Moscow subway with $3.16 \times 10^9$ passengers. The tenth busiest subway system was São Paulo's with $7.01 \times 10^8$ passengers in 1997. How many times greater was the Moscow subway volume than the São Paulo volume? Round to the nearest tenth. (*Source:* New York City Transit Authority)

**85.** In 1997, China had the largest armed forces in the world. Its fighting force numbered $2.93 \times 10^6$ soldiers. Taiwan's fighting force numbered only $4.25 \times 10^5$. How many times greater was China's armed forces than Taiwan's? Round to the nearest whole number. (*Source:* Russell Ash, *The Top 10 of Everything 1997*)

*Simplify the following. Assume that variables in the exponents represent integers and that all other variables are not 0. See Example 5.*

**86.** $(x^{3a+6})^3$

**87.** $(x^{2b+7})^2$

**88.** $\dfrac{x^{4a}(x^{4a})^3}{x^{4a-2}}$

**89.** $\dfrac{x^{-5y+2}x^{2y}}{x}$

**90.** $(b^{5x-2})^{2x}$

**91.** $(c^{2a+3})^3$

**92.** $\dfrac{(y^{2a})^8}{y^{a-3}}$

**93.** $\dfrac{(y^{4a})^7}{y^{2a-1}}$

**94.** $\left(\dfrac{2x^{3t}}{x^{2t-1}}\right)^4$

**95.** $\left(\dfrac{3y^{5a}}{y^{-a+1}}\right)^2$

**96.** $\dfrac{(z^{a+2})^b}{(z^{b-1})^a}$

**97.** $\dfrac{(y^{3-a})^b}{(y^{1-b})^a}$

**98.** $\dfrac{x^{2a+1}y^{a-1}}{x^{3a+1}y^{2a-3}}$

**99.** $\dfrac{x^{-5-3a}y^{-2a-b}}{x^{-5+3b}y^{-2b-a}}$

## REVIEW EXERCISES

*Simplify each expression. See Section 1.4.*

**100.** $-5y + 4y - 18 - y$

**101.** $12m - 14 - 15m - 1$

**102.** $-3x - (4x - 2)$

**103.** $-9y - (5 - 6y)$

**104.** $3(z - 4) - 2(3z + 1)$

**105.** $5(x - 3) - 4(2x - 5)$

## 5.3 POLYNOMIALS AND POLYNOMIAL FUNCTIONS

CD-ROM SSM SSG Video

**OBJECTIVES**

1. Identify term, constant, polynomial, monomial, binomial, trinomial, and the degree of a term and of a polynomial.
2. Define polynomial functions.
3. Review combining like terms.
4. Add polynomials.
5. Subtract polynomials.
6. Recognize the graph of a polynomial function from the degree of the polynomial.

**1** A **term** is a number or the product of a number and one or more variables raised to powers. The **numerical coefficient**, or simply the **coefficient**, is the numerical factor of a term.

| *Term* | *Numerical Coefficient* |
|---|---|
| $-12x^5$ | $-12$ |
| $x^3y$ | $1$ |
| $-z$ | $-1$ |
| $2$ | $2$ |

If a term contains only a number, it is called a **constant term**, or simply a **constant**.

A **polynomial** is a finite sum of terms in which all variables are raised to nonnegative integer powers and no variables appear in any denominator.

| *Polynomials* | *Not Polynomials* | |
|---|---|---|
| $4x^5y + 7xz$ | $5x^{-3} + 2x$ | Negative integer exponent |
| $-5x^3 + 2x + \dfrac{2}{3}$ | $\dfrac{6}{x^2} - 5x + 1$ | Variable in denominator |

A polynomial that contains only one variable is called a **polynomial in one variable**. For example, $3x^2 - 2x + 7$ is a **polynomial in $x$**. This polynomial in $x$ is written in *descending order* since the terms are listed in descending order of the variable's exponents. (The term 7 can be thought of as $7x^0$.) The following examples are polynomials in one variable written in **descending order**.

$$4x^3 - 7x^2 + 5 \qquad y^2 - 4 \qquad 8a^4 - 7a^2 + 4a$$

A **monomial** is a polynomial consisting of one term. A **binomial** is a polynomial consisting of two terms. A **trinomial** is a polynomial consisting of three terms.

| *Monomials* | *Binomials* | *Trinomials* |
|---|---|---|
| $ax^2$ | $x + y$ | $x^2 + 4xy + y^2$ |
| $-3x$ | $6y^2 - 2$ | $-x^4 + 3x^3 + 1$ |
| 4 | $\frac{5}{7}z^3 - 2z$ | $8y^2 - 2y - 10$ |

By definition, all monomials, binomials, and trinomials are also polynomials.

Each term of a polynomial has a **degree**.

### DEGREE OF A TERM

The **degree of a term** is the sum of the exponents on the *variables* contained in the term.

**Example 1** Find the degree of each term.

**a.** $3x^2$ **b.** $-2^3x^5$ **c.** $y$ **d.** $12x^2yz^3$ **e.** 5

*Solution*

**a.** The exponent on $x$ is 2, so the degree of the term is 2.
**b.** The exponent on $x$ is 5, so the degree of the term is 5. (Recall that the degree is the sum of the exponents on only the *variables*.)
**c.** The degree of $y$, or $y^1$, is 1.
**d.** The degree is the sum of the exponents on the variables, or $2 + 1 + 3 = 6$.
**c.** The degree of 5, which can be written as $5x^0$, is 0.

From the preceding example, we can say that the degree of a constant is 0. Also, the term 0 has no degree.

Each polynomial also has a degree.

### DEGREE OF A POLYNOMIAL

The **degree of a polynomial** is the largest degree of all its terms.

**Example 2** Find the degree of each polynomial and also indicate whether the polynomial is a monomial, binomial, or trinomial.

| | *Polynomial* | *Degree* | *Classification* |
|---|---|---|---|
| **a.** | $7x^3 - 3x + 2$ | 3 | Trinomial |
| **b.** | $-xyz$ | $1 + 1 + 1 = 3$ | Monomial |
| **c.** | $x^4 - 16$ | 4 | Binomial |

## Example 3

Find the degree of the polynomial

$$3xy + x^2y^2 - 5x^2 - 6.$$

*Solution* The degree of each term is

$$\begin{array}{lcccc} & 3xy & + x^2y^2 & - 5x^2 & - 6 \\ & \downarrow & \downarrow & \downarrow & \downarrow \\ \text{Degree:} & 2 & 4 & 2 & 0 \end{array}$$

The largest degree of any term is 4, so the degree of this polynomial is 4.

2 At times, it is convenient to use function notation to represent polynomials. For example, we may write $P(x)$ to represent the polynomial $3x^2 - 2x - 5$. In symbols, this is

$$P(x) = 3x^2 - 2x - 5$$

This function is called a **polynomial function** because the expression $3x^2 - 2x - 5$ is a polynomial.

**HELPFUL HINT**
Recall that the symbol $P(x)$ **does not mean** $P$ times $x$. It is a special symbol used to denote a function.

## Example 4

If $P(x) = 3x^2 - 2x - 5$, find the following.

**a.** $P(1)$ **b.** $P(-2)$

*Solution* **a.** Substitute 1 for $x$ in $P(x) = 3x^2 - 2x - 5$ and simplify.

$$P(x) = 3x^2 - 2x - 5$$
$$P(1) = 3(1)^2 - 2(1) - 5 = -4$$

**b.** Substitute $-2$ for $x$ in $P(x) = 3x^2 - 2x - 5$ and simplify.

$$P(x) = 3x^2 - 2x - 5$$
$$P(-2) = 3(-2)^2 - 2(-2) - 5 = 11$$

Many real-world phenomena are modeled by polynomial functions. If the polynomial function model is given, we can often find the solution of a problem by evaluating the function at a certain value.

**Example 5** FINDING THE HEIGHT OF AN OBJECT

The world's highest bridge, Royal Gorge suspension bridge in Colorado, is 1053 feet above the Arkansas River. An object is dropped from the top of this bridge. Neglecting air resistance, the height of the object at time $t$ seconds is given by the polynomial function $P(t) = -16t^2 + 1053$. Find the height of the object when $t = 1$ second and when $t = 8$ seconds.

*Solution* To find the height of the object at 1 second, we find $P(1)$.

$$P(t) = -16t^2 + 1053$$
$$P(1) = -16(1)^2 + 1053$$
$$P(1) = 1037$$

When $t = 1$ second, the height of the object is 1037 feet.
To find the height of the object at 8 seconds, we find $P(8)$.

$$P(t) = -16t^2 + 1053$$
$$P(8) = -16(8)^2 + 1053$$
$$P(8) = -1024 + 1053$$
$$P(8) = 29$$

When $t = 8$ seconds, the height of the object is 29 feet. Notice that as time $t$ increases, the height of the object decreases.

3 Before we add polynomials, recall that terms are considered to be **like terms** if they contain exactly the same variables raised to exactly the same powers.

| *Like Terms* | *Unlike Terms* |
|---|---|
| $-5x^2, -x^2$ | $4x^2, 3x$ |
| $7xy^3z, -2xzy^3$ | $12x^2y^3, -2xy^3$ |

To simplify a polynomial, **combine like terms** by using the distributive property. For example, by the distributive property,

$$5x + 7x = (5 + 7)x = 12x$$

**Example 6** Simplify by combining like terms.

**a.** $-12x^2 + 7x^2 - 6x$ **b.** $3xy - 2x + 5xy - x$

*Solution* By the distributive property,

**a.** $-12x^2 + 7x^2 - 6x = (-12 + 7)x^2 - 6x = -5x^2 - 6x$

**b.** Use the associative and commutative properties to group together like terms; then combine.

$$\begin{aligned} 3xy - 2x + 5xy - x &= 3xy + 5xy - 2x - x \\ &= (3 + 5)xy + (-2 - 1)x \\ &= 8xy - 3x \end{aligned}$$

4 Now we have reviewed the necessary skills to add polynomials.

**ADDING POLYNOMIALS**

Combine all like terms.

**Example 7** Add.

**a.** $(7x^3y - xy^3 + 11) + (6x^3y - 4)$ **b.** $(3a^3 - b + 2a - 5) + (a + b + 5)$

*Solution* **a.** To add, remove the parentheses and group like terms.

$$\begin{aligned}&(7x^3y - xy^3 + 11) + (6x^3y - 4)\\ &= 7x^3y - xy^3 + 11 + 6x^3y - 4\\ &= 7x^3y + 6x^3y - xy^3 + 11 - 4 && \text{Group like terms.}\\ &= 13x^3y - xy^3 + 7 && \text{Combine like terms.}\end{aligned}$$

**b.** $(3a^3 - b + 2a - 5) + (a + b + 5)$

$$\begin{aligned}&= 3a^3 - b + 2a - 5 + a + b + 5\\ &= 3a^3 - b + b + 2a + a - 5 + 5 && \text{Group like terms.}\\ &= 3a^3 + 3a && \text{Combine like terms.}\end{aligned}$$

**Example 8** Add $11x^3 - 12x^2 + x - 3$ and $x^3 - 10x + 5$.

*Solution* $(11x^3 - 12x^2 + x - 3) + (x^3 - 10x + 5)$

$$\begin{aligned}&= 11x^3 + x^3 - 12x^2 + x - 10x - 3 + 5 && \text{Group like terms.}\\ &= 12x^3 - 12x^2 - 9x + 2 && \text{Combine like terms.}\end{aligned}$$

Sometimes it is more convenient to add polynomials vertically. To do this, line up like terms beneath one another and add like terms.

5 The definition of subtraction of real numbers can be extended to apply to polynomials. To subtract a number, we add its opposite.

$$a - b = a + (-b)$$

Likewise, to subtract a polynomial, we add its opposite. In other words, if $P$ and $Q$ are polynomials, then

$$P - Q = P + (-Q)$$

The polynomial $-Q$ is the **opposite**, or **additive inverse**, of the polynomial $Q$. We can find $-Q$ by writing the opposite of each term of $Q$.

**SUBTRACTING POLYNOMIALS**

To subtract a polynomial, add its opposite.

For example,

To subtract, add its opposite (found by writing the opposite of each term).

$$(3x^2 + 4x - 7) - (3x^2 - 2x - 5) = (3x^2 + 4x - 7) + (-3x^2 + 2x + 5)$$
$$= 3x^2 + 4x - 7 - 3x^2 + 2x + 5$$
$$= 6x - 2 \quad \text{Combine like terms.}$$

**Example 9** Subtract $(12z^5 - 12z^3 + z) - (-3z^4 + z^3 + 12z)$.

*Solution* To subtract, add the opposite of the second polynomial to the first polynomial.

$$(12z^5 - 12z^3 + z) - (-3z^4 + z^3 + 12z)$$
$$= 12z^5 - 12z^3 + z + 3z^4 - z^3 - 12z) \quad \text{Add the opposite of the polynomial being subtracted.}$$
$$= 12z^5 + 3z^4 - 12z^3 - z^3 + z - 12z \quad \text{Group like terms.}$$
$$= 12z^5 + 3z^4 - 13z^3 - 11z \quad \text{Combine like terms.}$$

**Example 10** Subtract $4x^3y^2 - 3x^2y^2 + 2y^2$ from $10x^3y^2 - 7x^2y^2$.

*Solution* If we subtract 2 from 8, the difference is $8 - 2 = 6$. Notice the order of the numbers, and then write "Subtract $4x^3y^2 - 3x^2y^2 + 2y^2$ from $10x^3y^2 - 7x^2y^2$" as a mathematical expression.

$$(10x^3y^2 - 7x^2y^2) - (4x^3y^2 - 3x^2y^2 + 2y^2)$$
$$= 10x^3y^2 - 7x^2y^2 - 4x^3y^2 + 3x^2y^2 - 2y^2 \quad \text{Remove parentheses.}$$
$$= 6x^3y^2 - 4x^2y^2 - 2y^2 \quad \text{Combine like terms.}$$

Polynomials can also be added or subtracted vertically. Just remember to line up like terms. For example, perform the subtraction $(10x^3y^2 - 7x^2y^2) - (4x^3y^2 - 3x^2y^2 + 2y^2)$ vertically.

Add the opposite of the second polynomial.

$$\begin{array}{r} 10x^3y^2 - 7x^2y^2 \\ \underline{-(4x^3y^2 - 3x^2y^2 + 2y^2)} \end{array} \quad \text{is equivalent to} \quad \begin{array}{r} 10x^3y^2 - 7x^2y^2 \\ \underline{-4x^3y^2 + 3x^2y^2 - 2y^2} \\ 6x^3y^2 - 4x^2y^2 - 2y^2 \end{array}$$

Polynomial functions, like polynomials, can be added, subtracted, multiplied, and divided. For example, if

$$P(x) = x^2 + x + 1$$

then

$$2P(x) = 2(x^2 + x + 1) = 2x^2 + 2x + 2 \quad \text{Use the distributive property.}$$

Also, if $Q(x) = 5x^2 - 1$, then $P(x) + Q(x) = (x^2 + x + 1) + (5x^2 - 1) = 6x^2 + x$.

A useful business and economics application of subtracting polynomial functions is finding the profit function $P(x)$ when given a revenue function $R(x)$ and a cost function $C(x)$. In business, it is true that

$$\text{profit} = \text{revenue} - \text{cost, or}$$
$$P(x) = R(x) - C(x)$$

For example, if the revenue function is $R(x) = 7x$ and the cost function is $C(x) = 2x + 5000$, then the profit function is

$$P(x) = R(x) - C(x)$$

or

$$P(x) = 7x - (2x + 5000) \quad \text{Substitute } R(x) = 7x \text{ and } C(x) = 2x + 5000.$$
$$P(x) = 5x - 5000$$

Problem-solving exercises involving profit are in the exercise set.

6 In this section, we reviewed how to find the degree of a polynomial. Knowing the degree of a polynomial can help us recognize the graph of the related polynomial function. For example, we know from Section 3.1 that the graph of the polynomial function $f(x) = x^2$ is a parabola as shown to the left.

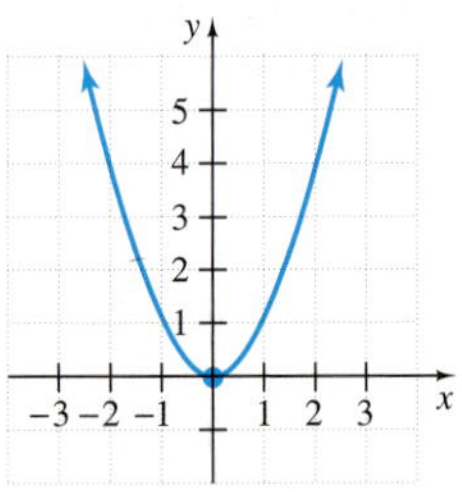


The polynomial $x^2$ has degree 2. The graphs of all polynomial functions of degree 2 will have this same general shape—opening upward, as shown, or downward. Graphs of polynomial functions of degree 2 or 3 will, in general, resemble one of the graphs shown next.

***Degree 2***

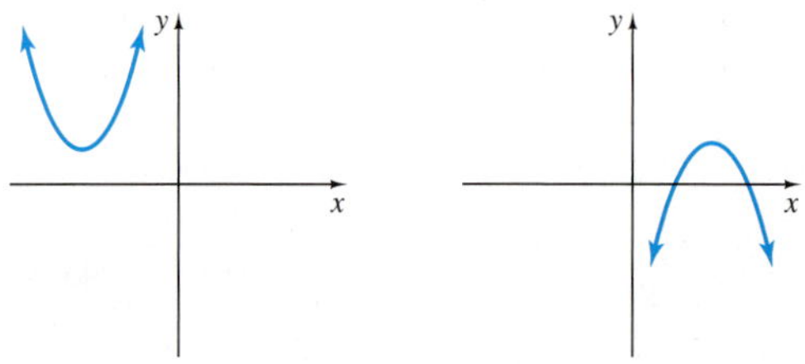


Coefficient of $x^2$ is a positive number.

Coefficient of $x^2$ is a negative number.

***Degree 3***

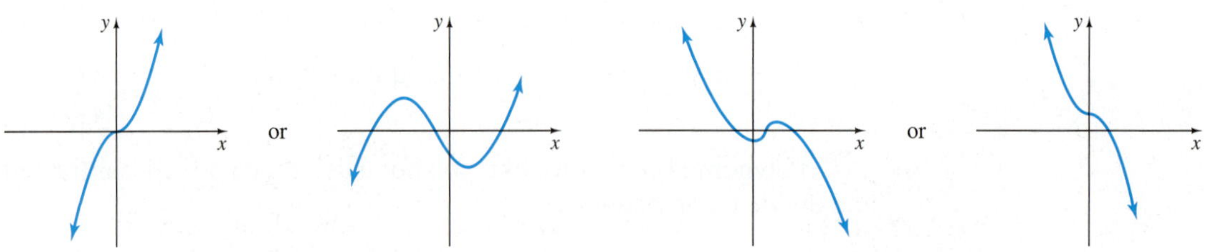


Coefficient of $x^3$ is a positive number.

Coefficient of $x^3$ is a negative number.

General Shapes of Graphs of Polynomial Functions

**Example 11** Determine which of the following graphs is the graph of $f(x) = 5x^3 - 6x^2 + 2x + 3$

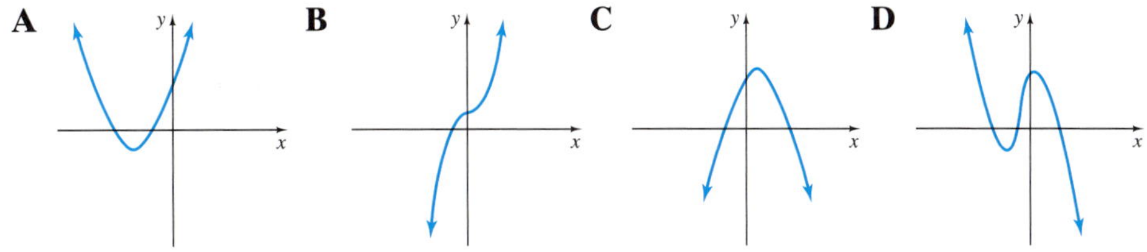


*Solution* The degree of $f(x)$ is 3, which means that its graph has the shape of B or D. The coefficient of $x^3$ is 5, a positive number, so the graph has the shape of B.

## GRAPHING CALCULATOR EXPLORATIONS

A graphing calculator may be used to visualize addition and subtraction of polynomials in one variable. For example, to visualize the following polynomial subtraction statement

$$(3x^2 - 6x + 9) - (x^2 - 5x + 6) = 2x^2 - x + 3$$

graph both

$$Y_1 = (3x^2 - 6x + 9) - (x^2 - 5x + 6) \quad \text{Left side of equation}$$

and

$$Y_2 = 2x^2 - x + 3 \quad \text{Right side of equation}$$

on the same screen and see that their graphs coincide. (*Note:* If the graphs do not coincide, we can be sure that a mistake has been made in combining polynomials or in calculator keystrokes. If the graphs appear to coincide, we cannot be sure that our work is correct. This is because it is possible for the graphs to differ so slightly that we do not notice it.)

10

−10 10

−10

The graphs of $Y_1$ and $Y_2$ are shown. The graphs appear to coincide, so the subtraction statement

$$(3x^2 - 6x + 9) - (x^2 - 5x + 6) = 2x^2 - x + 3$$

appears to be correct.

*Perform the indicated operations. Then visualize by using the procedure described above.*

**1.** $(2x^2 + 7x + 6) + (x^3 - 6x^2 - 14)$
**2.** $(-14x^3 - x + 2) + (-x^3 + 3x^2 + 4x)$
**3.** $(1.8x^2 - 6.8x - 1.7) - (3.9x^2 - 3.6x)$
**4.** $(-4.8x^2 + 12.5x - 7.8) - (3.1x^2 - 7.8x)$
**5.** $(1.29x - 5.68) + (7.69x^2 - 2.55x + 10.98)$
**6.** $(-0.98x^2 - 1.56x + 5.57) + (4.36x - 3.71)$

# Exercise Set 5.3

*Find the degree of each term. See Example 1.*

1. $4$
2. $7$
3. $5x^2$
4. $-z^3$
5. $-3xy^2$
6. $12x^3z$

*Find the degree of each polynomial and indicate whether the polynomial is a monomial, binomial, trinomial, or none of these. See Examples 2 and 3.*

7. $6x + 3$
8. $7x - 8$
9. $3x^2 - 2x + 5$
10. $5x^2 - 3x^2y - 2x^3$
11. $-xyz$
12. $-9$
13. $x^2y - 4xy^2 + 5x + y$
14. $-2x^2y - 3y^2 + 4x + y^5$
15. In your own words, describe how to find the degree of a term.
16. In your own words, describe how to find the degree of a polynomial.

*If $P(x) = x^2 + x + 1$ and $Q(x) = 5x^2 - 1$, find the following. See Example 4.*

17. $P(7)$
18. $Q(4)$
19. $Q(-10)$
20. $P(-4)$
21. $P(0)$
22. $Q(0)$

*Refer to Example 5 for Exercises 23 through 26.*

23. Find the height of the object at $t = 2$ seconds.
24. Find the height of the object at $t = 4$ seconds.
25. Find the height of the object at $t = 6$ seconds.
26. Approximate (to the nearest second) how long it takes before the object hits the ground. (*Hint:* The object hits the ground when $P(x) = 0$.)

*Simplify by combining like terms. See Example 6.*

27. $5y + y$
28. $-x + 3x$
29. $4x + 7x - 3$
30. $-8y + 9y + 4y^2$
31. $4xy + 2x - 3xy - 1$
32. $-8xy^2 + 4x - x + 2xy^2$

*Perform the indicated operations. See Examples 7 through 10.*

33. $(9y^2 - 8) + (9y^2 - 9)$
34. $(x^2 + 4x - 7) + (8x^2 + 9x - 7)$
35. Add $(x^2 + xy - y^2)$ and $(2x^2 - 4xy + 7y^2)$.
36. Add $(4x^3 - 6x^2 + 5x + 7)$ and $(2x^2 + 6x - 3)$.
37. $\begin{array}{r} x^2 - 6x + 3 \\ +\quad (2x + 5) \\ \hline \end{array}$
38. $\begin{array}{r} -2x^2 + 3x - 9 \\ +\quad (2x - 3) \\ \hline \end{array}$
39. $(9y^2 - 7y + 5) - (8y^2 - 7y + 2)$
40. $(2x^2 + 3x + 12) - (5x - 7)$
41. Subtract $(6x^2 - 3x)$ from $(4x^2 + 2x)$.
42. Subtract $(xy + x - y)$ from $(xy + x - 3)$.
43. $\begin{array}{r} 3x^2 - 4x + 8 \\ -\quad (5x^2 - 7) \\ \hline \end{array}$
44. $\begin{array}{r} -3x^2 - 4x + 8 \\ -\quad (5x + 12) \\ \hline \end{array}$
45. $(5x - 11) + (-x - 2)$
46. $(3x^2 - 2x) + (5x^2 - 9x)$
47. $(7x^2 + x + 1) - (6x^2 + x - 1)$
48. $(4x - 4) - (-x - 4)$
49. $(7x^3 - 4x + 8) + (5x^3 + 4x + 8x)$
50. $(9xyz + 4x - y) + (-9xyz - 3x + y + 2)$
51. $(9x^3 - 2x^2 + 4x - 7) - (2x^3 - 6x^2 - 4x + 3)$
52. $(3x^2 + 6xy + 3y^2) - (8x^2 - 6xy - y^2)$
53. Add $(y^2 + 4yx + 7)$ and $(-19y^2 + 7yx + 7)$.
54. Subtract $(x - 4)$ from $(3x^2 - 4x + 5)$.
55. $(3x^3 - b + 2a - 6) + (-4x^3 + b + 6a - 6)$
56. $(5x^2 - 6) + (2x^2 - 4x + 8)$
57. $(4x^2 - 6x + 2) - (-x^2 + 3x + 5)$
58. $(5x^2 + x + 9) - (2x^2 - 9)$
59. $(-3x + 8) + (-3x^2 + 3x - 5)$
60. $(5y^2 - 2y + 4) + (3y + 7)$
61. $(-3 + 4x^2 + 7xy^2) + (2x^3 - x^2 + xy^2)$
62. $(-3x^2y + 4) - (-7x^2y - 8y)$
63. $\begin{array}{r} 6y^2 - 6y + 4 \\ -(-y^2 - 6y + 7) \\ \hline \end{array}$
64. $\begin{array}{r} -4x^3 + 4x^2 - 4x \\ -(2x^3 - 2x^2 + 3x) \\ \hline \end{array}$
65. $\begin{array}{r} 3x^2 + 15x + 8 \\ +(2x^2 + 7x + 8) \\ \hline \end{array}$
66. $\begin{array}{r} 9x^2 + 9x - 4 \\ +(7x^2 - 3x - 4) \\ \hline \end{array}$
67. Find the sum of $(5q^4 - 2q^2 - 3q)$ and $(-6q^4 + 3q^2 + 5)$.
68. Find the sum of $(5y^4 - 7y^2 + x^2 - 3)$ and $(-3y^4 + 2y^2 + 4)$.
69. Subtract $(3x + 7)$ from the sum of $(7x^2 + 4x + 9)$ and $(8x^2 + 7x - 8)$.
70. Subtract $(9x + 8)$ from the sum of $(3x^2 - 2x - x^3 + 2)$ and $(5x^2 - 8x - x^3 + 4)$.
71. Find the sum of $(4x^4 - 7x^2 + 3)$ and $(2 - 3x^4)$.
72. Find the sum of $(8x^4 - 14x^2 + 6)$ and $(-12x^6 - 21x^4 - 9x^2)$.
73. $(8x^{2y} - 7x^y + 3) + (-4x^{2y} + 9x^y - 14)$

**74.** $(14z^{5x} + 3z^{2x} + z) - (2z^{5x} - 10z^{2x} + 3z)$

*Solve.*

**75.** The polynomial $P(t) = -32t + 500$ models the relationship between the length of time $t$ in seconds a particle flies through space, beginning at a velocity of 500 feet per second, and its accrued velocity, $P(t)$. Find $P(3)$, the accrued velocity after 3 seconds.

**76.** The polynomial function $P(x) = 45x - 100{,}000$ models the relationship between the number of lamps $x$ that Sherry's Lamp Shop sells and the profit the shop makes, $P(x)$. Find $P(4000)$, the profit from selling 4000 lamps.

**77.** The function $f(x) = 0.43x^2 + 164.6x + 949.3$ can be used to approximate spending for health care in the United States, where $x$ is the number of years since 1980 and $f(x)$ is the amount of money spent per capita. (*Source:* U.S. Health Care Financing Administration)

**a.** Approximate the amount of money spent on health care per capita in the year 1985.

**b.** Approximate the amount of money spent on health care per capita in the year 1995.

**c.** Use the given function to predict the amount of money that will be spent on health care per capita in the year 2010.

**d.** From parts **a**, **b**, and **c**, is the amount of money spent rising at a steady rate? Why or why not?

**78.** An object is thrown upward with an initial velocity of 50 feet per second from the top of the 350-foot high City Hall in Milwaukee, Wisconsin. The height of the object at any time $t$ can be described by the polynomial function $P(t) = -16t^2 + 50t + 350$. Find the height of the projectile when $t = 1$ second, $t = 2$ seconds, and $t = 3$ seconds. (*Source: World Almanac*)

**79.** The total cost (in dollars) for MCD, Inc., Manufacturing Company to produce $x$ blank audiocassette tapes per week is given by the polynomial function $C(x) = 0.8x + 10{,}000$. Find the total cost in producing 20,000 tapes per week.

**80.** The total revenues (in dollars) for MCD, Inc., Manufacturing Company to sell $x$ blank audiocassette tapes per week is given by the polynomial function $R(x) = 2x$. Find the total revenue in selling 20,000 tapes per week.

*A projectile is fired upward from the ground with an initial velocity of 300 feet per second. Neglecting air resistance, the height of the projectile at any time t can be described by the polynomial function*

$$P(t) = -16t^2 + 300t$$

*Use this in Exercises 81.–83.*

**81.** Find the height of the projectile at the given times.

**a.** $t = 1$ second

**b.** $t = 2$ seconds

**c.** $t = 3$ seconds

**d.** $t = 4$ seconds

**82.** Explain why the height increases and then decreases as time passes.

**83.** Approximate (to the nearest second) how long before the object hits the ground.

**84.** An object is thrown upward with an initial velocity of 25 feet per second from the top of the 984-foot-high Eiffel Tower in Paris, France. The height of the object at any time $t$ can be described by the polynomial function $P(t) = -16t^2 + 25t + 984$. Find the height of the projectile when $t = 1$ second, $t = 3$ seconds, and $t = 5$ seconds. (*Source:* Council on Tall Buildings and Urban Habitat, Lehigh University)

**85.** The function $f(x) = -0.85x^3 + 14.28x^2 - 49.38x + 574.16$ can be used to approximate the number of health maintenance organizations (HMO's) in the United States $x$ years after 1990. (Round each answer to the nearest whole.) (*Source:* Based on data from Interstudy, Minneapolis, MN)

**a.** Approximate the number of HMO's in 1995.

**b.** Approximate the number of HMO's in 1998.

**c.** Use the given model to predict the number of HMO's in the United States in 2003.

**86.** The function $f(x) = -1869x^3 + 30{,}581x^2 - 169{,}111x + 384{,}559$ can be used to approximate the number of AIDS cases reported in the United States from 1993 to 1997, where $x$ is the number of years since 1990. (*Source:* Based on data from the U.S. Centers for Disease Control and Prevention)

**a.** Approximate the number of AIDS cases reported in the United States in 1993.

**b.** Approximate the number of AIDS cases reported in the United States in 1995.

**c.** Approximate the number of AIDS cases reported in the United States in 1997.

**d.** Describe the trend in the number of AIDS cases reported during the period covered by the model.

*If* $P(x) = 3x + 3$, $Q(x) = 4x^2 - 6x + 3$, *and* $R(x) = 5x^2 - 7$, *find the following.*

**87.** $P(x) + Q(x)$

**88.** $R(x) + P(x)$

**89.** $Q(x) - R(x)$

**90.** $P(x) - Q(x)$

**91.** $2[Q(x)] - R(x)$

**92.** $-5[P(x)] - Q(x)$

**93.** $3[R(x)] + 4[P(x)]$

**94.** $2[Q(x)] + 7[R(x)]$

**95.** If the revenue function of a certain company is given by $R(x) = 5.5x$, where $x$ is the number of packages of personalized stationery sold and the cost function is given by $C(x) = 3x + 3000$,

**a.** Find the profit function. (Recall that revenue − cost = profit.)

**b.** Find the profit when 2000 packages of stationery are sold.

*Match each equation with its graph. See Example 11.*

**96.** $f(x) = 3x^2 - 2$

**97.** $h(x) = 5x^3 - 6x + 2$

**98.** $g(x) = -2x^3 - 3x^2 + 3x - 2$

**99.** $g(x) = -2x^2 - 6x + 2$

**A**

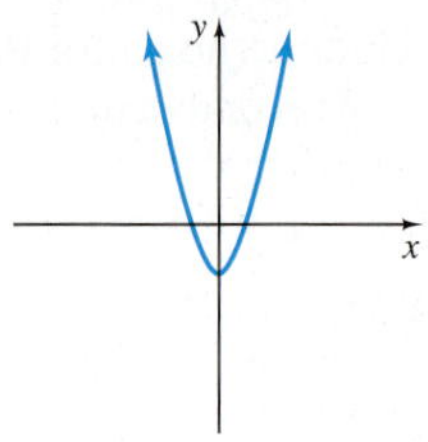


**B**

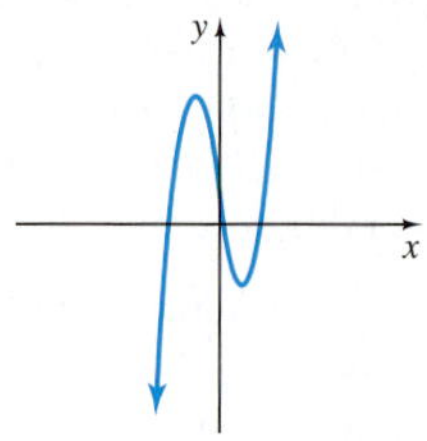


**C**

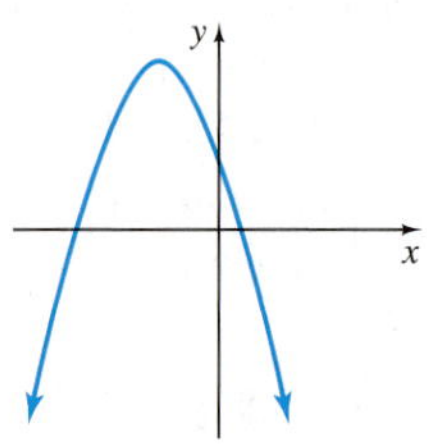


**D**

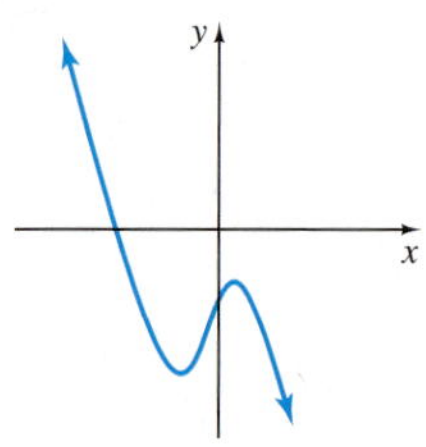


*Find each perimeter.*

△ **100.**

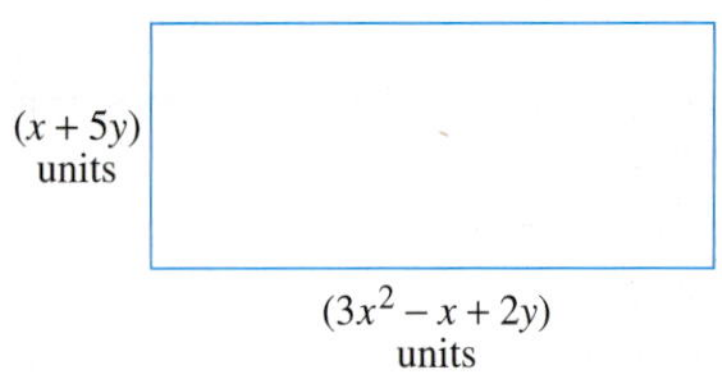


△ **101.**

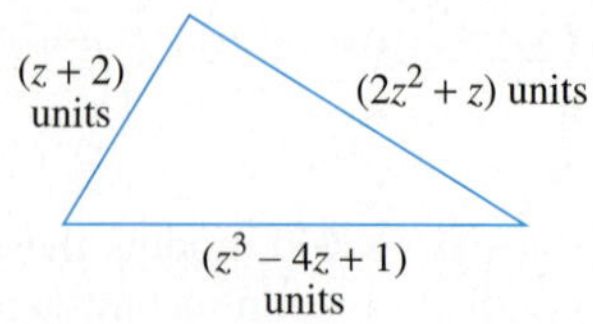


## REVIEW EXERCISES

*Multiply. See Section 1.4.*

**102.** $5(3x - 2)$

**103.** $-7(2z - 6y)$

**104.** $-2(x^2 - 5x + 6)$

**105.** $5(-3y^2 - 2y + 7)$

## *A Look Ahead*

**Example**

If $P(x) = -3x + 5$, find the following.

**a.** $P(a)$ **b.** $P(-x)$ **c.** $P(x + h)$

*Solution:*

**a.** $P(x) = -3x + 5$

$P(a) = -3a + 5$

**b.** $P(x) = -3x + 5$

$P(-x) = -3(-x) + 5$

$= 3x + 5$

**c.** $P(x) = -3x + 5$

$P(x + h) = -3(x + h) + 5$

$= -3x - 3h + 5$

*If $P(x)$ is the polynomial given, find* **a.** *$P(a)$,* **b.** *$P(-x)$, and* **c.** *$P(x + h)$. See the preceding example.*

**106.** $P(x) = 2x - 3$

**107.** $P(x) = 8x + 3$

**108.** $P(x) = 4x$

**109.** $P(x) = -4x$

**110.** $P(x) = 4x - 1$

**111.** $P(x) = 3x - 2$

## 5.4 MULTIPLYING POLYNOMIALS

CD-ROM SSM SSG Video

▶ **OBJECTIVES**

1. Multiply two polynomials.
2. Multiply binomials.
3. Square binomials.
4. Multiply the sum and difference of two terms.
5. Evaluate polynomial functions.

**1** Properties of real numbers and exponents are used continually in the process of multiplying polynomials. To multiply monomials, for example, we apply the commutative and associative properties of real numbers and the product rule for *exponents*.

**Example 1** Multiply.

**a.** $(2x^3)(5x^6)$ **b.** $(7y^4z^4)(-xy^{11}z^5)$

*Solution* Group like bases and apply the product rule for exponents.

**a.** $(2x^3)(5x^6) = 2(5)(x^3)(x^6) = 10x^9$

**b.** $(7y^4z^4)(-xy^{11}z^5) = 7(-1)x(y^4y^{11})(z^4z^5) = -7xy^{15}z^9$

**HELPFUL HINT**

See Sections 5.1 and 5.2 to review exponential expressions further.

To multiply a monomial by a polynomial other than a monomial, we use an expanded form of the distributive property.

$$a(b + c + d + \cdots + z) = ab + ac + ad + \cdots + az$$

Notice that the monomial $a$ is multiplied by each term of the polynomial.

**Example 2** Multiply.

**a.** $2x(5x - 4)$ **b.** $-3x^2(4x^2 - 6x + 1)$ **c.** $-xy(7x^2y + 3xy - 11)$

*Solution* Apply the distributive property.

**a.** $2x(5x - 4) = 2x(5x) + 2x(-4)$ Use the distributive property.

$= 10x^2 - 8x$ Multiply.

**b.** $-3x^2(4x^2 - 6x + 1) = -3x^2(4x^2) + (-3x^2)(-6x) + (-3x^2)(1)$

$= -12x^4 + 18x^3 - 3x^2$

**c.** $-xy(7x^2y + 3xy - 11) = -xy(7x^2y) + (-xy)(3xy) + (-xy)(-11)$

$= -7x^3y^2 - 3x^2y^2 + 11xy$

To multiply any two polynomials, we can use the following.

**MULTIPLYING TWO POLYNOMIALS**

To multiply any two polynomials, use the distributive property and multiply each term of one polynomial by each term of the other polynomial. Then combine any like terms.

**Example 3** Multiply and simplify the product if possible.

**a.** $(x + 3)(2x + 5)$ **b.** $(2x - 3)(5x^2 - 6x + 7)$

*Solution* **a.** Multiply each term of $(x + 3)$ by $(2x + 5)$.

$$\begin{aligned}(x + 3)(2x + 5) &= x(2x + 5) + 3(2x + 5) && \text{Apply the distributive property.}\\ &= 2x^2 + 5x + 6x + 15 && \text{Apply the distributive property again.}\\ &= 2x^2 + 11x + 15 && \text{Combine like terms.}\end{aligned}$$

**b.** Multiply each term of $(2x - 3)$ by each term of $(5x^2 - 6x + 7)$.

$$\begin{aligned}(2x - 3)(5x^2 - 6x + 7) &= 2x(5x^2 - 6x + 7) + (-3)(5x^2 - 6x + 7)\\ &= 10x^3 - 12x^2 + 14x - 15x^2 + 18x - 21\\ &= 10x^3 - 27x^2 + 32x - 21 && \text{Combine like terms.}\end{aligned}$$

Sometimes polynomials are easier to multiply vertically, in the same way we multiply real numbers. When multiplying vertically, we line up like terms in the **partial products** vertically. This makes combining like terms easier.

**Example 4** Multiply vertically $(4x^2 + 7)(x^2 + 2x + 8)$.

*Solution*

$$\begin{array}{rrrrrl} & & x^2 & +\ 2x & +\ 8 & \\ & & & 4x^2 & +\ 7 & \\ \hline & & 7x^2 & +\ 14x & +\ 56 & 7(x^2 + 2x + 8)\\ 4x^4 & +\ 8x^3 & +\ 32x^2 & & & 4x^2(x^2 + 2x + 8)\\ \hline 4x^4 & +\ 8x^3 & +\ 39x^2 & +\ 14x & +\ 56 & \text{Combine like terms.}\end{array}$$

2 When multiplying a binomial by a binomial, we can use a special order of multiplying terms, called the **FOIL** order. The letters of FOIL stand for "**F**irst-**O**uter-**I**nner-**L**ast." To illustrate this method, let's multiply $(2x - 3)$ by $(3x + 1)$.

Multiply the **F**irst terms of each binomial. $(2x - 3)(3x + 1)$ **F** $2x(3x) = 6x^2$

Multiply the **O**uter terms of each binomial. $(2x - 3)(3x + 1)$ **O** $2x(1) = 2x$

Multiply the **I**nner terms of each binomial. $(2x - 3)(3x + 1)$ **I** $-3(3x) = -9x$

Multiply the **L**ast terms of each binomial. $(2x - 3)(3x + 1)$ **L** $-3(1) = -3$

Combine like terms.

$$6x^2 + 2x - 9x - 3 = 6x^2 - 7x - 3$$

**Example 5** Use the FOIL order to multiply $(x - 1)(x + 2)$.

*Solution*

$$\begin{aligned}(x-1)(x+2) &= \overset{\text{First}}{x \cdot x} + \overset{\text{Outer}}{2 \cdot x} + \overset{\text{Inner}}{(-1)x} + \overset{\text{Last}}{(-1)(2)} \\ &= x^2 + 2x - x - 2 \\ &= x^2 + x - 2 \quad \text{Combine like terms.}\end{aligned}$$

**Example 6** Multiply. **a.** $(2x - 7)(3x - 4)$ **b.** $(3x + y)(5x - 2y)$

*Solution*

$$\textbf{a. } \begin{aligned}(2x-7)(3x-4) &= \overset{\text{First}}{2x(3x)} + \overset{\text{Outer}}{2x(-4)} + \overset{\text{Inner}}{(-7)(3x)} + \overset{\text{Last}}{(-7)(-4)} \\ &= 6x^2 - 8x - 21x + 28 \\ &= 6x^2 - 29x + 28\end{aligned}$$

$$\textbf{b. } \begin{aligned}(3x+y)(5x-2y) &= \overset{\text{F}}{15x^2} - \overset{\text{O}}{6xy} + \overset{\text{I}}{5xy} - \overset{\text{L}}{2y^2} \\ &= 15x^2 - xy + 2y^2\end{aligned}$$

3 The **square of a binomial** is a special case of the product of two binomials. By the FOIL order for multiplying two binomials, we have

$$\begin{aligned}(a+b)^2 &= (a+b)(a+b) \\ &= \overset{\text{F}}{a^2} + \overset{\text{O}}{ab} + \overset{\text{I}}{ba} + \overset{\text{L}}{b^2} \\ &= a^2 + 2ab + b^2\end{aligned}$$

This product can be visualized geometrically by analyzing areas.

Area of larger square: $(a + b)^2$

Sum of areas of smaller rectangles: $a^2 + 2ab + b^2$

Thus, $(a + b)^2 = a^2 + 2ab + b^2$

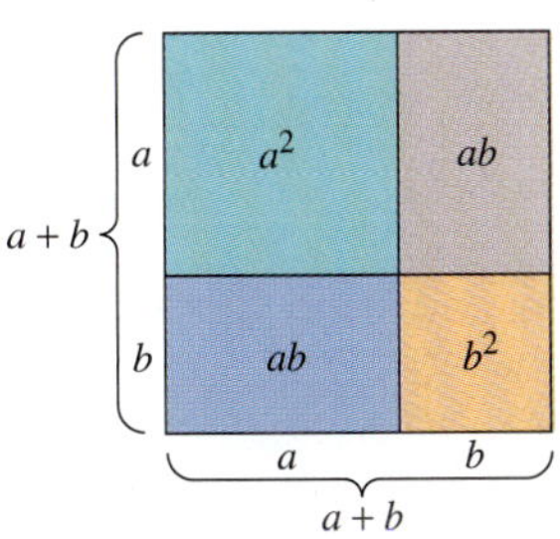


The same pattern occurs for the square of a difference. In general,

**SQUARE OF A BINOMIAL**

$$(a + b)^2 = a^2 + 2ab + b^2 \qquad (a - b)^2 = a^2 - 2ab + b^2$$

In other words, a binomial squared is the sum of the first term squared, twice the product of both terms, and the second term squared.

**Example 7** Multiply.

**a.** $(x + 5)^2$ **b.** $(x - 9)^2$ **c.** $(3x + 2z)^2$ **d.** $(4m^2 - 3n)^2$

$(a + b)^2 = a^2 + 2 \cdot a \cdot b + b^2$

*Solution* **a.** $(x + 5)^2 = x^2 + 2 \cdot x \cdot 5 + 5^2 = x^2 + 10x + 25$

**b.** $(x - 9)^2 = x^2 - 2 \cdot x \cdot 9 + 9^2 = x^2 - 18x + 81$

**c.** $(3x + 2z)^2 = (3x)^2 + 2(3x)(2z) + (2z)^2 = 9x^2 + 12xz + 4z^2$

**d.** $(4m^2 - 3n)^2 = (4m^2)^2 - 2(4m^2)(3n) + (3n)^2 = 16m^4 - 24m^2n + 9n^2$

**HELPFUL HINT**

Note that $(a + b)^2 = a^2 + 2ab + b^2$, **not** $a^2 + b^2$. Also, $(a - b)^2 = a^2 - 2ab + b^2$, **not** $a^2 - b^2$.

4 Another special product applies to the sum and difference of the same two terms. Multiply $(a + b)(a - b)$ to see a pattern.

$$(a + b)(a - b) = a^2 - ab + ba - b^2$$
$$= a^2 - b^2$$

**PRODUCT OF THE SUM AND DIFFERENCE OF TWO TERMS**

$$(a + b)(a - b) = a^2 - b^2$$

The product of the sum and difference of the same two terms is the difference of the first term squared and the second term squared.

**Example 8** Multiply.

**a.** $(x - 3)(x + 3)$ **b.** $(4y + 1)(4y - 1)$ **c.** $\left(3m^2 - \frac{1}{2}\right)\left(3m^2 + \frac{1}{2}\right)$

*Solution* $(a + b)(a - b) = a^2 - b^2$

**a.** $(x + 3)(x - 3) = x^2 - 3^2 = x^2 - 9$

**b.** $(4y + 1)(4y - 1) = (4y)^2 - 1^2 = 16y^2 - 1$

**c.** $\left(3m^2 - \frac{1}{2}\right)\left(3m^2 + \frac{1}{2}\right) = (3m^2)^2 - \left(\frac{1}{2}\right)^2 = 9m^4 - \frac{1}{4}$

**Example 9** Multiply $[3 + (2a + b)]^2$.

*Solution* Think of 3 as the first term and $(2a + b)$ as the second term, and apply the method for squaring a binomial.

$$[a + b]^2 = a^2 + 2(a) \cdot b + b^2$$

$$[3 + (2a + b)]^2 = 3^2 + 2(3)(2a + b) + (2a + b)^2$$
$$= 9 + 6(2a + b) + (2a + b)^2$$
$$= 9 + 12a + 6b + (2a)^2 + 2(2a)(b) + b^2 \quad \text{Square } (2a + b).$$
$$= 9 + 12a + 6b + 4a^2 + 4ab + b^2$$

**Example 10** Multiply $[(5x - 2y) - 1][(5x - 2y) + 1]$.

*Solution* Think of $(5x - 2y)$ as the first term and 1 as the second term, and apply the method for the product of the sum and difference of two terms.

$$\underbrace{[(5x - 2y)}_{(a} - \underbrace{1]}_{b)}\underbrace{[(5x - 2y)}_{(a} + \underbrace{1]}_{b)} = \underbrace{(5x - 2y)^2}_{a^2} - \underbrace{1^2}_{b^2}$$
$$= (5x)^2 - 2(5x)(2y) + (2y)^2 - 1 \quad \text{Square } (5x - 2y).$$
$$= 25x^2 - 20xy + 4y^2 - 1$$

**5** Our work in multiplying polynomials is often useful in evaluating polynomial functions.

**Example 11** If $f(x) = x^2 + 5x - 2$, find $f(a + 1)$.

*Solution* To find $f(a + 1)$, replace $x$ with the expression $a + 1$ in the polynomial function $f(x)$.

$$f(x) = x^2 + 5x - 2$$
$$f(a + 1) = (a + 1)^2 + 5(a + 1) - 2$$
$$= a^2 + 2a + 1 + 5a + 5 - 2$$
$$= a^2 + 7a + 4$$

## GRAPHING CALCULATOR EXPLORATIONS

In the previous section, we used a graphing calculator to visualize addition and subtraction of polynomials in one variable. In this section, the same method is used to visualize multiplication of polynomials in one variable. For example, to see that

$$(x - 2)(x + 1) = x^2 - x - 2,$$

graph both $Y_1 = (x - 2)(x + 1)$ and $Y_2 = x^2 - x - 2$ on the same screen and see whether their graphs coincide.

10

−10 10

−10

By tracing along both graphs, we see that the graphs of $Y_1$ and $Y_2$ appear to coincide, and thus $(x - 2)(x + 1) = x^2 - x - 2$ appears to be correct.

*Multiply. Then use a graphing calculator to visualize the results.*

**1.** $(x + 4)(x - 4)$

**2.** $(x + 3)(x + 3)$

**3.** $(3x - 7)^2$

**4.** $(5x - 2)^2$

**5.** $(5x + 1)(x^2 - 3x - 2)$

**6.** $(7x + 4)(2x^2 + 3x - 5)$

## SPOTLIGHT ON DECISION MAKING

Suppose you would like to sign up for an online service and have received the following advertisements from internet service providers in the mail.

**US Online**

Try our Internet service FREE for 30 days! See why everyone is talking about us:

- Unlimited Internet access and e-mail for a low monthly fee of $21.95
- No set-up fee
- Your own 6 MB Web site
- Thousands of local access numbers across the nation
- Around-the-clock help with our toll-free 888 number

Give us a call to set up your service today!

**Interconnect**

When you sign up with Interconnect as your Internet service provider, you get:

- Unlimited access to the Internet and e-mail
- Local access numbers in major metropolitan areas
- Online expert help

all for just $11.95 per month*! And, for a limited time, you can try Interconnect for a full month for FREE!

*A one-time $20 set-up fee applies, however.

**e-Link**

The sky's the limit with e-Link! If you act now, you can get your first 60 days for FREE, as well as a waiver of the $25 set-up fee. For just $19.95 per month, you get:

- 150 hours of Internet access and e-mail*
- a personal 3 MB Web page
- toll-free help, 24 hours a day, 7 days a week
- local access numbers around the country

*Each additional hour costs $2.95.

You construct a decision grid to help make your choice. In the decision grid, give each of the decision criteria a rank reflecting its importance to you, with 1 being not important to 10 being very important. Then for each online service, decide how well the criteria are supported, assigning a 1 in the rating column for poor support to a rating of 10 for excellent support. For US Online, fill in the Score column by multiplying rank by rating for each criteria. Repeat for each online service. Finally, total the scores in each column for each online service. The service with the highest score is likely to be the best choice for you.

Based on your decision-grid analysis, which online service would you choose? Explain.

| | | US Online | | Interconnect | | e-Link | |
|---|---|---|---|---|---|---|---|
| *Criteria* | *Rank* | *Rating* | *Score* | *Rating* | *Score* | *Rating* | *Score* |
| ***FREE TRIAL PERIOD*** | | | | | | | |
| ***MONTHLY FEE*** | | | | | | | |
| ***UNLIMITED ACCESS*** | | | | | | | |
| ***INCLUDES E-MAIL*** | | | | | | | |
| ***INCLUDES PERSONAL WEB PAGE*** | | | | | | | |
| ***SET-UP FEE*** | | | | | | | |
| ***TOLL-FREE HELP*** | | | | | | | |
| ***LOCAL ACCESS NUMBERS*** | | | | | | | |
| ***TOTAL*** | | | | | | | |

## Exercise Set 5.4

*Multiply. See Examples 1 through 4.*

1. $(-4x^3)(3x^2)$
2. $(-6a)(4a)$
3. $3x(4x + 7)$
4. $5x(6x - 4)$
5. $-6xy(4x + y)$
6. $-8y(6xy + 4x)$
7. $-4ab(xa^2 + ya^2 - 3)$
8. $-6b^2z(z^2a + baz - 3b)$
9. $(x - 3)(2x + 4)$
10. $(y + 5)(3y - 2)$
11. $(2x + 3)(x^3 - x + 2)$
12. $(a + 2)(3a^2 - a + 5)$
13. $\begin{array}{r} 3x - 2 \\ \times\ 5x + 1 \\ \hline \end{array}$
14. $\begin{array}{r} 2z - 4 \\ \times\ 6z - 2 \\ \hline \end{array}$
15. $\begin{array}{r} 3m^2 + 2m - 1 \\ \times \qquad 5m + 2 \\ \hline \end{array}$
16. $\begin{array}{r} 2x^2 - 3x - 4 \\ \times \qquad x + 5 \\ \hline \end{array}$
17. Explain how to multiply a polynomial by a polynomial.
18. Explain why $(3x + 2)^2$ does not equal $9x^2 + 4$.

*Multiply the binomials. See Examples 5 and 6.*

19. $(x - 3)(x + 4)$
20. $(c - 3)(c + 1)$
21. $(5x + 8y)(2x - y)$
22. $(2n - 9m)(n - 7m)$
23. $(3x - 1)(x + 3)$
24. $(5d - 3)(d + 6)$
25. $\left(3x + \frac{1}{2}\right)\left(3x - \frac{1}{2}\right)$
26. $\left(2x - \frac{1}{3}\right)\left(2x + \frac{1}{3}\right)$

*Multiply, using special product methods. See Examples 7 and 8.*

27. $(x + 4)^2$
28. $(x - 5)^2$
29. $(6y - 1)(6y + 1)$
30. $(x - 9)(x + 9)$
31. $(3x - y)^2$
32. $(4x - z)^2$
33. $(3b - 6y)(3b + 6y)$
34. $(2x - 4y)(2x + 4y)$

*Multiply, using special product methods. See Examples 9 and 10.*

35. $[3 + (4b + 1)]^2$
36. $[5 - (3b - 3)]^2$
37. $[(2s - 3) - 1][(2s - 3) + 1]$
38. $[(2y + 5) + 6][(2y + 5) - 6]$
39. $[(xy + 4) - 6]^2$
40. $[(2a^2 + 4a) + 1]^2$
41. Explain when the FOIL method can be used to multiply polynomials.
42. Explain why the product of $(a + b)$ and $(a - b)$ is not a trinomial.

*Multiply.*

43. $(3x + 1)(3x + 5)$
44. $(4x - 5)(5x + 6)$
45. $(2x^3 + 5)(5x^2 + 4x + 1)$
46. $(3y^3 - 1)(3y^3 - 6y + 1)$
47. $(7x - 3)(7x + 3)$
48. $(4x + 1)(4x - 1)$
49. $\begin{array}{r} 3x^2 + 4x - 4 \\ \times \qquad 3x + 6 \\ \hline \end{array}$
50. $\begin{array}{r} 6x^2 + 2x - 1 \\ \times \qquad 3x - 6 \\ \hline \end{array}$
51. $\left(4x + \frac{1}{3}\right)\left(4x - \frac{1}{2}\right)$
52. $\left(4y - \frac{1}{3}\right)\left(3y - \frac{1}{8}\right)$
53. $(6x + 1)^2$
54. $(4x + 7)^2$
55. $(x^2 + 2y)(x^2 - 2y)$
56. $(3x + 2y)(3x - 2y)$
57. $-6a^2b^2[5a^2b^2 - 6a - 6b]$
58. $7x^2y^3(-3ax - 4xy + z)$
59. $(a - 4)(2a - 4)$
60. $(2x - 3)(x + 1)$
61. $(7ab + 3c)(7ab - 3c)$
62. $(3xy - 2b)(3xy + 2b)$
63. $(m - 4)^2$
64. $(x + 2)^2$
65. $(3x + 1)^2$
66. $(4x + 6)^2$
67. $(y - 4)(y - 3)$
68. $(c - 8)(c + 2)$
69. $(x + y)(2x - 1)(x + 1)$
70. $(z + 2)(z - 3)(2z + 1)$
71. $(3x^2 + 2x - 1)^2$
72. $(4x^2 + 4x - 4)^2$
73. $(3x + 1)(4x^2 - 2x + 5)$
74. $(2x - 1)(5x^2 - x - 2)$

*If $R(x) = x + 5$, $Q(x) = x^2 - 2$, and $P(x) = 5x$, find the following.*

75. $P(x) \cdot R(x)$
76. $P(x) \cdot Q(x)$
77. $[Q(x)]^2$
78. $[R(x)]^2$
79. $R(x) \cdot Q(x)$
80. $P(x) \cdot R(x) \cdot Q(x)$
81. Perform the indicated operations. Explain the difference between the two problems.
    a. $(3x + 5) + (3x + 7)$
    b. $(3x + 5)(3x + 7)$

*Explain where the error occurs.*

82. $4x(x - 5) + 2x$
$$= 4x(x) + 4x(-5) + 4x(2x)$$
$$= 4x^2 - 20x + 8x^2$$
$$= 12x^2 - 20x$$

83. Find the area of the circle. Do not approximate $\pi$.

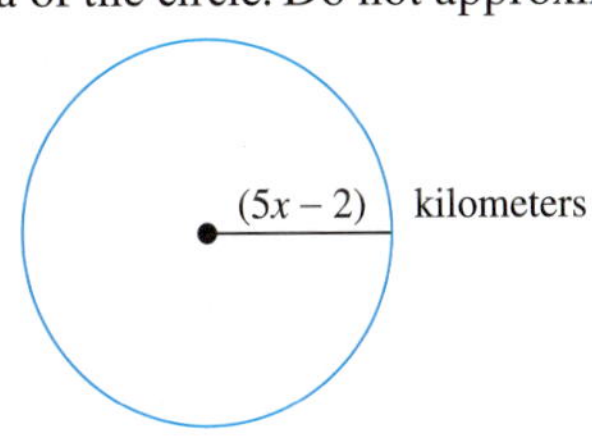

△ **84.** Find the volume of the cylinder. Do not approximate $\pi$.

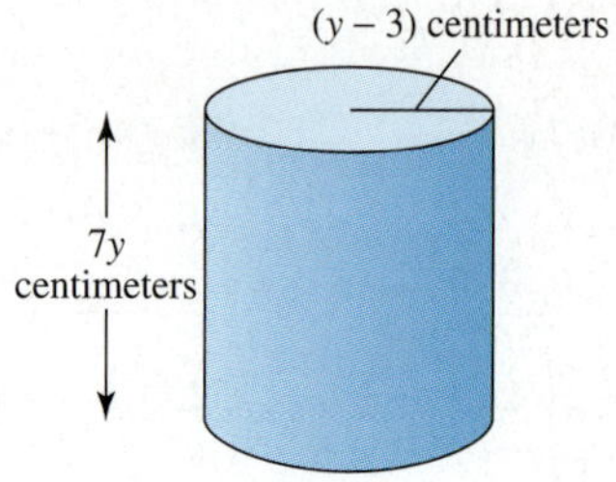


*If $f(x) = x^2 - 3x$, find the following. See Example 11.*

**85.** $f(a)$

**86.** $f(c)$

**87.** $f(a + h)$

**88.** $f(a + 5)$

**89.** $f(b - 2)$

**90.** $f(a - b)$

**91.** If $F(x) = x^2 + 3x + 2$, find
 **a.** $F(a + h)$
 **b.** $F(a)$
 **c.** $F(a + h) - F(a)$

**92.** If $g(x) = x^2 + 2x + 1$, find
 **a.** $g(a + h)$
 **b.** $g(a)$
 **c.** $g(a + h) - g(a)$

*Multiply. Assume that variables represent positive integers.*

**93.** $5x^2y^n(6y^{n+1} - 2)$

**94.** $-3yz^n(2y^3z^{2n} - 1)$

**95.** $(x^a + 5)(x^{2a} - 3)$

**96.** $(x^a + y^{2b})(x^a - y^{2b})$

## REVIEW EXERCISES

*Use the slope–intercept form of a line, $y = mx + b$, to find the slope of each line. See Section 3.4.*

**97.** $y = -2x + 7$

**98.** $y = \frac{3}{2}x - 1$

**99.** $3x - 5y = 14$

**100.** $x + 7y = 2$

*Use the vertical line test to determine which of the following are graphs of functions. See Section 3.2.*

**101.**

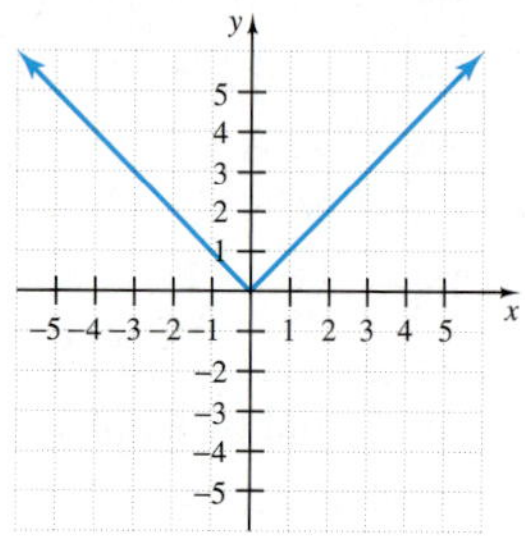


**102.**

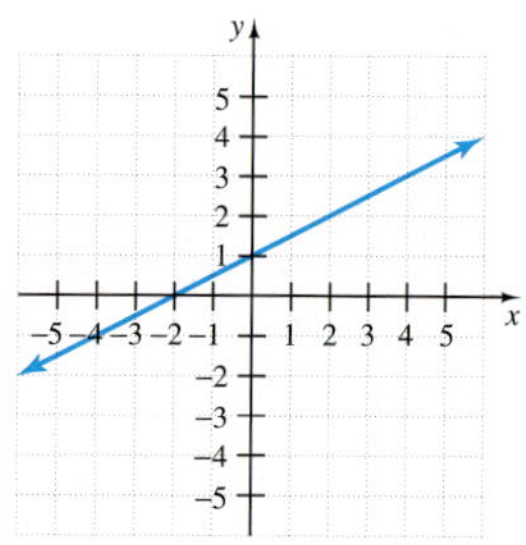


## 5.5 THE GREATEST COMMON FACTOR AND FACTORING BY GROUPING

▶ **OBJECTIVES**

1. Identify the GCF.
2. Factor out the GCF of a polynomial's terms.
3. Factor polynomials by grouping.

**1** **Factoring** is the reverse process of multiplying. It is the process of writing a polynomial as a product.

factoring →

$$6x^2 + 13x - 5 = (3x - 1)(2x + 5)$$

← multiplying

In the next few sections, we review techniques for factoring polynomials. These techniques are used at the end of this chapter to solve polynomial equations and to graph polynomial functions.

To factor a polynomial, we first factor out the greatest common factor (GCF) of its terms, using the distributive property. The GCF of a list of terms or monomials is the product of the GCF of the numerical coefficients and each GCF of the powers of a common variable.

**FINDING THE GCF OF A LIST OF MONOMIALS**

**Step 1:** Find the GCF of the numerical coefficients.
**Step 2:** Find the GCF of the variable factors.
**Step 3:** The product of the factors found in Steps 1 and 2 is the GCF of the monomials.

**Example 1** Find the GCF of $20x^3y$, $10x^2y^2$, and $35x^3$.

*Solution* The GCF of the numerical coefficients 20, 10, and 35 is 5, the largest integer that is a factor of each integer. The GCF of the variable factors $x^3$, $x^2$, and $x^3$ is $x^2$ because $x^2$ is the largest factor common to all three powers of $x$. The variable $y$ is not a common factor because it does not appear in all three monomials. The GCF is thus

$$5 \cdot x^2, \quad \text{or} \quad 5x^2$$

2 A first step in factoring polynomials is to use the distributive property and write the polynomial as a product of the GCF of its monomial terms and a simpler polynomial. This is called **factoring out** the GCF.

**Example 2** Factor.

**a.** $8x^2 + 4$ **b.** $5y - 2z^4$ **c.** $6x^2 - 3x^3$

*Solution* **a.** The GCF of terms $8x^2$ and 4 is 4.

$$8x^2 + 4 = 4 \cdot 2x^2 + 4 \cdot 1 \quad \text{Factor out 4 from each term.}$$
$$= 4(2x^2 + 1) \quad \text{Apply the distributive property.}$$

The factored form of $8x^2 + 4$ is $4(2x^2 + 1)$. To check, multiply $4(2x^2 + 1)$ to see that the product is $8x^2 + 4$.

**b.** There is no common factor of the terms $5y$ and $-2z^4$ other than 1 (or $-1$).

**c.** The greatest common factor of $6x^2$ and $-3x^3$ is $3x^2$. Thus,

$$6x^2 - 3x^3 = 3x^2 \cdot 2 - 3x^2 \cdot x$$
$$= 3x^2(2 - x)$$

**HELPFUL HINT**

To verity that the GCF has been factored out correctly, multiply the factors together and see that their product is the original polynomial.

**Example 3** Factor $17x^3y^2 - 34x^4y^2$.

*Solution* The GCF of the two terms is $17x^3y^2$, which we factor out of each term.

$$\begin{aligned} 17x^3y^2 - 34x^4y^2 &= 17x^3y^2 \cdot 1 - 17x^3y^2 \cdot 2x \\ &= 17x^3y^2(1 - 2x) \end{aligned}$$

**HELPFUL HINT**
If the GCF happens to be one of the terms in the polynomial, a factor of 1 will remain for this term when the GCF is factored out. For example, in the polynomial $21x^2 + 7x$, the, GCF of $21x^2$ and $7x$ is $7x$, so

$$21x^2 + 7x = 7x \cdot 3x + 7x \cdot 1 = 7x(3x + 1)$$

**Example 4** Factor $-3x^3y + 2x^2y - 5xy$.

*Solution* Two possibilities are shown for factoring this polynomial. First, the common factor $xy$ is factored out.

$$-3x^3y + 2x^2y - 5xy = xy(-3x^2 + 2x - 5)$$

Also, the common factor $-xy$ can be factored out as shown.

$$\begin{aligned} -3x^3y + 2x^2y - 5xy &= -xy(3x^2) + (-xy)(-2x) + (-xy)(5) \\ &= -xy(3x^2 - 2x + 5) \end{aligned}$$

Both of these alternatives are correct.

**Example 5** Factor $2(x - 5) + 3a(x - 5)$.

*Solution* The greatest common factor is the binomial factor $(x - 5)$.

$$2(x - 5) + 3a(x - 5) = (x - 5)(2 + 3a)$$

**Example 6** Factor $7x(x^2 + 5y) - (x^2 + 5y)$.

*Solution*

$$\begin{aligned} 7x(x^2 + 5y) - (x^2 + 5y) &= 7x(x^2 + 5y) - 1(x^2 + 5y) \\ &= (x^2 + 5y)(7x - 1) \end{aligned}$$

**HELPFUL HINT**
Notice that we wrote $-(x^2 + 5y)$ as $-1(x^2 + 5y)$ to aid in factoring.

3 Sometimes it is possible to factor a polynomial by grouping the terms of the polynomial and looking for common factors in each group. This method of factoring is called **factoring by grouping**.

**Example 7** Factor $ab - 6a + 2b - 12$.

*Solution* First look for the GCF of all four terms. The GCF of all four terms is 1. Next group the first two terms and the last two terms and factor out common factors from each group.

$$ab - 6a + 2b - 12 = (ab - 6a) + (2b - 12)$$

Factor $a$ from the first group and 2 from the second group.

$$= a(b - 6) + 2(b - 6)$$

Now we see a GCF of $(b - 6)$. Factor out $(b - 6)$ to get

$$a(b - 6) + 2(b - 6) = (b - 6)(a + 2)$$

**Check:** To check, multiply $(b - 6)$ and $(a + 2)$ to see that the product is $ab - 6a + 2b - 12$.

**HELPFUL HINT**

Notice that the polynomial $a(b - 6) + 2(b - 6)$ is *not* in factored form. It is a *sum*, not a *product*. The factored form is $(b - 6)(a + 2)$.

**Example 8** Factor $x^3 + 5x^2 + 3x + 15$.

*Solution*

$$\begin{aligned} x^3 + 5x^2 + 3x + 15 &= (x^3 + 5x^2) + (3x + 15) && \text{Group pairs of terms.} \\ &= x^2(x + 5) + 3(x + 5) && \text{Factor each binomial.} \\ &= (x + 5)(x^2 + 3) && \text{Factor out the common factor, } (x + 5). \end{aligned}$$

**Example 9** Factor $m^2n^2 + m^2 - 2n^2 - 2$.

*Solution*

$$\begin{aligned} m^2n^2 + m^2 - 2n^2 - 2 &= (m^2n^2 + m^2) + (-2n^2 - 2) && \text{Group pairs of terms.} \\ &= m^2(n^2 + 1) - 2(n^2 + 1) && \text{Factor each binomial.} \\ &= (n^2 + 1)(m^2 - 2) && \text{Factor out the common factor, } (n^2 + 1). \end{aligned}$$

**Example 10** Factor $xy + 2x - y - 2$.

*Solution*

$$\begin{aligned} xy + 2x - y - 2 &= (xy + 2x) + (-y - 2) && \text{Group pairs of terms.} \\ &= x(y + 2) - 1(y + 2) && \text{Factor each binomial.} \\ &= (y + 2)(x - 1) && \text{Factor out the common factor, } (y + 2). \end{aligned}$$

## MENTAL MATH

*Find the GCF of each list of monomials.*

**1.** 6, 12 **2.** 9, 27 **3.** $15x, 10$ **4.** $9x, 12$
**5.** $13x, 2x$ **6.** $4y, 5y$ **7.** $7x, 14x$ **8.** $8z, 4z$

## Exercise Set 5.5

*Find the GCF of each list of monomials. See Example 1.*

**1.** $a^8, a^5, a^3$
**2.** $b^9, b^2, b^5$
**3.** $x^2y^3z^3, y^2z^3, xy^2z^2$
**4.** $xy^2z^3, x^2y^2z^2, x^2y^3$
**5.** $6x^3y, 9x^2y^2, 12x^2y$
**6.** $4xy^2, 16xy^3, 8x^2y^2$
**7.** $10x^3yz^3, 20x^2z^5, 45xz^3$
**8.** $12y^2z^4, 9xy^3z^4, 15x^2y^2z^3$

*Factor out the GCF in each polynomial. See Examples 2 through 6.*

**9.** $18x - 12$
**10.** $21x + 14$
**11.** $4y^2 - 16xy^3$
**12.** $3z - 21xz^4$
**13.** $6x^5 - 8x^4 + 2x^3$
**14.** $9x + 3x^2 - 6x^3$
**15.** $8a^3b^3 - 4a^2b^2 + 4ab + 16ab^2$
**16.** $12a^3b - 6ab + 18ab^2 - 18a^2b$
**17.** $6(x + 3) + 5a(x + 3)$
**18.** $2(x - 4) + 3y(x - 4)$
**19.** $2x(z + 7) + (z + 7)$
**20.** $x(y - 2) + (y - 2)$
**21.** $3x(x^2 + 5) - 2(x^2 + 5)$
**22.** $4x(2y + 3) - 5(2y + 3)$
**23.** When $3x^2 - 9x + 3$ is factored, the result is $3(x^2 - 3x + 1)$. Explain why it is necessary to include the term 1 in this factored form.
**24.** Construct a trinomial whose GCF is $5x^2y^3$.

*Factor each polynomial by grouping. See Examples 7 through 10.*

**25.** $ab + 3a + 2b + 6$
**26.** $ab + 2a + 5b + 10$
**27.** $ac + 4a - 2c - 8$
**28.** $bc + 8b - 3c - 24$
**29.** $2xy - 3x - 4y + 6$
**30.** $12xy - 18x - 10y + 15$
**31.** $12xy - 8x - 3y + 2$
**32.** $20xy - 15x - 4y + 3$

*Factor each polynomial*

**33.** $6x^3 + 9$
**34.** $6x^2 - 8$
**35.** $x^3 + 3x^2$
**36.** $x^4 - 4x^3$
**37.** $8a^3 - 4a$
**38.** $12b^4 + 3b^2$
**39.** $-20x^2y + 16xy^3$
**40.** $-18xy^3 + 27x^4y$
**41.** $10a^2b^3 + 5ab^2 - 15ab^3$
**42.** $10ef - 20e^2f^3 + 30e^3f$
**43.** $9abc^2 + 6a^2bc - 6ab + 3bc$
**44.** $4a^2b^2c - 6ab^2c - 4ac + 8a$
**45.** $4x(y - 2) - 3(y - 2)$
**46.** $8y(z + 8) - 3(z + 8)$
**47.** $6xy + 10x + 9y + 15$
**48.** $15xy + 20x + 6y + 8$
**49.** $xy + 3y - 5x - 15$
**50.** $xy + 4y - 3x - 12$
**51.** $6ab - 2a - 9b + 3$
**52.** $16ab - 8a - 6b + 3$
**53.** $12xy + 18x + 2y + 3$
**54.** $20xy + 8x + 5y + 2$
**55.** $2m(n - 8) - (n - 8)$
**56.** $3a(b - 4) - (b - 4)$
**57.** $15x^3y^2 - 18x^2y^2$
**58.** $12x^4y^2 - 16x^3y^3$
**59.** $2x^2 + 3xy + 4x + 6y$
**60.** $3x^2 + 12x + 4xy + 16y$
**61.** $5x^2 + 5xy - 3x - 3y$
**62.** $4x^2 + 2xy - 10x - 5y$
**63.** $x^3 + 3x^2 + 4x + 12$
**64.** $x^3 + 4x^2 + 3x + 12$
**65.** $x^3 - x^2 - 2x + 2$
**66.** $x^3 - 2x^2 - 3x + 6$

*Solve.*

△ **67.** The material needed to manufacture a tin can is given by the polynomial

$$2\pi r^2 + 2\pi rh$$

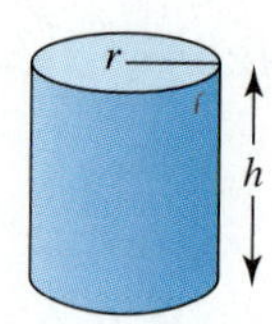


where the radius is $r$ and height is $h$. Factor this expression.

**68.** The amount $E$ of voltage in an electrical circuit is given by the formula

$$IR_1 + IR_2 = E$$

Write an equivalent equation by factoring the expression $IR_1 + IR_2$.

**69.** At the end of $T$ years, the amount of money $A$ in a savings account earning simple interest from an initial investment of $P$ dollars at rate $R$ is given by the formula

$$A = P + PRT$$

Write an equivalent equation by factoring the expression $P + PRT$.

△ **70.** An open-topped box has a square base and a height of 10 inches. If each of the bottom edges of the box has length $x$ inches, find the amount of material needed to construct the box. Write the answer in factored form.

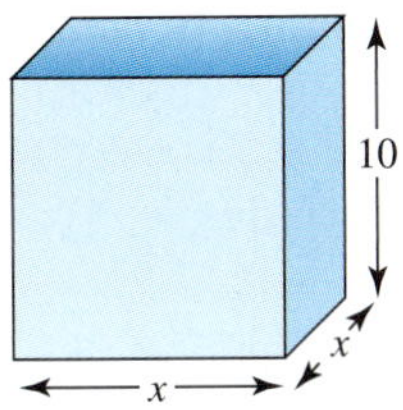


**71.** An object is thrown upward from the ground with an initial velocity of 64 feet per second. The height $h(t)$ of the object after $t$ seconds is given by the polynomial function

$$h(t) = -16t^2 + 64t$$

**a.** Write an equivalent factored expression for the function $h(t)$ by factoring $-16t^2 + 64t$.

**b.** Find $h(1)$ by using $h(t) = -16t^2 + 64t$ and then by using the factored form of $h(t)$.

**c.** Explain why the values found in part **b** are the same.

**72.** An object is dropped from the gondola of a hot-air balloon at a height of 224 feet. The height $h(t)$ of the object after $t$ seconds is given by the polynomial function

$$h(t) = -16t^2 + 224$$

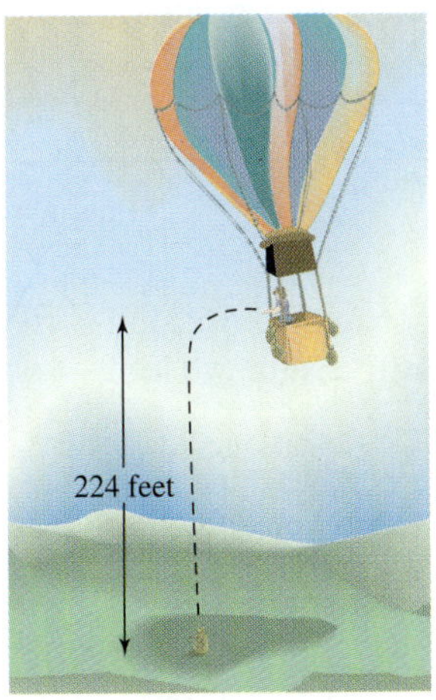


**a.** Write an equivalent factored expression for the function $h(t)$ by factoring $-16t^2 + 224$.

**b.** Find $h(2)$ by using $h(t) = -16t^2 + 224$ and then by using the factored form of the function.

**c.** Explain why the values found in part **b** are the same.

**73.** A factored polynomial can be in many forms. For example, a factored form of $xy - 3x - 2y + 6$ is $(x - 2)(y - 3)$. Which of the following is not a factored form of $xy - 3x - 2y + 6$?

**a.** $(2 - x)(3 - y)$ **b.** $(-2 + x)(-3 + y)$

**c.** $(x - 2)(y - 3)$ **d.** $(-x + 2)(-y + 3)$

**74.** Consider the following sequence of algebraic steps:

$$x^3 - 6x^2 + 2x - 10 = (x^3 - 6x^2) + (2x - 10)$$
$$= x^2(x - 6) + 2(x - 5)$$

Explain whether the final result is the factored form of the original polynomial.

**75.** Which factorization of $12x^2 + 9x + 3$ is correct?

**a.** $3(4x^2 + 3x + 1)$ **b.** $3(4x^2 + 3x - 1)$

**c.** $3(4x^2 + 3x - 3)$ **d.** $3(4x^2 + 3x)$

## REVIEW EXERCISES

*Simplify the following. See Section 5.1.*

**76.** $(5x^2)(11x^5)$ **77.** $(7y)(-2y^3)$

**78.** $(5x^2)^3$ **79.** $(-2y^3)^4$

*Find each product by using the FOIL order of multiplying binomials. See Section 5.4.*

**80.** $(x + 2)(x - 5)$ **81.** $(x - 7)(x - 1)$

**82.** $(x + 3)(x + 2)$ **83.** $(x - 4)(x + 2)$

**84.** $(y - 3)(y - 1)$ **85.** $(s + 8)(s + 10)$

## *A Look Ahead*

**Example**

Factor $x^{5a} - x^{3a} + x^{7a}$.

*Solution*

*The variable x is common to all three terms, and the power 3a is the smallest of the exponents. So factor out the common factor* $x^{3a}$.

$$x^{5a} - x^{3a} + x^{7a} = x^{3a}(x^{2a}) - x^{3a}(1) + x^{3a}(x^{4a})$$
$$= x^{3a}(x^{2a} - 1 + x^{4a})$$

*Factor. Assume that variables used as exponents represent positive integers.*

**86.** $x^{3n} - 2x^{2n} + 5x^n$ **87.** $3y^n + 3y^{2n} + 5y^{8n}$

**88.** $6x^{8a} - 2x^{5a} - 4x^{3a}$ **89.** $3x^{5a} - 6x^{3a} + 9x^{2a}$

## 5.6 FACTORING TRINOMIALS

CD-ROM SSM SSG Video

### ▶ OBJECTIVES

1. Factor trinomials of the form $x^2 + bx + c$.
2. Factor trinomials of the form $ax^2 + bx + c$.
3. Factor by substitution.
4. Factor trinomials by grouping.

**1** In the previous section, we used factoring by grouping to factor four-term polynomials. In this section, we present techniques for factoring trinomials. Since $(x - 2)(x + 5) = x^2 + 3x - 10$, we say that $(x - 2)(x + 5)$ is a factored form of $x^2 + 3x - 10$. Taking a close look at how $(x - 2)$ and $(x + 5)$ are multiplied suggests a pattern for factoring trinomials of the form

$$x^2 + bx + c$$

$$x \cdot x$$

$$(x - 2)(x + 5) = x^2 + 3x - 10$$

$$-2 + 5$$

$$-2 \cdot 5$$

The pattern for factoring is summarized next.

### FACTORING A TRINOMIAL OF THE FORM $x^2 + bx + c$

Find two numbers whose product is $c$ and whose sum is $b$. The factored form of $x^2 + bx + c$ is

$$(x + \text{one number})(x + \text{other number})$$

**Example 1** Factor $x^2 + 10x + 16$.

*Solution* We look for two integers whose product is 16 and whose sum is 10. Since our integers must have a positive product and a positive sum, we look at only positive factors of 16.

| *Positive Factors of 16* | *Sum of Factors* | |
|---|---|---|
| 1, 16 | $1 + 16 = 17$ | |
| 4, 4 | $4 + 4 = 8$ | |
| 2, 8 | $2 + 8 = 10$ | Correct pair |

The correct pair of numbers is 2 and 8 because their product is 16 and their sum is 10. Thus,

$$x^2 + 10x + 16 = (x + 2)(x + 8)$$

**Check:** To check, see that $(x + 2)(x + 8) = x^2 + 10x + 16$.

**Example 2** Factor $x^2 - 12x + 35$.

*Solution* We need to find two integers whose product is 35 and whose sum is −12. Since our integers must have a positive product and a negative sum, we consider only negative factors of 35. The numbers are −5 and −7.

$$x^2 - 12x + 35 = [x + (-5)][x + (-7)]$$
$$= (x - 5)(x - 7)$$

**Check:** To check, see that $(x - 5)(x - 7) = x^2 - 12x + 35$.

**Example 3** Factor $5x^3 - 30x^2 - 35x$.

*Solution* First we factor out the greatest common factor, $5x$.

$$5x^3 - 30x^2 - 35x = 5x(x^2 - 6x - 7)$$

Next we factor $x^2 - 6x - 7$ by finding two numbers whose product is −7 and whose sum is −6. The numbers are 1 and −7.

$$5x^3 - 30x^2 - 35x = 5x(x^2 - 6x - 7)$$
$$= 5x(x + 1)(x - 7)$$

**HELPFUL HINT**

If the polynomial to be factored contains a common factor that is factored out, don't forget to include that common factor in the final factored form of the original polynomial.

**Example 4** Factor $2n^2 - 38n + 80$.

*Solution* The terms of this polynomial have a greatest common factor of 2, which we factor out first.

$$2n^2 - 38n + 80 = 2(n^2 - 19n + 40)$$

Next we factor $n^2 - 19n + 40$ by finding two numbers whose product is 40 and whose sum is −19. Both numbers must be negative since their sum is −19. Possibilities are

−1 and −40, −2 and −20, −4 and −10, −5 and −8

None of the pairs has a sum of −19, so no further factoring with integers is possible. The factored form of $2n^2 - 38n + 80$ is

$$2n^2 - 38n + 80 = 2(n^2 - 19n + 40)$$

We call a polynomial such as $n^2 - 19n + 40$ that cannot be factored further, a **prime polynomial**.

2 Next, we factor trinomials of the form $ax^2 + bx + c$, where the coefficient $a$ of $x^2$ is not 1. Don't forget that the first step in factoring any polynomial is to factor out the greatest common factor of its terms.

**Example 5** Factor $2x^2 + 11x + 15$.

*Solution* Factors of $2x^2$ are $2x$ and $x$. Let's try these factors as first terms of the binomials.

$$2x^2 + 11x + 15 = (2x + \quad)(x + \quad)$$

Next we try combinations of factors of 15 until the correct middle term, $11x$, is obtained. We will try only positive factors of 15 since the coefficient of the middle term, 11, is positive. Positive factors of 15 are 1 and 15 and 3 and 5.

$(2x + 1)(x + 15)$: $1x + 30x = 31x$ Incorrect middle term

$(2x + 15)(x + 1)$: $15x + 2x = 17x$ Incorrect middle term

$(2x + 3)(x + 5)$: $3x + 10x = 13x$ Incorrect middle term

$(2x + 5)(x + 3)$: $5x + 6x = 11x$ Correct middle term

Thus, the factored form of $2x^2 + 11x + 15$ is $(2x + 5)(x + 3)$.

**FACTORING A TRINOMIAL OF THE FORM $ax^2 + bx + c$**

**Step 1:** Write all pairs of factors of $ax^2$.
**Step 2:** Write all pairs of factors of $c$, the constant term.
**Step 3:** Try various combinations of these factors until the correct middle term $bx$ is found.
**Step 4:** If no combination exists, the polynomial is **prime**.

**Example 6** Factor $3x^2 - x - 4$.

*Solution* Factors of $3x^2$: $3x \cdot x$
Factors of $-4$: $-1 \cdot 4$, $1 \cdot -4$, $-2 \cdot 2$, $2 \cdot -2$
Let's try possible combinations of these factors.

$(3x - 1)(x + 4)$: $-1x + 12x = 11x$ Incorrect middle term

$(3x + 4)(x - 1)$: $4x + (-3x) = 1x$ Incorrect middle term

$(3x - 4)(x + 1)$: $-4x + 3x = -1$ Correct middle term

Thus, $3x^2 - x - 4 = (3x - 4)(x + 1)$.

**HELPFUL HINT—SIGN PATTERNS**

A positive constant in a trinomial tells us to look for two numbers with the same sign. The sign of the coefficient of the middle term tells us whether the signs are both positive or both negative.

| both positive ↓ | same sign ↓ | | both negative ↓ | same sign ↓ |
|---|---|---|---|---|

$$2x^2 + 7x + 3 = (2x + 1)(x + 3) \qquad 2x^2 - 7x + 3 = (2x - 1)(x - 3)$$

A negative constant in a trinomial tells us to look for two numbers with opposite signs.

| opposite signs ↓ | | opposite signs ↓ |
|---|---|---|

$$2x^2 - 5x - 3 = (2x + 1)(x - 3) \qquad 2x^2 + 5x - 3 = (2x - 1)(x + 3)$$

**Example 7** Factor $12x^3y - 22x^2y + 8xy$.

*Solution* First we factor out the greatest common factor of the terms of this trinomial, $2xy$.

$$12x^3y - 22x^2y + 8xy = 2xy(6x^2 - 11x + 4)$$

Now we try to factor the trinomial $6x^2 - 11x + 4$.

Factors of $6x^2$: $2x \cdot 3x, \qquad 6x \cdot x$

Let's try $2x$ and $3x$.

$$2xy(6x^2 - 11x + 4) = 2xy(2x + \quad)(3x + \quad)$$

The constant term, 4, is positive and the coefficient of the middle term, −11, is negative, so we factor 4 into negative factors only.

Negative factors of 4: $-4(-1), \qquad -2(-2)$

Let's try −4 and −1.

$$2xy(2x - 4)(3x - 1)$$

$$\begin{array}{r} -12x \\ -2x \\ \hline -14x \end{array}$$ Incorrect middle term

This combination cannot be correct, because one of the factors, $(2x - 4)$, has a common factor of 2. This cannot happen if the polynomial $6x^2 - 11x + 4$ has no common factors.

Now let's try −1 and −4.

$$2xy(2x - 1)(3x - 4)$$

$$\begin{array}{r} -3x \\ -8x \\ \hline -11x \end{array}$$ Correct middle term

Thus,

$$12x^3y - 22x^2y + 8xy = 2xy(2x - 1)(3x - 4)$$

If this combination had not worked, we would have tried $-2$ and $-2$ as factors of 4 and then $6x$ and $x$ as factors of $6x^2$.

> **HELPFUL HINT**
> If a trinomial has no common factor (other than 1), then none of its binomial factors will contain a common factor (other than 1).

**Example 8** Factor $16x^2 + 24xy + 9y^2$.

*Solution* No greatest common factor can be factored out of this trinomial.

Factors of $16x^2$: $16x \cdot x$, $8x \cdot 2x$, $4x \cdot 4x$
Factors of $9y^2$: $y \cdot 9y$, $3y \cdot 3y$

We try possible combinations until the correct factorization is found.

$$16x^2 + 24xy + 9y^2 = (4x + 3y)(4x + 3y) \quad \text{or} \quad (4x + 3y)^2$$

The trinomial $16x^2 + 24xy + 9y^2$ in Example 8 is an example of a **perfect square trinomial** since its factors are two identical binomials. In the next section, we examine a special method for factoring perfect square trinomials.

**3** A complicated looking polynomial may be a simpler trinomial "in disguise." Revealing the simpler trinomial is possible by substitution.

**Example 9** Factor $2(a + 3)^2 - 5(a + 3) - 7$.

*Solution* The quantity $(a + 3)$ is in two of the terms of this polynomial. ***Substitute*** $x$ for $(a + 3)$, and the result is the following simpler trinomial.

$$2(a + 3)^2 - 5(a + 3) - 7 \quad \text{Original trinomial.}$$
$$= 2(x)^2 - 5(x) - 7 \quad \text{Substitute } x \text{ for } (a + 3).$$

Now factor $2x^2 - 5x - 7$.

$$2x^2 - 5x - 7 = (2x - 7)(x + 1)$$

But the quantity in the original polynomial was $(a + 3)$, not $x$. Thus, we need to reverse the substitution and replace $x$ with $(a + 3)$.

$$(2x - 7)(x + 1) \quad \text{Factored expression.}$$
$$= [2(a + 3) - 7][(a + 3) + 1] \quad \text{Substitute } (a + 3) \text{ for } x.$$
$$= (2a + 6 - 7)(a + 3 + 1) \quad \text{Remove inside parentheses.}$$
$$= (2a - 1)(a + 4) \quad \text{Simplify.}$$

Thus, $2(a + 3)^2 - 5(a + 3) - 7 = (2a - 1)(a + 4)$.

**Example 10** Factor $5x^4 + 29x^2 - 42$.

*Solution* Again, substitution may help us factor this polynomial more easily. Since this polynomial contains the variable $x$, we will choose a different substitution variable. Let $y = x^2$, so $y^2 = (x^2)^2$, or $x^4$. Then

$$5x^4 + 29x^2 - 42$$

becomes

$$5y^2 + 29y - 42$$

which factors as

$$5y^2 + 29y - 42 = (5y - 6)(y + 7)$$

Next, replace $y$ with $x^2$ to get

$$(5x^2 - 6)(x^2 + 7)$$

4 There is another method we can use when factoring trinomials of the form $ax^2 + bx + c$: Write the trinomial as a four-term polynomial, and then factor by grouping.

**FACTORING A TRINOMIAL OF THE FORM $ax^2 + bx + c$ BY GROUPING**

**Step 1:** Find two numbers whose product is $a \cdot c$ and whose sum is $b$.
**Step 2:** Write the term $bx$ as a sum by using the factors found in Step 1.
**Step 3:** Factor by grouping.

**Example 11** Factor $6x^2 + 13x + 6$.

*Solution* In this trinomial, $a = 6$, $b = 13$, and $c = 6$.

**Step 1:** Find two numbers whose product is $a \cdot c$, or $6 \cdot 6 = 36$, and whose sum is $b$, 13. The two numbers are 4 and 9.

**Step 2:** Write the middle term, $13x$, as the sum $4x + 9x$.

$$6x^2 + 13x + 6 = 6x^2 + 4x + 9x + 6$$

**Step 3:** Factor $6x^2 + 4x + 9x + 6$ by grouping.

$$(6x^2 + 4x) + (9x + 6) = 2x(3x + 2) + 3(3x + 2)$$
$$= (3x + 2)(2x + 3)$$

## MENTAL MATH

1. Find two numbers whose product is 10 and whose sum is 7.
2. Find two numbers whose product is 12 and whose sum is 8.
3. Find two numbers whose product is 24 and whose sum is 11.
4. Find two numbers whose product is 30 and whose sum is 13.

## Exercise Set 5.6

*Factor each trinomial. See Examples 1 through 4.*

1. $x^2 + 9x + 18$
2. $x^2 + 9x + 20$
3. $x^2 - 12x + 32$
4. $x^2 - 12x + 27$
5. $x^2 + 10x - 24$
6. $x^2 + 3x - 54$
7. $x^2 - 2x - 24$
8. $x^2 - 9x - 36$
9. $3x^2 - 18x + 24$
10. $x^2y^2 + 4xy^2 + 3y^2$
11. $4x^2z + 28xz + 40z$
12. $5x^2 - 45x + 70$
13. $2x^2 + 30x - 108$
14. $3x^2 + 12x - 96$
15. Find all positive and negative integers $b$ such that $x^2 + bx + 6$ factors.
16. Find all positive and negative integers $b$ such that $x^2 + bx - 10$ factors.

*Factor each trinomial. See Examples 5 through 8 and 11.*

17. $5x^2 + 16x + 3$
18. $3x^2 + 8x + 4$
19. $2x^2 - 11x + 12$
20. $3x^2 - 19x + 20$
21. $2x^2 + 25x - 20$
22. $6x^2 - 13x - 8$
23. $4x^2 - 12x + 9$
24. $25x^2 - 30x + 9$
25. $12x^2 + 10x - 50$
26. $12y^2 - 48y + 45$
27. $3y^4 - y^3 - 10y^2$
28. $2x^2z + 5xz - 12z$
29. $6x^3 + 8x^2 + 24x$
30. $18y^3 + 12y^2 + 2y$
31. $x^2 + 8xz + 7z^2$
32. $a^2 - 2ab - 15b^2$
33. $2x^2 - 5xy - 3y^2$
34. $6x^2 + 11xy + 4y^2$
35. $x^2 - x - 12$
36. $x^2 + 4x - 5$
37. $28y^2 + 22y + 4$
38. $24y^3 - 2y^2 - y$
39. $2x^2 + 15x - 27$
40. $3x^2 + 14x + 15$
41. Find all positive and negative integers $b$ such that $3x^2 + bx + 5$ factors.
42. Find all positive and negative integers $b$ such that $2x^2 + bx + 7$ factors.

*Use substitution to factor each polynomial completely. See Examples 9 and 10.*

43. $x^4 + x^2 - 6$
44. $x^4 - x^2 - 20$
45. $(5x + 1)^2 + 8(5x + 1) + 7$
46. $(3x - 1)^2 + 5(3x - 1) + 6$
47. $x^6 - 7x^3 + 12$
48. $x^6 - 4x^3 - 12$
49. $(a + 5)^2 - 5(a + 5) - 24$
50. $(3c + 6)^2 + 12(3c + 6) - 28$

*Solve.*

51. The volume $V(x)$ of a box in terms of its height $x$ is given by the function $V(x) = 3x^3 - 2x^2 - 8x$. Factor this expression for $V(x)$.

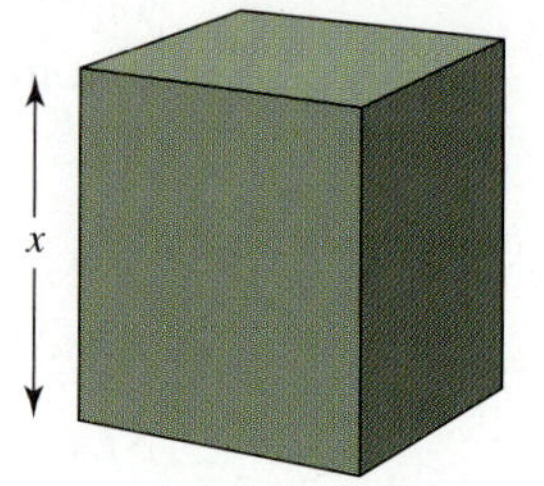


52. Based on your results from Exercise 51, find the length and width of the box if the height is 5 inches and the dimensions of the box are whole numbers.

*Factor each polynomial completely.*

53. $x^2 - 24x - 81$
54. $x^2 - 48x - 100$
55. $x^2 - 15x - 54$
56. $x^2 - 15x + 54$
57. $3x^2 - 6x + 3$
58. $8x^2 - 8x + 2$
59. $3x^2 - 5x - 2$
60. $5x^2 - 14x - 3$
61. $8x^2 - 26x + 15$
62. $12x^2 - 17x + 6$
63. $18x^4 + 21x^3 + 6x^2$
64. $20x^5 + 54x^4 + 10x^3$
65. $3a^2 + 12ab + 12b^2$
66. $2x^2 + 16xy + 32y^2$
67. $x^2 + 4x + 5$
68. $x^2 + 6x + 8$
69. $2(x + 4)^2 + 3(x + 4) - 5$
70. $3(x + 3)^2 + 2(x + 3) - 5$
71. $6x^2 - 49x + 30$
72. $4x^2 - 39x + 27$
73. $x^4 - 5x^2 - 6$
74. $x^4 - 5x^2 + 6$
75. $6x^3 - x^2 - x$
76. $12x^3 + x^2 - x$
77. $12a^2 - 29ab + 15b^2$
78. $16y^2 + 6yx - 27x^2$
79. $9x^2 + 30x + 25$
80. $4x^2 + 6x + 9$
81. $3x^2y - 11xy + 8y$
82. $5xy^2 - 9xy + 4x$
83. $2x^2 + 2x - 12$
84. $3x^2 + 6x - 45$
85. $(x - 4)^2 + 3(x - 4) - 18$
86. $(x - 3)^2 - 2(x - 3) - 8$
87. $2x^6 + 3x^3 - 9$
88. $3x^6 - 14x^3 + 8$
89. $72xy^4 - 24xy^2z + 2xz^2$
90. $36xy^2 - 48xyz^2 + 16xz^4$

*Recall that a graphing calculator may be used to visualize addition, subtraction, and multiplication of polynomials. In the same manner, a graphing calculator may be used to visualize factoring of polynomials in one variable. For example, to see that*

$$2x^3 - 9x^2 - 5x = x(2x + 1)(x - 5)$$

*graph* $Y_1 = 2x^3 - 9x^2 - 5x$ *and* $Y_2 = x(2x + 1)(x - 5)$. *Then trace along both graphs to see that they coincide. Factor the following and use this method to check your results.*

**91.** $x^4 + 6x^3 + 5x^2$

**92.** $x^3 + 6x^2 + 8x$

**93.** $30x^3 + 9x^2 - 3x$

**94.** $-6x^4 + 10x^3 - 4x^2$

### REVIEW EXERCISES

*Multiply the following. See Section 5.4.*

**95.** $(x - 2)(x^2 + 2x + 4)$

**96.** $(y + 1)(y^2 - y + 1)$

*If* $P(x) = 3x^2 + 2x - 9$, *find the following. See Section 5.3.*

**97.** $P(0)$

**98.** $P(1)$

**99.** $P(-1)$

**100.** $P(-2)$

### *A Look Ahead*

**Example**
Factor $x^{2n} + 7x^n + 12$.

*Solution*
Factors of $x^{2n}$ are $x^n$ and $x_n$ so $x^{2n} + 7x^n + 12 = (x^n + \text{one number})(x^n + \text{other number})$. Factors of 12 whose sum is 7 are 3 and 4. Thus

$$x^{2n} + 7x^n + 12 = (x^n + 4)(x^n + 3)$$

*Factor. Assume that variables used as exponents represent positive integers. See the preceding example.*

**101.** $x^{2n} + 10x^n + 16$

**102.** $x^{2n} - 7x^n + 12$

**103.** $x^{2n} - 3x^n - 18$

**104.** $x^{2n} + 7x^n - 18$

**105.** $2x^{2n} + 11x^n + 5$

**106.** $3x^{2n} - 8x^n + 4$

**107.** $4x^{2n} - 12x^n + 9$

**108.** $9x^{2n} + 24x^n + 16$

## 5.7 FACTORING BY SPECIAL PRODUCTS AND FACTORING STRATEGIES

CD-ROM SSM SSG Video

**OBJECTIVES**

1. Factor a perfect square trinomial.
2. Factor the difference of two squares.
3. Factor the sum or difference of two cubes.
4. Practice techniques for factoring polynomials.

**1** In the previous section, we considered a variety of ways to factor trinomials of the form $ax^2 + bx + c$. In one particular example, we factored $16x^2 + 24xy + 9y^2$ as

$$16x^2 + 24xy + 9y^2 = (4x + 3y)^2$$

Recall that $16x^2 + 24xy + 9y^2$ is a perfect square trinomial because its factors are two identical binomials. A perfect square trinomial can be factored quickly if you recognize the trinomial as a perfect square.

A trinomial is a perfect square trinomial if it can be written so that its first term is the square of some quantity $a$, its last term is the square of some quantity $b$, and its middle term is twice the product of the quantities $a$ and $b$. The following special formulas can be used to factor perfect square trinomials.

**PERFECT SQUARE TRINOMIALS**

$$a^2 + 2ab + b^2 = (a + b)^2$$
$$a^2 - 2ab + b^2 = (a - b)^2$$

Notice that these formulas above are the same special products from Section 5.4 for the square of a binomial.

From $a^2 + 2ab + b^2 = (a + b)^2$, we see that

$$16x^2 + 24xy + 9y^2 = (4x)^2 + 2(4x)(3y) + (3y)^2 = (4x + 3y)^2$$

**Example 1** Factor $m^2 + 10m + 25$.

*Solution* Notice that the first term is a square: $m^2 = (m)^2$, the last term is a square: $25 = 5^2$; and $10m = 2 \cdot 5 \cdot m$.
Thus,

$$m^2 + 10m + 25 = m^2 + 2(m)(5) + 5^2 = (m + 5)^2$$

**Example 2** Factor $3a^2x - 12abx + 12b^2x$.

*Solution* The terms of this trinomial have a GCF of $3x$, which we factor out first.

$$3a^2x - 12abx + 12b^2x = 3x(a^2 - 4ab + 4b^2)$$

Now, the polynomial $a^2 - 4ab + 4b^2$ is a perfect square trinomial. Notice that the first term is a square: $a^2 = (a)^2$; the last term is a square: $4b^2 = (2b)^2$; and $4ab = 2(a)(2b)$.
The factoring can now be completed as

$$3x(a^2 - 4ab + 4b^2) = 3x(a - 2b)^2$$

**HELPFUL HINT**

If you recognize a trinomial as a perfect square trinomial, use the special formulas to factor. However, methods for factoring trinomials in general from Section 5.6 will also result in the correct factored form.

2 We now factor special types of binomials, beginning with the **difference of two squares.** The special product pattern presented in Section 5.4 for the product of a sum and a difference of two terms is used again here. However, the emphasis is now on factoring rather than on multiplying.

**DIFFERENCE OF TWO SQUARES**

$$a^2 - b^2 = (a + b)(a - b)$$

Notice that a binomial is a difference of two squares when it is the difference of the square of some quantity $a$ and the square of some quantity $b$.

**Example 3** Factor the following.

**a.** $x^2 - 9$ **b.** $16y^2 - 9$ **c.** $50 - 8y^2$ **d.** $x^2 - \frac{1}{4}$

*Solution* **a.** $x^2 - 9 = x^2 - 3^2$
$= (x + 3)(x - 3)$

**b.** $16y^2 - 9 = (4y)^2 - 3^2$
$= (4y + 3)(4y - 3)$

**c.** First factor out the common factor of 2.
$50 - 8y^2 = 2(25 - 4y^2)$
$= 2(5 + 2y)(5 - 2y)$

**d.** $x^2 - \frac{1}{4} = x^2 - \left(\frac{1}{2}\right)^2 = \left(x + \frac{1}{2}\right)\left(x - \frac{1}{2}\right)$

The binomial $x^2 + 9$ is a **sum of two squares** and cannot be factored by using real numbers. **In general, except for factoring out a GCF, the sum of two squares usually cannot be factored by using real numbers.**

HELPFUL HINT
The sum of two squares whose GCF is 1 usually cannot be factored by using real numbers.

**Example 4** Factor the following.

**a.** $p^4 - 16$ **b.** $(x + 3)^2 - 36$

*Solution* **a.** $p^4 - 16 = (p^2)^2 - 4^2$
$= (p^2 + 4)(p^2 - 4)$

The binomial factor $p^2 + 4$ cannot be factored by using real numbers, but the binomial factor $p^2 - 4$ is a difference of squares.

$$(p^2 + 4)(p^2 - 4) = (p^2 + 4)(p + 2)(p - 2)$$

**b.** Factor $(x + 3)^2 - 36$ as the difference of squares.

$(x + 3)^2 - 36 = (x + 3)^2 - 6^2$
$= [(x + 3) + 6][(x + 3) - 6]$ Factor.
$= [x + 3 + 6][x + 3 - 6]$ Remove parentheses.
$= (x + 9)(x - 3)$ Simplify.

**Example 5** Factor $x^2 + 4x + 4 - y^2$.

*Solution* Factoring by grouping comes to mind since the sum of the first three terms of this polynomial is a perfect square trinomial.

$x^2 + 4x + 4 - y^2 = (x^2 + 4x + 4) - y^2$ Group the first three terms.
$= (x + 2)^2 - y^2$ Factor the perfect square trinomial.

This is not completely factored yet since we have a *difference*, not a *product*. Since $(x + 2)^2 - y^2$ is a difference of squares, we have

$$\begin{aligned}(x + 2)^2 - y^2 &= [(x + 2) + y][(x + 2) - y]\\ &= (x + 2 + y)(x + 2 - y)\end{aligned}$$

3 Although the sum of two squares usually cannot be factored, the sum of two cubes, as well as the difference of two cubes, can be factored as follows.

**SUM AND DIFFERENCE OF TWO CUBES**

$$a^3 + b^3 = (a + b)(a^2 - ab + b^2)$$
$$a^3 - b^3 = (a - b)(a^2 + ab + b^2)$$

To check the first pattern, let's find the product of $(a + b)$ and $(a^2 - ab + b^2)$.

$$\begin{aligned}(a + b)(a^2 - ab + b^2) &= a(a^2 - ab + b^2) + b(a^2 - ab + b^2)\\ &= a^3 - a^2b + ab^2 + a^2b - ab^2 + b^3\\ &= a^3 + b^3\end{aligned}$$

**Example 6** Factor $x^3 + 8$.

*Solution* First we write the binomial in the form $a^3 + b^3$. Then we use the formula

$$a^3 + b^3 = (a + b)(a^2 - a \cdot b + b^2), \quad \text{where } a \text{ is } x \text{ and } b \text{ is } 2.$$
$$\downarrow \quad \downarrow \quad \downarrow \quad \downarrow \quad \downarrow \quad \downarrow \quad \downarrow \quad \downarrow$$
$$x^3 + 8 = x^3 + 2^3 = (x + 2)(x^2 - x \cdot 2 + 2^2)$$

Thus, $x^3 + 8 = (x + 2)(x^2 - 2x + 4)$

**Example 7** Factor $p^3 + 27q^3$.

*Solution*

$$\begin{aligned}p^3 + 27q^3 &= p^3 + (3q)^3\\ &= (p + 3q)[p^2 - (p)(3q) + (3q)^2]\\ &= (p + 3q)(p^2 - 3pq + 9q^2)\end{aligned}$$

**Example 8** Factor $y^3 - 64$.

*Solution* This is a difference of cubes since $y^3 - 64 = y^3 - 4^3$.

From $a^3 - b^3 = (a - b)(a^2 + a \cdot b + b^2)$ we have that

$$\downarrow \quad \downarrow \quad \downarrow \quad \downarrow \quad \downarrow \quad \downarrow \quad \downarrow \quad \downarrow$$

$$\begin{aligned}y^3 - 4^3 &= (y - 4)(y^2 + y \cdot 4 + 4^2)\\ &= (y - 4)(y^2 + 4y + 16)\end{aligned}$$

**HELPFUL HINT**

When factoring sums or differences of cubes, be sure to notice the sign patterns.

same sign

$$x^3 + y^3 = (x + y)(x^2 - xy + y^2)$$

opposite sign   always positive

same sign

$$x^3 - y^3 = (x - y)(x^2 + xy + y^2)$$

opposite sign   always positive

**Example 9** Factor $125q^2 - n^3q^2$

*Solution* First we factor out a common factor of $q^2$.

$$\begin{aligned} 125q^2 - n^3q^2 &= q^2(125 - n^3) \\ &= q^2(5^3 - n^3) \end{aligned}$$

opposite sign   positive

$$\begin{aligned} &= q^2(5 - n)[5^2 + (5)(n) + n^2] \\ &= q^2(5 - n)(25 + 5n + n^2) \end{aligned}$$

Thus, $125q^2 - n^3q^2 = q^2(5 - n)(25 + 5n + n^2)$. The trinomial $25 + 5n + n^2$ cannot be factored further.

4 The key to proficiency in factoring polynomials is to practice until you are comfortable with each technique. A strategy for factoring polynomials is given next.

**FACTORING A POLYNOMIAL**

**Step 1:** Are there any common factors? If so, factor out the GCF.

**Step 2:** How many terms are in the polynomial?

- **a.** If there are **two** terms, decide if one of the following formulas may be applied.
  - **i.** Difference of two squares: $a^2 - b^2 = (a + b)(a - b)$.
  - **ii.** Difference of two cubes: $a^3 - b^3 = (a - b)(a^2 + ab + b^2)$.
  - **iii.** Sum of two cubes: $a^3 + b^3 = (a + b)(a^2 - ab + b^2)$.
- **b.** If there are **three** terms, try one of the following.
  - **i.** Perfect square trinomial: $a^2 + 2ab + b^2 = (a + b)^2$; $a^2 - 2ab + b^2 = (a - b)^2$
  - **ii.** If not a perfect square trinomial, factor by using the methods presented in Section 5.6.
- **c.** If there are **four** or more terms, try factoring by grouping.

**Step 3:** See if any factors in the factored polynomial can be factored further.

**Example 10** Factor each polynomial completely.

**a.** $8a^2b - 4ab$ **b.** $36x^2 - 9$ **c.** $2x^2 - 5x - 7$

*Solution* **a. Step 1:** The terms have a common factor of $4ab$, which we factor out.

$$8a^2b - 4ab = 4ab(2a - 1)$$

**Step 2:** There are two terms, but the binomial $2a - 1$ is not the difference of two squares or the sum or difference of two cubes.

**Step 3:** The factor $2a - 1$ cannot be factored further.

**b. Step 1:** Factor out a common factor of 9.

$$36x^2 - 9 = 9(4x^2 - 1)$$

**Step 2:** The factor $4x^2 - 1$ has two terms, and it is the difference of two squares.

$$9(4x^2 - 1) = 9(2x + 1)(2x - 1)$$

**Step 3:** No factor with more than one term can be factored further.

**c. Step 1:** The terms of $2x^2 - 5x - 7$ contain no common factor other than 1 or $-1$.

**Step 2.** There are three terms. The trinomial is not a perfect square, so we factor by methods from Section 5.6.

$$2x^2 - 5x - 7 = (2x - 7)(x + 1)$$

**Step 3:** No factor with more than one term can be factored further.

**Example 11** Factor each polynomial completely.

**a.** $5p^2 + 5 + qp^2 + q$ **b.** $9x^2 + 24x + 16$ **c.** $y^2 + 25$

*Solution* **a. Step 1:** There is no common factor of all terms of $5p^2 + 5 + qp^2 + q$.

**Step 2:** The polynomial has four terms, so try factoring by grouping.

$$\begin{aligned} 5p^2 + 5 + qp^2 + q &= (5p^2 + 5) + (qp^2 + q) && \text{Group the terms.} \\ &= 5(p^2 + 1) + q(p^2 + 1) \\ &= (p^2 + 1)(5 + q) \end{aligned}$$

**Step 3:** No factor can be factored further.

**b. Step 1:** The terms of $9x^2 + 24x + 16$ contain no common factor other than 1 or $-1$.

**Step 2:** The trinomial $9x^2 + 24x + 16$ is a perfect square trinomial, and $9x^2 + 24x + 16 = (3x + 4)^2$.

**Step 3:** No factor can be factored further.

**c. Step 1:** There is no common factor of $y^2 + 25$ other than 1.

**Step 2:** This binomial is the sum of two squares and is prime.

**Step 3:** The binomial $y^2 + 25$ cannot be factored further.

**Example 12** Factor each completely.

**a.** $27a^3 - b^3$ **b.** $3n^2m^4 - 48m^6$ **c.** $2x^2 - 12x + 18 - 2z^2$
**d.** $8x^4y^2 + 125xy^2$ **e.** $(x - 5)^2 - 49y^2$

*Solution* **a.** This binomial is the difference of two cubes.

$$\begin{aligned} 27a^3 - b^3 &= (3a)^3 - b^3 \\ &= (3a - b)[(3a)^2 + (3a)(b) + b^2] \\ &= (3a - b)(9a^2 + 3ab + b^2) \end{aligned}$$

**b.**
$$\begin{aligned} 3n^2m^4 - 48m^6 &= 3m^4(n^2 - 16m^2) && \text{Factor out the GCF, } 3m^4. \\ &= 3m^4(n + 4m)(n - 4m) && \text{Factor the difference of squares.} \end{aligned}$$

**c.**
$$\begin{aligned} 2x^2 - 12x + 18 - 2z^2 &= 2(x^2 - 6x + 9 - z^2) && \text{The GCF is 2.} \\ &= 2[(x^2 - 6x + 9) - z^2] && \text{Group the first three terms together.} \\ &= 2[(x - 3)^2 - z^2] && \text{Factor the perfect square trinomial.} \\ &= 2[(x - 3) + z][(x - 3) - z] && \text{Factor the difference of squares.} \\ &= 2(x - 3 + z)(x - 3 - z) \end{aligned}$$

**d.**
$$\begin{aligned} 8x^4y^2 + 125xy^2 &= xy^2(8x^3 + 125) && \text{The GCF is } xy^2. \\ &= xy^2[(2x)^3 + 5^3] \\ &= xy^2(2x + 5)[(2x)^2 - (2x)(5) + 5^2] && \text{Factor the sum of cubes.} \\ &= xy^2(2x + 5)(4x^2 - 10x + 25) \end{aligned}$$

**e.** This binomial is the difference of squares.

$$\begin{aligned} (x - 5)^2 - 49y^2 &= (x - 5)^2 - (7y)^2 \\ &= [(x - 5) + 7y][(x - 5) - 7y] \\ &= (x - 5 + 7y)(x - 5 - 7y) \end{aligned}$$

## Exercise Set 5.7

*Factor the following. See Examples 1 and 2.*

**1.** $x^2 + 6x + 9$
**2.** $x^2 - 10x + 25$
**3.** $4x^2 - 12x + 9$
**4.** $25x^2 + 10x + 1$
**5.** $3x^2 - 24x + 48$
**6.** $x^3 + 14x^2 + 49x$
**7.** $9y^2x^2 + 12yx^2 + 4x^2$
**8.** $32x^2 - 16xy + 2y^2$

*Factor the following. See Examples 3 through 5.*

**9.** $x^2 - 25$
**10.** $y^2 - 100$
**11.** $9 - 4z^2$
**12.** $16x^2 - y^2$
**13.** $(y + 2)^2 - 49$
**14.** $(x - 1)^2 - z^2$
**15.** $64x^2 - 100$
**16.** $4x^2 - 36$

*Factor the following. See Examples 6 through 8.*

**17.** $x^3 + 27$
**18.** $y^3 + 1$
**19.** $z^3 - 1$
**20.** $x^3 - 8$
**21.** $m^3 + n^3$
**22.** $r^3 + 125$
**23.** $x^3y^2 - 27y^2$
**24.** $64 - p^3$
**25.** $a^3b + 8b^4$
**26.** $8ab^3 + 27a^4$
**27.** $125y^3 - 8x^3$
**28.** $54y^3 - 128$

*Factor the following. See Example 9.*

**29.** $x^2 + 6x + 9 - y^2$
**30.** $x^2 + 12x + 36 - y^2$
**31.** $x^2 - 10x + 25 - y^2$
**32.** $x^2 - 18x + 81 - y^2$
**33.** $4x^2 + 4x + 1 - z^2$
**34.** $9y^2 + 12y + 4 - x^2$

*Factor each polynomial completely.*

**35.** $9x^2 - 49$
**36.** $25x^2 - 4$
**37.** $x^2 - 12x + 36$
**38.** $x^2 - 18x + 81$
**39.** $x^4 - 81$
**40.** $x^4 - 256$
**41.** $x^2 + 8x + 16 - 4y^2$
**42.** $x^2 + 14x + 49 - 9y^2$
**43.** $(x + 2y)^2 - 9$
**44.** $(3x + y)^2 - 25$

**45.** $x^3 - 216$

**46.** $8 - a^3$

**47.** $x^3 + 125$

**48.** $x^3 + 216$

**49.** $4x^2 + 25$

**50.** $16x^2 + 25$

**51.** $4a^2 + 12a + 9$

**52.** $9a^2 - 30a + 25$

**53.** $18x^2y - 2y$

**54.** $12xy^2 - 108x$

**55.** $8x^3 + y^3$

**56.** $27x^3 - y^3$

**57.** $x^6 - y^3$

**58.** $x^3 - y^6$

**59.** $x^2 + 16x + 64 - x^4$

**60.** $x^2 + 20x + 100 - x^4$

**61.** $3x^6y^2 + 81y^2$

**62.** $x^2y^9 + x^2y^3$

**63.** $(x + y)^3 + 125$

**64.** $(x + y)^3 + 27$

**65.** $(2x + 3)^3 - 64$

**66.** $(4x + 2)^3 - 125$

*Solve.*

△ **67.** The manufacturer of Antonio's Metal Washers needs to determine the cross-sectional area of each washer. If the outer radius of the washer is $R$ and the radius of the hole is $r$, express the area of the washer as a polynomial. Factor this polynomial completely.

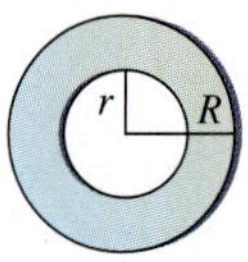


△ **68.** Express the area of the shaded region as a polynomial. Factor the polynomial completely.

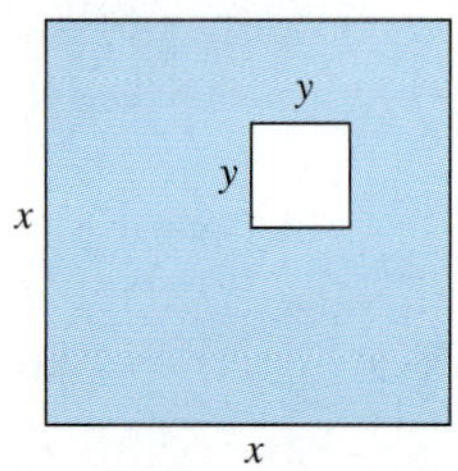


△ **69.** The manufacturer of Tootsie Roll Pops plans to change the size of its candy. To compute the new cost, the company needs a formula for the volume of the candy coating without the Tootsie Roll center. Given the diagram, express the volume as a polynomial. Factor this polynomial completely.

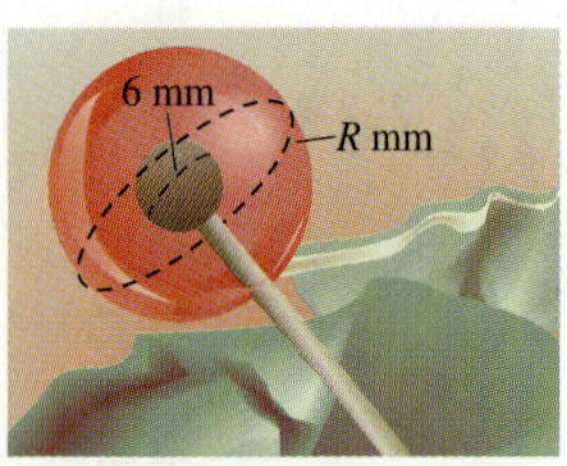


△ **70.** Express the area of the shaded region as a polynomial. Factor the polynomial completely.

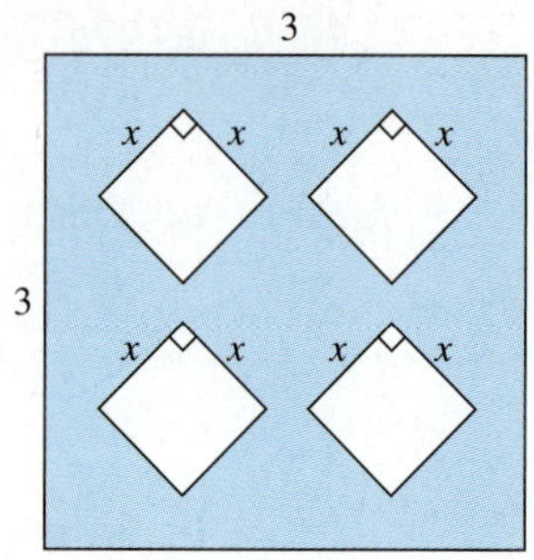


*Factor completely. See Examples 10 through 12.*

**71.** $1 - y^3$

**72.** $8 - a^3$

**73.** $9x^2 + 6x + 1$

**74.** $16y^2 - 24y + 9$

**75.** $x^2 - 8x + 16 - y^2$

**76.** $12x^2 - 22x - 20$

**77.** $x^4 - x$

**78.** $(2x + 1)^2 - 3(2x + 1) + 2$

**79.** $14x^2y - 2xy$

**80.** $24ab^2 - 6ab$

**81.** $4x^2 - 16$

**82.** $9x^2 - 81$

**83.** $128a^3 - 2b^3$

**84.** $32x^3 - 4y^3$

**85.** $3x^2 - 8x - 11$

**86.** $5x^2 - 2x - 3$

**87.** $4x^2 + 8x - 12$

**88.** $6x^2 - 6x - 12$

**89.** $4x^2 + 36x + 81$

**90.** $25x^2 + 40x + 16$

**91.** $8x^3 + 27y^3$

**92.** $125x^3 + 8y^3$

**93.** $64x^2y^3 - 8x^2$

**94.** $27x^5y^4 - 216x^2y$

**95.** $(x + 5)^3 + y^3$

**96.** $(y - 1)^3 + 27x^3$

**97.** $(5a - 3)^2 - 6(5a - 3) + 9$

**98.** $(4r + 1)^2 + 8(4r + 1) + 16$

*Find a value of c that makes each trinomial a perfect square trinomial.*

**99.** $x^2 + 6x + c$

**100.** $y^2 + 10y + c$

**101.** $m^2 - 14m + c$

**102.** $n^2 - 2n + c$

**103.** $x^2 + cx + 16$

**104.** $x^2 + cx + 36$

**105.** Factor $x^6 - 1$ completely, using the following methods from this chapter.

**a.** Factor the expression by treating it as the difference of two squares, $(x^3)^2 - 1^2$.

**b.** Factor the expression by treating it as the difference of two cubes, $(x^2)^3 - 1^3$.

**c.** Are the answers to parts **a** and **b** the same? Why or why not?

## REVIEW EXERCISES

*Solve the following equations. See Section 2.1.*

**106.** $x - 5 = 0$

**107.** $x + 7 = 0$

**108.** $3x + 1 = 0$

**109.** $5x - 15 = 0$

**110.** $-2x = 0$

**111.** $3x = 0$

**112.** $-5x + 25 = 0$

**113.** $-4x - 16 = 0$

## *A Look Ahead*

Example

Factor $x^{2n} - 100$.

*Solution*

This binomial is the difference of squares.

$$x^{2n} - 100 = (x^n)^2 - 10^2$$
$$= (x^n + 10)(x^n - 10)$$

*Factor each expression. Assume that variables used as exponents represent positive integers. See the preceding example.*

**114.** $x^{2n} - 25$

**115.** $x^{2n} - 36$

**116.** $36x^{2n} - 49$

**117.** $25x^{2n} - 81$

**118.** $x^{4n} - 16$

**119.** $x^{4n} - 625$

# 5.8 SOLVING EQUATIONS BY FACTORING AND PROBLEM SOLVING

CD-ROM SSM SSG Video

▶ **OBJECTIVES**

1. Solve polynomial equations by factoring.
2. Solve problems that can be modeled by polynomial equations.
3. Find the $x$-intercepts of a polynomial function.

**1** In this section, your efforts to learn factoring start to pay off. We use factoring to solve polynomial equations, which in turn helps us solve problems that can be modeled by polynomial equations and also helps us sketch the graph of polynomial functions.

A **polynomial equation** is the result of setting two polynomials equal to each other. Examples of polynomial equations are

$$3x^3 - 2x^2 = x^2 + 2x - 1 \qquad 2.6x + 7 = -1.3 \qquad -5x^2 - 5 = -9x^2 - 2x + 1$$

A polynomial equation is in **standard form** if one side of the equation is 0. In standard form the polynomial equations above are

$$3x^3 - 3x^2 - 2x + 1 = 0 \qquad 2.6x + 8.3 = 0 \qquad 4x^2 + 2x - 6 = 0$$

The degree of a simplified polynomial equation in standard form is the same as the highest degree of any of its terms. A polynomial equation of degree 2 is also called a **quadratic equation.**

A solution of a polynomial equation in one variable is a value of the variable that makes the equation true. The method presented in this section for solving polynomial equations is called the **factoring method.** This method is based on the **zero-factor property.**

**ZERO-FACTOR PROPERTY**

If $a$ and $b$ are real numbers and $a \cdot b = 0$, then $a = 0$ or $b = 0$.
This property is true for three or more factors also.

In other words, if the product of two or more real numbers is zero, then at least one number must be zero.

**Example 1** Solve $(x + 2)(x - 6) = 0$.

*Solution* By the zero-factor property, $(x + 2)(x - 6) = 0$ only if $x + 2 = 0$ or $x - 6 = 0$.

$x + 2 = 0$ or $x - 6 = 0$ Apply the zero-factor property.
$x = -2$ or $x = 6$ Solve each linear equation.

To check, let $x = -2$ and then let $x = 6$ in the original equation.

Let $x = -2$.
Then $(x + 2)(x - 6) = 0$
becomes $(-2 + 2)(-2 - 6) = 0$
$(0)(-8) = 0$
$0 = 0$ True.

Let $x = 6$.
Then $(x + 2)(x - 6) = 0$
becomes $(6 + 2)(6 - 6) = 0$
$(8)(0) = 0$
$0 = 0$ True.

Both −2 and 6 check, so they are both solutions.

**Example 2** Solve $2x^2 + 9x - 5 = 0$.

*Solution* To use the zero-factor property, one side of the equation must be 0, and the other side must be in factored form.

$2x^2 + 9x - 5 = 0$
$(2x - 1)(x + 5) = 0$ Factor.
$2x - 1 = 0$ or $x + 5 = 0$ Set each factor equal to zero.
$2x = 1$
$x = \frac{1}{2}$ or $x = -5$ Solve each linear equation.

The solutions are −5 and $\frac{1}{2}$. To check, let $x = \frac{1}{2}$ in the original equation; then let $x = -5$ in the original equation.

**SOLVING POLYNOMIAL EQUATIONS BY FACTORING**

**Step 1:** Write the equation in standard form so that one side of the equation is 0.
**Step 2:** Factor the polynomial completely.
**Step 3:** Set each factor containing a variable equal to 0.
**Step 4:** Solve the resulting equations.
**Step 5:** Check each solution in the original equation.

Since it is not always possible to factor a polynomial, not all polynomial equations can be solved by factoring. Other methods of solving polynomial equations are presented in Chapter 8.

**Example 3** Solve $x(2x - 7) = 4$.

*Solution* First write the equation in standard form; then factor.

$$\begin{aligned} x(2x - 7) &= 4 \\ 2x^2 - 7x &= 4 && \text{Multiply.} \\ 2x^2 - 7x - 4 &= 0 && \text{Write in standard form.} \\ (2x + 1)(x - 4) &= 0 && \text{Factor.} \\ 2x + 1 &= 0 \quad \text{or} \quad x - 4 = 0 && \text{Set each factor equal to zero.} \\ 2x &= -1 && \text{Solve.} \\ x &= -\frac{1}{2} \quad \text{or} \quad x = 4 \end{aligned}$$

The solutions are $-\frac{1}{2}$ and 4. Check both solutions in the original equation.

> **HELPFUL HINT**
> To apply the zero-factor property, one side of the equation must be 0, and the other side of the equation must be factored. To solve the equation $x(2x - 7) = 4$, for example, you may **not** set each factor equal to 4.

**Example 4** Solve $3(x^2 + 4) + 5 = -6(x^2 + 2x) + 13$.

*Solution* Rewrite the equation so that one side is 0.

$$\begin{aligned} 3(x^2 + 4) + 5 &= -6(x^2 + 2x) + 13 \\ 3x^2 + 12 + 5 &= -6x^2 - 12x + 13 && \text{Apply the distributive property.} \\ 9x^2 + 12x + 4 &= 0 && \text{Rewrite the equation so that one side is 0.} \\ (3x + 2)(3x + 2) &= 0 && \text{Factor.} \\ 3x + 2 &= 0 \quad \text{or} \quad 3x + 2 = 0 && \text{Set each factor equal to 0.} \\ 3x &= -2 \quad \text{or} \quad 3x = -2 \\ x &= -\frac{2}{3} \quad \text{or} \quad x = -\frac{2}{3} && \text{Solve each equation.} \end{aligned}$$

The solution is $-\frac{2}{3}$. Check by substituting $-\frac{2}{3}$ into the original equation.

If the equation contains fractions, we clear the equation of fractions as a first step.

**Example 5** Solve $2x^2 = \frac{17}{3}x + 1$.

*Solution*

$$2x^2 = \frac{17}{3}x + 1$$

$$3(2x^2) = 3\left(\frac{17}{3}x + 1\right) \quad \text{Clear the equation of fractions.}$$

$$6x^2 = 17x + 3 \quad \text{Apply the distributive property.}$$

$$6x^2 - 17x - 3 = 0 \quad \text{Rewrite the equation in standard form.}$$

$$(6x + 1)(x - 3) = 0 \quad \text{Factor.}$$

$$6x + 1 = 0 \quad \text{or} \quad x - 3 = 0 \quad \text{Set each factor equal to zero.}$$

$$6x = -1$$

$$x = -\frac{1}{6} \quad \text{or} \quad x = 3 \quad \text{Solve each equation.}$$

The solutions are $-\frac{1}{6}$ and 3.

**Example 6** Solve $x^3 = 4x$.

*Solution*

$$x^3 = 4x$$

$$x^3 - 4x = 0 \quad \text{Rewrite the equation so that one side is 0.}$$

$$x(x^2 - 4) = 0 \quad \text{Factor out the GCF, } x.$$

$$x(x + 2)(x - 2) = 0 \quad \text{Factor the difference of squares.}$$

$$x = 0 \quad \text{or} \quad x + 2 = 0 \quad \text{or} \quad x - 2 = 0 \quad \text{Set each factor equal to 0.}$$

$$x = 0 \quad \text{or} \quad x = -2 \quad \text{or} \quad x = 2 \quad \text{Solve each equation.}$$

The solutions are −2, 0, and 2. Check by substituting into the original equation.

Notice that the *third*-degree equation of Example 6 yielded *three* solutions.

**Example 7** Solve $x^3 + 5x^2 = x + 5$.

*Solution* First write the equation so that one side is 0.

$$x^3 + 5x^2 - x - 5 = 0$$

$$(x^3 - x) + (5x^2 - 5) = 0 \quad \text{Factor by grouping.}$$

$$x(x^2 - 1) + 5(x^2 - 1) = 0$$

$$(x^2 - 1)(x + 5) = 0$$

$$(x + 1)(x - 1)(x + 5) = 0 \quad \text{Factor the difference of squares.}$$

$$x + 1 = 0 \quad \text{or} \quad x - 1 = 0 \quad \text{or} \quad x + 5 = 0 \quad \text{Set each factor equal to 0.}$$

$$x = -1 \quad \text{or} \quad x = 1 \quad \text{or} \quad x = -5 \quad \text{Solve each equation.}$$

The solutions are −5, −1, and 1. Check in the original equation.

2 Some problems may be modeled by polynomial equations. To solve these problems, we use the same problem-solving steps that were introduced in Section 2.2. When solving these problems, keep in mind that a solution of an equation that models a problem is not always a solution to the problem. For example, a person's weight or

the length of a side of a geometric figure is always a positive number. Discard solutions that do not make sense as solutions of the problem.

## Example 8 FINDING THE RETURN TIME OF A ROCKET

An Alpha III model rocket is launched from the ground with an A8–3 engine. Without a parachute the height of the rocket $h$ at time $t$ seconds is approximated by the equation.

$$h = -16t^2 + 144t$$

Find how long it takes the rocket to return to the ground.

*Solution*

1. UNDERSTAND. Read and reread the problem. The equation $h = -16t^2 + 144t$ models the height of the rocket. Familiarize yourself with this equation by finding a few values.

   When $t = 1$ second, the height of the rocket is

   $$h = -16(1)^2 + 144(1) = 128 \text{ feet}$$

   When $t = 2$ seconds, the height of the rocket is

   $$h = -16(2)^2 + 144(2) = 224 \text{ feet}$$

2. TRANSLATE. To find how long it takes the rocket to return to the ground, we want to know what value of $t$ makes the height $h$ equal to 0. That is, we want to solve $h = 0$.

   $$-16t^2 + 144t = 0$$

3. SOLVE the quadratic equation by factoring.

   $$-16t^2 + 144t = 0$$
   $$-16t(t - 9) = 0$$
   $$-16t = 0 \quad \text{or} \quad t - 9 = 0$$
   $$t = 0 \qquad\qquad t = 9$$

4. INTERPRET. The height $h$ is 0 feet at time 0 seconds (when the rocket is launched) and at time 9 seconds.

   **Check:** See that the height of the rocket at 9 seconds equals 0.

   $$h = -16(9)^2 + 144(9) = -1296 + 1296 = 0$$

   **State:** The rocket returns to the ground 9 seconds after it is launched.

Some of the exercises at the end of this section make use of the **Pythagorean theorem.** Before we review this theorem, recall that a **right triangle** is a triangle that contains a 90° angle, or right angle. The **hypotenuse** of a right triangle is the side opposite the right angle and is the longest side of the triangle. The **legs** of a right triangle are the other sides of the triangle.

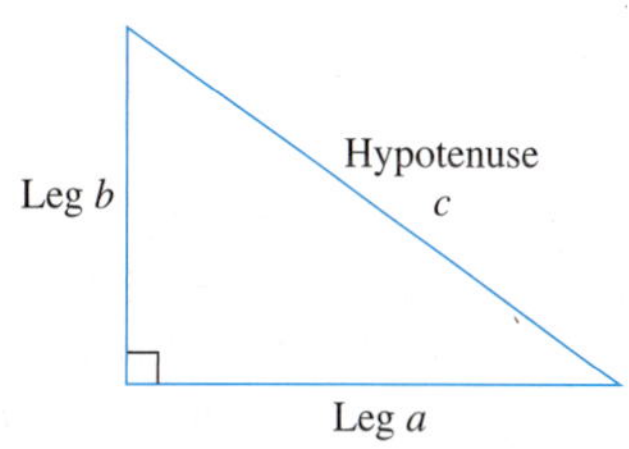

**PYTHAGOREAN THEOREM**

In a right triangle, the sum of the squares of the lengths of the two legs is equal to the square of the length of the hypotenuse.

$$(\text{leg})^2 + (\text{leg})^2 = (\text{hypotenuse})^2 \quad \text{or} \quad a^2 + b^2 = c^2$$

## △ Example 9 USING THE PYTHAGOREAN THEOREM

While framing an addition to an existing home, Kim Menzies, a carpenter, used the Pythagorean theorem to determine whether a wall was "square"—that is, whether the wall formed a right angle with the floor. He used a triangle whose sides are three consecutive integers. Find a right triangle whose sides are three consecutive integers.

*Solution* 1. UNDERSTAND. Read and reread the problem.

Let $x$, $x + 1$, and $x + 2$ be three consecutive integers. Since these integers represent lengths of the sides of a right triangle, we have

$$x = \text{one leg}$$
$$x + 1 = \text{other leg}$$
$$x + 2 = \text{hypotenuse (longest side)}$$

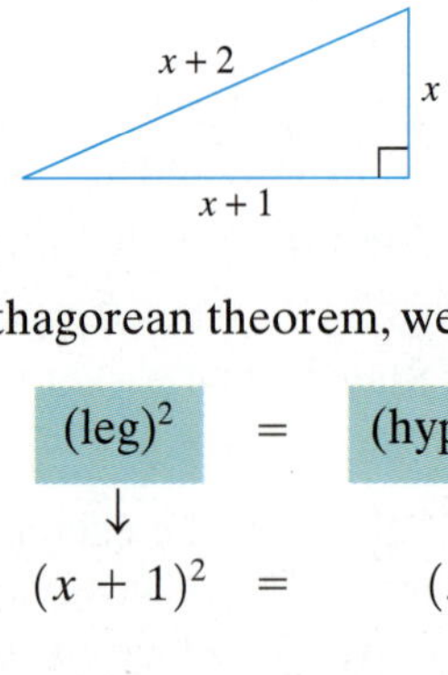


2. TRANSLATE. By the Pythagorean theorem, we have

| | | | | | |
|---|---|---|---|---|---|
| In words: | $(\text{leg})^2$ | + | $(\text{leg})^2$ | = | $(\text{hypotenuse})^2$ |
| | ↓ | | ↓ | | ↓ |
| Translate: | $(x)^2$ | + | $(x + 1)^2$ | = | $(x + 2)^2$ |

3. SOLVE the equation.

$$x^2 + (x + 1)^2 = (x + 2)^2$$
$$x^2 + x^2 + 2x + 1 = x^2 + 4x + 4 \quad \text{Multiply.}$$
$$2x^2 + 2x + 1 = x^2 + 4x + 4$$
$$x^2 - 2x - 3 = 0 \quad \text{Write in standard form.}$$
$$(x - 3)(x + 1) = 0$$
$$x - 3 = 0 \quad \text{or} \quad x + 1 = 0$$
$$x = 3 \qquad\qquad x = -1$$

4. INTERPRET. Discard $x = -1$ since length cannot be negative. If $x = 3$, then $x + 1 = 4$ and $x + 2 = 5$.

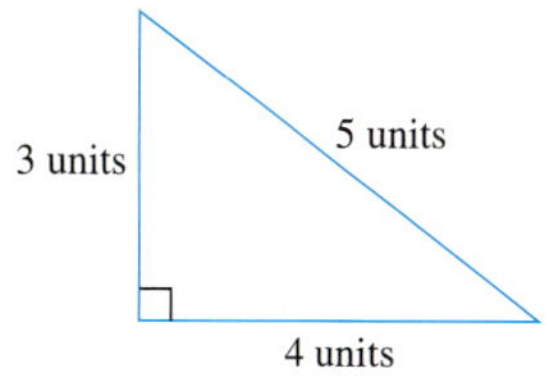


**Check:** To check, see that $(\text{leg})^2 + (\text{leg})^2 = (\text{hypotenuse})^2$

$$3^2 + 4^2 = 5^2$$
$$9 + 16 = 25 \quad \text{True.}$$

**State:** The lengths of the sides of the right triangle are 3, 4, and 5 units. Kim used this information, for example, by marking off lengths of 3 and 4 feet on the floor and framing respectively. If the diagonal length between these marks was 5 feet, the wall was "square." If not, adjustments were made.

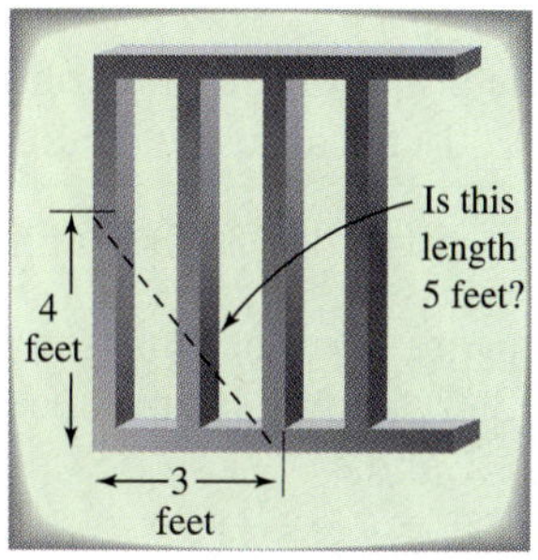


**3** Recall that to find the $x$-intercepts of the graph of a function, let $f(x) = 0$, or $y = 0$, and solve for $x$. This fact gives us a visual interpretation of the results of this section.

From Example 1, we know that the solutions of the equation $(x + 2)(x - 6) = 0$ are $-2$ and $6$. These solutions give us important information about the related polynomial function $p(x) = (x + 2)(x - 6)$. We know that when $x$ is $-2$ or when $x$ is 6, the value of $p(x)$ is 0.

$$p(x) = (x + 2)(x - 6)$$
$$p(-2) = (-2 + 2)(-2 - 6) = (0)(-8) = 0$$
$$p(6) = (6 + 2)(6 - 6) = (8)(0) = 0$$

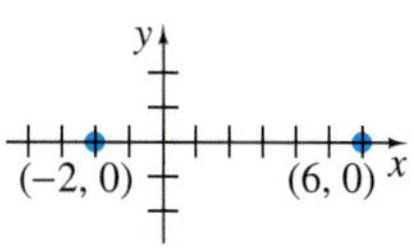


Thus, we know that $(-2, 0)$ and $(6, 0)$ are the $x$-intercepts of the graph of $p(x)$.

We also know that the graph of $p(x)$ does not cross the $x$-axis at any other point. For this reason, and the fact that $p(x) = (x + 2)(x - 6) = x^2 - 4x - 12$ has degree 2, we conclude that the graph of $p$ must look something like one of these two graphs:

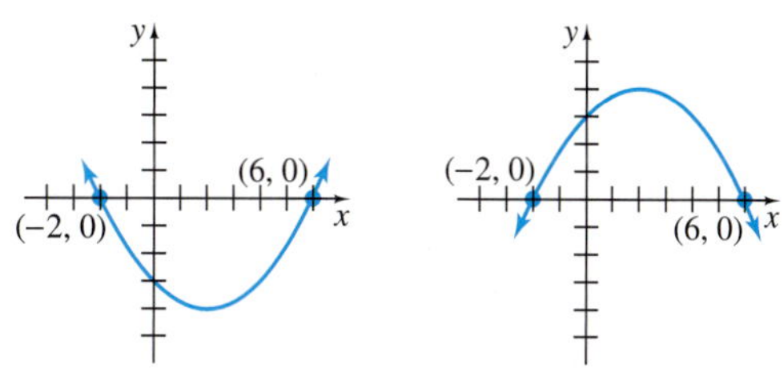

In the following section and in a later chapter, we explore these graphs more fully. For the moment, know that the solutions of a polynomial equation are the $x$-intercepts of the graph of the related function and that the $x$-intercepts of the graph of a polynomial function are the solutions of the related polynomial equation. These values are also called **roots**, or **zeros**, of a polynomial function.

**Example 10** Match each function with its graph.

$f(x) = (x - 3)(x + 2)$ $\quad g(x) = x(x + 2)(x - 2)$ $\quad h(x) = (x - 2)(x + 2)(x - 1)$

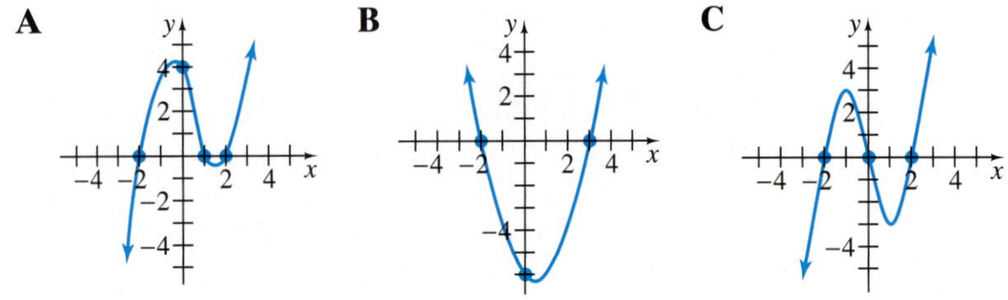


*Solution* The graph of the function $f(x) = (x - 3)(x + 2)$ has two $x$-intercepts, $(3, 0)$ and $(-2, 0)$, because the equation $0 = (x - 3)(x + 2)$ has two solutions, 3 and $-2$.

The graph of $f(x)$ is graph B.

The graph of the function $g(x) = x(x + 2)(x - 2)$ has three $x$-intercepts $(0, 0)$, $(-2, 0)$, and $(2, 0)$, because the equation $0 = x(x + 2)(x - 2)$ has three solutions, $0, -2$, and 2.

The graph of $g(x)$ is graph C.

The graph of the function $h(x) = (x - 2)(x + 2)(x - 1)$ has three $x$-intercepts, $(-2, 0)$, $(1, 0)$, and $(2, 0)$, because the equation $0 = (x - 2)(x + 2)(x - 1)$ has three solutions, $-2$, 1, and 2.

The graph of $h(x)$ is graph A.

## GRAPHING CALCULATOR EXPLORATIONS

We can use a graphing calculator to approximate real number solutions of any quadratic equation in standard form, whether the associated polynomial is factorable or not. For example, let's solve the quadratic equation $x^2 - 2x - 4 = 0$. The solutions of this equation will be the $x$-intercepts of the graph of the function $f(x) = x^2 - 2x - 4$. (Recall that to find $x$-intercepts, we let $f(x) = 0$, or $y = 0$.) When we use a standard window, the graph of this function looks like this.

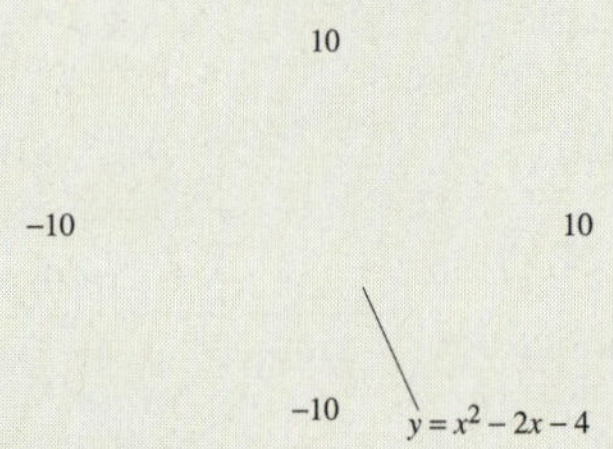

The graph appears to have one $x$-intercept between $-2$ and $-1$ and one between 3 and 4. To find the $x$-intercept between 3 and 4 to the nearest hundredth, we can use a zero feature, a Zoom feature, which magnifies a portion of the graph around the cursor, or we can redefine our window. If we redefine our window to

$$\begin{aligned} \text{Xmin} &= 2 & \text{Ymin} &= -1 \\ \text{Xmax} &= 5 & \text{Ymax} &= 1 \\ \text{Xscl} &= 1 & \text{Yscl} &= 1 \end{aligned}$$

the resulting screen is

2 5

By using the Trace feature, we can now see that one of the intercepts is between 3.21 and 3.25. To approximate to the nearest hundredth, Zoom again or redefine the window to

$$\begin{aligned} \text{Xmin} &= 3.2 & \text{Ymin} &= -0.1 \\ \text{Xmax} &= 3.3 & \text{Ymax} &= 0.1 \\ \text{Xscl} &= 1 & \text{Yscl} &= 1 \end{aligned}$$

If we use the Trace feature again, we see that, to the nearest hundredth, the $x$-intercept is 3.24. By repeating this process, we can approximate the other $x$-intercept to be $-1.24$.

To check, find $f(3.24)$ and $f(-1.24)$. Both of these values should be close to 0. (They will not be exactly 0 since we approximated these solutions.)

$$f(3.24) = 0.0176 \quad \text{and} \quad f(-1.24) = 0.0176$$

*Solve each of these quadratic equations by graphing a related function and approximating the $x$-intercepts to the nearest thousandth.*

**1.** $x^2 + 3x - 2 = 0$

**2.** $5x^2 - 7x + 1 = 0$

**3.** $2.3x^2 - 4.4x - 5.6 = 0$

**4.** $0.2x^2 + 6.2x + 2.1 = 0$

**5.** $0.09x^2 - 0.13x - 0.08 = 0$

**6.** $x^2 + 0.08x - 0.01 = 0$

## MENTAL MATH

*Solve each equation for the variable. See Example 1.*

**1.** $(x - 3)(x + 5) = 0$

**2.** $(y + 5)(y + 3) = 0$

**3.** $(z - 3)(z + 7) = 0$

**4.** $(c - 2)(c - 4) = 0$

**5.** $x(x - 9) = 0$

**6.** $w(w + 7) = 0$

## *Exercise Set 5.8*

*Solve each equation. See Example 1.*

**1.** $(x + 3)(3x - 4) = 0$

**2.** $(5x + 1)(x - 2) = 0$

**3.** $3(2x - 5)(4x + 3) = 0$

**4.** $8(3x - 4)(2x - 7) = 0$

*Solve each equation. See Examples 2 through 5.*

**5.** $x^2 + 11x + 24 = 0$

**6.** $y^2 - 10y + 24 = 0$

**7.** $12x^2 + 5x - 2 = 0$

**8.** $3y^2 - y - 14 = 0$

**9.** $z^2 + 9 = 10z$

**10.** $n^2 + n = 72$

**11.** $x(5x + 2) = 3$

**12.** $n(2n - 3) = 2$

**13.** $x^2 - 6x = x(8 + x)$

**14.** $n(3 + n) = n^2 + 4n$

**15.** $\frac{z^2}{6} - \frac{z}{2} - 3 = 0$

**16.** $\frac{c^2}{20} - \frac{c}{4} + \frac{1}{5} = 0$

**17.** $\frac{x^2}{2} + \frac{x}{20} = \frac{1}{10}$

**18.** $\frac{y^2}{30} = \frac{y}{15} + \frac{1}{2}$

**19.** $\frac{4t^2}{5} = \frac{t}{5} + \frac{3}{10}$

**20.** $\frac{5x^2}{6} - \frac{7x}{2} + \frac{2}{3} = 0$

*Solve each equation. See Examples 6 and 7.*

**21.** $(x + 2)(x - 7)(3x - 8) = 0$

**22.** $(4x + 9)(x - 4)(x + 1) = 0$

**23.** $y^3 = 9y$

**24.** $n^3 = 16n$

**25.** $x^3 - x = 2x^2 - 2$

**26.** $m^3 = m^2 + 12m$

**27.** Explain how solving $2(x - 3)(x - 1) = 0$ differs from solving $2x(x - 3)(x - 1) = 0$.

**28.** Explain why the zero-factor property works for more than two numbers whose product is 0.

*Solve each equation.*

**29.** $(2x + 7)(x - 10) = 0$

**30.** $(x + 4)(5x - 1) = 0$

**31.** $3x(x - 5) = 0$

**32.** $4x(2x + 3) = 0$

**33.** $x^2 - 2x - 15 = 0$

**34.** $x^2 + 6x - 7 = 0$

**35.** $12x^2 + 2x - 2 = 0$

**36.** $8x^2 + 13x + 5 = 0$

**37.** $w^2 - 5w = 36$

**38.** $x^2 + 32 = 12x$

**39.** $25x^2 - 40x + 16 = 0$

**40.** $9n^2 + 30n + 25 = 0$

**41.** $2r^3 + 6r^2 = 20r$

**42.** $-2t^3 = 108t - 30t^2$

**43.** $z(5z - 4)(z + 3) = 0$

**44.** $2r(r + 3)(5r - 4) = 0$

**45.** $2z(z + 6) = 2z^2 + 12z - 8$

**46.** $3c^2 - 8c + 2 = c(3c - 8)$

**47.** $(x - 1)(x + 4) = 24$

**48.** $(2x - 1)(x + 2) = -3$

**49.** $\frac{x^2}{4} - \frac{5}{2}x + 6 = 0$

**50.** $\frac{x^2}{18} + \frac{x}{2} + 1 = 0$

**51.** $y^2 + \frac{1}{4} = -y$

**52.** $\frac{x^2}{10} + \frac{5}{2} = x$

**53.** $y^3 + 4y^2 = 9y + 36$

**54.** $x^3 + 5x^2 = x + 5$

**55.** $2x^3 = 50x$

**56.** $m^5 = 36m^3$

**57.** $x^2 + (x + 1)^2 = 61$

**58.** $y^2 + (y + 2)^2 = 34$

**59.** $m^2(3m - 2) = m$

**60.** $x^2(5x + 3) = 26x$

**61.** $3x^2 = -x$

**62.** $y^2 = -5y$

**63.** $x(x - 3) = x^2 + 5x + 7$

**64.** $z^2 - 4z + 10 = z(z - 5)$

**65.** $3(t - 8) + 2t = 7 + t$

**66.** $7c - 2(3c + 1) = 5(4 - 2c)$

**67.** $-3(x - 4) + x = 5(3 - x)$

**68.** $-4(a + 1) - 3a = -7(2a - 3)$

**69.** Which solution strategies are incorrect? Why?

**a.** Solve $(y - 2)(y + 2) = 4$ by setting each factor equal to 4.

**b.** Solve $(x + 1)(x + 3) = 0$ by setting each factor equal to 0.

**c.** Solve $z^2 + 5z + 6 = 0$ by factoring $z^2 + 5z + 6$ and setting each factor equal to 0.

**d.** Solve $x^2 + 6x + 8 = 10$ by factoring $x^2 + 6x + 8$ and setting each factor equal to 0.

**70.** Describe two ways a linear equation differs from a quadratic equation.

*Solve. See Examples 8 and 9.*

**71.** One number exceeds another by five, and their product is 66. Find the numbers.

**72.** If the sum of two numbers is 4 and their product is $\frac{15}{4}$, find the numbers.

**73.** An electrician needs to run a cable from the top of a 60-foot tower to a transmitter box located 45 feet away from the base of the tower. Find how long he should make the cable.

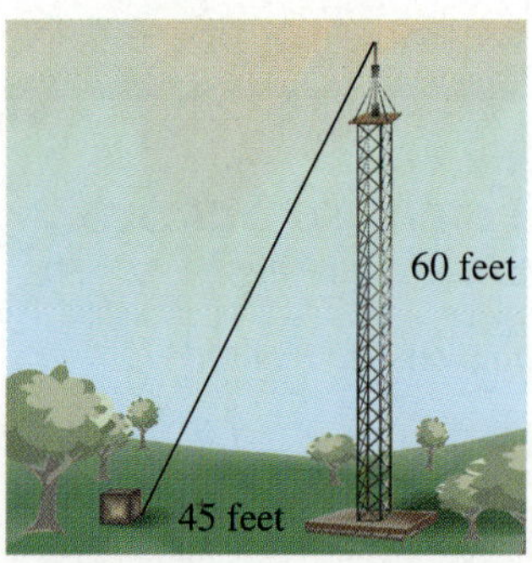

△ **74.** A stereo system installer needs to run speaker wire along the two diagonals of a rectangular room whose dimensions are 40 feet by 75 feet. Find how much speaker wire she needs.

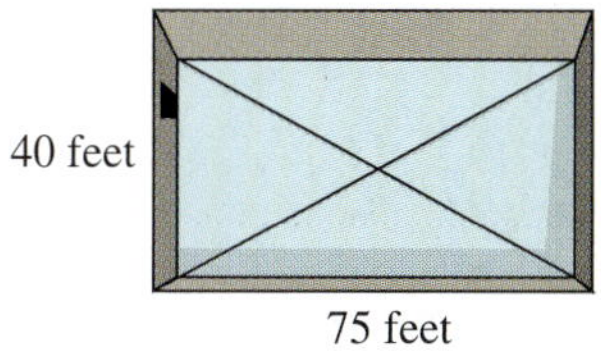


**75.** If the cost, $C(x)$, for manufacturing $x$ units of a certain product is given by $C(x) = x^2 - 15x + 50$, find the number of units manufactured at a cost of \$9500.

△ **76.** Determine whether any three consecutive integers represent the lengths of the sides of a right triangle.

△ **77.** The shorter leg of a right triangle is 3 centimeters less than the other leg. Find the length of the two legs if the hypotenuse is 15 centimeters.

△ **78.** The longer leg of a right triangle is 4 feet longer than the other leg. Find the length of the two legs if the hypotenuse is 20 feet.

△ **79.** Marie Mulroney has a rectangular board 12 inches by 16 inches around which she wants to put a uniform border of shells. If she has enough shells for a border whose area is 128 square inches, determine the width of the border.

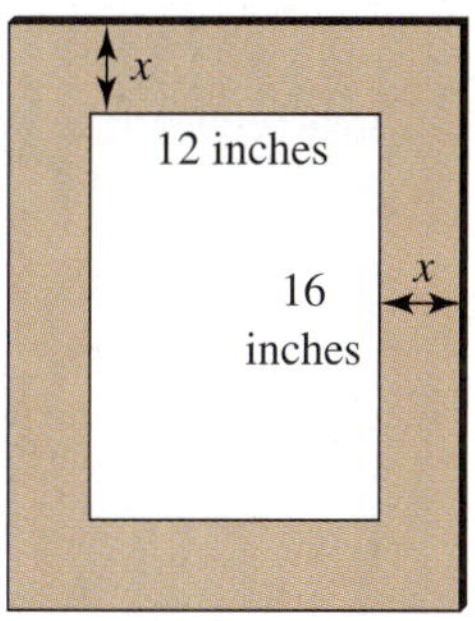


△ **80.** A gardener has a rose garden that measures 30 feet by 20 feet. He wants to put a uniform border of pine bark around the outside of the garden. Find how wide the border should be if he has enough pine bark to cover 336 square feet.

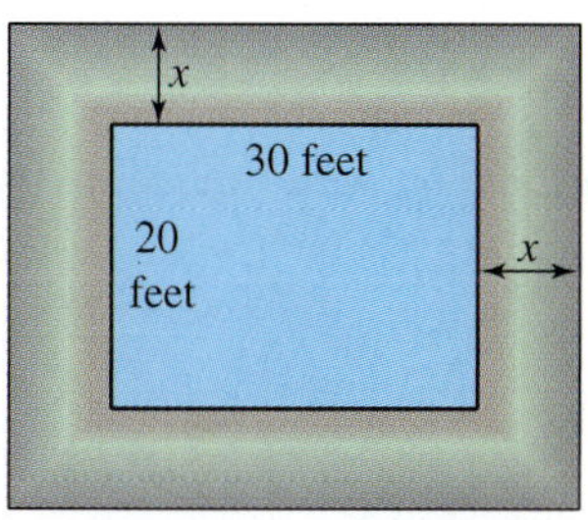


**81.** While hovering near the top of Ribbon Falls in Yosemite National Park at 1600 feet, a helicopter pilot accidentally drops his sunglasses. The height $h(t)$ of the sunglasses after $t$ seconds is given by the polynomial function

$$h(t) = -16t^2 + 1600$$

When will the sunglasses hit the ground?

**82.** After $t$ seconds, the height $h(t)$ of a model rocket launched from the ground into the air is given by the function

$$h(t) = -16t^2 + 80t$$

Find how long it takes the rocket to reach a height of 96 feet.

△ **83.** The floor of a shed has an area of 91 square feet. The floor is in the shape of a rectangle whose length is 6 feet more than the width. Find the length and the width of the floor of the shed.

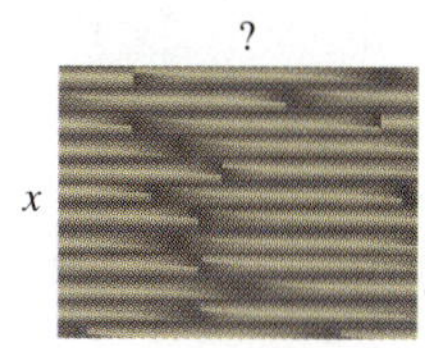


△ **84.** A vegetable garden with an area of 143 square feet is to be fertilized. If the width of the garden is 2 feet less than the length, find the dimensions of the garden.

**85.** The function $W(x) = 0.5x^2$ gives the number of servings of wedding cake that can be obtained from a two-layer $x$-inch square wedding cake tier. What size square wedding cake tier is needed to serve 50 people? (*Source:* Based on data from the *Wilton 2000 Yearbook of Cake Decorating*)

**86.** Use the function in Exercise 85 to determine what size wedding cake tier is needed to serve 200 people.

*Match each polynomial function with its graph (A–F). See Example 10.*

**87.** $f(x) = (x - 2)(x + 5)$

**88.** $g(x) = (x + 1)(x - 6)$

**89.** $h(x) = x(x + 3)(x - 3)$

**90.** $F(x) = (x + 1)(x - 2)(x + 5)$

**91.** $G(x) = 2x^2 + 9x + 4$

**92.** $H(x) = 2x^2 - 7x - 4$

**A**

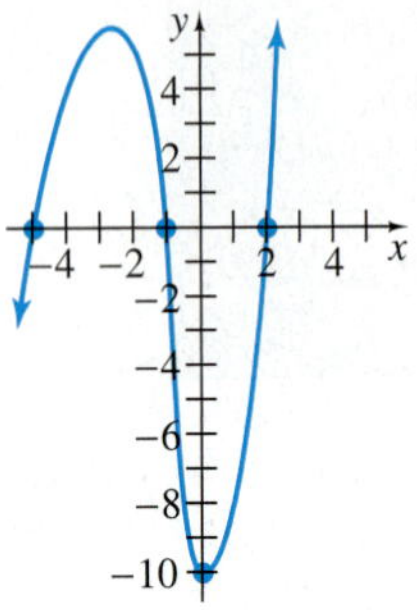


**B**

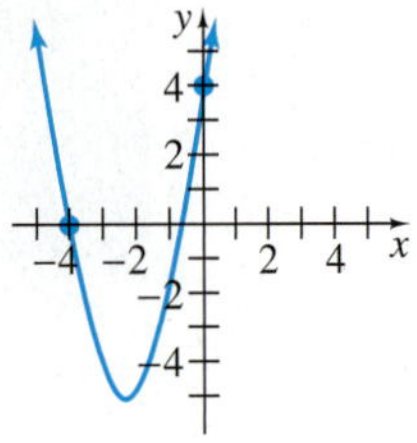


**C**

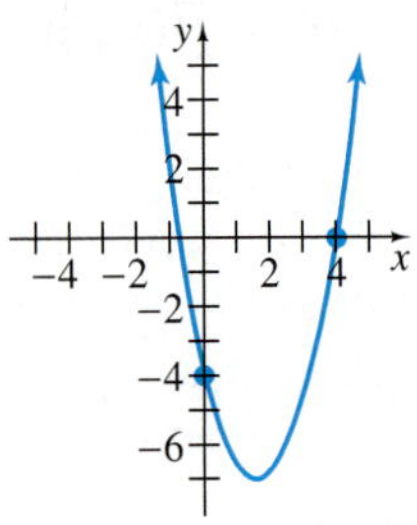


**D**

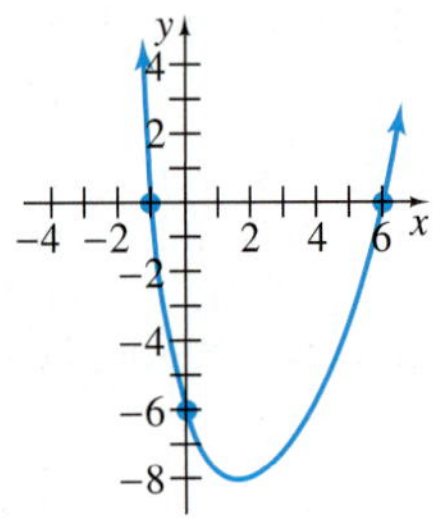


**E**

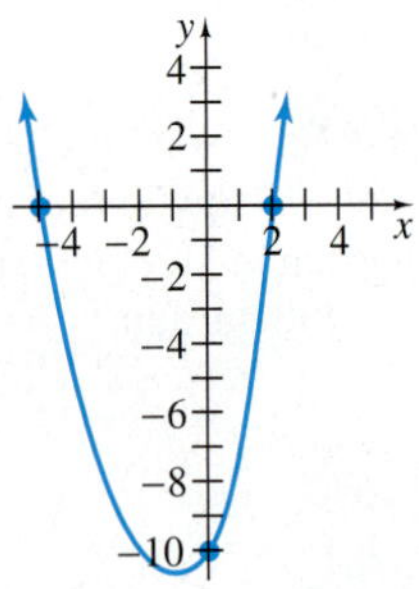


**F**

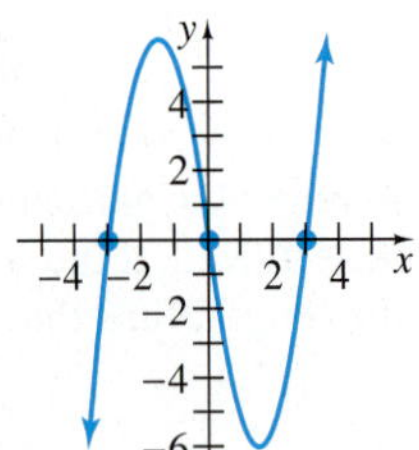


*Write a quadratic equation that has the given numbers as solutions.*

**93.** 5, 3

**94.** 6, 7

**95.** −1, 2

**96.** 4, −3

## REVIEW EXERCISES

*Write the x- and y-intercepts for each graph and determine whether the graph is the graph of a function. See Sections 3.1 and 3.2.*

**97.**

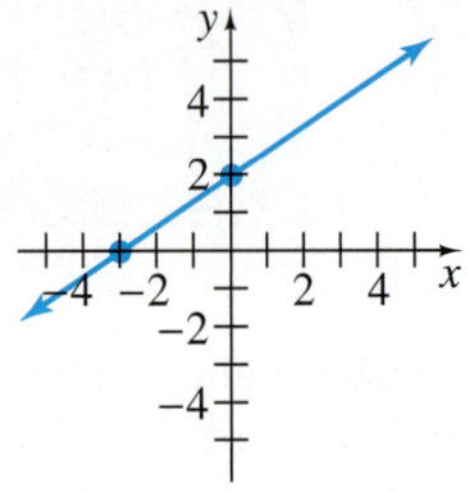


**98.**

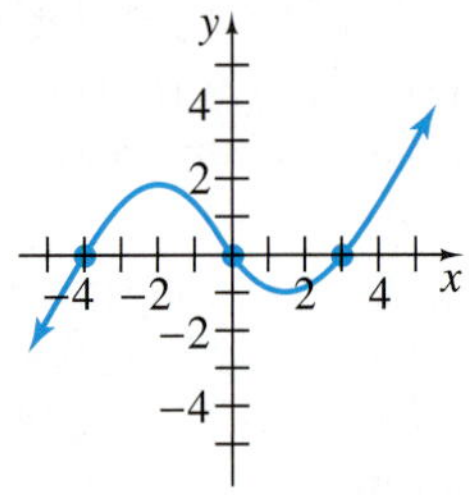


**99.**

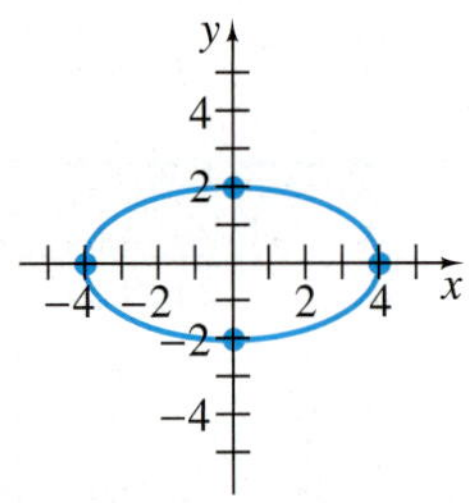


**100.**

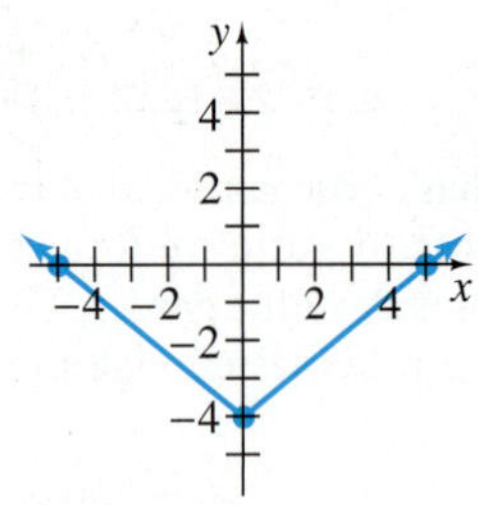


**101.** Draw a function with intercepts (−3, 0), (5, 0), and (0, 4).

**102.** Draw a function with intercepts (−7, 0), $\left(-\frac{1}{2}, 0\right)$, (4, 0), and (0, −1).

## 5.9 AN INTRODUCTION TO GRAPHING POLYNOMIAL FUNCTIONS

CD-ROM SSM SSG Video

**▶ OBJECTIVES**

1. Analyze the graph of a polynomial function.
2. Graph quadratic functions.
3. Find the vertex of a parabola by using the vertex formula.
4. Graph cubic functions.

**1** We discussed linear functions of the form $f(x) = mx + b$ in Chapter 3. In this chapter, we have thus far briefly discussed polynomial functions. In this section, we further discuss polynomial functions. As mentioned earlier, some polynomial functions are given special names according to their degree. For example,

$f(x) = 2x - 6$ is called a **linear function**; its **degree is one.**
$f(x) = 5x^2 - x + 3$ is called a **quadratic function;** its **degree is two.**
$f(x) = 7x^3 + 3x^2 - 1$ is called a **cubic function**; its **degree is three.**
$f(x) = -8x^4 - 3x^3 + 2x^2 + 20$ is called a **quartic function**; its **degree is four.**

All the above functions are also polynomial functions.

Before we practice graphing polynomial functions, let's analyze the graph of a polynomial function.

**Example 1** Given the graph of the function $g(x)$

**a.** Find the domain and the range of the function.
**b.** List the $x$- and $y$-intercepts
**c.** Find the coordinates of the point with the greatest $y$-value.
**d.** Find the coordinates of the point with the least $y$-value.
**e.** List the $x$-values whose $y$-values are equal to 0.
**f.** List the $x$-values whose $y$-values are greater than 0.
**g.** Find the solutions of $g(x) = 0$.

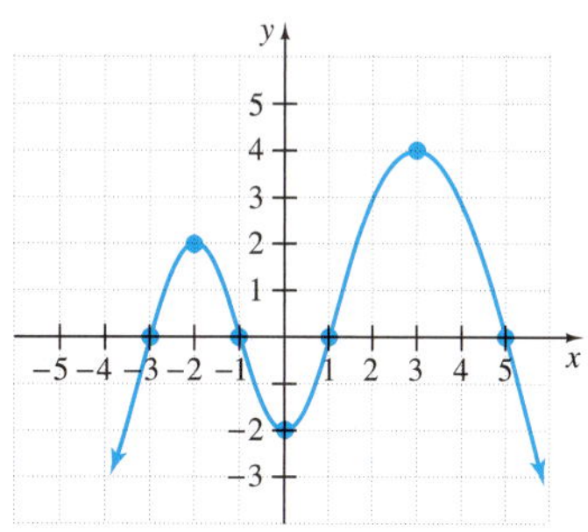


*Solution* **a.** The domain is the set of all real numbers, or in interval notation, $(-\infty, \infty)$. The range is $(-\infty, 4]$.
**b.** The $x$-intercepts are $(-3, 0)$, $(-1, 0)$, $(1, 0)$, and $(5, 0)$. The $y$-intercept is $(0, -2)$.
**c.** The point with the greatest $y$-value corresponds to the "highest" point. This is the point with coordinates $(3, 4)$. (This means that for all real number values for $x$, the greatest $y$-value, or $f(x)$ value, is 4.)

**d.** The point with the least $y$-value corresponds to the "lowest" point. This graph contains no "lowest" point, so there is no point with the least $y$-value.

**e.** The $y$-values are equal to 0 when the graph lies on the $x$-axis. The $x$-values when this occurs are the $x$-intercepts $(-3, 0)$, $(-1, 0)$, $(1, 0)$, and $(5, 0)$. Notice that this tells us that $g(-3) = 0$, $g(-1) = 0$, $g(1) = 0$, and $g(5) = 0$.

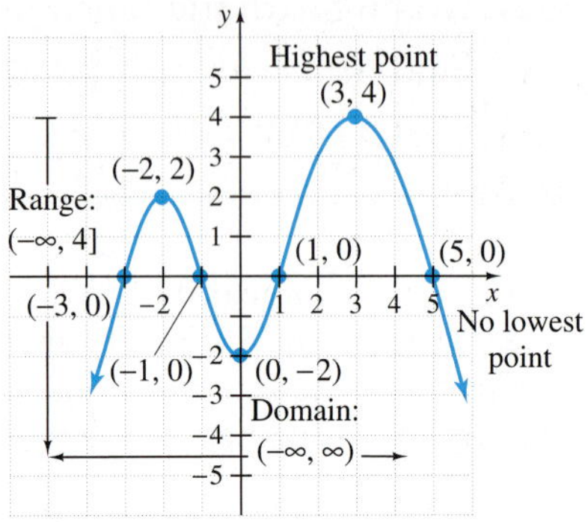


**f.** The $y$-values are greater than 0 when the graph lies above the $x$-axis. The $x$-values when this occurs are between $x = -3$ and $x = -1$ and between $x = 1$ and $x = 5$.

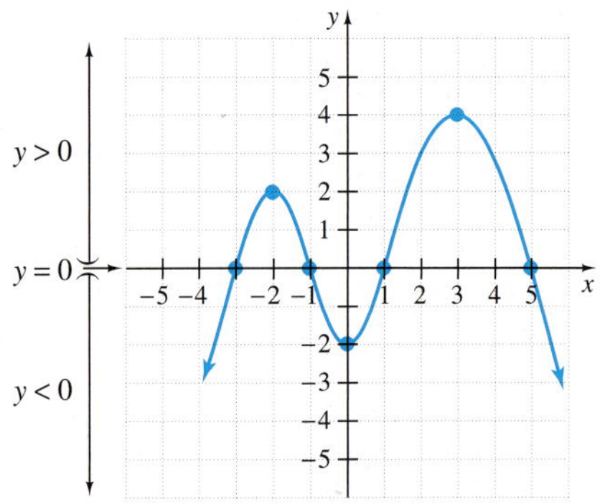


**g.** The solutions of $g(x) = 0$ are the $x$-intercepts of the graph. The $x$-intercepts are $(-3, 0)$, $(-1, 0)$, $(1, 0)$, and $(5, 0)$. This means that when $x = -3, -1, 1,$ or $5$, $y$ or $g(x) = 0$. The solutions are $-3$, $-1$, $1$, and $5$.

The graph of any polynomial function (linear, quadratic, cubic, and so on) can be sketched by plotting a sufficient number of ordered pairs that satisfy the function and connecting them to form a smooth curve. The graph of all polynomial functions will pass the vertical line test since they are graphs of functions. To graph a linear function defined by $f(x) = mx + b$, recall that two ordered pair solutions will suffice since its graph is a line. To graph other polynomial functions, we need to find and plot more ordered pair solutions to ensure a reasonable picture of its graph.

2 Since we know how to graph linear functions (see Chapter 3), we will now graph quadratic functions and discuss special characteristics of their graphs.

**QUADRATIC FUNCTION**

A quadratic function is a function that can be written in the form

$$f(x) = ax^2 + bx + c$$

where $a$, $b$, and $c$ are real numbers and $a \neq 0$.

We know that an equation of the form $f(x) = ax^2 + bx + c$ may be written as $y = ax^2 + bx + c$. Thus, both $f(x) = ax^2 + bx + c$ and $y = ax^2 + bx + c$ define quadratic functions as long as $a$ is not 0.

Recall the graph of the quadratic function defined by $f(x) = x^2$ by plotting points. Choose −3, −2, −1, 0, 1, 2, and 3 as $x$-values, and find corresponding $f(x)$ or $y$-values.

| $x$ | $y = f(x)$ |
|---|---|
| −3 | 9 |
| −2 | 4 |
| −1 | 1 |
| 0 | 0 |
| 1 | 1 |
| 2 | 4 |
| 3 | 9 |

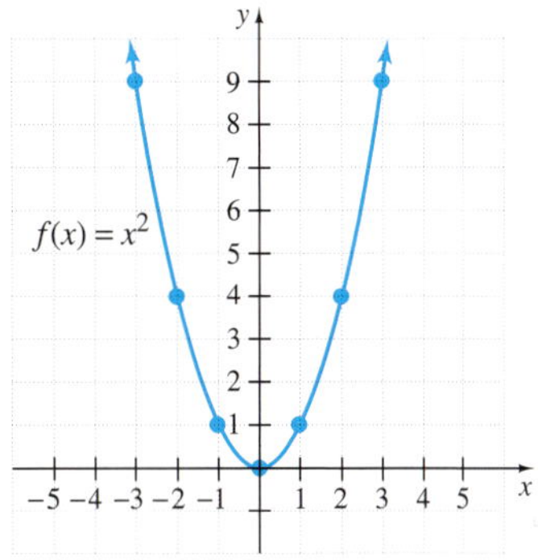


Notice that the graph passes the vertical line test, as it should since it is a function. Recall that this curve is called a **parabola.** The highest point on a parabola that opens downward or the lowest point on a parabola that opens upward is called the **vertex** of the parabola. The vertex of this parabola is $(0, 0)$, the lowest point on the graph. If we fold the graph along the $y$-axis, we can see that the two sides of the graph coincide. This means that this curve is symmetric about the $y$-axis, and the $y$-axis, or the line $x = 0$, is called the **axis of symmetry.** The graph of every quadratic function is a parabola and has an axis of symmetry: the vertical line that passes through the vertex of the parabola.

**Example 2** Graph the quadratic function $f(x) = -x^2 + 2x - 3$ by plotting points.

*Solution* To graph, choose values for $x$ and find corresponding $f(x)$ or $y$-values.

| $x$ | $y = f(x)$ |
|---|---|
| −2 | −11 |
| −1 | −6 |
| 0 | −3 |
| 1 | −2 |
| 2 | −3 |
| 3 | −6 |

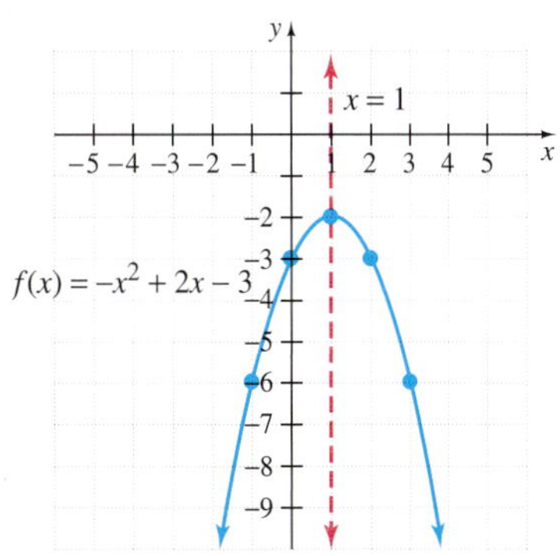


The vertex of this parabola is $(1, -2)$, the highest point on the graph. The vertical line $x = 1$ is the axis of symmetry. Recall that to find the $x$-intercepts of a graph, let $f(x)$ or $y = 0$. Since this graph has no $x$-intercepts, it means that $0 = -x^2 + 2x - 3$ has no real number solutions.

Notice that the parabola $f(x) = -x^2 + 2x - 3$ opens downward, whereas $f(x) = x^2$ opens upward. When the equation of a quadratic function is written in the form $f(x) = ax^2 + bx + c$, recall that the coefficient of the squared variable $a$ determines whether the parabola opens downward or upward. If $a > 0$, the parabola opens upward, and if $a < 0$, the parabola opens downward.

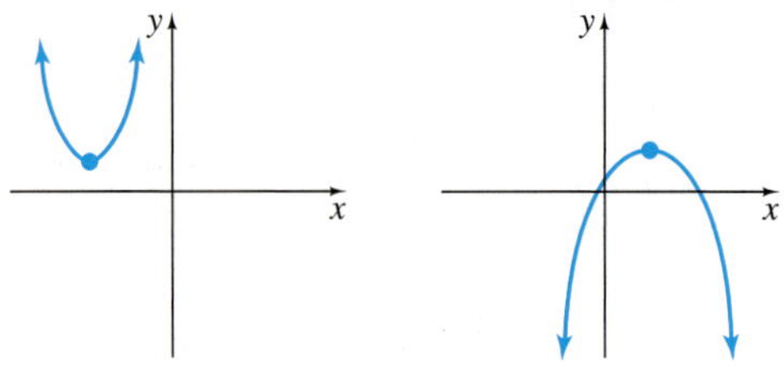


$f(x) = ax^2 + bx + c$; $a > 0$; opens upward

$f(x) = ax^2 + bx + c$; $a < 0$; opens downward

3 In both $f(x) = x^2$ and $f(x) = -x^2 + 2x - 3$, the vertex happens to be one of the points we chose to plot. Since this is not always the case, and since plotting the vertex allows us to draw the graph quickly, we need a consistent method for finding the vertex. One method is to use the following formula, which we shall derive in Chapter 8.

### VERTEX FORMULA

The graph of $f(x) = ax^2 + bx + c, a \neq 0$, is a parabola with vertex

$$\left(\frac{-b}{2a}, f\left(\frac{-b}{2a}\right)\right)$$

We can also find the $x$- and $y$-intercepts of a parabola to aid in graphing. Recall that $x$-intercepts of the graph of any equation may be found by letting $y = 0$ in the equation and solving for $x$. Also, $y$-intercepts may be found by letting $x = 0$ in the equation and solving for $y$ or $f(x)$.

**Example 3** Graph $f(x) = x^2 + 2x - 3$. Find the vertex and any intercepts.

*Solution* To find the vertex, use the vertex formula. For the function $f(x) = x^2 + 2x - 3$, $a = 1$ and $b = 2$. Thus,

$$x = \frac{-b}{2a} = \frac{-2}{2(1)} = -1$$

Next find $f(-1)$.

$$\begin{aligned} f(-1) &= (-1)^2 + 2(-1) - 3 \\ &= 1 - 2 - 3 \\ &= -4 \end{aligned}$$

The vertex is $(-1, -4)$, and since $a = 1$ is greater than 0, this parabola opens upward. This parabola will have two $x$-intercepts because its vertex lies below the $x$-axis and it opens upward. To find the $x$-intercepts, let $y$ or $f(x) = 0$ and solve for $x$.

$$f(x) = x^2 + 2x - 3$$
$$0 = x^2 + 2x - 3 \quad \text{Let } f(x) = 0.$$
$$0 = (x + 3)(x - 1) \quad \text{Factor.}$$
$$x + 3 = 0 \quad \text{or} \quad x - 1 = 0 \quad \text{Set each factor equal to 0.}$$
$$x = -3 \quad \text{or} \quad x = 1 \quad \text{Solve.}$$

The $x$-intercepts are $(-3, 0)$ and $(1, 0)$.

To find the $y$-intercept, let $x = 0$.

$$f(x) = x^2 + 2x - 3$$
$$f(0) = 0^2 + 2(0) - 3$$
$$f(0) = -3$$

The $y$-intercept is $(0, -3)$

Now plot these points and connect them with a smooth curve.

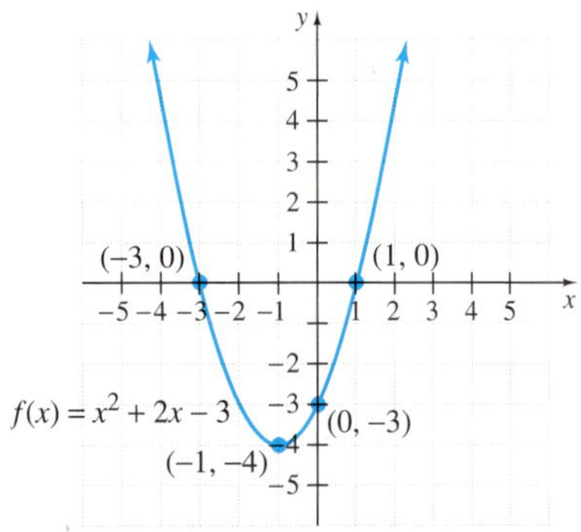


**HELPFUL HINT**

Not all graphs of parabolas have $x$-intercepts. To see this, first plot the vertex of the parabola and decide whether the parabola opens upward or downward. Then use this information to decide whether the graph of the parabola has $x$-intercepts.

**Example 4** Graph $f(x) = 3x^2 - 12x + 13$. Find the vertex and any intercepts.

*Solution* To find the vertex, use the vertex formula. For the function $y = 3x^2 - 12x + 13$, $a = 3$ and $b = -12$. Thus,

$$x = \frac{-b}{2a} = \frac{-(-12)}{2(3)} = \frac{12}{6} = 2$$

Next find $f(2)$.

$$f(2) = 3(2)^2 - 12(2) + 13$$
$$= 3(4) - 24 + 13$$
$$= 1$$

The vertex is $(2, 1)$. Also, this parabola opens upward, since $a = 3$, which is greater than 0. Notice that this parabola has no $x$-intercepts: Its vertex lies above the $x$-axis, and it opens upward.

To find the $y$-intercept, let $x = 0$.

$$\begin{aligned} f(0) &= 3(0)^2 - 12(0) + 13 \\ &= 0 - 0 + 13 \\ &= 13 \end{aligned}$$

The $y$-intercept is $(0, 13)$. Use this information along with symmetry of a parabola to sketch the graph of $f(x) = 3x^2 - 12x + 13$.

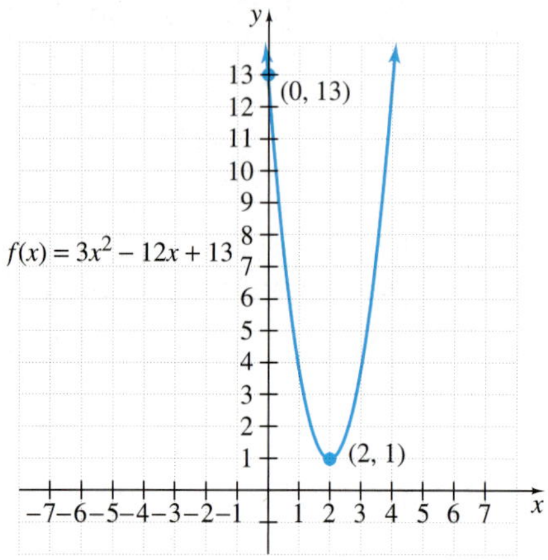


In Section 8.5, we study the graphing of quadratic functions further.

4 To sketch the graph of a cubic function, we again plot points and then connect the points with a smooth curve.

When graphing cubic functions, keep in mind their general shape from Section 5.3.

***Graph of a Polynomial Function of Degree 3***

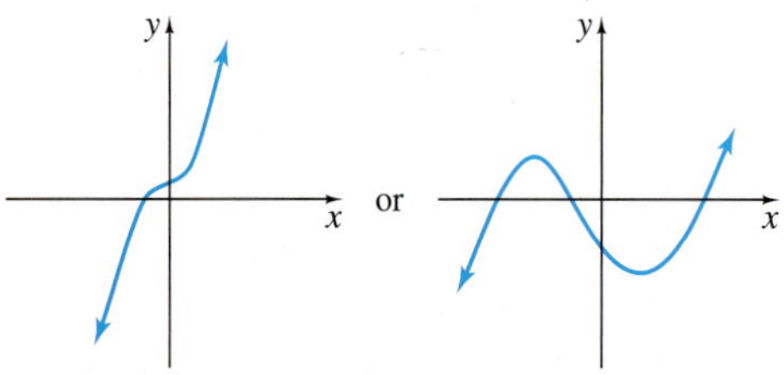


Coefficient of $x^3$ is a positive number.

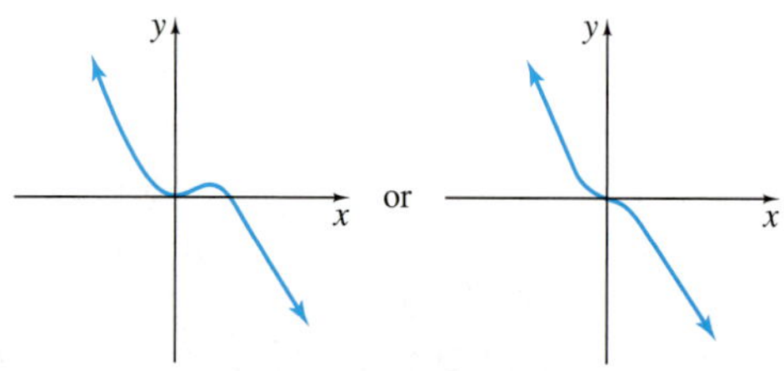


Coefficient of $x^3$ is a negative number.

**Example 5** Graph $f(x) = x^3 - 4x$. Find any intercepts.

*Solution* To find $x$-intercepts, let $y$ or $f(x) = 0$ and solve for $x$.

$$f(x) = x^3 - 4x$$

$$0 = x^3 - 4x \quad \text{Let } f(x) = 0.$$

$$0 = x(x^2 - 4)$$

$$0 = x(x + 2)(x - 2) \quad \text{Factor.}$$

$$x = 0 \quad \text{or} \quad x + 2 = 0 \quad \text{or} \quad x - 2 = 0 \quad \text{Set each factor equal to 0.}$$

$$x = 0 \quad \text{or} \quad x = -2 \quad \text{or} \quad x = 2 \quad \text{Solve.}$$

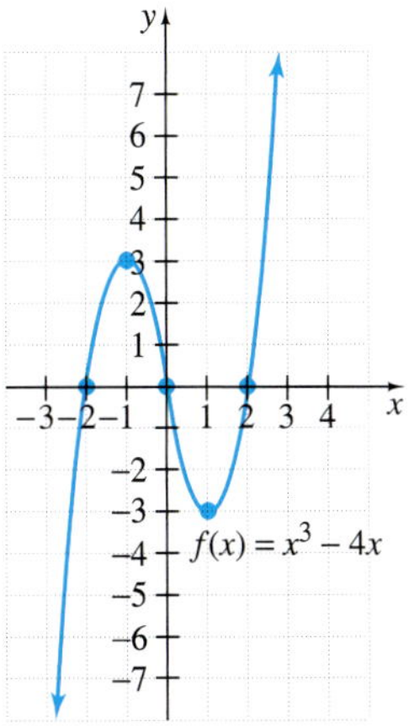


This graph has three $x$-intercepts. They are $(0, 0)$, $(-2, 0)$ and $(2, 0)$.

To find the $y$-intercept, let $x = 0$.

$$f(0) = 0^3 - 4(0) = 0$$

The $y$-intercept is $(0, 0)$. Next select some $x$-values and find their corresponding $f(x)$ or $y$-values.

$$f(x) = x^3 - 4x$$
$$f(-3) = (-3)^3 - 4(-3) = -27 + 12 = -15$$
$$f(-1) = (-1)^3 - 4(-1) = -1 + 4 = 3$$
$$f(1) = 1^3 - 4(1) = 1 - 4 = -3$$
$$f(3) = 3^3 - 4(3) = 27 - 12 = 15$$

| $x$ | $f(x)$ |
|---|---|
| −3 | −15 |
| −1 | 3 |
| 1 | −3 |
| 3 | 15 |

Plot the intercepts and points and connect them with a smooth curve. ■

**HELPFUL HINT**
When a graph has an $x$-intercept of $(0, 0)$, notice that the $y$-intercept will also be $(0, 0)$.

**HELPFUL HINT**
If unsure about the graph of a function, plot more points.

**Example 6** Graph $f(x) = -x^3$. Find any intercepts.

*Solution* To find $x$-intercepts, let $y$ or $f(x) = 0$ and solve for $x$.

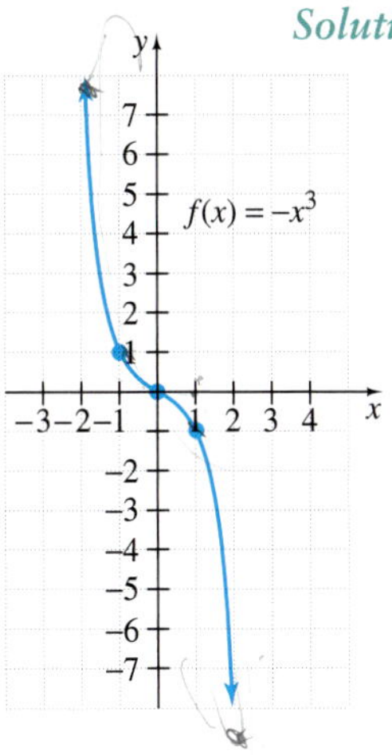


$$f(x) = -x^3$$
$$0 = -x^3$$
$$0 = x$$

The only $x$-intercept is $(0, 0)$. This means that the $y$-intercept is 0 also.

Next choose some $x$-values and find corresponding $y$-values.

$$f(x) = -x^3$$
$$f(-2) = -(-2)^3 = 8$$
$$f(-1) = -(-1)^3 = 1$$
$$f(1) = -(1)^3 = -1$$
$$f(2) = -2^3 = -8$$

| $x$ | $f(x)$ |
|---|---|
| −2 | 8 |
| −1 | 1 |
| 1 | −1 |
| 2 | −8 |

Plot the points and sketch the graph of $f(x) = -x^3$. ■

## MENTAL MATH

*State whether the graph of each quadratic function, a parabola, opens upward or downward.*

**1.** $f(x) = 2x^2 + 7x + 10$
**2.** $f(x) = -3x^2 - 5x$
**3.** $f(x) = -x^2 + 5$
**4.** $f(x) = x^2 + 3x + 7$

## Exercise Set 5.9

*For the graph of each function $f(x)$, answer the following. See Example 1.*

*a.* *Find the domain and the range of the function.*
*b.* *List the x- and y-intercepts.*
*c.* *Find the coordinates of the point with the greatest y-value.*
*d.* *Find the coordinates of the point with the least y-value.*
*e.* *List the x-values whose y-values are equal to 0.*
*f.* *List the x-values whose y-values are greater than 0.*
*g.* *Find the solutions of $f(x) = 0$.*

**1.**

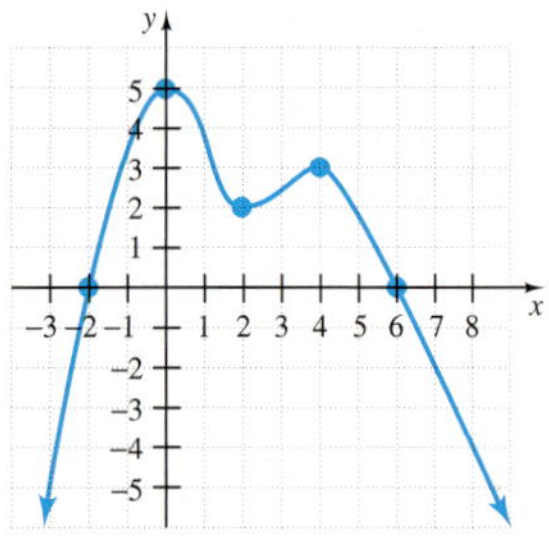


**2.**

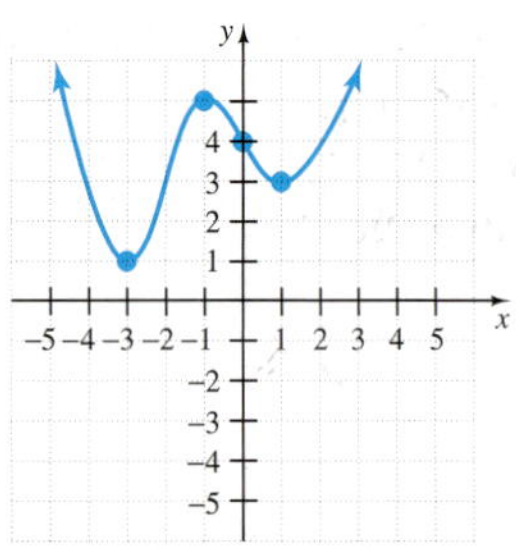


**3.** The graph in Example 3 of this section.

**4.** The graph in Example 4 of this section.

**5.** The graph in Example 5 of this section.

**6.** The graph in Example 6 of this section.

*Graph each quadratic function by plotting points. See Example 2.*

**7.** $f(x) = 2x^2$ **8.** $f(x) = -3x^2$

**9.** $f(x) = x^2 + 1$ **10.** $f(x) = x^2 - 2$

**11.** $f(x) = -x^2$ **12.** $f(x) = \frac{1}{2}x^2$

*Find the vertex of the graph of each function. See Examples 3 and 4.*

**13.** $f(x) = x^2 + 8x + 7$ **14.** $f(x) = x^2 + 6x + 5$

**15.** $f(x) = 3x^2 + 6x + 4$ **16.** $f(x) = -2x^2 + 2x + 1$

**17.** $f(x) = -x^2 + 10x + 5$ **18.** $f(x) = -x^2 - 8x + 2$

**19.** If the vertex of a parabola lies below the $x$-axis and the parabola opens upward, how many $x$-intercepts will the graph have?

**20.** If the vertex of a parabola lies below the $x$-axis and the parabola opens downward, how many $x$-intercepts will the graph have?

**21.** If the vertex of a parabola lies above the $x$-axis and the parabola opens upward, how many $x$-intercepts will the graph have?

**22.** If the vertex of a parabola lies above the $x$-axis and the parabola opens downward, how many $x$-intercepts will the graph have?

**23.** If the vertex of a parabola is the origin, how many $x$-intercepts and how many $y$-intercepts will the graph have?

*Graph each quadratic function. Find and label the vertex and intercepts. See Examples 3 and 4.*

**24.** $f(x) = x^2 + 8x + 7$ **25.** $f(x) = x^2 + 6x + 5$

**26.** $f(x) = x^2 - 2x - 24$ **27.** $f(x) = x^2 - 12x + 35$

**28.** $f(x) = 2x^2 - 6x$ **29.** $f(x) = -3x^2 + 6x$

*Graph each cubic function. Find any intercepts. See Examples 5 and 6.*

**30.** $f(x) = 4x^3 - 9x$ **31.** $f(x) = 2x^3 - 5x^2 - 3x$

**32.** $f(x) = x^3 + 3x^2 - x - 3$

**33.** $f(x) = x^3 + x^2 - 4x - 4$

**34.** Can the graph of a function ever have more than one $y$-intercept point? Why?

**35.** In general, is there a limit to the number of $x$-intercepts for the graph of a function?

*Graph each function. Find intercepts. If the function is a quadratic function, find the vertex.*

**36.** $f(x) = x^2 + 4x - 5$ **37.** $f(x) = x^2 + 2x - 3$

**38.** $f(x) = (x - 2)(x + 2)(x + 1)$

**39.** $f(x) = x^3 - 4x^2 + 3x$

**40.** $f(x) = x^2 + 1$ **41.** $f(x) = x^2 + 4$

**42.** $f(x) = -5x^2 + 5x$ **43.** $f(x) = 3x^2 - 12x$

**44.** $f(x) = x^3 - 9x$ **45.** $f(x) = x^3 + x^2 - 12x$

**46.** $f(x) = -x^3 - x^2 + 2x$

**47.** $f(x) = x^3 + x^2 - 9x - 9$

**48.** $f(x) = x^2 - 4x + 4$ **49.** $f(x) = x^2 - 2x + 1$

**50.** $f(x) = -x^3 + x$

**51.** $f(x) = x^2 + 6x$

**52.** $f(x) = 2x^2 - x - 3$

**53.** $f(x) = (x + 2)(x - 2)$

**54.** $f(x) = -x^3 + 3x^2 + x - 3$

**55.** $f(x) = -x^3 + 25x$

**56.** $f(x) = x^2 - 10x + 26$

**57.** $f(x) = x^2 + 2x + 4$

**58.** $f(x) = x(x - 4)(x + 2)$

**59.** $f(x) = 3x(x - 3)(x + 5)$

**60.** $g(x) = x(x - 2)(x + 3)(x + 5)$

**61.** $h(x) = (x - 4)(x - 2)(2x + 1)(x + 3)$

*Use a graphing calculator to verify the graph in each exercise.*

**62.** Exercise 36

**63.** Exercise 37

**64.** Exercise 54

**65.** Exercise 55

*Use a graphing calculator to approximate all x-intercepts to the nearest tenth.*

**66.** $F(x) = -x^4 + 2.1x^2 + 5.6$

**67.** $G(x) = x^4 - 6.2x^2 - 6.2$

## REVIEW EXERCISES

*Simplify each fraction. See Sections 5.1 and 5.2.*

**68.** $-\dfrac{8}{10}$

**69.** $-\dfrac{45}{100}$

**70.** $\dfrac{x^7y^{10}}{x^3y^{15}}$

**71.** $\dfrac{a^{14}b^2}{ab^4}$

**72.** $\dfrac{7n^{-9}m^{-2}}{14nm^{-5}}$

**73.** $\dfrac{20x^{-3}y^5}{25y^{-2}x}$

For additional Chapter Projects, visit the Real World Activities Website by going to http://www.prenhall.com/martin-gay.

# 5 CHAPTER PROJECT

## *Investigating Earth's Water*

Earth is covered by water. In fact, oceans cover nearly three-fourths of the surface of Earth. However, oceans aren't the only source of Earth's water. The melting of one of the other main sources of Earth's water, icecaps and glaciers, is expected to contribute to a global rise in ocean level due to global warming over the next 100 years. In this project, you will have the opportunity to investigate where Earth's water exists and how the ocean level will change. This project may be completed by working in groups or individually.

1. Refer to Table 1. Which accounts for more of Earth's water: groundwater or icecaps and glaciers?
2. Find the total volume of water that exists on planet Earth. Add this figure to the table.
3. Using the total you computed in Question 2, complete the percent column of the table. Discuss your findings.

Widespread industrialization during the nineteenth and twentieth centuries has led to an increase in the presence of carbon dioxide, methane, nitrous oxide, and chlorofluorocarbons in Earth's atmosphere. These so-called greenhouse gases are believed by some scientists to be responsible for an increase in the average global temperature of 0.2°C to 0.3°C in the last half of the twentieth century. If this global warming trend continues, one of its consequences may be a global increase in the level of the oceans. An overall increase in ocean level will be due to changes in icecaps and glaciers, as well as thermal expansion. Higher global temperatures lead to warming in the top layers of the ocean, causing the water to expand and elevate the sea level.

**TABLE 1. WHERE EARTH'S WATER EXISTS**

| | *Water Volume (cubic kilometers)* | *Percent* |
|---|---|---|
| Atmosphere | $1.3 \times 10^4$ | |
| Average in stream channels | $1.0 \times 10^3$ | |
| Freshwater lakes | $1.2 \times 10^5$ | |
| Groundwater | $8.3 \times 10^6$ | |
| Icecaps and glaciers | $2.9 \times 10^7$ | |
| Oceans | $1.32 \times 10^9$ | |
| Saline lakes and inland seas | $1.0 \times 10^5$ | |
| Water in soil above groundwater | $6.7 \times 10^5$ | |
| Total | | |

(*Source:* Data from B.J. Skinner, *Earth Resources*, 2nd Ed., Prentice Hall, 1976)

Table 2 lists each contributor to overall changes in ocean level along with a polynomial model describing the projected rise $y$ (in centimeters) each is expected to contribute $x$ years after 2000.

4. By 2020, how much will the ocean level have risen due to thermal expansion?
5. Using the polynomial models given in Table 2, find a single polynomial model that gives the overall rise in ocean level from 1990 to 2100.
6. Using your model from Question 5, find the projected overall increase in ocean level for
   **a.** 2025 **b.** 2050
   **c.** 2075 **d.** 2100
7. Discuss the impact of your findings in Question 6.

**TABLE 2. PROJECTIONS OF GLOBAL OCEAN LEVEL RISE BY CONTRIBUTOR, 1990–2100**

Alpine glaciers:
$$y = 0.0006x^2 + 0.0936x + 0.8788$$
Greenland ice sheet:
$$y = 0.0004x^2 + 0.0164x + 0.1212$$
Antarctica ice sheet:
$$y = -0.0001x^2 - 0.0002x + 0.0076$$
Thermal expansion:
$$y = 0.0011x^2 + 0.1564x + 1.4545$$

(*Source:* Based on data from Frederick K. Lutgens, Edward J. Tarbuck, *The Atmosphere: An Introduction to Meteorology*, 7th Ed., Prentice Hall, 1998)

## CHAPTER 5 VOCABULARY CHECK

*Fill in each blank with one of the words or phrases listed below.*

| | | | | | | |
|---|---|---|---|---|---|---|
| quadratic equation | scientific notation | polynomial | exponents | 1 | 0 | monomial |
| binomial | trinomial | degree of a polynomial | degree of a term | | | factoring |

1. A __________ is a finite sum of terms in which all variables are raised to nonnegative integer powers and no variables appear in any denominator.
2. __________ is the process of writing a polynomial as a product.
3. __________ are used to write repeated factors in a more compact form.
4. The __________ is the sum of the exponents on the variables contained in the term.
5. A __________ is a polynomial with one term.
6. If $a$ is not 0, $a^0 =$ ___.
7. A __________ is a polynomial with three terms.
8. A polynomial equation of degree 2 is also called a __________.
9. A positive number is written in __________ if it is written as the product of a number $a$, such that $1 \leq a < 10$ and a power of 10.
10. The __________ is the largest degree of all of its terms.
11. A __________ is a polynomial with two terms.
12. If $a$ and $b$ are real numbers and $a \cdot b =$ ___, then $a = 0$ or $b = 0$.

# CHAPTER 5 HIGHLIGHTS

| DEFINITIONS AND CONCEPTS | EXAMPLES |
|---|---|
| **Section 5.1 Exponents and Scientific Notation** | |
| Product rule: $a^m \cdot a^n = a^{m+n}$ | $x^2 \cdot x^3 = x^5$ |
| Zero exponent: $a^0 = 1, a \neq 0$ | $7^0 = 1, (-10)^0 = 1$ |
| Quotient rule: $\frac{a^m}{a^n} = a^{m-n}$ | $\frac{y^{10}}{y^4} = y^{10-4} = y^6$ |
| Negative exponent: $a^{-n} = \frac{1}{a^n}$ | $3^{-2} = \frac{1}{3^2} = \frac{1}{9}, \frac{x^{-5}}{x^{-7}} = x^{-5-(-7)} = x^2$ |
| A positive number is written in **scientific notation** if it is written as the product of a number $a$, where $1 \leq a < 10$, and an integer power of 10: $a \times 10^r$. | ***Numbers written in scientific notation*** <br> $568{,}000 = 5.68 \times 10^5$ <br> $0.0002117 = 2.117 \times 10^{-4}$ |
| **Section 5.2 More Work with Exponents and Scientific Notation** | |
| Power rules: <br> $(a^m)^n = a^{m \cdot n}$ <br> $(ab)^m = a^m b^m$ <br> $\left(\frac{a}{b}\right)^n = \frac{a^n}{b^n}$ | $(7^8)^2 = 7^{16}$ <br> $(2y)^3 = 2^3y^3 = 8y^3$ <br> $\left(\frac{5x^{-3}}{x^2}\right)^{-2} = \frac{5^{-2}x^6}{x^{-4}}$ <br> $= 5^{-2} \cdot x^{6-(-4)}$ <br> $= \frac{x^{10}}{5^2}$, or $\frac{x^{10}}{25}$ |
| **Section 5.3 Polynomials and Polynomial Functions** | |
| A **polynomial** is a finite sum of terms in which all variables have exponents raised to nonnegative integer powers and no variables appear in the denominator. | ***Polynomials*** <br> $1.3x^2$ (monomial) <br> $-\frac{1}{3}y + 5$ (binomial) <br> $6z^2 - 5z + 7$ (trinomial) |
| A function $P$ is a **polynomial function** if $P(x)$ is a polynomial. | For the polynomial function <br> $P(x) = -x^2 + 6x - 12$, find $P(-2)$ <br> $P(-2) = -(-2)^2 + 6(-2) - 12 = -28.$ |
| To add polynomials, combine all like terms. | ***Add*** <br> $(3y^2x - 2yx + 11) + (-5y^2x - 7)$ <br> $= -2y^2x - 2yx + 4$ |
| To subtract polynomials, change the signs of the terms of the polynomial being subtracted, then add. | ***Subtract*** <br> $(-2z^3 - z + 1) - (3z^3 + z - 6)$ <br> $= -2z^3 - z + 1 - 3z^3 - z + 6$ <br> $= -5z^3 - 2z + 7$ |

| DEFINITIONS AND CONCEPTS | EXAMPLES |
|---|---|
| **Section 5.4 Multiplying Polynomials** | |
| To multiply two polynomials, use the distributive property and multiply each term of one polynomial by each term of the other polynomial; then combine like terms. | ***Multiply*** $(x^2 - 2x)(3x^2 - 5x + 1)$ $= 3x^4 - 5x^3 + x^2 - 6x^3 + 10x^2 - 2x$ $= 3x^4 - 11x^3 + 11x^2 - 2x$ |
| ***Special products*** $(a + b)^2 = a^2 + 2ab + b^2$ $(a - b)^2 = a^2 - 2ab + b^2$ $(a + b)(a - b) = a^2 - b^2$ | $(3m + 2n)^2 = 9m^2 + 12mn + 4n^2$ $(z^2 - 5)^2 = z^4 - 10z^2 + 25$ $(7y + 1)(7y - 1) = 49y^2 - 1$ |
| The FOIL method may be used when multiplying two binomials. | ***Multiply*** $(x^2 + 5)(2x^2 - 9)$ F O I L $= x^2(2x^2) + x^2(-9) + 5(2x^2) + 5(-9)$ $= 2x^4 - 9x^2 + 10x^2 - 45$ $= 2x^4 + x^2 - 45$ |
| **Section 5.5 The Greatest Common Factor and Factoring by Grouping** | |
| The greatest common factor (GCF) of the terms of a polynomial is the product of the GCF of the numerical coefficients and the GCF of the variable factors. | Factor: $14xy^3 - 2xy^2 = 2 \cdot 7 \cdot x \cdot y^3 - 2 \cdot x \cdot y^2$. The GCF is $2 \cdot x \cdot y^2$, or $2xy^2$. $14xy^3 - 2xy^2 = 2xy^2(7y - 1)$ |
| To factor a polynomial by grouping, group the terms so that each group has a common factor. Factor out these common factors. Then see if the new groups have a common factor. | Factor $x^4y - 5x^3 + 2xy - 10$. $x^4y - 5x^3 + 2xy - 10 = x^3(xy - 5) + 2(xy - 5)$ $= (xy - 5)(x^3 + 2)$ |
| **Section 5.6 Factoring Trinomials** | |
| To factor $ax^2 + bx + c$, | Factor $28x^2 - 27x - 10$. |
| **Step 1:** Write all pairs of factors of $ax^2$. | Factors of $28x^2$: $28x$ and $x$, $2x$ and $14x$, $4x$ and $7x$. |
| **Step 2:** Write all pairs of factors of $c$. | Factors of $-10$: $-2$ and 5, 2 and $-5$, $-10$ and 1, 10 and $-1$. |
| **Step 3:** Try combinations of these factors until the middle term $bx$ is found. | $28x^2 - 27x - 10 = (7x + 2)(4x - 5)$ |
| **Section 5.7 Factoring by Special Products and Factoring Strategies** | |
| ***Perfect square trinomial*** | ***Factor*** |
| $a^2 + 2ab + b^2 = (a + b)^2$ | $25x^2 + 30x + 9 = (5x + 3)^2$ |
| $a^2 - 2ab + b^2 = (a - b)^2$ | $49z^2 - 28z + 4 = (7z - 2)^2$ |

*(continued)*

| DEFINITIONS AND CONCEPTS | EXAMPLES |
|---|---|
| **Section 5.7 Factoring by Special Products and Factoring Strategies** | |
| Difference of two squares<br>$a^2 - b^2 = (a + b)(a - b)$ | $36x^2 - y^2 = (6x + y)(6x - y)$ |
| Sum and difference of two cubes<br>$a^3 + b^3 = (a + b)(a^2 - ab + b^2)$<br>$a^3 - b^3 = (a - b)(a^2 + ab + b^2)$ | $8y^3 + 1 = (2y + 1)(4y^2 - 2y + 1)$<br>$27p^3 - 64q^3 = (3p - 4q)(9p^2 + 12pq + 16q^2)$ |
| ***To factor a polynomial***<br>**Step 1:** Factor out the GCF.<br>**Step 2:** If the polynomial is a binomial, see if it is a difference of two squares or a sum or difference of two cubes. If it is a trinomial, see if it is a perfect square trinomial. If not, try factoring by methods of Section 5.6. If it is a polynomial with 4 or more terms, try factoring by grouping.<br>**Step 3:** See if any factors can be factored further. | Factor $10x^4y + 5x^2y - 15y$.<br>$10x^4y + 5x^2y - 15y = 5y(2x^4 + x^2 - 3)$<br>$= 5y(2x^2 + 3)(x^2 - 1)$<br>$= 5y(2x^2 + 3)(x + 1)(x - 1)$ |
| **Section 5.8 Solving Equations by Factoring and Problem Solving** | |
| ***To solve polynomial equations by factoring:***<br>**Step 1:** Write the equation so that one side is 0.<br>**Step 2:** Factor the polynomial completely.<br>**Step 3:** Set each factor equal to 0.<br>**Step 4:** Solve the resulting equations.<br>**Step 5:** Check each solution. | Solve<br>$2x^3 - 5x^2 = 3x$<br>$2x^3 - 5x^2 - 3x = 0$<br>$x(2x + 1)(x - 3) = 0$<br>$x = 0$ or $2x + 1 = 0$ or $x - 3 = 0$<br>$x = 0$ or $x = -\frac{1}{2}$ or $x = 3$<br>The solutions are $0, -\frac{1}{2}$, and 3. |
| **Section 5.9 An Introduction to Graphing Polynomial Functions** | |
| To graph a polynomial function, find and plot $x$- and $y$-intercepts and a sufficient number of ordered pair solutions. Then connect the plotted points with a smooth curve. | Graph $f(x) = x^3 + 2x^2 - 3x$.<br>$0 = x^3 + 2x^2 - 3x$<br>$0 = x(x - 1)(x + 3)$<br>$x = 0$ or $x = 1$ or $x = -3$<br>The $x$-intercepts are $(0, 0)$, $(1, 0)$, and $(-3, 0)$.<br>$f(0) = 0^3 + 2 \cdot 0^2 - 3 \cdot 0 = 0.$<br>The $y$-intercept is $(0, 0)$ |

*(continued)*

| DEFINITIONS AND CONCEPTS | EXAMPLES |
|---|---|

### Section 5.9 An Introduction to Graphing Polynomial Functions

| $x$ | $f(x)$ |
|---|---|
| $-4$ | $-20$ |
| $-2$ | $6$ |
| $-1$ | $4$ |
| $\frac{1}{2}$ | $-\frac{7}{8}$ |
| $2$ | $10$ |

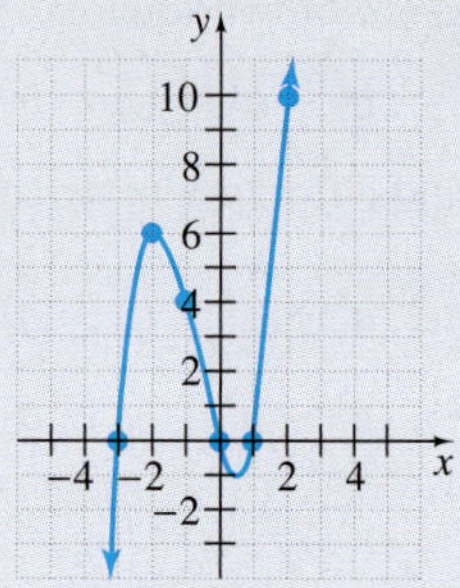


A quadratic function is a function that can be written in the form

$$f(x) = ax^2 + bx + c, a \neq 0$$

The graph of this quadratic function is a parabola with vertex $\left(\frac{-b}{2a}, f\left(\frac{-b}{2a}\right)\right)$.

Find the vertex of the graph of the quadratic function

$$f(x) = 2x^2 - 8x + 1$$

Here $a = 2$ and $b = -8$.

$$\frac{-b}{2a} = \frac{-(-8)}{2 \cdot 2} = 2$$

$$f(2) = 2 \cdot 2^2 - 8 \cdot (2) + 1 = -7$$

The vertex has coordinates $(2, -7)$.

## CHAPTER 5 REVIEW

**(5.1)** *Evaluate.*

**1.** $(-2)^2$ **2.** $(-3)^4$

**3.** $-2^2$ **4.** $-3^4$

**5.** $8^0$ **6.** $-9^0$

**7.** $-4^{-2}$ **8.** $(-4)^{-2}$

*Simplify each expression. Use only positive exponents.*

**9.** $-xy^2 \cdot y^3 \cdot xy^2z$ **10.** $(-4xy)(-3xy^2b)$

**11.** $a^{-14} \cdot a^5$ **12.** $\frac{a^{16}}{a^{17}}$

**13.** $\frac{x^{-7}}{x^4}$ **14.** $\frac{9a(a^{-3})}{18a^{15}}$

**15.** $\frac{y^{6p-3}}{y^{6p+2}}$

*Write in scientific notation.*

**16.** 36,890,000 **17.** $-0.000362$

*Write each number without exponents.*

**18.** $1.678 \times 10^{-6}$ **19.** $4.1 \times 10^5$

**(5.2)** *Simplify. Use only positive exponents.*

**20.** $(8^5)^3$ **21.** $\left(\frac{a}{4}\right)^2$

**22.** $(3x)^3$ **23.** $(-4x)^{-2}$

**24.** $\left(\frac{6x}{5}\right)^2$ **25.** $(8^6)^{-3}$

**26.** $\left(\frac{4}{3}\right)^{-2}$ **27.** $(-2x^3)^{-3}$

**28.** $\left(\frac{8p^6}{4p^4}\right)^{-2}$ **29.** $(-3x^{-2}y^2)^3$

**30.** $\left(\dfrac{x^{-5}y^{-3}}{z^3}\right)^{-5}$

**31.** $\dfrac{4^{-1}x^3yz}{x^{-2}yx^4}$

**32.** $(5xyz)^{-4}(x^{-2})^{-3}$

**33.** $\dfrac{2(3yz)^{-3}}{y^{-3}}$

*Simplify each expression.*

**34.** $x^{4a}(3x^{5a})^3$

**35.** $\dfrac{4y^{3x-3}}{2y^{2x+4}}$

*Use scientific notation to find the quotient. Express each quotient in scientific notation.*

**36.** $\dfrac{(0.00012)(144{,}000)}{0.0003}$

**37.** $\dfrac{(-0.00017)(0.00039)}{3000}$

*Simplify. Use only positive exponents.*

**38.** $\dfrac{27x^{-5}y^5}{18x^{-6}y^2} \cdot \dfrac{x^4y^{-2}}{x^{-2}y^3}$

**39.** $\dfrac{3x^5}{y^{-4}} \cdot \dfrac{(3xy^{-3})^{-2}}{(z^{-3})^{-4}}$

**40.** $\dfrac{(x^w)^2}{(x^{w-4})^{-2}}$

***(5.3)*** *Find the degree of each polynomial.*

**41.** $x^2y - 3xy^3z + 5x + 7y$

**42.** $3x + 2$

*Simplify by combining like terms.*

**43.** $4x + 8x - 6x^2 - 6x^2y$

**44.** $-8xy^3 + 4xy^3 - 3x^3y$

*Add or subtract as indicated.*

**45.** $(3x + 7y) + (4x^2 - 3x + 7) + (y - 1)$

**46.** $(4x^2 - 6xy + 9y^2) - (8x^2 - 6xy - y^2)$

**47.** $(3x^2 - 4b + 28) + (9x^2 - 30) - (4x^2 - 6b + 20)$

**48.** Add $(9xy + 4x^2 + 18)$ and $(7xy - 4x^3 - 9x)$.

**49.** Subtract $(x - 7)$ from the sum of $(3x^2y - 7xy - 4)$ and $(9x^2y + x)$.

**50.**
$$\begin{array}{r} x^2 - 5x + 7 \\ -\,(\;\;\; x + 4) \\ \hline \end{array}$$

**51.**
$$\begin{array}{l} x^3 \quad + 2xy^2 - y \\ +\,(x - 4xy^2 \qquad - 7) \\ \hline \end{array}$$

*If* $P(x) = 9x^2 - 7x + 8$, *find the following.*

**52.** $P(6)$

**53.** $P(-2)$

**54.** $P(-3)$

*If* $P(x) = 2x - 1$ *and* $Q(x) = x^2 + 2x - 5$, *find the following.*

**55.** $P(x) + Q(x)$

**56.** $2[P(x)] - Q(x)$

△ **57.** Find the perimeter of the rectangle.

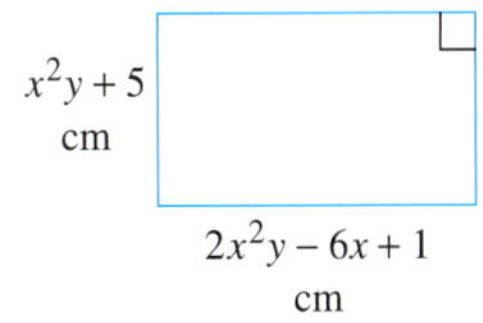


***(5.4)*** *Multiply.*

**58.** $-6x(4x^2 - 6x + 1)$

**59.** $-4ab^2(3ab^3 + 7ab + 1)$

**60.** $(x - 4)(2x + 9)$

**61.** $(-3xa + 4b)^2$

**62.** $(9x^2 + 4x + 1)(4x - 3)$

**63.** $(5x - 9y)(3x + 9y)$

**64.** $\left(x - \dfrac{1}{3}\right)\left(x + \dfrac{2}{3}\right)$

**65.** $(x^2 + 9x + 1)^2$

*Multiply, using special products.*

**66.** $(3x - y)^2$

**67.** $(4x + 9)^2$

**68.** $(x + 3y)(x - 3y)$

**69.** $[4 + (3a - b)][4 - (3a - b)]$

**70.** If $P(x) = 2x - 1$ and $Q(x) = x^2 + 2x - 5$, find $P(x) \cdot Q(x)$.

△ **71.** Find the area of the rectangle.

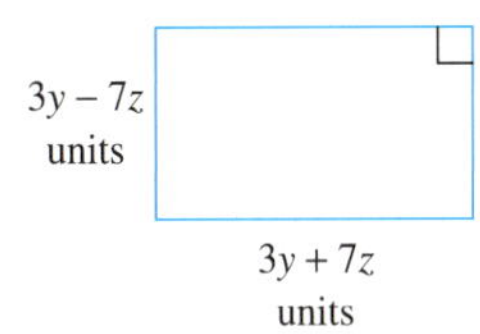


*Multiply. Assume that all variable exponents represent integers.*

**72.** $4a^b(3a^{b+2} - 7)$

**73.** $(4xy^z - b)^2$

**74.** $(3x^a - 4)(3x^a + 4)$

***(5.5)*** *Factor out the greatest common factor.*

**75.** $16x^3 - 24x^2$

**76.** $36y - 24y^2$

**77.** $6ab^2 + 8ab - 4a^2b^2$

**78.** $14a^2b^2 - 21ab^2 + 7ab$

**79.** $6a(a + 3b) - 5(a + 3b)$

**80.** $4x(x - 2y) - 5(x - 2y)$

**81.** $xy - 6y + 3x - 18$

**82.** $ab - 8b + 4a - 32$

**83.** $pq - 3p - 5q + 15$ **84.** $x^3 - x^2 - 2x + 2$

△ **85.** A smaller square is cut from a larger rectangle. Write the area of the shaded region as a factored polynomial.

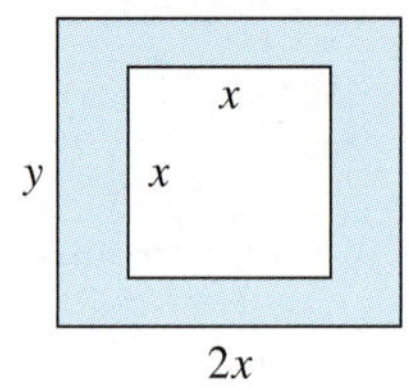


*(5.6) Completely factor each polynomial.*

**86.** $x^2 - 14x - 72$ **87.** $x^2 + 16x - 80$

**88.** $2x^2 - 18x + 28$ **89.** $3x^2 + 33x + 54$

**90.** $2x^3 - 7x^2 - 9x$ **91.** $3x^2 + 2x - 16$

**92.** $6x^2 + 17x + 10$ **93.** $15x^2 - 91x + 6$

**94.** $4x^2 + 2x - 12$ **95.** $9x^2 - 12x - 12$

**96.** $y^2(x + 6)^2 - 2y(x + 6)^2 - 3(x + 6)^2$

**97.** $(x + 5)^2 + 6(x + 5) + 8$

**98.** $x^4 - 6x^2 - 16$ **99.** $x^4 + 8x^2 - 20$

*(5.7) Factor each polynomial completely.*

**100.** $x^2 - 100$ **101.** $x^2 - 81$

**102.** $2x^2 - 32$ **103.** $6x^2 - 54$

**104.** $81 - x^4$ **105.** $16 - y^4$

**106.** $(y + 2)^2 - 25$ **107.** $(x - 3)^2 - 16$

**108.** $x^3 + 216$ **109.** $y^3 + 512$

**110.** $8 - 27y^3$ **111.** $1 - 64y^3$

**112.** $6x^4y + 48xy$ **113.** $2x^5 + 16x^2y^3$

**114.** $x^2 - 2x + 1 - y^2$ **115.** $x^2 - 6x + 9 - 4y^2$

**116.** $4x^2 + 12x + 9$ **117.** $16a^2 - 40ab + 25b^2$

△ **118.** The volume of the cylindrical shell is $\pi R^2h - \pi r^2h$ cubic units. Write this volume as a factored expression.

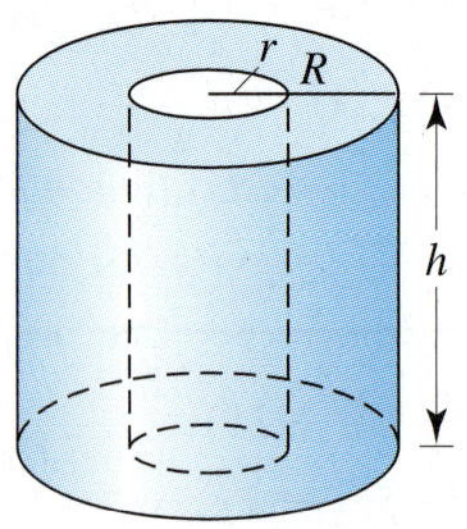


*(5.8) Solve each polynomial equation for the variable.*

**119.** $(3x - 1)(x + 7) = 0$

**120.** $3(x + 5)(8x - 3) = 0$

**121.** $5x(x - 4)(2x - 9) = 0$

**122.** $6(x + 3)(x - 4)(5x + 1) = 0$

**123.** $2x^2 = 12x$

**124.** $4x^3 - 36x = 0$

**125.** $(1 - x)(3x + 2) = -4x$

**126.** $2x(x - 12) = -40$

**127.** $3x^2 + 2x = 12 - 7x$

**128.** $2x^2 + 3x = 35$

**129.** $x^3 - 18x = 3x^2$

**130.** $19x^2 - 42x = -x^3$

**131.** $12x = 6x^3 + 6x^2$

**132.** $8x^3 + 10x^2 = 3x$

**133.** The sum of a number and twice its square is 105. Find the number.

△ **134.** The length of a rectangular piece of carpet is 2 meters less than 5 times its width. Find the dimensions of the carpet if its area is 16 square meters.

**135.** A scene from an adventure film calls for a stunt dummy to be dropped from above the second-story platform of the Eiffel Tower, a distance of 400 feet. Its height $h(t)$ at time $t$ seconds is given by

$$h(t) = -16t^2 + 400$$

Determine when the stunt dummy will reach the ground.

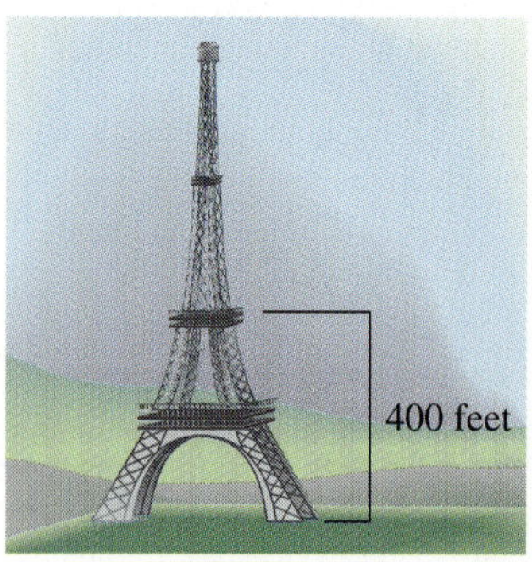

*(5.9) Exercises 136–139 refer to the following graph.*

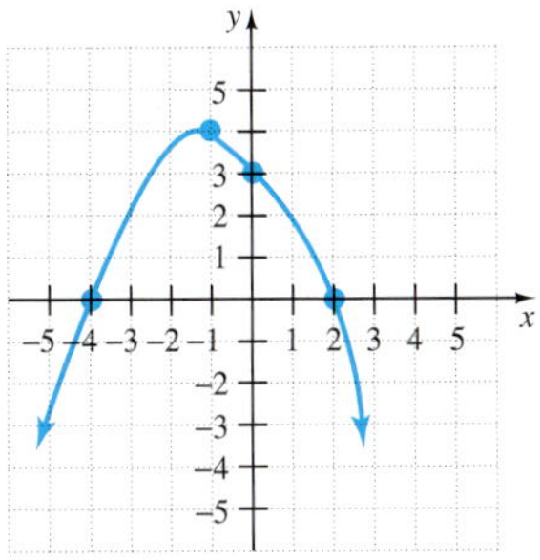


**136.** Find the domain and the range of the function.

**137.** List the $x$- and $y$-intercepts.

**138.** Find the coordinates of the point with the greatest $y$-value.

**139.** List the $x$-values for which the $y$-values are greater than 0.

*Graph each polynomial function defined by the equation. Find all intercepts. If the function is a quadratic function, find the vertex.*

**140.** $f(x) = x^2 + 6x + 9$

**141.** $f(x) = x^2 - 5x + 4$

**142.** $f(x) = (x - 1)(x^2 - 2x - 3)$

**143.** $f(x) = (x + 3)(x^2 - 4x + 3)$

**144.** $f(x) = 2x^2 - 4x + 5$

**145.** $f(x) = x^2 - 2x + 3$

**146.** $f(x) = x^3 - 16x$

**147.** $f(x) = x^3 + 5x^2 + 6x$

## CHAPTER 5 TEST

*Simplify. Use positive exponents to write the answers.*

**1.** $(-9x)^{-2}$

**2.** $-3xy^{-2}(4xy^2)z$

**3.** $\dfrac{6^{-1}a^2b^{-3}}{3^{-2}a^{-5}b^2}$

**4.** $\left(\dfrac{-xy^{-5}z}{xy^3}\right)^{-5}$

*Write in scientific notation.*

**5.** 630,000,000

**6.** 0.01200

**7.** Write $5 \times 10^{-6}$ without exponents.

**8.** Use scientific notation to find the quotient.

$$\frac{(0.0024)(0.00012)}{0.00032}$$

*Perform the indicated operations.*

**9.** $(4x^3 - 3x - 4) - (9x^3 + 8x + 5)$

**10.** $-3xy(4x + y)$

**11.** $(3x + 4)(4x - 7)$

**12.** $(5a - 2b)(5a + 2b)$

**13.** $(6m + n)^2$

**14.** $(2x - 1)(x^2 - 6x + 4)$

*Factor each polynomial completely.*

**15.** $16x^3y - 12x^2y^4$

**16.** $x^2 - 13x - 30$

**17.** $4y^2 + 20y + 25$

**18.** $6x^2 - 15x - 9$

**19.** $4x^2 - 25$

**20.** $x^3 + 64$

**21.** $3x^2y - 27y^3$

**22.** $6x^2 + 24$

**23.** $16y^3 - 2$

**24.** $x^2y - 9y - 3x^2 + 27$

*Solve the equation for the variable.*

**25.** $3(n - 4)(7n + 8) = 0$

**26.** $(x + 2)(x - 2) = 5(x + 4)$

**27.** $2x^3 + 5x^2 - 8x - 20 = 0$

△ **28.** Write the area of the shaded region as a factored polynomial.

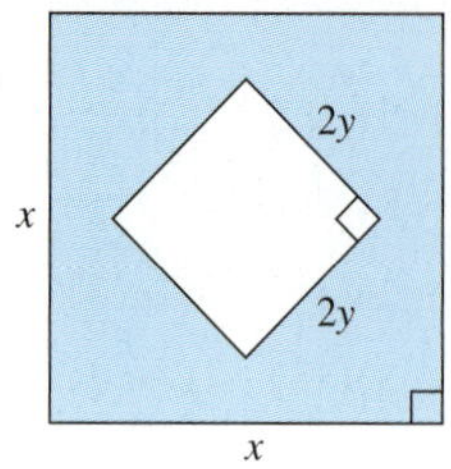


**29.** A pebble is hurled upward from the top of the Canada Trust Tower, which is 880 feet tall, with an initial velocity of 96 feet per second. Neglecting air resistance, the height $h(t)$ of the pebble after $t$ seconds is given by the polynomial function

$$h(t) = -16t^2 + 96t + 880$$

**a.** Find the height of the pebble when $t = 1$.

**b.** Find the height of the pebble when $t = 5.1$.

**c.** When will the pebble hit the ground?

*Graph. Find and label x- and y-intercepts. If the graph is a parabola, find its vertex.*

**30.** $f(x) = x^2 - 4x - 5$

**31.** $f(x) = x^3 - 1$

## CHAPTER 5 CUMULATIVE REVIEW

**1.** Find the roots.

**a.** $\sqrt[3]{27}$

**b.** $\sqrt[5]{1}$

**c.** $\sqrt[4]{16}$

**2.** Solve: $2(x - 3) = 5x - 9$.

**3.** Marial Callier just received an inheritance of \$10,000 and plans to place all the money in a savings account that pays 5% compounded quarterly to help her son go to college in 3 years. How much money will be in the account in 3 years?

**4.** Solve and graph the solution set.

**a.** $\frac{1}{4}x \leq \frac{3}{8}$

**b.** $-2.3x < 6.9$

*Solve.*

**5.** $-1 \leq \frac{2x}{3} + 5 \leq 2$

**6.** $|y| = 0$

**7.** $\left|2x - \frac{1}{10}\right| < -13$

**8.** Graph the linear equation $y = \frac{1}{3}x$.

**9.** Is the relation $x = y^2$ also a function?

**10.** Graph $x = 2$.

**11.** Find the slope of the line $y = 2$.

**12.** Find the equation of the horizontal line containing the point (2, 3).

**13.** Graph the union of $x + \frac{1}{2}y \geq -4$ or $y \leq -2$.

**14.** Use the substitution method to solve the system.

$$\begin{cases} 2x + 4y = -6 \\ x = 2y - 5 \end{cases}$$

**15.** Solve the system.

$$\begin{cases} 2x + 4y = 1 \\ 4x - 4z = -1 \\ y - 4z = -3 \end{cases}$$

**16.** A first number is 4 less than a second number. Four times the first number is 6 more than twice the second. Find the numbers.

**17.** Use matrices to solve the system.

$$\begin{cases} 2x - y = 3 \\ 4x - 2y = 5 \end{cases}$$

**18.** Use Cramer's rule to solve the system.

$$\begin{cases} x - 2y + z = 4 \\ 3x + y - 2z = 3 \\ 5x + 5y + 3z = -8 \end{cases}$$

**19.** Write each number in scientific notation.

**a.** 730,000

**b.** 0.00000104

**20.** Simplify each expression. Use positive exponents to write the answers.

**a.** $(2x^0y^{-3})^{-2}$

**b.** $\left(\dfrac{x^{-5}}{x^{-2}}\right)^{-3}$

**c.** $\left(\dfrac{2}{7}\right)^{-2}$

**d.** $\dfrac{5^{-2}x^{-3}y^{11}}{x^2y^{-5}}$

**21.** Find the degree of the polynomial $3xy + x^2y^2 - 5x^2 - 6$.

**22.** Multiply.

**a.** $(2x^3)(5x^6)$

**b.** $(7y^4z^4)(-xy^{11}z^5)$

*Factor.*

**23.** $17x^3y^2 - 34x^4y^2$

**24.** $x^2 + 10x + 16$

**25.** Solve: $2x^2 + 9x - 5 = 0$.

Have you ever thought about how many feet, or even miles, of wiring are needed in your house, dormitory, or apartment building to make all of your lights and electrical appliances work? Without electricians to wire our homes and buildings, we all would probably be in the dark right now.

In addition to installing wiring and coaxial or fiber-optic cable, electricians also may repair or maintain electrical components. Most electricians learn their trade through a four-or five-year apprenticeship program that includes both on-the-job training and classes such as electrical theory and mathematics. Electricians use math and problem-solving skills in tasks such as estimating job costs, testing circuits, and reading blueprints.

*For more information about a career as an electrician, visit the National Electrical Contractors Association Website by first going to www.prenhall.com/martin-gay.*

*In the Spotlight on Decision Making feature on page 337, you will have the opportunity to make a decision as an electrician about which resistor to use to repair a power supply.*

# 6

# RATIONAL EXPRESSIONS


6.1 Rational Functions and Multiplying and Dividing Rational Expressions

6.2 Adding and Subtracting Rational Expressions

6.3 Simplifying Complex Fractions

6.4 Dividing Polynomials

6.5 Synthetic Division and the Remainder Theorem

6.6 Solving Equations Containing Rational Expressions

6.7 Rational Equations and Problem Solving

6.8 Variation and Problem Solving


Polynomials are to algebra what integers are to arithmetic. We have added, subtracted, multiplied, and raised polynomials to powers, each operation yielding another polynomial, just as these operations on integers yield another integer. But when we divide one integer by another, the result may or may not be another integer. Likewise, when we divide one polynomial by another, we may or may not get a polynomial in return. The quotient $x \div (x + 1)$ is not a polynomial; it is a *rational expression* that can be written as $\frac{x}{x + 1}$.

In this chapter, we study these new algebraic forms known as rational expressions and the *rational functions* they generate.

## 6.1 RATIONAL FUNCTIONS AND MULTIPLYING AND DIVIDING RATIONAL EXPRESSIONS

CD-ROM SSM SSG Video

▶ **OBJECTIVES**

1. Define a rational expression and a rational function.
2. Find values for which a rational expression is undefined.
3. Simplify rational expressions.
4. Multiply rational expressions.
5. Divide rational expressions.

1 Recall that a *rational number*, or *fraction*, is a number that can be written as the quotient $\frac{p}{q}$ of two integers $p$ and $q$ as long as $q$ is not 0. A **rational expression** is an expression that can be written as the quotient $\frac{P}{Q}$ of two polynomials $P$ and $Q$ as long as $Q$ is not 0.

***Examples of Rational Expressions***

$$\frac{3x+7}{2} \quad \frac{5x^2-3}{x-1} \quad \frac{7x-2}{2x^2+7x+6}$$

Rational expressions are sometimes used to describe functions. For example, we call the function $f(x) = \frac{x^2+2}{x-3}$ a **rational function** since $\frac{x^2+2}{x-3}$ is a rational expression.

### Example 1 COST FOR PRESSING COMPACT DISCS

For the ICL Production Company, the rational function $C(x) = \frac{2.6x + 10{,}000}{x}$ describes the company's cost per disc of pressing $x$ compact discs. Find the cost per disc for pressing

**a.** 100 compact discs
**b.** 1000 compact discs

*Solution* **a.** $C(100) = \frac{2.6(100) + 10{,}000}{100} = \frac{10{,}260}{100} = 102.6$

The cost per disc for pressing 100 compact discs is \$102.60.

**b.** $C(1000) = \frac{2.6(1000) + 10{,}000}{1000} = \frac{12{,}600}{1000} = 12.6$

The cost per disc for pressing 1000 compact discs is \$12.60. Notice that as more compact discs are produced, the cost per disc decreases.

2 As with fractions, a rational expression is **undefined** if the denominator is 0. If a variable in a rational expression is replaced with a number that makes the denominator 0, we say that the rational expression is **undefined** for this value of the variable. For

example, the rational expression $\frac{x^2+2}{x-3}$ is undefined when $x$ is 3, because replacing $x$ with 3 results in a denominator of 0. For this reason, we must exclude 3 from the domain of the function defined by $f(x) = \frac{x^2+2}{x-3}$.

The domain of $f$ is then

$$\{x \mid x \text{ is a real number and } x \neq 3\}$$

"The set of all $x$ such that $x$ is a real number and $x$ is not equal to 3."

Unless told otherwise, we assume that the domain of a function described by an equation is the set of all real numbers for which the equation is defined.

**Example 2** Find the domain of each rational function.

**a.** $f(x) = \frac{8x^3 + 7x^2 + 20}{2}$ **b.** $g(x) = \frac{5x^2 - 3}{x - 1}$ **c.** $f(x) = \frac{7x - 2}{x^2 - 2x - 15}$

*Solution* The domain of each function will contain all real numbers except those values that make the denominator 0.

**a.** No matter what the value of $x$, the denominator of $f(x) = \frac{8x^3 + 7x^2 + 20}{2}$ is never 0, so the domain of $f$ is $\{x \mid x \text{ is a real number}\}$.

**b.** To find the values of $x$ that make the denominator of $g(x)$ equal to 0, we solve the equation "denominator = 0":

$$x - 1 = 0, \quad \text{or} \quad x = 1$$

The domain of $g(x)$ must exclude 1 since the rational expression is undefined when $x$ is 1. The domain of $g$ is $\{x \mid x \text{ is a real number and } x \neq 1\}$.

**c.** We find the domain by setting the denominator equal to 0.

$$x^2 - 2x - 15 = 0 \quad \text{Set the denominator equal to 0 and solve.}$$

$$(x - 5)(x + 3) = 0$$

$$x - 5 = 0 \quad \text{or} \quad x + 3 = 0$$

$$x = 5 \quad \text{or} \quad x = -3$$

If $x$ is replaced with 5 or with $-3$, the rational expression is undefined.

The domain of $f$ is $\{x \mid x \text{ is a real number and } x \neq 5 \text{ and } x \neq -3\}$.

3 Recall that a fraction is in lowest terms or simplest form if the numerator and denominator have no common factors other than 1 (or $-1$). For example, $\frac{3}{13}$ is in lowest terms since 3 and 13 have no common factors other than 1 (or $-1$).

To **simplify** a rational expression, or to write it in lowest terms, we use the fundamental principle of rational expressions.

**FUNDAMENTAL PRINCIPLE OF RATIONAL EXPRESSIONS**

For any rational expression $\frac{P}{Q}$ and any polynomial $R$, where $R \neq 0$,

$$\frac{PR}{QR} = \frac{P}{Q}$$

Thus, the fundamental principle says that multiplying or dividing the numerator and denominator of a rational expression by the same nonzero polynomial yields an equivalent rational expression.

To simplify a rational expression such as $\frac{(x + 2)^2}{x^2 - 4}$, factor the numerator and the denominator and then use the fundamental principle of rational expressions to divide out common factors.

$$\frac{(x + 2)^2}{x^2 - 4} = \frac{(x + 2)(x + 2)}{(x + 2)(x - 2)} = \frac{x + 2}{x - 2}$$

This means that the rational expression $\frac{(x + 2)^2}{x^2 - 4}$ has the same value as the rational expression $\frac{x + 2}{x - 2}$ for all values of $x$ except 2 and $-2$. (Remember that when $x$ is 2, the denominators of both rational expressions are 0 and that when $x$ is $-2$, the original rational expression has a denominator of 0.)

As we simplify rational expressions, we will assume that the simplified rational expression is equivalent to the original rational expression for all real numbers except those for which either denominator is 0.

In general, the following steps may be used to simplify rational expressions or to write a rational expression in lowest terms.

**SIMPLIFYING OR WRITING A RATIONAL EXPRESSION IN LOWEST TERMS**

**Step 1:** Completely factor the numerator and denominator of the rational expression.

**Step 2:** Apply the fundamental principle of rational expressions to divide out factors common to both the numerator and denominator.

For now, we assume that variables in a rational expression do not represent values that make the denominator 0.

**Example 3** Simplify $\frac{2x^2}{10x^3 - 2x^2}$.

*Solution* Factor out $2x^2$ from the denominator. Then divide numerator and denominator by their GCF, $2x^2$.

$$\frac{2x^2}{10x^3 - 2x^2} = \frac{2x^2 \cdot 1}{2x^2(5x - 1)} = \frac{1}{5x - 1}$$

When the terms in the numerator of a rational expression differ by sign from the terms of the denominator, the polynomials are opposites of each other and the expression simplifies to $-1$. To see this, factor out $-1$ from the numerator or the denominator. For example,

$$\frac{2-x}{x-2}=\frac{-1(-2+x)}{x-2}=\frac{-1(x-2)}{x-2}=-1$$

If $-1$ is factored from the denominator of the same rational expression, the result is the same.

$$\frac{2-x}{x-2}=\frac{2-x}{-1(-x+2)}=\frac{2-x}{-1(2-x)}=\frac{1}{-1}=-1$$

**HELPFUL HINT**

When the numerator and the denominator of a rational expression are opposites of each other, the expression simplifies to $-1$.

**Example 4** Simplify $\dfrac{18-2x^2}{x^2-2x-3}$.

*Solution*

$$\frac{18-2x^2}{x^2-2x-3}=\frac{2(9-x^2)}{(x+1)(x-3)} \quad \text{Factor.}$$

$$=\frac{2(3+x)(3-x)}{(x+1)(x-3)} \quad \text{Factor completely.}$$

$$=\frac{2(3+x)\cdot -1(x-3)}{(x+1)(x-3)}$$

Notice the opposites $3-x$ and $x-3$. Write $3-x$ as $-1(x-3)$ and simplify.

$$=-\frac{2(3+x)}{x+1}$$

**HELPFUL HINT**

Recall that for a fraction $\dfrac{a}{b}$,

$$\frac{a}{-b}=\frac{-a}{b}=-\frac{a}{b}$$

For example

$$\frac{-(x+1)}{(x+2)}=\frac{(x+1)}{-(x+2)}=-\frac{x+1}{x+2}$$

**Example 5** Simplify each rational expression.

**a.** $\dfrac{x^3+8}{2+x}$

**b.** $\dfrac{2y^2+2}{y^3-5y^2+y-5}$

*Solution* **a.**

$$\frac{x^3+8}{2+x} = \frac{(x+2)(x^2-2x+4)}{x+2}$$ Factor the sum of the two cubes.

$$= x^2 - 2x + 4$$ Divide out common factors.

**b.**

$$\frac{2y^2+2}{y^3-5y^2+y-5} = \frac{2(y^2+1)}{(y^3-5y^2)+(y-5)}$$ Factor the numerator.

$$= \frac{2(y^2+1)}{y^2(y-5)+1(y-5)}$$ Factor the denominator by grouping.

$$= \frac{2(y^2+1)}{(y-5)(y^2+1)}$$

$$= \frac{2}{y-5}$$ Divide out common factors.

**4** Arithmetic operations on rational expressions are performed in the same way as they are on rational numbers.

### MULTIPLYING RATIONAL EXPRESSIONS

The rule for multiplying rational expressions is

$$\frac{P}{Q}\cdot\frac{R}{S} = \frac{PR}{QS} \quad \text{as long as } Q \neq 0 \text{ and } S \neq 0.$$

To multiply rational expressions, you may use these steps:

**Step 1:** Completely factor each numerator and denominator.
**Step 2:** Use the rule above and multiply the numerators and the denominators.
**Step 3:** Simplify the product by dividing the numerator and denominator by their common factors.

**Example 6** Multiply.

**a.** $\dfrac{2x^3}{9y}\cdot\dfrac{y^2}{4x^3}$ **b.** $\dfrac{1+3n}{2n}\cdot\dfrac{2n-4}{3n^2-2n-1}$

*Solution* **a.** $\dfrac{2x^3}{9y}\cdot\dfrac{y^2}{4x^3} = \dfrac{2x^3y^2}{36x^3y}$

To simplify, divide the numerator and the denominator by the common factor, $2x^3y$.

$$\frac{2x^3y^2}{36x^3y} = \frac{y(2x^3y)}{18(2x^3y)} = \frac{y}{18}$$

**b.** $\frac{1+3n}{2n}\cdot\frac{2n-4}{3n^2-2n-1} = \frac{1+3n}{2n}\cdot\frac{2(n-2)}{(3n+1)(n-1)}$ Factor.

$= \frac{(1+3n)\cdot 2(n-2)}{2n(3n+1)(n-1)}$ Multiply.

$= \frac{n-2}{n(n-1)}$ Divide out common factors.

When we multiply rational expressions, notice that we factor each numerator and denominator first. This helps when we apply the fundamental principle to write the product in lowest terms.

## Example 7

Multiply.

**a.** $\frac{2x^2+3x-2}{-4x-8}\cdot\frac{16x^2}{4x^2-1}$ **b.** $\frac{x^3-1}{-3x+3}\cdot\frac{15x^2}{x^2+x+1}$

*Solution* **a.** $\frac{2x^2+3x-2}{-4x-8}\cdot\frac{16x^2}{4x^2-1} = \frac{(2x-1)(x+2)}{-4(x+2)}\cdot\frac{16x^2}{(2x+1)(2x-1)}$ Factor.

$= \frac{4\cdot 4x^2(2x-1)(x+2)}{-1\cdot 4(x+2)(2x+1)(2x-1)}$ Multiply.

$= -\frac{4x^2}{2x+1}$ Divide out common factors.

**b.** $\frac{x^3-1}{-3x+3}\cdot\frac{15x^2}{x^2+x+1} = \frac{(x-1)(x^2+x+1)}{-3(x-1)}\cdot\frac{15x^2}{x^2+x+1}$ Factor.

$= \frac{(x-1)(x^2+x+1)\cdot 3\cdot 5x^2}{-1\cdot 3(x-1)(x^2+x+1)}$ Factor.

$= \frac{5x^2}{-1}$ Divide out common factors.

$= -5x^2$

5 Recall that two numbers are reciprocals of each other if their product is 1. Similarly, if $\frac{P}{Q}$ is a rational expression, then $\frac{Q}{P}$ is its **reciprocal**, since

$$\frac{P}{Q}\cdot\frac{Q}{P} = \frac{P\cdot Q}{Q\cdot P} = 1$$

The following are examples of expressions and their reciprocals.

| ***Expression*** | ***Reciprocal*** |
|---|---|
| $\frac{3}{x}$ | $\frac{x}{3}$ |
| $\frac{2+x^2}{4x-3}$ | $\frac{4x-3}{2+x^2}$ |
| $x^3$ | $\frac{1}{x^3}$ |
| 0 | no reciprocal |

## DIVIDING RATIONAL EXPRESSIONS

The rule for dividing rational expressions is

$$\frac{P}{Q} \div \frac{R}{S} = \frac{P}{Q} \cdot \frac{S}{R} = \frac{PS}{QR} \quad \text{as long as } Q \neq 0, S \neq 0, \text{and } R \neq 0.$$

To divide by a rational expression, use the rule above and multiply by its reciprocal. Then simplify if possible.

Notice that division of rational expressions is the same as for rational numbers.

**Example 8** Divide.

**a.** $\frac{3x}{5y} \div \frac{9y}{x^5}$ **b.** $\frac{8m^2}{3m^2-12} \div \frac{40}{2-m}$

*Solution* **a.** $\frac{3x}{5y} \div \frac{9y}{x^5} = \frac{3x}{5y} \cdot \frac{x^5}{9y}$ Multiply by the reciprocal of the divisor.

$= \frac{x^6}{15y^2}$ Simplify.

**b.** $\frac{8m^2}{3m^2-12} \div \frac{40}{2-m} = \frac{8m^2}{3m^2-12} \cdot \frac{2-m}{40}$ Multiply by the reciprocal of the divisor.

$= \frac{8m^2(2-m)}{3(m+2)(m-2)\cdot 40}$ Factor and multiply.

$= \frac{8\,m^2 \cdot -1(m-2)}{3(m+2)(m-2)\cdot 8 \cdot 5}$ Write $(2-m)$ as $-1(m-2)$.

$= -\frac{m^2}{15(m+2)}$ Simplify.

**HELPFUL HINT**

When dividing rational expressions, do not divide out common factors until the division problem is rewritten as a multiplication problem.

**Example 9** Perform each indicated operation.

$$\frac{x^2 - 25}{(x + 5)^2} \cdot \frac{3x + 15}{4x} \div \frac{x^2 - 3x - 10}{x}$$

*Solution*

$$\frac{x^2 - 25}{(x + 5)^2} \cdot \frac{3x + 15}{4x} \div \frac{x^2 - 3x - 10}{x}$$

$$= \frac{x^2 - 25}{(x + 5)^2} \cdot \frac{3x + 15}{4x} \cdot \frac{x}{x^2 - 3x - 10}$$

To divide, multiply by the reciprocal

$$= \frac{(x + 5)(x - 5)}{(x + 5)(x + 5)} \cdot \frac{3(x + 5)}{4x} \cdot \frac{x}{(x - 5)(x + 2)}$$

$$= \frac{3}{4(x + 2)}$$

## SPOTLIGHT ON DECISION MAKING

Suppose you are an electrician at a small packaging plant. You are repairing machinery that heats the hot glue gun used for sealing boxes. You have determined that a resistor in the machinery's 50-volt direct current power supply must be replaced. To keep the glue warm, the power supply must dissipate about 2000 watts of power.

You know that the power $P$(in watts) dissipated by a resistor in a direct current circuit is given by the formula $P = \frac{V^2}{R}$, where $V$ is the voltage (in volts) and $R$ is the resistance (in ohms). Which of the three resistors shown in the parts list would you use to replace the faulty resistor? Why?

**PARTS LIST RESISTORS**

| *Part Number* | *Material* | *Resistance (in ohms)* |
|---|---|---|
| 1298 | Aluminum | 0.95 |
| 3169 | Nickel | 1.81 |
| 4203 | Tungsten | 1.22 |

## GRAPHING CALCULATOR EXPLORATIONS

Recall that since the rational expression $\frac{7x - 2}{(x - 2)(x + 5)}$ is not defined when $x = 2$ or when $x = -5$, we say that the domain of the rational function $f(x) = \frac{7x - 2}{(x - 2)(x + 5)}$ is all real numbers except 2 and $-5$. This domain can be written as $\{x \mid x \text{ is a real number and } x \neq 2, x \neq -5\}$. This means that the graph of $f(x)$ should not cross the vertical lines $x = 2$ and $x = -5$. The graph of $f(x)$ in *connected* mode follows. In connected mode the graphing calculator tries to connect all dots of the graph so that the result is a smooth curve. This is what has happened in the graph. Notice that the graph appears to contain vertical lines at $x = 2$ and at $x = -5$. We know that this cannot happen because the function is not defined at $x = 2$ and at $x = -5$. We also know that this cannot happen because the graph of this function would not pass the vertical line test.

10

−10 10

−10

If we graph $f(x)$ in *dot* mode, the graph appears as follows. In dot mode the graphing calculator will not connect dots with a smooth curve. Notice that the vertical lines have disappeared, and we have a better picture of the graph. The graph, however, actually appears more like the hand-drawn graph to its right. By using a Table feature, a Calculate Value feature, or by tracing, we can see that the function is not defined at $x = 2$ and at $x = -5$.

10

−10 10

−10

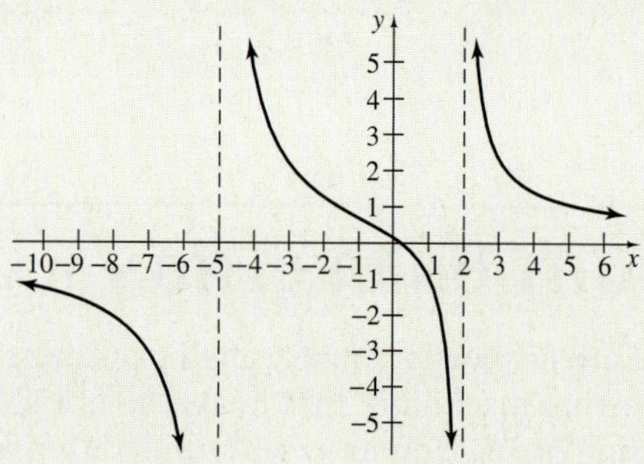


*Find the domain of each rational function. Then graph each rational function and use the graph to confirm the domain.*

**1.** $f(x) = \frac{x + 1}{x^2 - 4}$

**2.** $g(x) = \frac{5x}{x^2 - 9}$

**3.** $h(x) = \frac{x^2}{2x^2 + 7x - 4}$

**4.** $f(x) = \frac{3x + 2}{4x^2 - 19x - 5}$

## Exercise Set 6.1

*Find each function value. See Example 1.*

1. $f(x) = \dfrac{x+8}{2x-1}$; $f(2), f(0), f(-1)$
2. $f(y) = \dfrac{y-2}{-5+y}$; $f(-5), f(0), f(10)$
3. $g(x) = \dfrac{x^2+8}{x^3-25x}$; $g(3), g(-2), g(1)$
4. $s(t) = \dfrac{t^3+1}{t^2+1}$; $s(-1), s(1), s(2)$

*Find the domain of each rational function. See Example 2.*

5. $f(x) = \dfrac{5x-7}{4}$
6. $g(x) = \dfrac{4-3x}{2}$
7. $s(t) = \dfrac{t^2+1}{2t}$
8. $v(t) = -\dfrac{5t+t^2}{3t}$
9. $f(x) = \dfrac{3x}{7-x}$
10. $f(x) = \dfrac{-4x}{-2+x}$
11. $R(x) = \dfrac{3+2x}{x^3+x^2-2x}$
12. $h(x) = \dfrac{5-3x}{2x^2-14x+20}$
13. $C(x) = \dfrac{x+3}{x^2-4}$
14. $R(x) = \dfrac{5}{x^2-7x}$
15. In your own words, explain how to find the domain of a rational function.
16. In your own words, explain how to simplify a rational expression or to write it in lowest terms.

*Write each rational expression in lowest terms. See Examples 3 through 5.*

17. $\dfrac{4x-8}{3x-6}$
18. $\dfrac{12-6x}{30-15x}$
19. $\dfrac{2x-14}{7-x}$
20. $\dfrac{9-x}{5x-45}$
21. $\dfrac{x^2-2x-3}{x^2-6x+9}$
22. $\dfrac{x^2+10x+25}{x^2+8x+15}$
23. $\dfrac{2x^2+12x+18}{x^2-9}$
24. $\dfrac{x^2-4}{2x^2+8x+8}$
25. $\dfrac{3x+6}{x^2+2x}$
26. $\dfrac{3x+4}{9x^2+4}$
27. $\dfrac{2x^2-x-3}{2x^3-3x^2+2x-3}$
28. $\dfrac{3x^2-5x-2}{6x^3+2x^2+3x+1}$
29. $\dfrac{8q^2}{16q^3-16q^2}$
30. $\dfrac{3y}{6y^2-30y}$
31. $\dfrac{x^2+6x-40}{10+x}$
32. $\dfrac{x^2-8x+16}{4-x}$
33. $\dfrac{x^3-125}{5-x}$
34. $\dfrac{4x+4}{2x^3+2}$
35. $\dfrac{8x^3-27}{4x-6}$
36. $\dfrac{9x^2-15x+25}{27x^3+125}$

*Multiply or divide as indicated. Simplify all answers. See Examples 6 through 9.*

37. $\dfrac{3xy^3}{4x^3y^2}\cdot\dfrac{-8x^3y^4}{9x^4y^7}$
38. $-\dfrac{2xyz^3}{5x^2z^2}\cdot\dfrac{10xy}{x^3}$
39. $\dfrac{8a}{3a^4b^2}\div\dfrac{4b^5}{6a^2b}$
40. $\dfrac{3y^3}{14x^4}\div\dfrac{8y^3}{7x}$
41. $\dfrac{a^2b}{a^2-b^2}\cdot\dfrac{a+b}{4a^3b}$
42. $\dfrac{3ab^2}{a^2-4}\cdot\dfrac{a-2}{6a^2b^2}$
43. $\dfrac{x^2-9}{4}\div\dfrac{x^2-6x+9}{x^2-x-6}$
44. $\dfrac{a-5b}{a^2+ab}\div\dfrac{15b-3a}{b^2-a^2}$
45. $\dfrac{9x+9}{4x+8}\cdot\dfrac{2x+4}{3x^2-3}$
46. $\dfrac{x^2-1}{10x+30}\cdot\dfrac{12x+36}{3x-3}$
47. $\dfrac{a+b}{ab}\div\dfrac{a^2-b^2}{4a^3b}$
48. $\dfrac{6a^2b^2}{a^2-4}\div\dfrac{3ab^2}{a-2}$
49. $\dfrac{2x^2-4x-30}{5x^2-40x-75}\div\dfrac{x^2-8x+15}{x^2-6x+9}$
50. $\dfrac{4a+36}{a^2-7a-18}\div\dfrac{a^2-a-6}{a^2-81}$
51. $\dfrac{2x^3-16}{6x^2+6x-36}\cdot\dfrac{9x+18}{3x^2+6x+12}$
52. $\dfrac{x^2-3x+9}{5x^2-20x-105}\cdot\dfrac{x^2-49}{x^3+27}$
53. $\dfrac{15b-3a}{b^2-a^2}\div\dfrac{a-5b}{ab+b^2}$
54. $\dfrac{4x+4}{x-1}\div\dfrac{x^2-4x-5}{x^2-1}$
55. $\dfrac{a^3+a^2b+a+b}{a^3+a}\cdot\dfrac{6a^2}{2a^2-2b^2}$
56. $\dfrac{a^2-2a}{ab-2b+3a-6}\cdot\dfrac{8b+24}{3a+6}$
57. $\dfrac{5a}{12}\cdot\dfrac{2}{25a^2}\cdot\dfrac{15a}{2}$
58. $\dfrac{4a}{7}\div\dfrac{a^2}{14}\cdot\dfrac{3}{a}$

**59.** $\dfrac{3x - x^2}{x^3 - 27} \div \dfrac{x}{x^2 + 3x + 9}$

**60.** $\dfrac{x^2 - 3x}{x^3 - 27} \div \dfrac{2x}{2x^2 + 6x + 18}$

**61.** $\dfrac{4a}{7} \div \left(\dfrac{a^2}{14} \cdot \dfrac{3}{a}\right)$

**62.** $\dfrac{a^2}{14} \cdot \dfrac{3}{a} \div \dfrac{4a}{7}$

**63.** $\dfrac{8b + 24}{3a + 6} \div \dfrac{ab - 2b + 3a - 6}{a^2 - 4a + 4}$

**64.** $\dfrac{2a^2 - 2b^2}{a^3 + a^2b + a + b} \div \dfrac{6a^2}{a^3 + a}$

**65.** $\dfrac{4}{x} \div \dfrac{3xy}{x^2} \cdot \dfrac{6x^2}{x^4}$

**66.** $\dfrac{4}{x} \cdot \dfrac{3xy}{x^2} \div \dfrac{6x^2}{x^4}$

**67.** $\dfrac{3x^2 - 5x - 2}{y^2 + y - 2} \cdot \dfrac{y^2 + 4y - 5}{12x^2 + 7x + 1} \div \dfrac{5x^2 - 9x - 2}{8x^2 - 2x - 1}$

**68.** $\dfrac{x^2 + x - 2}{3y^2 - 5y - 2} \cdot \dfrac{12y^2 + y - 1}{x^2 + 4x - 5} \div \dfrac{8y^2 - 6y + 1}{5y^2 - 9y - 2}$

**69.** $\dfrac{5a^2 - 20}{3a^2 - 12a} \div \dfrac{a^3 + 2a^2}{2a^2 - 8a} \cdot \dfrac{9a^3 + 6a^2}{2a^2 - 4a}$

**70.** $\dfrac{5a^2 - 20}{3a^2 - 12a} \div \left(\dfrac{a^3 + 2a^2}{2a^2 - 8a} \cdot \dfrac{9a^3 + 6a^2}{2a^2 - 4a}\right)$

**71.** $\dfrac{5x^4 + 3x^2 - 2}{x - 1} \cdot \dfrac{x + 1}{x^4 - 1}$

**72.** $\dfrac{3x^4 - 10x^2 - 8}{x - 2} \cdot \dfrac{3x + 6}{15x^2 + 10}$

△ **73.** Find the area of the rectangle.

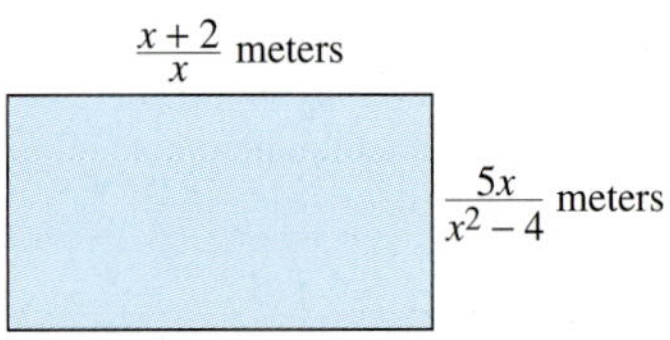


△ **74.** Find the area of the triangle.

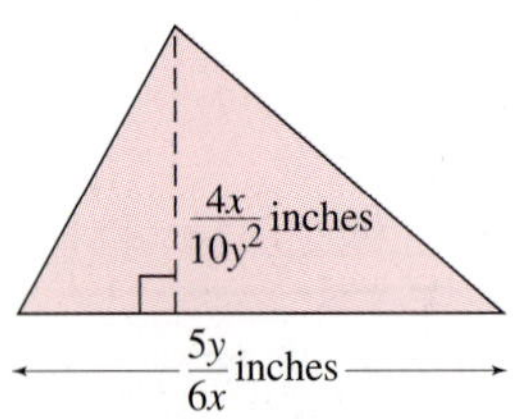


**75.** In our definition of division for

$$\frac{P}{Q} \div \frac{R}{S}$$

we stated that $Q \neq 0, S \neq 0$, and $R \neq 0$. Explain why $R$ cannot equal 0.

**76.** Find the polynomial in the second numerator such that the following statement is true.

$$\frac{x^2 - 4}{x^2 - 7x + 10} \cdot \frac{?}{2x^2 + 11x + 14} = 1$$

△ **77.** A parallelogram has area $\dfrac{x^2 + x - 2}{x^3}$ square feet and height $\dfrac{x^2}{x - 1}$ feet. Express the length of its base as a rational expression in $x$. (*Hint:* Since $A = b \cdot h$, then $b = \dfrac{A}{h}$ or $b = A \div h$.)

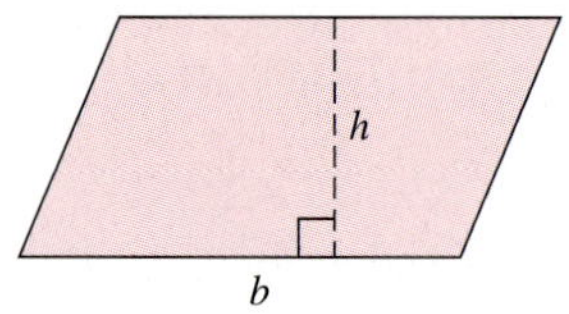


**78.** A lottery prize of $\dfrac{15x^3}{y^2}$ dollars is to be divided among $5x$ people. Express the amount of money each person is to receive as a rational expression in $x$ and $y$.

**79.** Graph a portion of the function $f(x) = \dfrac{20x}{100 - x}$. To do so, complete the given table, plot the points, and then connect the plotted points with a smooth curve.

| $x$ | 0 | 10 | 30 | 50 | 70 | 90 | 95 | 99 |
|---|---|---|---|---|---|---|---|---|
| $y$ or $f(x)$ | | | | | | | | |

**80.** The domain of the function $f(x) = \dfrac{1}{x}$ is all real numbers except 0. This means that the graph of this function will be in two pieces: one piece corresponding to $x$ values less than 0 and one piece corresponding to $x$ values greater than 0. Graph the function by completing the following tables, separately plotting the points, and connecting each set of plotted points with a smooth curve.

| $x$ | $\frac{1}{4}$ | $\frac{1}{2}$ | 1 | 2 | 4 |
|---|---|---|---|---|---|
| $y$ or $f(x)$ | | | | | |

| $x$ | $-4$ | $-2$ | $-1$ | $-\frac{1}{2}$ | $-\frac{1}{4}$ |
|---|---|---|---|---|---|
| $y$ or $f(x)$ | | | | | |

**81.** The function $f(x) = \dfrac{100{,}000x}{100 - x}$ models the cost in dollars for removing $x$ percent of the pollutants from a bayou in which a nearby company dumped creosote.

**a.** What is the domain of $f(x)$?

**b.** Find the cost of removing 30% of the pollutants from the bayou. (*Hint:* Find $f(30)$.)

**c.** Find the cost of removing 60% of the pollutants and then 80% of the pollutants.

**d.** Find $f(90)$, then $f(95)$, and then $f(99)$. What happens to the cost as $x$ approaches 100%?

**82.** The total revenue from the sale of a popular book is approximated by the rational function $R(x) = \dfrac{1000x^2}{x^2 + 4}$ where $x$ is the number of years since publication and $R(x)$ is the total revenue in millions of dollars.

**a.** Find the total revenue at the end of the first year.

**b.** Find the total revenue at the end of the second year.

**c.** Find the revenue during the second year only.

## REVIEW EXERCISES

*Perform the indicated operations. See Section 1.3.*

**83.** $\dfrac{4}{5} + \dfrac{3}{5}$

**84.** $\dfrac{4}{10} - \dfrac{7}{10}$

**85.** $\dfrac{5}{28} - \dfrac{2}{21}$

**86.** $\dfrac{5}{13} + \dfrac{2}{7}$

**87.** $\dfrac{3}{8} + \dfrac{1}{2} - \dfrac{3}{16}$

**88.** $\dfrac{2}{9} - \dfrac{1}{6} + \dfrac{2}{3}$

## *A Look Ahead*

Example

Perform the following operation.

$$\frac{x^{2n} - 3x^n - 18}{x^{2n} - 9} \cdot \frac{3x^n + 9}{x^{2n}}$$

*Solution* $\dfrac{x^{2n} - 3x^n - 18}{x^{2n} - 9} \cdot \dfrac{3x^n + 9}{x^{2n}}$

$$= \frac{(x^n + 3)(x^n - 6) \cdot 3(x^n + 3)}{(x^n + 3)(x^n - 3) \cdot x^{2n}}$$

$$= \frac{3(x^n - 6)(x^n + 3)}{x^{2n}(x^n - 3)}$$

*Perform the indicated operation. Write all answers in lowest terms. See the preceding example.*

**89.** $\dfrac{x^{2n} - 4}{7x} \cdot \dfrac{14x^3}{x^n - 2}$

**90.** $\dfrac{x^{2n} + 4x^n + 4}{4x - 3} \cdot \dfrac{8x^2 - 6x}{x^n + 2}$

**91.** $\dfrac{y^{2n} + 9}{10y} \cdot \dfrac{y^n - 3}{y^{4n} - 81}$

**92.** $\dfrac{y^{4n} - 16}{y^{2n} + 4} \cdot \dfrac{6y}{y^n + 2}$

**93.** $\dfrac{y^{2n} - y^n - 2}{2y^n - 4} \div \dfrac{y^{2n} - 1}{1 + y^n}$

**94.** $\dfrac{y^{2n} + 7y^n + 10}{10} \div \dfrac{y^{2n} + 4y^n + 4}{5y^n + 25}$

## 6.2 ADDING AND SUBTRACTING RATIONAL EXPRESSIONS

CD-ROM SSM SSG Video

▶ **OBJECTIVES**

1. Add or subtract rational expressions with common denominators.
2. Identify the least common denominator of two or more rational expressions.
3. Add or subtract rational expressions with unlike denominators.

**1** Rational expressions, like rational numbers, can be added or subtracted. We define the sum or difference of rational expressions in the same way that we defined the sum or difference of rational numbers (fractions).

### ADDING OR SUBTRACTING RATIONAL EXPRESSIONS WITH COMMON DENOMINATORS

If $\dfrac{P}{Q}$ and $\dfrac{R}{Q}$ are rational expressions, then

$$\frac{P}{Q} + \frac{R}{Q} = \frac{P + R}{Q} \quad \text{and} \quad \frac{P}{Q} - \frac{R}{Q} = \frac{P - R}{Q}$$

To add or subtract rational expressions with common denominators, add or subtract the numerators and write the sum or difference over the common denominator.

**Example 1** Add or subtract.

a. $\frac{x}{4} + \frac{5x}{4}$   b. $\frac{x^2}{x+7} - \frac{49}{x+7}$   c. $\frac{x}{3y^2} - \frac{x+1}{3y^2}$

*Solution* The rational expressions have common denominators, so add or subtract their numerators and place the sum or difference over their common denominator.

a. $\frac{x}{4} + \frac{5x}{4} = \frac{x+5x}{4} = \frac{6x}{4} = \frac{3x}{2}$   Add the numerators and write the result over the common denominator.

b. $\frac{x^2}{x+7} - \frac{49}{x+7} = \frac{x^2-49}{x+7}$   Subtract the numerators and write the result over the common denominator.

$= \frac{(x+7)(x-7)}{x+7}$   Factor the numerator.

$= x - 7$   Simplify.

**HELPFUL HINT**
Be sure to insert parentheses here so that the entire numerator is subtracted.

c. $\frac{x}{3y^2} - \frac{x+1}{3y^2} = \frac{x-(x+1)}{3y^2}$   Subtract the numerators.

$= \frac{x-x-1}{3y^2}$   Use the distributive property.

$= -\frac{1}{3y^2}$   Simplify.

2 To add or subtract rational expressions with unlike denominators, first write the rational expressions as equivalent rational expressions with common denominators.

The **least common denominator (LCD)** is usually the easiest common denominator to work with. The LCD of a list of rational expressions is a polynomial of least degree whose factors include the denominator factors in the list.

Use the following steps to find the LCD.

**FINDING THE LEAST COMMON DENOMINATOR (LCD)**

**Step 1:** Factor each denominator completely.
**Step 2:** The LCD is the product of all unique factors each raised to the greatest power that appears in any factored denominator.

**Example 2** Find the LCD of the rational expressions in each list.

a. $\frac{2}{3x^5y^2}, \frac{3z}{5xy^3}$   b. $\frac{7}{z+1}, \frac{z}{z-1}$

c. $\frac{m-1}{m^2-25}, \frac{2m}{2m^2-9m-5}, \frac{7}{m^2-10m+25}$   d. $\frac{x}{x^2-4}, \frac{11}{6-3x}$

*Solution*

**HELPFUL HINT**
The greatest power of $x$ is 5, so we have a factor of $x^5$. The greatest power of $y$ is 3, so we have a factor of $y^3$.

**a.** First we factor each denominator.

$$3x^5y^2 = 3 \cdot x^5 \cdot y^2$$
$$5xy^3 = 5 \cdot x \cdot y^3$$
$$\text{LCD} = 3 \cdot 5 \cdot x^5 \cdot y^3 = 15x^5y^3$$

**b.** The denominators $z + 1$ and $z - 1$ do not factor further. Thus,

$$\text{LCD} = (z + 1)(z - 1)$$

**c.** We first factor each denominator.

$$m^2 - 25 = (m + 5)(m - 5)$$
$$2m^2 - 9m - 5 = (2m + 1)(m - 5)$$
$$m^2 - 10m + 25 = (m - 5)(m - 5)$$
$$\text{LCD} = (m + 5)(2m + 1)(m - 5)^2$$

**HELPFUL HINT**
$(x - 2)$ and $(2 - x)$ are opposite factors. Notice that $-1$ was factored from $(2 - x)$ so that the factors are identical.

**d.** Factor each denominator.

$$x^2 - 4 = (x + 2)(x - 2)$$
$$6 - 3x = 3(2 - x) = 3(-1)(x - 2)$$
$$\text{LCD} = 3(-1)(x + 2)(x - 2)$$
$$= -3(x + 2)(x - 2)$$

**HELPFUL HINT**
If opposite factors occur, do not use both in the LCD. Instead, factor $-1$ from one of the opposite factors so that the factors are then identical.

3 To add or subtract rational expressions with unlike denominators, we write each rational expression as an equivalent rational expression so that their denominators are alike.

**ADDING OR SUBTRACTING RATIONAL EXPRESSIONS WITH UNLIKE DENOMINATORS**

**Step 1:** Find the LCD of the rational expressions.
**Step 2:** Write each rational expression as an equivalent rational expression whose denominator is the LCD found in Step 1.
**Step 3:** Add or subtract numerators, and write the result over the common denominator.
**Step 4:** Simplify the resulting rational expression.

**Example 3** Perform the indicated operation.

**a.** $\dfrac{2}{x^2y} + \dfrac{5}{3x^3y}$ **b.** $\dfrac{3x}{x + 2} + \dfrac{2x}{x - 2}$ **c.** $\dfrac{x}{x - 1} - \dfrac{4}{1 - x}$

*Solution* **a.** The LCD is $3x^3y$. Write each fraction as an equivalent fraction with denominator $3x^3y$. To do this, we multiply both the numerator and denominator of each fraction by the factors needed to obtain the LCD as denominator.

The first fraction is multiplied by $\frac{3x}{3x}$ so that the new denominator is the LCD.

$$\frac{2}{x^2y}+\frac{5}{3x^3y}=\frac{2\cdot 3x}{x^2y\cdot 3x}+\frac{5}{3x^3y}$$ The second expression already has a denominator of $3x^3y$.

$$=\frac{6x}{3x^3y}+\frac{5}{3x^3y}$$

$$=\frac{6x+5}{3x^3y}$$ Add the numerators.

**b.** The LCD is the product of the two denominators: $(x+2)(x-2)$.

$$\frac{3x}{x+2}+\frac{2x}{x-2}=\frac{3x\cdot(x-2)}{(x+2)\cdot(x-2)}+\frac{2x\cdot(x+2)}{(x-2)\cdot(x+2)}$$ Write equivalent rational expressions.

$$=\frac{3x(x-2)+2x(x+2)}{(x+2)(x-2)}$$ Add the numerators.

$$=\frac{3x^2-6x+2x^2+4x}{(x+2)(x-2)}$$ Apply the distributive property.

$$=\frac{5x^2-2x}{(x+2)(x-2)}$$ Simplify the numerator.

**c.** The LCD is either $x-1$ or $1-x$. To get a common denominator of $x-1$, we factor $-1$ from the denominator of the second rational expression.

$$\frac{x}{x-1}-\frac{4}{1-x}=\frac{x}{x-1}-\frac{4}{-1(x-1)}$$ Write $1-x$ as $-1(x-1)$.

$$=\frac{x}{x-1}-\frac{-1\cdot 4}{x-1}$$ Write $\frac{4}{-1(x-1)}$ as $\frac{-1\cdot 4}{x-1}$.

$$=\frac{x-(-4)}{x-1}$$

$$=\frac{x+4}{x-1}$$ Simplify.

**Example 4** Subtract $\frac{5k}{k^2-4}-\frac{2}{k^2+k-2}$.

*Solution*

$$\frac{5k}{k^2-4}-\frac{2}{k^2+k-2}=\frac{5k}{(k+2)(k-2)}-\frac{2}{(k+2)(k-1)}$$ Factor each denominator to find the LCD.

The LCD is $(k+2)(k-2)(k-1)$. We write equivalent rational expressions with the LCD as denominators.

$$\frac{5k}{(k+2)(k-2)}-\frac{2}{(k+2)(k-1)}=\frac{5k\cdot(k-1)}{(k+2)(k-2)\cdot(k-1)}-\frac{2\cdot(k-2)}{(k+2)(k-1)\cdot(k-2)}$$

$$=\frac{5k(k-1)-2(k-2)}{(k+2)(k-2)(k-1)}$$ Subtract the numerators.

$$= \frac{5k^2 - 5k - 2k + 4}{(k + 2)(k - 2)(k - 1)}$$ Multiply in the numerator.

$$= \frac{5k^2 - 7k + 4}{(k + 2)(k - 2)(k - 1)}$$ Simplify.

**Example 5** Add $\frac{2x - 1}{2x^2 - 9x - 5} + \frac{x + 3}{6x^2 - x - 2}$.

*Solution*

$$\frac{2x - 1}{2x^2 - 9x - 5} + \frac{x + 3}{6x^2 - x - 2} = \frac{2x - 1}{(2x + 1)(x - 5)} + \frac{x + 3}{(2x + 1)(3x - 2)}$$ Factor the denominators.

The LCD is $(2x + 1)(x - 5)(3x - 2)$.

$$= \frac{(2x - 1) \cdot (3x - 2)}{(2x + 1)(x - 5) \cdot (3x - 2)} + \frac{(x + 3) \cdot (x - 5)}{(2x + 1)(3x - 2) \cdot (x - 5)}$$

$$= \frac{(2x - 1)(3x - 2) + (x + 3)(x - 5)}{(2x + 1)(x - 5)(3x - 2)}$$ Add the numerators.

$$= \frac{6x^2 - 7x + 2 + x^2 - 2x - 15}{(2x + 1)(x - 5)(3x - 2)}$$ Multiply in the numerator.

$$= \frac{7x^2 - 9x - 13}{(2x + 1)(x - 5)(3x - 2)}$$ Simplify.

**Example 6** Perform each indicated operation.

$$\frac{7}{x - 1} + \frac{10x}{x^2 - 1} - \frac{5}{x + 1}$$

*Solution*

$$\frac{7}{x - 1} + \frac{10x}{x^2 - 1} - \frac{5}{x + 1} = \frac{7}{x - 1} + \frac{10x}{(x - 1)(x + 1)} - \frac{5}{x + 1}$$ Factor the denominators.

The LCD is $(x - 1)(x + 1)$.

$$= \frac{7 \cdot (x + 1)}{(x - 1) \cdot (x + 1)} + \frac{10x}{(x - 1)(x + 1)} - \frac{5 \cdot (x - 1)}{(x + 1) \cdot (x - 1)}$$

$$= \frac{7(x + 1) + 10x - 5(x - 1)}{(x - 1)(x + 1)}$$ Add and subtract the numerators.

$$= \frac{7x + 7 + 10x - 5x + 5}{(x - 1)(x + 1)}$$ Multiply in the numerator.

$$= \frac{12x + 12}{(x - 1)(x + 1)}$$ Simplify.

$$= \frac{12(x + 1)}{(x - 1)(x + 1)}$$ Factor the numerator.

$$= \frac{12}{x - 1}$$ Divide out common factors.

## GRAPHING CALCULATOR EXPLORATIONS

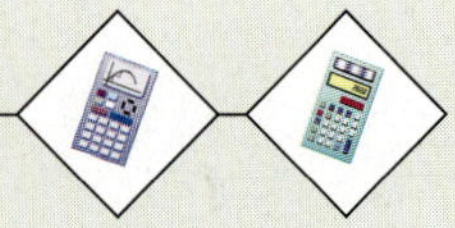

A graphing calculator can be used to support the results of operations on rational expressions. For example, to verify the result of Example 3b, graph

$$Y_1 = \frac{3x}{x+2} + \frac{2x}{x-2} \quad \text{and} \quad Y_2 = \frac{5x^2 - 2x}{(x+2)(x-2)}$$

on the same set of axes. The graphs should be the same. Use a Table feature or a Trace feature to see that this is true.

## Exercise Set 6.2

*Perform the indicated operation. If possible, simplify your answer. See Example 1.*

1. $\frac{2}{x} - \frac{5}{x}$
2. $\frac{4}{x^2} + \frac{2}{x^2}$
3. $\frac{2}{x-2} + \frac{x}{x-2}$
4. $\frac{x}{5-x} + \frac{2}{5-x}$
5. $\frac{x^2}{x+2} - \frac{4}{x+2}$
6. $\frac{4}{x-2} - \frac{x^2}{x-2}$
7. $\frac{2x-6}{x^2+x-6} + \frac{3-3x}{x^2+x-6}$
8. $\frac{5x+2}{x^2+2x-8} + \frac{2-4x}{x^2+2x-8}$

△ 9. Find the perimeter and the area of the square.

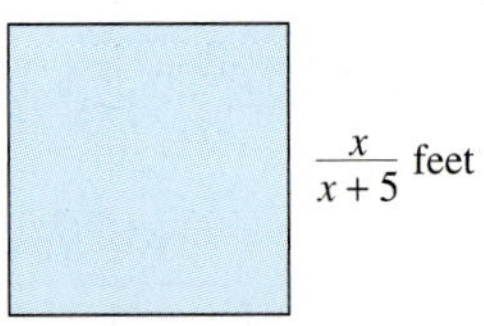


△ 10. Find the perimeter of the quadrilateral.

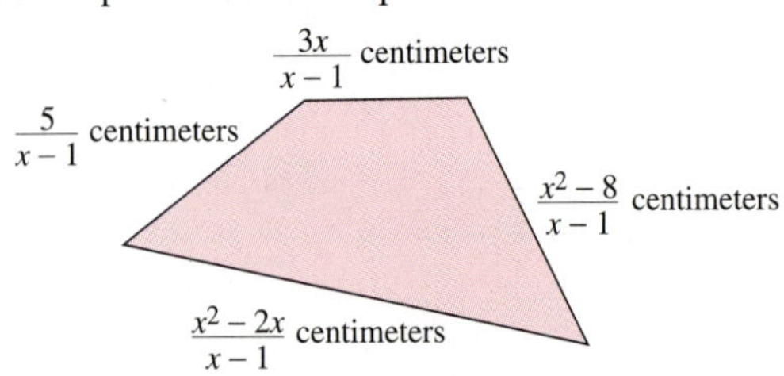


*Find the LCD of the rational expressions in each list. See Example 2.*

11. $\frac{2}{7}, \frac{3}{5x}$
12. $\frac{4}{5y}, \frac{3}{4y^2}$
13. $\frac{3}{x}, \frac{2}{x+1}$
14. $\frac{5}{2x}, \frac{7}{2+x}$
15. $\frac{12}{x+7}, \frac{8}{x-7}$
16. $\frac{1}{2x-1}, \frac{x}{2x+1}$
17. $\frac{5}{3x+6}, \frac{2x}{2x-4}$
18. $\frac{2}{3a+9}, \frac{5}{5a-15}$
19. $\frac{5+x}{(3x-1)(x+2)}, \frac{2}{3x-1}$
20. $\frac{6-x}{(x+3)(x-3)}, \frac{9}{x+3}$
21. $\frac{2a}{a^2-b^2}, \frac{1}{a^2-2ab+b^2}$
22. $\frac{2a}{a^2+8a+16}, \frac{7a}{a^2+a-12}$
23. $\frac{x}{x^2-9}, \frac{5x}{x}, \frac{7}{12-4x}$
24. $\frac{9}{x^2-25}, \frac{1}{50-10x}, \frac{6}{x}$
25. When is the LCD of two rational expressions equal to the product of their denominators? (*Hint:* What is the LCD of $\frac{1}{x}$ and $\frac{7}{x+5}$?)
26. When is the LCD of two rational expressions with different denominators equal to one of the denominators? (*Hint:* What is the LCD of $\frac{3x}{x+2}$ and $\frac{7x+1}{(x+2)^3}$?)

*Perform the indicated operation. If possible, simplify your answer. Write each answer in lowest terms. See Example 3a and 3b.*

27. $\frac{4}{3x} + \frac{3}{2x}$
28. $\frac{10}{7x} - \frac{5}{2x}$
29. $\frac{5}{2y^2} - \frac{2}{7y}$
30. $\frac{4}{11x^4y} - \frac{1}{4x^2y^3}$
31. $\frac{x-3}{x+4} - \frac{x+2}{x-4}$
32. $\frac{x-1}{x-5} - \frac{x+2}{x+5}$

**33.** $\frac{1}{x-5}+\frac{x}{x^2-x-20}$

**34.** $\frac{x+1}{x^2-x-20}-\frac{2}{x+4}$

*Perform the indicated operation. If possible, simplify your answer. See Example 3c.*

**35.** $\frac{1}{a-b}+\frac{1}{b-a}$

**36.** $\frac{1}{a-3}-\frac{1}{3-a}$

**37.** $\frac{x+1}{1-x}+\frac{1}{x-1}$

**38.** $\frac{5}{1-x}-\frac{1}{x-1}$

**39.** $\frac{5}{x-2}+\frac{x+4}{2-x}$

**40.** $\frac{3}{5-x}+\frac{x+2}{x-5}$

*Perform each indicated operation. If possible, simplify your answer. Write each answer in lowest terms. See Examples 4 through 6.*

**41.** $\frac{y+1}{y^2-6y+8}-\frac{3}{y^2-16}$

**42.** $\frac{x+2}{x^2-36}-\frac{x}{x^2+9x+18}$

**43.** $\frac{x+4}{3x^2+11x+6}+\frac{x}{2x^2+x-15}$

**44.** $\frac{x+3}{5x^2+12x+4}+\frac{6}{x^2-x-6}$

**45.** $\frac{7}{x^2-x-2}+\frac{x}{x^2+4x+3}$

**46.** $\frac{a}{a^2+10a+25}+\frac{4}{a^2+6a+5}$

**47.** $\frac{2}{x+1}-\frac{3x}{3x+3}+\frac{1}{2x+2}$

**48.** $\frac{5}{3x-6}-\frac{x}{x-2}+\frac{3+2x}{5x-10}$

**49.** $\frac{3}{x+3}+\frac{5}{x^2+6x+9}-\frac{x}{x^2-9}$

**50.** $\frac{x+2}{x^2-2x-3}+\frac{x}{x-3}-\frac{4}{x+1}$

*Add or subtract as indicated. If possible, simplify your answer.*

**51.** $\frac{4}{3x^2y^3}+\frac{5}{3x^2y^3}$

**52.** $\frac{7}{2xy^4}+\frac{1}{2xy^4}$

**53.** $\frac{x-5}{2x}-\frac{x+5}{2x}$

**54.** $\frac{x+4}{4x}-\frac{x-4}{4x}$

**55.** $\frac{3}{2x+10}+\frac{8}{3x+15}$

**56.** $\frac{10}{3x-3}+\frac{1}{7x-7}$

**57.** $\frac{-2}{x^2-3x}-\frac{1}{x^3-3x^2}$

**58.** $\frac{-3}{2a+8}-\frac{8}{a^2+4a}$

**59.** $\frac{ab}{a^2-b^2}+\frac{b}{a+b}$

**60.** $\frac{x}{25-x^2}+\frac{2}{3x-15}$

**61.** $\frac{5}{x^2-4}-\frac{3}{x^2+4x+4}$

**62.** $\frac{3z}{z^2-9}-\frac{2}{3-z}$

**63.** $\frac{2}{a^2+2a+1}+\frac{3}{a^2-1}$

**64.** $\frac{9x+2}{3x^2-2x-8}+\frac{7}{3x^2+x-4}$

**65.** In your own words, explain how to add rational expressions with different denominators.

**66.** In your own words, explain how to multiply rational expressions.

**67.** In your own words, explain how to divide rational expressions.

**68.** In your own words, explain how to subtract rational expressions with different denominators.

*Perform the indicated operation. If possible, simplify your answer.*

**69.** $\left(\frac{2}{3}-\frac{1}{x}\right)\cdot\left(\frac{3}{x}+\frac{1}{2}\right)$

**70.** $\left(\frac{2}{3}-\frac{1}{x}\right)\div\left(\frac{3}{x}+\frac{1}{2}\right)$

**71.** $\left(\frac{1}{x}+\frac{2}{3}\right)-\left(\frac{1}{x}-\frac{2}{3}\right)$

**72.** $\left(\frac{1}{2}+\frac{2}{x}\right)-\left(\frac{1}{2}-\frac{1}{x}\right)$

**73.** $\left(\frac{2a}{3}\right)^2\div\left(\frac{a^2}{a+1}-\frac{1}{a+1}\right)$

**74.** $\left(\frac{x+2}{2x}-\frac{x-2}{2x}\right)\cdot\left(\frac{5x}{4}\right)^2$

**75.** $\left(\frac{2x}{3}\right)^2\div\left(\frac{x}{3}\right)^2$

**76.** $\left(\frac{2x}{3}\right)^2\cdot\left(\frac{3}{x}\right)^2$

**77.** $\frac{x}{x^2-9}+\frac{3}{x^2-6x+9}-\frac{1}{x+3}$

**78.** $\frac{3}{x^2-9}-\frac{x}{x^2-6x+9}+\frac{1}{x+3}$

**79.** $\left(\frac{x}{x+1}-\frac{x}{x-1}\right)\div\frac{x}{2x+2}$

**80.** $\frac{x}{2x+2}\div\left(\frac{x}{x+1}+\frac{x}{x-1}\right)$

**81.** $\frac{4}{x}\cdot\left(\frac{2}{x+2}-\frac{2}{x-2}\right)$

**82.** $\frac{1}{x+1}\cdot\left(\frac{5}{x}+\frac{2}{x-3}\right)$

*Use a graphing calculator to support the results of each exercise.*

**83.** Exercise 3

**84.** Exercise 4

**85.** Exercise 31

**86.** Exercise 32

## REVIEW EXERCISES

*Use the distributive property to multiply the following. See Section 1.4*

**87.** $12\left(\dfrac{2}{3} + \dfrac{1}{6}\right)$

**88.** $14\left(\dfrac{1}{7} + \dfrac{3}{14}\right)$

**89.** $x^2\left(\dfrac{4}{x^2} + 1\right)$

**90.** $5y^2\left(\dfrac{1}{y^2} - \dfrac{1}{5}\right)$

*Find each root. See Section 1.3.*

**91.** $\sqrt{100}$

**92.** $\sqrt{25}$

**93.** $\sqrt[3]{8}$

**94.** $\sqrt[3]{27}$

**95.** $\sqrt[4]{81}$

**96.** $\sqrt[4]{16}$

*Use the Pythagorean theorem to find each unknown length of a right triangle. See Section 5.8.*

△ **97.**

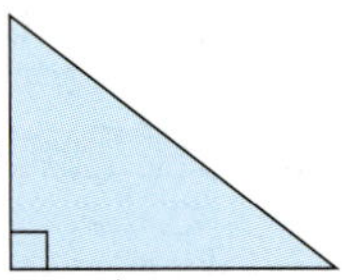


△ **98.**

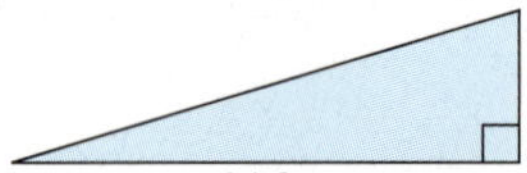


### *A Look Ahead*

**Example**

Add $x^{-1} + 3x^{-2}$.

*Solution*

$$\begin{aligned} x^{-1} + 3x^{-2} &= \frac{1}{x} + \frac{3}{x^2} \\ &= \frac{1 \cdot x}{x \cdot x} + \frac{3}{x^2} \\ &= \frac{x}{x^2} + \frac{3}{x^2} \\ &= \frac{x+3}{x^2} \end{aligned}$$

*Perform the indicated operation. See the preceding example.*

**99.** $x^{-1} + (2x)^{-1}$

**100.** $3y^{-1} + (4y)^{-1}$

**101.** $4x^{-2} - 3x^{-1}$

**102.** $(4x)^{-2} - (3x)^{-1}$

**103.** $x^{-3}(2x + 1) - 5x^{-2}$

**104.** $4x^{-3} + x^{-4}(5x + 7)$

## 6.3 SIMPLIFYING COMPLEX FRACTIONS

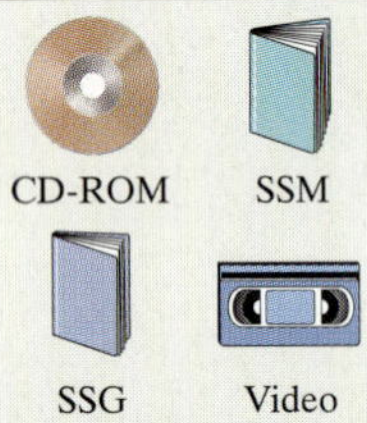


▶ **OBJECTIVES**

1. Simplify complex fractions by simplifying the numerator and denominator and then dividing.
2. Simplify complex fractions by multiplying by a common denominator.
3. Simplify expressions with negative exponents.

**1** A rational expression whose numerator, denominator, or both contain one or more rational expressions is called a **complex rational expression** or a **complex fraction**.

***Complex Fractions***

$$\frac{\frac{1}{a}}{\frac{b}{2}} \qquad \frac{\frac{x}{2y^2}}{\frac{6x-2}{9y}} \qquad \frac{x + \frac{1}{y}}{y+1}$$

The parts of a complex fraction are

$$\frac{\left.\frac{x}{y+2}\right\}}{\left.7 + \frac{1}{y}\right\}}$$

← Numerator of complex fraction.

← Main fraction bar.

← Denominator of complex fraction.

Our goal in this section is to simplify complex fractions. A complex fraction is simplified when it is in the form $\frac{P}{Q}$, where $P$ and $Q$ are polynomials that have no common factors. Two methods of simplifying complex fractions are introduced. The first method evolves from the definition of a fraction as a quotient.

### SIMPLIFYING A COMPLEX FRACTION: METHOD I

**Step 1:** Simplify the numerator and the denominator of the complex fraction so that each is a single fraction.
**Step 2:** Perform the indicated division by multiplying the numerator of the complex fraction by the reciprocal of the denominator of the complex fraction.
**Step 3:** Simplify if possible.

**Example 1** Simplify each complex fraction.

**a.** $\dfrac{\frac{2x}{27y^2}}{\frac{6x^2}{9}}$ **b.** $\dfrac{\frac{5x}{x+2}}{\frac{10}{x-2}}$ **c.** $\dfrac{\frac{x}{y^2}+\frac{1}{y}}{\frac{y}{x^2}+\frac{1}{x}}$

*Solution* **a.** The numerator of the complex fraction is already a single fraction, and so is the denominator. Perform the indicated division by multiplying the numerator, $\frac{2x}{27y^2}$, by the reciprocal of the denominator, $\frac{6x^2}{9}$. Then simplify.

$$\frac{\frac{2x}{27y^2}}{\frac{6x^2}{9}} = \frac{2x}{27y^2} \div \frac{6x^2}{9}$$

$$= \frac{2x}{27y^2} \cdot \frac{9}{6x^2} \qquad \text{Multiply by the reciprocal of } \frac{6x^2}{9}.$$

$$= \frac{2x \cdot 9}{27y^2 \cdot 6x^2}$$

$$= \frac{1}{9xy^2}$$

**HELPFUL HINT**
Both the numerator and denominator are single fractions, so we perform the indicated division.

**b.** $$\frac{\frac{5x}{x+2}}{\frac{10}{x-2}} = \frac{5x}{x+2} \cdot \frac{x-2}{10}$$ Multiply by the reciprocal of $\frac{10}{x-2}$.

$$= \frac{5x(x-2)}{2 \cdot 5(x+2)}$$

$$= \frac{x(x-2)}{2(x+2)}$$ Simplify.

**c.** First simplify the numerator and the denominator of the complex fraction separately so that each is a single fraction. Then perform the indicated division.

$$\frac{\frac{x}{y^2}+\frac{1}{y}}{\frac{y}{x^2}+\frac{1}{x}} = \frac{\frac{x}{y^2}+\frac{1 \cdot y}{y \cdot y}}{\frac{y}{x^2}+\frac{1 \cdot x}{x \cdot x}}$$ The LCD is $y^2$. The LCD is $x^2$.

$$= \frac{\frac{x+y}{y^2}}{\frac{y+x}{x^2}}$$ Add. Add.

$$= \frac{x+y}{y^2} \cdot \frac{x^2}{y+x}$$ Multiply by the reciprocal of $\frac{y+x}{x^2}$.

$$= \frac{x^2(x+y)}{y^2(y+x)}$$

$$= \frac{x^2}{y^2}$$ Simplify.

2 Next we look at another method of simplifying complex fractions. With this method we multiply the numerator and the denominator of the complex fraction by the LCD of all fractions in the complex fraction.

**SIMPLIFYING A COMPLEX FRACTION: METHOD II**

**Step 1:** Multiply the numerator and the denominator of the complex fraction by the LCD of the fractions in both the numerator and the denominator.

**Step 2:** Simplify.

**Example 2** Simplify each complex fraction.

**a.** $\dfrac{\frac{5x}{x+2}}{\frac{10}{x-2}}$

**b.** $\dfrac{\frac{x}{y^2}+\frac{1}{y}}{\frac{y}{x^2}+\frac{1}{x}}$

*Solution* **a.** The least common denominator of $\frac{5x}{x+2}$ and $\frac{10}{x-2}$ is $(x+2)(x-2)$.

Multiply both the numerator, $\frac{5x}{x+2}$, and the denominator, $\frac{10}{x-2}$, by the LCD.

$$\frac{\frac{5x}{x+2}}{\frac{10}{x-2}} = \frac{\left(\frac{5x}{x+2}\right)\cdot(x+2)(x-2)}{\left(\frac{10}{x-2}\right)\cdot(x+2)(x-2)}$$ Multiply numerator and denominator by the LCD.

$$= \frac{5x\cdot(x-2)}{2\cdot 5\cdot(x+2)}$$ Simplify.

$$= \frac{x(x-2)}{2(x+2)}$$ Simplify.

**b.** The least common denominator of $\frac{x}{y^2}, \frac{1}{y}, \frac{y}{x^2}$, and $\frac{1}{x}$ is $x^2y^2$.

$$\frac{\frac{x}{y^2}+\frac{1}{y}}{\frac{y}{x^2}+\frac{1}{x}} = \frac{\left(\frac{x}{y^2}+\frac{1}{y}\right)\cdot x^2y^2}{\left(\frac{y}{x^2}+\frac{1}{x}\right)\cdot x^2y^2}$$ Multiply the numerator and denominator by the LCD.

$$= \frac{\frac{x}{y^2}\cdot x^2y^2+\frac{1}{y}\cdot x^2y^2}{\frac{y}{x^2}\cdot x^2y^2+\frac{1}{x}\cdot x^2y^2}$$ Use the distributive property.

$$= \frac{x^3+x^2y}{y^3+xy^2}$$ Simplify.

$$= \frac{x^2(x+y)}{y^2(y+x)}$$ Factor.

$$= \frac{x^2}{y^2}$$ Simplify.

3 If an expression contains negative exponents, write the expression as an equivalent expression with positive exponents.

## Example 3 Simplify.

$$\frac{x^{-1}+2xy^{-1}}{x^{-2}-x^{-2}y^{-1}}$$

*Solution* This fraction does not appear to be a complex fraction. If we write it by using only positive exponents, however, we see that it is a complex fraction.

$$\frac{x^{-1}+2xy^{-1}}{x^{-2}-x^{-2}y^{-1}} = \frac{\frac{1}{x}+\frac{2x}{y}}{\frac{1}{x^2}-\frac{1}{x^2y}}$$

The LCD of $\frac{1}{x}, \frac{2x}{y}, \frac{1}{x^2}$, and $\frac{1}{x^2y}$ is $x^2y$. Multiply both the numerator and denominator by $x^2y$.

$$= \frac{\left(\frac{1}{x} + \frac{2x}{y}\right) \cdot x^2y}{\left(\frac{1}{x^2} - \frac{1}{x^2y}\right) \cdot x^2y}$$

$$= \frac{\frac{1}{x} \cdot x^2y + \frac{2x}{y} \cdot x^2y}{\frac{1}{x^2} \cdot x^2y - \frac{1}{x^2y} \cdot x^2y}$$ Apply the distributive property.

$$= \frac{xy + 2x^3}{y - 1}$$ Simplify.

## Exercise Set 6.3

*Simplify each complex fraction. See Examples 1 and 2.*

1. $\dfrac{\frac{1}{3}}{\frac{2}{5}}$

2. $\dfrac{\frac{3}{5}}{\frac{4}{5}}$

3. $\dfrac{\frac{4}{x}}{\frac{5}{2x}}$

4. $\dfrac{\frac{5}{2x}}{\frac{4}{x}}$

5. $\dfrac{\frac{10}{3x}}{\frac{5}{6x}}$

6. $\dfrac{\frac{15}{2x}}{\frac{5}{6x}}$

7. $\dfrac{1 + \frac{2}{5}}{2 + \frac{3}{5}}$

8. $\dfrac{2 + \frac{1}{7}}{3 - \frac{4}{7}}$

9. $\dfrac{\frac{4}{x-1}}{\frac{x}{x-1}}$

10. $\dfrac{\frac{x}{x+2}}{\frac{2}{x+2}}$

11. $\dfrac{1 - \frac{2}{x}}{x - \frac{4}{9x}}$

12. $\dfrac{5 - \frac{3}{x}}{x + \frac{2}{3x}}$

13. $\dfrac{\frac{1}{x+1} - 1}{\frac{1}{x-1} + 1}$

14. $\dfrac{1 + \frac{1}{x-1}}{1 - \frac{1}{x+1}}$

*Simplify. See Example 3.*

15. $\dfrac{x^{-1}}{x^{-2} + y^{-2}}$

16. $\dfrac{a^{-3} + b^{-1}}{a^{-2}}$

17. $\dfrac{2a^{-1} + 3b^{-2}}{a^{-1} - b^{-1}}$

18. $\dfrac{x^{-1} + y^{-1}}{3x^{-2} + 5y^{-2}}$

19. $\dfrac{1}{x - x^{-1}}$

20. $\dfrac{x^{-2}}{x + 3x^{-1}}$

*Simplify.*

21. $\dfrac{\frac{x+1}{7}}{\frac{x+2}{7}}$

22. $\dfrac{\frac{y}{10}}{\frac{x+1}{10}}$

23. $\dfrac{\frac{1}{2} - \frac{1}{3}}{\frac{3}{4} + \frac{2}{5}}$

24. $\dfrac{\frac{5}{6} - \frac{1}{2}}{\frac{1}{3} + \frac{1}{8}}$

25. $\dfrac{\frac{x+1}{3}}{\frac{2x-1}{6}}$

26. $\dfrac{\frac{x+3}{12}}{\frac{4x-5}{15}}$

27. $\dfrac{\frac{x}{3}}{\frac{2}{x+1}}$

28. $\dfrac{\frac{x-1}{5}}{\frac{3}{x}}$

29. $\dfrac{\frac{2}{x} + 3}{\frac{4}{x^2} - 9}$

30. $\dfrac{2 + \frac{1}{x}}{4x - \frac{1}{x}}$

**31.** $\dfrac{1 - \dfrac{x}{y}}{\dfrac{x^2}{y^2} - 1}$

**32.** $\dfrac{1 - \dfrac{2}{x}}{x - \dfrac{4}{x}}$

**33.** $\dfrac{\dfrac{-2x}{x - y}}{\dfrac{y}{x^2}}$

**34.** $\dfrac{\dfrac{7y}{x^2 + xy}}{\dfrac{y^2}{x^2}}$

**35.** $\dfrac{\dfrac{2}{x} + \dfrac{1}{x^2}}{\dfrac{y}{x^2}}$

**36.** $\dfrac{\dfrac{5}{x^2} - \dfrac{2}{x}}{\dfrac{1}{x} + 2}$

**37.** $\dfrac{\dfrac{x}{9} - \dfrac{1}{x}}{1 + \dfrac{3}{x}}$

**38.** $\dfrac{\dfrac{x}{4} - \dfrac{4}{x}}{1 - \dfrac{4}{x}}$

**39.** $\dfrac{\dfrac{x - 1}{x^2 - 4}}{1 + \dfrac{1}{x - 2}}$

**40.** $\dfrac{\dfrac{2}{x + 5} + \dfrac{4}{x + 3}}{\dfrac{3x + 13}{x^2 + 8x + 15}}$

**41.** $\dfrac{\dfrac{4}{5 - x} + \dfrac{5}{x - 5}}{\dfrac{2}{x} + \dfrac{3}{x - 5}}$

**42.** $\dfrac{\dfrac{3}{x - 4} - \dfrac{2}{4 - x}}{\dfrac{2}{x - 4} - \dfrac{2}{x}}$

**43.** $\dfrac{\dfrac{x + 2}{x} - \dfrac{2}{x - 1}}{\dfrac{x + 1}{x} + \dfrac{x + 1}{x - 1}}$

**44.** $\dfrac{\dfrac{5}{a + 2} - \dfrac{1}{a - 2}}{\dfrac{3}{2 + a} + \dfrac{6}{2 - a}}$

**45.** $\dfrac{\dfrac{x - 2}{x + 2} + \dfrac{x + 2}{x - 2}}{\dfrac{x - 2}{x + 2} - \dfrac{x + 2}{x - 2}}$

**46.** $\dfrac{\dfrac{x - 1}{x + 1} - \dfrac{x + 1}{x - 1}}{\dfrac{x - 1}{x + 1} + \dfrac{x + 1}{x - 1}}$

**47.** $\dfrac{\dfrac{2}{y^2} - \dfrac{5}{xy} - \dfrac{3}{x^2}}{\dfrac{2}{y^2} + \dfrac{7}{xy} + \dfrac{3}{x^2}}$

**48.** $\dfrac{\dfrac{2}{x^2} - \dfrac{1}{xy} - \dfrac{1}{y^2}}{\dfrac{1}{x^2} - \dfrac{3}{xy} + \dfrac{2}{y^2}}$

**49.** $\dfrac{a^{-1} + 1}{a^{-1} - 1}$

**50.** $\dfrac{a^{-1} - 4}{4 + a^{-1}}$

**51.** $\dfrac{3x^{-1} + (2y)^{-1}}{x^{-2}}$

**52.** $\dfrac{5x^{-2} - 3y^{-1}}{x^{-1} + y^{-1}}$

**53.** $\dfrac{2a^{-1} + (2a)^{-1}}{a^{-1} + 2a^{-2}}$

**54.** $\dfrac{a^{-1} + 2a^{-2}}{2a^{-1} + (2a)^{-1}}$

**55.** $\dfrac{5x^{-1} + 2y^{-1}}{x^{-2}y^{-2}}$

**56.** $\dfrac{x^{-2}y^{-2}}{5x^{-1} + 2y^{-1}}$

**57.** $\dfrac{5x^{-1} - 2y^{-1}}{25x^{-2} - 4y^{-2}}$

**58.** $\dfrac{3x^{-1} + 3y^{-1}}{4x^{-2} - 9y^{-2}}$

**59.** $(x^{-1} + y^{-1})^{-1}$

**60.** $\dfrac{xy}{x^{-1} + y^{-1}}$

**61.** $\dfrac{x}{1 - \dfrac{1}{1 + \dfrac{1}{x}}}$

**62.** $\dfrac{1}{1 - \dfrac{1}{1 - \dfrac{1}{x}}}$

**63.** When the source of a sound is traveling toward a listener, the pitch that the listener hears due to the Doppler effect is given by the complex rational compression $\dfrac{a}{1 - \dfrac{s}{770}}$, where $a$ is the actual pitch of the sound and $s$ is the speed of the sound source. Simplify this expression.

**64.** Which of the following are equivalent to $\dfrac{\dfrac{1}{x}}{\dfrac{3}{y}}$?

**a.** $\dfrac{1}{x} \div \dfrac{3}{y}$

**b.** $\dfrac{1}{x} \cdot \dfrac{y}{3}$

**c.** $\dfrac{1}{x} \div \dfrac{y}{3}$

*In the study of calculus, the difference quotient* $\dfrac{f(a + h) - f(a)}{h}$ *is often found and simplified. Find and simplify this quotient for each function* $f(x)$ *by following steps* **a** *through* **d**.

**a.** *Find* $(a + h)$.

**b.** *Find* $f(a)$.

**c.** *Use steps* **a** *and* **b** *to find* $\dfrac{f(a + h) - f(a)}{h}$

**d.** *Simplify the result of step* **c**.

**65.** $f(x) = \dfrac{1}{x}$

**66.** $f(x) = \dfrac{5}{x}$

**67.** $\dfrac{3}{x+1}$ **68.** $\dfrac{2}{x^2}$

## REVIEW EXERCISES

*Simplify. See Sections 5.1 and 5.2.*

**69.** $\dfrac{3x^3y^2}{12x}$ **70.** $\dfrac{-36xb^3}{9xb^2}$

**71.** $\dfrac{144x^5y^5}{-16x^2y}$ **72.** $\dfrac{48x^3y^2}{-4xy}$

*Solve the following. See Sections 2.6 and 2.7*

**73.** $|x-5|=9$ **74.** $|2y+1|=1$

**75.** $|x-5|<9$ **76.** $|2x+1|\geq 1$

## *A Look Ahead*

**Example**

Simplify $\dfrac{2(a+b)^{-1}-5(a-b)^{-1}}{4(a^2-b^2)^{-1}}$

*Solution*

$$\frac{2(a+b)^{-1}-5(a-b)^{-1}}{4(a^2-b^2)^{-1}}=\frac{\dfrac{2}{a+b}-\dfrac{5}{a-b}}{\dfrac{4}{a^2-b^2}}$$

$$=\frac{\left(\dfrac{2}{a+b}-\dfrac{5}{a-b}\right)\cdot(a+b)(a-b)}{\left[\dfrac{4}{(a+b)(a-b)}\right]\cdot(a+b)(a-b)}$$

$$=\frac{\dfrac{2}{a+b}\cdot(a+b)(a-b)-\dfrac{5}{a-b}\cdot(a+b)(a-b)}{\dfrac{4(a+b)(a-b)}{(a+b)(a-b)}}$$

$$=\frac{2(a-b)-5(a+b)}{4}$$

$$=\frac{-3a-7b}{4},\text{ or }-\frac{3a+7b}{4}$$

*Simplify. See the preceding example.*

**77.** $\dfrac{1}{1-(1-x)^{-1}}$ **78.** $\dfrac{1}{1+(1+x)^{-1}}$

**79.** $\dfrac{(x+2)^{-1}+(x-2)^{-1}}{(x^2-4)^{-1}}$ **80.** $\dfrac{(y-1)^{-1}-(y+4)^{-1}}{(y^2+3y-4)^{-1}}$

**81.** $\dfrac{3(a+1)^{-1}+4a^{-2}}{(a^3+a^2)^{-1}}$ **82.** $\dfrac{9x^{-1}-5(x-y)^{-1}}{4(x-y)^{-1}}$

## 6.4 DIVIDING POLYNOMIALS

CD-ROM SSM SSG Video

**▶ OBJECTIVES**

1. Divide a polynomial by a monomial.
2. Divide by a polynomial.

**1** Recall that a rational expression is a quotient of polynomials. An equivalent form of a rational expression can be obtained by performing the indicated division. For example, the rational expression $\dfrac{10x^3-5x^2+20x}{5x}$ can be thought of as the polynomial $10x^3-5x^2+20x$ divided by the monomial $5x$. To perform this division of a polynomial by a monomial (which we do below) recall the following addition fact for fractions with a common denominator.

$$\frac{a}{c}+\frac{b}{c}=\frac{a+b}{c}$$

If $a$, $b$, and $c$ are monomials, we might read this equation from right to left and gain insight into dividing a polynomial by a monomial.

**DIVIDING A POLYNOMIAL BY A MONOMIAL**

Divide each term in the polynomial by the monomial.

$$\frac{a+b}{c} = \frac{a}{c} + \frac{b}{c}, \text{ where } c \neq 0$$

**Example 1** Divide $10x^3 - 5x^2 + 20x$ by $5x$.

*Solution* We divide each term of $10x^3 - 5x^2 + 20x$ by $5x$ and simplify.

$$\frac{10x^3 - 5x^2 + 20x}{5x} = \frac{10x^3}{5x} - \frac{5x^2}{5x} + \frac{20x}{5x} = 2x^2 - x + 4$$

**Check:** To check, see that (quotient) (divisor) = dividend, or

$$(2x^2 - x + 4)(5x) = 10x^3 - 5x^2 + 20x.$$

**Example 2** Divide $\dfrac{3x^5y^2 - 15x^3y - x^2y - 6x}{x^2y}$

*Solution* We divide each term in the numerator by $x^2y$.

$$\frac{3x^5y^2 - 15x^3y - x^2y - 6x}{x^2y} = \frac{3x^5y^2}{x^2y} - \frac{15x^3y}{x^2y} - \frac{x^2y}{x^2y} - \frac{6x}{x^2y}$$

$$= 3x^3y - 15x - 1 - \frac{6}{xy}$$

2 To divide a polynomial by a polynomial other than a monomial, we use **long division.** Polynomial long division is similar to long division of real numbers. We review long division of real numbers by dividing 7 into 296.

$$\begin{array}{r} 42 \\ \text{Divisor: } 7\overline{)296} \\ -28 \\ \hline 16 \\ -14 \\ \hline 2 \end{array}$$

- $4(7) = 28.$
- Subtract and bring down the next digit in the dividend.
- $2(7) = 14.$
- Subtract. The remainder is 2.

The quotient is $42\dfrac{2}{7}$ (remainder) / (divisor).

**Check:** To check, notice that

$$42(7) + 2 = 296, \text{ the dividend.}$$

This same division process can be applied to polynomials, as shown next.

**Example 3** Divide $2x^2 - x - 10$ by $x + 2$.

*Solution* $2x^2 - x - 10$ is the dividend, and $x + 2$ is the divisor.

**Step 1:** Divide $2x^2$ by $x$.

$$x+2\overline{)2x^2 - x - 10}\quad\text{quotient: } 2x$$

$\frac{2x^2}{x} = 2x$, so $2x$ is the first term of the quotient.

**Step 2:** Multiply $2x(x + 2)$.

$$\begin{array}{r} 2x \phantom{-10}\\ x+2\overline{)2x^2 - x - 10} \\ 2x^2 + 4x \phantom{-10} \end{array}$$

$2x(x + 2)$
Like terms are lined up vertically.

**Step 3:** Subtract $(2x^2 + 4x)$ from $(2x^2 - x - 10)$ by changing the signs of $(2x^2 + 4x)$ and adding.

$$\begin{array}{r} 2x \phantom{-10}\\ x+2\overline{)2x^2 - x - 10} \\ \underline{-2x^2 - 4x} \phantom{-10} \\ -5x \phantom{-10} \end{array}$$

**Step 4:** Bring down the next term, $-10$, and start the process over.

$$\begin{array}{r} 2x \phantom{-10}\\ x+2\overline{)2x^2 - x - 10} \\ \underline{-2x^2 - 4x} \phantom{-10} \\ -5x - 10 \end{array}$$

**Step 5:** Divide $-5x$ by $x$.

$$\begin{array}{r} 2x - 5\\ x+2\overline{)2x^2 - x - 10} \\ \underline{-2x^2 - 4x} \phantom{-10} \\ -5x - 10 \end{array}$$

$\frac{-5x}{x} = -5$, so $-5$ is the second term of the quotient.

**Step 6:** Multiply $-5(x + 2)$.

$$\begin{array}{r} 2x - 5\\ x+2\overline{)2x^2 - x - 10} \\ \underline{-2x^2 - 4x} \phantom{-10} \\ -5x - 10 \\ -5x - 10 \end{array}$$

$-5(x + 2)$
Like terms are lined up vertically.

**Step 7:** Subtract $(-5x - 10)$ from $(-5x - 10)$.

$$\begin{array}{r} 2x - 5\\ x+2\overline{)2x^2 - x - 10} \\ \underline{-2x^2 - 4x} \phantom{-10} \\ -5x - 10 \\ \underline{+5x + 10} \\ 0 \end{array}$$

Then $\dfrac{2x^2 - x - 10}{x + 2} = 2x - 5$. There is no remainder.

**Check:** Check this result by multiplying $2x - 5$ by $x + 2$. Their product is

$(2x - 5)(x + 2) = 2x^2 - x - 10$, the dividend.

**Example 4** Divide: $(6x^2 - 19x + 12) \div (3x - 5)$

*Solution*

$$
\begin{array}{r}
2x \phantom{{}-19x+12} \\
3x - 5\overline{)6x^2 - 19x + 12} \\
\underline{-6x^2 \overset{+}{\not -} 10x} \phantom{{}+12} \\
-9x + 12
\end{array}
$$

Divide $\frac{6x^2}{3x} = 2x$.
Multiply $2x(3x - 5)$.
Subtract by adding the opposite.
Bring down the next term, $+12$.

$$
\begin{array}{r}
2x - 3 \\
3x - 5\overline{)6x^2 - 19x + 12} \\
\underline{-6x^2 \overset{+}{\not -} 10x} \phantom{{}+12} \\
-9x + 12 \\
\underline{\overset{+}{\not -}9x \overset{-}{\not +} 15} \\
-3
\end{array}
$$

Divide $\frac{-9x}{3x} = -3$.

Multiply $-3(3x - 5)$.
Subtract by adding the opposite.

**Check:** divisor · quotient + remainder

$(3x - 5)\quad(2x - 3) + (-3) = 6x^2 - 19x + 15 - 3$

$= 6x^2 - 19x + 12$ The dividend

The division checks, so

$$\frac{6x^2 - 19x + 12}{3x - 5} = 2x - 3 - \frac{3}{3x - 5}$$

**HELPFUL HINT**
This fraction is the remainder over the divisor.

**Example 5** Divide $3x^4 + 2x^3 - 8x + 6$ by $x^2 - 1$.

*Solution* Before dividing, we represent any "missing powers" by the product of 0 and the variable raised to the missing power. There is no $x^2$ term in the dividend, so we include $0x^2$ to represent the missing term. Also, there is no $x$ term in the divisor, so we include $0x$ in the divisor.

$$
\begin{array}{r}
3x^2 + 2x + 3 \\
x^2 + 0x - 1\overline{)3x^4 + 2x^3 + 0x^2 - 8x + 6} \\
\underline{-3x^4 \overset{-}{\not +} 0x^3 \overset{+}{\not -} 3x^2} \phantom{{}-8x+6} \\
2x^3 + 3x^2 - 8x \phantom{{}+6} \\
\underline{-2x^3 \overset{-}{\not +} 0x^2 \overset{+}{\not -} 2x} \phantom{{}+6} \\
3x^2 - 6x + 6 \\
\underline{-3x^2 \overset{-}{\not +} 0x \overset{+}{\not -} 3} \\
-6x + 9
\end{array}
$$

$\frac{3x^4}{x^2} = 3x^2$

$3x^2(x^2 + 0x - 1)$
Subtract. Bring down $-8x$.
$\frac{2x^3}{x^2} = 2x$, a term of the quotient.
$2x(x^2 + 0x - 1)$
Subtract. Bring down 6.
$\frac{3x^2}{x^2} = 3$, a term of the quotient.
$3(x^2 + 0x - 1)$
Subtract.

*The division process is finished when the degree of the remainder polynomial is less than the degree of the divisor.* Thus,

$$\frac{3x^4 + 2x^3 - 8x + 6}{x^2 - 1} = 3x^2 + 2x + 3 + \frac{-6x + 9}{x^2 - 1}$$

**Example 6** Divide $27x^3 + 8$ by $3x + 2$.

*Solution* We replace the missing terms in the dividend with $0x^2$ and $0x$.

$$\begin{array}{r|l}
 & 9x^2 - 6x + 4 \\
3x + 2 & 27x^3 + 0x^2 + 0x + 8 \\
 & \underline{-27x^3 \mp 18x^2} \quad 9x^2(3x + 2) \\
 & -18x^2 + 0x \quad \text{Subtract. Bring down } 0x. \\
 & \underline{\mp 18x^2 \mp 12x} \quad -6x(3x + 2) \\
 & 12x + 8 \quad \text{Subtract. Bring down } 8. \\
 & \underline{-12x \mp 8} \quad 4(3x + 2)
\end{array}$$

Thus, $\frac{27x^3 + 8}{3x + 2} = 9x^2 - 6x + 4$.

## Exercise Set 6.4

*Divide. See Examples 1 and 2*

1. Divide $4a^2 + 8a$ by $2a$.
2. Divide $6x^4 - 3x^3$ by $3x^2$.
3. $\frac{12a^5b^2 + 16a^4b}{4a^4b}$
4. $\frac{4x^3y + 12x^2y^2 - 4xy^3}{4xy}$
5. $\frac{4x^2y^2 + 6xy^2 - 4y^2}{2x^2y}$
6. $\frac{6x^5 + 74x^4 + 24x^3}{2x^3}$
7. $\frac{4x^2 + 8x + 4}{4}$
8. $\frac{15x^3 - 5x^2 + 10x}{5x^2}$
9. A board of length $(3x^4 + 6x^2 - 18)$ meters is to be cut into three pieces of the same length. Find the length of each piece.

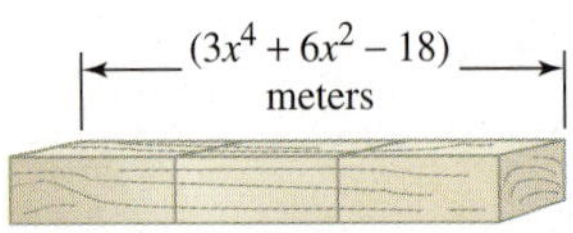


△ 10. The perimeter of a regular hexagon is given to be $12x^5 - 48x^3 + 3$ miles. Find the length of each side.

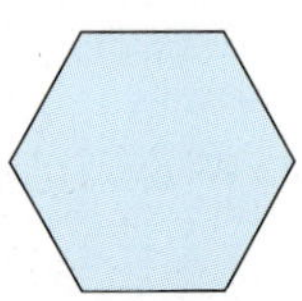

*Divide. See Examples 3 through 6.*

11. $(x^2 + 3x + 2) \div (x + 2)$
12. $(y^2 + 7y + 10) \div (y + 5)$
13. $(2x^2 - 6x - 8) \div (x + 1)$
14. $(3x^2 + 19x + 20) \div (x + 5)$
15. $2x^2 + 3x - 2$ by $2x + 4$
16. $6x^2 - 17x - 3$ by $3x - 9$
17. $(4x^3 + 7x^2 + 8x + 20) \div (2x + 4)$
18. $(18x^3 + x^2 - 90x - 5) \div (9x^2 - 45)$

△ 19. If the area of the rectangle is $(15x^2 - 29x - 14)$ square inches and its length is $(5x + 2)$ inches, find its width.

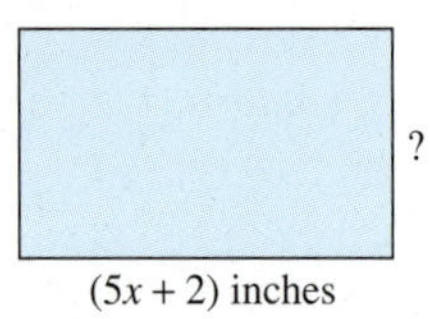


△ 20. If the area of a parallelogram is $(2x^2 - 17x + 35)$ square centimeters and its base is $(2x - 7)$ centimeters, find its height.

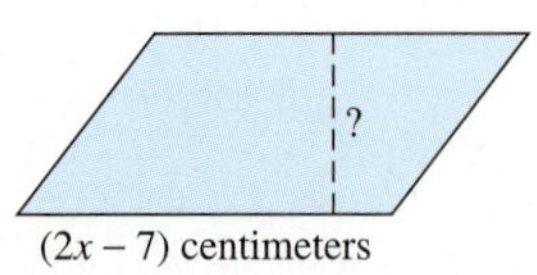

*Divide.*

**21.** $25a^2b^{12}$ by $10a^5b^7$

**22.** $12a^2b^3$ by $8a^7b$

**23.** $(x^6y^6 - x^3y^3) \div x^3y^3$

**24.** $(25xy^2 + 75xyz + 125x^2yz) \div -5x^2y$

**25.** $(a^2 + 4a + 3) \div (a + 1)$

**26.** $(3x^2 - 14x + 16) \div (x - 2)$

**27.** $(2x^2 + x - 10) \div (x - 2)$

**28.** $(x^2 - 7x + 12) \div (x - 5)$

**29.** $-16y^3 + 24y^4$ by $-4y^2$

**30.** $-20a^2b + 12ab^2$ by $-4ab$

**31.** $(2x^2 + 13x + 15) \div (x - 5)$

**32.** $(2x^2 + 13x + 5) \div (2x + 3)$

**33.** $(20x^2y^3 + 6xy^4 - 12x^3y^5) \div 2xy^3$

**34.** $(3x^2y + 6x^2y^2 + 3xy) \div 3xy$

**35.** $(6x^2 + 16x + 8) \div (3x + 2)$

**36.** $(x^2 - 25) \div (x + 5)$

**37.** $(2y^2 + 7y - 15) \div (2y - 3)$

**38.** $(3x^2 - 4x + 6) \div (x - 2)$

**39.** $4x^2 - 9$ by $2x - 3$

**40.** $8x^2 + 6x - 27$ by $4x + 9$

**41.** $2x^3 + 6x - 4$ by $x + 4$

**42.** $4x^3 - 5x$ by $2x - 1$

**43.** $3x^2 - 4$ by $x - 1$

**44.** $x^2 - 9$ by $x + 4$

**45.** $(-13x^3 + 2x^4 + 16x^2 - 9x + 20) \div (5 - x)$

**46.** $(5x^2 - 5x + 2x^3 + 20) \div (4 + x)$

**47.** $3x^5 - x^3 + 4x^2 - 12x - 8$ by $x^2 - 2$

**48.** $-8x^3 + 2x^4 + 19x^2 - 33x + 15$ by $x^2 - x + 5$

**49.** $(3x^3 - 5) \div 3x^2$

**50.** $(14x^3 - 2) \div (7x - 1)$

**51.** Find $P(1)$ for the polynomial function $P(x) = 3x^3 + 2x^2 - 4x + 3$. Next, divide $3x^3 + 2x^2 - 4x + 3$ by $x - 1$. Compare the remainder with $P(1)$.

**52.** Find $P(-2)$ for the polynomial function $P(x) = x^3 - 4x^2 - 3x + 5$. Next, divide $x^3 - 4x^2 - 3x + 5$ by $x + 2$. Compare the remainder with $P(-2)$.

**53.** Find $P(-3)$ for the polynomial $P(x) = 5x^4 - 2x^2 + 3x - 6$. Next, divide $5x^4 - 2x^2 + 3x - 6$ by $x + 3$. Compare the remainder with $P(-3)$.

**54.** Find $P(2)$ for the polynomial function $P(x) = -4x^4 + 2x^3 - 6x + 3$. Next, divide $-4x^4 + 2x^3 - 6x + 3$ by $x - 2$. Compare the remainder with $P(2)$.

**55.** Write down any patterns you noticed from Exercises 51–54.

**56.** Explain how to check polynomial long division.

**57.** Try performing the following division without changing the order of the terms. Describe why this makes the process more complicated. Then perform the division again after putting the terms in the dividend in descending order of exponents.

$$\frac{4x^2 - 12x - 12 + 3x^3}{x - 2}$$

**58.** Gateway is a leading direct marketer of personal computers. Gateway's annual net profit can be modeled by the polynomial function

$$P(x) = -88.5x^3 + 454x^2 - 506.5x + 251,$$

where $P(x)$ is net profit in millions of dollars in the year $x$. Gateway's annual revenue can be modeled by the function $R(x) = 1200.7x + 5059.7$, where $R(x)$ is revenue in millions of dollars in the year $x$. In both models, $x = 0$ represents the year 1996. (*Source:* Gateway, Inc., 1996–1999)

**a.** Suppose that a market analyst has found the model $P(x)$, and another analyst at the same firm has found the model $R(x)$. The analysts have been asked by their manager to work together to find a model for Gateway's net profit margin. The analysts know that a company's net profit margin is the ratio of its net profit to its revenue. Describe how these two analysts could collaborate to find a function $m(x)$ that models Gateway's net profit margin based on the work they have done independently.

**b.** Without actually finding $m(x)$, give a general description of what you would expect the form of the result to be.

## REVIEW EXERCISES

*Insert <, >, or = to make each statement true. See Section 1.3.*

**59.** $3^2$ ____ $(-3)^2$

**60.** $(-5)^2$ ____ $5^2$

**61.** $-2^3$ ____ $(-2)^3$

**62.** $3^4$ ____ $(-3)^4$

*Solve each inequality. See Section 2.7.*

**63.** $|x + 5| < 4$

**64.** $|x - 1| \leq 8$

**65.** $|2x + 7| \geq 9$

**66.** $|4x + 2| > 10$

## A Look Ahead

**Example**

$\left(x^2 - \frac{7}{2}x + 4\right) \div (x + 2)$

*Solution*

$$
\begin{array}{r|l}
 & x - \frac{11}{2} \\
\hline
x + 2 & x^2 - \frac{7}{2}x + 4 \\
 & \underline{x^2 + 2x\phantom{+4}} \\
 & -\frac{11}{2}x + 4 \\
 & \underline{-\frac{11}{2}x - 11} \\
 & 15
\end{array}
$$

The quotient is $x - \frac{11}{2} + \frac{15}{x + 2}$.

*Divide. See the preceding example.*

**67.** $\left(x^4 + \frac{2}{3}x^3 + x\right) \div (x - 1)$

**68.** $\left(2x^3 + \frac{9}{2}x^2 - 4x - 10\right) \div (x + 2)$

**69.** $\left(3x^4 - x - x^3 + \frac{1}{2}\right) \div (2x - 1)$

**70.** $\left(2x^4 + \frac{1}{2}x^3 + x^2 + x\right) \div (x - 2)$

**71.** $(5x^4 - 2x^2 + 10x^3 - 4x) \div (5x + 10)$

**72.** $(9x^5 + 6x^4 - 6x^2 - 4x) \div (3x + 2)$

## 6.5 SYNTHETIC DIVISION AND THE REMAINDER THEOREM

CD-ROM SSM SSG Video

▶ **OBJECTIVES**

1. Use synthetic division to divide a polynomial by a binomial.
2. Use the remainder theorem to evaluate polynomials.

**1** When a polynomial is to be divided by a binomial of the form $x - c$, a shortcut process called **synthetic division** may be used. On the left is an example of long division, and on the right, the same example showing the coefficients of the variables only.

$$
\begin{array}{r|l}
 & 2x^2 + 5x + 2 \\
\hline
x - 3 & 2x^3 - x^2 - 13x + 1 \\
 & \underline{2x^3 - 6x^2} \\
 & 5x^2 - 13x \\
 & \underline{5x^2 - 15x} \\
 & 2x + 1 \\
 & \underline{2x - 6} \\
 & 7
\end{array}
\qquad
\begin{array}{r|l}
 & 2 \quad 5 \quad 2 \\
\hline
1 - 3 & 2 - 1 - 13 + 1 \\
 & \underline{2 - 6} \\
 & 5 - 13 \\
 & \underline{5 - 15} \\
 & 2 + 1 \\
 & \underline{2 - 6} \\
 & 7
\end{array}
$$

Notice that as long as we keep coefficients of powers of $x$ in the same column, we can perform division of polynomials by performing algebraic operations on the coefficients only. This shortcut process of dividing with coefficients only in a special format is called synthetic division. To find $(2x^3 - x^2 - 13x + 1) \div (x - 3)$ by synthetic division, follow the next example.

**Example 1** Use synthetic division to divide $2x^3 - x^2 - 13x + 1$ by $x - 3$.

*Solution* To use synthetic division, the divisor must be in the form $x - c$. Since we are dividing by $x - 3$, $c$ is 3. Write down 3 and the coefficients of the dividend.

$c$

$$\begin{array}{r|rrrr} 3 & 2 & -1 & -13 & 1 \\ & \downarrow & & & \\ \hline & 2 & & & \end{array}$$

Next, draw a line and bring down the first coefficient of the dividend.

$$\begin{array}{r|rrrr} 3 & 2 & -1 & -13 & 1 \\ & & 6 & & \\ \hline & 2 & & & \end{array}$$

Multiply $3 \cdot 2$ and write down the product, 6.

$$\begin{array}{r|rrrr} 3 & 2 & -1 & -13 & 1 \\ & & 6 & & \\ \hline & 2 & 5 & & \end{array}$$

Add $-1 + 6$. Write down the sum, 5.

$$\begin{array}{r|rrrr} 3 & 2 & -1 & -13 & 1 \\ & & 6 & 15 & \\ \hline & 2 & 5 & 2 & \end{array}$$

$3 \cdot 5 = 15$.
$-13 + 15 = 2$.

$$\begin{array}{r|rrrr} 3 & 2 & -1 & -13 & 1 \\ & & 6 & 15 & 6 \\ \hline & 2 & 5 & 2 & 7 \end{array}$$

$3 \cdot 2 = 6$.
$1 + 6 = 7$.

The quotient is found in the bottom row. The numbers 2, 5, and 2 are the coefficients of the quotient polynomial, and the number 7 is the remainder. The degree of the quotient polynomial is one less than the degree of the dividend. In our example, the degree of the dividend is 3, so the degree of the quotient polynomial is 2. As we found when we performed the long division, the quotient is

$$2x^2 + 5x + 2, \qquad \text{remainder } 7$$

or

$$2x^2 + 5x + 2 + \frac{7}{x - 3}$$

**Example 2** Use synthetic division to divide $x^4 - 2x^3 - 11x^2 + 5x + 34$ by $x + 2$.

*Solution* The divisor is $x + 2$, which we write in the form $x - c$ as $x - (-2)$. Thus, $c$ is $-2$. The dividend coefficients are 1, $-2$, $-11$, 5, and 34.

$c$

$$\begin{array}{r|rrrrr} -2 & 1 & -2 & -11 & 5 & 34 \\ & & -2 & 8 & 6 & -22 \\ \hline & 1 & -4 & -3 & 11 & 12 \end{array}$$

The dividend is a fourth-degree polynomial, so the quotient polynomial is a third-degree polynomial. The quotient is $x^3 - 4x^2 - 3x + 11$ with a remainder of 12. Thus,

$$\frac{x^4 - 2x^3 - 11x^2 + 5x + 34}{x + 2} = x^3 - 4x^2 - 3x + 11 + \frac{12}{x + 2}$$

**HELPFUL HINT**

Before dividing by synthetic division, write the dividend in descending order of variable exponents. Any "missing powers" of the variable should be represented by 0 times the variable raised to the missing power.

**Example 3** If $P(x) = 2x^3 - 4x^2 + 5$

**a.** Find $P(2)$ by substitution.
**b.** Use synthetic division to find the remainder when $P(x)$ is divided by $x - 2$.

*Solution* **a.** $P(x) = 2x^3 - 4x^2 + 5$

$P(2) = 2(2)^3 - 4(2)^2 + 5$

$= 2(8) - 4(4) + 5 = 16 - 16 + 5 = 5$

Thus, $P(2) = 5$.

**b.** The coefficients of $P(x)$ are 2, −4, 0, and 5. The number 0 is a coefficient of the missing power of $x^1$. The divisor is $x - 2$, so $c$ is 2.

| c → 2 | 2 | −4 | 0 | 5 | |
|---|---|---|---|---|---|
| | | 4 | 0 | 0 | |
| | 2 | 0 | 0 | 5 | remainder |

The remainder when $P(x)$ is divided by $x - 2$ is 5.

2 Notice in the preceding example that $P(2) = 5$ and that the remainder when $P(x)$ is divided by $x - 2$ is 5. This is no accident. This illustrates the **remainder theorem.**

**REMAINDER THEOREM**

If a polynomial $P(x)$ is divided by $x - c$, then the remainder is $P(c)$.

**Example 4** Use the remainder theorem and synthetic division to find $P(4)$ if

$$P(x) = 4x^6 - 25x^5 + 35x^4 + 17x^2.$$

*Solution* To find $P(4)$ by the remainder theorem, we divide $P(x)$ by $x - 4$. The coefficients of $P(x)$ are 4, −25, 35, 0, 17, 0, and 0. Also, $c$ is 4.

| c → 4 | 4 | −25 | 35 | 0 | 17 | 0 | 0 | |
|---|---|---|---|---|---|---|---|---|
| | | 16 | −36 | −4 | −16 | 4 | 16 | |
| | 4 | −9 | −1 | −4 | 1 | 4 | 16 | remainder |

Thus, $P(4) = 16$, the remainder.

# Exercise Set 6.5

*Use synthetic division to divide. See Examples 1 and 2.*

1. $(x^2 + 3x - 40) \div (x - 5)$
2. $(x^2 - 14x + 24) \div (x - 2)$
3. $(x^2 + 5x - 6) \div (x + 6)$
4. $(x^2 + 12x + 32) \div (x + 4)$
5. $(x^3 - 7x^2 - 13x + 5) \div (x - 2)$
6. $(x^3 + 6x^2 + 4x - 7) \div (x + 5)$
7. $(4x^2 - 9) \div (x - 2)$
8. $(3x^2 - 4) \div (x - 1)$

*For the given polynomial $P(x)$ and the given c, find $P(c)$ by (a) direct substitution and (b) the remainder theorem. See Examples 3 and 4.*

9. $P(x) = 3x^2 - 4x - 1; P(2)$
10. $P(x) = x^2 - x + 3; P(5)$
11. $P(x) = 4x^4 + 7x^2 + 9x - 1; P(-2)$
12. $P(x) = 8x^5 + 7x + 4; P(-3)$
13. $P(x) = x^5 + 3x^4 + 3x - 7; P(-1)$
14. $P(x) = 5x^4 - 4x^3 + 2x - 1; P(-1)$

*Use synthetic division to divide.*

15. $(x^3 - 3x^2 + 2) \div (x - 3)$
16. $(x^2 + 12) \div (x + 2)$
17. $(6x^2 + 13x + 8) \div (x + 1)$
18. $(x^3 - 5x^2 + 7x - 4) \div (x - 3)$
19. $(2x^4 - 13x^3 + 16x^2 - 9x + 20) \div (x - 5)$
20. $(3x^4 + 5x^3 - x^2 + x - 2) \div (x + 2)$
21. $(3x^2 - 15) \div (x + 3)$
22. $(3x^2 + 7x - 6) \div (x + 4)$
23. $(3x^3 - 6x^2 + 4x + 5) \div \left(x - \frac{1}{2}\right)$
24. $(8x^3 - 6x^2 - 5x + 3) \div \left(x + \frac{3}{4}\right)$
25. $(3x^3 + 2x^2 - 4x + 1) \div \left(x - \frac{1}{3}\right)$
26. $(9y^3 + 9y^2 - y + 2) \div \left(y + \frac{2}{3}\right)$
27. $(7x^2 - 4x + 12 + 3x^3) \div (x + 1)$
28. $(x^4 + 4x^3 - x^2 - 16x - 4) \div (x - 2)$
29. $(x^3 - 1) \div (x - 1)$
30. $(y^3 - 8) \div (y - 2)$
31. $(x^2 - 36) \div (x + 6)$
32. $(4x^3 + 12x^2 + x - 12) \div (x + 3)$

*For the given polynomial $P(x)$ and the given c, use the remainder theorem to find $P(c)$.*

33. $P(x) = x^3 + 3x^2 - 7x + 4; 1$
34. $P(x) = x^3 + 5x^2 - 4x - 6; 2$
35. $P(x) = 3x^3 - 7x^2 - 2x + 5; -3$
36. $P(x) = 4x^3 + 5x^2 - 6x - 4; -2$
37. $P(x) = 4x^4 + x^2 - 2; -1$
38. $P(x) = x^4 - 3x^2 - 2x + 5; -2$
39. $P(x) = 2x^4 - 3x^2 - 2; \frac{1}{3}$
40. $P(x) = 4x^4 - 2x^3 + x^2 - x - 4; \frac{1}{2}$
41. $P(x) = x^5 + x^4 - x^3 + 3; \frac{1}{2}$
42. $P(x) = x^5 - 2x^3 + 4x^2 - 5x + 6; \frac{2}{3}$
43. Explain an advantage of using the remainder theorem instead of direct substitution.
44. Explain an advantage of using synthetic division instead of long division.

*We say that 2 is a factor of 8 because 2 divides 8 evenly, or with a remainder of 0. In the same manner, the polynomial $x - 2$ is a factor of the polynomial $x^3 - 14x^2 + 24x$ because the remainder is 0 when $x^3 - 14x^2 + 24x$ is divided by $x - 2$. Use this information for Exercises 45 through 47.*

45. Use synthetic division to show that $x + 3$ is a factor of $x^3 + 3x^2 + 4x + 12$.
46. Use synthetic division to show that $x - 2$ is a factor of $x^3 - 2x^2 - 3x + 6$.
47. From the remainder theorem, the polynomial $x - c$ is a factor of a polynomial function $P(x)$ if $P(c)$ is what value?
48. If a polynomial is divided by $x - 5$, the quotient is $2x^2 + 5x - 6$ and the remainder is 3. Find the original polynomial.
49. If a polynomial is divided by $x + 3$, the quotient is $x^2 - x + 10$ and the remainder is $-2$. Find the original polynomial.

△ **50.** If the area of a parallelogram is $(x^4 - 23x^2 + 9x - 5)$ square centimeters and its base is $(x + 5)$ centimeters, find its height.

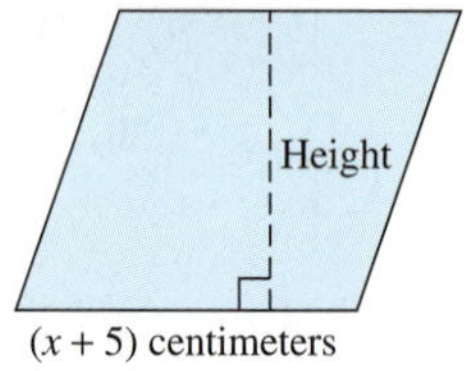


△ **51.** If the volume of a box is $(x^4 + 6x^3 - 7x^2)$ cubic meters, its height is $x^2$ meters, and its length is $(x + 7)$ meters, find its width.

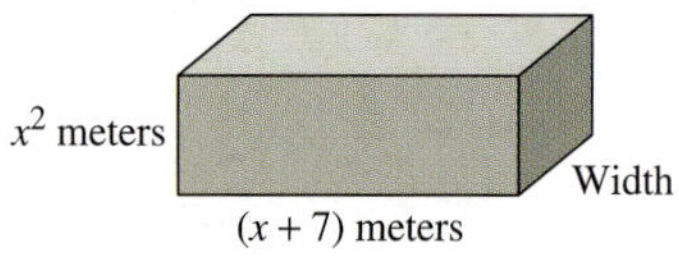


## REVIEW EXERCISES

*Solve each equation for x. See Sections 2.1 and 5.8.*

**52.** $7x + 2 = x - 3$

**53.** $4 - 2x = 17 - 5x$

**54.** $x^2 = 4x - 4$

**55.** $5x^2 + 10x = 15$

**56.** $\frac{x}{3} - 5 = 13$

**57.** $\frac{2x}{9} + 1 = \frac{7}{9}$

*Factor the following. See Sections 5.5 and 5.7.*

**58.** $x^3 - 1$

**59.** $8y^3 + 1$

**60.** $125z^3 + 8$

**61.** $a^3 - 27$

**62.** $xy + 2x + 3y + 6$

**63.** $x^2 - x + xy - y$

**64.** $x^3 - 9x$

**65.** $2x^3 - 32x$

# 6.6 SOLVING EQUATIONS CONTAINING RATIONAL EXPRESSIONS

CD-ROM SSM SSG Video

▶ **OBJECTIVE**

1. Solve equations containing rational expressions.

**1** In this section, we solve equations containing rational expressions. Before beginning this section, make sure that you understand the difference between an *equation* and an *expression*. An **equation** contains an equal sign and an **expression** does not.

| ***Equation*** | ***Expression*** |
|---|---|
| $\frac{x}{2} + \frac{x}{6} = \frac{2}{3}$ | $\frac{x}{2} + \frac{x}{6}$ |

### SOLVING EQUATIONS CONTAINING RATIONAL EXPRESSIONS

To solve *equations* containing rational expressions, first clear the equation of fractions by multiplying both sides of the equation by the LCD of all rational expressions. Then solve as usual.

**HELPFUL HINT**

The method described above is for equations only. It may *not* be used for performing operations on expressions.

**Example 1** Solve: $\dfrac{4x}{5} + \dfrac{3}{2} = \dfrac{3x}{10}$

*Solution* The LCD of $\dfrac{4x}{5}, \dfrac{3}{2}$, and $\dfrac{3x}{10}$ is 10. We multiply both sides of the equation by 10.

$$\frac{4x}{5} + \frac{3}{2} = \frac{3x}{10}$$

$$10\left(\frac{4x}{5} + \frac{3}{2}\right) = 10\left(\frac{3x}{10}\right) \quad \text{Multiply both sides by the LCD.}$$

$$10 \cdot \frac{4x}{5} + 10 \cdot \frac{3}{2} = 10 \cdot \frac{3x}{10} \quad \text{Use the distributive property.}$$

$$8x + 15 = 3x \quad \text{Simplify.}$$

$$15 = -5x \quad \text{Subtract } 8x \text{ from both sides.}$$

$$-3 = x \quad \text{Solve.}$$

Verify this solution by replacing $x$ with $-3$ in the original equation.

**Check:**

$$\frac{4x}{5} + \frac{3}{2} = \frac{3x}{10}$$

$$\frac{4(-3)}{5} + \frac{3}{2} \stackrel{?}{=} \frac{3(-3)}{10}$$

$$\frac{-12}{5} + \frac{3}{2} \stackrel{?}{=} \frac{-9}{10}$$

$$-\frac{24}{10} + \frac{15}{10} \stackrel{?}{=} -\frac{9}{10}$$

$$-\frac{9}{10} = -\frac{9}{10} \quad \text{True.}$$

The solution is $-3$.

The important difference of the equations in this section is that the denominator of a rational expression may contain a variable. Recall that a rational expression is undefined for values of the variable that make the denominator 0. If a proposed solution makes the denominator 0, then it must be rejected as a solution of the original equation. Such proposed solutions are called **extraneous solutions**.

**Example 2** Solve: $\dfrac{3}{x} - \dfrac{x + 21}{3x} = \dfrac{5}{3}$

*Solution* The LCD of the denominators $x$, $3x$, and 3 is $3x$. We multiply both sides by $3x$.

$$\frac{3}{x} - \frac{x + 21}{3x} = \frac{5}{3}$$

$$3x\left(\frac{3}{x} - \frac{x + 21}{3x}\right) = 3x\left(\frac{5}{3}\right) \quad \text{Multiply both sides by the LCD.}$$

$$3x \cdot \frac{3}{x} - 3x \cdot \frac{x + 21}{3x} = 3x \cdot \frac{5}{3} \quad \text{Use the distributive property.}$$

$$9 - (x + 21) = 5x \quad \text{Simplify.}$$

$$9 - x - 21 = 5x$$

$$-12 = 6x$$

$$-2 = x \quad \text{Solve.}$$

The proposed solution is −2.

**Check:** Check the proposed solution in the original equation.

$$\frac{3}{x} - \frac{x + 21}{3x} = \frac{5}{3}$$

$$\frac{3}{-2} - \frac{-2 + 21}{3(-2)} \stackrel{?}{=} \frac{5}{3}$$

$$-\frac{9}{6} + \frac{19}{6} \stackrel{?}{=} \frac{5}{3}$$

$$\frac{10}{6} \stackrel{?}{=} \frac{5}{3} \quad \text{True.}$$

The solution is −2.

The following steps may be used to solve equations containing rational expressions.

**SOLVING AN EQUATION CONTAINING RATIONAL EXPRESSIONS**

**Step 1:** Multiply both sides of the equation by the LCD of all rational expressions in the equation.
**Step 2:** Simplify both sides.
**Step 3:** Determine whether the equation is linear, quadratic, or higher degree and solve accordingly.
**Step 4:** Check the solution in the original equation.

**Example 3** Solve: $\frac{x + 6}{x - 2} = \frac{2(x + 2)}{x - 2}$

*Solution* First we multiply both sides of the equation by the LCD, $x - 2$.

$$\frac{x + 6}{x - 2} = \frac{2(x + 2)}{x - 2}$$

$$(x - 2) \cdot \frac{x + 6}{x - 2} = (x - 2) \cdot \frac{2(x + 2)}{x - 2} \quad \text{Multiply both sides by } x - 2.$$

$$x + 6 = 2(x + 2) \quad \text{Simplify.}$$

$$x + 6 = 2x + 4 \quad \text{Use the distributive property.}$$

$$2 = x \quad \text{Solve.}$$

**Check:** The proposed solution is 2. Notice that 2 makes a denominator 0 in the original equation. This can also be seen in a check. Check the proposed solution 2 in the original equation.

$$\frac{x + 6}{x - 2} = \frac{2(x + 2)}{x - 2}$$

$$\frac{2 + 6}{2 - 2} = \frac{2(2 + 2)}{2 - 2}$$

$$\frac{8}{0} = \frac{2(4)}{0}$$

The denominators are 0, so 2 is not a solution of the original equation. The solution is { } or ∅.

**Example 4** Solve: $\frac{2x}{2x-1} + \frac{1}{x} = \frac{1}{2x-1}$

*Solution* The LCD is $x(2x-1)$. Multiply both sides by $x(2x-1)$. By the distributive property, this is the same as multiplying each term by $x(2x-1)$.

$$x(2x-1) \cdot \frac{2x}{2x-1} + x(2x-1) \cdot \frac{1}{x} = x(2x-1) \cdot \frac{1}{2x-1}$$

$$x(2x) + (2x-1) = x \quad \text{Simplify.}$$

$$2x^2 + 2x - 1 - x = 0$$

$$2x^2 + x - 1 = 0$$

$$(x+1)(2x-1) = 0$$

$$x + 1 = 0 \quad \text{or} \quad 2x - 1 = 0$$

$$x = -1 \qquad x = \frac{1}{2}$$

The number $\frac{1}{2}$ makes the denominator $2x - 1$ equal 0, so it is not a solution. The solution is −1.

**Example 5** Solve: $\frac{2x}{x-3} + \frac{6-2x}{x^2-9} = \frac{x}{x+3}$

*Solution* We factor the second denominator to find that the LCD is $(x+3)(x-3)$. We multiply both sides of the equation by $(x+3)(x-3)$. By the distributive property, this is the same as multiplying each term by $(x+3)(x-3)$.

$$\frac{2x}{x-3} + \frac{6-2x}{x^2-9} = \frac{x}{x+3}$$

$$(x+3)(x-3) \cdot \frac{2x}{x-3} + (x+3)(x-3) \cdot \frac{6-2x}{(x+3)(x-3)}$$

$$= (x+3)(x-3)\left(\frac{x}{x+3}\right)$$

$$2x(x+3) + (6-2x) = x(x-3) \quad \text{Simplify.}$$

$$2x^2 + 6x + 6 - 2x = x^2 - 3x \quad \text{Use the distributive property.}$$

Next we solve this quadratic equation by the factoring method. To do so, we first write the equation so that one side is 0.

$$x^2 + 7x + 6 = 0$$

$$(x+6)(x+1) = 0 \quad \text{Factor.}$$

$$x = -6 \text{ or } x = -1 \quad \text{Set each factor equal to 0.}$$

Neither −6 nor −1 makes any denominator 0 so they are both solutions. The solutions are −6 and −1.

**Example 6** Solve: $\frac{z}{2z^2 + 3z - 2} - \frac{1}{2z} = \frac{3}{z^2 + 2z}$

*Solution* Factor the denominators to find that the LCD is $2z(z + 2)(2z - 1)$. Multiply both sides by the LCD. Remember, by using the distributive property, this is the same as multiplying each term by $2z(z + 2)(2z - 1)$.

$$\frac{z}{2z^2 + 3z - 2} - \frac{1}{2z} = \frac{3}{z^2 + 2z}$$

$$\frac{z}{(2z - 1)(z + 2)} - \frac{1}{2z} = \frac{3}{z(z + 2)}$$

$$2z(z + 2)(2z - 1) \cdot \frac{z}{(2z - 1)(z + 2)} - 2z(z + 2)(2z - 1) \cdot \frac{1}{2z}$$

$$= 2z(z + 2)(2z - 1) \cdot \frac{3}{z(z + 2)}$$ Apply the distributive property.

$$2z(z) - (z + 2)(2z - 1) = 3 \cdot 2(2z - 1)$$ Simplify.

$$2z^2 - (2z^2 + 3z - 2) = 12z - 6$$

$$2z^2 - 2z^2 - 3z + 2 = 12z - 6$$

$$-3z + 2 = 12z - 6$$

$$-15z = -8$$

$$z = \frac{8}{15}$$ Solve.

The proposed solution $\frac{8}{15}$ does not make any denominator 0; the solution is $\frac{8}{15}$.

A graph can be helpful in visualizing solutions of equations. For example, to visualize the solution of the equation $\frac{3}{x} - \frac{x + 21}{3x} = \frac{5}{3}$ in Example 2, the graph of the related rational function $f(x) = \frac{3}{x} - \frac{x + 21}{3x}$ is shown. A solution of the equation is an $x$-value that corresponds to a $y$-value of $\frac{5}{3}$.

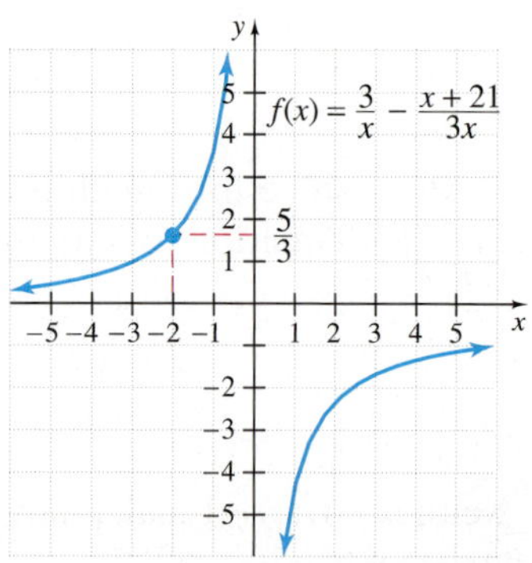

Notice that an $x$-value of $-2$ corresponds to a $y$-value of $\frac{5}{3}$. The solution of the equation is indeed $-2$ as shown in Example 2.

## Exercise Set 6.6

*Solve each equation. See Examples 1 and 2.*

**1.** $\frac{x}{2} - \frac{x}{3} = 12$

**2.** $x = \frac{x}{2} - 4$

**3.** $\frac{x}{3} = \frac{1}{6} + \frac{x}{4}$

**4.** $\frac{x}{2} = \frac{21}{10} - \frac{x}{5}$

**5.** $\frac{2}{x} + \frac{1}{2} = \frac{5}{x}$

**6.** $\frac{5}{3x} + 1 = \frac{7}{6}$

**7.** $\frac{x+3}{x} = \frac{5}{x}$

**8.** $\frac{4-3x}{2x} = -\frac{8}{2x}$

*Solve each equation. See Examples 3 through 6.*

**9.** $\frac{x+5}{x+3} = \frac{8}{x+3}$

**10.** $\frac{5}{x-2} - \frac{2}{x+4} = -\frac{4}{x^2+2x-8}$

**11.** $\frac{1}{x-1} + \frac{1}{x+1} = \frac{2}{x^2-1}$

**12.** $\frac{1}{x-1} = \frac{2}{x+1}$

**13.** $\frac{6}{x+3} = \frac{4}{x-3}$

**14.** $\frac{1}{x-4} - \frac{3x}{x^2-16} = \frac{2}{x+4}$

**15.** $\frac{3}{2x+3} - \frac{1}{2x-3} = \frac{4}{4x^2-9}$

**16.** $\frac{1}{x-4} = \frac{8}{x^2-16}$

**17.** $\frac{2}{x^2-4} = \frac{1}{2x-4}$

**18.** $\frac{1}{x-2} - \frac{2}{x^2-2x} = 1$

**19.** $\frac{12}{3x^2+12x} = 1 - \frac{1}{x+4}$

*Solve each equation.*

**20.** $\frac{5}{x} = \frac{20}{12}$

**21.** $\frac{2}{x} = \frac{10}{5}$

**22.** $1 - \frac{4}{a} = 5$

**23.** $7 + \frac{6}{a} = 5$

**24.** $\frac{1}{2x} - \frac{1}{x+1} = \frac{1}{3x^2+3x}$

**25.** $\frac{2}{x-5} + \frac{1}{2x} = \frac{5}{3x^2-15x}$

**26.** $\frac{1}{x} - \frac{x}{25} = 0$

**27.** $\frac{x}{4} + \frac{5}{x} = 3$

**28.** $5 - \frac{2}{2y-5} = \frac{3}{2y-5}$

**29.** $1 - \frac{5}{y+7} = \frac{4}{y+7}$

**30.** $\frac{x-1}{x+2} = \frac{2}{3}$

**31.** $\frac{6x+7}{2x+9} = \frac{5}{3}$

**32.** $\frac{x+3}{x+2} = \frac{1}{x+2}$

**33.** $\frac{2x+1}{4-x} = \frac{9}{4-x}$

**34.** $\frac{1}{a-3} + \frac{2}{a+3} = \frac{1}{a^2-9}$

**35.** $\frac{12}{9-a^2} + \frac{3}{3+a} = \frac{2}{3-a}$

**36.** $\frac{64}{x^2-16} + 1 = \frac{2x}{x-4}$

**37.** $2 + \frac{3}{x} = \frac{2x}{x+3}$

**38.** $\frac{-15}{4y+1} + 4 = y$

**39.** $\frac{36}{x^2-9} + 1 = \frac{2x}{x+3}$

**40.** $\frac{28}{x^2-9} + \frac{2x}{x-3} + \frac{6}{x+3} = 0$

**41.** $\frac{x^2-20}{x^2-7x+12} = \frac{3}{x-3} + \frac{5}{x-4}$

**42.** $\frac{x+2}{x^2+7x+10} = \frac{1}{3x+6} - \frac{1}{x+5}$

**43.** $\frac{3}{2x-5} + \frac{2}{2x+3} = 0$

**44.** The average cost of producing $x$ game disks for a computer is given by the function $f(x) = 3.3 + \frac{5400}{x}$. Find the number of game disks that must be produced for the average cost to be \$5.10.

**45.** The average cost of producing $x$ electric pencil sharpeners is given by the function $f(x) = 20 + \frac{4000}{x}$. Find the number of electric pencil sharpeners that must be produced for the average cost to be \$25.

*Solve each equation. Begin by writing each equation with positive exponents only.*

**46.** $x^{-2} - 19x^{-1} + 48 = 0$

**47.** $x^{-2} - 5x^{-1} - 36 = 0$

**48.** $p^{-2} + 4p^{-1} - 5 = 0$

**49.** $6p^{-2} - 5p^{-1} + 1 = 0$

*Solve each equation. Round solutions to two decimal places.*

**50.** $\dfrac{1.4}{x - 2.6} = \dfrac{-3.5}{x + 7.1}$

**51.** $\dfrac{-8.5}{x + 1.9} = \dfrac{5.7}{x - 3.6}$

**52.** $\dfrac{10.6}{y} - 14.7 = \dfrac{9.92}{3.2} + 7.6$

**53.** $\dfrac{12.2}{x} + 17.3 = \dfrac{9.6}{x} - 14.7$

*Use a graphing calculator to verify the solution of each given exercise.*

**54.** Exercise 20

**55.** Exercise 21

**56.** Exercise 30

**57.** Exercise 31

## REVIEW EXERCISES

*Write each sentence as an equation and solve. See Section 2.2.*

**58.** Four more than 3 times a number is 19.

**59.** The sum of two consecutive integers is 147.

**60.** The length of a rectangle is 5 inches more than the width. Its perimeter is 50 inches. Find the length and width.

**61.** The sum of a number and its reciprocal is $\frac{5}{2}$.

*The following graph is from a survey of state and federal prisons. Use this histogram to answer Exercises 62–66.*

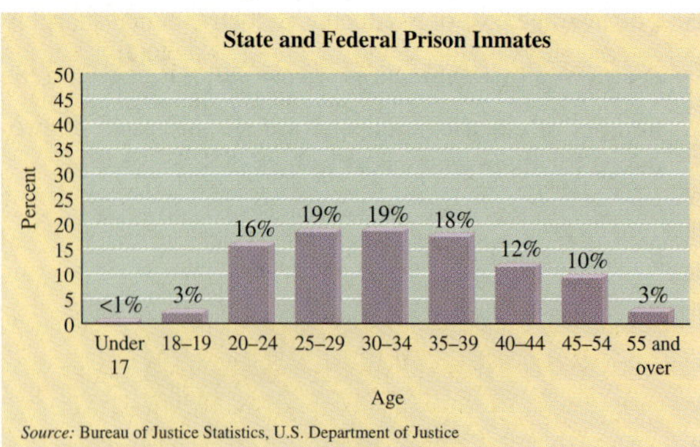


**62.** What percent of state and federal prison inmates are age 45 to 54?

**63.** What percent of state and federal prison inmates are 55 years old or older?

**64.** What age category shows the highest percent of prison inmates?

**65.** What percent of state and federal prison inmates are 20 to 34 years old?

**66.** At the end of 1998, there were 30,907 inmates under the jurisdiction of state and federal correctional authorities in the state of Louisiana. Approximately how many 25- to 29-year-old inmates would you expect to have been held in Louisiana at the end of 1998? Round to the nearest whole. (*Source:* Bureau of Justice Statistics)

## A Look Ahead

**Example**

Solve $\left(\dfrac{x}{x+1}\right)^2 - 7\left(\dfrac{x}{x+1}\right) + 10 = 0.$

*Solution*

Let $u = \dfrac{x}{x+1}$ and solve for $u$. Then substitute back and solve for $x$.

$$\left(\frac{x}{x+1}\right)^2 - 7\left(\frac{x}{x+1}\right) + 10 = 0$$

$$u^2 - 7u + 10 = 0 \quad \text{Let } u = \frac{x}{x+1}.$$

$$(u - 5)(u - 2) = 0 \quad \text{Factor.}$$

$$u = 5 \quad \text{or} \quad u = 2 \quad \text{Solve.}$$

Since $u = \dfrac{x}{x+1}$, we have that $5 = \dfrac{x}{x+1}$ or $2 = \dfrac{x}{x+1}$. Thus, there are two rational equations to solve.

**1.**
$$5 = \frac{x}{x+1}$$
$$5 \cdot (x + 1) = x$$
$$5x + 5 = x$$
$$5 = -4x$$
$$x = -\frac{5}{4}$$

**2.**
$$2 = \frac{x}{x+1}$$
$$2 \cdot (x + 1) = x$$
$$2x + 2 = x$$
$$2 = -x$$
$$x = -2$$

Since neither $-\frac{5}{4}$ nor $-2$ makes the denominator 0, the solutions are $-\frac{5}{4}$ and $-2$.

*Solve each equation by substitution. See the preceding example.*

**67.** $(x - 1)^2 + 3(x - 1) + 2 = 0$

**68.** $(4 - x)^2 - 5(4 - x) + 6 = 0$

**69.** $\left(\dfrac{3}{x-1}\right)^2 + 2\left(\dfrac{3}{x-1}\right) + 1 = 0$

**70.** $\left(\dfrac{5}{2+x}\right)^2 + \left(\dfrac{5}{2+x}\right) - 20 = 0$

## *Supplementary Exercises on Expressions and Equations*

It is very important that you understand the difference between an expression and an equation containing rational expressions. An equation contains an equal sign; an expression does not.

***Expression to be Simplified***

$$\frac{x}{2} + \frac{x}{6}$$

Write both rational expressions with the LCD, 6, as the denominator.

$$\frac{x}{2} + \frac{x}{6} = \frac{x \cdot 3}{2 \cdot 3} + \frac{x}{6}$$

$$= \frac{3x}{6} + \frac{x}{6}$$

$$= \frac{4x}{6} = \frac{2x}{3}$$

***Equation to be Solved***

$$\frac{x}{2} + \frac{x}{6} = \frac{2}{3}$$

Multiply both sides by the LCD, 6.

$$6\left(\frac{x}{2} + \frac{x}{6}\right) = 6\left(\frac{2}{3}\right)$$

$$3x + x = 4$$

$$4x = 4$$

$$x = 1$$

Check to see that the solution is 1.

**HELPFUL HINT**

Remember: Equations can be cleared of fractions; expressions cannot.

*Perform each indicated operation and simplify, or solve the equation for the variable.*

1. $\frac{x}{2} = \frac{1}{8} + \frac{x}{4}$

2. $\frac{x}{4} = \frac{3}{2} + \frac{x}{10}$

3. $\frac{1}{8} + \frac{x}{4}$

4. $\frac{3}{2} + \frac{x}{10}$

5. $\frac{4}{x+2} - \frac{2}{x-1}$

6. $\frac{5}{x-2} - \frac{10}{x+4}$

7. $\frac{4}{x+2} = \frac{2}{x-1}$

8. $\frac{5}{x-2} = \frac{10}{x+4}$

9. $\frac{2}{x^2-4} = \frac{1}{x+2} - \frac{3}{x-2}$

10. $\frac{3}{x^2-25} = \frac{1}{x+5} + \frac{2}{x-5}$

11. $\frac{5}{x^2-3x} + \frac{4}{2x-6}$

12. $\frac{5}{x^2-3x} \div \frac{4}{2x-6}$

13. $\frac{x-1}{x+1} + \frac{x+7}{x-1} = \frac{4}{x^2-1}$

14. $\left(1 - \frac{y}{x}\right) \div \left(1 - \frac{x}{y}\right)$

15. $\frac{a^2-9}{a-6} \cdot \frac{a^2-5a-6}{a^2-a-6}$

16. $\frac{2}{a-6} + \frac{3a}{a^2-5a-6} - \frac{a}{5a+5}$

17. $\frac{2x+3}{3x-2} = \frac{4x+1}{6x+1}$

18. $\frac{5x-3}{2x} = \frac{10x+3}{4x+1}$

19. $\frac{a}{9a^2-1} + \frac{2}{6a-2}$

20. $\frac{3}{4a-8} - \frac{a+2}{a^2-2a}$

21. $-\frac{3}{x^2} - \frac{1}{x} + 2 = 0$

22. $\frac{x}{2x+6} + \frac{5}{x^2-9}$

23. $\frac{x-8}{x^2-x-2} + \frac{2}{x-2}$

24. $\frac{x-8}{x^2-x-2} + \frac{2}{x-2} = \frac{3}{x+1}$

25. $\frac{3}{a} - 5 = \frac{7}{a} - 1$

26. $\frac{7}{3z-9} + \frac{5}{z}$

## 6.7 RATIONAL EQUATIONS AND PROBLEM SOLVING

CD-ROM SSM SSG Video

**OBJECTIVES**

1. Solve an equation containing rational expressions for a specified variable.
2. Solve problems by writing equations containing rational expressions.

**1** In Section 2.3 we solved equations for a specified variable. In this section, we continue practicing this skill by solving equations containing rational expressions for a specified variable. The steps given in Section 2.3 for solving equations for a specified variable are repeated here.

### SOLVING EQUATIONS FOR A SPECIFIED VARIABLE

**Step 1:** Clear the equation of fractions or rational expressions by multiplying each side of the equation by the least common denominator (LCD) of all denominators in the equation.

**Step 2:** Use the distributive property to remove grouping symbols such as parentheses.

**Step 3:** Combine like terms on each side of the equation.

**Step 4:** Use the addition property of equality to rewrite the equation as an equivalent equation with terms containing the specified variable on one side and all other terms on the other side.

**Step 5:** Use the distributive property and the multiplication property of equality to get the specified variable alone.

**Example 1** Solve $\frac{1}{x} + \frac{1}{y} = \frac{1}{z}$ for $x$.

*Solution* To clear this equation of fractions, we multiply both sides of the equation by $xyz$, the LCD of $\frac{1}{x}, \frac{1}{y}$, and $\frac{1}{z}$.

$$\frac{1}{x} + \frac{1}{y} = \frac{1}{z}$$

$$xyz\left(\frac{1}{x} + \frac{1}{y}\right) = xyz\left(\frac{1}{z}\right) \quad \text{Multiply both sides by } xyz.$$

$$xyz\left(\frac{1}{x}\right) + xyz\left(\frac{1}{y}\right) = xyz\left(\frac{1}{z}\right) \quad \text{Use the distributive property.}$$

$$yz + xz = xy \quad \text{Simplify.}$$

Notice the two terms that contain the specified variable $x$.

Next, we subtract $xz$ from both sides so that all terms containing the specified variable $x$ are on one side of the equation and all other terms are on the other side.

$$yz = xy - xz$$

Now we use the distributive property to factor $x$ from $xy - xz$ and then the multiplication property of equality to solve for $x$.

$$yz = x(y - z)$$

$$\frac{yz}{y - z} = x \quad \text{or} \quad x = \frac{yz}{y - z} \quad \text{Divide both sides by } y - z.$$

**2** Problem solving sometimes involves modeling a described situation with an equation containing rational expressions. In Examples 2 through 5, we practice solving such problems and use the problem-solving steps first introduced in Section 2.2.

**Example 2** FINDING AN UNKNOWN NUMBER

If a certain number is subtracted from the numerator and added to the denominator of $\frac{9}{19}$, the new fraction is equivalent to $\frac{1}{3}$. Find the number.

*Solution* 1. UNDERSTAND the problem. Read and reread the problem and try guessing the solution. For example, if the unknown number is 3, we have

$$\frac{9 - 3}{19 + 3} = \frac{1}{3}$$

To see if this is a true statement, we simplify the fraction on the left side.

$$\frac{6}{22} = \frac{1}{3} \quad \text{or} \quad \frac{3}{11} = \frac{1}{3} \quad \text{False.}$$

Since this is not a true statement, 3 is not the correct number. Remember that the purpose of this step is not to guess the correct solution but to gain an understanding of the problem posed.

We will let $n$ = the number to be subtracted from the numerator and added to the denominator.

2. TRANSLATE the problem.

| | | | |
|---|---|---|---|
| In words: | when the number is subtracted from the numerator and added to the denominator of the fraction $\frac{9}{19}$ | this is equivalent to | $\frac{1}{3}$ |
| | ↓ | ↓ | ↓ |
| Translate: | $\frac{9-n}{19+n}$ | $=$ | $\frac{1}{3}$ |

3. SOLVE the equation for $n$.

$$\frac{9-n}{19+n} = \frac{1}{3}$$

To solve for $n$, we begin by multiplying both sides by the LCD $3(19 + n)$.

$$3(19+n) \cdot \frac{9-n}{19+n} = 3(19+n) \cdot \frac{1}{3} \quad \text{Multiply both sides by the LCD.}$$

$$3(9-n) = 19+n \quad \text{Simplify.}$$

$$27 - 3n = 19 + n$$

$$8 = 4n$$

$$2 = n \quad \text{Solve.}$$

4. INTERPRET the results.

**Check:** If we subtract 2 from the numerator and add 2 to the denominator of $\frac{9}{19}$, we have $\frac{9-2}{19+2} = \frac{7}{21} = \frac{1}{3}$, and the problem checks.

**State:** The unknown number is 2.

## Example 3 FINDING THE DISTANCE OF A LIGHT SOURCE

The intensity $I$ of light, as measured in foot-candles, $x$ feet from its source is given by the rational equation

$$I = \frac{320}{x^2}$$

How far away is the source if the intensity of light is 5 foot-candles?

*Solution*

1. UNDERSTAND. Read and reread the problem, and guess a solution. Since an equation has been given that describes the relationship between $I$ and $x$, we replace $x$ with a few values to help us become familiar with the equation.

   To find the intensity $I$ of light 1 foot from the source, we let $x = 1$.

$$I = \frac{320}{1^2} = \frac{320}{1} = 320 \text{ foot-candles}$$

   To find the intensity $I$ of light 3 feet from the source, we let $x = 3$.

$$I = \frac{320}{3^2} = \frac{320}{9} = 35\frac{5}{9} \text{ foot-candles}$$

   Notice that as $x$ increases, $I$ decreases. That is, as the number of feet from the light source increases, the intensity decreases, as expected.

2. TRANSLATE. We are given that the intensity $I$ is 5 foot-candles, and we are asked to find how far away the light source, $x$ is. To do so, we let $I = 5$.

$$I = \frac{320}{x^2}$$

$$5 = \frac{320}{x^2} \quad \text{Let } I = 5.$$

3. SOLVE the equation for $x$.

$$5 = \frac{320}{x^2}$$

$$x^2 \cdot 5 = x^2 \cdot \frac{320}{x^2} \quad \text{Multiply both sides by } x^2.$$

$$5x^2 = 320 \quad \text{Simplify.}$$

$$5x^2 - 320 = 0 \quad \text{Subtract 320.}$$

$$5(x^2 - 64) = 0 \quad \text{Factor.}$$

$$5(x + 8)(x - 8) = 0 \quad \text{Factor.}$$

$$x = -8 \quad \text{or} \quad x = 8$$

4. INTERPRET. Since $x$ represents distance and distance cannot be negative, the proposed solution $-8$ must be rejected. *Check* the solution 8 feet in the given formula. Then *state* the conclusion: The source of light is 8 feet away when the intensity is 5 foot-candles.

The following work example leads to an equation containing rational expressions.

## Example 4 CALCULATING WORK HOURS

Melissa Scarlatti can clean the house in 4 hours, whereas her husband, Zack, can do the same job in 5 hours. They have agreed to clean together so that they can finish in time to watch a movie on TV that starts in 2 hours. How long will it take them to clean the house together? Can they finish before the movie starts?

*Solution*

1. UNDERSTAND. Read and reread the problem. The key idea here is the relationship between the *time* (in hours) it takes to complete the job and the *part of the job* completed in 1 unit of time (1 hour). For example, if the *time* it takes Melissa to

complete the job is 4 hours, the *part of the job* she can complete in 1 hour is $\frac{1}{4}$. Similarly, Zack can complete $\frac{1}{5}$ of the job in 1 hour.

We will let $t$ = *the time* in hours it takes Melissa and Zack to clean the house together. Then $\frac{1}{t}$ represents the *part of the job* they complete in 1 hour. We summarize the given information in a chart.

| | *Hours to Complete the Job* | *Part of Job Completed in 1 Hour* |
|---|---|---|
| ***MELISSA ALONE*** | 4 | $\frac{1}{4}$ |
| ***ZACK ALONE*** | 5 | $\frac{1}{5}$ |
| ***TOGETHER*** | $t$ | $\frac{1}{t}$ |

2. TRANSLATE.

In words:

| part of job Melissa can complete in 1 hour | added to | part of job Zack can complete in 1 hour | is equal to | part of job they can complete together in 1 hour |
|---|---|---|---|---|
| ↓ | ↓ | ↓ | ↓ | ↓ |
| Translate: $\frac{1}{4}$ | $+$ | $\frac{1}{5}$ | $=$ | $\frac{1}{t}$ |

3. SOLVE.

$$\frac{1}{4} + \frac{1}{5} = \frac{1}{t}$$

$$20t\left(\frac{1}{4} + \frac{1}{5}\right) = 20t\left(\frac{1}{t}\right) \quad \text{Multiply both sides by the LCD, } 20t.$$

$$5t + 4t = 20$$

$$9t = 20$$

$$t = \frac{20}{9} \text{ or } 2\frac{2}{9} \quad \text{Solve.}$$

4. INTERPRET.

**Check:** The proposed solution is $2\frac{2}{9}$. That is, Melissa and Zack would take $2\frac{2}{9}$ hours to clean the house together. This proposed solution is reasonable since $2\frac{2}{9}$ hours is more than half of Melissa's time and less than half of Zack's time. Check this solution in the originally stated problem.

**State:** Melissa and Zack can clean the house together in $2\frac{2}{9}$ hours. They cannot complete the job before the movie starts.

## Example 5 FINDING THE SPEED OF A CURRENT

Steve Deitmer takes $1\frac{1}{2}$ times as long to go 72 miles upstream in his boat as he does to return. If the boat cruises at 30 mph in still water, what is the speed of the current?

*Solution*

1. UNDERSTAND. Read and reread the problem. Guess a solution. Suppose that the current is 4 mph. The speed of the boat upstream is slowed down by the current: $30 - 4$, or 26 mph, and the speed of the boat downstream is speeded up by the current: $30 + 4$, or 34 mph. Next let's find out how long it takes to travel 72 miles upstream and 72 miles downstream. To do so, we use the formula $d = rt$, or $\frac{d}{r} = t$.

| ***Upstream*** | ***Downstream*** |
|---|---|
| $\frac{d}{r} = t$ | $\frac{d}{r} = t$ |
| $\frac{72}{26} = t$ | $\frac{72}{34} = t$ |
| $2\frac{10}{13} = t$ | $2\frac{2}{17} = t$ |

Since the time upstream $\left(2\frac{10}{13}\text{ hours}\right)$ is not $1\frac{1}{2}$ times the time downstream $\left(2\frac{2}{17}\text{ hours}\right)$, our guess is not correct. We do, however, have a better understanding of the problem.

We will let

$$x = \text{the speed of the current}$$
$$30 + x = \text{the speed of the boat downstream}$$
$$30 - x = \text{the speed of the boat upstream}$$

This information is summarized in the following chart, where we use the formula $\frac{d}{r} = t$.

| | *Distance* | *Rate* | *Time* $\left(\frac{d}{r}\right)$ |
|---|---|---|---|
| *UPSTREAM* | 72 | $30 - x$ | $\frac{72}{30 - x}$ |
| *DOWNSTREAM* | 72 | $30 + x$ | $\frac{72}{30 + x}$ |

2. TRANSLATE. Since the time spent traveling upstream is $1\frac{1}{2}$ times the time spent traveling downstream, we have

In words:

| time upstream | is | $1\frac{1}{2}$ | times | time downstream |
|---|---|---|---|---|
| ↓ | ↓ | ↓ | ↓ | ↓ |

Translate: $\frac{72}{30 - x} = \frac{3}{2} \cdot \frac{72}{30 + x}$

3. SOLVE. $\frac{72}{30 - x} = \frac{3}{2} \cdot \frac{72}{30 + x}$

First we multiply both sides by the LCD, $2(30 + x)(30 - x)$.

$$2(30 + x)(30 - x) \cdot \frac{72}{30 - x} = 2(30 + x)(30 - x)\left(\frac{3}{2} \cdot \frac{72}{30 + x}\right)$$

$$72 \cdot 2(30 + x) = 3 \cdot 72 \cdot (30 - x) \quad \text{Simplify.}$$

$$2(30 + x) = 3(30 - x) \quad \text{Divide both sides by 72.}$$

$$60 + 2x = 90 - 3x \quad \text{Use the distributive property.}$$

$$5x = 30$$

$$x = 6 \quad \text{Solve.}$$

4. INTERPRET.

**Check:** Check the proposed solution of 6 mph in the originally stated problem.

**State:** The current's speed is 6 mph.

## SPOTLIGHT ON DECISION MAKING

Suppose you are an aviation safety inspector. You are testing the accuracy of the radioaltimeter on an airplane. To be acceptable, the altitude reading given by the altimeter must be within 3% of the actual altitude. You know that you can check the altimeter's altitude reading with the equation $t = \frac{2a}{c}$, where $t$ is the time it takes a radar pulse aimed downward from the airplane to bounce off Earth's surface and return to the radioaltimeter, $a$ is the altitude, and $c$ is the speed of light, $3 \times 10^8$ meters per second.

During your test, you find that it takes $5 \times 10^{-5}$ second for the radar pulse emitted by the altimeter to be returned. The altimeter reads an altitude of 7420 meters. Is the altimeter reading acceptable? Explain.

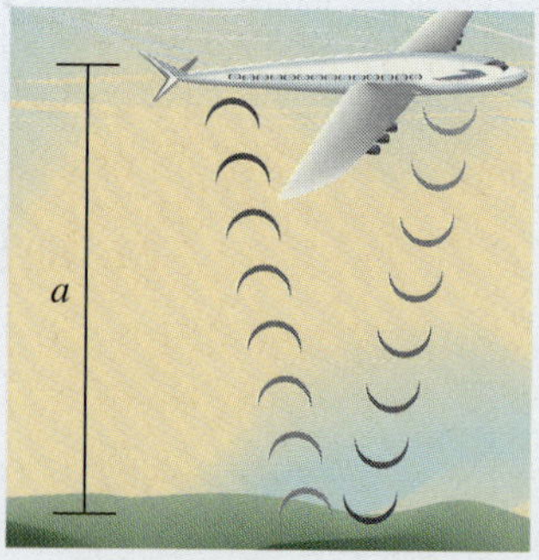

## Exercise Set 6.7

*Solve each equation for the specified variable. See Example 1.*

1. $F = \frac{9}{5}C + 32$ for $C$ (Meteorology)
2. $V = \frac{1}{3}\pi r^2 h$ for $h$ (Volume)
3. $Q = \frac{A - I}{L}$ for $I$ (Finance)
4. $P = 1 - \frac{C}{S}$ for $S$ (Finance)
5. $\frac{1}{R} = \frac{1}{R_1} + \frac{1}{R_2}$ for $R$ (Electronics)
6. $\frac{1}{R} = \frac{1}{R_1} + \frac{1}{R_2}$ for $R_1$ (Electronics)
7. $S = \frac{n(a + L)}{2}$ for $n$ (Sequences)
8. $S = \frac{n(a + L)}{2}$ for $a$ (Sequences)
9. $A = \frac{h(a + b)}{2}$ for $b$ (Geometry)
10. $A = \frac{h(a + b)}{2}$ for $h$ (Geometry)
11. $\frac{P_1V_1}{T_1} = \frac{P_2V_2}{T_2}$ for $T_2$ (Chemistry)
12. $H = \frac{kA(T_1 - T_2)}{L}$ for $T_2$ (Physics)
13. $f = \frac{f_1 f_2}{f_1 + f_2}$ for $f_2$ (Optics)
14. $I = \frac{E}{R + r}$ for $r$ (Electronics)
15. $\lambda = \frac{2L}{n}$ for $L$ (Physics)
16. $S = \frac{a_1 - a_n r}{1 - r}$ for $a_1$ (Sequences)
17. $\frac{\theta}{\omega} = \frac{2L}{c}$ for $c$
18. $F = \frac{-GMm}{r^2}$ for $M$ (Physics)

*Solve. See Example 2.*

19. The sum of a number and 5 times its reciprocal is 6. Find the number(s).
20. The quotient of a number and 9 times its reciprocal is 1. Find the number(s).
21. If a number is added to the numerator of $\frac{12}{41}$ and twice the number is added to the denominator of $\frac{12}{41}$, the resulting fraction is equivalent to $\frac{1}{3}$. Find the number.
22. If a number is subtracted from the numerator of $\frac{13}{8}$ and added to the denominator of $\frac{13}{8}$, the resulting fraction is equivalent to $\frac{2}{5}$. Find the number.

*In electronics, the relationship among the resistances $R_1$ and $R_2$ of two resistors wired in a parallel circuit and their combined resistance $R$ is described by the formula $\frac{1}{R} = \frac{1}{R_1} + \frac{1}{R_2}$. Use this formula to solve Exercises 23 through 26. See Example 3.*

23. If the combined resistance is 2 ohms and one of the two resistances is 3 ohms, find the other resistance.
24. Find the combined resistance of two resistors of 12 ohms each when they are wired in a parallel circuit.
25. The relationship among resistance of two resistors wired in a parallel circuit and their combined resistance may be extended to three resistors of resistances $R_1$, $R_2$, and $R_3$. Write an equation you believe may describe the relationship, and use it to find the combined resistance if $R_1$, is 5 $R_2$ is 6, and $R_3$ is 2.
26. Use your formula from Exercise 25 to find the combined resistance if $R_1$ is 3, $R_2$ is 5, and $R_3$ is 15.

*Solve. See Example 4.*

27. Alan Cantrell can word process a research paper in 6 hours. With Steve Isaac's help, the paper can be processed in 4 hours. Find how long it takes Steve to word process the paper alone.
28. An experienced roofer can roof a house in 26 hours. A beginning roofer needs 39 hours to complete the same job. Find how long it takes for the two to do the job together.
29. A new printing press can print newspapers twice as fast as the old one can. The old one can print the afternoon edition in 4 hours. Find how long it takes to print the afternoon edition if both printers are operating.
30. Three postal workers can sort a stack of mail in 20 minutes, 30 minutes, and 60 minutes, respectively. Find how long it takes to sort the mail if all three work together.

*Solve. See Example 5*

31. An F-100 plane and a Toyota truck leave the same town at sunrise and head for a town 450 miles away. The speed of the plane is three times the speed of the truck, and the plane arrives 6 hours ahead of the truck. Find the speed of the truck.
32. Mattie Evans drove 150 miles in the same amount of time that it took a turbo propeller plane to travel 600 miles. The speed of the plane was 150 mph faster than the speed of the car. Find the speed of the plane.

**33.** The speed of a boat in still water is 24 mph. If the boat travels 54 miles upstream in the same time that it takes to travel 90 miles downstream, find the speed of the current.

**34.** The speed of Lazy River's current is 5 mph. If a boat travels 20 miles downstream in the same time that it takes to travel 10 miles upstream, find the speed of the boat in still water.

*Solve.*

**35.** The sum of the reciprocals of two consecutive odd integers is $\frac{20}{99}$. Find the two integers.

**36.** The sum of the reciprocals of two consecutive integers is $-\frac{15}{56}$. Find the two integers.

**37.** If Sarah Clark can do a job in 5 hours and Dick Belli and Sarah working together can do the same job in 2 hours, find how long it takes Dick to do the job alone.

**38.** One hose can fill a goldfish pond in 45 minutes, and two hoses can fill the same pond in 20 minutes. Find how long it takes the second hose alone to fill the pond.

**39.** The speed of a bicyclist is 10 mph faster than the speed of a walker. If the bicyclist travels 26 miles in the same amount of time that the walker travels 6 miles, find the speed of the bicyclist.

**40.** Two trains going in opposite directions leave at the same time. One train travels 15 mph faster than the other. In 6 hours the trains are 630 miles apart. Find the speed of each.

**41.** The numerator of a fraction is 4 less than the denominator. If both the numerator and the denominator are increased by 2, the resulting fraction is equivalent to $\frac{2}{3}$. Find the fraction.

**42.** The denominator of a fraction is 1 more than the numerator. If both the numerator and the denominator are decreased by 3, the resulting fraction is equivalent to $\frac{4}{5}$. Find the fraction.

**43.** Cyclist Lance Armstrong of the United States won the 1999 Tour de France. An amateur cyclist training for a road race rode the first 20-mile portion of his workout at a constant rate. For the 16-mile cool-down portion of his workout, he reduced his speed by 2 miles per hour. Each portion of the workout took equal time. Find the cyclist's rate during the first portion and his rate during the cool-down portion.

**44.** Moo Dairy has three machines to fill gallon milk cartons. The machines can fill the daily quota in 5 hours, 6 hours, and 7.5 hours, respectively. Find how long it takes to fill the daily quota if all three machines are running.

**45.** The inlet pipe of an oil tank can fill the tank in 1 hour 30 minutes. The outlet pipe can empty the tank in 1 hour. Find how long it takes to empty a full tank if both pipes are open.

**46.** A plane flies 465 miles with the wind and 345 miles against the wind in the same length of time. If the speed of the wind is 20 mph, find the speed of the plane in still air.

**47.** Two rockets are launched. The first travels at 9000 mph. Fifteen minutes later the second is launched at 10,000 mph. Find the distance at which both rockets are an equal distance from Earth.

**48.** Two joggers, one averaging 8 mph and one averaging 6 mph, start from a designated initial point. The slower jogger arrives at the end of the run a half hour after the other jogger. Find the distance of the run.

**49.** Smith Engineering is in the process of reviewing the salaries of their surveyors. During this review, the company has found that an experienced surveyor surveys a roadbed in 4 hours. An apprentice surveyor needs 5 hours to survey the same stretch of road. If the two work together, find how long it takes them to complete the job.

**50.** A semi truck travels 300 miles through the flatland in the same amount of time that it travels 180 miles through the Great Smoky mountains. The rate of the truck is 20 miles per hour slower in the mountains than in the flatland. Find both the flatland rate and mountain rate.

**51.** An experienced bricklayer constructs a small wall in 3 hours. An apprentice completes the job in 6 hours. Find how long it takes if they work together.

**52.** A marketing manager travels 1080 miles in a corporate jet and then an additional 240 miles by car. If the car ride takes 1 hour longer, and if the rate of the jet is 6 times the rate of the car, find the time the manager travels by jet and find the time she travels by car.

**53.** Gary Marcus and Tony Alva work at Lombardo's Pipe and Concrete. Mr. Lombardo is preparing an estimate for a customer. He knows that Gary lays a slab of concrete in 6 hours. Tony lays the same size slab in 4 hours. If both work on the job and the cost of labor is \$45 per hour, determine what the labor estimate should be.

**54.** In 2 minutes, a conveyor belt moves 300 pounds of recyclable aluminum from the delivery truck to a storage area. A smaller belt moves the same quantity of cans the same distance in 6 minutes. If both belts are used, find how long it takes to move the cans to the storage area.

**55.** Mr. Dodson can paint his house by himself in four days. His son needs an additional day to complete the job if he works by himself. If they work together, find how long it takes to paint the house.

**56.** While road testing a new make of car, the editor of a consumer magazine finds that she can go 10 miles into a 3-mile-per-hour wind in the same amount of time that she can go 11 miles with a 3-mile-per-hour wind behind her. Find the speed of the car in still air.

**57.** The world record for the largest white bass caught is held by Ronald Sprouse of Virginia. The bass weighed 6 pounds 13 ounces. If Ronald rows to his favorite fishing spot 9 miles downstream in the same amount of time that he rows 3 miles upstream, and if the current is 6 miles per hour, find how long it takes him to cover the 12 miles.

*Calculating body-mass index is a way to gauge whether a person should lose weight. Doctors recommend that body-mass index values fall between 19 and 25. The formula for body-mass index B is* $B = \frac{705w}{h^2}$, *where w is weight in pounds and h is height in inches. Use this formula to answer Exercises 58 and 59.*

**58.** A patient is 5 ft 8 in. tall. What should his or her weight be to have a bodymass index of 25? Round to the nearest whole pound.

**59.** A doctor recorded a body-mass index of 47 on a patient's chart. Later, a nurse notices that the doctor recorded the patient's weight as 240 pounds but neglected to record the patient's height. Explain how the nurse can use the information from the chart to find the patient's height. Then find the height.

*In physics, when the source of a sound is traveling toward an observer, the relationship between the actual pitch a of the sound and the pitch h that the observer hears due to the Doppler effect is described by the formula* $h = \frac{a}{1 - \frac{s}{770}}$, *where s is the speed of the sound source in miles per hour. Use this formula to answer Exercise 60.*

**60.** An emergency vehicle has a single-tone siren with the pitch of the musical note E. As it approaches an observer standing by the road, the vehicle is traveling 50 miles per hour. Is the pitch that the observer hears due to the Doppler effect lower or higher than the actual pitch? To which musical note is the pitch that the observer hears closest?

Pitch of an Octave of Musical Notes in Hertz (Hz)

| Note | Pitch |
|---|---|
| Middle C | 261.63 |
| D | 293.66 |
| E | 329.63 |
| F | 349.23 |
| G | 392.00 |
| A | 440.00 |
| B | 493.88 |

*Note:* Greater numbers indicate higher pitches (acoustically).
(*Source:* American Standards Association)

## REVIEW EXERCISES

*Solve each equation for x. See Section 2.1.*

**61.** $\frac{x}{5} = \frac{x + 2}{3}$

**62.** $\frac{x}{4} = \frac{x + 3}{6}$

**63.** $\frac{x - 3}{2} = \frac{x - 5}{6}$

**64.** $\frac{x - 6}{4} = \frac{x - 2}{5}$

## 6.8 VARIATION AND PROBLEM SOLVING

CD-ROM SSM SSG Video

▶ OBJECTIVES

1. Solve problems involving direct variation.
2. Solve problems involving inverse variation.
3. Solve problems involving joint variation.
4. Solve problems involving combined variation.

1 A very familiar example of direct variation is the relationship of the circumference $C$ of a circle to its radius $r$. The formula $C = 2\pi r$ expresses that the circumference is always $2\pi$ times the radius. In other words, $C$ is always a constant multiple $(2\pi)$ of $r$. Because it is, we say that **$C$ varies directly as $r$**, that **$C$ varies directly with $r$**, or that **$C$ is directly proportional to $r$.**

### DIRECT VARIATION

***y* varies directly as *x*,** or ***y* is directly proportional to *x*,** if there is a nonzero constant $k$ such that

$$y = kx$$

The number $k$ is called the **constant of variation** or the **constant of proportionality.**

In the above definition, the relationship described between $x$ and $y$ is a linear one. In other words, the graph of $y = kx$ is a line. The slope of the line is $k$, and the line passes through the origin.

For example, the graph of the direct variation equation $C = 2\pi r$ is shown. The horizontal axis represents the radius $r$, and the vertical axis is the circumference $C$. From the graph we can read that when the radius is 6 units, the circumference is approximately 38 units. Also, when the circumference is 45 units, the radius is between 7 and 8 units. Notice that as the radius increases, the circumference increases.

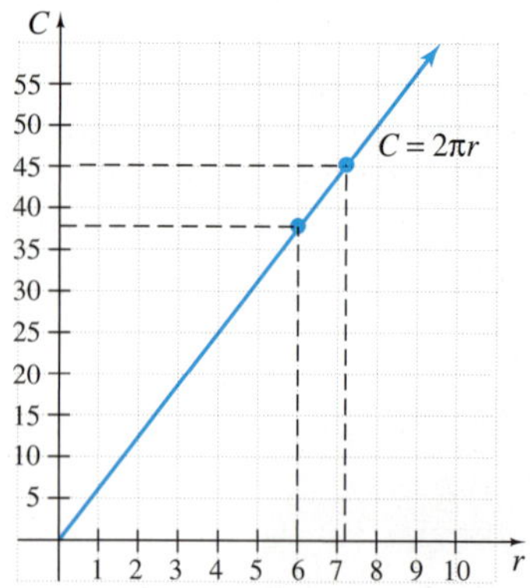


**Example 1** Suppose that $y$ varies directly as $x$. If $y$ is 5 when $x$ is 30, find the constant of variation and the direct variation equation.

*Solution* Since $y$ varies directly as $x$, we write $y = kx$. If $y = 5$ when $x = 30$, we have that

$$y = kx$$

$$5 = k(30) \quad \text{Replace } y \text{ with 5 and } x \text{ with 30.}$$

$$\frac{1}{6} = k \quad \text{Solve for } k.$$

The constant of variation is $\frac{1}{6}$.

After finding the constant of variation $k$, the direct variation equation can be written as $y = \frac{1}{6}x$.

## Example 2 USING DIRECT VARIATION AND HOOKE'S LAW

Hooke's law states that the distance a spring stretches is directly proportional to the weight attached to the spring. If a 40-pound weight attached to the spring stretches the spring 5 inches, find the distance that a 65-pound weight attached to the spring stretches the spring.

*Solution*

1. UNDERSTAND. Read and reread the problem. Notice that we are given that the distance a spring stretches is **directly proportional** to the weight attached. We let

$d$ = the distance stretched

$w$ = the weight attached

The constant of variation is represented by $k$.

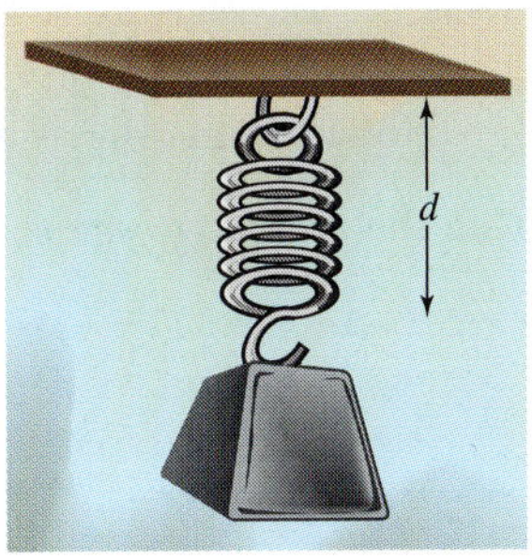


2. TRANSLATE. Because $d$ is directly proportional to $w$, we write

$$d = kw$$

3. SOLVE. When a weight of 40 pounds is attached, the spring stretches 5 inches. That is, when $w = 40$, $d = 50$.

$$d = kw$$

$$5 = k(40) \quad \text{Replace } d \text{ with 5 and } w \text{ with 40.}$$

$$\frac{1}{8} = k \quad \text{Solve for } k.$$

Now when we replace $k$ with $\frac{1}{8}$ in the equation

$d = kw$, we have

$$d = \frac{1}{8}w$$

To find the stretch when a weight of 65 pounds is attached, we replace $w$ with 65 to find $d$.

$$d = \frac{1}{8}(65)$$

$$= \frac{65}{8} = 8\frac{1}{8} \quad \text{or} \quad 8.125$$

4. INTERPRET.

**Check:** Check the proposed solution of 8.125 inches in the original problem.

**State:** The spring stetches 8.125 inches when a 65-pound weight is attached.

2 When $y$ is proportional to the **reciprocal** of another variable $x$, we say that ***y* varies inversely as *x***, or that ***y* is inversely proportional to *x***. An example of the inverse variation relationship is the relationship between the pressure that a gas exerts and the volume of its container. As the volume of a container decreases, the pressure of the gas it contains increases.

### INVERSE VARIATION

***y* varies inversely as *x***, or ***y* is inversely proportional to *x***, if there is a nonzero constant $k$ such that

$$y = \frac{k}{x}$$

The number $k$ is called the **constant of variation** or the **constant of proportionality.**

Notice that $y = \dfrac{k}{x}$ is a rational equation. Its graph for $k > 0$ and $x > 0$ is shown. From the graph, we can see that as $x$ increases, $y$ decreases.

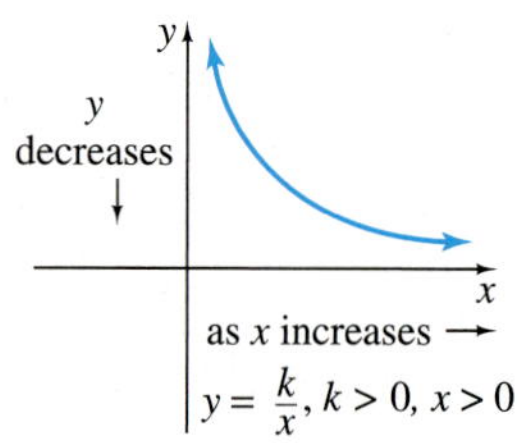


**Example 3** Suppose that $u$ varies inversely as $w$. If $u$ is 3 when $w$ is 5, find the constant of variation and the inverse variation equation.

*Solution* Since $u$ varies inversely as $w$, we have $u = \dfrac{k}{w}$. We let $u = 3$ and $w = 5$, and we solve for $k$.

$$u = \frac{k}{w}$$

$$3 = \frac{k}{5} \quad \text{Let } u = 3 \text{ and } w = 5.$$

$$15 = k \quad \text{Multiply both sides by 5.}$$

The constant of variation $k$ is 15. This gives the inverse variation equation

$$u = \frac{15}{w}$$

## Example 4 USING INVERSE VARIATION AND BOYLE'S LAW

Boyle's law says that if the temperature stays the same, the pressure $P$ of a gas is inversely proportional to the volume $V$. If a cylinder in a steam engine has a pressure of 960 kilopascals when the volume is 1.4 cubic meters, find the pressure when the volume increases to 2.5 cubic meters.

*Solution*

1. UNDERSTAND. Read and reread the problem. Notice that we are given that the pressure of a gas is *inversely proportional* to the volume. We will let $P$ = the pressure and $V$ = the volume. The constant of variation is represented by $k$.
2. TRANSLATE. Because $P$ is inversely proportional to $V$, we write

$$P = \frac{k}{V}$$

When $P = 960$ kilopascals, the volume $V = 1.4$ cubic meters. We use this information to find $k$.

$$960 = \frac{k}{1.4} \quad \text{Let } P = 960 \text{ and } V = 1.4.$$

$$1344 = k \quad \text{Multiply both sides by 1.4.}$$

Thus, the value of $k$ is 1344. Replacing $k$ with 1344 in the variation equation, we have

$$P = \frac{1344}{V}$$

Next we find $P$ when $V$ is 2.5 cubic meters.

3. SOLVE.

$$P = \frac{1344}{2.5} \quad \text{Let } V = 2.5.$$

$$= 537.6$$

4. INTERPRET. *Check* the proposed solution in the original problem.

**State:** When the volume is 2.5 cubic meters, the pressure is 537.6 kilopascals.

3 Sometimes the ratio of a variable to the product of many other variables is constant. For example, the ratio of distance traveled to the product of speed and time traveled is always 1.

$$\frac{d}{rt} = 1 \qquad \text{or} \qquad d = rt$$

Such a relationship is called **joint variation.**

### JOINT VARIATION

If the ratio of a variable $y$ to the product of two or more variables is constant, then $y$ **varies jointly as,** or **is jointly proportional to,** the other variables. If

$$y = kxz$$

then the number $k$ is the **constant of variation** or the **constant of proportionality.**

## Example 5 EXPRESSING SURFACE AREA

The surface area of a cylinder varies jointly as its radius and height. Express surface area $S$ in terms of radius $r$ and height $h$.

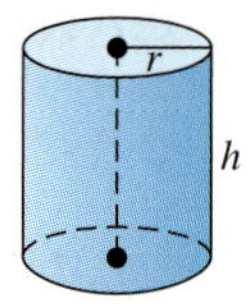


*Solution* Because the surface area varies jointly as the radius $r$ and the height $h$, we equate $S$ to a constant multiple of $r$ and $h$.

$$S = krh$$

In the equation, $S = krh$, it can be determined that the constant $k$ is $2\pi$, and we then have the formula $S = 2\pi rh$. (This formula does not include the areas of the two circular bases.)

4 Some examples of variation involve combinations of direct, inverse, and joint variation. We will call these variations **combined variation.**

## Example 6 FINDING COLUMN WEIGHT

The maximum weight that a circular column can support is directly proportional to the fourth power of its diameter and is inversely proportional to the square of its height. A 2-meter-diameter column that is 8 meters in height can support 1 ton. Find the weight that a 1-meter-diameter column that is 4 meters in height can support.

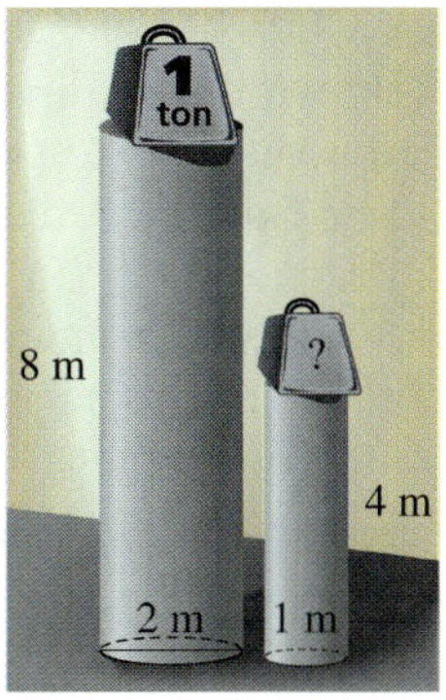


*Solution*

1. UNDERSTAND. Read and reread the problem. Let $w$ = weight, $d$ = diameter, $h$ = height, and $k$ = the constant of variation.
2. TRANSLATE. Since $w$ is directly proportional to $d^4$ and inversely proportional to $h^2$, we have

$$w = \frac{kd^4}{h^2}$$

3. SOLVE. To find $k$, we are given that a 2-meter-diameter column that is 8 meters in height can support 1 ton. That is, $w = 1$ when $d = 2$ and $h = 8$, or

$$1 = \frac{k \cdot 2^4}{8^2} \quad \text{Let } w = 1, d = 2, \text{ and } h = 8.$$

$$1 = \frac{k \cdot 16}{64}$$

$$4 = k \quad \text{Solve for } k.$$

Now replace $k$ with 4 in the equation $w = \frac{kd^4}{h^2}$ and we have

$$w = \frac{4d^4}{h^2}$$

To find weight $w$ for a 1-meter-diameter column that is 4 meters in height, let $d = 1$ and $h = 4$.

$$w = \frac{4 \cdot 1^4}{4^2}$$

$$w = \frac{4}{16} = \frac{1}{4}$$

4. INTERPRET. *Check* the proposed solution in the original problem. *State:* The 1-meter-diameter column that is 4 meters in height can hold $\frac{1}{4}$ ton of weight. ■

## SPOTLIGHT ON DECISION MAKING

Suppose you are painting the ceilings of your one-story home, whose layout is shown in the figure. The amount of paint you need is directly proportional to the area of what is to be painted. A clerk at the paint store says that 450 square feet can be painted with 4 quarts of paint. Quarts of paint cost \$5.95 each and gallons of paint cost \$21.50 each. You have brought only \$50 with you to the store. Can you get all the paint you need for the project with the money you have brought? If so, explain how. If not, how much more money will you need?

## Exercise Set 6.8

*If y varies directly as x, find the constant of variation k and the direct variation equation for each situation. See Example 1.*

1. $y = 4$ when $x = 20$
2. $y = 5$ when $x = 30$
3. $y = 6$ when $x = 4$
4. $y = 12$ when $x = 8$
5. $y = 7$ when $x = \frac{1}{2}$
6. $y = 11$ when $x = \frac{1}{3}$
7. $y = 0.2$ when $x = 0.8$
8. $y = 0.4$ when $x = 2.5$

*Solve. See Example 2.*

9. The weight of a synthetic ball varies directly with the cube of its radius. A ball with a radius of 2 inches weighs 1.20 pounds. Find the weight of a ball of the same material with a 3-inch radius.

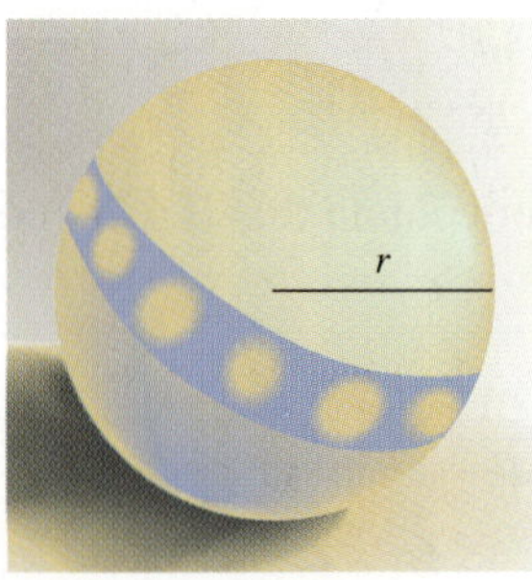


10. At sea, the distance to the horizon is directly proportional to the square root of the elevation of the observer. If a person who is 36 feet above the water can see 7.4 miles, find how far a person 64 feet above the water can see. Round answer to one decimal place.

11. The amount $P$ of pollution varies directly with the population $N$ of people. Kansas City has a population of 450,000 and produces 260,000 tons of pollutants. Find how many tons of pollution we should expect St. Louis to produce, if we know that its population is 980,000. Round answer to the nearest whole ton.

12. Charles' law states that if the pressure $P$ stays the same, the volume $V$ of a gas is directly proportional to its temperature $T$. If a balloon is filled with 20 cubic meters of a gas at a temperature of 300 $K$, find the new volume if the temperature rises to 360 $K$ while the pressure stays the same.

*If y varies inversely as x, find the constant of variation k and the inverse variation equation for each situation. See Example 3.*

13. $y = 6$ when $x = 5$
14. $y = 20$ when $x = 9$
15. $y = 100$ when $x = 7$
16. $y = 63$ when $x = 3$
17. $y = \frac{1}{8}$ when $x = 16$
18. $y = \frac{1}{10}$ when $x = 40$
19. $y = 0.2$ when $x = 0.7$
20. $y = 0.6$ when $x = 0.3$

*Solve. See Example 4.*

21. Pairs of markings a set distance apart are made on highways so that police can detect drivers exceeding the speed limit. Over a fixed distance, the speed $R$ varies inversely with the time $T$. In one particular pair of markings, $R$ is 45 mph when $T$ is 6 seconds. Find the speed of a car that travels the given distance in 5 seconds.

**22.** The weight of an object on or above the surface of Earth varies inversely as the square of the distance between the object and Earth's center. If a person weighs 160 pounds on Earth's surface, find the individual's weight if he moves 200 miles above Earth. Round answer to the nearest pound. (Assume that Earth's radius is 4000 miles.)

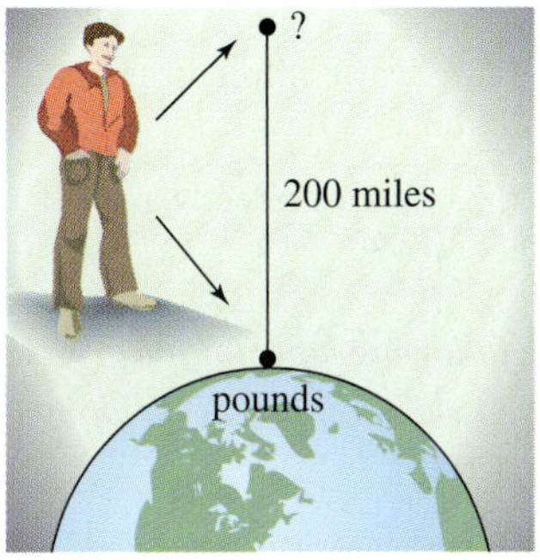


**23.** If the voltage $V$ in an electric circuit is held constant, the current $I$ is inversely proportional to the resistance $R$. If the current is 40 amperes when the resistance is 270 ohms, find the current when the resistance is 150 ohms.

**24.** Because it is more efficient to produce larger numbers of items, the cost of producing Dysan computer disks is inversely proportional to the number produced. If 4000 can be produced at a cost of \$1.20 each, find the cost per disk when 6000 are produced.

**25.** The intensity $I$ of light varies inversely as the square of the distance $d$ from the light source. If the distance from the light source is doubled (see the figure at the bottom of this column and the figure at the top of the next column), determine what happens to the intensity of light at the new location.

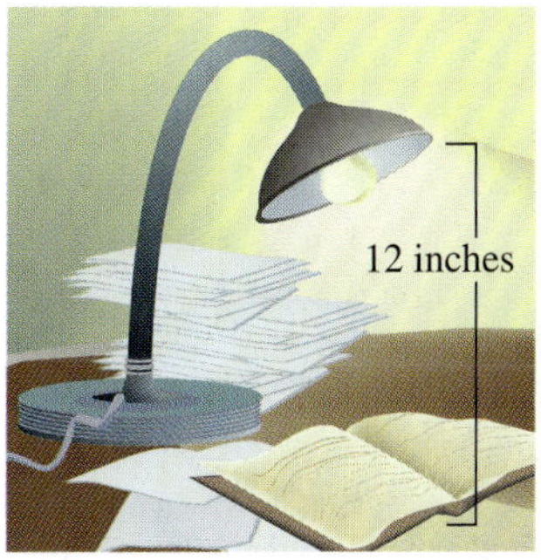


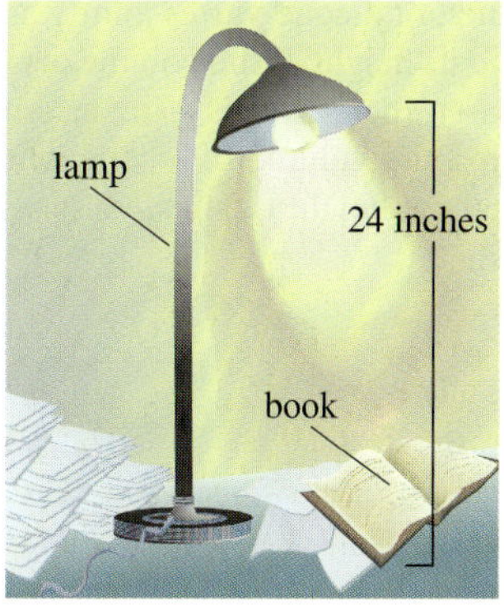


△ **26.** The maximum weight that a circular column can hold is inversely proportional to the square of its height. If an 8-foot column can hold 2 tons, find how much weight a 10-foot column can hold.

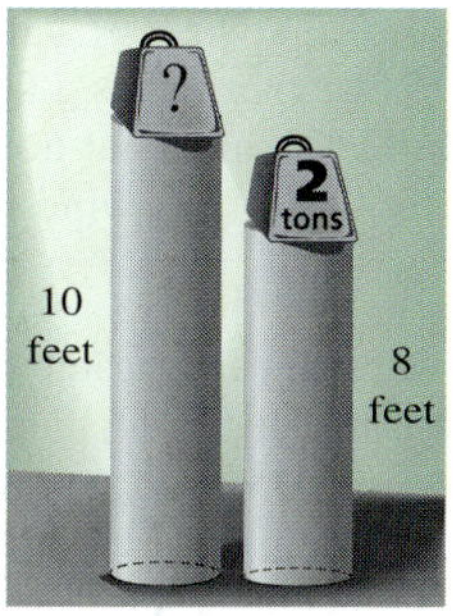


*Write each statement as an equation. See Example 5.*

**27.** $x$ varies jointly as $y$ and $z$.

**28.** $P$ varies jointly as $R$ and the square of $S$.

**29.** $r$ varies jointly as $s$ and the cube of $t$.

**30.** $a$ varies jointly as $b$ and $c$.

*Solve. See Examples 5 and 6.*

△ **31.** The maximum weight that a rectangular beam can support varies jointly as its width and the square of its height and inversely as its length. If a beam $\frac{1}{2}$ foot wide, $\frac{1}{3}$ foot high, and 10 feet long can support 12 tons, find how much a similar beam can support if the beam is $\frac{2}{3}$ foot wide, $\frac{1}{2}$ foot high, and 16 feet long.

**32.** The number of cars manufactured on an assembly line at a General Motors plant varies jointly as the number of workers and the time they work. If 200 workers can produce 60 cars in 2 hours, find how many cars 240 workers should be able to make in 3 hours.

△ **33.** The volume of a cone varies jointly as the square of its radius and its height. If the volume of a cone is $32\pi$ cubic inches when the radius is 4 inches and the height is 6 inches, find the volume of a cone when the radius is 3 inches and the height is 5 inches.

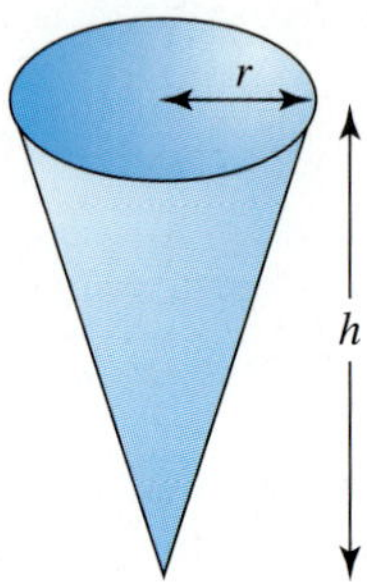


**34.** When a wind blows perpendicularly against a flat surface, its force is jointly proportional to the surface area and the speed of the wind. A sail whose surface area is 12 square feet experiences a 20-pound force when the wind speed is 10 miles per hour. Find the force on an 8-square-foot sail if the wind speed is 12 miles per hour.

**35.** The horsepower that can be safely transmitted to a shaft varies jointly as the shaft's angular speed of rotation (in revolutions per minute) and the cube of its diameter. A 2-inch shaft making 120 revolutions per minute safely transmits 40 horsepower. Find how much horsepower can be safely transmitted by a 3-inch shaft making 80 revolutions per minute.

△ **36.** The maximum weight that a rectangular beam can support varies jointly as its width and the square of its height and inversely as its length. If a beam $\frac{1}{3}$ foot wide, 1 foot high, and 10 feet long can support 3 tons, find how much weight a similar beam can support if it is 1 foot wide, $\frac{1}{3}$ foot high, and 9 feet long.

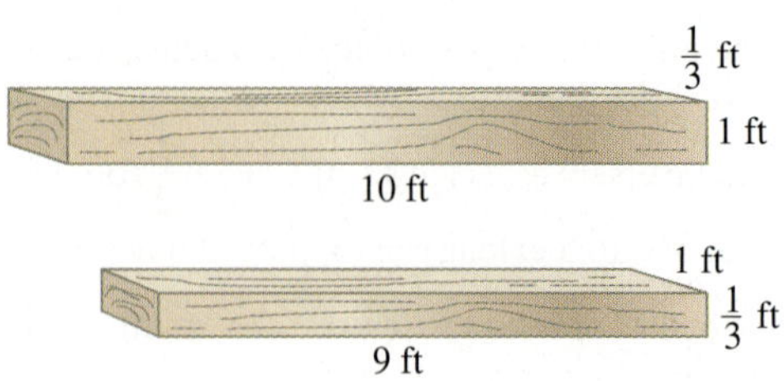


**37.** The atmospheric pressure $y$ (in millibars) is inversely proportional to the altitude $x$ (in kilometers). If the atmospheric pressure is 400 millibars at an altitude of 8 kilometers, find the atmospheric pressure at an altitude of 4 kilometers.

**38.** The horsepower to drive a boat varies directly as the cube of the speed of the boat. If the speed of the boat is to double, determine the corresponding increase in horsepower required.

△ **39.** The volume of a cylinder varies jointly as the height and the square of the radius. If the height is halved and the radius is doubled, determine what happens to the volume.

**40.** Suppose that $y$ varies directly as $x$. If $x$ is doubled, what is the effect on $y$?

**41.** Suppose that $y$ varies directly as $x^2$. If $x$ is doubled, what is the effect on $y$?

*Complete the following table for the inverse variation $y = \frac{k}{x}$ over each given value of k. Plot the points on a rectangular coordinate system.*

| $x$ | $\frac{1}{4}$ | $\frac{1}{2}$ | 1 | 2 | 4 |
|---|---|---|---|---|---|
| $y = \frac{k}{x}$ | | | | | |

**42.** $k = 1$

**43.** $k = 3$

**44.** $k = 5$

**45.** $k = \frac{1}{2}$

## REVIEW EXERCISES

*Find the exact circumference and area of each circle. See the inside cover for a list of geometric formulas.*

△ **46.**

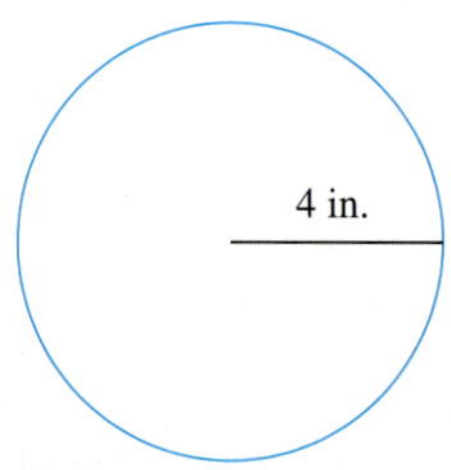


△ **47.**

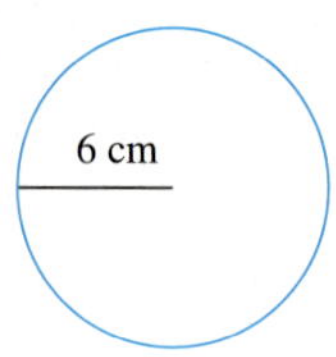


△ **48.**

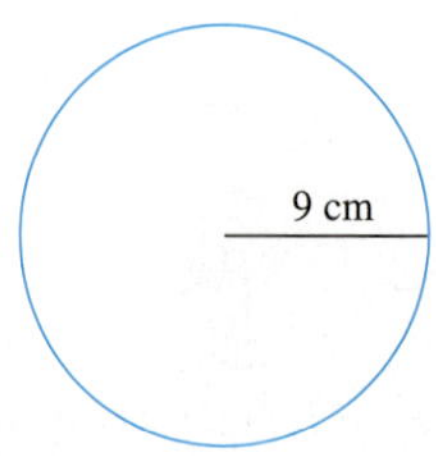

△ **49.**

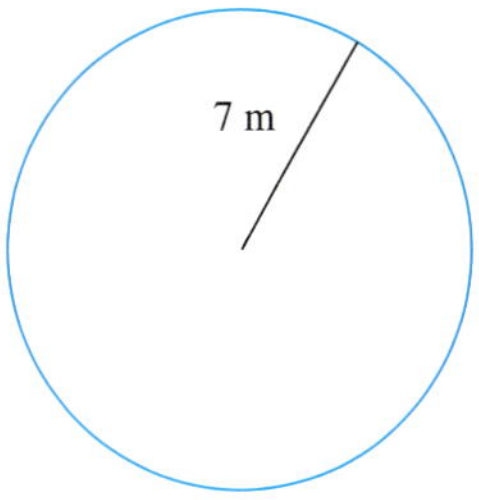


*Find each square root. See Section 1.3*

**50.** $\sqrt{81}$

**51.** $\sqrt{36}$

**52.** $\sqrt{1}$

**53.** $\sqrt{4}$

**54.** $\sqrt{\frac{1}{4}}$

**55.** $\sqrt{\frac{1}{25}}$

**56.** $\sqrt{\frac{4}{9}}$

**57.** $\sqrt{\frac{25}{121}}$

6

For additional Chapter Projects, visit the Real World Activities Website by going to http://www.prenhall.com/martin-gay.

# CHAPTER PROJECT

## *Modeling Electricity Production*

According to the U.S. Department of Energy, energy produced by renewable sources (including hydroelectric, geothermal, wind, and solar powers) accounted for nearly 7% of the United States' total production of energy in 1997. Wind energy can be harnessed by windmills to produce electricity, but it is the least utilized of these renewable energy sources. However, progressive communities are experimenting with fields of windmills for communal electricity needs, examining exactly how wind speed affects the amount of electricity produced.

A community in California is experimenting with electricity generated by windmills. City engineers are analyzing data they have gathered about their field of windmills. The engineers are familiar with other research demonstrating that the amount of electricity that a windmill generates hourly (in watt-hours) is directly proportional to the cube of the wind speed (in miles per hour). In this project, you will use this fact to help the engineers analyze the amount of electricity generated by the windmill field. This project may be completed by working in groups or individually.

1. The city engineers have documented that when the wind speed is exactly 10 miles per hour, a windmill generates electricity at a rate of 15 watt-hours. Find a formula that models the relationship between the wind speed and the amount of electricity generated hourly by a windmill.
2. Complete the following table for the given wind speeds. Then use the table to estimate the wind speed required to obtain: **a.** 100 watt-hours, and **b.** 400 watt-hours.

| *Wind Speed (miles per hour)* | *Electricity Generated (watt-hours)* |
|---|---|
| 15 | |
| 17 | |
| 19 | |
| 21 | |
| 23 | |
| 25 | |
| 27 | |
| 29 | |
| 31 | |
| 33 | |
| 35 | |

3. Plot the ordered pairs from the table. Describe the trend shown by the graph.
4. The engineers' data show that for several days the wind speed was more or less steady at 20 miles per hour, and the windmills generated the expected 120 watt-hours. According to the weather forecast for the coming few days, wind speed will fluctuate wildly, but will still average 20 miles per hour. Should the engineers still expect the windmills to generate 120 watt-hours? Demonstrate your reasoning with a numerical example.
5. During one three-day period, each windmill generated 150 watt-hours. The forecast predicts that the wind speed for the coming few days will drop by half. How many watt-hours should the engineers now expect each windmill to generate? In general, if the wind speed yields $c$ watt-hours, how many watt-hours does half the wind speed yield?

## CHAPTER 6 VOCABULARY CHECK

*Fill in each blank with one of the words or phrases listed below.*

| | | | | | |
|---|---|---|---|---|---|
| rational expression | equation | complex fraction | opposites | synthetic division | |
| least common denominator | expression | long division | jointly | directly | inversely |

1. A rational expression whose numerator, denominator, or both contain one or more rational expressions is called a ________________.
2. To divide a polynomial by a polynomial other than a monomial, we use ____________.
3. In the equation $y = kx$, $y$ varies ________ as $x$.
4. In the equation $y = \frac{k}{x}$, $y$ varies _________ as $x$.
5. The ______________________ of a list of rational expressions is a polynomial of least degree whose factors include the denominator factors in the list.
6. When a polynomial is to be divided by a binomial of the form $x - c$, a shortcut process called ________________ may be used.
7. In the equation $y = kxz$, $y$ varies _______ as $x$ and $z$.
8. The expressions $(x - 5)$ and $(5 - x)$ are called _________.
9. A ________________ is an expression that can be written as the quotient $\frac{P}{Q}$ of two polynomials $P$ and $Q$ as long as $Q$ is not 0.
10. Which is an expression and which is an equation? An example of an _________ is $\frac{2}{x} + \frac{2}{x^2} = 7$ and an example of an _________ is $\frac{2}{x} + \frac{5}{x^2}$.

# CHAPTER 6 HIGHLIGHTS

| DEFINITIONS AND CONCEPTS | EXAMPLES |
|---|---|
| **Section 6.1 Rational Functions and Multiplying and Dividing Rational Expressions** | |
| A **rational expression** is the quotient $\frac{P}{Q}$ of two polynomials $P$ and $Q$, as long as Q is not 0. | $\frac{2x-6}{7}$, $\frac{t^2-3t+5}{t-1}$ |
| ***To Simplify a Rational Expression***<br>**Step 1:** Completely factor the numerator and the denominator.<br>**Step 2:** Apply the fundamental principle of rational expressions. | Simplify.<br>$\frac{2x^2+9x-5}{x^2-25} = \frac{(2x-1)(x+5)}{(x-5)(x+5)}$<br>$= \frac{2x-1}{x-5}$ |
| ***To Multiply Rational Expressions***<br>**Step 1:** Completely factor numerators and denominators.<br>**Step 2:** Multiply the numerators and multiply the denominators.<br>**Step 3:** Apply the fundamental principle of rational expressions. | Multiply $\frac{x^3+8}{12x-18} \cdot \frac{14x^2-21x}{x^2+2x}$.<br>$= \frac{(x+2)(x^2-2x+4)}{6(2x-3)} \cdot \frac{7x(2x-3)}{x(x+2)}$<br>$= \frac{7(x^2-2x+4)}{6}$ |
| ***To Divide Rational Expressions***<br>Multiply the first rational expression by the reciprocal of the second rational expression. | Divide $\frac{x^2+6x+9}{5xy-5y} \div \frac{x+3}{10y}$.<br>$= \frac{(x+3)(x+3)}{5y(x-1)} \cdot \frac{2 \cdot 5y}{x+3}$<br>$= \frac{2(x+3)}{x-1}$ |
| A **rational function** is a function described by a rational expression. | $f(x) = \frac{2x-6}{7}$, $h(t) = \frac{t^2-3t+5}{t-1}$ |
| **Section 6.2 Adding and Subtracting Rational Expressions** | |
| ***To Add or Subtract Rational Expressions***<br>**Step 1:** Find the LCD.<br>**Step 2:** Write each rational expression as an equivalent rational expression whose denominator is the LCD.<br>**Step 3:** Add or subtract numerators and write the result over the common denominator.<br>**Step 4:** Simplify the resulting rational expression. | Subtract $\frac{3}{x+2} - \frac{x+1}{x-3}$.<br>$= \frac{3 \cdot (x-3)}{(x+2) \cdot (x-3)} - \frac{(x+1) \cdot (x+2)}{(x-3) \cdot (x+2)}$<br>$= \frac{3(x-3)-(x+1)(x+2)}{(x+2)(x-3)}$<br>$= \frac{3x-9-(x^2+3x+2)}{(x+2)(x-3)}$<br>$= \frac{3x-9-x^2-3x-2}{(x+2)(x-3)}$<br>$= \frac{-x^2-11}{(x+2)(x-3)}$ |

| DEFINITIONS AND CONCEPTS | EXAMPLES |
|---|---|

## Section 6.3 Simplifying Complex Fractions

**Method 1:** Simplify the numerator and the denominator so that each is a single fraction. Then perform the indicated division and simplify if possible.

Simplify $\dfrac{\dfrac{x+2}{x}}{x-\dfrac{4}{x}}$.

Method 1: $\dfrac{\dfrac{x+2}{x}}{\dfrac{x \cdot x}{1 \cdot x}-\dfrac{4}{x}} = \dfrac{\dfrac{x+2}{x}}{\dfrac{x^2-4}{x}}$

$$= \frac{x+2}{x} \cdot \frac{x}{(x+2)(x-2)} = \frac{1}{x-2}$$

**Method 2:** Multiply the numerator and the denominator of the complex fraction by the LCD of the fractions in both the numerator and the denominator. Then simplify if possible.

Method 2: $\dfrac{\left(\dfrac{x+2}{x}\right) \cdot x}{\left(x-\dfrac{4}{x}\right) \cdot x} = \dfrac{x+2}{x \cdot x - \dfrac{4}{x} \cdot x}$

$$= \frac{x+2}{x^2-4} = \frac{x+2}{(x+2)(x-2)} = \frac{1}{x-2}$$

## Section 6.4 Dividing Polynomials

To divide a polynomial by a monomial:
Divide each term in the polynomial by the monomial.

Divide $\dfrac{12a^5b^3 - 6a^2b^2 + ab}{6a^2b^2}$.

$$= \frac{12a^5b^3}{6a^2b^2} - \frac{6a^2b^2}{6a^2b^2} + \frac{ab}{6a^2b^2}$$

$$= 2a^3b - 1 + \frac{1}{6ab}$$

To divide a polynomial by a polynomial, other than a monomial:
Use **long division.**

Divide $2x^3 - x^2 - 8x - 1$ by $x - 2$.

$$\begin{array}{r} 2x^2 + 3x - 2 \\ x-2 \overline{)\,2x^3 - x^2 - 8x - 1} \\ \underline{2x^3 - 4x^2 \qquad\qquad} \\ 3x^2 - 8x \qquad \\ \underline{3x^2 - 6x \qquad} \\ -2x - 1 \\ \underline{-2x + 4} \\ -5 \end{array}$$

The quotient is $2x^2 + 3x - 2 - \dfrac{5}{x-2}$.

## Section 6.5 Synthetic Division and the Remainder Theorem

A shortcut method called **synthetic division** may be used to divide a polynomial by a binomial of the form $x - c$.

Use synthetic division to divide $2x^3 - x^2 - 8x - 1$ by $x - 2$.

$$\begin{array}{r|rrrr} 2 & 2 & -1 & -8 & -1 \\ & \downarrow & 4 & 6 & -4 \\ \hline & 2 & 3 & -2 & -5 \end{array}$$

The quotient is $2x^2 + 3x - 2 - \dfrac{5}{x-2}$.

| DEFINITIONS AND CONCEPTS | EXAMPLES |
|---|---|

## Section 6.6 Solving Equations Containing Rational Expressions

To solve an equation containing rational expressions: Multiply both sides of the equation by the LCD of all rational expressions. Then apply the distributive property and simplify. Solve the resulting equation and then check each proposed solution to see whether it makes the denominator 0. If so, it is an **extraneous solution.**

Solve $x - \frac{3}{x} = \frac{1}{2}$.

$$2x\left(x - \frac{3}{x}\right) = 2x\left(\frac{1}{2}\right) \quad \text{The LCD is } 2x.$$

$$2x \cdot x - 2x\left(\frac{3}{x}\right) = 2x\left(\frac{1}{2}\right) \quad \text{Distribute.}$$

$$2x^2 - 6 = x$$

$$2x^2 - x - 6 = 0 \quad \text{Subtract } x.$$

$$(2x + 3)(x - 2) = 0 \quad \text{Factor.}$$

$$x = -\frac{3}{2} \quad \text{or} \quad x = 2 \quad \text{Solve.}$$

Both $-\frac{3}{2}$ and 2 check. The solutions are 2 and $-\frac{3}{2}$.

## Section 6.7 Rational Equations and Problem Solving

***Solving an Equation for a Specified Variable***

Treat the specified variable as the only variable of the equation and solve as usual.

Solve for $x$.

$$A = \frac{2x + 3y}{5}$$

$$5A = 2x + 3y \quad \text{Multiply both sides by 5.}$$

$$5A - 3y = 2x \quad \text{Subtract } 3y \text{ from both sides.}$$

$$\frac{5A - 3y}{2} = x \quad \text{Divide both sides by 2.}$$

***Problem-Solving Steps to Follow***

Jeanee and David Dillon volunteer every year to clean a strip of Lake Ponchartrain beach. Jeanee can clean all the trash in this area of beach in 6 hours; David takes 5 hours. Find how long it will take them to clean the area of beach together.

1. UNDERSTAND.

1. Read and reread the problem.
   Let $x$ = time in hours that it takes Jeanee and David to clean the beach together.

| | *Hours to Complete* | *Part Completed in 1 Hour* |
|---|---|---|
| ***JEANEE ALONE*** | 6 | $\frac{1}{6}$ |
| ***DAVID ALONE*** | 5 | $\frac{1}{5}$ |
| ***TOGETHER*** | $x$ | $\frac{1}{x}$ |

*(continued)*

| DEFINITIONS AND CONCEPTS | EXAMPLES |
|---|---|
| **Section 6.7 Rational Equations and Problem Solving** | |
| **2.** TRANSLATE. | **2.** In words: part Jeanee can complete in 1 hour + part David can complete in 1 hour = part they can complete together in 1 hour<br>Translate: $\frac{1}{6} + \frac{1}{5} = \frac{1}{x}$ |
| **3.** SOLVE. | **3.** $\frac{1}{6} + \frac{1}{5} = \frac{1}{x}$ Multiply both sides by $30x$.<br>$5x + 6x = 30$<br>$11x = 30$<br>$x = \frac{30}{11}$ or $2\frac{8}{11}$ |
| **4.** INTERPRET. | **4.** *Check* and then *state.* Together, they can clean the beach in $2\frac{8}{11}$ hours. |
| **Section 6.8 Variation and Problem Solving** | |
| $y$ **varies directly as** $x$, or $y$ is **directly proportional to** $x$, if there is a nonzero constant $k$ such that $$y = kx$$ | The circumference of a circle $C$ varies directly as its radius $r$. $$C = \underbrace{2\pi}_{k} r$$ |
| $y$ **varies inversely as** $x$, or $y$ is **inversely proportional to** x, if there is a nonzero constant $k$ such that $$y = \frac{k}{x}$$ | Pressure $P$ varies inversely with volume $V$. $$P = \frac{k}{V}$$ |
| $y$ **varies jointly as** $x$ and $z$ or $y$ is **jointly proportional to** $x$ and $z$ if there is a nonzero constant $k$ such that $$y = kxz$$ | The lateral surface area $S$ of a cylinder varies jointly as its radius $r$ and height $h$. $$S = \underbrace{2\pi}_{k} rh$$ |

## CHAPTER 6 REVIEW

*Find the domain for each rational function.*

**1.** $f(x) = \dfrac{3 - 5x}{7}$

**2.** $g(x) = \dfrac{2x + 4}{11}$

**3.** $F(x) = \dfrac{-3x^2}{x - 5}$

**4.** $h(x) = \dfrac{4x}{3x - 12}$

**5.** $f(x) = \dfrac{x^3 + 2}{x^2 + 8x}$

**6.** $G(x) = \dfrac{20}{3x^2 - 48}$

*Write each rational expression in lowest terms.*

**7.** $\dfrac{15x^4}{45x^2}$

**8.** $\dfrac{x + 2}{2 + x}$

**9.** $\dfrac{18m^6p^2}{10m^4p}$

**10.** $\dfrac{x - 12}{12 - x}$

**11.** $\dfrac{5x - 15}{25x - 75}$

**12.** $\dfrac{22x + 8}{11x + 4}$

**13.** $\dfrac{2x}{2x^2 - 2x}$

**14.** $\dfrac{x + 7}{x^2 - 49}$

**15.** $\dfrac{2x^2 + 4x - 30}{x^2 + x - 20}$

**16.** $\dfrac{xy - 3x + 2y - 6}{x^2 + 4x + 4}$

**17.** The average cost of manufacturing $x$ bookcases is given by the rational function

$$C(x) = \frac{35x + 4200}{x}$$

**a.** Find the average cost per bookcase of manufacturing 50 bookcases.

**b.** Find the average cost per bookcase of manufacturing 100 bookcases.

**c.** As the number of bookcases increases, does the average cost per bookcase increase or decrease? (See parts **a** and **b**.)

*Perform the indicated operation. If possible, simplify your answer.*

**18.** $\dfrac{5}{x^3} \cdot \dfrac{x^2}{15}$

**19.** $\dfrac{3x^4yz^3}{15x^2y^2} \cdot \dfrac{10xy}{z^6}$

**20.** $\dfrac{4 - x}{5} \cdot \dfrac{15}{2x - 8}$

**21.** $\dfrac{x^2 - 6x + 9}{2x^2 - 18} \cdot \dfrac{4x + 12}{5x - 15}$

**22.** $\dfrac{a - 4b}{a^2 + ab} \cdot \dfrac{b^2 - a^2}{8b - 2a}$

**23.** $\dfrac{x^2 - x - 12}{2x^2 - 32} \cdot \dfrac{x^2 + 8x + 16}{3x^2 + 21x + 36}$

**24.** $\dfrac{2x^3 + 54}{5x^2 + 5x - 30} \cdot \dfrac{6x + 12}{3x^2 - 9x + 27}$

**25.** $\dfrac{3}{4x} \div \dfrac{8}{2x^2}$

**26.** $\dfrac{4x + 8y}{3} \div \dfrac{5x + 10y}{9}$

**27.** $\dfrac{5ab}{14c^3} \div \dfrac{10a^4b^2}{6ac^5}$

**28.** $\dfrac{2}{5x} \div \dfrac{4 - 18x}{6 - 27x}$

**29.** $\dfrac{x^2 - 25}{3} \div \dfrac{x^2 - 10x + 25}{x^2 - x - 20}$

**30.** $\dfrac{a - 4b}{a^2 + ab} \div \dfrac{20b - 5a}{b^2 - a^2}$

**31.** $\dfrac{7x + 28}{2x + 4} \div \dfrac{x^2 + 2x - 8}{x^2 - 2x - 8}$

**32.** $\dfrac{3x + 3}{x - 1} \div \dfrac{x^2 - 6x - 7}{x^2 - 1}$

**33.** $\dfrac{2x - x^2}{x^3 - 8} \div \dfrac{x^2}{x^2 + 2x + 4}$

**34.** $\dfrac{5a^2 - 20}{a^3 + 2a^2 + a + 2} \div \dfrac{7a}{a^3 + a}$

**35.** $\dfrac{2a}{21} \div \dfrac{3a^2}{7} \cdot \dfrac{4}{a}$

**36.** $\dfrac{5x - 15}{3 - x} \cdot \dfrac{x + 2}{10x + 20} \cdot \dfrac{x^2 - 9}{x^2 - x - 6}$

**37.** $\dfrac{4a + 8}{5a^2 - 20} \cdot \dfrac{3a^2 - 6a}{a + 3} \div \dfrac{2a^2}{5a + 15}$

***(6.2)*** *Find the LCD of the rational expressions in the list.*

**38.** $\dfrac{4}{9}, \dfrac{5}{2}$

**39.** $\dfrac{5}{4x^2y^5}, \dfrac{3}{10x^2y^4}, \dfrac{x}{6y^4}$

**40.** $\dfrac{5}{2x}, \dfrac{7}{x - 2}$

**41.** $\dfrac{3}{5x}, \dfrac{2}{x - 5}$

**42.** $\dfrac{1}{5x^3}, \dfrac{4}{x^2 + 3x - 28}, \dfrac{11}{10x^2 - 30x}$

*Perform the indicated operation. If possible, simplify your answer.*

**43.** $\dfrac{2}{15} + \dfrac{4}{15}$

**44.** $\dfrac{4}{x-4} + \dfrac{x}{x-4}$

**45.** $\dfrac{4}{3x^2} + \dfrac{2}{3x^2}$

**46.** $\dfrac{1}{x-2} - \dfrac{1}{4-2x}$

**47.** $\dfrac{2x+1}{x^2+x-6} + \dfrac{2-x}{x^2+x-6}$

**48.** $\dfrac{7}{2x} + \dfrac{5}{6x}$

**49.** $\dfrac{1}{3x^2y^3} - \dfrac{1}{5x^4y}$

**50.** $\dfrac{1}{10-x} + \dfrac{x-1}{x-10}$

**51.** $\dfrac{x-2}{x+1} - \dfrac{x-3}{x-1}$

**52.** $\dfrac{x}{9-x^2} - \dfrac{2}{5x-15}$

**53.** $2x + 1 - \dfrac{1}{x-3}$

**54.** $\dfrac{2}{a^2-2a+1} + \dfrac{3}{a^2-1}$

**55.** $\dfrac{x}{9x^2+12x+16} - \dfrac{3x+4}{27x^3-64}$

*Perform the indicated operation. If possible, simplify your answer.*

**56.** $\dfrac{2}{x-1} - \dfrac{3x}{3x-3} + \dfrac{1}{2x-2}$

**57.** $\dfrac{3}{2x} \cdot \left(\dfrac{2}{x+1} - \dfrac{2}{x-3}\right)$

**58.** $\left(\dfrac{2}{x} - \dfrac{1}{5}\right) \cdot \left(\dfrac{2}{x} + \dfrac{1}{3}\right)$

**59.** $\dfrac{2}{x^2-16} - \dfrac{3x}{x^2+8x+16} + \dfrac{3}{x+4}$

△ **60.** Find the perimeter of the heptagon (polygon with 7 sides).

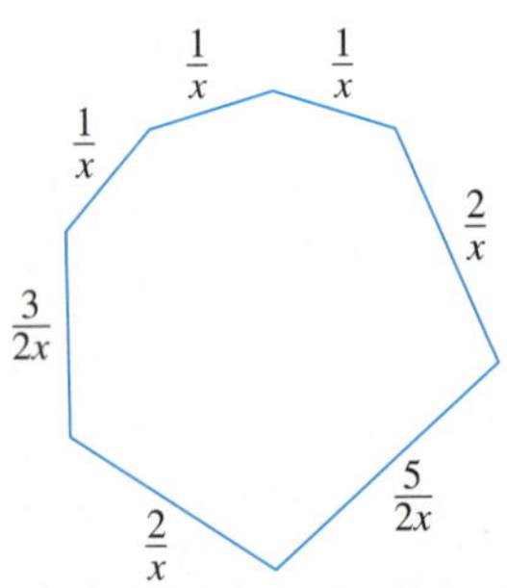


**(6.3)** *Simplify each complex fraction.*

**61.** $\dfrac{\frac{2}{5}}{\frac{3}{5}}$

**62.** $\dfrac{1 - \frac{3}{4}}{2 + \frac{1}{4}}$

**63.** $\dfrac{\frac{1}{x} - \frac{2}{3x}}{\frac{5}{2x} - \frac{1}{3}}$

**64.** $\dfrac{\frac{x^2}{15}}{\frac{x+1}{5x}}$

**65.** $\dfrac{\frac{3}{y^2}}{\frac{6}{y^3}}$

**66.** $\dfrac{\frac{x+2}{3}}{\frac{5}{x-2}}$

**67.** $\dfrac{2 - \frac{3}{2x}}{x - \frac{2}{5x}}$

**68.** $\dfrac{1 + \frac{x}{y}}{\frac{x^2}{y^2} - 1}$

**69.** $\dfrac{\frac{5}{x} + \frac{1}{xy}}{\frac{3}{x^2}}$

**70.** $\dfrac{\frac{x}{3} - \frac{3}{x}}{1 + \frac{3}{x}}$

**71.** $\dfrac{\frac{1}{x-1} + 1}{\frac{1}{x+1} - 1}$

**72.** $\dfrac{2}{1 - \frac{2}{x}}$

**73.** $\dfrac{1}{1 + \frac{2}{1 - \frac{1}{x}}}$

**74.** $\dfrac{\frac{x^2+5x-6}{4x+3}}{\frac{(x+6)^2}{8x+6}}$

**75.** $\dfrac{\frac{x-3}{x+3} + \frac{x+3}{x-3}}{\frac{x-3}{x+3} - \frac{x+3}{x-3}}$

**76.** $\dfrac{\frac{3}{x-1} - \frac{2}{1-x}}{\frac{2}{x-1} - \frac{2}{x}}$

**77.** If $f(x) = \dfrac{3}{x}$, find each of the following:

**a.** $f(a+h)$

**b.** $f(a)$

**c.** Use parts **a** and **b** to find $\dfrac{f(a+h) - f(a)}{h}$.

**d.** Simplify the results of part **c.**

***(6.4)*** *Divide.*

**78.** Divide $3x^5yb^9$ by $9xy^7$.

**79.** Divide $-9xb^4z^3$ by $-4axb^2$.

**80.** $(4xy + 2x^2 - 9) \div 4xy$

**81.** Divide $12xb^2 + 16xb^4$ by $4xb^3$.

**82.** $(3x^4 - 25x^2 - 20) \div (x - 3)$

**83.** $(-x^2 + 2x^4 + 5x - 12) \div (x + 2)$

**84.** $(2x^4 - x^3 + 2x^2 - 3x + 1) \div \left(x - \frac{1}{2}\right)$

**85.** $(2x^3 + 3x^2 - 2x + 2) \div \left(x + \frac{3}{2}\right)$

**86.** $(3x^4 + 5x^3 + 7x^2 + 3x - 2) \div (x^2 + x + 2)$

**87.** $(9x^4 - 6x^3 + 3x^2 - 12x - 30) \div (3x^2 - 2x - 5)$

***(6.5)*** *Use synthetic division to find each quotient.*

**88.** $(3x^3 + 12x - 4) \div (x - 2)$

**89.** $(3x^3 + 2x^2 - 4x - 1) \div \left(x + \frac{3}{2}\right)$

**90.** $(x^5 - 1) \div (x + 1)$

**91.** $(x^3 - 81) \div (x - 3)$

**92.** $(x^3 - x^2 + 3x^4 - 2) \div (x - 4)$

**93.** $(3x^4 - 2x^2 + 10) \div (x + 2)$

*If $P(x) = 3x^5 - 9x + 7$, use the remainder theorem to find the following.*

**94.** $P(4)$

**95.** $P(-5)$

**96.** $P\left(\frac{2}{3}\right)$

**97.** $P\left(-\frac{1}{2}\right)$

△ **98.** If the area of the rectangle is $(x^4 - x^3 - 6x^2 - 6x + 18)$ square miles and its width is $(x - 3)$ miles, find the length.

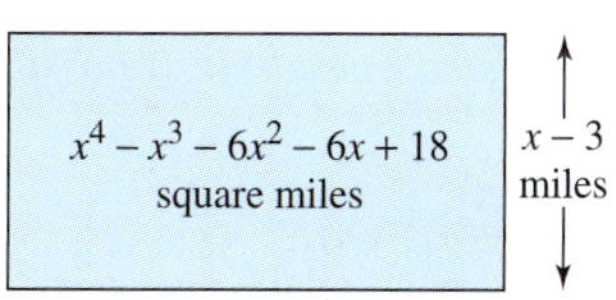


***(6.6)*** *Solve each equation for x.*

**99.** $\frac{2}{5} = \frac{x}{15}$

**100.** $\frac{3}{x} + \frac{1}{3} = \frac{5}{x}$

**101.** $4 + \frac{8}{x} = 8$

**102.** $\frac{2x + 3}{5x - 9} = \frac{3}{2}$

**103.** $\frac{1}{x-2} - \frac{3x}{x^2-4} = \frac{2}{x+2}$

**104.** $\frac{7}{x} - \frac{x}{7} = 0$

**105.** $\frac{x - 2}{x^2 - 7x + 10} = \frac{1}{5x - 10} - \frac{1}{x - 5}$

*Solve the equations for x or perform the indicated operation. Simplify.*

**106.** $\frac{5}{x^2 - 7x} + \frac{4}{2x - 14}$

**107.** $3 - \frac{5}{x} - \frac{2}{x^2} = 0$

**108.** $\frac{4}{3 - x} - \frac{7}{2x - 6} + \frac{5}{x}$

***(6.7)*** *Solve the equation for the specified variable.*

△ **109.** $A = \frac{h(a + b)}{2}$, $a$

**110.** $\frac{1}{R} = \frac{1}{R_1} + \frac{1}{R_2}$, $R_2$

**111.** $I = \frac{E}{R + r}$, $R$

**112.** $A = P + Prt$, $r$

**113.** $H = \frac{kA(T_1 - T_2)}{L}$, $A$

*Solve.*

**114.** The sum of a number and twice its reciprocal is 3. Find the number(s).

**115.** If a number is added to the numerator of $\frac{3}{7}$, and twice that number is added to the denominator of $\frac{3}{7}$, the result is equivalent to $\frac{10}{21}$. Find the number.

**116.** The denominator of a fraction is 2 more than the numerator. If the numerator is decreased by 3 and the denominator is increased by 5, the resulting fraction is equivalent to $\frac{2}{3}$. Find the fraction.

**117.** The sum of the reciprocals of two consecutive even integers is $-\frac{9}{40}$. Find the two integers.

**118.** Three boys can paint a fence in 4 hours, 5 hours, and 6 hours, respectively. Find how long it will take all three boys to paint the fence.

**119.** If Sue Katz can type a certain number of mailing labels in 6 hours and Tom Neilson and Sue working together can type the same number of mailing labels in 4 hours, find how long it takes Tom alone to type the mailing labels.

**120.** The inlet pipe of a water tank can fill the tank in 2 hours and 30 minutes. The outlet pipe can empty the tank in 2 hours. Find how long it takes to empty a full tank if both pipes are open.

**121.** Timmy Garnica drove 210 miles in the same amount of time that it took a DC-10 jet to travel 1715 miles. The speed of the jet was 430 mph faster than the speed of the car. Find the speed of the jet.

**122.** The combined resistance $R$ of two resistors in parallel with resistances $R_1$ and $R_2$ is given by the formula $\frac{1}{R} = \frac{1}{R_1} + \frac{1}{R_2}$. If the combined resistance is $\frac{30}{11}$ ohms and the resistance of one of the two resistors is 5 ohms, find the resistance of the other resistor.

**123.** The speed of a Ranger boat in still water is 32 mph. If the boat travels 72 miles upstream in the same time that it takes to travel 120 miles downstream, find the speed of the current.

**124.** A B737 jet flies 445 miles with the wind and 355 miles against the wind in the same length of time. If the speed of the jet in still air is 400 mph, find the speed of the wind.

**125.** The speed of a jogger is 3 mph faster than the speed of a walker. If the jogger travels 14 miles in the same amount of time that the walker travels 8 miles, find the speed of the walker.

**126.** Two Amtrak trains traveling on parallel tracks leave Tucson at the same time. The speed of one train is 18 mph faster than the other. If the faster train travels 378 miles in the same time that the other train travels 270 miles, find the speed of each train.

*(6.8) Solve each variation problem.*

**127.** A is directly proportional to $B$. If $A = 6$ when $B = 14$, find $A$ when $B = 21$.

**128.** $C$ is inversely proportional to $D$. If $C = 12$ when $D = 8$, find $C$ when $D = 24$.

**129.** According to Boyle's law, the pressure exerted by a gas is inversely proportional to the volume, as long as the temperature stays the same. If a gas exerts a pressure of 1250 pounds per square inch when the volume is 2 cubic feet, find the volume when the pressure is 800 pounds per square inch.

△ **130.** The surface area of a sphere varies directly as the square of its radius. If the surface area is $36\pi$ square inches when the radius is 3 inches, find the surface area when the radius is 4 inches.

## CHAPTER 6 TEST

*Find the domain of each rational function.*

**1.** $f(x) = \frac{5x^2}{1 - x}$

**2.** $g(x) = \frac{9x^2 - 9}{x^2 + 4x + 3}$

*Write each rational expression in lowest terms.*

**3.** $\frac{5x^7}{3x^4}$

**4.** $\frac{7x - 21}{24 - 8x}$

**5.** $\frac{x^2 - 4x}{x^2 + 5x - 36}$

*Perform the indicated operation. If possible, simplify your answer.*

**6.** $\frac{x}{x - 2} \cdot \frac{x^2 - 4}{5x}$

**7.** $\dfrac{2x^3 + 16}{6x^2 + 12x} \cdot \dfrac{5}{x^2 - 2x + 4}$

**8.** $\dfrac{26ab}{7c} \div \dfrac{13a^2c^5}{14a^4b^3}$

**9.** $\dfrac{3x^2 - 12}{x^2 + 2x - 8} \div \dfrac{6x + 18}{x + 4}$

**10.** $\dfrac{4x - 12}{2x - 9} \div \dfrac{3 - x}{4x^2 - 81} \cdot \dfrac{x + 3}{5x + 15}$

**11.** $\dfrac{5}{4x^3} + \dfrac{7}{4x^3}$

**12.** $\dfrac{3 + 2x}{10 - x} + \dfrac{13 + x}{x - 10}$

**13.** $\dfrac{3}{x^2 - x - 6} + \dfrac{2}{x^2 - 5x + 6}$

**14.** $\dfrac{5}{x - 7} - \dfrac{2x}{3x - 21} + \dfrac{x}{2x - 14}$

**15.** $\dfrac{3x}{5} \cdot \left(\dfrac{5}{x} - \dfrac{5}{2x}\right)$

*Simplify each complex fraction.*

**16.** $\dfrac{\dfrac{4x}{13}}{\dfrac{20x}{13}}$

**17.** $\dfrac{\dfrac{5}{x} - \dfrac{7}{3x}}{\dfrac{9}{8x} - \dfrac{1}{x}}$

**18.** $\dfrac{\dfrac{x^2 - 5x + 6}{x + 3}}{\dfrac{x^2 - 4x + 4}{x^2 - 9}}$

*Divide.*

**19.** $(4x^2y + 9x + z) \div 3xz$

**20.** $(x^6 + 3x^5 - 2x^4 + x^2 - 3x + 2) \div (x - 2)$

**21.** Use synthetic division to divide $(4x^4 - 3x^3 + 2x^2 - x - 1)$ by $(x + 3)$.

**22.** If $P(x) = 4x^4 + 7x^2 - 2x - 5$, use the remainder theorem to find $P(-2)$.

*Solve each equation for x.*

**23.** $\dfrac{5x + 3}{3x - 7} = \dfrac{19}{7}$

**24.** $\dfrac{5}{x - 5} + \dfrac{x}{x + 5} = -\dfrac{29}{21}$

**25.** $\dfrac{x}{x - 4} = 3 - \dfrac{4}{x - 4}$

**26.** Solve for $x$: $\dfrac{x + b}{a} = \dfrac{4x - 7a}{b}$

**27.** The product of one more than a number and twice the reciprocal of the number is $\dfrac{12}{5}$. Find the number.

**28.** If Jan can weed the garden in 2 hours and her husband can weed it in 1 hour and 30 minutes, find how long, it takes them to weed the garden together.

**29.** Suppose that $W$ is inversely proportional to $V$. If $W = 20$ when $V = 12$, find $W$ when $V = 15$.

**30.** Suppose that $Q$ is jointly proportional to $R$ and the square of $S$. If $Q = 24$ when $R = 3$ and $S = 4$, find $Q$ when $R = 2$ and $S = 3$.

**31.** When an anvil is dropped into a gorge, the speed with which it strikes the ground is directly proportional to the square root of the distance it falls. An anvil that falls 400 feet hits the ground at a speed of 160 feet per second. Find the height of a cliff over the gorge if a dropped anvil hits the ground at a speed of 128 feet per second.

## CHAPTER 6 CUMULATIVE REVIEW

1. Translate each phrase to an algebraic expression. Use the variable $x$ to represent each unknown number.
   - **a.** Eight times a number
   - **b.** Three more than eight times a number
   - **c.** The quotient of a number and $-7$
   - **d.** One and six-tenths subtracted from twice a number

2. Solve for $y$: $\frac{y}{3} - \frac{y}{4} = \frac{1}{6}$

3. In the United States, the annual consumption of cigarettes is declining. The consumption $c$ in billions of cigarettes per year since the year 1985 can be approximated by the formula

$$c = -14.25t + 598.69$$

where $t$ is the number of years after 1985. Use this formula to predict the years that the consumption of cigarettes will be less than 200 billion per year.

4. Solve: $\left|\frac{3x + 1}{2}\right| = -2$

5. Solve for $x$: $\left|\frac{2(x + 1)}{3}\right| \le 0$

6. Graph the equation $y = -2x + 3$.

7. Which of the following relations are also functions?
   - **a.** $\{(-2, 5), (2, 7), (-3, 5), (9, 9)\}$
   - **b.**

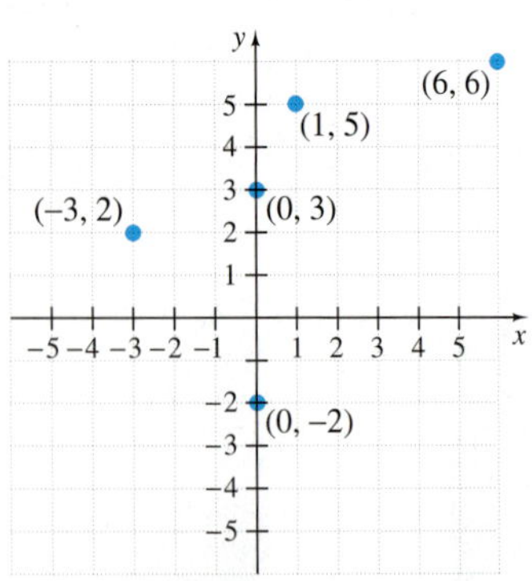


   - **c.**

| Input | Correspondence | Output |
|---|---|---|
| People in a certain city | Each person's age | The set of nonnegative integers |

8. Graph $x - 3y = 6$ by plotting intercept points.

9. Find an equation of the line with slope $-3$ containing the point $(1, -5)$. Write the equation in slope–intercept form $y = mx + b$.

10. Graph the intersection of $x \ge 1$ and $y \ge 2x - 1$.

11. Use the elimination method to solve the system.

$$\begin{cases} 3x - 2y = 10 \\ 4x - 3y = 15 \end{cases}$$

12. Solve the system.

$$\begin{cases} 2x - 4y + 8z = 2 \\ -x - 3y + z = 11 \\ x - 2y + 4z = 0 \end{cases}$$

13. The measure of the largest angle of a triangle is 80° more than the measure of the smallest angle, and the measure of the remaining angle is 10° more than the measure of the smallest angle. Find the measure of each angle.

14. Use matrices to solve the system.

$$\begin{cases} x + 2y + z = 2 \\ -2x - y + 2z = 5 \\ x + 3y - 2z = -8 \end{cases}$$

15. Evaluate the following.
   - **a.** $7^0$
   - **b.** $-7^0$
   - **c.** $(2x + 5)^0$
   - **d.** $2x^0$

16. Simplify each. Assume that $a$ and $b$ are integers and that $x$ and $y$ are not 0.
   - **a.** $x^{-b}(2x^b)^2$
   - **b.** $\frac{(y^{3a})^2}{y^{a-6}}$

17. Find the degree of each term.
   - **a.** $3x^2$
   - **b.** $-2^3x^5$
   - **c.** $y$
   - **d.** $12x^2yz^3$
   - **e.** $5$

**18.** Multiply $[3 + (2a + b)]^2$.

**19.** Factor $ab - 6a + 2b - 12$

**20.** $2n^2 - 38n + 80$

**21.** Factor $x^2 + 4x + 4 - y^2$

**22.** Solve $(x + 2)(x - 6) = 0$

**23.** Graph $f(x) = -x^3$. Find any intercepts.

**24.** Subtract $\dfrac{5k}{k^2 - 4} - \dfrac{2}{k^2 + k - 2}$.

**25.** Solve: $\dfrac{3}{x} - \dfrac{x + 21}{3x} = \dfrac{5}{3}$

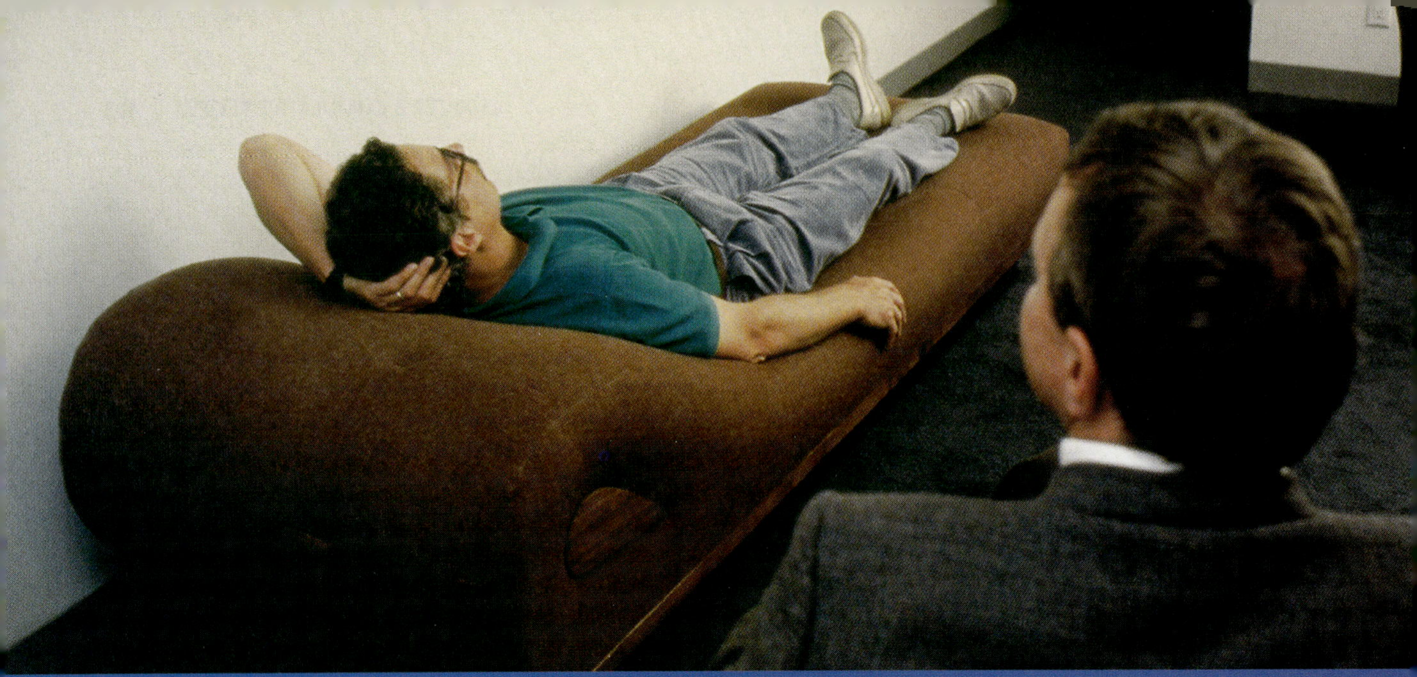

## Studying Human Behavior

Psychology is the scientific study of the human mind and behavior. Over 150,000 psychologists practice in the United States in such diverse areas as experimental psychology, clinical psychology, industrial psychology, educational psychology, counseling psychology, psychotherapy, military psychology, consumer psychology, family psychology, and sports psychology.

Although psychology careers in teaching, research, and counseling frequently require advanced degrees, there are many career paths in which a two- or four-year degree is useful. Employment counselors, child protection workers, corrections officers, social service directors, day-care-center supervisors, and hospital patient service representatives are all examples of positions for which advanced psychology degrees are not necessarily required. No matter which educational path is chosen, psychologists must have good communication, interpersonal, research, and analytical skills, including the ability to reason numerically, interpret statistics, read tables and graphs, and solve problems.

*For more information about psychology careers, visit the American Psychological Association Website by first going to www.prenhall.com/martin-gay.*

*In the Spotlight on Decision Making feature on page 441, you will have the opportunity to make a decision about assigning subjects to the appropriate test groups for a psychology experiment.*

# RATIONAL EXPONENTS, RADICALS, AND COMPLEX NUMBERS

# 7


7.1 Radicals and Radical Functions

7.2 Rational Exponents

7.3 Simplifying Radical Expressions

7.4 Adding, Subtracting, and Multiplying Radical Expressions

7.5 Rationalizing Denominators and Numerators of Radical Expressions

7.6 Radical Equations and Problem Solving

7.7 Complex Numbers


In this chapter, radical notation is reviewed, and then rational exponents are introduced. As the name implies, rational exponents are exponents that are rational numbers. We present an interpretation of rational exponents that is consistent with the meaning and rules already established for integer exponents, and we present two forms of notation for roots: radical and exponent. We conclude this chapter with complex numbers, a natural extension of the real number system.

## 7.1 RADICALS AND RADICAL FUNCTIONS

CD-ROM SSM SSG Video

**OBJECTIVES**

1. Find square roots.
2. Approximate roots using a calculator.
3. Find cube roots.
4. Find $n$th roots.
5. Find $\sqrt[n]{a^n}$ where $a$ is a real number.
6. Graph square and cube root functions.

**1** Recall from Section 1.3 that to find a **square root** of a number $a$, we find a number that was squared to get $a$.

Thus, because

$$5^2 = 25 \quad \text{and} \quad (-5)^2 = 25, \text{ then}$$

both 5 and $-5$ are square roots of 25.

Recall that we denote the **nonnegative**, or **principal, square root** with the **radical sign**.

$$\sqrt{25} = 5$$

We denote the **negative square root** with the **negative radical sign**.

$$-\sqrt{25} = -5$$

An expression containing a radical sign is called a **radical expression**. An expression within, or "under," a radical sign is called a **radicand**.

radical sign

radical expression: $\sqrt{a}$

radicand

**PRINCIPAL AND NEGATIVE SQUARE ROOTS**

The **principal square root** of a nonnegative number $a$ is its nonnegative square root. The principal square root is written as $\sqrt{a}$. The **negative square root** of $a$ is written as $-\sqrt{a}$.

**Example 1** Simplify. Assume that all variables represent positive numbers.

**a.** $\sqrt{36}$ **b.** $\sqrt{0}$ **c.** $\sqrt{\frac{4}{49}}$ **d.** $\sqrt{0.25}$ **e.** $\sqrt{x^6}$ **f.** $\sqrt{9x^{10}}$ **g.** $-\sqrt{81}$

*Solution*

**a.** $\sqrt{36} = 6$ because $6^2 = 36$ and 6 is not negative.

**b.** $\sqrt{0} = 0$ because $0^2 = 0$ and 0 is not negative.

**c.** $\sqrt{\frac{4}{49}} = \frac{2}{7}$ because $\left(\frac{2}{7}\right)^2 = \frac{4}{49}$ and $\frac{2}{7}$ is not negative.

**d.** $\sqrt{0.25} = 0.5$ because $(0.5)^2 = 0.25$.

**e.** $\sqrt{x^6} = x^3$ because $(x^3)^2 = x^6$.

**f.** $\sqrt{9x^{10}} = 3x^5$ because $(3x^5)^2 = 9x^{10}$.

**g.** $-\sqrt{81} = -9$. The negative in front of the radical indicates the negative square root of 81.

Can we find the square root of a negative number, say $\sqrt{-4}$? That is, can we find a real number whose square is $-4$? No, there is no real number whose square is $-4$, and we say that $\sqrt{-4}$ is not a real number. In general:

*The square root of a negative number is not a real number.*

> **HELPFUL HINT**
> Don't forget, the square root of a negative number, such as $\sqrt{-9}$, is not a real number. In Section 7.7, we will see what kind of a number $\sqrt{-9}$ is.

2 Recall that numbers such as 1, 4, 9, and 25 are called **perfect squares**, since $1 = 1^2$, $4 = 2^2$, $9 = 3^2$, and $25 = 5^2$. Square roots of perfect square radicands simplify to rational numbers. What happens when we try to simplify a root such as $\sqrt{3}$? Since 3 is not a perfect square, $\sqrt{3}$ is not a rational number. It is called an **irrational number**, and we can find a decimal **approximation** of it. To find decimal approximations, use a calculator. For example, an approximation for $\sqrt{3}$ is

$$\sqrt{3} \approx 1.732$$

$\uparrow$ approximation symbol

To see if the approximation is reasonable, notice that since

$$1 < 3 < 4, \text{ then}$$
$$\sqrt{1} < \sqrt{3} < \sqrt{4}, \text{ or}$$
$$1 < \sqrt{3} < 2.$$

We found $\sqrt{3} \approx 1.732$, a number between 1 and 2, so our result is reasonable.

**Example 2** Use a calculator to approximate $\sqrt{20}$. Round the approximation to 3 decimal places and check to see that your approximation is reasonable.

*Solution*

$$\sqrt{20} \approx 4.472$$

Is this reasonable? Since $16 < 20 < 25$, then $\sqrt{16} < \sqrt{20} < \sqrt{25}$, or $4 < \sqrt{20} < 5$. The approximation is between 4 and 5 and thus is reasonable.

3 Finding roots can be extended to other roots such as cube roots. For example, since $2^3 = 8$, we call 2 the **cube root** of 8. In symbols, we write

$$\sqrt[3]{8} = 2$$

**CUBE ROOT**

The **cube root** of a real number $a$ is written as $\sqrt[3]{a}$, and

$$\sqrt[3]{a} = b \text{ only if } b^3 = a$$

From this definition, we have

$$\sqrt[3]{64} = 4 \text{ since } 4^3 = 64$$
$$\sqrt[3]{-27} = -3 \text{ since } (-3)^3 = -27$$
$$\sqrt[3]{x^3} = x \text{ since } x^3 = x^3$$

Notice that, unlike with square roots, *it is possible to have a negative radicand when finding a cube root.* This is so because the *cube* of a negative number is a negative number. Therefore, the *cube root* of a negative number is a negative number.

**Example 3** Find the cube roots.

**a.** $\sqrt[3]{1}$ **b.** $\sqrt[3]{-64}$ **c.** $\sqrt[3]{\frac{8}{125}}$ **d.** $\sqrt[3]{x^6}$ **e.** $\sqrt[3]{-8x^9}$

*Solution*
**a.** $\sqrt[3]{1} = 1$ because $1^3 = 1$.
**b.** $\sqrt[3]{-64} = -4$ because $(-4)^3 = -64$.
**c.** $\sqrt[3]{\frac{8}{125}} = \frac{2}{5}$ because $\left(\frac{2}{5}\right)^3 = \frac{8}{125}$.
**d.** $\sqrt[3]{x^6} = x^2$ because $(x^2)^3 = x^6$.
**e.** $\sqrt[3]{-8x^9} = -2x^3$ because $(-2x^3)^3 = -8x^9$.

4 Just as we can raise a real number to powers other than 2 or 3, we can find roots other than square roots and cube roots. In fact, we can find the ***n*th root** of a number, where $n$ is any natural number. In symbols, the $n$th root of $a$ is written as $\sqrt[n]{a}$, where $n$ is called the **index**. The index 2 is usually omitted for square roots.

**HELPFUL HINT**

If the index is even, such as $\sqrt{\ }$, $\sqrt[4]{\ }$, $\sqrt[6]{\ }$, and so on, the radicand must be nonnegative for the root to be a real number. For example,

$$\sqrt[4]{16} = 2, \text{ but } \sqrt[4]{-16} \text{ is not a real number.}$$
$$\sqrt[6]{64} = 2, \text{ but } \sqrt[6]{-64} \text{ is not a real number.}$$

If the index is odd, such as $\sqrt[3]{\ }$, $\sqrt[5]{\ }$, and so on, the radicand may be any real number. For example,

$$\sqrt[3]{64} = 4 \text{ and } \sqrt[3]{-64} = -4$$
$$\sqrt[5]{32} = 2 \text{ and } \sqrt[5]{-32} = -2$$

**Example 4** Simplify the following expressions.

**a.** $\sqrt[4]{81}$ **b.** $\sqrt[5]{-243}$ **c.** $-\sqrt{25}$ **d.** $\sqrt[4]{-81}$ **e.** $\sqrt[3]{64x^3}$

*Solution*
**a.** $\sqrt[4]{81} = 3$ because $3^4 = 81$ and 3 is positive.
**b.** $\sqrt[5]{-243} = -3$ because $(-3)^5 = -243$.

**c.** $-\sqrt{25} = -5$ because $-5$ is the opposite of $\sqrt{25}$.

**d.** $\sqrt[4]{-81}$ is not a real number. There is no real number that, when raised to the fourth power, is $-81$.

**e.** $\sqrt[3]{64x^3} = 4x$ because $(4x)^3 = 64x^3$.

5 Recall that the notation $\sqrt{a^2}$ indicates the positive square root of $a^2$ only. For example,

$$\sqrt{(-5)^2} = \sqrt{25} = 5$$

When variables are present in the radicand and it is unclear whether the variable represents a positive number or a negative number, absolute value bars are sometimes needed to ensure that the result is a positive number. For example,

$$\sqrt{x^2} = |x|$$

This ensures that the result is positive. This same situation may occur when the index is any *even* positive integer. When the index is any *odd* positive integer, absolute value bars are not necessary.

**FINDING $\sqrt[n]{a^n}$**

If $n$ is an *even* positive integer, then $\sqrt[n]{a^n} = |a|$.

If $n$ is an *odd* positive integer, then $\sqrt[n]{a^n} = a$.

**Example 5** Simplify.

**a.** $\sqrt{(-3)^2}$ **b.** $\sqrt{x^2}$ **c.** $\sqrt[4]{(x-2)^4}$ **d.** $\sqrt[3]{(-5)^3}$ **e.** $\sqrt[5]{(2x-7)^5}$

*Solution* **a.** $\sqrt{(-3)^2} = |-3| = 3$ When the index is even, the absolute value bars ensure us that our result is not negative.

**b.** $\sqrt{x^2} = |x|$

**c.** $\sqrt[4]{(x-2)^4} = |x-2|$

**d.** $\sqrt[3]{(-5)^3} = -5$

**e.** $\sqrt[5]{(2x-7)^5} = 2x - 7$ Absolute value bars are not needed when the index is odd.

6 Recall that an equation in $x$ and $y$ describes a function if each $x$-value is paired with exactly one $y$-value. With this in mind, does the equation

$$y = \sqrt{x}$$

describe a function? First, notice that replacement values for $x$ must be nonnegative real numbers, since $\sqrt{x}$ is not a real number if $x < 0$. The notation $\sqrt{x}$ denotes the principal square root of $x$, so for every nonnegative number $x$, there is exactly one number, $\sqrt{x}$. Therefore, $y = \sqrt{x}$ describes a function, and we may write it as

$$f(x) = \sqrt{x}$$

Recall that the domain of a function in $x$ is the set of all possible replacement values for $x$. This means that the domain of this function is the set of all nonnegative numbers, or $\{x|x \geq 0\}$.

We find function values for $f(x)$ as usual. For example,

$$f(0) = \sqrt{0} = 0$$
$$f(1) = \sqrt{1} = 1$$
$$f(4) = \sqrt{4} = 2$$
$$f(9) = \sqrt{9} = 3$$

Choosing perfect squares for $x$ ensures us that $f(x)$ is a rational number, but it is important to stress that $f(x) = \sqrt{x}$ is defined for all nonnegative real numbers. For example,

$$f(3) = \sqrt{3} \approx 1.732$$

**Example 6** Graph the square root function $f(x) = \sqrt{x}$.

*Solution* To graph, we identify the domain, evaluate the function for several values of $x$, plot the resulting points, and connect the points with a smooth curve. The domain of this function is the set of all nonnegative numbers or $\{x|x \geq 0\}$. The table comes from the function values obtained earlier.

| $x$ | $f(x) = \sqrt{x}$ |
|---|---|
| 0 | 0 |
| 1 | 1 |
| 3 | $\sqrt{3} \approx 1.7$ |
| 4 | 2 |
| 9 | 3 |

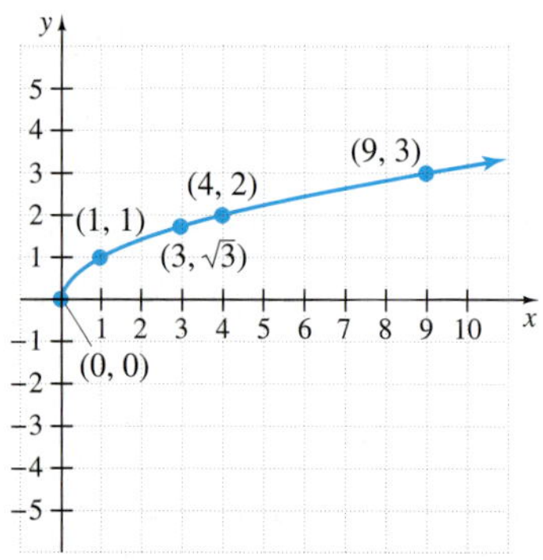


Notice that the graph of this function passes the vertical line test, as expected.

The equation $f(x) = \sqrt[3]{x}$ also describes a function. Here $x$ may be any real number, so the domain of this function is the set of all real numbers. A few function values are given next.

$$f(0) = \sqrt[3]{0} = 0$$
$$f(1) = \sqrt[3]{1} = 1$$
$$f(-1) = \sqrt[3]{-1} = -1$$
$$f(6) = \sqrt[3]{6}$$
$$f(-6) = \sqrt[3]{-6}$$
$$f(8) = \sqrt[3]{8} = 2$$
$$f(-8) = \sqrt[3]{-8} = -2$$

Here, the radicands are not perfect cubes. The radicals do not simplify to rational numbers.

**Example 7** Graph the function $f(x) = \sqrt[3]{x}$.

*Solution* To graph, we identify the domain, plot points, and connect the points with a smooth curve. The domain of this function is the set of all real numbers. The table comes from the function values obtained earlier. We have approximated $\sqrt[3]{6}$ and $\sqrt[3]{-6}$ for graphing purposes.

| $x$ | $f(x) = \sqrt[3]{x}$ |
|---|---|
| 0 | 0 |
| 1 | 1 |
| −1 | −1 |
| 6 | $\sqrt[3]{6} \approx 1.8$ |
| −6 | $\sqrt[3]{-6} \approx -1.8$ |
| 8 | 2 |
| −8 | −2 |

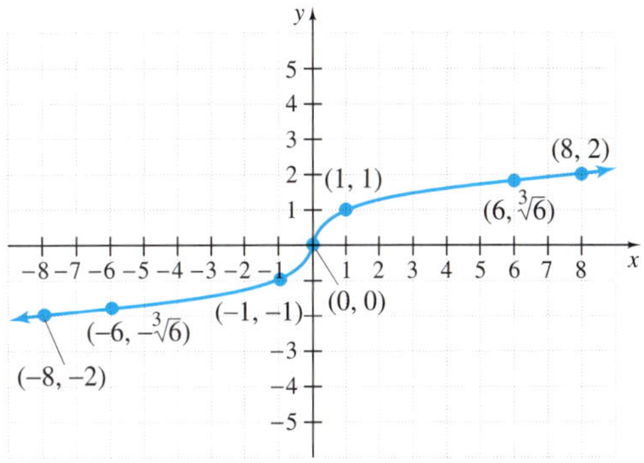


The graph of this function passes the vertical line test, as expected.

## SPOTLIGHT ON DECISION MAKING

Suppose you are a scientist working for NASA. A new moon, S/2001U1, has been discovered orbiting the planet Uranus in our outer solar system. You have been asked to check whether it is possible for this moon to have an oxygen atmosphere. You can do so by comparing the average speed of an oxygen molecule (480 meters per second) to the moon's **escape velocity**, the speed an object must travel to permanently leave the moon's gravitational pull. If the moon's escape velocity is greater than the average speed of oxygen molecules, then it is possible for the moon to retain oxygen in its atmosphere—that is, if oxygen exists on the moon at all.

Data about the new moon are listed in the table. Use that along with the escape velocity formula given below to decide whether it is possible for S/2001U1 to have an oxygen atmosphere.

$$v = \sqrt{\frac{2GM}{r}}, \text{ where}$$

$v$ is the escape velocity (in meters per second, m/s),

$M$ is the mass of the moon (in kilograms, kg),

$r$ is the radius of the moon (in meters, m), and

$G$ is the universal constant of gravitation where

$$\left(G = 6.67 \times 10^{-11} \frac{\text{m}^3}{\text{kg} \cdot \text{s}^2}\right).$$

*S/2001U1 Parameters*

| | |
|---|---|
| Mass | $9.07 \times 10^{20}$ kg |
| Radius | 620,000 m |
| Visual geometric albedo | 0.07 |
| Orbital period | 7.2 days |

# Exercise Set 7.1

*Simplify. Assume that variables represent positive real numbers. See Example 1.*

1. $\sqrt{100}$
2. $\sqrt{400}$
3. $\sqrt{\frac{1}{4}}$
4. $\sqrt{\frac{9}{25}}$
5. $\sqrt{0.0001}$
6. $\sqrt{0.04}$
7. $-\sqrt{36}$
8. $-\sqrt{9}$
9. $\sqrt{x^{10}}$
10. $\sqrt{x^{16}}$
11. $\sqrt{16y^6}$
12. $\sqrt{64y^{20}}$

*Use a calculator to approximate each square root to 3 decimal places. Check to see that each approximation is reasonable. See Example 2.*

13. $\sqrt{7}$
14. $\sqrt{11}$
15. $\sqrt{38}$
16. $\sqrt{56}$
17. $\sqrt{200}$
18. $\sqrt{300}$

*Find each cube root. See Example 3.*

19. $\sqrt[3]{64}$
20. $\sqrt[3]{27}$
21. $\sqrt[3]{\frac{1}{8}}$
22. $\sqrt[3]{\frac{27}{64}}$
23. $\sqrt[3]{-1}$
24. $\sqrt[3]{-125}$
25. $\sqrt[3]{x^{12}}$
26. $\sqrt[3]{x^{15}}$
27. $\sqrt[3]{-27x^9}$
28. $\sqrt[3]{-64x^6}$

*Find each root. Assume that all variables represent nonnegative real numbers. See Example 4.*

29. $-\sqrt[4]{16}$
30. $\sqrt[5]{-243}$
31. $\sqrt[4]{-16}$
32. $\sqrt{-16}$
33. $\sqrt[5]{-32}$
34. $\sqrt[5]{-1}$
35. $\sqrt[5]{x^{20}}$
36. $\sqrt[4]{x^{20}}$
37. $\sqrt[6]{64x^{12}}$
38. $\sqrt[5]{-32x^{15}}$
39. $\sqrt{81x^4}$
40. $\sqrt[4]{81x^4}$
41. $\sqrt[4]{256x^8}$
42. $\sqrt{256x^8}$

*Simplify. Assume that the variables represent any real number. See Example 5.*

43. $\sqrt{(-8)^2}$
44. $\sqrt{(-7)^2}$
45. $\sqrt[3]{(-8)^3}$
46. $\sqrt[5]{(-7)^5}$
47. $\sqrt{4x^2}$
48. $\sqrt[4]{16x^4}$
49. $\sqrt[3]{x^3}$
50. $\sqrt[5]{x^5}$
51. $\sqrt{(x-5)^2}$
52. $\sqrt{(y-6)^2}$
53. $\sqrt{x^2+4x+4}$
 (*Hint:* Factor the polynomial first.)
54. $\sqrt{x^2-8x+16}$
 (*Hint:* Factor the polynomial first.)

*Simplify each radical. Assume that all variables represent positive real numbers.*

55. $-\sqrt{121}$
56. $-\sqrt[3]{125}$
57. $\sqrt[3]{8x^3}$
58. $\sqrt{16x^8}$
59. $\sqrt{y^{12}}$
60. $\sqrt[3]{y^{12}}$
61. $\sqrt{25a^2b^{20}}$
62. $\sqrt{9x^4y^6}$
63. $\sqrt[3]{-27x^{12}y^9}$
64. $\sqrt[3]{-8a^{21}b^6}$
65. $\sqrt[4]{a^{16}b^4}$
66. $\sqrt[4]{x^8y^{12}}$
67. $\sqrt[5]{-32x^{10}y^5}$
68. $\sqrt[5]{-243z^{15}}$
69. $\sqrt{\frac{25}{49}}$
70. $\sqrt{\frac{4}{81}}$
71. $\sqrt{\frac{x^2}{4y^2}}$
72. $\sqrt{\frac{y^{10}}{9x^6}}$
73. $-\sqrt[3]{\frac{z^{21}}{27x^3}}$
74. $-\sqrt[3]{\frac{64a^3}{b^9}}$
75. $\sqrt[4]{\frac{x^4}{16}}$
76. $\sqrt[4]{\frac{y^4}{81x^4}}$

*If $f(x) = \sqrt{2x+3}$ and $g(x) = \sqrt[3]{x-8}$, find the following function values. See Examples 6 and 7.*

77. $f(0)$
78. $g(0)$
79. $g(7)$
80. $f(-1)$
81. $g(-19)$
82. $f(3)$
83. $f(2)$
84. $g(1)$

*Identify the domain and then graph each function. See Example 6.*

85. $f(x) = \sqrt{x} + 2$
86. $f(x) = \sqrt{x} - 2$
87. $f(x) = \sqrt{x-3}$; use the following table.

| $x$ | $f(x)$ |
|---|---|
| 3 | |
| 4 | |
| 7 | |
| 12 | |

88. $f(x) = \sqrt{x+1}$; use the following table.

| $x$ | $f(x)$ |
|---|---|
| −1 | |
| 0 | |
| 3 | |
| 8 | |

*Identify the domain and then graph each function. See Example 7.*

89. $f(x) = \sqrt[3]{x} + 1$
90. $f(x) = \sqrt[3]{x} - 2$

**91.** $g(x) = \sqrt[3]{x - 1}$; use the following table.

| $x$ | $g(x)$ |
|---|---|
| 1 | 0 |
| 2 | 1 |
| 0 | −1 |
| 9 | 2 |
| −7 | −2 |

**92.** $g(x) = \sqrt[3]{x + 1}$; use the following table.

| $x$ | $g(x)$ |
|---|---|
| −1 | 0 |
| 0 | 1 |
| −2 | −1 |
| 7 | 2 |
| −9 | −2 |

**93.** Suppose that a friend tells you that $\sqrt{13} \approx 5.7$. Without a calculator, how can you convince your friend that he must have made an error?

**94.** Escape velocity is the minimum speed that an object must reach to escape a planet's pull of gravity. Escape velocity $v$ is given by the equation $v = \sqrt{\frac{2GM}{r}}$, where $M$ is the mass of the planet, $r$ is its radius, and $G$ is the universal gravitational constant, which has a value of $G = 6.67 \times 10^{-11}$ $\text{m}^3/\text{kg} \cdot \text{s}^2$. The mass of Earth is $5.97 \times 10^{24}$ kg and its radius is $6.37 \times 10^6$ m. Use this information to find the escape velocity for Earth. Round to the nearest whole number. (*Source:* National Space Science Data Center)

*Use a graphing calculator to verify the domain of each function and its graph.*

**95.** Exercise 85 **96.** Exercise 86

**97.** Exercise 89 **98.** Exercise 90

## REVIEW EXERCISES

*Simplify each exponential expression. See Sections 5.1 and 5.2.*

**99.** $(-2x^3y^2)^5$ **100.** $(4y^6z^7)^3$

**101.** $(-3x^2y^3z^5)(20x^5y^7)$ **102.** $(-14a^5bc^2)(2abc^4)$

**103.** $\dfrac{7x^{-1}y}{14(x^5y^2)^{-2}}$ **104.** $\dfrac{(2a^{-1}b^2)^3}{(8a^2b)^{-2}}$

# 7.2 RATIONAL EXPONENTS

CD-ROM SSM SSG Video

**▶ OBJECTIVES**

1. Understand the meaning of $a^{1/n}$.
2. Understand the meaning of $a^{m/n}$.
3. Understand the meaning of $a^{-m/n}$.
4. Use rules for exponents to simplify expressions that contain rational exponents.
5. Use rational exponents to simplify radical expressions.

**1** So far in this text, we have not defined expressions with rational exponents such as $3^{1/2}$, $x^{2/3}$, and $-9^{-1/4}$. We will define these expressions so that the rules for exponents will apply to these rational exponents as well.

Suppose that $x = 5^{1/3}$. Then

$$x^3 = (5^{1/3})^3 = 5^{1/3 \cdot 3} = 5^1 \text{ or } 5$$

↳ using rules for exponents ↑

Since $x^3 = 5$, then $x$ is the number whose cube is 5, or $x = \sqrt[3]{5}$. Notice that we also know that $x = 5^{1/3}$. This means

$$5^{1/3} = \sqrt[3]{5}$$

**DEFINITION OF $a^{1/n}$**

If $n$ is a positive integer greater than 1 and $\sqrt[n]{a}$ is a real number, then

$$a^{1/n} = \sqrt[n]{a}$$

Notice that the denominator of the rational exponent corresponds to the index of the radical.

**Example 1** Use radical notation to write the following. Simplify if possible.

**a.** $4^{1/2}$ **b.** $64^{1/3}$ **c.** $x^{1/4}$ **d.** $0^{1/6}$ **e.** $-9^{1/2}$ **f.** $(81x^8)^{1/4}$ **g.** $(5y)^{1/3}$

*Solution*

**a.** $4^{1/2} = \sqrt{4} = 2$

**b.** $64^{1/3} = \sqrt[3]{64} = 4$

**c.** $x^{1/4} = \sqrt[4]{x}$

**d.** $0^{1/6} = \sqrt[6]{0} = 0$

**e.** $-9^{1/2} = -\sqrt{9} = -3$

**f.** $(81x^8)^{1/4} = \sqrt[4]{81x^8} = 3x^2$

**g.** $(5y)^{1/3} = \sqrt[3]{5y}$

2 As we expand our use of exponents to include $\frac{m}{n}$, we define their meaning so that rules for exponents still hold true. For example, by properties of exponents,

$$8^{2/3} = (8^{1/3})^2 = (\sqrt[3]{8})^2 \qquad \text{or}$$
$$8^{2/3} = (8^2)^{1/3} = \sqrt[3]{8^2}$$

**DEFINITION OF $a^{m/n}$**

If $m$ and $n$ are positive integers greater than 1 with $\frac{m}{n}$ in lowest terms, then

$$a^{m/n} = \sqrt[n]{a^m} = (\sqrt[n]{a})^m$$

as long as $\sqrt[n]{a}$ is a real number.

Notice that the denominator $n$ of the rational exponent corresponds to the index of the radical. The numerator $m$ of the rational exponent indicates that the base is to be raised to the $m$th power. This means

$$8^{2/3} = \sqrt[3]{8^2} = \sqrt[3]{64} = 4 \qquad \text{or}$$
$$8^{2/3} = (\sqrt[3]{8})^2 = 2^2 = 4$$

**HELPFUL HINT**
Most of the time, $(\sqrt[n]{a})^m$ will be easier to calculate than $\sqrt[n]{a^m}$.

**Example 2** Use radical notation to write the following. Then simplify if possible.

**a.** $4^{3/2}$ **b.** $-16^{3/4}$ **c.** $(-27)^{2/3}$ **d.** $\left(\frac{1}{9}\right)^{3/2}$ **e.** $(4x - 1)^{3/5}$

*Solution* **a.** $4^{3/2} = (\sqrt{4})^3 = 2^3 = 8$

**b.** $-16^{3/4} = -(\sqrt[4]{16})^3 = -(2)^3 = -8$

**c.** $(-27)^{2/3} = (\sqrt[3]{-27})^2 = (-3)^2 = 9$

**d.** $\left(\frac{1}{9}\right)^{3/2} = \left(\sqrt{\frac{1}{9}}\right)^3 = \left(\frac{1}{3}\right)^3 = \frac{1}{27}$

**e.** $(4x - 1)^{3/5} = \sqrt[5]{(4x - 1)^3}$

**HELPFUL HINT**
The *denominator* of a rational exponent is the index of the corresponding radical. For example, $x^{1/5} = \sqrt[5]{x}$ and $z^{2/3} = \sqrt[3]{z^2}$, or $z^{2/3} = (\sqrt[3]{z})^2$.

3 The rational exponents we have given meaning to exclude negative rational numbers. To complete the set of definitions, we define $a^{-m/n}$.

**DEFINITION OF $a^{-m/n}$**

$$a^{-m/n} = \frac{1}{a^{m/n}}$$

as long as $a^{m/n}$ is a nonzero real number.

**Example 3** Write each expression with a positive exponent, and then simplify.

**a.** $16^{-3/4}$ **b.** $(-27)^{-2/3}$

*Solution* **a.** $16^{-3/4} = \frac{1}{16^{3/4}} = \frac{1}{(\sqrt[4]{16})^3} = \frac{1}{2^3} = \frac{1}{8}$

**b.** $(-27)^{-2/3} = \frac{1}{(-27)^{2/3}} = \frac{1}{(\sqrt[3]{-27})^2} = \frac{1}{(-3)^2} = \frac{1}{9}$

**HELPFUL HINT**

If an expression contains a negative rational exponent, such as $9^{-3/2}$, you may want to first write the expression with a positive exponent and then interpret the rational exponent. Notice that the sign of the base is not affected by the sign of its exponent. For example,

$$9^{-3/2} = \frac{1}{9^{3/2}} = \frac{1}{(\sqrt{9})^3} = \frac{1}{27}$$

Also,

$$(-27)^{-1/3} = \frac{1}{(-27)^{1/3}} = -\frac{1}{3}$$

4 It can be shown that the properties of integer exponents hold for rational exponents. By using these properties and definitions, we can now simplify expressions that contain rational exponents.

These rules are repeated here for review.

### SUMMARY OF EXPONENT RULES

If $m$ and $n$ are rational numbers, and $a$, $b$, and $c$ are numbers for which the expressions below exist, then

Product rule for exponents: $a^m \cdot a^n = a^{m+n}$

Power rule for exponents: $(a^m)^n = a^{m \cdot n}$

Power rules for products and quotients: $(ab)^n = a^n b^n$ and $\left(\frac{a}{c}\right)^n = \frac{a^n}{c^n}, c \neq 0$

Quotient rule for exponents: $\frac{a^m}{a^n} = a^{m-n}, a \neq 0$

Zero exponent: $a^0 = 1, a \neq 0$

Negative exponent: $a^{-n} = \frac{1}{a^n}, a \neq 0$

**Example 4** Use properties of exponents to simplify. Write results with only positive exponents.

**a.** $x^{1/2}x^{1/3}$ **b.** $\frac{7^{1/3}}{7^{4/3}}$ **c.** $\frac{(2x^{2/5}y^{-1/3})^5}{x^2y}$

*Solution* **a.** $x^{1/2}x^{1/3} = x^{(1/2+1/3)} = x^{3/6+2/6} = x^{5/6}$

**b.** $\frac{7^{1/3}}{7^{4/3}} = 7^{1/3-4/3} = 7^{-3/3} = 7^{-1} = \frac{1}{7}$

**c.** We begin by using the power rule $(ab)^m = a^m b^m$ to simplify the numerator.

$$\frac{(2x^{2/5}y^{-1/3})^5}{x^2y} = \frac{2^5(x^{2/5})^5(y^{-1/3})^5}{x^2y} = \frac{32x^2y^{-5/3}}{x^2y}$$

$$= 32x^{2-2}y^{-5/3-3/3} \quad \text{Apply the quotient rule.}$$

$$= 32x^0y^{-8/3}$$

$$= \frac{32}{y^{8/3}}$$

## Example 5

Multiply.

**a.** $z^{2/3}(z^{1/3} - z^5)$ **b.** $(x^{1/3} - 5)(x^{1/3} + 2)$

*Solution*

**a.** $z^{2/3}(z^{1/3} - z^5) = z^{2/3}z^{1/3} - z^{2/3}z^5$ Apply the distributive property.

$= z^{(2/3+1/3)} - z^{(2/3+5)}$ Use the product rule.

$= z^{3/3} - z^{(2/3+15/3)}$

$= z - z^{17/3}$

**b.** $(x^{1/3} - 5)(x^{1/3} + 2) = x^{2/3} + 2x^{1/3} - 5x^{1/3} - 10$ Think of $(x^{1/3} - 5)$ and $(x^{1/3} + 2)$ as 2 binomials, and FOIL.

$= x^{2/3} - 3x^{1/3} - 10$

## Example 6

Factor $x^{-1/2}$ from the expression $3x^{-1/2} - 7x^{5/2}$. Assume that all variables represent positive numbers.

*Solution*

$$3x^{-1/2} - 7x^{5/2} = (x^{-1/2})(3) - (x^{-1/2})(7x^{6/2})$$
$$= x^{-1/2}(3 - 7x^3)$$

To check, multiply $x^{-1/2}(3 - 7x^3)$ to see that the product is $3x^{-1/2} - 7x^{5/2}$.

5 Some radical expressions are easier to simplify when we first write them with rational exponents. We can simplify some radical expressions by first writing the expression with rational exponents. Use properties of exponents to simplify, and then convert back to radical notation.

## Example 7

Use rational exponents to simplify. Assume that variables represent positive numbers.

**a.** $\sqrt[6]{25}$ **b.** $\sqrt[8]{x^4}$ **c.** $\sqrt[4]{r^2s^6}$

*Solution*

**a.** $\sqrt[6]{25} = 25^{1/6} = (5^2)^{1/6} = 5^{2/6} = 5^{1/3} = \sqrt[3]{5}$

**b.** $\sqrt[8]{x^4} = x^{4/8} = x^{1/2} = \sqrt{x}$

**c.** $\sqrt[4]{r^2s^6} = (r^2s^6)^{1/4} = r^{2/4}s^{6/4} = r^{1/2}s^{3/2} = (rs^3)^{1/2} = \sqrt{rs^3}$

**Example 8** Use rational exponents to write as a single radical.

**a.** $\sqrt{x} \cdot \sqrt[4]{x}$ **b.** $\dfrac{\sqrt{x}}{\sqrt[3]{x}}$ **c.** $\sqrt[3]{3} \cdot \sqrt{2}$

*Solution* **a.** $\sqrt{x} \cdot \sqrt[4]{x} = x^{1/2} \cdot x^{1/4} = x^{1/2+1/4}$

$= x^{3/4} = \sqrt[4]{x^3}$

**b.** $\dfrac{\sqrt{x}}{\sqrt[3]{x}} = \dfrac{x^{1/2}}{x^{1/3}} = x^{1/2-1/3} = x^{3/6-2/6}$

$= x^{1/6} = \sqrt[6]{x}$

**c.** $\sqrt[3]{3} \cdot \sqrt{2} = 3^{1/3} \cdot 2^{1/2}$ Write with rational exponents.

$= 3^{2/6} \cdot 2^{3/6}$ Write the exponents so that they have the same denominator.

$= (3^2 \cdot 2^3)^{1/6}$ Use $a^nb^n = (ab)^n$.

$= \sqrt[6]{3^2 \cdot 2^3}$ Write with radical notation.

$= \sqrt[6]{72}$ Multiply $3^2 \cdot 2^3$.

## SPOTLIGHT ON DECISION MAKING

Suppose you are a telecommunications industry analyst. A colleague has just formulated a mathematical model for the number of cellular telephone subscriptions in the United States from 1985 to 1998. The model is $y = 341.8x^{21/5}$, where $y$ is the number of cellular telephone subscriptions $x$ years after 1980. The actual data from 1985 to 1998 are listed in the table.

Your colleague has asked your help in evaluating whether this model represents the actual data well. By comparing the numbers of subscriptions given by the model to the actual data given in the table, decide whether this mathematical model is acceptable. Explain your reasoning.

**U.S. CELLULAR TELEPHONE SUBSCRIPTIONS, 1985–1998**

| *Year* | *Subscriptions* |
|---|---|
| 1985 | 340,213 |
| 1986 | 681,825 |
| 1987 | 1,230,855 |
| 1988 | 2,069,441 |
| 1989 | 3,508,944 |
| 1990 | 5,283,055 |
| 1991 | 7,557,148 |
| 1992 | 11,032,753 |
| 1993 | 16,009,461 |
| 1994 | 24,134,421 |
| 1995 | 33,785,661 |
| 1996 | 44,042,992 |
| 1997 | 55,312,293 |
| 1998 | 69,209,321 |

(*Source:* The CTIA Semi-Annual Wireless Survey)

## Exercise Set 7.2

*Use radical notation to write each expression. Simplify if possible. See Example 1.*

1. $49^{1/2}$
2. $64^{1/3}$
3. $27^{1/3}$
4. $8^{1/3}$
5. $\left(\frac{1}{16}\right)^{1/4}$
6. $\left(\frac{1}{64}\right)^{1/2}$
7. $169^{1/2}$
8. $81^{1/4}$
9. $2m^{1/3}$
10. $(2m)^{1/3}$
11. $(9x^4)^{1/2}$
12. $(16x^8)^{1/2}$
13. $(-27)^{1/3}$
14. $-64^{1/2}$
15. $-16^{1/4}$
16. $(-32)^{1/5}$

*Use radical notation to write each expression. Simplify if possible. See Example 2.*

17. $16^{3/4}$
18. $4^{5/2}$
19. $(-64)^{2/3}$
20. $(-8)^{4/3}$
21. $(-16)^{3/4}$
22. $(-9)^{3/2}$
23. $(2x)^{3/5}$
24. $2x^{3/5}$
25. $(7x + 2)^{2/3}$
26. $(x - 4)^{3/4}$
27. $\left(\frac{16}{9}\right)^{3/2}$
28. $\left(\frac{49}{25}\right)^{3/2}$

*Write with positive exponents. Simplify if possible. See Example 3.*

29. $8^{-4/3}$
30. $64^{-2/3}$
31. $(-64)^{-2/3}$
32. $(-8)^{-4/3}$
33. $(-4)^{-3/2}$
34. $(-16)^{-5/4}$
35. $x^{-1/4}$
36. $y^{-1/6}$
37. $\frac{1}{a^{-2/3}}$
38. $\frac{1}{n^{-8/9}}$
39. $\frac{5}{7x^{-3/4}}$
40. $\frac{2}{3y^{-5/7}}$
41. Explain how writing $x^{-7}$ with positive exponents is similar to writing $x^{-1/4}$ with positive exponents.
42. Explain how writing $2x^{-5}$ with positive exponents is similar to writing $2x^{-3/4}$ with positive exponents.

*Use the properties of exponents to simplify each expression. Write with positive exponents. See Example 4.*

43. $a^{2/3}a^{5/3}$
44. $b^{9/5}b^{8/5}$
45. $x^{-2/5} \cdot x^{7/5}$
46. $y^{4/3} \cdot y^{-1/3}$
47. $3^{1/4} \cdot 3^{3/8}$
48. $5^{1/2} \cdot 5^{1/6}$
49. $\frac{y^{1/3}}{y^{1/6}}$
50. $\frac{x^{3/4}}{x^{1/8}}$
51. $(4u^2)^{3/2}$
52. $(32^{1/5}x^{2/3})^3$
53. $\frac{b^{1/2}b^{3/4}}{-b^{1/4}}$
54. $\frac{a^{1/4}a^{-1/2}}{a^{2/3}}$
55. $\frac{(3x^{1/4})^3}{x^{1/12}}$
56. $\frac{(2x^{1/5})^4}{x^{3/10}}$

*Multiply. See Example 5.*

57. $y^{1/2}(y^{1/2} - y^{2/3})$
58. $x^{1/2}(x^{1/2} + x^{3/2})$
59. $x^{2/3}(2x - 2)$
60. $3x^{1/2}(x + y)$
61. $(2x^{1/3} + 3)(2x^{1/3} - 3)$
62. $(y^{1/2} + 5)(y^{1/2} + 5)$

*Factor the common factor from the given expression. See Example 6.*

63. $x^{8/3};\ x^{8/3} + x^{10/3}$
64. $x^{3/2};\ x^{5/2} - x^{3/2}$
65. $x^{1/5};\ x^{2/5} - 3x^{1/5}$
66. $x^{2/7};\ x^{3/7} - 2x^{2/7}$
67. $x^{-1/3};\ 5x^{-1/3} + x^{2/3}$
68. $x^{-3/4};\ x^{-3/4} + 3x^{1/4}$

*Use rational exponents to simplify each radical. Assume that all variables represent positive numbers. See Example 7.*

69. $\sqrt[6]{x^3}$
70. $\sqrt[9]{a^3}$
71. $\sqrt[6]{4}$
72. $\sqrt[4]{36}$
73. $\sqrt[4]{16x^2}$
74. $\sqrt[8]{4y^2}$
75. $\sqrt[8]{x^4y^4}$
76. $\sqrt[9]{y^6z^3}$

*Use rational expressions to write as a single radical expression. See Example 8.*

77. $\sqrt[3]{y} \cdot \sqrt[5]{y^2}$
78. $\sqrt[3]{y^2} \cdot \sqrt[6]{y}$
79. $\frac{\sqrt[3]{b^2}}{\sqrt[4]{b}}$
80. $\frac{\sqrt[4]{a}}{\sqrt[5]{a}}$
81. $\frac{\sqrt[3]{a^2}}{\sqrt[6]{a}}$
82. $\frac{\sqrt[5]{b^2}}{\sqrt[10]{b^3}}$
83. $\sqrt{3} \cdot \sqrt[3]{4}$
84. $\sqrt[3]{5} \cdot \sqrt{2}$
85. $\sqrt[5]{7} \cdot \sqrt[3]{y}$
86. $\sqrt[4]{5} \cdot \sqrt[3]{x}$
87. In physics, the speed of a wave traveling over a stretched string with tension $t$ and density $u$ is given by the expression $\frac{\sqrt{t}}{\sqrt{u}}$. Write this expression with rational exponents.
88. In electronics, the angular frequency of oscillations in a certain type of circuit is given by the expression $(LC)^{-1/2}$. Use radical notation to write this expression.

*Basal metabolic rate (BMR) is the number of calories per day a person needs to maintain life. A person's basal metabolic rate $B(w)$ in calories per day can be estimated with the function $B(w) = 70w^{3/4}$, where $w$ is the person's weight in kilograms.*

89. Estimate the BMR for a person who weighs 50 kilograms. Round to the nearest calorie. (Note: 50 kilograms is approximately 110 pounds.)

**90.** Estimate the BMR for a person who weighs 85 kilograms. Round to the nearest calorie. (Note: 85 kilograms is approximately 187 pounds.)

**91.** Hewlett-Packard (HP) is a global leader in computing and imaging products. HP's annual net revenue can be modeled by the function $f(x) = 6550x^{43/50}$, where $f(x)$ is net revenue in millions of dollars in the year $x$, and $x = 0$ represents the year 1990. (*Source:* Hewlett-Packard Company, 1995–1999)

**a.** Use this model to find HP's net revenue in 1999.

**b.** Predict HP's net revenue in 2004.

*Fill in the box with the correct expression.*

**92.** $\square \cdot a^{2/3} = a^{3/3}$, or $a$

**93.** $\square \cdot x^{1/8} = x^{4/8}$, or $x^{1/2}$

**94.** $\dfrac{\square}{x^{-2/5}} = x^{3/5}$

**95.** $\dfrac{\square}{y^{-3/4}} = y^{4/4}$, or $y$

*Use a calculator to write a four-decimal-place approximation of each.*

**96.** $8^{1/4}$

**97.** $20^{1/5}$

**98.** $18^{3/5}$

**99.** $76^{5/7}$

## REVIEW EXERCISES

*Write each integer as a product of two integers such that one of the factors is a perfect square. For example, write 18 as $9 \cdot 2$, because 9 is a perfect square.*

**100.** 75

**101.** 20

**102.** 48

**103.** 45

*Write each integer as a product of two integers such that one of the factors is a perfect cube. For example, write 24 as $8 \cdot 3$, because 8 is a perfect cube.*

**104.** 16

**105.** 56

**106.** 54

**107.** 80

## 7.3 SIMPLIFYING RADICAL EXPRESSIONS

CD-ROM SSM SSG Video

▶ **OBJECTIVES**

1. Use the product rule for radicals.
2. Use the quotient rule for radicals.
3. Simplify radicals.

**1** It is possible to simplify some radicals that do not evaluate to rational numbers. To do so, we use a product rule and a quotient rule for radicals. To discover the product rule, notice the following pattern.

$$\sqrt{9} \cdot \sqrt{4} = 3 \cdot 2 = 6$$
$$\sqrt{9 \cdot 4} = \sqrt{36} = 6$$

Since both expressions simplify to 6, it is true that

$$\sqrt{9} \cdot \sqrt{4} = \sqrt{9 \cdot 4}$$

This pattern suggests the following product rule for radicals.

### PRODUCT RULE FOR RADICALS

If $\sqrt[n]{a}$ and $\sqrt[n]{b}$ are real numbers, then

$$\sqrt[n]{a} \cdot \sqrt[n]{b} = \sqrt[n]{ab}$$

Notice that the product rule is the relationship $a^{1/n} \cdot b^{1/n} = (ab)^{1/n}$ stated in radical notation.

**Example 1** Multiply.

**a.** $\sqrt{3} \cdot \sqrt{5}$ **b.** $\sqrt{21} \cdot \sqrt{x}$ **c.** $\sqrt[3]{4} \cdot \sqrt[3]{2}$ **d.** $\sqrt[4]{5y^2} \cdot \sqrt[4]{2x^3}$ **e.** $\sqrt{\frac{2}{a}} \cdot \sqrt{\frac{b}{3}}$

*Solution*

**a.** $\sqrt{3} \cdot \sqrt{5} = \sqrt{3 \cdot 5} = \sqrt{15}$

**b.** $\sqrt{21} \cdot \sqrt{x} = \sqrt{21x}$

**c.** $\sqrt[3]{4} \cdot \sqrt[3]{2} = \sqrt[3]{4 \cdot 2} = \sqrt[3]{8} = 2$

**d.** $\sqrt[4]{5y^2} \cdot \sqrt[4]{2x^3} = \sqrt[4]{5y^2 \cdot 2x^3} = \sqrt[4]{10y^2x^3}$

**e.** $\sqrt{\frac{2}{a}} \cdot \sqrt{\frac{b}{3}} = \sqrt{\frac{2}{a} \cdot \frac{b}{3}} = \sqrt{\frac{2b}{3a}}$

2 To discover a quotient rule for radicals, notice the following pattern.

$$\sqrt{\frac{4}{9}} = \frac{2}{3}$$

$$\frac{\sqrt{4}}{\sqrt{9}} = \frac{2}{3}$$

Since both expressions simplify to $\frac{2}{3}$, it is true that

$$\sqrt{\frac{4}{9}} = \frac{\sqrt{4}}{\sqrt{9}}$$

This pattern suggests the following quotient rule for radicals.

**QUOTIENT RULE FOR RADICALS**

If $\sqrt[n]{a}$ and $\sqrt[n]{b}$ are real numbers and $\sqrt[n]{b}$ is not zero, then

$$\sqrt[n]{\frac{a}{b}} = \frac{\sqrt[n]{a}}{\sqrt[n]{b}}$$

Notice that the quotient rule is the relationship $\left(\frac{a}{b}\right)^{1/n} = \frac{a^{1/n}}{b^{1/n}}$ stated in radical notation. We can use the quotient rule to simplify radical expressions by reading the rule from left to right, or to divide radicals by reading the rule from right to left.

For example,

$$\sqrt{\frac{x}{16}} = \frac{\sqrt{x}}{\sqrt{16}} = \frac{\sqrt{x}}{4} \qquad \text{Using } \sqrt[n]{\frac{a}{b}} = \frac{\sqrt[n]{a}}{\sqrt[n]{b}}$$

$$\frac{\sqrt{75}}{\sqrt{3}} = \sqrt{\frac{75}{3}} = \sqrt{25} = 5 \qquad \text{Using } \frac{\sqrt[n]{a}}{\sqrt[n]{b}} = \sqrt[n]{\frac{a}{b}}$$

Note: *For the remainder of this chapter, we will assume that variables represent positive real numbers. Since this is so, we need not insert absolute value bars when we simplify even roots.*

**Example 2** Use the quotient rule to simplify.

**a.** $\sqrt{\frac{25}{49}}$ **b.** $\sqrt{\frac{x}{9}}$ **c.** $\sqrt[3]{\frac{8}{27}}$ **d.** $\sqrt[4]{\frac{3}{16y^4}}$

*Solution*

**a.** $\sqrt{\frac{25}{49}} = \frac{\sqrt{25}}{\sqrt{49}} = \frac{5}{7}$

**b.** $\sqrt{\frac{x}{9}} = \frac{\sqrt{x}}{\sqrt{9}} = \frac{\sqrt{x}}{3}$

**c.** $\sqrt[3]{\frac{8}{27}} = \frac{\sqrt[3]{8}}{\sqrt[3]{27}} = \frac{2}{3}$

**d.** $\sqrt[4]{\frac{3}{16y^4}} = \frac{\sqrt[4]{3}}{\sqrt[4]{16y^4}} = \frac{\sqrt[4]{3}}{2y}$

3 Both the product and quotient rules can be used to simplify a radical. If the product rule is read from right to left, we have that $\sqrt[n]{ab} = \sqrt[n]{a} \cdot \sqrt[n]{b}$. This is used to simplify the following radicals.

**Example 3** Simplify the following.

**a.** $\sqrt{50}$ **b.** $\sqrt[3]{24}$ **c.** $\sqrt{26}$ **d.** $\sqrt[4]{32}$

*Solution*

**a.** Factor 50 such that one factor is the largest perfect square that divides 50. The largest perfect square factor of 50 is 25, so we write 50 as $25 \cdot 2$ and use the product rule for radicals to simplify.

$$\sqrt{50} = \sqrt{25 \cdot 2} = \sqrt{25} \cdot \sqrt{2} = 5\sqrt{2}$$

The largest perfect square factor of 50.

**HELPFUL HINT**
Don't forget that, for example, $5\sqrt{2}$ means $5 \cdot \sqrt{2}$.

**b.** $\sqrt[3]{24} = \sqrt[3]{8 \cdot 3} = \sqrt[3]{8} \cdot \sqrt[3]{3} = 2\sqrt[3]{3}$

The largest perfect cube factor of 24.

**c.** $\sqrt{26}$ The largest perfect square factor of 26 is 1, so $\sqrt{26}$ cannot be simplified further.

**d.** $\sqrt[4]{32} = \sqrt[4]{16 \cdot 2} = \sqrt[4]{16} \cdot \sqrt[4]{2} = 2\sqrt[4]{2}$

The largest fourth power factor of 32.

After simplifying a radical such as a square root, always check the radicand to see that it contains no other perfect square factors. It may, if the largest perfect square factor of the radicand was not originally recognized. For example,

$$\sqrt{200} = \sqrt{4 \cdot 50} = \sqrt{4} \cdot \sqrt{50} = 2\sqrt{50}$$

Notice that the radicand 50 still contains the perfect square factor 25. This is because 4 is not the largest perfect square factor of 200. We continue as follows.

$$2\sqrt{50} = 2\sqrt{25 \cdot 2} = 2 \cdot \sqrt{25} \cdot \sqrt{2} = 2 \cdot 5 \cdot \sqrt{2} = 10\sqrt{2}$$

The radical is now simplified since 2 contains no perfect square factors (other than 1).

**HELPFUL HINT**

To help you recognize largest perfect power factors of a radicand, it will help if you are familiar with some perfect powers. A few are listed below.

| | | | | | | | | | | | | |
|---|---|---|---|---|---|---|---|---|---|---|---|---|
| Perfect Squares | 1, | 4, | 9, | 16, | 25, | 36, | 49, | 64, | 81, | 100, | 121, | 144 |
| | $1^2$ | $2^2$ | $3^2$ | $4^2$ | $5^2$ | $6^2$ | $7^2$ | $8^2$ | $9^2$ | $10^2$ | $11^2$ | $12^2$ |
| Perfect Cubes | 1, | 8, | 27, | 64, | 125 | | | | | | | |
| | $1^3$ | $2^3$ | $3^3$ | $4^3$ | $5^3$ | | | | | | | |
| Perfect Fourth Powers | 1, | 16, | 81, | 256 | | | | | | | | |
| | $1^4$ | $2^4$ | $3^4$ | $4^4$ | | | | | | | | |

In general, we say that a radicand of the form $\sqrt[n]{a}$ is simplified when $a$ contains no factors that are perfect $n$th powers (other than 1 or $-1$).

**Example 4** Use the product rule to simplify.

**a.** $\sqrt{25x^3}$ **b.** $\sqrt[3]{54x^6y^8}$ **c.** $\sqrt[4]{81z^{11}}$

*Solution*

**a.** $\sqrt{25x^3} = \sqrt{25x^2 \cdot x}$ Find the largest perfect square factor.

$= \sqrt{25x^2} \cdot \sqrt{x}$ Apply the product rule.

$= 5x\sqrt{x}$ Simplify.

**b.** $\sqrt[3]{54x^6y^8} = \sqrt[3]{27 \cdot 2 \cdot x^6 \cdot y^6 \cdot y^2}$ Factor the radicand and identify perfect cube factors.

$= \sqrt[3]{27x^6y^6 \cdot 2y^2}$

$= \sqrt[3]{27x^6y^6} \cdot \sqrt[3]{2y^2}$ Apply the product rule.

$= 3x^2y^2\sqrt[3]{2y^2}$ Simplify.

**c.** $\sqrt[4]{81z^{11}} = \sqrt[4]{81 \cdot z^8 \cdot z^3}$ Factor the radicand and identify perfect fourth power factors.

$= \sqrt[4]{81z^8} \cdot \sqrt[4]{z^3}$ Apply the product rule.

$= 3z^2\sqrt[4]{z^3}$ Simplify.

**Example 5** Use the quotient rule to divide, and simplify if possible.

**a.** $\dfrac{\sqrt{20}}{\sqrt{5}}$ **b.** $\dfrac{\sqrt{50x}}{2\sqrt{2}}$ **c.** $\dfrac{7\sqrt[3]{48x^4y^8}}{\sqrt[3]{6y^2}}$

*Solution* **a.** $\dfrac{\sqrt{20}}{\sqrt{5}} = \sqrt{\dfrac{20}{5}}$ Apply the quotient rule.

$= \sqrt{4}$ Simplify.

$= 2$ Simplify.

**b.** $\dfrac{\sqrt{50x}}{2\sqrt{2}} = \dfrac{1}{2} \cdot \sqrt{\dfrac{50x}{2}}$ Apply the quotient rule.

$= \dfrac{1}{2} \cdot \sqrt{25x}$ Simplify.

$= \dfrac{1}{2} \cdot \sqrt{25} \cdot \sqrt{x}$ Factor $25x$.

$= \dfrac{1}{2} \cdot 5 \cdot \sqrt{x}$ Simplify.

$= \dfrac{5}{2}\sqrt{x}$

**c.** $\dfrac{7\sqrt[3]{48x^4y^8}}{\sqrt[3]{6y^2}} = 7 \cdot \sqrt[3]{\dfrac{48x^4y^8}{6y^2}}$ Apply the quotient rule.

$= 7 \cdot \sqrt[3]{8x^4y^6}$ Simplify.

$= 7\sqrt[3]{8x^3y^6 \cdot x}$ Factor.

$= 7 \cdot \sqrt[3]{8x^3y^6} \cdot \sqrt[3]{x}$ Apply the product rule.

$= 7 \cdot 2xy^2 \cdot \sqrt[3]{x}$ Simplify.

$= 14xy^2\sqrt[3]{x}$

## Exercise Set 7.3

*Use the product rule to multiply. See Example 1.*

1. $\sqrt{7} \cdot \sqrt{2}$
2. $\sqrt{11} \cdot \sqrt{10}$
3. $\sqrt[4]{8} \cdot \sqrt[4]{2}$
4. $\sqrt[4]{27} \cdot \sqrt[4]{3}$
5. $\sqrt[3]{4} \cdot \sqrt[3]{9}$
6. $\sqrt[3]{10} \cdot \sqrt[3]{5}$
7. $\sqrt{2} \cdot \sqrt{3x}$
8. $\sqrt{3y} \cdot \sqrt{5x}$
9. $\sqrt{\dfrac{7}{x}} \cdot \sqrt{\dfrac{2}{y}}$
10. $\sqrt{\dfrac{6}{m}} \cdot \sqrt{\dfrac{n}{5}}$
11. $\sqrt[4]{4x^3} \cdot \sqrt[4]{5}$
12. $\sqrt[4]{ab^2} \cdot \sqrt[4]{27ab}$

*Use the quotient rule to simplify. See Examples 2 and 3.*

13. $\sqrt{\dfrac{6}{49}}$
14. $\sqrt{\dfrac{8}{81}}$
15. $\sqrt{\dfrac{2}{49}}$
16. $\sqrt{\dfrac{5}{121}}$
17. $\sqrt[4]{\dfrac{x^3}{16}}$
18. $\sqrt[4]{\dfrac{y}{81x^4}}$
19. $\sqrt[3]{\dfrac{4}{27}}$
20. $\sqrt[3]{\dfrac{3}{64}}$

**21.** $\sqrt[4]{\frac{8}{x^8}}$

**22.** $\sqrt[4]{\frac{a^3}{81}}$

**23.** $\sqrt[3]{\frac{2x}{81y^{12}}}$

**24.** $\sqrt[3]{\frac{3}{8x^6}}$

**25.** $\sqrt{\frac{x^2y}{100}}$

**26.** $\sqrt{\frac{y^2z}{36}}$

**27.** $\sqrt{\frac{5x^2}{4y^2}}$

**28.** $\sqrt{\frac{y^{10}}{9x^6}}$

**29.** $-\sqrt[3]{\frac{z^7}{27x^3}}$

**30.** $-\sqrt[3]{\frac{64a}{b^9}}$

*Simplify. See Examples 3 and 4.*

**31.** $\sqrt{32}$

**32.** $\sqrt{27}$

**33.** $\sqrt[3]{192}$

**34.** $\sqrt[3]{108}$

**35.** $5\sqrt{75}$

**36.** $3\sqrt{8}$

**37.** $\sqrt{24}$

**38.** $\sqrt{20}$

**39.** $\sqrt{100x^5}$

**40.** $\sqrt{64y^9}$

**41.** $\sqrt[3]{16y^7}$

**42.** $\sqrt[3]{64y^9}$

**43.** $\sqrt[4]{a^8b^7}$

**44.** $\sqrt[5]{32z^{12}}$

**45.** $\sqrt{y^5}$

**46.** $\sqrt[3]{y^5}$

**47.** $\sqrt{25a^2b^3}$

**48.** $\sqrt{9x^5y^7}$

**49.** $\sqrt[5]{-32x^{10}y}$

**50.** $\sqrt[5]{-243z^9}$

**51.** $\sqrt[3]{50x^{14}}$

**52.** $\sqrt[3]{40y^{10}}$

**53.** $-\sqrt{32a^8b^7}$

**54.** $-\sqrt{20ab^6}$

**55.** $\sqrt{9x^7y^9}$

**56.** $\sqrt{12r^9s^{12}}$

**57.** $\sqrt[3]{125r^9s^{12}}$

**58.** $\sqrt[3]{8a^6b^9}$

*Use the quotient rule to divide. Then simplify if possible. See Example 5.*

**59.** $\frac{\sqrt{14}}{\sqrt{7}}$

**60.** $\frac{\sqrt{45}}{\sqrt{9}}$

**61.** $\frac{\sqrt[3]{24}}{\sqrt[3]{3}}$

**62.** $\frac{\sqrt[3]{10}}{\sqrt[3]{2}}$

**63.** $\frac{5\sqrt[4]{48}}{\sqrt[4]{3}}$

**64.** $\frac{7\sqrt[4]{162}}{\sqrt[4]{2}}$

**65.** $\frac{\sqrt{x^5y^3}}{\sqrt{xy}}$

**66.** $\frac{\sqrt{a^7b^6}}{\sqrt{a^3b^2}}$

**67.** $\frac{8\sqrt[3]{54m^7}}{\sqrt[3]{2m}}$

**68.** $\frac{\sqrt[3]{128x^3}}{-3\sqrt[3]{2x}}$

**69.** $\frac{3\sqrt{100x^2}}{2\sqrt{2x^{-1}}}$

**70.** $\frac{\sqrt{270y^2}}{5\sqrt{3y^{-4}}}$

**71.** $\frac{\sqrt[4]{96a^{10}b^3}}{\sqrt[4]{3a^2b^3}}$

**72.** $\frac{\sqrt[5]{64x^{10}y^3}}{\sqrt[5]{2x^3y^{-7}}}$

**73.** The formula for the surface area $A$ of a cone with height $h$ and radius $r$ is given by

$$A = \pi r\sqrt{r^2 + h^2}$$

**a.** Find the surface area of a cone whose height is 3 centimeters and whose radius is 4 centimeters.

**b.** Approximate to two decimal places the surface area of a cone whose height is 7.2 feet and whose radius is 6.8 feet.

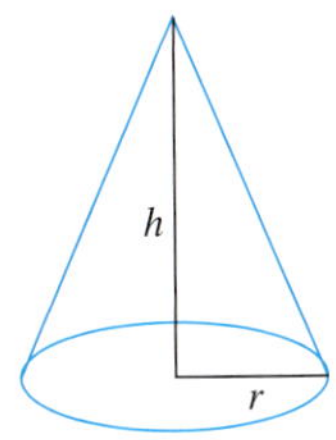


**74.** Before Mount Vesuvius, a volcano in Italy, erupted violently in 79 A.D., its height was 4190 feet. Vesuvius was roughly cone-shaped, and its base had a radius of approximately 25,200 feet. Use the formula for the surface area of a cone, given in Exercise 73, to approximate the surface area this volcano had before it erupted. (*Source:* Global Volcanism Network)

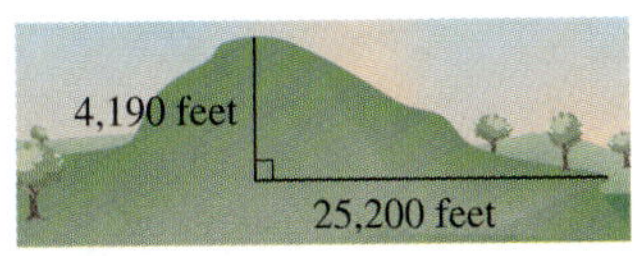


**75.** The owner of Knightime Video has determined that the demand equation for renting older releases is given by the equation $F(x) = 0.6\sqrt{49 - x^2}$, where $x$ is the price in dollars per two-day rental and $F(x)$ is the number of times the video is demanded per week.

**a.** Approximate to one decimal place the demand per week of an older release if the rental price is \$3 per two-day rental.

**b.** Approximate to one decimal place the demand per week of an older release if the rental price is \$5 per two-day rental.

**c.** Explain how the owner of the video store can use this equation to predict the number of copies of each tape that should be in stock.

## REVIEW EXERCISES

*Perform each indicated operation. See Sections 1.4 and 5.4.*

**76.** $6x + 8x$

**77.** $(6x)(8x)$

**78.** $(2x + 3)(x - 5)$

**79.** $(2x + 3) + (x - 5)$

**80.** $9y^2 - 8y^2$

**81.** $(9y^2)(-8y^2)$

**82.** $-3(x + 5)$

**83.** $-3 + x + 5$

**84.** $(x - 4)^2$

**85.** $(2x + 1)^2$

## 7.4 ADDING, SUBTRACTING, AND MULTIPLYING RADICAL EXPRESSIONS

CD-ROM SSM SSG Video

**OBJECTIVES**

1. Add or subtract radical expressions.
2. Multiply radical expressions.

**1** We have learned that sums or differences of like terms can be simplified. To simplify these sums or differences, we use the distributive property. For example,

$$2x + 3x = (2 + 3)x = 5x \quad \text{and} \quad 7x^2y - 4x^2y = (7 - 4)x^2y = 3x^2y$$

The distributive property can also be used to add **like radicals.**

### LIKE RADICALS

Radicals with the same index and the same radicand are like radicals.

For example, $2\sqrt{7} + 3\sqrt{7} = (2 + 3)\sqrt{7} = 5\sqrt{7}$. Also,

$$5\sqrt{3x} - 7\sqrt{3x} = (5 - 7)\sqrt{3x} = -2\sqrt{3x}$$

The expression $2\sqrt{7} + 2\sqrt[3]{7}$ cannot be simplified further since $2\sqrt{7}$ and $2\sqrt[3]{7}$ are not like radicals.

**Example 1** Add or subtract. Assume that variables represent positive real numbers.

**a.** $\sqrt{20} + 2\sqrt{45}$ **b.** $\sqrt[3]{54} - 5\sqrt[3]{16} + \sqrt[3]{2}$ **c.** $\sqrt{27x} - 2\sqrt{9x} + \sqrt{72x}$

**d.** $\sqrt[3]{98} + \sqrt{98}$ **e.** $\sqrt[3]{48y^4} + \sqrt[3]{6y^4}$

*Solution* First, simplify each radical. Then add or subtract any like radicals.

**a.** $\sqrt{20} + 2\sqrt{45} = \sqrt{4 \cdot 5} + 2\sqrt{9 \cdot 5}$ Factor 20 and 45.

$= \sqrt{4} \cdot \sqrt{5} + 2 \cdot \sqrt{9} \cdot \sqrt{5}$ Use the product rule.

$= 2 \cdot \sqrt{5} + 2 \cdot 3 \cdot \sqrt{5}$ Simplify $\sqrt{4}$ and $\sqrt{9}$.

$= 2\sqrt{5} + 6\sqrt{5}$

$= 8\sqrt{5}$ Add like radicals.

**b.** $\sqrt[3]{54} - 5\sqrt[3]{16} + \sqrt[3]{2}$

$= \sqrt[3]{27} \cdot \sqrt[3]{2} - 5 \cdot \sqrt[3]{8} \cdot \sqrt[3]{2} + \sqrt[3]{2}$ Factor and use the product rule.

$= 3 \cdot \sqrt[3]{2} - 5 \cdot 2 \cdot \sqrt[3]{2} + \sqrt[3]{2}$ Simplify $\sqrt[3]{27}$ and $\sqrt[3]{8}$.

$= 3\sqrt[3]{2} - 10\sqrt[3]{2} + \sqrt[3]{2}$ Write $5 \cdot 2$ as 10.

$= -6\sqrt[3]{2}$ Combine like radicals.

**c.** $\sqrt{27x} - 2\sqrt{9x} + \sqrt{72x}$

$= \sqrt{9} \cdot \sqrt{3x} - 2 \cdot \sqrt{9} \cdot \sqrt{x} + \sqrt{36} \cdot \sqrt{2x}$ Factor and use the product rule.

$= 3 \cdot \sqrt{3x} - 2 \cdot 3 \cdot \sqrt{x} + 6 \cdot \sqrt{2x}$ Simplify $\sqrt{9}$ and $\sqrt{36}$.

$= 3\sqrt{3x} - 6\sqrt{x} + 6\sqrt{2x}$ Write $2 \cdot 3$ as 6.

**HELPFUL HINT**

None of these terms contain like radicals. We can simplify no further.

**d.** $\sqrt[3]{98} + \sqrt{98} = \sqrt[3]{98} + \sqrt{49} \cdot \sqrt{2}$ Factor and use the product rule.

$= \sqrt[3]{98} + 7\sqrt{2}$ No further simplification is possible.

**e.** $\sqrt[3]{48y^4} + \sqrt[3]{6y^4} = \sqrt[3]{8y^3} \cdot \sqrt[3]{6y} + \sqrt[3]{y^3} \cdot \sqrt[3]{6y}$ Factor and use the product rule.

$= 2y\sqrt[3]{6y} + y\sqrt[3]{6y}$ Simplify $\sqrt[3]{8y^3}$ and $\sqrt[3]{y^3}$.

$= 3y\sqrt[3]{6y}$ Combine like radicals.

**Example 2** Add or subtract as indicated.

**a.** $\dfrac{\sqrt{45}}{4} - \dfrac{\sqrt{5}}{3}$ **b.** $\sqrt[3]{\dfrac{7x}{8}} + 2\sqrt[3]{7x}$

*Solution* **a.** $\dfrac{\sqrt{45}}{4} - \dfrac{\sqrt{5}}{3} = \dfrac{3\sqrt{5}}{4} - \dfrac{\sqrt{5}}{3}$ To subtract, notice that the LCD is 12.

$= \dfrac{3\sqrt{5} \cdot 3}{4 \cdot 3} - \dfrac{\sqrt{5} \cdot 4}{3 \cdot 4}$ Write each expression as an equivalent expression with a denominator of 12.

$= \dfrac{9\sqrt{5}}{12} - \dfrac{4\sqrt{5}}{12}$ Multiply factors in the numerator and the denominator.

$= \dfrac{5\sqrt{5}}{12}$ Subtract.

**b.** $\sqrt[3]{\dfrac{7x}{8}} + 2\sqrt[3]{7x} = \dfrac{\sqrt[3]{7x}}{\sqrt[3]{8}} + 2\sqrt[3]{7x}$ Apply the quotient rule for radicals.

$= \dfrac{\sqrt[3]{7x}}{2} + 2\sqrt[3]{7x}$ Simplify.

$= \dfrac{\sqrt[3]{7x}}{2} + \dfrac{2\sqrt[3]{7x} \cdot 2}{2}$ Write each expression as an equivalent expression with a denominator of 2.

$= \dfrac{\sqrt[3]{7x}}{2} + \dfrac{4\sqrt[3]{7x}}{2}$

$= \dfrac{5\sqrt[3]{7x}}{2}$ Add.

**2** We can multiply radical expressions by using many of the same properties used to multiply polynomial expressions. For instance, to multiply $\sqrt{2}(\sqrt{6} - 3\sqrt{2})$, we use the distributive property and multiply $\sqrt{2}$ by each term inside the parentheses.

$$\sqrt{2}(\sqrt{6} - 3\sqrt{2}) = \sqrt{2}(\sqrt{6}) - \sqrt{2}(3\sqrt{2}) \quad \text{Use the distributive property.}$$
$$= \sqrt{2 \cdot 6} - 3\sqrt{2 \cdot 2}$$
$$= \sqrt{2 \cdot 2 \cdot 3} - 3 \cdot 2 \quad \text{Use the product rule for radicals.}$$
$$= 2\sqrt{3} - 6$$

**Example 3** Multiply.

**a.** $\sqrt{3}(5 + \sqrt{30})$ **b.** $(\sqrt{5} - \sqrt{6})(\sqrt{7} + 1)$ **c.** $(7\sqrt{x} + 5)(3\sqrt{x} - \sqrt{5})$
**d.** $(4\sqrt{3} - 1)^2$ **e.** $(\sqrt{2x} - 5)(\sqrt{2x} + 5)$

*Solution* **a.** 
$$\sqrt{3}(5 + \sqrt{30}) = \sqrt{3}(5) + \sqrt{3}(\sqrt{30})$$
$$= 5\sqrt{3} + \sqrt{3 \cdot 30}$$
$$= 5\sqrt{3} + \sqrt{3 \cdot 3 \cdot 10}$$
$$= 5\sqrt{3} + 3\sqrt{10}$$

**b.** To multiply, we can use the FOIL method.

$$(\sqrt{5} - \sqrt{6})(\sqrt{7} + 1) = \underbrace{\sqrt{5} \cdot \sqrt{7}}_{\text{First}} + \underbrace{\sqrt{5} \cdot 1}_{\text{Outer}} - \underbrace{\sqrt{6} \cdot \sqrt{7}}_{\text{Inner}} - \underbrace{\sqrt{6} \cdot 1}_{\text{Last}}$$
$$= \sqrt{35} + \sqrt{5} - \sqrt{42} - \sqrt{6}$$

**c.** 
$$(7\sqrt{x} + 5)(3\sqrt{x} - \sqrt{5}) = 7\sqrt{x}(3\sqrt{x}) - 7\sqrt{x}(\sqrt{5}) + 5(3\sqrt{x}) - 5(\sqrt{5})$$
$$= 21x - 7\sqrt{5x} + 15\sqrt{x} - 5\sqrt{5}$$

**d.** 
$$(4\sqrt{3} - 1)^2 = (4\sqrt{3} - 1)(4\sqrt{3} - 1)$$
$$= 4\sqrt{3}(4\sqrt{3}) - 4\sqrt{3}(1) - 1(4\sqrt{3}) - 1(-1)$$
$$= 16 \cdot 3 - 4\sqrt{3} - 4\sqrt{3} + 1$$
$$= 48 - 8\sqrt{3} + 1$$
$$= 49 - 8\sqrt{3}$$

**e.** 
$$(\sqrt{2x} - 5)(\sqrt{2x} + 5) = \sqrt{2x} \cdot \sqrt{2x} + 5\sqrt{2x} - 5\sqrt{2x} - 5 \cdot 5$$
$$= 2x - 25$$

## MENTAL MATH

*Simplify. Assume that all variables represent positive real numbers.*

**1.** $2\sqrt{3} + 4\sqrt{3}$ **2.** $5\sqrt{7} + 3\sqrt{7}$ **3.** $8\sqrt{x} - 5\sqrt{x}$
**4.** $3\sqrt{y} + 10\sqrt{y}$ **5.** $7\sqrt[3]{x} + 5\sqrt[3]{x}$ **6.** $8\sqrt[3]{z} - 2\sqrt[3]{z}$

## Exercise Set 7.4

*Add or subtract. See Examples 1 and 2.*

**1.** $\sqrt{8} - \sqrt{32}$

**2.** $\sqrt{27} - \sqrt{75}$

**3.** $2\sqrt{2x^3} + 4x\sqrt{8x}$

**4.** $3\sqrt{45x^3} + x\sqrt{5x}$

**5.** $2\sqrt{50} - 3\sqrt{125} + \sqrt{98}$

**6.** $4\sqrt{32} - \sqrt{18} + 2\sqrt{128}$

**7.** $\sqrt[3]{16x} - \sqrt[3]{54x}$

**8.** $2\sqrt[3]{3a^4} - 3a\sqrt[3]{81a}$

**9.** $\sqrt{9b^3} - \sqrt{25b^3} + \sqrt{49b^3}$

**10.** $\sqrt{4x^7} + 9x^2\sqrt{x^3} - 5x\sqrt{x^5}$

**11.** $\dfrac{5\sqrt{2}}{3} + \dfrac{2\sqrt{2}}{5}$

**12.** $\dfrac{\sqrt{3}}{2} + \dfrac{4\sqrt{3}}{3}$

**13.** $\sqrt[3]{\dfrac{11}{8}} - \dfrac{\sqrt[3]{11}}{6}$

**14.** $\dfrac{2\sqrt[3]{4}}{7} - \dfrac{\sqrt[3]{4}}{14}$

**15.** $\dfrac{\sqrt{20x}}{9} + \sqrt{\dfrac{5x}{9}}$

**16.** $\dfrac{3x\sqrt{7}}{5} + \sqrt{\dfrac{7x^2}{100}}$

**17.** $7\sqrt{9} - 7 + \sqrt{3}$

**18.** $\sqrt{16} - 5\sqrt{10} + 7$

**19.** $2 + 3\sqrt{y^2} - 6\sqrt{y^2} + 5$

**20.** $3\sqrt{7} - \sqrt[3]{x} + 4\sqrt{7} - 3\sqrt[3]{x}$

**21.** $3\sqrt{108} - 2\sqrt{18} - 3\sqrt{48}$

**22.** $-\sqrt{75} + \sqrt{12} - 3\sqrt{3}$

**23.** $-5\sqrt[3]{625} + \sqrt[3]{40}$

**24.** $-2\sqrt[3]{108} - \sqrt[3]{32}$

**25.** $\sqrt{9b^3} - \sqrt{25b^3} + \sqrt{16b^3}$

**26.** $\sqrt{4x^7y^5} + 9x^2\sqrt{x^3y^5} - 5xy\sqrt{x^5y^3}$

**27.** $5y\sqrt{8y} + 2\sqrt{50y^3}$

**28.** $3\sqrt{8x^2y^3} - 2x\sqrt{32y^3}$

**29.** $\sqrt[3]{54xy^3} - 5\sqrt[3]{2xy^3} + y\sqrt[3]{128x}$

**30.** $2\sqrt[3]{24x^3y^4} + 4x\sqrt[3]{81y^4}$

**31.** $6\sqrt[3]{11} + 8\sqrt{11} - 12\sqrt{11}$

**32.** $3\sqrt[3]{5} + 4\sqrt{5}$

**33.** $-2\sqrt[4]{x^7} + 3\sqrt[4]{16x^7}$

**34.** $6\sqrt[3]{24x^3} - 2\sqrt[3]{81x^3} - x\sqrt[3]{3}$

**35.** $\dfrac{4\sqrt{3}}{3} - \dfrac{\sqrt{12}}{3}$

**36.** $\dfrac{\sqrt{45}}{10} + \dfrac{7\sqrt{5}}{10}$

**37.** $\dfrac{\sqrt[3]{8x^4}}{7} + \dfrac{3x\sqrt[3]{x}}{7}$

**38.** $\dfrac{\sqrt[4]{48}}{5x} - \dfrac{2\sqrt[4]{3}}{10x}$

**39.** $\sqrt{\dfrac{28}{x^2}} + \sqrt{\dfrac{7}{4x^2}}$

**40.** $\dfrac{\sqrt{99}}{5x} - \sqrt{\dfrac{44}{x^2}}$

**41.** $\sqrt[3]{\dfrac{16}{27}} - \dfrac{\sqrt[3]{54}}{6}$

**42.** $\dfrac{\sqrt[3]{3}}{10} + \sqrt[3]{\dfrac{24}{125}}$

**43.** $-\dfrac{\sqrt[3]{2x^4}}{9} + \sqrt[3]{\dfrac{250x^4}{27}}$

**44.** $\dfrac{\sqrt[3]{y^5}}{8} + \dfrac{5y\sqrt[3]{y^2}}{4}$

△ **45.** Find the perimeter of the trapezoid.

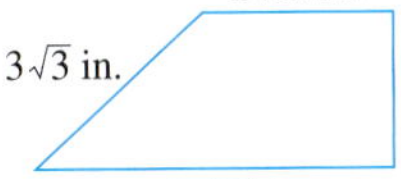


△ **46.** Find the perimeter of the triangle.

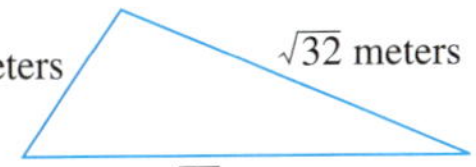


*Multiply, and then simplify if possible. See Example 3.*

**47.** $\sqrt{7}(\sqrt{5} + \sqrt{3})$

**48.** $\sqrt{5}(\sqrt{15} - \sqrt{35})$

**49.** $(\sqrt{5} - \sqrt{2})^2$

**50.** $(3x - \sqrt{2})(3x - \sqrt{2})$

**51.** $\sqrt{3x}(\sqrt{3} - \sqrt{x})$

**52.** $\sqrt{5y}(\sqrt{y} + \sqrt{5})$

**53.** $(2\sqrt{x} - 5)(3\sqrt{x} + 1)$

**54.** $(8\sqrt{y} + z)(4\sqrt{y} - 1)$

**55.** $(\sqrt[3]{a} - 4)(\sqrt[3]{a} + 5)$

**56.** $(\sqrt[3]{a} + 2)(\sqrt[3]{a} + 7)$

**57.** $6(\sqrt{2} - 2)$

**58.** $\sqrt{5}(6 - \sqrt{5})$

**59.** $\sqrt{2}(\sqrt{2} + x\sqrt{6})$

**60.** $\sqrt{3}(\sqrt{3} - 2\sqrt{5x})$

**61.** $(2\sqrt{7} + 3\sqrt{5})(\sqrt{7} - 2\sqrt{5})$

**62.** $(\sqrt{6} - 4\sqrt{2})(3\sqrt{6} + 1)$

**63.** $(\sqrt{x} - y)(\sqrt{x} + y)$

**64.** $(3\sqrt{x} + 2)(\sqrt{3x} - 2)$

**65.** $(\sqrt{3} + x)^2$

**66.** $(\sqrt{y} - 3x)^2$

**67.** $(\sqrt{5x} - 3\sqrt{2})(\sqrt{5x} - 3\sqrt{3})$

**68.** $(5\sqrt{3x} - \sqrt{y})(4\sqrt{x} + 1)$

**69.** $(\sqrt[3]{4} + 2)(\sqrt[3]{2} - 1)$

**70.** $(\sqrt[3]{3} + \sqrt[3]{2})(\sqrt[3]{9} - \sqrt[3]{4})$

**71.** $(\sqrt[3]{x} + 1)(\sqrt[3]{x} - 4\sqrt{x} + 7)$

**72.** $(\sqrt[3]{3x} + 3)(\sqrt[3]{2x} - 3x - 1)$

△ **73.** Baseboard needs to be installed around the perimeter of a rectangular room.

**a.** Find how much baseboard should be ordered by finding the perimeter of the room.

**b.** Find the area of the room.

△ **74.** A border of wallpaper is to be used around the perimeter of the odd-shaped room shown.

**a.** Find how much wallpaper border is needed by finding the perimeter of the room.

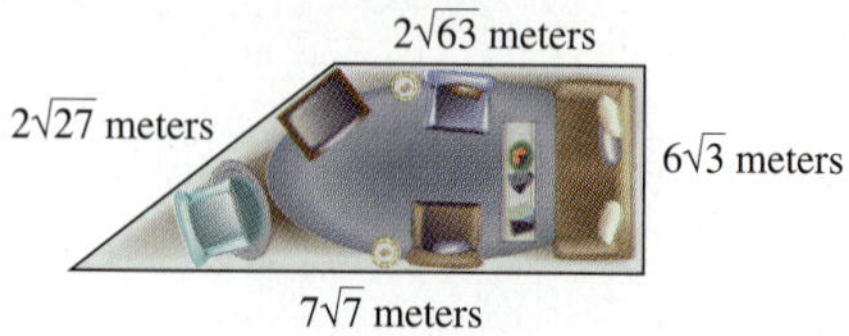


**b.** Find the area of the room. (*Hint:* The area of a trapezoid is the product of half the height $6\sqrt{3}$ meters and the sum of the bases $2\sqrt{63}$ and $7\sqrt{7}$ meters.)

**75.** Explain how simplifying $2x + 3x$ is similar to simplifying $2\sqrt{x} + 3\sqrt{x}$.

**76.** Explain how multiplying $(x - 2)(x + 3)$ is similar to multiplying $(\sqrt{x} - \sqrt{2})(\sqrt{x} + 3)$.

## REVIEW EXERCISES

*Factor each numerator and denominator. Then simplify if possible. See Section 6.1.*

**77.** $\dfrac{2x - 14}{2}$

**78.** $\dfrac{8x - 24y}{4}$

**79.** $\dfrac{7x - 7y}{x^2 - y^2}$

**80.** $\dfrac{x^3 - 8}{4x - 8}$

**81.** $\dfrac{6a^2b - 9ab}{3ab}$

**82.** $\dfrac{14r - 28r^2s^2}{7rs}$

**83.** $\dfrac{-4 + 2\sqrt{3}}{6}$

**84.** $\dfrac{-5 + 10\sqrt{7}}{5}$

# 7.5 RATIONALIZING DENOMINATORS AND NUMERATORS OF RADICAL EXPRESSIONS

CD-ROM SSM SSG Video

▶ **OBJECTIVES**

1. Rationalize denominators.
2. Rationalize numerators.
3. Rationalize denominators or numerators having two terms.

**1** Often in mathematics, it is helpful to write a radical expression such as $\dfrac{\sqrt{3}}{\sqrt{2}}$ either without a radical in the denominator or without a radical in the numerator. The process of writing this expression as an equivalent expression but without a radical in the denominator is called **rationalizing the denominator**. To rationalize the denominator of $\dfrac{\sqrt{3}}{\sqrt{2}}$, we use the fundamental principle of fractions and multiply the numerator and the denominator by $\sqrt{2}$. Recall that this is the same as multiplying by $\dfrac{\sqrt{2}}{\sqrt{2}}$, which simplifies to 1.

$$\frac{\sqrt{3}}{\sqrt{2}} = \frac{\sqrt{3} \cdot \sqrt{2}}{\sqrt{2} \cdot \sqrt{2}} = \frac{\sqrt{6}}{\sqrt{4}} = \frac{\sqrt{6}}{2}$$

**Example 1** Rationalize the denominator of each expression.

**a.** $\dfrac{2}{\sqrt{5}}$ **b.** $\dfrac{2\sqrt{16}}{\sqrt{9x}}$ **c.** $\sqrt[3]{\dfrac{1}{2}}$

*Solution* **a.** To rationalize the denominator, we multiply the numerator and denominator by a factor that makes the radicand in the denominator a perfect square.

$$\frac{2}{\sqrt{5}} = \frac{2 \cdot \sqrt{5}}{\sqrt{5} \cdot \sqrt{5}} = \frac{2\sqrt{5}}{5}$$

The denominator is now rationalized.

**b.** First, we simplify the radicals and then rationalize the denominator.

$$\frac{2\sqrt{16}}{\sqrt{9x}} = \frac{2(4)}{3\sqrt{x}} = \frac{8}{3\sqrt{x}}$$

To rationalize the denominator, multiply the numerator and denominator by $\sqrt{x}$. Then

$$\frac{8}{3\sqrt{x}} = \frac{8 \cdot \sqrt{x}}{3\sqrt{x} \cdot \sqrt{x}} = \frac{8\sqrt{x}}{3x}$$

**c.** $\sqrt[3]{\frac{1}{2}} = \frac{\sqrt[3]{1}}{\sqrt[3]{2}} = \frac{1}{\sqrt[3]{2}}$. Now we rationalize the denominator. Since $\sqrt[3]{2}$ is a cube root, we want to multiply by a value that will make the radicand 2 a perfect cube. If we multiply by $\sqrt[3]{2^2}$, we get $\sqrt[3]{2^3} = \sqrt[3]{8} = 2$.

$$\frac{1 \cdot \sqrt[3]{2^2}}{\sqrt[3]{2} \cdot \sqrt[3]{2^2}} = \frac{\sqrt[3]{4}}{\sqrt[3]{2^3}} = \frac{\sqrt[3]{4}}{2}$$

Multiply numerator and denominator by $\sqrt[3]{2^2}$ and then simplify.

**Example 2** Rationalize the denominator of $\sqrt{\frac{7x}{3y}}$.

*Solution*

$$\sqrt{\frac{7x}{3y}} = \frac{\sqrt{7x}}{\sqrt{3y}}$$

Use the quotient rule. No radical may be simplified further.

$$= \frac{\sqrt{7x} \cdot \sqrt{3y}}{\sqrt{3y} \cdot \sqrt{3y}}$$

Multiply numerator and denominator by $\sqrt{3y}$ so that the radicand in the denominator is a perfect square.

$$= \frac{\sqrt{21xy}}{3y}$$

Use the product rule in the numerator and denominator. Remember that $\sqrt{3y} \cdot \sqrt{3y} = 3y$.

**Example 3** Rationalize the denominator of $\frac{\sqrt[4]{x}}{\sqrt[4]{81y^5}}$.

*Solution* First, simplify each radical if possible.

$$\frac{\sqrt[4]{x}}{\sqrt[4]{81y^5}} = \frac{\sqrt[4]{x}}{\sqrt[4]{81y^4} \cdot \sqrt[4]{y}}$$

Use the product rule in the denominator.

$$= \frac{\sqrt[4]{x}}{3y\sqrt[4]{y}}$$

Write $\sqrt[4]{81y^4}$ as $3y$.

$$= \frac{\sqrt[4]{x} \cdot \sqrt[4]{y^3}}{3y\sqrt[4]{y} \cdot \sqrt[4]{y^3}}$$

Multiply numerator and denominator by $\sqrt[4]{y^3}$ so that the radicand in the denominator is a perfect fourth power.

$$= \frac{\sqrt[4]{xy^3}}{3y\sqrt[4]{y^4}}$$

Use the product rule in the numerator and denominator.

$$= \frac{\sqrt[4]{xy^3}}{3y^2}$$

In the denominator, $\sqrt[4]{y^4} = y$ and $3y \cdot y = 3y^2$.

2 As mentioned earlier, it is also often helpful to write an expression such as $\frac{\sqrt{3}}{\sqrt{2}}$ as an equivalent expression without a radical in the numerator. This process is called **rationalizing the numerator.** To rationalize the numerator of $\frac{\sqrt{3}}{\sqrt{2}}$, we multiply the numerator and the denominator by $\sqrt{3}$.

$$\frac{\sqrt{3}}{\sqrt{2}} = \frac{\sqrt{3} \cdot \sqrt{3}}{\sqrt{2} \cdot \sqrt{3}} = \frac{\sqrt{9}}{\sqrt{6}} = \frac{3}{\sqrt{6}}$$

**Example 4** Rationalize the numerator of $\frac{\sqrt{7}}{\sqrt{45}}$.

*Solution* First we simplify $\sqrt{45}$.

$$\frac{\sqrt{7}}{\sqrt{45}} = \frac{\sqrt{7}}{\sqrt{9 \cdot 5}} = \frac{\sqrt{7}}{3\sqrt{5}}$$

Next we rationalize the numerator by multiplying the numerator and the denominator by $\sqrt{7}$.

$$\frac{\sqrt{7}}{3\sqrt{5}} = \frac{\sqrt{7} \cdot \sqrt{7}}{3\sqrt{5} \cdot \sqrt{7}} = \frac{7}{3\sqrt{5 \cdot 7}} = \frac{7}{3\sqrt{35}}$$

**Example 5** Rationalize the numerator of $\frac{\sqrt[3]{2x^2}}{\sqrt[3]{5y}}$.

*Solution* The numerator and the denominator of this expression are already simplified. To rationalize the numerator, $\sqrt[3]{2x^2}$, we multiply the numerator and denominator by a factor that will make the radicand a perfect cube. If we multiply $\sqrt[3]{2x^2}$ by $\sqrt[3]{4x}$, we get $\sqrt[3]{8x^3} = 2x$.

$$\frac{\sqrt[3]{2x^2}}{\sqrt[3]{5y}} = \frac{\sqrt[3]{2x^2} \cdot \sqrt[3]{4x}}{\sqrt[3]{5y} \cdot \sqrt[3]{4x}} = \frac{\sqrt[3]{8x^3}}{\sqrt[3]{20xy}} = \frac{2x}{\sqrt[3]{20xy}}$$

3 Remember the product of the sum and difference of two terms?

$$(a + b)(a - b) = a^2 - b^2$$

These two expressions are called conjugates of each other.

To rationalize a numerator or denominator that is a sum or difference of two terms, we use conjugates. To see how and why this works, let's rationalize the denominator of the expression $\frac{5}{\sqrt{3} - 2}$. To do so, we multiply both the numerator and the denominator by $\sqrt{3} + 2$, the **conjugate** of the denominator $\sqrt{3} - 2$, and see what happens.

$$\frac{5}{\sqrt{3} - 2} = \frac{5(\sqrt{3} + 2)}{(\sqrt{3} - 2)(\sqrt{3} + 2)}$$

$$= \frac{5(\sqrt{3} + 2)}{(\sqrt{3})^2 - 2^2}$$ Multiply the sum and difference of two terms: $(a + b)(a - b) = a^2 - b^2$.

$$= \frac{5(\sqrt{3} + 2)}{3 - 4}$$

$$= \frac{5(\sqrt{3} + 2)}{-1}$$

$$= -5(\sqrt{3} + 2) \quad \text{or} \quad -5\sqrt{3} - 10$$

Notice in the denominator that the product of $(\sqrt{3} - 2)$ and its conjugate, $(\sqrt{3} + 2)$, is $-1$. In general, the product of an expression and its conjugate will contain no radical terms. This is why, when rationalizing a denominator or a numerator containing two terms, we multiply by its conjugate. Examples of conjugates are

$$\sqrt{a} - \sqrt{b} \quad \text{and} \quad \sqrt{a} + \sqrt{b}$$

$$x + \sqrt{y} \quad \text{and} \quad x - \sqrt{y}$$

**Example 6** Rationalize each denominator.

**a.** $\frac{2}{3\sqrt{2} + 4}$ **b.** $\frac{\sqrt{6} + 2}{\sqrt{5} - \sqrt{3}}$ **c.** $\frac{2\sqrt{m}}{3\sqrt{x} + \sqrt{m}}$

*Solution* **a.** Multiply the numerator and denominator by the conjugate of the denominator, $3\sqrt{2} + 4$.

$$\frac{2}{3\sqrt{2} + 4} = \frac{2(3\sqrt{2} - 4)}{(3\sqrt{2} + 4)(3\sqrt{2} - 4)}$$

$$= \frac{2(3\sqrt{2} - 4)}{(3\sqrt{2})^2 - 4^2}$$

$$= \frac{2(3\sqrt{2} - 4)}{18 - 16}$$

$$= \frac{2(3\sqrt{2} - 4)}{2}, \quad \text{or} \quad 3\sqrt{2} - 4$$

It is often helpful to leave a numerator in factored form to help determine whether the expression can be simplified.

**b.** Multiply the numerator and denominator by the conjugate of $\sqrt{5} - \sqrt{3}$.

$$\frac{\sqrt{6}+2}{\sqrt{5}-\sqrt{3}} = \frac{(\sqrt{6}+2)(\sqrt{5}+\sqrt{3})}{(\sqrt{5}-\sqrt{3})(\sqrt{5}+\sqrt{3})}$$
$$= \frac{\sqrt{6}\sqrt{5}+\sqrt{6}\sqrt{3}+2\sqrt{5}+2\sqrt{3}}{(\sqrt{5})^2-(\sqrt{3})^2}$$
$$= \frac{\sqrt{30}+\sqrt{18}+2\sqrt{5}+2\sqrt{3}}{5-3}$$
$$= \frac{\sqrt{30}+3\sqrt{2}+2\sqrt{5}+2\sqrt{3}}{2}$$

**c.** Multiply by the conjugate of $3\sqrt{x} + \sqrt{m}$ to eliminate the radicals from the denominator.

$$\frac{2\sqrt{m}}{3\sqrt{x}+\sqrt{m}} = \frac{2\sqrt{m}(3\sqrt{x}-\sqrt{m})}{(3\sqrt{x}+\sqrt{m})(3\sqrt{x}-\sqrt{m})} = \frac{6\sqrt{mx}-2m}{(3\sqrt{x})^2-(\sqrt{m})^2}$$
$$= \frac{6\sqrt{mx}-2m}{9x-m}$$

**Example 7** Rationalize the numerator of $\frac{\sqrt{x}+2}{5}$.

*Solution* We multiply the numerator and the denominator by the conjugate of the numerator, $\sqrt{x} + 2$.

$$\frac{\sqrt{x}+2}{5} = \frac{(\sqrt{x}+2)(\sqrt{x}-2)}{5(\sqrt{x}-2)}$$ Multiply by $\sqrt{x} - 2$, the conjugate of $\sqrt{x} + 2$.

$$= \frac{(\sqrt{x})^2-2^2}{5(\sqrt{x}-2)}$$ $(a+b)(a-b) = a^2 - b^2$.

$$= \frac{x-4}{5(\sqrt{x}-2)}$$

## MENTAL MATH

*Find the conjugate of each expression.*

**1.** $\sqrt{2} + x$
**2.** $\sqrt{3} + y$
**3.** $5 - \sqrt{a}$
**4.** $6 - \sqrt{b}$
**5.** $7\sqrt{5} + 8\sqrt{x}$
**6.** $9\sqrt{2} - 6\sqrt{y}$

## Exercise Set 7.5

*Rationalize each denominator. See Examples 1 through 3.*

**1.** $\frac{\sqrt{2}}{\sqrt{7}}$
**2.** $\frac{\sqrt{3}}{\sqrt{2}}$
**3.** $\sqrt{\frac{1}{5}}$
**4.** $\sqrt{\frac{1}{2}}$

5. $\sqrt[3]{\frac{3}{4}}$ 6. $\sqrt[3]{\frac{2}{9}}$

7. $\frac{4}{\sqrt[3]{3}}$ 8. $\frac{6}{\sqrt[3]{9}}$

9. $\frac{3}{\sqrt{8x}}$ 10. $\frac{5}{\sqrt{27a}}$

11. $\frac{3}{\sqrt[3]{4x^2}}$ 12. $\frac{5}{\sqrt[3]{3y}}$

13. $\sqrt{\frac{4}{x}}$ 14. $\sqrt{\frac{25}{y}}$

15. $\frac{9}{\sqrt{3a}}$ 16. $\frac{x}{\sqrt{5}}$

17. $\frac{3}{\sqrt[3]{2}}$ 18. $\frac{5}{\sqrt[3]{9}}$

19. $\frac{2\sqrt{3}}{\sqrt{7}}$ 20. $\frac{-5\sqrt{2}}{\sqrt{11}}$

21. $\sqrt{\frac{2x}{5y}}$ 22. $\sqrt{\frac{13a}{2b}}$

23. $\sqrt[4]{\frac{81}{8}}$ 24. $\sqrt[4]{\frac{1}{9}}$

25. $\sqrt[4]{\frac{16}{9x^7}}$ 26. $\sqrt[5]{\frac{32}{m^6n^{13}}}$

27. $\frac{5a}{\sqrt[5]{8a^9b^{11}}}$ 28. $\frac{9y}{\sqrt[4]{4y^9}}$

*Rationalize each numerator. See Examples 4 and 5.*

29. $\sqrt{\frac{5}{3}}$ 30. $\sqrt{\frac{3}{2}}$

31. $\sqrt{\frac{18}{5}}$ 32. $\sqrt{\frac{12}{7}}$

33. $\frac{\sqrt{4x}}{7}$ $\sqrt{x}$ 34. $\frac{\sqrt{3x^5}}{6}$

35. $\frac{\sqrt[3]{5y^2}}{\sqrt[3]{4x}}$ 36. $\frac{\sqrt[3]{4x}}{\sqrt[3]{z^4}}$

37. $\sqrt{\frac{2}{5}}$ 38. $\sqrt{\frac{3}{7}}$

39. $\frac{\sqrt{2x}}{11}$ 40. $\frac{\sqrt{y}}{7}$

41. $\sqrt[3]{\frac{7}{8}}$ 42. $\sqrt[3]{\frac{25}{2}}$

43. $\frac{\sqrt[3]{3x^5}}{10}$ 44. $\sqrt[3]{\frac{9y}{7}}$

45. $\sqrt{\frac{18x^4y^6}{3z}}$ 46. $\sqrt{\frac{8x^5y}{2z}}$

47. When rationalizing the denominator of $\frac{\sqrt{5}}{\sqrt{7}}$, explain why both the numerator and the denominator must be multiplied by $\sqrt{7}$.

48. When rationalizing the numerator of $\frac{\sqrt{5}}{\sqrt{7}}$, explain why both the numerator and the denominator must be multiplied by $\sqrt{5}$.

*Rationalize each denominator. See Example 6.*

49. $\frac{6}{2-\sqrt{7}}$ 50. $\frac{3}{\sqrt{7}-4}$

51. $\frac{-7}{\sqrt{x}-3}$ 52. $\frac{-8}{\sqrt{y}+4}$

53. $\frac{\sqrt{2}-\sqrt{3}}{\sqrt{2}+\sqrt{3}}$ 54. $\frac{\sqrt{3}+\sqrt{4}}{\sqrt{2}+\sqrt{3}}$

55. $\frac{\sqrt{a}+1}{2\sqrt{a}-\sqrt{b}}$ 56. $\frac{2\sqrt{a}-3}{2\sqrt{a}-\sqrt{b}}$

57. $\frac{8}{1+\sqrt{10}}$ 58. $\frac{-3}{\sqrt{6}-2}$

59. $\frac{\sqrt{x}}{\sqrt{x}+\sqrt{y}}$ 60. $\frac{2\sqrt{a}}{2\sqrt{x}-\sqrt{y}}$

61. $\frac{2\sqrt{3}+\sqrt{6}}{4\sqrt{3}-\sqrt{6}}$ 62. $\frac{4\sqrt{5}+\sqrt{2}}{2\sqrt{5}-\sqrt{2}}$

*Rationalize each numerator. See Example 7.*

63. $\frac{2-\sqrt{11}}{6}$ 64. $\frac{\sqrt{15}+1}{2}$

65. $\frac{2-\sqrt{7}}{-5}$ 66. $\frac{\sqrt{5}+2}{\sqrt{2}}$

67. $\frac{\sqrt{x}+3}{\sqrt{x}}$ 68. $\frac{5+\sqrt{2}}{\sqrt{2x}}$

69. $\frac{\sqrt{2}-1}{\sqrt{2}+1}$ 70. $\frac{\sqrt{8}-\sqrt{3}}{\sqrt{2}+\sqrt{3}}$

71. $\frac{\sqrt{x}+1}{\sqrt{x}-1}$ 72. $\frac{\sqrt{x}+\sqrt{y}}{\sqrt{x}-\sqrt{y}}$

△ 73. The formula of the radius of a sphere $r$ with surface area $A$ is given by the formula

$$r = \sqrt{\frac{A}{4\pi}}$$

Rationalize the denominator of the radical expression in this formula.

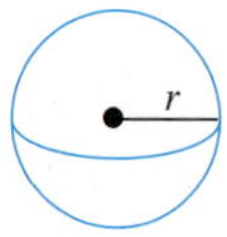


△ **74.** The formula for the radius of a cone $r$ with height 7 centimeters and volume $V$ is given by the formula

$$r = \sqrt{\frac{3V}{7\pi}}$$

Rationalize the numerator of the radical expression in this formula.

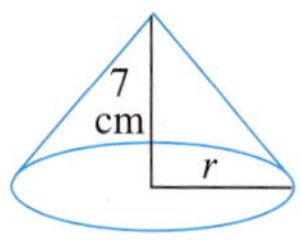


**75.** Explain why rationalizing the denominator does not change the value of the original expression.

**76.** Explain why rationalizing the numerator does not change the value of the original expression.

## REVIEW EXERCISES

*Solve each equation. See Sections 2.1 and 5.8.*

**77.** $2x - 7 = 3(x - 4)$

**78.** $9x - 4 = 7(x - 2)$

**79.** $(x - 6)(2x + 1) = 0$

**80.** $(y + 2)(5y + 4) = 0$

**81.** $x^2 - 8x = -12$

**82.** $x^3 = x$

# 7.6 RADICAL EQUATIONS AND PROBLEM SOLVING

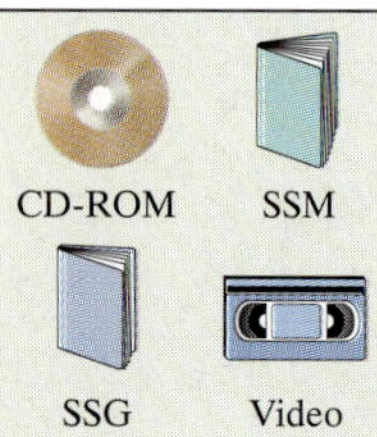


**OBJECTIVES**

1. Solve equations that contain radical expressions.
2. Use the Pythagorean theorem to model problems.

**1** In this section, we present techniques to solve equations containing radical expressions such as

$$\sqrt{2x - 3} = 9$$

We use the power rule to help us solve these radical equations.

**POWER RULE**

If both sides of an equation are raised to the same power, **all** solutions of the original equation are **among** the solutions of the new equation.

This property *does not* say that raising both sides of an equation to a power yields an equivalent equation. A solution of the new equation *may or may not* be a solution of the original equation. Thus, *each solution of the new equation must be checked* to make sure it is a solution of the original equation. Recall that a proposed solution that is not a solution of the original equation is called an **extraneous solution**.

**Example 1** Solve $\sqrt{2x - 3} = 9$.

*Solution* We use the power rule to square both sides of the equation to eliminate the radical.

$$\begin{aligned} \sqrt{2x - 3} &= 9 \\ (\sqrt{2x - 3})^2 &= 9^2 \\ 2x - 3 &= 81 \\ 2x &= 84 \\ x &= 42 \end{aligned}$$

Now we, check the solution in the original equation.

**Check:**

$$\sqrt{2x - 3} = 9$$
$$\sqrt{2(42) - 3} \stackrel{?}{=} 9 \quad \text{Let } x = 42.$$
$$\sqrt{84 - 3} \stackrel{?}{=} 9$$
$$\sqrt{81} \stackrel{?}{=} 9$$
$$9 = 9 \quad \text{True.}$$

The solution checks, so we conclude that the solution is 42.

To solve a radical equation, first isolate a radical on one side of the equation.

**Example 2** Solve $\sqrt{-10x - 1} + 3x = 0$.

*Solution* First, isolate the radical on one side of the equation. To do this, we subtract $3x$ from both sides.

$$\sqrt{-10x - 1} + 3x = 0$$
$$\sqrt{-10x - 1} + 3x - 3x = 0 - 3x$$
$$\sqrt{-10x - 1} = -3x$$

Next we use the power rule to eliminate the radical.

$$(\sqrt{-10x - 1})^2 = (-3x)^2$$
$$-10x - 1 = 9x^2$$

Since this is a quadratic equation, we can set the equation equal to 0 and try to solve by factoring.

$$9x^2 + 10x + 1 = 0$$
$$(9x + 1)(x + 1) = 0 \quad \text{Factor.}$$
$$9x + 1 = 0 \quad \text{or} \quad x + 1 = 0 \quad \text{Set each factor equal to 0.}$$
$$x = -\frac{1}{9} \qquad x = -1$$

**Check:** Let $x = -\frac{1}{9}$.

$$\sqrt{-10x - 1} + 3x = 0$$
$$\sqrt{-10\left(-\frac{1}{9}\right) - 1} + 3\left(-\frac{1}{9}\right) \stackrel{?}{=} 0$$
$$\sqrt{\frac{10}{9} - \frac{9}{9}} - \frac{3}{9} \stackrel{?}{=} 0$$
$$\sqrt{\frac{1}{9}} - \frac{1}{3} \stackrel{?}{=} 0$$
$$\frac{1}{3} - \frac{1}{3} = 0 \quad \text{True.}$$

Let $x = -1$.

$$\sqrt{-10x - 1} + 3x = 0$$
$$\sqrt{-10(-1) - 1} + 3(-1) \stackrel{?}{=} 0$$
$$\sqrt{10 - 1} - 3 \stackrel{?}{=} 0$$
$$\sqrt{9} - 3 \stackrel{?}{=} 0$$
$$3 - 3 = 0 \quad \text{True.}$$

Both solutions check. The solutions are $-\frac{1}{9}$ and $-1$.

The following steps may be used to solve a radical equation.

**SOLVING A RADICAL EQUATION**

**Step 1:** Isolate one radical on one side of the equation.
**Step 2:** Raise each side of the equation to a power equal to the index of the radical and simplify.
**Step 3:** If the equation still contains a radical term, repeat Steps 1 and 2. If not, solve the equation.
**Step 4:** Check all proposed solutions in the original equation.

**Example 3** Solve $\sqrt[3]{x+1} + 5 = 3$.

*Solution* First we isolate the radical by subtracting 5 from both sides of the equation.

$$\sqrt[3]{x+1} + 5 = 3$$
$$\sqrt[3]{x+1} = -2$$

Next we raise both sides of the equation to the third power to eliminate the radical.

$$(\sqrt[3]{x+1})^3 = (-2)^3$$
$$x + 1 = -8$$
$$x = -9$$

The solution checks in the original equation, so the solution is −9.

**Example 4** Solve $\sqrt{4-x} = x - 2$.

*Solution*

$$\sqrt{4-x} = x - 2$$
$$(\sqrt{4-x})^2 = (x-2)^2$$
$$4 - x = x^2 - 4x + 4$$
$$x^2 - 3x = 0 \quad \text{Write the quadratic equation in standard form.}$$
$$x(x-3) = 0 \quad \text{Factor.}$$
$$x = 0 \quad \text{or} \quad x - 3 = 0 \quad \text{Set each factor equal to 0.}$$
$$x = 3$$

**Check:**

$$\sqrt{4-x} = x - 2 \qquad\qquad \sqrt{4-x} = x - 2$$
$$\sqrt{4-0} \stackrel{?}{=} 0 - 2 \quad \text{Let } x = 0. \qquad \sqrt{4-3} \stackrel{?}{=} 3 - 2 \quad \text{Let } x = 3.$$
$$2 = -2 \quad \text{False.} \qquad 1 = 1 \quad \text{True.}$$

The proposed solution 3 checks, but 0 does not. Since 0 is an extraneous solution, the only solution is 3.

**HELPFUL HINT**

In Example 4, notice that $(x-2)^2 = x^2 - 4x + 4$. Make sure binomials are squared correctly.

**Example 5** Solve $\sqrt{2x + 5} + \sqrt{2x} = 3$.

*Solution* We get one radical alone by subtracting $\sqrt{2x}$ from both sides.

$$\sqrt{2x + 5} + \sqrt{2x} = 3$$
$$\sqrt{2x + 5} = 3 - \sqrt{2x}$$

Now we use the power rule to begin eliminating the radicals. First we square both sides.

$$(\sqrt{2x + 5})^2 = (3 - \sqrt{2x})^2$$

$$2x + 5 = 9 - 6\sqrt{2x} + 2x \quad \text{Multiply } (3 - \sqrt{2x})(3 - \sqrt{2x}).$$

There is still a radical in the equation, so we get the radical alone again. Then we square both sides.

$$2x + 5 = 9 - 6\sqrt{2x} + 2x \quad \text{Get the radical alone.}$$
$$6\sqrt{2x} = 4$$
$$36(2x) = 16 \quad \text{Square both sides of the equation to eliminate the radical.}$$
$$72x = 16 \quad \text{Multiply.}$$
$$x = \frac{16}{72} \quad \text{Solve.}$$
$$x = \frac{2}{9} \quad \text{Simplify.}$$

The proposed solution, $\frac{2}{9}$, checks in the original equation. The solution is $\frac{2}{9}$.

**HELPFUL HINT**

Make sure expressions are squared correctly. In Example 5, we squared $(3 - \sqrt{2x})$ as

$$(3 - \sqrt{2x})^2 = (3 - \sqrt{2x})(3 - \sqrt{2x})$$
$$= 3 \cdot 3 - 3\sqrt{2x} - 3\sqrt{2x} + \sqrt{2x} \cdot \sqrt{2x}$$
$$= 9 - 6\sqrt{2x} + 2x$$

2 Recall that the Pythagorean theorem states that in a right triangle, the length of the hypotenuse squared equals the sum of the lengths of each of the legs squared.

**PYTHAGOREAN THEOREM**

If $a$ and $b$ are the lengths of the legs of a right triangle and $c$ is the length of the hypotenuse, then $a^2 + b^2 = c^2$.

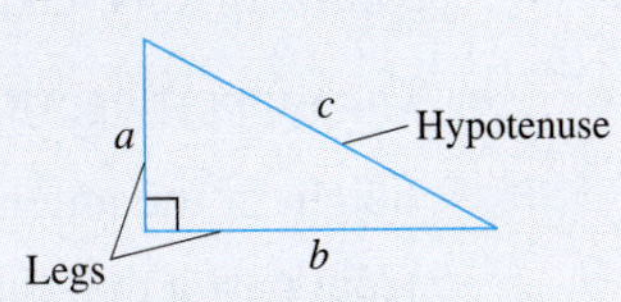

**△ Example 6** Find the length of the unknown leg of the right triangle.

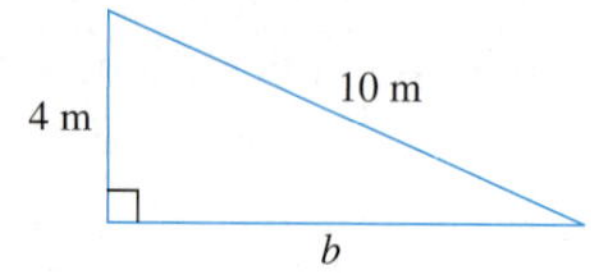


*Solution* In the formula $a^2 + b^2 = c^2$, $c$ is the hypotenuse. Here, $c = 10$, the length of the hypotenuse, and $a = 4$. We solve for $b$. Then $a^2 + b^2 = c^2$ becomes

$$4^2 + b^2 = 10^2$$
$$16 + b^2 = 100$$
$$b^2 = 84 \quad \text{Subtract 16 from both sides.}$$

Since $b$ is a length and thus is positive, we have that

$$b = \sqrt{84} = \sqrt{4 \cdot 21} = 2\sqrt{21}$$

The unknown leg of the triangle is $2\sqrt{21}$ meters long.

**△ Example 7** **CALCULATING PLACEMENT OF A WIRE**

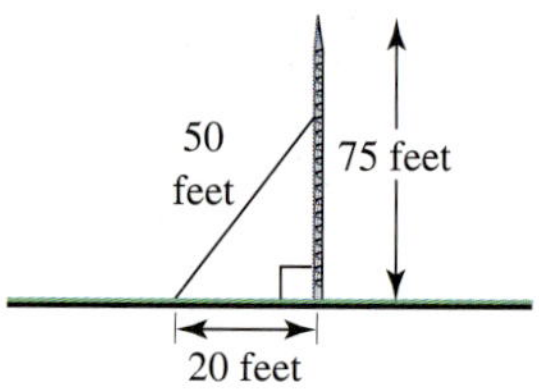


A 50-foot supporting wire is to be attached to a 75-foot antenna. Because of surrounding buildings, sidewalks, and roadways, the wire must be anchored exactly 20 feet from the base of the antenna.

**a.** How high from the base of the antenna is the wire attached?

**b.** Local regulations require that a supporting wire be attached at a height no less than $\frac{3}{5}$ of the total height of the antenna. From part **a**, have local regulations been met?

*Solution*

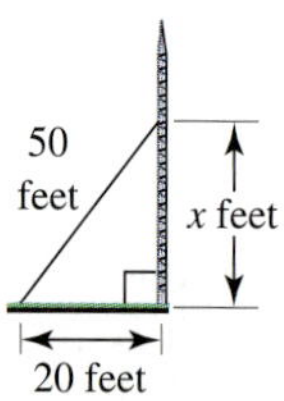


1. UNDERSTAND. Read and reread the problem. From the diagram we notice that a right triangle is formed with hypotenuse 50 feet and one leg 20 feet. Let $x$ be the height from the base of the antenna to the attached wire.

2. TRANSLATE. Use the Pythagorean theorem.

$$a^2 + b^2 = c^2$$
$$20^2 + x^2 = 50^2 \quad a = 20, c = 50$$

3. SOLVE.

$$20^2 + x^2 = 50^2$$
$$400 + x^2 = 2500$$
$$x^2 = 2100 \quad \text{Subtract 400 from both sides.}$$
$$x = \sqrt{2100}$$
$$= 10\sqrt{21}$$

4. INTERPRET. *Check* the work and *state* the solution.

**a.** The wire is attached exactly $10\sqrt{21}$ feet from the base of the pole, or approximately 45.8 feet.

**b.** The supporting wire must be attached at a height no less than $\frac{3}{5}$ of the total height of the antenna. This height is $\frac{3}{5}$(75 feet), or 45 feet. Since we know from part **a** that the wire is to be attached at a height of approximately 45.8 feet, local regulations have been met.

## GRAPHING CALCULATOR EXPLORATIONS

We can use a graphing calculator to solve radical equations. For example, to use a graphing calculator to approximate the solutions of the equation solved in Example 4, we graph the following.

$$Y_1 = \sqrt{4 - x} \quad \text{and} \quad Y_2 = x - 2$$

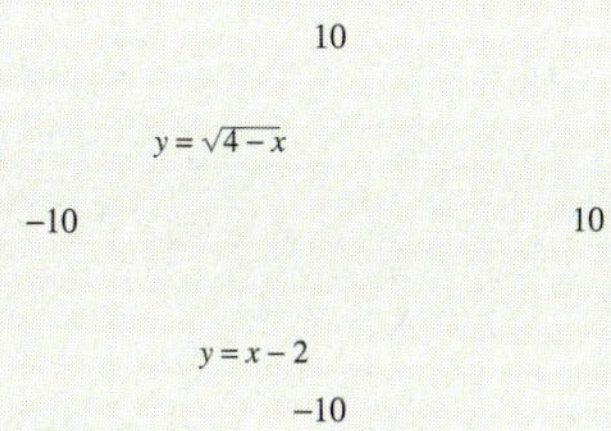


The $x$-value of the point of intersection is the solution. Use the Intersect feature or the Zoom and Trace features of your graphing calculator to see that the solution is 3.

*Use a graphing calculator to solve each radical equation. Round all solutions to the nearest hundredth.*

**1.** $\sqrt{x + 7} = x$

**2.** $\sqrt{3x + 5} = 2x$

**3.** $\sqrt{2x + 1} = \sqrt{2x + 2}$

**4.** $\sqrt{10x - 1} = \sqrt{-10x + 10} - 1$

**5.** $1.2x = \sqrt{3.1x + 5}$

**6.** $\sqrt{1.9x^2 - 2.2} = -0.8x + 3$

## SPOTLIGHT ON DECISION MAKING

Suppose you are a psychologist studying a person's ability to recognize patterns. You theorize that IQ is linked to pattern-recognition ability and design a research study using human subjects to test your theory. As part of your study, you have decided to form three groups of test subjects based roughly on IQ. Group $A$ will consist of subjects having IQ's under 90, Group $B$ will consist of subjects having IQ's from 90 to 105, and Group $C$ will consist of subjects having IQ's over 105.

While preparing for your research study, you came across the findings of another psychologist suggesting that the number $S$ of nonsense syllables that a person can repeat consecutively depends on his or her IQ score $I$ according to the equation $S = 2\sqrt{I} - 9$. Because administering IQ tests can be time-consuming and because your groupings by IQ need only be approximate, you decide to use this equation as a quick way to assign test subjects to Groups $A$, $B$, and $C$. Each subject is individually screened by listening to a string of 20 random nonsense syllables and then repeating as many as possible. The results for the first 5 test subjects are listed in the table. For each subject, decide to which group—$A$, $B$, or $C$—the subject should be assigned.

TEST SUBJECT SCREENING

| *Subject* | *S—the number of Nonsense Syllables Successfully Repeated* |
|---|---|
| 1 | 11 |
| 2 | 13 |
| 3 | 9 |
| 4 | 12 |
| 5 | 10 |

## Exercise Set 7.6

*Solve. See Examples 1 and 2.*

1. $\sqrt{2x} = 4$
2. $\sqrt{3x} = 3$
3. $\sqrt{x-3} = 2$
4. $\sqrt{x+1} = 5$
5. $\sqrt{2x} = -4$
6. $\sqrt{5x} = -5$
7. $\sqrt{4x-3} - 5 = 0$
8. $\sqrt{x-3} - 1 = 0$
9. $\sqrt{2x-3} - 2 = 1$
10. $\sqrt{3x+3} - 4 = 8$

*Solve. See Example 3.*

11. $\sqrt[3]{6x} = -3$
12. $\sqrt[3]{4x} = -2$
13. $\sqrt[3]{x-2} - 3 = 0$
14. $\sqrt[3]{2x-6} - 4 = 0$

*Solve. See Examples 4 and 5.*

15. $\sqrt{13-x} = x - 1$
16. $\sqrt{2x-3} = 3 - x$
17. $x - \sqrt{4-3x} = -8$
18. $2x + \sqrt{x+1} = 8$
19. $\sqrt{y+5} = 2 - \sqrt{y-4}$
20. $\sqrt{x+3} + \sqrt{x-5} = 3$
21. $\sqrt{x-3} + \sqrt{x+2} = 5$
22. $\sqrt{2x-4} - \sqrt{3x+4} = -2$

*Solve. See Examples 1 through 5.*

23. $\sqrt{3x-2} = 5$
24. $\sqrt{5x-4} = 9$
25. $-\sqrt{2x} + 4 = -6$
26. $-\sqrt{3x+9} = -12$
27. $\sqrt{3x+1} + 2 = 0$
28. $\sqrt{3x+1} - 2 = 0$
29. $\sqrt[4]{4x+1} - 2 = 0$
30. $\sqrt[4]{2x-9} - 3 = 0$
31. $\sqrt{4x-3} = 7$
32. $\sqrt{3x+9} = 6$
33. $\sqrt[3]{6x-3} - 3 = 0$
34. $\sqrt[3]{3x} + 4 = 7$
35. $\sqrt[3]{2x-3} - 2 = -5$
36. $\sqrt[3]{x-4} - 5 = -7$
37. $\sqrt{x+4} = \sqrt{2x-5}$
38. $\sqrt{3y+6} = \sqrt{7y-6}$
39. $x - \sqrt{1-x} = -5$
40. $x - \sqrt{x-2} = 4$
41. $\sqrt[3]{-6x-1} = \sqrt[3]{-2x-5}$
42. $x + \sqrt{x+5} = 7$
43. $\sqrt{5x-1} - \sqrt{x} + 2 = 3$
44. $\sqrt{2x-1} - 4 = -\sqrt{x-4}$
45. $\sqrt{2x-1} = \sqrt{1-2x}$
46. $\sqrt{7x-4} = \sqrt{4-7x}$
47. $\sqrt{3x+4} - 1 = \sqrt{2x+1}$
48. $\sqrt{x-2} + 3 = \sqrt{4x+1}$
49. $\sqrt{y+3} - \sqrt{y-3} = 1$
50. $\sqrt{x+1} - \sqrt{x-1} = 2$
51. What is wrong with the following steps?

$$\sqrt{2x+5} + \sqrt{4-x} = 8$$
$$(\sqrt{2x+5} + \sqrt{4-x})^2 = 8^2$$
$$(2x+5) + (4-x) = 64$$
$$x + 9 = 64$$
$$x = 55$$

52. How can you immediately tell that the equation $\sqrt{2y+3} = -4$ has no real solution?

*Find the length of the unknown side of each triangle. See Example 6.*

53.
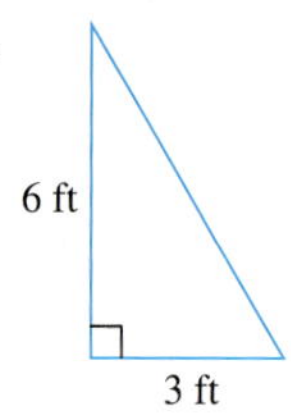


54.
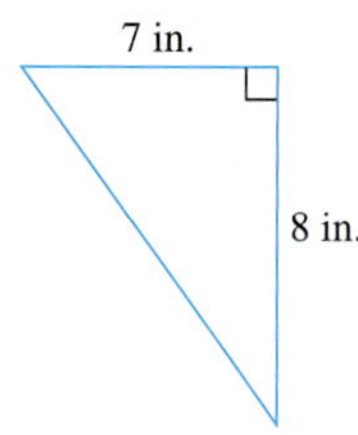


55.
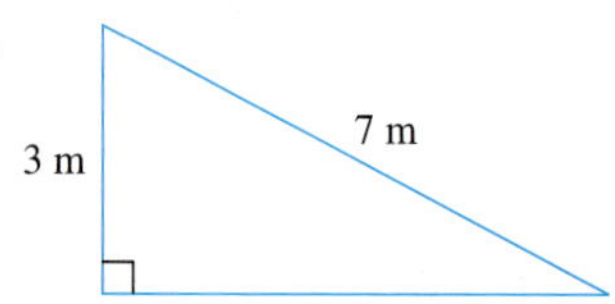


56.
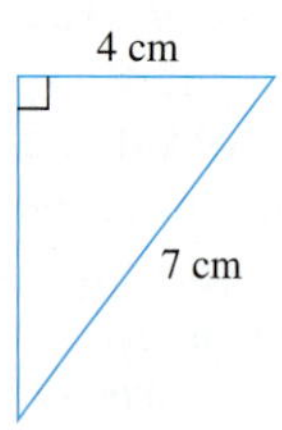


*Find the length of the unknown side of each triangle. Give the exact length and a one-decimal-place approximation. See Example 6.*

57.
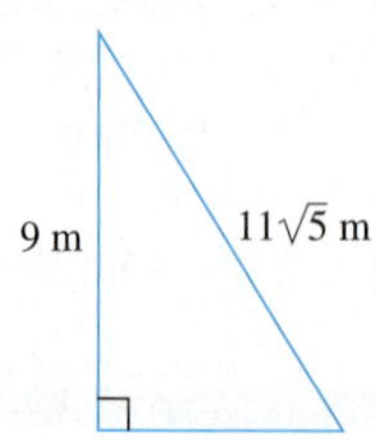

△ **58.**

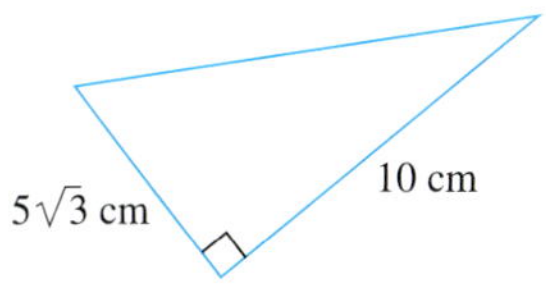


△ **59.**

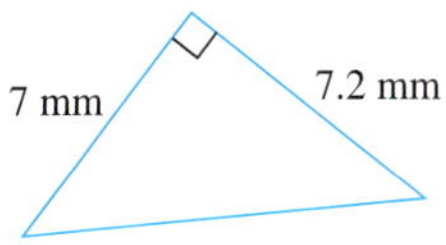


△ **60.**

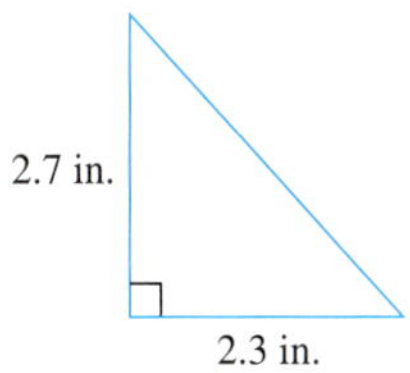


*Solve. See Example 7. Give exact answers and two-decimal-place approximations where appropriate.*

△ **61.** A wire is needed to support a vertical pole 15 feet high. The cable will be anchored to a stake 8 feet from the base of the pole. How much cable is needed?

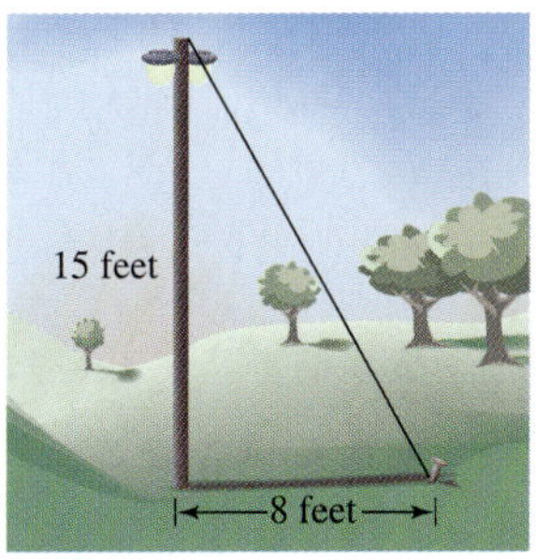


△ **62.** The tallest structure in the United States is a TV tower in Blanchard, North Dakota. Its height is 2063 feet. A 2382-foot length of wire is to be used as a guy wire attached to the top of the tower. Approximate to the nearest foot how far from the base of the tower the guy wire must be anchored. (*Source:* U.S. Geological Survey)

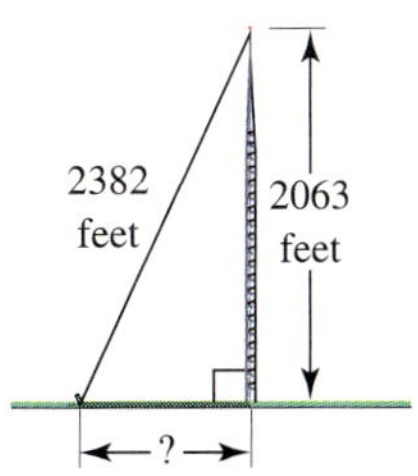


△ **63.** A spotlight is mounted on the eaves of a house 12 feet above the ground. A flower bed runs between the house and the sidewalk, so the closest the ladder can be placed to the house is 5 feet. How long a ladder is needed so that an electrician can reach the place where the light is mounted?

△ **64.** A wire is to be attached to support a telephone pole. Because of surrounding buildings, sidewalks, and roadways, the wire must be anchored exactly 15 feet from the base of the pole. Telephone company workers have only 30 feet of cable, and 2 feet of that must be used to attach the cable to the pole and to the stake on the ground. How high from the base of the pole can the wire be attached?

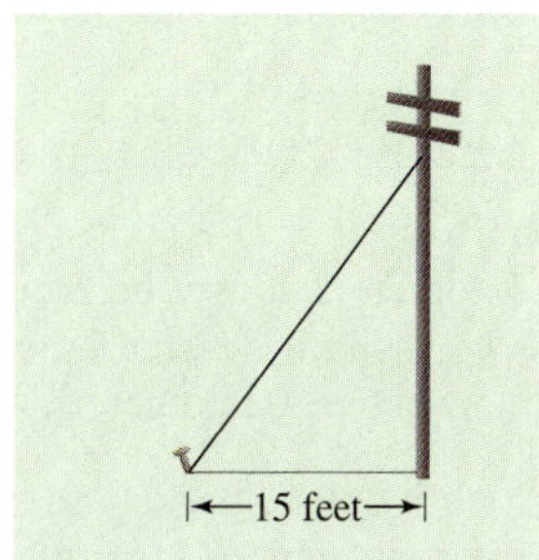


△ **65.** The radius of the Moon is 1080 miles. Use the formula for the radius $r$ of a sphere given its surface area $A$,

$$r = \sqrt{\frac{A}{4\pi}}$$

to find the surface area of the Moon. Round to the nearest square mile. (*Source:* National Space Science Data Center)

**66.** Police departments find it very useful to be able to approximate the speed of a car when they are given the distance that the car skidded before it came to a stop. If the road surface is wet concrete, the function $S(x) = \sqrt{10.5x}$ is used, where $S(x)$ is the speed of the car in miles per hour and $x$ is the distance skidded in feet. Find how fast a car was moving if it skidded 280 feet on wet concrete.

**67.** The formula $v = \sqrt{2gh}$ gives the velocity $v$, in feet per second, of an object when it falls $h$ feet accelerated by gravity $g$, in feet per second squared. If $g$ is approximately 32 feet per second squared, find how far an object has fallen if its velocity is 80 feet per second.

**68.** Two tractors are pulling a tree stump from a field. If two forces $A$ and $B$ pull at right angles (90°) to each other, the size of the resulting force $R$ is given by the formula $R = \sqrt{A^2 + B^2}$. If tractor $A$ is exerting 600 pounds of force and the resulting force is 850 pounds, find how much force tractor $B$ is exerting.

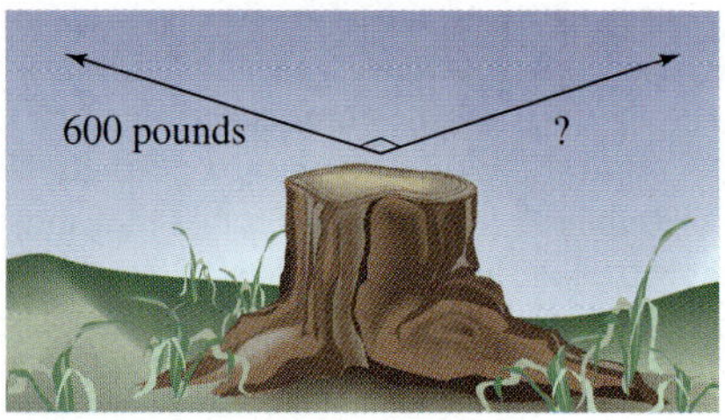


**69.** Solve: $\sqrt{\sqrt{x+3} + \sqrt{x}} = \sqrt{3}$

**70.** The maximum distance $D(h)$ that a person can see from a height $h$ kilometers above the ground is given by the function $D(h) = 111.7\sqrt{h}$. Find the height that would allow a person to see 80 kilometers.

**71.** The cost $C(x)$ in dollars per day to operate a small delivery service is given by $C(x) = 80\sqrt[3]{x} + 500$, where $x$ is the number of deliveries per day. In July, the manager decides that it is necessary to keep delivery costs below \$1620. Find the greatest number of deliveries this company can make per day and still keep overhead below \$1620.

**72.** Explain why proposed solutions of radical equations must be checked.

**73.** Consider the equations $\sqrt{2x} = 4$ and $\sqrt[3]{2x} = 4$.

**a.** Explain the difference in solving these equations.

**b.** Explain the similarity in solving these equations.

## REVIEW EXERCISES

*Use the vertical line test to determine whether each graph represents the graph of a function. See Section 3.2.*

**74.**

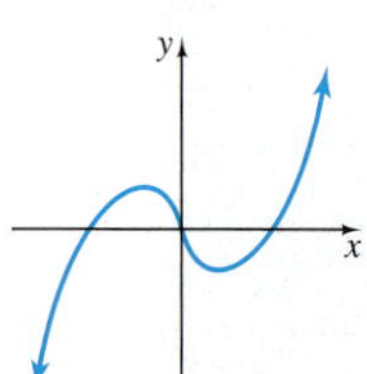


**75.**

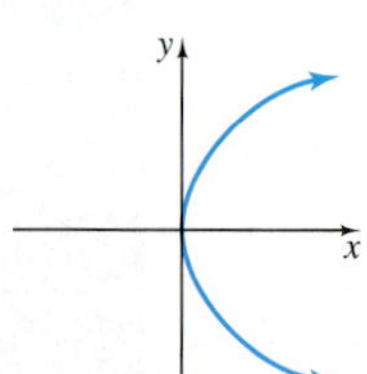


**76.**

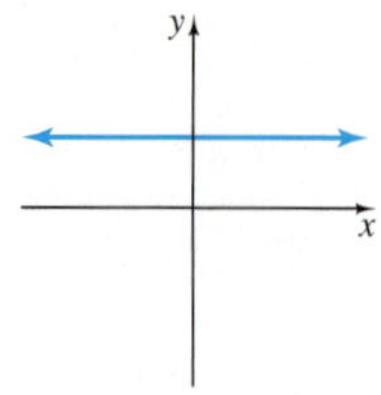


**77.**

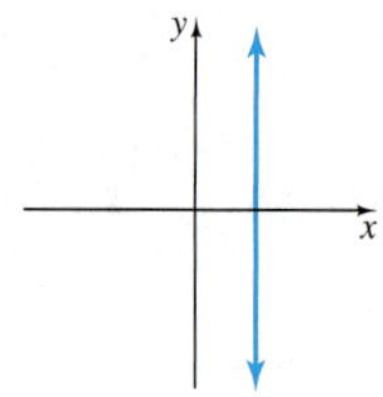


**78.**

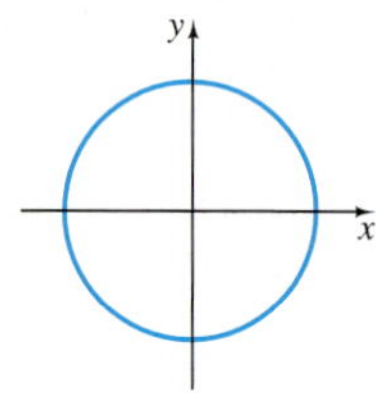


**79.**

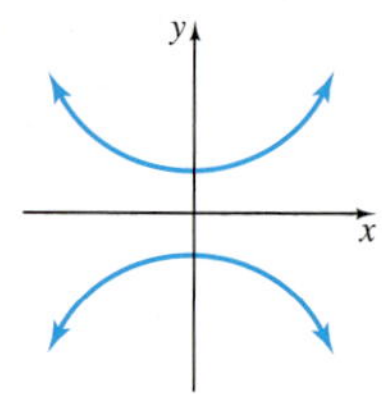


*Simplify. See Section 6.4.*

**80.** $\dfrac{\frac{x}{6}}{\frac{2x}{3} + \frac{1}{2}}$

**81.** $\dfrac{\frac{1}{y} + \frac{4}{5}}{\frac{-3}{20}}$

**82.** $\dfrac{\frac{z}{5} + \frac{1}{10}}{\frac{z}{20} - \frac{z}{5}}$

**83.** $\dfrac{\frac{1}{y} + \frac{1}{x}}{\frac{1}{y} - \frac{1}{x}}$

## *A Look Ahead*

**Example**

Solve $(t^2 - 3t) - 2\sqrt{t^2 - 3t} = 0$.

*Solution*

Substitution can be used to make this problem somewhat simpler. Since $t^2 - 3t$ occurs more than once, let $x = t^2 - 3t$.

$$(t^2 - 3t) - 2\sqrt{t^2 - 3t} = 0$$
$$x - 2\sqrt{x} = 0 \quad \text{Let } x = t^2 - 3t.$$
$$x = 2\sqrt{x}$$
$$x^2 = (2\sqrt{x})^2$$
$$x^2 = 4x$$
$$x^2 - 4x = 0$$
$$x(x - 4) = 0$$
$$x = 0 \quad \text{or} \quad x - 4 = 0$$
$$x = 4$$

Now we "undo" the substitution.

$x = 0$ Replace $x$ with $t^2 - 3t$.

$$t^2 - 3t = 0$$
$$t(t - 3) = 0$$
$$t = 0 \quad \text{or} \quad t - 3 = 0$$
$$t = 3$$

$x = 4$ Replace $x$ with $t^2 - 3t$.

$$\begin{aligned} t^2 - 3t &= 4 \\ t^2 - 3t - 4 &= 0 \\ (t - 4)(t + 1) &= 0 \\ t - 4 = 0 \quad &\text{or} \quad t + 1 = 0 \\ t = 4 \quad &\qquad\quad t = -1 \end{aligned}$$

In this problem, we have four possible solutions: 0, 3, 4, and −1. All four solutions check in the original equation, so the solutions are −1, 0, 3, 4.

*Solve. See the preceding example.*

**84.** $3\sqrt{x^2 - 8x} = x^2 - 8x$

**85.** $\sqrt{(x^2 - x) + 7} = 2(x^2 - x) - 1$

**86.** $7 - (x^2 - 3x) = \sqrt{(x^2 - 3x) + 5}$

**87.** $x^2 + 6x = 4\sqrt{x^2 + 6x}$

## 7.7 COMPLEX NUMBERS

CD-ROM SSM SSG Video

▶ **OBJECTIVES**

1. Define imaginary and complex numbers.
2. Add or subtract complex numbers.
3. Multiply complex numbers.
4. Divide complex numbers.
5. Raise $i$ to powers.

**1** Our work with radical expressions has excluded expressions such as $\sqrt{-16}$ because $\sqrt{-16}$ is not a real number; there is no real number whose square is −16. In this section, we discuss a number system that includes roots of negative numbers. This number system is the **complex number system**, and it includes the set of real numbers as a subset. The complex number system allows us to solve equations such as $x^2 + 1 = 0$ that have no real number solutions. The set of complex numbers includes the **imaginary unit**.

### IMAGINARY UNIT

The imaginary unit, written $i$, is the number whose square is −1. That is,

$$i^2 = -1 \quad \text{and} \quad i = \sqrt{-1}$$

To write the square root of a negative number in terms of $i$, use the property that if $a$ is a positive number, then

$$\begin{aligned} \sqrt{-a} &= \sqrt{-1} \cdot \sqrt{a} \\ &= i \cdot \sqrt{a} \end{aligned}$$

Using $i$, we can write $\sqrt{-16}$ as

$$\sqrt{-16} = \sqrt{-1 \cdot 16} = \sqrt{-1} \cdot \sqrt{16} = i \cdot 4, \text{ or } 4i$$

**Example 1** Write with $i$ notation.

**a.** $\sqrt{-36}$ **b.** $\sqrt{-5}$ **c.** $-\sqrt{-20}$

*Solution*

**a.** $\sqrt{-36} = \sqrt{-1 \cdot 36} = \sqrt{-1} \cdot \sqrt{36} = i \cdot 6$, or $6i$

**b.** $\sqrt{-5} = \sqrt{-1(5)} = \sqrt{-1} \cdot \sqrt{5} = i\sqrt{5}$. Since $\sqrt{5}i$ can easily be confused with $\sqrt{5i}$, we write $\sqrt{5}i$ as $i\sqrt{5}$.

**c.** $-\sqrt{-20} = -\sqrt{-1 \cdot 20} = -\sqrt{-1} \cdot \sqrt{4 \cdot 5} = -i \cdot 2\sqrt{5} = -2i\sqrt{5}$

The product rule for radicals does not necessarily hold true for imaginary numbers. *To multiply square roots of negative numbers, first we write each number in terms of the imaginary unit i.* For example, to multiply $\sqrt{-4}$ and $\sqrt{-9}$, we first write each number in the form $bi$.

$$\sqrt{-4}\sqrt{-9} = 2i(3i) = 6i^2 = 6(-1) = -6$$

We will also use this method to simplify quotients of square roots of negative numbers.

**Example 2** Multiply or divide as indicated.

**a.** $\sqrt{-3} \cdot \sqrt{-5}$ **b.** $\sqrt{-36} \cdot \sqrt{-1}$ **c.** $\sqrt{8} \cdot \sqrt{-2}$ **d.** $\dfrac{\sqrt{-125}}{\sqrt{5}}$

*Solution* **a.** $\sqrt{-3} \cdot \sqrt{-5} = i\sqrt{3}(i\sqrt{5}) = i^2\sqrt{15} = -1\sqrt{15} = -\sqrt{15}$

**b.** $\sqrt{-36} \cdot \sqrt{-1} = 6i(i) = 6i^2 = 6(-1) = -6$

**c.** $\sqrt{8} \cdot \sqrt{-2} = 2\sqrt{2}(i\sqrt{2}) = 2i(\sqrt{2}\sqrt{2}) = 2i(2) = 4i$

**d.** $\dfrac{\sqrt{-125}}{\sqrt{5}} = \dfrac{i\sqrt{125}}{\sqrt{5}} = i\sqrt{25} = 5i$

Now that we have practiced working with the imaginary unit, we define complex numbers.

**COMPLEX NUMBERS**

A **complex number** is a number that can be written in the form $a + bi$, where $a$ and $b$ are real numbers.

Notice that the set of real numbers is a subset of the complex numbers since any real number can be written in the form of a complex number. For example,

$$16 = 16 + 0i$$

In general, a complex number $a + bi$ is a real number if $b = 0$. Also, a complex number is called an **imaginary number** if $a = 0$. For example,

$$3i = 0 + 3i \quad \text{and} \quad i\sqrt{7} = 0 + i\sqrt{7}$$

are imaginary numbers.

The following diagram shows the relationship between complex numbers and their subsets.

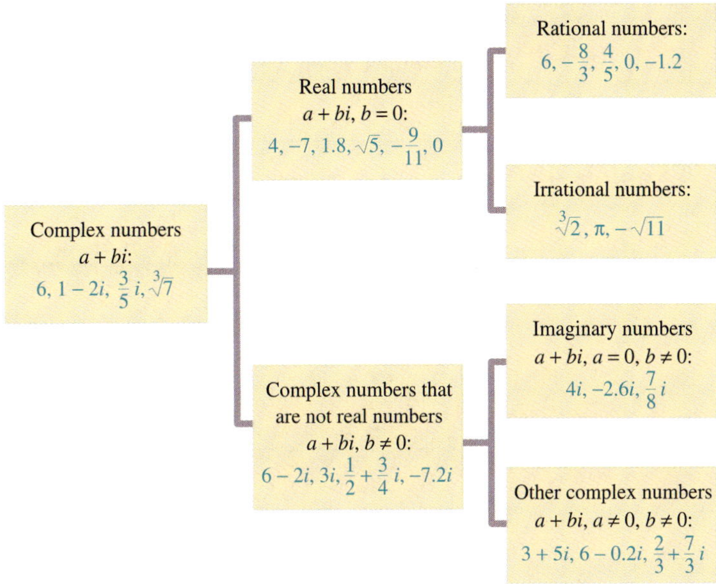


2 Two complex numbers $a + bi$ and $c + di$ are equal if and only if $a = c$ and $b = d$. Complex numbers can be added or subtracted by adding or subtracting their real parts and then adding or subtracting their imaginary parts.

### SUM OR DIFFERENCE OF COMPLEX NUMBERS

If $a + bi$ and $c + di$ are complex numbers, then their sum is

$$(a + bi) + (c + di) = (a + c) + (b + d)i$$

Their difference is

$$(a + bi) - (c + di) = a + bi - c - di = (a - c) + (b - d)i$$

**Example 3** Add or subtract the complex numbers. Write the sum or difference in the form $a + bi$.

**a.** $(2 + 3i) + (-3 + 2i)$ **b.** $(5i) - (1 - i)$ **c.** $(-3 - 7i) - (-6)$

*Solution* **a.** $(2 + 3i) + (-3 + 2i) = (2 - 3) + (3 + 2)i = -1 + 5i$

**b.** $5i - (1 - i) = 5i - 1 + i$

$= -1 + (5 + 1)i$

$= -1 + 6i$

**c.** $(-3 - 7i) - (-6) = -3 - 7i + 6$

$= (-3 + 6) - 7i$

$= 3 - 7i$

**3** To multiply two complex numbers of the form $a + bi$, we multiply as though they are binomials. Then we use the relationship $i^2 = -1$ to simplify.

**Example 4** Multiply the complex numbers. Write the product in the form $a + bi$.

**a.** $-7i \cdot 3i$ **b.** $3i(2 - i)$ **c.** $(2 - 5i)(4 + i)$

**d.** $(2 - i)^2$ **e.** $(7 + 3i)(7 - 3i)$

*Solution*

**a.**
$$\begin{aligned} -7i \cdot 3i &= -21i^2 \\ &= -21(-1) \quad \text{Replace } i^2 \text{ with } -1. \\ &= 21 \end{aligned}$$

**b.**
$$\begin{aligned} 3i(2 - i) &= 3i \cdot 2 - 3i \cdot i && \text{Use the distributive property.} \\ &= 6i - 3i^2 && \text{Multiply.} \\ &= 6i - 3(-1) && \text{Replace } i^2 \text{ with } -1. \\ &= 6i + 3 \\ &= 3 + 6i \end{aligned}$$

Use the FOIL method. (First, Outer, Inner, Last)

**c.**
$$\begin{aligned} (2 - 5i)(4 + i) &= \underset{F}{2(4)} + \underset{O}{2(i)} - \underset{I}{5i(4)} - \underset{L}{5i(i)} \\ &= 8 + 2i - 20i - 5i^2 \\ &= 8 - 18i - 5(-1) \quad i^2 = -1. \\ &= 8 - 18i + 5 \\ &= 13 - 18i \end{aligned}$$

**d.**
$$\begin{aligned} (2 - i)^2 &= (2 - i)(2 - i) \\ &= 2(2) - 2(i) - 2(i) + i^2 \\ &= 4 - 4i + (-1) \quad i^2 = -1. \\ &= 3 - 4i \end{aligned}$$

**e.**
$$\begin{aligned} (7 + 3i)(7 - 3i) &= 7(7) - 7(3i) + 3i(7) - 3i(3i) \\ &= 49 - 21i + 21i - 9i^2 \\ &= 49 - 9(-1) \quad i^2 = -1. \\ &= 49 + 9 \\ &= 58 \end{aligned}$$

Notice that if you add, subtract, or multiply two complex numbers, just like real numbers, the result is a complex number.

**4** From Example 4e, notice that the product of $7 + 3i$ and $7 - 3i$ is a real number. These two complex numbers are called **complex conjugates** of one another. In general, we have the following definition.

### COMPLEX CONJUGATES

The complex numbers $(a + bi)$ and $(a - bi)$ are called **complex conjugates** of each other, and $(a + bi)(a - bi) = a^2 + b^2$.

To see that the product of a complex number $a + bi$ and its conjugate $a - bi$ is the real number $a^2 + b^2$ we multiply.

$$\begin{aligned}(a + bi)(a - bi) &= a^2 - abi + abi - b^2i^2 \\ &= a^2 - b^2(-1) \\ &= a^2 + b^2\end{aligned}$$

We use complex conjugates to divide by a complex number.

**Example 5** Find each quotient. Write in the form $a + bi$.

**a.** $\dfrac{2 + i}{1 - i}$ **b.** $\dfrac{7}{3i}$

*Solution* **a.** Multiply the numerator and denominator by the complex conjugate of $1 - i$ to eliminate the imaginary number in the denominator.

$$\begin{aligned}\frac{2 + i}{1 - i} &= \frac{(2 + i)(1 + i)}{(1 - i)(1 + i)} \\ &= \frac{2(1) + 2(i) + 1(i) + i^2}{1^2 - i^2} \\ &= \frac{2 + 3i - 1}{1 + 1} \\ &= \frac{1 + 3i}{2} \quad \text{or} \quad \frac{1}{2} + \frac{3}{2}i\end{aligned}$$

**HELPFUL HINT**

Recall that division can be checked by multiplication.

To check that $\dfrac{2 + i}{1 - i} = \dfrac{1}{2} + \dfrac{3}{2}i$, in Example 5**a**, multiply $\left(\dfrac{1}{2} + \dfrac{3}{2}i\right)(1 - i)$ to verify that the product is $2 + i$.

**b.** Multiply the numerator and denominator by the conjugate of $3i$. Note that $3i = 0 + 3i$, so its conjugate is $0 - 3i$ or $-3i$.

$$\frac{7}{3i} = \frac{7(-3i)}{(3i)(-3i)} = \frac{-21i}{-9i^2} = \frac{-21i}{-9(-1)} = \frac{-21i}{9} = \frac{-7i}{3} \quad \text{or} \quad 0 - \frac{7}{3}i$$

**5** We can use the fact that $i^2 = -1$ to find higher powers of $i$. To find $i^3$, we rewrite it as the product of $i^2$ and $i$.

$$i^3 = i^2 \cdot i = (-1)i = -i$$
$$i^4 = i^2 \cdot i^2 = (-1) \cdot (-1) = 1$$

We continue this process and use the fact that $i^4 = 1$ and $i^2 = -1$ to simplify $i^5$ and $i^6$.

$$i^5 = i^4 \cdot i = 1 \cdot i = i$$
$$i^6 = i^4 \cdot i^2 = 1 \cdot (-1) = -1$$

If we continue finding powers of $i$, we generate the following pattern. Notice that the values $i$, $-1$, $-i$, and 1 repeat as $i$ is raised to higher and higher powers.

| | | |
|---|---|---|
| $i^1 = i$ | $i^5 = i$ | $i^9 = i$ |
| $i^2 = -1$ | $i^6 = -1$ | $i^{10} = -1$ |
| $i^3 = -i$ | $i^7 = -i$ | $i^{11} = -i$ |
| $i^4 = 1$ | $i^8 = 1$ | $i^{12} = 1$ |

This pattern allows us to find other powers of $i$. To do so, we will use the fact that $i^4 = 1$ and rewrite a power of $i$ in terms of $i^4$.
For example, $i^{22} = i^{20} \cdot i^2 = (i^4)^5 \cdot i^2 = 1^5 \cdot (-1) = 1 \cdot (-1) = -1$.

**Example 6** Find the following powers of $i$.

**a.** $i^7$ **b.** $i^{20}$ **c.** $i^{46}$ **d.** $i^{-12}$

*Solution*

**a.** $i^7 = i^4 \cdot i^3 = 1(-i) = -i$

**b.** $i^{20} = (i^4)^5 = 1^5 = 1$

**c.** $i^{46} = i^{44} \cdot i^2 = (i^4)^{11} \cdot i^2 = 1^{11}(-1) = -1$

**d.** $i^{-12} = \dfrac{1}{i^{12}} = \dfrac{1}{(i^4)^3} = \dfrac{1}{(1)^3} = \dfrac{1}{1} = 1$

## MENTAL MATH

*Simplify. See Example 1.*

1. $\sqrt{-81}$
2. $\sqrt{-49}$
3. $\sqrt{-7}$
4. $\sqrt{-3}$
5. $-\sqrt{16}$
6. $-\sqrt{4}$
7. $\sqrt{-64}$
8. $\sqrt{-100}$

## Exercise Set 7.7

*Write in terms of i. See Example 1.*

1. $\sqrt{-24}$
2. $\sqrt{-32}$
3. $-\sqrt{-36}$
4. $-\sqrt{-121}$
5. $8\sqrt{-63}$
6. $4\sqrt{-20}$
7. $-\sqrt{54}$
8. $\sqrt{-63}$

*Multiply or divide. See Example 2.*

9. $\sqrt{-2} \cdot \sqrt{-7}$
10. $\sqrt{-11} \cdot \sqrt{-3}$
11. $\sqrt{-5} \cdot \sqrt{-10}$
12. $\sqrt{-2} \cdot \sqrt{-6}$
13. $\sqrt{16} \cdot \sqrt{-1}$
14. $\sqrt{3} \cdot \sqrt{-27}$
15. $\dfrac{\sqrt{-9}}{\sqrt{3}}$
16. $\dfrac{\sqrt{49}}{\sqrt{-10}}$

17. $\frac{\sqrt{-80}}{\sqrt{-10}}$

18. $\frac{\sqrt{-40}}{\sqrt{-8}}$

*Add or subtract. Write the sum or difference in the form $a + bi$. See Example 3.*

19. $(4 - 7i) + (2 + 3i)$

20. $(2 - 4i) - (2 - i)$

21. $(6 + 5i) - (8 - i)$

22. $(8 - 3i) + (-8 + 3i)$

23. $6 - (8 + 4i)$

24. $(9 - 4i) - 9$

*Multiply. Write the product in the form $a + bi$. See Example 4.*

25. $6i(2 - 3i)$

26. $5i(4 - 7i)$

27. $(\sqrt{3} + 2i)(\sqrt{3} - 2i)$

28. $(\sqrt{5} - 5i)(\sqrt{5} + 5i)$

29. $(4 - 2i)^2$

30. $(6 - 3i)^2$

*Write each quotient in the form $a + bi$. See Example 5.*

31. $\frac{4}{i}$

32. $\frac{5}{6i}$

33. $\frac{7}{4 + 3i}$

34. $\frac{9}{1 - 2i}$

35. $\frac{3 + 5i}{1 + i}$

36. $\frac{6 + 2i}{4 - 3i}$

37. $\frac{5 - i}{3 - 2i}$

38. $\frac{6 - i}{2 + i}$

*Perform the indicated operation. Write the result in the form $a + bi$.*

39. $(7i)(-9i)$

40. $(-6i)(-4i)$

41. $(6 - 3i) - (4 - 2i)$

42. $(-2 - 4i) - (6 - 8i)$

43. $(6 - 2i)(3 + i)$

44. $(2 - 4i)(2 - i)$

45. $(8 - 3i) + (2 + 3i)$

46. $(7 + 4i) + (4 - 4i)$

47. $(1 - i)(1 + i)$

48. $(6 + 2i)(6 - 2i)$

49. $\frac{16 + 15i}{-3i}$

50. $\frac{2 - 3i}{-7i}$

51. $(9 + 8i)^2$

52. $(4 - 7i)^2$

53. $\frac{2}{3 + i}$

54. $\frac{5}{3 - 2i}$

55. $(5 - 6i) - 4i$

56. $(6 - 2i) + 7i$

57. $\frac{2 - 3i}{2 + i}$

58. $\frac{6 + 5i}{6 - 5i}$

59. $(2 + 4i) + (6 - 5i)$

60. $(5 - 3i) + (7 - 8i)$

*Find each power of i. See Example 6.*

61. $i^8$

62. $i^{10}$

63. $i^{21}$

64. $i^{15}$

65. $i^{11}$

66. $i^{40}$

67. $i^{-6}$

68. $i^{-9}$

*Write in the form $a + bi$.*

69. $i^3 + i^4$

70. $i^8 - i^7$

71. $i^6 + i^8$

72. $i^4 + i^{12}$

73. $2 + \sqrt{-9}$

74. $5 - \sqrt{-16}$

75. $\frac{6 + \sqrt{-18}}{3}$

76. $\frac{4 - \sqrt{-8}}{2}$

77. $\frac{5 - \sqrt{-75}}{10}$

78. Describe how to find the conjugate of a complex number.

79. Explain why the product of a complex number and its complex conjugate is a real number.

*Simplify.*

80. $(8 - \sqrt{-3}) - (2 + \sqrt{-12})$

81. $(8 - \sqrt{-4}) - (2 + \sqrt{-16})$

82. Determine whether $2i$ is a solution of $x^2 + 4 = 0$.

83. Determine whether $-1 + i$ is a solution of $x^2 + 2x = -2$.

## REVIEW EXERCISES

*Recall that the sum of the measures of the angles of a triangle is 180°. Find the unknown angle in each triangle. See Section 4.3.*

84.

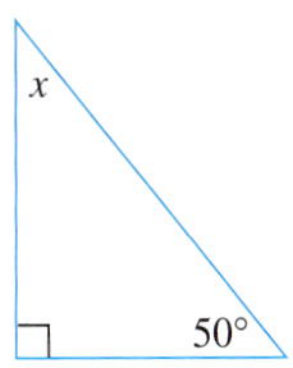


85.

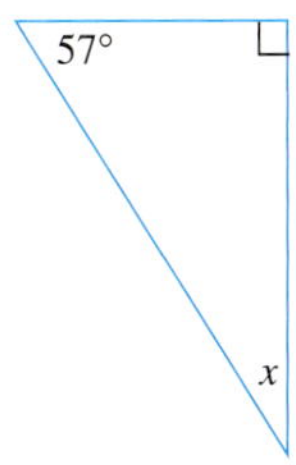


*Use synthetic division to divide the following. See Section 6.5.*

86. $(x^3 - 6x^2 + 3x - 4) \div (x - 1)$

87. $(5x^4 - 3x^2 + 2) \div (x + 2)$

*Thirty people were recently polled about their average monthly balance in their checking accounts. The results of this poll are shown in the following histogram. Use this graph to answer Exercises 88 through 93. See Section 1.2.*

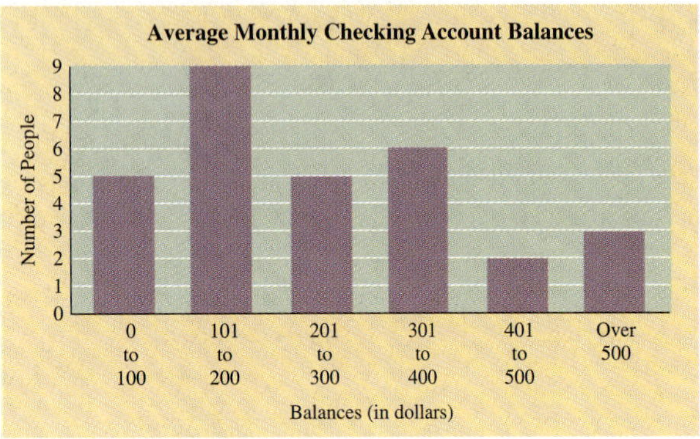


**88.** How many people polled reported an average checking balance of \$201 to \$300?

**89.** How many people polled reported an average checking balance of \$0 to \$100?

**90.** How many people polled reported an average checking balance of \$200 or less?

**91.** How many people polled reported an average checking balance of \$301 or more?

**92.** What percent of people polled reported an average checking balance of \$201 to \$300?

**93.** What percent of people polled reported an average checking balance of \$0 to \$100?

7

For additional Chapter Projects, visit the Real World Activities Website by going to http://www.prenhall.com/martin-gay.

## CHAPTER PROJECT

### *Calculating the Length and Period of a Pendulum*

A simple pendulum of a given length, like the kind found in a clock, has a unique property. The time required to complete one full back-and-forth swing (called the **period**) is the same regardless of the mass of the pendulum or the distance it travels. The time to complete one full swing *does*, however, depend on the pendulum's length. In this project, you will have the opportunity to investigate the relationship between the length of a pendulum and its period. You will need at least 1 meter of string, a weight of some sort, a meter stick, a stopwatch, and a calculator. This project may be completed by working in groups or individually.

Make a simple pendulum by securely tying the string to the weight.

The formula relating a pendulum's period $T$ (in seconds) to its length $l$ (in centimeters) is

$$T = 2\pi\sqrt{\frac{l}{980}}$$

The period of a pendulum is defined as the time it takes the pendulum to complete one full back-and-forth swing. In this project, you will be measuring your simple pendulum's period with a stopwatch. Because the periods will be only a few seconds long, it will be more accurate for you to time a total of 5 complete swings and then find the average time of one complete swing.

1. For each of the pendulum (string) lengths $l$ given in Table 1, measure the time required for 5 complete swings and record it in the appropriate column. Next, divide this value by 5 to find the measured period of the pendulum for the given length and record it in the Measured Period $T_m$ column in the table. Use the given formula to calculate the theoretical period $T$ for the same pendulum length and record it in the appropriate column. (Round to two decimal places.) Find and record in the last column the difference between the measured period and the theoretical period.
2. For each of the periods $T$ given in Table 2, use the given formula to calculate the theoretical pendulum length $l$ required to yield the given period. Record $l$ in the appropriate column; round to one decimal place. Next, using this length $l$, measure and record the time for 5 complete swings. Divide this time by 5 to find the measured period $T_m$, and record it. Then find and record in the last column the difference between the theoretical period and the measured period.
3. Use the general trends you find in the tables to describe the relationship between a pendulum's period and its length.

4. Discuss the differences you found between the values of the theoretical period and the measured period. What factors contributed to these differences?

TABLE 1

| *Length l* (*in centimeters*) | *Time for 5 Swings* (*in seconds*) | *Measured Period* $T_m$ (*in seconds*) | *Theoretical Period T* (*in seconds*) | *Difference* $\|T - T_m\|$ |
|---|---|---|---|---|
| 30 | | | | |
| 55 | | | | |
| 70 | | | | |

TABLE 2

| *Period I* (*in seconds*) | *Theoretical Length l* (*in centimeters*) | *Time for 5 swings* (*in seconds*) | *Measured Periods* $T_m$ | *Difference* $\|T - T_m\|$ |
|---|---|---|---|---|
| 1 | | | | |
| 1.25 | | | | |
| 2 | | | | |

## CHAPTER 7 VOCABULARY CHECK

*Fill in each blank with one of the words or phrases listed below.*

| | | | | |
|---|---|---|---|---|
| index | rationalizing | conjugate | principal square root | cube root |
| complex number | like radicals | radicand | imaginary unit | |

1. The ________ of $\sqrt{3} + 2$ is $\sqrt{3} - 2$.
2. The ________________ of a nonnegative number $a$ is written as $\sqrt{a}$.
3. The process of writing a radical expression as an equivalent expression but without a radical in the denominator is called __________ the denominator.
4. The ____________, written $i$, is the number whose square is $-1$.
5. The ________ of a number is written as $\sqrt[3]{a}$.
6. In the notation $\sqrt[n]{a}$, $n$ is called the ______ and $a$ is called the _________.
7. Radicals with the same index and the same radicand are called __________.
8. A ____________ is a number that can be written in the form $a + bi$ where $a$ and $b$ are real numbers.

# CHAPTER 7 HIGHLIGHTS

| DEFINITIONS AND CONCEPTS | EXAMPLES |
|---|---|
| **Section 7.1 Radicals and Radical Functions** | |
| The **positive**, or **principal**, **square root** of a nonnegative number $a$ is written as $\sqrt{a}$. $\sqrt{a} = b$ only if $b^2 = a$ and $b \geq 0$ | $\sqrt{36} = 6$ $\quad \sqrt{\frac{9}{100}} = \frac{3}{10}$ <br> $-\sqrt{36} = -6$ $\quad \sqrt{0.04} = 0.2$ |
| The **negative square root** of $a$ is written as $-\sqrt{a}$. | |
| The **cube root** of a real number $a$ is written as $\sqrt[3]{a}$. $\sqrt[3]{a} = b$ only if $b^3 = a$ | $\sqrt[3]{27} = 3$ $\quad \sqrt[3]{-\frac{1}{8}} = -\frac{1}{2}$ <br> $\sqrt[3]{y^6} = y^2$ $\quad \sqrt[3]{64x^9} = 4x^3$ |
| If $n$ is an even positive integer, then $\sqrt[n]{a^n} = \lvert a \rvert$. | $\sqrt{(-3)^2} = \lvert -3 \rvert = 3$ |
| If $n$ is an odd positive integer, then $\sqrt[n]{a^n} = a$. | $\sqrt[3]{(-7)^3} = -7$ |
| A **radical function** in $x$ is a function defined by an expression containing a root of $x$. | If $f(x) = \sqrt{x} + 2$, <br> $f(1) = \sqrt{1} + 2 = 1 + 2 = 3$ <br> $f(3) = \sqrt{3} + 2 \approx 3.73$ |
| **Section 7.2 Rational Exponents** | |
| $a^{1/n} = \sqrt[n]{a}$ if $\sqrt[n]{a}$ is a real number. | $81^{1/2} = \sqrt{81} = 9$ <br> $(-8x^3)^{1/3} = \sqrt[3]{-8x^3} = -2x$ |
| If $m$ and $n$ are positive integers greater than 1 with $\frac{m}{n}$ in lowest terms and $\sqrt[n]{a}$ is a real number, then $a^{m/n} = (a^{1/n})^m = (\sqrt[n]{a})^m$ | $4^{5/2} = (\sqrt{4})^5 = 2^5 = 32$ <br> $27^{2/3} = (\sqrt[3]{27})^2 = 3^2 = 9$ |
| $a^{-m/n} = \frac{1}{a^{m/n}}$ as long as $a^{m/n}$ is a nonzero number. | $16^{-3/4} = \frac{1}{16^{3/4}} = \frac{1}{(\sqrt[4]{16})^3} = \frac{1}{2^3} = \frac{1}{8}$ |
| **Exponent rules** are true for rational exponents. | $x^{2/3} \cdot x^{-5/6} = x^{2/3-5/6} = x^{-1/6} = \frac{1}{x^{1/6}}$ <br> $(8^{14})^{1/7} = 8^2 = 64$ <br> $\frac{a^{4/5}}{a^{-2/5}} = a^{4/5-(-2/5)} = a^{6/5}$ |
| **Section 7.3 Simplifying Radical Expressions** | |
| ***Product and Quotient Rules*** <br> If $\sqrt[n]{a}$ and $\sqrt[n]{b}$ are real numbers, <br> $\sqrt[n]{a} \cdot \sqrt[n]{b} = \sqrt[n]{a \cdot b}$ <br> $\frac{\sqrt[n]{a}}{\sqrt[n]{b}} = \sqrt[n]{\frac{a}{b}}$, provided $\sqrt[n]{b} \neq 0$ | Multiply or divide as indicated: <br> $\sqrt{11} \cdot \sqrt{3} = \sqrt{33}$ <br> $\frac{\sqrt[3]{40x}}{\sqrt[3]{5x}} = \sqrt[3]{8} = 2$ |
| A radical of the form $\sqrt[n]{a}$ is **simplified** when $a$ contains no factors that are perfect $n$th powers. | $\sqrt{40} = \sqrt{4 \cdot 10} = 2\sqrt{10}$ <br> $\sqrt{36x^5} = \sqrt{36x^4 \cdot x} = 6x^2\sqrt{x}$ <br> $\sqrt[3]{24x^7y^3} = \sqrt[3]{8x^6y^3 \cdot 3x} = 2x^2y\sqrt[3]{3x}$ <br> $\sqrt{36x^4 \cdot x} = 6x^2\sqrt{x}$ |

| DEFINITIONS AND CONCEPTS | EXAMPLES |
|---|---|
| **Section 7.4 Adding, Subtracting, and Multiplying Radical Expressions** | |
| Radicals with the same index and the same radicand are **like radicals**. | $5\sqrt{6} + 2\sqrt{6} = (5 + 2)\sqrt{6} = 7\sqrt{6}$ |
| The distributive property can be used to add like radicals. | $-\sqrt[3]{3x} - 10\sqrt[3]{3x} + 3\sqrt[3]{10x}$ <br> $= (-1 - 10)\sqrt[3]{3x} + 3\sqrt[3]{10x}$ <br> $= -11\sqrt[3]{3x} + 3\sqrt[3]{10x}$ |
| Radical expressions are multiplied by using many of the same properties used to multiply polynomials. | Multiply. <br> $(\sqrt{5} - \sqrt{2x})(\sqrt{2} + \sqrt{2x})$ <br> $= \sqrt{10} + \sqrt{10x} - \sqrt{4x} - 2x$ <br> $= \sqrt{10} + \sqrt{10x} - 2\sqrt{x} - 2x$ <br> $(2\sqrt{3} - \sqrt{8x})(2\sqrt{3} + \sqrt{8x})$ <br> $= 4(3) - 8x = 12 - 8x$ |
| **Section 7.5 Rationalizing Denominators and Numerators of Radical Expressions** | |
| The **conjugate** of $a + b$ is $a - b$. | The conjugate of $\sqrt{7} + \sqrt{3}$ is $\sqrt{7} - \sqrt{3}$ |
| The process of writing the denominator of a radical expression without a radical is called **rationalizing the denominator.** | Rationalize each denominator. <br> $\frac{\sqrt{5}}{\sqrt{3}} = \frac{\sqrt{5} \cdot \sqrt{3}}{\sqrt{3} \cdot \sqrt{3}} = \frac{\sqrt{15}}{3}$ <br> $\frac{6}{\sqrt{7} + \sqrt{3}} = \frac{6(\sqrt{7} - \sqrt{3})}{(\sqrt{7} + \sqrt{3})(\sqrt{7} - \sqrt{3})}$ <br> $= \frac{6(\sqrt{7} - \sqrt{3})}{7 - 3}$ <br> $= \frac{6(\sqrt{7} - \sqrt{3})}{4} = \frac{3(\sqrt{7} - \sqrt{3})}{2}$ |
| The process of writing the numerator of a radical expression without a radical is called **rationalizing the numerator.** | Rationalize each numerator. <br> $\frac{\sqrt[3]{9}}{\sqrt[3]{5}} = \frac{\sqrt[3]{9} \cdot \sqrt[3]{3}}{\sqrt[3]{5} \cdot \sqrt[3]{3}} = \frac{\sqrt[3]{27}}{\sqrt[3]{15}} = \frac{3}{\sqrt[3]{15}}$ <br> $\frac{\sqrt{9} + \sqrt{3x}}{12} = \frac{(\sqrt{9} + \sqrt{3x})(\sqrt{9} - \sqrt{3x})}{12(\sqrt{9} - \sqrt{3x})}$ <br> $= \frac{9 - 3x}{12(\sqrt{9} - \sqrt{3x})}$ <br> $= \frac{3(3 - x)}{3 \cdot 4(3 - \sqrt{3x})} = \frac{3 - x}{4(3 - \sqrt{3x})}$ |
| **Section 7.6 Radical Equations and Problem Solving** | |
| ***To Solve a Radical Equation*** | Solve $x = \sqrt{4x + 9} + 3$. |
| **Step 1:** Write the equation so that one radical is by itself on one side of the equation. | **1.** $x - 3 = \sqrt{4x + 9}$ |
| **Step 2:** Raise each side of the equation to a power equal to the index of the radical and simplify. | **2.** $(x - 3)^2 = (\sqrt{4x + 9})^2$ <br> $x^2 - 6x + 9 = 4x + 9$ |
| **Step 3:** If the equation still contains a radical, repeat Steps 1 and 2. If not, solve the equation. | **3.** $x^2 - 10x = 0$ <br> $x(x - 10) = 0$ <br> $x = 0$ or $x = 10$ |
| **Step 4:** Check all proposed solutions in the original equation. | **4.** The proposed solution 10 checks, but 0 does not. The solution is 10. |

| DEFINITIONS AND CONCEPTS | EXAMPLES |
| --- | --- |
| **Section 7.7 Complex Numbers** | |
| $i^2 = -1$ and $i = \sqrt{-1}$ | Simplify $\sqrt{-9}$.<br>$\sqrt{-9} = \sqrt{-1 \cdot 9} = \sqrt{-1} \cdot \sqrt{9} = i \cdot 3$ or $3i$ |
| A **complex number** is a number that can be written in the form $a + bi$, where $a$ and $b$ are real numbers. | ***Complex Numbers*** — ***Written in form a + bi***<br>$12$ — $12 + 0i$<br>$-5i$ — $0 + (-5)i$<br>$-2 - 3i$ — $-2 + (-3)i$<br>Multiply.<br>$\sqrt{-3} \cdot \sqrt{-7} = i\sqrt{3} \cdot i\sqrt{7}$<br>$= i^2\sqrt{21}$<br>$= -\sqrt{21}$ |
| **To add or subtract complex numbers,** add or subtract their real parts and then add or subtract their imaginary parts. | Perform each indicated operation.<br>$(-3 + 2i) - (7 - 4i) = -3 + 2i - 7 + 4i$<br>$= -10 + 6i$ |
| **To multiply complex numbers,** multiply as though they are binomials. | $(-7 - 2i)(6 + i) = -42 - 7i - 12i - 2i^2$<br>$= -42 - 19i - 2(-1)$<br>$= -42 - 19i + 2$<br>$= -40 - 19i$ |
| The complex numbers $(a + bi)$ and $(a - bi)$ are called **complex conjugates.** | The complex conjugate of<br>$(3 + 6i)$ is $(3 - 6i)$.<br>Their product is a real number.<br>$(3 - 6i)(3 + 6i) = 9 - 36i^2$<br>$= 9 - 36(-1) = 9 + 36 = 45$ |
| **To divide complex numbers,** multiply the numerator and the denominator by the conjugate of the denominator. | Divide.<br>$\frac{4}{2 - i} = \frac{4(2 + i)}{(2 - i)(2 + i)}$<br>$= \frac{4(2 + i)}{4 - i^2}$<br>$= \frac{4(2 + i)}{5}$<br>$= \frac{8 + 4i}{5} = \frac{8}{5} + \frac{4}{5}i$ |

## CHAPTER 7 REVIEW

**(7.1)** *Find the root. Assume that all variables represent positive numbers.*

**1.** $\sqrt{81}$

**2.** $\sqrt[4]{81}$

**3.** $\sqrt[3]{-8}$

**4.** $\sqrt[4]{-16}$

**5.** $-\sqrt{\frac{1}{49}}$

**6.** $\sqrt{x^{64}}$

**7.** $-\sqrt{36}$

**8.** $\sqrt[3]{64}$

**9.** $\sqrt[3]{-a^6b^9}$

**10.** $\sqrt{16a^4b^{12}}$

**11.** $\sqrt[5]{32a^5b^{10}}$

**12.** $\sqrt[5]{-32x^{15}y^{20}}$

**13.** $\sqrt{\frac{x^{12}}{36y^2}}$

**14.** $\sqrt[3]{\frac{27y^3}{z^{12}}}$

*Simplify. Use absolute value bars when necessary.*

**15.** $\sqrt{(-x)^2}$

**16.** $\sqrt[4]{(x^2-4)^4}$

**17.** $\sqrt[3]{(-27)^3}$

**18.** $\sqrt[5]{(-5)^5}$

**19.** $-\sqrt[5]{x^5}$

**20.** $\sqrt[4]{16(2y+z)^{12}}$

**21.** $\sqrt{25(x-y)^{10}}$

**22.** $\sqrt[5]{-y^5}$

**23.** $\sqrt[9]{-x^9}$

*Identify the domain and then graph each function.*

**24.** $f(x) = \sqrt{x} + 3$

**25.** $g(x) = \sqrt[3]{x-3}$; use the accompanying table.

| $x$ | $-5$ | 2 | 3 | 4 | 11 |
|---|---|---|---|---|---|
| $g(x)$ | | | | | |

**(7.2)** *Evaluate the following.*

**26.** $\left(\frac{1}{81}\right)^{1/4}$

**27.** $\left(-\frac{1}{27}\right)^{1/3}$

**28.** $(-27)^{-1/3}$

**29.** $(-64)^{-1/3}$

**30.** $-9^{3/2}$

**31.** $64^{-1/3}$

**32.** $(-25)^{5/2}$

**33.** $\left(\frac{25}{49}\right)^{-3/2}$

**34.** $\left(\frac{8}{27}\right)^{-2/3}$

**35.** $\left(-\frac{1}{36}\right)^{-1/4}$

*Write with rational exponents.*

**36.** $\sqrt[3]{x^2}$

**37.** $\sqrt[5]{5x^2y^3}$

*Write with radical notation.*

**38.** $y^{4/5}$

**39.** $5(xy^2z^5)^{1/3}$

**40.** $(x+2y)^{-1/2}$

*Simplify each expression. Assume that all variables represent positive numbers. Write with only positive exponents.*

**41.** $a^{1/3}a^{4/3}a^{1/2}$

**42.** $\frac{b^{1/3}}{b^{4/3}}$

**43.** $(a^{1/2}a^{-2})^3$

**44.** $(x^{-3}y^6)^{1/3}$

**45.** $\left(\frac{b^{3/4}}{a^{-1/2}}\right)^8$

**46.** $\frac{x^{1/4}x^{-1/2}}{x^{2/3}}$

**47.** $\left(\frac{49c^{5/3}}{a^{-1/4}b^{5/6}}\right)^{-1}$

**48.** $a^{-1/4}(a^{5/4} - a^{9/4})$

*Use a calculator and write a three-decimal-place approximation.*

**49.** $\sqrt{20}$

**50.** $\sqrt[3]{-39}$

**51.** $\sqrt[4]{726}$

**52.** $56^{1/3}$

**53.** $-78^{3/4}$

**54.** $105^{-2/3}$

*Use rational exponents to write each radical with the same index. Then multiply.*

**55.** $\sqrt[3]{2} \cdot \sqrt{7}$

**56.** $\sqrt[3]{3} \cdot \sqrt[4]{x}$

**(7.3)** *Perform the indicated operations and then simplify if possible. For the remainder of this review, assume that variables represent positive numbers only.*

**57.** $\sqrt{3} \cdot \sqrt{8}$

**58.** $\sqrt[3]{7y} \cdot \sqrt[3]{x^2z}$

**59.** $\frac{\sqrt{44x^3}}{\sqrt{11x}}$

**60.** $\frac{\sqrt[4]{a^6b^{13}}}{\sqrt[4]{a^2b}}$

*Simplify.*

**61.** $\sqrt{60}$

**62.** $-\sqrt{75}$

**63.** $\sqrt[3]{162}$

**64.** $\sqrt[3]{-32}$

**65.** $\sqrt{36x^7}$

**66.** $\sqrt[3]{24a^5b^7}$

**67.** $\sqrt{\frac{p^{17}}{121}}$

**68.** $\sqrt[3]{\frac{y^5}{27x^6}}$

**69.** $\sqrt[4]{\frac{xy^6}{81}}$

**70.** $\sqrt{\frac{2x^3}{49y^4}}$

△ **71.** The formula for the radius $r$ of a circle of area $A$ is

$$r = \sqrt{\frac{A}{\pi}}$$

**a.** Find the exact radius of a circle whose area is 25 square meters.

**b.** Approximate to two decimal places the radius of a circle whose area is 104 square inches.

**(7.4)** *Perform the indicated operation.*

**72.** $x\sqrt{75xy} - \sqrt{27x^3y}$

**73.** $2\sqrt{32x^2y^3} - xy\sqrt{98y}$

**74.** $\sqrt[3]{128} + \sqrt[3]{250}$

**75.** $3\sqrt[4]{32a^5} - a\sqrt[4]{162a}$

**76.** $\dfrac{5}{\sqrt{4}} + \dfrac{\sqrt{3}}{3}$

**77.** $\sqrt{\dfrac{8}{x^2}} - \sqrt{\dfrac{50}{16x^2}}$

**78.** $2\sqrt{50} - 3\sqrt{125} + \sqrt{98}$

**79.** $2a\sqrt[4]{32b^5} - 3b\sqrt[4]{162a^4b} + \sqrt[4]{2a^4b^5}$

*Multiply and then simplify if possible.*

**80.** $\sqrt{3}(\sqrt{27} - \sqrt{3})$

**81.** $(\sqrt{x} - 3)^2$

**82.** $(\sqrt{5} - 5)(2\sqrt{5} + 2)$

**83.** $(2\sqrt{x} - 3\sqrt{y})(2\sqrt{x} + 3\sqrt{y})$

**84.** $(\sqrt{a} + 3)(\sqrt{a} - 3)$

**85.** $(\sqrt[3]{a} + 2)^2$

**86.** $(\sqrt[3]{5x} + 9)(\sqrt[3]{5x} - 9)$

**87.** $(\sqrt[3]{a} + 4)(\sqrt[3]{a^2} - 4\sqrt[3]{a} + 16)$

***(7.5)*** *Rationalize each denominator.*

**88.** $\dfrac{3}{\sqrt{7}}$

**89.** $\sqrt{\dfrac{x}{12}}$

**90.** $\dfrac{5}{\sqrt[3]{4}}$

**91.** $\sqrt{\dfrac{24x^5}{3y^2}}$

**92.** $\sqrt[3]{\dfrac{15x^6y^7}{z^2}}$

**93.** $\dfrac{5}{2 - \sqrt{7}}$

**94.** $\dfrac{3}{\sqrt{y} - 2}$

**95.** $\dfrac{\sqrt{2} - \sqrt{3}}{\sqrt{2} + \sqrt{3}}$

*Rationalize each numerator.*

**96.** $\dfrac{\sqrt{11}}{3}$

**97.** $\sqrt{\dfrac{18}{y}}$

**98.** $\dfrac{\sqrt[3]{9}}{7}$

**99.** $\sqrt{\dfrac{24x^5}{3y^2}}$

**100.** $\sqrt[3]{\dfrac{xy^2}{10z}}$

**101.** $\dfrac{\sqrt{x} + 5}{-3}$

***(7.6)*** *Solve each equation for the variable.*

**102.** $\sqrt{y - 7} = 5$

**103.** $\sqrt{2x} + 10 = 4$

**104.** $\sqrt[3]{2x - 6} = 4$

**105.** $\sqrt{x + 6} = \sqrt{x + 2}$

**106.** $2x - 5\sqrt{x} = 3$

**107.** $\sqrt{x + 9} = 2 + \sqrt{x - 7}$

*Find each unknown length.*

△ **108.**

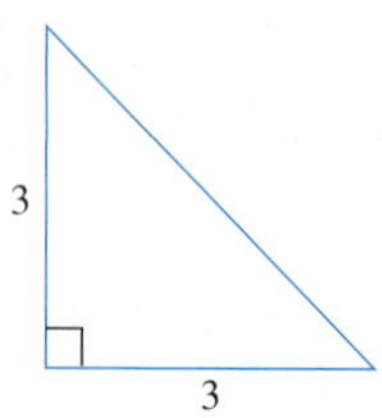


△ **109.**

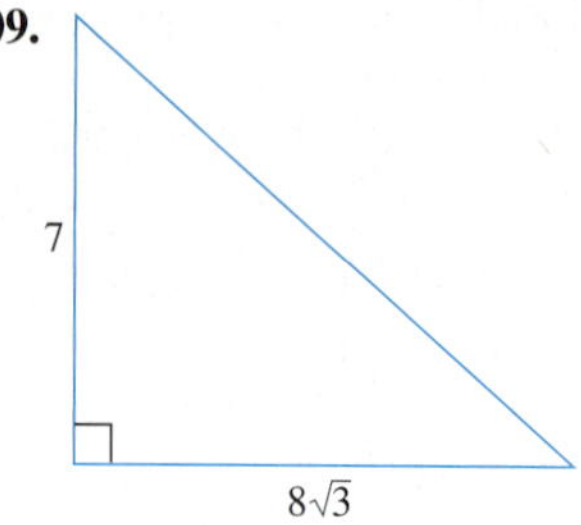


△ **110.** Beverly Hillis wants to determine the distance $x$ across a pond on her property. She is able to measure the distances shown on the following diagram. Find how wide the lake is at the crossing point, indicated by the triangle, to the nearest tenth of a foot.

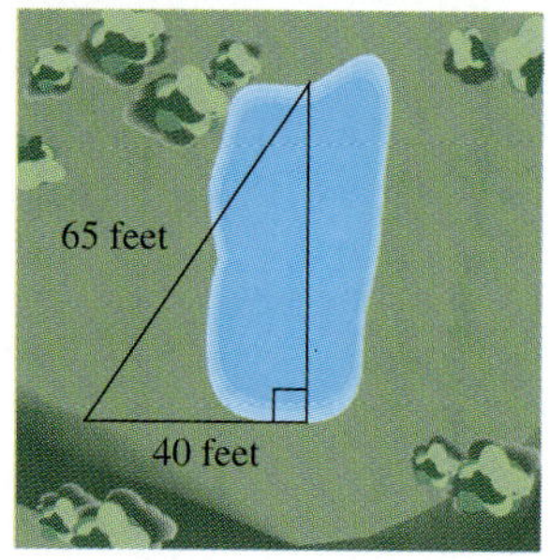


△ **111.** A pipe fitter needs to connect two underground pipelines that are offset by 3 feet, as pictured in the diagram. Neglecting the joints needed to join the pipes, find the length of the shortest possible connecting pipe rounded to the nearest hundredth of a foot.

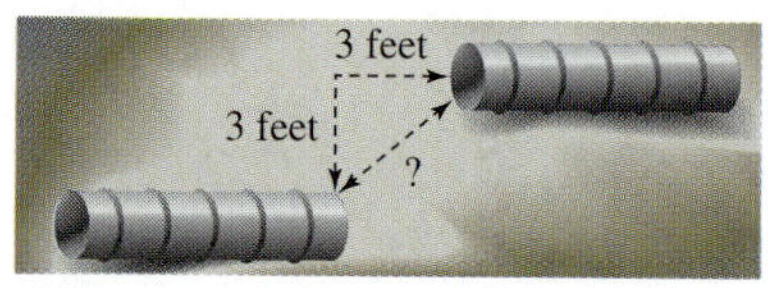


***(7.7)*** *Perform the indicated operation and simplify. Write the result in the form $a + bi$.*

**112.** $\sqrt{-8}$

**113.** $-\sqrt{-6}$

**114.** $\sqrt{-4} + \sqrt{-16}$

**115.** $\sqrt{-2} \cdot \sqrt{-5}$

**116.** $(12 - 6i) + (3 + 2i)$

**117.** $(-8 - 7i) - (5 - 4i)$

**118.** $(\sqrt{3} + \sqrt{2}) + (3\sqrt{2} - \sqrt{-8})$

**119.** $2i(2 - 5i)$

**120.** $-3i(6 - 4i)$

**121.** $(3 + 2i)(1 + i)$

**122.** $(2 - 3i)^2$

**123.** $(\sqrt{6} - 9i)(\sqrt{6} + 9i)$

**124.** $\dfrac{2 + 3i}{2i}$

**125.** $\dfrac{1 + i}{-3i}$

## CHAPTER 7 TEST

*Raise to the power or find the root. Assume that all variables represent positive numbers. Write with only positive exponents.*

**1.** $\sqrt{216}$

**2.** $-\sqrt[4]{x^{64}}$

**3.** $\left(\frac{1}{125}\right)^{1/3}$

**4.** $\left(\frac{1}{125}\right)^{-1/3}$

**5.** $\left(\frac{8x^3}{27}\right)^{2/3}$

**6.** $\sqrt[3]{-a^{18}b^9}$

**7.** $\left(\frac{64c^{4/3}}{a^{-2/3}b^{5/6}}\right)^{1/2}$

**8.** $a^{-2/3}(a^{5/4} - a^3)$

*Find the root. Use absolute value bars when necessary.*

**9.** $\sqrt[4]{(4xy)^4}$

**10.** $\sqrt[3]{(-27)^3}$

*Rationalize the denominator. Assume that all variables represent positive numbers.*

**11.** $\sqrt{\frac{9}{y}}$

**12.** $\frac{4 - \sqrt{x}}{4 + 2\sqrt{x}}$

**13.** $\frac{\sqrt[3]{ab}}{\sqrt[3]{ab^2}}$

**14.** Rationalize the numerator of $\frac{\sqrt{6} + x}{8}$ and simplify.

*Perform the indicated operations. Assume that all variables represent positive numbers.*

**15.** $\sqrt{125x^3} - 3\sqrt{20x^3}$

**16.** $\sqrt{3}(\sqrt{16} - \sqrt{2})$

**17.** $(\sqrt{x} + 1)^2$

**18.** $(\sqrt{2} - 4)(\sqrt{3} + 1)$

**19.** $(\sqrt{5} + 5)(\sqrt{5} - 5)$

*Use a calculator to approximate each to three decimal places.*

**20.** $\sqrt{561}$

**21.** $386^{-2/3}$

*Solve.*

**22.** $x = \sqrt{x - 2} + 2$

**23.** $\sqrt{x^2 - 7} + 3 = 0$

**24.** $\sqrt{x + 5} = \sqrt{2x - 1}$

*Perform the indicated operation and simplify. Write the result in the form $a + bi$.*

**25.** $\sqrt{-2}$

**26.** $-\sqrt{-8}$

**27.** $(12 - 6i) - (12 - 3i)$

**28.** $(6 - 2i)(6 + 2i)$

**29.** $(4 + 3i)^2$

**30.** $\frac{1 + 4i}{1 - i}$

△ **31.** Find $x$.

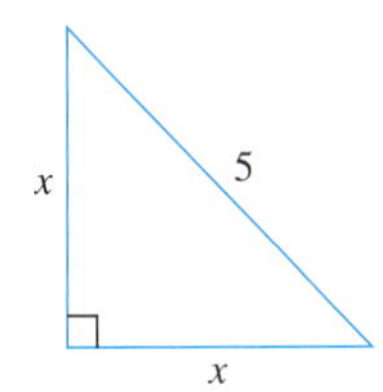


**32.** Identify the domain of $g(x)$. Then complete the accompanying table and graph $g(x)$.

$$g(x) = \sqrt{x + 2}$$

| $x$ | $-2$ | $-1$ | $2$ | $7$ |
|---|---|---|---|---|
| $g(x)$ | | | | |

*Solve.*

**33.** The function $V(r) = \sqrt{2.5r}$ can be used to estimate the maximum safe velocity $V$ in miles per hour at which a car can travel if it is driven along a curved road with a *radius of curvature r* in feet. To the nearest whole number, find the maximum safe speed if a cloverleaf exit on an expressway has a radius of curvature of 300 feet.

**34.** Use the formula from Exercise 33 to find the radius of curvature if the safe velocity is 30 mph.

## CHAPTER 7 CUMULATIVE REVIEW

**1.** Simplify each expression.

**a.** $3xy - 2xy + 5 - 7 + xy$

**b.** $7x^2 + 3 - 5(x^2 - 4)$

**c.** $(2.1x - 5.6) - (-x - 5.3)$

**2.** Solve for $x$: $\dfrac{x+5}{2} + \dfrac{1}{2} = 2x - \dfrac{x-3}{8}$

**3.** A salesperson earns \$600 per month plus a commission of 20% of sales. Find the minimum amount of sales needed to receive a total income of at least \$1500 per month.

**4.** Solve: $2|x| + 25 = 23$

**5.** Solve: $\left|\dfrac{x}{3} - 1\right| - 7 \geq -5$

**6.** Graph the equation $y = |x|$.

**7.** Determine the domain and range of each relation.

**a.** $\{(2, 3), (2, 4), (0, -1), (3, -1)\}$

**b.**

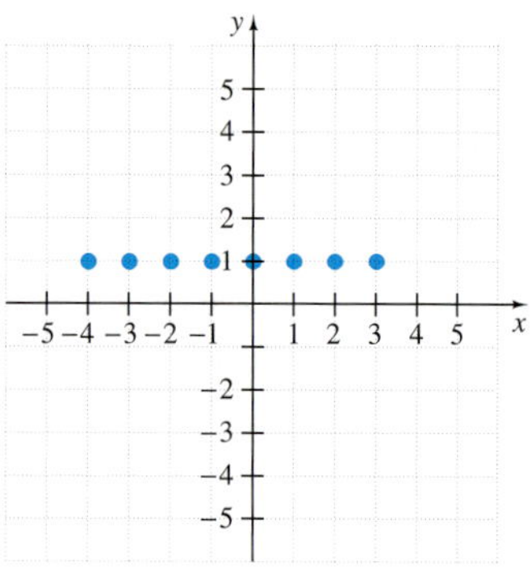


**c.** Input: Cities Output: Population (in thousands)

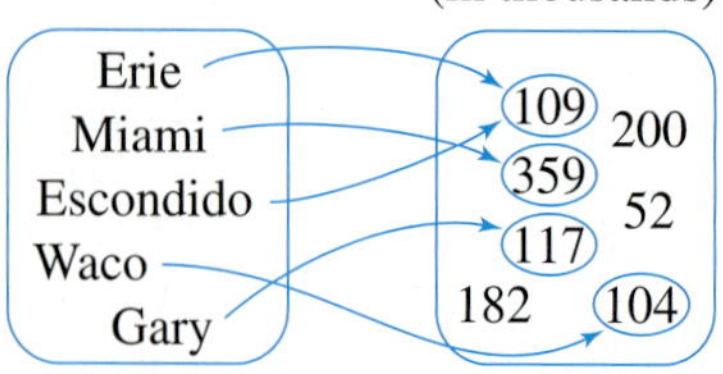


**8.** Graph $y = -3$.

**9.** Find the slope of the line $x = -5$.

**10.** Use the substitution method to solve the system.

$$\begin{cases} -\dfrac{x}{6} + \dfrac{y}{2} = \dfrac{1}{2} \\ \dfrac{x}{3} - \dfrac{y}{6} = -\dfrac{3}{4} \end{cases}$$

**11.** Use the product rule to simplify.

**a.** $2^2 \cdot 2^5$

**b.** $x^7x^3$

**c.** $y \cdot y^2 \cdot y^4$

**12.** Use scientific notation to simplify $\dfrac{2000 \times 0.000021}{700}$.

**13.** If $P(x) = 3x^2 - 2x - 5$, find the following.

**a.** $P(1)$

**b.** $P(-2)$

**14.** Multiply and simplify the product if possible.

**a.** $(x + 3)(2x + 5)$

**b.** $(2x - 3)(5x^2 - 6x + 7)$

**15.** Find the GCF of $20x^3y$, $10x^2y^2$, and $35x^3$.

**16.** Simplify each rational expression.

**a.** $\dfrac{x^3 + 8}{2 + x}$

**b.** $\dfrac{2y^2 + 2}{y^3 - 5y^2 + y - 5}$

**17.** Perform the indicated operation.

**a.** $\dfrac{2}{x^2y} + \dfrac{5}{3x^3y}$

**b.** $\dfrac{3x}{x+2} + \dfrac{2x}{x-2}$

**c.** $\dfrac{x}{x-1} - \dfrac{4}{1-x}$

**18.** Simplify each complex fraction.

**a.** $\dfrac{\dfrac{5x}{x+2}}{\dfrac{10}{x-2}}$

**b.** $\dfrac{\dfrac{x}{y^2} + \dfrac{1}{y}}{\dfrac{y}{x^2} + \dfrac{1}{x}}$

**19.** Divide $10x^3 - 5x^2 + 20x$ by $5x$.

**20.** Use synthetic division to divide $2x^3 - x^2 - 13x + 1$ by $x - 3$.

**21.** Solve: $\dfrac{x+6}{x-2} = \dfrac{2(x+2)}{x-2}$

**22.** Solve $\frac{1}{x} + \frac{1}{y} = \frac{1}{z}$ for $x$.

**23.** Suppose that $u$ varies inversely as $w$. If $u$ is 3 when $w$ is 5, find the constant of variation and the inverse variation equation.

**24.** Write each expression with a positive exponent, and then simplify.

**a.** $16^{-3/4}$ **b.** $(-27)^{-2/3}$

**25.** Rationalize the numerator of $\frac{\sqrt{x} + 2}{5}$.

## Allies of Good Nutrition

Diet and nutrition play a major role in good health. Diets low in saturated fats and dietary cholesterol tend to lower the risk of cardiovascular disease. High sodium intakes have been associated with high blood pressure and stroke. Antioxidants, such as beta carotene and vitamin C, can help protect against heart disease and some cancers. It is no wonder, then, that public interest in healthy eating habits is soaring.

Registered dieticians, food and nutrition experts, are valuable allies in an effort to eat right. They plan nutrition programs, supervise food preparation, and educate about the health benefits of good nutrition. Registered dieticians work in diverse environments: hospitals, schools, day-care centers, nursing homes, government or university laboratories, private practice, corporate wellness programs, food producing companies, pharmaceutical companies, restaurant management, and community health settings. Dieticians use math and problem-solving skills in tasks such as analyzing the nutritional content of a recipe or food product, assessing a client's diet, and determining an individual's nutritional requirements.

*For more information about careers in dietetics, visit the American Dietetic Association Web site by first going to www.prenhall.com/martin-gay.*

*In the Spotlight on Decision Making feature on page 480, you will have the opportunity to make a decision about the adequacy of vitamin A intake as a registered dietician.*

# QUADRATIC EQUATIONS AND FUNCTIONS

# 8

8.1 Solving Quadratic Equations by Completing the Square

8.2 Solving Quadratic Equations by the Quadratic Formula

8.3 Solving Equations by Using Quadratic Methods

8.4 Nonlinear Inequalities in One Variable

8.5 Quadratic Functions and Their Graphs

8.6 Further Graphing of Quadratic Functions

An important part of the study of algebra is learning to model and solve problems. Often, the model of a problem is a quadratic equation or a function containing a second-degree polynomial. In this chapter, we continue the work begun in Chapter 5, when we solved polynomial equations in one variable by factoring. Two additional methods of solving quadratic equations are analyzed, as well as methods of solving nonlinear inequalities in one variable.

## 8.1 SOLVING QUADRATIC EQUATIONS BY COMPLETING THE SQUARE

### OBJECTIVES

CD-ROM SSM SSG Video

1. Use the square root property to solve quadratic equations.
2. Solve quadratic equations by completing the square.
3. Use quadratic equations to solve problems.

1 In Chapter 5, we solved quadratic equations by factoring. Recall that a **quadratic,** or **second-degree, equation** is an equation that can be written in the form $ax^2 + bx + c = 0$, where $a$, $b$, and $c$ are real numbers and $a$ is not 0. To solve a quadratic equation such as $x^2 = 9$ by factoring, we use the zero-factor theorem. To use the zero-factor theorem, the equation must first be written in standard form, $ax^2 + bx + c = 0$.

$$x^2 = 9$$

$$x^2 - 9 = 0 \qquad \text{Subtract 9 from both sides.}$$

$$(x + 3)(x - 3) = 0 \qquad \text{Factor.}$$

$$x + 3 = 0 \quad \text{or} \quad x - 3 = 0 \qquad \text{Set each factor equal to 0.}$$

$$x = -3 \qquad\qquad x = 3 \qquad \text{Solve.}$$

The solution set is $\{-3, 3\}$, the positive and negative square roots of 9. Not all quadratic equations can be solved by factoring, so we need to explore other methods. Notice that the solutions of the equation $x^2 = 9$ are two numbers whose square is 9.

$$3^2 = 9 \quad \text{and} \quad (-3)^2 = 9$$

Thus, we can solve the equation $x^2 = 9$ by taking the square root of both sides. Be sure to include both $\sqrt{9}$ and $-\sqrt{9}$ as solutions since both $\sqrt{9}$ and $-\sqrt{9}$ are numbers whose square is 9.

$$x^2 = 9$$

$$\sqrt{x^2} = \pm\sqrt{9}$$

$$x = \pm 3$$

The notation $\pm\sqrt{9}$ (read as "plus or minus $\sqrt{9}$") indicates the pair of numbers $+\sqrt{9}$ and $-\sqrt{9}$.

This illustrates the square root property.

**HELPFUL HINT**

The notation $\pm 3$, for example, is read as "plus or minus 3." It is a shorthand notation for the pair of numbers $+3$ and $-3$.

**SQUARE ROOT PROPERTY**

If $b$ is a real number and if $a^2 = b$, then $a = \pm\sqrt{b}$.

**Example 1** Use the square root property to solve $x^2 = 50$.

*Solution*

$$x^2 = 50$$

$$x = \pm\sqrt{50} \quad \text{Use the square root property.}$$

$$x = \pm 5\sqrt{2} \quad \text{Simplify the radical.}$$

*Check* Let $x = 5\sqrt{2}$. Let $x = -5\sqrt{2}$.

$$x^2 = 50 \qquad\qquad x^2 = 50$$

$$(5\sqrt{2})^2 \stackrel{?}{=} 50 \qquad\qquad (-5\sqrt{2})^2 \stackrel{?}{=} 50$$

$$25 \cdot 2 \stackrel{?}{=} 50 \qquad\qquad 25 \cdot 2 \stackrel{?}{=} 50$$

$$50 = 50 \quad \text{True.} \qquad\qquad 50 = 50 \quad \text{True.}$$

The solutions are $5\sqrt{2}$ and $-5\sqrt{2}$.

**Example 2** Use the square root property to solve $2x^2 = 14$.

*Solution* First we get the squared variable alone on one side of the equation.

$$2x^2 = 14$$

$$x^2 = 7 \quad \text{Divide both sides by 2.}$$

$$x = \pm\sqrt{7} \quad \text{Use the square root property.}$$

Check to see that the solutions are $\sqrt{7}$ and $-\sqrt{7}$.

**Example 3** Use the square root property to solve $(x + 1)^2 = 12$.

*Solution*

$$(x + 1)^2 = 12$$

$$x + 1 = \pm\sqrt{12} \quad \text{Use the square root property.}$$

$$x + 1 = \pm 2\sqrt{3} \quad \text{Simplify the radical.}$$

$$x = -1 \pm 2\sqrt{3} \quad \text{Subtract 1 from both sides.}$$

*Check* Below is a check for $-1 + 2\sqrt{3}$. The check for $-1 - 2\sqrt{3}$ is almost the same and is left for you to do on your own.

$$(x + 1)^2 = 12$$

$$(-1 + 2\sqrt{3} + 1)^2 \stackrel{?}{=} 12$$

$$(2\sqrt{3})^2 \stackrel{?}{=} 12$$

$$4 \cdot 3 \stackrel{?}{=} 12$$

$$12 = 12 \quad \text{True.}$$

The solutions are $-1 + 2\sqrt{3}$, $-1 - 2\sqrt{3}$.

**Example 4** Use the square root property to solve $(2x - 5)^2 = -16$.

*Solution* $(2x - 5)^2 = -16$

$$2x - 5 = \pm\sqrt{-16} \quad \text{Use the square root property.}$$

$$2x - 5 = \pm 4i \quad \text{Simplify the radical.}$$

$$2x = 5 \pm 4i \quad \text{Add 5 to both sides.}$$

$$x = \frac{5 \pm 4i}{2} \quad \text{Divide both sides by 2.}$$

The solutions are $\frac{5 + 4i}{2}$ and $\frac{5 - 4i}{2}$.

Notice from Examples 3 and 4 that, if we write a quadratic equation so that one side is the square of a binomial, we can solve by using the square root property. To write the square of a binomial, we write perfect square trinomials. Recall that a perfect square trinomial is a trinomial that can be factored into two identical binomial factors.

| ***Perfect Square Trinomials*** | ***Factored Form*** |
|---|---|
| $x^2 + 8x + 16$ | $(x + 4)^2$ |
| $x^2 - 6x + 9$ | $(x - 3)^2$ |
| $x^2 + 3x + \frac{9}{4}$ | $\left(x + \frac{3}{2}\right)^2$ |

Notice that for each perfect square trinomial, **the constant term of the trinomial is the square of half the coefficient of the *x*-term.** For example,

$$x^2 + 8x + 16 \qquad\qquad x^2 - 6x + 9$$

$$\frac{1}{2}(8) = 4 \text{ and } 4^2 = 16 \qquad \frac{1}{2}(-6) = -3 \text{ and } (-3)^2 = 9$$

2 The process of writing a quadratic equation so that one side is a perfect square trinomial is called **completing the square.**

**Example 5** Solve $p^2 + 2p = 4$ by completing the square.

*Solution* First, add the square of half the coefficient of $p$ to both sides so that the resulting trinomial will be a perfect square trinomial. The coefficient of $p$ is 2.

$$\frac{1}{2}(2) = 1 \quad \text{and} \quad 1^2 = 1$$

Add 1 to both sides of the original equation.

$$p^2 + 2p = 4$$

$$p^2 + 2p + 1 = 4 + 1 \quad \text{Add 1 to both sides.}$$

$$(p + 1)^2 = 5 \quad \text{Factor the trinomial; simplify the right side.}$$

We may now use the square root property and solve for $p$.

$$p + 1 = \pm\sqrt{5} \quad \text{Use the square root property.}$$

$$p = -1 \pm \sqrt{5} \quad \text{Subtract 1 from both sides.}$$

Notice that there are two solutions: $-1 + \sqrt{5}$ and $-1 - \sqrt{5}$.

**Example 6** Solve $m^2 - 7m - 1 = 0$ for $m$ by completing the square.

*Solution* First, add 1 to both sides of the equation so that the left side has no constant term.

$$m^2 - 7m - 1 = 0$$

$$m^2 - 7m = 1$$

Now find the constant term that makes the left side a perfect square trinomial by squaring half the coefficient of $m$. Add this constant to both sides of the equation.

$$\frac{1}{2}(-7) = -\frac{7}{2} \quad \text{and} \quad \left(-\frac{7}{2}\right)^2 = \frac{49}{4}$$

$$m^2 - 7m + \frac{49}{4} = 1 + \frac{49}{4} \quad \text{Add } \frac{49}{4} \text{ to both sides of the equation.}$$

$$\left(m - \frac{7}{2}\right)^2 = \frac{53}{4} \quad \text{Factor the perfect square trinomial and simplify the right side.}$$

$$m - \frac{7}{2} = \pm\sqrt{\frac{53}{4}} \quad \text{Apply the square root property.}$$

$$m = \frac{7}{2} \pm \frac{\sqrt{53}}{2} \quad \text{Add } \frac{7}{2} \text{ to both sides and simplify } \sqrt{\frac{53}{4}}.$$

$$m = \frac{7 \pm \sqrt{53}}{2} \quad \text{Simplify.}$$

The solutions are $\dfrac{7 + \sqrt{53}}{2}$ and $\dfrac{7 - \sqrt{53}}{2}$.

**Example 7** Solve $2x^2 - 8x + 3 = 0$.

*Solution* Our procedure for finding the constant term to complete the square works only if the coefficient of the squared variable term is 1. Therefore, to solve this equation, the first step is to divide both sides by 2, the coefficient of $x^2$.

$$2x^2 - 8x + 3 = 0$$

$$x^2 - 4x + \frac{3}{2} = 0 \qquad \text{Divide both sides by 2.}$$

$$x^2 - 4x = -\frac{3}{2} \qquad \text{Subtract } \frac{3}{2} \text{ from both sides.}$$

Next find the square of half of $-4$.

$$\frac{1}{2}(-4) = -2 \quad \text{and} \quad (-2)^2 = 4$$

Add 4 to both sides of the equation to complete the square.

$$x^2 - 4x + 4 = -\frac{3}{2} + 4$$

$$(x - 2)^2 = \frac{5}{2} \qquad \text{Factor the perfect square and simplify the right side.}$$

$$x - 2 = \pm\sqrt{\frac{5}{2}} \qquad \text{Apply the square root property.}$$

$$x - 2 = \pm\frac{\sqrt{10}}{2} \qquad \text{Rationalize the denominator.}$$

$$x = 2 \pm \frac{\sqrt{10}}{2} \qquad \text{Add 2 to both sides.}$$

$$= \frac{4}{2} \pm \frac{\sqrt{10}}{2} \qquad \text{Find the common denominator.}$$

$$= \frac{4 \pm \sqrt{10}}{2} \qquad \text{Simplify.}$$

The solutions are $\frac{4 + \sqrt{10}}{2}$ and $\frac{4 - \sqrt{10}}{2}$.

The following steps may be used to solve a quadratic equation such as $ax^2 + bx + c = 0$ by completing the square. This method may be used whether or not the polynomial $ax^2 + bx + c$ is factorable.

### SOLVING A QUADRATIC EQUATION IN $x$ BY COMPLETING THE SQUARE

**Step 1:** If the coefficient of $x^2$ is 1, go to Step 2. Otherwise, divide both sides of the equation by the coefficient of $x^2$.
**Step 2:** Isolate all variable terms on one side of the equation.
**Step 3:** Complete the square for the resulting binomial by adding the square of half of the coefficient of $x$ to both sides of the equation.
**Step 4:** Factor the resulting perfect square trinomial and write it as the square of a binomial.
**Step 5:** Use the square root property to solve for $x$.

**Example 8** Solve $3x^2 - 9x + 8 = 0$ by completing the square.

*Solution* $3x^2 - 9x + 8 = 0$

**Step 1:** $x^2 - 3x + \frac{8}{3} = 0$ Divide both sides of the equation by 3.

**Step 2:** $x^2 - 3x = -\frac{8}{3}$ Subtract $\frac{8}{3}$ from both sides.

Since $\frac{1}{2}(-3) = -\frac{3}{2}$ and $\left(-\frac{3}{2}\right)^2 = \frac{9}{4}$, we add $\frac{9}{4}$ to both sides of the equation.

**Step 3:** $x^2 - 3x + \frac{9}{4} = -\frac{8}{3} + \frac{9}{4}$

**Step 4:** $\left(x - \frac{3}{2}\right)^2 = -\frac{5}{12}$ Factor the perfect square trinomial.

**Step 5:** $x - \frac{3}{2} = \pm\sqrt{-\frac{5}{12}}$ Apply the square root property.

$x - \frac{3}{2} = \pm\frac{i\sqrt{5}}{2\sqrt{3}}$ Simplify the radical.

$x - \frac{3}{2} = \pm\frac{i\sqrt{15}}{6}$ Rationalize the denominator.

$x = \frac{3}{2} \pm \frac{i\sqrt{15}}{6}$ Add $\frac{3}{2}$ to both sides.

$= \frac{9}{6} \pm \frac{i\sqrt{15}}{6}$ Find a common denominator.

$= \frac{9 \pm i\sqrt{15}}{6}$ Simplify.

The solutions are $\frac{9 + i\sqrt{15}}{6}$ and $\frac{9 - i\sqrt{15}}{6}$.

3 Recall the **simple interest** formula $I = Prt$, where $I$ is the interest earned, $P$ is the principal, $r$ is the rate of interest, and $t$ is time. If \$100 is invested at a simple interest rate of 5% annually, at the end of 3 years the total interest $I$ earned is

$$I = P \cdot r \cdot t$$

or

$$I = 100 \cdot 0.05 \cdot 3 = \$15$$

and the new principal is

$$\$100 + \$15 = \$115$$

Most of the time, the interest computed on money borrowed or money deposited is **compound interest.** Compound interest, unlike simple interest, is computed on original principal *and* on interest already earned. To see the difference between simple interest and compound interest, suppose that \$100 is invested at a rate of 5% compounded annually. To find the total amount of money at the end of 3 years, we calculate as follows.

$$I = P \cdot r \cdot t$$

| | |
|---|---|
| First year: | Interest $= \$100 \cdot 0.05 \cdot 1 = \$5.00$ |
| | New principal $= \$100.00 + \$5.00 = \$105.00$ |
| Second year: | Interest $= \$105.00 \cdot 0.05 \cdot 1 = \$5.25$ |
| | New principal $= \$105.00 + \$5.25 = \$110.25$ |
| Third year: | Interest $= \$110.25 \cdot 0.05 \cdot 1 \approx \$5.51$ |
| | New principal $= \$110.25 + \$5.51 = \$115.76$ |

At the end of the third year, the total compound interest earned is \$15.76, whereas the total simple interest earned is \$15.

It is tedious to calculate compound interest as we did above, so we use a compound interest formula. The formula for calculating the total amount of money when interest is compounded annually is

$$A = P(1 + r)^t$$

where $P$ is the original investment, $r$ is the interest rate per compounding period, and $t$ is the number of periods. For example, the amount of money $A$ at the end of 3 years if \$100 is invested at 5% compounded annually is

$$A = \$100(1 + 0.05)^3 \approx \$100(1.1576) = \$115.76$$

as we previously calculated.

## Example 9 FINDING INTEREST RATES

Find the interest rate $r$ if \$2000 compounded annually grows to \$2420 in 2 years.

*Solution*

1. UNDERSTAND the problem. Since the \$2000 is compounded annually, we use the compound interest formula. For this example, make sure that you understand the formula for compounding interest annually.

2. TRANSLATE. We substitute the given values into the formula.

$$A = P(1 + r)^t$$

$$2420 = 2000(1 + r)^2 \quad \text{Let } A = 2420, P = 2000, \text{ and } t = 2.$$

3. SOLVE. Solve the equation for $r$.

$$2420 = 2000(1 + r)^2$$

$$\frac{2420}{2000} = (1 + r)^2 \quad \text{Divide both sides by 2000.}$$

$$\frac{121}{100} = (1 + r)^2 \quad \text{Simplify the fraction.}$$

$$\pm\sqrt{\frac{121}{100}} = 1 + r \quad \text{Use the square root property.}$$

$$\pm\frac{11}{10} = 1 + r \quad \text{Simplify.}$$

$$-1 \pm \frac{11}{10} = r$$

$$-\frac{10}{10} \pm \frac{11}{10} = r$$

$$\frac{1}{10} = r \quad \text{or} \quad -\frac{21}{10} = r$$

4. INTERPRET. The rate cannot be negative, so we reject $-\frac{21}{10}$.

**Check:** $\frac{1}{10} = 0.10 = 10\%$ per year. If we invest \$2000 at 10% compounded annually, in 2 years the amount in the account would be $2000(1 + 0.10)^2 = 2420$ dollars, the desired amount.

**State:** The interest rate is 10% compounded annually.

## GRAPHING CALCULATOR EXPLORATIONS

In Section 5.8, we showed how we can use a grapher to approximate real number solutions of a quadratic equation written in standard form. We can also use a grapher to solve a quadratic equation when it is not written in standard form. For example, to solve $(x + 1)^2 = 12$, the quadratic equation in Example 3, we graph the following on the same set of axes. Use Xmin $= -10$, Xmax $= 10$, Ymin $= -13$, and Ymax $= 13$.

$$Y_1 = (x + 1)^2 \quad \text{and} \quad Y_2 = 12$$

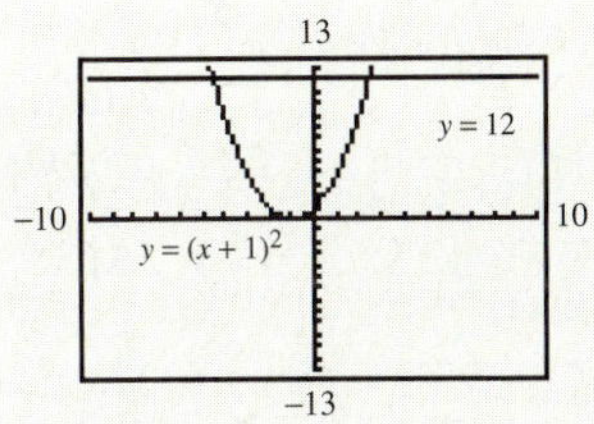

Use the Intersect feature or the Zoom and Trace features to locate the points of intersection of the graphs. The $x$-values of these points are the solutions of $(x + 1)^2 = 12$. The solutions, rounded to two decimal places, are 2.46 and −4.46.

Check to see that these numbers are approximations of the exact solutions $-1 \pm 2\sqrt{3}$.

*Use a graphing calculator to solve each quadratic equation. Round all solutions to the nearest hundredth.*

**1.** $x(x - 5) = 8$

**2.** $x(x + 2) = 5$

**3.** $x^2 + 0.5x = 0.3x + 1$

**4.** $x^2 - 2.6x = -2.2x + 3$

**5.** Use a grapher and solve $(2x - 5)^2 = -16$, Example 4 in this section, using the window

$$\text{Xmin} = -20$$
$$\text{Xmax} = 20$$
$$\text{Xscl} = 1$$
$$\text{Ymin} = -20$$
$$\text{Ymax} = 20$$
$$\text{Yscl} = 1$$

Explain the results. Compare your results with the solution found in Example 4.

**6.** What are the advantages and disadvantages of using a grapher to solve quadratic equations?

## Exercise Set 8.1

*Use the square root property to solve each equation. These equations have real-number solutions. See Examples 1 through 3.*

**1.** $x^2 = 16$

**2.** $x^2 = 49$

**3.** $x^2 - 7 = 0$

**4.** $x^2 - 11 = 0$

**5.** $x^2 = 18$

**6.** $y^2 = 20$

**7.** $3z^2 - 30 = 0$

**8.** $2x^2 = 4$

**9.** $(x + 5)^2 = 9$

**10.** $(y - 3)^2 = 4$

**11.** $(z - 6)^2 = 18$

**12.** $(y + 4)^2 = 27$

**13.** $(2x - 3)^2 = 8$

**14.** $(4x + 9)^2 = 6$

*Use the square root property to solve each equation. See Examples 1 through 4.*

**15.** $x^2 + 9 = 0$

**16.** $x^2 + 4 = 0$

**17.** $x^2 - 6 = 0$

**18.** $y^2 - 10 = 0$

**19.** $2z^2 + 16 = 0$

**20.** $3p^2 + 36 = 0$

**21.** $(x - 1)^2 = -16$

**22.** $(y + 2)^2 = -25$

**23.** $(z + 7)^2 = 5$

**24.** $(x + 10)^2 = 11$

**25.** $(x + 3)^2 = -8$

**26.** $(y - 4)^2 = -18$

*Add the proper constant to each binomial so that the resulting trinomial is a perfect square trinomial. Then factor the trinomial.*

**27.** $x^2 + 16x$

**28.** $y^2 + 2y$

**29.** $z^2 - 12z$

**30.** $x^2 - 8x$

**31.** $p^2 + 9p$

**32.** $n^2 + 5n$

**33.** $x^2 + x$

**34.** $y^2 - y$

*Find two possible missing terms so that each is a perfect square trinomial.*

**35.** $x^2 + \square + 16$

**36.** $y^2 + \square + 9$

**37.** $z^2 + \square + \frac{25}{4}$

**38.** $x^2 + \square + \frac{1}{4}$

*Solve each equation by completing the square. These equations have real number solutions. See Examples 5 through 7.*

**39.** $x^2 + 8x = -15$

**40.** $y^2 + 6y = -8$

**41.** $x^2 + 6x + 2 = 0$

**42.** $x^2 - 2x - 2 = 0$

**43.** $x^2 + x - 1 = 0$

**44.** $x^2 + 3x - 2 = 0$

**45.** $x^2 + 2x - 5 = 0$

**46.** $y^2 + y - 7 = 0$

**47.** $3p^2 - 12p + 2 = 0$

**48.** $2x^2 + 14x - 1 = 0$

**49.** $4y^2 - 12y - 2 = 0$

**50.** $6x^2 - 3 = 6x$

**51.** $2x^2 + 7x = 4$

**52.** $3x^2 - 4x = 4$

**53.** $x^2 - 4x - 5 = 0$

**54.** $y^2 + 6y - 8 = 0$

**55.** $x^2 + 8x + 1 = 0$

**56.** $x^2 - 10x + 2 = 0$

**57.** $3y^2 + 6y - 4 = 0$

**58.** $2y^2 + 12y + 3 = 0$

**59.** $2x^2 - 3x - 5 = 0$

**60.** $5x^2 + 3x - 2 = 0$

*Solve each equation by completing the square. See Examples 5 through 8.*

**61.** $y^2 + 2y + 2 = 0$

**62.** $x^2 + 4x + 6 = 0$

**63.** $x^2 - 6x + 3 = 0$

**64.** $x^2 - 7x - 1 = 0$

**65.** $2a^2 + 8a = -12$

**66.** $3x^2 + 12x = -14$

**67.** $5x^2 + 15x - 1 = 0$

**68.** $16y^2 + 16y - 1 = 0$

**69.** $2x^2 - x + 6 = 0$

**70.** $4x^2 - 2x + 5 = 0$

**71.** $x^2 + 10x + 28 = 0$

**72.** $y^2 + 8y + 18 = 0$

**73.** $z^2 + 3z - 4 = 0$

**74.** $y^2 + y - 2 = 0$

**75.** $2x^2 - 4x + 3 = 0$

**76.** $9x^2 - 36x = -40$

**77.** $3x^2 + 3x = 5$

**78.** $5y^2 - 15y = 1$

*Use the formula $A = P(1 + r)^t$ to solve Exercises 79–82. See Example 9.*

**79.** Find the rate $r$ at which \$3000 grows to \$4320 in 2 years.

**80.** Find the rate $r$ at which \$800 grows to \$882 in 2 years.

**81.** Find the rate at which \$810 grows to \$1000 in 2 years.

**82.** Find the rate at which \$2000 grows to \$2880 in 2 years.

**83.** In your own words, what is the difference between simple interest and compound interest?

**84.** If you are depositing money in an account that pays 4%, would you prefer the interest to be simple or compound? Explain why.

**85.** If you are borrowing money at a rate of 10%, would you prefer the interest to be simple or compound? Explain why.

*Neglecting air resistance, the distance $s(t)$ in feet traveled by a freely falling object is given by the function $s(t) = 16t^2$, where $t$ is time in seconds. Use this formula to solve Exercises 86 through 89. Round answers to two decimal places.*

**86.** The Petronas Towers in Kuala Lumpur, built in 1997, are the tallest buildings in Malaysia. Each tower is 1483 feet tall. How long would it take an object to fall to the ground from the top of one of the towers? (*Source:* Council on Tall Buildings and Urban Habitat, Lehigh University)

**87.** The height of the Chicago Beach Tower Hotel, built in 1998 in Dubai, United Arab Emirates, is 1053 feet. How long would it take an object to fall to the ground from the top of the building? (*Source:* Council on Tall Buildings and Urban Habitat, Lehigh University)

**88.** The height of the Nurek Dam in Tajikistan (part of the former USSR that borders Afghanistan) is 984 feet. How long would it take an object to fall from the top to the base of the dam? (*Source:* U.S. Committee on Large Dams of the International Commission on Large Dams)

**89.** The Hoover Dam, located on the Colorado River on the border of Nevada and Arizona near Las Vegas, is 725 feet tall. How long would it take an object to fall from the top to the base of the dam? (*Source:* U.S. Committee on Large Dams of the International Commission on Large Dams)

*Solve.*

△ **90.** The area of a square room is 225 square feet. Find the dimensions of the room.

△ **91.** The area of a circle is $36\pi$ square inches. Find the radius of the circle.

△ **92.** An isosceles right triangle has legs of equal length. If the hypotenuse is 20 centimeters long, find the length of each leg.

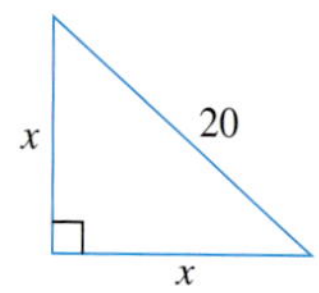


△ **93.** A 27-inch TV is advertised in the *Daily Sentry* newspaper. If 27 inches is the measure of the diagonal of the picture tube, find the measure of the side of the picture tube.

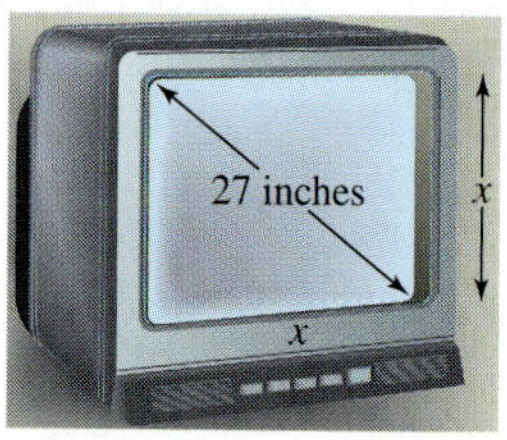


*A common equation used in business is a demand equation. It expresses the relationship between the unit price of some commodity and the quantity demanded. For Exercises 94 and 95, p represents the unit price and x represents the quantity demanded in thousands.*

**94.** A manufacturing company has found that the demand equation for a certain type of scissors is given by the equation $p = -x^2 + 47$. Find the demand for the scissors if the price is \$11 per pair.

**95.** Acme, Inc., sells desk lamps and has found that the demand equation for a certain style of desk lamp is given by the equation $p = -x^2 + 15$. Find the demand for the desk lamp if the price is \$7 per lamp.

### REVIEW EXERCISES

*Simplify each expression. See Section 7.1.*

**96.** $\frac{3}{4} - \sqrt{\frac{25}{16}}$

**97.** $\frac{3}{5} + \sqrt{\frac{16}{25}}$

**98.** $\frac{1}{2} - \sqrt{\frac{9}{4}}$

**99.** $\frac{9}{10} - \sqrt{\frac{49}{100}}$

*Simplify each expression. See Section 7.5.*

**100.** $\frac{6 + 4\sqrt{5}}{2}$

**101.** $\frac{10 - 20\sqrt{3}}{2}$

**102.** $\frac{3 - 9\sqrt{5}}{6}$

**103.** $\frac{12 - 8\sqrt{7}}{16}$

*Evaluate $\sqrt{b^2 - 4ac}$ for each set of values. See Section 7.3.*

**104.** $a = 2, b = 4, c = -1$

**105.** $a = 1, b = 6, c = 2$

**106.** $a = 3, b = -1, c = -2$

**107.** $a = 1, b = -3, c = -1$

## 8.2 SOLVING QUADRATIC EQUATIONS BY THE QUADRATIC FORMULA

**OBJECTIVES**

CD-ROM SSM SSG Video

1. Solve quadratic equations by using the quadratic formula.
2. Determine the number and type of solutions of a quadratic equation by using the discriminant.
3. Solve geometric pr\oblems modeled by quadratic equations.

**1** Any quadratic equation can be solved by completing the square. Since the same sequence of steps is repeated each time we complete the square, let's complete the square for a general quadratic equation, $ax^2 + bx + c = 0$. By doing so, we find a pattern for the solutions of a quadratic equation known as the **quadratic formula.**

Recall that to complete the square for an equation such as $ax^2 + bx + c = 0$, we first divide both sides by the coefficient of $x^2$.

$$ax^2 + bx + c = 0$$

$$x^2 + \frac{b}{a}x + \frac{c}{a} = 0 \qquad \text{Divide both sides by } a\text{, the coefficient of } x^2.$$

$$x^2 + \frac{b}{a}x = -\frac{c}{a} \qquad \text{Subtract the constant } \frac{c}{a} \text{ from both sides.}$$

Next, find the square of half $\frac{b}{a}$, the coefficient of $x$.

$$\frac{1}{2}\left(\frac{b}{a}\right) = \frac{b}{2a} \quad \text{and} \quad \left(\frac{b}{2a}\right)^2 = \frac{b^2}{4a^2}$$

Add this result to both sides of the equation.

$$x^2 + \frac{b}{a}x + \frac{b^2}{4a^2} = -\frac{c}{a} + \frac{b^2}{4a^2} \qquad \text{Add } \frac{b^2}{4a^2} \text{ to both sides.}$$

$$x^2 + \frac{b}{a}x + \frac{b^2}{4a^2} = \frac{-c \cdot 4a}{a \cdot 4a} + \frac{b^2}{4a^2} \qquad \text{Find a common denominator on the right side.}$$

$$x^2 + \frac{b}{a}x + \frac{b^2}{4a^2} = \frac{b^2 - 4ac}{4a^2} \qquad \text{Simplify the right side.}$$

$$\left(x + \frac{b}{2a}\right)^2 = \frac{b^2 - 4ac}{4a^2} \qquad \text{Factor the perfect square trinomial on the left side.}$$

$$x + \frac{b}{2a} = \pm\sqrt{\frac{b^2 - 4ac}{4a^2}} \qquad \text{Apply the square root property.}$$

$$x + \frac{b}{2a} = \pm\frac{\sqrt{b^2 - 4ac}}{2a} \qquad \text{Simplify the radical.}$$

$$x = -\frac{b}{2a} \pm \frac{\sqrt{b^2 - 4ac}}{2a} \qquad \text{Subtract } \frac{b}{2a} \text{ from both sides.}$$

$$x = \frac{-b \pm \sqrt{b^2 - 4ac}}{2a} \qquad \text{Simplify.}$$

This equation identifies the solutions of the general quadratic equation in standard form and is called the quadratic formula. It can be used to solve any equation written in standard form $ax^2 + bx + c = 0$ as long as $a$ is not 0.

**QUADRATIC FORMULA**

A quadratic equation written in the form $ax^2 + bx + c = 0$ has the solutions

$$x = \frac{-b \pm \sqrt{b^2 - 4ac}}{2a}$$

**Example 1** Solve $3x^2 + 16x + 5 = 0$ for $x$.

*Solution* This equation is in standard form, so $a = 3, b = 16$, and $c = 5$. Substitute these values into the quadratic formula.

$$x = \frac{-b \pm \sqrt{b^2 - 4ac}}{2a} \quad \text{Quadratic formula.}$$

$$= \frac{-16 \pm \sqrt{16^2 - 4(3)(5)}}{2 \cdot 3} \quad \text{Use } a = 3, b = 16, \text{ and } c = 5.$$

$$= \frac{-16 \pm \sqrt{256 - 60}}{6}$$

$$= \frac{-16 \pm \sqrt{196}}{6} = \frac{-16 \pm 14}{6}$$

$$x = \frac{-16 + 14}{6} = -\frac{1}{3} \quad \text{or} \quad x = \frac{-16 - 14}{6} = -\frac{30}{6} = -5$$

The solutions are $-\frac{1}{3}$ and $-5$.

**Example 2** Solve $2x^2 - 4x = 3$.

*Solution* First write the equation in standard form by subtracting 3 from both sides.

$$2x^2 - 4x - 3 = 0$$

> **HELPFUL HINT**
> To replace $a$, $b$, and $c$ correctly in the quadratic formula, write the quadratic equation in standard form $ax^2 + bx + c = 0$.

Now $a = 2, b = -4$, and $c = -3$. Substitute these values into the quadratic formula.

$$x = \frac{-b \pm \sqrt{b^2 - 4ac}}{2a}$$

$$= \frac{-(-4) \pm \sqrt{(-4)^2 - 4(2)(-3)}}{2 \cdot 2}$$

$$= \frac{4 \pm \sqrt{16 + 24}}{4}$$

$$= \frac{4 \pm \sqrt{40}}{4} = \frac{4 \pm 2\sqrt{10}}{4}$$

$$= \frac{2(2 \pm \sqrt{10})}{2 \cdot 2} = \frac{2 \pm \sqrt{10}}{2}$$

The solutions are $\frac{2 + \sqrt{10}}{2}$ and $\frac{2 - \sqrt{10}}{2}$.

> **HELPFUL HINT**
> To simplify the expression $\frac{4 \pm 2\sqrt{10}}{4}$ in the preceding example, note that 2 is factored out of both terms of the numerator *before* simplifying.
>
> $$\frac{4 \pm 2\sqrt{10}}{4} = \frac{2(2 \pm \sqrt{10})}{2 \cdot 2} = \frac{2 \pm \sqrt{10}}{2}$$

**Example 3** Solve $\frac{1}{4}m^2 - m + \frac{1}{2} = 0$.

*Solution* We could use the quadratic formula with $a = \frac{1}{4}$, $b = -1$, and $c = \frac{1}{2}$. Instead, we find a simpler, equivalent standard form equation whose coefficients are not fractions.

Multiply both sides of the equation by 4 to clear fractions.

$$4\left(\frac{1}{4}m^2 - m + \frac{1}{2}\right) = 4 \cdot 0$$

$$m^2 - 4m + 2 = 0 \qquad \text{Simplify.}$$

Substitute $a = 1$, $b = -4$, and $c = 2$ into the quadratic formula and simplify.

$$m = \frac{-(-4) \pm \sqrt{(-4)^2 - 4(1)(2)}}{2 \cdot 1} = \frac{4 \pm \sqrt{16 - 8}}{2}$$

$$= \frac{4 \pm \sqrt{8}}{2} = \frac{4 \pm 2\sqrt{2}}{2} = \frac{2(2 \pm \sqrt{2})}{2}$$

$$= 2 \pm \sqrt{2}$$

The solutions are $2 + \sqrt{2}$ and $2 - \sqrt{2}$.

**Example 4** Solve $x = -3x^2 - 3$.

*Solution* The equation in standard form is $3x^2 + x + 3 = 0$. Thus, let $a = 3$, $b = 1$, and $c = 3$ in the quadratic formula.

$$x = \frac{-1 \pm \sqrt{1^2 - 4(3)(3)}}{2 \cdot 3} = \frac{-1 \pm \sqrt{1 - 36}}{6} = \frac{-1 \pm \sqrt{-35}}{6} = \frac{-1 \pm i\sqrt{35}}{6}$$

The solutions are $\frac{-1 + i\sqrt{35}}{6}$ and $\frac{-1 - i\sqrt{35}}{6}$.

In Example 1, the equation $3x^2 + 16x + 5 = 0$ had 2 real roots, $-\frac{1}{3}$ and $-5$. In Example 4, the equation $3x^2 + x + 3 = 0$ (written in standard form) had no real roots. How do their related graphs compare? Recall that the $x$-intercepts of $f(x) = 3x^2 + 16x + 5$ occur where $f(x) = 0$ or where $3x^2 + 16x + 5 = 0$. Since this equation has 2 real roots, the graph has 2 $x$-intercepts. Similarly, since the equation $3x^2 + x + 3 = 0$ has no real roots, the graph of $f(x) = 3x^2 + x + 3$ has no $x$-intercepts.

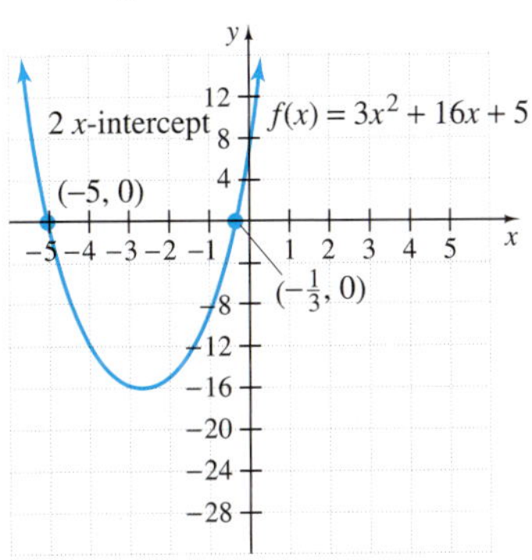


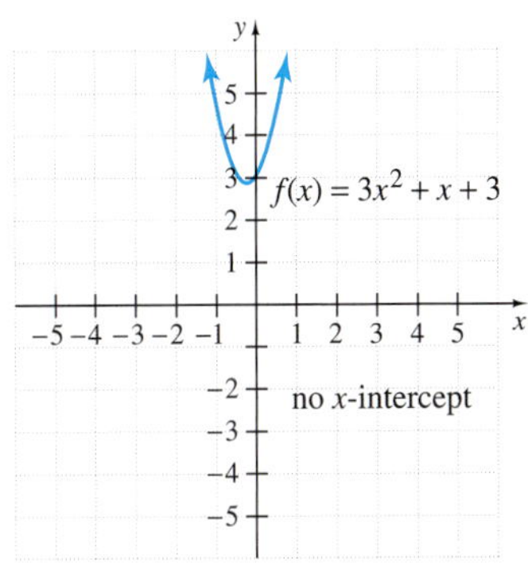

2 In the quadratic formula, $x = \dfrac{-b \pm \sqrt{b^2 - 4ac}}{2a}$, the radicand $b^2 - 4ac$ is called the **discriminant** because, by knowing its value, we can **discriminate** among the possible number and type of solutions of a quadratic equation. Possible values of the discriminant and their meanings are summarized next.

### DISCRIMINANT

The following table corresponds the discriminant $b^2 - 4ac$ of a quadratic equation of the form $ax^2 + bx + c = 0$ with the number and type of solutions of the equation.

| $b^2 - 4ac$ | *Number and Type of Solutions* |
|---|---|
| Positive | Two real solutions |
| Zero | One real solution |
| Negative | Two complex but not real solutions |

**Example 5** Use the discriminant to determine the number and type of solutions of each quadratic equation.

**a.** $x^2 + 2x + 1 = 0$ **b.** $3x^2 + 2 = 0$ **c.** $2x^2 - 7x - 4 = 0$

*Solution* **a.** In $x^2 + 2x + 1 = 0$, $a = 1$, $b = 2$, and $c = 1$. Thus,

$$b^2 - 4ac = 2^2 - 4(1)(1) = 0$$

Since $b^2 - 4ac = 0$, this quadratic equation has one real solution.

**b.** In this equation, $a = 3$, $b = 0$, $c = 2$. Then $b^2 - 4ac = 0 - 4(3)(2) = -24$. Since $b^2 - 4ac$ is negative, the quadratic equation has two complex but not real solutions.

**c.** In this equation, $a = 2$, $b = -7$, and $c = -4$. Then

$$b^2 - 4ac = (-7)^2 - 4(2)(-4) = 81$$

Since $b^2 - 4ac$ is positive, the quadratic equation has two real solutions.

The discriminant helps us determine the number and type of solutions of a quadratic equation, $ax^2 + bx + c = 0$. Recall that the solutions of this equation are the same as the $x$-intercepts of its related graph $f(x) = ax^2 + bx + c$. This means that the discriminant of $ax^2 + bx + c = 0$ also tells us the number of $x$-intercepts for the graph of $f(x) = ax^2 + bx + c$.

## GRAPH OF $f(x) = ax^2 + bx + c$

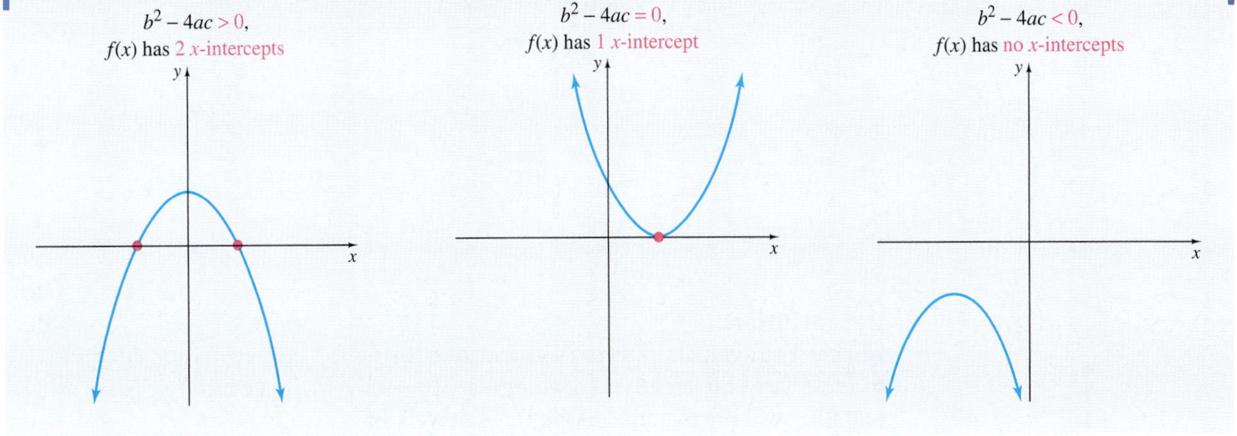


**3** The quadratic formula is useful in solving problems that are modeled by quadratic equations.

## Example 6 CALCULATING DISTANCE SAVED

At a local university, students often leave the sidewalk and cut across the lawn to save walking distance. Given the diagram below of a favorite place to cut across the lawn, approximate how many feet of walking distance a student saves by cutting across the lawn instead of walking on the sidewalk.

*Solution*

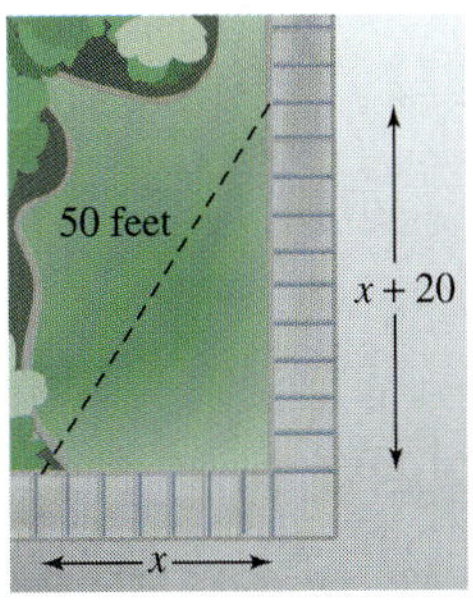


1. UNDERSTAND. Read and reread the problem. In the diagram, notice that a triangle is formed. Since the corner of the block forms a right angle, we use the Pythagorean theorem for right triangles. You may want to review this theorem.

2. TRANSLATE. By the Pythagorean theorem, we have

   In words: $(\text{leg})^2 + (\text{leg})^2 = (\text{hypotenuse})^2$

   Translate: $x^2 + (x + 20)^2 = 50^2$

3. SOLVE. Use the quadratic formula to solve.

$$x^2 + x^2 + 40x + 400 = 2500 \qquad \text{Square } (x + 20) \text{ and } 50.$$

$$2x^2 + 40x - 2100 = 0 \qquad \text{Set the equation equal to } 0.$$

$$x^2 + 20x - 1050 = 0 \qquad \text{Divide by } 2.$$

Here, $a = 1, b = 20, c = -1050$. By the quadratic formula,

$$x = \frac{-20 \pm \sqrt{20^2 - 4(1)(-1050)}}{2 \cdot 1}$$

$$= \frac{-20 \pm \sqrt{400 + 4200}}{2} = \frac{-20 \pm \sqrt{4600}}{2}$$

$$= \frac{-20 \pm \sqrt{100 \cdot 46}}{2} = \frac{-20 \pm 10\sqrt{46}}{2}$$

$$= -10 \pm 5\sqrt{46} \quad \text{Simplify.}$$

4. INTERPRET.

**Check:** Your calculations in the quadratic formula. The length of a side of a triangle can't be negative, so we reject $-10 - 5\sqrt{46}$. Since $-10 + 5\sqrt{46} \approx 24$ feet, the walking distance along the sidewalk is

$$x + (x + 20) \approx 24 + (24 + 20) = 68 \text{ feet.}$$

**State:** A student saves $68 - 50$ or 18 feet of walking distance by cutting across the lawn.

## SPOTLIGHT ON DECISION MAKING

Suppose you are a registered dietician. Recently, you read an article in a nutrition journal that described a relationship between weight and the Recommended Dietary Allowance (RDA) for vitamin A in children up to age 10. The relationship is $y = 0.149x^2 - 4.475x + 406.478$, where $y$ is the RDA for vitamin A in micrograms for a child whose weight is $x$ pounds. (*Source:* Food and Nutrition Board, National Academy of Sciences—Institute of Medicine, 1989)

You are working with a 4-year-old patient who weighs 40 pounds. After analyzing her diet, you are able to determine that she is currently getting an average of 400 micrograms of vitamin A daily. Decide whether her current vitamin A intake is adequate. If not, how much more is needed each day? In either case, determine how much weight she will need to gain before a daily intake of 500 micrograms of vitamin A is appropriate.

## Exercise Set 8.2

*Use the quadratic formula to solve each equation. These equations have real number solutions. See Examples 1 through 3.*

**1.** $m^2 + 5m - 6 = 0$

**2.** $p^2 + 11p - 12 = 0$

**3.** $2y = 5y^2 - 3$

**4.** $5x^2 - 3 = 14x$

**5.** $x^2 - 6x + 9 = 0$

**6.** $y^2 + 10y + 25 = 0$

**7.** $x^2 + 7x + 4 = 0$

**8.** $y^2 + 5y + 3 = 0$

**9.** $8m^2 - 2m = 7$

**10.** $11n^2 - 9n = 1$

**11.** $3m^2 - 7m = 3$

**12.** $x^2 - 13 = 5x$

**13.** $\frac{1}{2}x^2 - x - 1 = 0$

**14.** $\frac{1}{6}x^2 + x + \frac{1}{3} = 0$

**15.** $\frac{2}{5}y^2 + \frac{1}{5}y = \frac{3}{5}$

**16.** $\frac{1}{8}x^2 + x = \frac{5}{2}$

**17.** $\frac{1}{3}y^2 - y - \frac{1}{6} = 0$

**18.** $\frac{1}{2}y^2 = y + \frac{1}{2}$

**19.** Solve Exercise 1 by factoring. Explain the result.

**20.** Solve Exercise 2 by factoring. Explain the result.

*Use the quadratic formula to solve each equation. See Example 4.*

**21.** $6 = -4x^2 + 3x$
**22.** $9x^2 + x + 2 = 0$
**23.** $(x + 5)(x - 1) = 2$
**24.** $x(x + 6) = 2$
**25.** $10y^2 + 10y + 3 = 0$
**26.** $3y^2 + 6y + 5 = 0$

*The solutions of the quadratic equation* $ax^2 + bx + c = 0$ *are*

$$\frac{-b + \sqrt{b^2 - 4ac}}{2a} \quad \text{and} \quad \frac{-b - \sqrt{b^2 - 4ac}}{2a}$$

**27.** Show that the sum of these solutions is $\frac{-b}{a}$.

**28.** Show that the product of these solutions is $\frac{c}{a}$.

*Use the discriminant to determine the number and type of solutions of each equation. See Example 5.*

**29.** $9x - 2x^2 + 5 = 0$
**30.** $5 - 4x + 12x^2 = 0$
**31.** $4x^2 + 12x = -9$
**32.** $9x^2 + 1 = 6x$
**33.** $3x = -2x^2 + 7$
**34.** $3x^2 = 5 - 7x$
**35.** $6 = 4x - 5x^2$
**36.** $8x = 3 - 9x^2$

*Use the quadratic formula to solve each equation. These equations have real number solutions.*

**37.** $x^2 + 5x = -2$
**38.** $y^2 - 8 = 4y$
**39.** $(m + 2)(2m - 6) = 5(m - 1) - 12$
**40.** $7p(p - 2) + 2(p + 4) = 3$
**41.** $\frac{x^2}{3} - x = \frac{5}{3}$
**42.** $\frac{x^2}{2} - 3 = -\frac{9}{2}x$
**43.** $x(6x + 2) - 3 = 0$
**44.** $x(7x + 1) = 2$

*Use the quadratic formula to solve each equation.*

**45.** $x^2 + 6x + 13 = 0$
**46.** $x^2 + 2x + 2 = 0$
**47.** $\frac{2}{5}y^2 + \frac{1}{5}y + \frac{3}{5} = 0$
**48.** $\frac{1}{8}x^2 + x + \frac{5}{2} = 0$
**49.** $\frac{1}{2}y^2 = y - \frac{1}{2}$
**50.** $\frac{2}{3}x^2 - \frac{20}{3}x = -\frac{100}{6}$
**51.** $(n - 2)^2 = 15n$
**52.** $\left(p - \frac{1}{2}\right)^2 = \frac{p}{2}$

*Solve. See Example 6.*

△ **53.** Nancy, Thelma, and John Varner live on a corner lot. Often, neighborhood children cut across their lot to save walking distance. Given the diagram below, approximate to the nearest foot how many feet of walking distance is saved by cutting across their property instead of walking around the lot.

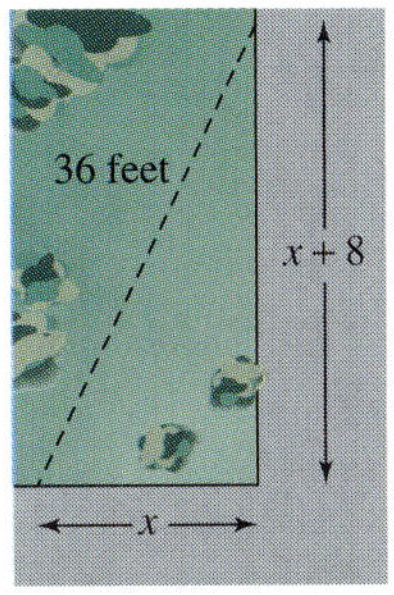


△ **54.** Given the diagram below, approximate to the nearest foot how many feet of walking distance a person saves by cutting across the lawn instead of walking on the sidewalk.

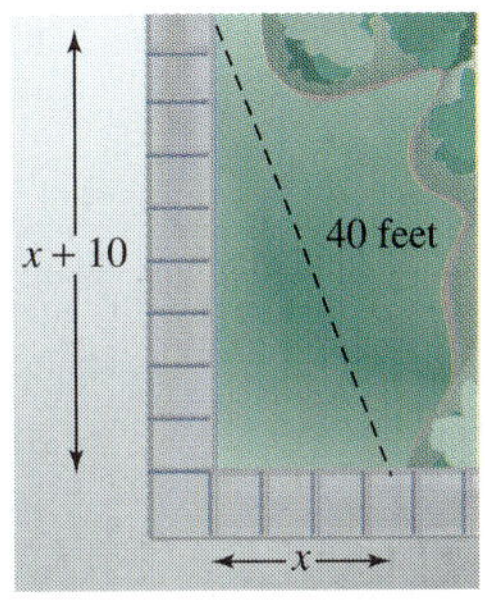


△ **55.** The hypotenuse of an isosceles right triangle is 2 centimeters longer than either of its legs. Find the exact length of each side. (*Hint:* An isosceles right triangle is a right triangle whose legs are the same length.)

△ **56.** The hypotenuse of an isosceles right triangle is one meter longer than either of its legs. Find the length of each side.

△ **57.** Uri Chechov's rectangular dog pen for his Irish setter must have an area of 400 square feet. Also, the length must be 10 feet longer than the width. Find the dimensions of the pen.

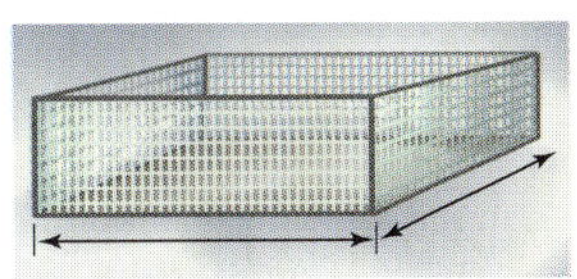

△ **58.** An entry in the Peach Festival Poster Contest must be rectangular and have an area of 1200 square inches. Furthermore, its length must be 20 inches longer than its width. Find the dimensions each entry must have.

△ **59.** A holding pen for cattle must be square and have a diagonal length of 100 meters.

**a.** Find the length of a side of the pen.

**b.** Find the area of the pen.

△ **60.** A rectangle is three times longer than it is wide. It has a diagonal of length 50 centimeters.

**a.** Find the dimensions of the rectangle.
**b.** Find the perimeter of the rectangle.

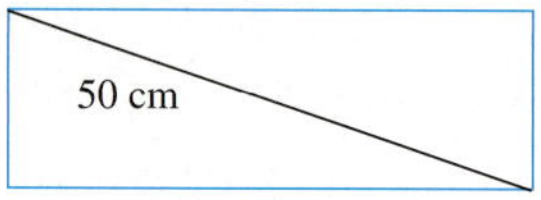


**61.** If a point $B$ divides a line segment such that the smaller portion is to the larger portion as the larger is to the whole, the whole is the length of the *golden ratio.*

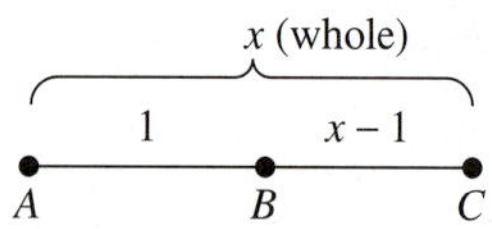


The golden ratio was thought by the Greeks to be the most pleasing to the eye, and many of their buildings contained numerous examples of the golden ratio. The value of the golden ratio is the positive solution of

$$\frac{x-1}{1} = \frac{1}{x}$$

(smaller)/(larger) = (larger)/(whole)

Find this value.

△ **62.** The base of a triangle is four more than twice its height. If the area of the triangle is 42 square centimeters, find its base and height.

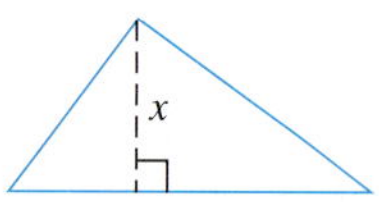


*The Wollomombi Falls in Australia have a height of 1100 feet. A pebble is thrown upward from the top of the falls with an initial velocity of 20 feet per second. The height of the pebble $h$ after $t$ seconds is given by the equation $h = -16t^2 + 20t + 1100$. Use this equation for Exercises 63 and 64.*

**63.** How long after the pebble is thrown will it hit the ground? Round to the nearest tenth of a second.

**64.** How long after the pebble is thrown will it be 550 feet from the ground? Round to the nearest tenth of a second.

*A ball is thrown downward from the top of a 180-foot building with an initial velocity of 20 feet per second. The height of the ball $h$ after $t$ seconds is given by the equation $h = -16t^2 - 20t + 180$. Use this equation to answer Exercises 65 and 66.*

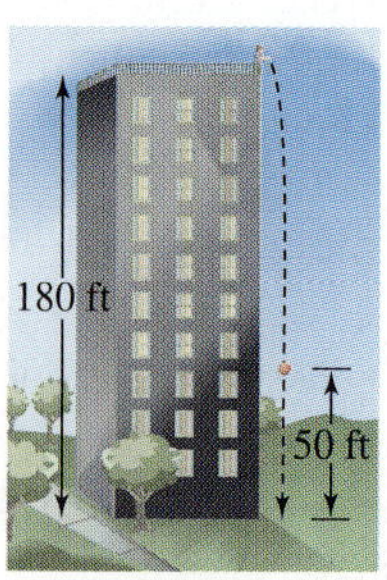


**65.** How long after the ball is thrown will it strike the ground? Round the result to the nearest tenth of a second.

**66.** How long after the ball is thrown will it be 50 feet from the ground? Round the result to the nearest tenth of a second.

*The accompanying graph shows the daily low temperatures for one week in New Orleans, Louisiana.*

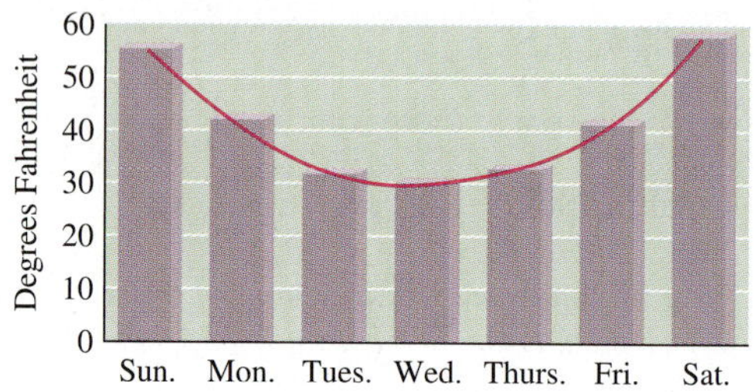


**67.** Which day of the week shows the greatest decrease in low temperature?

**68.** Which day of the week shows the greatest increase in low temperature?

**69.** Which day of the week had the lowest low temperature?

**70.** Use the graph to estimate the low temperature on Thursday.

*Notice that the shape of the temperature graph is similar to a parabola (see Section 5.9). In fact, this graph can be modeled by the quadratic function $f(x) = 3x^2 - 18x + 56$, where $f(x)$ is the temperature in degrees Fahrenheit and $x$ is the number of days from Sunday. (This graph is shown in red above.) Use this function to answer Exercises 71 and 72.*

**71.** Use the quadratic function given to approximate the temperature on Thursday. Does your answer agree with the graph above?

**72.** Use the function given and the quadratic formula to find when the temperature was 35°F. [*Hint:* Let $f(x) = 35$ and solve for $x$.] Round your answer to one decimal place and interpret your result. Does your answer agree with the graph above?

**73.** Wal-Mart Stores' net income can be modeled by the quadratic function $f(x) = 128.5x^2 - 69.5x + 2681$, where $f(x)$ is net income in millions of dollars and $x$ is the number of years after 1995. (*Source:* Based on data from Wal-Mart Stores, Inc.)

**a.** Find Wal-Mart's net income in 1997.

**b.** If the trend described by the model continues, predict the year after 1995 in which Wal-Mart's net income will be $15,000 million. Round to the nearest whole year.

**74.** The number of inmates in custody in U.S. prisons and jails can be modeled by the quadratic function

$p(x) = -716.2x^2 + 87{,}453.7x + 1{,}148{,}702$ where $p(x)$ is the number of inmates and $x$ is the number of years after 1990. (*Source:* Based on data from the Bureau of Justice Statistics, U.S. Department of Justice, 1990–1998) Round **a** and **b** to the nearest ten thousand.

**a.** Find the number of prison inmates in the United States in 1992.

**b.** Find the number of prison inmates in the United States in 1998.

**75.** Use a graphing calculator to solve Exercises 63 and 65.

**76.** Use a graphing calculator to solve Exercises 64 and 66.

*Recall that the discriminant also tells us the number of x-intercepts of the related function.*

**77.** Check the results of Exercise 29 by graphing $y = 9x - 2x^2 + 5$.

**78.** Check the results of Exercise 30 by graphing $y = 5 - 4x + 12x^2$.

## REVIEW EXERCISES

*Solve each equation. See Sections 6.6 and 7.6.*

**79.** $\sqrt{5x - 2} = 3$

**80.** $\sqrt{y + 2} + 7 = 12$

**81.** $\frac{1}{x} + \frac{2}{5} = \frac{7}{x}$

**82.** $\frac{10}{z} = \frac{5}{z} - \frac{1}{3}$

*Factor. See Section 5.7.*

**83.** $x^4 + x^2 - 20$

**84.** $2y^4 + 11y^2 - 6$

**85.** $z^4 - 13z^2 + 36$

**86.** $x^4 - 1$

## *A Look Ahead*

**Example**

Solve $x^2 - 3\sqrt{2}x + 2 = 0$.

*Solution*

In this equation, $a = 1$, $b = -3\sqrt{2}$, and $c = 2$. By the quadratic formula, we have

$$x = \frac{-b \pm \sqrt{b^2 - 4ac}}{2a}$$

$$= \frac{3\sqrt{2} \pm \sqrt{(-3\sqrt{2})^2 - 4(1)(2)}}{2(1)}$$

$$= \frac{3\sqrt{2} \pm \sqrt{18 - 8}}{2} = \frac{3\sqrt{2} \pm \sqrt{10}}{2}$$

The solution set is $\left\{\frac{3\sqrt{2} + \sqrt{10}}{2}, \frac{3\sqrt{2} - \sqrt{10}}{2}\right\}$.

*Use the quadratic formula to solve each quadratic equation. See the preceding example.*

**87.** $3x^2 - \sqrt{12}x + 1 = 0$

**88.** $5x^2 + \sqrt{20}x + 1 = 0$

**89.** $x^2 + \sqrt{2}x + 1 = 0$

**90.** $x^2 - \sqrt{2}x + 1 = 0$

**91.** $2x^2 - \sqrt{3}x - 1 = 0$

**92.** $7x^2 + \sqrt{7}x - 2 = 0$

# 8.3 SOLVING EQUATIONS BY USING QUADRATIC METHODS

▶ OBJECTIVES

1. Solve various equations that are quadratic in form.
2. Solve problems that lead to quadratic equations.

**1** In this section, we discuss various types of equations that can be solved in part by using the methods for solving quadratic equations.

Once each equation is simplified, you may want to use these steps when deciding what method to use to solve the quadratic equation.

**SOLVING A QUADRATIC EQUATION**

**Step 1:** If the equation is in the form $(ax + b)^2 = c$, use the square root property and solve. If not, go to Step 2.
**Step 2:** Write the equation in standard form: $ax^2 + bx + c = 0$.
**Step 3:** Try to solve the equation by the factoring method. If not possible, go to Step 4.
**Step 4:** Solve the equation by the quadratic formula.

The first example is a radical equation that becomes a quadratic equation once we square both sides.

**Example 1** Solve $x - \sqrt{x} - 6 = 0$.

*Solution* Recall that to solve a radical equation, first get the radical alone on one side of the equation. Then square both sides.

$$x - 6 = \sqrt{x} \quad \text{Add } \sqrt{x} \text{ to both sides.}$$

$$(x - 6)^2 = (\sqrt{x})^2 \quad \text{Square both sides.}$$

$$x^2 - 12x + 36 = x$$

$$x^2 - 13x + 36 = 0 \quad \text{Set the equation equal to 0.}$$

$$(x - 9)(x - 4) = 0$$

$$x - 9 = 0 \quad \text{or} \quad x - 4 = 0$$

$$x = 9 \qquad\qquad x = 4$$

*Check*

| Let $x = 9$ | | Let $x = 4$ | |
|---|---|---|---|
| $x - \sqrt{x} - 6 = 0$ | | $x - \sqrt{x} - 6 = 0$ | |
| $9 - \sqrt{9} - 6 \stackrel{?}{=} 0$ | | $4 - \sqrt{4} - 6 \stackrel{?}{=} 0$ | |
| $9 - 3 - 6 \stackrel{?}{=} 0$ | | $4 - 2 - 6 \stackrel{?}{=} 0$ | |
| $0 = 0$ | True. | $-4 = 0$ | False. |

The solution is 9.

**Example 2** Solve $\dfrac{3x}{x - 2} - \dfrac{x + 1}{x} = \dfrac{6}{x(x - 2)}$.

*Solution* In this equation, $x$ cannot be either 2 or 0, because these values cause denominators to equal zero. To solve for $x$, we first multiply both sides of the equation by $x(x - 2)$ to clear the fractions. By the distributive property, this means that we multiply each term by $x(x-2)$.

$$x(x-2)\left(\frac{3x}{x-2}\right) - x(x-2)\left(\frac{x+1}{x}\right) = x(x-2)\left[\frac{6}{x(x-2)}\right]$$

$$3x^2 - (x-2)(x+1) = 6 \quad \text{Simplify.}$$

$$3x^2 - (x^2 - x - 2) = 6 \quad \text{Multiply.}$$

$$3x^2 - x^2 + x + 2 = 6$$

$$2x^2 + x - 4 = 0 \quad \text{Simplify.}$$

This equation cannot be factored using integers, so we solve by the quadratic formula.

$$x = \frac{-1 \pm \sqrt{1^2 - 4(2)(-4)}}{2 \cdot 2}$$ Use $a = 2, b = 1,$ and $c = -4$ in the quadratic formula.

$$= \frac{-1 \pm \sqrt{1 + 32}}{4}$$ Simplify.

$$= \frac{-1 \pm \sqrt{33}}{4}$$

Neither proposed solution will make the denominators 0.

The solutions are $\frac{-1 + \sqrt{33}}{4}$ and $\frac{-1 - \sqrt{33}}{4}$.

**Example 3** Solve $p^4 - 3p^2 - 4 = 0$.

*Solution* First we factor the trinomial.

$$p^4 - 3p^2 - 4 = 0$$

$$(p^2 - 4)(p^2 + 1) = 0 \quad \text{Factor.}$$

$$(p-2)(p+2)(p^2+1) = 0 \quad \text{Factor further.}$$

$$p - 2 = 0 \quad \text{or} \quad p + 2 = 0 \quad \text{or} \quad p^2 + 1 = 0$$ Set each factor equal to 0 and solve.

$$p = 2 \qquad p = -2 \qquad p^2 = -1$$

$$p = \pm\sqrt{-1} = \pm i$$

The solutions are 2, $-2$, $i$ and $-i$.

**HELPFUL HINT**

Example 3 can be solved using substitution also. Think of

$p^4 - 3p^2 - 4 = 0$ as

$(p^2)^2 - 3p^2 - 4 = 0$ Then let $x = p^2$, and solve and substitute back. The solutions will be the same.

$x^2 - 3x - 4 = 0$

**Example 4** Solve $(x-3)^2 - 3(x-3) - 4 = 0$.

*Solution* Notice that the quantity $(x-3)$ is repeated in this equation. Sometimes it is help-ful to substitute a variable (in this case other than $x$) for the repeated quantity. We will let $y = x - 3$. Then

$$(x - 3)^2 - 3(x - 3) - 4 = 0$$

becomes

$$y^2 - 3y - 4 = 0 \quad \text{Let } x - 3 = y.$$

$$(y - 4)(y + 1) = 0 \quad \text{Factor.}$$

To solve, we use the zero factor property.

$$y - 4 = 0 \quad \text{or} \quad y + 1 = 0 \quad \text{Set each factor equal to 0.}$$

$$y = 4 \qquad\qquad y = -1 \quad \text{Solve.}$$

**HELPFUL HINT**
When using substitution, don't forget to substitute back to the original variable.

To find values of $x$, we substitute back. That is, we substitute $x - 3$ for $y$.

$$x - 3 = 4 \quad \text{or} \quad x - 3 = -1$$

$$x = 7 \qquad\qquad x = 2$$

Both 2 and 7 check. The solutions are 2 and 7.

**Example 5** Solve $x^{2/3} - 5x^{1/3} + 6 = 0$.

*Solution* The key to solving this equation is recognizing that $x^{2/3} = (x^{1/3})^2$. We replace $x^{1/3}$ with $m$ so that

$$(x^{1/3})^2 - 5x^{1/3} + 6 = 0$$

becomes

$$m^2 - 5m + 6 = 0$$

Now we solve by factoring.

$$m^2 - 5m + 6 = 0$$

$$(m - 3)(m - 2) = 0 \quad \text{Factor.}$$

$$m - 3 = 0 \quad \text{or} \quad m - 2 = 0 \quad \text{Set each factor equal to 0.}$$

$$m = 3 \qquad\qquad m = 2$$

Since $m = x^{1/3}$, we have

$$x^{1/3} = 3 \quad \text{or} \quad x^{1/3} = 2$$

$$x = 3^3 = 27 \quad \text{or} \quad x = 2^3 = 8$$

Both 8 and 27 check. The solutions are 8 and 27.

2 The next example is a work problem. This problem is modeled by a rational equation that simplifies to a quadratic equation.

## Example 6 FINDING WORK TIME

Together, an experienced typist and an apprentice typist can process a document in 6 hours. Alone, the experienced typist can process the document 2 hours faster than the apprentice typist can. Find the time in which each person can process the document alone.

*Solution* 1. UNDERSTAND. Read and reread the problem. The key idea here is the relationship between the *time* (hours) it takes to complete the job and the *part of the job* completed in one unit of time (hour). For example, because they can complete the job together in 6 hours, the *part of the job* they can complete in 1 hour is $\frac{1}{6}$. Let

$x$ = the *time* in hours it takes the apprentice typist to complete the job alone

$x - 2$ = the *time* in hours it takes the experienced typist to complete the job alone

We can summarize in a chart the information discussed

| | *Total Hours to Complete Job* | *Part of Job Completed in 1 Hour* |
|---|---|---|
| *APPRENTICE TYPIST* | $x$ | $\frac{1}{x}$ |
| *EXPERIENCED TYPIST* | $x - 2$ | $\frac{1}{x-2}$ |
| *TOGETHER* | 6 | $\frac{1}{6}$ |

2. TRANSLATE.

| | | | | | |
|---|---|---|---|---|---|
| In words: | part of job completed by apprentice typist in 1 hour | added to | part of job completed by experienced typist in 1 hour | is equal to | part of job completed together in 1 hour |
| | ↓ | ↓ | ↓ | ↓ | ↓ |
| Translate: | $\frac{1}{x}$ | $+$ | $\frac{1}{x-2}$ | $=$ | $\frac{1}{6}$ |

3. SOLVE.

$$\frac{1}{x} + \frac{1}{x-2} = \frac{1}{6}$$

$$6x(x-2)\left(\frac{1}{x} + \frac{1}{x-2}\right) = 6x(x-2) \cdot \frac{1}{6}$$ Multiply both sides by the LCD, $6x(x-2)$.

$$6x(x-2) \cdot \frac{1}{x} + 6x(x-2) \cdot \frac{1}{x-2} = 6x(x-2) \cdot \frac{1}{6}$$ Use the distributive property.

$$6(x-2) + 6x = x(x-2)$$

$$6x - 12 + 6x = x^2 - 2x$$

$$0 = x^2 - 14x + 12$$

Now we can substitute $a = 1$, $b = -14$, and $c = 12$ into the quadratic formula and simplify.

$$x = \frac{-(-14) \pm \sqrt{(-14)^2 - 4(1)(12)}}{2 \cdot 1} = \frac{14 \pm \sqrt{148}}{2}$$

Using a calculator or a square root table, we see that $\sqrt{148} \approx 12.2$ rounded to one decimal place. Thus,

$$x \approx \frac{14 \pm 12.2}{2}$$

$$x \approx \frac{14 + 12.2}{2} = 13.1 \quad \text{or} \quad x \approx \frac{14 - 12.2}{2} = 0.9$$

4. INTERPRET.

**Check:** If the apprentice typist completes the job alone in 0.9 hours, the experienced typist completes the job alone in $x - 2 = 0.9 - 2 = -1.1$ hours. Since this is not possible, we reject the solution of 0.9. The approximate solution thus is 13.1 hours.

**State:** The apprentice typist can complete the job alone in approximately 13.1 hours, and the experienced typist can complete the job alone in approximately

$$x - 2 = 13.1 - 2 = 11.1 \text{ hours.}$$

## Example 7 FINDING SPEED

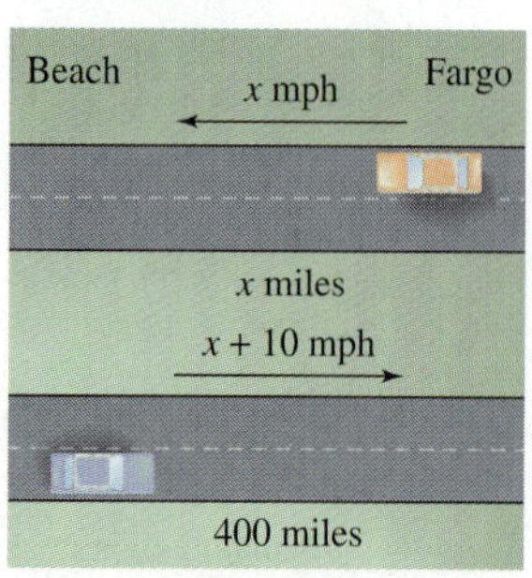


Beach and Fargo are about 400 miles apart. A salesperson travels from Fargo to Beach one day at a certain speed. She returns to Fargo the next day and drives 10 mph faster. Her total travel time was $14\frac{2}{3}$ hours. Find her speed to Beach and the return speed to Fargo.

*Solution* 1. UNDERSTAND. Read and reread the problem. Let

$x$ = the speed to Beach, so

$x + 10$ = the return speed to Fargo.

Then organize the given information in a table.

**HELPFUL HINT**
Since $d = rt$, then $t = \frac{d}{r}$. The time column was completed using $\frac{d}{r}$.

| | distance | = | rate | · | time |
|---|---|---|---|---|---|
| TO BEACH | 400 | | $x$ | | $\frac{400}{x}$ ← distance / ← rate |
| RETURN TO FARGO | 400 | | $x + 10$ | | $\frac{400}{x+10}$ ← distance / ← rate |

2. TRANSLATE.

In words: time to Beach + return time to Fargo = $14\frac{2}{3}$ hours

Translate: $\frac{400}{x} + \frac{400}{x+10} = \frac{44}{3}$

3. SOLVE.

$$\frac{400}{x} + \frac{400}{x+10} = \frac{44}{3}$$

$$\frac{100}{x} + \frac{100}{x+10} = \frac{11}{3} \quad \text{Divide both sides by 4.}$$

$$3x(x+10)\left(\frac{100}{x} + \frac{100}{x+10}\right) = 3x(x+10)\cdot\frac{11}{3} \quad \text{Multiply both sides by the LCD } 3x(x+10).$$

$$3x(x+10)\cdot\frac{100}{x} + 3x(x+10)\cdot\frac{100}{x+10} = 3x(x+10)\cdot\frac{11}{3} \quad \text{Use the distributive property.}$$

$$3(x+10)\cdot 100 + 3x\cdot 100 = x(x+10)\cdot 11$$

$$300x + 3000 + 300x = 11x^2 + 110x$$

$$0 = 11x^2 - 490x - 3000 \quad \text{Set equation equal to 0.}$$

$$0 = (11x + 60)(x - 50) \quad \text{Factor.}$$

$$11x + 60 = 0 \quad \text{or} \quad x - 50 = 0 \quad \text{Set each factor equal to 0.}$$

$$x = -\frac{60}{11} \text{ or } -5\frac{5}{11} \qquad x = 50$$

4. INTERPRET.

*Check* The speed is not negative, so it's not $-5\frac{5}{11}$. The number 50 does check.

*State* The speed to Beach was 50 mph and her return speed to Fargo was 60 mph.

## Exercise Set 8.3

*Solve. See Example 1.*

1. $2x = \sqrt{10 + 3x}$
2. $3x = \sqrt{8x + 1}$
3. $x - 2\sqrt{x} = 8$
4. $x - \sqrt{2x} = 4$
5. $\sqrt{9x} = x + 2$
6. $\sqrt{16x} = x + 3$

*Solve. See Example 2.*

7. $\dfrac{2}{x} + \dfrac{3}{x - 1} = 1$
8. $\dfrac{6}{x^2} = \dfrac{3}{x + 1}$
9. $\dfrac{3}{x} + \dfrac{4}{x + 2} = 2$
10. $\dfrac{5}{x - 2} + \dfrac{4}{x + 2} = 1$
11. $\dfrac{7}{x^2 - 5x + 6} = \dfrac{2x}{x - 3} - \dfrac{x}{x - 2}$
12. $\dfrac{11}{2x^2 + x - 15} = \dfrac{5}{2x - 5} - \dfrac{x}{x + 3}$

*Solve. See Example 3.*

13. $p^4 - 16 = 0$
14. $x^4 + 2x^2 - 3 = 0$
15. $4x^4 + 11x^2 = 3$
16. $z^4 = 81$
17. $z^4 - 13z^2 + 36 = 0$
18. $9x^4 + 5x^2 - 4 = 0$

*Solve. See Examples 4 and 5.*

19. $x^{2/3} - 3x^{1/3} - 10 = 0$
20. $x^{2/3} + 2x^{1/3} + 1 = 0$
21. $(5n + 1)^2 + 2(5n + 1) - 3 = 0$
22. $(m - 6)^2 + 5(m - 6) + 4 = 0$
23. $2x^{2/3} - 5x^{1/3} = 3$
24. $3x^{2/3} + 11x^{1/3} = 4$
25. $1 + \dfrac{2}{3t - 2} = \dfrac{8}{(3t - 2)^2}$
26. $2 - \dfrac{7}{x + 6} = \dfrac{15}{(x + 6)^2}$
27. $20x^{2/3} - 6x^{1/3} - 2 = 0$
28. $4x^{2/3} + 16x^{1/3} = -15$

*Solve. See Examples 1 through 5.*

29. $a^4 - 5a^2 + 6 = 0$
30. $x^4 - 12x^2 + 11 = 0$
31. $\dfrac{2x}{x - 2} + \dfrac{x}{x + 3} = -\dfrac{5}{x + 3}$
32. $\dfrac{5}{x - 3} + \dfrac{x}{x + 3} = \dfrac{19}{x^2 - 9}$
33. $(p + 2)^2 = 9(p + 2) - 20$
34. $2(4m - 3)^2 - 9(4m - 3) = 5$
35. $2x = \sqrt{11x + 3}$
36. $4x = \sqrt{2x + 3}$
37. $x^{2/3} - 8x^{1/3} + 15 = 0$
38. $x^{2/3} - 2x^{1/3} - 8 = 0$
39. $y^3 + 9y - y^2 - 9 = 0$
40. $x^3 + x - 3x^2 - 3 = 0$
41. $2x^{2/3} + 3x^{1/3} - 2 = 0$
42. $6x^{2/3} - 25x^{1/3} - 25 = 0$
43. $x^{-2} - x^{-1} - 6 = 0$
44. $y^{-2} - 8y^{-1} + 7 = 0$
45. $x - \sqrt{x} = 2$
46. $x - \sqrt{3x} = 6$
47. $\dfrac{x}{x - 1} + \dfrac{1}{x + 1} = \dfrac{2}{x^2 - 1}$
48. $\dfrac{x}{x - 5} + \dfrac{5}{x + 5} = -\dfrac{1}{x^2 - 25}$
49. $p^4 - p^2 - 20 = 0$
50. $x^4 - 10x^2 + 9 = 0$
51. $2x^3 = -54$
52. $y^3 - 216 = 0$
53. $1 = \dfrac{4}{x - 7} + \dfrac{5}{(x - 7)^2}$
54. $3 + \dfrac{1}{2p + 4} = \dfrac{10}{(2p + 4)^2}$
55. $27y^4 + 15y^2 = 2$
56. $8z^4 + 14z^2 = -5$

*Solve. See Examples 6 and 7.*

57. A jogger ran 3 miles, decreased her speed by 1 mile per hour, and then ran another 4 miles. If her total time jogging was $1\frac{3}{5}$ hours, find her speed for each part of her run.

58. Mark Keaton's workout consists of jogging for 3 miles, and then riding his bike for 5 miles at a speed 4 miles per hour faster than he jogs. If his total workout time is 1 hour, find his jogging speed and his biking speed.

**59.** A Chinese restaurant in Mandeville, Louisiana, has a large goldfish pond around the restaurant. Suppose that an inlet pipe and a hose together can fill the pond in 8 hours. The inlet pipe alone can complete the job in one hour less time than the hose alone. Find the time that the hose can complete the job alone and the time that the inlet pipe can complete the job alone. Round each to the nearest tenth of an hour.

**60.** A water tank on a farm in Flatonia, Texas, can be filled with a large inlet pipe and a small inlet pipe in 3 hours. The large inlet pipe alone can fill the tank in 2 hours less time than the small inlet pipe alone. Find the time to the nearest tenth of an hour each pipe can fill the tank alone.

**61.** Roma Sherry drove 330 miles from her hometown to Tucson. During her return trip, she was able to increase her speed by 11 mph. If her return trip took 1 hour less time, find her original speed and her speed returning home.

**62.** A salesperson drove to Portland, a distance of 300 miles. During the last 80 miles of his trip, heavy rainfall forced him to decrease his speed by 15 mph. If his total driving time was 6 hours, find his original speed and his speed during the rainfall.

**63.** Bill Shaughnessy and his son Billy can clean the house together in 4 hours. When the son works alone, it takes him an hour longer to clean than it takes his dad alone. Find how long to the nearest tenth of an hour it takes the son to clean alone.

**64.** Together, Noodles and Freckles eat a 50-pound bag of dog food in 30 days. Noodles by himself eats a 50-pound bag in 2 weeks less time than Freckles does by himself. How many days to the nearest whole day would a 50-pound bag of dog food last Freckles?

**65.** The product of a number and 4 less than the number is 96. Find the number.

**66.** A whole number increased by its square is two more than twice itself. Find the number.

△ **67.** Suppose that an open box is to be made from a square sheet of cardboard by cutting out squares from each corner as shown and then folding along the dotted lines. If the box is to have a volume of 300 cubic centimeters, find the original dimensions of the sheet of cardboard.

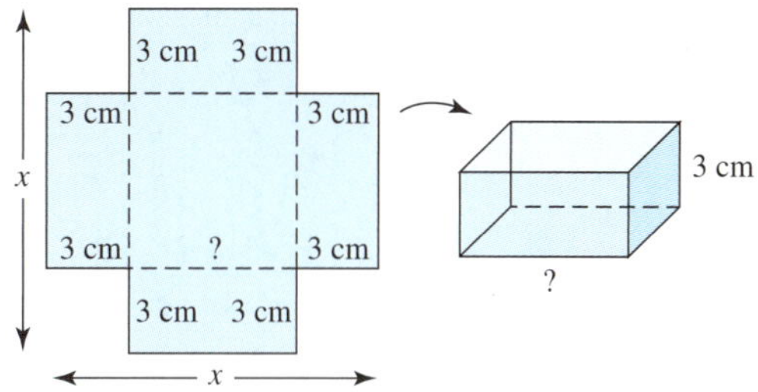


**a.** The ? in the drawing to the left will be the length (and also the width) of the box as shown in the drawing above. Represent this length in terms of $x$.

**b.** Use the formula for volume of a box, $V = l \cdot w \cdot h$, to write an equation in $x$.

**c.** Solve the equation for $x$ and give the dimensions of the sheet of cardboard. Check your solution.

△ **68.** Suppose that an open box is to be made from a square sheet of cardboard by cutting out squares from each corner as shown and then folding along the dotted lines. If the box is to have a volume of 128 cubic inches, find the original dimensions of the sheet of cardboard. (*Hint:* Use Exercise 67 Parts **a, b,** and **c** to help you.)

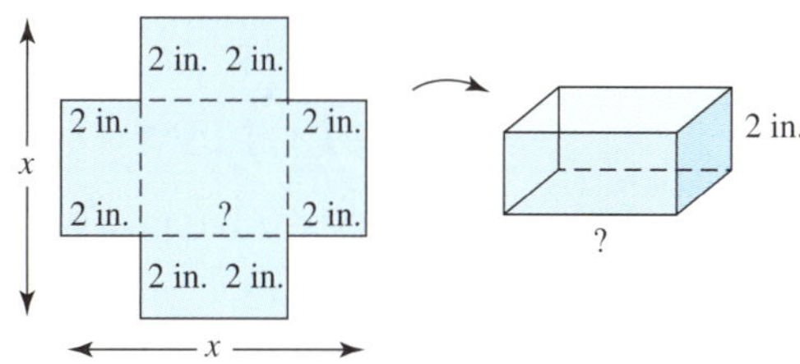


**69.** During the 2000 Grand Prix of Miami auto race, Juan Montoya posted the fastest lap speed but Max Papis won the race. The track is 7920 feet (1.5 miles) long. Montoya's fastest lap speed was 3.8 feet per second faster than Papis' fastest lap speed. Traveling at these fastest speeds, Papis would have taken 0.376 seconds longer than Montoya to complete a lap. (*Source:* Championship Auto Racing Teams, Inc.)

**a.** Find Max Papis' fastest lap speed during the race. Round to one decimal place.

**b.** Find Juan Montoya's fastest lap speed during the race. Round to one decimal place.

**c.** Convert each speed to miles per hour. Round to one decimal place.

**70.** Use a graphing calculator to solve Exercise 29. Compare the solution with the solution from Exercise 29. Explain any differences.

**71.** Write a polynomial equation that has three solutions: 2, 5, and −7.

**72.** Write a polynomial equation that has three solutions: 0, $2i$, and $-2i$.

## REVIEW EXERCISES

*Solve each inequality. See Section 2.4.*

**73.** $\frac{5x}{3} + 2 \le 7$

**74.** $\frac{2x}{3} + \frac{1}{6} \ge 2$

**75.** $\frac{y-1}{15} > -\frac{2}{5}$

**76.** $\frac{z-2}{12} < \frac{1}{4}$

*Find the domain and range of each graphed relation. Decide which relations are also functions. See Section 3.2.*

**77.**

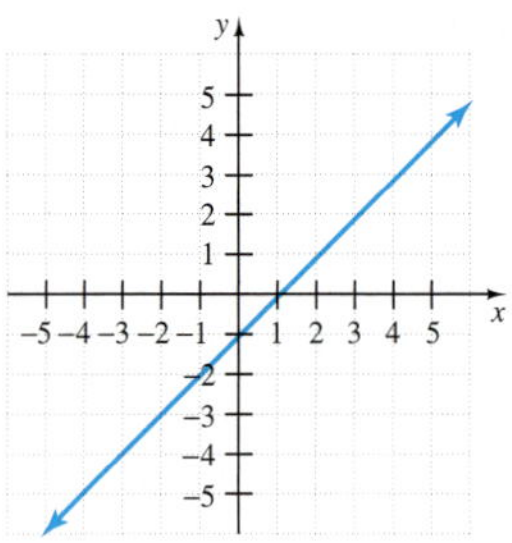

**78.**

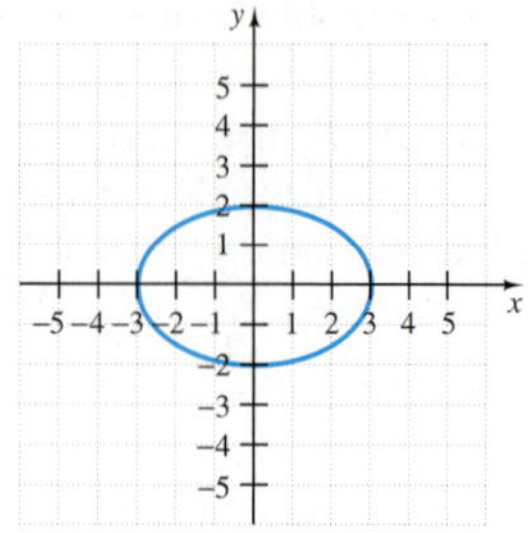

**79.**

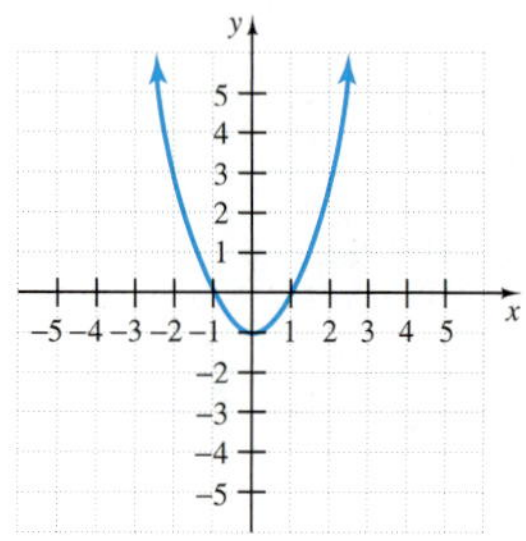

**80.**

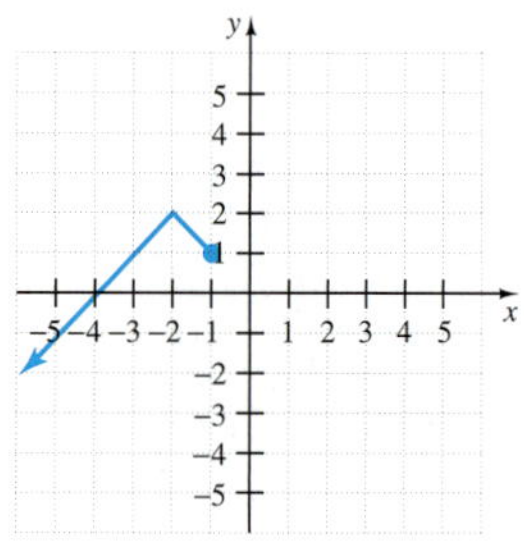

## 8.4 NONLINEAR INEQUALITIES IN ONE VARIABLE

**OBJECTIVES**

1. Solve polynomial inequalities of degree 2 or greater.
2. Solve inequalities that contain rational expressions with variables in the denominator.

**1** Just as we can solve linear inequalities in one variable, so can we also solve quadratic inequalities in one variable. A **quadratic inequality** is an inequality that can be written so that one side is a quadratic expression and the other side is 0. Here are examples of quadratic inequalities in one variable. Each is written in **standard form.**

$$x^2 - 10x + 7 \le 0 \qquad 3x^2 + 2x - 6 > 0$$

$$2x^2 + 9x - 2 < 0 \qquad x^2 - 3x + 11 \ge 0$$

A solution of a quadratic inequality in one variable is a value of the variable that makes the inequality a true statement.

The value of an expression such as $x^2 - 3x - 10$ will sometimes be positive, sometimes negative, and sometimes 0, depending on the value substituted for $x$. To solve the inequality $x^2 - 3x - 10 < 0$, we are looking for all values of $x$ that make the expression $x^2 - 3x - 10$ **less than 0,** or **negative.** To understand how we find these values, we'll study the graph of the quadratic function $y = x^2 - 3x - 10$.

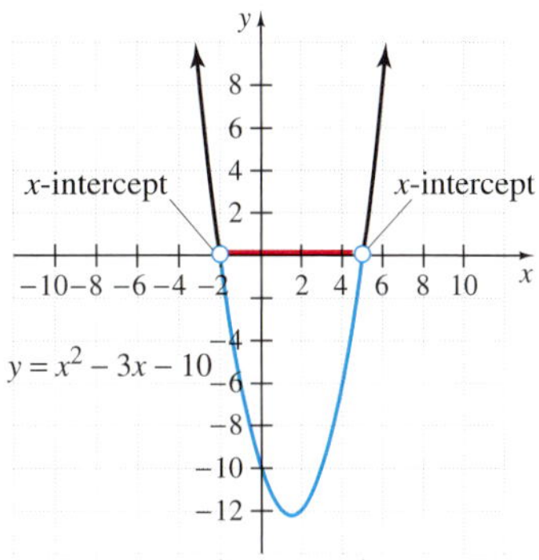


Notice that the $x$-values for which $y$ is positive are separated from the $x$ values for which $y$ is negative by the $x$-intercepts. (Recall that the $x$-intercepts correspond to values of $x$ for which $y = 0$.) Thus, the solution set of $x^2 - 3x - 10 < 0$ consists of all real numbers from $-2$ to 5, or in interval notation, $(-2, 5)$.

It is not necessary to graph $y = x^2 - 3x - 10$ to solve the related inequality $x^2 - 3x - 10 < 0$. Instead, we can draw a number line representing the $x$-axis and keep the following in mind: *A region on the number line for which the value of* $x^2 - 3x - 10$ *is positive is separated from a region on the number line for which the value of* $x^2 - 3x - 10$ *is negative by a value for which the expression is* 0.

Let's find these values for which the expression is 0 by solving the related equation:

$$x^2 - 3x - 10 = 0$$

$$(x - 5)(x + 2) = 0 \quad \text{Factor.}$$

$$x - 5 = 0 \quad \text{or} \quad x + 2 = 0 \quad \text{Set each factor equal to 0.}$$

$$x = 5 \qquad\qquad x = -2 \quad \text{Solve.}$$

These two numbers $-2$ and 5, divide the number line into three regions. We will call the regions $A$, $B$, and $C$. These regions are important because, if the value of $x^2 - 3x - 10$ is negative when a number from a region is substituted for $x$, then $x^2 - 3x - 10$ is negative when any number in that region is substituted for $x$. The same is true if the value of $x^2 - 3x - 10$ is positive for a particular value of $x$ in a region.

To see whether the inequality $x^2 - 3x - 10 < 0$ is true or false in each region, we choose a test point from each region and substitute its value for $x$ in the inequality

$x^2 - 3x - 10 < 0$. If the resulting inequality is true, the region containing the test point is a solution region.

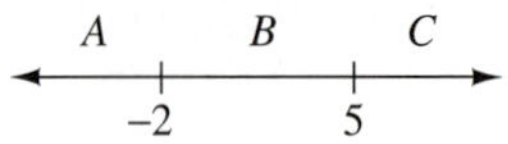


| *Region* | *Test Point Value* | *Factored Form* $(x - 5)(x + 2) < 0$ | *Result* |
|---|---|---|---|
| *A* | −3 | $(-8)(-1) < 0$ | False |
| *B* | 0 | $(-5)(2) < 0$ | True |
| *C* | 6 | $(1)(8) < 0$ | False |

The values in region *B* satisfy the inequality. The numbers −2 and 5 are not included in the solution set since the inequality symbol is <. The solution set is $(-2, 5)$, and its graph is shown.

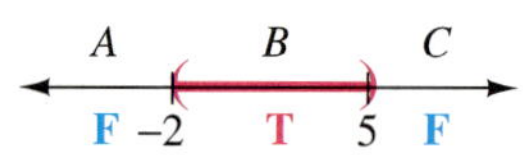


**Example 1** Solve $(x + 3)(x - 3) > 0$.

*Solution* First we solve the related equation $(x + 3)(x - 3) = 0$.

$$(x + 3)(x - 3) = 0$$

$$x + 3 = 0 \quad \text{or} \quad x - 3 = 0$$

$$x = -3 \qquad\qquad x = 3$$

The two numbers −3 and 3 separate the number line into three regions, *A*, *B*, and *C*.

Now we substitute the value of a test point from each region. If the test value satisfies the inequality, every value in the region containing the test value is a solution.

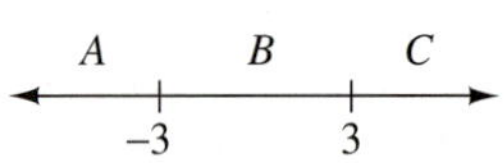


| *Region* | *Test Point Value* | $(x + 3)(x - 3) > 0$ | *Result* |
|---|---|---|---|
| *A* | −4 | $(-1)(-7) > 0$ | True |
| *B* | 0 | $(3)(-3) > 0$ | False |
| *C* | 4 | $(7)(1) > 0$ | True |

The points in regions *A* and *C* satisfy the inequality. The numbers −3 and 3 are not included in the solution since the inequality symbol is >. The solution set is $(-\infty, -3) \cup (3, \infty)$, and its graph is shown.

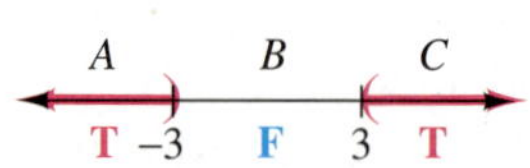

The following steps may be used to solve a polynomial inequality.

**SOLVING A POLYNOMIAL INEQUALITY**

**Step 1:** Write the inequality in standard form and then solve the related equation.
**Step 2:** Separate the number line into regions with the solutions from Step 1.
**Step 3:** For each region, choose a test point and determine whether its value satisfies the *original inequality.*
**Step 4:** The solution set includes the regions whose test point value is a solution. If the inequality symbol is $\leq$ or $\geq$, the values from Step 1 are solutions; if $<$ or $>$, they are not.

## Example 2

Solve $x^2 - 4x \leq 0$.

*Solution* First we solve the related equation $x^2 - 4x = 0$.

$$x^2 - 4x = 0$$
$$x(x - 4) = 0$$
$$x = 0 \quad \text{or} \quad x = 4$$

The numbers 0 and 4 separate the number line into three regions, $A$, $B$, and $C$.

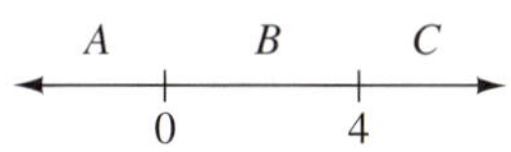


Check a test value in each region in the original inequality. Values in region $B$ satisfy the inequality. The numbers 0 and 4 are included in the solution since the inequality symbol is $\leq$. The solution set is $[0, 4]$, and its graph is shown.

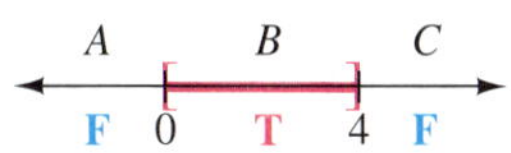


## Example 3

Solve $(x + 2)(x - 1)(x - 5) \leq 0$.

*Solution* First we solve $(x + 2)(x - 1)(x - 5) = 0$. By inspection, we see that the solutions are $-2, 1$, and $5$. They separate the number line into four regions, $A$, $B$, $C$, and $D$. Next we check test points from each region.

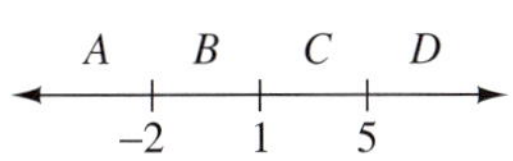


| *Region* | *Test Point Value* | $(x + 2)(x - 1)(x - 5) \leq 0$ | *Result* |
|---|---|---|---|
| $A$ | $-3$ | $(-1)(-4)(-8) \leq 0$ | True |
| $B$ | $0$ | $(2)(-1)(-5) \leq 0$ | False |
| $C$ | $2$ | $(4)(1)(-3) \leq 0$ | True |
| $D$ | $6$ | $(8)(5)(1) \leq 0$ | False |

The solution set is $(-\infty, -2] \cup [1, 5]$, and its graph is shown. We include the numbers $-2$, 1, and 5 because the inequality symbol is $\leq$.

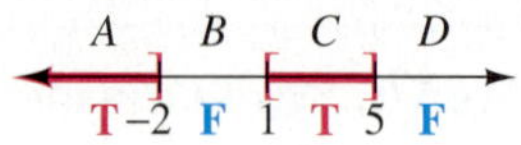


**2** Inequalities containing rational expressions with variables in the denominator are solved by using a similar procedure.

**Example 4** Solve $\dfrac{x+2}{x-3} \leq 0$.

*Solution* First we find all values that make the denominator equal to 0. To do this, we solve $x - 3 = 0$ and find that $x = 3$.

Next, we solve the related equation $\dfrac{x+2}{x-3} = 0$.

$$\frac{x+2}{x-3} = 0$$

$$x + 2 = 0 \quad \text{Multiply both sides by the LCD, } x - 3.$$

$$x = -2$$

Now we place these numbers on a number line and proceed as before, checking test point values in the original inequality.

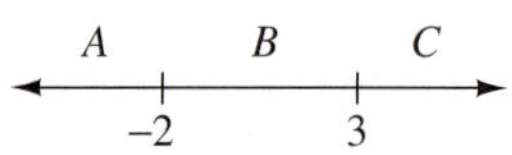


***Choose −3 from region A.***

$$\frac{x+2}{x-3} \leq 0$$

$$\frac{-3+2}{-3-3} \leq 0$$

$$\frac{-1}{-6} \leq 0$$

$$\frac{1}{6} \leq 0 \quad \text{False.}$$

***Choose 0 from region B.***

$$\frac{x+2}{x-3} \leq 0$$

$$\frac{0+2}{0-3} \leq 0$$

$$-\frac{2}{3} \leq 0 \quad \text{True.}$$

***Choose 4 from region C.***

$$\frac{x+2}{x-3} \leq 0$$

$$\frac{4+2}{4-3} \leq 0$$

$$6 \leq 0 \quad \text{False.}$$

The solution set is $[-2, 3)$. This interval includes $-2$ because $-2$ satisfies the original inequality. This interval does not include 3, because 3 would make the denominator 0.

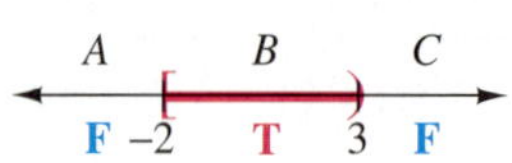


The following steps may be used to solve a rational inequality with variables in the denominator.

**SOLVING A RATIONAL INEQUALITY**

**Step 1:** Solve for values that make all denominators 0.
**Step 2:** Solve the related equation.
**Step 3:** Separate the number line into regions with the solutions from Steps 1 and 2.
**Step 4:** For each region, choose a test point and determine whether its value satisfies the *original inequality.*
**Step 5:** The solution set includes the regions whose test point value is a solution. Check whether to include values from Step 2. Be sure *not* to include values that make any denominator 0.

**Example 5** Solve $\frac{5}{x+1} < -2$.

*Solution* First we find values for $x$ that make the denominator equal to 0.

$$x + 1 = 0$$
$$x = -1$$

Next we solve $\frac{5}{x+1} = -2$.

$$(x+1) \cdot \frac{5}{x+1} = (x+1) \cdot -2 \quad \text{Multiply both sides by the LCD, } x + 1.$$
$$5 = -2x - 2 \quad \text{Simplify.}$$
$$7 = -2x$$
$$-\frac{7}{2} = x$$

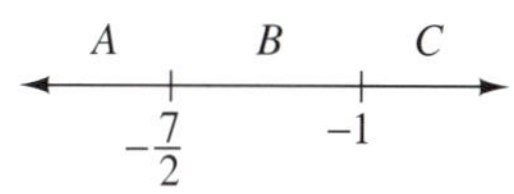


We use these two solutions to divide a number line into three regions and choose test points. Only a test point value from region $B$ satisfies the *original inequality.* The solution set is $\left(-\frac{7}{2}, -1\right)$, and its graph is shown.

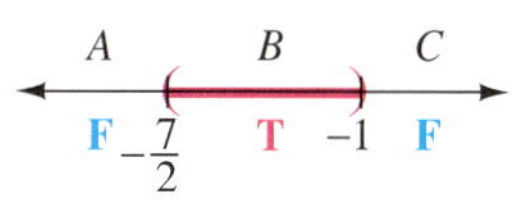


## *Exercise Set 8.4*

*Solve each quadratic inequality. Graph the solution set and write the solution set in interval notation. See Examples 1 through 3.*

**1.** $(x + 1)(x + 5) > 0$
**2.** $(x + 1)(x + 5) \le 0$
**3.** $(x - 3)(x + 4) \le 0$
**4.** $(x + 4)(x - 1) > 0$
**5.** $x^2 - 7x + 10 \le 0$
**6.** $x^2 + 8x + 15 \ge 0$
**7.** $3x^2 + 16x < -5$
**8.** $2x^2 - 5x < 7$
**9.** $(x - 6)(x - 4)(x - 2) > 0$
**10.** $(x - 6)(x - 4)(x - 2) \le 0$
**11.** $x(x - 1)(x + 4) \le 0$
**12.** $x(x - 6)(x + 2) > 0$
**13.** $(x^2 - 9)(x^2 - 4) > 0$
**14.** $(x^2 - 16)(x^2 - 1) \le 0$

*Solve each inequality. Graph the solution set and write the solution set in interval notation. See Example 4.*

**15.** $\frac{x+7}{x-2} < 0$ **16.** $\frac{x-5}{x-6} > 0$ **17.** $\frac{5}{x+1} > 0$

**18.** $\frac{3}{y-5} < 0$ **19.** $\frac{x+1}{x-4} \geq 0$ **20.** $\frac{x+1}{x-4} \leq 0$

**21.** Explain why $\frac{x+2}{x-3} > 0$ and $(x+2)(x-3) > 0$ have the same solutions.

**22.** Explain why $\frac{x+2}{x-3} \geq 0$ and $(x+2)(x-3) \geq 0$ do not have the same solutions.

*Solve each inequality. Graph the solution set and write the solution set in interval notation. See Example 5.*

**23.** $\frac{3}{x-2} < 4$ **24.** $\frac{-2}{y+3} > 2$ **25.** $\frac{x^2+6}{5x} \geq 1$

**26.** $\frac{y^2+15}{8y} \leq 1$

*Solve each inequality. Graph the solution set and write the solution set in interval notation.*

**27.** $(x-8)(x+7) > 0$ **28.** $(x-5)(x+1) < 0$

**29.** $(2x-3)(4x+5) \leq 0$ **30.** $(6x+7)(7x-12) > 0$

**31.** $x^2 > x$

**32.** $x^2 < 25$

**33.** $(2x-8)(x+4)(x-6) \leq 0$

**34.** $(3x-12)(x+5)(2x-3) \geq 0$

**35.** $6x^2 - 5x \geq 6$

**36.** $12x^2 + 11x \leq 15$

**37.** $4x^3 + 16x^2 - 9x - 36 > 0$

**38.** $x^3 + 2x^2 - 4x - 8 < 0$

**39.** $x^4 - 26x^2 + 25 \geq 0$

**40.** $16x^4 - 40x^2 + 9 \leq 0$

**41.** $(2x-7)(3x+5) > 0$

**42.** $(4x-9)(2x+5) < 0$

**43.** $\frac{x}{x-10} < 0$ **44.** $\frac{x+10}{x-10} > 0$

**45.** $\frac{x-5}{x+4} \geq 0$ **46.** $\frac{x-3}{x+2} \leq 0$

**47.** $\frac{x(x+6)}{(x-7)(x+1)} \geq 0$

**48.** $\frac{(x-2)(x+2)}{(x+1)(x-4)} \leq 0$

**49.** $\frac{-1}{x-1} > -1$ **50.** $\frac{4}{y+2} < -2$

**51.** $\frac{x}{x+4} \leq 2$ **52.** $\frac{4x}{x-3} \geq 5$

**53.** $\frac{z}{z-5} \geq 2z$ **54.** $\frac{p}{p+4} \leq 3p$

**55.** $\frac{(x+1)^2}{5x} > 0$ **56.** $\frac{(2x-3)^2}{x} < 0$

*Find all numbers that satisfy each of the following.*

**57.** A number minus its reciprocal is less than zero. Find the numbers.

**58.** Twice a number added to its reciprocal is nonnegative. Find the numbers.

**59.** The total profit function $P(x)$ for a company producing $x$ thousand units is given by

$$P(x) = -2x^2 + 26x - 44$$

Find the values of $x$ for which the company makes a profit. [*Hint:* The company makes a profit when $P(x) > 0$.]

**60.** A projectile is fired straight up from the ground with an initial velocity of 80 feet per second. Its height $s(t)$ in feet at any time $t$ is given by the function

$$s(t) = -16t^2 + 80t$$

Find the interval of time for which the height of the projectile is greater than 96 feet.

*Use a graphing calculator to check each exercise.*

**61.** Exercise 27. **62.** Exercise 28.

**63.** Exercise 39. **64.** Exercise 40.

## REVIEW EXERCISES

*Recall that the graph of $f(x) + K$ is the same as the graph of $f(x)$ shifted $K$ units upward if $K > 0$ and $|K|$ units downward if $K < 0$. Use the graph of $f(x) = |x|$ below to sketch the graph of each function. (See Sections 3.1 and 3.3.)*

**65.** $g(x) = |x| + 2$

**66.** $H(x) = |x| - 2$

**67.** $F(x) = |x| - 1$

**68.** $h(x) = |x| + 5$

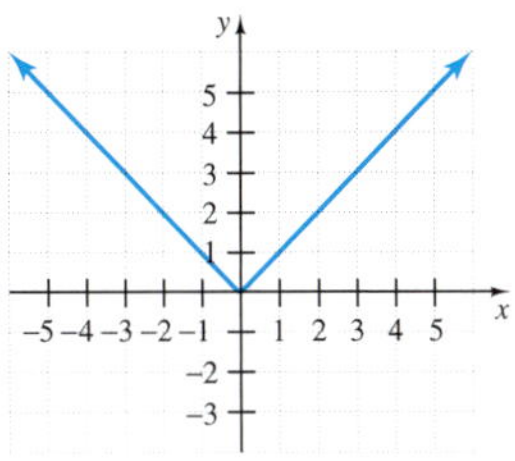


*Use the graph of $f(x) = x^2$ below to sketch the graph of each function.*

**69.** $F(x) = x^2 - 3$

**70.** $h(x) = x^2 - 4$

**71.** $H(x) = x^2 + 1$

**72.** $g(x) = x^2 + 3$

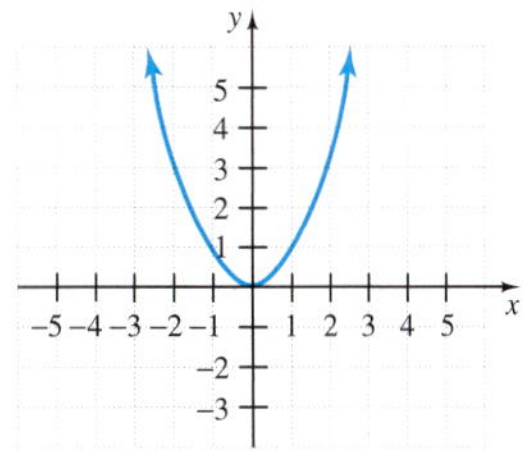


## 8.5 QUADRATIC FUNCTIONS AND THEIR GRAPHS

▶ **OBJECTIVES**

CD-ROM SSM SSG Video

1. Graph quadratic functions of the form $f(x) = x^2 + k$.
2. Graph quadratic functions of the form $f(x) = (x - h)^2$.
3. Graph quadratic functions of the form $f(x) = (x - h)^2 + k$.
4. Graph quadratic functions of the form $f(x) = ax^2$.
5. Graph quadratic functions of the form $f(x) = a(x - h)^2 + k$.

**1** We first graphed the quadratic equation $y = x^2$ in Section 3.1. In Section 3.2, we learned that this graph defines a function, and we wrote $y = x^2$ as $f(x) = x^2$. Quadratic functions and their graphs were studied further in Section 5.9. In these sections, we discovered that the graph of a quadratic function is a parabola opening upward or downward. In this section, we continue our study of quadratic functions and their graphs.

First, let's recall the definition of a quadratic function.

### QUADRATIC FUNCTION

A quadratic function is a function that can be written in the form $f(x) = ax^2 + bx + c$, where $a$, $b$, and $c$ are real numbers and $a \neq 0$.

Notice that equations of the form $y = ax^2 + bx + c$, where $a \neq 0$, define quadratic functions, since $y$ is a function of $x$ or $y = f(x)$.

Recall that if $a > 0$, the parabola opens upward and if $a < 0$, the parabola opens downward. Also, the vertex of a parabola is the lowest point if the parabola opens upward and the highest point if the parabola opens downward. The axis of symmetry is the vertical line that passes through the vertex.

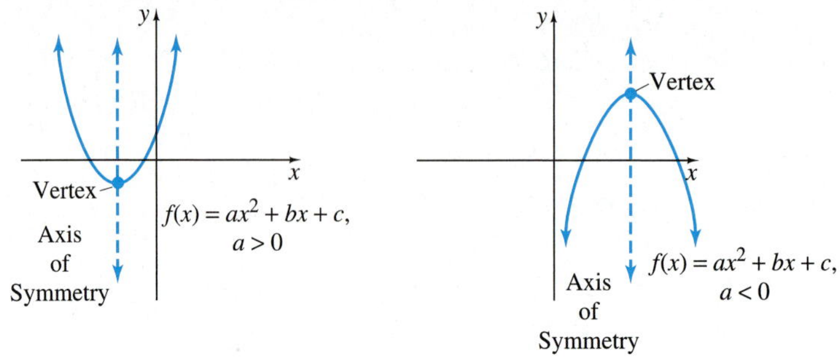


**Example 1** Graph $f(x) = x^2$ and $g(x) = x^2 + 3$ on the same set of axes.

*Solution* First we construct a table of values for $f(x)$ and plot the points. Notice that for each $x$-value, the corresponding value of $g(x)$ must be 3 more than the corresponding value of $f(x)$ since $f(x) = x^2$ and $g(x) = x^2 + 3$. In other words, the graph of $g(x) = x^2 + 3$ is the same as the graph of $f(x) = x^2$ shifted upward 3 units. The axis of symmetry for both graphs is the $y$-axis.

| $x$ | $f(x) = x^2$ | $g(x) = x^2 + 3$ |
|---|---|---|
| −2 | 4 | 7 |
| −1 | 1 | 4 |
| 0 | 0 | 3 |
| 1 | 1 | 4 |
| 2 | 4 | 7 |

Each $y$-value is increased by 3.

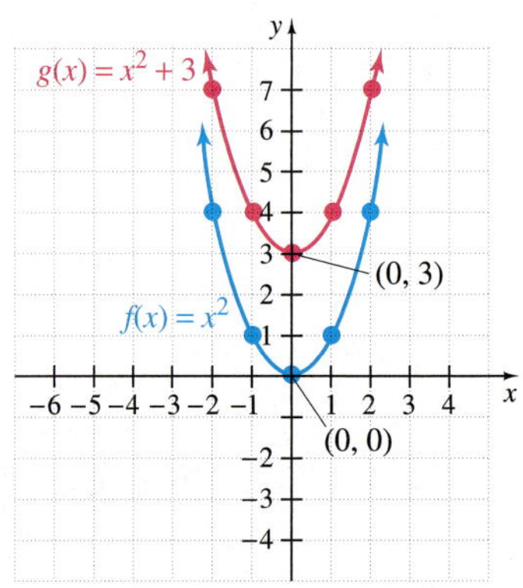


In general, we have the following properties.

**GRAPHING THE PARABOLA DEFINED BY $f(x) = x^2 + k$**

If $k$ is positive, the graph of $f(x) = x^2 + k$ is the graph of $y = x^2$ shifted upward $k$ units.

If $k$ is negative, the graph of $f(x) = x^2 + k$ is the graph of $y = x^2$ shifted downward $|k|$ units.

The vertex is $(0, k)$, and the axis of symmetry is the $y$-axis.

**Example 2** Graph each function.

**a.** $F(x) = x^2 + 2$ **b.** $g(x) = x^2 - 3$

*Solution* **a.** $F(x) = x^2 + 2$

The graph of $F(x) = x^2 + 2$ is obtained by shifting the graph of $y = x^2$ upward 2 units.

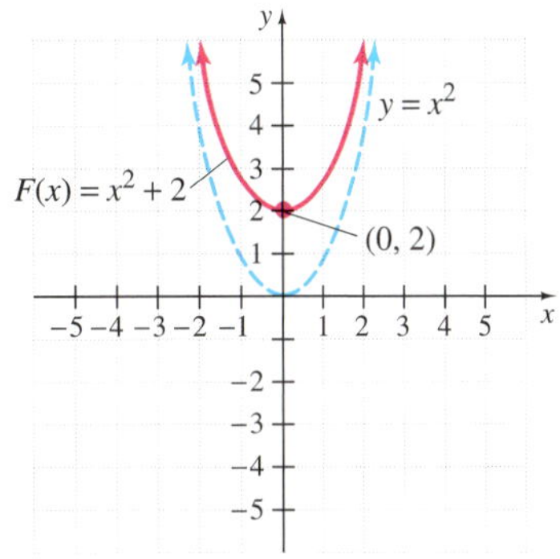


**b.** $g(x) = x^2 - 3$

The graph of $g(x) = x^2 - 3$ is obtained by shifting the graph of $y = x^2$ downward 3 units.

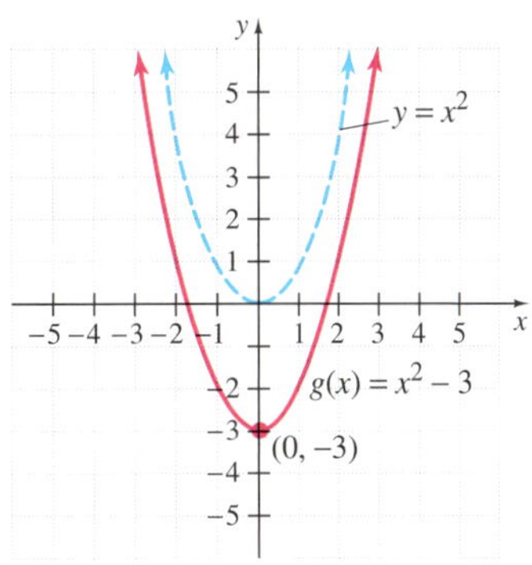


**2** Now we will graph functions of the form $f(x) = (x - h)^2$.

**Example 3** Graph $f(x) = x^2$ and $g(x) = (x - 2)^2$ on the same set of axes.

*Solution* By plotting points, we see that for each $x$-value, the corresponding value of $g(x)$ is the same as the value of $f(x)$ when the $x$-value is increased by 2. Thus, the graph of $g(x) = (x - 2)^2$ is the graph of $f(x) = x^2$ shifted to the right 2 units. The axis of symmetry for the graph of $g(x) = (x - 2)^2$ is also shifted 2 units to the right and is the line $x = 2$.

| $x$ | $f(x) = x^2$ | $x$ | $g(x) = (x - 2)^2$ |
|---|---|---|---|
| −2 | 4 | 0 | 4 |
| −1 | 1 | 1 | 1 |
| 0 | 0 | 2 | 0 |
| 1 | 1 | 3 | 1 |
| 2 | 4 | 4 | 4 |

Each $x$-value increased by 2 corresponds to same $y$-value.

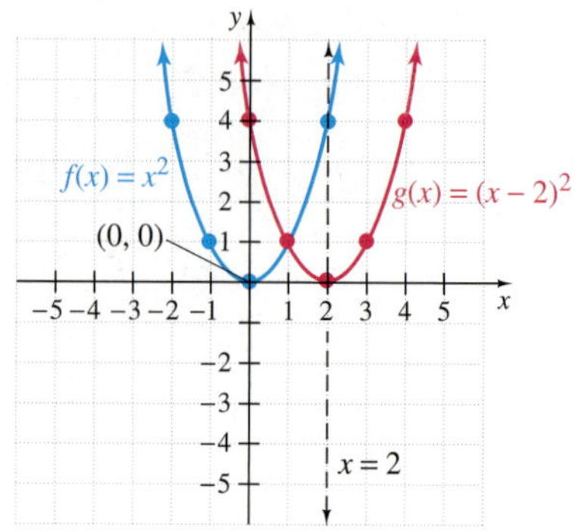


In general, we have the following properties.

## GRAPHING THE PARABOLA DEFINED BY $f(x) = (x - h)^2$

If $h$ is positive, the graph of $f(x) = (x - h)^2$ is the graph of $y = x^2$ shifted to the right $h$ units.

If $h$ is negative, the graph of $f(x) = (x - h)^2$ is the graph of $y = x^2$ shifted to the left $|h|$ units.

The vertex is $(h, 0)$, and the axis of symmetry is the vertical line $x = h$.

**Example 4** Graph each function.

**a.** $G(x) = (x - 3)^2$ **b.** $F(x) = (x + 1)^2$

*Solution* **a.** The graph of $G(x) = (x - 3)^2$ is obtained by shifting the graph of $y = x^2$ to the right 3 units. The graph of $G(x)$ is below on the left.

**b.** The equation $F(x) = (x + 1)^2$ can be written as $F(x) = [x - (-1)]^2$. The graph of $F(x) = [x - (-1)]^2$ is obtained by shifting the graph of $y = x^2$ to the left 1 unit. The graph of $F(x)$ is below on the right.

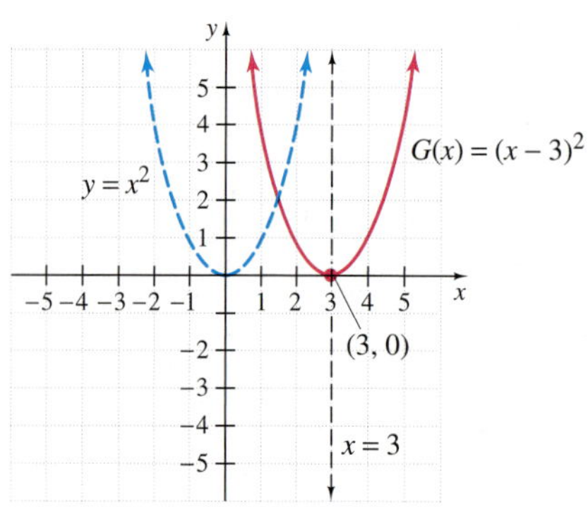


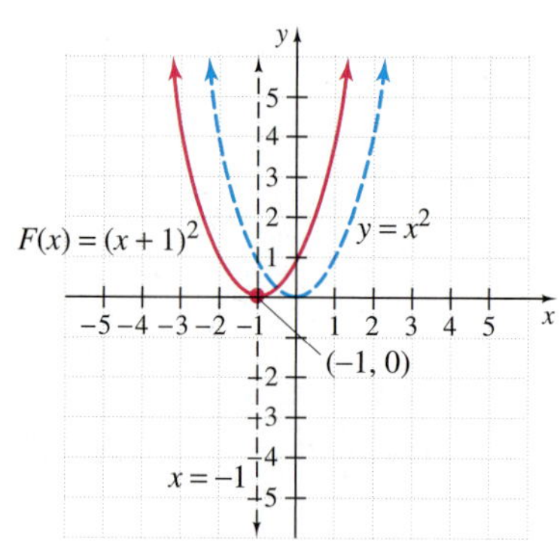


**3** As we will see in graphing functions of the form $f(x) = (x - h)^2 + k$, it is possible to combine vertical and horizontal shifts.

**GRAPHING THE PARABOLA DEFINED BY $f(x) = (x - h)^2 + k$**

The parabola has the same shape as $y = x^2$.
The vertex is $(h, k)$, and the axis of symmetry is the vertical line $x = h$.

**Example 5** Graph $F(x) = (x - 3)^2 + 1$.

*Solution* The graph of $F(x) = (x - 3)^2 + 1$ is the graph of $y = x^2$ shifted 3 units to the right and 1 unit up. The vertex is then $(3, 1)$, and the axis of symmetry is $x = 3$. A few ordered pair solutions are plotted to aid in graphing.

| $x$ | $F(x) = (x - 3)^2 + 1$ |
|---|---|
| 1 | 5 |
| 2 | 2 |
| 4 | 2 |
| 5 | 5 |

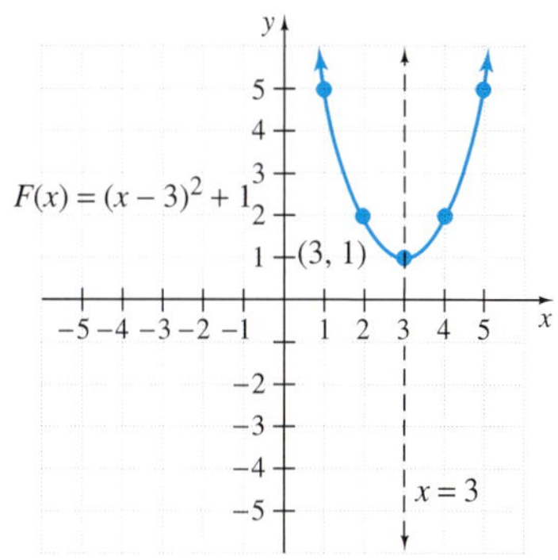


4 Next, we discover the change in the shape of the graph when the coefficient of $x^2$ is not 1.

**Example 6** Graph $f(x) = x^2, g(x) = 3x^2$, and $h(x) = \frac{1}{2}x^2$ on the same set of axes.

*Solution* Comparing the tables of values, we see that for each $x$-value, the corresponding value of $g(x)$ is triple the corresponding value of $f(x)$. Similarly, the value of $h(x)$ is half the value of $f(x)$.

| $x$ | $f(x) = x^2$ |
|---|---|
| −2 | 4 |
| −1 | 1 |
| 0 | 0 |
| 1 | 1 |
| 2 | 4 |

| $x$ | $g(x) = 3x^2$ |
|---|---|
| −2 | 12 |
| −1 | 3 |
| 0 | 0 |
| 1 | 3 |
| 2 | 12 |

| $x$ | $h(x) = \frac{1}{2}x^2$ |
|---|---|
| −2 | 2 |
| −1 | $\frac{1}{2}$ |
| 0 | 0 |
| 1 | $\frac{1}{2}$ |
| 2 | 2 |

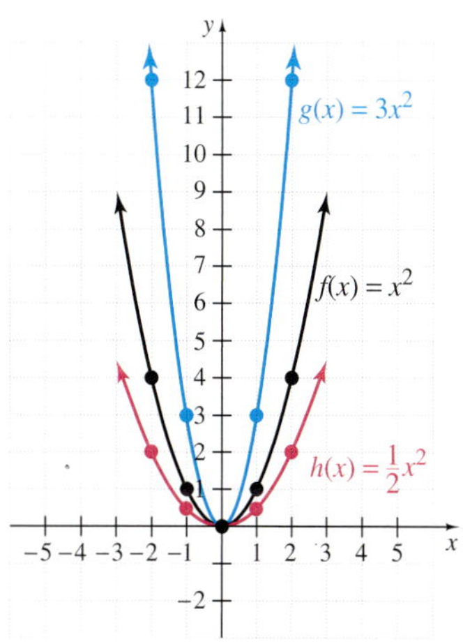

The result is that the graph of $g(x) = 3x^2$ is narrower than the graph of $f(x) = x^2$ and the graph of $h(x) = \frac{1}{2}x^2$ is wider. The vertex for each graph is (0, 0), and the axis of symmetry is the $y$-axis.

**GRAPHING THE PARABOLA DEFINED BY $f(x) = ax^2$**

If $a$ is positive, the parabola opens upward, and if $a$ is negative, the parabola opens downward.
If $|a| > 1$, the graph of the parabola is narrower than the graph of $y = x^2$.
If $|a| < 1$, the graph of the parabola is wider than the graph of $y = x^2$.

**Example 7** Graph $f(x) = -2x^2$.

*Solution* Because $a = -2$, a negative value, this parabola opens downward. Since $|-2| = 2$ and $2 > 1$, the parabola is narrower than the graph of $y = x^2$. The vertex is (0, 0), and the axis of symmetry is the $y$-axis. We verify this by plotting a few points.

| $x$ | $f(x) = -2x^2$ |
|---|---|
| −2 | −8 |
| −1 | −2 |
| 0 | 0 |
| 1 | −2 |
| 2 | −8 |

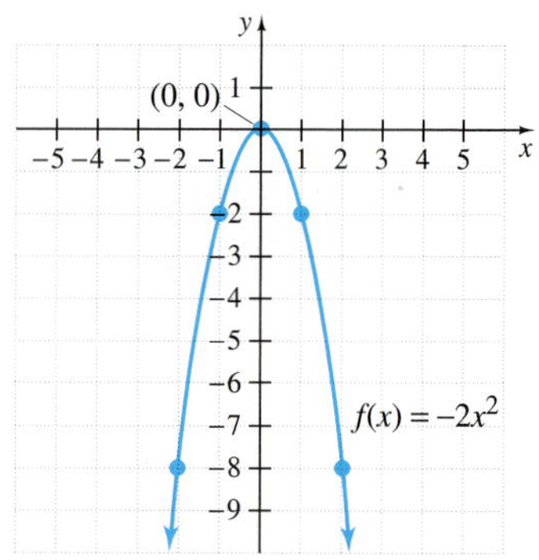


5 Now we will see the shape of the graph of a quadratic function of the form $f(x) = a(x - h)^2 + k$.

**Example 8** Graph $g(x) = \frac{1}{2}(x + 2)^2 + 5$. Find the vertex and the axis of symmetry.

*Solution* The function $g(x) = \frac{1}{2}(x + 2)^2 + 5$ may be written as $g(x) = \frac{1}{2}[x - (-2)]^2 + 5$. Thus, this graph is the same as the graph of $y = x^2$ shifted 2 units to the left and 5 units up, and it is wider because $a$ is $\frac{1}{2}$. The vertex is (−2, 5), and the axis of symmetry is $x = -2$. We plot a few points to verify.

| $x$ | $g(x) = \frac{1}{2}(x + 2)^2 + 5$ |
|---|---|
| $-4$ | $7$ |
| $-3$ | $5\frac{1}{2}$ |
| $-2$ | $5$ |
| $-1$ | $5\frac{1}{2}$ |
| $0$ | $7$ |

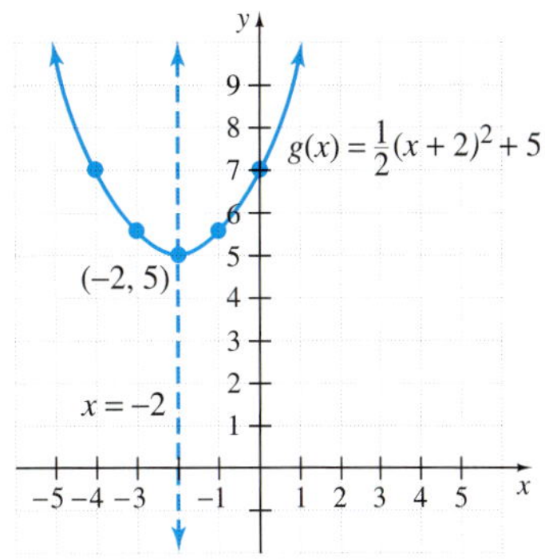


In general, the following holds.

## GRAPH OF A QUADRATIC FUNCTION

The graph of a quadratic function written in the form $f(x) = a(x - h)^2 + k$ is a parabola with vertex $(h, k)$. If $a > 0$, the parabola opens upward, and if $a < 0$, the parabola opens downward. The axis of symmetry is the line whose equation is $x = h$.

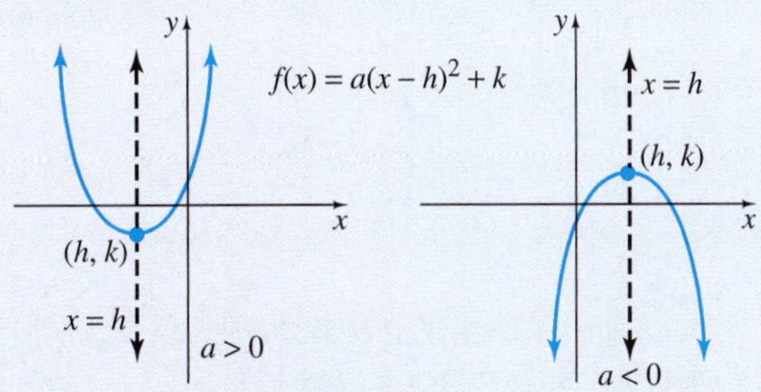


## GRAPHING CALCULATOR EXPLORATIONS

*Use a graphing calculator to graph the first function of each pair that follows. Then use its graph to predict the graph of the second function. Check your prediction by graphing both on the same set of axes.*

**1.** $F(x) = \sqrt{x}$; $G(x) = \sqrt{x} + 1$

**2.** $g(x) = x^3$; $H(x) = x^3 - 2$

**3.** $H(x) = |x|$; $f(x) = |x - 5|$

**4.** $h(x) = x^3 + 2$; $g(x) = (x - 3)^3 + 2$

**5.** $f(x) = |x + 4|$; $F(x) = |x + 4| + 3$

**6.** $G(x) = \sqrt{x} - 2$; $g(x) = \sqrt{x - 4} - 2$

## MENTAL MATH

*State the vertex of the graph of each quadratic function.*

**1.** $f(x) = x^2$

**2.** $f(x) = -5x^2$

**3.** $g(x) = (x - 2)^2$

**4.** $g(x) = (x + 5)^2$

**5.** $f(x) = 2x^2 + 3$

**6.** $h(x) = x^2 - 1$

**7.** $g(x) = (x + 1)^2 + 5$

**8.** $h(x) = (x - 10)^2 - 7$

## Exercise Set 8.5

*Sketch the graph of each quadratic function. Label the vertex, and sketch and label the axis of symmetry. See Examples 1 through 5.*

**1.** $f(x) = x^2 - 1$
**2.** $g(x) = x^2 + 3$
**3.** $h(x) = x^2 + 5$
**4.** $h(x) = x^2 - 4$
**5.** $g(x) = x^2 + 7$
**6.** $f(x) = x^2 - 2$
**7.** $f(x) = (x - 5)^2$
**8.** $g(x) = (x + 5)^2$
**9.** $h(x) = (x + 2)^2$
**10.** $H(x) = (x - 1)^2$
**11.** $G(x) = (x + 3)^2$
**12.** $f(x) = (x - 6)^2$
**13.** $f(x) = (x - 2)^2 + 5$
**14.** $g(x) = (x - 6)^2 + 1$
**15.** $h(x) = (x + 1)^2 + 4$
**16.** $G(x) = (x + 3)^2 + 3$
**17.** $g(x) = (x + 2)^2 - 5$
**18.** $h(x) = (x + 4)^2 - 6$

*Sketch the graph of each quadratic function. Label the vertex, and sketch and label the axis of symmetry. See Examples 6 and 7.*

**19.** $g(x) = -x^2$
**20.** $f(x) = 5x^2$
**21.** $h(x) = \frac{1}{3}x^2$
**22.** $g(x) = -3x^2$
**23.** $H(x) = 2x^2$
**24.** $f(x) = -\frac{1}{4}x^2$

*Sketch the graph of each quadratic function. Label the vertex, and sketch and label the axis of symmetry. See Example 8.*

**25.** $f(x) = 2(x - 1)^2 + 3$
**26.** $g(x) = 4(x - 4)^2 + 2$
**27.** $h(x) = -3(x + 3)^2 + 1$
**28.** $f(x) = -(x - 2)^2 - 6$
**29.** $H(x) = \frac{1}{2}(x - 6)^2 - 3$
**30.** $G(x) = \frac{1}{5}(x + 4)^2 + 3$

*Sketch the graph of each quadratic function. Label the vertex, and sketch and label the axis of symmetry.*

**31.** $f(x) = -(x - 2)^2$
**32.** $g(x) = -(x + 6)^2$
**33.** $F(x) = -x^2 + 4$
**34.** $H(x) = -x^2 + 10$
**35.** $F(x) = 2x^2 - 5$
**36.** $g(x) = \frac{1}{2}x^2 - 2$
**37.** $h(x) = (x - 6)^2 + 4$
**38.** $f(x) = (x - 5)^2 + 2$
**39.** $F(x) = \left(x + \frac{1}{2}\right)^2 - 2$
**40.** $H(x) = \left(x + \frac{1}{2}\right)^2 - 3$
**41.** $F(x) = \frac{3}{2}(x + 7)^2 + 1$
**42.** $g(x) = -\frac{3}{2}(x - 1)^2 - 5$
**43.** $f(x) = \frac{1}{4}x^2 - 9$
**44.** $H(x) = \frac{3}{4}x^2 - 2$
**45.** $G(x) = 5\left(x + \frac{1}{2}\right)^2$
**46.** $F(x) = 3\left(x - \frac{3}{2}\right)^2$
**47.** $h(x) = -(x - 1)^2 - 1$
**48.** $f(x) = -3(x + 2)^2 + 2$
**49.** $g(x) = \sqrt{3}(x + 5)^2 + \frac{3}{4}$
**50.** $G(x) = \sqrt{5}(x - 7)^2 - \frac{1}{2}$
**51.** $h(x) = 10(x + 4)^2 - 6$
**52.** $h(x) = 8(x + 1)^2 + 9$
**53.** $f(x) = -2(x - 4)^2 + 5$
**54.** $G(x) = -4(x + 9)^2 - 1$

*Write the equation of the parabola that has the same shape as $f(x) = 5x^2$ but with the following vertex.*

**55.** $(2, 3)$
**56.** $(1, 6)$
**57.** $(-3, 6)$
**58.** $(4, -1)$

*The shifting properties covered in this section apply to the graphs of all functions. Given the accompanying graph of $y = f(x)$, sketch the graph of each of the following.*

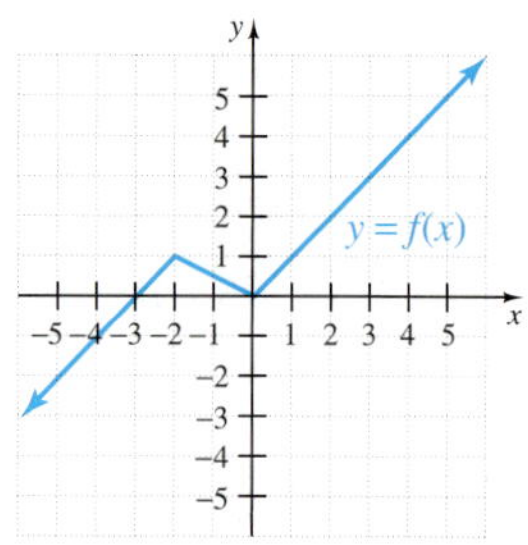


**59.** $y = f(x) + 1$
**60.** $y = f(x) - 2$
**61.** $y = f(x - 3)$
**62.** $y = f(x + 3)$
**63.** $y = f(x + 2) + 2$
**64.** $y = f(x - 1) + 1$

### REVIEW EXERCISES

*Add the proper constant to each binomial so that the resulting trinomial is a perfect square trinomial. See Section 8.1.*

**65.** $x^2 + 8x$
**66.** $y^2 + 4y$
**67.** $z^2 - 16z$
**68.** $x^2 - 10x$
**69.** $y^2 + y$
**70.** $z^2 - 3z$

*Solve by completing the square. See Section 8.1.*

**71.** $x^2 + 4x = 12$
**72.** $y^2 + 6y = -5$
**73.** $z^2 + 10z - 1 = 0$
**74.** $x^2 + 14x + 20 = 0$
**75.** $z^2 - 8z = 2$
**76.** $y^2 - 10y = 3$

## 8.6 FURTHER GRAPHING OF QUADRATIC FUNCTIONS

**OBJECTIVES**

1. Write quadratic functions in the form $y = a(x - h)^2 + k$.
2. Derive a formula for finding the vertex of a parabola.
3. Find the minimum or maximum value of a quadratic function.

1 We know that the graph of a quadratic function is a parabola. If a quadratic function is written in the form

$$f(x) = a(x - h)^2 + k$$

we can easily find the vertex $(h, k)$ and graph the parabola. To write a quadratic function in this form, complete the square. (See Section 8.1 for a review of completing the square.)

**Example 1** Graph $f(x) = x^2 - 4x - 12$. Find the vertex and any intercepts.

*Solution* The graph of this quadratic function is a parabola. To find the vertex of the parabola, we will write the function in the form $y = (x - h)^2 + k$. To do this, we complete the square on the binomial $x^2 - 4x$. To simplify our work, we let $f(x) = y$.

$$y = x^2 - 4x - 12 \quad \text{Let } f(x) = y.$$

$$y + 12 = x^2 - 4x \quad \text{Add 12 to both sides to get the } x\text{-variable terms alone.}$$

Now we add the square of half of −4 to both sides.

$$\frac{1}{2}(-4) = -2 \quad \text{and} \quad (-2)^2 = 4$$

$$y + 12 + 4 = x^2 - 4x + 4 \quad \text{Add 4 to both sides.}$$

$$y + 16 = (x - 2)^2 \quad \text{Factor the trinomial.}$$

$$y = (x - 2)^2 - 16 \quad \text{Subtract 16 from both sides.}$$

$$f(x) = (x - 2)^2 - 16 \quad \text{Replace } y \text{ with } f(x).$$

From this equation, we can see that the vertex of the parabola is $(2, -16)$, a point in quadrant IV, and the axis of symmetry is the line $x = 2$.

Notice that $a = 1$. Since $a > 0$, the parabola opens upward. This parabola opening upward with vertex $(2, -16)$ will have two $x$-intercepts and one $y$-intercept. (See the Helpful Hint after this example.)

$x$-intercepts: let $y$ or $f(x) = 0$

$$f(x) = x^2 - 4x - 12$$
$$0 = x^2 - 4x - 12$$
$$0 = (x - 6)(x + 2)$$
$$0 = x - 6 \quad \text{or} \quad 0 = x + 2$$
$$6 = x \quad \quad -2 = x$$

$y$-intercept: let $x = 0$

$$f(x) = x^2 - 4x - 12$$
$$f(0) = 0^2 - 4 \cdot 0 - 12$$
$$= -12$$

The two $x$-intercepts are $(6, 0)$ and $(-2, 0)$. The $y$-intercept is $(0, -12)$. The sketch of $f(x) = x^2 - 4x - 12$ is shown.

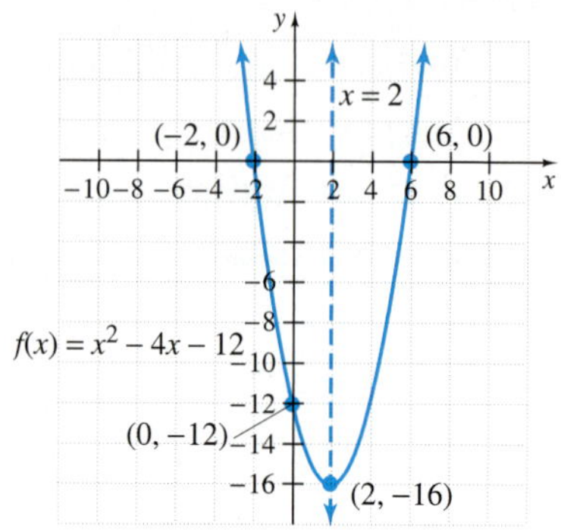


### HELPFUL HINT

Parabola Opens Upward
Vertex in I or II: no $x$-intercept
Vertex in III or IV: 2 $x$-intercepts

Parabola Opens Downward
Vertex in I or II: 2 $x$-intercept
Vertex in III or IV: no $x$-intercepts.

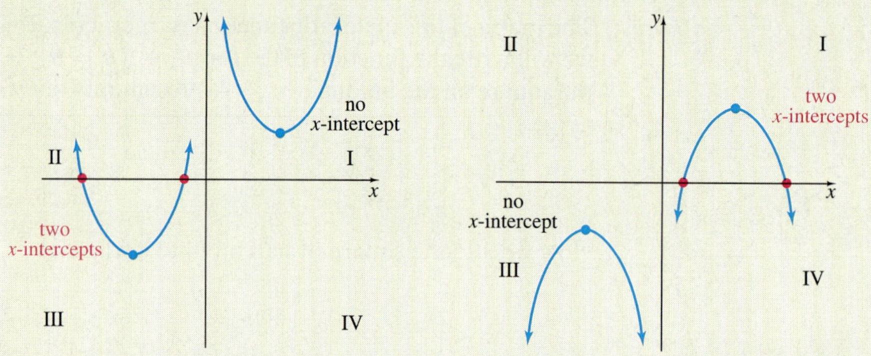


**Example 2** Graph $f(x) = 3x^2 + 3x + 1$. Find the vertex and any intercepts.

*Solution* Replace $f(x)$ with $y$ and complete the square on $x$ to write the equation in the form $y = a(x - h)^2 + k$.

$$y = 3x^2 + 3x + 1 \qquad \text{Replace } f(x) \text{ with } y.$$

$$y - 1 = 3x^2 + 3x \qquad \text{Isolate } x\text{-variable terms.}$$

Factor 3 from the terms $3x^2 + 3x$ so that the coefficient of $x^2$ is 1.

$$y - 1 = 3(x^2 + x) \qquad \text{Factor out 3.}$$

The coefficient of $x$ in the parentheses above is 1. Then $\frac{1}{2}(1) = \frac{1}{2}$ and $\left(\frac{1}{2}\right)^2 = \frac{1}{4}$. Since we are adding $\frac{1}{4}$ inside the parentheses, we are really adding $3\left(\frac{1}{4}\right)$, so we *must* add $3\left(\frac{1}{4}\right)$ to the left side.

$$y - 1 + 3\left(\frac{1}{4}\right) = 3\left(x^2 + x + \frac{1}{4}\right)$$

$$y - \frac{1}{4} = 3\left(x + \frac{1}{2}\right)^2$$ Simplify the left side and factor the right side.

$$y = 3\left(x + \frac{1}{2}\right)^2 + \frac{1}{4}$$ Add $\frac{1}{4}$ to both sides.

$$f(x) = 3\left(x + \frac{1}{2}\right)^2 + \frac{1}{4}$$ Replace $y$ with $f(x)$.

Then $a = 3$, $h = -\frac{1}{2}$, and $k = \frac{1}{4}$. This means that the parabola opens upward with vertex $\left(-\frac{1}{2}, \frac{1}{4}\right)$ and that the axis of symmetry is the line $x = -\frac{1}{2}$.

To find the $y$-intercept, let $x = 0$. Then

$$f(0) = 3(0)^2 + 3(0) + 1 = 1$$

Thus the $y$-intercept is (0, 1).

This parabola has no $x$-intercepts since the vertex is in the second quadrant and opens upward. Use the vertex, axis of symmetry, and $y$-intercept to sketch the parabola.

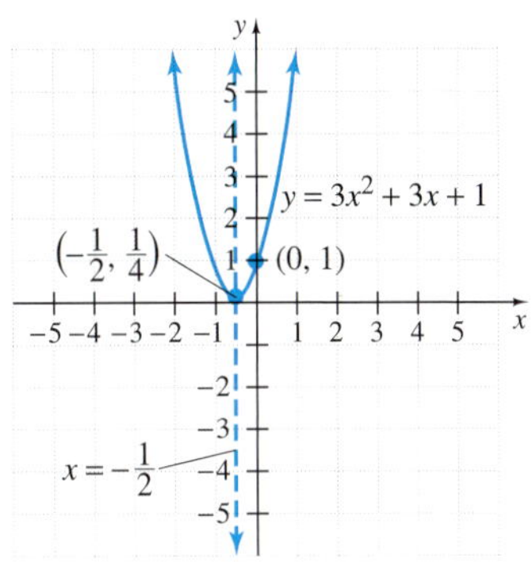


**Example 3** Graph $f(x) = -x^2 - 2x + 3$. Find the vertex and any intercepts.

*Solution* We write $f(x)$ in the form $a(x - h)^2 + k$ by completing the square. First we replace $f(x)$ with $y$.

$$f(x) = -x^2 - 2x + 3$$

$$y = -x^2 - 2x + 3$$

$$y - 3 = -x^2 - 2x$$ Subtract 3 from both sides to get the $x$-variable terms alone.

$$y - 3 = -1(x^2 + 2x)$$ Factor −1 from the terms $-x^2 - 2x$.

The coefficient of $x$ is 2. Then $\frac{1}{2}(2) = 1$ and $1^2 = 1$. We add 1 to the right side inside the parentheses and add $-1(1)$ to the left side.

$$y - 3 - 1(1) = -1(x^2 + 2x + 1)$$

$$y - 4 = -1(x + 1)^2 \quad \text{Simplify the left side and factor the right side.}$$

$$y = -1(x + 1)^2 + 4 \quad \text{Add 4 to both sides.}$$

$$f(x) = -1(x + 1)^2 + 4 \quad \text{Replace } y \text{ with } f(x).$$

**HELPFUL HINT**
This can be written as $f(x) = -1[x - (-1)]^2 + 4$. Notice that the vertex is $(-1, 4)$.

Since $a = -1$, the parabola opens downward with vertex $(-1, 4)$ and axis of symmetry $x = -1$.

To find the $y$-intercept, we let $x = 0$ and solve for $y$. Then

$$f(0) = -0^2 - 2(0) + 3 = 3$$

Thus, $(0, 3)$ is the $y$-intercept.

To find the $x$-intercepts, we let $y$ or $f(x) = 0$ and solve for $x$.

$$f(x) = -x^2 - 2x + 3$$

$$0 = -x^2 - 2x + 3 \quad \text{Let } f(x) = 0.$$

Now we divide both sides by $-1$ so that the coefficient of $x^2$ is 1.

$$\frac{0}{-1} = \frac{-x^2}{-1} - \frac{2x}{-1} + \frac{3}{-1} \quad \text{Divide both sides by } -1.$$

$$0 = x^2 + 2x - 3 \quad \text{Simplify.}$$

$$0 = (x + 3)(x - 1) \quad \text{Factor.}$$

$$x + 3 = 0 \quad \text{or} \quad x - 1 = 0 \quad \text{Set each factor equal to 0.}$$

$$x = -3 \qquad x = 1 \quad \text{Solve.}$$

The $x$-intercepts are $(-3, 0)$ and $(1, 0)$. Use these points to sketch the parabola.

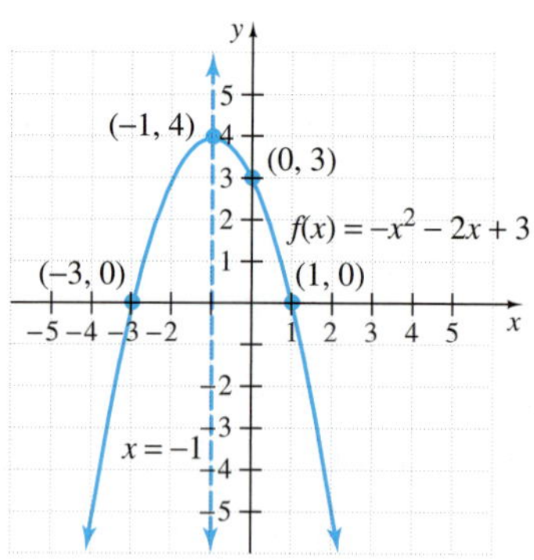


**2** Recall from Section 5.9 that we introduced a formula for finding the vertex of a parabola. Now that we have practiced completing the square, we will show that the $x$-coordinate of the vertex of the graph of $f(x)$ or $y = ax^2 + bx + c$ can be found by the formula $x = \frac{-b}{2a}$. To do so, we complete the square on $x$ and write the equation in the form $y = a(x - h)^2 + k$.

First, isolate the $x$-variable terms by subtracting $c$ from both sides.

$$y = ax^2 + bx + c$$

$$y - c = ax^2 + bx$$

Next, factor $a$ from the terms $ax^2 + bx$.

$$y - c = a\left(x^2 + \frac{b}{a}x\right)$$

Next, add the square of half of $\frac{b}{a}$, or $\left(\frac{b}{2a}\right)^2 = \frac{b^2}{4a^2}$, to the right side inside the parentheses. Because of the factor $a$, what we really added was $a\left(\frac{b^2}{4a^2}\right)$ and this must be added to the left side.

$$y - c + a\left(\frac{b^2}{4a^2}\right) = a\left(x^2 + \frac{b}{a}x + \frac{b^2}{4a^2}\right)$$

$$y - c + \frac{b^2}{4a} = a\left(x + \frac{b}{2a}\right)^2 \quad \text{Simplify the left side and factor the right side.}$$

$$y = a\left(x + \frac{b}{2a}\right)^2 + c - \frac{b^2}{4a} \quad \text{Add } c \text{ to both sides and subtract } \frac{b^2}{4a} \text{ from both sides.}$$

Compare this form with $f(x)$ or $y = a(x - h)^2 + k$ and see that $h$ is $\frac{-b}{2a}$, which means that the $x$-coordinate of the vertex of the graph of $f(x) = ax^2 + bx + c$ is $\frac{-b}{2a}$.

**VERTEX FORMULA**

The graph of $f(x) = ax^2 + bx + c$, when $a \neq 0$, is a parabola with vertex

$$\left(\frac{-b}{2a}, f\left(\frac{-b}{2a}\right)\right)$$

Let's use this formula to find the vertex of the parabola we graphed in Example 1.

**Example 4** Find the vertex of the graph of $f(x) = x^2 - 4x - 12$.

*Solution* In the quadratic function $f(x) = x^2 - 4x - 12$, notice that $a = 1$, $b = -4$, and $c = -12$. Then

$$\frac{-b}{2a} = \frac{-(-4)}{2(1)} = 2$$

The $x$-value of the vertex is 2. To find the corresponding $f(x)$ or $y$-value, find $f(2)$. Then

$$f(2) = 2^2 - 4(2) - 12 = 4 - 8 - 12 = -16$$

The vertex is $(2, -16)$. These results agree with our findings in Example 1.

**3** The vertex of a parabola gives us some important information about its corresponding quadratic function. The quadratic function whose graph is a parabola that opens upward has a minimum value, and the quadratic function whose graph is a parabola that opens downward has a maximum value. The $f(x)$ or $y$-value of the vertex is the minimum or maximum value of the function.

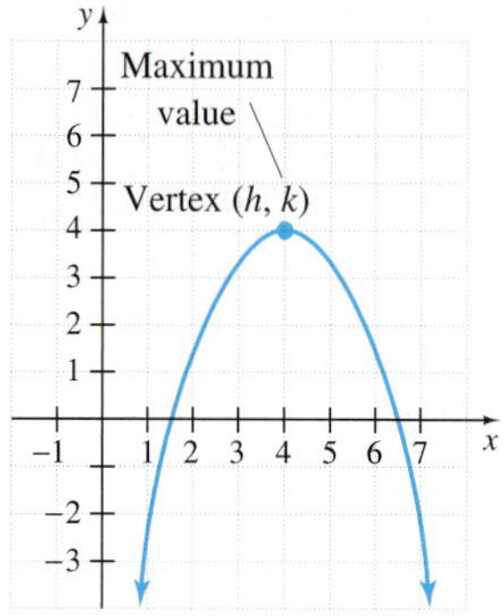


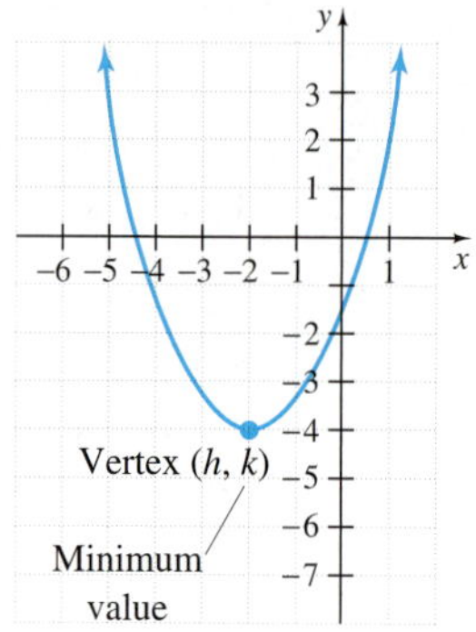


## Example 5 FINDING MAXIMUM HEIGHT

A rock is thrown upward from the ground. Its height in feet above ground after $t$ seconds is given by the function $f(t) = -16t^2 + 20t$. Find the maximum height of the rock and the number of seconds it took for the rock to reach its maximum height.

*Solution*

1. UNDERSTAND. The maximum height of the rock is the largest value of $f(t)$. Since the function $f(t) = -16t^2 + 20t$ is a quadratic function, its graph is a parabola. It opens downward since $-16 < 0$. Thus, the maximum value of $f(t)$ is the $f(t)$ or $y$-value of the vertex of its graph.
2. TRANSLATE. To find the vertex $(h, k)$, notice that for $f(t) = -16t^2 + 20t$, $a = -16$, $b = 20$, and $c = 0$. We will use these values and the vertex formula

$$\left(\frac{-b}{2a}, f\left(\frac{-b}{2a}\right)\right)$$

3. SOLVE.

$$h = \frac{-b}{2a} = \frac{-20}{-32} = \frac{5}{8}$$

$$f\left(\frac{5}{8}\right) = -16\left(\frac{5}{8}\right)^2 + 20\left(\frac{5}{8}\right)$$

$$= -16\left(\frac{25}{64}\right) + \frac{25}{2}$$

$$= -\frac{25}{4} + \frac{50}{4} = \frac{25}{4}$$

4. INTERPRET. The graph of $f(t)$ is a parabola opening downward with vertex $\left(\frac{5}{8}, \frac{25}{4}\right)$. This means that the rock's maximum height is $\frac{25}{4}$ feet, or $6\frac{1}{4}$ feet, which was reached in $\frac{5}{8}$ second.

## SPOTLIGHT ON DECISION MAKING

Suppose you are a member of a community theater group, the Slidell Players. For an upcoming performance of *Grease*, your group must decide on a ticket price. The graph shows the relationship between ticket price and box office receipts for past Slidell Players performances.

What ticket price would you suggest that the Slidell Players charge for its performance of *Grease*? Explain your reasoning.

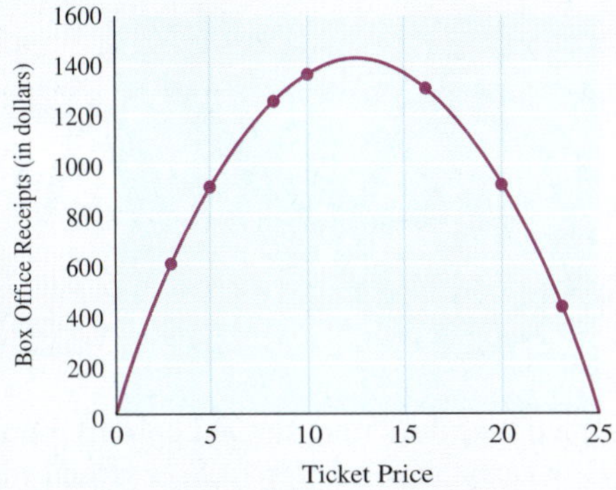


## Exercise Set 8.6

*Find the vertex of the graph of each quadratic function. See Examples 1 through 4.*

**1.** $f(x) = x^2 + 8x + 7$

**2.** $f(x) = x^2 + 6x + 5$

**3.** $f(x) = -x^2 + 10x + 5$

**4.** $f(x) = -x^2 - 8x + 2$

**5.** $f(x) = 5x^2 - 10x + 3$

**6.** $f(x) = -3x^2 + 6x + 4$

**7.** $f(x) = -x^2 + x + 1$

**8.** $f(x) = x^2 - 9x + 8$

*Match each function with its graph. See Examples 1 through 4.*

A

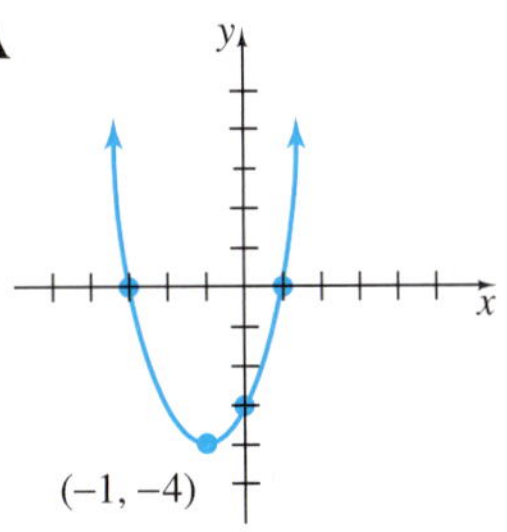


B

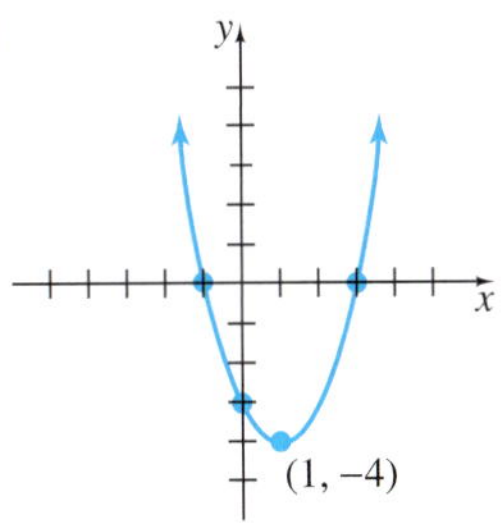


C

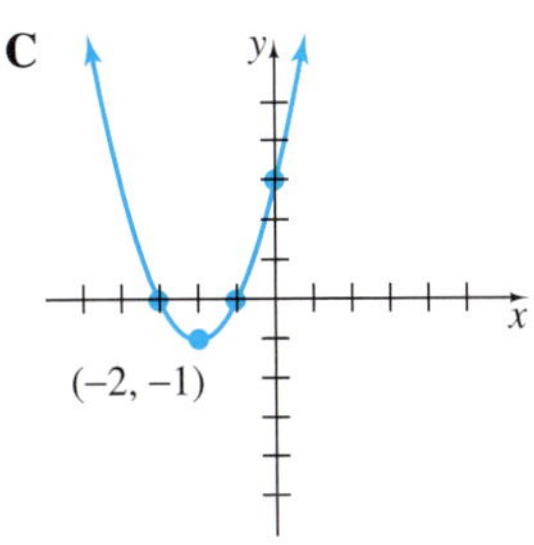


D

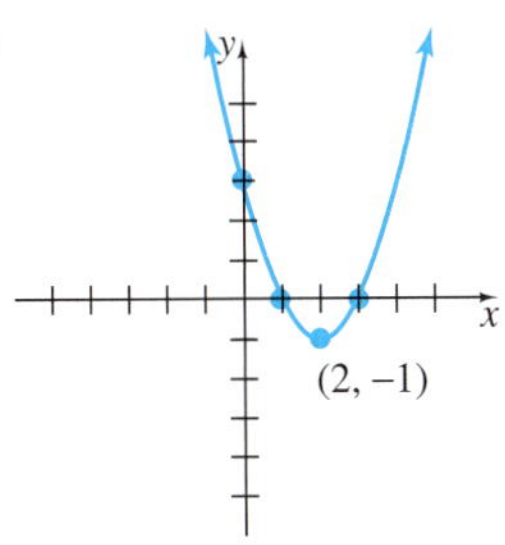


**9.** $f(x) = x^2 - 4x + 3$

**10.** $f(x) = x^2 + 2x - 3$

**11.** $f(x) = x^2 - 2x - 3$

**12.** $f(x) = x^2 + 4x + 3$

*Find the vertex of the graph of each quadratic function. Determine whether the graph opens upward or downward, find any intercepts, and sketch the graph. See Examples 1 through 3.*

**13.** $f(x) = x^2 + 4x - 5$

**14.** $f(x) = x^2 + 2x - 3$

**15.** $f(x) = -x^2 + 2x - 1$

**16.** $f(x) = -x^2 + 4x - 4$

**17.** $f(x) = x^2 - 4$

**18.** $f(x) = x^2 - 1$

**19.** $f(x) = 4x^2 + 4x - 3$

**20.** $f(x) = 2x^2 - x - 3$

**21.** $f(x) = x^2 + 8x + 15$

**22.** $f(x) = x^2 + 10x + 9$

**23.** $f(x) = x^2 - 6x + 5$

**24.** $f(x) = x^2 - 4x + 3$

**25.** $f(x) = x^2 - 4x + 5$

**26.** $f(x) = x^2 - 6x + 11$

**27.** $f(x) = 2x^2 + 4x + 5$

**28.** $f(x) = 3x^2 + 12x + 16$

**29.** $f(x) = -2x^2 + 12x$

**30.** $f(x) = -4x^2 + 8x$

**31.** $f(x) = x^2 + 1$

**32.** $f(x) = x^2 + 4$

**33.** $f(x) = x^2 - 2x - 15$

**34.** $f(x) = x^2 - 4x + 3$

**35.** $f(x) = -5x^2 + 5x$

**36.** $f(x) = 3x^2 - 12x$

**37.** $f(x) = -x^2 + 2x - 12$

**38.** $f(x) = -x^2 + 8x - 17$

**39.** $f(x) = 3x^2 - 12x + 15$

**40.** $f(x) = 2x^2 - 8x + 11$

**41.** $f(x) = x^2 + x - 6$

**42.** $f(x) = x^2 + 3x - 18$

**43.** $f(x) = -2x^2 - 3x + 35$

**44.** $f(x) = 3x^2 - 13x - 10$

*Solve. See Example 5.*

**45.** The cost $C$ in dollars of manufacturing $x$ bicycles at Holladay's Production Plant is given by the function $C(x) = 2x^2 - 800x + 92{,}000$.

**a.** Find the number of bicycles that must be manufactured to minimize the cost.

**b.** Find the minimum cost.

**46.** If a projectile is fired straight upward from the ground with an initial speed of 96 feet per second, then its height $h$ in feet after $t$ seconds is given by the equation

$$h(t) = -16t^2 + 96t$$

Find the maximum height of the projectile.

**47.** If Rheam Gaspar throws a ball upward with an initial speed of 32 feet per second, then its height $h$ in feet after $t$ seconds is given by the equation

$$h(t) = -16t^2 + 32t$$

Find the maximum height of the ball.

**48.** The Utah Ski Club sells calendars to raise money. The profit $P$, in cents, from selling $x$ calendars is given by the equation $P(x) = 360x - x^2$.

**a.** Find how many calendars must be sold to maximize profit.

**b.** Find the maximum profit.

**49.** Find two numbers whose sum is 60 and whose product is as large as possible. [*Hint:* Let $x$ and $60 - x$ be the two positive numbers. Their product can be described by the function $f(x) = x(60 - x)$.]

**50.** Find two numbers whose sum is 11 and whose product is as large as possible. (Use the hint for Exercise 49.)

**51.** Find two numbers whose difference is 10 and whose product is as small as possible. (Use the hint for Exercise 49.)

**52.** Find two numbers whose difference is 8 and whose product is as small as possible.

△ **53.** The length and width of a rectangle must have a sum of 40. Find the dimensions of the rectangle that will have the maximum area. (Use the hint for Exercise 49.)

△ **54.** The length and width of a rectangle must have a sum of 50. Find the dimensions of the rectangle that will have maximum area.

**55.** Methane is a gas produced by landfills, natural gas systems, and coal mining that contributes to the greenhouse effect and global warming. Methane emissions in the United States can be modeled by the quadratic function $f(x) = -0.74x^2 + 8.66x + 159.07$, where $f(x)$ is the amount of methane produced in million metric tons and $x$ is the number of years after 1990. (*Source:* based on data from the U.S. Environmental Protection Agency, 1993–1998)

**a.** If this trend continues, what will U.S. emissions of methane be in 2004?

**b.** In what year were methane emissions in the United States at their maximum? Round to the nearest whole year.

**c.** Use the result of part b. to determine the maximum methane emissions level.

**56.** The number of inmates in custody in U.S. prisons and jails can be modeled by the quadratic function $p(x) = -716.2x^2 + 87{,}453.7x + 1{,}148{,}702$, where $p(x)$ is the number of inmates and $x$ is the number of years after 1990. (*Source:* based on data from the Bureau of Justice Statistics, U.S. Department of Justice, 1990–1998)

**a.** Will this function have a maximum or a minimum? How can you tell?

**b.** According to this model, when will the number of prison inmates in custody in the United States be at its maximum/minimum?

**c.** What is the number of inmates predicted for that year? Round answer to the nearest hundred inmates.

*Find the vertex of the graph of each quadratic function. Determine whether the graph opens upward or downward, find the y-intercept, approximate the x-intercepts to one decimal place, and sketch the graph.*

**57.** $f(x) = x^2 + 10x + 15$

**58.** $f(x) = x^2 - 6x + 4$

**59.** $f(x) = 3x^2 - 6x + 7$

**60.** $f(x) = 2x^2 + 4x - 1$

*Use a graphing calculator to check each exercise.*

**61.** Exercise 27

**62.** Exercise 28

**63.** Exercise 37

**64.** Exercise 38

*Find the maximum or minimum value of each function. Approximate to two decimal places.*

**65.** $f(x) = 2.3x^2 - 6.1x + 3.2$

**66.** $f(x) = 7.6x^2 + 9.8x - 2.1$

**67.** $f(x) = -1.9x^2 + 5.6x - 2.7$

**68.** $f(x) = -5.2x^2 - 3.8x + 5.1$

## REVIEW EXERCISES

*Sketch the graph of each function. See Section 8.5.*

**69.** $f(x) = x^2 + 2$

**70.** $f(x) = (x - 3)^2$

**71.** $g(x) = x + 2$

**72.** $h(x) = x - 3$

**73.** $f(x) = (x + 5)^2 + 2$

**74.** $f(x) = 2(x - 3)^2 + 2$

**75.** $f(x) = 3(x - 4)^2 + 1$

**76.** $f(x) = (x + 1)^2 + 4$

**77.** $f(x) = -(x - 4)^2 + \frac{3}{2}$

**78.** $f(x) = -2(x + 7)^2 + \frac{1}{2}$

For additional Chapter Projects, visit the Real World Activities Website by going to http://www.prenhall.com/martin-gay.

# 8 CHAPTER PROJECT

## *Fitting a Quadratic Model to Data*

Throughout the twentieth century, the eating habits of Americans changed noticeably. Americans started consuming less whole milk and butter, and started consuming more skim and low-fat milk and margarine. We also started eating more poultry and fish. In this project, you will have the opportunity to investigate trends in per capita consumption of poultry during the twentieth century. This project may be completed by working in groups or individually.

We will start by finding a quadratic model, $y = ax^2 + bx + c$, that has ordered pair solutions that correspond to the data for U.S. per capita consumption of poultry given in the table. To do so, substitute each data pair into the equation. Each time, the result is an equation in three unknowns: $a$, $b$, and $c$. Because there are three pairs of data, we can form a system of three linear equations in three unknowns. Solving for the values of $a$, $b$, and $c$ gives a quadratic model that represents the given data.

U.S. PER CAPITA CONSUMPTION OF POULTRY (IN POUNDS)

| *Year* | $x$ | *Poultry Consumption, y (in pounds)* |
|---|---|---|
| 1909 | 9 | 11 |
| 1969 | 69 | 33 |
| 1998 | 98 | 68 |

(*Source:* Economic research service, U.S. Department of Agriculture)

1. Write the system of equations that must be solved to find the values of $a$, $b$, and $c$ needed for a quadratic model of the given data.
2. Solve the system of equations for $a$, $b$, and $c$. Recall the various methods of solving linear systems used in Chapter 4. You might consider using matrices, Cramer's rule, or a graphing calculator to do so. Round to the nearest thousandth.
3. Write the quadratic model for the data. Note that the variable $x$ represents the number of years after 1900.
4. In 1939, the actual U.S. per capita consumption of poultry was 12 pounds per person. Based on this information, how accurate do you think this model is for years other than those given in the table?
5. Use your model to estimate the per capita consumption of poultry in 1950.
6. According to the model, in what year was per capita consumption of poultry 50 pounds per person?
7. In what year was the per capita consumption of poultry at its lowest level? What was that level?
8. Who might be interested in a model like this and how would it be helpful?

## CHAPTER 8 VOCABULARY CHECK

*Fill in each blank with one of the words or phrases listed below.*

| | | | |
|---|---|---|---|
| quadratic formula | quadratic | discriminant | $\pm\sqrt{b}$ |
| completing the square | quadratic inequality | | |

$(h, k)$ $(0, k)$ $(h, 0)$ $\frac{-b}{2a}$

1. The ___________ helps us know find the number and type of solutions of a quadratic equation.
2. If $a^2 = b$, then $a =$ ______.
3. The graph of $f(x) = ax^2 + bx + c$ where $a$ is not 0 is a parabola whose vertex has $x$-value of ____.
4. A(n) ________________ is an inequality that can be written so that one side is a quadratic expression and the other side is 0.
5. The process of writing a quadratic equation so that one side is a perfect square trinomial is called ________________.
6. The graph of $f(x) = x^2 + k$ has vertex ______.
7. The graph of $f(x) = (x - h)^2$ has vertex ______.
8. The graph of $f(x) = (x - h)^2 + k$ has vertex ______.
9. The formula $x = \frac{-b \pm \sqrt{b^2 - 4ac}}{2a}$ is called the ________________.
10. A _________ equation is one that can be written in the form $ax^2 + bx + c = 0$ where $a$, $b$, and $c$ are real numbers and $a$ is not 0.

# CHAPTER 8 HIGHLIGHTS

| DEFINITIONS AND CONCEPTS | EXAMPLES |
|---|---|
| **Section 8.1 Solving Quadratic Equations by Completing the Square** | |
| ***Square root property*** <br> If $b$ is a real number and if $a^2 = b$, then $a = \pm\sqrt{b}$. | Solve $(x + 3)^2 = 14$. <br> $x + 3 = \pm\sqrt{14}$ <br> $x = -3 \pm \sqrt{14}$ |
| ***To solve a quadratic equation in x by completing the square*** <br> **Step 1:** If the coefficient of $x^2$ is not 1, divide both sides of the equation by the coefficient of $x^2$. <br> **Step 2:** Isolate the variable terms. <br> **Step 3:** Complete the square by adding the square of half of the coefficient of $x$ to both sides. <br> **Step 4:** Write the resulting trinomial as the square of a binomial. <br> **Step 5:** Apply the square root property and solve for $x$. | Solve $3x^2 - 12x - 18 = 0$. <br> 1. $x^2 - 4x - 6 = 0$ <br> 2. $x^2 - 4x = 6$ <br> 3. $\frac{1}{2}(-4) = -2$ and $(-2)^2 = 4$ <br> $x^2 - 4x + 4 = 6 + 4$ <br> 4. $(x - 2)^2 = 10$ <br> 5. $x - 2 = \pm\sqrt{10}$ <br> $x = 2 \pm \sqrt{10}$ |

| DEFINITIONS AND CONCEPTS | EXAMPLES |
|---|---|

## Section 8.2 Solving Quadratic Equations by the Quadratic Formula

A quadratic equation written in the form $ax^2 + bx + c = 0$ has solutions

$$x = \frac{-b \pm \sqrt{b^2 - 4ac}}{2a}$$

Solve $x^2 - x - 3 = 0$.

$$a = 1, b = -1, c = -3$$

$$x = \frac{-(-1) \pm \sqrt{(-1)^2 - 4(1)(-3)}}{2 \cdot 1}$$

$$x = \frac{1 \pm \sqrt{13}}{2}$$

## Section 8.3 Solving Equations by Using Quadratic Methods

Substitution is often helpful in solving an equation that contains a repeated variable expression.

Solve $(2x + 1)^2 - 5(2x + 1) + 6 = 0$.
Let $m = 2x + 1$. Then

$m^2 - 5m + 6 = 0$ Let $m = 2x + 1$.

$(m - 3)(m - 2) = 0$

$m = 3$ or $m = 2$

$2x + 1 = 3$ or $2x + 1 = 2$ Substitute back.

$x = 1$ or $x = \frac{1}{2}$

## Section 8.4 Nonlinear Inequalities in One Variable

***To solve a polynomial inequality***

**Step 1:** Write the inequality in standard form.
**Step 2:** Solve the related equation.
**Step 3:** Use solutions from Step 2 to separate the number line into regions.
**Step 4:** Use test points to determine whether values in each region satisfy the original inequality.
**Step 5:** Write the solution set as the union of regions whose test point value is a solution.

Solve $x^2 \geq 6x$.

1. $x^2 - 6x \geq 0$
2. $x^2 - 6x = 0$

   $x(x - 6) = 0$

   $x = 0$ or $x = 6$
3. 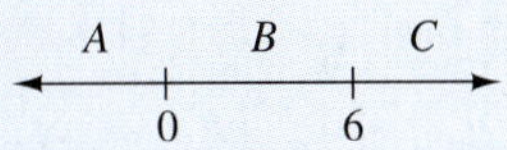


4.

| Region | Test Point Value | $x^2 \geq 6x$ | Result |
|---|---|---|---|
| A | −2 | $(-2)^2 \geq 6(-2)$ | True |
| B | 1 | $1^2 \geq 6(1)$ | False |
| C | 7 | $7^2 \geq 6(7)$ | True |

5. 

The solution set is $(-\infty, 0] \cup [6, \infty)$.

*(continued)*

| DEFINITIONS AND CONCEPTS | EXAMPLES |
|---|---|

## Section 8.4 Nonlinear Inequalities in One Variable

***To solve a rational inequality***

**Step 1:** Solve for values that make all denominators 0.
**Step 2:** Solve the related equation.
**Step 3:** Use solutions from Steps 1 and 2 to separate the number line into regions.
**Step 4:** Use test points to determine whether values in each region satisfy the original inequality.
**Step 5:** Write the solution set as the union of regions whose test point value is a solution.

Solve $\dfrac{6}{x-1} < -2.$

1. $x - 1 = 0$    Set denominator equal to 0.
   $x = 1$
2. $\dfrac{6}{x-1} = -2$
   $6 = -2(x - 1)$    Multiply by $(x - 1)$.
   $6 = -2x + 2$
   $4 = -2x$
   $-2 = x$
3. 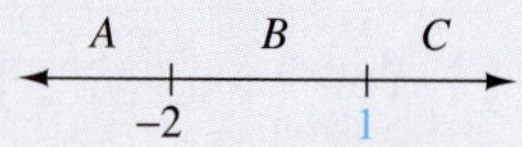

4. Only a test value from region $B$ satisfies the original inequality.
5. 

The solution set is $(-2, 1)$.

## Section 8.5 Quadratic Functions and Their Graphs

***Graph of a quadratic function***

The graph of a quadratic function written in the form $f(x) = a(x - h)^2 + k$ is a parabola with vertex $(h, k)$. If $a > 0$, the parabola opens upward; if $a < 0$, the parabola opens downward. The axis of symmetry is the line whose equation is $x = h$.

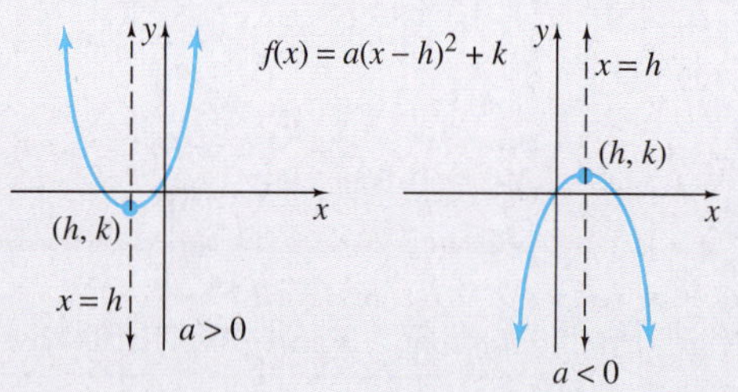


Graph $g(x) = 3(x - 1)^2 + 4$.

The graph is a parabola with vertex $(1, 4)$ and axis of symmetry $x = 1$. Since $a = 3$ is positive, the graph opens upward.

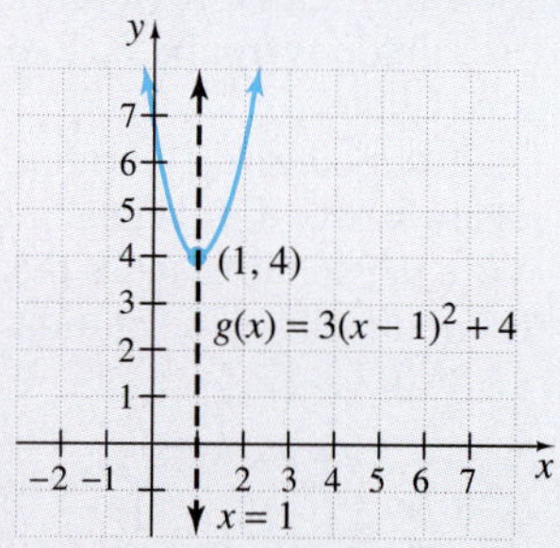


## Section 8.6 Further Graphing of Quadratic Functions

The graph of $f(x) = ax^2 + bx + c$, where $a \neq 0$, is a parabola with vertex

$$\left(\frac{-b}{2a}, f\left(\frac{-b}{2a}\right)\right)$$

Graph $f(x) = x^2 - 2x - 8$. Find the vertex and $x$- and $y$-intercepts.

$$\frac{-b}{2a} = \frac{-(-2)}{2 \cdot 1} = 1$$

*(continued)*

| DEFINITIONS AND CONCEPTS | EXAMPLES |
|---|---|

**Section 8.6 Further Graphing of Quadratic Functions**

$$f(1) = 1^2 - 2(1) - 8 = -9$$

The vertex is $(1, -9)$.

$$0 = x^2 - 2x - 8$$
$$0 = (x - 4)(x + 2)$$
$$x = 4 \quad \text{or} \quad x = -2$$

The $x$-intercepts are $(4, 0)$ and $(-2, 0)$.

$$f(0) = 0^2 - 2 \cdot 0 - 8 = -8$$

The $y$-intercept is $(0, -8)$.

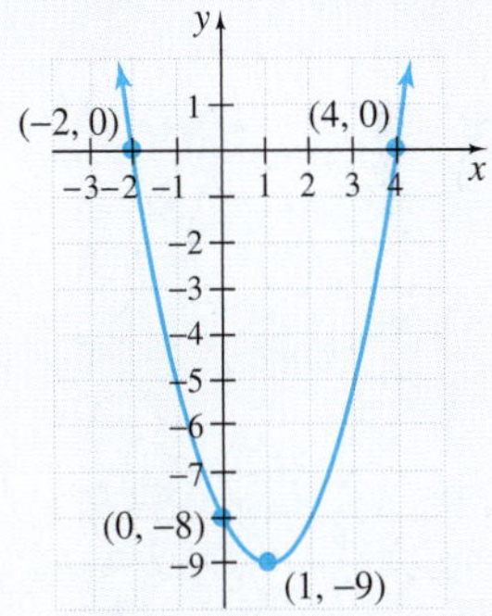


## CHAPTER 8 REVIEW

*(8.1) Solve by factoring.*

**1.** $x^2 - 15x + 14 = 0$

**2.** $x^2 - x - 30 = 0$

**3.** $10x^2 = 3x + 4$

**4.** $7a^2 = 29a + 30$

*Solve by using the square root property.*

**5.** $4m^2 = 196$

**6.** $9y^2 = 36$

**7.** $(9n + 1)^2 = 9$

**8.** $(5x - 2)^2 = 2$

*Solve by completing the square.*

**9.** $z^2 + 3z + 1 = 0$

**10.** $x^2 + x + 7 = 0$

**11.** $(2x + 1)^2 = x$

**12.** $(3x - 4)^2 = 10x$

**13.** If $P$ dollars are originally invested, the formula $A = P(1 + r)^2$ gives the amount $A$ in an account paying interest rate $r$ compounded annually after 2 years. Find the interest rate $r$ such that \$2500 increases to \$2717 in 2 years. Round the result to the nearest hundredth of a percent.

**14.** Two ships leave a port at the same time and travel at the same speed. One ship is traveling due north and the other due east. In a few hours, the ships are 150 miles apart. How many miles has each ship traveled? Give an exact answer and a one-decimal-place approximation.

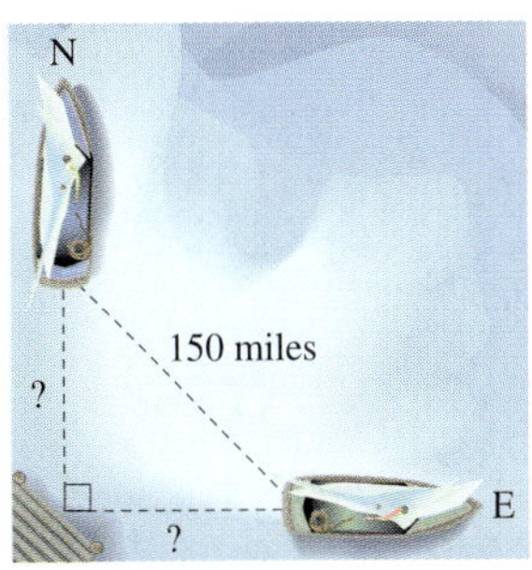

***(8.2)*** *If the discriminant of a quadratic equation has the given value, determine the number and type of solutions of the equation.*

**15.** $-8$ **16.** $48$

**17.** $100$ **18.** $0$

*Solve by using the quadratic formula.*

**19.** $x^2 - 16x + 64 = 0$ **20.** $x^2 + 5x = 0$

**21.** $x^2 + 11 = 0$ **22.** $2x^2 + 3x = 5$

**23.** $6x^2 + 7 = 5x$ **24.** $9a^2 + 4 = 2a$

**25.** $(5a - 2)^2 - a = 0$ **26.** $(2x - 3)^2 = x$

**27.** Cadets graduating from military school usually toss their hats high into the air at the end of the ceremony. One cadet threw his hat so that its distance $d(t)$ in feet above the ground $t$ seconds after it was thrown was $d(t) = -16t^2 + 30t + 6$.

**a.** Find the distance above the ground of the hat 1 second after it was thrown.

**b.** Find the time it takes the hat to hit the ground. Give an exact time and a one-decimal-place approximation.

△ **28.** The hypotenuse of an isosceles right triangle is 6 centimeters longer than either of the legs. Find the length of the legs.

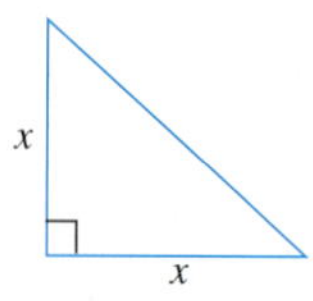


***(8.3)*** *Solve each equation for the variable.*

**29.** $x^3 = 27$ **30.** $y^3 = -64$

**31.** $\frac{5}{x} + \frac{6}{x-2} = 3$ **32.** $\frac{7}{8} = \frac{8}{x^2}$

**33.** $x^4 - 21x^2 - 100 = 0$

**34.** $5(x + 3)^2 - 19(x + 3) = 4$ $-\frac{16}{5}, 1$

**35.** $x^{2/3} - 6x^{1/3} + 5 = 0$ **36.** $x^{2/3} - 6x^{1/3} = -8$

**37.** $a^6 - a^2 = a^4 - 1$ **38.** $y^{-2} + y^{-1} = 20$ $-\frac{1}{5}, \frac{1}{4}$

**39.** Two postal workers, Jerome Grant and Tim Bozik, can sort a stack of mail in 5 hours. Working alone, Tim can sort the mail in 1 hour less time than Jerome can. Find the time that each postal worker can sort the mail alone. Round the result to one decimal place.

**40.** A negative number decreased by its reciprocal is $-\frac{24}{5}$. Find the number.

***(8.4)*** *Solve each inequality for x. Graph the solution set and write each solution set in interval notation.*

**41.** $2x^2 - 50 \le 0$

**42.** $\frac{1}{4}x^2 < \frac{1}{16}$

**43.** $(2x - 3)(4x + 5) \ge 0$

**44.** $(x^2 - 16)(x^2 - 1) > 0$

**45.** $\frac{x-5}{x-6} < 0$

**46.** $\frac{x(x+5)}{4x-3} \ge 0$

**47.** $\frac{(4x+3)(x-5)}{x(x+6)} > 0$

**48.** $(x + 5)(x - 6)(x + 2) \le 0$

**49.** $x^3 + 3x^2 - 25x - 75 > 0$

**50.** $\frac{x^2+4}{3x} \le 1$

**51.** $\frac{(5x+6)(x-3)}{x(6x-5)} < 0$

**52.** $\frac{3}{x-2} > 2$

*(8.5) Sketch the graph of each function. Label the vertex and the axis of symmetry.*

**53.** $f(x) = x^2 - 4$

**54.** $g(x) = x^2 + 7$

**55.** $H(x) = 2x^2$

**56.** $h(x) = -\frac{1}{3}x^2$

**57.** $F(x) = (x - 1)^2$

**58.** $G(x) = (x + 5)^2$

**59.** $f(x) = (x - 4)^2 - 2$

**60.** $f(x) = -3(x - 1)^2 + 1$

*(8.6) Sketch the graph of each function. Find the vertex and the intercepts.*

**61.** $f(x) = x^2 + 10x + 25$

**62.** $f(x) = -x^2 + 6x - 9$

**63.** $f(x) = 4x^2 - 1$

**64.** $f(x) = -5x^2 + 5$

**65.** Find the vertex of the graph of $f(x) = -3x^2 - 5x + 4$. Determine whether the graph opens upward or downward, find the $y$-intercept, approximate the $x$-intercepts to one decimal place, and sketch the graph.

**66.** The function $h(t) = -16t^2 + 120t + 300$ gives the height in feet of a projectile fired from the top of a building in $t$ seconds.

**a.** When will the object reach a height of 350 feet? Round your answer to one decimal place.

**b.** Explain why Part **a** has two answers.

**67.** Find two numbers whose product is as large as possible, given that their sum is 420.

**68.** Write an equation of a quadratic function whose graph is a parabola that has vertex $(-3, 7)$ and that passes through the origin.

## CHAPTER 8 TEST

*Solve each equation for the variable.*

**1.** $5x^2 - 2x = 7$

**2.** $(x + 1)^2 = 10$

**3.** $m^2 - m + 8 = 0$

**4.** $u^2 - 6u + 2 = 0$

**5.** $7x^2 + 8x + 1 = 0$

**6.** $a^2 - 3a = 5$

**7.** $\dfrac{4}{x + 2} + \dfrac{2x}{x - 2} = \dfrac{6}{x^2 - 4}$

**8.** $x^4 - 8x^2 - 9 = 0$

**9.** $x^6 + 1 = x^4 + x^2$

**10.** $(x + 1)^2 - 15(x + 1) + 56 = 0$

*Solve the equation for the variable by completing the square.*

**11.** $x^2 - 6x = -2$

**12.** $2a^2 + 5 = 4a$

*Solve each inequality for x. Graph the solution set and then write the solution set in interval notation.*

**13.** $2x^2 - 7x > 15$

**14.** $(x^2 - 16)(x^2 - 25) > 0$

**15.** $\dfrac{5}{x + 3} < 1$

**16.** $\dfrac{7x - 14}{x^2 - 9} \le 0$

*Graph each function. Label the vertex.*

**17.** $f(x) = 3x^2$

**18.** $G(x) = -2(x - 1)^2 + 5$

*Graph each function. Find and label the vertex, y-intercept, and x-intercepts (if any).*

**19.** $h(x) = x^2 - 4x + 4$

**20.** $F(x) = 2x^2 - 8x + 9$

△ **21.** A 10-foot ladder is leaning against a house. The distance from the bottom of the ladder to the house is 4 feet less than the distance from the top of the ladder to the ground. Find how far the top of the ladder is from the ground. Give an exact answer and a one-decimal-place approximation.

**22.** Dave and Sandy Hartranft can paint a room together in 4 hours. Working alone, Dave can paint the room in 2 hours less time than Sandy can. Find how long it takes Sandy to paint the room alone.

**23.** A stone is thrown upward from a bridge. The stone's height in feet, $s(t)$, above the water $t$ seconds after the stone is thrown is a function given by the equation $s(t) = -16t^2 + 32t + 256$.

**a.** Find the maximum height of the stone.

**b.** Find the time it takes the stone to hit the water. Round the answer to two decimal places.

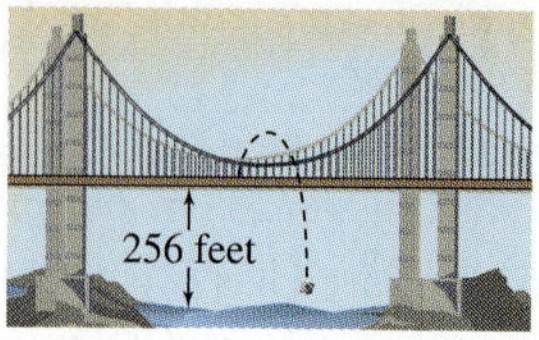


△ **24.** Given the diagram shown, approximate to the nearest foot how many feet of walking distance a person saves by cutting across the lawn instead of walking on the sidewalk.

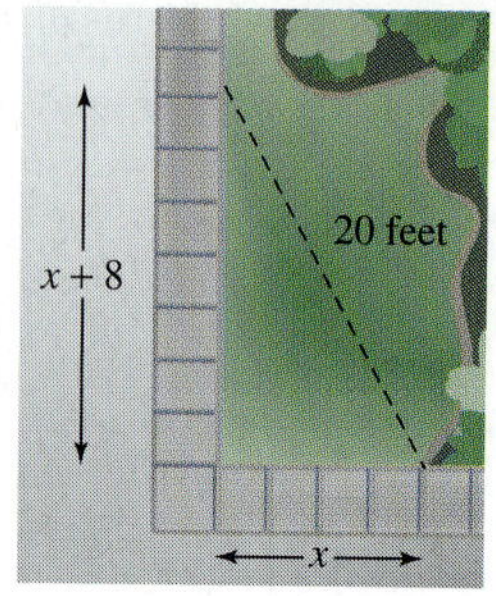


## CHAPTER 8 CUMULATIVE REVIEW

**1.** Write each sentence using mathematical symbols.
  **a.** The sum of 5 and $y$ is greater than or equal to 7.
  **b.** 11 is not equal to $z$.
  **c.** 20 is less than the difference of 5 and twice $x$.

**2.** Find the slope of the line containing the points $(0, 3)$ and $(2, 5)$. Graph the line.

**3.** Use the elimination method to solve the system:
$$\begin{cases} x - 5y = -12 \\ -x + y = 4 \end{cases}$$

**4.** Multiply.
  **a.** $(2x - 7)(3x - 4)$
  **b.** $(3x + y)(5x - 2y)$

**5.** Factor.
  **a.** $8x^2 + 4$
  **b.** $5x - 2z^4$
  **c.** $6x^2 - 3x^3$

Factor.

**6.** $x^2 - 12x + 35$

**7.** $3a^2x - 12abx + 12b^2x$

**8.** Solve $3(x^2 + 4) + 5 = -6(x^2 + 2x) + 13$.

**9.** Graph the quadratic function $f(x) = -x^2 + 2x - 3$ by plotting points.

**10.** Simplify.
$$\frac{2x^2}{10x^3 - 2x^2}$$

**11.** Add $\dfrac{2x - 1}{2x^2 - 9x - 5} + \dfrac{x + 3}{6x^2 - x - 2}$.

**12.** Simplify.
$$\frac{x^{-1} + 2xy^{-1}}{x^{-2} - x^{-2}y^{-1}}$$

**13.** Divide: $\dfrac{3x^5y^2 - 15x^3y - x^2y - 6x}{x^2y}$

**14.** If $P(x) = 2x^3 - 4x^2 + 5$
  **a.** Find $P(2)$ by substitution.
  **b.** Use synthetic division to find the remainder when $P(x)$ is divided by $x - 2$.

**15.** Solve: $\dfrac{4x}{5} + \dfrac{3}{2} = \dfrac{3x}{10}$

**16.** If a certain number is subtracted from the numerator and added to the denominator of $\dfrac{9}{19}$, the new fraction is equivalent to $\dfrac{1}{3}$. Find the number.

**17.** Suppose that $y$ varies directly as $x$. If $y$ is 5 when $x$ is 30, find the constant of variation and the direct variation equation.

**18.** Simplify.
  **a.** $\sqrt{(-3)^2}$
  **b.** $\sqrt{x^2}$
  **c.** $\sqrt[4]{(x - 2)^4}$
  **d.** $\sqrt[3]{(-5)^3}$
  **e.** $\sqrt[5]{(2x - 7)^5}$

**19.** Use rational exponents to simplify. Assume that variables represent positive numbers.

**a.** $\sqrt[6]{25}$ **b.** $\sqrt[8]{x^4}$

**c.** $\sqrt[4]{r^2s^6}$

**20.** Use the product rule to simplify.

**a.** $\sqrt{25x^3}$ **b.** $\sqrt[3]{54x^6y^8}$

**c.** $\sqrt[4]{81z^{11}}$

**21.** Rationalize the denominator of each expression.

**a.** $\dfrac{2}{\sqrt{5}}$ **b.** $\dfrac{2\sqrt{16}}{\sqrt{9x}}$

**c.** $\sqrt[3]{\dfrac{1}{2}}$

**22.** Solve $\sqrt{2x+5} + \sqrt{2x} = 3$.

**23.** Find each quotient. Write in the form $a + bi$.

**a.** $\dfrac{2+i}{1-i}$

**b.** $\dfrac{7}{3i}$

**24.** Use the square root property to solve $(x+1)^2 = 12$.

**25.** Solve $x - \sqrt{x} - 6 = 0$.

## Designing Your World

Schools, homes, hospitals, airports, auditoriums, community centers, jails, theaters, day-care centers, and office buildings are just some of the types of structures designed by architects to be safe, economical, and functional.

To become a licensed architect, a person must have a professional degree in architecture, complete a three-year internship, and pass the Architect Registration Examination. An architecture degree typically includes courses in building design, computer-aided design and drafting (CADD), physics and other physical sciences, architectural history, and mathematics. Architects need solid computer and communication skills. They also need a good understanding of geometry and spatial relationships to visualize a building during the design process.

*For more information about a career in architecture, visit The American Institute of Architects Web site by first going to* ***www.prenhall.com/martin-gay.***

*In the Spotlight on Decision Making feature on page 553, you will have the opportunity to make a decision as an architect about redesigning a bridge arch in the shape of a half-ellipse.*

# CONIC SECTIONS

# 9

9.1 The Parabola and the Circle

9.2 The Ellipse and the Hyperbola

9.3 Solving Nonlinear Systems of Equations

9.4 Nonlinear Inequalities and Systems of Inequalities

In Chapter 8, we analyzed some of the important connections between a parabola and its equation. Parabolas are interesting in their own right but are more interesting still because they are part of a collection of curves known as conic sections. This chapter is devoted to quadratic equations in two variables and their conic section graphs: the parabola, circle, ellipse, and hyperbola.

## 9.1 THE PARABOLA AND THE CIRCLE

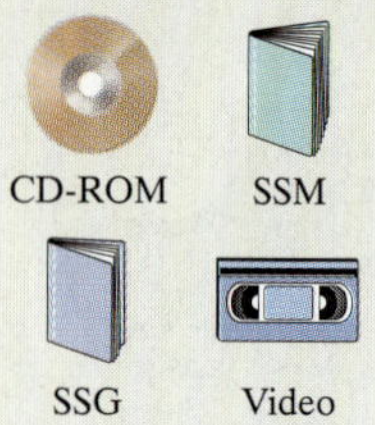


**OBJECTIVES**

1. Graph parabolas of the form $x = a(y - k)^2 + h$ and $y = a(x - h)^2 + k$.
2. Use the distance formula and the midpoint formula.
3. Graph circles of the form $(x - h)^2 + (y - k)^2 = r^2$.
4. Write the equation of a circle, given its center and radius.
5. Find the center and the radius of a circle, given its equation.

**Conic sections** derive their name because each conic section is the intersection of a right circular cone and a plane. The circle, parabola, ellipse, and hyperbola are the conic sections.

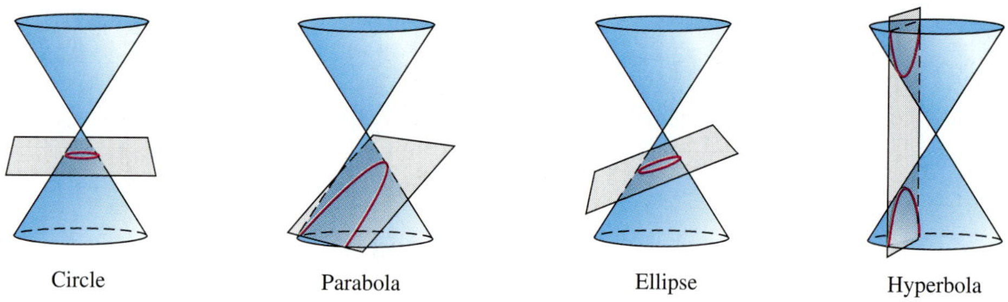


1 Thus far, we have seen that $f(x)$ or $y = a(x - h)^2 + k$ is the equation of a parabola that opens upward if $a > 0$ or downward if $a < 0$. Parabolas can also open left or right, or even on a slant. Equations of these parabolas are not functions of $x$, of course, since a parabola opening any way other than upward or downward fails the vertical line test. In this section, we introduce parabolas that open to the left and to the right. Parabolas opening on a slant will not be developed in this book.

Just as $y = a(x - h)^2 + k$ is the equation of a parabola that opens upward or downward, $x = a(y - k)^2 + h$ is the equation of a parabola that opens to the right or to the left. The parabola opens to the right if $a > 0$ and to the left if $a < 0$. The parabola has vertex $(h, k)$, and its axis of symmetry is the line $y = k$.

### PARABOLAS

$$y = a(x - h)^2 + k$$

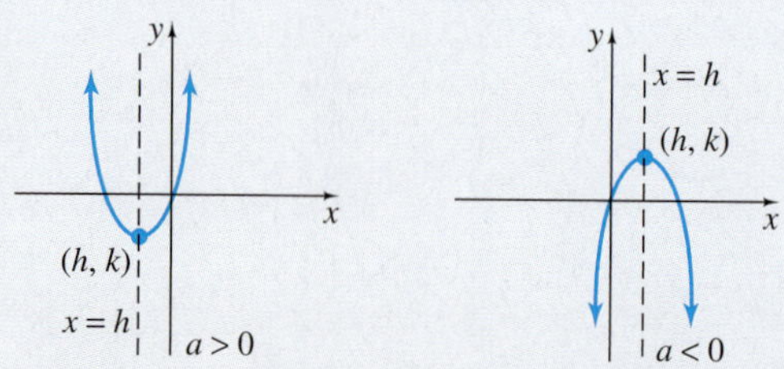

$$x = a(y - k)^2 + h$$

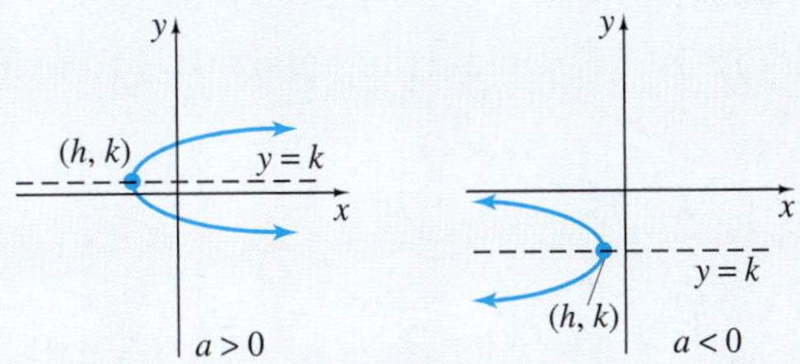


The equations $y = a(x - h)^2 + k$ and $x = a(y - k)^2 + h$ are called **standard forms.**

**Example 1** Graph the parabola $x = 2y^2$.

*Solution* Written in standard form, the equation $x = 2y^2$ is $x = 2(y - 0)^2 + 0$ with $a = 2$, $h = 0$, and $k = 0$. Its graph is a parabola with vertex $(0, 0)$, and its axis of symmetry is the line $y = 0$. Since $a > 0$, this parabola opens to the right. The table shows a few more ordered pair solutions of $x = 2y^2$. Its graph is also shown.

| $x$ | $y$ |
|---|---|
| 8 | −2 |
| 2 | −1 |
| 0 | 0 |
| 2 | 1 |
| 8 | 2 |

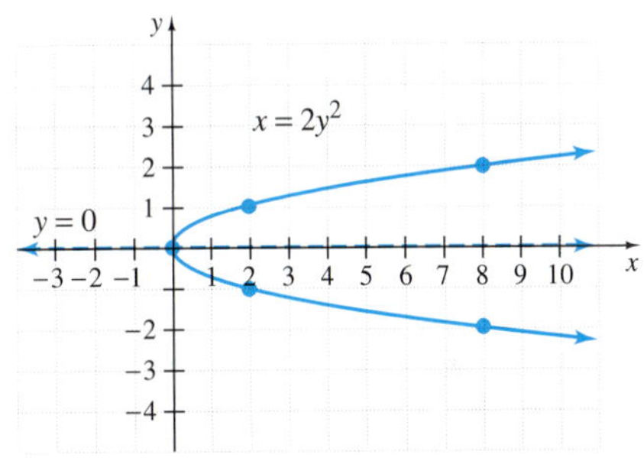


**Example 2** Graph the parabola $x = -3(y - 1)^2 + 2$.

*Solution* The equation $x = -3(y - 1)^2 + 2$ is in the form $x = a(y - k)^2 + h$ with $a = -3$, $k = 1$, and $h = 2$. Since $a < 0$, the parabola opens to the left. The vertex $(h, k)$ is $(2, 1)$, and the axis of symmetry is the line $y = 1$. When $y = 0$, $x = -1$, so the $x$-intercept is $(-1, 0)$. Again, we obtain a few ordered pair solutions and then graph the parabola.

| $x$ | $y$ |
|---|---|
| 2 | 1 |
| −1 | 0 |
| −1 | 2 |
| −10 | 3 |
| −10 | −1 |

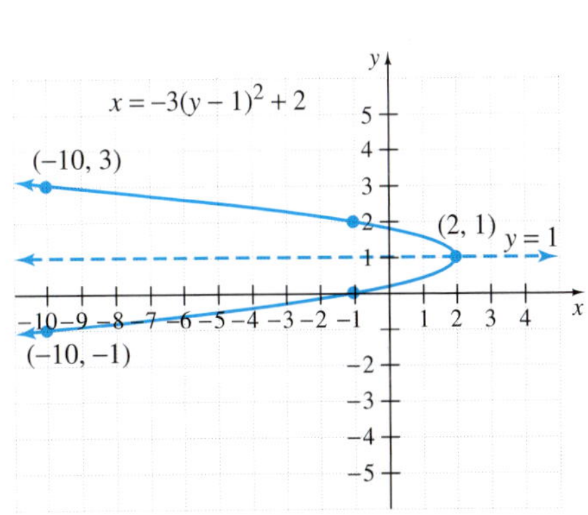

**Example 3** Graph $y = -x^2 - 2x + 15$.

*Solution* Complete the square on $x$ to write the equation in standard form.

$$y - 15 = -x^2 - 2x \qquad \text{Subtract 15 from both sides.}$$

$$y - 15 = -1(x^2 + 2x) \qquad \text{Factor } -1 \text{ from the terms } -x^2 - 2x.$$

The coefficient of $x$ is 2. Find the square of half of 2.

$$\frac{1}{2}(2) = 1 \quad \text{and} \quad 1^2 = 1$$

$$y - 15 - 1(1) = -1(x^2 + 2x + 1) \qquad \text{Add } -1(1) \text{ to both sides.}$$

$$y - 16 = -1(x + 1)^2 \qquad \text{Simplify the left side and factor the right side.}$$

$$y = -(x + 1)^2 + 16 \qquad \text{Add 16 to both sides.}$$

The equation is now in standard form $y = a(x - h)^2 + k$ with $a = -1$, $h = -1$, and $k = 16$.

The vertex is then $(h, k)$, or $(-1, 16)$.

A second method for finding the vertex is by using the formula $\frac{-b}{2a}$.

$$x = \frac{-(-2)}{2(-1)} = \frac{2}{-2} = -1$$

$$y = -(-1)^2 - 2(-1) + 15 = -1 + 2 + 15 = 16$$

Again, we see that the vertex is $(-1, 16)$, and the axis of symmetry is the vertical line $x = -1$. The $y$-intercept is $(0, 15)$. Now we can use a few more ordered pair solutions to graph the parabola.

| $x$ | $y$ |
|---|---|
| −1 | 16 |
| 0 | 15 |
| −2 | 15 |
| 1 | 12 |
| −3 | 12 |
| 3 | 0 |
| −5 | 0 |

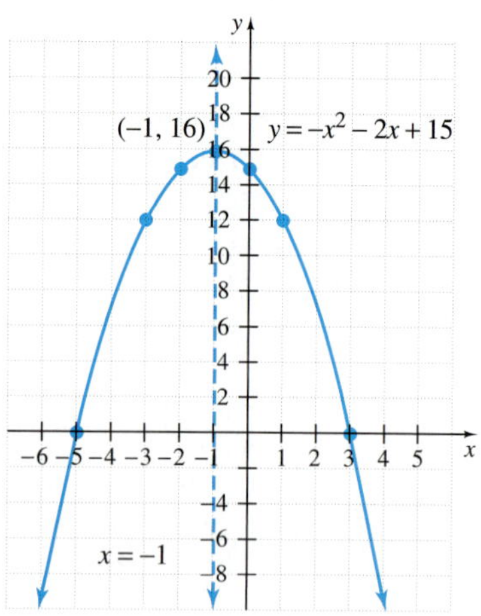

**Example 4** Graph $x = 2y^2 + 4y + 5$.

*Solution* Notice that this equation is quadratic in $y$, so its graph is a parabola that opens to the left or the right. We can complete the square on $y$ or we can use the formula $\dfrac{-b}{2a}$ to find the vertex.

Since the equation is quadratic in $y$, the formula gives us the $y$-value of the vertex.

$$y = \frac{-4}{2 \cdot 2} = \frac{-4}{4} = -1$$

$$x = 2(-1)^2 + 4(-1) + 5 = 2 \cdot 1 - 4 + 5 = 3$$

The vertex is $(3, -1)$, and the axis of symmetry is the line $y = -1$. The parabola opens to the right since $a > 0$. The $x$-intercept is $(5, 0)$.

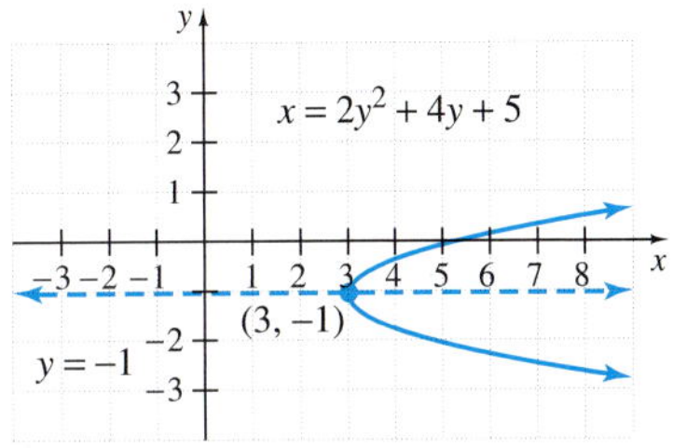


2 Another conic section is the circle. Before we review the circle, we need a formula to calculate the distance between points of the Cartesian coordinate system. To find the distance between two points, we use the distance formula, which is derived from the Pythagorean theorem.

To find the distance $d$ between two points $(x_1, y_1)$ and $(x_2, y_2)$ as shown to the left, notice that the length of leg $a$ is $x_2 - x_1$ and that the length of leg $b$ is $y_2 - y_1$.

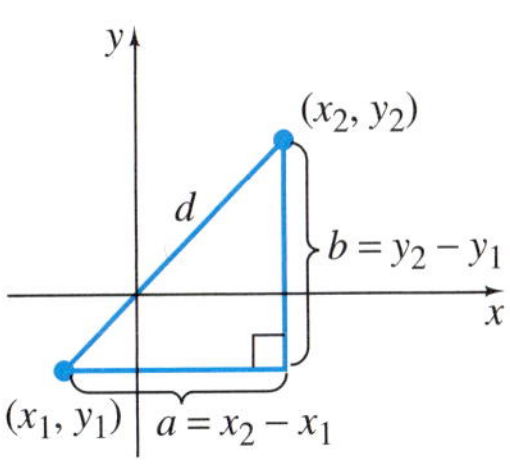


Thus, the Pythagorean theorem tells us that

$$d^2 = a^2 + b^2$$

or

$$d^2 = (x_2 - x_1)^2 + (y_2 - y_1)^2$$

or

$$d = \sqrt{(x_2 - x_1)^2 + (y_2 - y_1)^2}$$

This formula gives us the distance between any two points on the real plane.

**DISTANCE FORMULA**

The distance $d$ between two points $(x_1, y_1)$ and $(x_2, y_2)$ is given by

$$d = \sqrt{(x_2 - x_1)^2 + (y_2 - y_1)^2}$$

**Example 5** Find the distance between $(2, -5)$ and $(1, -4)$. Give an exact distance and a three-decimal-place approximation.

*Solution* To use the distance formula, it makes no difference which point we call $(x_1, y_1)$ and which point we call $(x_2, y_2)$. We will let $(x_1, y_1) = (2, -5)$ and $(x_2, y_2) = (1, -4)$.

$$\begin{aligned} d &= \sqrt{(x_2 - x_1)^2 + (y_2 - y_1)^2} \\ &= \sqrt{(1 - 2)^2 + [-4 - (-5)]^2} \\ &= \sqrt{(-1)^2 + (1)^2} \\ &= \sqrt{1 + 1} \\ &= \sqrt{2} \approx 1.414 \end{aligned}$$

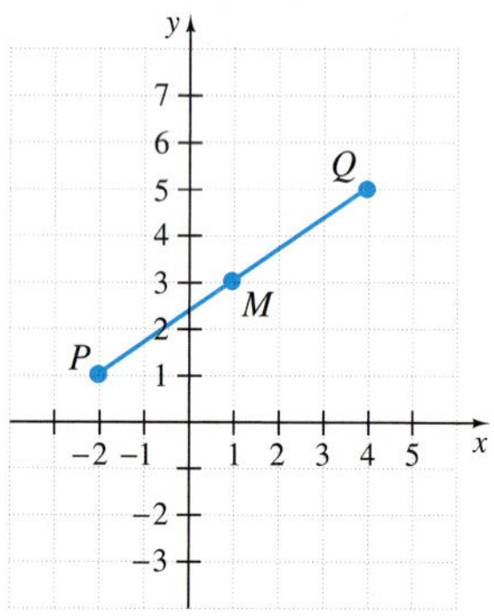


The distance between the two points is exactly $\sqrt{2}$ units, or approximately 1.414 uniits.

The **midpoint** of a line segment is the **point** located exactly halfway between the two endpoints of the line segment. On the graph to the left, the point $M$ is the midpoint of line segment $PQ$. Thus, the distance between $M$ and $P$ equals the distance between $M$ and $Q$.

The $x$-coordinate of $M$ is at half the distance between the $x$-coordinates of $P$ and $Q$, and the $y$-coordinate of $M$ is at half the distance between the $y$-coordinates of $P$ and $Q$. That is, the $x$-coordinate of $M$ is the average of the $x$-coordinates of $P$ and $Q$; the $y$-coordinate of $M$ is the average of the $y$-coordinates of $P$ and $Q$.

### MIDPOINT FORMULA

The midpoint of the line segment whose endpoints are $(x_1, y_1)$ and $(x_2, y_2)$ is the point with coordinates

$$\left(\frac{x_1 + x_2}{2}, \frac{y_1 + y_2}{2}\right)$$

**Example 6** Find the midpoint of the line segment that joins points $P(-3, 3)$ and $Q(1, 0)$.

*Solution* Use the midpoint formula. It makes no difference which point we call $(x_1, y_1)$ or which point we call $(x_2, y_2)$. Let $(x_1, y_1) = (-3, 3)$ and $(x_2, y_2) = (1, 0)$.

$$\begin{aligned} \text{midpoint} &= \left(\frac{x_1 + x_2}{2}, \frac{y_1 + y_2}{2}\right) \\ &= \left(\frac{-3 + 1}{2}, \frac{3 + 0}{2}\right) \\ &= \left(\frac{-2}{2}, \frac{3}{2}\right) \\ &= \left(-1, \frac{3}{2}\right) \end{aligned}$$

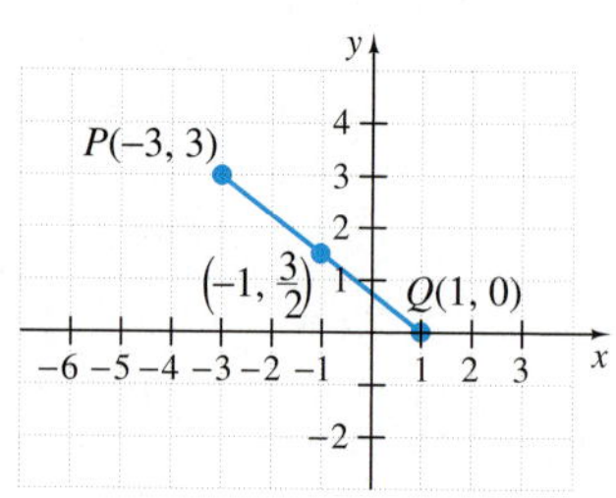


The midpoint of the segment is $\left(-1, \frac{3}{2}\right)$.

**3** Another conic section is the **circle.** A circle is the set of all points in a plane that are the same distance from a fixed point called the **center.** The distance is called the **radius** of the circle. To find a standard equation for a circle, let $(h, k)$ represent the center of the circle, and let $(x, y)$ represent any point on the circle. The distance between $(h, k)$ and $(x, y)$ is defined to be the circle's radius, $r$ units. We can find this distance $r$ by using the distance formula.

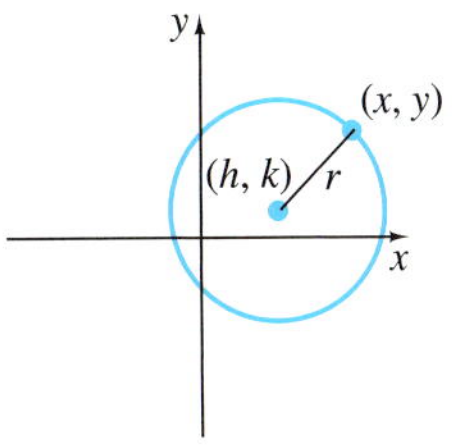


$$r = \sqrt{(x - h)^2 + (y - k)^2}$$

$$r^2 = (x - h)^2 + (y - k)^2 \quad \text{Square both sides.}$$

### CIRCLE

The graph of $(x - h)^2 + (y - k)^2 = r^2$ is a circle with center $(h, k)$ and radius $r$.

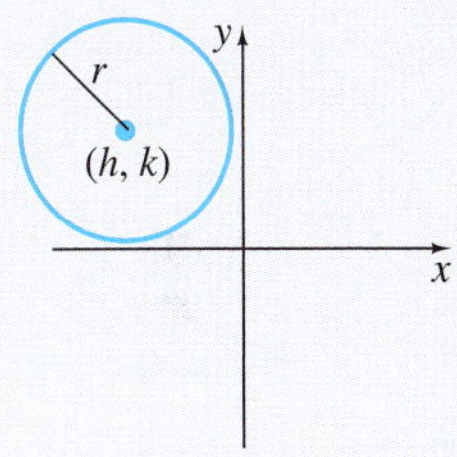


The equation $(x - h)^2 + (y - k)^2 = r^2$ is called **standard form.**

If an equation can be written in the standard form

$$(x - h)^2 + (y - k)^2 = r^2$$

then its graph is a circle, which we can draw by graphing the center $(h, k)$ and using the radius $r$.

**Example 7** Graph $x^2 + y^2 = 4$.

*Solution* The equation can be written in standard form as

$$(x - 0)^2 + (y - 0)^2 = 2^2$$

The center of the circle is $(0, 0)$, and the radius is 2. Its graph is shown.

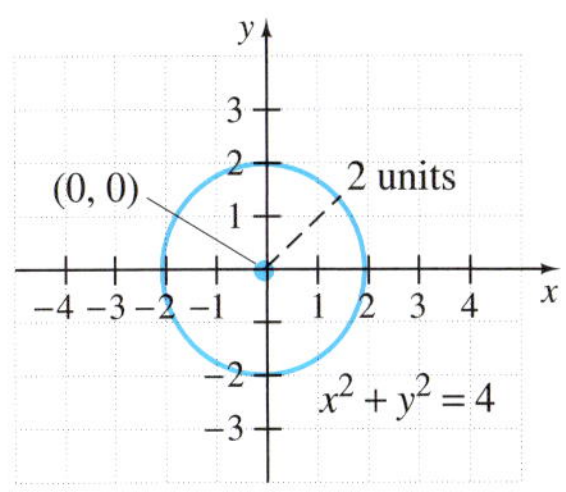


**HELPFUL HINT**

Notice the difference between the equation of a circle and the equation of a parabola. The equation of a circle contains both $x^2$ and $y^2$ terms on the same side of the equation with equal coefficients. The equation of a parabola has either an $x^2$ term or a $y^2$ term but not both.

**Example 8** Graph $(x + 1)^2 + y^2 = 8$.

*Solution* The equation can be written as $(x + 1)^2 + (y - 0)^2 = 8$ with $h = -1$, $k = 0$, and $r = \sqrt{8}$. The center is $(-1, 0)$, and the radius is $\sqrt{8} = 2\sqrt{2} \approx 2.8$.

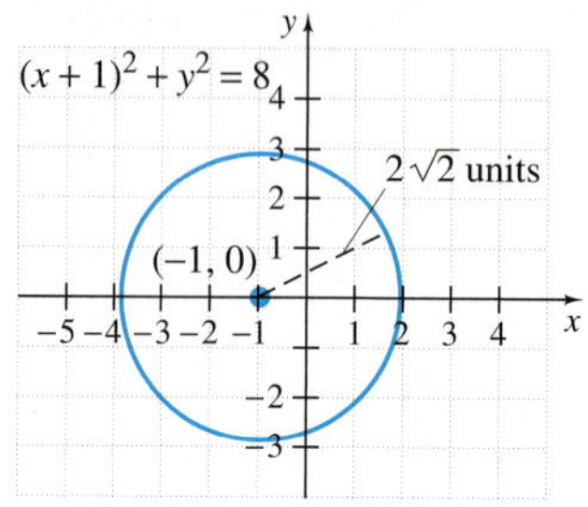


**4** Since a circle is determined entirely by its center and radius, this information is all we need to write the equation of a circle.

**Example 9** Find an equation of the circle with center $(-7, 3)$ and radius 10.

*Solution* Using the given values $h = -7$, $k = 3$, and $r = 10$, we write the equation

$$(x - h)^2 + (y - k)^2 = r^2$$

or

$$[x - (-7)]^2 + (y - 3)^2 = 10^2 \quad \text{Substitute the given values.}$$

or

$$(x + 7)^2 + (y - 3)^2 = 100$$

**5** To find the center and the radius of a circle from its equation, write the equation in standard form. To write the equation of a circle in standard form, we complete the square on both $x$ and $y$.

**Example 10** Graph $x^2 + y^2 + 4x - 8y = 16$.

*Solution* Since this equation contains $x^2$ and $y^2$ terms on the same side of the equation with equal coefficients, its graph is a circle. To write the equation in standard form, group the terms involving $x$ and the terms involving $y$, and then complete the square on each variable.

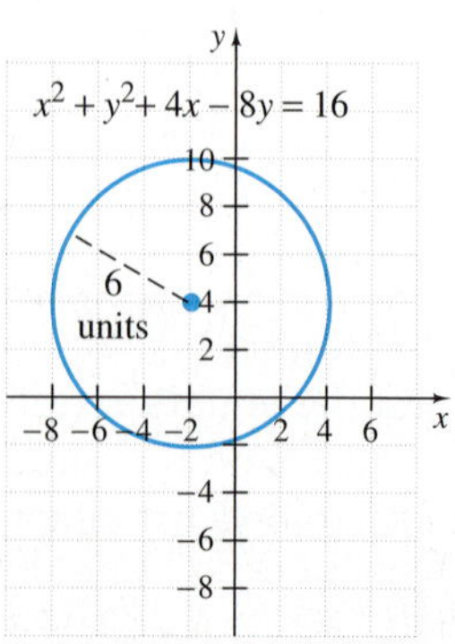


$$(x^2 + 4x) + (y^2 - 8y) = 16$$

Thus, $\frac{1}{2}(4) = 2$ and $2^2 = 4$. Also, $\frac{1}{2}(-8) = -4$ and $(-4)^2 = 16$. Add 4 and then 16 to both sides.

$$(x^2 + 4x + 4) + (y^2 - 8y + 16) = 16 + 4 + 16$$

$$(x + 2)^2 + (y - 4)^2 = 36 \quad \text{Factor.}$$

This circle has the center $(-2, 4)$ and radius 6, as shown.

## GRAPHING CALCULATOR EXPLORATIONS

To graph an equation such as $x^2 + y^2 = 25$ with a graphing calculator, we first solve the equation for $y$.

$$x^2 + y^2 = 25$$
$$y^2 = 25 - x^2$$
$$y = \pm\sqrt{25 - x^2}$$

The graph of $y = \sqrt{25 - x^2}$ will be the top half of the circle, and the graph of $y = -\sqrt{25 - x^2}$ will be the bottom half of the circle.

To graph, press [Y =] and enter $Y_1 = \sqrt{25 - x^2}$ and $Y_2 = -\sqrt{25 - x^2}$. Insert parentheses around $25 - x^2$ so that $\sqrt{25 - x^2}$ and not $\sqrt{25} - x^2$ is graphed.

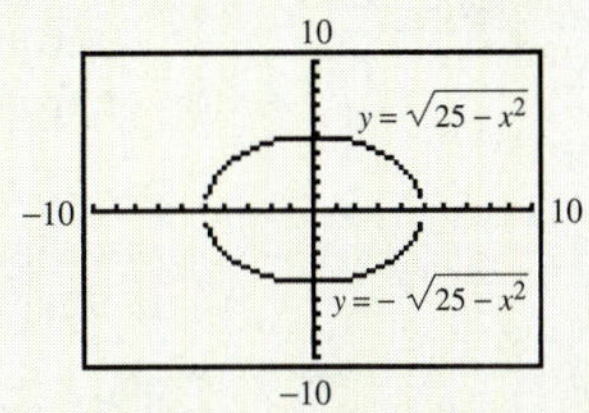


The graph does not appear to be a circle because we are currently using a standard window and the screen is rectangular. This causes the tick marks on the $x$-axis to be farther apart than the tick marks on the $y$-axis and, thus, creates the distorted circle. If we want the graph to appear circular, we must define a square window by using a feature of the graphing calculator or by redefining the window to show the $x$-axis from −15 to 15 and the $y$-axis from −10 to 10. Using a square window, the graph appears as follows.

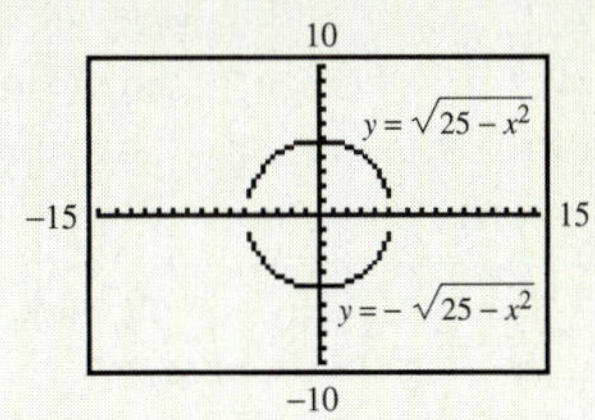


*Use a graphing calculator to graph each circle.*

**1.** $x^2 + y^2 = 55$
**2.** $x^2 + y^2 = 20$
**3.** $5x^2 + 5y^2 = 50$
**4.** $6x^2 + 6y^2 = 105$
**5.** $2x^2 + 2y^2 - 34 = 0$
**6.** $4x^2 + 4y^2 - 48 = 0$
**7.** $7x^2 + 7y^2 - 89 = 0$
**8.** $3x^2 + 3y^2 - 35 = 0$

## MENTAL MATH

*The graph of each equation is a parabola. Determine whether the parabola opens upward, downward, to the left, or to the right.*

**1.** $y = x^2 - 7x + 5$
**2.** $y = -x^2 + 16$
**3.** $x = -y^2 - y + 2$
**4.** $x = 3y^2 + 2y - 5$
**5.** $y = -x^2 + 2x + 1$
**6.** $x = -y^2 + 2y - 6$

## Exercise Set 9.1

*The graph of each equation is a parabola. Find the vertex of the parabola and sketch its graph. See Examples 1 through 4.*

**1.** $x = 3y^2$
**2.** $x = -2y^2$
**3.** $x = (y - 2)^2 + 3$
**4.** $x = (y - 4)^2 - 1$
**5.** $y = 3(x - 1)^2 + 5$
**6.** $x = -4(y - 2)^2 + 2$
**7.** $x = y^2 + 6y + 8$
**8.** $x = y^2 - 6y + 6$
**9.** $y = x^2 + 10x + 20$
**10.** $y = x^2 + 4x - 5$
**11.** $x = -2y^2 + 4y + 6$
**12.** $x = 3y^2 + 6y + 7$

*Find the distance between each pair of points. Approximate the distance in Exercises 21 and 22 to two decimal places. See Example 5.*

**13.** $(5, 1)$ and $(8, 5)$
**14.** $(2,3)$ and $(14,8)$
**15.** $(-3, 2)$ and $(1, -3)$
**16.** $(3, -2)$ and $(-4, 1)$
**17.** $(-9, 4)$ and $(-8, 1)$
**18.** $(-5, -2)$ and $(-6, -6)$
**19.** $(0, -\sqrt{2})$ and $(\sqrt{3}, 0)$
**20.** $(-\sqrt{5}, 0)$ and $(0, \sqrt{7})$
**21.** $(1.7, -3.6)$ and $(-8.6, 5.7)$
**22.** $(9.6, 2.5)$ and $(-1.9, -3.7)$
**23.** $(2\sqrt{3}, \sqrt{6})$ and $(-\sqrt{3}, 4\sqrt{6})$
**24.** $(5\sqrt{2}, -4)$ and $(-3\sqrt{2}, -8)$

*Find the midpoint of the line segment whose endpoints are given. See Example 6.*

**25.** $(6, -8), (2, 4)$
**26.** $(3, 9), (7, 11)$
**27.** $(-2, -1), (-8, 6)$
**28.** $(-3, -4), (6, -8)$
**29.** $(7, 3), (-1, -3)$
**30.** $(-2, 5), (-1, 6)$
**31.** $\left(\frac{1}{2}, \frac{3}{8}\right), \left(-\frac{3}{2}, \frac{5}{8}\right)$
**32.** $\left(-\frac{2}{5}, \frac{7}{15}\right), \left(-\frac{2}{5}, -\frac{4}{15}\right)$
**33.** $(\sqrt{2}, 3\sqrt{5}), (\sqrt{2}, -2\sqrt{5})$
**34.** $(\sqrt{8}, -\sqrt{12}), (3\sqrt{2}, 7\sqrt{3})$
**35.** $(4.6, -3.5), (7.8, -9.8)$
**36.** $(-4.6, 2.1), (-6.7, 1.9)$

*The graph of each equation is a circle. Find the center and the radius, and then sketch. See Examples 7, 8, and 10.*

**37.** $x^2 + y^2 = 9$
**38.** $x^2 + y^2 = 25$
**39.** $x^2 + (y - 2)^2 = 1$
**40.** $(x - 3)^2 + y^2 = 9$
**41.** $(x - 5)^2 + (y + 2)^2 = 1$
**42.** $(x + 3)^2 + (y + 3)^2 = 4$
**43.** $x^2 + y^2 + 6y = 0$
**44.** $x^2 + 10x + y^2 = 0$
**45.** $x^2 + y^2 + 2x - 4y = 4$
**46.** $x^2 + 6x - 4y + y^2 = 3$
**47.** $x^2 + y^2 - 4x - 8y - 2 = 0$
**48.** $x^2 + y^2 - 2x - 6y - 5 = 0$

*Write an equation of the circle with the given center and radius. See Example 9.*

**49.** $(2, 3); 6$
**50.** $(-7, 6); 2$
**51.** $(0, 0); \sqrt{3}$
**52.** $(0, -6); \sqrt{2}$
**53.** $(-5, 4); 3\sqrt{5}$
**54.** the origin; $4\sqrt{7}$

**55.** If you are given a list of equations of circles and parabolas and none are in standard form, explain how you would determine which is an equation of a circle and which is an equation of a parabola. Explain also how you would distinguish the upward or downward parabolas from the left-opening or right-opening parabolas.

*Sketch the graph of each equation. If the graph is a parabola, find its vertex. If the graph is a circle, find its center and radius.*

**56.** $x = y^2 + 2$
**57.** $x = y^2 - 3$
**58.** $y = (x + 3)^2 + 3$
**59.** $y = (x - 2)^2 - 2$
**60.** $x^2 + y^2 = 49$
**61.** $x^2 + y^2 = 1$
**62.** $x = (y - 1)^2 + 4$
**63.** $x = (y + 3)^2 - 1$
**64.** $(x + 3)^2 + (y - 1)^2 = 9$
**65.** $(x - 2)^2 + (y - 2)^2 = 16$
**66.** $x = -2(y + 5)^2$
**67.** $x = -(y - 1)^2$
**68.** $x^2 + (y + 5)^2 = 5$
**69.** $(x - 4)^2 + y^2 = 7$
**70.** $y = 3(x - 4)^2 + 2$
**71.** $y = 5(x + 5)^2 + 3$
**72.** $2x^2 + 2y^2 = \frac{1}{2}$
**73.** $\frac{x^2}{8} + \frac{y^2}{8} = 2$
**74.** $y = x^2 - 2x - 15$
**75.** $y = x^2 + 7x + 6$
**76.** $x^2 + y^2 + 6x + 10y - 2 = 0$
**77.** $x^2 + y^2 + 2x + 12y - 12 = 0$
**78.** $x = y^2 + 6y + 2$
**79.** $x = y^2 + 8y - 4$
**80.** $x^2 + y^2 - 8y + 5 = 0$
**81.** $x^2 - 10y + y^2 + 4 = 0$
**82.** $x = -2y^2 - 4y$
**83.** $x = -3y^2 + 30y$
**84.** $\frac{x^2}{3} + \frac{y^2}{3} = 2$
**85.** $5x^2 + 5y^2 = 25$
**86.** $y = 4x^2 - 40x + 105$
**87.** $y = 5x^2 - 20x + 16$

*Solve.*

**88.** Two surveyors need to find the distance across a lake. They place a reference pole at point $A$ in the diagram. Point $B$ is 3 meters east and 1 meter north of the reference point $A$. Point $C$ is 19 meters east and 13 meters north of point $A$. Find the distance across the lake, from $B$ to $C$.

△ **89.** Determine whether the triangle with vertices (2, 6), (0, −2), and (5, 1) is an isosceles triangle.

△ **90.** Cindy Brown, an architect, is drawing plans on grid paper for a circular pool with a fountain in the middle. The paper is marked off in centimeters, and each centimeter represents 1 foot. On the paper, the diameter of the "pool" is 20 centimeters, and "fountain" is the point (0, 0).

**a.** Sketch the architect's drawing. Be sure to label the axes.

**b.** Write an equation that describes the circular pool.

**c.** Cindy plans to place a circle of lights around the fountain such that each light is 5 feet from the fountain. Write an equation for the circle of lights and sketch the circle on your drawing.

**91.** A bridge constructed over a bayou has a supporting arch in the shape of a parabola. Find an equation of the parabolic arch if the length of the road over the arch is 100 meters and the maximum height of the arch is 40 meters.

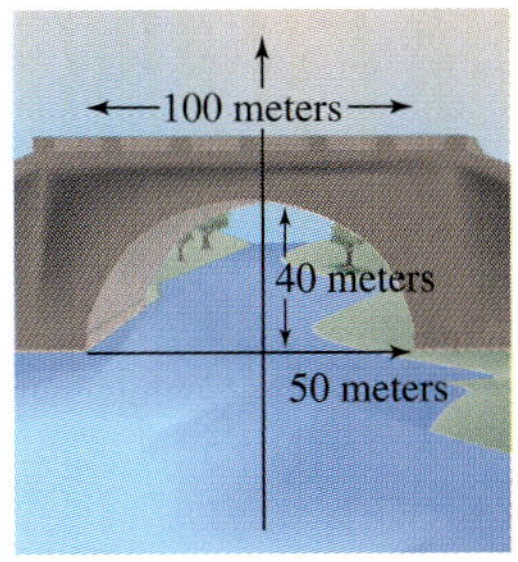


*Use a graphing calculator to verify each exercise. Use a square viewing window.*

**92.** Exercise 84. **93.** Exercise 85.

**94.** Exercise 86. **95.** Exercise 87.

## REVIEW EXERCISES

*Graph each equation. See Section 3.3.*

**96.** $y = 2x + 5$ **97.** $y = -3x + 3$

**98.** $y = 3$ **99.** $x = -2$

*Rationalize each denominator and simplify if possible. See Section 7.5.*

**100.** $\frac{1}{\sqrt{3}}$ **101.** $\frac{\sqrt{5}}{\sqrt{8}}$

**102.** $\frac{4\sqrt{7}}{\sqrt{6}}$ **103.** $\frac{10}{\sqrt{5}}$

# 9.2 THE ELLIPSE AND THE HYPERBOLA

**OBJECTIVES**

1. Define and graph an ellipse.
2. Define and graph a hyperbola.

**1** An **ellipse** can be thought of as the set of points in a plane such that the sum of the distances of those points from two fixed points is constant. Each of the two fixed points is called a **focus.** (The plural of focus is **foci.**) The point midway between the foci is called the **center.**

An ellipse may be drawn by hand by using two thumbtacks, a piece of string, and a pencil. Secure the two thumbtacks in a piece of cardboard, for example, and tie each end of the string to a tack. Use your pencil to pull the string tight and draw the ellipse. The two thumbtacks are the foci of the drawn ellipse.

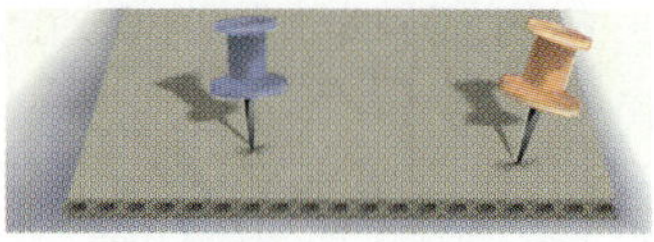

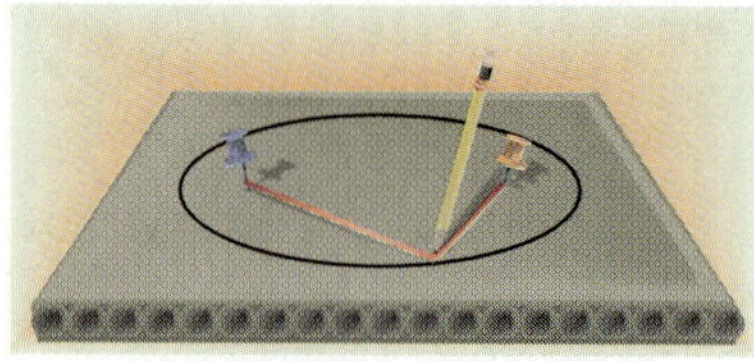

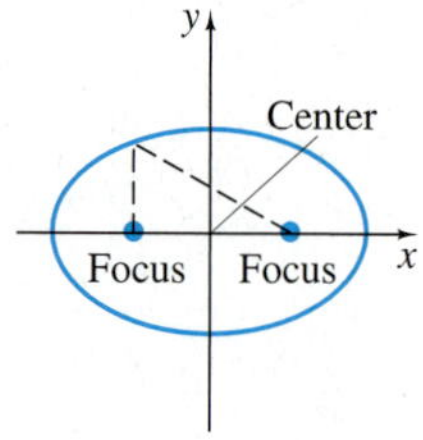


## ELLIPSE WITH CENTER (0, 0)

The graph of an equation of the form $\frac{x^2}{a^2} + \frac{y^2}{b^2} = 1$ is an ellipse with center $(0, 0)$. The $x$-intercepts are $(a, 0)$ and $(-a, 0)$, and the $y$-intercepts are $(0, b)$, and $(0, -b)$.

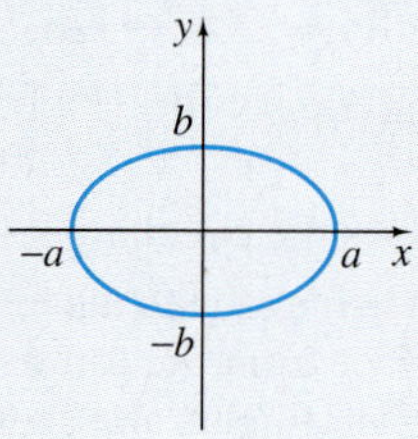


The **standard form** of an ellipse with center $(0, 0)$ is $\frac{x^2}{a^2} + \frac{y^2}{b^2} = 1$.

**Example 1** Graph $\frac{x^2}{9} + \frac{y^2}{16} = 1$.

*Solution* The equation is of the form $\frac{x^2}{a^2} + \frac{y^2}{b^2} = 1$, with $a = 3$ and $b = 4$, so its graph is an ellipse with center $(0, 0)$, $x$-intercepts $(3, 0)$ and $(-3, 0)$, and $y$-intercepts $(0, 4)$ and $(0, -4)$.

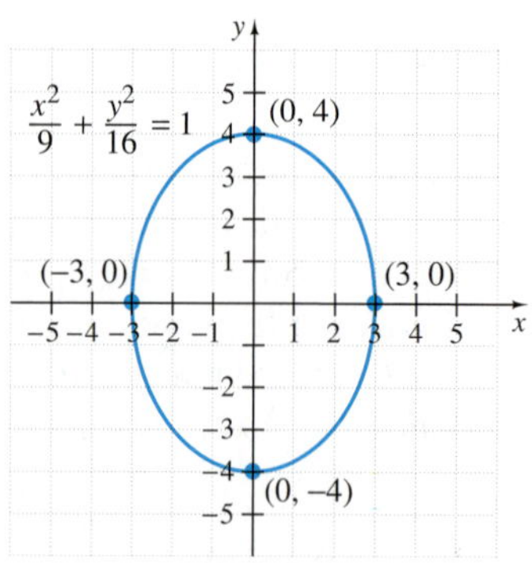

**Example 2** Graph the equation $4x^2 + 16y^2 = 64$.

*Solution* Although this equation contains a sum of squared terms in $x$ and $y$ on the same side of an equation, this is not the equation of a circle since the coefficients of $x^2$ and $y^2$ are not the same. The graph of this equation is an ellipse. Since the standard form of the equation of an ellipse has 1 on one side, divide both sides of this equation by 64.

$$4x^2 + 16y^2 = 64$$

$$\frac{4x^2}{64} + \frac{16y^2}{64} = \frac{64}{64} \qquad \text{Divide both sides by 64.}$$

$$\frac{x^2}{16} + \frac{y^2}{4} = 1 \qquad \text{Simplify.}$$

We now recognize the equation of an ellipse with $a = 4$ and $b = 2$. This ellipse has center $(0, 0)$, $x$-intercepts $(4, 0)$ and $(-4, 0)$, and $y$-intercepts $(0, 2)$ and $(0, -2)$.

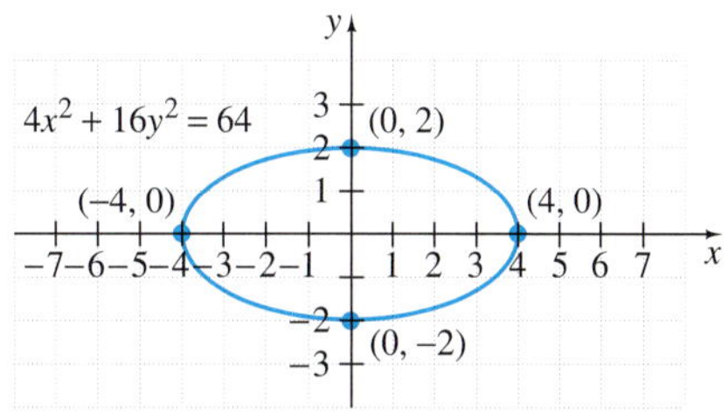


The center of an ellipse is not always $(0, 0)$, as shown in the next example.

**Example 3** Graph $\frac{(x + 3)^2}{25} + \frac{(y - 2)^2}{36} = 1$.

*Solution* The center of this ellipse is found in a way that is similar to finding the center of a circle. This ellipse has center $(-3, 2)$. Notice that $a = 5$ and $b = 6$. To find four points on the graph of the ellipse, first graph the center, $(-3, 2)$. Since $a = 5$, count 5 units right and then 5 units left of the point with coordinates $(-3, 2)$. Next, since $b = 6$, start at $(-3, 2)$ and count 6 units up and then 6 units down to find two more points on the ellipse.

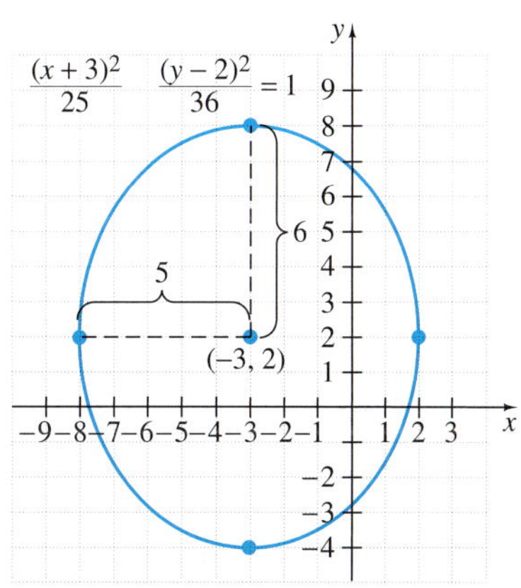

**2** The final conic section is the **hyperbola.** A hyperbola is the set of points in a plane such that the absolute value of the difference of the distances from two fixed points is constant. Each of the two fixed points is called a **focus.** The point midway between the foci is called the **center.**

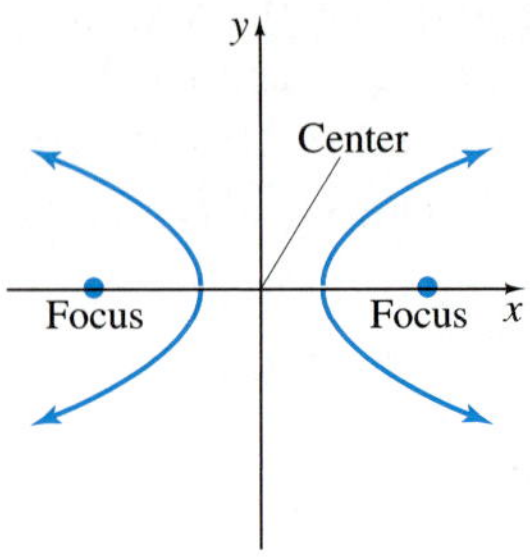


Using the distance formula, we can show that the graph of $\frac{x^2}{a^2} - \frac{y^2}{b^2} = 1$ is a hyperbola with center $(0, 0)$ and $x$-intercepts $(a, 0)$ and $(-a, 0)$. Also, the graph of $\frac{y^2}{b^2} - \frac{x^2}{a^2} = 1$ is a hyperbola with center $(0, 0)$ and $y$-intercepts $(0, b)$ and $(0, -b)$.

### HYPERBOLA WITH CENTER (0, 0)

The graph of an equation of the form $\frac{x^2}{a^2} - \frac{y^2}{b^2} = 1$ is a hyperbola with center $(0, 0)$ and $x$-intercepts $(a, 0)$ and $(-a, 0)$.

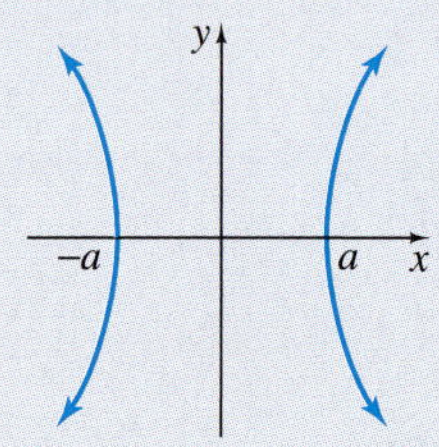


The graph of an equation of the form $\frac{y^2}{b^2} - \frac{x^2}{a^2} = 1$ is a hyperbola with center $(0, 0)$ and $y$-intercepts $(0, b)$ and $(0, -b)$.

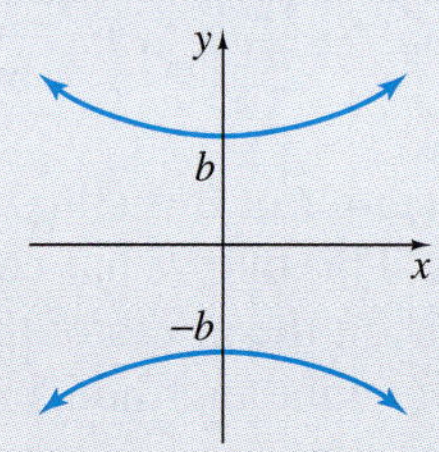


The equations $\frac{x^2}{a^2} - \frac{y^2}{b^2} = 1$ and $\frac{y^2}{b^2} - \frac{x^2}{a^2} = 1$ are the **standard forms** for the equation of a hyperbola.

**HELPFUL HINT**

Notice the difference between the equation of an ellipse and a hyperbola. The equation of the ellipse contains $x^2$ and $y^2$ terms on the same side of the equation with same-sign coefficients. For a hyperbola, the coefficients on the same side of the equation have different signs.

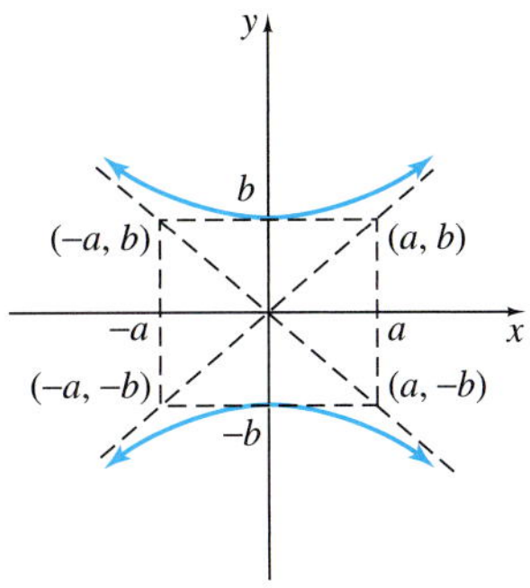


Graphing a hyperbola such as $\frac{y^2}{b^2} - \frac{x^2}{a^2} = 1$ is made easier by recognizing one of its important characteristics. Examining the figure to the left, notice how the sides of the branches of the hyperbola extend indefinitely and seem to approach the dashed lines in the figure. These dashed lines are called the **asymptotes** of the hyperbola.

To sketch these lines, or asymptotes, draw a rectangle with vertices $(a, b)$, $(-a, b)$, $(a, -b)$, and $(-a, -b)$. The asymptotes of the hyperbola are the extended diagonals of this rectangle.

## Example 4

Sketch the graph of $\frac{x^2}{16} - \frac{y^2}{25} = 1$.

*Solution* This equation has the form $\frac{x^2}{a^2} - \frac{y^2}{b^2} = 1$, with $a = 4$ and $b = 5$. Thus, its graph is a hyperbola that opens to the left and right. It has center $(0, 0)$ and $x$-intercepts $(4, 0)$ and $(-4, 0)$. To aid in graphing the hyperbola, we first sketch its asymptotes. The extended diagonals of the rectangle with corners $(4, 5)$, $(4, -5)$, $(-4, 5)$, and $(-4, -5)$ are the asymptotes of the hyperbola. Then we use the asymptotes to aid in sketching the hyperbola.

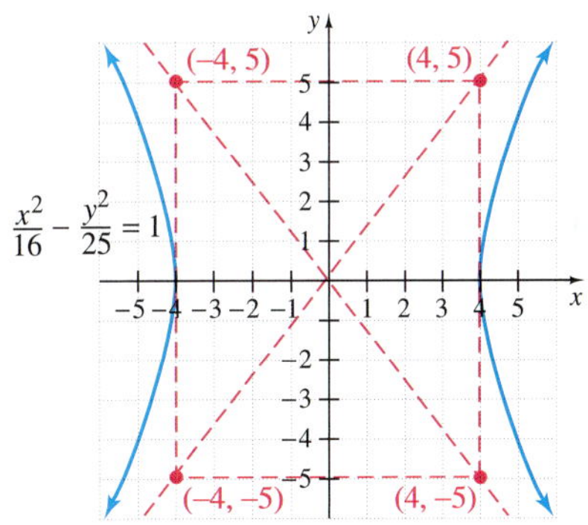


## Example 5

Sketch the graph of the equation $4y^2 - 9x^2 = 36$.

*Solution* Since this is a difference of squared terms in $x$ and $y$ on the same side of the equation, its graph is a hyperbola, as opposed to an ellipse or a circle. The standard form of the equation of a hyperbola has a 1 on one side, so divide both sides of the equation by 36.

$$4y^2 - 9x^2 = 36$$

$$\frac{4y^2}{36} - \frac{9x^2}{36} = \frac{36}{36} \qquad \text{Divide both sides by 36.}$$

$$\frac{y^2}{9} - \frac{x^2}{4} = 1 \qquad \text{Simplify.}$$

The equation is of the form $\frac{y^2}{b^2} - \frac{x^2}{a^2} = 1$, with $a = 2$ and $b = 3$, so the hyperbola is centered at $(0, 0)$ with $y$-intercepts $(0, 3)$ and $(0, -3)$. The sketch of the hyperbola is shown.

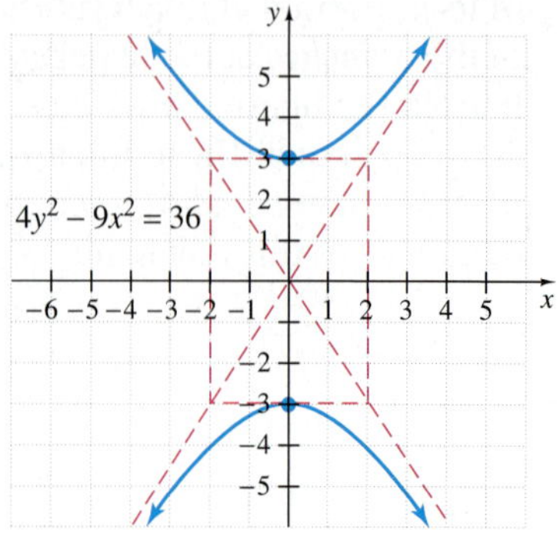


Following is a summary of conic sections.

## CONIC SECTIONS

| | ***Standard Form*** | ***Graph*** |
|---|---|---|
| ***Parabola*** | $y = a(x - h)^2 + k$ | |
| ***Parabola*** | $x = a(y - k)^2 + h$ | |
| ***Circle*** | $(x - h)^2 + (y - k)^2 = r^2$ | |
| ***Ellipse*** | $\frac{x^2}{a^2} + \frac{y^2}{b^2} = 1$ | |

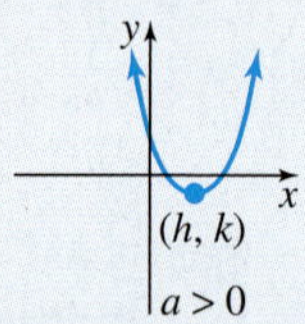


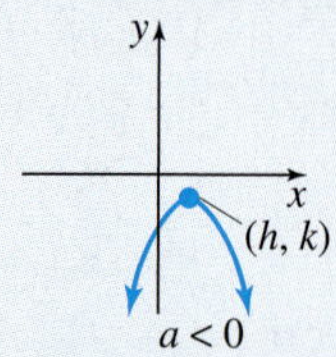


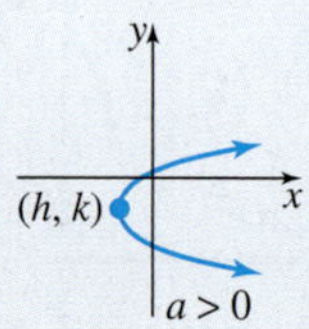


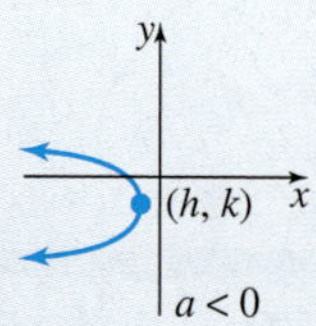


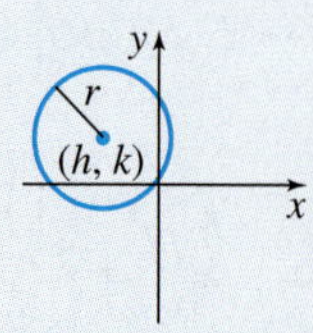


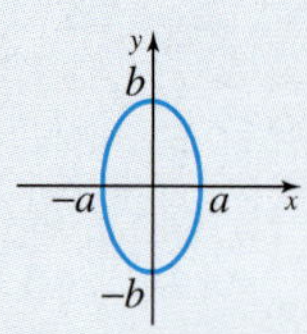

| | | |
|---|---|---|
| ***Hyperbola*** | $\dfrac{x^2}{a^2} - \dfrac{y^2}{b^2} = 1$ | 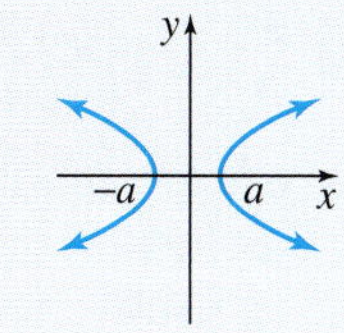 |
| ***Hyperbola*** | $\dfrac{y^2}{b^2} - \dfrac{x^2}{a^2} = 1$ | 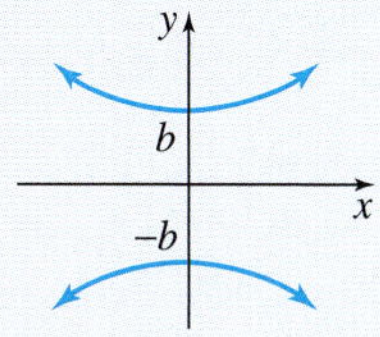 |

## GRAPHING CALCULATOR EXPLORATIONS

To find the graph of an ellipse by using a graphing calculator, use the same procedure as for graphing a circle. For example, to graph $x^2 + 3y^2 = 22$, first solve for $y$.

$$3y^2 = 22 - x^2$$

$$y^2 = \frac{22 - x^2}{3}$$

$$y = \pm\sqrt{\frac{22 - x^2}{3}}$$

Next press the [Y =] key and enter $Y_1 = \sqrt{\dfrac{22 - x^2}{3}}$ and $Y_2 = -\sqrt{\dfrac{22 - x^2}{3}}$. (Insert two sets of parentheses in the radicand as $\sqrt{((22 - x^2)/3)}$ so that the desired graph is obtained.) The graph appears as follows.

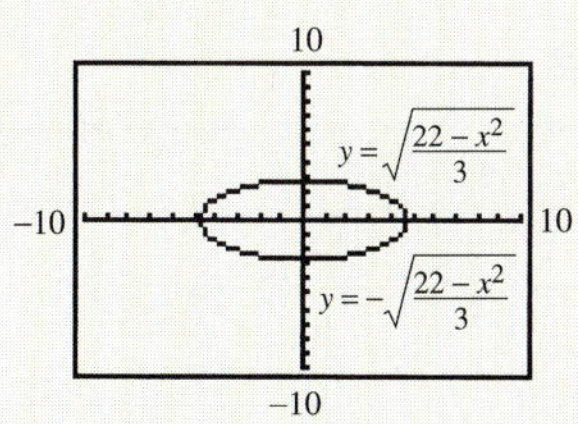


*Use a graphing calculator to graph each ellipse.*

**1.** $10x^2 + y^2 = 32$

**2.** $x^2 + 6y^2 = 35$

**3.** $20x^2 + 5y^2 = 100$

**4.** $4y^2 + 12x^2 = 48$

**5.** $7.3x^2 + 15.5y^2 = 95.2$

**6.** $18.8x^2 + 36.1y^2 = 205.8$

## SPOTLIGHT ON DECISION MAKING

Suppose you are an astronomer. You know that the orbits of stars, planets, comets, asteroids, and satellites all have the shape of one of the conic sections. *Eccentricity* is a measure used to describe the shape and elongation of an orbital path. The table shows ranges of eccentricities for the different types of conic sections.

For each of the following comets known to pass through our solar system, decide what type of orbit the comet has based on its eccentricity $e$. Describe how the shape of the comet's orbit affects how often it passes through our solar system. (For more exercises on eccentricity, see Exercise Set 9.2.)

**a.** Spacewatch (1997 P2), $e = 1.02851919$

**b.** Whipple, $e = 0.25871336$

**c.** Lee (1999 H1), $e = 0.99973749$

**d.** Giacobini-Zinner, $e = 0.70647162$

**e.** Tabur (1997 N1), $e = 1.00004712$

| *Conic Section* | *Eccentricity e* |
|---|---|
| Circle | $e = 0$ |
| Ellipse | $0 < e < 1$ |
| Parabola | $e = 1$ |
| Hyperbola | $e > 1$ |

## Exercise Set 9.2

*Sketch the graph of each equation. See Examples 1 and 2.*

**1.** $\frac{x^2}{4} + \frac{y^2}{25} = 1$

**2.** $\frac{x^2}{9} + y^2 = 1$

**3.** $\frac{x^2}{16} + \frac{y^2}{9} = 1$

**4.** $x^2 + \frac{y^2}{4} = 1$

**5.** $9x^2 + 4y^2 = 36$

**6.** $x^2 + 4y^2 = 16$

**7.** $4x^2 + 25y^2 = 100$

**8.** $36x^2 + y^2 = 36$

*Sketch the graph of each equation. See Example 3.*

**9.** $\frac{(x + 1)^2}{36} + \frac{(y - 2)^2}{49} = 1$

**10.** $\frac{(x - 3)^2}{9} + \frac{(y + 3)^2}{16} = 1$

**11.** $\frac{(x - 1)^2}{4} + \frac{(y - 1)^2}{25} = 1$

**12.** $\frac{(x + 3)^2}{16} + \frac{(y + 2)^2}{4} = 1$

*Sketch the graph of each equation. See Examples 4 and 5.*

**13.** $\frac{x^2}{4} - \frac{y^2}{9} = 1$

**14.** $\frac{x^2}{36} - \frac{y^2}{36} = 1$

**15.** $\frac{y^2}{25} - \frac{x^2}{16} = 1$

**16.** $\frac{y^2}{25} - \frac{x^2}{49} = 1$

*Sketch the graph of each equation. See Example 5.*

**17.** $x^2 - 4y^2 = 16$

**18.** $4x^2 - y^2 = 36$

**19.** $16y^2 - x^2 = 16$

**20.** $4y^2 - 25x^2 = 100$

**21.** If you are given a list of equations of circles, parabolas, ellipses, and hyperbolas, explain how you could distinguish the different conic sections from their equations.

*Identify whether each equation, when graphed, will be a parabola, circle, ellipse, or hyperbola. Sketch the graph of each equation.*

**22.** $(x - 7)^2 + (y - 2)^2 = 4$

**23.** $y = x^2 + 4$

**24.** $y = x^2 + 12x + 36$

**25.** $\dfrac{x^2}{4} + \dfrac{y^2}{9} = 1$

**26.** $\dfrac{y^2}{9} - \dfrac{x^2}{9} = 1$

**27.** $\dfrac{x^2}{16} - \dfrac{y^2}{4} = 1$

**28.** $\dfrac{x^2}{16} + \dfrac{y^2}{4} = 1$

**29.** $x^2 + y^2 = 16$

**30.** $x = y^2 + 4y - 1$

**31.** $x = -y^2 + 6y$

**32.** $9x^2 - 4y^2 = 36$

**33.** $9x^2 + 4y^2 = 36$

**34.** $\dfrac{(x-1)^2}{49} + \dfrac{(y+2)^2}{25} = 1$

**35.** $y^2 = x^2 + 16$

**36.** $\left(x + \dfrac{1}{2}\right)^2 + \left(y - \dfrac{1}{2}\right)^2 = 1$

**37.** $y = -2x^2 + 4x - 3$

*The orbits of stars, planets, comets, asteroids, and satellites all have the shape of one of the conic sections. Astronomers use a measure called* eccentricity *to describe the shape and elongation of an orbital path. For the circle and ellipse, eccentricity e is calculated with the formula* $e = \dfrac{c}{d}$, *where* $c^2 = |a^2 - b^2|$ *and d is the larger value of a or b. For a hyperbola, eccentricity e is calculated with the formula* $e = \dfrac{c}{d}$, *where* $c^2 = a^2 + b^2$ *and the value of d is equal to a if the hyperbola has x-intercepts or equal to b if the hyperbola has y-intercepts. (For more information about eccentricity, see the Spotlight on Decision Making in this section.)*

**A** $\dfrac{x^2}{36} - \dfrac{y^2}{13} = 1$

**B** $\dfrac{x^2}{4} + \dfrac{y^2}{4} = 1$

**C** $\dfrac{x^2}{25} + \dfrac{y^2}{16} = 1$

**D** $\dfrac{y^2}{25} - \dfrac{x^2}{39} = 1$

**E** $\dfrac{x^2}{17} + \dfrac{y^2}{81} = 1$

**F** $\dfrac{x^2}{36} + \dfrac{y^2}{36} = 1$

**G** $\dfrac{x^2}{16} - \dfrac{y^2}{65} = 1$

**H** $\dfrac{x^2}{144} + \dfrac{y^2}{140} = 1$

**38.** Identify the type of conic section represented by each of the equations A–H.

**39.** For each of the equations A–H, identify the values of $a^2$ and $b^2$.

**40.** For each of the equations A–H, calculate the value of $c^2$ and $c$.

**41.** For each of the equations A–H, find the value of $d$.

**42.** For each of the equations A–H, calculate the eccentricity $e$.

**43.** What do you notice about the values of $e$ for the equations you identified as ellipses?

**44.** What do you notice about the values of $e$ for the equations you identified as circles?

**45.** What do you notice about the values of $e$ for the equations you identified as hyperbolas?

*Solve.*

**46.** A planet's orbit about the Sun can be described as an ellipse. Consider the Sun as the origin of a rectangular coordinate system. Suppose that the $x$-intercepts of the elliptical path of the planet are $\pm 130{,}000{,}000$ and that the $y$-intercepts are $\pm 125{,}000{,}000$. Write the equation of the elliptical path of the planet.

**47.** Comets orbit the Sun in elongated ellipses. Consider the Sun as the origin of a rectangular coordinate system. Suppose that the equation of the path of the comet is

$$\frac{(x - 1{,}782{,}000{,}000)^2}{3.42 \cdot 10^{23}} + \frac{(y - 356{,}400{,}000)^2}{1.368 \cdot 10^{22}} = 1$$

Find the center of the path of the comet.

**48.** Use a graphing calculator to verify Exercise 5.

**49.** Use a graphing calculator to verify Exercise 6.

## REVIEW EXERCISES

*Solve each inequality. See Section 2.5.*

**50.** $x < 5$ and $x < 1$

**51.** $x < 5$ or $x < 1$

**52.** $2x - 1 \geq 7$ or $-3x \leq -6$

**53.** $2x - 1 \geq 7$ and $-3x \leq -6$

*Perform the indicated operations. See Sections 5.3 and 5.4.*

**54.** $(2x^3)(-4x^2)$

**55.** $2x^3 - 4x^3$

**56.** $-5x^2 + x^2$

**57.** $(-5x^2)(x^2)$

## *A Look Ahead*

**Example**

Sketch the graph of $\dfrac{(x-2)^2}{25} - \dfrac{(y-1)^2}{9} = 1$.

*Solution*
This hyperbola has center (2, 1). Notice that $a = 5$ and $b = 3$.

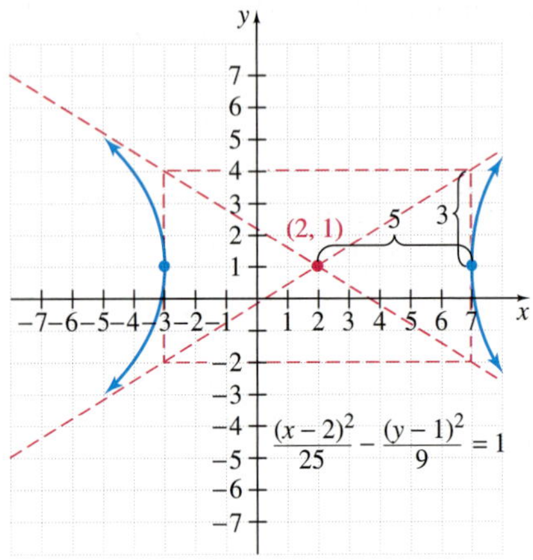


*Sketch the graph of each equation. See the preceding example.*

**59.** $\frac{(x-1)^2}{4} - \frac{(y+1)^2}{25} = 1$

**60.** $\frac{(x+2)^2}{9} - \frac{(y-1)^2}{4} = 1$

**61.** $\frac{y^2}{16} - \frac{(x+3)^2}{9} = 1$

**62.** $\frac{(y+4)^2}{4} - \frac{x^2}{25} = 1$

**63.** $\frac{(x+5)^2}{16} - \frac{(y+2)^2}{25} = 1$

**64.** $\frac{(x-3)^2}{9} - \frac{(y-2)^2}{4} = 1$

## 9.3 SOLVING NONLINEAR SYSTEMS OF EQUATIONS

▶ **OBJECTIVES**

CD-ROM SSM SSG Video

1. Solve a nonlinear system by substitution.
2. Solve a nonlinear system by elimination.

In Section 4.1, we used graphing, substitution, and elimination methods to find solutions of systems of linear equations in two variables. We now apply these same methods to nonlinear systems of equations in two variables. A **nonlinear system of equations** is a system of equations at least one of which is not linear. Since we will be graphing the equations in each system, we are interested in real number solutions only.

**1** First, nonlinear systems are solved by the substitution method.

**Example 1** Solve the system.

$$\begin{cases} x^2 - 3y = 1 \\ x - y = 1 \end{cases}$$

*Solution* We can solve this system by substitution if we solve one equation for one of the variables. Solving the first equation for $x$ is not the best choice since doing so introduces a radical. Also, solving for $y$ in the first equation introduces a fraction. We solve the second equation for $y$.

$$x - y = 1 \quad \text{Second equation}$$

$$x - 1 = y \quad \text{Solve for } y.$$

Replace $y$ with $x - 1$ in the first equation, and then solve for $x$.

$$x^2 - 3y = 1 \quad \text{First equation}$$
$$x^2 - 3(x - 1) = 1 \quad \text{Replace } y \text{ with } x - 1.$$
$$x^2 - 3x + 3 = 1$$
$$x^2 - 3x + 2 = 0$$
$$(x - 2)(x - 1) = 0$$
$$x = 2 \quad \text{or} \quad x = 1$$

Let $x = 2$ and then let $x = 1$ in the equation $y = x - 1$ to find corresponding $y$-values.

Let $x = 2$.
$$y = x - 1$$
$$y = 2 - 1 = 1$$

Let $x = 1$.
$$y = x - 1$$
$$y = 1 - 1 = 0$$

The solutions are $(2, 1)$ and $(1, 0)$. Check both solutions in both equations. Both solutions satisfy both equations, so both are solutions of the system. The graph of each equation in the system is shown next. Intersections of the graphs are at $(2, 1)$ and $(1, 0)$.

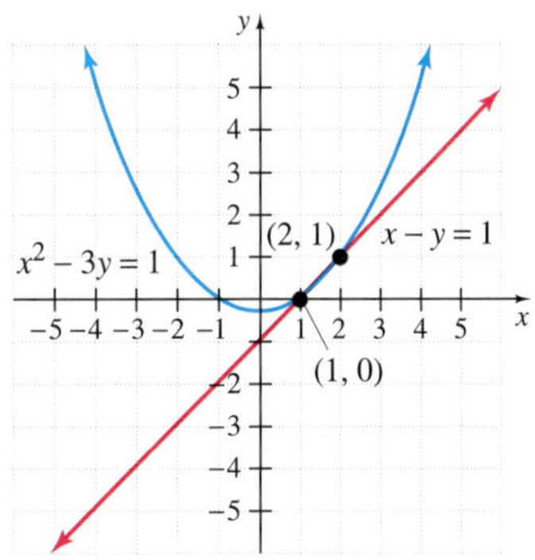


## Example 2

Solve the system.

$$\begin{cases} y = \sqrt{x} \\ x^2 + y^2 = 6 \end{cases}$$

*Solution* This system is ideal for substitution since $y$ is expressed in terms of $x$ in the first equation. Notice that if $y = \sqrt{x}$, then both $x$ and $y$ must be nonnegative if they are real numbers. Substitute $\sqrt{x}$ for $y$ in the second equation, and solve for $x$.

$$x^2 + y^2 = 6$$
$$x^2 + (\sqrt{x})^2 = 6 \quad \text{Let } y = \sqrt{x}.$$
$$x^2 + x = 6$$
$$x^2 + x - 6 = 0$$
$$(x + 3)(x - 2) = 0$$
$$x = -3 \quad \text{or} \quad x = 2$$

The solution $-3$ is discarded because we have noted that $x$ must be nonnegative. To see this, let $x = -3$ in the first equation. Then let $x = 2$ in the first equation to find a corresponding $y$-value.

Let $x = -3$.

$$y = \sqrt{x}$$

$$y = \sqrt{-3} \quad \text{Not a real number}$$

Let $x = 2$.

$$y = \sqrt{x}$$

$$y = \sqrt{2}$$

Since we are interested only in real number solutions, the only solution is $(2, \sqrt{2})$. Check to see that this solution satisfies both equations. The graph of each equation in the system is shown next.

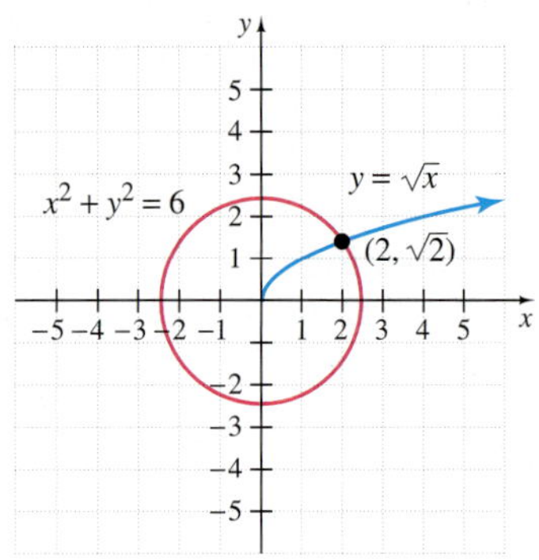


**Example 3** Solve the system.

$$\begin{cases} x^2 + y^2 = 4 \\ x + y = 3 \end{cases}$$

*Solution* We use the substitution method and solve the second equation for $x$.

$$x + y = 3 \quad \text{Second equation}$$

$$x = 3 - y$$

Now we let $x = 3 - y$ in the first equation.

$$x^2 + y^2 = 4 \quad \text{First equation}$$

$$(3 - y)^2 + y^2 = 4 \quad \text{Let } x = 3 - y.$$

$$9 - 6y + y^2 + y^2 = 4$$

$$2y^2 - 6y + 5 = 0$$

By the quadratic formula, where $a = 2$, $b = -6$, and $c = 5$, we have

$$y = \frac{6 \pm \sqrt{(-6)^2 - 4 \cdot 2 \cdot 5}}{2 \cdot 2} = \frac{6 \pm \sqrt{-4}}{4}$$

Since $\sqrt{-4}$ is not a real number, there is no real solution, or $\varnothing$. Graphically, the circle and the line do not intersect, as shown.

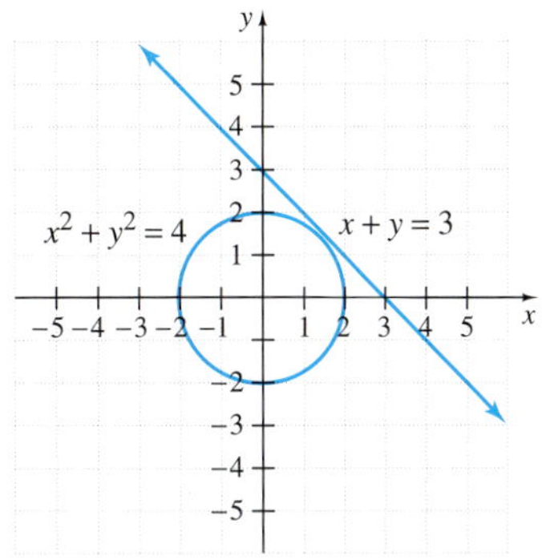


2 Some nonlinear systems may be solved by the elimination method.

**Example 4** Solve the system.

$$\begin{cases} x^2 + 2y^2 = 10 \\ x^2 - y^2 = 1 \end{cases}$$

*Solution* We will use the elimination, or addition, method to solve this system. To eliminate $x^2$ when we add the two equations, multiply both sides of the second equation by $-1$. Then

$$\begin{cases} x^2 + 2y^2 = 10 \\ (-1)(x^2 - y^2) = -1 \cdot 1 \end{cases} \quad \text{is equivalent to} \quad \begin{cases} x^2 + 2y^2 = 10 \\ -x^2 + y^2 = -1 \end{cases}$$

$$3y^2 = 9 \quad \text{Add.}$$

$$y^2 = 3 \quad \text{Divide both sides by 3.}$$

$$y = \pm\sqrt{3}$$

To find the corresponding $x$-values, we let $y = \sqrt{3}$ and $y = -\sqrt{3}$ in either original equation. We choose the second equation.

| Let $y = \sqrt{3}$. | Let $y = -\sqrt{3}$. |
|---|---|
| $x^2 - y^2 = 1$ | $x^2 - y^2 = 1$ |
| $x^2 - (\sqrt{3})^2 = 1$ | $x^2 - (-\sqrt{3})^2 = 1$ |
| $x^2 - 3 = 1$ | $x^2 - 3 = 1$ |
| $x^2 = 4$ | $x^2 = 4$ |
| $x = \pm\sqrt{4} = \pm 2$ | $x = \pm\sqrt{4} = \pm 2$ |

The solutions are $(2, \sqrt{3})$, $(-2, \sqrt{3})$, $(2, -\sqrt{3})$, and $(-2, -\sqrt{3})$. Check all four ordered pairs in both equations of the system. The graph of each equation in this system is shown.

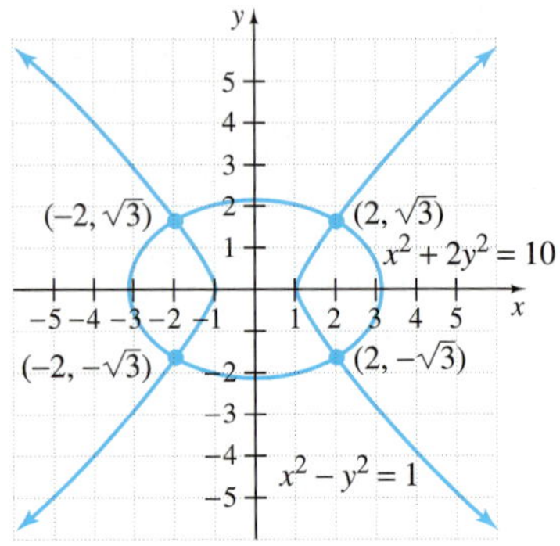


## Exercise Set 9.3

*Solve each nonlinear system of equations for real solutions. See Examples 1 through 4.*

1. $\begin{cases} x^2 + y^2 = 25 \\ 4x + 3y = 0 \end{cases}$

2. $\begin{cases} x^2 + y^2 = 25 \\ 3x + 4y = 0 \end{cases}$

3. $\begin{cases} x^2 + 4y^2 = 10 \\ y = x \end{cases}$

4. $\begin{cases} 4x^2 + y^2 = 10 \\ y = x \end{cases}$

5. $\begin{cases} y^2 = 4 - x \\ x - 2y = 4 \end{cases}$

6. $\begin{cases} x^2 + y^2 = 4 \\ x + y = -2 \end{cases}$

7. $\begin{cases} x^2 + y^2 = 9 \\ 16x^2 - 4y^2 = 64 \end{cases}$

8. $\begin{cases} 4x^2 + 3y^2 = 35 \\ 5x^2 + 2y^2 = 42 \end{cases}$

9. $\begin{cases} x^2 + 2y^2 = 2 \\ x - y = 2 \end{cases}$

10. $\begin{cases} x^2 + 2y^2 = 2 \\ x^2 - 2y^2 = 6 \end{cases}$

11. $\begin{cases} y = x^2 - 3 \\ 4x - y = 6 \end{cases}$

12. $\begin{cases} y = x + 1 \\ x^2 - y^2 = 1 \end{cases}$

13. $\begin{cases} y = x^2 \\ 3x + y = 10 \end{cases}$

14. $\begin{cases} 6x - y = 5 \\ xy = 1 \end{cases}$

15. $\begin{cases} y = 2x^2 + 1 \\ x + y = -1 \end{cases}$

16. $\begin{cases} x^2 + y^2 = 9 \\ x + y = 5 \end{cases}$

17. $\begin{cases} y = x^2 - 4 \\ y = x^2 - 4x \end{cases}$

18. $\begin{cases} x = y^2 - 3 \\ x = y^2 - 3y \end{cases}$

19. $\begin{cases} 2x^2 + 3y^2 = 14 \\ -x^2 + y^2 = 3 \end{cases}$

20. $\begin{cases} 4x^2 - 2y^2 = 2 \\ -x^2 + y^2 = 2 \end{cases}$

21. $\begin{cases} x^2 + y^2 = 1 \\ x^2 + (y + 3)^2 = 4 \end{cases}$

22. $\begin{cases} x^2 + 2y^2 = 4 \\ x^2 - y^2 = 4 \end{cases}$

23. $\begin{cases} y = x^2 + 2 \\ y = -x^2 + 4 \end{cases}$

24. $\begin{cases} x = -y^2 - 3 \\ x = y^2 - 5 \end{cases}$

25. $\begin{cases} 3x^2 + y^2 = 9 \\ 3x^2 - y^2 = 9 \end{cases}$

26. $\begin{cases} x^2 + y^2 = 25 \\ x = y^2 - 5 \end{cases}$

27. $\begin{cases} x^2 + 3y^2 = 6 \\ x^2 - 3y^2 = 10 \end{cases}$

28. $\begin{cases} x^2 + y^2 = 1 \\ y = x^2 - 9 \end{cases}$

29. $\begin{cases} x^2 + y^2 = 36 \\ y = \frac{1}{6}x^2 - 6 \end{cases}$

30. $\begin{cases} x^2 + y^2 = 16 \\ y = -\frac{1}{4}x^2 + 4 \end{cases}$

31. How many real solutions are possible for a system of equations whose graphs are a circle and a parabola? Draw diagrams to illustrate each possibility.

32. How many real solutions are possible for a system of equations whose graphs are an ellipse and a line? Draw diagrams to illustrate each possibility.

33. The sum of the squares of two numbers is 130. The difference of the squares of the two numbers is 32. Find the two numbers.

**34.** The sum of the squares of two numbers is 20. Their product is 8. Find the two numbers.

△ **35.** During the development stage of a new rectangular keypad for a security system, it was decided that the area of the rectangle should be 285 square centimeters and the perimeter should be 68 centimeters. Find the dimensions of the keypad.

△ **36.** A rectangular holding pen for cattle is to be designed so that its perimeter is 92 feet and its area is 525 feet. Find the dimensions of the holding pen.

*Recall that in business, a demand function expresses the quantity of a commodity demanded as a function of the commodity's unit price. A supply function expresses the quantity of a commodity supplied as a function of the commodity's unit price. When the quantity produced and supplied is equal to the quantity demanded, then we have what is called* ***market equilibrium.***

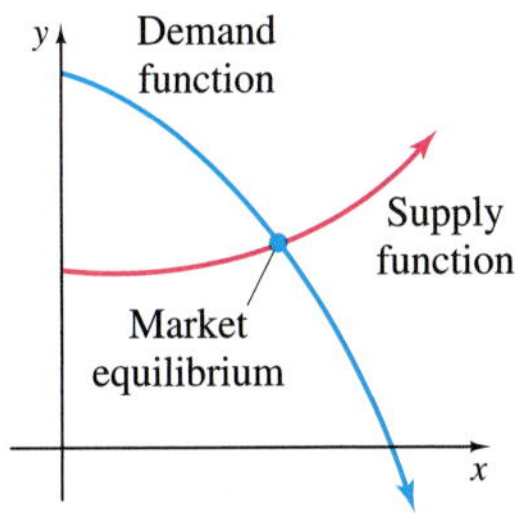


**37.** The demand function for a certain compact disc is given by the function

$$p = -0.01x^2 - 0.2x + 9$$

and the corresponding supply function is given by

$$p = 0.01x^2 - 0.1x + 3$$

where $p$ is in dollars and $x$ is in thousands of units. Find the equilibrium quantity and the corresponding price by solving the system consisting of the two given equations.

**38.** The demand function for a certain style of picture frame is given by the function

$$p = -2x^2 + 90$$

and the corresponding supply function is given by

$$p = 9x + 34$$

where $p$ is in dollars and $x$ is in thousands of units. Find the equilibrium quantity and the corresponding price by solving the system consisting of the two given equations.

*Use a graphing calculator to verify the results of each exercise.*

**39.** Exercise 3.

**40.** Exercise 4.

**41.** Exercise 23.

**42.** Exercise 24.

## REVIEW EXERCISES

*Graph each inequality in two variables. See Section 3.6.*

**43.** $x > -3$

**44.** $y \leq 1$

**45.** $y < 2x - 1$

**46.** $3x - y \leq 4$

*Find the perimeter of each geometric figure. See Section 5.3.*

△ **47.**

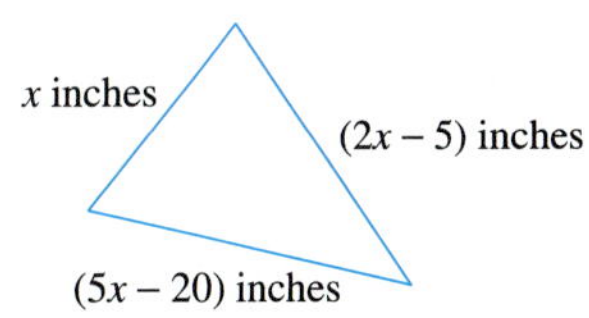


△ **48.**

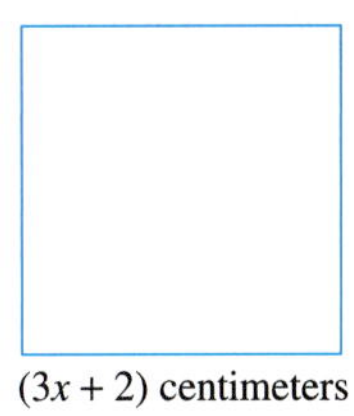


△ **49.**

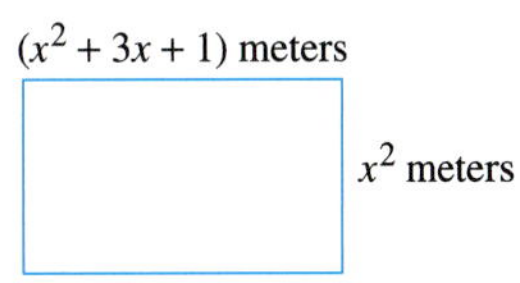


△ **50.**

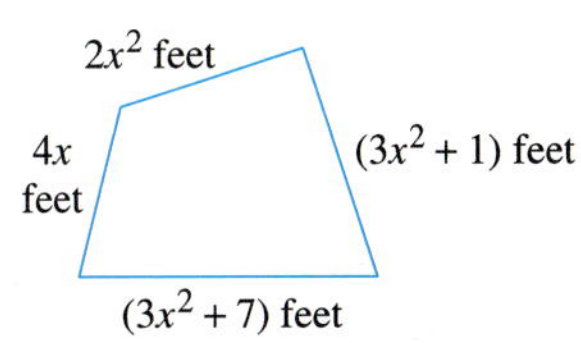

## 9.4 NONLINEAR INEQUALITIES AND SYSTEMS OF INEQUALITIES

▶ **OBJECTIVES**

CD-ROM SSM SSG Video

1. Graph a nonlinear inequality.
2. Graph a system of nonlinear inequalities.

**1** We can graph a nonlinear inequality in two variables such as $\frac{x^2}{9} + \frac{y^2}{16} \le 1$ in a way similar to the way we graphed a linear inequality in two variables in Section 3.6. First, we graph the related equation $\frac{x^2}{9} + \frac{y^2}{16} = 1$. The graph of the equation is our boundary. Then, using test points, we determine and shade the region whose points satisfy the inequality.

**Example 1** Graph $\frac{x^2}{9} + \frac{y^2}{16} \le 1$.

*Solution* First, graph the equation $\frac{x^2}{9} + \frac{y^2}{16} = 1$. Sketch a solid curve since the graph of $\frac{x^2}{9} + \frac{y^2}{16} \le 1$ includes the graph of $\frac{x^2}{9} + \frac{y^2}{16} = 1$. The graph is an ellipse, and it divides the plane into two regions, the "inside" and the "outside" of the ellipse. To determine which region contains the solutions, select a test point in either region and determine whether the coordinates of the point satisfy the inequality. We choose (0, 0) as the test point.

$$\frac{x^2}{9} + \frac{y^2}{16} \le 1$$

$$\frac{0^2}{9} + \frac{0^2}{16} \le 1 \quad \text{Let } x = 0 \text{ and } y = 0.$$

$$0 \le 1 \quad \text{True.}$$

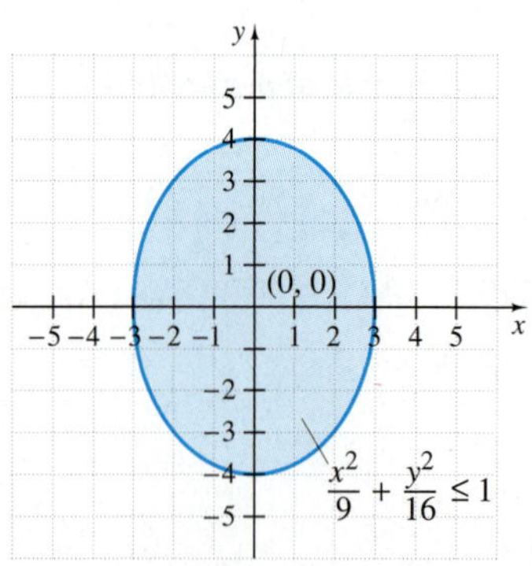

Since this statement is true, the solution set is the region containing $(0, 0)$. The graph of the solution set includes the points on and inside the ellipse, as shaded in the figure.

**Example 2** Graph $4y^2 > x^2 + 16$.

*Solution* The related equation is $4y^2 = x^2 + 16$. Subtract $x^2$ from both sides and divide both sides by 16, and we have $\frac{y^2}{4} - \frac{x^2}{16} = 1$, which is a hyperbola. Graph the hyperbola as a dashed curve since the graph of $4y^2 > x^2 + 16$ does *not* include the graph of $4y^2 = x^2 + 16$. The hyperbola divides the plane into three regions. Select a test point in each region—not on a boundary line—to determine whether that region contains solutions of the inequality.

| ***Test region A with* (0, 4)** | ***Test region B with* (0, 0)** | ***Test region C with* (0, −4)** |
|---|---|---|
| $4y^2 > x^2 + 16$ | $4y^2 > x^2 + 16$ | $4y^2 > x^2 + 16$ |
| $4(4)^2 > 0^2 + 16$ | $4(0)^2 > 0^2 + 16$ | $4(-4)^2 > 0^2 + 16$ |
| $64 > 16$ True. | $0 > 16$ False. | $64 > 16$ True. |

The graph of the solution set includes the shaded regions $A$ and $C$ only, not the boundary.

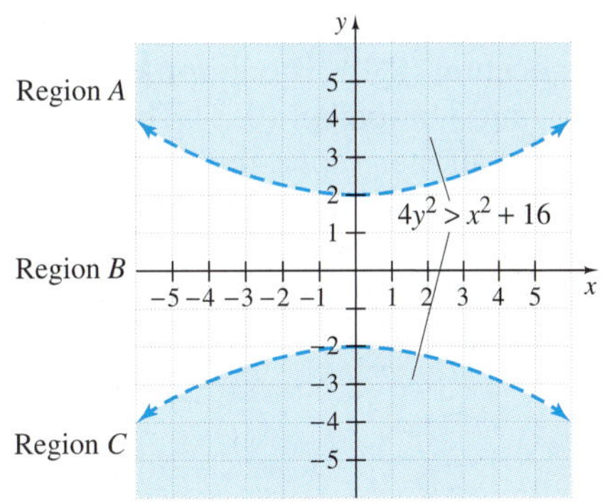


2 In Section 3.6 we graphed systems of linear inequalities. Recall that the graph of a system of inequalities is the intersection of the graphs of the inequalities.

**Example 3** Graph the system.

$$\begin{cases} x \le 1 - 2y \\ y \le x^2 \end{cases}$$

*Solution* We graph each inequality on the same set of axes. The intersection is shown in the third graph. It is the darkest shaded region along with its boundary lines. The coordinates of the points of intersection can be found by solving the related system.

$$\begin{cases} x = 1 - 2y \\ y = x^2 \end{cases}$$

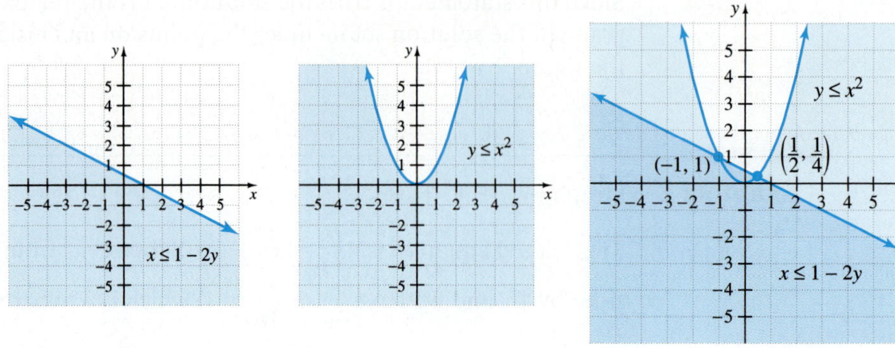


**Example 4** Graph the system.

$$\begin{cases} x^2 + y^2 < 25 \\ \dfrac{x^2}{9} - \dfrac{y^2}{25} < 1 \\ y < x + 3 \end{cases}$$

*Solution* We graph each inequality. The graph of $x^2 + y^2 < 25$ contains points "inside" the circle that has center $(0, 0)$ and radius 5. The graph of $\frac{x^2}{9} - \frac{y^2}{25} < 1$ is the region between the two branches of the hyperbola with $x$-intercepts $-3$ and 3 and center $(0, 0)$. The graph of $y < x + 3$ is the region "below" the line with slope 1 and $y$-intercept $(0, 3)$. The graph of the solution set of the system is the intersection of all the graphs, the darkest shaded region shown. The boundary of this region is not part of the solution.

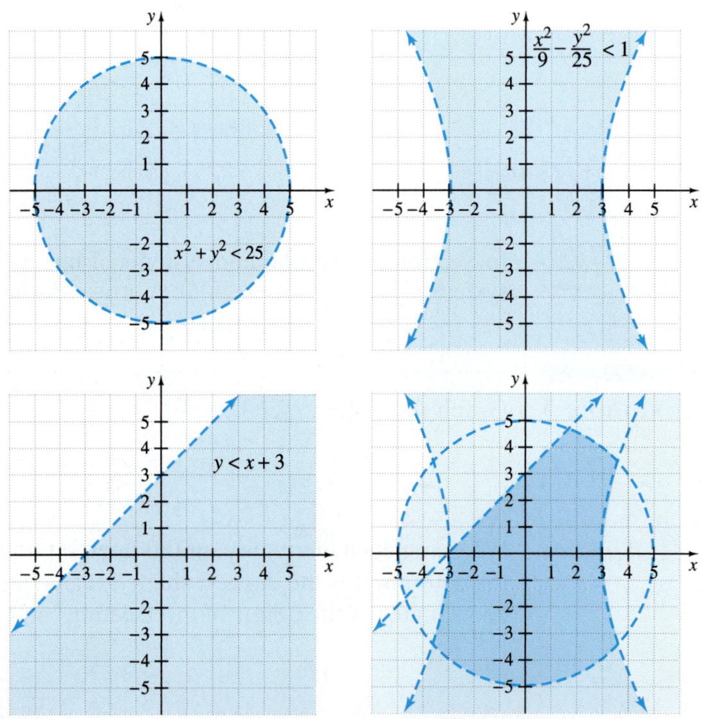

## SPOTLIGHT ON DECISION MAKING

Suppose you are an architect. You have just designed a bridge over a two-lane road. Spanning the road is an arch in the shape of a half-ellipse that is 40 feet wide at the base of the arch and is 15 feet tall at the center of the arch. A colleague has just pointed out that the bridge must have a 13-foot clearance for vehicles on the road. The road is 22 feet wide, and its center line falls directly beneath the highest point of the arch. Decide whether your current bridge design will allow 13-foot-tall vehicles to pass on the road beneath it or if your bridge must be redesigned. Explain your reasoning. (*Hint:* Envision a 13-foot-tall semi truck passing under the bridge in the right-hand lane of the road. Try checking whether points along the top of the truck would fall within the ellipse that defines the arch of the bridge. Remember that the truck can drive within any portion of the right-hand lane.)

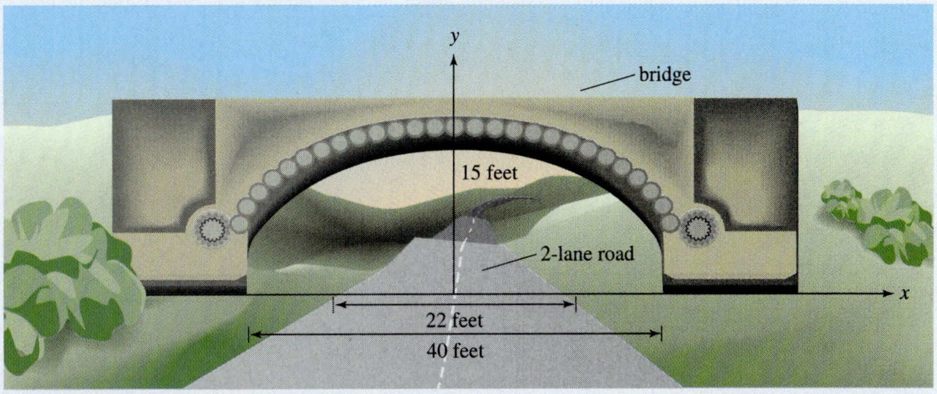


## *Exercise Set 9.4*

*Graph each inequality. See Examples 1 and 2.*

1. $y < x^2$
2. $y < -x^2$
3. $x^2 + y^2 \geq 16$
4. $x^2 + y^2 < 36$
5. $\frac{x^2}{4} - y^2 < 1$
6. $x^2 - \frac{y^2}{9} \geq 1$
7. $y > (x - 1)^2 - 3$
8. $y > (x + 3)^2 + 2$
9. $x^2 + y^2 \leq 9$
10. $x^2 + y^2 > 4$
11. $y > -x^2 + 5$
12. $y < -x^2 + 5$
13. $\frac{x^2}{4} + \frac{y^2}{9} \leq 1$
14. $\frac{x^2}{25} + \frac{y^2}{4} \geq 1$
15. $\frac{y^2}{4} - x^2 \leq 1$
16. $\frac{y^2}{16} - \frac{x^2}{9} > 1$
17. $y < (x - 2)^2 + 1$
18. $y > (x - 2)^2 + 1$
19. $y \leq x^2 + x - 2$
20. $y > x^2 + x - 2$
21. Discuss how graphing a linear inequality such as $x + y < 9$ is similar to graphing a nonlinear inequality such as $x^2 + y^2 < 9$.
22. Discuss how graphing a linear inequality such as $x + y < 9$ is different from graphing a nonlinear inequality such as $x^2 + y^2 < 9$.

*Graph each system. See Examples 3 and 4.*

23. $\begin{cases} 2x - y < 2 \\ y \leq -x^2 \end{cases}$
24. $\begin{cases} x - 2y > 4 \\ y > -x^2 \end{cases}$
25. $\begin{cases} 4x + 3y \geq 12 \\ x^2 + y^2 < 16 \end{cases}$
26. $\begin{cases} 3x - 4y \leq 12 \\ x^2 + y^2 < 16 \end{cases}$
27. $\begin{cases} x^2 + y^2 \leq 9 \\ x^2 + y^2 \geq 1 \end{cases}$
28. $\begin{cases} x^2 + y^2 \geq 9 \\ x^2 + y^2 \geq 16 \end{cases}$
29. $\begin{cases} y > x^2 \\ y \geq 2x + 1 \end{cases}$
30. $\begin{cases} y \leq -x^2 + 3 \\ y \leq 2x - 1 \end{cases}$

**31.** $\begin{cases} x > y^2 \\ y > 0 \end{cases}$

**32.** $\begin{cases} x < (y+1)^2 + 2 \\ x + y \geq 3 \end{cases}$

**33.** $\begin{cases} x^2 + y^2 > 9 \\ y > x^2 \end{cases}$

**34.** $\begin{cases} x^2 + y^2 \leq 9 \\ y < x^2 \end{cases}$

**35.** $\begin{cases} \frac{x^2}{4} + \frac{y^2}{9} \geq 1 \\ x^2 + y^2 \geq 4 \end{cases}$

**36.** $\begin{cases} x^2 + (y-2)^2 \geq 9 \\ \frac{x^2}{4} + \frac{y^2}{25} < 1 \end{cases}$

**37.** $\begin{cases} x^2 - y^2 \geq 1 \\ y \geq 0 \end{cases}$

**38.** $\begin{cases} x^2 - y^2 \geq 1 \\ x \geq 0 \end{cases}$

**39.** $\begin{cases} x + y \geq 1 \\ 2x + 3y < 1 \\ x > -3 \end{cases}$

**40.** $\begin{cases} x - y < -1 \\ 4x - 3y > 0 \\ y > 0 \end{cases}$

**41.** $\begin{cases} x^2 - y^2 < 1 \\ \frac{x^2}{16} + y^2 \leq 1 \\ x \geq -2 \end{cases}$

**42.** $\begin{cases} x^2 - y^2 \geq 1 \\ \frac{x^2}{16} + \frac{y^2}{4} \leq 1 \\ y \geq 1 \end{cases}$

**43.** Graph the system.

$$\begin{cases} y \leq x^2 \\ y \geq x + 2 \\ x \geq 0 \\ y \geq 0 \end{cases}$$

## REVIEW EXERCISES

*Determine which graph is the graph of a function. See Section 3.2.*

**44.**

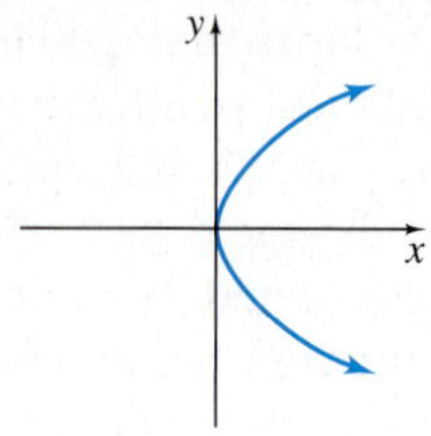


**45.**

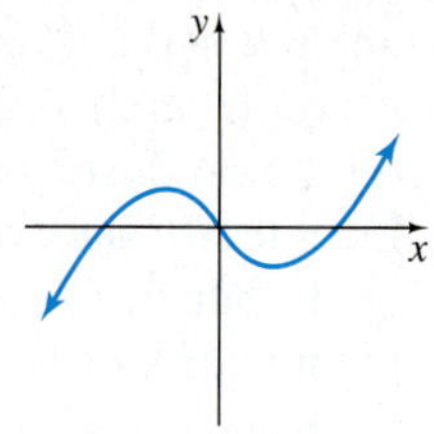


**46.**

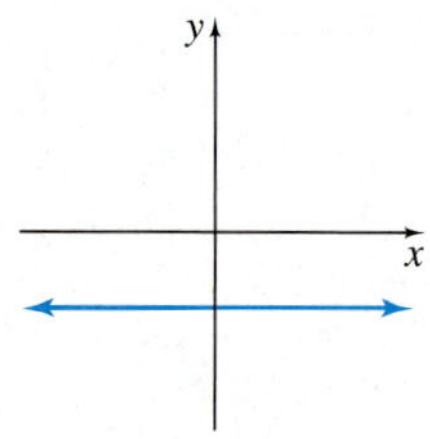


**47.**

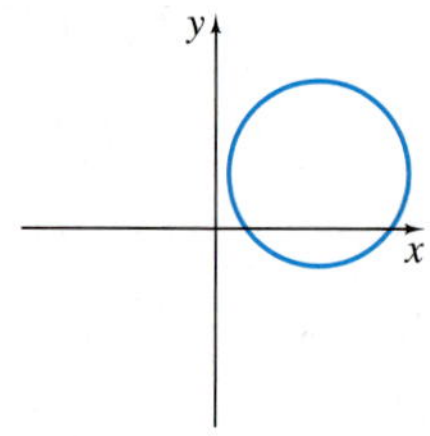


*Find each function value if* $f(x) = 3x^2 - 2$. *See Section 3.2.*

**48.** $f(-1)$

**49.** $f(-3)$

**50.** $f(a)$

**51.** $f(b)$

For additional Chapter Projects, visit the Real World Activities Website by going to http://www.prenhall.com/martin-gay.

# 9 CHAPTER PROJECT

## *Modeling Conic Sections*

In this project, you will have the opportunity to construct and investigate a model of an ellipse. You will need two thumbtacks or nails, graph paper, cardboard, tape, string, a pencil, and a ruler. This project may be completed by working in groups or individually.

Follow these steps, answering any questions as you go.

1. Draw an $x$-axis and a $y$-axis on the graph paper as shown in Figure 1.
2. Place the graph paper on the cardboard and attach it with tape.
3. Locate two points on the $x$-axis, each about $1\frac{1}{2}$ inches from the origin and on opposite sides of the origin (see Figure 1). Insert thumbtacks (or nails) at each of these locations.
4. Fasten a 9-inch piece of string to the thumbtacks as shown in Figure 2. Use your pencil to draw and keep the string taut while you carefully move the pencil in a path all around the thumbtacks.
5. Using the grid of the graph paper as a guide, find an approximate equation of the ellipse you drew.
6. Experiment by moving the tacks closer together or farther apart and drawing new ellipses. What do you observe?

7. Write a paragraph explaining why the figure drawn by the pencil is an ellipse. How might you use the same materials to draw a circle?
8. (Optional) Choose one of the ellipses you drew with the string and pencil. Use a ruler to draw any six tangent lines to the ellipse. (A line is tangent to the ellipse if it intersects, or just touches, the ellipse at only one point. See Figure 3.) Extend the tangent lines to yield six points of intersection among the tangents. Use a straightedge to draw a line connecting each pair of opposite points of intersection. What do you observe? Repeat with a different ellipse. Can you make a conjecture about the relationship among the lines that connect opposite points of intersection?

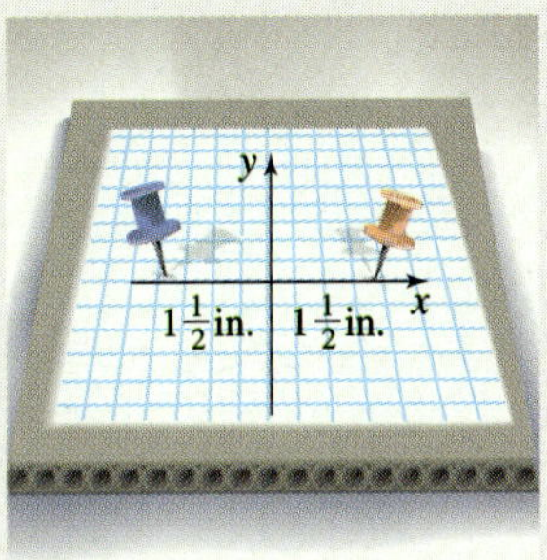

Figure 1

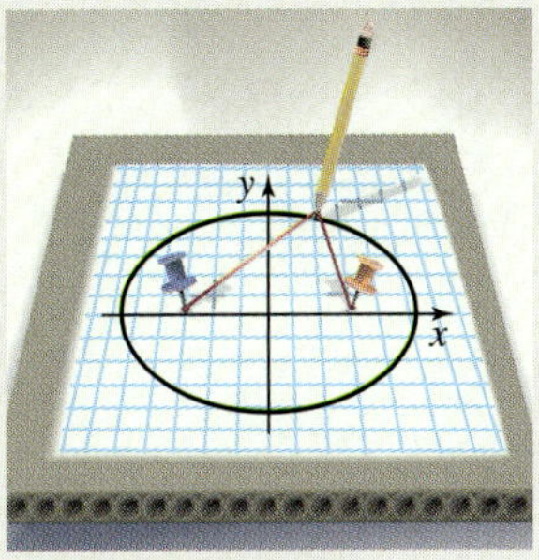

Figure 2

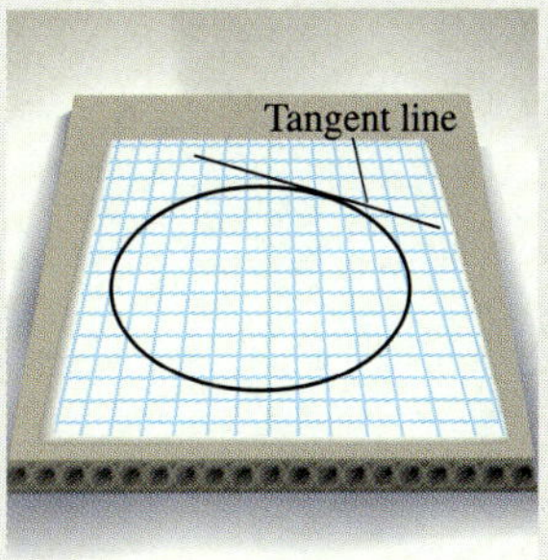

Figure 3

## CHAPTER 9 VOCABULARY CHECK

*Fill in each blank with one of the words or phrases listed below.*

| | | | |
|---|---|---|---|
| circle | midpoint | radius | distance |
| center | ellipse | hyperbola | nonlinear system of equations |

1. The ________ formula is $d = \sqrt{(x_2 - x_1)^2 + (y_2 - y_1)^2}$.
2. A(n) ________ is the set of all points in a plane that are the same distance from a fixed point, called the ________.
3. A ______________________________ is a system of equations at least one of which is not linear.
4. A(n) ________ is the set of points on a plane such that the sum of the distances of those points from two fixed points is a constant.
5. In a circle, the distance from the center to a point of the circle is called its ________.
6. A(n) __________ is the set of points in a plane such that the absolute value of the difference of the distance from two fixed points is constant.
7. The __________ formula is $\left(\frac{x_1 + x_2}{2}, \frac{y_1 + y_2}{2}\right)$.

# CHAPTER 9 HIGHLIGHTS

**DEFINITIONS AND CONCEPTS** | **EXAMPLES**

## Section 9.1 The Parabola and the Circle

***Parabolas***

$$y = a(x - h)^2 + k$$

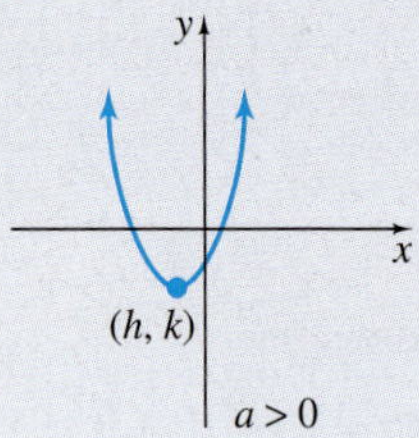


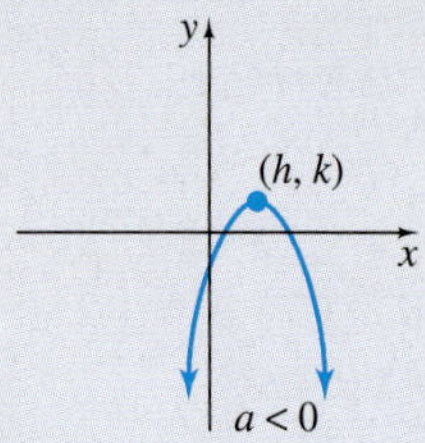


$$x = a(y - k)^2 + h$$

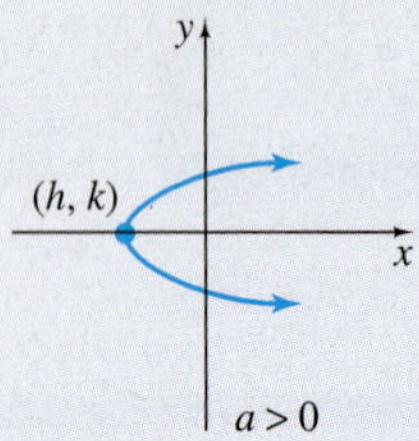


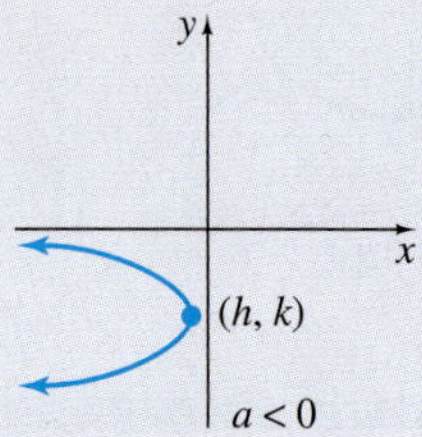


Graph $x = 3y^2 - 12y + 13$.

$$x - 13 = 3y^2 - 12y$$

$$x - 13 + 3(4) = 3(y^2 - 4y + 4) \quad \text{Add 3(4) to both sides.}$$

$$x = 3(y - 2)^2 + 1$$

Since $a = 3$, this parabola opens to the right with vertex $(1, 2)$. Its axis of symmetry is $y = 2$. The $x$-intercept is $(13, 0)$.

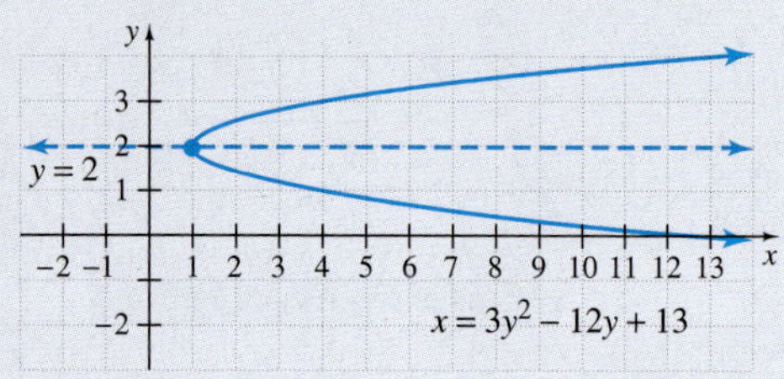


***Distance formula***

The distance $d$ between two points $(x_1, y_1)$ and $(x_2, y_2)$ is given by

$$d = \sqrt{(x_2 - x_1)^2 + (y_2 - y_1)^2}$$

Find the distance between points $(-1, 6)$ and $(-2, -4)$. Let $(x_1, y_1) = (-1, 6)$ and $(x_2, y_2) = (-2, -4)$.

$$d = \sqrt{(x_2 - x_1)^2 + (y_2 - y_1)^2}$$

$$= \sqrt{(-2 - (-1))^2 + (-4 - 6)^2}$$

$$= \sqrt{1 + 100} = \sqrt{101}$$

***Midpoint formula***

The midpoint of the line segment whose endpoints are $(x_1, y_1)$ and $(x_2, y_2)$ is the point with coordinates

$$\left(\frac{x_1 + x_2}{2}, \frac{y_1 + y_2}{2}\right)$$

Find the midpoint of the line segment whose endpoints are $(-1, 6)$ and $(-2, -4)$.

$$\left(\frac{-1 + (-2)}{2}, \frac{6 + (-4)}{2}\right)$$

The midpoint is

$$\left(-\frac{3}{2}, 1\right)$$

***Circle***

The graph of $(x - h)^2 + (y - k)^2 = r^2$ is a circle with center $(h, k)$ and radius $r$.

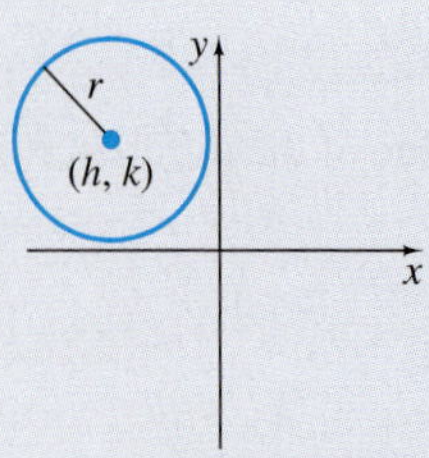


Graph $x^2 + (y + 3)^2 = 5$.

This equation can be written as

$(x - 0)^2 + (y + 3)^2 = 5$ with $h = 0$, $k = -3$, and $r = \sqrt{5}$.

The center of this circle is $(0, -3)$, and the radius is $\sqrt{5}$.

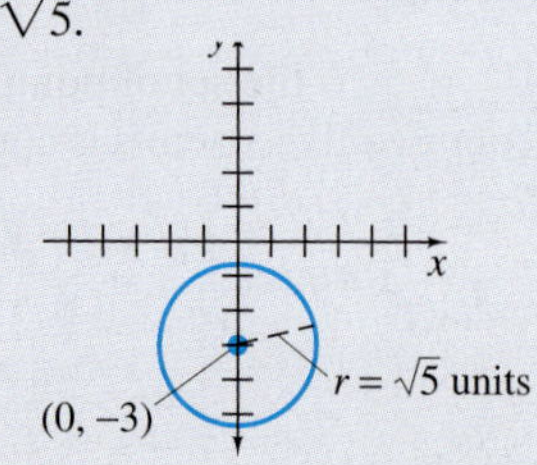

| DEFINITIONS AND CONCEPTS | EXAMPLES |
|---|---|

## Section 9.2 The Ellipse and the Hyperbola

***Ellipse with center* (0, 0)**

The graph of an equation of the form $\frac{x^2}{a^2} + \frac{y^2}{b^2} = 1$ is an ellipse with center $(0, 0)$. The $x$-intercepts are $(a, 0)$ and $(-a, 0)$, and the $y$-intercepts are $(0, b)$ and $(0, -b)$.

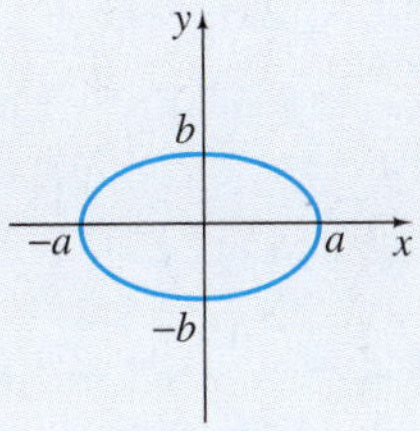


Graph $4x^2 + 9y^2 = 36$.

$$\frac{x^2}{9} + \frac{y^2}{4} = 1 \quad \text{Divide by 36.}$$

$$\frac{x^2}{3^2} + \frac{y^2}{2^2} = 1$$

The ellipse has center $(0, 0)$, $x$-intercepts $(3, 0)$ and $(-3, 0)$, and $y$-intercepts $(0, 2)$ and $(0, -2)$.

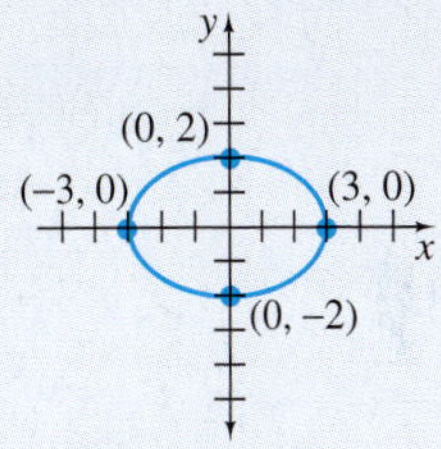


***Hyperbola with center* (0, 0)**

The graph of an equation of the form $\frac{x^2}{a^2} - \frac{y^2}{b^2} = 1$ is a hyperbola with center $(0, 0)$ and $x$-intercepts $(a, 0)$ and $(-a, 0)$.

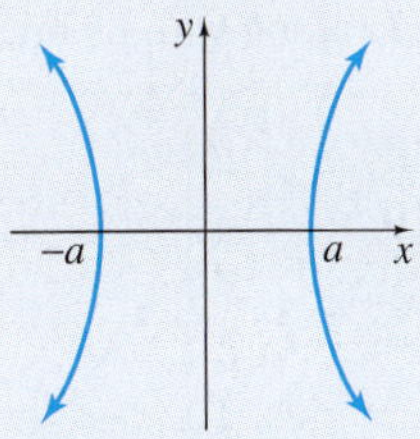


Graph $\frac{x^2}{9} - \frac{y^2}{4} = 1$. Here $a = 3$ and $b = 2$.

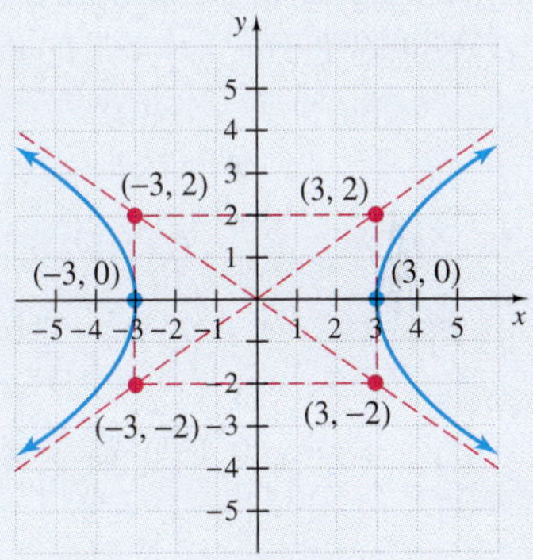


The graph of an equation of the form $\frac{y^2}{b^2} - \frac{x^2}{a^2} = 1$ is a hyperbola with center $(0, 0)$ and $y$-intercepts $(0, b)$ and $(0, -b)$.

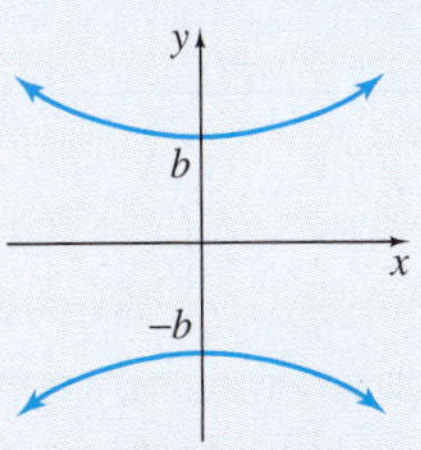


## Section 9.3 Solving Nonlinear Systems of Equations

A **nonlinear system of equations** is a system of equations at least one of which is not linear. Both the substitution method and the elimination method may be used to solve a nonlinear system of equations.

Solve the nonlinear system $\begin{cases} y = x + 2 \\ 2x^2 + y^2 = 3 \end{cases}$

*(continued)*

| DEFINITIONS AND CONCEPTS | EXAMPLES |
|---|---|
| **Section 9.3 Solving Nonlinear Systems of Equations** | |
| | Substitute $x + 2$ for $y$ in the second equation. $$2x^2 + y^2 = 3$$ $$2x^2 + (x + 2)^2 = 3$$ $$2x^2 + x^2 + 4x + 4 = 3$$ $$3x^2 + 4x + 1 = 0$$ $$(3x + 1)(x + 1) = 0$$ $$x = -\frac{1}{3}, x = -1$$ If $x = -\frac{1}{3}, y = x + 2 = -\frac{1}{3} + 2 = \frac{5}{3}$. If $x = -1, y = x + 2 = -1 + 2 = 1$. The solutions are $\left(-\frac{1}{3}, \frac{5}{3}\right)$ and $(-1, 1)$. |
| **Section 9.4 Nonlinear Inequalities and Systems of Inequalities** | |
| The graph of a system of inequalities is the intersection of the graphs of the inequalities. | Graph the system $\begin{cases} x \geq y^2 \\ x + y \leq 4 \end{cases}$ The graph of the system is the darkest shaded region along with its boundary lines. 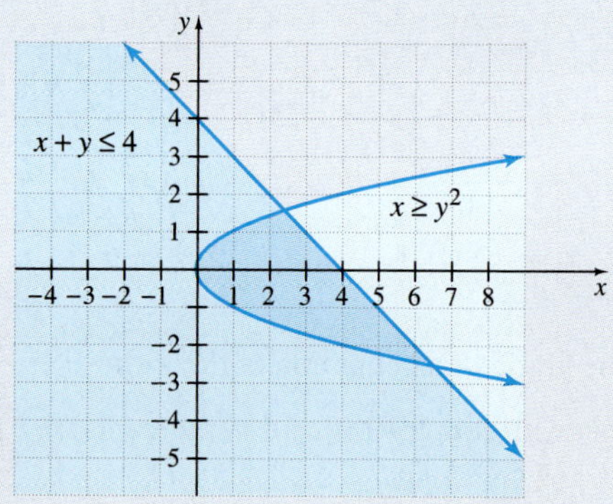 |



## CHAPTER 9 REVIEW

***(9.1)*** *Find the distance between each pair of points. For Exercises 7 and 8, round the distance to two decimal places.*

**1.** $(-6, 3)$ and $(8, 4)$

**2.** $(3, 5)$ and $(8, 9)$

**3.** $(-4, -6)$ and $(-1, 5)$

**4.** $(-1, 5)$ and $(2, -3)$

**5.** $(-\sqrt{2}, 0)$ and $(0, -4\sqrt{6})$

**6.** $(-\sqrt{5}, -\sqrt{11})$ and $(-\sqrt{5}, -3\sqrt{11})$

**7.** $(7.4, -8.6)$ and $(-1.2, 5.6)$

**8.** $(2.3, 1.8)$ and $(10.7, -9.2)$

*Find the midpoint of the line segment whose endpoints are given.*

**9.** $(2, 6)$ and $(-12, 4)$

**10.** $(-3, 8)$ and $(11, 24)$

**11.** $(-6, -5)$ and $(-9, 7)$

**12.** $(4, -6)$ and $(-15, 2)$

**13.** $\left(0, -\frac{3}{8}\right)$ and $\left(\frac{1}{10}, 0\right)$

**14.** $\left(\frac{3}{4}, -\frac{1}{7}\right)$ and $\left(-\frac{1}{4}, -\frac{3}{7}\right)$

**15.** $(\sqrt{3}, -2\sqrt{6})$ and $(\sqrt{3}, -4\sqrt{6})$

**16.** $(-5\sqrt{3}, 2\sqrt{7})$ and $(-3\sqrt{3}, 10\sqrt{7})$

*Write an equation of the circle with the given center and radius.*

**17.** center $(-4, 4)$, radius 3

**18.** center $(5, 0)$, radius 5

**19.** center $(-7, -9)$, radius $\sqrt{11}$

**20.** center $(0, 0)$, radius $\frac{7}{2}$

*Sketch the graph of the equation. If the graph is a circle, find its center. If the graph is a parabola, find its vertex.*

**21.** $x^2 + y^2 = 7$

**22.** $x = 2(y - 5)^2 + 4$

**23.** $x = -(y + 2)^2 + 3$

**24.** $(x - 1)^2 + (y - 2)^2 = 4$

**25.** $y = -x^2 + 4x + 10$

**26.** $x = -y^2 - 4y + 6$

**27.** $x = \frac{1}{2}y^2 + 2y + 1$

**28.** $y = -3x^2 + \frac{1}{2}x + 4$

**29.** $x^2 + y^2 + 2x + y = \frac{3}{4}$

**30.** $x^2 + y^2 + 3y = \frac{7}{4}$

**31.** $4x^2 + 4y^2 + 16x + 8y = 1$

**32.** $3x^2 + 6x + 3y^2 = 9$

**33.** $y = x^2 + 6x + 9$

**34.** $x = y^2 + 6y + 9$

**35.** Write an equation of the circle centered at $(5.6, -2.4)$ with diameter 6.2.

***(9.2)*** *Sketch the graph of each equation.*

**36.** $x^2 + \frac{y^2}{4} = 1$

**37.** $x^2 - \frac{y^2}{4} = 1$

**38.** $\frac{y^2}{4} - \frac{x^2}{16} = 1$

**39.** $\frac{y^2}{4} + \frac{x^2}{16} = 1$

**40.** $\frac{x^2}{5} + \frac{y^2}{5} = 1$

**41.** $\frac{x^2}{5} - \frac{y^2}{5} = 1$

**42.** $-5x^2 + 25y^2 = 125$

**43.** $4y^2 + 9x^2 = 36$

**44.** $\frac{(x - 2)^2}{4} + (y - 1)^2 = 1$

**45.** $\frac{(x + 3)^2}{9} + \frac{(y - 4)^2}{25} = 1$

**46.** $x^2 - y^2 = 1$

**47.** $36y^2 - 49x^2 = 1764$

**48.** $y^2 = x^2 + 9$

**49.** $x^2 = 4y^2 - 16$

**50.** $100 - 25x^2 = 4y^2$

*Sketch the graph of each equation. Identify whether each equation, when graphed, will be a parabola, circle, ellipse, or hyperbola.*

**51.** $y = x^2 + 4x + 6$

**52.** $y^2 = x^2 + 6$

**53.** $y^2 + x^2 = 4x + 6$

**54.** $y^2 + 2x^2 = 4x + 6$

**55.** $x^2 + y^2 - 8y = 0$

**56.** $x - 4y = y^2$

**57.** $x^2 - 4 = y^2$

**58.** $x^2 = 4 - y^2$

**59.** $6(x - 2)^2 + 9(y + 5)^2 = 36$

**60.** $36y^2 = 576 + 16x^2$

**61.** $\frac{x^2}{16} - \frac{y^2}{25} = 1$

**62.** $3(x - 7)^2 + 3(y + 4)^2 = 1$

*Use a graphing calculator to verify the results of each exercise.*

**63.** Exercise 39.

**64.** Exercise 40.

**65.** Exercise 51.

**66.** Exercise 58.

***(9.3)*** *Solve each system of equations.*

**67.** $\begin{cases} y = 2x - 4 \\ y^2 = 4x \end{cases}$

**68.** $\begin{cases} x^2 + y^2 = 4 \\ x - y = 4 \end{cases}$

**69.** $\begin{cases} y = x + 2 \\ y = x^2 \end{cases}$

**70.** $\begin{cases} y = x^2 - 5x + 1 \\ y = -x + 6 \end{cases}$

**71.** $\begin{cases} 4x - y^2 = 0 \\ 2x^2 + y^2 = 16 \end{cases}$

**72.** $\begin{cases} x^2 + 4y^2 = 16 \\ x^2 + y^2 = 4 \end{cases}$

**73.** $\begin{cases} x^2 + y^2 = 10 \\ 9x^2 + y^2 = 18 \end{cases}$

**74.** $\begin{cases} x^2 + 2y = 9 \\ 5x - 2y = 5 \end{cases}$

**75.** $\begin{cases} y = 3x^2 + 5x - 4 \\ y = 3x^2 - x + 2 \end{cases}$

**76.** $\begin{cases} x^2 - 3y^2 = 1 \\ 4x^2 + 5y^2 = 21 \end{cases}$

△ **77.** Find the length and the width of a room whose area is 150 square feet and whose perimeter is 50 feet.

**78.** What is the greatest number of real solutions possible for a system of two equations whose graphs are an ellipse and a hyperbola?

***(9.4)*** *Graph the inequality or system of inequalities.*

**79.** $y \le -x^2 + 3$

**80.** $x^2 + y^2 < 9$

**81.** $x^2 - y^2 < 1$

**82.** $\dfrac{x^2}{4} + \dfrac{y^2}{9} \ge 1$

**83.** $\begin{cases} 2x \le 4 \\ x + y \ge 1 \end{cases}$

**84.** $\begin{cases} 3x + 4y \le 12 \\ x - 2y > 6 \end{cases}$

**85.** $\begin{cases} y > x^2 \\ x + y \ge 3 \end{cases}$

**86.** $\begin{cases} x^2 + y^2 \le 16 \\ x^2 + y^2 \ge 4 \end{cases}$

**87.** $\begin{cases} x^2 + y^2 < 4 \\ x^2 - y^2 \le 1 \end{cases}$

**88.** $\begin{cases} x^2 + y^2 < 4 \\ y \ge x^2 - 1 \\ x \ge 0 \end{cases}$

## CHAPTER 9 TEST

**1.** Find the distance between the points $(-6, 3)$ and $(-8, -7)$.

**2.** Find the distance between the points $(-2\sqrt{5}, \sqrt{10})$ and $(-\sqrt{5}, 4\sqrt{10})$.

**3.** Find the midpoint of the line segment whose endpoints are $(-2, -5)$ and $(-6, 12)$.

**4.** Find the midpoint of the line segment whose endpoints are $\left(-\dfrac{2}{3}, -\dfrac{1}{5}\right)$ and $\left(-\dfrac{1}{3}, \dfrac{4}{5}\right)$.

*Sketch the graph of each equation.*

**5.** $x^2 + y^2 = 36$

**6.** $x^2 - y^2 = 36$

**7.** $16x^2 + 9y^2 = 144$

**8.** $y = x^2 - 8x + 16$

**9.** $x^2 + y^2 + 6x = 16$

**10.** $x = y^2 + 8y - 3$

**11.** $\dfrac{(x - 4)^2}{16} + \dfrac{(y - 3)^2}{9} = 1$

**12.** $y^2 - x^2 = 1$

*Solve each system.*

**13.** $\begin{cases} x^2 + y^2 = 169 \\ 5x + 12y = 0 \end{cases}$

**14.** $\begin{cases} x^2 + y^2 = 26 \\ x^2 - y^2 = 24 \end{cases}$

**15.** $\begin{cases} y = x^2 - 5x + 6 \\ y = 2x \end{cases}$

**16.** $\begin{cases} x^2 + 4y^2 = 5 \\ y = x \end{cases}$

*Graph the solution of each system.*

**17.** $\begin{cases} 2x + 5y \geq 10 \\ y \geq x^2 + 1 \end{cases}$

**18.** $\begin{cases} \dfrac{x^2}{4} + y^2 \leq 1 \\ x + y > 1 \end{cases}$

**19.** $\begin{cases} x^2 + y^2 > 1 \\ \dfrac{x^2}{4} - y^2 \geq 1 \end{cases}$

**20.** $\begin{cases} x^2 + y^2 \geq 4 \\ x^2 + y^2 < 16 \\ y \geq 0 \end{cases}$

**21.** Which graph best resembles the graph of $x = a(y - k)^2 + h$ if $a > 0$, $h < 0$, and $k > 0$?

**A**

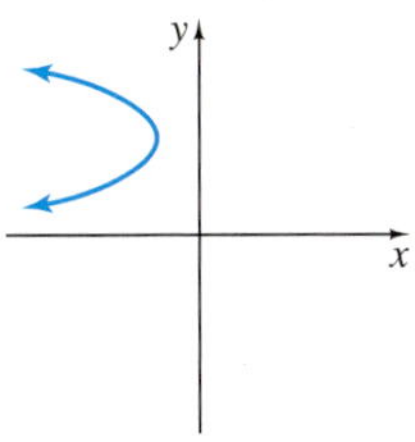


**B**

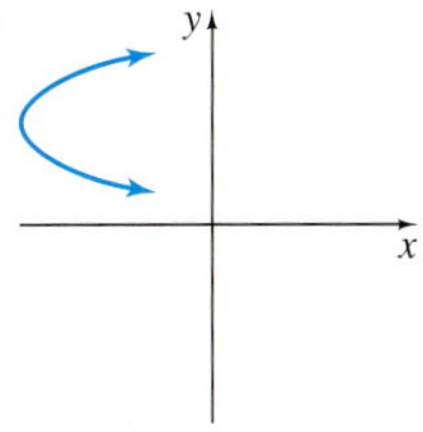


**C**

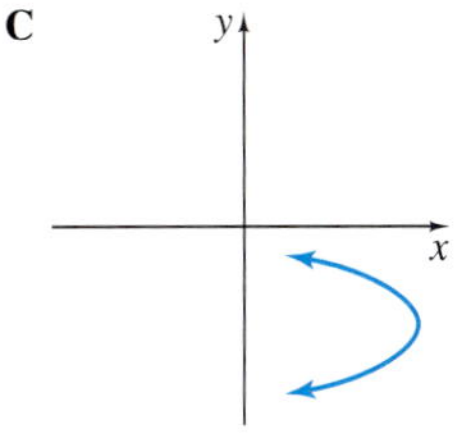


**D**

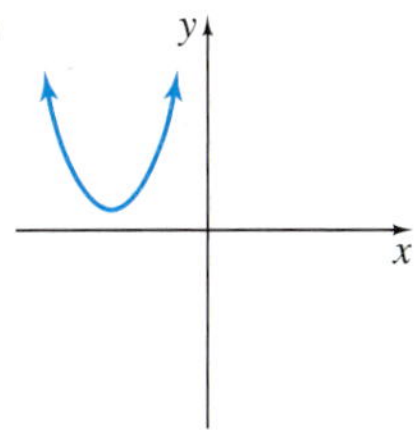


**22.** A bridge has an arch in the shape of a half-ellipse. If the equation of the ellipse, measured in feet, is $100x^2 + 225y^2 = 22{,}500$, find the height of the arch from the road and the width of the arch.

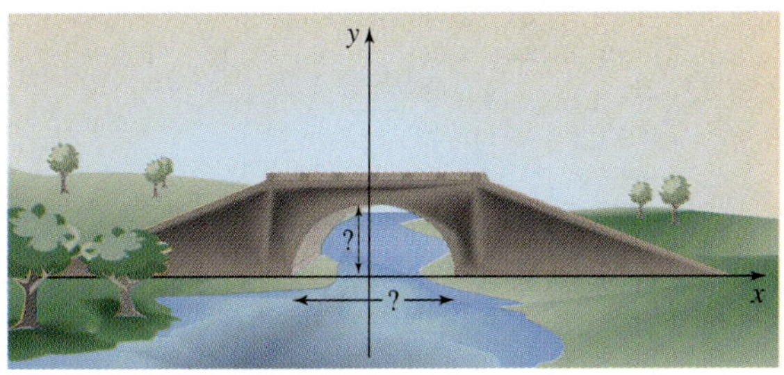


## CHAPTER 9 CUMULATIVE REVIEW

**1.** Use the associative property of multiplication to write an expression equivalent to $4 \cdot (9y)$. Then simplify the equivalent expression.

**2.** Graph $x = -2y$ by plotting intercepts.

**3.** Use the elimination method to solve the system:

$$\begin{cases} 3x + \dfrac{y}{2} = 2 \\ 6x + y = 5 \end{cases}$$

**4.** Use the power rules to simplify the following. Use positive exponents to write all results.

**a.** $(5x^2)^3$

**b.** $\left(\frac{2}{3}\right)^3$

**c.** $\left(\frac{3p^4}{q^5}\right)^2$

**d.** $\left(\frac{2^{-3}}{y}\right)^{-2}$

**e.** $(x^{-5}y^2z^{-1})^7$

**5.** Solve $2x^2 = \frac{17}{3}x + 1$.

**6.** Perform each indicated operation.

$$\frac{7}{x-1} + \frac{10x}{x^2-1} - \frac{5}{x+1}$$

**7.** Simplify each complex fraction.

**a.** $\dfrac{\frac{2x}{27y^2}}{\frac{6x^2}{9}}$

**b.** $\dfrac{\frac{5x}{x+2}}{\frac{10}{x-2}}$

**c.** $\dfrac{\frac{x}{y^2}+\frac{1}{y}}{\frac{y}{x^2}+\frac{1}{x}}$

**8.** Divide $2x^2 - x - 10$ by $x + 2$.

**9.** Use the remainder theorem and synthetic division to find $P(4)$ if

$P(x) = 4x^6 - 25x^5 + 35x^4 + 17x^2$.

**10.** Solve: $\frac{2x}{x-3} + \frac{6-2x}{x^2-9} = \frac{x}{x+3}$.

**11.** Melissa Scarlatti can clean the house in 4 hours, whereas her husband, Zack, can do the same job in 5 hours. They have agreed to clean together so that they can finish in time to watch a movie on TV that starts in 2 hours. How long will it take them to clean the house together? Can they finish before the movie starts?

**12.** Find the cube roots.

**a.** $\sqrt[3]{1}$

**b.** $\sqrt[3]{-64}$

**c.** $\sqrt[3]{\frac{8}{125}}$

**d.** $\sqrt[3]{x^6}$

**e.** $\sqrt[3]{-8x^9}$

**13.** Multiply.

**a.** $z^{2/3}(z^{1/3} - z^5)$

**b.** $(x^{1/3} - 5)(x^{1/3} + 2)$

**14.** Use the quotient rule to divide, and simplify if possible.

**a.** $\frac{\sqrt{20}}{\sqrt{5}}$

**b.** $\frac{\sqrt{50x}}{2\sqrt{2}}$

**c.** $\frac{7\sqrt[3]{48x^4y^8}}{\sqrt[3]{6y^2}}$

**15.** Add or subtract as indicated.

**a.** $\frac{\sqrt{45}}{4} - \frac{\sqrt{5}}{3}$

**b.** $\sqrt[3]{\frac{7x}{8}} + 2\sqrt[3]{7x}$

**16.** Rationalize the denominator of $\sqrt{\frac{7x}{3y}}$.

**17.** Solve $\sqrt{2x-3} = 9$.

**18.** Find the following powers of $i$.

**a.** $i^7$

**b.** $i^{20}$

**c.** $i^{46}$

**d.** $i^{-12}$

**19.** Solve $p^2 + 2p = 4$ by completing the square.

**20.** Solve $\frac{1}{4}m^2 - m + \frac{1}{2} = 0$.

**21.** Solve $p^4 - 3p^2 - 4 = 0$.

**22.** Solve $\frac{x + 2}{x - 3} \leq 0$.

**23.** Graph $g(x) = \frac{1}{2}(x + 2)^2 + 5$. Find the vertex and the axis of symmetry.

**24.** Find the vertex of the graph of $f(x) = x^2 - 4x - 12$.

**25.** Find the distance between $(2, -5)$ and $(1, -4)$. Give an exact distance and a three-decimal-place approximation.

## The Number One Job in the United States

A Webmaster is responsible for creating and managing Web sites on the World Wide Web. According to the 1999 *Jobs Rated Almanac,* Webmaster is rated as the number one job in the United States based on factors such as income, stress, physical demands, potential growth, job security, and work environment.

Webmasters have technical, business, and visual design skills or understanding. Some Webmasters may focus on the computer programming aspects of developing Web sites, including graphics and Web site security issues. Others may focus on Web site content, such as company or organization history, product information, or other marketing topics. Webmasters should be comfortable with people management, project management, and strategic planning. Webmasters use math and problem-solving skills in tasks such as creating budgets, making hardware or software purchasing decisions, and compiling and analyzing Web site usage statistics.

*For more information about a career as a Webmaster or other Web professional, visit the World Organization of Webmasters Web site by first going to www.prenhall.com/martin-gay.*

*In the Spotlight on Decision Making feature on page 595, you will have the opportunity to make a decision about modeling Web site usage statistics as a Webmaster.*

# EXPONENTIAL AND LOGARITHMIC FUNCTIONS

# 10

10.1 The Algebra of Functions; Composite Functions

10.2 Inverse Functions

10.3 Exponential Functions

10.4 Logarithmic Functions

10.5 Properties of Logarithms

10.6 Common Logarithms, Natural Logarithms, and Change of Base

10.7 Exponential and Logarithmic Equations and Applications

In this chapter, we discuss two closely related functions: exponential and logarithmic functions. These functions are vital to applications in economics, finance, engineering, the sciences, education, and other fields. Models of tumor growth and learning curves are two examples of the uses of exponential and logarithmic functions.

## 10.1 THE ALGEBRA OF FUNCTIONS; COMPOSITE FUNCTIONS

▶ **OBJECTIVES**

CD-ROM SSM SSG Video

1. Add, subtract, multiply, and divide functions.
2. Construct composite functions.

**1** As we have seen in earlier chapters, it is possible to add, subtract, multiply, and divide functions. Although we have not stated it as such, the sums, differences, products, and quotients of functions are themselves functions. For example, if $f(x) = 3x$ and $g(x) = x + 1$, their product, $f(x) \cdot g(x) = 3x(x + 1) = 3x^2 + 3x$, is a new function. We can use the notation $(f \cdot g)(x)$ to denote this new function. Finding the sum, difference, product, and quotient of functions to generate new functions is called the **algebra of functions.**

### ALGEBRA OF FUNCTIONS

Let $f$ and $g$ be functions. New functions from $f$ and $g$ are defined as follows.

***Sum*** $(f + g)(x) = f(x) + g(x)$

***Difference*** $(f - g)(x) = f(x) - g(x)$

***Product*** $(f \cdot g)(x) = f(x) \cdot g(x)$

***Quotient*** $\left(\frac{f}{g}\right)(x) = \frac{f(x)}{g(x)}, g(x) \neq 0$

**Example 1** If $f(x) = x - 1$ and $g(x) = 2x - 3$, find

**a.** $(f + g)(x)$

**b.** $(f - g)(x)$

**c.** $(f \cdot g)(x)$

**d.** $\left(\frac{f}{g}\right)(x)$

*Solution* Use the algebra of functions and replace $f(x)$ by $x - 1$ and $g(x)$ by $2x - 3$. Then we simplify.

**a.** 
$$\begin{aligned}(f + g)(x) &= f(x) + g(x)\\ &= (x - 1) + (2x - 3)\\ &= 3x - 4\end{aligned}$$

**b.** $(f - g)(x) = f(x) - g(x)$
$$= (x - 1) - (2x - 3)$$
$$= x - 1 - 2x + 3$$
$$= -x + 2$$

**c.** $(f \cdot g)(x) = f(x) \cdot g(x)$
$$= (x - 1)(2x - 3)$$
$$= 2x^2 - 5x + 3$$

**d.** $\left(\frac{f}{g}\right)(x) = \frac{f(x)}{g(x)} = \frac{x - 1}{2x - 3}$, where $x \neq \frac{3}{2}$

There is an interesting but not surprising relationship between the graphs of functions and the graphs of their sum, difference, product, and quotient. For example, the graph of $(f + g)(x)$ can be found by adding the graph of $f(x)$ to the graph of $g(x)$. We add two graphs by adding corresponding $y$-values.

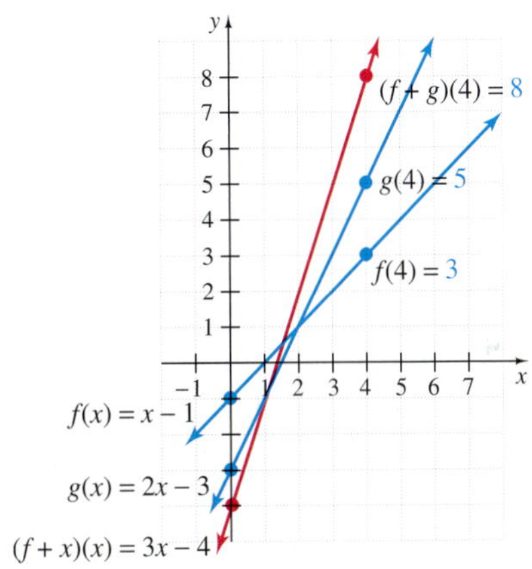


2 Another way to combine functions is called **function composition.** To understand this new way of combining functions, study the tables below. The first table shows degrees Celsius $C(x)$ as a function of degrees Fahrenheit $x$. The second table shows Kelvins $K(C)$ as a function of degrees Celsius $C$. (The Kelvin scale is a temperature scale devised by Lord Kelvin in 1848.) The function represented by the first table we will call $C$, and the second function we will call $K$.

| | | | | | | |
|---|---|---|---|---|---|---|
| $x$ = Degrees Fahrenheit (Input) | −31 | −13 | 32 | 68 | 149 | 212 |
| $C(x)$ = Degrees Celsius (Output) | −35 | −25 | 0 | 20 | 65 | 100 |

| | | | | | | |
|---|---|---|---|---|---|---|
| $C$ = Degrees Celsius (Input) | −35 | −25 | 0 | 20 | 65 | 100 |
| $K(C)$ = Kelvins (Output) | 238.15 | 248.15 | 273.15 | 293.15 | 338.15 | 373.15 |

Suppose that we want a table that shows a direct conversion from degrees Fahrenheit to kelvins. In other words, suppose that a table is needed that shows kelvins as a function of degrees Fahrenheit. This can easily be done because in the tables, the output of the first table $C(x)$ is the same as the input of the second table. If we use $C(x)$ to represent this, then we get the following table.

| $x$ = Degrees Fahrenheit (Input) | −31 | −13 | 32 | 68 | 149 | 212 |
|---|---|---|---|---|---|---|
| $K(C(x))$ = Kelvins (Output) | 238.15 | 248.15 | 273.15 | 293.15 | 338.15 | 373.15 |

Since the output of the first table is used as the input of the second table, we write the new function as $K(C(x))$. The new function is formed from the composition of the other two functions. The mathematical symbol for this composition is $(K \circ C)(x)$. Thus, $(K \circ C)(x) = K(C(x))$.

It is possible to find an equation for the composition of the two functions $C$ and $K$. In other words, we can find a function that converts degrees Fahrenheit directly to kelvins. The function $C(x) = \frac{5}{9}(x - 32)$ converts degrees Fahrenheit to degrees Celsius, and the function $K(C) = C + 273.15$ converts degrees Celsius to kelvins. Thus,

$$(K \circ C)(x) = K(C(x)) = K\left(\frac{5}{9}(x - 32)\right) = \frac{5}{9}(x - 32) + 273.15$$

In general, the notation $\boldsymbol{f(g(x))}$ means "$f$ composed with $g$" and can be written as $\boldsymbol{(f \circ g)(x)}$. Also $g(f(x))$, or $(g \circ f)(x)$, means "$g$ composed with $f$."

### COMPOSITION OF FUNCTIONS

The composition of functions $f$ and $g$ is

$$(f \circ g)(x) = f(g(x))$$

**HELPFUL HINT**

$(f \circ g)(x)$ does not mean the same as $(f \cdot g)(x)$.

$$(f \circ g)(x) = f(g(x)) \text{ while } (f \cdot g)(x) = f(x) \cdot g(x)$$

**Example 2** If $f(x) = x^2$ and $g(x) = x + 3$, find each composition.

**a.** $(f \circ g)(2)$ and $(g \circ f)(2)$

**b.** $(f \circ g)(x)$ and $(g \circ f)(x)$

*Solution* **a.** $(f \circ g)(2) = f(g(2))$

$= f(5)$ Replace $g(2)$ with 5. [Since $g(x) = x + 3$, then $g(2) = 2 + 3 = 5$.]

$= 5^2 = 25$

$(g \circ f)(2) = g(f(2))$

$= g(4)$ Since $f(x) = x^2$, then $f(2) = 2^2 = 4$.

$= 4 + 3 = 7$

**b.** $(f \circ g)(x) = f(g(x))$

$= f(x + 3)$ — Replace $g(x)$ with $x + 3$.

$= (x + 3)^2$ — $f(x + 3) = (x + 3)^2$

$= x^2 + 6x + 9$ — Square $(x + 3)$.

$(g \circ f)(x) = g(f(x))$

$= g(x^2)$ — Replace $f(x)$ with $x^2$.

$= x^2 + 3$ — $g(x^2) = x^2 + 3$

**Example 3** If $f(x) = |x|$ and $g(x) = x - 2$, find each composition.

**a.** $(f \circ g)(x)$

**b.** $(g \circ f)(x)$

*Solution* **a.** $(f \circ g)(x) = f(g(x)) = f(x - 2) = |x - 2|$

**b.** $(g \circ f)(x) = g(f(x)) = g(|x|) = |x| - 2$

**HELPFUL HINT**

In Examples 2 and 3, notice that $(g \circ f)(x) \neq (f \circ g)(x)$. In general, $(g \circ f)(x)$ *may* or *may not* equal $(f \circ g)(x)$.

**Example 4** If $f(x) = 5x$, $g(x) = x - 2$, and $h(x) = \sqrt{x}$, write each function as a composition using two of the given functions.

**a.** $F(x) = \sqrt{x - 2}$

**b.** $G(x) = 5x - 2$

*Solution* **a.** Notice the order in which the function $F$ operates on an input value $x$. First, 2 is subtracted from $x$. This is the function $g(x) = x - 2$. Then the square root *of that result* is taken. The square root function is $h(x) = \sqrt{x}$. This means that $F = h \circ g$. To check, we find $h \circ g$.

$$(h \circ g)(x) = h(g(x)) = h(x - 2) = \sqrt{x - 2}$$

**b.** Notice the order in which the function $G$ operates on an input value $x$. First, $x$ is multiplied by 5, and then 2 is subtracted from the result. This means that $G = g \circ f$. To check, we find $g \circ f$.

$$(g \circ f)(x) = g(f(x)) = g(5x) = 5x - 2$$

## GRAPHING CALCULATOR EXPLORATIONS

If $f(x) = \frac{1}{2}x + 2$ and $g(x) = \frac{1}{3}x^2 + 4$, then

$$\begin{aligned}(f + g)(x) &= f(x) + g(x)\\ &= \left(\frac{1}{2}x + 2\right) + \left(\frac{1}{3}x^2 + 4\right)\\ &= \frac{1}{3}x^2 + \frac{1}{2}x + 6.\end{aligned}$$

To visualize this addition of functions with a graphing calculator, graph

$$Y_1 = \frac{1}{2}x + 2, \qquad Y_2 = \frac{1}{3}x^2 + 4, \qquad Y_3 = \frac{1}{3}x^2 + \frac{1}{2}x + 6$$

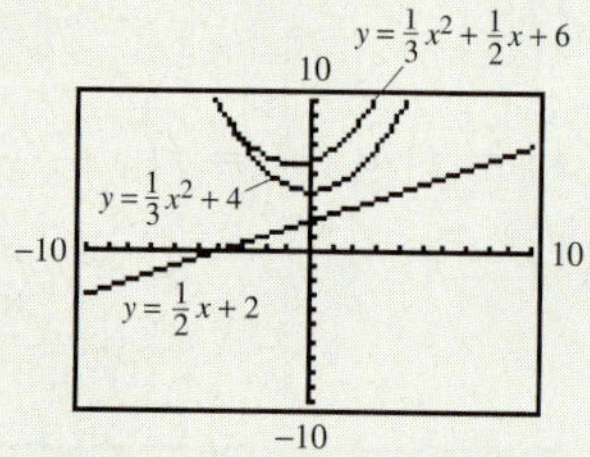


Use a TABLE feature to verify that for a given $x$ value, $Y_1 + Y_2 = Y_3$. For example, verify that when $x = 0$, $Y_1 = 2$, $Y_2 = 4$, and $Y_3 = 2 + 4 = 6$.

## Exercise Set 10.1

*For the functions f and g, find* **a.** $(f + g)(x)$, **b.** $(f - g)(x)$, **c.** $(f \cdot g)(x)$, *and* **d.** $\left(\frac{f}{g}\right)(x)$. *See Example 1.*

**1.** $f(x) = x - 7, g(x) = 2x + 1$

**2.** $f(x) = x + 4, g(x) = 5x - 2$

**3.** $f(x) = x^2 + 1, g(x) = 5x$

**4.** $f(x) = x^2 - 2, g(x) = 3x$

**5.** $f(x) = \sqrt{x}, g(x) = x + 5$

**6.** $f(x) = \sqrt[3]{x}, g(x) = x - 3$

**7.** $f(x) = -3x, g(x) = 5x^2$

**8.** $f(x) = 4x^3, g(x) = -6x$

*If* $f(x) = x^2 - 6x + 2$, $g(x) = -2x$, *and* $h(x) = \sqrt{x}$, *find each composition. See Example 2.*

**9.** $(f \circ g)(2)$

**10.** $(h \circ f)(-2)$

**11.** $(g \circ f)(-1)$

**12.** $(f \circ h)(1)$

**13.** $(g \circ h)(0)$

**14.** $(h \circ g)(0)$

*Find $(f \circ g)(x)$ and $(g \circ f)(x)$. See Examples 2 and 3.*

**15.** $f(x) = x^2 + 1, g(x) = 5x$

**16.** $f(x) = x - 3, g(x) = x^2$

**17.** $f(x) = 2x - 3, g(x) = x + 7$

**18.** $f(x) = x + 10, g(x) = 3x + 1$

**19.** $f(x) = x^3 + x - 2, g(x) = -2x$

**20.** $f(x) = -4x, g(x) = x^3 + x^2 - 6$

**21.** $f(x) = \sqrt{x}, g(x) = -5x + 2$

**22.** $f(x) = 7x - 1, g(x) = \sqrt[3]{x}$

*If $f(x) = 3x$, $g(x) = \sqrt{x}$, and $h(x) = x^2 + 2$, write each function as a composition using two of the given functions. See Example 4.*

**23.** $H(x) = \sqrt{x^2 + 2}$

**24.** $G(x) = \sqrt{3x}$

**25.** $F(x) = 9x^2 + 2$

**26.** $H(x) = 3x^2 + 6$

**27.** $G(x) = 3\sqrt{x}$

**28.** $F(x) = x + 2$

*Find $f(x)$ and $g(x)$ so that the given function $h(x) = (f \circ g)(x)$.*

**29.** $h(x) = (x + 2)^2$

**30.** $h(x) = |x - 1|$

**31.** $h(x) = \sqrt{x + 5} + 2$

**32.** $h(x) = (3x + 4)^2 + 3$

**33.** $h(x) = \dfrac{1}{2x - 3}$

**34.** $h(x) = \dfrac{1}{x + 10}$

*Given that* $f(-1) = 4 \quad g(-1) = -4$
$f(0) = 5 \quad g(0) = -3$
$f(2) = 7 \quad g(2) = -1$
$f(7) = 1 \quad g(7) = 4$

*Find each function value.*

**35.** $(f + g)(2)$

**36.** $(f - g)(7)$

**37.** $(f \circ g)(2)$

**38.** $(g \circ f)(2)$

**39.** $(f \cdot g)(7)$

**40.** $(f \cdot g)(0)$

**41.** $\left(\dfrac{f}{g}\right)(-1)$

**42.** $\left(\dfrac{g}{f}\right)(-1)$

*Solve.*

**43.** Business people are concerned with cost functions, revenue functions, and profit functions. Recall that the profit $P(x)$ obtained from $x$ units of a product is equal to the revenue $R(x)$ from selling the $x$ units minus the cost $C(x)$ of manufacturing the $x$ units. Write an equation expressing this relationship among $C(x)$, $R(x)$, and $P(x)$.

**44.** Suppose the revenue $R(x)$ for $x$ units of a product can be described by $R(x) = 25x$, and the cost $C(x)$ can be described by $C(x) = 50 + x^2 + 4x$. Find the profit $P(x)$ for $x$ units.

### REVIEW EXERCISES

*Solve each equation for y. See Section 2.3.*

**45.** $x = y + 2$

**46.** $x = y - 5$

**47.** $x = 3y$

**48.** $x = -6y$

**49.** $x = -2y - 7$

**50.** $x = 4y + 7$

## 10.2 INVERSE FUNCTIONS

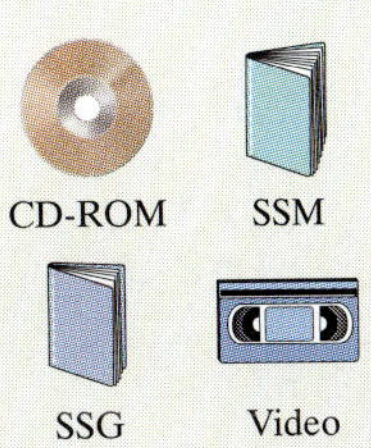


▶ **OBJECTIVES**

1. Determine whether a function is a one-to-one function.
2. Use the horizontal line test to decide whether a function is a one-to-one function.
3. Find the inverse of a function.
4. Find the equation of the inverse of a function.
5. Graph functions and their inverses.
6. Determine whether two functions are inverses of each other.

1 In the next section, we begin a study of two new functions: exponential and logarithmic functions. As we learn more about these functions, we will discover that they share a special relation to each other: They are inverses of each other.

Before we study these functions, we need to learn about inverses. We begin by defining one-to-one functions.

Study the following table.

| Degrees Fahrenheit (Input) | −31 | −13 | 32 | 68 | 149 | 212 |
|---|---|---|---|---|---|---|
| Degrees Celsius (Output) | −35 | −25 | 0 | 20 | 65 | 100 |

Recall that since each degrees Fahrenheit (input) corresponds to exactly one degrees Celsius (output), this table of inputs and outputs does describe a function. Also notice that each output corresponds to a different input. This type of function is given a special name—a one-to-one function.

Does the set $f = \{(0, 1), (2, 2), (-3, 5), (7, 6)\}$ describe a one-to-one function? It is a function since each $x$-value corresponds to a unique $y$-value. For this particular function $f$, each $y$-value corresponds to a unique $x$-value. Thus, this function is also a **one-to-one function.**

### ONE-TO-ONE FUNCTION

For a **one-to-one function,** each $x$-value (input) corresponds to only one $y$-value (output), and each $y$-value (output) corresponds to only one $x$-value (input).

**Example 1** Determine whether each function described is one-to-one.

**a.** $f = \{(6, 2), (5, 4), (-1, 0), (7, 3)\}$

**b.** $g = \{(3, 9), (-4, 2), (-3, 9), (0, 0)\}$

**c.** $h = \{(1, 1), (2, 2), (10, 10), (-5, -5)\}$

**d.**

| MINERAL (INPUT) | Talc | Gypsum | Diamond | Topaz | Stibnite |
|---|---|---|---|---|---|
| HARDNESS ON THE MOHS SCALE (OUTPUT) | 1 | 2 | 10 | 8 | 2 |

**e.**

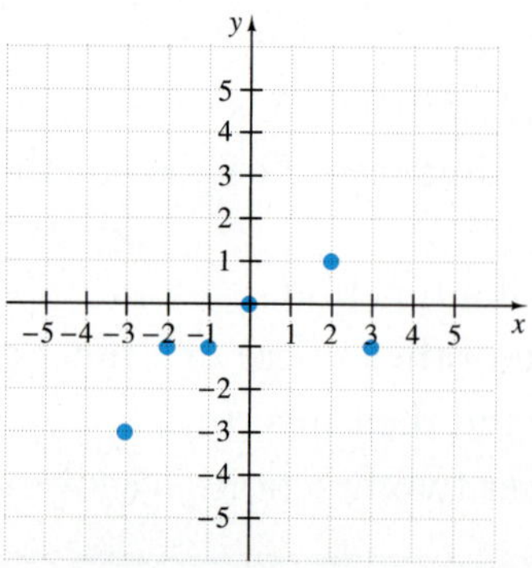

*Solution*

**a.** $f$ is one-to-one since each $y$-value corresponds to only one $x$-value.

**b.** $g$ is not one-to-one because the $y$-value 9 in $(3, 9)$ and $(-3, 9)$ corresponds to two different $x$-values.

**c.** $h$ is a one-to-one function since each $y$-value corresponds to only one $x$-value.

**d.** This table does not describe a one-to-one function since the output 2 corresponds to two different inputs, gypsum and stibnite.

**e.** This graph does not describe a one-to-one function since the $y$-value $-1$ corresponds to three different $x$-values, $-2$, $-1$, and 3.

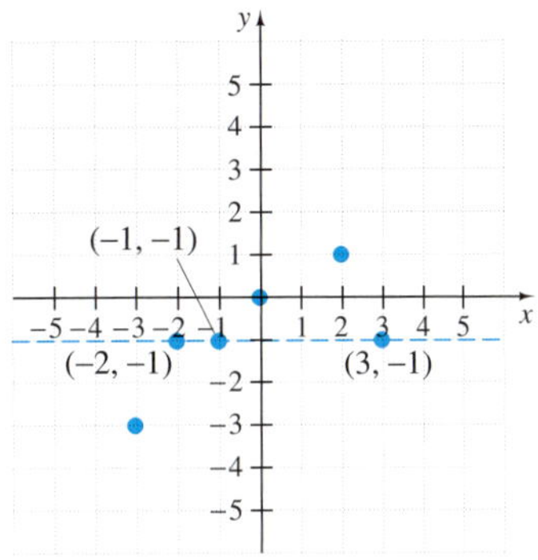


2 Recall that we recognize the graph of a function when it passes the vertical line test. Since every $x$-value of the function corresponds to exactly one $y$-value, each vertical line intersects the function's graph at most once. The graph shown next, for instance, is the graph of a function.

Is this function a *one-to-one* function? The answer is no. To see why not, notice that the $y$-value of the ordered pair $(-3, 3)$, for example, is the same as the $y$-value of the ordered pair $(3, 3)$. This function is therefore not one-to-one.

To test whether a graph is the graph of a one-to-one function, apply the vertical line test to see if it is a function, and then apply a similar **horizontal line test** to see if it is a one-to-one function.

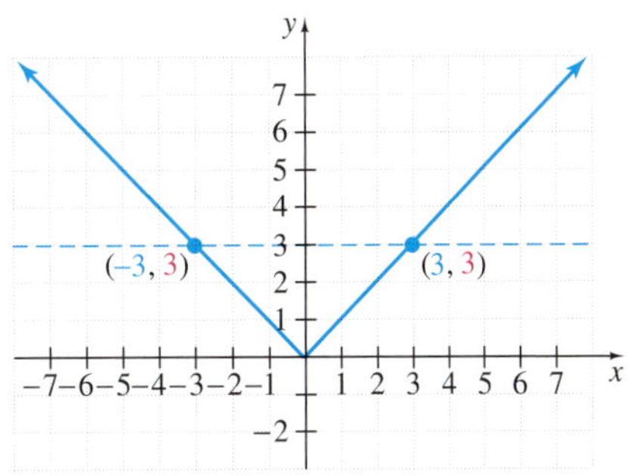


**HORIZONTAL LINE TEST**

If every horizontal line intersects the graph of a function at most once, then the function is a one-to-one function.

**Example 2** Determine whether each graph is the graph of a one-to-one function.

**a.**

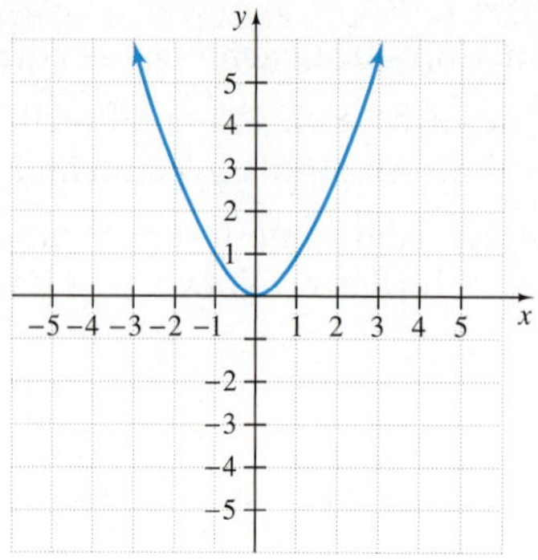

**b.**

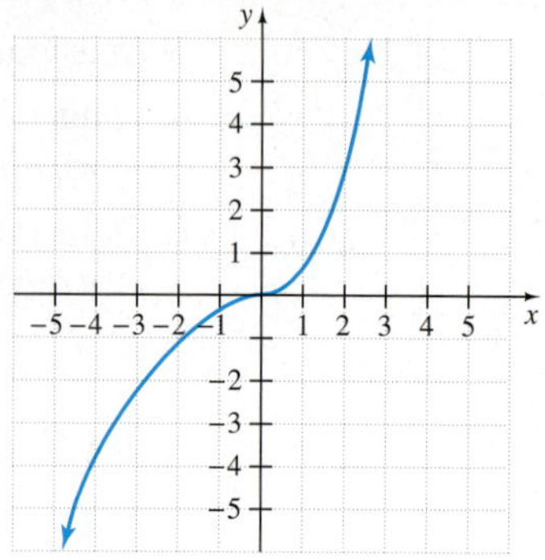

**c.**

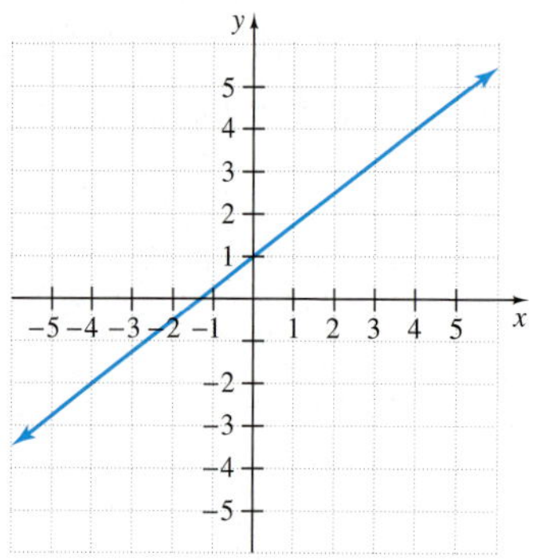

**d.**

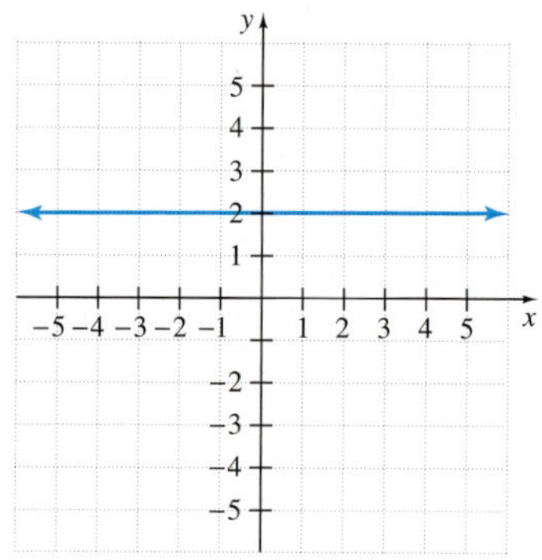

**e.**

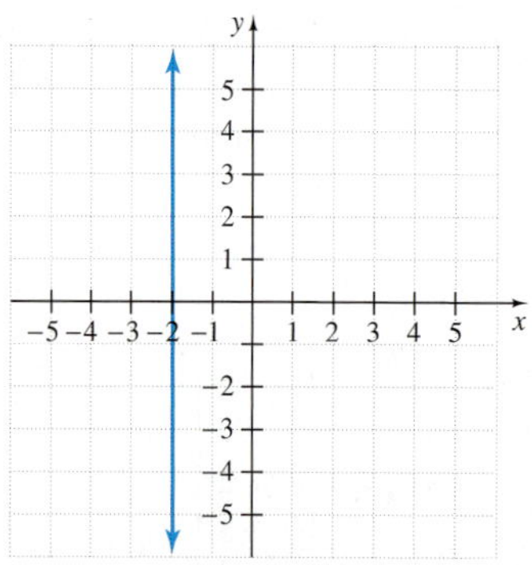

*Solution* Graphs **a**, **b**, **c**, and **d** all pass the vertical line test, so only these graphs are graphs of functions. But, of these, only **b** and **c** pass the horizontal line test, so only **b** and **c** are graphs of one-to-one functions.

**HELPFUL HINT**

All linear equations are one-to-one functions except those whose graphs are horizontal or vertical lines. A vertical line does not pass the vertical line test and hence is not the graph of a function. A horizontal line is the graph of a function but does not pass the horizontal line test and hence is not the graph of a one-to-one function.

3 One-to-one functions are special in that their graphs pass both the vertical and horizontal line tests. They are special, too, in another sense: For each one-to-one function, we can find its **inverse function** by switching the coordinates of the ordered pairs of the function, or the inputs and the outputs. For example, the inverse of the one-to-one function

| | | | | | | |
|---|---|---|---|---|---|---|
| Degrees Fahrenheit (Input) | −31 | −13 | 32 | 68 | 149 | 212 |
| Degrees Celsius (Output) | −35 | −25 | 0 | 20 | 65 | 100 |

is the function

| | | | | | | |
|---|---|---|---|---|---|---|
| Degrees Celsius (Input) | −35 | −25 | 0 | 20 | 65 | 100 |
| Degrees Fahrenheit (Output) | −31 | −13 | 32 | 68 | 149 | 212 |

Notice that the ordered pair $(-31, -35)$ of the function, for example, becomes the ordered pair $(-35, -31)$ of its inverse.

Also, the inverse of the one-to-one function $f = \{(2, -3), (5, 10), (9, 1)\}$ is $\{(-3, 2), (10, 5), (1, 9)\}$. For a function $f$, we use the notation $f^{-1}$, read "$f$ inverse," to denote its inverse function. Notice that since the coordinates of each ordered pair have been switched, the domain (set of inputs) of $f$ is the range (set of outputs) of $f^{-1}$, and the range of $f$ is the domain of $f^{-1}$. See the definition of inverse function.

### INVERSE FUNCTION

The inverse of a one-to-one function $f$ is the one-to-one function $f^{-1}$ that consists of the set of all ordered pairs $(y, x)$ where $(x, y)$ belongs to $f$.

**Example 3** Find the inverse of the one-to-one function.

$$f = \{(0, 1), (-2, 7), (3, -6), (4, 4)\}$$

*Solution* $f^{-1} = \{(1, 0), (7, -2), (-6, 3), (4, 4)\}$

Switch coordinates of each ordered pair.

**HELPFUL HINT**

The symbol $f^{-1}$ is the single symbol used to denote the inverse of the function $f$. It is read as "$f$ inverse." This symbol *does not mean* $\frac{1}{f}$.

4 If a one-to-one function $f$ is defined as a set of ordered pairs, we can find $f^{-1}$ by interchanging the $x$- and $y$-coordinates of the ordered pairs. If a one-to-one

function $f$ is given in the form of an equation, we can find $f^{-1}$ by using a similar procedure.

**FINDING THE INVERSE OF A ONE-TO-ONE FUNCTION $f(x)$**

**Step 1:** Replace $f(x)$ with $y$.
**Step 2:** Interchange $x$ and $y$.
**Step 3:** Solve the equation for $y$.
**Step 4:** Replace $y$ with the notation $f^{-1}(x)$.

**Example 4** Find an equation of the inverse of $f(x) = x + 3$.

*Solution* $f(x) = x + 3$

| | | |
|---|---|---|
| **Step 1:** | $y = x + 3$ | Replace $f(x)$ with $y$. |
| **Step 2:** | $x = y + 3$ | Interchange $x$ and $y$. |
| **Step 3:** | $x - 3 = y$ | Solve for $y$. |
| **Step 4:** | $f^{-1}(x) = x - 3$ | Replace $y$ with $f^{-1}(x)$. |

The inverse of $f(x) = x + 3$ is $f^{-1}(x) = x - 3$. Notice that, for example,

$f(1) = 1 + 3 = 4$ and $f^{-1}(4) = 4 - 3 = 1$

Ordered pair: (1, 4) Ordered pair: (4, 1)

The coordinates are switched, as expected.

**Example 5** Find the equation of the inverse of $f(x) = 3x - 5$. Graph $f$ and $f^{-1}$ on the same set of axes.

*Solution* $f(x) = 3x - 5$

| | | |
|---|---|---|
| **Step 1:** | $y = 3x - 5$ | Replace $f(x)$ with $y$. |
| **Step 2:** | $x = 3y - 5$ | Interchange $x$ and $y$. |
| **Step 3:** | $3y = x + 5$ | Solve for $y$. |
| | $y = \dfrac{x + 5}{3}$ | |
| **Step 4:** | $f^{-1}(x) = \dfrac{x + 5}{3}$ | Replace $y$ with $f^{-1}(x)$. |

Now we graph $f(x)$ and $f^{-1}(x)$ on the same set of axes. Both $f(x) = 3x - 5$ and

$f^{-1}(x) = \dfrac{x + 5}{3}$ are linear functions, so each graph is a line.

$f(x) = 3x - 5$

| $x$ | $y = f(x)$ |
|---|---|
| 1 | −2 |
| 0 | −5 |
| $\frac{5}{3}$ | 0 |

$f^{-1}(x) = \dfrac{x + 5}{3}$

| $x$ | $y = f^{-1}(x)$ |
|---|---|
| −2 | 1 |
| −5 | 0 |
| 0 | $\frac{5}{3}$ |

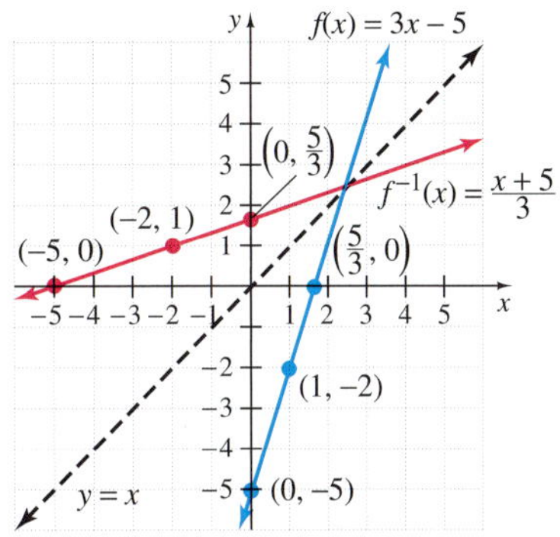


5 Notice that the graphs of $f$ and $f^{-1}$ in Example 5 are mirror images of each other, and the "mirror" is the dashed line $y = x$. This is true for every function and its inverse. For this reason, we say that *the graphs of $f$ and $f^{-1}$ are symmetric about the line $y = x$.*

To see why this happens, study the graph of a few ordered pairs and their switched coordinates.

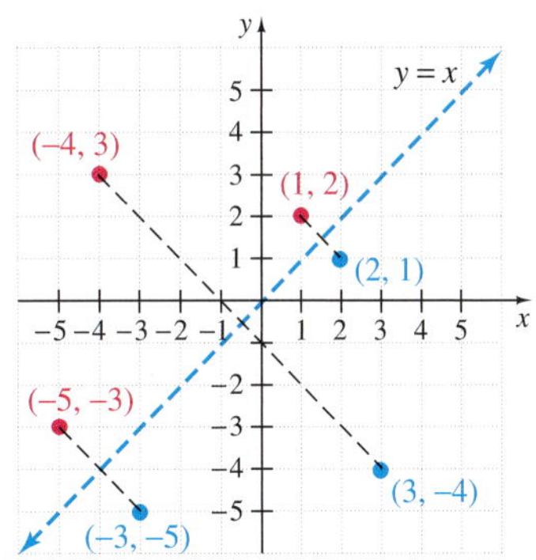


6 Notice also in the table of values in Example 5 that $f(0) = -5$ and $f^{-1}(-5) = 0$, as expected. Also, for example, $f(1) = -2$ and $f^{-1}(-2) = 1$. In words, we say that for some input $x$, the function $f^{-1}$ takes the output of $x$, called $f(x)$, back to $x$.

$$x \rightarrow f(x) \quad \text{and} \quad f^{-1}(f(x)) \rightarrow x$$
$$\downarrow \quad \downarrow \qquad\qquad \downarrow \quad \downarrow$$
$$f(0) = -5 \quad \text{and} \quad f^{-1}(-5) = 0$$
$$f(1) = -2 \quad \text{and} \quad f^{-1}(-2) = 1$$

In general,

> If $f$ is a one-to-one function, then the inverse of $f$ is the function $f^{-1}$ such that
>
> $$(f^{-1} \circ f)(x) = x \quad \text{and} \quad (f \circ f^{-1})(x) = x$$

**Example 6** Graph the inverse of each function.

*Solution* The function is graphed in blue and the inverse is graphed in red.

**a.**

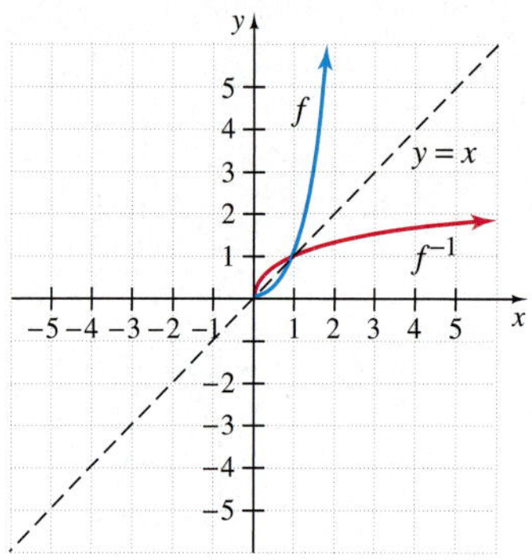


**b.**

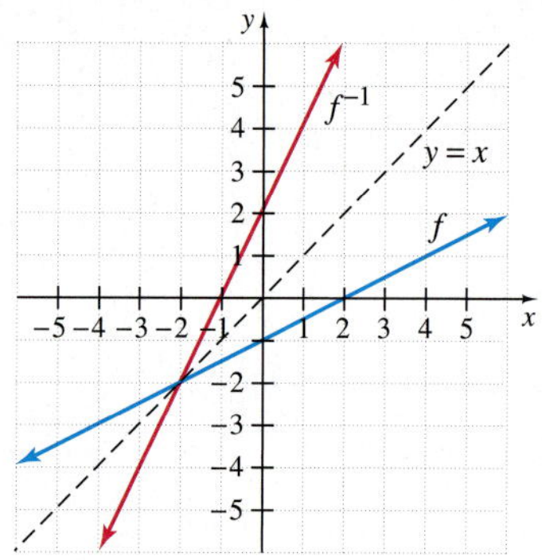


**Example 7** Show that if $f(x) = 3x + 2$, then $f^{-1}(x) = \dfrac{x-2}{3}$.

*Solution* See that $f^{-1}(f(x)) = x$ and $f(f^{-1}(x)) = x$.

$$
\begin{aligned}
(f^{-1} \circ f)(x) &= f^{-1}(f(x)) \\
&= f^{-1}(3x + 2) && \text{Replace } f(x) \text{ with } 3x + 2. \\
&= \frac{3x + 2 - 2}{3} \\
&= \frac{3x}{3} \\
&= x
\end{aligned}
$$

$$
\begin{aligned}
(f \circ f^{-1})(x) &= f(f^{-1}(x)) \\
&= f\left(\frac{x-2}{3}\right) && \text{Replace } f^{-1}(x) \text{ with } \frac{x-2}{3}. \\
&= 3\left(\frac{x-2}{3}\right) + 2 \\
&= x - 2 + 2 \\
&= x
\end{aligned}
$$

## GRAPHING CALCULATOR EXPLORATIONS

A graphing calculator can be used to visualize the results of Example 7. Recall that the graph of a function $f$ and its inverse $f^{-1}$ are mirror images of each other across the line $y = x$. To see this for the function from Example 7, use a square window and graph

the given function: $Y_1 = 3x + 2$

its inverse: $Y_2 = \dfrac{x - 2}{3}$

and the line: $Y_3 = x$

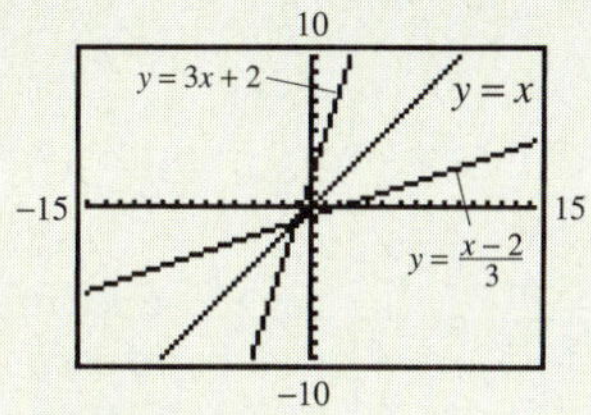


Exercises will follow in Exercise Set 10.2.

## *Exercise Set 10.2*

*Determine whether each function is a one-to-one function. If it is one-to-one, list the inverse function by switching coordinates, or inputs and outputs. See Examples 1 and 3.*

**1.** $f = \{(-1, -1), (1, 1), (0, 2), (2, 0)\}$

**2.** $g = \{(8, 6), (9, 6), (3, 4), (-4, 4)\}$

**3.** $h = \{(10, 10)\}$

**4.** $r = \{(1, 2), (3, 4), (5, 6), (6, 7)\}$

**5.** $f = \{(11, 12), (4, 3), (3, 4), (6, 6)\}$

**6.** $g = \{(0, 3), (3, 7), (6, 7), (-2, -2)\}$

**7.**

| Month of 1998 (Input) | January | February | March | April | May | June |
|---|---|---|---|---|---|---|
| Thousands of Houses on Sale at Month's End (Output) | 282 | 277 | 281 | 285 | 282 | 287 |

(*Source:* U.S. Department of Housing and Urban Development)

**8.**

| State (Input) | Washington | Ohio | Georgia | Colorado | California | Arizona |
|---|---|---|---|---|---|---|
| Electoral Votes (Output) | 11 | 21 | 13 | 8 | 54 | 8 |

(*Source:* U.S. Bureau of the Census)

**9.**

| State (Input) | California | Vermont | Virginia | Texas | South Dakota |
|---|---|---|---|---|---|
| Rank in Population (Output) | 1 | 49 | 12 | 2 | 45 |

(*Source:* U.S. Bureau of the Census)

△ **10.**

| Shape (Input) | Triangle | Pentagon | Quadrilateral | Hexagon | Decagon |
|---|---|---|---|---|---|
| Number of Sides (Output) | 3 | 5 | 4 | 6 | 10 |

*Given the one-to-one function $f(x) = x^3 + 2$, find the following. [Hint: You do not need to find the equation for $f^{-1}(x)$.]*

**11. a.** $f(1)$
**b.** $f^{-1}(3)$

**12. a.** $f(0)$
**b.** $f^{-1}(2)$

**13. a.** $f(-1)$
**b.** $f^{-1}(1)$

**14. a.** $f(-2)$
**b.** $f^{-1}(-6)$

*Determine whether the graph of each function is the graph of a one-to-one function. See Example 2.*

**15.**

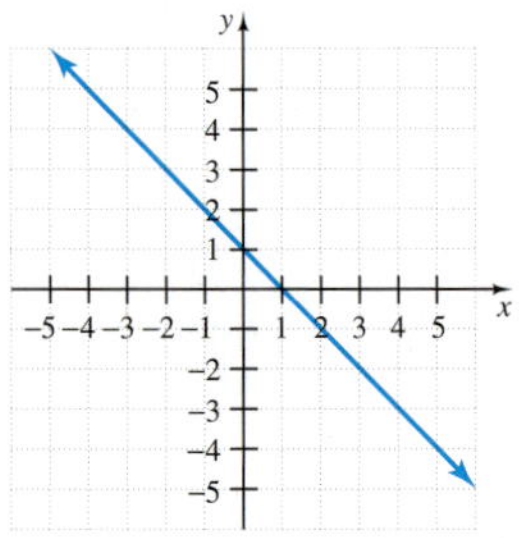

**16.**

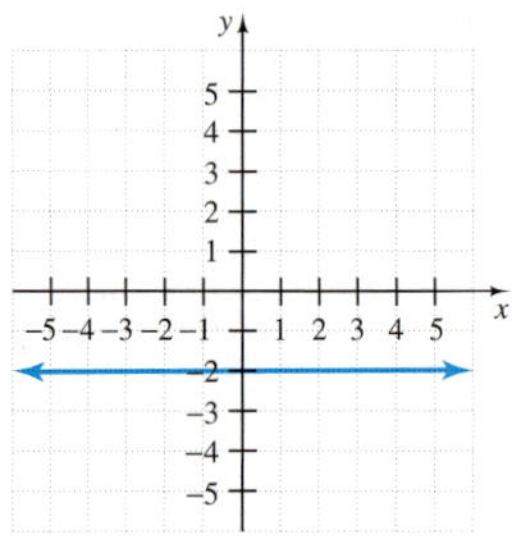

**17.**

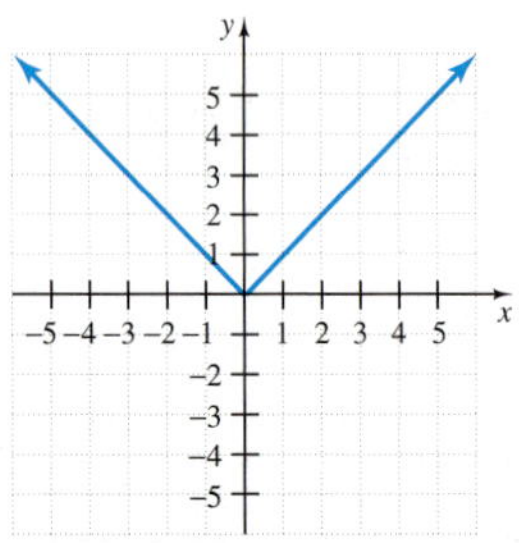

**18.**

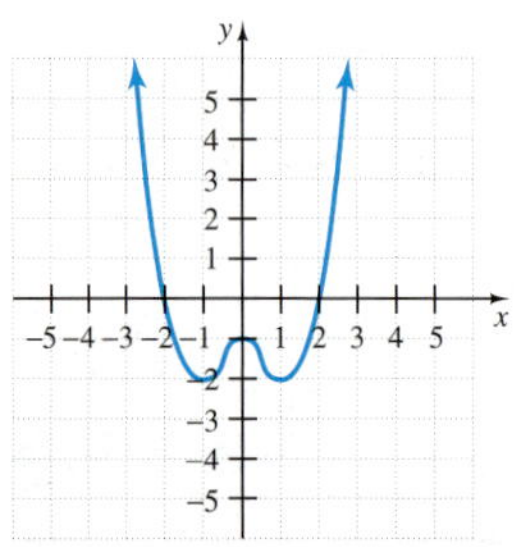

**19.**

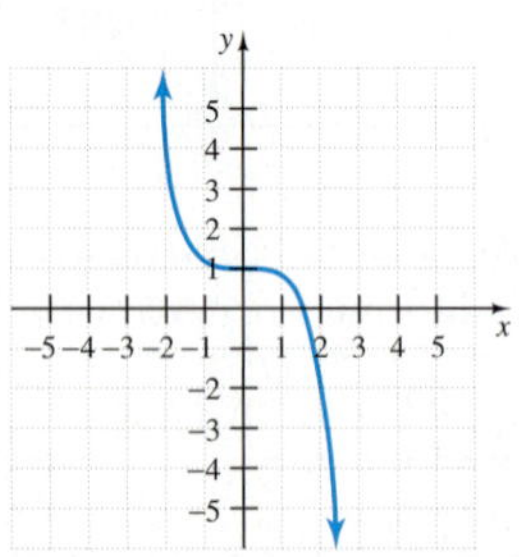

**20.**

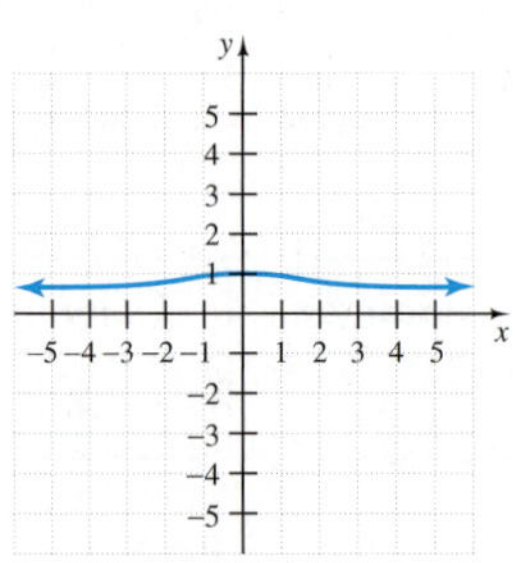

**21.**

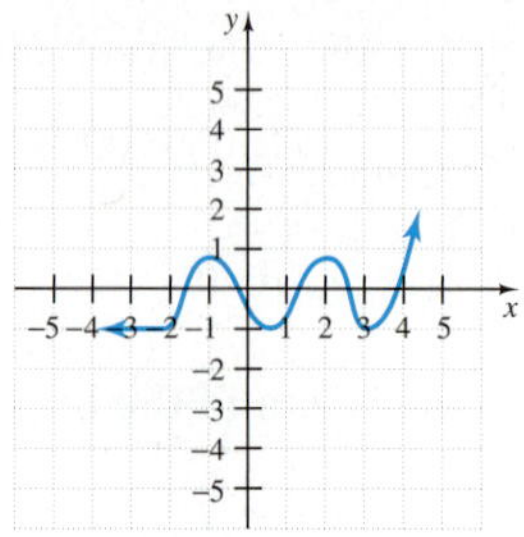

**22.**

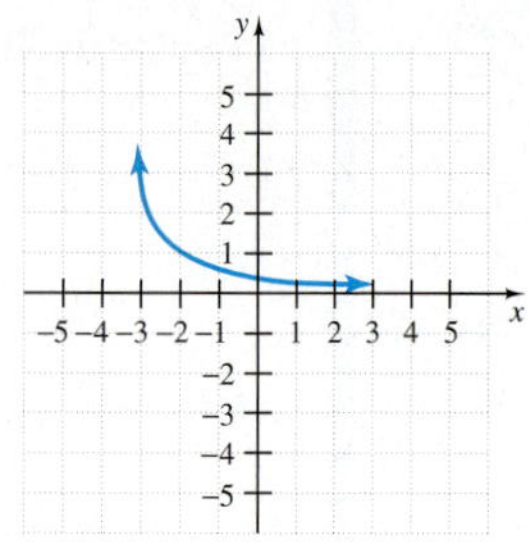

*Each of the following functions is one-to-one. Find the inverse of each function and graph the function and its inverse on the same set of axes. See Examples 4 and 5.*

**23.** $f(x) = x + 4$

**24.** $f(x) = x - 5$

**25.** $f(x) = 2x - 3$

**26.** $f(x) = 4x + 9$

**27.** $f(x) = \frac{1}{2}x - 1$

**28.** $f(x) = -\frac{1}{2}x + 2$

**29.** $f(x) = x^3$

**30.** $f(x) = x^3 - 1$

*Find the inverse of each one-to-one function. See Examples 4 and 5.*

**31.** $f(x) = 5x + 2$

**32.** $f(x) = 6x - 1$

**33.** $f(x) = \dfrac{x - 2}{5}$

**34.** $f(x) = \dfrac{4x - 3}{2}$

**35.** $f(x) = \sqrt[3]{x}$

**36.** $f(x) = \sqrt[3]{x + 1}$

**37.** $f(x) = \dfrac{5}{3x + 1}$

**38.** $f(x) = \dfrac{7}{2x + 4}$

**39.** $f(x) = (x + 2)^3$

**40.** $f(x) = (x - 5)^3$

*Graph the inverse of each function on the same set of axes. See Example 6.*

**41.**

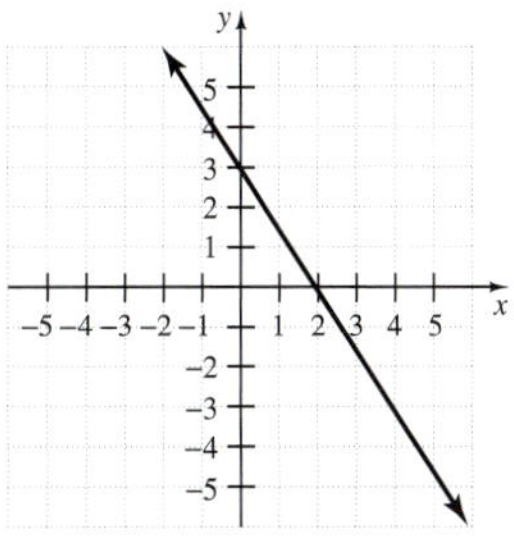

**42.**

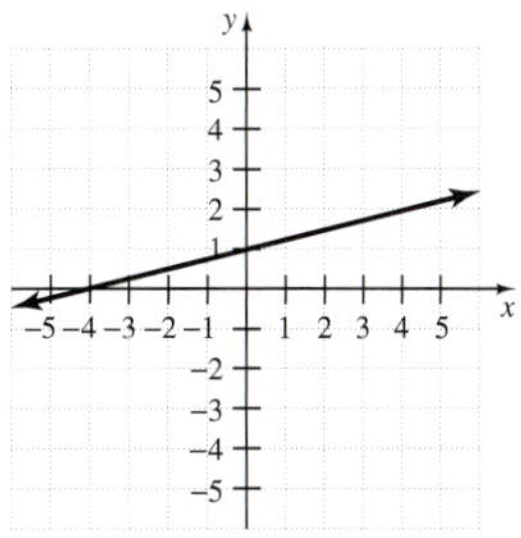

**43.**

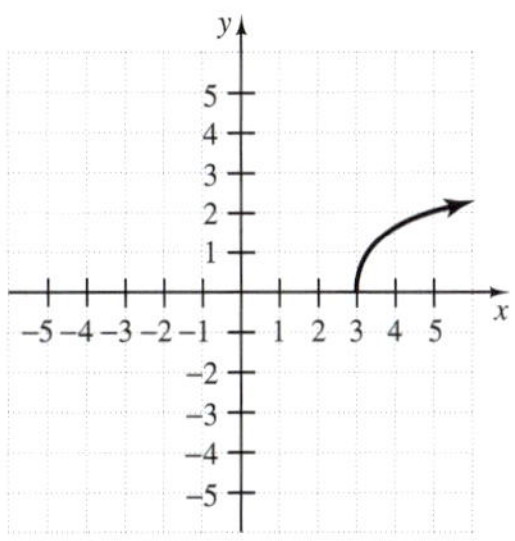

**44.**

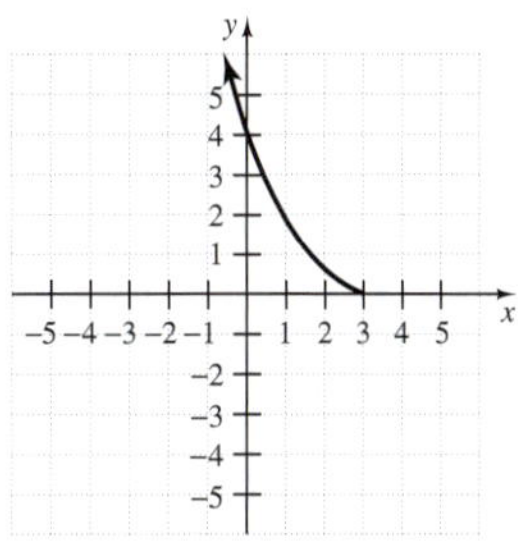

**45.**

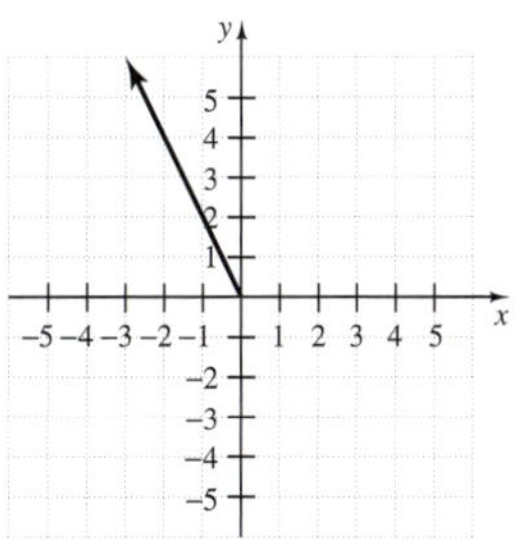

**46.**

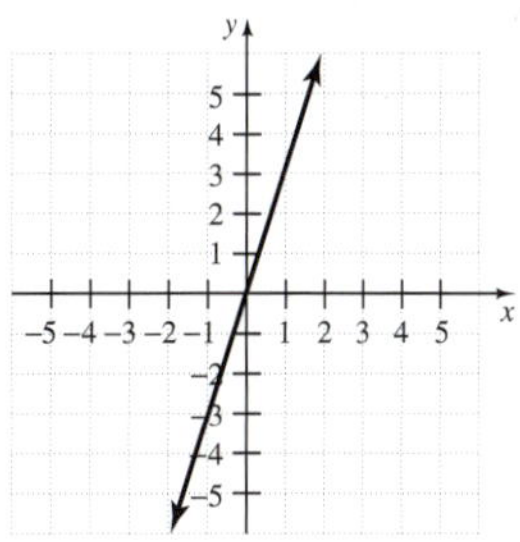

*Solve. See Example 7.*

**47.** If $f(x) = 2x + 1$, show that $f^{-1}(x) = \dfrac{x-1}{2}$.

**48.** If $f(x) = 3x - 10$, show that $f^{-1}(x) = \dfrac{x+10}{3}$.

**49.** If $f(x) = x^3 + 6$, show that $f^{-1}(x) = \sqrt[3]{x-6}$.

**50.** If $f(x) = x^3 - 5$, show that $f^{-1}(x) = \sqrt[3]{x+5}$.

*For Exercises 51 and 52,*

**a.** Write the ordered pairs for $f(x)$ whose points are highlighted. (Include the points whose coordinates are given.)

**b.** Write the corresponding ordered pairs for the inverse of $f$, $f^{-1}$.

**c.** Graph the ordered pairs for $f^{-1}$ found in Part **b.**

**d.** Graph $f^{-1}(x)$ by drawing a smooth curve through the plotted points.

**51.**

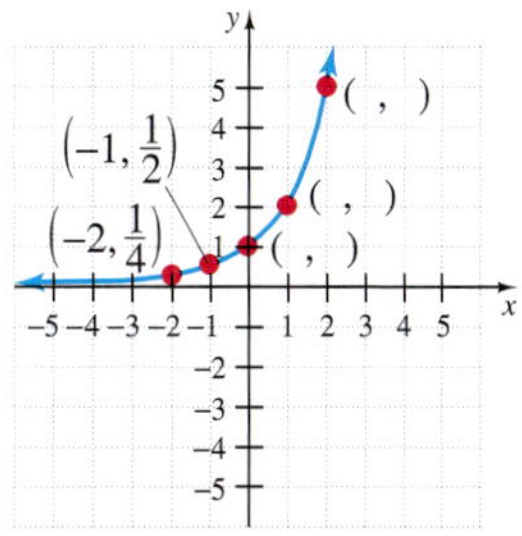


**52.**

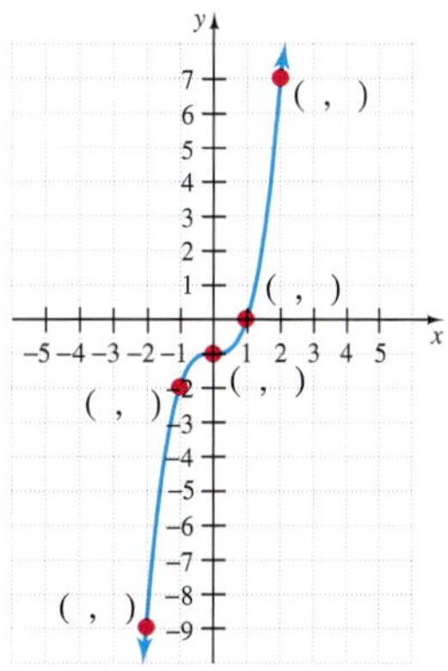

*Find the inverse of each given one-to-one function. Then graph the function and its inverse on a square window.*

**53.** $f(x) = 3x + 1$

**54.** $f(x) = -2x - 6$

**55.** $f(x) = \sqrt[3]{x+1}$

**56.** $f(x) = x^3 - 3$

## REVIEW EXERCISES

*Evaluate each of the following. See Section 7.2.*

**57.** $25^{1/2}$

**58.** $49^{1/2}$

**59.** $16^{3/4}$

**60.** $27^{2/3}$

**61.** $9^{-3/2}$

**62.** $81^{-3/4}$

*If $f(x) = 3^x$, find the following. In Exercises 65 and 66, give an exact answer and a two-decimal-place approximation. See Sections 3.2 and 5.1.*

**63.** $f(2)$

**64.** $f(0)$

**65.** $f(\frac{1}{2})$

**66.** $f(\frac{2}{3})$

## 10.3 EXPONENTIAL FUNCTIONS

CD-ROM SSM SSG Video

**OBJECTIVES**

1. Graph exponential functions.
2. Solve equations of the form $b^x = b^y$.
3. Solve problems modeled by exponential equations.

1 In earlier chapters, we gave meaning to exponential expressions such as $2^x$, where $x$ is a rational number. For example,

$$2^3 = 2 \cdot 2 \cdot 2 \qquad \text{Three factors, each factor is 2}$$

$$2^{3/2} = (2^{1/2})^3 = \sqrt{2} \cdot \sqrt{2} \cdot \sqrt{2} \qquad \text{Three factors, each factor is } \sqrt{2}$$

When $x$ is an irrational number (for example, $\sqrt{3}$), what meaning can we give to $2^{\sqrt{3}}$?

It is beyond the scope of this book to give precise meaning to $2^x$ if $x$ is irrational. We can confirm your intuition and say that $2^{\sqrt{3}}$ is a real number, and since $1 < \sqrt{3} < 2$, then $2^1 < 2^{\sqrt{3}} < 2^2$. We can also use a calculator and approximate $2^{\sqrt{3}}$: $2^{\sqrt{3}} \approx 3.321997$. In fact, as long as the base $b$ is positive, $b^x$ is a real number for all real numbers $x$. Finally, the rules of exponents apply whether $x$ is rational or irrational, as long as $b$ is positive. In this section, we are interested in functions of the form $f(x) = b^x$, where $b > 0$. A function of this form is called an **exponential function.**

**EXPONENTIAL FUNCTION**

A function of the form

$$f(x) = b^x$$

is called an **exponential function** if $b > 0$, $b$ is not 1, and $x$ is a real number.

Next, we practice graphing exponential functions.

**Example 1** Graph the exponential functions defined by $f(x) = 2^x$ and $g(x) = 3^x$ on the same set of axes.

*Solution* Graph each function by plotting points. Set up a table of values for each of the two functions.

$f(x) = 2^x$

| $x$ | 0 | 1 | 2 | 3 | −1 | −2 |
|---|---|---|---|---|---|---|
| $f(x)$ | 1 | 2 | 4 | 8 | $\frac{1}{2}$ | $\frac{1}{4}$ |

$g(x) = 3^x$

| $x$ | 0 | 1 | 2 | 3 | −1 | −2 |
|---|---|---|---|---|---|---|
| $g(x)$ | 1 | 3 | 9 | 27 | $\frac{1}{3}$ | $\frac{1}{9}$ |

If each set of points is plotted and connected with a smooth curve, the following graphs result.

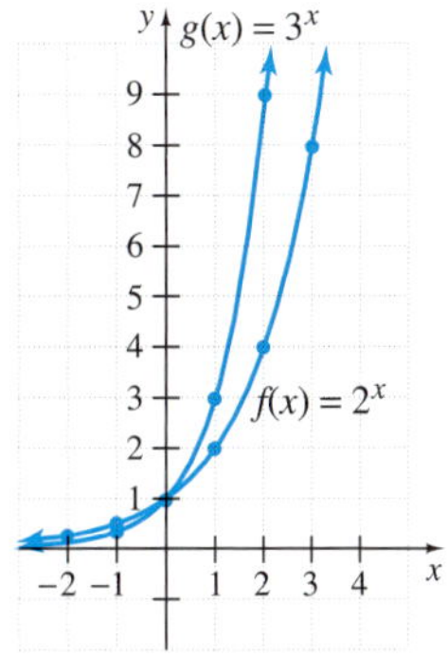


A number of things should be noted about the two graphs of exponential functions in Example 1. First, the graphs show that $f(x) = 2^x$ and $g(x) = 3^x$ are one-to-one functions since each graph passes the vertical and horizontal line tests. The $y$-intercept of each graph is $(0, 1)$, but neither graph has an $x$-intercept. From the graph, we can also see that the domain of each function is all real numbers and that the range is $(0, \infty)$. We can also see that as $x$-values are increasing, $y$-values are increasing also.

**Example 2** Graph the exponential functions $y = \left(\frac{1}{2}\right)^x$ and $y = \left(\frac{1}{3}\right)^x$ on the same set of axes.

*Solution* As before, plot points and connect them with a smooth curve.

$y = \left(\frac{1}{2}\right)^x$

| $x$ | 0 | 1 | 2 | 3 | −1 | −2 |
|---|---|---|---|---|---|---|
| $y$ | 1 | $\frac{1}{2}$ | $\frac{1}{4}$ | $\frac{1}{8}$ | 2 | 4 |

$y = \left(\frac{1}{3}\right)^x$

| $x$ | 0 | 1 | 2 | 3 | −1 | −2 |
|---|---|---|---|---|---|---|
| $y$ | 1 | $\frac{1}{3}$ | $\frac{1}{9}$ | $\frac{1}{27}$ | 3 | 9 |

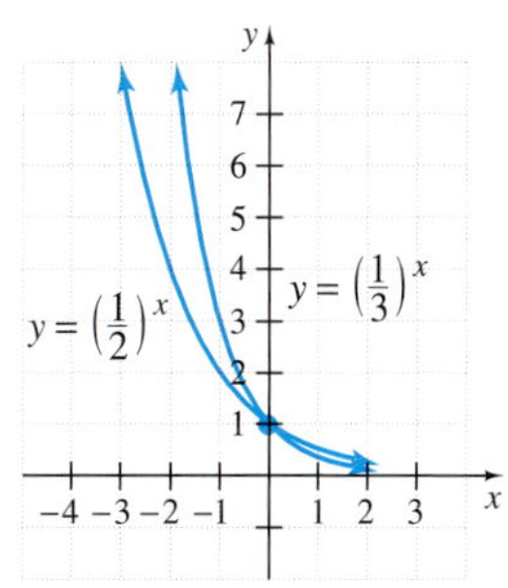


Each function in Example 2 again is a one-to-one function. The $y$-intercept of both is $(0, 1)$. The domain is the set of all real numbers, and the range is $(0, \infty)$.

Notice the difference between the graphs of Example 1 and the graphs of Example 2. An exponential function is always increasing if the base is greater than 1.

When the base is between 0 and 1, the graph is always decreasing. The following figures summarize these characteristics of exponential functions.

$$f(x) = b^x, \quad b > 0, \quad b \neq 1$$

- one-to-one function
- $y$-intercept $(0, 1)$
- no $x$-intercept
- domain: $(-\infty, \infty)$
- range: $(0, \infty)$

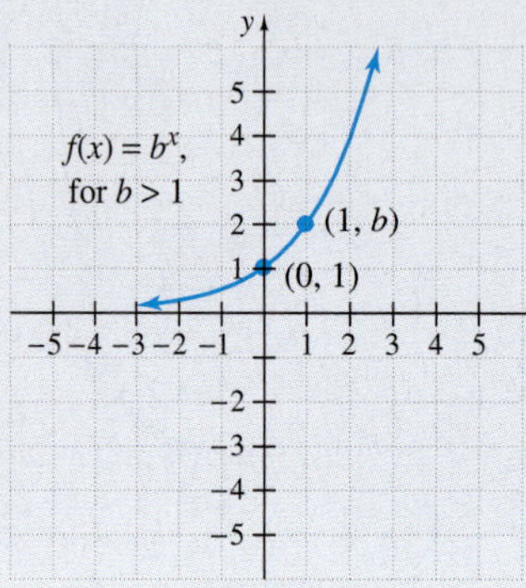


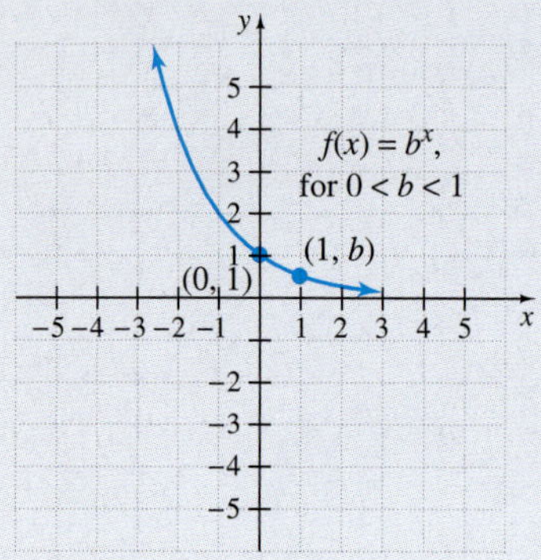


**Example 3** Graph the exponential function $f(x) = 3^{x+2}$.

*Solution* As before, we find and plot a few ordered pair solutions. Then we connect the points with a smooth curve.

$y = 3^{x+2}$

| $x$ | 0 | −1 | −2 | −3 | −4 |
|---|---|---|---|---|---|
| $y$ | 9 | 3 | 1 | $\frac{1}{3}$ | $\frac{1}{9}$ |

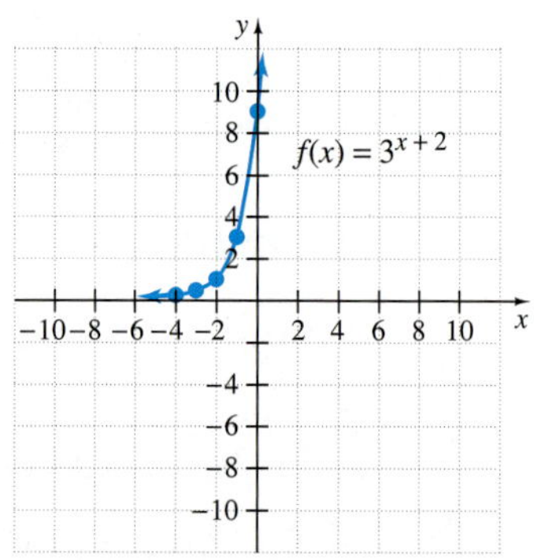


2 We have seen that an exponential function $y = b^x$ is a one-to-one function. Another way of stating this fact is a property that we can use to solve exponential equations.

**UNIQUENESS OF $b^x$**

Let $b > 0$ and $b \neq 1$. Then $b^x = b^y$ is equivalent to $x = y$.

**Example 4** Solve each equation for $x$.

**a.** $2^x = 16$ **b.** $9^x = 27$ **c.** $4^{x+3} = 8^x$

*Solution* **a.** We write 16 as a power of 2 and then use the uniqueness of $b^x$ to solve.

$$2^x = 16$$
$$2^x = 2^4$$

Since the bases are the same and are nonnegative, by the uniqueness of $b^x$, we then have that the exponents are equal. Thus,

$$x = 4$$

The solution is 4.

**b.** Notice that both 9 and 27 are powers of 3.

$$9^x = 27$$

$$(3^2)^x = 3^3 \quad \text{Write 9 and 27 as powers of 3.}$$

$$3^{2x} = 3^3$$

$$2x = 3 \quad \text{Apply the uniqueness of } b^x.$$

$$x = \frac{3}{2} \quad \text{Divide by 2.}$$

To check, replace $x$ with $\frac{3}{2}$ in the original expression, $9^x = 27$. The solution is $\frac{3}{2}$.

**c.** Write both 4 and 8 as powers of 2.

$$4^{x+3} = 8^x$$

$$(2^2)^{x+3} = (2^3)^x$$

$$2^{2x+6} = 2^{3x}$$

$$2x + 6 = 3x \quad \text{Apply the uniqueness of } b^x.$$

$$6 = x \quad \text{Subtract } 2x \text{ from both sides.}$$

The solution is 6.

There is one major problem with the preceding technique. Often the two sides of an equation cannot easily be written as powers of a common base. We explore how to solve an equation such as $4 = 3^x$ with the help of **logarithms** later.

3 The bar graph here shows the increase in the number of cellular phone users. Notice that the graph of the exponential function $y = 6.052(1.378)^x$ approximates the heights of the bars.

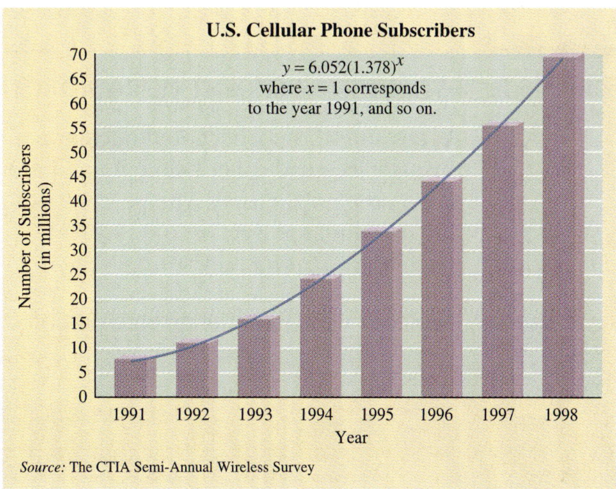


The graph above shows just one example of how the world abounds with patterns that can be modeled by exponential functions. To make these applications realistic, we use numbers that warrant a calculator. Another application of an exponential function has to do with interest rates on loans.

The exponential function defined by $A = P\left(1 + \frac{r}{n}\right)^{nt}$ models the dollars $A$ accrued (or owed) after $P$ dollars are invested (or loaned) at an annual rate of interest $r$ compounded $n$ times each year for $t$ years. This function is known as the compound interest formula.

## Example 5 USING THE COMPOUND INTEREST FORMULA

Find the amount owed at the end of 5 years if \$1600 is loaned at a rate of 9% compounded monthly.

*Solution* We use the formula $A = P\left(1 + \frac{r}{n}\right)^{nt}$, with the following values.

$$P = \$1600 \text{ (the amount of the loan)}$$
$$r = 9\% = 0.09 \text{ (the annual rate of interest)}$$
$$n = 12 \text{ (the number of times interest is compounded each year)}$$
$$t = 5 \text{ (the duration of the loan, in years)}$$

$$A = P\left(1 + \frac{r}{n}\right)^{nt} \quad \text{Compound interest formula}$$
$$= 1600\left(1 + \frac{0.09}{12}\right)^{12(5)} \quad \text{Substitute known values.}$$
$$= 1600(1.0075)^{60}$$

To approximate $A$, use the $\boxed{y^x}$ or $\boxed{\wedge}$ key on your calculator.

$\boxed{2505.0896}$

Thus, the amount $A$ owed is approximately \$2505.09.

## Example 6 ESTIMATING PERCENT OF RADIOACTIVE MATERIAL

As a result of the Chernobyl nuclear accident, radioactive debris was carried through the atmosphere. One immediate concern was the impact that the debris had on the milk supply. The percent $y$ of radioactive material in raw milk after $t$ days is estimated by $y = 100\,(2.7)^{-0.1t}$. Estimate the expected percent of radioactive material in the milk after 30 days.

*Solution* Replace $t$ with 30 in the given equation.

$$y = 100(2.7)^{-0.1t}$$
$$= 100(2.7)^{-0.1(30)} \quad \text{Let } t = 30.$$
$$= 100(2.7)^{-3}$$

To approximate the percent $y$, the following keystrokes may be used on a scientific calculator.

$\boxed{2.7}\ \boxed{y^x}\ \boxed{3}\ \boxed{+/-}\ \boxed{=}\ \boxed{\times}\ \boxed{100}\ \boxed{=}$

The display should read

| |
|---|
| 5.0805263 |

Thus, approximately 5% of the radioactive material still remained in the milk supply after 30 days.

## GRAPHING CALCULATOR EXPLORATIONS

We can use a graphing calculator and its TRACE feature to solve Example 6 graphically.

To estimate the expected percent of radioactive material in the milk after 30 days, enter $Y_1 = 100(2.7)^{-0.1x}$. (The variable $t$ in Example 6 is changed to $x$ here to better accommodate our work on the graphing calculator.) The graph does not appear on a standard viewing window, so we need to determine an appropriate viewing window. Because it doesn't make sense to look at radioactivity *before* the Chernobyl nuclear accident, we use Xmin $= 0$. We are interested in finding the percent of radioactive material in the milk when $x = 30$, so we choose Xmax $= 35$ to leave enough space to see the graph at $x = 30$. Because the values of $y$ are percents, it seems appropriate that $0 \le y \le 100$. (We also use Xscl $= 1$ and Yscl $= 10$.) Now we graph the function.

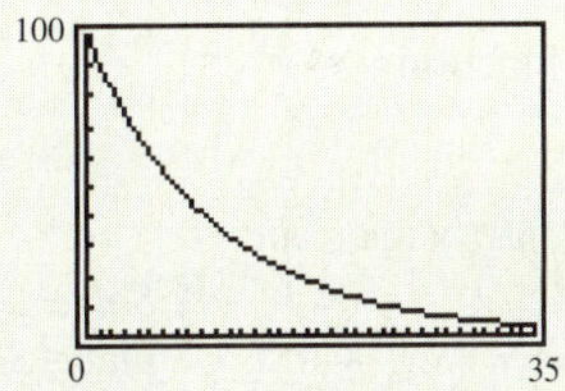


We can use the TRACE feature to obtain an approximation of the expected percent of radioactive material in the milk when $x = 30$. (A TABLE feature may also be used to approximate the percent.) To obtain a better approximation, let's use the ZOOM feature several times to zoom in near $x = 30$.

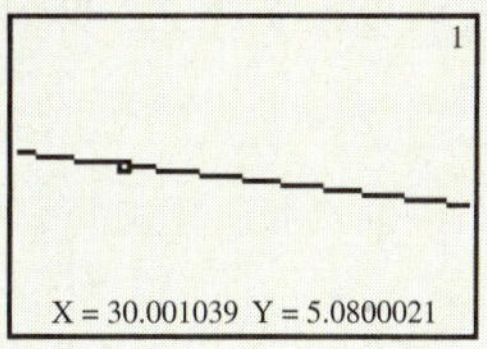


The percent of radioactive material in the milk 30 days after the Chernobyl accident was 5.08%, accurate to two decimal places.

*Use a graphing calculator to find each percent. Approximate your solutions so that they are accurate to two decimal places.*

**1.** Estimate the expected percent of radioactive material in the milk 2 days after the Chernobyl nuclear accident.

**2.** Estimate the expected percent of radioactive material in the milk 10 days after the Chernobyl nuclear accident.

**3.** Estimate the expected percent of radioactive material in the milk 15 days after the Chernobyl nuclear accident.

**4.** Estimate the expected percent of radioactive material in the milk 25 days after the Chernobyl nuclear accident.

## Exercise Set 10.3

*Graph each exponential function. See Examples 1 through 3.*

**1.** $y = 4^x$

**2.** $y = 5^x$

**3.** $y = 2^x + 1$

**4.** $y = 3^x - 1$

**5.** $y = \left(\frac{1}{4}\right)^x$

**6.** $y = \left(\frac{1}{5}\right)^x$

**7.** $y = \left(\frac{1}{2}\right)^x - 2$

**8.** $y = \left(\frac{1}{3}\right)^x + 2$

**9.** $y = -2^x$

**10.** $y = -3^x$

**11.** $y = -\left(\frac{1}{4}\right)^x$

**12.** $y = -\left(\frac{1}{5}\right)^x$

**13.** $f(x) = 2^{x+1}$

**14.** $f(x) = 3^{x-1}$

**15.** $f(x) = 4^{x-2}$

**16.** $f(x) = 2^{x+3}$

**17.** Explain why the graph of an exponential function $y = b^x$ contains the point $(1, b)$.

**18.** Explain why an exponential function $y = b^x$ has a $y$-intercept of $(0, 1)$.

*Solve each equation for x. See Example 4.*

**19.** $3^x = 27$

**20.** $6^x = 36$

**21.** $16^x = 8$

**22.** $64^x = 16$

**23.** $32^{2x-3} = 2$

**24.** $9^{2x+1} = 81$

**25.** $\frac{1}{4} = 2^{3x}$

**26.** $\frac{1}{27} = 3^{2x}$

**27.** $5^x = 625$

**28.** $2^x = 64$

**29.** $4^x = 8$

**30.** $32^x = 4$

**31.** $27^{x+1} = 9$

**32.** $125^{x-2} = 25$

**33.** $81^{x-1} = 27^{2x}$

**34.** $4^{3x-7} = 32^{2x}$

*Match each exponential equation with its graph.*

**35.** $f(x) = \left(\frac{1}{2}\right)^x$

**36.** $f(x) = 2^x$

**37.** $f(x) = \left(\frac{1}{4}\right)^x$

**38.** $f(x) = 3^x$

**A**

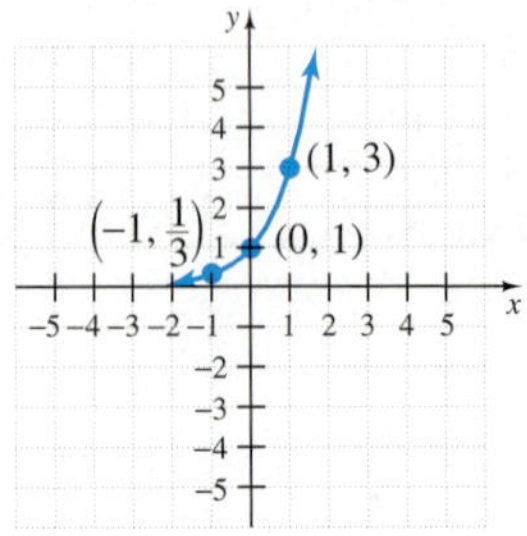


**B**

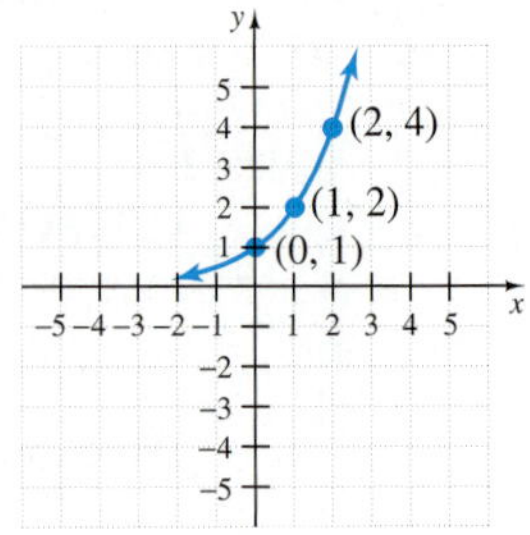


**C**

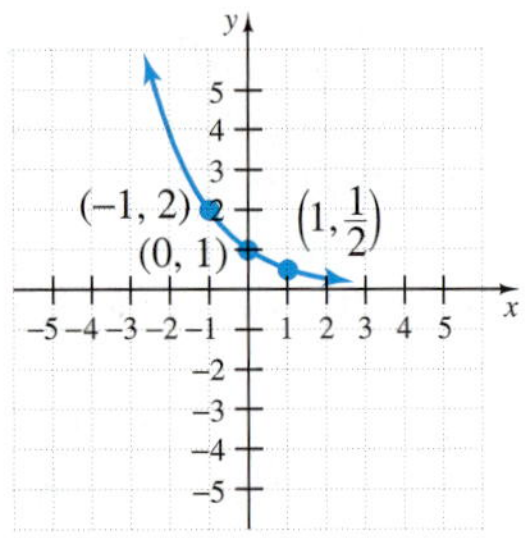


**D**

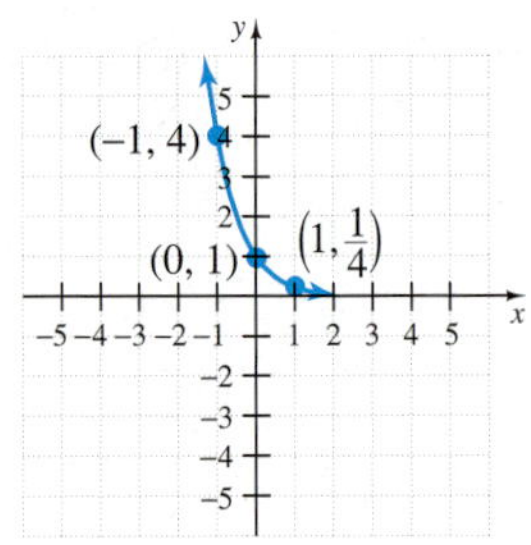


*Solve. Unless otherwise indicated, round results to one decimal place. See Example 6.*

**39.** One type of uranium has a daily radioactive decay rate of 0.4%. If 30 pounds of this uranium is available today, find how much will still remain after 50 days. Use $y = 30(2.7)^{-0.004t}$, and let $t$ be 50.

**40.** The nuclear waste from an atomic energy plant decays at a rate of 3% each century. If 150 pounds of nuclear waste is disposed of, find how much of it will still remain after 10 centuries. Use $y = 150(2.7)^{-0.03t}$, and let $t$ be 10.

**41.** The size of the rat population of a wharf area grows at a rate of 8% monthly. If there are 200 rats in January, find how many rats (rounded to the nearest whole) should be expected by next January. Use $y = 200(2.7)^{0.08t}$.

**42.** National Park Service personnel are trying to increase the size of the bison population of Theodore Roosevelt National Park. If 260 bison currently live in the park, and if the population's rate of growth is 2.5% annually, find how many bison (rounded to the nearest whole) there should be in 10 years. Use $y = 260(2.7)^{0.025t}$.

**43.** A rare isotope of a nuclear material is very unstable, decaying at a rate of 15% each second. Find how much isotope remains 10 seconds after 5 grams of the isotope is created. Use $y = 5(2.7)^{-0.15t}$.

**44.** An accidental spill of 75 grams of radioactive material in a local stream has led to the presence of radioactive debris decaying at a rate of 4% each day. Find how much debris still remains after 14 days. Use $y = 75(2.7)^{-0.04t}$.

**45.** Mexico City is growing at a rate of 0.7% annually. If there were 15,525,000 residents of Mexico City in 1994, find how many (to the nearest ten-thousand) are living in the city in 2000. Use $y = 15{,}525{,}000(2.7)^{0.007t}$.

**46.** An unusually wet spring has caused the size of the Cape Cod mosquito population to increase by 8% each day. If an estimated 200,000 mosquitoes are on Cape Cod on May 12, find how many thousands of mosquitoes will inhabit the Cape on May 25. Use $y = 200{,}000(2.7)^{0.08t}$.

*Solve. Use* $A = P\left(1 + \frac{r}{n}\right)^{nt}$. *Round answers to two decimal places. See Example 5.*

**47.** Find the amount Erica owes at the end of 3 years if \$6000 is loaned to her at a rate of 8% compounded monthly.

**48.** Find the amount owed at the end of 5 years if \$3000 is loaned at a rate of 10% compounded quarterly.

**49.** Find the total amount Janina has in a college savings account if \$2000 was invested and earned 6% compounded semiannually for 12 years.

**50.** Find the amount accrued if \$500 is invested and earns 7% compounded monthly for 4 years.

*Use a graphing calculator to solve. Estimate each result to two decimal places.*

**51.** Verify the results of Exercise 39.

**52.** From Exercise 39, estimate the number of pounds of uranium that will be available after 100 days.

**53.** From Exercise 39, estimate the number of pounds of uranium that will be available after 120 days.

**54.** Verify the results of Exercise 44.

**55.** From Exercise 44, estimate the amount of debris that remains after 10 days.

**56.** From Exercise 44, estimate the amount of debris that remains after 20 days.

*Solve.*

**57.** The world population is currently growing at a rate of 1.32% annually. In 1998, the midyear population of the world was 5,926,466,814 people. Predict the midyear world population (to the nearest million) in 2005. Use $y = 5{,}926{,}466{,}814(2.7)^{0.0132t}$, where $t$ is the number of years after 1998. (*Source:* Based on data from the U.S. Bureau of the Census, International Data Base)

**58.** Retail revenue from shopping on the Internet is expected to grow at a rate of 64% per year. In 1997, a total of \$2.4 billion in revenue was collected through Internet retail sales. To make the following predictions, use $y = 2.4(1.64)^t$, where $t$ is the number of years after 1997. (*Source:* Based on data from Forrester Research Inc.)

**a.** What level of retail revenues from Internet shopping is expected in 2001?

**b.** Predict the level of Internet shopping revenues in 2010.

**59.** Carbon dioxide ($CO_2$) is a greenhouse gas that contributes to global warming. Due to the combustion of fossil fuels, the amount of $CO_2$ in Earth's atmosphere has been increasing by 0.4% annually over the past century. In 1994, the concentration of $CO_2$ in the atmosphere was 358 parts per million by volume. To make the following predictions, use $y = 358(1.004)^t$, where $t$ is the number of years after 1994. (*Source:* Based on data from the United Nations Environment Programme's Information Unit for Conventions)

**a.** Predict the concentration of $CO_2$ in the atmosphere in the year 2004.

**b.** Predict the concentration of $CO_2$ in the atmosphere in the year 2025.

*The formula* $y = 6.052(1.378)^x$ *gives the number of cellular phone users y (in millions) in the United States for the years 1991 through 1998. In this formula, x = 0 corresponds to 1991, x = 1 corresponds to 1992, and so on. Use this formula to solve exercises 60 and 61.*

**60.** Use this model to predict the number of cellular phone users in the year 2005.

**61.** Use this model to predict the number of cellular phone users in the year 2008.

## REVIEW EXERCISES

*Solve each equation. See Sections 2.1 and 5.8.*

**62.** $5x - 2 = 18$

**63.** $3x - 7 = 11$

**64.** $3x - 4 = 3(x + 1)$

**65.** $2 - 6x = 6(1 - x)$

**66.** $x^2 + 6 = 5x$

**67.** $18 = 11x - x^2$

*By inspection, find the value for x that makes each statement true.*

**68.** $2^x = 8$

**69.** $3^x = 9$

**70.** $5^x = \frac{1}{5}$

**71.** $4^x = 1$

## 10.4 LOGARITHMIC FUNCTIONS

**OBJECTIVES**

1. Write exponential equations with logarithmic notation and write logarithmic equations with exponential notation.
2. Solve logarithmic equations by using exponential notation.
3. Identify and graph logarithmic functions.

**1** Since the exponential function $f(x) = 2^x$ is a one-to-one function, it has an inverse. We can create a table of values for $f^{-1}$ by switching the coordinates in the accompanying table of values for $f(x) = 2^x$.

| $x$ | $y = f(x)$ |
|---|---|
| $-3$ | $\frac{1}{8}$ |
| $-2$ | $\frac{1}{4}$ |
| $-1$ | $\frac{1}{2}$ |
| 0 | 1 |
| 1 | 2 |
| 2 | 4 |
| 3 | 8 |

| $x$ | $y = f^{-1}(x)$ |
|---|---|
| $\frac{1}{8}$ | $-3$ |
| $\frac{1}{4}$ | $-2$ |
| $\frac{1}{2}$ | $-1$ |
| 1 | 0 |
| 2 | 1 |
| 4 | 2 |
| 8 | 3 |

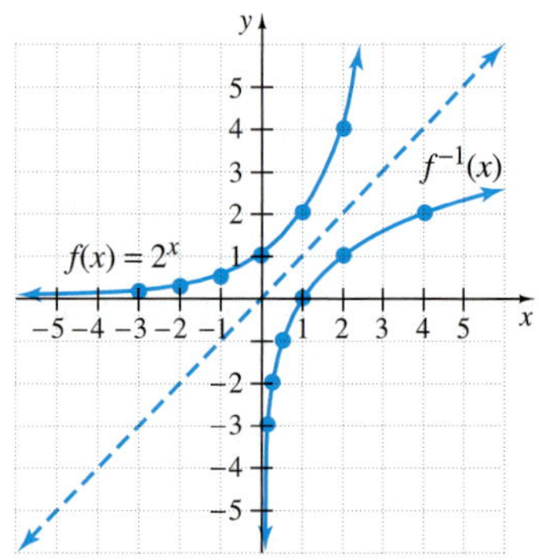


The graphs of $f(x)$ and its inverse are shown above. Notice that the graphs of $f$ and $f^{-1}$ are symmetric about the line $y = x$, as expected.

Now we would like to be able to write an equation for $f^{-1}$. To do so, we follow the steps for finding an inverse.

$$f(x) = 2^x$$

**Step 1:** Replace $f(x)$ by $y$. $\quad y = 2^x$

**Step 2:** Interchange $x$ and $y$. $\quad x = 2^y$

**Step 3:** Solve for $y$.

At this point, we are stuck. To solve this equation for $y$, a new notation, the **logarithmic notation,** is needed. The symbol $\log_b x$ means "the power to which $b$ is raised in order to produce a result of $x$."

$$\log_b x = y \quad \text{means} \quad b^y = x$$

We say that $\log_b x$ is "the logarithm of $x$ to the base $b$" or "the log of $x$ to the base $b$."

### LOGARITHMIC DEFINITION

If $b > 0$ and $b \neq 1$, then

$$y = \log_b x \text{ means } x = b^y$$

for every $x > 0$ and every real number $y$.

Before returning to the function $x = 2^y$ and solving it for $y$ in terms of $x$, let's practice using the new notation $\log_b x$.

It is important to be able to write exponential equations with logarithmic notation, and vice versa. The following table shows examples of both forms.

**HELPFUL HINT**
Notice that a *logarithm* is an *exponent*. In other words, $\log_3 9$ is the *power* that we raise 3 to in order to get 9.

| *Logarithmic Equation* | *Corresponding Exponential Equation* |
|---|---|
| $\log_3 9 = 2$ | $3^2 = 9$ |
| $\log_6 1 = 0$ | $6^0 = 1$ |
| $\log_2 8 = 3$ | $2^3 = 8$ |
| $\log_4 \frac{1}{16} = -2$ | $4^{-2} = \frac{1}{16}$ |
| $\log_8 2 = \frac{1}{3}$ | $8^{1/3} = 2$ |

**Example 1** Write as an exponential equation.

**a.** $\log_5 25 = 2$ **b.** $\log_6 \frac{1}{6} = -1$ **c.** $\log_2 \sqrt{2} = \frac{1}{2}$

*Solution* **a.** $\log_5 25 = 2$ means $5^2 = 25$

**b.** $\log_6 \frac{1}{6} = -1$ means $6^{-1} = \frac{1}{6}$

**c.** $\log_2 \sqrt{2} = \frac{1}{2}$ means $2^{1/2} = \sqrt{2}$

**Example 2** Write as a logarithmic equation.

**a.** $9^3 = 729$ **b.** $6^{-2} = \frac{1}{36}$ **c.** $5^{1/3} = \sqrt[3]{5}$

*Solution* **a.** $9^3 = 729$ means $\log_9 729 = 3$

**b.** $6^{-2} = \frac{1}{36}$ means $\log_6 \frac{1}{36} = -2$

**c.** $5^{1/3} = \sqrt[3]{5}$ means $\log_5 \sqrt[3]{5} = \frac{1}{3}$

**Example 3** Find the value of each logarithmic expression.

**a.** $\log_4 16$ **b.** $\log_{10} \frac{1}{10}$ **c.** $\log_9 3$

*Solution* **a.** $\log_4 16 = 2$ because $4^2 = 16$ **b.** $\log_{10} \frac{1}{10} = -1$ because $10^{-1} = \frac{1}{10}$

**c.** $\log_9 3 = \frac{1}{2}$ because $9^{1/2} = \sqrt{9} = 3$

**HELPFUL HINT**

Another method for evaluating logarithms such as those in Example 3 is to set the expression equal to $x$ and then write them in exponential form to find $x$. For example:

**a.** $\log_4 16 = x$ means $4^x = 16$. Since $4^2 = 16$, $x = 2$ or $\log_4 16 = 2$.

**b.** $\log_{10} \frac{1}{10} = x$ means $10^x = \frac{1}{10}$. Since $10^{-1} = \frac{1}{10}$, $x = -1$ or $\log_{10} \frac{1}{10} = -1$.

**c.** $\log_9 3 = x$ means $9^x = 3$. Since $9^{1/2} = 3$, $x = \frac{1}{2}$ or $\log_9 3 = \frac{1}{2}$.

2 The ability to interchange the logarithmic and exponential forms of a statement is often the key to solving logarithmic equations.

**Example 4** Solve each equation for $x$.

**a.** $\log_4 \frac{1}{4} = x$ **b.** $\log_5 x = 3$ **c.** $\log_x 25 = 2$

**d.** $\log_3 1 = x$ **e.** $\log_b 1 = x$

*Solution* **a.** $\log_4 \frac{1}{4} = x$ means $4^x = \frac{1}{4}$. Solve $4^x = \frac{1}{4}$ for $x$.

$$4^x = \frac{1}{4}$$

$$4^x = 4^{-1}$$

Since the bases are the same, by the uniqueness of $b^x$, we have that

$$x = -1$$

The solution is $-1$. To check, see that $\log_4 \frac{1}{4} = -1$, since $4^{-1} = \frac{1}{4}$.

**b.** $\log_5 x = 3$ means $5^3 = x$ or

$$x = 125$$

The solution is 125.

**c.** $\log_x 25 = 2$ means $x^2 = 25$ and $x > 0$ and $x \neq 1$.

$$x = 5$$

Even though $(-5)^2 = 25$, the base $b$ of a logarithm must be positive. The solution is 5.

**d.** $\log_3 1 = x$ means $3^x = 1$. Either solve this equation by inspection or solve by writing 1 as $3^0$ as shown.

$$3^x = 3^0 \quad \text{Write 1 as } 3^0.$$

$$x = 0 \quad \text{Apply the uniqueness of } b^x.$$

The solution is 0.

**e.** $\log_b 1 = x$ means $b^x = 1$ and $b > 0$ and $b \neq 1$.

$$b^x = b^0 \quad \text{Write 1 as } b^0.$$

$$x = 0 \quad \text{Apply the uniqueness of } b^x.$$

The solution is 0.

In Example 4**e** we proved an important property of logarithms. That is, $\log_b 1$ is always 0. This property as well as two important others are given next.

**PROPERTIES OF LOGARITHMS**

If $b$ is a real number, $b > 0$, and $b \neq 1$, then

1. $\log_b 1 = 0$
2. $\log_b b^x = x$
3. $b^{\log_b x} = x$

To see that $\log_b b^x = x$, change the logarithmic form to exponential form. Then, $\log_b b^x = x$ means $b^x = b^x$. In exponential form, the statement is true, so in logarithmic form, the statement is also true.

**Example 5** Simplify.

**a.** $\log_3 3^2$ **b.** $\log_7 7^{-1}$ **c.** $5^{\log_5 3}$ **d.** $2^{\log_2 6}$

*Solution*

**a.** From Property 2, $\log_3 3^2 = 2$.

**b.** From Property 2, $\log_7 7^{-1} = -1$.

**c.** From Property 3, $5^{\log_5 3} = 3$.

**d.** From Property 3, $2^{\log_2 6} = 6$.

3 Let us now return to the function $f(x) = 2^x$ and write an equation for its inverse, $f^{-1}(x)$. Recall our earlier work.

$$f(x) = 2^x$$

**Step 1:** Replace $f(x)$ by $y$. $\quad y = 2^x$

**Step 2:** Interchange $x$ and $y$. $\quad x = 2^y$

Having gained proficiency with the notation $\log_b x$, we can now complete the steps for writing the inverse equation.

**Step 3:** Solve for $y$. $\quad y = \log_2 x$

**Step 4:** Replace $y$ with $f^{-1}(x)$. $\quad f^{-1}(x) = \log_2 x$

Thus, $f^{-1}(x) = \log_2 x$ defines a function that is the inverse function of the function $f(x) = 2^x$. The function $f^{-1}(x)$ or $y = \log_2 x$ is called a **logarithmic function.**

**LOGARITHMIC FUNCTION**

If $x$ is a positive real number, $b$ is a constant positive real number, and $b$ is not 1, then a **logarithmic function** is a function that can be defined by

$$f(x) = \log_b x$$

The domain of $f$ is the set of positive real numbers, and the range of $f$ is the set of real numbers.

We can explore logarithmic functions by graphing them.

**Example 6** Graph the logarithmic function $y = \log_2 x$.

*Solution* First we write the equation with exponential notation as $2^y = x$. Then we find some ordered pair solutions that satisfy this equation. Finally, we plot the points and connect them with a smooth curve. The domain of this function is $(0, \infty)$, and the range is all real numbers.

Since $x = 2^y$ is solved for $x$, we choose $y$-values and compute corresponding $x$-values.

If $y = 0$, $x = 2^0 = 1$

If $y = 1$, $x = 2^1 = 2$

If $y = 2$, $x = 2^2 = 4$

If $y = -1$, $x = 2^{-1} = \frac{1}{2}$

| $x = 2^y$ | $y$ |
|---|---|
| 1 | 0 |
| 2 | 1 |
| 4 | 2 |
| $\frac{1}{2}$ | $-1$ |

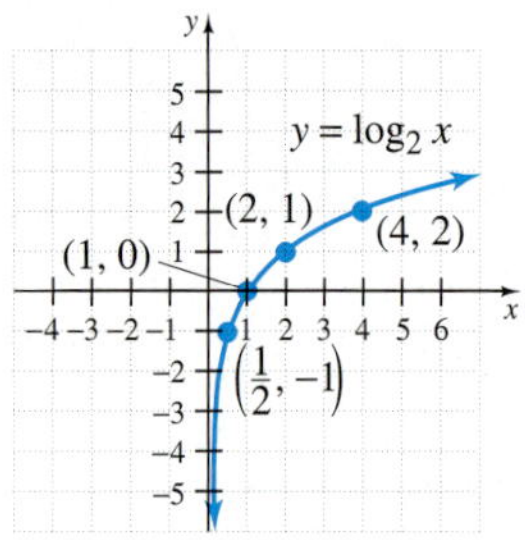


**Example 7** Graph the logarithmic function $f(x) = \log_{1/3} x$.

*Solution* Replace $f(x)$ with $y$, and write the result with exponential notation.

$$f(x) = \log_{1/3} x$$

$$y = \log_{1/3} x \quad \text{Replace } f(x) \text{ with } y.$$

$$\left(\frac{1}{3}\right)^y = x \quad \text{Write in exponential form.}$$

Now we can find ordered pair solutions that satisfy $\left(\frac{1}{3}\right)^y = x$, plot these points, and connect them with a smooth curve.

If $y = 0$, $x = \left(\frac{1}{3}\right)^0 = 1$

If $y = 1$, $x = \left(\frac{1}{3}\right)^1 = \frac{1}{3}$

If $y = -1$, $x = \left(\frac{1}{3}\right)^{-1} = 3$

If $y = -2$, $x = \left(\frac{1}{3}\right)^{-2} = 9$

| $x = \left(\frac{1}{3}\right)^y$ | $y$ |
|---|---|
| 1 | 0 |
| $\frac{1}{3}$ | 1 |
| 3 | $-1$ |
| 9 | $-2$ |

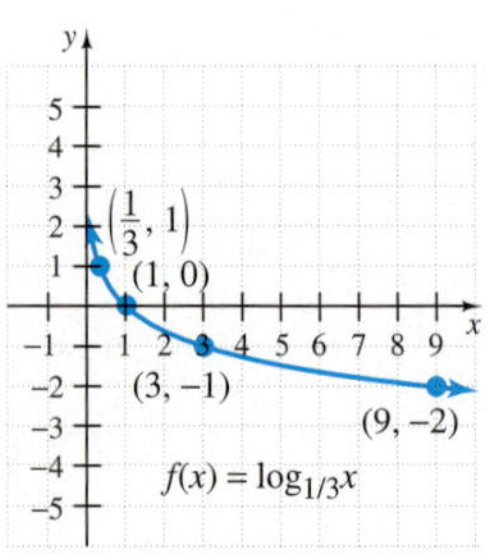


The domain of this function is $(0, \infty)$, and the range is the set of all real numbers.

The following figures summarize characteristics of logarithmic functions.

$$f(x) = \log_b x,\ b > 0,\ b \neq 1$$

- one-to-one function
- $x$-intercept $(1, 0)$
- no $y$-intercept
- domain: $(0, \infty)$
- range: $(-\infty, \infty)$

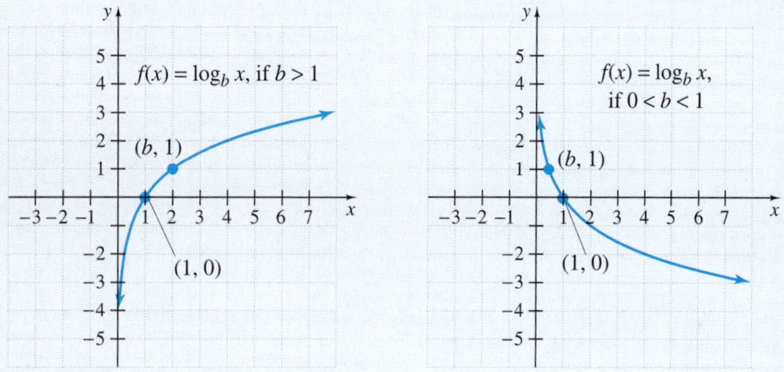


## SPOTLIGHT ON DECISION MAKING

Suppose you are the Webmaster for a small but growing company. One of your duties is to ensure that your company's newly established Web site can adequately handle the number of visitors to it. You decide to find a mathematical model for recent Web site usage statistics to help predict future numbers of visitors. Ultimately, you would like to use this model to predict when your Web site's server capacity must be expanded.

The first step in finding a model for the usage statistics is to decide what type of mathematical model to use: linear, quadratic, exponential, or logarithmic. The graph shows the number of visitors to your company's Web site in each of the first five months since it was established. Use the graph to decide which type of mathematical model to use. Explain your reasoning.

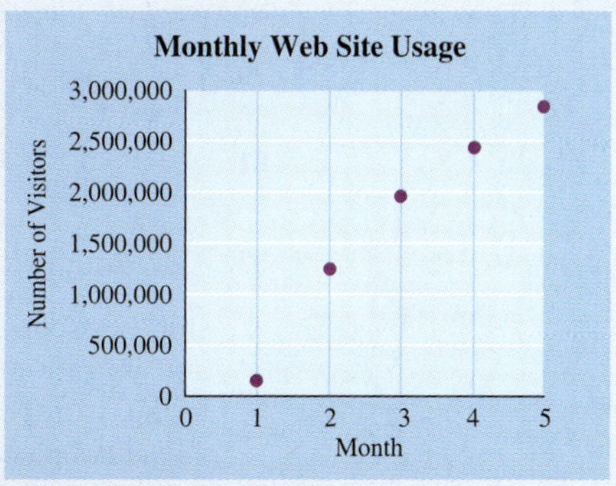


| Month | 1 | 2 | 3 | 4 | 5 |
|---|---|---|---|---|---|
| Visitors | 166,511 | 1,320,978 | 1,996,298 | 2,475,445 | 2,847,100 |

## Exercise Set 10.4

*Write each as an exponential equation. See Example 1.*

1. $\log_6 36 = 2$
2. $\log_2 32 = 5$
3. $\log_3 \frac{1}{27} = -3$
4. $\log_5 \frac{1}{25} = -2$
5. $\log_{10} 1000 = 3$
6. $\log_{10} 10 = 1$
7. $\log_e x = 4$
8. $\log_e \frac{1}{e} = -1$
9. $\log_e \frac{1}{e^2} = -2$
10. $\log_e y = 7$
11. $\log_7 \sqrt{7} = \frac{1}{2}$
12. $\log_{11} \sqrt[4]{11} = \frac{1}{4}$

*Write each as a logarithmic equation. See Example 2.*

13. $2^4 = 16$
14. $5^3 = 125$
15. $10^2 = 100$
16. $10^4 = 10{,}000$
17. $e^3 = x$
18. $e^5 = y$
19. $10^{-1} = \frac{1}{10}$
20. $10^{-2} = \frac{1}{100}$
21. $4^{-2} = \frac{1}{16}$
22. $3^{-4} = \frac{1}{81}$
23. $5^{1/2} = \sqrt{5}$
24. $4^{1/3} = \sqrt[3]{4}$

*Find the value of each logarithmic expression. See Example 3.*

25. $\log_2 8$
26. $\log_3 9$
27. $\log_3 \frac{1}{9}$
28. $\log_2 \frac{1}{32}$
29. $\log_{25} 5$
30. $\log_8 \frac{1}{2}$
31. $\log_{1/2} 2$
32. $\log_{2/3} \frac{4}{9}$
33. $\log_7 1$
34. $\log_9 9$
35. $\log_2 2^4$
36. $\log_6 6^{-2}$
37. $\log_{10} 100$
38. $\log_{10} \frac{1}{10}$
39. $3^{\log_3 5}$
40. $5^{\log_5 7}$
41. $\log_3 81$
42. $\log_2 16$
43. $\log_4 \frac{1}{64}$
44. $\log_3 \frac{1}{9}$
45. Explain why negative numbers are not included as logarithmic bases.
46. Explain why 1 is not included as a logarithmic base.

*Solve each equation for x. See Example 4.*

47. $\log_3 9 = x$
48. $\log_2 8 = x$
49. $\log_3 x = 4$
50. $\log_2 x = 3$
51. $\log_x 49 = 2$
52. $\log_x 8 = 3$
53. $\log_2 \frac{1}{8} = x$
54. $\log_3 \frac{1}{81} = x$
55. $\log_3 \frac{1}{27} = x$
56. $\log_5 \frac{1}{125} = x$
57. $\log_8 x = \frac{1}{3}$
58. $\log_9 x = \frac{1}{2}$
59. $\log_4 16 = x$
60. $\log_2 16 = x$
61. $\log_{3/4} x = 3$
62. $\log_{2/3} x = 2$
63. $\log_x 100 = 2$
64. $\log_x 27 = 3$

*Simplify. See Example 5.*

65. $\log_5 5^3$
66. $\log_6 6^2$
67. $2^{\log_2 3}$
68. $7^{\log_7 4}$
69. $\log_9 9$
70. $\log_8 (8)^{-1}$

*Graph each logarithmic function. Label any intercepts. See Examples 6 and 7.*

71. $y = \log_3 x$
72. $y = \log_2 x$
73. $f(x) = \log_{1/4} x$
74. $f(x) = \log_{1/2} x$
75. $f(x) = \log_5 x$
76. $f(x) = \log_6 x$
77. $f(x) = \log_{1/6} x$
78. $f(x) = \log_{1/5} x$

*Graph each function and its inverse function on the same set of axes. Label any intercepts.*

79. $y = 4^x$; $y = \log_4 x$
80. $y = 3^x$; $y = \log_3 x$
81. $y = \left(\frac{1}{3}\right)^x$; $y = \log_{1/3} x$
82. $y = \left(\frac{1}{2}\right)^x$; $y = \log_{1/2} x$
83. The formula $\log_{10}(1 - k) = \frac{-0.3}{H}$ models the relationship between the half-life $H$ of a radioactive material and its rate of decay $k$. Find the rate of decay of the iodine isotope I-131 if its half-life is 8 days. Round to 4 decimal places.
84. Explain why the graph of the function $y = \log_b x$ contains the point $(1, 0)$ no matter what $b$ is.
85. $\text{Log}_3 10$ is between which two integers? Explain your answer.

## REVIEW EXERCISES

*Simplify each rational expression. See Section 6.1.*

**86.** $\dfrac{x+3}{3+x}$ **87.** $\dfrac{x-5}{5-x}$

**88.** $\dfrac{x^2-8x+16}{2x-8}$ **89.** $\dfrac{x^2-3x-10}{2+x}$

*Add or subtract as indicated. See Section 6.2.*

**90.** $\dfrac{2}{x}+\dfrac{3}{x^2}$ **91.** $\dfrac{3x}{x+3}+\dfrac{9}{x+3}$

**92.** $\dfrac{m^2}{m+1}-\dfrac{1}{m+1}$ **93.** $\dfrac{5}{y+1}-\dfrac{4}{y-1}$

# 10.5 PROPERTIES OF LOGARITHMS

▶ **OBJECTIVES**

CD-ROM SSM SSG Video

1. Use the product property of logarithms.
2. Use the quotient property of logarithms.
3. Use the power property of logarithms.
4. Use the properties of logarithms together.

In the previous section we explored some basic properties of logarithms. We now introduce and explore additional properties. Because a logarithm is an exponent, logarithmic properties are just restatements of exponential properties.

**1** The first of these properties is called the **product property of logarithms,** because it deals with the logarithm of a product.

### PRODUCT PROPERTY OF LOGARITHMS

If $x$, $y$, and $b$ are positive real numbers and $b \neq 1$, then

$$\log_b xy = \log_b x + \log_b y$$

To prove this, let $\log_b x = M$ and $\log_b y = N$. Now write each logarithm with exponential notation.

$$\log_b x = M \quad \text{is equivalent to} \quad b^M = x$$

$$\log_b y = N \quad \text{is equivalent to} \quad b^N = y$$

Multiply the left sides and the right sides of the exponential equations, and we have that

$$xy = (b^M)(b^N) = b^{M+N}$$

If we write the equation $xy = b^{M+N}$ in equivalent logarithmic form, we have

$$\log_b xy = M + N$$

But since $M = \log_b x$ and $N = \log_b y$, we can write

$$\log_b xy = \log_b x + \log_b y \qquad \text{Let } M = \log_b x \text{ and } N = \log_b y.$$

In other words, the logarithm of a product is the sum of the logarithms of the factors. This property is sometimes used to simplify logarithmic expressions.

In the examples that follow, assume that variables represent positive numbers.

**Example 1** Write each sum as a single logarithm.

**a.** $\log_{11} 10 + \log_{11} 3$ **b.** $\log_3 \frac{1}{2} + \log_3 12$ **c.** $\log_2 (x + 2) + \log_2 x$

*Solution* In each case, both terms have a common logarithmic base.

**a.** $\log_{11} 10 + \log_{11} 3 = \log_{11} (10 \cdot 3)$ Apply the product property.

$= \log_{11} 30$

**b.** $\log_3 \frac{1}{2} + \log_3 12 = \log_3 \left(\frac{1}{2} \cdot 12\right) = \log_3 6$

**c.** $\log_2 (x + 2) + \log_2 x = \log_2 [(x + 2) \cdot x] = \log_2 (x^2 + 2x)$

**HELPFUL HINT**
Check your logarithm properties. Make sure you understand that $\log_2 (x + 2)$ *is not* $\log_2 x + \log_2 2$.

**2** The second property is the **quotient property of logarithms.**

**QUOTIENT PROPERTY OF LOGARITHMS**

If $x$, $y$, and $b$ are positive real numbers and $b \neq 1$, then

$$\log_b \frac{x}{y} = \log_b x - \log_b y$$

The proof of the quotient property of logarithms is similar to the proof of the product property. Notice that the quotient property says that the logarithm of a quotient is the difference of the logarithms of the dividend and divisor.

**Example 2** Write each difference as a single logarithm.

**a.** $\log_{10} 27 - \log_{10} 3$ **b.** $\log_5 8 - \log_5 x$ **c.** $\log_3 (x^2 + 5) - \log_3 (x^2 + 1)$

*Solution* All terms have a common logarithmic base.

**a.** $\log_{10} 27 - \log_{10} 3 = \log_{10} \frac{27}{3} = \log_{10} 9$

**b.** $\log_5 8 - \log_5 x = \log_5 \frac{8}{x}$

**c.** $\log_3 (x^2 + 5) - \log_3 (x^2 + 1) = \log_3 \frac{x^2 + 5}{x^2 + 1}$ Apply the quotient property.

3 The third and final property we introduce is the **power property of logarithms.**

**POWER PROPERTY OF LOGARITHMS**

If $x$ and $b$ are positive real numbers, $b \neq 1$, and $r$ is a real number, then

$$\log_b x^r = r \log_b x$$

**Example 3** Use the power property to rewrite each expression.

**a.** $\log_5 x^3$ **b.** $\log_4 \sqrt{2}$

*Solution* **a.** $\log_5 x^3 = 3 \log_5 x$

**b.** $\log_4 \sqrt{2} = \log_4 2^{1/2} = \frac{1}{2} \log_4 2$

4 Many times we must use more than one property of logarithms to simplify a logarithmic expression.

**Example 4** Write as a single logarithm.

**a.** $2 \log_5 3 + 3 \log_5 2$ **b.** $3 \log_9 x - \log_9 (x + 1)$ **c.** $\log_4 25 + \log_4 3 - \log_4 5$

*Solution* In each case, both terms have a common logarithmic base.

**a.**
$$\begin{aligned} 2 \log_5 3 + 3 \log_5 2 &= \log_5 3^2 + \log_5 2^3 && \text{Apply the power property.} \\ &= \log_5 9 + \log_5 8 \\ &= \log_5 (9 \cdot 8) && \text{Apply the product property.} \\ &= \log_5 72 \end{aligned}$$

**b.**
$$\begin{aligned} 3 \log_9 x - \log_9 (x + 1) &= \log_9 x^3 - \log_9 (x + 1) && \text{Apply the power property.} \\ &= \log_9 \frac{x^3}{x + 1} && \text{Apply the quotient property.} \end{aligned}$$

**c.** Use both the product and quotient properties.

$$\begin{aligned} \log_4 25 + \log_4 3 - \log_4 5 &= \log_4 (25 \cdot 3) - \log_4 5 && \text{Apply the product property.} \\ &= \log_4 75 - \log_4 5 && \text{Simplify.} \\ &= \log_4 \frac{75}{5} && \text{Apply the quotient property.} \\ &= \log_4 15 && \text{Simplify.} \end{aligned}$$

**Example 5** Write each expression as sums or differences of multiples of logarithms.

**a.** $\log_3 \frac{5 \cdot 7}{4}$ **b.** $\log_2 \frac{x^5}{y^2}$

*Solution* **a.** $\log_3 \frac{5 \cdot 7}{4} = \log_3 (5 \cdot 7) - \log_3 4$ Apply the quotient property.

$= \log_3 5 + \log_3 7 - \log_3 4$ Apply the product property.

**b.** $\log_2 \frac{x^5}{y^2} = \log_2 (x^5) - \log_2 (y^2)$ Apply the quotient property.

$= 5 \log_2 x - 2 \log_2 y$ Apply the power property.

**HELPFUL HINT**

Notice that we are not able to simplify further a logarithmic expression such as $\log_5 (2x - 1)$. None of the basic properties gives a way to write the logarithm of a difference in some equivalent form.

**Example 6** If $\log_b 2 = 0.43$ and $\log_b 3 = 0.68$, use the properties of logarithms to evaluate.

**a.** $\log_b 6$ **b.** $\log_b 9$ **c.** $\log_b \sqrt{2}$

*Solution* **a.** $\log_b 6 = \log_b (2 \cdot 3)$ Write 6 as $2 \cdot 3$.

$= \log_b 2 + \log_b 3$ Apply the product property.

$= 0.43 + 0.68$ Substitute given values.

$= 1.11$ Simplify.

**b.** $\log_b 9 = \log_b 3^2$ Write 9 as $3^2$.

$= 2 \log_b 3$

$= 2(0.68)$ Substitute 0.68 for $\log_b 3$.

$= 1.36$ Simplify.

**c.** First, recall that $\sqrt{2} = 2^{1/2}$. Then

$\log_b \sqrt{2} = \log_b 2^{1/2}$ Write $\sqrt{2}$ as $2^{1/2}$.

$= \frac{1}{2} \log_b 2$ Apply the power property.

$= \frac{1}{2}(0.43)$ Substitute the given value.

$= 0.215$ Simplify.

A summary of the basic properties of logarithms that we have developed so far is given next.

### PROPERTIES OF LOGARITHMS

If $x$, $y$, and $b$ are positive real numbers, $b \neq 1$, and $r$ is a real number, then

1. $\log_b 1 = 0$
2. $\log_b b^x = x$
3. $b^{\log_b x} = x$
4. $\log_b xy = \log_b x + \log_b y$ — Product property.
5. $\log_b \dfrac{x}{y} = \log_b x - \log_b y$ — Quotient property.
6. $\log_b x^r = r \log_b x$ — Power property.

## SPOTLIGHT ON DECISION MAKING

Suppose you are a quality assurance inspector for an electronics manufacturer. Your department has conducted reliability studies of a new model of CD player. Your studies show that the CD player's reliability can be described by the exponential function $R(t) = 2.7^{-(1/3)t}$, where the reliability $R$ is the probability that the CD player is still working $t$ years after it is manufactured.

The marketing department asks for your input in choosing a warranty period for the CD player. Popular warranty periods for similar competing CD players are 1 year, 2 years, and 3 years. Using the graph of the reliability for this CD player, which warranty period would you recommend? Explain your reasoning. What other factors would you want to consider?

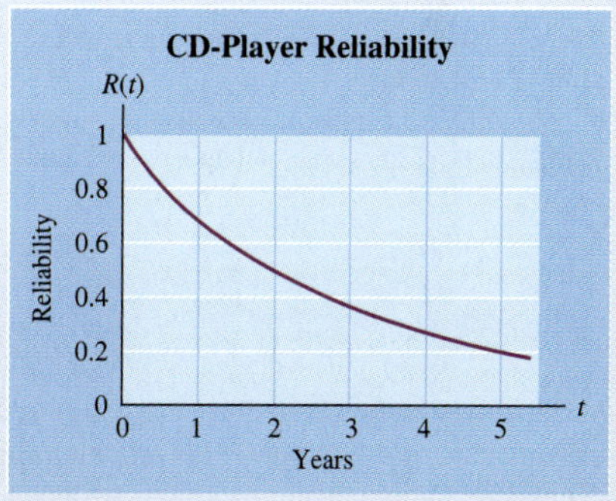


## Exercise Set 10.5

*Write each sum as the logarithm of a single expression. Assume that variables represent positive numbers. See Example 1.*

**1.** $\log_5 2 + \log_5 7$

**2.** $\log_3 8 + \log_3 4$

**3.** $\log_4 9 + \log_4 x$

**4.** $\log_2 x + \log_2 y$

**5.** $\log_{10} 5 + \log_{10} 2 + \log_{10} (x^2 + 2)$

**6.** $\log_6 3 + \log_6 (x + 4) + \log_6 5$

*Write each as the logarithm of a single expression. Assume that variables represent positive numbers. See Examples 2 and 4.*

**7.** $\log_5 12 - \log_5 4$

**8.** $\log_7 20 - \log_7 4$

**9.** $\log_2 x - \log_2 y$

**10.** $\log_3 12 - \log_3 z$

**11.** $\log_4 2 + \log_4 10 - \log_4 5$

**12.** $\log_6 18 + \log_6 2 - \log_6 9$

*Use the power property to rewrite each expression. See Example 3.*

**13.** $\log_3 x^2$

**14.** $\log_2 x^5$

**15.** $\log_4 5^{-1}$

**16.** $\log_6 7^{-2}$

**17.** $\log_5 \sqrt{y}$

**18.** $\log_5 \sqrt[3]{x}$

*Write each as a single logarithm. Assume that variables represent positive numbers. See Example 4.*

**19.** $2 \log_2 5$

**20.** $3 \log_5 2$

**21.** $3 \log_5 x + 6 \log_5 z$

**22.** $2 \log_7 y + 6 \log_7 z$

**23.** $\log_{10} x - \log_{10} (x + 1) + \log_{10} (x^2 - 2)$

**24.** $\log_9 (4x) - \log_9 (x - 3) + \log_9 (x^3 + 1)$

**25.** $\log_4 5 + \log_4 7$

**26.** $\log_3 2 + \log_3 5$

**27.** $\log_3 8 - \log_3 2$

**28.** $\log_5 12 - \log_5 3$

**29.** $\log_7 6 + \log_7 3 - \log_7 4$

**30.** $\log_8 5 + \log_8 15 - \log_8 20$

**31.** $3 \log_4 2 + \log_4 6$

**32.** $2 \log_3 5 + \log_3 2$

**33.** $3 \log_2 x + \frac{1}{2} \log_2 x - 2 \log_2 (x + 1)$

**34.** $2 \log_5 x + \frac{1}{3} \log_5 x - 3 \log_5 (x + 5)$

**35.** $2 \log_8 x - \frac{2}{3} \log_8 x + 4 \log_8 x$

**36.** $5 \log_6 x - \frac{3}{4} \log_6 x + 3 \log_6 x$

*Write each expression as a sum or difference of multiples of logarithms. Assume that variables represent positive numbers. See Example 5.*

**37.** $\log_2 \frac{7 \cdot 11}{3}$

**38.** $\log_5 \frac{2 \cdot 9}{13}$

**39.** $\log_3 \frac{4y}{5}$

**40.** $\log_4 \frac{2}{9z}$

**41.** $\log_2 \frac{x^3}{y}$

**42.** $\log_5 \frac{x}{y^4}$

**43.** $\log_b \sqrt{7x}$

**44.** $\log_b \sqrt{\frac{3}{y}}$

**45.** $\log_7 \frac{5x}{4}$

**46.** $\log_9 \frac{7}{y}$

**47.** $\log_5 x^3(x + 1)$

**48.** $\log_2 y^3 z$

**49.** $\log_6 \frac{x^2}{x + 3}$

**50.** $\log_3 \frac{(x + 5)^2}{x}$

*If $\log_b 3 = 0.5$ and $\log_b 5 = 0.7$, evaluate the following. See Example 6. If necessary, round to three decimal places.*

**51.** $\log_b \frac{5}{3}$

**52.** $\log_b 25$

**53.** $\log_b 15$

**54.** $\log_b \frac{3}{5}$

**55.** $\log_b \sqrt[3]{5}$

**56.** $\log_b \sqrt[4]{3}$

*Answer the following true or false.*

**57.** $\log_2 x^3 = 3 \log_2 x$

**58.** $\log_3 (x + y) = \log_3 x + \log_3 y$

**59.** $\frac{\log_7 10}{\log_7 5} = \log_7 2$

**60.** $\log_7 \frac{14}{8} = \log_7 14 - \log_7 8$

**61.** $\frac{\log_7 x}{\log_7 y} = (\log_7 x) - (\log_7 y)$

**62.** $(\log_3 6) \cdot (\log_3 4) = \log_3 24$

*If $\log_b 2 = 0.43$ and $\log_b 3 = 0.68$, evaluate the following.*

**63.** $\log_b 8$

**64.** $\log_b 81$

**65.** $\log_b \frac{3}{9}$

**66.** $\log_b \frac{4}{32}$

**67.** $\log_b \sqrt{\frac{2}{3}}$

**68.** $\log_b \sqrt{\frac{3}{2}}$

## REVIEW EXERCISES

**69.** Graph the functions $y = 10^x$ and $y = \log_{10} x$ on the same set of axes. See Section 10.4.

*Evaluate each expression. See Section 10.4.*

**70.** $\log_{10} 100$

**71.** $\log_{10} \frac{1}{10}$

**72.** $\log_7 7^2$

**73.** $\log_7 \sqrt{7}$

## 10.6 COMMON LOGARITHMS, NATURAL LOGARITHMS, AND CHANGE OF BASE

CD-ROM SSM SSG Video

**OBJECTIVES**

1. Identify common logarithms and approximate them by calculator.
2. Evaluate common logarithms of powers of 10.
3. Identify natural logarithms and approximate them by calculator.
4. Evaluate natural logarithms of powers of $e$.
5. Use the change of base formula.

In this section we look closely at two particular logarithmic bases. These two logarithmic bases are used so frequently that logarithms to their bases are given special names. **Common logarithms** are logarithms to base 10. **Natural logarithms** are logarithms to base $e$, which we introduce in this section. The work in this section is based on the use of the calculator, which has both the common "log" [LOG] and the natural "log" [LN] keys.

1 Logarithms to base 10, common logarithms, are used frequently because our number system is a base 10 decimal system. The notation $\log x$ means the same as $\log_{10} x$.

**COMMON LOGARITHMS**

$$\log x \text{ means } \log_{10} x$$

**Example 1** Use a calculator to approximate log 7 to four decimal places.

*Solution* Press the following sequence of keys.

[7] [LOG] or [LOG] [7] [ENTER]

To four decimal places,

$$\log 7 \approx 0.8451$$

2 To evaluate the common log of a power of 10, a calculator is not needed. According to the property of logarithms,

$$\log_b b^x = x$$

It follows that if $b$ is replaced with 10, we have

$$\log 10^x = x$$

> **HELPFUL HINT**
> Remember that $\log 10^x$ means $\log_{10} 10^x = x$.

**Example 2** Find the exact value of each logarithm.

**a.** $\log 10$ **b.** $\log 1000$ **c.** $\log \frac{1}{10}$ **d.** $\log \sqrt{10}$

*Solution*

**a.** $\log 10 = \log 10^1 = 1$

**b.** $\log 1000 = \log 10^3 = 3$

**c.** $\log \frac{1}{10} = \log 10^{-1} = -1$

**d.** $\log \sqrt{10} = \log 10^{1/2} = \frac{1}{2}$

As we will soon see, equations containing common logs are useful models of many natural phenomena.

**Example 3** Solve $\log x = 1.2$ for $x$. Give an exact solution, and then approximate the solution to four decimal places.

*Solution* Remember that the base of a common log is understood to be 10.

> **HELPFUL HINT**
> The understood base is 10.

$$\log x = 1.2$$

$$10^{1.2} = x \quad \text{Write with exponential notation.}$$

The exact solution is $10^{1.2}$. To four decimal places, $x \approx 15.8489$.

The Richter scale measures the intensity, or magnitude, of an earthquake. The formula for the magnitude $R$ of an earthquake is $R = \log\left(\frac{a}{T}\right) + B$, where $a$ is the amplitude in micrometers of the vertical motion of the ground at the recording station, $T$ is the number of seconds between successive seismic waves, and $B$ is an adjustment factor that takes into account the weakening of the seismic wave as the distance increases from the epicenter of the earthquake.

**Example 4** **FINDING THE MAGNITUDE OF AN EARTHQUAKE**

Find an earthquake's magnitude on the Richter scale if a recording station measures an amplitude of 300 micrometers and 2.5 seconds between waves. Assume that $B$ is 4.2. Approximate the solution to the nearest tenth.

*Solution* Substitute the known values into the formula for earthquake intensity.

$$R = \log\left(\frac{a}{T}\right) + B \qquad \text{Richter scale formula}$$

$$= \log\left(\frac{300}{2.5}\right) + 4.2 \qquad \text{Let } a = 300, T = 2.5, \text{ and } B = 4.2.$$

$$= \log(120) + 4.2$$

$$\approx 2.1 + 4.2 \qquad \text{Approximate } \log 120 \text{ by } 2.1.$$

$$= 6.3$$

This earthquake had a magnitude of 6.3 on the Richter scale.

3 **Natural logarithms** are also frequently used, especially to describe natural events; hence the label "natural logarithm." Natural logarithms are logarithms to the base $e$, which is a constant approximately equal to 2.7183. The number $e$ is an irrational number, as is $\pi$. The notation $\log_e x$ is usually abbreviated to $\ln x$. (The abbreviation ln is read "el en.")

### NATURAL LOGARITHMS

$\ln x$ means $\log_e x$

The graph of $y = \ln x$ is shown to the right.

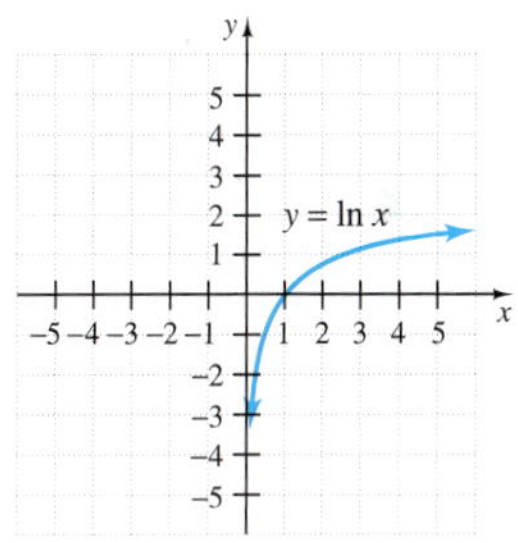


**Example 5** Use a calculator to approximate ln 8 to four decimal places.

*Solution* Press the following sequence of keys.

[8] [LN] or [LN] [8] [ENTER]

To four decimal places,

$$\ln 8 \approx 2.0794$$

4 As a result of the property $\log_b b^x = x$, we know that $\log_e e^x = x$, or $\ln e^x = x$.

**Example 6** Find the exact value of each natural logarithm.

**a.** $\ln e^3$ **b.** $\ln \sqrt[5]{e}$

*Solution* **a.** $\ln e^3 = 3$ **b.** $\ln \sqrt[5]{e} = \ln e^{1/5} = \frac{1}{5}$

**Example 7** Solve $\ln 3x = 5$. Give an exact solution, and then approximate the solution to four decimal places.

*Solution* Remember that the base of a natural logarithm is understood to be $e$.

**HELPFUL HINT**
The understood base is $e$.

$$\ln 3x = 5$$

$$e^5 = 3x \quad \text{Write with exponential notation.}$$

$$\frac{e^5}{3} = x \quad \text{Solve for } x.$$

The exact solution is $\frac{e^5}{3}$. To four decimal places,

$$x \approx 49.4711.$$

Recall from Section 10.3 the formula $A = P\left(1 + \frac{r}{n}\right)^{nt}$ for compound interest, where $n$ represents the number of compoundings per year. When interest is compounded continuously, the formula $A = Pe^{rt}$ is used, where $r$ is the annual interest rate and interest is compounded continuously for $t$ years.

**Example 8** **FINDING FINAL LOAN PAYMENT**

Find the amount owed at the end of 5 years if \$1600 is loaned at a rate of 9% compounded continuously.

*Solution* Use the formula $A = Pe^{rt}$, where

$$P = \$1600 \text{ (the size of the loan)}$$

$$r = 9\% = 0.09 \text{ (the rate of interest)}$$

$$t = 5 \text{ (the 5-year duration of the loan)}$$

$$A = Pe^{rt}$$

$$= 1600e^{0.09(5)} \quad \text{Substitute in known values.}$$

$$= 1600e^{0.45}$$

Now we can use a calculator to approximate the solution.

$$A \approx 2509.30$$

The total amount of money owed is \$2509.30.

5 Calculators are handy tools for approximating natural and common logarithms. Unfortunately, some calculators cannot be used to approximate logarithms to bases

other than $e$ or 10—at least not directly. In such cases, we use the change of base formula.

**CHANGE OF BASE**

If $a$, $b$, and $c$ are positive real numbers and neither $b$ nor $c$ is 1, then

$$\log_b a = \frac{\log_c a}{\log_c b}$$

**Example 9** Approximate $\log_5 3$ to four decimal places.

*Solution* Use the change of base property to write $\log_5 3$ as a quotient of logarithms to base 10.

$$\log_5 3 = \frac{\log 3}{\log 5}$$ Use the change of base property. In the change of base property, we let $a = 3$, $b = 5$, and $c = 10$.

$$\approx \frac{0.4771213}{0.69897}$$ Approximate logarithms by calculator.

$$\approx 0.6826062$$ Simplify by calculator.

To four decimal places, $\log_5 3 \approx 0.6826$.

## Exercise Set 10.6

*Use a calculator to approximate each logarithm to four decimal places. See Examples 1 and 5.*

1. $\log 8$
2. $\log 6$
3. $\log 2.31$
4. $\log 4.86$
5. $\ln 2$
6. $\ln 3$
7. $\ln 0.0716$
8. $\ln 0.0032$
9. $\log 12.6$
10. $\log 25.9$
11. $\ln 5$
12. $\ln 7$
13. $\log 41.5$
14. $\ln 41.5$
15. Use a calculator and try to approximate $\log 0$. Describe what happens and explain why.
16. Use a calculator and try to approximate $\ln 0$. Describe what happens and explain why.

*Find the exact value. See Examples 2 and 6.*

17. $\log 100$
18. $\log 10{,}000$
19. $\log\left(\frac{1}{1000}\right)$
20. $\log\left(\frac{1}{100}\right)$
21. $\ln e^2$
22. $\ln e^4$
23. $\ln \sqrt[4]{e}$
24. $\ln \sqrt[5]{e}$
25. $\log 10^3$
26. $\ln e^5$
27. $\ln e^2$
28. $\log 10^7$
29. $\log 0.0001$
30. $\log 0.001$
31. $\ln \sqrt{e}$
32. $\log \sqrt{10}$
33. Without using a calculator, explain which of $\log 50$ or $\ln 50$ must be larger.
34. Without using a calculator, explain which of $\log 50^{-1}$ or $\ln 50^{-1}$ must be larger.

*Solve each equation for x. Give an exact solution and a four-decimal-place approximation. See Examples 3 and 7.*

35. $\log x = 1.3$
36. $\log x = 2.1$
37. $\log 2x = 1.1$
38. $\log 3x = 1.3$
39. $\ln x = 1.4$
40. $\ln x = 2.1$
41. $\ln(3x - 4) = 2.3$
42. $\ln(2x + 5) = 3.4$
43. $\log x = 2.3$
44. $\log x = 3.1$
45. $\ln x = -2.3$
46. $\ln x = -3.7$
47. $\log(2x + 1) = -0.5$
48. $\log(3x - 2) = -0.8$
49. $\ln 4x = 0.18$
50. $\ln 3x = 0.76$

*Approximate each logarithm to four decimal places. See Example 9.*

51. $\log_2 3$
52. $\log_3 2$
53. $\log_{1/2} 5$
54. $\log_{1/3} 2$
55. $\log_4 9$
56. $\log_9 4$

**57.** $\log_3 \frac{1}{6}$

**58.** $\log_6 \frac{2}{3}$

**59.** $\log_8 6$

**60.** $\log_6 8$

*Use the formula* $R = \log\left(\frac{a}{T}\right) + B$ *to find the intensity R on the Richter scale of the earthquakes that fit the descriptions given. Round answers to one decimal place. See Example 4.*

**61.** Amplitude $a$ is 200 micrometers, time $T$ between waves is 1.6 seconds, and $B$ is 2.1.

**62.** Amplitude $a$ is 150 micrometers, time $T$ between waves is 3.6 seconds, and $B$ is 1.9.

**63.** Amplitude $a$ is 400 micrometers, time $T$ between waves is 2.6 seconds, and $B$ is 3.1.

**64.** Amplitude $a$ is 450 micrometers, time $T$ between waves is 4.2 seconds, and $B$ is 2.7.

*Use the formula* $A = Pe^{rt}$ *to solve. See Example 8.*

**65.** Find how much money Dana Jones has after 12 years if \$1400 is invested at 8% interest compounded continuously.

**66.** Determine the size of an account in which \$3500 earns 6% interest compounded continuously for 1 year.

**67.** Find the amount of money Barbara Mack owes at the end of 4 years if 6% interest is compounded continuously on her \$2000 debt.

**68.** Find the amount of money for which a \$2500 certificate of deposit is redeemable if it has been paying 10% interest compounded continuously for 3 years.

*Graph each function by finding ordered pair solutions, plotting the solutions, and then drawing a smooth curve through the plotted points.*

**69.** $f(x) = e^x$

**70.** $f(x) = e^{2x}$

**71.** $f(x) = e^{-3x}$

**72.** $f(x) = e^{-x}$

**73.** $f(x) = e^x + 2$

**74.** $f(x) = e^x - 3$

**75.** $f(x) = e^{x-1}$

**76.** $f(x) = e^{x+4}$

**77.** $f(x) = 3e^x$

**78.** $f(x) = -2e^x$

**79.** $f(x) = \ln x$

**80.** $f(x) = \log x$

**81.** $f(x) = -2\log x$

**82.** $f(x) = 3\ln x$

**83.** $f(x) = \log(x + 2)$

**84.** $f(x) = \log(x - 2)$

**85.** $f(x) = \ln x - 3$

**86.** $f(x) = \ln x + 3$

**87.** Graph $f(x) = e^x$ (Exercise 69), $f(x) = e^x + 2$ (Exercise 73), and $f(x) = e^x - 3$ (Exercise 74) on the same screen. Discuss any trends shown on the graphs.

**88.** Graph $f(x) = \ln x$ (Exercise 79), $f(x) = \ln x - 3$ (Exercise 85), and $f(x) = \ln x + 3$ (Exercise 86). Discuss any trends shown on the graphs.

## REVIEW EXERCISES

*Solve each equation for x. See Sections 2.1 and 5.8.*

**89.** $6x - 3(2 - 5x) = 6$

**90.** $2x + 3 = 5 - 2(3x - 1)$

**91.** $2x + 3y = 6x$

**92.** $4x - 8y = 10x$

**93.** $x^2 + 7x = -6$

**94.** $x^2 + 4x = 12$

*Solve each system of equations. See Section 4.1.*

**95.** $\begin{cases} x + 2y = -4 \\ 3x - y = 9 \end{cases}$

**96.** $\begin{cases} 5x + y = 5 \\ -3x - 2y = -10 \end{cases}$

# 10.7 EXPONENTIAL AND LOGARITHMIC EQUATIONS AND APPLICATIONS

**OBJECTIVES**

CD-ROM SSM SSG Video

1. Solve exponential equations.
2. Solve logarithmic equations.
3. Solve problems that can be modeled by exponential and logarithmic equations.

1 In Section 10.3 we solved exponential equations such as $2^x = 16$ by writing 16 as a power of 2 and applying the uniqueness of $b^x$.

$$2^x = 16$$

$$2^x = 2^4 \qquad \text{Write 16 as } 2^4.$$

$$x = 4 \qquad \text{Use the uniqueness of } b^x.$$

Solving the equation in this manner is possible since 16 is a power of 2. If solving an equation such as $2^x = a\ number$, where the number is not a power of 2, we use logarithms. For example, to solve an equation such as $3^x = 7$, we use the fact that $f(x) = \log_b x$ is a one-to-one function. Another way of stating this fact is as a property of equality.

**LOGARITHM PROPERTY OF EQUALITY**

Let $a$, $b$, and $c$ be real numbers such that $\log_b a$ and $\log_b c$ are real numbers and $b$ is not 1. Then

$$\log_b a = \log_b c \text{ is equivalent to } a = c$$

**Example 1** Solve $3^x = 7$.

*Solution* To solve, we use the logarithm property of equality and take the logarithm of both sides. For this example, we use the common logarithm.

$$3^x = 7$$

$$\log 3^x = \log 7 \qquad \text{Take the common log of both sides.}$$

$$x \log 3 = \log 7 \qquad \text{Apply the power property of logarithms.}$$

$$x = \frac{\log 7}{\log 3} \qquad \text{Divide both sides by log 3.}$$

The exact solution is $\frac{\log 7}{\log 3}$. If a decimal approximation is preferred,

$$\frac{\log 7}{\log 3} \approx \frac{0.845098}{0.4771213} \approx 1.7712 \text{ to four decimal places.}$$

The solution is $\frac{\log 7}{\log 3}$, or *approximately* 1.7712.

2 By applying the appropriate properties of logarithms, we can solve a broad variety of logarithmic equations.

**Example 2** Solve $\log_4(x-2) = 2$.

*Solution* Notice that $x - 2$ must be positive, so $x$ must be greater than 2. With this in mind, we first write the equation with exponential notation.

$$\log_4(x-2) = 2$$
$$4^2 = x - 2$$
$$16 = x - 2$$
$$18 = x \quad \text{Add 2 to both sides.}$$

*Check* To check, we replace $x$ with 18 in the original equation.

$$\log_4(x-2) = 2$$
$$\log_4(18-2) \stackrel{?}{=} 2 \quad \text{Let } x = 18.$$
$$\log_4 16 \stackrel{?}{=} 2$$
$$4^2 = 16 \quad \text{True.}$$

The solution is 18.

**Example 3** Solve $\log_2 x + \log_2(x-1) = 1$.

*Solution* Notice that $x - 1$ must be positive, so $x$ must be greater than 1. We use the product property on the left side of the equation.

$$\log_2 x + \log_2(x-1) = 1$$
$$\log_2 x(x-1) = 1 \quad \text{Apply the product property.}$$
$$\log_2(x^2 - x) = 1$$

Next we write the equation with exponential notation and solve for $x$.

$$2^1 = x^2 - x$$
$$0 = x^2 - x - 2 \quad \text{Subtract 2 from both sides.}$$
$$0 = (x-2)(x+1) \quad \text{Factor.}$$
$$0 = x - 2 \quad \text{or} \quad 0 = x + 1 \quad \text{Set each factor equal to 0.}$$
$$2 = x \qquad -1 = x$$

Recall that $-1$ cannot be a solution because $x$ must be greater than 1. If we forgot this, we would still reject $-1$ after checking. To see this, we replace $x$ with $-1$ in the original equation.

$$\log_2 x + \log_2(x-1) = 1$$
$$\log_2(-1) + \log_2(-1-1) \stackrel{?}{=} 1 \quad \text{Let } x = -1.$$

Because the logarithm of a negative number is undefined, $-1$ is rejected. Check to see that the solution is 2.

**Example 4** Solve $\log(x + 2) - \log x = 2$.

*Solution* We use the quotient property of logarithms on the left side of the equation.

$$\log(x + 2) - \log x = 2$$

$$\log \frac{x + 2}{x} = 2 \qquad \text{Apply the quotient property.}$$

$$10^2 = \frac{x + 2}{x} \qquad \text{Write using exponential notation.}$$

$$100 = \frac{x + 2}{x} \qquad \text{Simplify.}$$

$$100x = x + 2 \qquad \text{Multiply both sides by } x.$$

$$99x = 2 \qquad \text{Subtract } x \text{ from both sides.}$$

$$x = \frac{2}{99} \qquad \text{Divide both sides by 99.}$$

Verify that the solution is $\frac{2}{99}$.

3 Logarithmic and exponential functions are used in a variety of scientific, technical, and business settings. A few examples follow.

**Example 5** **ESTIMATING POPULATION SIZE**

The population size $y$ of a community of lemmings varies according to the relationship $y = y_0e^{0.15t}$. In this formula, $t$ is time in months, and $y_0$ is the initial population at time 0. Estimate the population after 6 months if there were originally 5000 lemmings.

*Solution* We substitute 5000 for $y_0$ and 6 for $t$.

$$y = y_0e^{0.15t}$$

$$= 5000e^{0.15(6)} \qquad \text{Let } t = 6 \text{ and } y_0 = 5000.$$

$$= 5000e^{0.9} \qquad \text{Multiply.}$$

Using a calculator, we find that $y \approx 12{,}298.016$. In 6 months the population will be approximately 12,300 lemmings.

**Example 6** **DOUBLING AN INVESTMENT**

How long does it take an investment of \$2000 to double if it is invested at 5% interest compounded quarterly? The necessary formula is $A = P\left(1 + \frac{r}{n}\right)^{nt}$, where $A$ is the accrued (or owed) amount, $P$ is the principal invested, $r$ is the annual rate of interest, $n$ is the number of compounding periods per year, and $t$ is the number of years.

*Solution* We are given that $P = \$2000$ and $r = 5\% = 0.05$. Compounding quarterly means 4 times a year, so $n = 4$. The investment is to double, so $A$ must be \$4000. Substitute these values and solve for $t$.

$$A = P\left(1 + \frac{r}{n}\right)^{nt}$$

$$4000 = 2000\left(1 + \frac{0.05}{4}\right)^{4t} \quad \text{Substitute in known values.}$$

$$4000 = 2000(1.0125)^{4t} \quad \text{Simplify } 1 + \frac{0.05}{4}.$$

$$2 = (1.0125)^{4t} \quad \text{Divide both sides by 2000.}$$

$$\log 2 = \log 1.0125^{4t} \quad \text{Take the logarithm of both sides.}$$

$$\log 2 = 4t(\log 1.0125) \quad \text{Apply the power property.}$$

$$\frac{\log 2}{4 \log 1.0125} = t \quad \text{Divide both sides by } 4 \log 1.0125.$$

$$13.949408 \approx t \quad \text{Approximate by calculator.}$$

Thus, it takes nearly 14 years for the money to double in value.

## GRAPHING CALCULATOR EXPLORATIONS

Use a graphing calculator to find how long it takes an investment of \$1500 to triple if it is invested at 8% interest compounded monthly.

First, let $P = \$1500$, $r = 0.08$, and $n = 12$ (for 12 months) in the formula

$$A = P\left(1 + \frac{r}{n}\right)^{nt}$$

Notice that when the investment has tripled, the accrued amount $A$ is \$4500. Thus,

$$4500 = 1500\left(1 + \frac{0.08}{12}\right)^{12t}$$

Determine an appropriate viewing window and enter and graph the equations

$$Y_1 = 1500\left(1 + \frac{0.08}{12}\right)^{12x}$$

and

$$Y_2 = 4500$$

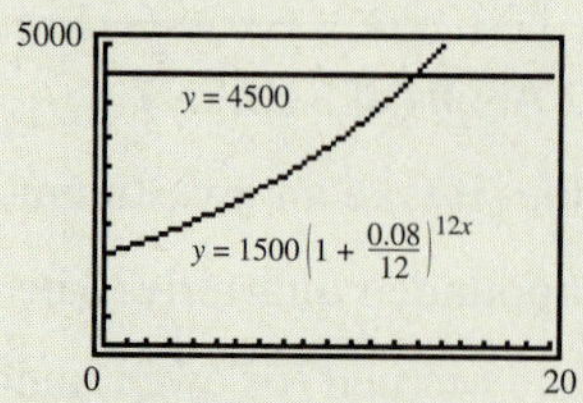

The point of intersection of the two curves is the solution. The $x$-coordinate tells how long it takes for the investment to triple.

Use a TRACE feature or an INTERSECT feature to approximate the coordinates of the point of intersection of the two curves. It takes approximately 13.78 years, or 13 years and 9 months, for the investment to triple in value to \$4500.

*Use this graphical solution method to solve each problem. Round each answer to the nearest hundredth.*

1. Find how long it takes an investment of \$5000 to grow to \$6000 if it is invested at 5% interest compounded quarterly.
2. Find how long it takes an investment of \$1000 to double if it is invested at 4.5% interest compounded daily. (Use 365 days in a year.)
3. Find how long it takes an investment of \$10,000 to quadruple if it is invested at 6% interest compounded monthly.
4. Find how long it takes \$500 to grow to \$800 if it is invested at 4% interest compounded semiannually.

## Exercise Set 10.7

*Solve each equation. Give an exact solution, and also approximate the solution to four decimal places. See Example 1.*

**1.** $3^x = 6$

**2.** $4^x = 7$

**3.** $3^{2x} = 3.8$

**4.** $5^{3x} = 5.6$

**5.** $2^{x-3} = 5$

**6.** $8^{x-2} = 12$

**7.** $9^x = 5$

**8.** $3^x = 11$

**9.** $4^{x+7} = 3$

**10.** $6^{x+3} = 2$

**11.** $7^{3x-4} = 11$

**12.** $5^{2x-6} = 12$

**13.** $e^{6x} = 5$

**14.** $e^{2x} = 8$

*Solve each equation. See Examples 2 through 4.*

**15.** $\log_2(x+5) = 4$

**16.** $\log_6(x^2 - x) = 1$

**17.** $\log_3 x^2 = 4$

**18.** $\log_2 x^2 = 6$

**19.** $\log_4 2 + \log_4 x = 0$

**20.** $\log_3 5 + \log_3 x = 1$

**21.** $\log_2 6 - \log_2 x = 3$

**22.** $\log_4 10 - \log_4 x = 2$

**23.** $\log_4 x + \log_4(x+6) = 2$

**24.** $\log_3 x + \log_3(x+6) = 3$

**25.** $\log_5(x+3) - \log_5 x = 2$

**26.** $\log_6(x+2) - \log_6 x = 2$

**27.** $\log_3(x-2) = 2$

**28.** $\log_2(x-5) = 3$

**29.** $\log_4(x^2 - 3x) = 1$

**30.** $\log_8(x^2 - 2x) = 1$

**31.** $\ln 5 + \ln x = 0$

**32.** $\ln 3 + \ln(x-1) = 0$

**33.** $3 \log x - \log x^2 = 2$

**34.** $2 \log x - \log x = 3$

**35.** $\log_2 x + \log_2(x+5) = 1$

**36.** $\log_4 x + \log_4(x+7) = 1$

**37.** $\log_4 x - \log_4(2x-3) = 3$

**38.** $\log_2 x - \log_2(3x+5) = 4$

**39.** $\log_2 x + \log_2(3x+1) = 1$

**40.** $\log_3 x + \log_3(x-8) = 2$

*Solve. See Example 5.*

**41.** The size of the wolf population at Isle Royale National Park increases at a rate of 4.3% per year. If the size of the current population is 83 wolves, find how many there should be in 5 years. Use $y = y_0 e^{0.043t}$ and round to the nearest whole.

**42.** The number of victims of a flu epidemic is increasing at a rate of 7.5% per week. If 20,000 persons are currently infected, find in how many days we can expect 45,000 to have the flu. Use $y = y_0 e^{0.075t}$ and round to the nearest whole. (Hint: Don't forget to convert your answer to days.)

**43.** The size of the population of Senegal is increasing at a rate of 2.6% per year. If 10,052,000 people lived in Senegal in 1999, find how many inhabitants there will be by 2005. Round to the nearest ten-thousand. Use $y = y_0 e^{0.026t}$.

**44.** In 1999, 1001 million people were citizens of India. Find how long it will take India's population to reach a size of 1500 million (that is, 1.5 billion) if the population size is growing at a rate of 1.7% per year. Use $y = y_0 e^{0.017t}$ and round to the nearest tenth. (*Source:* U.S. Bureau of the Census, International Data Base)

**45.** In 1999, Russia had a population of 146,394 thousand. At that time, Russia's population was declining at a rate of 0.5% per year. How long will it take for Russia's population to reach 120,000 thousand? Use $y = y_0e^{-0.005t}$ and round to the nearest tenth. (*Source:* U.S. Bureau of the Census, International Data Base)

**46.** The population of Italy has been decreasing at a rate of 0.1% per year. If there were 56,735,000 people living in Italy in 1999, how many inhabitants will there be by 2020? Use $y = y_0e^{-0.001t}$ and round to the nearest whole number. (*Source:* U.S. Bureau of the Census, International Data Base)

*Use the formula* $A = P\left(1 + \frac{r}{n}\right)^{nt}$ *to solve these compound interest problems. Round to the nearest tenth. See Example 6.*

**47.** Find how long it takes $600 to double if it is invested at 7% interest compounded monthly.

**48.** Find how long it takes $600 to double if it is invested at 12% interest compounded monthly.

**49.** Find how long it takes a $1200 investment to earn $200 interest if it is invested at 9% interest compounded quarterly.

**50.** Find how long it takes a $1500 investment to earn $200 interest if it is invested at 10% compounded semiannually.

**51.** Find how long it takes $1000 to double if it is invested at 8% interest compounded semiannually.

**52.** Find how long it takes $1000 to double if it is invested at 8% interest compounded monthly.

*The formula* $w = 0.00185h^{2.67}$ *is used to estimate the normal weight w of a boy h inches tall. Use this formula to solve the height–weight problems. Round to the nearest tenth.*

**53.** Find the expected weight of a boy who is 35 inches tall.

**54.** Find the expected weight of a boy who is 43 inches tall.

**55.** Find the expected height of a boy who weighs 85 pounds.

**56.** Find the expected height of a boy who weighs 140 pounds.

*The formula* $P = 14.7e^{-0.21x}$ *gives the average atmospheric pressure P, in pounds per square inch, at an altitude x, in miles above sea level. Use this formula to solve these pressure problems. Round answers to the nearest tenth.*

**57.** Find the average atmospheric pressure of Denver, which is 1 mile above sea level.

**58.** Find the average atmospheric pressure of Pikes Peak, which is 2.7 miles above sea level.

**59.** Find the elevation of a Delta jet if the atmospheric pressure outside the jet is 7.5 lb/in.$^2$.

**60.** Find the elevation of a remote Himalayan peak if the atmospheric pressure atop the peak is 6.5 lb/in.$^2$.

*Psychologists call the graph of the formula* $t = \frac{1}{c}\ln\left(\frac{A}{A - N}\right)$ *the learning curve, since the formula relates time t passed, in weeks, to a measure N of learning achieved, to a measure A of maximum learning possible, and to a measure c of an individual's learning style. Round to the nearest week.*

**61.** Norman is learning to type. If he wants to type at a rate of 50 words per minute ($N$ is 50) and his expected maximum rate is 75 words per minute ($A$ is 75), find how many weeks it should take him to achieve his goal. Assume that $c$ is 0.09.

**62.** An experiment with teaching chimpanzees sign language shows that a typical chimp can master a maximum of 65 signs. Find how many weeks it should take a chimpanzee to master 30 signs if $c$ is 0.03.

**63.** Janine is working on her dictation skills. She wants to take dictation at a rate of 150 words per minute and believes that the maximum rate she can hope for is 210 words per minute. Find how many weeks it should take her to achieve the 150 words per minute level if $c$ is 0.07.

**64.** A psychologist is measuring human capability to memorize nonsense syllables. Find how many weeks it should take a subject to learn 15 nonsense syllables if the maximum possible to learn is 24 syllables and $c$ is 0.17.

*Use a graphing calculator to solve each equation. For example, to solve Exercise 65, let* $Y_1 = e^{0.3x}$ *and* $Y_2 = 8$, *and graph the equations. The x-value of the point of intersection is the solution. Round all solutions to two decimal places.*

**65.** $e^{0.3x} = 8$

**66.** $10^{0.5x} = 7$

**67.** $2\log(-5.6x + 1.3) = -x - 1$

**68.** $\ln(1.3x - 2.1) = -3.5x + 5$

**69.** Check Exercise 11.

**70.** Check Exercise 12.

**71.** Check Exercise 31.

**72.** Check Exercise 32.

## REVIEW EXERCISES

*If* $x = -2$, $y = 0$, *and* $z = 3$, *find the value of each expression. See Section 1.4.*

**73.** $\dfrac{x^2 - y + 2z}{3x}$

**74.** $\dfrac{x^3 - 2y + z}{2z}$

**75.** $\dfrac{3z - 4x + y}{x + 2z}$

**76.** $\dfrac{4y - 3x + z}{2x + y}$

*Find the inverse function of each one-to-one function. See Section 10.2.*

**77.** $f(x) = 5x + 2$

**78.** $f(x) = \dfrac{x - 3}{4}$

For additional Chapter Projects, visit the Real World Activities Website by going to http://www.prenhall.com/martin-gay.

# 10 Chapter Project

## Modeling Temperature

When a cold object is placed in a warm room, the object's temperature gradually rises until it becomes, or nearly becomes, room temperature. Similarly, if a hot object is placed in a cooler room, the object's temperature gradually falls to room temperature. The way in which a cold or hot object warms up or cools off is modeled by an exact mathematical relationship, known as Newton's law of cooling. This law relates the temperature of an object to the time elapsed since its warming or cooling began. In this project, you will have the opportunity to investigate this model of cooling and warming. This project may be completed by working in groups or individually.

To investigate Newton's law of cooling in this project, you will collect experimental data in one of two methods: Method 1, using a stopwatch and thermometer, or Method 2, using Texas Instruments' Calculator-Based Laboratory (CBL™) or Second Generation Calculator-Based Laboratory (CBL 2™).

***Method 1 Materials***

- Container of either cold or hot liquid
- Thermometer
- Stopwatch
- Graphing calculator with regression capabilities

***Method 2 Materials***

- Container of either cold or hot liquid
- A TI-82, TI-83, or TI-85 graphing calculator with unit-to-unit link cable
- CBL™ or CBL 2™ unit with temperature probe

DATA TABLE

| *Time, t* | *Temperature, T* |
|---|---|
| 0 | |
| | |
| | |
| | |

***Steps for Collecting Data with Method 1:***

**a.** Insert the thermometer into the liquid and allow a thermometer reading to register. Take a temperature reading $T$ as you start the stopwatch (at $t = 0$) and record it in the accompanying data table.

**b.** Continue taking temperature readings at uniform intervals anywhere between 5 and 10 minutes long. At each reading use the stopwatch to measure the length of time that has elapsed since the temperature readings started with your first reading at $t = 0$. Record your time $t$ and liquid temperature $T$ in the data table. Gather data for six to twelve readings.

**c.** Plot the data from the data table. Plot $t$ on the horizontal axis and $T$ on the vertical axis.

***Steps for Collecting Data with Method 2:***

**a.** Enter the HEAT program appropriate for your calculator.

**b.** Prepare the CBL or CBL 2 and the graphing calculator. Insert the temperature probe into the liquid.

**c.** Start the HEAT program on the graphing calculator and follow its instructions to begin collecting data. The program will collect 36 temperature readings in degrees Celsius and plot them in real time with $t$ on the horizontal axis and $T$ on the vertical axis.

1. Which of the following mathematical models best fits the data you collected? Explain your reasoning. (Assume $a > 0$.)

   **a.** $T = ab^t + c$
   **b.** $T = ab^{-t} + c$
   **c.** $T = -ab^{-t} + c$
   **d.** $T = \ln(-ax + b) + c$
   **e.** $T = -\ln(-ax + b) + c$

2. What does the constant $c$ represent in the model you chose? What is the value of $c$ in this activity?

3. (Optional) Subtract the value of $c$ from each of your observations of $T$. Enter the new

ordered pairs $(t, T - c)$ into a graphing calculator. Use the exponential or logarithmic regression feature to find a model for your experimental data. Graph the ordered pairs $(t, T - c)$ with the model you found. How well does the model fit the data? How does the model compare with your selection from Question 1?

### *Graphing Calculator Programs*

#### *TI-82 or TI-83 Program*

```
PROGRAM:HEAT82
:PlotsOff
:Func
:FnOff
:AxesOn
:ClrDraw
:ClrList L3, L4
:-10→Ymin
:90→Ymax
:10→Yscl
:ClrHome
:{1, 0}→L1
:Send (L1)
:{1, 1, 1}→L1
:Send (L1)
:36→dim L3
:36→dim L4
:Disp "HOW MUCH TIME"
:Disp "BETWEEN POINTS"
:Disp "IN SECONDS?"
:Input T
:-2*T→Xmin
:36*T→Xmax
:T→Xscl
:seq(K, K, T, 36*T, T)→L3
:ClrHome
:Disp "PRESS ENTER"
:Disp "TO START"
:Pause
:ClrHome
:{3, T, -1, 0}→L1
:Send(L1)
:For (K, 1, 36, 1)
:Get (L4 (K))
:Pt-On(L3 (K), L4 (K))
:End
:ClrHome
:Plot1(Scatter, L3, L4,·)
:DispGraph
:Stop
```

#### *TI-85 Program*

```
PROGRAM:HEAT85
:Func
:FnOff
:AxesOn
:ClDrw
:1→dimL L3:1→dimL L4
:-10→yMin
:90→yMax
:10→yScl
:Cl LCD
:{1, 0}→L1
:Outpt("CBLSEND", L1)
:{1, 1, 1}→L1
:Outpt ("CBLSEND", L1)
:36→dimL L3
:36→dimL L4
:Disp "HOW MUCH TIME"
:Disp "BETWEEN POINTS"
:Disp "IN SECONDS"
:Input T
:-2*T→xMin
:36*T→xMax
:T→xScl
:seq(K, K, T, 36*T, T)→L3
:ClLCD
:Disp "PRESS ENTER"
:Disp "TO START"
:Pause
:ClLCD
:{3, T, -1, 0}→L1
:Outpt ("CBLSEND", L1)
:For (K, 1, 36, 1)
:Input "CBLGET", L4(K)
:PtOn (L3(K), L4(K))
:End
:ClLCD
:Scatter L3, L4
:DispG
:Stop
```

## CHAPTER 10 VOCABULARY CHECK

*Fill in each blank with one of the words or phrases listed below.*

| | | | | |
|---|---|---|---|---|
| inverse | common | composition | symmetric | exponential |
| vertical | logarithmic | natural | horizontal | |

1. For each one-to-one function, we can find its ________ function by switching the coordinates of the ordered pairs of the function.
2. The ____________ of functions $f$ and $g$ is $(f \circ g)(x) = f(g(x))$.
3. A function of the form $f(x) = b^x$ is called an ____________ function if $b > 0$, $b$ is not 1, and $x$ is a real number.
4. The graphs of $f$ and $f^{-1}$ are __________ about the line $y = x$.
5. ________ logarithms are logarithms to base $e$.
6. __________ logarithms are logarithms to base 10.
7. To see whether a graph is the graph of a one-to-one function, apply the ________ line test to see if it is a function, and then apply the ___________ line test to see if it is a one-to-one function.
8. A ___________ function is a function that can be defined by $f(x) = \log_b x$ where $x$ is a positive real number, $b$ is a constant positive real number, and $b$ is not 1.

# CHAPTER 10 HIGHLIGHTS

| DEFINITIONS AND CONCEPTS | EXAMPLES |
|---|---|
| **Section 10.1 The Algebra of Functions; Composite Functions** | |
| ***Algebra of Functions***<br>Sum $(f + g)(x) = f(x) + g(x)$<br>Difference $(f - g)(x) = f(x) - g(x)$<br>Product $(f \cdot g)(x) = f(x) \cdot g(x)$<br>Quotient $\left(\frac{f}{g}\right)(x) = \frac{f(x)}{g(x)}, g(x) \neq 0$ | If $f(x) = 7x$ and $g(x) = x^2 + 1$,<br>$(f + g)(x) = f(x) + g(x) = 7x + x^2 + 1$<br>$(f - g)(x) = f(x) - g(x) = 7x - (x^2 + 1)$<br>$= 7x - x^2 - 1$<br>$(f \cdot g)(x) = f(x) \cdot g(x) = 7x(x^2 + 1)$<br>$= 7x^3 + 7x$<br>$\left(\frac{f}{g}\right)(x) = \frac{f(x)}{g(x)} = \frac{7x}{x^2 + 1}$ |
| ***Composite Functions***<br>The notation $(f \circ g)(x)$ means "$f$ composed with $g$."<br>$(f \circ g)(x) = f(g(x))$<br>$(g \circ f)(x) = g(f(x))$ | If $f(x) = x^2 + 1$ and $g(x) = x - 5$, find $(f \circ g)(x)$.<br>$(f \circ g)(x) = f(g(x))$<br>$= f(x - 5)$<br>$= (x - 5)^2 + 1$<br>$= x^2 - 10x + 26$ |

| DEFINITIONS AND CONCEPTS | EXAMPLES |
|---|---|

## Section 10.2 Inverse Functions

If $f$ is a function, then $f$ is a **one-to-one function** only if each $y$-value (output) corresponds to only one $x$-value (input).

***Horizontal Line Test***

If every horizontal line intersects the graph of a function at most once, then the function is a one-to-one function.

Determine whether each graph is a one-to-one function.

**A**

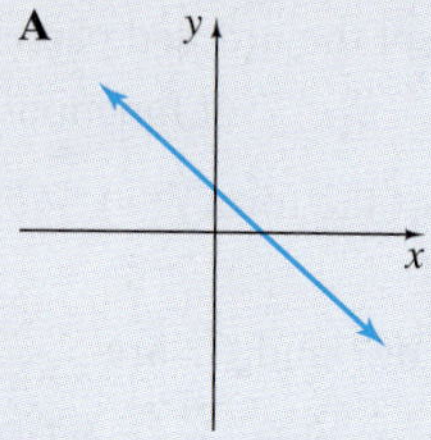


**B**

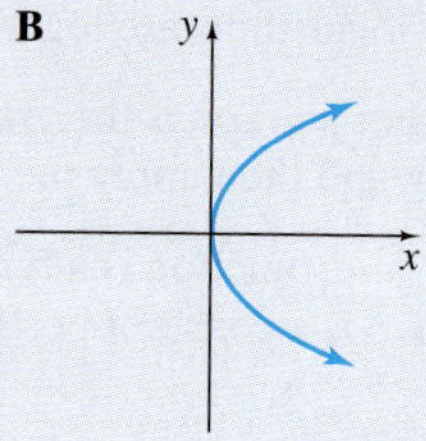


**C**

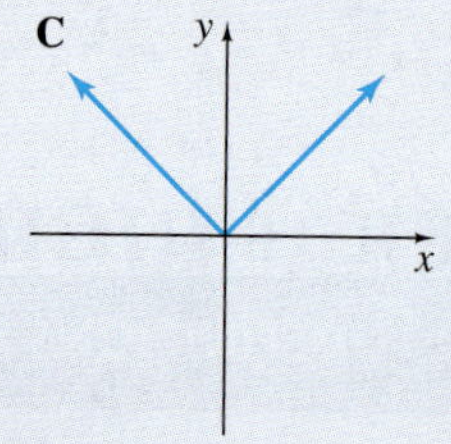


Graphs A and C pass the vertical line test, so only these are graphs of functions. Of graphs A and C, only graph A passes the horizontal line test, so only graph A is the graph of a one-to-one function.

The **inverse** of a one-to-one function $f$ is the one-to-one function $f^{-1}$ that is the set of all ordered pairs $(b, a)$ such that $(a, b)$ belongs to $f$.

***To Find the Inverse of a One-to-One Function f(x)***

**Step 1:** Replace $f(x)$ with $y$.
**Step 2:** Interchange $x$ and $y$.
**Step 3:** Solve for $y$.
**Step 4:** Replace $y$ with $f^{-1}(x)$.

Find the inverse of $f(x) = 2x + 7$.

$y = 2x + 7$ Replace $f(x)$ with $y$.

$x = 2y + 7$ Interchange $x$ and $y$.

$2y = x - 7$ Solve for $y$.

$$y = \frac{x-7}{2}$$

$$f^{-1}(x) = \frac{x-7}{2}$$ Replace $y$ with $f^{-1}(x)$.

The inverse of $f(x) = 2x + 7$ is

$$f^{-1}(x) = \frac{x-7}{2}.$$

| DEFINITIONS AND CONCEPTS | EXAMPLES |
|---|---|

## Section 10.3 Exponential Functions

A function of the form $f(x) = b^x$ is an **exponential function,** where $b > 0, b \neq 1$, and $x$ is a real number.

Graph the exponential function $y = 4^x$.

| $x$ | $y$ |
|---|---|
| $-2$ | $\frac{1}{16}$ |
| $-1$ | $\frac{1}{4}$ |
| 0 | 1 |
| 1 | 4 |
| 2 | 16 |

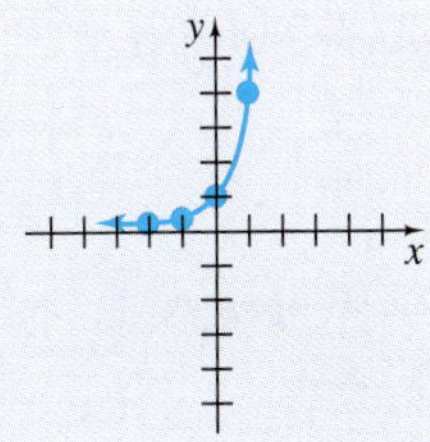


***Uniqueness of $b^x$***

If $b > 0$ and $b \neq 1$, then $b^x = b^y$ is equivalent to $x = y$.

Solve $2^{x+5} = 8$.

$2^{x+5} = 2^3$ Write 8 as $2^3$.

$x + 5 = 3$ Use the uniqueness of $b^x$.

$x = -2$ Subtract 5 from both sides.

## Section 10.4 Logarithmic Functions

***Logarithmic Definition***

If $b > 0$ and $b \neq 1$, then

$$y = \log_b x \quad \text{means} \quad x = b^y$$

for any positive number $x$ and real number $y$.

| *Logarithmic Form* | *Corresponding Exponential Statement* |
|---|---|
| $\log_5 25 = 2$ | $5^2 = 25$ |
| $\log_9 3 = \frac{1}{2}$ | $9^{1/2} = 3$ |

***Properties of Logarithms***

If $b$ is a real number, $b > 0$ and $b \neq 1$, then

$$\log_b 1 = 0, \quad \log_b b^x = x, \quad b^{\log_b x} = x$$

$\log_5 1 = 0, \quad \log_7 7^2 = 2, \quad 3^{\log_3 6} = 6$

***Logarithmic Function***

If $b > 0$ and $b \neq 1$, then a **logarithmic function** is a function that can be defined as

$$f(x) = \log_b x$$

The domain of $f$ is the set of positive real numbers, and the range of $f$ is the set of real numbers.

Graph $y = \log_3 x$.
Write $y = \log_3 x$ as $3^y = x$. Plot the ordered pair solutions listed in the table, and connect them with a smooth curve.

| $x$ | $y$ |
|---|---|
| 3 | 1 |
| 1 | 0 |
| $\frac{1}{3}$ | $-1$ |
| $\frac{1}{9}$ | $-2$ |

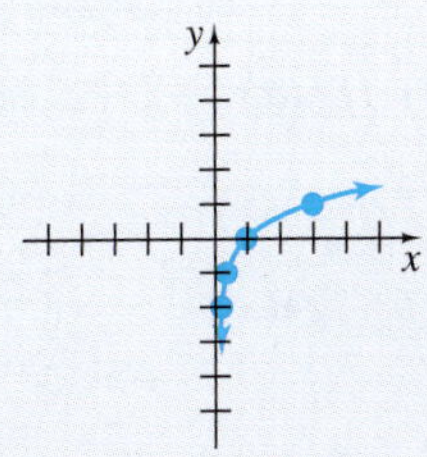

| DEFINITIONS AND CONCEPTS | EXAMPLES |
|---|---|
| **Section 10.5 Properties of Logarithms** | |
| Let $x$, $y$, and $b$ be positive numbers and $b \neq 1$.<br>***Product Property***<br>$\log_b xy = \log_b x + \log_b y$<br>***Quotient Property***<br>$\log_b \frac{x}{y} = \log_b x - \log_b y$<br>***Power Property***<br>$\log_b x^r = r \log_b x$ | Write as a single logarithm.<br>$2 \log_5 6 + \log_5 x - \log_5 (y + 2)$<br>$= \log_5 6^2 + \log_5 x - \log_5 (y + 2)$ Power property<br>$= \log_5 36 \cdot x - \log_5 (y + 2)$ Product property<br>$= \log_5 \frac{36x}{y + 2}$ Quotient property |
| **Section 10.6 Common Logarithms, Natural Logarithms, and Change of Base** | |
| ***Common Logarithms***<br>$\log x$ means $\log_{10} x$<br>***Natural Logarithms***<br>$\ln x$ means $\log_e x$<br>***Continuously Compounded Interest Formula***<br>$A = Pe^{rt}$<br>where $r$ is the annual interest rate for $P$ dollars invested for $t$ years. | $\log 5 = \log_{10} 5 \approx 0.69897$<br>$\ln 7 = \log_e 7 \approx 1.94591$<br>Find the amount in an account at the end of 3 years if \$1000 is invested at an interest rate of 4% compounded continuously.<br>Here, $t = 3$ years, $P = \$1000$, and $r = 0.04$.<br>$A = Pe^{rt}$<br>$= 1000e^{0.04(3)}$<br>$\approx \$1127.50$ |
| **Section 10.7 Exponential and Logarithmic Equations and Applications** | |
| ***Logarithm Property of Equality***<br>Let $\log_b a$ and $\log_b c$ be real numbers and $b \neq 1$. Then<br>$\log_b a = \log_b c$ is equivalent to $a = c$ | Solve $2^x = 5$.<br>$\log 2^x = \log 5$ Log property of equality<br>$x \log 2 = \log 5$ Power property<br>$x = \frac{\log 5}{\log 2}$ Divide both sides by log 2.<br>$x \approx 2.3219$ Use a calculator. |

## CHAPTER 10 REVIEW

***(10.1)*** *If* $f(x) = x - 5$ *and* $g(x) = 2x + 1$, *find*

**1.** $(f + g)(x)$

**2.** $(f - g)(x)$

**3.** $(f \cdot g)(x)$

**4.** $\left(\frac{g}{f}\right)(x)$

*If $f(x) = x^2 - 2$, $g(x) = x + 1$, and $h(x) = x^3 - x^2$, find each composition.*

**5.** $(f \circ g)(x)$

**6.** $(g \circ f)(x)$

**7.** $(h \circ g)(2)$

**8.** $(f \circ f)(x)$

**9.** $(f \circ g)(-1)$

**10.** $(h \circ h)(2)$

***(10.2)*** *Determine whether each function is a one-to-one function. If it is one-to-one, list the elements of its inverse.*

**11.** $h = \{(-9, 14), (6, 8), (-11, 12), (15, 15)\}$

**12.** $f = \{(-5, 5), (0, 4), (13, 5), (11, -6)\}$

**13.**

| U.S. Region (Input) | West | Midwest | South | Northeast |
|---|---|---|---|---|
| Rank in Automobile Thefts (Output) | 2 | 4 | 1 | 3 |

△ **14.**

| Shape (Input) | Square | Triangle | Parallelogram | Rectangle |
|---|---|---|---|---|
| Number of Sides (Output) | 4 | 3 | 4 | 4 |

*Given that $f(x) = \sqrt{x + 2}$ is a one-to-one function, find the following.*

**15. a.** $f(7)$

**b.** $f^{-1}(3)$

**16. a.** $f(-1)$

**b.** $f^{-1}(1)$

*Determine whether each function is a one-to-one function.*

**17.**

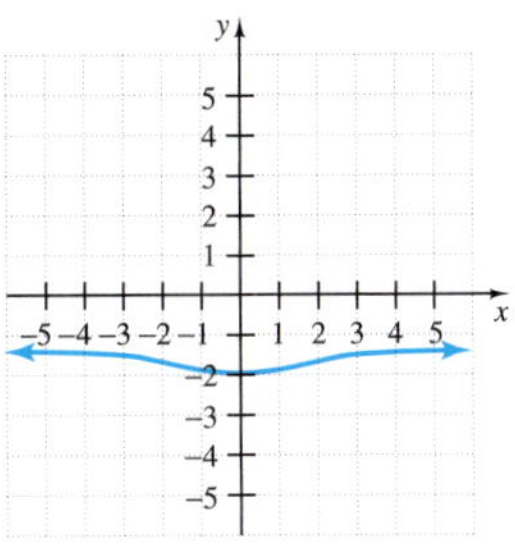


**18.**

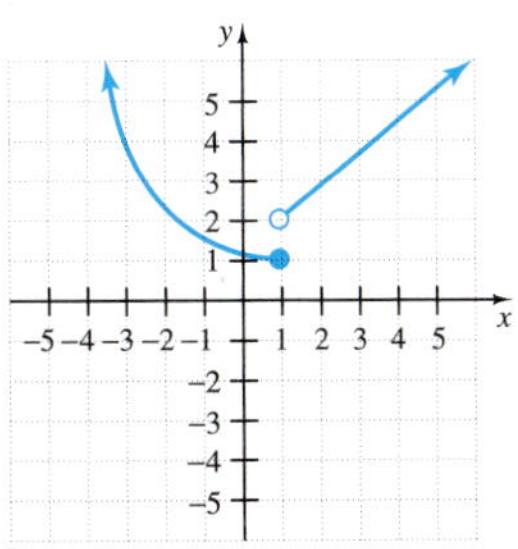


**19.**

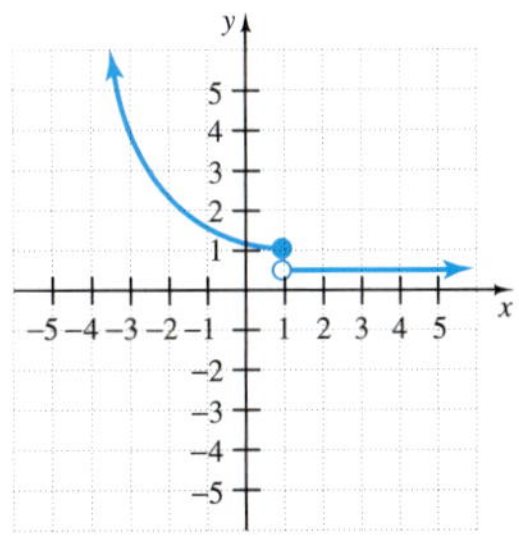


**20.**

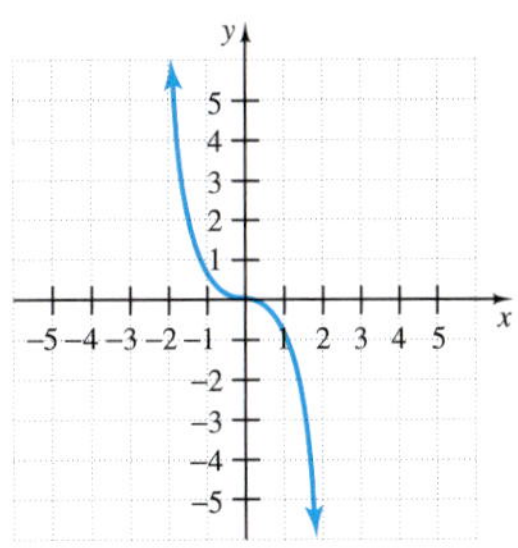


*Find an equation defining the inverse function of the given one-to-one function.*

**21.** $f(x) = x - 9$

**22.** $f(x) = x + 8$

**23.** $f(x) = 6x + 11$

**24.** $f(x) = 12x$

**25.** $f(x) = x^3 - 5$

**26.** $f(x) = \sqrt[3]{x + 2}$

**27.** $g(x) = \dfrac{12x - 7}{6}$

**28.** $r(x) = \dfrac{13}{2}x - 4$

*On the same set of axes, graph the given one-to-one function and its inverse.*

**29.** $g(x) = \sqrt{x}$

**30.** $h(x) = 5x - 5$

**31.** Find the inverse of the one-to-one function $f(x) = 2x - 3$. Then graph both $f(x)$ and $f^{-1}(x)$ with a square window.

***(10.3)*** *Solve each equation for x.*

**32.** $4^x = 64$

**33.** $3^x = \dfrac{1}{9}$

**34.** $2^{3x} = \dfrac{1}{16}$

**35.** $5^{2x} = 125$

**36.** $9^{x+1} = 243$

**37.** $8^{3x-2} = 4$

*Graph each exponential function.*

**38.** $y = 3^x$

**39.** $y = \left(\dfrac{1}{3}\right)^x$

**40.** $y = 4 \cdot 2^x$

**41.** $y = 2^x + 4$

*Use the formula* $A = P\left(1 + \dfrac{r}{n}\right)^{nt}$ *to solve the interest problems. In this formula,*

$A$ = amount accrued (or owed)

$P$ = principal invested (or loaned)

$r$ = rate of interest

$n$ = number of compounding periods per year

$t$ = time in years

**42.** Find the amount accrued if \$1600 is invested at 9% interest compounded semiannually for 7 years.

**43.** A total of \$800 is invested in a 7% certificate of deposit for which interest is compounded quarterly. Find the value that this certificate will have at the end of 5 years.

**44.** Use a graphing calculator to verify the results of Exercise 40.

***(10.4)*** *Write each equation with logarithmic notation.*

**45.** $49 = 7^2$

**46.** $2^{-4} = \dfrac{1}{16}$

*Write each logarithmic equation with exponential notation.*

**47.** $\log_{1/2} 16 = -4$

**48.** $\log_{0.4} 0.064 = 3$

*Solve for x.*

**49.** $\log_4 x = -3$

**50.** $\log_3 x = 2$

**51.** $\log_3 1 = x$

**52.** $\log_4 64 = x$

**53.** $\log_x 64 = 2$

**54.** $\log_x 81 = 4$

**55.** $\log_4 4^5 = x$

**56.** $\log_7 7^{-2} = x$

**57.** $5^{\log_5 4} = x$

**58.** $2^{\log_2 9} = x$

**59.** $\log_2 (3x - 1) = 4$

**60.** $\log_3 (2x + 5) = 2$

**61.** $\log_4 (x^2 - 3x) = 1$

**62.** $\log_8 (x^2 + 7x) = 1$

*Graph each pair of equations on the same coordinate system.*

**63.** $y = 2^x$ and $y = \log_2 x$

**64.** $y = \left(\dfrac{1}{2}\right)^x$ and $y = \log_{1/2} x$

***(10.5)*** *Write each of the following as single logarithms.*

**65.** $\log_3 8 + \log_3 4$

**66.** $\log_2 6 + \log_2 3$

**67.** $\log_7 15 - \log_7 20$

**68.** $\log 18 - \log 12$

**69.** $\log_{11} 8 + \log_{11} 3 - \log_{11} 6$

**70.** $\log_5 14 + \log_5 3 - \log_5 21$

**71.** $2 \log_5 x - 2 \log_5 (x + 1) + \log_5 x$

**72.** $4 \log_3 x - \log_3 x + \log_3 (x + 2)$

*Use properties of logarithms to write each expression as a sum or difference of multiples of logarithms.*

**73.** $\log_3 \dfrac{x^3}{x + 2}$

**74.** $\log_4 \dfrac{x + 5}{x^2}$

**75.** $\log_2 \dfrac{3x^2y}{z}$

**76.** $\log_7 \dfrac{yz^3}{x}$

*If $\log_b 2 = 0.36$ and $\log_b 5 = 0.83$, find the following.*

**77.** $\log_b 50$

**78.** $\log_b \dfrac{4}{5}$

***(10.6)*** *Use a calculator to approximate the logarithm to four decimal places.*

**79.** $\log 3.6$

**80.** $\log 0.15$

**81.** $\ln 1.25$

**82.** $\ln 4.63$

*Find the exact value.*

**83.** $\log 1000$

**84.** $\log \dfrac{1}{10}$

**85.** $\ln \dfrac{1}{e}$

**86.** $\ln e^4$

*Solve each equation for x.*

**87.** $\ln (2x) = 2$

**88.** $\ln (3x) = 1.6$

**89.** $\ln (2x - 3) = -1$

**90.** $\ln (3x + 1) = 2$

*Use the formula $\ln \dfrac{I}{I_0} = -kx$ to solve radiation problems. In this formula,*

$x$ = depth in millimeters
$I$ = intensity of radiation
$I_0$ = initial intensity
$k$ = a constant measure dependent on the material

*Round answers to two decimal places.*

**91.** Find the depth at which the intensity of the radiation passing through a lead shield is reduced to 3% of the original intensity if the value of $k$ is 2.1.

**92.** If $k$ is 3.2, find the depth at which 2% of the original radiation will penetrate.

*Approximate the logarithm to four decimal places.*

**93.** $\log_5 1.6$

**94.** $\log_3 4$

*Use the formula $A = Pe^{rt}$ to solve the interest problems in which interest is compounded continuously. In this formula,*

$A$ = amount accrued (or owed)
$P$ = principal invested (or loaned)
$r$ = rate of interest
$t$ = time in years

**95.** Bank of New York offers a 5-year, 6% continuously compounded investment option. Find the amount accrued if \$1450 is invested.

**96.** Find the amount to which a \$940 investment grows if it is invested at 11% compounded continuously for 3 years.

***(10.7)*** *Solve each exponential equation for x. Give an exact solution and also approximate the solution to four decimal places.*

**97.** $3^{2x} = 7$

**98.** $6^{3x} = 5$

**99.** $3^{2x+1} = 6$

**100.** $4^{3x+2} = 9$

**101.** $5^{3x-5} = 4$

**102.** $8^{4x-2} = 3$

**103.** $2 \cdot 5^{x-1} = 1$

**104.** $3 \cdot 4^{x+5} = 2$

*Solve the equation for x.*

**105.** $\log_5 2 + \log_5 x = 2$

**106.** $\log_3 x + \log_3 10 = 2$

**107.** $\log(5x) - \log(x + 1) = 4$

**108.** $\ln(3x) - \ln(x - 3) = 2$

**109.** $\log_2 x + \log_2 2x - 3 = 1$

**110.** $-\log_6(4x + 7) + \log_6 x = 1$

*Use the formula $y = y_0e^{kt}$ to solve the population growth problems. In this formula,*

$y$ = size of population
$y_0$ = initial count of population
$k$ = rate of growth
$t$ = time

*Round each answer to the nearest whole.*

**111.** The population of mallard ducks in Nova Scotia is expected to grow at a rate of 6% per week during the spring migration. If 155,000 ducks are already in Nova Scotia, find how many are expected by the end of 4 weeks.

**112.** The population of Indonesia is growing at a rate of 1.5% per year. If the population in 1998 was 212,942,000, find the expected population by the year 2006. (*Source:* U.S. Bureau of the Census, International Data Base)

**113.** Japan is experiencing an annual growth rate of 0.2%. In 1998, the population of Japan was 125,932,000. How long will it take for the population to be 140,000,000? (*Source:* U.S. Bureau of the Census, International Data Base)

**114.** In 1998, Canada had a population of 30,675,000. How long will it take Canada to double in population if its growth rate is 1.1% annually? (*Source:* U.S. Bureau of the Census, International Data Base)

**115.** Egypt's population is increasing at a rate of 1.9% per year. How long will it take for its 1998 population of 66,050,000 to double in size? (*Source:* U.S. Bureau of the Census, International Data Base)

*Use the compound interest equation $A = P\left(1 + \frac{r}{n}\right)^{nt}$ to solve the following. (See the directions for Exercises 42 and 43 for an explanation of this formula. Round answers to the nearest tenth.)*

**116.** Find how long it will take a \$5000 investment to grow to \$10,000 if it is invested at 8% interest compounded quarterly.

**117.** An investment of \$6000 has grown to \$10,000 while the money was invested at 6% interest compounded monthly. Find how long it was invested.

*Use a graphing calculator to solve each equation. Round all solutions to two decimal places.*

**118.** $e^x = 2$ **119.** $10^{0.3x} = 7$

## CHAPTER 10 TEST

*If $f(x) = x$, $g(x) = x - 7$, and $h(x) = x^2 - 6x + 5$, find the following.*

**1.** $(f \circ h)(0)$

**2.** $(g \circ f)(x)$

**3.** $(g \circ h)(x)$

*On the same set of axes, graph the given one-to-one function and its inverse.*

**4.** $f(x) = 7x - 14$

*Determine whether the given graph is the graph of a one-to-one function.*

**5.**

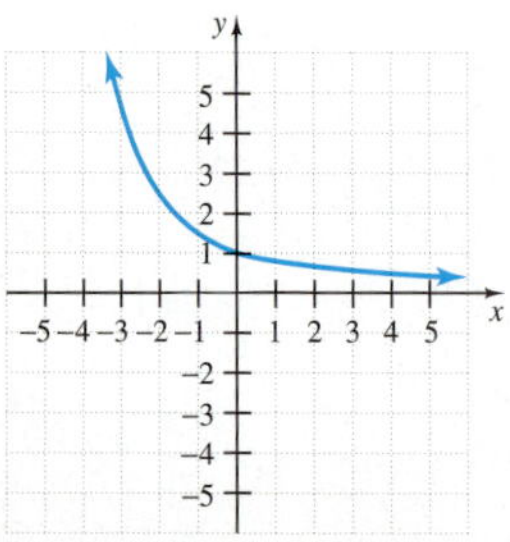


**6.**

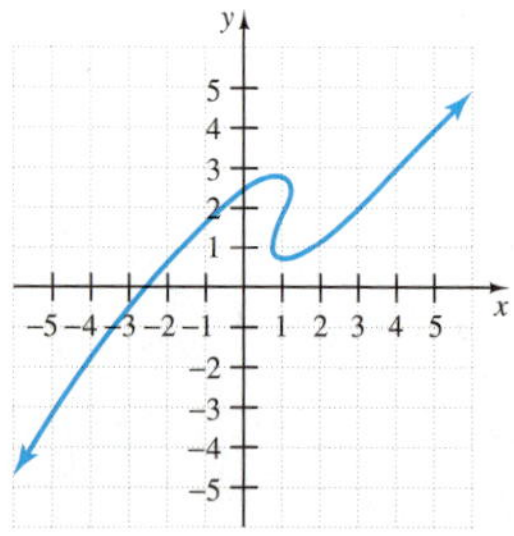


*Determine whether each function is one-to-one. If it is one-to-one, find an equation or a set of ordered pairs that defines the inverse function of the given function.*

**7.** $y = 6 - 2x$

**8.** $f = \{(0, 0), (2, 3), (-1, 5)\}$

**9.**

| Word (Input) | Dog | Cat | House | Desk | Circle |
|---|---|---|---|---|---|
| First Letter of Word (Output) | d | c | h | d | c |

*Use the properties of logarithms to write each expression as a single logarithm.*

**10.** $\log_3 6 + \log_3 4$

**11.** $\log_5 x + 3 \log_5 x - \log_5 (x + 1)$

**12.** Write the expression $\log_6 \frac{2x}{y^3}$ as the sum or difference of multiples of logarithms.

**13.** If $\log_b 3 = 0.79$ and $\log_b 5 = 1.16$, find the value of $\log_b \frac{3}{25}$.

**14.** Approximate $\log_7 8$ to four decimal places.

**15.** Solve $8^{x-1} = \frac{1}{64}$ for $x$. Give an exact solution.

**16.** Solve $3^{2x+5} = 4$ for $x$. Give an exact solution, and also approximate the solution to four decimal places.

*Solve each logarithmic equation for x. Give an exact solution.*

**17.** $\log_3 x = -2$

**18.** $\ln \sqrt{e} = x$

**19.** $\log_8 (3x - 2) = 2$

**20.** $\log_5 x + \log_5 3 = 2$

**21.** $\log_4 (x + 1) - \log_4 (x - 2) = 3$

**22.** Solve $\ln (3x + 7) = 1.31$ accurate to four decimal places.

**23.** Graph $y = \left(\frac{1}{2}\right)^x + 1$.

**24.** Graph the functions $y = 3^x$ and $y = \log_3 x$ on the same coordinate system.

*Use the formula $A = P\left(1 + \frac{r}{n}\right)^{nt}$ to solve Exercises 25 and 26.*

**25.** Find the amount in the account if \$4000 is invested for 3 years at 9% interest compounded monthly.

**26.** Find how long it will take \$2000 to grow to \$3000 if the money is invested at 7% interest compounded semiannually. Round to the nearest whole.

*Use the population growth formula $y = y_0 e^{kt}$ to solve Exercises 27 and 28.*

**27.** The prairie dog population of the Grand Rapids area now stands at 57,000 animals. If the population is growing at a rate of 2.6% annually, find how many prairie dogs there will be in that area 5 years from now.

**28.** In an attempt to save an endangered species of wood duck, naturalists would like to increase the wood duck population from 400 to 1000 ducks. If the annual population growth rate is 6.2%, find how long it will take the naturalists to reach their goal. Round to the nearest whole year.

**29.** The formula $\log (1 + k) = \frac{0.3}{D}$ relates the doubling time $D$, in days, and the growth rate $k$ for a population of mice. Find the rate at which the population is increasing if the doubling time is 56 days. Round to the nearest tenth of a percent.

**30.** Use a graphing calculator to approximate the solution of

$$e^{0.2x} = e^{-0.4x} + 2$$

to two decimal places.

## CHAPTER 10 CUMULATIVE REVIEW

**1.** Multiply.

**a.** $(-8)(-1)$

**b.** $(-2)\frac{1}{6}$

**c.** $3(-3)$

**d.** $0(11)$

**e.** $\left(\frac{1}{5}\right)\left(-\frac{10}{11}\right)$

**f.** $(7)(1)(-2)(-3)$

**g.** $8(-2)(0)$

**2.** Graph $y = x^2$.

**3.** Solve the system.

$$\begin{cases} x - 5y - 2z = 6 \\ -2x + 10y + 4z = -12 \\ \frac{1}{2}x - \frac{5}{2}y - z = 3 \end{cases}$$

**4.** Use the quotient rule to simplify.

**a.** $\dfrac{x^7}{x^4}$

**b.** $\dfrac{5^8}{5^2}$

**c.** $\dfrac{20x^6}{4x^5}$

**d.** $\dfrac{12y^{10}z^7}{14y^8z^7}$

**5.** For the ICL Production Company, the rational function $C(x) = \dfrac{2.6x + 10{,}000}{x}$ describes the company's cost per disc of pressing $x$ compact discs. Find the cost per disc for pressing:

**a.** 100 compact discs

**b.** 1000 compact discs

**6.** Add or subtract.

**a.** $\dfrac{x}{4} + \dfrac{5x}{4}$

**b.** $\dfrac{x^2}{x+7} - \dfrac{49}{x+7}$

**c.** $\dfrac{x}{3y^2} - \dfrac{x+1}{3y^2}$

**7.** Divide $3x^4 + 2x^3 - 8x + 6$ by $x^2 - 1$.

**8.** Solve: $\dfrac{2x}{2x-1} + \dfrac{1}{x} = \dfrac{1}{2x-1}$

**9.** Steve Deitmer takes $1\frac{1}{2}$ times as long to go 72 miles upstream in his boat as he does to return. If the boat cruises at 30 mph in still water, what is the speed of the current?

**10.** Simplify the following expressions.

**a.** $\sqrt[4]{81}$

**b.** $\sqrt[5]{-243}$

**c.** $-\sqrt{25}$

**d.** $\sqrt[4]{-81}$

**e.** $\sqrt[3]{64x^3}$

**11.** Use rational exponents to write as a single radical.

**a.** $\sqrt{x} \cdot \sqrt[4]{x}$

**b.** $\dfrac{\sqrt{x}}{\sqrt[3]{x}}$

**c.** $\sqrt[3]{3} \cdot \sqrt{2}$

**12.** Multiply.

**a.** $\sqrt{3}(5 + \sqrt{30})$

**b.** $(\sqrt{5} - \sqrt{6})(\sqrt{7} + 1)$

**c.** $(7\sqrt{x} + 5)(3\sqrt{x} - \sqrt{5})$

**d.** $(4\sqrt{3} - 1)^2$

**e.** $(\sqrt{2x} - 5)(\sqrt{2x} + 5)$

**13.** Rationalize the denominator of $\dfrac{\sqrt[4]{x}}{\sqrt[4]{81y^5}}$.

**14.** Solve $\sqrt{4 - x} = x - 2$.

**15.** Solve $3x^2 - 9x + 8 = 0$ by completing the square.

**16.** Solve $\dfrac{3x}{x-2} - \dfrac{x+1}{x} = \dfrac{6}{x(x-2)}$.

**17.** Solve $x^2 - 4x \le 0$.

**18.** Graph $F(x) = (x - 3)^2 + 1$.

**19.** Find the midpoint of the line segment that joins points $P(-3, 3)$ and $Q(1, 0)$.

**20.** Graph the equation $4x^2 + 16y^2 = 64$.

**21.** Solve the system:

$$\begin{cases} x^2 + y^2 = 4 \\ x + y = 3 \end{cases}$$

**22.** Graph the system:

$$\begin{cases} x \le 1 - 2y \\ y \le x^2 \end{cases}$$

**23.** If $f(x) = x - 1$ and $g(x) = 2x - 3$, find

**a.** $(f + g)(x)$

**b.** $(f - g)(x)$

**c.** $(f \cdot g)(x)$

**d.** $\left(\frac{f}{g}\right)(x)$

**24.** Find an equation of the inverse of $f(x) = x + 3$.

**25.** Find the value of each logarithmic expression.

**a.** $\log_4 16$

**b.** $\log_{10} \frac{1}{10}$

**c.** $\log_9 3$

A certified financial planner works with clients to develop a sound financial plan that helps the client reach his or her life goals. Financial planning is a growing field: As Baby Boomers approach retirement age and life spans lengthen overall, more and more people will seek professional assistance with financial management.

Financial planners work for investment firms, accounting firms, insurance companies, banks, credit counseling organizations, law firms, or in private practice. Through interviews and discussions, they assess clients' current financial positions. Planners then give advice on retirement planning, insurance needs, investment options, estate planning, tax strategies, and employee benefits. Financial planners may then help clients implement their new financial plans. Certified financial planners use math and problem-solving skills in such tasks as analyzing clients' current cash flow, estimating cash needs for future goals, and calculating investment returns.

*For more information about a career as a certified financial planner, visit the Certified Financial Planner Board of Standards Web site by first going to www.prenhall.com/martin-gay.*

*In the Spotlight on Decision Making feature on page 651, you will have the opportunity to make a decision about reaching a client's retirement goals as a certified financial planner.*

# 11 SEQUENCES, SERIES, AND THE BINOMIAL THEOREM


**11.1** Sequences

**11.2** Arithmetic and Geometric Sequences

**11.3** Series

**11.4** Partial Sums of Arithmetic and Geometric Sequences

**11.5** The Binomial Theorem


Having explored in some depth the concept of function, we turn now in this final chapter to *sequences.* In one sense, a sequence is simply an ordered list of numbers. In another sense, a sequence is itself a function. Phenomena modeled by such functions are everywhere around us. The starting place for all mathematics is the sequence of natural numbers: 1, 2, 3, 4, and so on.

Sequences lead us to *series,* which are a sum of ordered numbers. Through series we gain new insight, for example about the expansion of a binomial $(a + b)^n$, the concluding topic of this book.

## 11.1 SEQUENCES

▶ **OBJECTIVES**

CD-ROM SSM SSG Video

1. Write the terms of a sequence given its general term.
2. Find the general term of a sequence.
3. Solve applications that involve sequences.

Suppose that a town's present population of 100,000 is growing by 5% each year. After the first year, the town's population will be

$$100{,}000 + 0.05(100{,}000) = 105{,}000$$

After the second year, the town's population will be

$$105{,}000 + 0.05(105{,}000) = 110{,}250$$

After the third year, the town's population will be

$$110{,}250 + 0.05(110{,}250) \approx 115{,}763$$

If we continue to calculate, the town's yearly population can be written as the **infinite sequence** of numbers

$$105{,}000,\ 110{,}250,\ 115{,}763,\ \ldots$$

If we decide to stop calculating after a certain year (say, the fourth year), we obtain the **finite sequence**

$$105{,}000,\ 110{,}250,\ 115{,}763,\ 121{,}551$$

### SEQUENCES

An infinite sequence is a function whose domain is the set of natural numbers $\{1, 2, 3, 4, \ldots\}$.

A finite sequence is a function whose domain is the set of natural numbers $\{1, 2, 3, 4, \ldots, n\}$, where $n$ is some natural number.

1 Given the sequence $2, 4, 8, 16, \ldots$, we say that each number is a **term** of the sequence. Because a sequence is a function, we could describe it by writing $f(n) = 2^n$, where $n$ is a natural number. Instead, we use the notation

$$a_n = 2^n$$

Some function values are

| | |
|---|---|
| $a_1 = 2^1 = 2$ | First term of the sequence |
| $a_2 = 2^2 = 4$ | Second term |
| $a_3 = 2^3 = 8$ | Third term |
| $a_4 = 2^4 = 16$ | Fourth term |
| $a_{10} = 2^{10} = 1024$ | Tenth term |

The $n$th term of the sequence $a_n$ is called the **general term.**

**Example 1** Write the first five terms of the sequence whose general term is given by

$$a_n = n^2 - 1$$

*Solution* Evaluate $a_n$, where $n$ is 1, 2, 3, 4, and 5.

$$a_n = n^2 - 1$$

$$a_1 = 1^2 - 1 = 0 \quad \text{Replace } n \text{ with 1.}$$

$$a_2 = 2^2 - 1 = 3 \quad \text{Replace } n \text{ with 2.}$$

$$a_3 = 3^2 - 1 = 8 \quad \text{Replace } n \text{ with 3.}$$

$$a_4 = 4^2 - 1 = 15 \quad \text{Replace } n \text{ with 4.}$$

$$a_5 = 5^2 - 1 = 24 \quad \text{Replace } n \text{ with 5.}$$

Thus, the first five terms of the sequence $a_n = n^2 - 1$ are 0, 3, 8, 15, and 24.

**Example 2** If the general term of a sequence is given by $a_n = \dfrac{(-1)^n}{3n}$, find

**a.** the first term of the sequence
**b.** $a_8$
**c.** the one-hundredth term of the sequence
**d.** $a_{15}$

*Solution*

**a.** $a_1 = \dfrac{(-1)^1}{3(1)} = -\dfrac{1}{3}$ Replace $n$ with 1.

**b.** $a_8 = \dfrac{(-1)^8}{3(8)} = \dfrac{1}{24}$ Replace $n$ with 8.

**c.** $a_{100} = \dfrac{(-1)^{100}}{3(100)} = \dfrac{1}{300}$ Replace $n$ with 100.

**d.** $a_{15} = \dfrac{(-1)^{15}}{3(15)} = -\dfrac{1}{45}$ Replace $n$ with 15.

**2** Suppose we know the first few terms of a sequence and want to find a general term that fits the pattern of the first few terms.

**Example 3** Find a general term $a_n$ of the sequence whose first few terms are given.

**a.** $1, 4, 9, 16, \ldots$

**b.** $\dfrac{1}{1}, \dfrac{1}{2}, \dfrac{1}{3}, \dfrac{1}{4}, \dfrac{1}{5}, \ldots$

**c.** $-3, -6, -9, -12, \ldots$

**d.** $\dfrac{1}{2}, \dfrac{1}{4}, \dfrac{1}{8}, \dfrac{1}{16}, \ldots$

*Solution* **a.** These numbers are the squares of the first four natural numbers, so a general term might be $a_n = n^2$.

**b.** These numbers are the reciprocals of the first five natural numbers, so a general term might be $a_n = \frac{1}{n}$.

**c.** These numbers are the product of $-3$ and the first four natural numbers, so a general term might be $a_n = -3n$.

**d.** Notice that the denominators double each time.

$$\frac{1}{2}, \quad \frac{1}{2 \cdot 2}, \quad \frac{1}{2(2 \cdot 2)}, \quad \frac{1}{2(2 \cdot 2 \cdot 2)}$$

or

$$\frac{1}{2^1}, \quad \frac{1}{2^2}, \quad \frac{1}{2^3}, \quad \frac{1}{2^4}$$

We might then suppose that the general term is $a_n = \frac{1}{2^n}$.

3 Sequences model many phenomena of the physical world, as illustrated by the following example.

## Example 4 FINDING A PUPPY'S WEIGHT GAIN

The amount of weight, in pounds, a puppy gains in each month of its first year is modeled by a sequence whose general term is $a_n = n + 4$, where $n$ is the number of the month. Write the first five terms of the sequence, and find how much weight the puppy should gain in its fifth month.

*Solution* Evaluate $a_n = n + 4$ when $n$ is 1, 2, 3, 4, and 5.

$$a_1 = 1 + 4 = 5$$
$$a_2 = 2 + 4 = 6$$
$$a_3 = 3 + 4 = 7$$
$$a_4 = 4 + 4 = 8$$
$$a_5 = 5 + 4 = 9$$

The puppy should gain 9 pounds in its fifth month.

### SPOTLIGHT ON DECISION MAKING

Suppose you are considering two job offers. The first job offer pays \$11.50 per hour and guarantees a \$0.65-per-hour raise each year. The second job offer pays \$10.75 per hour and guarantees a \$1.10-per-hour raise each year. If one of your goals is to be earning at least \$15 per hour in 5 years, which job offer would you accept? Explain. What other factors would you want to consider?

## Exercise Set 11.1

*Write the first five terms of each sequence whose general term is given. See Example 1.*

1. $a_n = n + 4$
2. $a_n = 5 - n$
3. $a_n = (-1)^n$
4. $a_n = (-2)^n$
5. $a_n = \dfrac{1}{n + 3}$
6. $a_n = \dfrac{1}{7 - n}$
7. $a_n = 2n$
8. $a_n = -6n$
9. $a_n = -n^2$
10. $a_n = n^2 + 2$
11. $a_n = 2^n$
12. $a_n = 3^{n-2}$
13. $a_n = 2n + 5$
14. $a_n = 1 - 3n$
15. $a_n = (-1)^n n^2$
16. $a_n = (-1)^{n+1}(n - 1)$

*Find the indicated term for each sequence whose general term is given. See Example 2.*

17. $a_n = 3n^2;\ a_5$
18. $a_n = -n^2;\ a_{15}$
19. $a_n = 6n - 2;\ a_{20}$
20. $a_n = 100 - 7n;\ a_{50}$
21. $a_n = \dfrac{n + 3}{n};\ a_{15}$
22. $a_n = \dfrac{n}{n + 4};\ a_{24}$
23. $a_n = (-3)^n;\ a_6$
24. $a_n = 5^{n+1};\ a_3$
25. $a_n = \dfrac{n - 2}{n + 1};\ a_6$
26. $a_n = \dfrac{n + 3}{n + 4};\ a_8$
27. $a_n = \dfrac{(-1)^n}{n};\ a_8$
28. $a_n = \dfrac{(-1)^n}{2n};\ a_{100}$
29. $a_n = -n^2 + 5;\ a_{10}$
30. $a_n = 8 - n^2;\ a_{20}$
31. $a_n = \dfrac{(-1)^n}{n + 6};\ a_{19}$
32. $a_n = \dfrac{n - 4}{(-2)^n};\ a_6$

*Find a general term $a_n$ for each sequence whose first four terms are given. See Example 3.*

33. 3, 7, 11, 15
34. 2, 7, 12, 17
35. −2, −4, −8, −16
36. −4, 16, −64, 256
37. $\dfrac{1}{3}, \dfrac{1}{9}, \dfrac{1}{27}, \dfrac{1}{81}$
38. $\dfrac{2}{5}, \dfrac{2}{25}, \dfrac{2}{125}, \dfrac{2}{625}$

*Solve. See Example 4.*

39. The distance, in feet, that a Thermos dropped from a cliff falls in each consecutive second is modeled by a sequence whose general term is $a_n = 32n - 16$, where $n$ is the number of seconds. Find the distance the Thermos falls in the second, third, and fourth seconds.
40. The population size of a culture of bacteria triples every hour such that its size is modeled by the sequence $a_n = 50(3)^{n-1}$, where $n$ is the number of the hour just beginning. Find the size of the culture at the beginning of the fourth hour and the size of the culture at the beginning of the first hour.
41. Mrs. Laser agrees to give her son Mark an allowance of \$0.10 on the first day of his 14-day vacation, \$0.20 on the second day, \$0.40 on the third day, and so on. Write an equation of a sequence whose terms correspond to Mark's allowance. Find the allowance Mark will receive on the last day of his vacation.
42. A small theater has 10 rows with 12 seats in the first row, 15 seats in the second row, 18 seats in the third row, and so on. Write an equation of a sequence whose terms correspond to the seats in each row. Find the number of seats in the eighth row.
43. The number of cases of a new infectious disease is doubling every year such that the number of cases is modeled by a sequence whose general term is $a_n = 75(2)^{n-1}$, where $n$ is the number of the year just beginning. Find how many cases there will be at the beginning of the sixth year. Find how many cases there were at the beginning of the first year.
44. A new college had an initial enrollment of 2700 students in 2000, and each year the enrollment increases by 150 students. Find the enrollment for each of 5 years, beginning with 2000.
45. An endangered species of sparrow had an estimated population of 800 in 2000, and scientists predict that its population will decrease by half each year. Estimate the population in 2004. Estimate the year the sparrow will be extinct.
46. A **Fibonacci sequence** is a special type of sequence in which the first two terms are 1 and each term thereafter is the sum of the two previous terms: 1, 1, 2, 3, 5, 8, . . . . Many plants and animals seem to grow according to a Fibonacci sequence, including pine cones, pineapple scales, nautilus shells, and certain flowers. Write the first 15 terms of the Fibonacci sequence.

*Find the first five terms of each sequence. Round each term after the first to four decimal places.*

47. $a_n = \dfrac{1}{\sqrt{n}}$
48. $\dfrac{\sqrt{n}}{\sqrt{n} + 1}$
49. $a_n = \left(1 + \dfrac{1}{n}\right)^n$
50. $a_n = \left(1 + \dfrac{0.05}{n}\right)^n$

### REVIEW EXERCISES

*Sketch the graph of each quadratic function. See Section 8.5.*

51. $f(x) = (x - 1)^2 + 3$
52. $f(x) = (x - 2)^2 + 1$
53. $f(x) = 2(x + 4)^2 + 2$
54. $f(x) = 3(x - 3)^2 + 4$

*Find the distance between each pair of points. See Section 9.1.*

55. $(-4, -1)$ and $(-7, -3)$
56. $(-2, -1)$ and $(-1, 5)$
57. $(2, -7)$ and $(-3, -3)$
58. $(10, -14)$ and $(5, -11)$

## 11.2 ARITHMETIC AND GEOMETRIC SEQUENCES

▶ **OBJECTIVES**

CD-ROM SSM SSG Video

1. Identify arithmetic sequences and their common differences.
2. Identify geometric sequences and their common ratios.

**1** Find the first four terms of the sequence whose general term is $a_n = 5 + (n - 1)3$.

$$a_1 = 5 + (1 - 1)3 = 5 \quad \text{Replace } n \text{ with 1.}$$
$$a_2 = 5 + (2 - 1)3 = 8 \quad \text{Replace } n \text{ with 2.}$$
$$a_3 = 5 + (3 - 1)3 = 11 \quad \text{Replace } n \text{ with 3.}$$
$$a_4 = 5 + (4 - 1)3 = 14 \quad \text{Replace } n \text{ with 4.}$$

The first four terms are 5, 8, 11, and 14. Notice that the difference of any two successive terms is 3.

$$8 - 5 = 3$$
$$11 - 8 = 3$$
$$14 - 11 = 3$$
$$\vdots$$
$$a_n - a_{n-1} = 3$$

($a_n$: *n*th term; $a_{n-1}$: previous term)

Because the difference of any two successive terms is a constant, we call the sequence an **arithmetic sequence,** or an **arithmetic progression.** The constant difference $d$ in successive terms is called the **common difference.** In this example, $d$ is 3.

### ARITHMETIC SEQUENCE AND COMMON DIFFERENCE

An **arithmetic sequence** is a sequence in which each term (after the first) differs from the preceding term by a constant amount $d$. The constant $d$ is called the **common difference** of the sequence.

The sequence 2, 6, 10, 14, 18, . . . is an arithmetic sequence. Its common difference is 4. Given the first term $a_1$ and the common difference $d$ of an arithmetic sequence, we can find any term of the sequence.

**Example 1** Write the first five terms of the arithmetic sequence whose first term is 7 and whose common difference is 2.

*Solution*

$$a_1 = 7$$
$$a_2 = 7 + 2 = 9$$
$$a_3 = 9 + 2 = 11$$
$$a_4 = 11 + 2 = 13$$
$$a_5 = 13 + 2 = 15$$

The first five terms are 7, 9, 11, 13, 15.

Notice the general pattern of the terms in Example 1.

$$a_1 = 7$$
$$a_2 = 7 + 2 = 9 \quad \text{or} \quad a_2 = a_1 + d$$
$$a_3 = 9 + 2 = 11 \quad \text{or} \quad a_3 = a_2 + d = (a_1 + d) + d = a_1 + 2d$$
$$a_4 = 11 + 2 = 13 \quad \text{or} \quad a_4 = a_3 + d = (a_1 + 2d) + d = a_1 + 3d$$
$$a_5 = 13 + 2 = 15 \quad \text{or} \quad a_5 = a_4 + d = (a_1 + 3d) + d = a_1 + 4d$$

(subscript − 1) is multiplier

The pattern on the right suggests that the general term $a_n$ of an arithmetic sequence is given by

$$a_n = a_1 + (n - 1)d$$

### GENERAL TERM OF AN ARITHMETIC SEQUENCE

The general term $a_n$ of an arithmetic sequence is given by

$$a_n = a_1 + (n - 1)d$$

where $a_1$ is the first term and $d$ is the common difference.

**Example 2** Consider the arithmetic sequence whose first term is 3 and common difference is −5.

**a.** Write an expression for the general term $a_n$.

**b.** Find the twentieth term of this sequence.

*Solution* **a.** Since this is an arithmetic sequence, the general term $a_n$ is given by $a_n = a_1 + (n - 1)d$. Here, $a_1 = 3$ and $d = -5$, so

$$a_n = 3 + (n - 1)(-5) \quad \text{Let } a_1 = 3 \text{ and } d = -5.$$
$$= 3 - 5n + 5 \quad \text{Multiply.}$$
$$= 8 - 5n \quad \text{Simplify.}$$

**b.** $a_n = 8 - 5n$

$$a_{20} = 8 - 5 \cdot 20 \quad \text{Let } n = 20.$$
$$= 8 - 100 = -92$$

**Example 3** Find the eleventh term of the arithmetic sequence whose first three terms are 2, 9, and 16.

*Solution* Since the sequence is arithmetic, the eleventh term is

$$a_{11} = a_1 + (11 - 1)d = a_1 + 10d$$

We know $a_1$ is the first term of the sequence, so $a_1 = 2$. Also, $d$ is the constant difference of terms, so $d = a_2 - a_1 = 9 - 2 = 7$. Thus,

$$a_{11} = a_1 + 10d$$
$$= 2 + 10 \cdot 7 \quad \text{Let } a_1 = 2 \text{ and } d = 7.$$
$$= 72$$

**Example 4** If the third term of an arithmetic progression is 12 and the eighth term is 27, find the fifth term.

*Solution* We need to find $a_1$ and $d$ to write the general term, which then enables us to find $a_5$, the fifth term. The given facts about terms $a_3$ and $a_8$ lead to a system of linear equations.

$$\begin{cases} a_3 = a_1 + (3 - 1)d \\ a_8 = a_1 + (8 - 1)d \end{cases} \quad \text{or} \quad \begin{cases} 12 = a_1 + 2d \\ 27 = a_1 + 7d \end{cases}$$

Next, we solve the system $\begin{cases} 12 = a_1 + 2d \\ 27 = a_1 + 7d \end{cases}$ by elimination. Multiply both sides of the second equation by $-1$ so that

$$\begin{cases} 12 = a_1 + 2d \\ -1(27) = -1(a_1 + 7d) \end{cases} \quad \text{simplifies to} \quad \begin{cases} 12 = a_1 + 2d \\ -27 = -a_1 - 7d \end{cases}$$

$$-15 = -5d \qquad \text{Add the equations.}$$

$$3 = d \qquad \text{Divide both sides by } -5.$$

To find $a_1$, let $d = 3$ in $12 = a_1 + 2d$. Then

$$12 = a_1 + 2(3)$$

$$12 = a_1 + 6$$

$$6 = a_1$$

Thus, $a_1 = 6$ and $d = 3$, so

$$\begin{aligned} a_n &= 6 + (n - 1)(3) \\ &= 6 + 3n - 3 \\ &= 3 + 3n \end{aligned}$$

and

$$a_5 = 3 + 3 \cdot 5 = 18$$

**Example 5** **FINDING SALARY**

Donna Theime has an offer for a job starting at \$40,000 per year and guaranteeing her a raise of \$1600 per year for the next 5 years. Write the general term for the arithmetic sequence that models Donna's potential annual salaries, and find her salary for the fourth year.

*Solution* The first term, $a_1$, is 40,000, and $d$ is 1600. So

$$a_n = 40{,}000 + (n - 1)(1600) = 38{,}400 + 1600n$$

$$a_4 = 38{,}400 + 1600 \cdot 4 = 44{,}800$$

Her salary for the fourth year will be \$44,800.

2 We now investigate a **geometric sequence,** also called a **geometric progression.** In the sequence 5, 15, 45, 135, . . . , each term after the first is the *product* of 3 and the preceding term. This pattern of multiplying by a constant to get the next term defines a geometric sequence. The constant is called the **common ratio** because it is the ratio of any term (after the first) to its preceding term.

$$\frac{15}{5} = 3$$

$$\frac{45}{15} = 3$$

$$\frac{135}{45} = 3$$

$$\vdots$$

nth term ⟶ previous term ⟶

$$\frac{a_n}{a_{n-1}} = 3$$

## GEOMETRIC SEQUENCE AND COMMON RATIO

A **geometric sequence** is a sequence in which each term (after the first) is obtained by multiplying the preceding term by a constant $r$. The constant $r$ is called the **common ratio** of the sequence.

The sequence $12, 6, 3, \frac{3}{2}, \ldots$ is geometric since each term after the first is the product of the previous term and $\frac{1}{2}$.

**Example 6** Write the first five terms of a geometric sequence whose first term is 7 and whose common ratio is 2.

*Solution*

$$a_1 = 7$$

$$a_2 = 7(2) = 14$$

$$a_3 = 14(2) = 28$$

$$a_4 = 28(2) = 56$$

$$a_5 = 56(2) = 112$$

The first five terms are 7, 14, 28, 56, and 112. ■

Notice the general pattern of the terms in Example 6.

$a_1 = 7$

$a_2 = 7(2) = 14$ or $a_2 = a_1(r)$

$a_3 = 14(2) = 28$ or $a_3 = a_2(r) = (a_1 \cdot r) \cdot r = a_1r^2$

$a_4 = 28(2) = 56$ or $a_4 = a_3(r) = (a_1 \cdot r^2) \cdot r = a_1r^3$

$a_5 = 56(2) = 112$ or $a_5 = a_4(r) = (a_1 \cdot r^3) \cdot r = a_1r^4$

(subscript − 1) is power

The pattern on the right on the previous page suggests that the general term of a geometric sequence is given by $a_n = a_1r^{n-1}$.

**GENERAL TERM OF A GEOMETRIC SEQUENCE**

The general term $a_n$ of a geometric sequence is given by

$$a_n = a_1r^{n-1}$$

where $a_1$ is the first term and $r$ is the common ratio.

**Example 7** Find the eighth term of the geometric sequence whose first term is 12 and whose common ratio is $\frac{1}{2}$.

*Solution* Since this is a geometric sequence, the general term $a_n$ is given by

$$a_n = a_1r^{n-1}$$

Here $a_1 = 12$ and $r = \frac{1}{2}$, so $a_n = 12\left(\frac{1}{2}\right)^{n-1}$. Evaluate $a_n$ for $n = 8$.

$$a_8 = 12\left(\frac{1}{2}\right)^{8-1} = 12\left(\frac{1}{2}\right)^7 = 12\left(\frac{1}{128}\right) = \frac{3}{32}$$

**Example 8** Find the fifth term of the geometric sequence whose first three terms are 2, −6, and 18.

*Solution* Since the sequence is geometric and $a_1 = 2$, the fifth term must be $a_1r^{5-1}$, or $2r^4$. We know that $r$ is the common ratio of terms, so $r$ must be $\frac{-6}{2}$, or −3. Thus,

$$a_5 = 2r^4$$
$$a_5 = 2(-3)^4 = 162$$

**Example 9** If the second term of a geometric sequence is $\frac{5}{4}$ and the third term is $\frac{5}{16}$, find the first term and the common ratio.

*Solution* Notice that $\frac{5}{16} \div \frac{5}{4} = \frac{1}{4}$, so $r = \frac{1}{4}$. Then

$$a_2 = a_1\left(\frac{1}{4}\right)^{2-1}$$

$$\frac{5}{4} = a_1\left(\frac{1}{4}\right)^1, \text{ or } a_1 = 5 \qquad \text{Replace } a_2 \text{ with } \frac{5}{4}.$$

The first term is 5.

**Example 10** **PREDICTING POPULATION OF A BACTERIAL CULTURE**

The population size of a bacterial culture growing under controlled conditions is doubling each day. Predict how large the culture will be at the beginning of day 7 if it measures 10 units at the beginning of day 1.

*Solution* Since the culture doubles in size each day, the population sizes are modeled by a geometric sequence. Here $a_1 = 10$ and $r = 2$. Thus,

$$a_n = a_1 r^{n-1} = 10(2)^{n-1} \quad \text{and} \quad a_7 = 10(2)^{7-1} = 640$$

The bacterial culture should measure 640 units at the beginning of day 7.

## SPOTLIGHT ON DECISION MAKING

Suppose you are a research biologist studying a particular strain of bacteria that grows at a rate of 1.5 times per hour. For a particular experiment, you will start with a culture of 200 units of bacteria and will allow the culture to grow for 7 hours. Decide whether a culture dish that holds 5000 units will be large enough for this experiment. If not, would a dish that holds 10,000 units be a better choice?

## Exercise Set 11.2

*Write the first five terms of the arithmetic or geometric sequence whose first term, $a_1$, and common difference, d, or common ratio, r, are given. See Examples 1 and 6.*

**1.** $a_1 = 4; d = 2$

**2.** $a_1 = 3; d = 10$

**3.** $a_1 = 6; d = -2$

**4.** $a_1 = -20; d = 3$

**5.** $a_1 = 1; r = 3$

**6.** $a_1 = -2; r = 2$

**7.** $a_1 = 48; r = \frac{1}{2}$

**8.** $a_1 = 1; r = \frac{1}{3}$

*Find the indicated term of each sequence. See Examples 2 and 7.*

**9.** The eighth term of the arithmetic sequence whose first term is 12 and whose common difference is 3

**10.** The twelfth term of the arithmetic sequence whose first term is 32 and whose common difference is $-4$

**11.** The fourth term of the geometric sequence whose first term is 7 and whose common ratio is $-5$

**12.** The fifth term of the geometric sequence whose first term is 3 and whose common ratio is 3

**13.** The fifteenth term of the arithmetic sequence whose first term is $-4$ and whose common difference is $-4$

**14.** The sixth term of the geometric sequence whose first term is 5 and whose common ratio is $-4$

*Find the indicated term of each sequence. See Examples 3 and 8.*

**15.** The ninth term of the arithmetic sequence $0, 12, 24, \ldots$

**16.** The thirteenth term of the arithmetic sequence $-3, 0, 3, \ldots$

**17.** The twenty-fifth term of the arithmetic sequence $20, 18, 16, \ldots$

**18.** The ninth term of the geometric sequence $5, 10, 20, \ldots$

**19.** The fifth term of the geometric sequence $2, -10, 50, \ldots$

**20.** The sixth term of the geometric sequence $\frac{1}{2}, \frac{3}{2}, \frac{9}{2}, \ldots$

*Find the indicated term of each sequence. See Examples 4 and 9.*

**21.** The eighth term of the arithmetic sequence whose fourth term is 19 and whose fifteenth term is 52

**22.** If the second term of an arithmetic sequence is 6 and the tenth term is 30, find the twenty-fifth term.

**23.** If the second term of an arithmetic progression is $-1$ and the fourth term is 5, find the ninth term.

**24.** If the second term of a geometric progression is 15 and the third term is 3, find $a_1$ and $r$.

**25.** If the second term of a geometric progression is $-\frac{4}{3}$ and the third term is $\frac{8}{3}$, find $a_1$ and $r$.

**26.** If the third term of a geometric sequence is 4 and the fourth term is $-12$, find $a_1$ and $r$.

**27.** Explain why 14, 10, and 6 may be the first three terms of an arithmetic sequence when it appears we are subtracting instead of adding to get the next term.

**28.** Explain why 80, 20, and 5 may be the first three terms of a geometric sequence when it appears we are dividing instead of multiplying to get the next term.

*Given are the first three terms of a sequence that is either arithmetic or geometric. If the sequence is arithmetic, find $a_1$ and $d$. If a sequence is geometric, find $a_1$ and $r$.*

**29.** 2, 4, 6

**30.** 8, 16, 24

**31.** 5, 10, 20

**32.** 2, 6, 18

**33.** $\frac{1}{2}, \frac{1}{10}, \frac{1}{50}$

**34.** $\frac{2}{3}, \frac{4}{3}, 2$

**35.** $x, 5x, 25x$

**36.** $y, -3y, 9y$

**37.** $p, p + 4, p + 8$

**38.** $t, t - 1, t - 2$

*Find the indicated term of each sequence.*

**39.** The twenty-first term of the arithmetic sequence whose first term is 14 and whose common difference is $\frac{1}{4}$

**40.** The fifth term of the geometric sequence whose first term is 8 and whose common ratio is $-3$

**41.** The fourth term of the geometric sequence whose first term is 3 and whose common ratio is $-\frac{2}{3}$

**42.** The fourth term of the arithmetic sequence whose first term is 9 and whose common difference is 5

**43.** The fifteenth term of the arithmetic sequence $\frac{3}{2}, 2, \frac{5}{2}, \ldots$

**44.** The eleventh term of the arithmetic sequence $2, \frac{5}{3}, \frac{4}{3}, \ldots$

**45.** The sixth term of the geometric sequence $24, 8, \frac{8}{3}, \ldots$

**46.** The eighteenth term of the arithmetic sequence $5, 2, -1, \ldots$

**47.** If the third term of an arithmetic sequence is 2 and the seventeenth term is $-40$, find the tenth term.

**48.** If the third term of a geometric sequence is $-28$ and the fourth term is $-56$, find $a_1$ and $r$.

*Solve. See Examples 5 and 10.*

**49.** An auditorium has 54 seats in the first row, 58 seats in the second row, 62 seats in the third row, and so on. Find the general term of this arithmetic sequence and the number of seats in the twentieth row.

**50.** A triangular display of cans in a grocery store has 20 cans in the first row, 17 cans in the next row, and so on, in an arithmetic sequence. Find the general term and the number of cans in the fifth row. Find how many rows there are in the display and how many cans are in the top row.

**51.** The initial size of a virus culture is 6 units, and it triples its size every day. Find the general term of the geometric sequence that models the culture's size.

**52.** A real estate investment broker predicts that a certain property will increase in value 15% each year. Thus, the yearly property values can be modeled by a geometric sequence whose common ratio $r$ is 1.15. If the initial property value was \$500,000, write the first four terms of the sequence and predict the value at the end of the third year.

**53.** A rubber ball is dropped from a height of 486 feet, and it continues to bounce one-third the height from which it last fell. Write out the first five terms of this geometric sequence and find the general term. Find how many bounces it takes for the ball to rebound less than 1 foot.

**54.** On the first swing, the length of the arc through which a pendulum swings is 50 inches. The length of each successive swing is 80% of the preceding swing. Determine whether this sequence is arithmetic or geometric. Find the length of the fourth swing.

**55.** Jose takes a job that offers a monthly starting salary of \$4000 and guarantees him a monthly raise of \$125 during his first year of training. Find the general term of this arithmetic sequence and his monthly salary at the end of his training.

**56.** At the beginning of Claudia Schaffer's exercise program, she rides 15 minutes on the Lifecycle. Each week she increases her riding time by 5 minutes. Write the general term of this arithmetic sequence, and find her riding time after 7 weeks. Find how many weeks it takes her to reach a riding time of 1 hour.

**57.** If a radioactive element has a half-life of 3 hours, then $x$ grams of the element dwindles to $\frac{x}{2}$ grams after 3 hours. If a nuclear reactor has 400 grams of that radioactive element, find the amount of radioactive material after 12 hours.

*Write the first four terms of the arithmetic or geometric sequence whose first term, $a_1$, and common difference, $d$, or common ratio, $r$, are given.*

**58.** $a_1 = \$3720, d = -\$268.50$

**59.** $a_1 = \$11{,}782.40, r = 0.5$

**60.** $a_1 = 26.8, r = 2.5$

**61.** $a_1 = 19.652; d = -0.034$

**62.** Describe a situation in your life that can be modeled by a geometric sequence. Write an equation for the sequence.

**63.** Describe a situation in your life that can be modeled by an arithmetic sequence. Write an equation for the sequence.

### REVIEW EXERCISES

*Evaluate. See Section 1.3.*

**64.** $5(1) + 5(2) + 5(3) + 5(4)$

**65.** $\frac{1}{3(1)} + \frac{1}{3(2)} + \frac{1}{3(3)}$

**66.** $2(2 - 4) + 3(3 - 4) + 4(4 - 4)$

**67.** $3^0 + 3^1 + 3^2 + 3^3$

**68.** $\frac{1}{4(1)} + \frac{1}{4(2)} + \frac{1}{4(3)}$

**69.** $\frac{8-1}{8+1} + \frac{8-2}{8+2} + \frac{8-3}{8+3}$

## 11.3 SERIES

▶ **OBJECTIVES**

CD-ROM SSM SSG Video

1. Identify finite and infinite series and use summation notation.
2. Find partial sums.

**1** A person who conscientiously saves money by saving first \$100 and then saving \$10 more each month than he saved the preceding month is saving money according to the arithmetic sequence

$$a_n = 100 + 10(n - 1)$$

Following this sequence, he can predict how much money he should save for any particular month. But if he also wants to know how much money *in total* he has saved, say, by the fifth month, he must find the *sum* of the first five terms of the sequence

$$\underbrace{100}_{a_1} + \underbrace{100 + 10}_{a_2} + \underbrace{100 + 20}_{a_3} + \underbrace{100 + 30}_{a_4} + \underbrace{100 + 40}_{a_5}$$

A sum of the terms of a sequence is called a **series** (the plural is also "series"). As our example here suggests, series are frequently used to model financial and natural phenomena.

A series is a **finite series** if it is the sum of only the first $k$ terms of the sequence, for some natural number $k$. A series is an **infinite series** if it is the sum of all the terms of the sequence. For example,

| *Sequence* | *Series* | |
|---|---|---|
| 5, 9, 13 | $5 + 9 + 13$ | Finite; $k$ is 3. |
| 5, 9, 13, . . . | $5 + 9 + 13 + \cdots$ | Infinite |
| $4, -2, 1, -\frac{1}{2}, \frac{1}{4}$ | $4 + (-2) + 1 + \left(-\frac{1}{2}\right) + \left(\frac{1}{4}\right)$ | Finite; $k$ is 5. |
| 4, −2, 1, . . . | $4 + (-2) + 1 + \cdots$ | Infinite |
| 3, 6, . . . , 99 | $3 + 6 + \cdots + 99$ | Finite; $k$ is 33. |

A shorthand notation for denoting a series when the general term of the sequence is known is called **summation notation.** The Greek uppercase letter **sigma,** $\Sigma$, is used to mean "sum." The expression $\sum_{n=1}^{5}(3n + 1)$ is read "the sum of $3n + 1$ as $n$ goes from 1 to 5"; this expression means the sum of the first five terms of the sequence whose general term is $a_n = 3n + 1$. Often, the variable $i$ is used instead of $n$ in summation notation: $\sum_{i=1}^{5}(3i + 1)$. Whether we use $n$, $i$, $k$, or some other variable, the variable is called the **index of summation.** The notation $i = 1$ below the symbol $\Sigma$ indicates the beginning value of $i$, and the number 5 above the symbol $\Sigma$ indicates the ending value of $i$. Thus, the terms of the sequence are found by successively replacing $i$ with the natural numbers 1, 2, 3, 4, 5. To find the sum, we write out the terms and then add.

$$\sum_{i=1}^{5}(3i + 1) = (3 \cdot 1 + 1) + (3 \cdot 2 + 1) + (3 \cdot 3 + 1) + (3 \cdot 4 + 1) + (3 \cdot 5 + 1)$$
$$= 4 + 7 + 10 + 13 + 16 = 50$$

**Example 1** Evaluate.

**a.** $\sum_{i=0}^{6}\frac{i-2}{2}$ **b.** $\sum_{i=3}^{5}2^i$

*Solution* **a.**

$$\sum_{i=0}^{6}\frac{i-2}{2} = \frac{0-2}{2} + \frac{1-2}{2} + \frac{2-2}{2} + \frac{3-2}{2} + \frac{4-2}{2} + \frac{5-2}{2} + \frac{6-2}{2}$$
$$= (-1) + \left(-\frac{1}{2}\right) + 0 + \frac{1}{2} + 1 + \frac{3}{2} + 2$$
$$= \frac{7}{2}, \text{ or } 3\frac{1}{2}$$

**b.**

$$\sum_{i=3}^{5}2^i = 2^3 + 2^4 + 2^5$$
$$= 8 + 16 + 32$$
$$= 56$$

**Example 2** Write each series with summation notation.

**a.** $3 + 6 + 9 + 12 + 15$ **b.** $\frac{1}{2} + \frac{1}{4} + \frac{1}{8} + \frac{1}{16}$

*Solution* **a.** Since the *difference* of each term and the preceding term is 3, the terms correspond to the first five terms of the arithmetic sequence $a_n = a_1 + (n - 1)d$ with $a_1 = 3$ and $d = 3$. So $a_n = 3 + (n - 1)3$. Thus, in summation notation,

$$3 + 6 + 9 + 12 + 15 = \sum_{i=1}^{5} [3 + (i - 1)3]$$

**b.** Since each term is the *product* of the preceding term and $\frac{1}{2}$, these terms correspond to the first four terms of the geometric sequence $a_n = a_1 r^{n-1}$. Here $a_1 = \frac{1}{2}$ and $r = \frac{1}{2}$, so $a_n = \left(\frac{1}{2}\right)\left(\frac{1}{2}\right)^{n-1} = \left(\frac{1}{2}\right)^{1+(n-1)} = \left(\frac{1}{2}\right)^{n}$. In summation notation,

$$\frac{1}{2} + \frac{1}{4} + \frac{1}{8} + \frac{1}{16} = \sum_{i=1}^{4} \left(\frac{1}{2}\right)^{i}$$

2 The sum of the first $n$ terms of a sequence is a finite series known as a **partial sum,** $S_n$. Thus, for the sequence $a_1, a_2, \ldots, a_n$, the first three partial sums are

$$S_1 = a_1$$

$$S_2 = a_1 + a_2$$

$$S_3 = a_1 + a_2 + a_3$$

In general, $S_n$ is the sum of the first $n$ terms of a sequence.

$$S_n = \sum_{i=1}^{n} a_n$$

**Example 3** Find the sum of the first three terms of the sequence whose general term is $a_n = \frac{n + 3}{2n}$.

*Solution*

$$S_3 = \sum_{i=1}^{3} \frac{i + 3}{2i} = \frac{1 + 3}{2 \cdot 1} + \frac{2 + 3}{2 \cdot 2} + \frac{3 + 3}{2 \cdot 3}$$

$$= 2 + \frac{5}{4} + 1 = 4\frac{1}{4}$$

The next example illustrates how these sums model real-life phenomena.

**Example 4** **NUMBER OF BABY GORILLAS BORN**

The number of baby gorillas born at the San Diego Zoo is a sequence defined by $a_n = n(n - 1)$, where $n$ is the number of years the zoo has owned gorillas. Find the *total* number of baby gorillas born in the *first 4 years*.

*Solution* To solve, find the sum

$$S_4 = \sum_{i=1}^{4} i(i-1)$$
$$= 1(1-1) + 2(2-1) + 3(3-1) + 4(4-1)$$
$$= 0 + 2 + 6 + 12 = 20$$

There were 20 gorillas born in the first 4 years.

## Exercise Set 11.3

*Evaluate. See Example 1.*

1. $\sum_{i=1}^{4}(i-3)$
2. $\sum_{i=1}^{5}(i+6)$
3. $\sum_{i=4}^{7}(2i+4)$
4. $\sum_{i=2}^{3}(5i-1)$
5. $\sum_{i=2}^{4}(i^2-3)$
6. $\sum_{i=3}^{5} i^3$
7. $\sum_{i=1}^{3}\left(\frac{1}{i+5}\right)$
8. $\sum_{i=2}^{4}\left(\frac{2}{i+3}\right)$
9. $\sum_{i=1}^{3}\frac{1}{6i}$
10. $\sum_{i=1}^{3}\frac{1}{3i}$
11. $\sum_{i=2}^{6} 3i$
12. $\sum_{i=3}^{6} -4i$
13. $\sum_{i=3}^{5} i(i+2)$
14. $\sum_{i=2}^{4} i(i-3)$
15. $\sum_{i=1}^{5} 2^i$
16. $\sum_{i=1}^{4} 3^{i-1}$
17. $\sum_{i=1}^{4}\frac{4i}{i+3}$
18. $\sum_{i=2}^{5}\frac{6-i}{6+i}$

*Write each series with summation notation. See Example 2.*

19. $1 + 3 + 5 + 7 + 9$
20. $4 + 7 + 10 + 13$
21. $4 + 12 + 36 + 108$
22. $5 + 10 + 20 + 40 + 80 + 160$
23. $12 + 9 + 6 + 3 + 0 + (-3)$
24. $5 + 1 + (-3) + (-7)$
25. $12 + 4 + \frac{4}{3} + \frac{4}{9}$
26. $80 + 20 + 5 + \frac{5}{4} + \frac{5}{16}$
27. $1 + 4 + 9 + 16 + 25 + 36 + 49$
28. $1 + (-4) + 9 + (-16)$

*Find each partial sum. See Example 3.*

29. Find the sum of the first two terms of the sequence whose general term is $a_n = (n+2)(n-5)$.
30. Find the sum of the first six terms of the sequence whose general term is $a_n = (-1)^n$.
31. Find the sum of the first two terms of the sequence whose general term is $a_n = n(n-6)$.
32. Find the sum of the first seven terms of the sequence whose general term is $a_n = (-1)^{n-1}$.
33. Find the sum of the first four terms of the sequence whose general term is $a_n = (n+3)(n+1)$.
34. Find the sum of the first five terms of the sequence whose general term is $a_n = \frac{(-1)^n}{2n}$.
35. Find the sum of the first four terms of the sequence whose general term is $a_n = -2n$.
36. Find the sum of the first five terms of the sequence whose general term is $a_n = (n-1)^2$.
37. Find the sum of the first three terms of the sequence whose general term is $a_n = -\frac{n}{3}$.
38. Find the sum of the first three terms of the sequence whose general term is $a_n = (n+4)^2$.

*Solve. See Example 4.*

39. A gardener is making a triangular planting with 1 tree in the first row, 2 trees in the second row, 3 trees in the third row, and so on for 10 rows. Write the sequence that describes the number of trees in each row. Find the total number of trees planted.

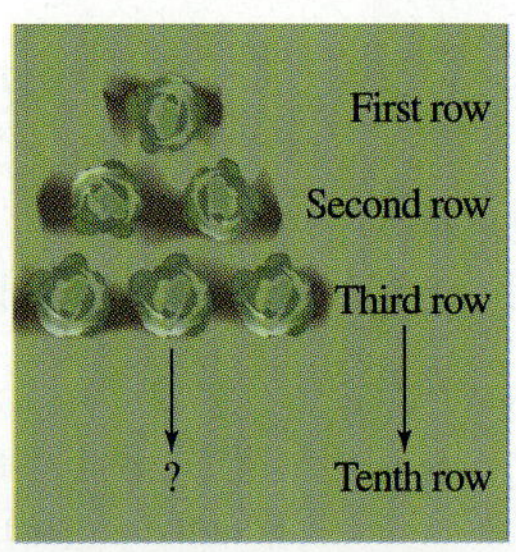

**40.** Some surfers at the beach form a human pyramid with 2 surfers in the top row, 3 surfers in the second row, 4 surfers in the third row, and so on. If there are 6 rows in the pyramid, write the sequence that describes the number of surfers in each row of the pyramid. Find the total number of surfers.

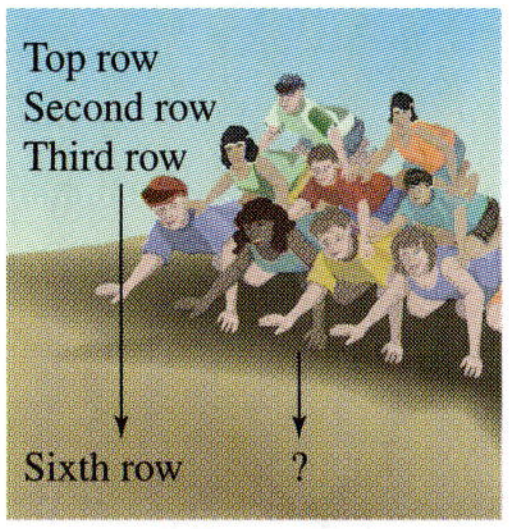


**41.** A culture of fungus starts with 6 units and doubles every day. Write the general term of the sequence that describes the growth of this fungus. Find the number of fungus units there will be at the beginning of the fifth day.

**42.** A bacterial colony begins with 100 bacteria and doubles every 6 hours. Write the general term of the sequence describing the growth of the bacteria. Find the number of bacteria there will be after 24 hours.

**43.** A bacterial colony begins with 50 bacteria and doubles every 12 hours. Write the sequence that describes the growth of the bacteria. Find the number of bacteria there will be after 48 hours.

**44.** The number of otters born each year in a new aquarium forms a sequence whose general term is $a_n = (n - 1)(n + 3)$. Find the number of otters born in the third year, and find the total number of otters born in the first three years.

**45.** The number of opossums killed each month on a new highway forms the sequence whose general term is $a_n = (n + 1)(n + 2)$, where $n$ is the number of the months. Find the number of opossums killed in the fourth month, and find the total number killed in the first four months.

**46.** In 1998 the population of an endangered fish was estimated by environmentalists to be decreasing each year. The size of the population in a given year is $24 - 4n$ thousand fish fewer than the previous year. Find the decrease in population in 2000, if year 1 is 1998. Find the total decrease in the fish population for the years 1998 through 2000.

**47.** The amount of decay in pounds of a radioactive isotope each year is given by the sequence whose general term is $a_n = 100(0.5)^n$, where $n$ is the number of the year. Find the amount of decay in the fourth year, and find the total amount of decay in the first four years.

**48.** Susan has a choice between two job offers. Job $A$ has an annual starting salary of \$20,000 with guaranteed annual raises of \$1200 for the next four years, whereas job $B$ has an annual starting salary of \$18,000 with guaranteed annual raises of \$2500 for the next four years. Compare the fifth partial sums for each sequence to determine which job would pay Susan more money over the next 5 years.

**49.** A pendulum swings a length of 40 inches on its first swing. Each successive swing is $\frac{4}{5}$ of the preceding swing. Find the length of the fifth swing and the total length swung during the first five swings. (Round to the nearest tenth of an inch.)

**50.** Explain the difference between a sequence and a series.

**51.** **a.** Write the sum $\sum_{i=1}^{7}(i + i^2)$ without summation notation.

**b.** Write the sum $\sum_{i=1}^{7} i + \sum_{i=1}^{7} i^2$ without summation notation.

**c.** Compare the results of Parts **a.** and **b.**

**d.** Do you think the following is true or false? Explain your answer.

$$\sum_{i=1}^{n}(a_n + b_n) = \sum_{i=1}^{n} a_n + \sum_{i=1}^{n} b_n$$

**52.** **a.** Write the sum $\sum_{i=1}^{6} 5i^3$ without summation notation.

**b.** Write the expression $5 \cdot \sum_{i=1}^{6} i^3$ without summation notation.

**c.** Compare the results of Parts **a.** and **b.**

**d.** Do you think the following is true or false? Explain your answer.

$$\sum_{i=1}^{n} c \cdot a_n = c \cdot \sum_{i=1}^{n} a_n \text{ where } c \text{ is a constant}$$

## REVIEW EXERCISES

*Evaluate. See Section 1.3.*

**53.** $\dfrac{5}{1 - \frac{1}{2}}$

**54.** $\dfrac{-3}{1 - \frac{1}{7}}$

**55.** $\dfrac{\frac{1}{3}}{1 - \frac{1}{10}}$

**56.** $\dfrac{\frac{6}{11}}{1 - \frac{1}{10}}$

**57.** $\dfrac{3(1 - 2^4)}{1 - 2}$

**58.** $\dfrac{2(1 - 5^3)}{1 - 5}$

**59.** $\frac{10}{2}(3 + 15)$

**60.** $\frac{12}{2}(2 + 19)$

## 11.4 PARTIAL SUMS OF ARITHMETIC AND GEOMETRIC SEQUENCES

CD-ROM SSM SSG Video

### ▶ OBJECTIVES

1. Find the partial sum of an arithmetic sequence.
2. Find the partial sum of a geometric sequence.
3. Find the sum of the terms of an infinite geometric sequence.

1 Partial sums $S_n$ are relatively easy to find when $n$ is small—that is, when the number of terms to add is small. But when $n$ is large, finding $S_n$ can be tedious. For a large $n$, $S_n$ is still relatively easy to find if the addends are terms of an arithmetic sequence or a geometric sequence.

For an arithmetic sequence, $a_n = a_1 + (n - 1)d$ for some first term $a_1$ and some common difference $d$. So $S_n$, the sum of the first $n$ terms, is

$$S_n = a_1 + (a_1 + d) + (a_1 + 2d) + \cdots + (a_1 + (n - 1)d)$$

We might also find $S_n$ by "working backward" from the $n$th term $a_n$, finding the preceding term $a_{n-1}$, by subtracting $d$ each time.

$$S_n = a_n + (a_n - d) + (a_n - 2d) + \cdots + (a_n - (n - 1)d)$$

Now add the left sides of these two equations and add the right sides.

$$2S_n = (a_1 + a_n) + (a_1 + a_n) + (a_1 + a_n) + \cdots + (a_1 + a_n)$$

The $d$ terms subtract out, leaving $n$ sums of the first term, $a_1$, and last term, $a_n$. Thus, we write

$$2S_n = n(a_1 + a_n)$$

or

$$S_n = \frac{n}{2}(a_1 + a_n)$$

**PARTIAL SUM $S_n$ OF AN ARITHMETIC SEQUENCE**

The partial sum $S_n$ of the first $n$ terms of an arithmetic sequence is given by

$$S_n = \frac{n}{2}(a_1 + a_n)$$

where $a_1$ is the first term of the sequence and $a_n$ is the $n$th term.

**Example 1** Use the partial sum formula to find the sum of the first six terms of the arithmetic sequence 2, 5, 8, 11, 14, 17, . . . .

*Solution* Use the formula for $S_n$ of an arithmetic sequence, replacing $n$ with 6, $a_1$ with 2, and $a_n$ with 17.

$$S_n = \frac{n}{2}(a_1 + a_n) = \frac{6}{2}(2 + 17) = 3(19) = 57$$

**Example 2** Find the sum of the first 30 positive integers.

*Solution* Because $1, 2, 3, \ldots, 30$ is an arithmetic sequence, use the formula for $S_n$ with $n = 30, a_1 = 1$, and $a_n = 30$. Thus,

$$S_n = \frac{n}{2}(a_1 + a_n) = \frac{30}{2}(1 + 30) = 15(31) = 465$$

**Example 3** **STACKING ROLLS OF CARPET**

Rolls of carpet are stacked in 20 rows with 3 rolls in the top row, 4 rolls in the next row, and so on, forming an arithmetic sequence. Find the total number of carpet rolls if there are 22 rolls in the bottom row.

*Solution* The list $3, 4, 5, \ldots, 22$ is the first 20 terms of an arithmetic sequence. Use the formula for $S_n$ with $a_1 = 3$, $a_n = 22$, and $n = 20$ terms. Thus,

$$S_{20} = \frac{20}{2}(3 + 22) = 10(25) = 250$$

There are a total of 250 rolls of carpet.

2 We can also derive a formula for the partial sum $S_n$ of the first $n$ terms of a geometric series. If $a_n = a_1 r^{n-1}$, then

$$S_n = \underset{\substack{\uparrow \\ \text{1st term}}}{a_1} + \underset{\substack{\uparrow \\ \text{2nd term}}}{a_1 r} + \underset{\substack{\uparrow \\ \text{3rd term}}}{a_1 r^2} + \cdots + \underset{\substack{\uparrow \\ n\text{th term}}}{a_1 r^{n-1}}$$

Multiply each side of the equation by $-r$.

$$-rS_n = -a_1 r - a_1 r^2 - a_1 r^3 - \cdots - a_1 r^n$$

Add the two equations.

$$S_n - rS_n = a_1 + (a_1 r - a_1 r) + (a_1 r^2 - a_1 r^2) + (a_1 r^3 - a_1 r^3) + \cdots - a_1 r^n$$
$$S_n - rS_n = a_1 - a_1 r^n$$

Now factor each side.

$$S_n(1 - r) = a_1(1 - r^n)$$

Solve for $S_n$ by dividing both sides by $1 - r$. Thus,

$$S_n = \frac{a_1(1 - r^n)}{1 - r}$$

as long as $r$ is not 1.

**PARTIAL SUM $S_n$ OF A GEOMETRIC SEQUENCE**

The partial sum $S_n$ of the first $n$ terms of a geometric sequence is given by

$$S_n = \frac{a_1(1 - r^n)}{1 - r}$$

where $a_1$ is the first term of the sequence, $r$ is the common ratio, and $r \neq 1$.

**Example 4** Find the sum of the first six terms of the geometric sequence 5, 10, 20, 40, 80, 160.

*Solution* Use the formula for the partial sum $S_n$ of the terms of a geometric sequence. Here, $n = 6$, the first term $a_1 = 5$, and the common ratio $r = 2$.

$$S_n = \frac{a_1(1 - r^n)}{1 - r}$$

$$S_6 = \frac{5(1 - 2^6)}{1 - 2} = \frac{5(-63)}{-1} = 315$$

**Example 5** **FINDING AMOUNT OF DONATION**

A grant from an alumnus to a university specified that the university was to receive \$800,000 during the first year and 75% of the preceding year's donation during each of the following 5 years. Find the total amount donated during the 6 years.

*Solution* The donations are modeled by the first six terms of a geometric sequence. Evaluate $S_n$ when $n = 6$, $a_1 = 800{,}000$, and $r = 0.75$.

$$S_6 = \frac{800{,}000[1 - (0.75)^6]}{1 - 0.75}$$

$$= \$2{,}630{,}468.75$$

The total amount donated during the 6 years is \$2,630,468.75.

3 Is it possible to find the sum of all the terms of an infinite sequence? Examine the partial sums of the geometric sequence $\frac{1}{2}, \frac{1}{4}, \frac{1}{8}, \ldots$.

$$S_1 = \frac{1}{2}$$

$$S_2 = \frac{1}{2} + \frac{1}{4} = \frac{3}{4}$$

$$S_3 = \frac{1}{2} + \frac{1}{4} + \frac{1}{8} = \frac{7}{8}$$

$$S_4 = \frac{1}{2} + \frac{1}{4} + \frac{1}{8} + \frac{1}{16} = \frac{15}{16}$$

$$S_5 = \frac{1}{2} + \frac{1}{4} + \frac{1}{8} + \frac{1}{16} + \frac{1}{32} = \frac{31}{32}$$

$$\vdots$$

$$S_{10} = \frac{1}{2} + \frac{1}{4} + \frac{1}{8} + \cdots + \frac{1}{2^{10}} = \frac{1023}{1024}$$

Even though each partial sum is larger than the preceding partial sum, we see that each partial sum is closer to 1 than the preceding partial sum. If $n$ gets larger and larger, then $S_n$ gets closer and closer to 1. We say that 1 is the **limit** of $S_n$ and also that 1 is the sum of the terms of this infinite sequence. In general, if $|r| < 1$, the following formula gives the sum of the terms of an infinite geometric sequence.

### SUM OF THE TERMS OF AN INFINITE GEOMETRIC SEQUENCE

The sum $S_\infty$ of the terms of an infinite geometric sequence is given by

$$S_\infty = \frac{a_1}{1 - r}$$

where $a_1$ is the first term of the sequence, $r$ is the common ratio, and $|r| < 1$. If $|r| \geq 1$, $S_\infty$ does not exist.

What happens for other values of $r$? For example, in the following geometric sequence, $r = 3$.

$$6, 18, 54, 162, \ldots$$

Here, as $n$ increases, the sum $S_n$ increases also. This time, though, $S_n$ does not get closer and closer to a fixed number but instead increases without bound.

**Example 6** Find the sum of the terms of the geometric sequence $2, \frac{2}{3}, \frac{2}{9}, \frac{2}{27}, \ldots$.

*Solution* For this geometric sequence, $r = \frac{1}{3}$. Since $|r| < 1$, we may use the formula for $S_\infty$ of a geometric sequence with $a_1 = 2$ and $r = \frac{1}{3}$.

$$S_\infty = \frac{a_1}{1 - r} = \frac{2}{1 - \frac{1}{3}} = \frac{2}{\frac{2}{3}} = 3$$

The formula for the sum of the terms of an infinite geometric sequence can be used to write a repeating decimal as a fraction. For example,

$$0.33\overline{3} = \frac{3}{10} + \frac{3}{100} + \frac{3}{1000} + \cdots$$

This sum is the sum of the terms of an infinite geometric sequence whose first term $a_1$ is $\frac{3}{10}$ and whose common ratio $r$ is $\frac{1}{10}$. Using the formula for $S_\infty$,

$$S_\infty = \frac{a_1}{1 - r} = \frac{\frac{3}{10}}{1 - \frac{1}{10}} = \frac{1}{3}$$

So $0.33\overline{3} = \frac{1}{3}$.

## Example 7 DISTANCE TRAVELED BY A PENDULUM

On its first pass, a pendulum swings through an arc whose length is 24 inches. On each pass thereafter, the arc length is 75% of the arc length on the preceding pass. Find the total distance the pendulum travels before it comes to rest.

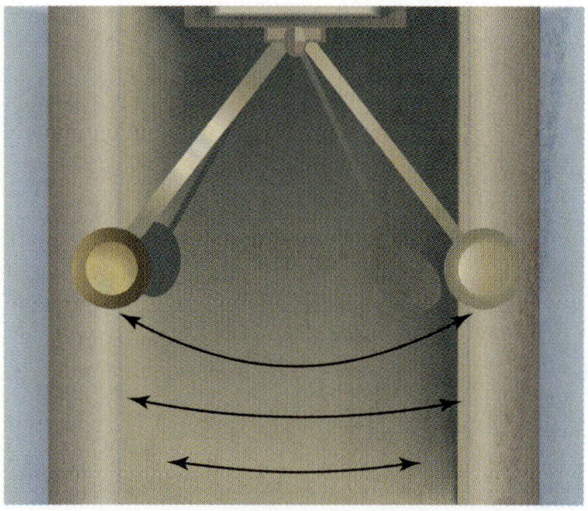

*Solution* We must find the sum of the terms of an infinite geometric sequence whose first term, $a_1$, is 24 and whose common ratio, $r$, is 0.75. Since $|r| < 1$, we may use the formula for $S_\infty$.

$$S_\infty = \frac{a_1}{1 - r} = \frac{24}{1 - 0.75} = \frac{24}{0.25} = 96$$

The pendulum travels a total distance of 96 inches before it comes to rest.

## SPOTLIGHT ON DECISION MAKING

Suppose you are a certified financial planner. You are working with a 30-year-old client whose goal is to retire at age 65 with a sum of $500,000 to live off. This year she just started making the maximum annual contribution of $2000 to a Roth IRA (Individual Retirement Account) that pays 8% interest compounded annually.

You know that if she continues to make a $2000 contribution at the beginning of each year, by the end of the $n$th year, her account increases in value by the $n$th term of the geometric sequence $a_n = 2000(1.08)^n$, which considers both the annual $2000 contribution and her earned interest. Decide whether your client will be able to reach her retirement goal by making only the maximum Roth IRA contribution each year until she retires, or if you should suggest an additional investment to help her reach her goal. (Note: Use a partial sum to find the value of the Roth IRA at the end of 35 years. To find $a_1$ of the geometric sequence, be sure to evaluate the equation for the $n$th term at $n = 1$.)

## Exercise Set 11.4

*Use the partial sum formula to find the partial sum of the given arithmetic or geometric sequence. See Examples 1 and 4.*

**1.** Find the sum of the first six terms of the arithmetic sequence $1, 3, 5, 7, \ldots$.

**2.** Find the sum of the first seven terms of the arithmetic sequence $-7, -11, -15, \ldots$.

**3.** Find the sum of the first five terms of the geometric sequence $4, 12, 36, \ldots$.

**4.** Find the sum of the first eight terms of the geometric sequence $-1, 2, -4, \ldots$.

**5.** Find the sum of the first six terms of the arithmetic sequence $3, 6, 9, \ldots$.

**6.** Find the sum of the first four terms of the arithmetic sequence $-4, -8, -12, \ldots$.

**7.** Find the sum of the first four terms of the geometric sequence $2, \frac{2}{5}, \frac{2}{25}, \ldots$.

**8.** Find the sum of the first five terms of the geometric sequence $\frac{1}{3}, -\frac{2}{3}, \frac{4}{3}, \ldots$.

*Solve. See Example 2.*

**9.** Find the sum of the first ten positive integers.

**10.** Find the sum of the first eight negative integers.

**11.** Find the sum of the first four positive odd integers.

**12.** Find the sum of the first five negative odd integers.

*Find the sum of the terms of each infinite geometric sequence. See Example 6.*

**13.** $12, 6, 3, \ldots$

**14.** $45, 15, 5, \ldots$

**15.** $\frac{1}{10}, \frac{1}{100}, \frac{1}{1000}, \ldots$

**16.** $\frac{3}{5}, \frac{3}{20}, \frac{3}{80}, \ldots$

**17.** $-10, -5, -\frac{5}{2}, \ldots$

**18.** $-16, -4, -1, \ldots$

**19.** $2, -\frac{1}{4}, \frac{1}{32}, \ldots$

**20.** $-3, \frac{3}{5}, -\frac{3}{25}, \ldots$

**21.** $\frac{2}{3}, -\frac{1}{3}, \frac{1}{6}, \ldots$

**22.** $6, -4, \frac{8}{3}, \ldots$

*Solve.*

**23.** Find the sum of the first ten terms of the sequence $-4, 1, 6, \ldots, 41$ where 41 is the tenth term.

**24.** Find the sum of the first twelve terms of the sequence $-3, -13, -23, \ldots, -113$ where $-113$ is the twelfth term.

**25.** Find the sum of the first seven terms of the sequence $3, \frac{3}{2}, \frac{3}{4}, \ldots$.

**26.** Find the sum of the first five terms of the sequence $-2, -6, -18, \ldots$.

**27.** Find the sum of the first five terms of the sequence $-12, 6, -3, \ldots$.

**28.** Find the sum of the first four terms of the sequence $-\frac{1}{4}, -\frac{3}{4}, -\frac{9}{4}, \ldots$.

**29.** Find the sum of the first twenty terms of the sequence $\frac{1}{2}, \frac{1}{4}, 0, \ldots, -\frac{17}{4}$ where $-\frac{17}{4}$ is the twentieth term.

**30.** Find the sum of the first fifteen terms of the sequence $-5, -9, -13, \ldots, -61$ where $-61$ is the fifteenth term.

**31.** If $a_1$ is 8 and $r$ is $-\frac{2}{3}$, find $S_3$.

**32.** If $a_1$ is 10, $a_{18}$ is $\frac{3}{2}$, and $d$ is $-\frac{1}{2}$, find $S_{18}$.

*Solve. See Example 3.*

**33.** Modern Car Company has come out with a new car model. Market analysts predict that 4000 cars will be sold in the first month and that sales will drop by 50 cars per month after that during the first year. Write out the first five terms of the sequence, and find the number of sold cars predicted for the twelfth month. Find the total predicted number of sold cars for the first year.

**34.** A company that sends faxes charges \$3 for the first page sent and \$0.10 less than the preceding page for each additional page sent. The cost per page forms an arithmetic sequence. Write the first five terms of this sequence, and use a partial sum to find the cost of sending a nine-page document.

**35.** Sal has two job offers: Firm *A* starts at \$22,000 per year and guarantees raises of \$1000 per year, whereas Firm *B* starts at \$20,000 and guarantees raises of \$1200 per year. Over a 10-year period, determine the more profitable offer.

**36.** The game of pool uses 15 balls numbered 1 to 15. In the variety called rotation, a player who sinks a ball receives as many points as the number on the ball. Use an arithmetic series to find the score of a player who sinks all 15 balls.

*Solve. See Example 5.*

**37.** A woman made \$30,000 during the first year she owned her business and made an additional 10% over the previous year in each subsequent year. Find how much she made during her fourth year of business. Find her total earnings during the first four years.

**38.** In free fall, a parachutist falls 16 feet during the first second, 48 feet during the second second, 80 feet during the third second, and so on. Find how far she falls during the eighth second. Find the total distance she falls during the first 8 seconds.

**39.** A trainee in a computer company takes 0.9 times as long to assemble each computer as he took to assemble the preceding computer. If it took him 30 minutes to assemble the first computer, find how long it takes him to assemble the fifth computer. Find the total time he takes to assemble the first five computers (round to the nearest minute).

**40.** On a gambling trip to Reno, Carol doubled her bet each time she lost. If her first losing bet was \$5 and she lost six consecutive bets, find how much she lost on the sixth bet. Find the total amount lost on these six bets.

*Solve. See Example 7.*

**41.** A ball is dropped from a height of 20 feet and repeatedly rebounds to a height that is $\frac{4}{5}$ of its previous height. Find the total distance the ball covers before it comes to rest.

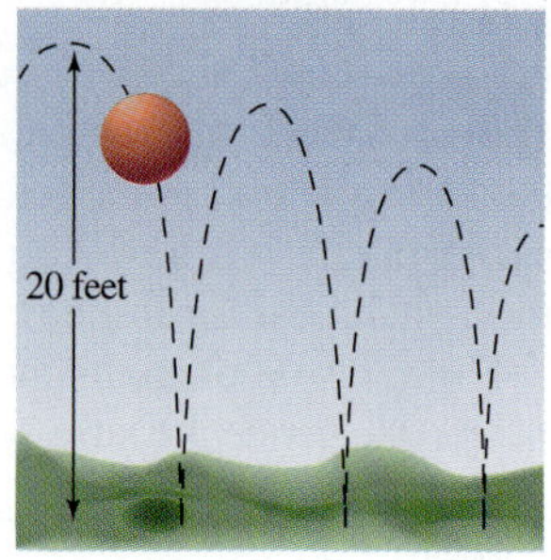


**42.** A rotating flywheel coming to rest makes 300 revolutions in the first minute and in each minute thereafter makes $\frac{2}{5}$ as many revolutions as in the preceding minute. Find how many revolutions the wheel makes before it comes to rest.

*Solve.*

**43.** In the pool game of rotation, player *A* sinks balls numbered 1 to 9, and player *B* sinks the rest of the balls. Use arithmetic series to find each player's score (see Exercise 36).

**44.** A godfather deposited \$250 in a savings account on the day his godchild was born. On each subsequent birthday he deposited \$50 more than he deposited the previous year. Find how much money he deposited on his godchild's twenty-first birthday. Find the total amount deposited over the 21 years.

**45.** During the holiday rush a business can rent a computer system for \$200 the first day, with the rental fee decreasing \$5 for each additional day. Find the fee paid for 20 days during the holiday rush.

**46.** The spraying of a field with insecticide killed 6400 weevils the first day, 1600 the second day, 400 the third day, and so on. Find the total number of weevils killed during the first 5 days.

**47.** A college student humorously asks his parents to charge him room and board according to this geometric sequence: \$0.01 for the first day of the month, \$0.02 for the second day, \$0.04 for the third day, and so on. Find the total room and board he would pay for 30 days.

**48.** Following its television advertising campaign, a bank attracted 80 new customers the first day, 120 the second day, 160 the third day, and so on, in an arithmetic sequence. Find how many new customers were attracted during the first 5 days following its television campaign.

**49.** Write $0.88\overline{8}$ as an infinite geometric series and use the formula for $S_\infty$ to write it as a rational number.

**50.** Write $0.54\overline{54}$ as an infinite geometric series and use the formula $S_\infty$ to write it as a rational number.

**51.** Explain whether the sequence 5, 5, 5, . . . is arithmetic, geometric, neither, or both.

**52.** Describe a situation in everyday life that can be modeled by an infinite geometric series.

### REVIEW EXERCISES

*Evaluate. See Section 1.3.*

**53.** $6 \cdot 5 \cdot 4 \cdot 3 \cdot 2 \cdot 1$

**54.** $8 \cdot 7 \cdot 6 \cdot 5 \cdot 4 \cdot 3 \cdot 2 \cdot 1$

**55.** $\dfrac{3 \cdot 2 \cdot 1}{2 \cdot 1}$

**56.** $\dfrac{5 \cdot 4 \cdot 3 \cdot 2 \cdot 1}{3 \cdot 2 \cdot 1}$

*Multiply. See Section 5.4.*

**57.** $(x + 5)^2$

**58.** $(x - 2)^2$

**59.** $(2x - 1)^3$

**60.** $(3x + 2)^3$

## 11.5 THE BINOMIAL THEOREM

CD-ROM SSM SSG Video

**OBJECTIVES**

1. Use Pascal's triangle to expand binomials.
2. Evaluate factorials.
3. Use the binomial theorem to expand binomials.
4. Find the $n$th term in the expansion of a binomial raised to a positive power.

In this section, we learn how to **expand** binomials of the form $(a + b)^n$ easily. Expanding a binomial such as $(a + b)^n$ means to write the factored form as a sum. First, we review the patterns in the expansions of $(a + b)^n$.

$$(a + b)^0 = 1 \qquad \text{1 term}$$
$$(a + b)^1 = a + b \qquad \text{2 terms}$$
$$(a + b)^2 = a^2 + 2ab + b^2 \qquad \text{3 terms}$$
$$(a + b)^3 = a^3 + 3a^2b + 3ab^2 + b^3 \qquad \text{4 terms}$$
$$(a + b)^4 = a^4 + 4a^3b + 6a^2b^2 + 4ab^3 + b^4 \qquad \text{5 terms}$$
$$(a + b)^5 = a^5 + 5a^4b + 10a^3b^2 + 10a^2b^3 + 5ab^4 + b^5 \qquad \text{6 terms}$$

Notice the following patterns.

1. The expansion of $(a + b)^n$ contains $n + 1$ terms. For example, for $(a + b)^3$, $n = 3$, and the expansion contains $3 + 1$ terms, or 4 terms.
2. The first term of the expansion of $(a + b)^n$ is $a^n$, and the last term is $b^n$.
3. The powers of $a$ decrease by 1 for each term, whereas the powers of $b$ increase by 1 for each term.
4. For each term of the expansion of $(a + b)^n$, the sum of the exponents of $a$ and $b$ is $n$. (For example, the sum of the exponents of $5a^4b$ is $4 + 1$, or 5, and the sum of the exponents of $10a^3b^2$ is $3 + 2$, or 5.)

1 There are patterns in the coefficients of the terms as well. Written in a triangular array, the coefficients are called **Pascal's triangle.**

$$
\begin{array}{lcccccccccccl}
(a+b)^0: & & & & & & 1 & & & & & & n=0 \\
(a+b)^1: & & & & & 1 & & 1 & & & & & n=1 \\
(a+b)^2: & & & & 1 & & 2 & & 1 & & & & n=2 \\
(a+b)^3: & & & 1 & & 3 & & 3 & & 1 & & & n=3 \\
(a+b)^4: & & 1 & & 4 & & 6 & & 4 & & 1 & & n=4 \\
(a+b)^5: & 1 & & 5 & & 10 & & 10 & & 5 & & 1 & n=5
\end{array}
$$

Each row in Pascal's triangle begins and ends with 1. Any other number in a row is the sum of the two closest numbers above it. Using this pattern, we can write the next row, for $n = 6$, by first writing the number 1. Then we can add the consecutive numbers in the row for $n = 5$ and write each sum "between and below" the pair. We complete the row by writing a 1.

$$
\begin{array}{ccccccccccccccl}
 & 1 & & 5 & & 10 & & 10 & & 5 & & 1 & & & n=5 \\
1 & & 6 & & 15 & & 20 & & 15 & & 6 & & 1 & & n=6
\end{array}
$$

We can use Pascal's triangle and the patterns noted to expand $(a + b)^n$ without actually multiplying any terms.

**Example 1** Expand $(a + b)^6$.

*Solution* Using the $n = 6$ row of Pascal's triangle as the coefficients and following the patterns noted, $(a + b)^6$ can be expanded as

$$a^6 + 6a^5b + 15a^4b^2 + 20a^3b^3 + 15a^2b^4 + 6ab^5 + b^6$$

2 For a large $n$, the use of Pascal's triangle to find coefficients for $(a + b)^n$ can be tedious. An alternative method for determining these coefficients is based on the concept of a **factorial.**

The **factorial of $n$,** written $n!$ (read "$n$ factorial"), is the product of the first $n$ consecutive natural numbers.

**FACTORIAL OF $n$: $n!$**

If $n$ is a natural number, then $n! = n(n - 1)(n - 2)(n - 3) \cdot \cdots \cdot 3 \cdot 2 \cdot 1$.
The factorial of 0, written $0!$, is defined to be 1.

For example, $3! = 3 \cdot 2 \cdot 1 = 6$, $5! = 5 \cdot 4 \cdot 3 \cdot 2 \cdot 1 = 120$, and $0! = 1$.

**Example 2** Evaluate each expression.

**a.** $\dfrac{5!}{6!}$  **b.** $\dfrac{10!}{7!3!}$

**c.** $\dfrac{3!}{2!1!}$  **d.** $\dfrac{7!}{7!0!}$

*Solution* **a.** $\frac{5!}{6!} = \frac{5 \cdot 4 \cdot 3 \cdot 2 \cdot 1}{6 \cdot 5 \cdot 4 \cdot 3 \cdot 2 \cdot 1} = \frac{1}{6}$

**b.** $\frac{10!}{7!3!} = \frac{10 \cdot 9 \cdot 8 \cdot 7!}{7! \cdot 3 \cdot 2 \cdot 1} = \frac{10 \cdot 9 \cdot 8}{3 \cdot 2 \cdot 1} = 10 \cdot 3 \cdot 4 = 120$

**c.** $\frac{3!}{2!1!} = \frac{3 \cdot 2 \cdot 1}{2 \cdot 1 \cdot 1} = 3$

**d.** $\frac{7!}{7!0!} = \frac{7!}{7! \cdot 1} = 1$

**HELPFUL HINT**

We can use a calculator with a factorial key to evaluate a factorial. A calculator uses scientific notation for large results.

3 It can be proved, although we won't do so here, that the coefficients of terms in the expansion of $(a + b)^n$ can be expressed in terms of factorials. Following patterns 1 through 4 given earlier and using the factorial expressions of the coefficients, we have what is known as the **binomial theorem.**

**BINOMIAL THEOREM**

If $n$ is a positive integer, then

$$(a + b)^n = a^n + \frac{n}{1!}a^{n-1}b^1 + \frac{n(n-1)}{2!}a^{n-2}b^2 + \frac{n(n-1)(n-2)}{3!}a^{n-3}b^3 + \cdots + b^n$$

We call the formula for $(a + b)^n$ given by the binomial theorem the **binomial formula.**

**Example 3** Use the binomial theorem to expand $(x + y)^{10}$.

*Solution* Let $a = x$, $b = y$, and $n = 10$ in the binomial formula.

$$\begin{aligned}(x + y)^{10} &= x^{10} + \frac{10}{1!}x^9y + \frac{10 \cdot 9}{2!}x^8y^2 + \frac{10 \cdot 9 \cdot 8}{3!}x^7y^3 + \frac{10 \cdot 9 \cdot 8 \cdot 7}{4!}x^6y^4 \\ &\quad + \frac{10 \cdot 9 \cdot 8 \cdot 7 \cdot 6}{5!}x^5y^5 + \frac{10 \cdot 9 \cdot 8 \cdot 7 \cdot 6 \cdot 5}{6!}x^4y^6 \\ &\quad + \frac{10 \cdot 9 \cdot 8 \cdot 7 \cdot 6 \cdot 5 \cdot 4}{7!}x^3y^7 \\ &\quad + \frac{10 \cdot 9 \cdot 8 \cdot 7 \cdot 6 \cdot 5 \cdot 4 \cdot 3}{8!}x^2y^8 \\ &\quad + \frac{10 \cdot 9 \cdot 8 \cdot 7 \cdot 6 \cdot 5 \cdot 4 \cdot 3 \cdot 2}{9!}xy^9 + y^{10} \\ &= x^{10} + 10x^9y + 45x^8y^2 + 120x^7y^3 + 210x^6y^4 + 252x^5y^5 + 210x^4y^6 \\ &\quad + 120x^3y^7 + 45x^2y^8 + 10xy^9 + y^{10}\end{aligned}$$

**Example 4** Use the binomial theorem to expand $(x + 2y)^5$.

*Solution* Let $a = x$ and $b = 2y$ in the binomial formula.

$$(x + 2y)^5 = x^5 + \frac{5}{1!}x^4(2y) + \frac{5 \cdot 4}{2!}x^3(2y)^2 + \frac{5 \cdot 4 \cdot 3}{3!}x^2(2y)^3 + \frac{5 \cdot 4 \cdot 3 \cdot 2}{4!}x(2y)^4 + (2y)^5$$
$$= x^5 + 10x^4y + 40x^3y^2 + 80x^2y^3 + 80xy^4 + 32y^5$$

**Example 5** Use the binomial theorem to expand $(3m - n)^4$.

*Solution* Let $a = 3m$ and $b = -n$ in the binomial formula.

$$(3m - n)^4 = (3m)^4 + \frac{4}{1!}(3m)^3(-n) + \frac{4 \cdot 3}{2!}(3m)^2(-n)^2 + \frac{4 \cdot 3 \cdot 2}{3!}(3m)(-n)^3 + (-n)^4$$
$$= 81m^4 - 108m^3n + 54m^2n^2 - 12mn^3 + n^4$$

4 Sometimes it is convenient to find a specific term of a binomial expansion without writing out the entire expansion. By studying the expansion of binomials, a pattern forms for each term. This pattern is most easily stated for the $(r + 1)$st term.

**($r$ + 1)ST TERM IN A BINOMIAL EXPANSION**

The $(r + 1)$st term of the expansion of $(a + b)^n$ is $\dfrac{n!}{r!(n - r)!}a^{n-r}b^r$.

**Example 6** Find the eighth term in the expansion of $(2x - y)^{10}$.

*Solution* Use the formula, with $n = 10$, $a = 2x$, $b = -y$, and $r + 1 = 8$. Notice that, since $r + 1 = 8$, $r = 7$.

$$\frac{n!}{r!(n - r)!}a^{n-r}b^r = \frac{10!}{7!3!}(2x)^3(-y)^7$$
$$= 120(8x^3)(-y^7)$$
$$= -960x^3y^7$$

## Exercise Set 11.5

*Use Pascal's triangle to expand the binomial. See Example 1.*

1. $(m + n)^3$
2. $(x + y)^4$
3. $(c + d)^5$
4. $(a + b)^6$
5. $(y - x)^5$
6. $(q - r)^7$
7. Explain how to generate a row of Pascal's triangle.
8. Write the $n = 8$ row of Pascal's triangle.

*Evaluate each expression. See Example 2.*

9. $\frac{8!}{7!}$
10. $\frac{6!}{0!}$
11. $\frac{7!}{5!}$
12. $\frac{8!}{5!}$
13. $\frac{10!}{7!2!}$
14. $\frac{9!}{5!3!}$
15. $\frac{8!}{6!0!}$
16. $\frac{10!}{4!6!}$

*Use the binomial formula to expand each binomial. See Examples 3 through 5.*

17. $(a + b)^7$
18. $(x + y)^8$
19. $(a + 2b)^5$
20. $(x + 3y)^6$
21. $(q + r)^9$
22. $(b + c)^6$
23. $(4a + b)^5$
24. $(3m + n)^4$
25. $(5a - 2b)^4$
26. $(m - 4)^6$
27. $(2a + 3b)^3$
28. $(4 - 3x)^5$
29. $(x + 2)^5$
30. $(3 + 2a)^4$

*Find the indicated term. See Example 6.*

31. The fifth term of the expansion of $(c - d)^5$
32. The fourth term of the expansion of $(x - y)^6$
33. The eighth term of the expansion of $(2c + d)^7$
34. The tenth term of the expansion of $(5x - y)^9$
35. The fourth term of the expansion of $(2r - s)^5$
36. The first term of the expansion of $(3q - 7r)^6$
37. The third term of the expansion of $(x + y)^4$
38. The fourth term of the expansion of $(a + b)^8$
39. The second term of the expansion of $(a + 3b)^{10}$
40. The third term of the expansion of $(m + 5n)^7$

### REVIEW EXERCISES

*Sketch the graph of each function. Decide whether each function is one-to-one. See Sections 8.5 and 10.2.*

41. $f(x) = |x|$
42. $g(x) = 3(x - 1)^2$
43. $H(x) = 2x + 3$
44. $F(x) = -2$
45. $f(x) = x^2 + 3$
46. $h(x) = -(x + 1)^2 - 4$

For additional Chapter Projects, visit the Real World Activities Website by going to http://www.prenhall.com/martin-gay.

# 11 Chapter Project

## Modeling College Tuition

Annual college tuition has steadily increased since 1970. According to the College Board, by the 1999–2000 academic year, the average annual tuition at a public four-year university had increased to \$3356. Similarly, average annual tuition at private four-year universities had grown to \$15,380.

Over the past few years, annual tuition at four-year public universities has been increasing at an average rate of 4.5% per year. Over the same time period, annual tuition at four-year private universities has been increasing at an average rate of \$804.30 per year. In this project, you will have the opportunity to model and investigate the trend in increasing tuition at public and private universities. This project may be completed by working in groups or individually.

1. Using the information given in the introductory paragraphs, decide whether the sequence of public university tuitions is arithmetic or geometric.
2. Using the information given in the introductory paragraphs, decide whether the sequence of private university tuitions is arithmetic or geometric.
3. Find the general term of the sequence that describes the pattern of average annual tuition for four-year public universities. Let $n = 1$ represent the 1999–2000 academic year.
4. Find the general term of the sequence that describes the pattern of average annual tuition for four-year private universities. Let $n = 1$ represent the 1999–2000 academic year.
5. Assuming that the rate of tuition increase remains the same, use the general term equation from Question 3 to find the average annual tuition at a four-year public university for the 2002–2003 academic year.
6. Assuming that the rate of tuition increase remains the same, use the general term equation from Question 4 to find the average annual tuition at a four-year private university for the 2003–2004 academic year.
7. Use partial sums to find the average cost of a four-year college education at a public university for a student who started college in the 1999–2000 academic year.
8. Use partial sums to find the average cost of a four-year college education at a private university for a student who starts college in the 2002–2003 academic year. (Hint: One way to do this is to find $S_7$ and subtract $S_3$ from it. If you use this method, explain why this gives the desired sum.)
9. (Optional) Use newspapers or news magazines to find a situation that can be modeled by a sequence. Briefly describe the situation, and decide whether it is an arithmetic or a geometric sequence. Find an equation of the general term of the sequence.

| *Academic Year* | ***n*** |
|---|---|
| 1999–2000 | $n = 1$ |
| 2000–2001 | $n = 2$ |
| 2001–2002 | $n = 3$ |
| 2002–2003 | $n = 4$ |
| 2003–2004 | $n = 5$ |
| 2004–2005 | $n = 6$ |
| 2005–2006 | $n = 7$ |

## CHAPTER 11 VOCABULARY CHECK

*Fill in each blank with one of the words or phrases listed below.*

| | | | |
|---|---|---|---|
| general term | common difference | infinite sequence | common ratio |
| Pascal's triangle | finite sequence | factorial of $n$ | arithmetic sequence |
| geometric sequence | series | | |

1. A(n) ____________ is a function whose domain is the set of natural numbers $\{1, 2, 3, \ldots, n\}$, where $n$ is some natural number.
2. The ____________, written $n!$, is the product of the first $n$ consecutive natural numbers.
3. A(n) ______________ is a function whose domain is the set of natural numbers.
4. A(n) ________________ is a sequence in which each term (after the first) is obtained by multiplying the preceding term by a constant amount $r$. The constant $r$ is called the ____________ of the sequence.
5. A sum of the terms of a sequence is called a _____.
6. The $n$th term of the sequence $a_n$ is called the ___________.
7. A(n) ________________ is a sequence in which each term (after the first) differs from the preceding term by a constant amount $d$. The constant $d$ is called the ________________ of the sequence.
8. A(n) triangle array of the coefficients of the terms of the expansions of $(a + b)^n$ is called ____________.

# CHAPTER 11 HIGHLIGHTS

| DEFINITIONS AND CONCEPTS | EXAMPLES |
|---|---|
| **Section 11.1 Sequences** | |
| An **infinite sequence** is a function whose domain is the set of natural numbers $\{1, 2, 3, 4, \ldots\}$. | ***Infinite Sequence*** <br> $2, 4, 6, 8, 10, \ldots$ |
| A **finite sequence** is a function whose domain is the set of natural numbers $\{1, 2, 3, 4, \ldots, n\}$, where $n$ is some natural number. | ***Finite Sequence*** <br> $1, -2, 3, -4, 5, -6$ |
| The notation $a_n$, where $n$ is a natural number, is used to denote a sequence. | Write the first four terms of the sequence whose general term is $a_n = n^2 + 1$. <br> $a_1 = 1^2 + 1 = 2$ <br> $a_2 = 2^2 + 1 = 5$ <br> $a_3 = 3^2 + 1 = 10$ <br> $a_4 = 4^2 + 1 = 17$ |
| **Section 11.2 Arithmetic and Geometric Sequences** | |
| An **arithmetic sequence** is a sequence in which each term differs from the preceding term by a constant amount $d$, called the **common difference.** | ***Arithmetic Sequence*** <br> $5, 8, 11, 14, 17, 20, \ldots$ <br> Here, $a_1 = 5$ and $d = 3$. |
| The **general term** $a_n$ of an arithmetic sequence is given by <br> $a_n = a_1 + (n - 1)d$ <br> where $a_1$ is the first term and $d$ is the common difference. | The general term is <br> $a_n = a_1 + (n - 1)d$ or <br> $a_n = 5 + (n - 1)3$ <br> *(continued)* |

| DEFINITIONS AND CONCEPTS | EXAMPLES |
|---|---|

## Section 11.2 Arithmetic and Geometric Sequences

A **geometric sequence** is a sequence in which each term is obtained by multiplying the preceding term by a constant $r$, called the **common ratio.**

***Geometric Sequence***

$$12, -6, 3, -\frac{3}{2}, \ldots$$

Here $a_1 = 12$ and $r = -\frac{1}{2}$.

The **general term** $a_n$ of a geometric sequence is given by

$$a_n = a_1 r^{n-1}$$

where $a_1$ is the first term and $r$ is the common ratio.

The general term is

$$a_n = a_1 r^{n-1} \text{ or}$$

$$a_n = 12\left(-\frac{1}{2}\right)^{n-1}$$

## Section 11.3 Series

A sum of the terms of a sequence is called a **series.**

A shorthand notation for denoting a series is called **summation notation:**

$$\sum_{i=1}^{4}$$

Greek letter sigma used to mean sum

index of summation → $i=1$

| *Sequence* | *Series* | |
|---|---|---|
| 3, 7, 11, 15 | 3 + 7 + 11 + 15 | finite |
| 3, 7, 11, 15, . . . | 3 + 7 + 11 + 15 + · · · | infinite |

$$\sum_{i=1}^{4} 3^i = 3^1 + 3^2 + 3^3 + 3^4$$

$$= 3 + 9 + 27 + 81$$

$$= 120$$

## Section 11.4 Partial Sums of Arithmetic and Geometric Sequences

**Partial sum,** $S_n$, of the first $n$ terms of an arithmetic sequence:

$$S_n = \frac{n}{2}(a_1 + a_n)$$

where $a_1$ is the first term and $a_n$ is the $n$th term.

The sum of the first five terms of the arithmetic sequence

$$12, 24, 36, 48, 60, \ldots \text{ is}$$

$$S_n = \frac{5}{2}(12 + 60) = 180$$

**Partial sum,** $S_n$, of the first $n$ terms of a geometric sequence:

$$S_n = \frac{a_1(1 - r^n)}{1 - r}$$

where $a_1$ is the first term, $r$ is the common ratio, and $r \neq 1$.

The sum of the first five terms of the geometric sequence

$$15, 30, 60, 120, 240, \ldots \text{ is}$$

$$S_5 = \frac{15(1 - 2^5)}{1 - 2} = 465$$

Sum of the terms of an infinite geometric sequence:

$$S_\infty = \frac{a_1}{1 - r}$$

where $a_1$ is the first term, $r$ is the common ratio, and $|r| < 1$. (If $|r| \geq 1$, $S_\infty$ does not exist.)

The sum of the terms of the infinite geometric sequence

$$1, \frac{1}{3}, \frac{1}{9}, \frac{1}{27}, \ldots \text{ is}$$

$$S_\infty = \frac{1}{1 - \frac{1}{3}} = \frac{3}{2}$$

| DEFINITIONS AND CONCEPTS | EXAMPLES |
|---|---|
| **Section 11.5 The Binomial Theorem** | |
| The **factorial of *n*,** written $n!$, is the product of the first $n$ consecutive natural numbers. | $5! = 5 \cdot 4 \cdot 3 \cdot 2 \cdot 1 = 120$ |
| ***Binomial Theorem*** If $n$ is a positive integer, then $(a+b)^n = a^n + \frac{n}{1!}a^{n-1}b^1 + \frac{n(n-1)}{2!}a^{n-2}b^2 + \frac{n(n-1)(n-2)}{3!}a^{n-3}b^3 + \cdots + b^n$ | Expand $(3x+y)^4$. $(3x+y)^4 = (3x)^4 + \frac{4}{1!}(3x)^3(y)^1 + \frac{4 \cdot 3}{2!}(3x)^2(y)^2 + \frac{4 \cdot 3 \cdot 2}{3!}(3x)^1y^3 + y^4$ $= 81x^4 + 108x^3y + 54x^2y^2 + 12xy^3 + y^4$ |

## CHAPTER 11 REVIEW

*(11.1) Find the indicated term(s) of the given sequence.*

**1.** The first five terms of the sequence $a_n = -3n^2$

**2.** The first five terms of the sequence $a_n = n^2 + 2n$

**3.** The one-hundredth term of the sequence $a_n = \frac{(-1)^n}{100}$

**4.** The fiftieth term of the sequence $a_n = \frac{2n}{(-1)^2}$

**5.** The general term $a_n$ of the sequence $\frac{1}{6}, \frac{1}{12}, \frac{1}{18}, \ldots$.

**6.** The general term $a_n$ of the sequence $-1, 4, -9, 16, \ldots$.

*Solve the following applications.*

**7.** The distance in feet that an olive falling from rest in a vacuum will travel during each second is given by an arithmetic sequence whose general term is $a_n = 32n - 16$, where $n$ is the number of the second. Find the distance the olive will fall during the fifth, sixth, and seventh seconds.

**8.** A culture of yeast doubles every day in a geometric progression whose general term is $a_n = 100(2)^{n-1}$, where $n$ is the number of the day just ending. Find how many days it takes the yeast culture to measure at least 10,000. Find the original measure of the yeast culture.

**9.** The Centers for Disease Control and Prevention (CDC) reported that a new type of virus infected approximately 450 people during 1999, the year it was first discovered. The CDC predicts that during the next decade the virus will infect three times as many people each year as the year before. Write out the first five terms of this geometric sequence, and predict the number of infected people there will be in 2003.

**10.** The first row of an amphitheater contains 50 seats, and each row thereafter contains 8 additional seats. Write the first ten terms of this arithmetic progression, and find the number of seats in the tenth row.

*(11.2)*

**11.** Find the first five terms of the geometric sequence whose first term is $-2$ and whose common ratio is $\frac{2}{3}$.

**12.** Find the first five terms of the arithmetic sequence whose first term is 12 and whose common difference is $-1.5$.

**13.** Find the thirtieth term of the arithmetic sequence whose first term is $-5$ and whose common difference is 4.

**14.** Find the eleventh term of the arithmetic sequence whose first term is 2 and whose common difference is $\frac{3}{4}$.

**15.** Find the twentieth term of the arithmetic sequence whose first three terms are 12, 7, and 2.

**16.** Find the sixth term of the geometric sequence whose first three terms are 4, 6, and 9.

**17.** If the fourth term of an arithmetic sequence is 18 and the twentieth term is 98, find the first term and the common difference.

**18.** If the third term of a geometric sequence is $-48$ and the fourth term is 192, find the first term and the common ratio.

**19.** Find the general term of the sequence $\frac{3}{10}, \frac{3}{100}, \frac{3}{1000}, \ldots$.

**20.** Find a general term that satisfies the terms shown for the sequence $50, 58, 66, \ldots$.

*Determine whether each of the following sequences is arithmetic, geometric, or neither. If a sequence is arithmetic, find $a_1$ and d. If a sequence is geometric, find $a_1$ and r.*

**21.** $\frac{8}{3}, 4, 6, \ldots$

**22.** $-10.5, -6.1, -1.7$

**23.** $7x, -14x, 28x$

**24.** $3x^2, 9x^4, 81x^8, \ldots$

*Solve the following applications.*

**25.** To test the bounce of a racquetball, the ball is dropped from a height of 8 feet. The ball is judged "good" if it rebounds at least 75% of its previous height with each bounce. Write out the first six terms of this geometric sequence (round to the nearest tenth). Determine if a ball is "good" that rebounds to a height of 2.5 feet after the fifth bounce.

**26.** A display of oil cans in an auto parts store has 25 cans in the bottom row, 21 cans in the next row, and so on, in an arithmetic progression. Find the general term and the number of cans in the top row.

**27.** Suppose that you save $1 the first day of a month, $2 the second day, $4 the third day, continuing to double your savings each day. Write the general term of this geometric sequence and find the amount you will save on the tenth day. Estimate the amount you will save on the thirtieth day of the month, and check your estimate with a calculator.

**28.** On the first swing, the length of an arc through which a pendulum swings is 30 inches. The length of the arc for each successive swing is 70% of the preceding swing. Find the length of the arc for the fifth swing.

**29.** Rosa takes a job that has a monthly starting salary of $900 and guarantees her a monthly raise of $150 during her 6-month training period. Find the general term of this sequence and her salary at the end of her training.

**30.** A sheet of paper is $\frac{1}{512}$-inch thick. By folding the sheet in half, the total thickness will be $\frac{1}{256}$ inch. A second fold produces a total thickness of $\frac{1}{128}$ inch. Estimate the thickness of the stack after 15 folds, and then check your estimate with a calculator.

***(11.3)*** *Write out the terms and find the sum for each of the following.*

**31.** $\sum_{i=1}^{5} (2i - 1)$

**32.** $\sum_{i=1}^{5} i(i + 2)$

**33.** $\sum_{i=2}^{4} \frac{(-1)^i}{2i}$

**34.** $\sum_{i=3}^{5} 5(-1)^{i-1}$

*Find the partial sum of the given sequence.*

**35.** $S_4$ of the sequence $a_n = (n - 3)(n + 2)$

**36.** $S_6$ of the sequence $a_n = n^2$

**37.** $S_5$ of the sequence $a_n = -8 + (n - 1)3$

**38.** $S_3$ of the sequence $a_n = 5(4)^{n-1}$

*Write the sum with $\Sigma$ notation.*

**39.** $1 + 3 + 9 + 27 + 81 + 243$

**40.** $6 + 2 + (-2) + (-6) + (-10) + (-14) + (-18)$

**41.** $\frac{1}{4} + \frac{1}{16} + \frac{1}{64} + \frac{1}{256}$

**42.** $1 + \left(-\frac{3}{2}\right) + \frac{9}{4}$

*Solve.*

**43.** A yeast colony begins with 20 yeast and doubles every 8 hours. Write the sequence that describes the growth of the yeast, and find the total yeast after 48 hours.

**44.** The number of cranes born each year in a new aviary forms a sequence whose general term is $a_n = n^2 + 2n - 1$. Find the number of cranes born in the fourth year and the total number of cranes born in the first four years.

**45.** Harold has a choice between two job offers. Job *A* has an annual starting salary of $39,500 with guaranteed annual raises of $2200 for the next four years, whereas job *B* has an annual starting salary of $41,000 with guaranteed annual raises of $1400 for the next four years. Compare the salaries for the fifth year under each job offer.

**46.** A sample of radioactive waste is decaying such that the amount decaying in kilograms during year *n* is $a_n = 200(0.5)^n$. Find the amount of decay in the third year, and the total amount of decay in the first three years.

***(11.4)*** *Find the partial sum of the given sequence.*

**47.** The sixth partial sum of the sequence $15, 19, 23, \ldots$.

**48.** The ninth partial sum of the sequence $5, -10, 20, \ldots$.

**49.** The sum of the first 30 odd positive integers

**50.** The sum of the first 20 positive multiples of 7

**51.** The sum of the first 20 terms of the sequence $8, 5, 2, \ldots$.

**52.** The sum of the first eight terms of the sequence $\frac{3}{4}, \frac{9}{4}, \frac{27}{4}, \ldots$

**53.** $S_4$ if $a_1 = 6$ and $r = 5$

**54.** $S_{100}$ if $a_1 = -3$ and $d = -6$

*Find the sum of each infinite geometric sequence.*

**55.** $5, \frac{5}{2}, \frac{5}{4}, \ldots$

**56.** $18, -2, \frac{2}{9}, \ldots$

**57.** $-20, -4, -\frac{4}{5}, \ldots$

**58.** $0.2, 0.02, 0.002, \ldots$

*Solve.*

**59.** A frozen yogurt store owner cleared \$20,000 the first year he owned his business and made an additional 15% over the previous year in each subsequent year. Find how much he made during his fourth year of business. Find his total earnings during the first 4 years (round to the nearest dollar).

**60.** On his first morning in a television assembly factory, a trainee takes 0.8 times as long to assemble each television as he took to assemble the one before. If it took him 40 minutes to assemble the first television, find how long it takes him to assemble the fourth television. Find the total time he takes to assemble the first four televisions (round to the nearest minute).

**61.** During the harvest season a farmer can rent a combine machine for \$100 the first day, with the rental fee decreasing \$7 for each additional day. Find how much the farmer pays for the rental on the seventh day. Find how much total rent the farmer pays for 7 days.

**62.** A rubber ball is dropped from a height of 15 feet and rebounds 80% of its previous height after each bounce. Find the total distance the ball travels before it comes to rest.

**63.** After a pond was sprayed once with insecticide, 1800 mosquitoes were killed the first day, 600 the second day, 200 the third day, and so on. Find the total number of mosquitoes killed during the first 6 days after the spraying (round to the nearest unit).

**64.** See Exercise 63. Find the day on which the insecticide is no longer effective, and find the total number of mosquitoes killed (round to the nearest mosquito).

**65.** Use the formula $S_\infty$ to write $0.5\overline{55}$ as a fraction.

**66.** A movie theater has 27 seats in the first row, 30 seats in the second row, 33 seats in the third row, and so on. Find the total number of seats in the theater if there are 20 rows.

***(11.5)*** *Use Pascal's triangle to expand each binomial.*

**67.** $(x + z)^5$ **68.** $(y - r)^6$

**69.** $(2x + y)^4$ **70.** $(3y - z)^4$

*Use the binomial formula to expand the following.*

**71.** $(b + c)^8$ **72.** $(x - w)^7$

**73.** $(4m - n)^4$ **74.** $(p - 2r)^5$

*Find the indicated term.*

**75.** The fourth term of the expansion of $(a + b)^7$

**76.** The eleventh term of the expansion of $(y + 2z)^{10}$

## CHAPTER 11 TEST

*Find the indicated term(s) of the given sequence.*

**1.** The first five terms of the sequence $a_n = \dfrac{(-1)^n}{n + 4}$

**2.** The first five terms of the sequence $a_n = \dfrac{3}{(-1)^n}$

**3.** The eightieth term of the sequence $a_n = 10 + 3(n - 1)$

**4.** The two-hundredth term of the sequence $a_n = (n + 1)(n - 1)(-1)^n$

**5.** The general term of the sequence $\dfrac{2}{5}, \dfrac{2}{25}, \dfrac{2}{125}, \ldots$

**6.** The general term of the sequence $-9, 18, -27, 36, \ldots$

*Find the partial sum of the given sequence.*

**7.** $S_5$ of the sequence $a_n = 5(2)^{n-1}$

**8.** $S_{30}$ of the sequence $a_n = 18 + (n - 1)(-2)$

**9.** $S_\infty$ of the sequence $a_1 = 24$ and $r = \dfrac{1}{6}$

**10.** $S_\infty$ of the sequence $\dfrac{3}{2}, -\dfrac{3}{4}, \dfrac{3}{8}, \ldots$

**11.** $\displaystyle\sum_{i=1}^{4} i(i - 2)$ **12.** $\displaystyle\sum_{i=2}^{4} 5(2)^i(-1)^{i-1}$

*Expand the binomial by using Pascal's triangle.*

**13.** $(a - b)^6$ **14.** $(2x + y)^5$

*Expand the binomial by using the binomial formula.*

**15.** $(y + z)^8$ **16.** $(2p + r)^7$

*Solve the following applications.*

**17.** The population of a small town is growing yearly according to the sequence defined by $a_n = 250 + 75(n - 1)$, where $n$ is the number of the year just beginning. Predict the population at the beginning of the tenth year. Find the town's initial population.

**18.** A gardener is making a triangular planting with one shrub in the first row, three shrubs in the second row, five shrubs in the third row, and so on, for eight rows. Write the finite series of this sequence, and find the total number of shrubs planted.

**19.** A pendulum swings through an arc of length 80 centimeters on its first swing. On each successive swing, the length of the arc is $\dfrac{3}{4}$ the length of the arc on the preceding swing. Find the length of the arc on the fourth swing, and find the total arc length for the first four swings.

**20.** See Exercise 19. Find the total arc length before the pendulum comes to rest.

**21.** A parachutist in free-fall falls 16 feet during the first second, 48 feet during the second second, 80 feet during the third second, and so on. Find how far he falls during the tenth second. Find the total distance he falls during the first 10 seconds.

**22.** Use the formula $S_\infty$ to write $0.42\overline{42}$ as a fraction.

## CHAPTER 11 CUMULATIVE REVIEW

**1.** Divide.

**a.** $\frac{20}{-4}$ **b.** $\frac{-9}{-3}$

**c.** $-\frac{3}{8} \div 3$ **d.** $\frac{-40}{10}$

**e.** $\frac{-1}{10} \div \frac{-2}{5}$ **f.** $\frac{8}{0}$

**2.** Suppose that The Digital Store just announced an 8% decrease in the price of their Compaq Presario computers. If one particular computer model sells for \$2162 after the decrease, find the original price of this computer.

**3.** Use Cramer's rule to solve each system.

**a.** $\begin{cases} 3x + 4y = -7 \\ x - 2y = -9 \end{cases}$

**b.** $\begin{cases} 5x + y = 5 \\ -7x - 2y = -7 \end{cases}$

**4.** Use the product rule to multiply.

**a.** $(3x^6)(5x)$ **b.** $(-2x^3p^2)(4xp^{10})$

**5.** Use synthetic division to divide $(x^4 - 2x^3 - 11x^2 + 5x + 34)$ by $(x + 2)$.

**6.** Simplify the following.

**a.** $\sqrt{50}$ **b.** $\sqrt[3]{24}$

**c.** $\sqrt{26}$ **d.** $\sqrt[4]{32}$

**7.** Find the interest rate $r$ if \$2000 compounded annually grows to \$2420 in 2 years.

**8.** Solve $(x - 3)^2 - 3(x - 3) - 4 = 0$.

**9.** Solve $\frac{5}{x + 1} < -2$.

**10.** A rock is thrown upward from the ground. Its height in feet above ground after $t$ seconds is given by the function $f(t) = -16t^2 + 20t$. Find the maximum height of the rock and the number of seconds it took for the rock to reach its maximum height.

**11.** Sketch the graph of $\frac{x^2}{16} - \frac{y^2}{25} = 1$.

**12.** Solve the system.

$\begin{cases} y = \sqrt{x} \\ x^2 + y^2 = 6 \end{cases}$

**13.** Graph $\frac{x^2}{9} + \frac{y^2}{16} \leq 1$.

**14.** If $f(x) = x^2$ and $g(x) = x + 3$, find each composition.

**a.** $(f \circ g)(2)$ and $(g \circ f)(2)$

**b.** $(f \circ g)(x)$ and $(g \circ f)(x)$

**15.** Find the inverse of the one-to-one function.

$f = \{(0, 1), (-2, 7), (3, -6), (4, 4)\}$

**16.** Solve each equation for $x$.

**a.** $2^x = 16$ **b.** $9^x = 27$

**c.** $4^{x+3} = 8^x$

**17.** Simplify.

**a.** $\log_3 3^2$ **b.** $\log_7 7^{-1}$

**c.** $5^{\log_5 3}$ **d.** $2^{\log_2 6}$

**18.** Write each sum as a single logarithm.

**a.** $\log_{11} 10 + \log_{11} 3$

**b.** $\log_3 \frac{1}{2} + \log_3 12$

**c.** $\log_2 (x + 2) + \log_2 x$

**19.** Find the amount owed at the end of 5 years if \$1600 is loaned at a rate of 9% compounded continuously.

**20.** Solve $3^x = 7$.

**21.** Solve $\log_4 (x - 2) = 2$.

**22.** Write the first five terms of the sequence whose general term is given by $a_n = n^2 - 1$.

**23.** Find the eleventh term of the arithmetic sequence whose first three terms are 2, 9, and 16.

**24.** Evaluate.

**a.** $\sum_{i=0}^{6} \frac{i - 2}{2}$ **b.** $\sum_{i=3}^{5} 2^i$

**25.** Find the sum of the first 30 positive integers.

# Appendix A REVIEW OF ANGLES, LINES, AND SPECIAL TRIANGLES

The word **geometry** is formed from the Greek words, **geo**, meaning earth, and **metron**, meaning measure. Geometry literally means to measure the earth.

This section contains a review of some basic geometric ideas. It will be assumed that fundamental ideas of geometry such as point, line, ray, and angle are known. In this appendix, the notation $\angle 1$ is read "angle 1" and the notation $m\angle 1$ is read "the measure of angle 1."

We first review types of angles.

## ANGLES

A **right angle** is an angle whose measure is 90°. A right angle can be indicated by a square drawn at the vertex of the angle, as shown below.
An angle whose measure is more than 0° but less than 90° is called an **acute angle**.
An angle whose measure is greater than 90° but less than 180° is called an **obtuse angle**.
An angle whose measure is 180° is called a **straight angle**.
Two angles are said to be **complementary** if the sum of their measures is 90°. Each angle is called the **complement** of the other.
Two angles are said to be **supplementary** if the sum of their measures is 180°. Each angle is called the **supplement** of the other.

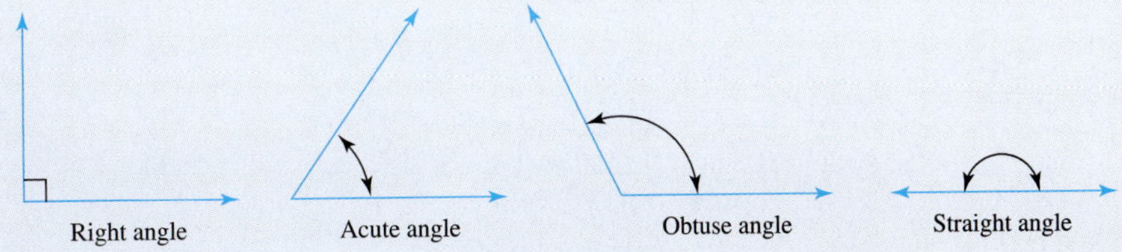


(*continued*)

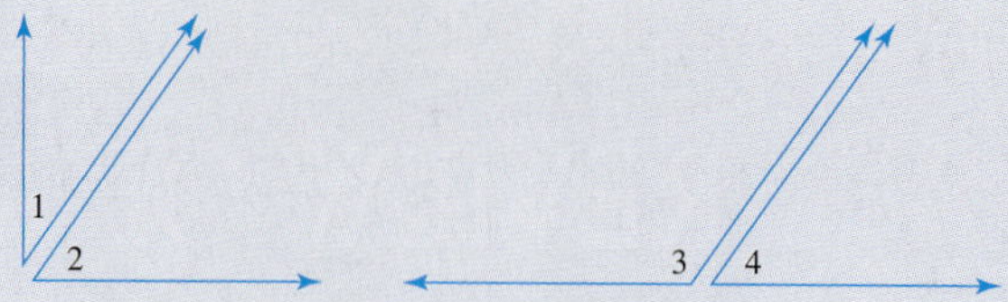


Complementary angles: $m\angle 1 + m\angle 2 = 90°$

Supplementary angles: $m\angle 3 + m\angle 4 = 180°$

**Example 1** If an angle measures 28°, find its complement.

*Solution* Two angles are complementary if the sum of their measures is 90°. The complement of a 28° angle is an angle whose measure is $90° - 28° = 62°$. To check, notice that $28° + 62° = 90°$.

**Plane** is an undefined term that we will describe. A plane can be thought of as a flat surface with infinite length and width, but no thickness. A plane is two dimensional. The arrows in the following diagram indicate that a plane extends indefinitely and has no boundaries.

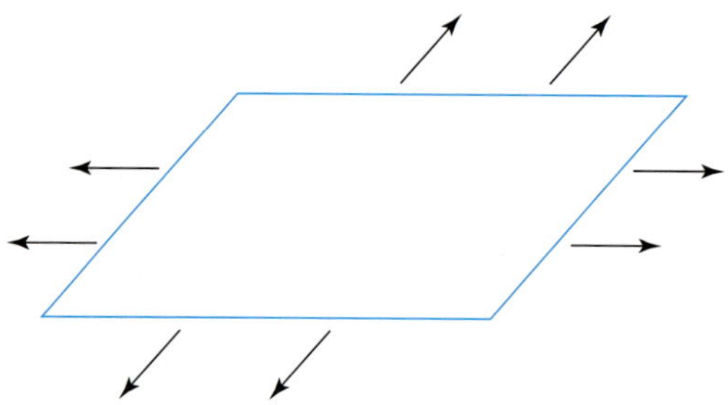

Figures that lie on a plane are called **plane figures**. (See the description of common plane figures in Appendix B.) Lines that lie in the same plane are called **coplanar**.

## LINES

Two lines are **parallel** if they lie in the same plane but never meet.
**Intersecting lines** meet or cross in one point.
Two lines that form right angles when they intersect are said to be **perpendicular**.

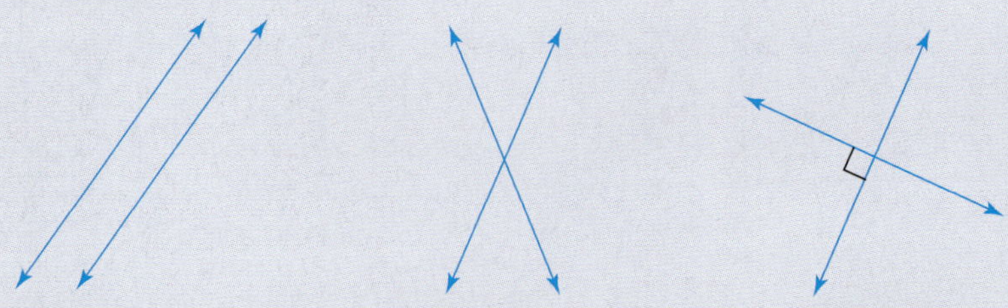

Parallel lines

Intersecting lines

Intersecting lines that are perpendicular

Two intersecting lines form **vertical angles**. Angles 1 and 3 are vertical angles. Also, angles 2 and 4 are vertical angles. It can be shown that **vertical angles have equal measures**.

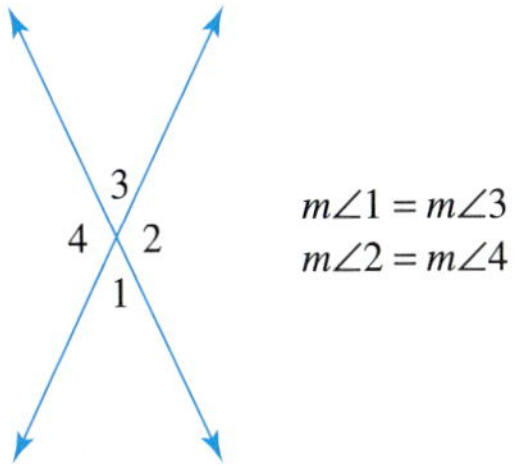


**Adjacent angles** have the same vertex and share a side but have no interior points in common. Angles 1 and 2 are adjacent angles. Other pairs of adjacent angles are angles 2 and 3, angles 3 and 4, and angles 4 and 1.

A **transversal** is a line that intersects two or more lines in the same plane. Line $l$ is a transversal that intersects lines $m$ and $n$. The eight angles formed are numbered and certain pairs of these angles are given special names.

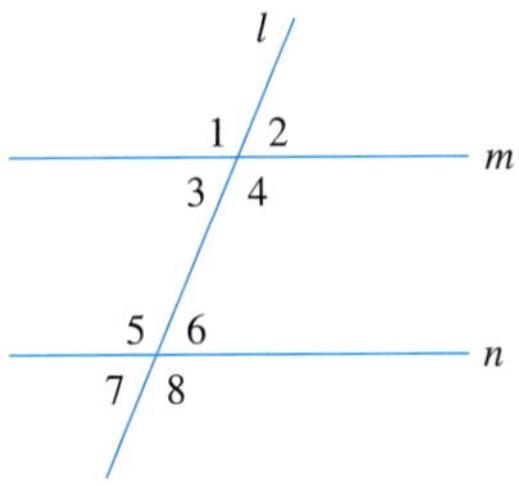


**Corresponding angles:** $\angle 1$ and $\angle 5$, $\angle 3$ and $\angle 7$, $\angle 2$ and $\angle 6$, $\angle 4$ and $\angle 8$.

**Exterior angles:** $\angle 1, \angle 2, \angle 7, \angle 8$.

**Interior angles:** $\angle 3, \angle 4, \angle 5, \angle 6$.

**Alternate interior angles:** $\angle 3$ and $\angle 6$, $\angle 4$ and $\angle 5$.

These angles and parallel lines are related in the following manner:

## PARALLEL LINES CUT BY A TRANSVERSAL

1. If two parallel lines are cut by a transversal, then
   a. **corresponding angles are equal** and
   b. **alternate interior angles are equal.**
2. If corresponding angles formed by two lines and a transversal are equal, then the lines are parallel.
3. If alternate interior angles formed by two lines and a transversal are equal, then the lines are parallel.

**Example 2** Given that lines *m* and *n* are parallel and that the measure of angle 1 is 100°, find the measures of angles 2, 3, and 4.

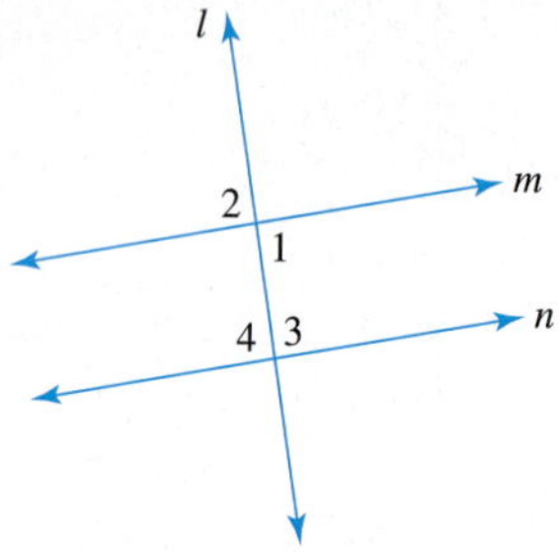


*Solution* $m\angle 2 = 100°$, since angles 1 and 2 are vertical angles.
$m\angle 4 = 100°$, since angles 1 and 4 are alternate interior angles.
$m\angle 3 = 180° - 100° = 80°$, since angles 4 and 3 are supplementary angles.

A **polygon** is the union of three or more coplanar line segments that intersect each other only at each end point, with each end point shared by exactly two segments.

A **triangle** is a polygon with three sides. The sum of the measures of the three angles of a triangle is 180°. In the following figure, $m\angle 1 + m\angle 2 + m\angle 3 = 180°$.

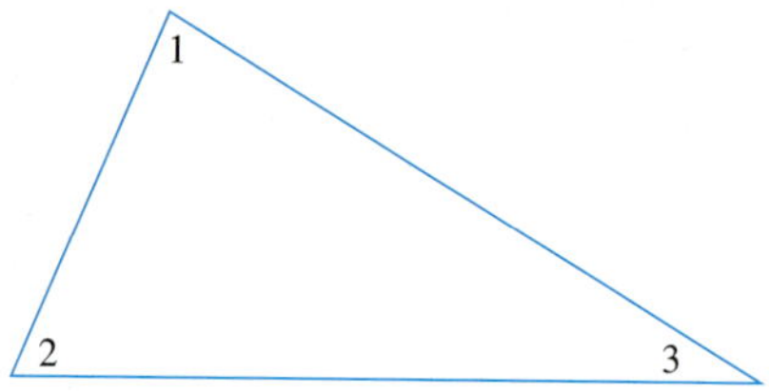


**Example 3** Find the measure of the third angle of the triangle shown.

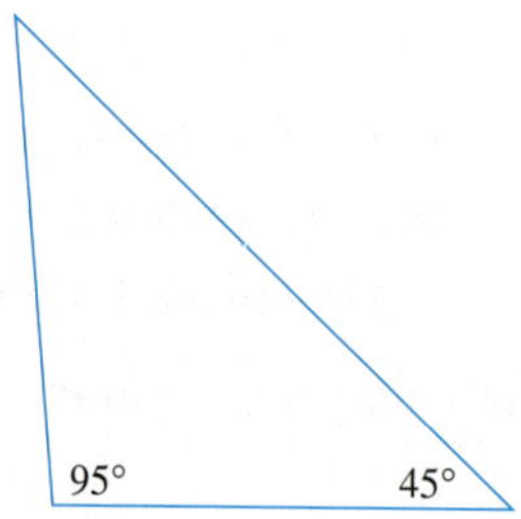


*Solution* The sum of the measures of the angles of a triangle is 180°. Since one angle measures 45° and the other angle measures 95°, the third angle measures $180° - 45° - 95° = 40°$.

Two triangles are **congruent** if they have the same size and the same shape. In congruent triangles, the measures of corresponding angles are equal and the lengths of corresponding sides are equal. The following triangles are congruent:

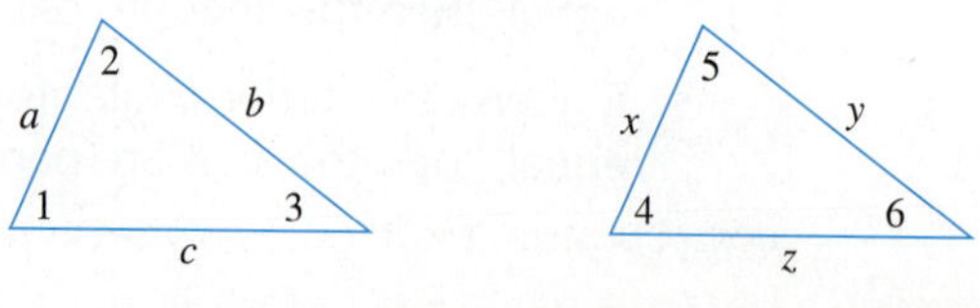

Corresponding angles are equal: $m\angle 1 = m\angle 4$, $m\angle 2 = m\angle 5$, and $m\angle 3 = m\angle 6$. Also, lengths of corresponding sides are equal: $a = x$, $b = y$, and $c = z$.

Any one of the following may be used to determine whether two triangles are congruent:

## CONGRUENT TRIANGLES

1. If the measures of two angles of a triangle equal the measures of two angles of another triangle and the lengths of the sides between each pair of angles are equal, the triangles are congruent.

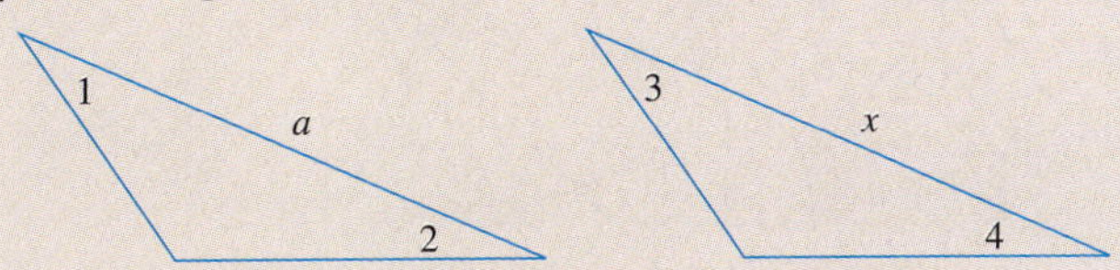


$m\angle 1 = m\angle 3$
$m\angle 2 = m\angle 4$
and
$a = x$

2. If the lengths of the three sides of a triangle equal the lengths of corresponding sides of another triangle, the triangles are congruent.

$a = x$
$b = y$
and
$c = z$

3. If the lengths of two sides of a triangle equal the lengths of corresponding sides of another triangle, and the measures of the angles between each pair of sides are equal, the triangles are congruent.

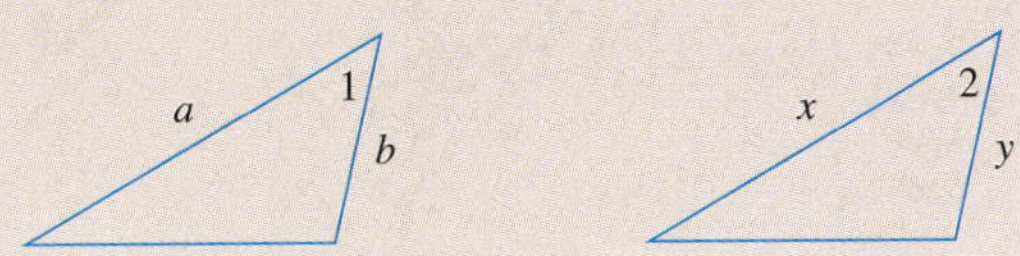


$a = x$
$b = y$
and
$m\angle 1 = m\angle 2$

Two triangles are **similar** if they have the same shape. In similar triangles, the measures of corresponding angles are equal and corresponding sides are in proportion. The following triangles are similar. (All similar triangles drawn in this appendix will be oriented the same.)

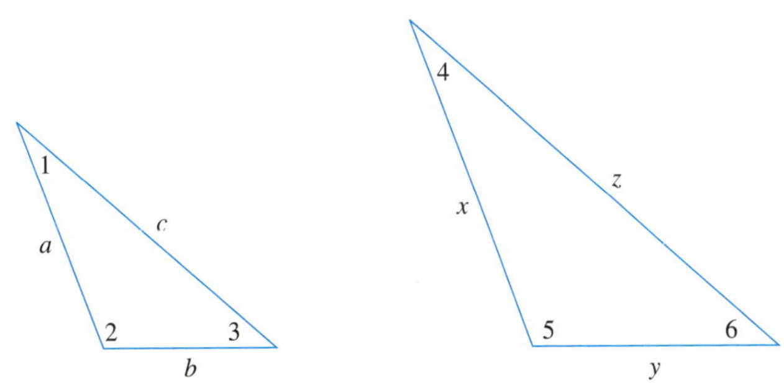


Corresponding angles are equal: $m\angle 1 = m\angle 4$, $m\angle 2 = m\angle 5$, and $m\angle 3 = m\angle 6$. Also, corresponding sides are proportional: $\frac{a}{x} = \frac{b}{y} = \frac{c}{z}$.

Any one of the following may be used to determine whether two triangles are similar:

## SIMILAR TRIANGLES

1. If the measures of two angles of a triangle equal the measures of two angles of another triangle, the triangles are similar.

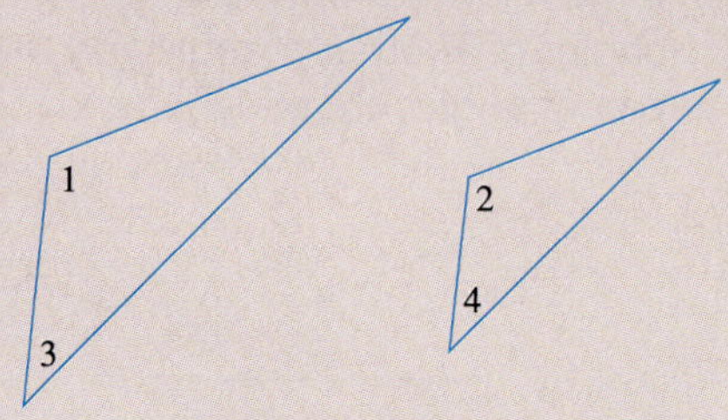


$m\angle 1 = m\angle 2$
and
$m\angle 3 = m\angle 4$

2. If three sides of one triangle are proportional to three sides of another triangle, the triangles are similar.

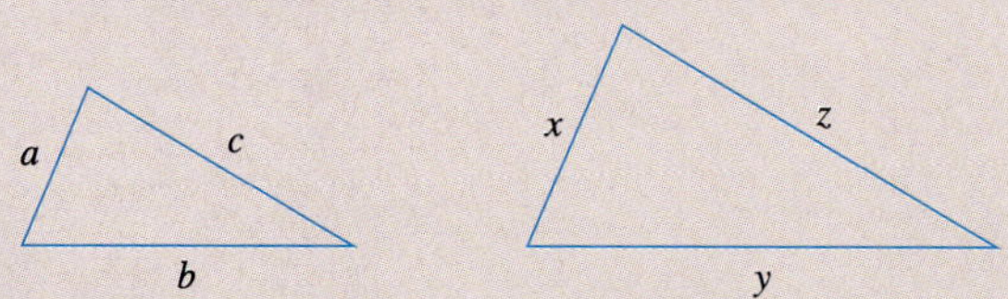


$\frac{a}{x} = \frac{b}{y} = \frac{c}{z}$

3. If two sides of a triangle are proportional to two sides of another triangle and the measures of the included angles are equal, the triangles are similar.

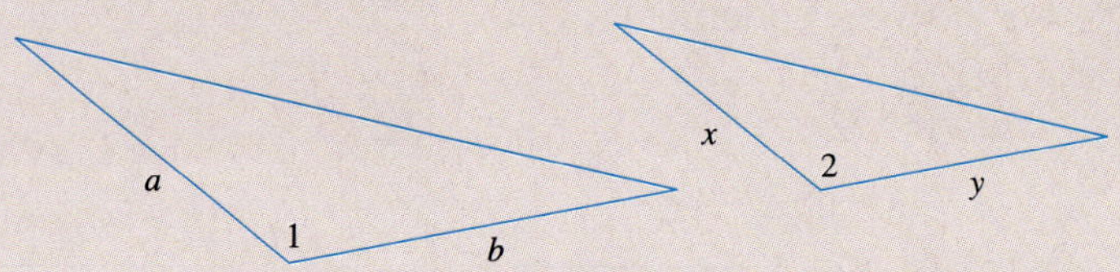


$m\angle 1 = m\angle 2$
and
$\frac{a}{x} = \frac{b}{y}$

**Example 4** Given that the following triangles are similar, find the missing length $x$.

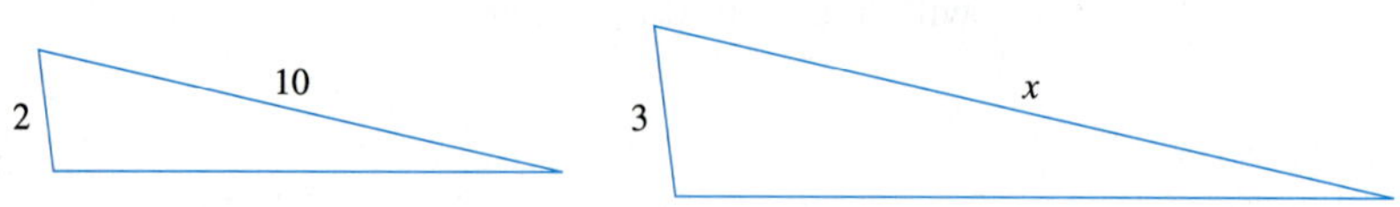


*Solution* Since the triangles are similar, corresponding sides are in proportion. Thus, $\frac{2}{3} = \frac{10}{x}$. To solve this equation for $x$, we multiply both sides by the LCD, $3x$.

$$3x\left(\frac{2}{3}\right) = 3x\left(\frac{10}{x}\right)$$
$$2x = 30$$
$$x = 15$$

The missing length is 15 units.

A **right triangle** contains a right angle. The side opposite the right angle is called the **hypotenuse**, and the other two sides are called the **legs**. The **Pythagorean theorem** gives a formula that relates the lengths of the three sides of a right triangle.

**THE PYTHAGOREAN THEOREM**

If $a$ and $b$ are the lengths of the legs of a right triangle, and $c$ is the length of the hypotenuse, then $a^2 + b^2 = c^2$.

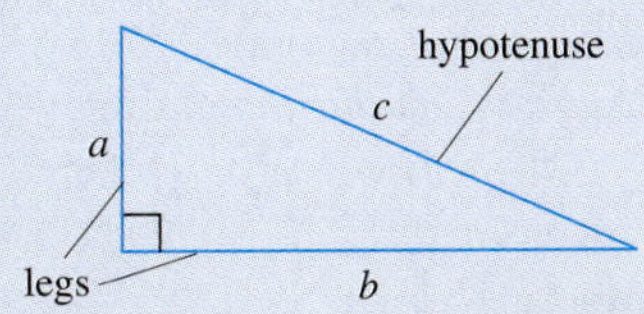


**Example 5** Find the length of the hypotenuse of a right triangle whose legs have lengths of 3 centimeters and 4 centimeters.

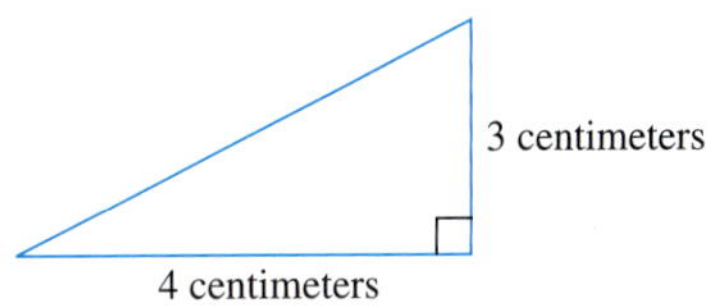


*Solution* Because we have a right triangle, we use the Pythagorean theorem. The legs are 3 centimeters and 4 centimeters, so let $a = 3$ and $b = 4$ in the formula.

$$a^2 + b^2 = c^2$$
$$3^2 + 4^2 = c^2$$
$$9 + 16 = c^2$$
$$25 = c^2$$

Since $c$ represents a length, we assume that $c$ is positive. Thus, if $c^2$ is 25, $c$ must be 5. The hypotenuse has a length of 5 centimeters.

## *Appendix A Exercise Set*

*Find the complement of each angle. See Example 1.*

**1.** 19°

**2.** 65°

**3.** 70.8°

**4.** $45\frac{2}{3}°$

**5.** $11\frac{1}{4}°$

**6.** 19.6°

*Find the supplement of each angle.*

**7.** 150°

**8.** 90°

**9.** 30.2°

**10.** 81.9°

**11.** $79\frac{1}{2}°$

**12.** $165\frac{8}{9}°$

**13.** If lines $m$ and $n$ are parallel, find the measures of angles 1 through 7. See Example 2.

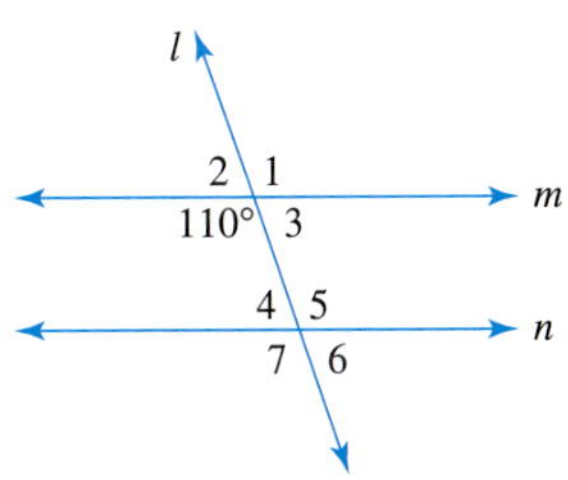

**14.** If lines $m$ and $n$ are parallel, find the measures of angles 1 through 5. See Example 2.

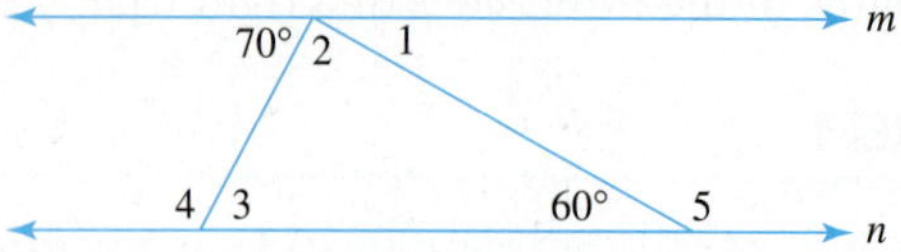


*In each of the following, the measures of two angles of a triangle are given. Find the measure of the third angle. See Example 3.*

**15.** 11°, 79°
**16.** 8°, 102°
**17.** 25°, 65°
**18.** 44°, 19°
**19.** 30°, 60°
**20.** 67°, 23°

*In each of the following, the measure of one angle of a right triangle is given. Find the measures of the other two angles.*

**21.** 45°
**22.** 60°
**23.** 17°
**24.** 30°
**25.** $39\frac{3}{4}°$
**26.** 72.6°

*Given that each of the following pairs of triangles is similar, find the missing lengths. See Example 4.*

**27.**

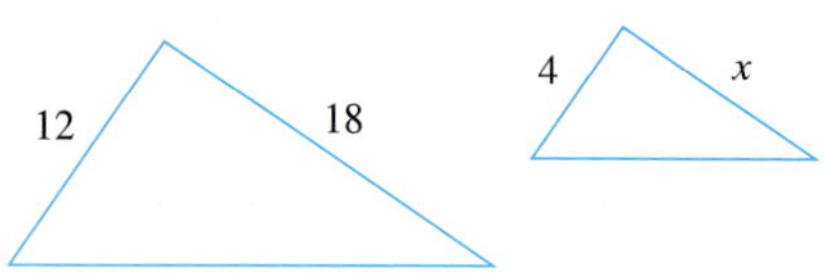


**28.**

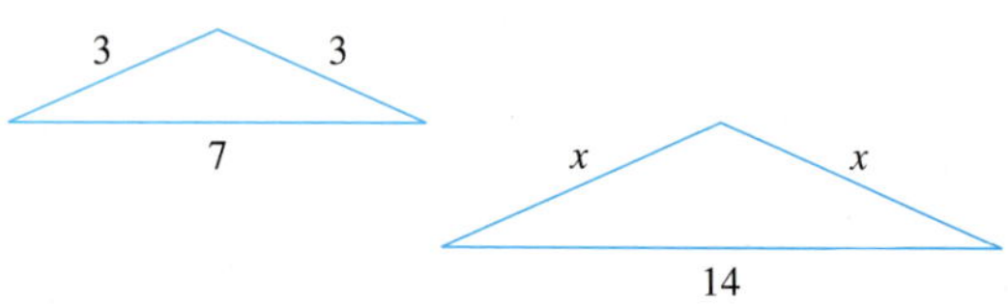


**29.**

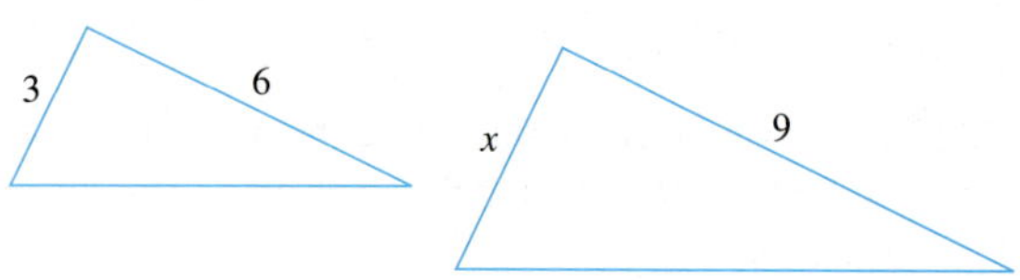


**30.**

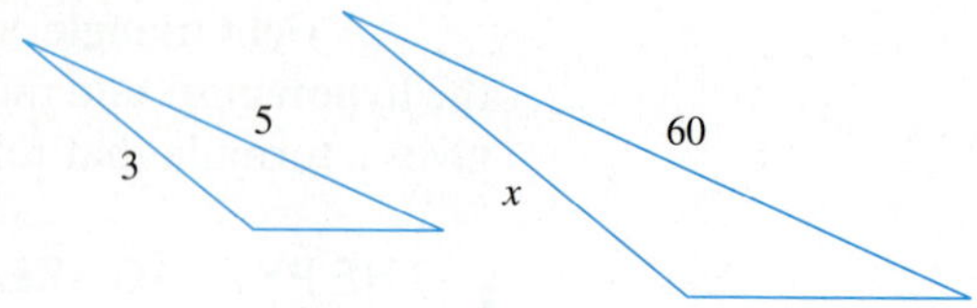


*Use the Pythagorean theorem to find the missing lengths in the right triangles. See Example 5.*

**31.**

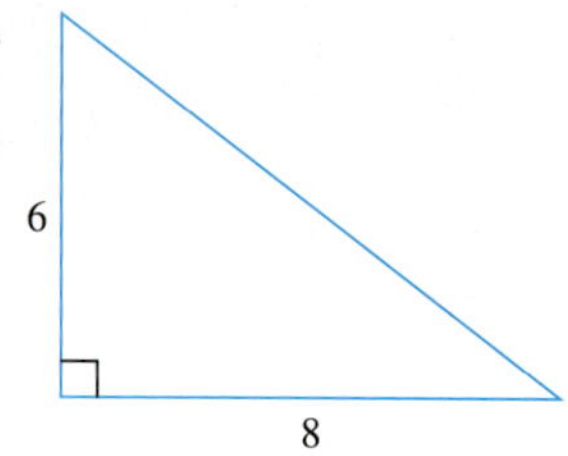


**32.**

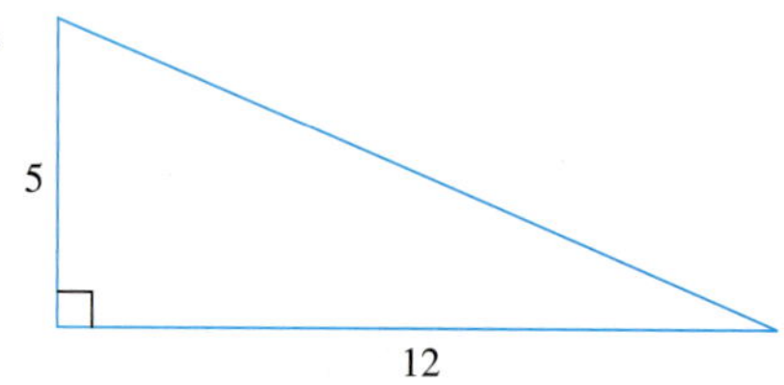


**33.**

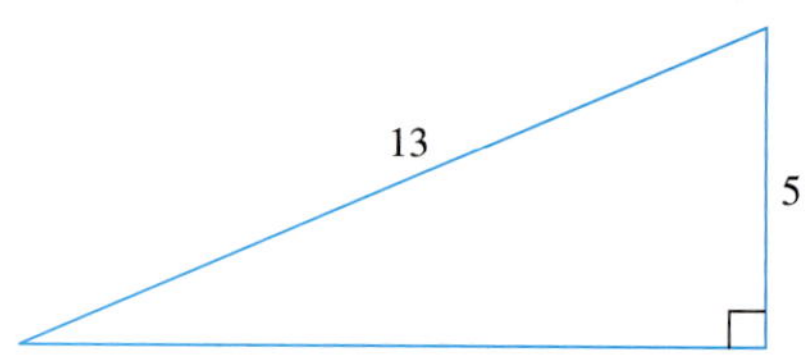


**34.**

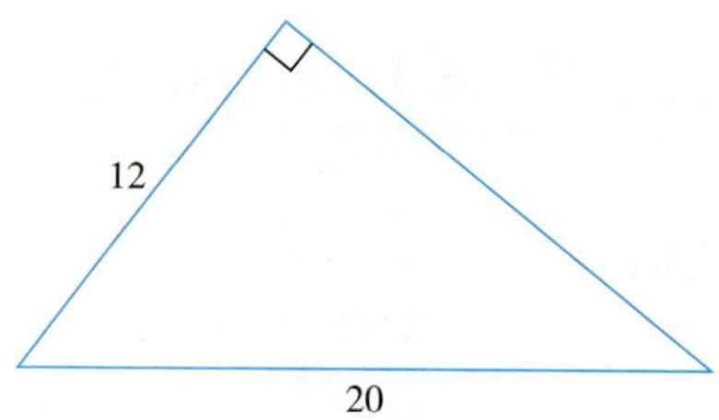

# Appendix B
# REVIEW OF GEOMETRIC FIGURES

| Plane figures have length and width but no thickness or depth. | | |
|---|---|---|
| *Name* | *Description* | *Figure* |
| ***POLYGON*** | Union of three or more coplanar line segments that intersect with each other only at each endpoint, with each endpoint shared by two segments. | 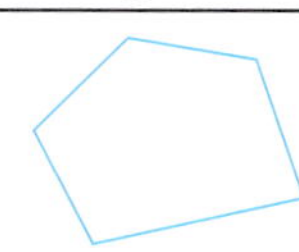 |
| ***TRIANGLE*** | Polygon with three sides (sum of measures of three angles is 180°). |  |
| ***SCALENE TRIANGLE*** | Triangle with no sides of equal length. | 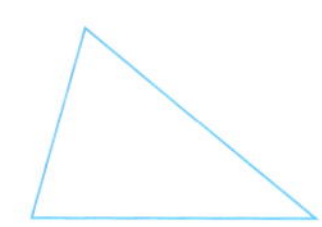 |
| ***ISOSCELES TRIANGLE*** | Triangle with two sides of equal length. | 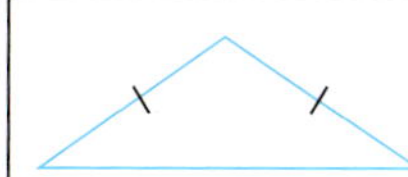 |
| ***EQUILATERAL TRIANGLE*** | Triangle with all sides of equal length. | 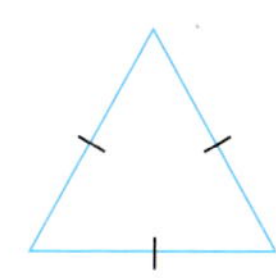 |
| ***RIGHT TRIANGLE*** | Triangle that contains a right angle. | 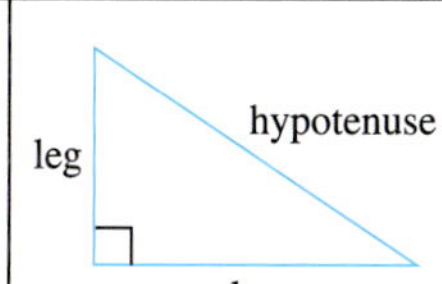  |

***Plane figures have length and width but no thickness or depth.***

| *Name* | *Description* | *Figure* |
|---|---|---|
| ***QUADRILATERAL*** | Polygon with four sides (sum of measures of four angles is 360°). | 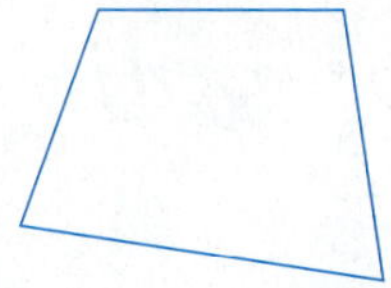 |
| ***TRAPEZOID*** | Quadrilateral with exactly one pair of opposite sides parallel. | 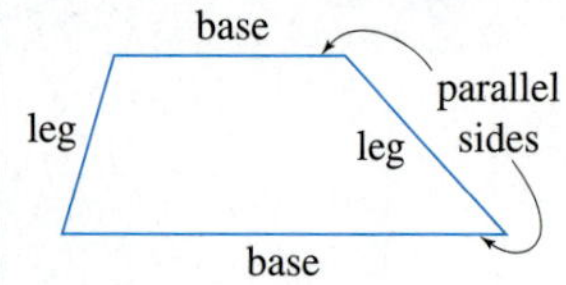 |
| ***ISOSCELES TRAPEZOID*** | Trapezoid with legs of equal length. |  |
| ***PARALLELOGRAM*** | Quadrilateral with both pairs of opposite sides parallel and equal in length. |  |
| ***RHOMBUS*** | Parallelogram with all sides of equal length. | 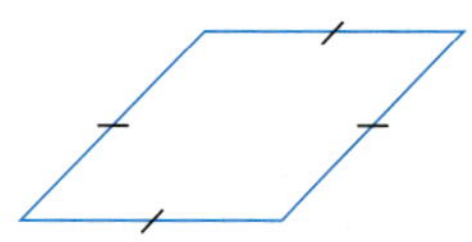 |
| ***RECTANGLE*** | Parallelogram with four right angles. | 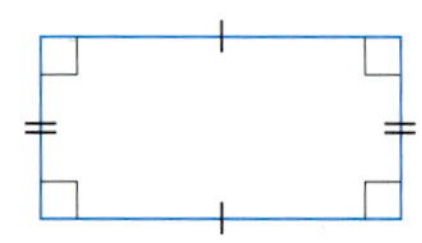 |
| ***SQUARE*** | Rectangle with all sides of equal length. | 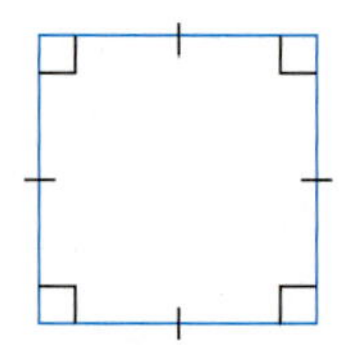 |
| ***CIRCLE*** | All points in a plane the same distance from a fixed point called the **center**. | 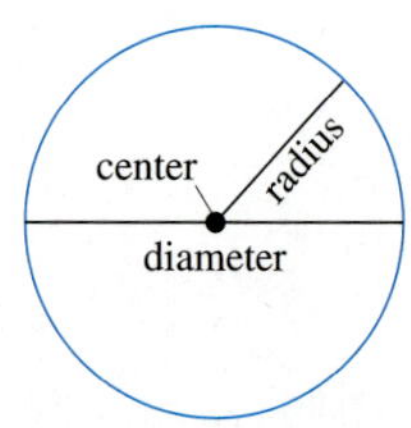 |

| *Solids have length, width, and depth.* | | |
|---|---|---|
| *Name* | *Description* | *Figure* |
| ***RECTANGULAR SOLID*** | A solid with six sides, all of which are rectangles. | 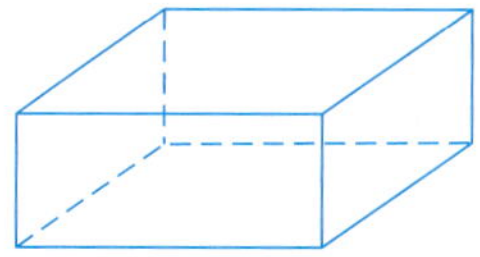 |
| ***CUBE*** | A rectangular solid whose six sides are squares. | 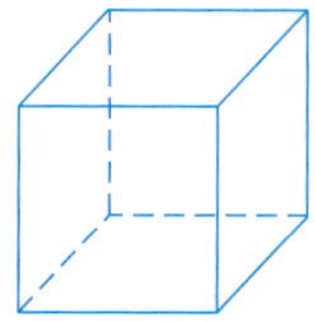 |
| ***SPHERE*** | All points the same distance from a fixed point, called the center. | 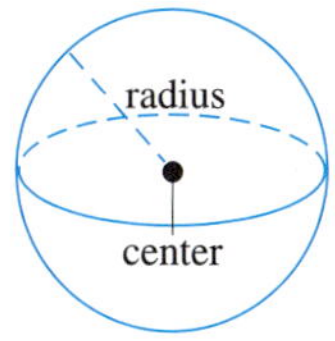  |
| ***RIGHT CIRCULAR CYLINDER*** | A cylinder with two circular bases that are perpendicular to its altitude. | 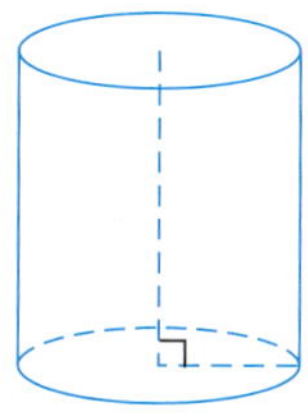 |
| ***RIGHT CIRCULAR CONE*** | A cone with a circular base that is perpendicular to its altitude. | 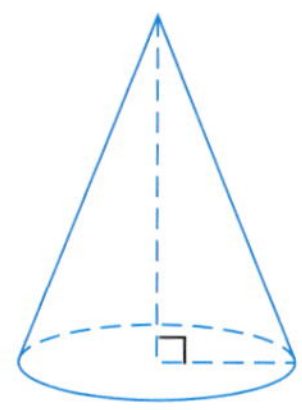 |

# Appendix C REVIEW OF VOLUME AND SURFACE AREA

A **convex solid** is a set of points, $S$, not all in one plane, such that for any two points $A$ and $B$ in $S$, all points between $A$ and $B$ are also in $S$. In this appendix, we will find the volume and surface area of special types of solids called polyhedrons. A solid formed by the intersection of a finite number of planes is called a **polyhedron**. The box below is an example of a polyhedron.

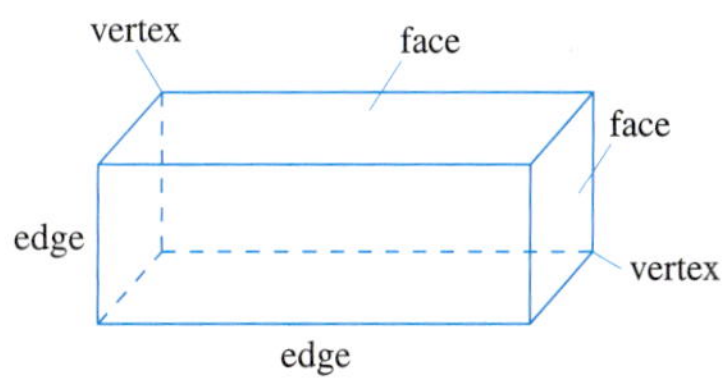

Polyhedron

Each of the plane regions of the polyhedron is called a **face** of the polyhedron. If the intersection of two faces is a line segment, this line segment is an **edge** of the polyhedron. The intersections of the edges are the **vertices** of the polyhedron.

**Volume** is a measure of the space of a solid. The volume of a box or can, for example, is the amount of space inside. Volume can be used to describe the amount of juice in a pitcher or the amount of concrete needed to pour a foundation for a house.

The volume of a solid is the number of **cubic units** in the solid. A cubic centimeter and a cubic inch are illustrated.

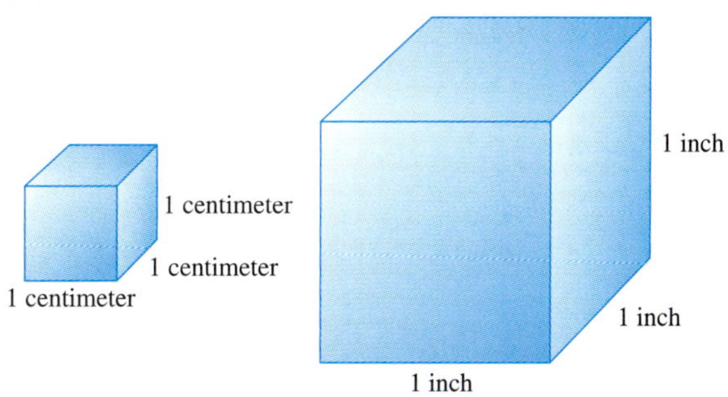

**1 cubic centimeter** **1 cubic inch**

The **surface area** of a polyhedron is the sum of the areas of the faces of the polyhedron. For example, each face of the cube to the left above has an area of 1 square centimeter. Since there are 6 faces of the cube, the sum of the areas of the faces is 6 square centimeters. Surface area can be used to describe the amount of material needed to cover or form a solid. Surface area is measured in square units.

Formulas for finding the volumes, $V$, and surface areas, $SA$, of some common solids are given next.

## VOLUME AND SURFACE AREA FORMULAS OF COMMON SOLIDS

***Solid*** — ***Formulas***

RECTANGULAR SOLID

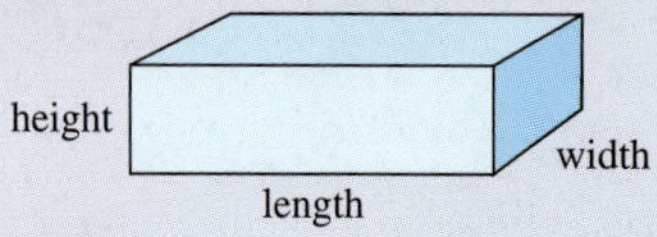


$V = lwh$

$SA = 2lh + 2wh + 2lw$

where $h$ = height, $w$ = width, $l$ = length

CUBE

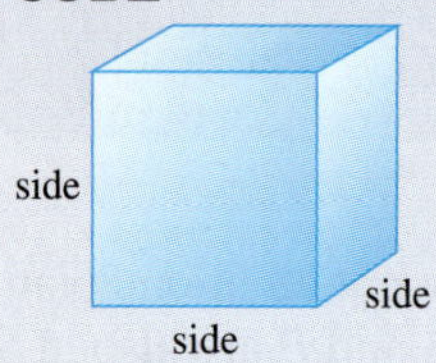


$V = s^3$

$SA = 6s^2$

where $s$ = side

SPHERE

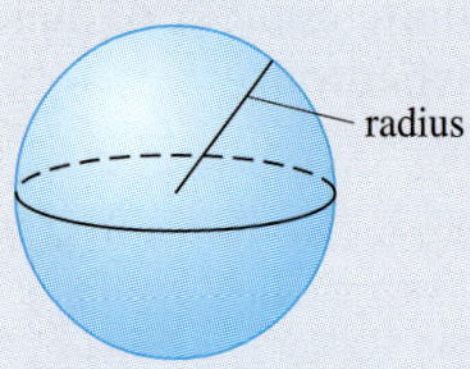


$V = \frac{4}{3}\pi r^3$

$SA = 4\pi r^2$

where $r$ = radius

CIRCULAR CYLINDER

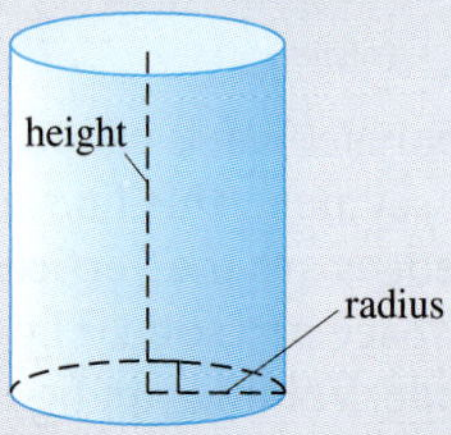


$V = \pi r^2 h$

$SA = 2\pi rh + 2\pi r^2$

where $h$ = height, $r$ = radius

CONE

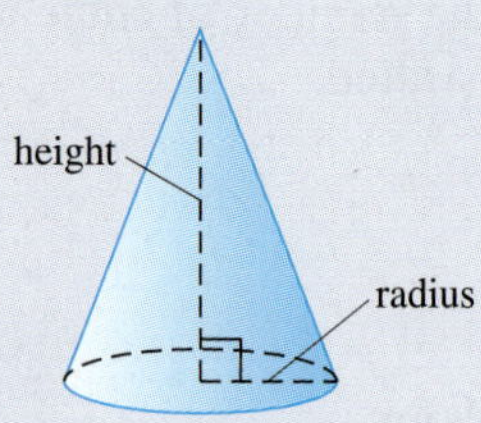


$V = \frac{1}{3}\pi r^2 h$

$SA = \pi r\sqrt{r^2 + h^2} + \pi r^2$

where $h$ = height, $r$ = radius

SQUARE-BASED PYRAMID

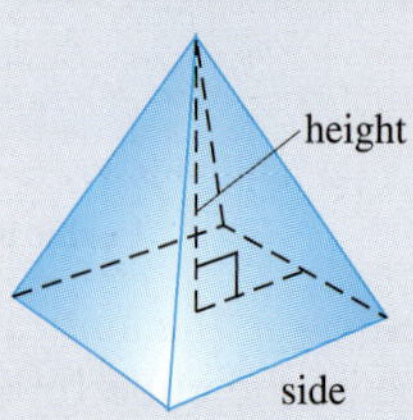


$V = \frac{1}{3}s^2 h$

$SA = B + \frac{1}{2}pl$

where $B$ = area of base, $p$ = perimeter of base, $h$ = height, $s$ = side, $l$ = slant height

**HELPFUL HINT**
Volume is measured in cubic units. Surface area is measured in square units.

**Example 1** Find the volume and surface area of a rectangular box that is 12 inches long, 6 inches wide, and 3 inches high.

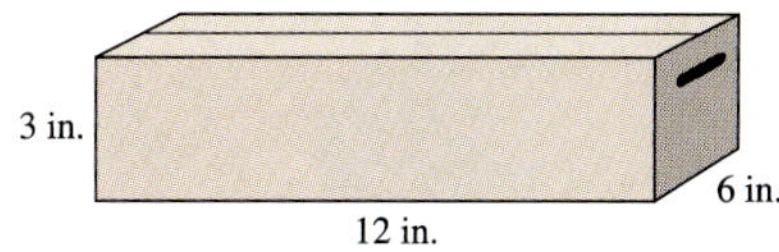


*Solution* Let $h = 3$ in., $l = 12$ in., and $w = 6$ in.

$$V = lwh$$
$$V = 12 \text{ inches} \cdot 6 \text{ inches} \cdot 3 \text{ inches} = 216 \text{ cubic inches}$$

The volume of the rectangular box is 216 cubic inches.

$$\begin{aligned} SA &= 2lh + 2wh + 2lw \\ &= 2(12 \text{ in.})(3 \text{ in.}) + 2(6 \text{ in.})(3 \text{ in.}) + 2(12 \text{ in.})(6 \text{ in.}) \\ &= 72 \text{ sq. in.} + 36 \text{ sq. in.} + 144 \text{ sq. in.} \\ &= 252 \text{ sq. in.} \end{aligned}$$

The surface area of the rectangular box is 252 square inches.

**Example 2** Find the volume and surface area of a ball of radius 2 inches. Give the exact volume and surface area and then use the approximation $\frac{22}{7}$ for $\pi$.

*Solution*

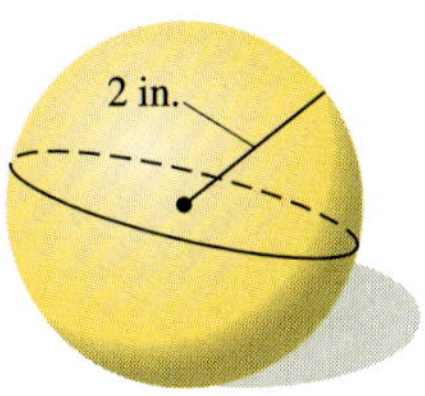


$$V = \frac{4}{3}\pi r^3 \quad \text{Formula for volume of a sphere.}$$
$$V = \frac{4}{3}\pi(2 \text{ in.})^3 \quad \text{Let } r = 2 \text{ inches.}$$
$$= \frac{32}{3}\pi \text{ cu. in.} \quad \text{Simplify.}$$
$$\approx \frac{32}{3} \cdot \frac{22}{7} \text{ cu. in.} \quad \text{Approximate } \pi \text{ with } \frac{22}{7}.$$
$$= \frac{704}{21} \text{ or } 33\frac{11}{21} \text{ cu. in.}$$

The volume of the sphere is exactly $\frac{32}{3}\pi$ cubic inches or approximately $33\frac{11}{21}$ cubic inches.

$$SA = 4\pi r^2 \quad \text{Formula for surface area.}$$
$$SA = 4\pi(2 \text{ in.})^2 \quad \text{Let } r = 2 \text{ inches.}$$
$$= 16\pi \text{ sq. in.} \quad \text{Simplify.}$$
$$\approx 16 \cdot \frac{22}{7} \text{ sq. in.} \quad \text{Approximate } \pi \text{ with } \frac{22}{7}.$$
$$= \frac{352}{7} \text{ or } 50\frac{2}{7} \text{ sq. in.}$$

The surface area of the sphere is exactly $16\pi$ square inches or approximately $50\frac{2}{7}$ square inches.

## Appendix C Exercise Set

*Find the volume and surface area of each solid. See Examples 1 and 2. For formulas that contain $\pi$, give an exact answer and then approximate using $\frac{22}{7}$ for $\pi$.*

**1.**

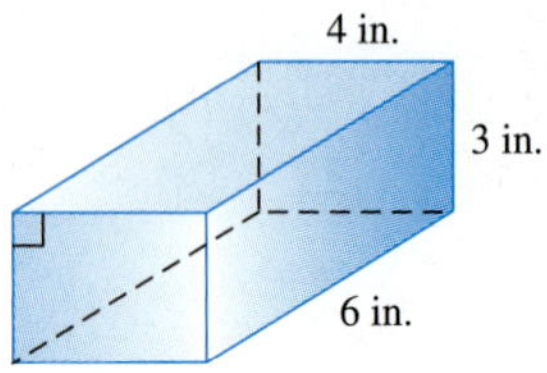


**2.**

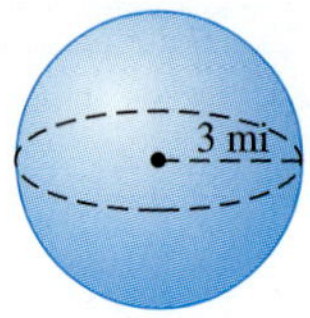


**3.**

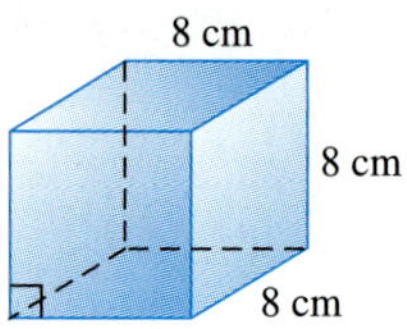


**4.**

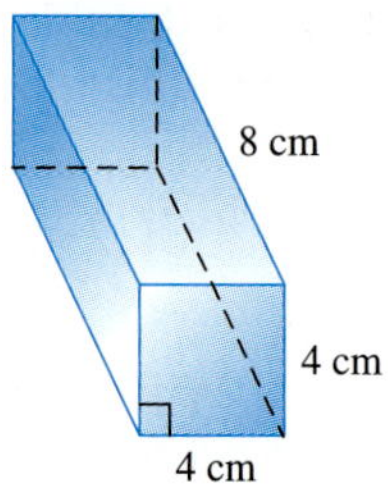


**5.** (For surface area, use 3.14 for $\pi$ and approximate to two decimal places.)

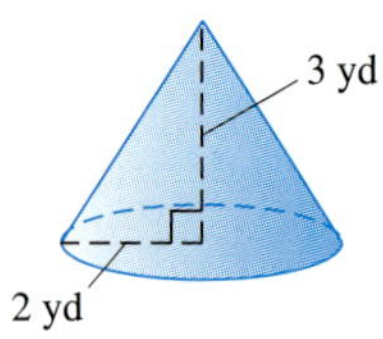


**6.**

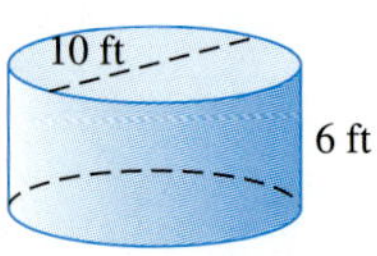


**7.**

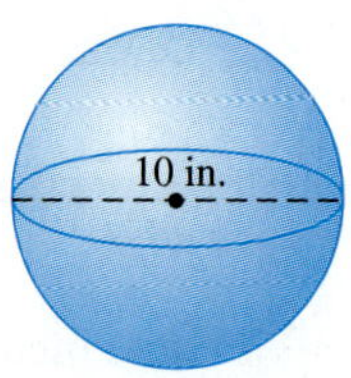


**8.** Find the volume only.

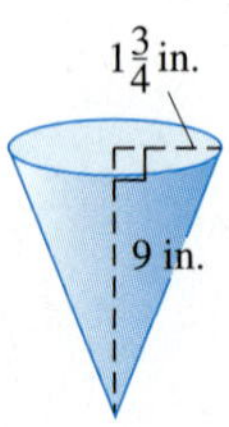


**9.**

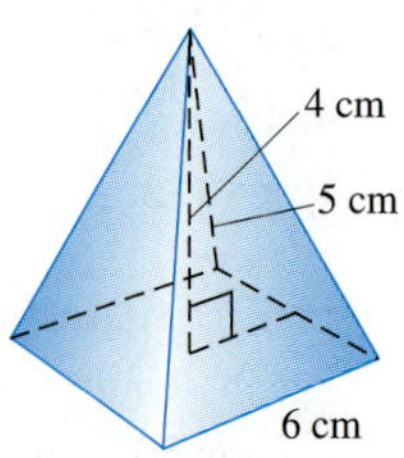


**10.**

*Solve.*

**11.** Find the volume of a cube with edges of $1\frac{1}{3}$ inches.

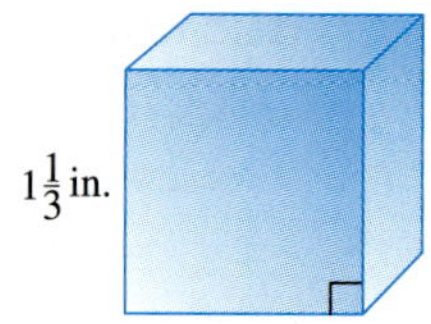


**12.** A water storage tank is in the shape of a cone with the pointed end down. If the radius is 14 ft and the depth of the tank is 15 ft, approximate the volume of the tank in cubic feet. Use $\frac{22}{7}$ for $\pi$.

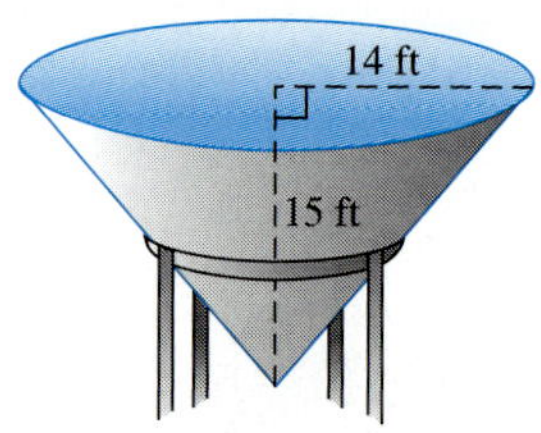


**13.** Find the surface area of a rectangular box 2 ft by 1.4 ft by 3 ft.

**14.** Find the surface area of a box in the shape of a cube that is 5 ft on each side.

**15.** Find the volume of a pyramid with a square base 5 in. on a side and a height of 1.3 in.

**16.** Approximate to the nearest hundredth the volume of a sphere with a radius of 2 cm. Use 3.14 for $\pi$.

**17.** A paperweight is in the shape of a square-based pyramid 20 cm tall. If an edge of the base is 12 cm, find the volume of the paperweight.

**18.** A bird bath is made in the shape of a hemisphere (half-sphere). If its radius is 10 in., approximate the volume. Use $\frac{22}{7}$ for $\pi$.

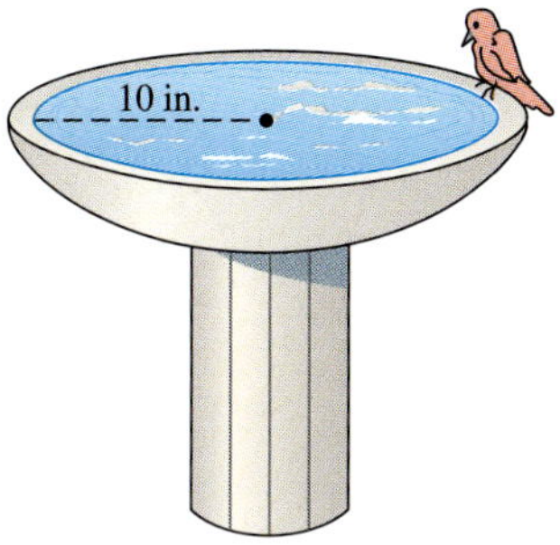


**19.** Find the exact surface area of a sphere with a radius of 7 in.

**20.** A tank is in the shape of a cylinder 8 ft tall and 3 ft in radius. Find the exact surface area of the tank.

**21.** Find the volume of a rectangular block of ice 2 ft by $2\frac{1}{2}$ ft by $1\frac{1}{2}$ ft.

**22.** Find the capacity (volume in cubic feet) of a rectangular ice chest with inside measurements of 3 ft by $1\frac{1}{2}$ ft by $1\frac{3}{4}$ ft.

**23.** An ice cream cone with a 4-cm diameter and 3-cm depth is filled exactly level with the top of the cone. Approximate how much ice cream (in cubic centimeters) is in the cone. Use $\frac{22}{7}$ for $\pi$.

**24.** A child's toy is in the shape of a square-based pyramid 10 in. tall. If an edge of the base is 7 in., find the volume of the toy.

Appendix D

# AN INTRODUCTION TO USING A GRAPHING UTILITY

## The Viewing Window and Interpreting Window Settings

In this appendix, we will use the term **graphing utility** to mean a graphing calculator or a computer software graphing package. All graphing utilities graph equations by plotting points on a screen. While plotting several points can be slow and sometimes tedious for us, a graphing utility can quickly and accurately plot hundreds of points. How does a graphing utility show plotted points? A computer or calculator screen is made up of a grid of small rectangular areas called **pixels**. If a pixel contains a point to be plotted, the pixel is turned "on"; otherwise, the pixel remains "off." The graph of an equation is then a collection of pixels turned "on." The graph of $y = 3x + 1$ from a graphing calculator is shown in Figure A-1. Notice the irregular shape of the line caused by the rectangular pixels.

Figure A-1

The portion of the coordinate plane shown on the screen in Figure A-1 is called the **viewing window** or the **viewing rectangle.** Notice the $x$-axis and the $y$-axis on the graph. While tick marks are shown on the axes, they are not labeled. This means that from this screen alone, we do not know how many units each tick mark represents. To see what each tick mark represents and the minimum and maximum values on the axes, check the *window setting* of the graphing utility. It defines the viewing window. The window of the graph of $y = 3x + 1$ shown in Figure A-1 has the following setting (Figure A-2):

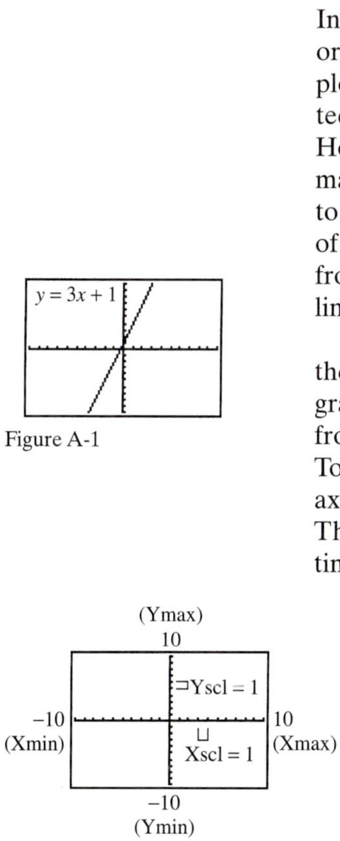

Figure A-2

| | |
|---|---|
| Xmin = −10 | The minimum $x$-value is −10. |
| Xmax = 10 | The maximum $x$-value is 10. |
| Xscl = 1 | The $x$-axis scale is 1 unit per tick mark. |
| Ymin = −10 | The minimum $y$-value is −10. |
| Ymax = 10 | The maximum $y$-value is 10. |
| Yscl = 1 | The $y$-axis scale is 1 unit per tick mark. |

By knowing the scale, we can find the minimum and the maximum values on the axes simply by counting tick marks. For example, if both the Xscl ($x$-axis scale) and the Yscl are 1 unit per tick mark on the graph in Figure A-3, we can count the tick marks and find that the minimum $x$-value is $-10$ and the maximum $x$-value is 10. Also, the minimum $y$-value is $-10$ and the maximum $y$-value is 10. If the Xscl ($x$-axis scale) changes to 2 units per tick mark (shown in Figure A-4), by counting tick marks, we see that the minimum $x$-value is now $-20$ and the maximum $x$-value is now 20.

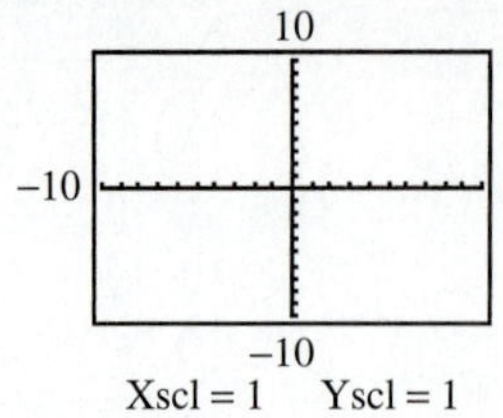


Figure A-3

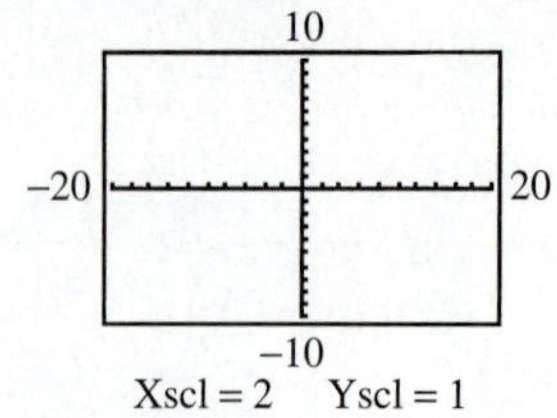


Figure A-4

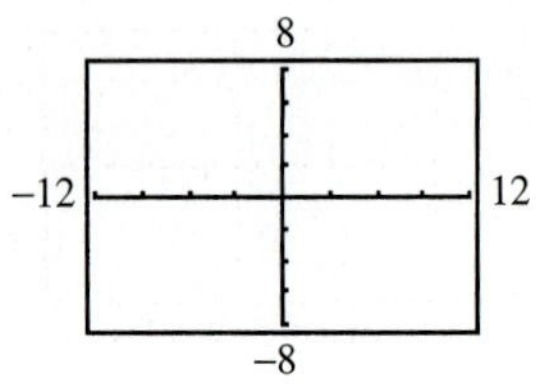


Figure A-5

It is also true that if we know the Xmin and the Xmax values, we can calculate the Xscl by the displayed axes. For example, the Xscl of the graph in Figure A-5 must be 3 units per tick mark for the maximum and minimum $x$-values to be as shown. Also, the Yscl of that graph must be 2 units per tick mark for the maximum and minimum $y$-values to be as shown.

We will call the viewing window in Figure A-3 a *standard* viewing window or rectangle. Although a standard viewing window is sufficient for much of this text, special care must be taken to ensure that all key features of a graph are shown. Figures A-6, A-7, and A-8 show the graph of $y = x^2 + 11x - 1$ on three different viewing windows. Note that certain viewing windows for this equation are misleading.

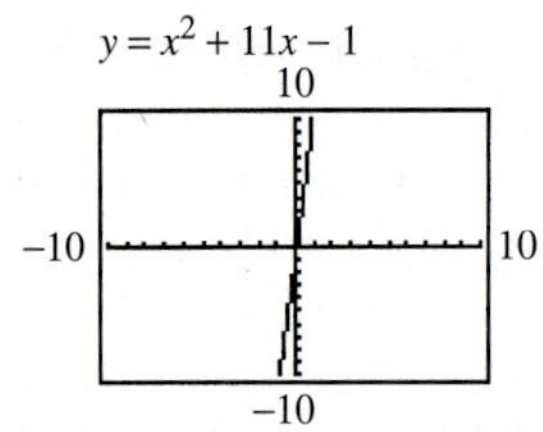


Figure A-6

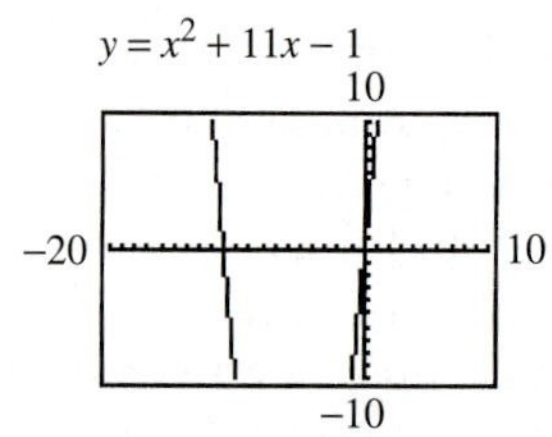


Figure A-7

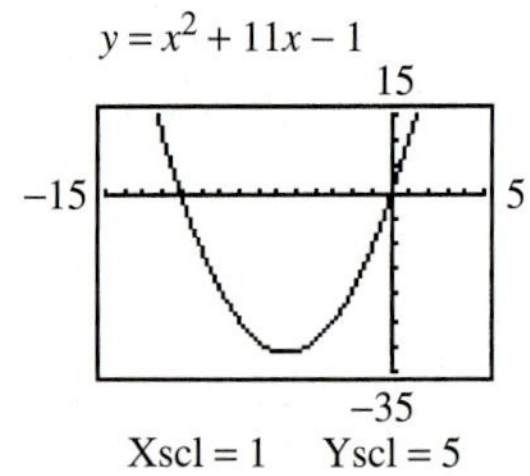


Figure A-8

How do we ensure that all distinguishing features of the graph of an equation are shown? It helps to know about the equation that is being graphed. For example, the equation $y = x^2 + 11x - 1$ is not a linear equation and its graph is not a line. This equation is a quadratic equation and, therefore, its graph is a parabola. By knowing this information, we know that the graph shown in Figure A-6, although correct, is misleading. Of the three viewing rectangles shown, the graph in Figure A-8 is best because it shows more of the distinguishing features of the parabola. Properties of equations needed for graphing will be studied in this text.

## *Viewing Window and Interpreting Window Settings Exercise Set*

*In Exercises 1–4, determine whether all ordered pairs listed will lie within a standard viewing rectangle.*

**1.** $(-9, 0), (5, 8), (1, -8)$

**2.** $(4, 7), (0, 0), (-8, 9)$

**3.** $(-11, 0), (2, 2), (7, -5)$

**4.** $(3, 5), (-3, -5), (15, 0)$

*In Exercises 5–10, choose an Xmin, Xmax, Ymin, and Ymax so that all ordered pairs listed will lie within the viewing rectangle.*

**5.** $(-90, 0), (55, 80), (0, -80)$

**6.** $(4, 70), (20, 20), (-18, 90)$

**7.** $(-11, 0), (2, 2), (7, -5)$

**8.** (3, 5), (−3, −5), (15, 0)
**9.** (200, 200), (50, −50), (70, −50)
**10.** (40, 800), (−30, 500), (15, 0).

*Write the window setting for each viewing window shown. Use the following format:*

| | |
|---|---|
| Xmin = | Ymin = |
| Xmax = | Ymax = |
| Xscl = | Yscl = |

**11.**
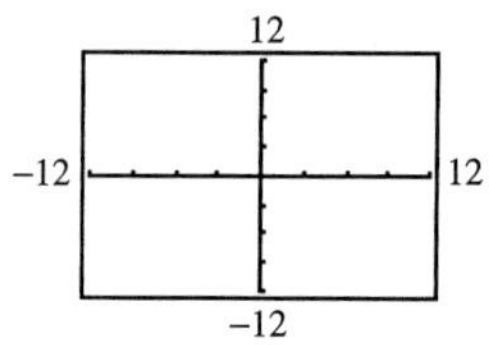


**12.**
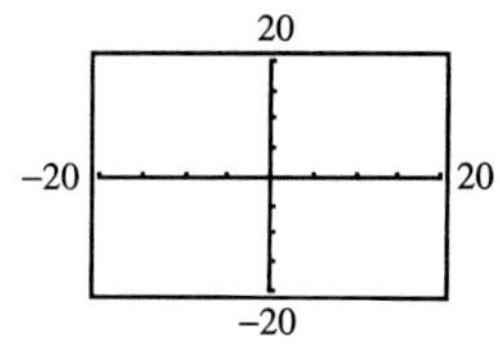


**13.**
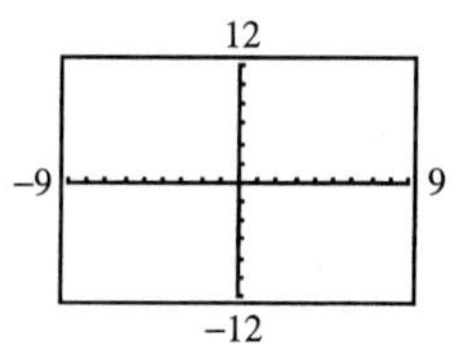


**14.**
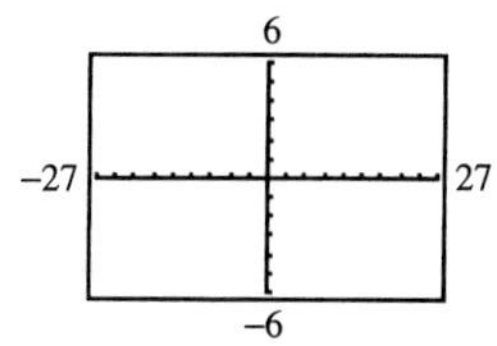


**15.**
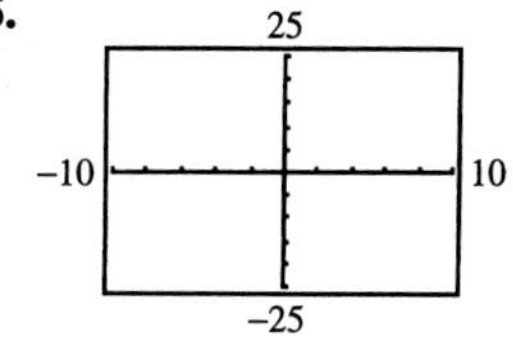


**16.**
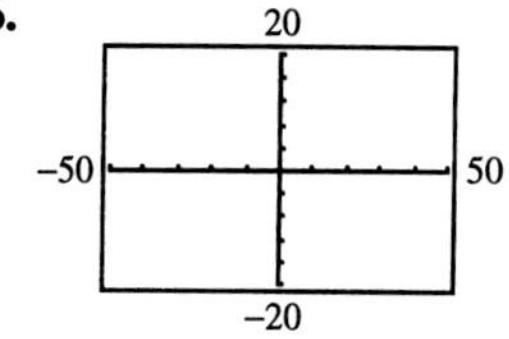


**17.**
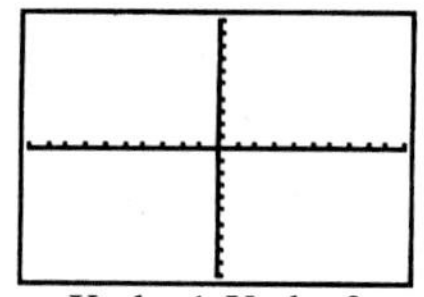
Xscl = 1, Yscl = 3

**18.**
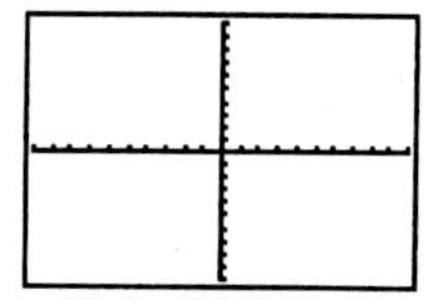
Xscl = 10, Yscl = 2

**19.**
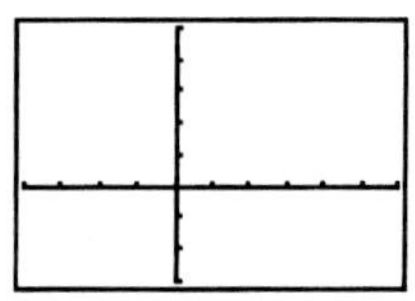
Xscl = 5, Yscl = 10

**20.**
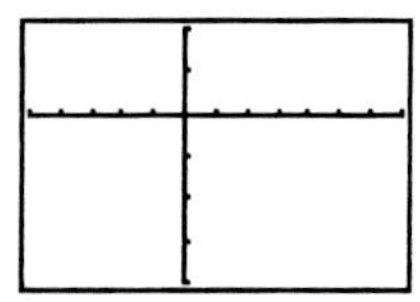
Xscl = 100, Yscl = 200

## Graphing Equations and Square Viewing Window

In general, the following steps may be used to graph an equation on a standard viewing window.

**GRAPHING AN EQUATION IN $x$ AND $y$ WITH A GRAPHIING UTILITY ON A STANDARD VIEWING WINDOW**

**Step 1:** Solve the equation for $y$.
**Step 2:** Use your graphing utility and enter the equation in the form
$\text{Y} = \textit{expression involving } x$
**Step 3:** Activate the graphing utility.

Special care must be taken when entering the *expression involving x* in Step 2. You must be sure that the graphing utility you are using interprets the expression as you want it to. For example, let's graph $3y = 4x$. To do so,

**Step 1:** Solve the equation for $y$.

$$3y = 4x$$

$$\frac{3y}{3} = \frac{4x}{3}$$

$$y = \frac{4}{3}x$$

**Step 2:** Using your graphing utility, enter the expression $\frac{4}{3}x$ after the Y = prompt. In order for your graphing utility to correctly interpret the expression, you may need to enter $(4/3)x$ or $(4 \div 3)x$.

**Step 3:** Activate the graphing utility. The graph should appear as in Figure A-9.

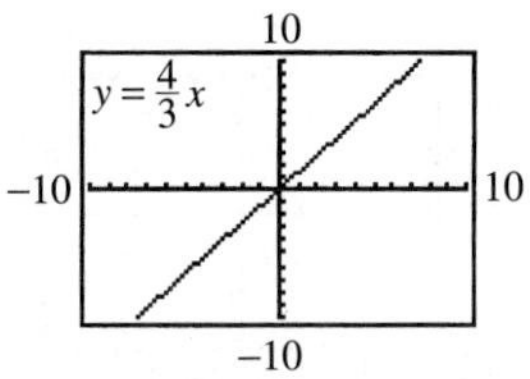


Figure A-9

Distinguishing features of the graph of a line include showing all the intercepts of the line. For example, the window of the graph of the line in Figure A-10 does not show both intercepts of the line, but the window of the graph of the same line in Figure A-11 does show both intercepts. Notice the notation below each graph. This is a shorthand notation of the range setting of the graph. This notation means [Xmin, Xmax] by [Ymin, Ymax].

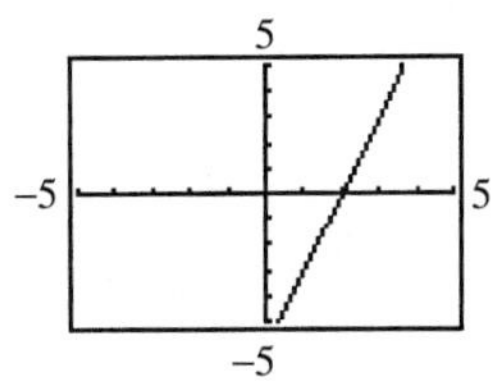


[−5, 5] by [−5, 5]

Figure A-10

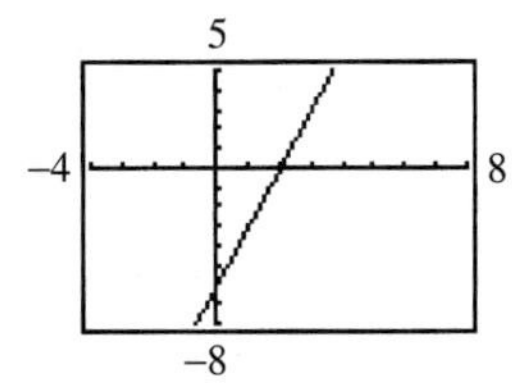


[−4, 8] by [−8, 5]

Figure A-11

On a standard viewing window, the tick marks on the $y$-axis are closer together than the tick marks on the $x$-axis. This happens because the viewing window is a rectangle, and so 10 equally spaced tick marks on the positive $y$-axis will be closer together than 10 equally spaced tick marks on the positive $x$-axis. This causes the appearance of graphs to be distorted.

For example, notice the different appearances of the same line graphed using different viewing windows. The line in Figure A-12 is distorted because the tick marks along the $x$-axis are farther apart than the tick marks along the $y$-axis. The graph of the same line in Figure A-13 is not distorted because the viewing rectangle has been selected so that there is equal spacing between tick marks on both axes.

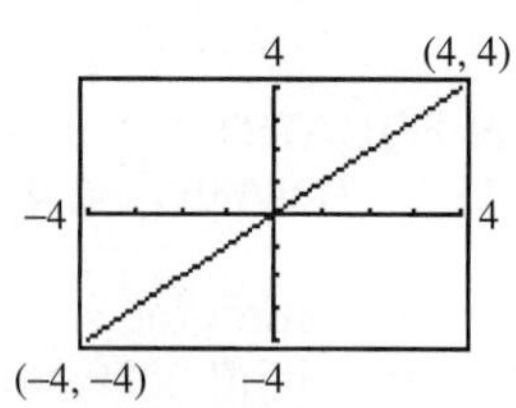


Figure A-12

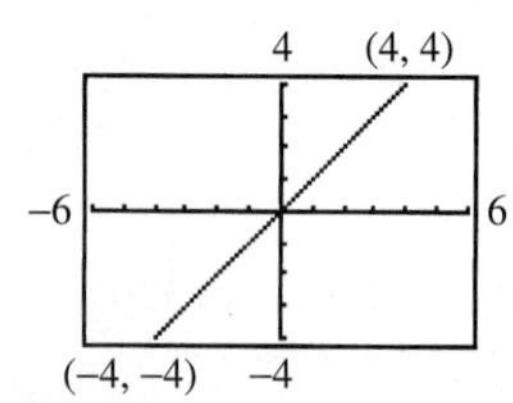


Figure A-13

We say that the line in Figure A-13 is graphed on a *square* setting. Some graphing utilities have a built-in program that, if activated, will automatically provide a square setting. A square setting is especially helpful when we are graphing perpendicular lines, circles, or when a true geometric perspective is desired. Some examples of square screens are shown in Figures A-14 and A-15.

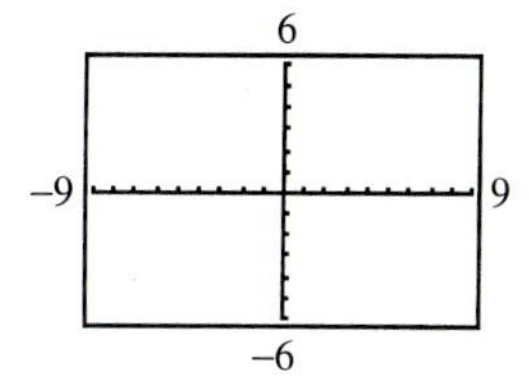


Figure A-14

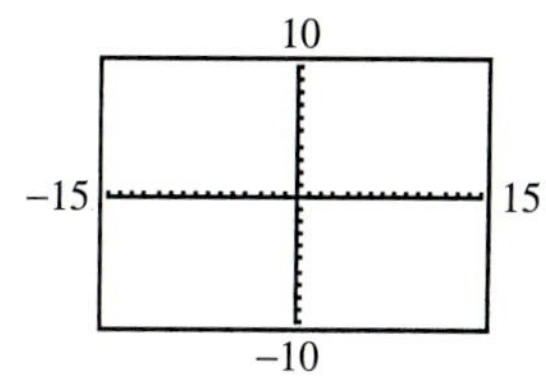


Figure A-15

Other features of a graphing utility such as Trace, Zoom, Intersect, and Table are discussed in appropriate Graphing Calculator Explorations in this text.

## *Graphing Equations and Square Viewing Window Exersise Set*

*Graph each linear equation in two variables, using the two different range settings given. Determine which setting shows all intercepts of a line.*

**1.** $y = 2x + 12$
Setting A: $[-10, 10]$ by $[-10, 10]$
Setting B: $[-10, 10]$ by $[-10, 15]$

**2.** $y = -3x + 25$
Setting A: $[-5, 5]$ by $[-30, 10]$
Setting B: $[-10, 10]$ by $[-10, 30]$

**3.** $y = -x - 41$
Setting A: $[-50, 10]$ by $[-10, 10]$
Setting B: $[-50, 10]$ by $[-50, 15]$

**4.** $y = 6x - 18$
Setting A: $[-10, 10]$ by $[-20, 10]$
Setting B: $[-10, 10]$ by $[-10, 10]$

**5.** $y = \frac{1}{2}x - 15$
Setting A: $[-10, 10]$ by $[-20, 10]$
Setting B: $[-10, 35]$ by $[-20, 15]$

**6.** $y = -\frac{2}{3}x - \frac{29}{3}$
Setting A: $[-10, 10]$ by $[-10, 10]$
Setting B: $[-15, 5]$ by $[-15, 5]$

*The graph of each equation is a line. Use a graphing utility and a standard viewing window to graph each equation.*

**7.** $3x = 5y$ **8.** $7y = -3x$ **9.** $9x - 5y = 30$
**10.** $4x + 6y = 20$ **11.** $y = -7$ **12.** $y = 2$
**13.** $x + 10y = -5$ **14.** $x - 5y = 9$

*Graph the following equations using the square setting given. Some keystrokes that may be helpful are given.*

**15.** $y = \sqrt{x}$ $[-12, 12]$ by $[-8, 8]$
Suggested keystrokes: $\sqrt{\phantom{x}}\, x$

**16.** $y = \sqrt{2x}$ $[-12, 12]$ by $[-8, 8]$
Suggested keystrokes: $\sqrt{\phantom{x}}\, (2x)$

**17.** $y = x^2 + 2x + 1$ $[-15, 15]$ by $[-10, 10]$
Suggested keystrokes: $x \wedge 2 + 2\,x + 1$

**18.** $y = x^2 - 5$ $[-15, 15]$ by $[-10, 10]$
Suggested keystrokes: $x \wedge 2 - 5$

**19.** $y = |x|$ $[-9, 9]$ by $[-6, 6]$
Suggested keystrokes: *ABS* $(x)$

**20.** $y = |x - 2|$ $[-9, 9]$ by $[-6, 6]$
Suggested keystrokes: *ABS* $(x - 2)$

*Graph each line. Use a standard viewing window; then, if necessary, change the viewing window so that all intercepts of each line show.*

**21.** $x + 2y = 30$ **22.** $1.5x - 3.7y = 40.3$

# ANSWERS TO SELECTED EXERCISES

## ■ CHAPTER 1 REAL NUMBERS AND ALGEBRAIC EXPRESSIONS

***Exercise Set 1.2*** **1.** 35 **3.** 30.38 **5.** $\frac{3}{8}$ **7.** 22 **9.** 2000 mi **11.** 20.4 sq. ft **13.** \$36,909.60 **15.** $\{1, 2, 3, 4, 5\}$ **17.** $\{11, 12, 13, 14, 15, 16\}$ **19.** $\{0\}$ **21.** $\{0, 2, 4, 6, 8\}$ **23.** 

**25.** 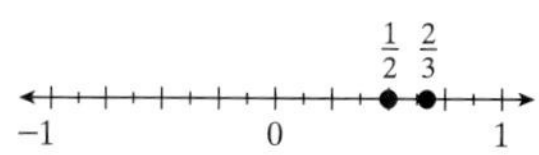

**27.** 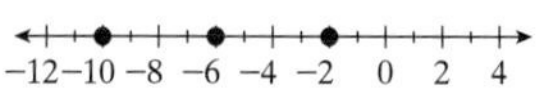

**29.** Answers may vary. **31.** $\{3, 0, \sqrt{36}\}$ **33.** $\{3, \sqrt{36}\}$ **35.** $\{\sqrt{7}\}$ **37.** $\in$ **39.** $\notin$ **41.** $\notin$ **43.** $\notin$ **45.** true **47.** true **49.** false **51.** false **53.** true **55.** false **57.** Answers may vary. **59.** $-2$ **61.** 4 **63.** 0 **65.** $-3$ **67.** Answers may vary. **69.** 6.2 **71.** $-\frac{4}{7}$ **73.** $\frac{2}{3}$ **75.** 0 **77.** $2x$ **79.** $2x + 5$ **81.** $x - 10$ **83.** $x + 2$ **85.** $\frac{x}{11}$ **87.** $3x + 12$ **89.** $x - 17$ **91.** $2(x + 3)$ **93.** $\frac{5}{4 - x}$ **95.** 137; 102; 93; 71; 59

***Exercise Set 1.3*** **1.** 5 **3.** $-24$ **5.** $-11$ **7.** $-4$ **9.** $\frac{4}{3}$ **11.** $-2$ **13.** $-60$ **15.** 80 **17.** 3 **19.** 0 **21.** $-8$ **23.** $-\frac{3}{7}$ **25.** $\frac{1}{21}$ **27.** $-49$ **29.** 36 **31.** $-8$ **33.** Answers may vary. **35.** 7 **37.** $\frac{1}{3}$ **39.** 4 **41.** 3 **43.** 48 **45.** $-1$ **47.** $-3$ **49.** 14.4 **51.** $-2.1$ **53.** $-\frac{1}{3}$ **55.** 17 **57.** $-\frac{79}{15}$ **59.** $-\frac{5}{14}$ **61.** $-0.5876$ **63.** 13 **65.** 65 **67.** $-2$ **69.** $\frac{5}{2}$ **71. a.** 18; 22; 28; 208 **b.** increase **73. a.** 600; 150; 105 **b.** decrease **75.** $\frac{13}{35}$ **77.** 4205 m **79.** b **81.** d **83.** Yes. Two players have 6 points each (the third player has 0 points) or two players have 5 points each (the third has 2 points). **85.** 16.5227 **87.** 4.4272 **89.** 13.2% **91.** 10.8%

***Exercise Set 1.4*** **1.** $4c = 7$ **3.** $3(x + 1) = 7$ **5.** $\frac{n}{5} = 4n$ **7.** $z - 2 = 2z$ **9.** $>$ **11.** $=$ **13.** $<$ **15.** $7x \leq -21$ **17.** $-2 + x \neq 10$ **19.** $2(x - 6) > \frac{1}{11}$ **21.** $y - 7 = 6$ **23.** $2(x - 6) = -27$ **25.** $-5; \frac{1}{5}$ **27.** $8; -\frac{1}{8}$ **29.** $\frac{1}{4}; -4$ **31.** 0; undefined **33.** $-\frac{7}{8}; \frac{8}{7}$ **35.** Zero. For every real number $x$, $0 \cdot x \neq 1$, so 0 has no reciprocal. It is the only real number that has no reciprocal because if $x \neq 0$, then $x \cdot \frac{1}{x} = 1$ by definition. **37.** $y + 7x$ **39.** $w \cdot z$ **41.** $\frac{x}{5} \cdot \frac{1}{3}$ **43.** no **45.** $(5 \cdot 7)x$ **47.** $x + (1.2 + y)$ **49.** $14(z \cdot y)$ **51.** 10 and 4. Subtraction is not associative. **53.** $3x + 15$ **55.** $-2a - b$ **57.** $12x + 10y + 4z$ **59.** $2y - 6 = \frac{1}{8}$ **61.** $\frac{n + 5}{2} > 2n$ **63.** $6 + 3x$ **65.** 0

**67.** 7 **69.** $(10 \cdot 2)y$ **71.** $a(b + c) = ab + ac$ **73.** $0.1d$ **75.** $112 - x$ **77.** $90 - 5x$ **79.** $\$35.61y$ **81.** $2x + 2$ **83.** $-8y - 14$ **85.** $-9c - 4$ **87.** $4 - 8y$ **89.** $-11y - 11$ **91.** $3t - 14$ **93.** 0 **95.** $13n - 20$ **97.** $-180.96y - 74.33$ **99.** $6.5y - 7.92x + 25.47$ **101.** no **103.** 70 million **105.** 35 million **107.** 12.3%

***Chapter 1 Review*** **1.** 21 **3.** 324,000 **5.** $\{-2, 0, 2, 4, 6\}$ **7.** $\varnothing$ **9.** $\{\ldots, -1, 0, 1, 2\}$ **11.** false **13.** true **15.** true **17.** true **19.** true **21.** true **23.** true **25.** true **27.** true **29.** $\left\{5, \frac{8}{2}, \sqrt{9}\right\}$ **31.** $\{\sqrt{7}, \pi\}$ **33.** $\left\{5, \frac{8}{2}, \sqrt{9}, -1\right\}$ **35.** $-0.6$ **37.** $-1$ **39.** $\frac{1}{0.6}$ **41.** 1 **43.** $-35$ **45.** 0.31 **47.** 13.3 **49.** 0 **51.** 0 **53.** $-5$ **55.** 4 **57.** 9 **59.** 3 **61.** $-\frac{32}{135}$ **63.** $-\frac{5}{4}$ **65.** $\frac{5}{8}$ **67.** $-1$ **69.** 1 **71.** $-4$ **73.** $\frac{5}{7}$ **75.** $\frac{1}{5}$ **77.** $-5$ **79.** 5 **81.** $-xy + 1$ **83.** $2x^2 - 2$ **85.** $-1.1x - 0.3$ **87. a.** 6.28; 62.8; 628 **b.** increase **89.** $n + 2n = -15$ **91.** $6(t - 5) = 4$ **93.** $9x - 10 = 5$ **95.** $-4 < 7y$ **97.** $t + 6 \leq -12$ **99.** distributive property **101.** commutative property of addition **103.** multiplicative inverse property **105.** associative property of multiplication **107.** multiplicative identity property **109.** $(3 + x) + (7 + y)$ **111.** $2 \cdot \frac{1}{2}$, for example **113.** $7 + 0$ **115.** $>$ **117.** $=$ **119.** $>$

***Chapter 1 Test*** **1.** true **2.** false **3.** false **4.** false **5.** true **6.** false **7.** $-3$ **8.** 43 **9.** $-225$ **10.** $-2$ **11.** 1 **12.** 12 **13.** 1 **14. a.** 5.75; 17.25; 57.50; 115.00 **b.** increase **15.** $2|x + 5| = 30$ **16.** $\frac{(6 - y)^2}{7} < -2$ **17.** $\frac{9z}{|-12|} \neq 10$ **18.** $3\left(\frac{n}{5}\right) = -n$ **19.** $20 = 2x - 6$ **20.** $-2 = \frac{x}{x + 5}$ **21.** distributive property **22.** associative property of addition **23.** additive inverse property **24.** multiplication property of zero **25.** $0.05n + 0.1d$ **26.** $2y^2 - 10$ **27.** $-1.3x + 1.9$

## CHAPTER 2 EQUATIONS, INEQUALITIES, AND PROBLEM SOLVING

***Mental Math*** **1.** $8x + 21$ **3.** $6n - 7$ **5.** $-4x - 1$

***Exercise Set 2.1*** **1.** $-12$ **3.** $-0.9$ **5.** 6 **7.** $-5$ **9.** $-1.1$ **11.** $-5$ **13.** 0 **15.** 2 **17.** $-9$ **19.** $-\frac{10}{7}$ **21. a.** $4x + 5$ **b.** $-3$ **c.** Answers may vary. **23.** $\frac{1}{6}$ **25.** 4 **27.** 1 **29.** 5 **31.** all real numbers **33.** $\varnothing$ **35.** Answers may vary. **37.** 8 **39.** $-5.9$ **41.** 7 **43.** 4.2 **45.** 2 **47.** $-2$ **49.** 0 **51.** 5 **53.** $\varnothing$ **55.** $\frac{1}{8}$ **57.** 0 **59.** 29 **61.** $-8$ **63.** all real numbers **65.** 4 **67.** 8 **69.** all real numbers **71.** $\frac{40}{3}$ **73.** 17 **75.** $\frac{3}{5}$ **77.** $\frac{4}{5}$ **79.** $K = -11$ **81.** $K = 24$ **83.** $-4.86$ **85.** 1.53 **87.** not a fair game **89.** $8 + x$ **91.** $8 - x$ **93.** $2x - 5$ **95.** $-\frac{3}{22}$ **97.** 4

***Exercise Set 2.2*** **1.** $4y$ **3.** $3z + 3$ **5.** $15x + 30$ **7.** $10x + 3$ **9.** $-5$ **11.** 45, 225 **13.** 78 **15.** 1.92 **17.** approximately 658.59 million acres **19.** 51,700 homes **21.** 20% **23.** 117 automobile loans **25.** Dallas/Ft. Worth, 60.5; Atlanta, 68.2; Chicago, 70.3 **27.** B767-300ER, 216 seats; B737-200, 112 seats **29.** \$430.00 **31. a.** 49,057 telephone company operators **b.** Answers may vary. **33.** 5 years **35.** 17 million returns **37.** square: each side, 18 cm; triangle: each side, 24 cm **39.** length, 14 cm; width, 6 cm **41.** width: 8.4 m; height: 47 m **43.** 64°, 32°, 84° **45.** 80°, 100° **47.** 15°, 75° **49.** 40°, 140° **51.** 75, 76, 77 **53.** Fallon's zip code is 89406; Fernley's zip code is 89408; Gardnerville Ranchos' zip code is 89410 **55.** any three consecutive integers **57.** 25 skateboards **59.** 800 books **61.** Answers may vary. **63.** 11 million trees **65.** 6 **67.** 208 **69.** $-55$ **71.** 3195 **73.** 121.8152

***Mental Math*** **1.** $y = 5 - 2x$ **3.** $a = 5b + 8$ **5.** $k = h - 5j + 6$

***Exercise Set 2.3*** **1.** $t = \frac{D}{r}$ **3.** $R = \frac{I}{PT}$ **5.** $y = \frac{9x - 16}{4}$ **7.** $W = \frac{P - 2L}{2}$ **9.** $A = \frac{J + 3}{C}$ **11.** $g = \frac{W}{h - 3t^2}$ **13.** $B = \frac{T - 2C}{AC}$ **15.** $r = \frac{C}{2\pi}$ **17.** $r = \frac{E - IR}{I}$ **19.** $L = \frac{2s - an}{n}$ **21.** $v = \frac{3st^4 - N}{5s}$ **23.** $H = \frac{S - 2LW}{2L + 2W}$ **25.** \$4703.71; \$4713.99; \$4719.22; \$4722.74; \$4724.45 **27.** 12 times a year; you earn more interest **29. a.** \$7313.97 **b.** \$7321.14 **c.** \$7325.98 **31.** 40°C **33.** 3.6 hr, or 3 hr and 36 min **35.** 171 packages **37.** 0.42 ft **39.** 0.25 sec **41.** $41.125\pi$ ft $\approx$ 129.1325 ft **43.** Estimates will vary; actual cost was \$6.80 each. **45.** \$1831.96 **47.** 2 gal **49. a.** 1174.86 cu. m **b.** 310.34 cu. m **c.** 1485.20 cu. m **51.** 2.25 hr, or 2 hr and 15 min

**53.** approximately 34,507 mph **55.** 0.388; 0.723; 1.00; 1.523; 5.202; 9.538; 19.193; 30.065; 39.505
**57.** $\frac{1}{4}$ **59.** $\frac{3}{8}$ **61.** $\frac{3}{8}$ **63.** $\frac{3}{4}$ **65.** 1 **67.** 1 **69.** $\{2, 3\}$ **71.** $\{-3, -2, -1, 0, 1, 2, 3\}$ **73.** Answers may vary.

***Mental Math*** **1.** $\{x|x < 6\}$ **3.** $\{x|x \geq 10\}$ **5.** $\{x|x > 4\}$ **7.** $\{x|x \leq 2\}$

***Exercise Set 2.4*** **1.** ; $(-\infty, -3)$ **3.** ; $[0.3, \infty)$ **5.** ; $(5, \infty)$
**7.** ; $(-2, 5)$ **9.** ; $(-1, 5)$ **11.** Answers may vary. **13.** 100 **15.** $-5, -1{,}000{,}000$
**17.** ; $[-2, \infty)$ **19.** ; $(-\infty, 1)$ **21.** ; $(-\infty, 2]$
**23.** ; $(-\infty, -4)$ **25.** ; $\left[\frac{8}{3}, \infty\right)$ **27.** ; $(-\infty, -4.7)$
**29.** ; $(-\infty, -3]$ **31.** ; $(4, \infty)$ **33.** $(-\infty, -1]$ **35.** $(-\infty, 11]$ **37.** $(-13, \infty)$ **39.** $(-\infty, 7]$
**41.** $(-\infty, \infty)$ **43.** $\varnothing$ **45.** $(0, \infty)$ **47.** $(-2, \infty)$ **49.** $\left[-\frac{3}{5}, \infty\right]$ **51.** $[-9.6, \infty)$ **53.** $(38, \infty)$ **55.** Answers may vary.
**57.** $[0, \infty)$ **59.** $(-\infty, -5]$ **61.** $\left(-\infty, \frac{1}{4}\right)$ **63.** $(-\infty, -1)$ **65.** $\left[-\frac{79}{3}, \infty\right)$ **67.** $(-\infty, -15)$ **69.** $[3, \infty)$
**71.** $\left[-\frac{37}{3}, \infty\right)$ **73.** $(-\infty, 5)$ **75.** 30 **77.** 1040 lb **79.** 16 oz **81.** more than 200 calls **83.** $F \geq 932°$
**85. a.** 2002 **b.** Answers may vary. **87.** decreasing **89.** 6.74 gal **91.** 2004 **93.** Answers may vary. **95.** 2001
**97.** 0, 1, 2, 3, 4, 5, 6, 7 **99.** $-6, -7, -8, \ldots$ **101.** ; $(-7, 1]$ **103.** ; $[-2.5, 5.3)$

***Exercise Set 2.5*** **1.** $\{2, 3, 4, 5, 6, 7\}$ **3.** $\{4, 6\}$ **5.** $\{\ldots, -2, -1, 0, 1, \ldots\}$ **7.** $\{5, 7\}$
**9.** $\{x|x$ is an odd integer or $x = 2$ or $x = 4\}$ **11.** $\{2, 4\}$ **13.** ; $(-2, 5)$ **15.** ; $[6, \infty)$
**17.** ; $(-\infty, -3]$ **19.** ; $(11, 17)$ **21.** ; $[1, 4]$
**23.** ; $\left[-3, \frac{3}{2}\right]$ **25.** ; $[-21, -9]$ **27.** ; $(-\infty, -1) \cup (0, \infty)$
**29.** ; $[2, \infty)$ **31.** ; $(-\infty, \infty)$ **33.** Answers may vary. **35.** ; $(-1, 2)$
**37.** ; $(-\infty, \infty)$ **39.** ; $[-1, \infty)$ **41.** ; $[-5, \infty)$ **43.** ; $\left[\frac{3}{2}, 6\right]$
**45.** ; $\left(\frac{5}{4}, \frac{11}{4}\right)$ **47.** ; $\varnothing$ **49.** ; $(-7, \infty)$ **51.** ; $\left(-5, \frac{5}{2}\right)$
**53.** ; $\left(0, \frac{14}{3}\right]$ **55.** ; $(-\infty, -3]$ **57.** ; $(-\infty, 1] \cup \left(\frac{29}{7}, \infty\right)$
**59.** ; $\varnothing$ **61.** ; $\left[-\frac{1}{2}, \frac{3}{2}\right)$ **63.** ; $\left(-\frac{4}{3}, \frac{7}{3}\right)$ **65.** ; $(6, 12)$
**67.** $-20.2° \leq F \leq 95°$ **69.** $67 \leq$ final score $\leq 94$ **71.** 1993, 1994, 1995 **73.** $-12$ **75.** $-4$ **77.** $-7, 7$ **79.** 0
**81.** ; $(6, \infty)$ **83.** ; $[3, 7]$ **85.** ; $(-\infty, -1)$

***Mental Math*** **1.** 7 **3.** $-5$ **5.** $-6$ **7.** 12

***Exercise Set 2.6*** **1.** $7, -7$ **3.** $4.2, -4.2$ **5.** $7; -2$ **7.** 8, 4 **9.** $5, -5$ **11.** $3, -3$ **13.** 0 **15.** $\varnothing$ **17.** $\frac{1}{5}$ **19.** $|x| = 5$
**21.** $9, -\frac{1}{2}$ **23.** $-\frac{5}{2}$ **25.** Answers may vary. **27.** $4, -4$ **29.** 0 **31.** $\varnothing$ **33.** $0, \frac{14}{3}$ **35.** $2, -2$ **37.** $\varnothing$ **39.** $7, -1$

**41.** ∅ **43.** ∅ **45.** $-\frac{1}{8}$ **47.** $\frac{1}{2}, -\frac{5}{6}$ **49.** $2, -\frac{12}{5}$ **51.** 3, −2 **53.** $-8, \frac{2}{3}$ **55.** ∅ **57.** 4 **59.** 13, −8 **61.** 3, −3 **63.** 8, −7 **65.** 2, 3 **67.** $2, -\frac{10}{3}$ **69.** $\frac{3}{2}$ **71.** ∅ **73.** Answers may vary. **75.** 13% **77.** \$1.088 billion **79.** Answers may vary. **81.** no solution

***Exercise Set 2.7*** **1.** ; [−4, 4] **3.** ; (1, 5) **5.** ; (−5, −1) **7.** ; [−10, 3] **9.** ; [−5, 5] **11.** ; ∅ **13.** ; [0, 12] **15.** ; $(-\infty, -3) \cup (3, \infty)$ **17.** ; $(-\infty, -24] \cup [4, \infty)$ **19.** ; $(-\infty, -4) \cup (4, \infty)$ **21.** ; $(-\infty, \infty)$ **23.** ; $\left(-\infty, \frac{2}{3}\right) \cup (2, \infty)$ **25.** ; {0} **27.** ; $\left(-\infty, -\frac{3}{8}\right) \cup \left(-\frac{3}{8}, \infty\right)$ **29.** $|x| < 7$ **31.** $|x| \le 5$ **33.** ; [−2, 2] **35.** ; $(-\infty, -1) \cup (1, \infty)$ **37.** ; (−5, 11) **39.** ; $(-\infty, 4) \cup (6, \infty)$ **41.** ; ∅ **43.** ; $(-\infty, \infty)$ **45.** ; [−2, 9] **47.** ; $(-\infty, -11] \cup [1, \infty)$ **49.** ; $(-\infty, 0) \cup (0, \infty)$ **51.** ; $(-\infty, \infty)$ **53.** ; $\left[-\frac{1}{2}, 1\right]$ **55.** ; $(-\infty, -3) \cup (0, \infty)$ **57.** ; ∅ **59.** ; $(-\infty, \infty)$ **61.** ; $\left(-\frac{2}{3}, 0\right)$ **63.** ; $(-\infty, \infty)$ **65.** ; $(-\infty, -1) \cup (1, \infty)$ **67.** ; $(-\infty, -12) \cup (0, \infty)$ **69.** ; $(-\infty, -6) \cup (0, \infty)$ **71.** ; $\left(-\frac{31}{5}, \frac{11}{5}\right)$ **73.** ; [−1, 8] **75.** ; $\left[-\frac{23}{8}, \frac{17}{8}\right]$ **77.** (−2, 5) **79.** 5, −2 **81.** $(-\infty, -7] \cup [17, \infty)$ **83.** $-\frac{9}{4}$ **85.** (−2, 1) **87.** $2, \frac{4}{3}$ **89.** ∅ **91.** $\frac{19}{2}, -\frac{17}{2}$ **93.** $\left(-\infty, -\frac{25}{3}\right) \cup \left(\frac{35}{3}, \infty\right)$ **95.** Answers may vary. **97.** $3.45 < x < 3.55$ **99.** $\frac{1}{6}$ **101.** 0 **103.** $\frac{1}{3}$ **105.** −1.5 **107.** 0

***Chapter 2 Review*** **1.** 3 **3.** $-\frac{45}{14}$ **5.** 0 **7.** 6 **9.** all real numbers **11.** ∅ **13.** −3 **15.** $\frac{96}{5}$ **17.** 32 **19.** 8 **21.** ∅ **23.** 2 **25.** −7 **27.** 52 **29.** \$22,896 **31.** No such odd integers exist. **33.** 358 mi **35.** 5 plants, \$200 **37.** $r = \frac{C}{2\pi}$ **39.** $x = \frac{4y - 12}{5}$ **41.** $x = \frac{y - y_1 + mx_1}{m}$ **43.** $g = \frac{s - vt}{t^2}$ **45.** $P = \frac{I}{1 + rt}$ **47.** $h = \frac{3V}{\pi r^2}$ **49.** $T_2 = \frac{T_1V_2}{V_1}$ **51.** $\left(\frac{290}{9}\right)$°C ≈ 32.2°C **53.** 16 packages **55.** 58 mph **57.** $(-\infty, -4]$ **59.** $(-17, \infty)$ **61.** $(-\infty, 4]$ **63.** $(-\infty, 1)$ **65.** $(2, \infty)$ **67.** 260° ≤ C ≤ 538°C **69.** \$1750 to \$3750 **71.** $\left[-2, -\frac{9}{5}\right)$ **73.** $\left(-\frac{3}{5}, 0\right)$ **75.** $\left[-\frac{4}{3}, \frac{7}{6}\right]$ **77.** $(-\infty, \infty)$

**79.** $(5, \infty)$ **81.** 5, 11 **83.** $-1, \frac{11}{3}$ **85.** $-\frac{1}{6}$ **87.** $\varnothing$ **89.** 1, 5 **91.** $\varnothing$ **93.** $-10, -\frac{4}{3}$

**95.** 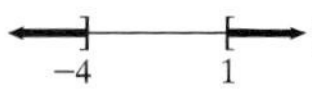; $(-\infty, -4] \cup [1, \infty)$ **97.** ; $(-3, 3)$ **99.** 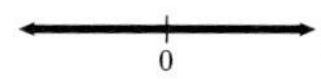; $(-\infty, \infty)$

**101.** 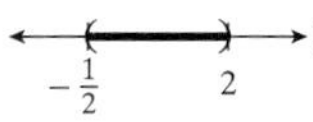; $\left(-\frac{1}{2}, 2\right)$ **103.** 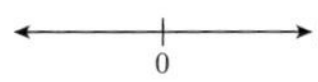; $\varnothing$

***Chapter 2 Test*** **1.** 10 **2.** 1 **3.** $\varnothing$ **4.** all real numbers **5.** 12 **6.** $-\frac{80}{29}$ **7.** $1, \frac{2}{3}$ **8.** $\varnothing$ **9.** $y = \frac{3x - 8}{4}$
**10.** $n = \frac{9}{7}m$ **11.** $g = \frac{S}{t^2 + vt}$ **12.** $C = \frac{5}{9}(F - 32)$ **13.** $(5, \infty)$ **14.** $[2, \infty)$ **15.** $\left(\frac{3}{2}, 5\right]$ **16.** $(-\infty, -2) \cup \left(\frac{4}{3}, \infty\right)$
**17.** $[5, \infty)$ **18.** $[4, \infty)$ **19.** $[-3, -1)$ **20.** $(-\infty, \infty)$ **21.** 9.6 **22.** 211,468 people **23.** approximately 8
**24.** more than 850 sunglasses **25.** \$3542.27 **26.** \$0.87 per lb **27.** 1402 watts

***Chapter 2 Cumulative Review*** **1. a.** $\{2, 3, 4, 5\}$ **b.** $\{101, 102, 103, \ldots\}$; Sec. 1.2, Ex. 3 **2. a.** 3 **b.** 5 **c.** $-2$ **d.** $-8$
**e.** 0; Sec. 1.2, Ex. 6 **3. a.** $-14$ **b.** $-4$ **c.** 5 **d.** $-10.2$ **e.** $-\frac{5}{21}$; Sec. 1.3, Ex. 1 **4. a.** 3 **b.** 5 **c.** $\frac{1}{2}$; Sec. 1.3, Ex. 7
**5. a.** $-2$ **b.** 9 **c.** $-1$; Sec. 1.3, Ex. 10 **6. a.** $x + 5 = 20$ **b.** $2(3 + y) = 4$ **c.** $x - 8 = 2x$
**d.** $\frac{z}{9} = 3(z - 5)$; Sec. 1.4, Ex. 1 **7.** $5 + 7x$; Sec. 1.4, Ex. 6 **8.** 2; Sec. 2.1, Ex. 1 **9.** all real numbers; Sec. 2.1, Ex. 9
**10. a.** $2x + 1$ **b.** $12x - 3$; Sec. 2.2, Ex. 1 **11.** 23, 49; Sec. 2.2, Ex. 3 **12.** $y = \frac{2x + 7}{3}$ or $y = \frac{2x}{3} + \frac{7}{3}$; Sec. 2.3, Ex. 2
**13.** $b = \frac{2A - Bh}{h}$; Sec. 2.3, Ex. 3 **14. a.** 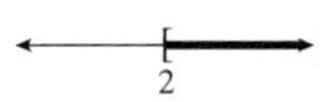; $[2, \infty)$ **b.** 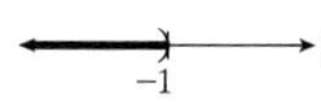; $(-\infty, -1)$
**c.** ; (0.5 3]; Sec. 2.4, Ex. 1 **15.** $\left[\frac{5}{2}, \infty\right)$; Sec. 2.4, Ex. 5 **16.** $(-\infty, \infty)$; Sec. 2.4, Ex. 7

**17.** $\{4, 6\}$; Sec. 2.5, Ex. 1 **18.** $(-\infty, 4)$; Sec. 2.5, Ex. 2 **19.** $\{2, 3, 4, 5, 6, 8\}$; Sec. 2.5, Ex. 6 **20.** $(-\infty, \infty)$; Sec. 2.5, Ex. 8
**21.** 2, $-2$; Sec. 2.6, Ex. 1 **22.** 24, $-20$; Sec. 2.6, Ex. 3 **23.** 4; Sec. 2.6, Ex. 9 **24.** $[-3, 3]$; Sec. 2.7, Ex. 1
**25.** $(-\infty, \infty)$; Sec. 2.7, Ex. 1

## ■ CHAPTER 3 GRAPHS AND FUNCTIONS

***Graphing Calculator Explorations***

**1.** 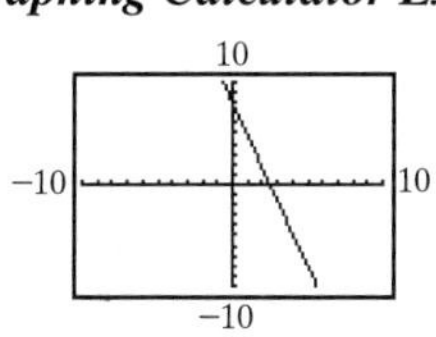

**3.** 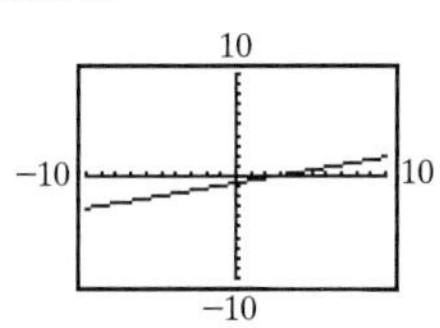

**5.** 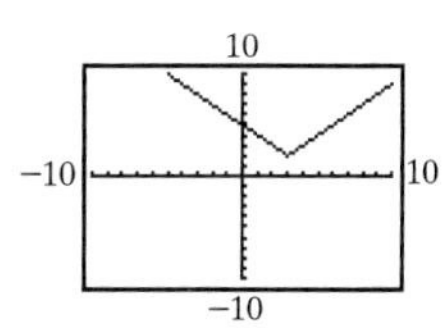

**7.** 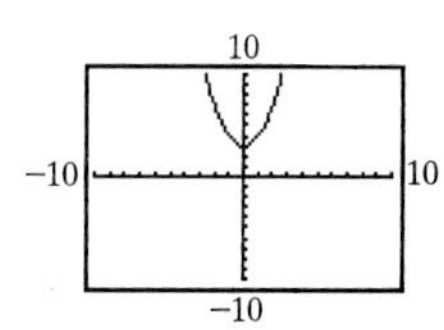


***Mental Math*** **1.** $(5, 2)$ **3.** $(3, -1)$ **5.** $(-5, -2)$ **7.** $(-1, 0)$

***Exercise Set 3.1***

**1.** Quadrant I
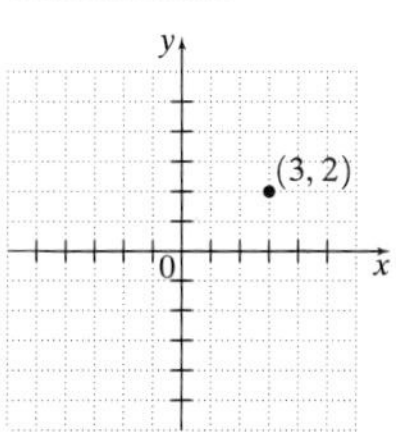

**3.** Quadrant II
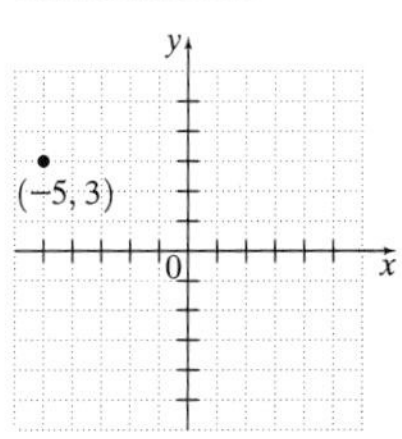

**5.** Quadrant IV
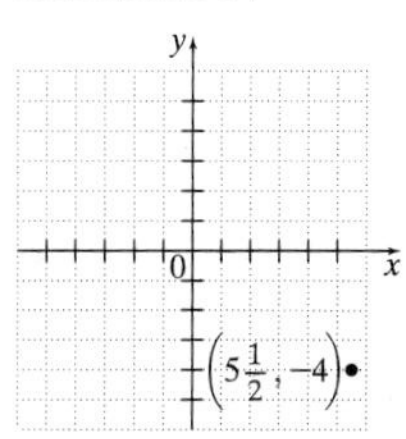

**7.** $y$-axis
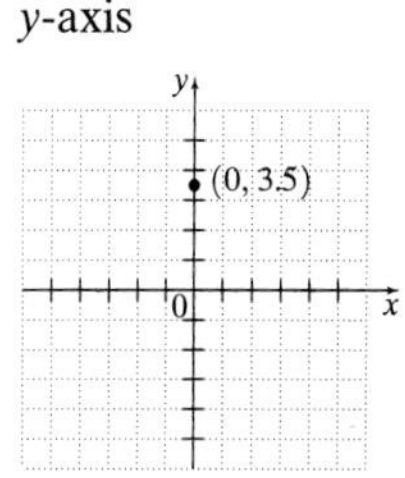

**9.** Quadrant III
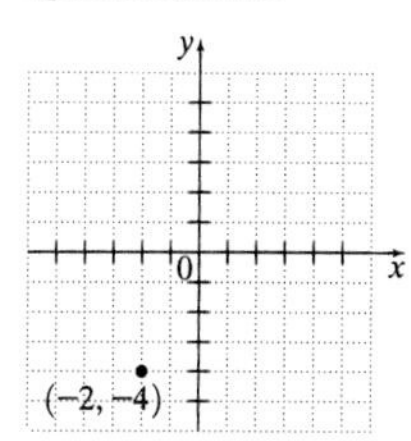


**11.** Quadrant IV **13.** $x$-axis **15.** Quadrant III **17.** no; yes **19.** yes; yes **21.** yes; yes **23.** yes; no **25.** yes; yes

**27.** linear
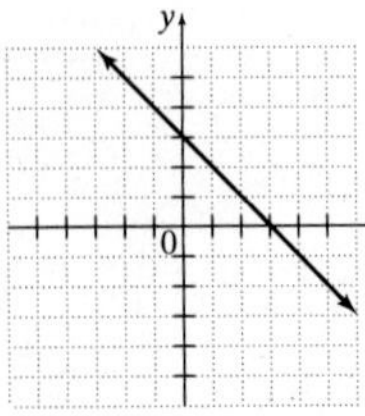

**29.** linear
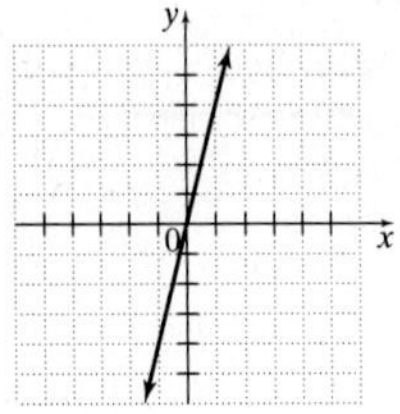

**31.** linear
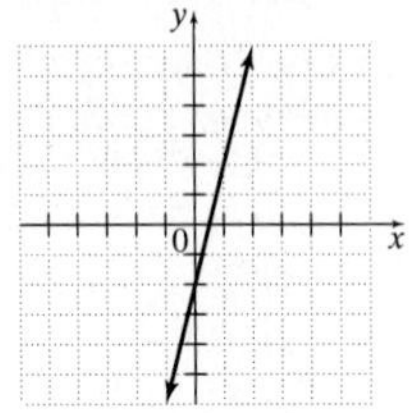

**33.** not linear
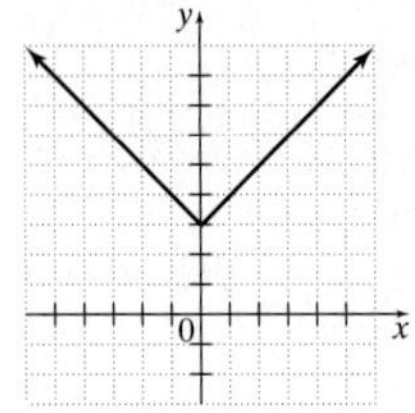

**35.** linear
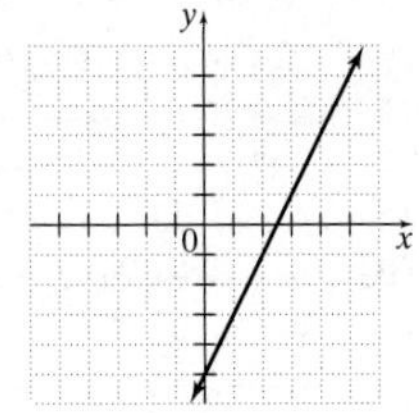

**37.** not linear
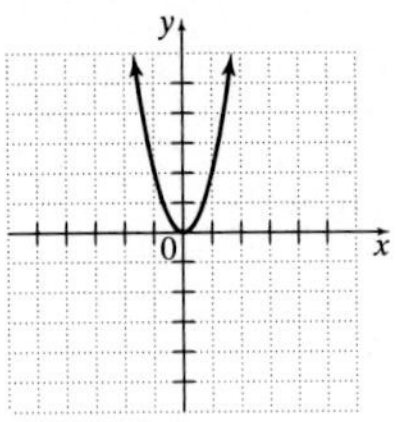

**39.** not linear
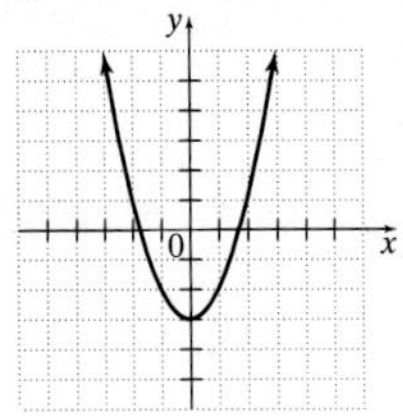

**41.** linear
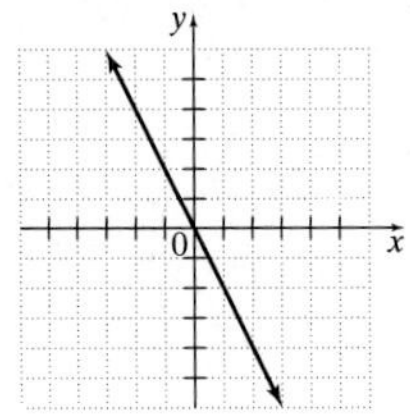

**43.** linear
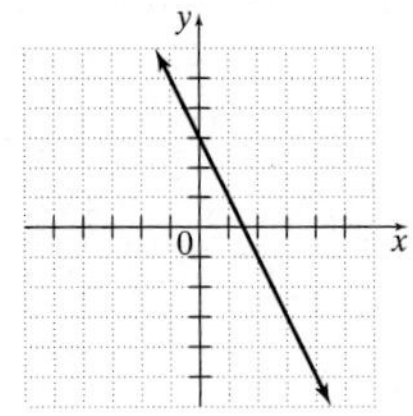

**45.** not linear
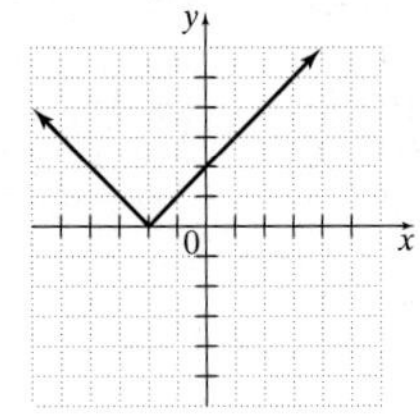

**47.** not linear
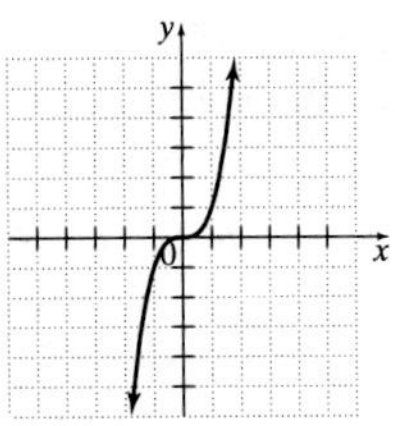

**49.** not linear
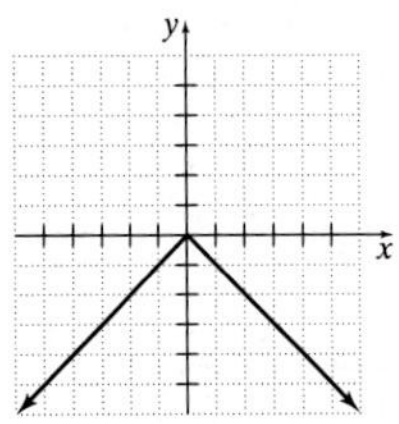

**51.** linear
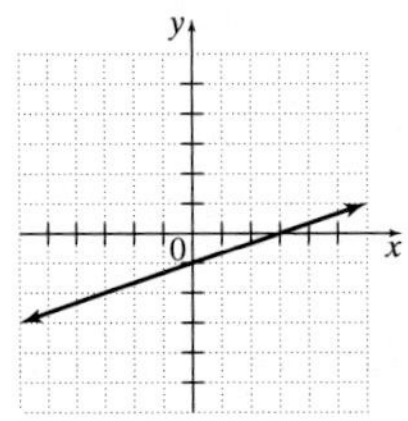

**53.** linear
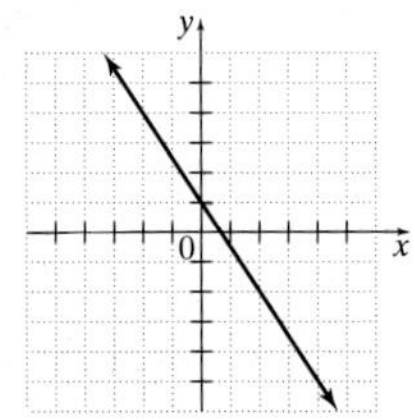

**55.**
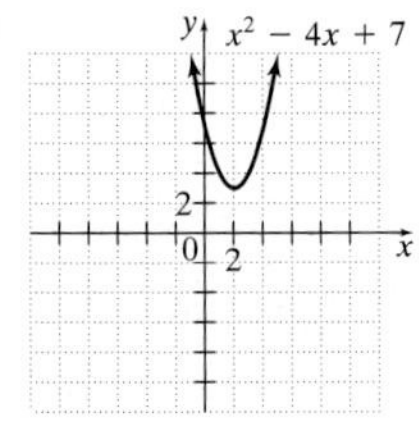


**57. a.**
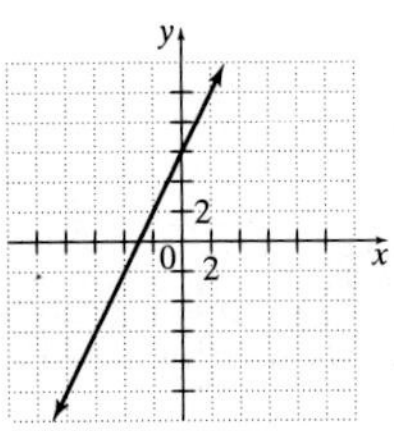

**b.** 14 in. **59.** 1991 **61.** Answers may vary. **63.** \$7000 **65.** \$500 **67.** Depreciation is the same from year to year. **69.** B **71.** C **73.**
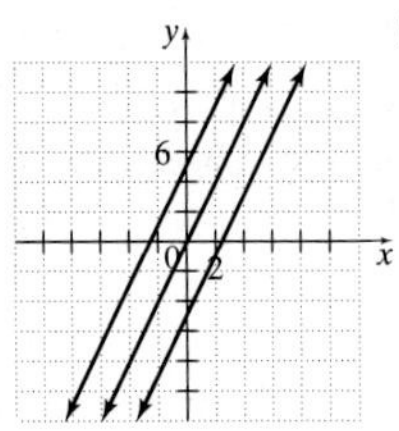
; Answers may vary. **75.** Answers may vary. **77.** $y = -3 - 2x$;
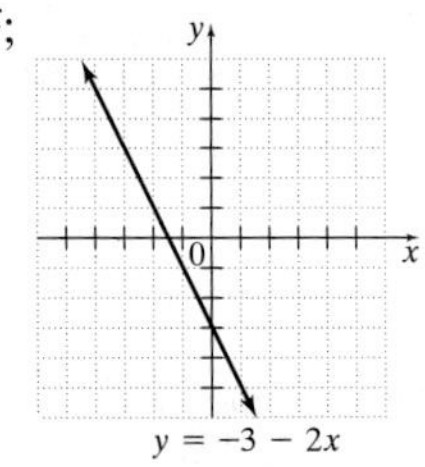


**79.** $y = 5 - x^2$;
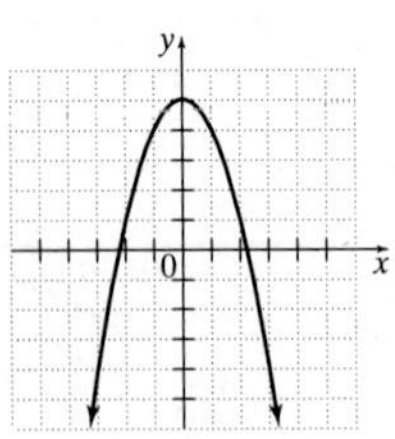

**81.**
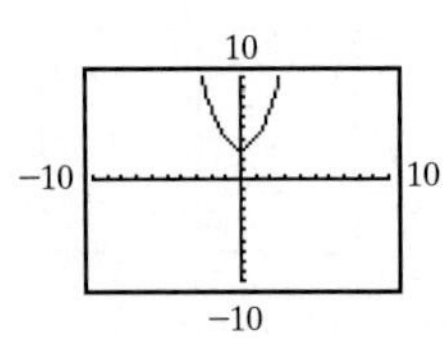

**83.**
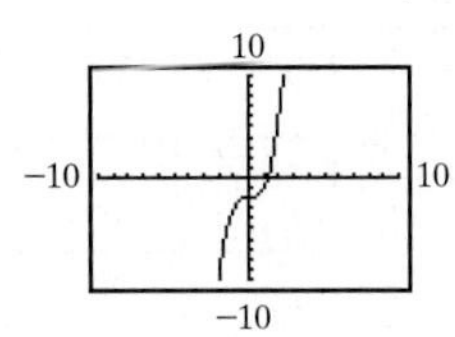

**85.** 0 **87.** $\frac{1}{4}$ **89.** $(-\infty, -6)$ **91.** $(-\infty, -4]$

***Graphing Calculator Explorations*** **1.**
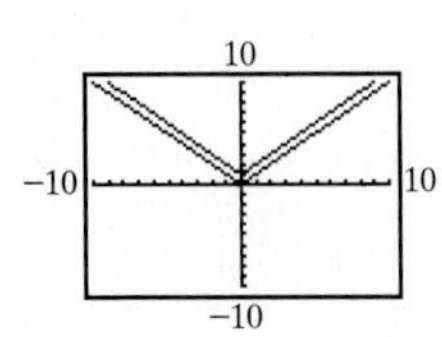

**3.**
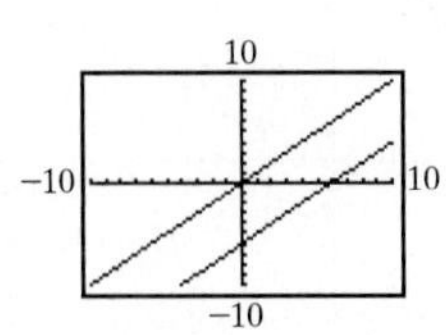

**5.**
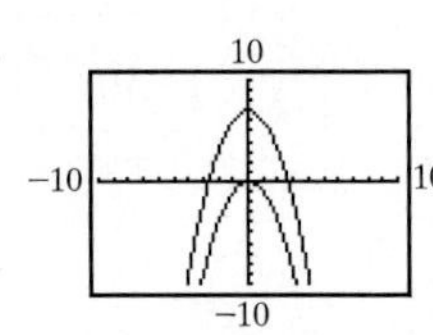

***Exercise Set 3.2*** **1.** domain: $\{-1, 0, -2, 5\}$; range: $\{7, 6, 2\}$; function **3.** domain: $\{-2, 6, -7\}$; range: $\{4, -3, -8\}$; not a function **5.** domain: $\{1\}$; range: $\{1, 2, 3, 4\}$; not a function **7.** domain: $\left\{\frac{3}{2}, 0\right\}$; range: $\left\{\frac{1}{2}, -7, \frac{4}{5}\right\}$; not a function
**9.** domain: $\{-3, 0, 3\}$; range: $\{-3, 0, 3\}$; function **11.** domain: $\{-1, 1, 2, 3\}$; range: $\{2, 1\}$; function
**13.** domain: {Colorado, Alaska, Delaware, Illinois, Connecticut, Texas}; range: $\{6, 1, 20, 30\}$; function
**15.** domain: $\{32°, 104°, 212°, 50°\}$; range: $\{0°, 40°, 10°, 100°\}$; function **17.** domain: $\{2, -1, 5, 100\}$; range: $\{0\}$; function
**19.** function **21.** Answers may vary. **23.** function **25.** not a function **27.** function
**29.** domain: $[0, \infty)$; range: $(-\infty, \infty)$; not a function **31.** domain: $[-1, 1]$; range: $(-\infty, \infty)$; not a function
**33.** domain: $(-\infty, \infty)$; range: $(-\infty, -3] \cup [3, \infty)$; not a function **35.** domain: $[2, 7]$; range $[1, 6]$; not a function
**37.** domain: $\{-2\}$; range: $(-\infty, \infty)$; not a function **39.** domain: $(-\infty, \infty)$; range: $(-\infty, 3]$; function
**41.** Answers may vary. **43.** yes **45.** no **47.** yes **49.** yes **51.** yes **53.** no **55.** 15 **57.** 38 **59.** 7 **61.** 3
**63. a.** 0 **b.** 1 **c.** $-1$ **65. a.** 246 **b.** 6 **c.** $\frac{9}{2}$ **67. a.** $-5$ **b.** $-5$ **c.** $-5$ **69. a.** 5.1 **b.** 15.5 **c.** 9.533 **71.** $(1, -10)$
**73.** $f(-1) = -2$ **75.** $-4, 0$ **77.** infinite number **79. a.** \$13.4 billion **b.** \$13.672 billion **81.** \$34.374 billion
**83.** $f(x) = x + 7$ **85.** $25\pi$ sq. cm **87.** 2744 cu. in. **89.** 166.38 cm **91.** 163.2 mg
**93. a.** 91.4; The per capita consumption of poultry was 91.4 pounds in 1997. **b.** 106.7 lb

**95.** $5, -5, 6$

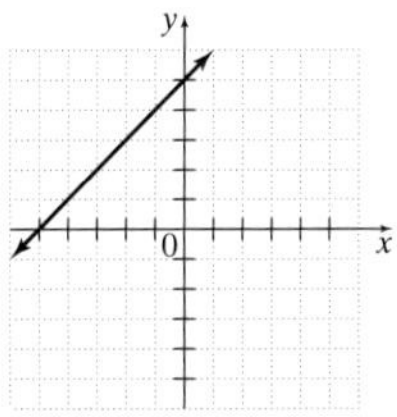

**97.** $2, \frac{8}{7}, \frac{12}{7}$

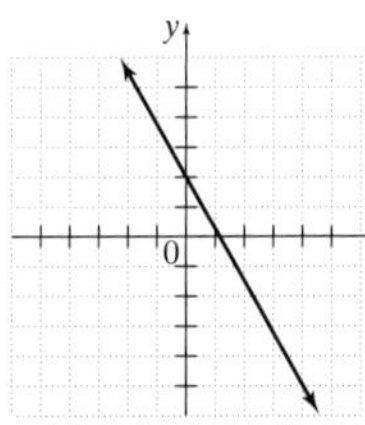

**99.** $0, 0, -6$

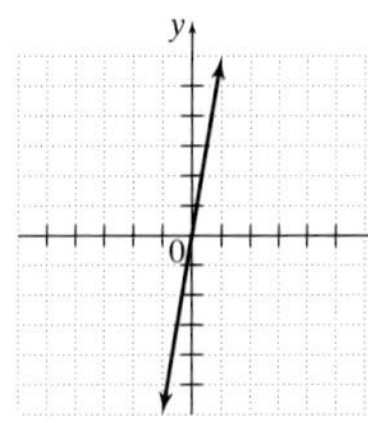

**101.** yes; 170 m

**103. a.** $-3s + 12$ **b.** $-3r + 12$

**105. a.** 132 **b.** $a^2 - 12$

***Graphing Calculator Explorations***

**1.** $y = \frac{x}{3.5}$

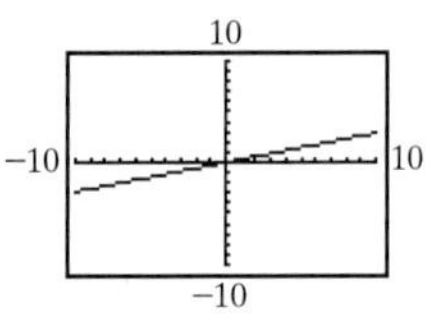

**3.** $y = -\frac{5.78}{2.31}x + \frac{10.98}{2.31}$

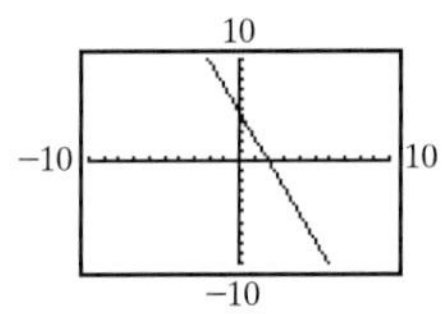

**5.** $y = |x| + 3.78$

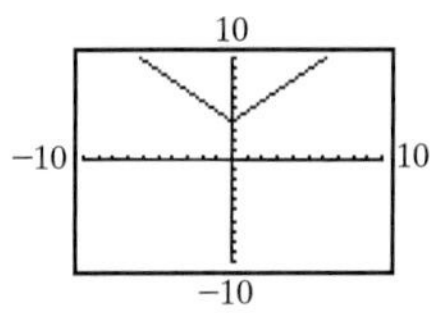

**7.** $y = 5.6x^2 + 7.7x + 1.5$

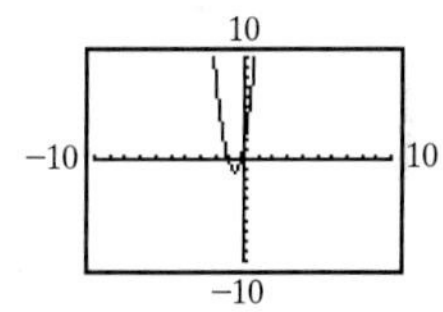

***Exercise Set 3.3*** **1.**

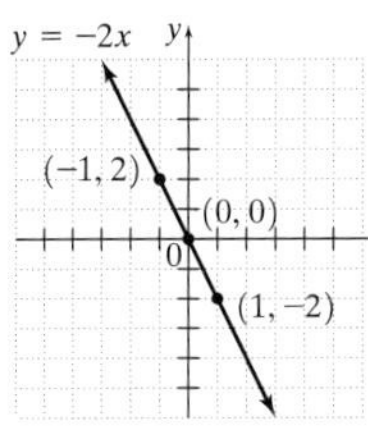


**3.**

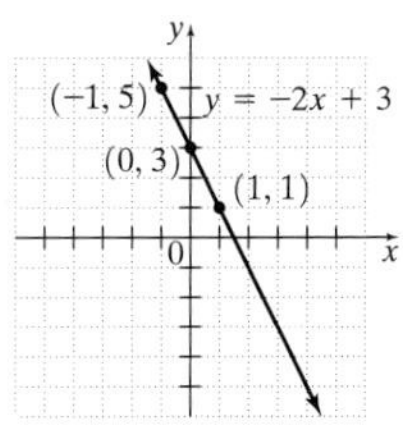


**5.**

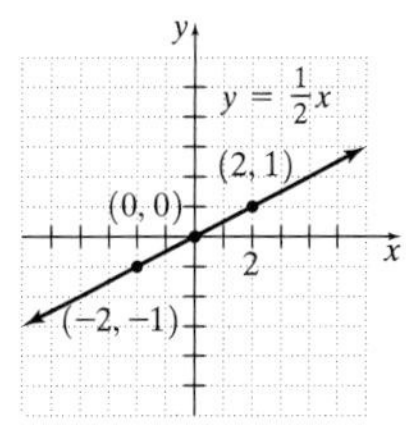


**7.**

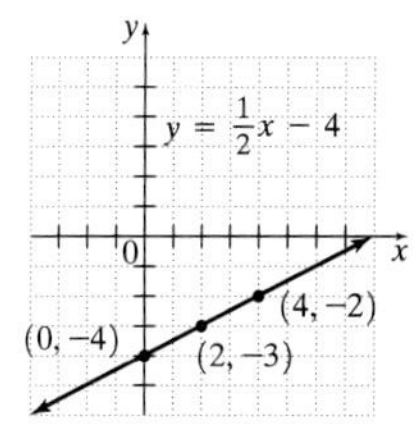


**9.** C **11.** D **13.**

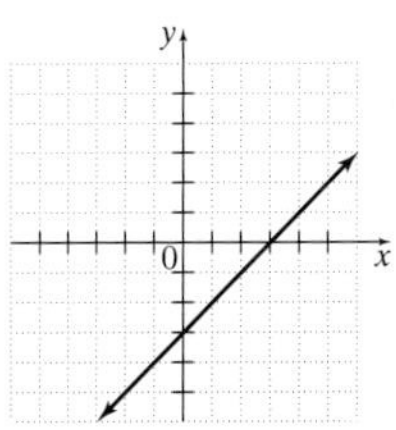

**15.**

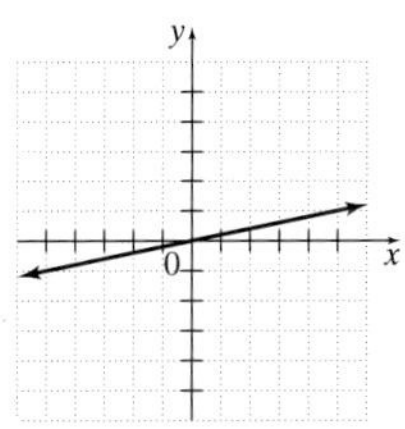

**17.**

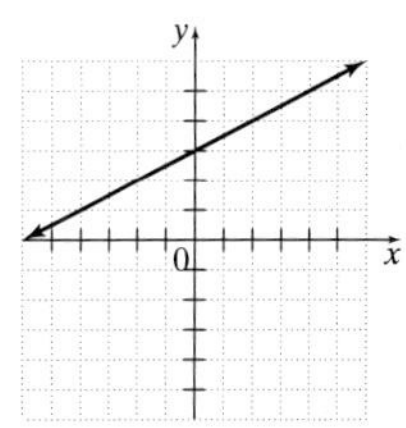

**19.**

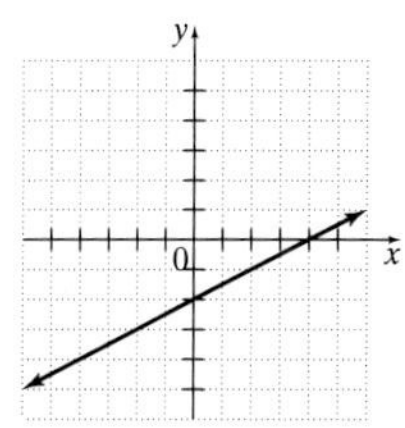

**21.** Answers may vary. **23.**

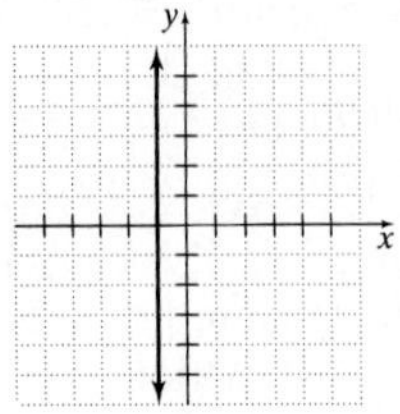

**25.**

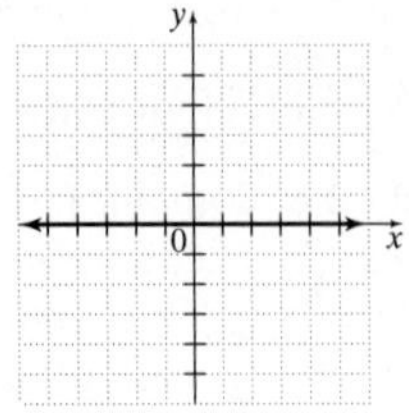

**27.**

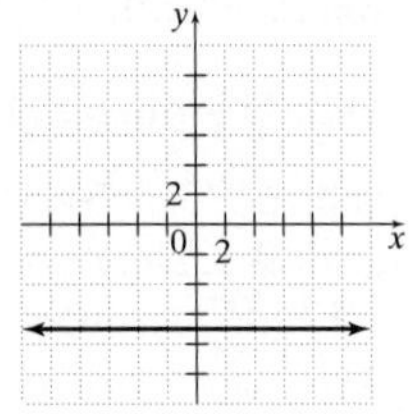


**29.** C **31.** A **33.** The vertical line $x = 0$ has $y$-intercepts.

**35.**

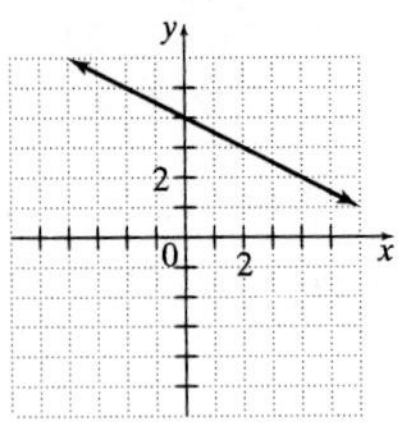


**37.**

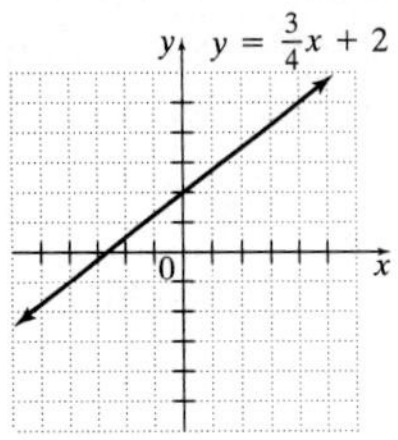


**39.**

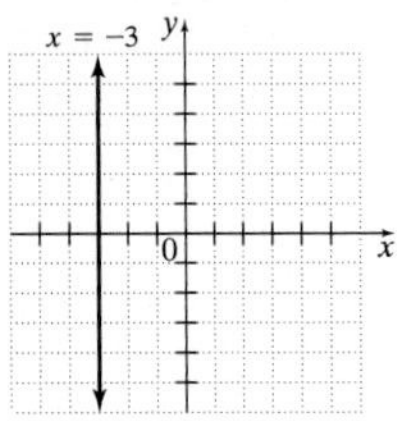


**41.**

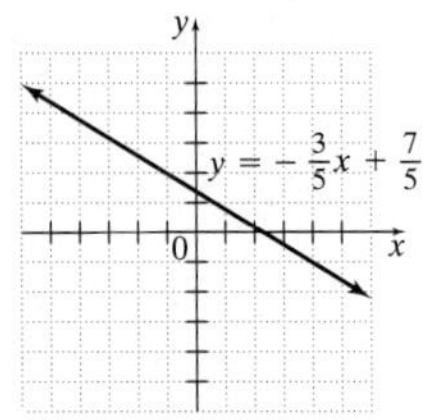


**43.**

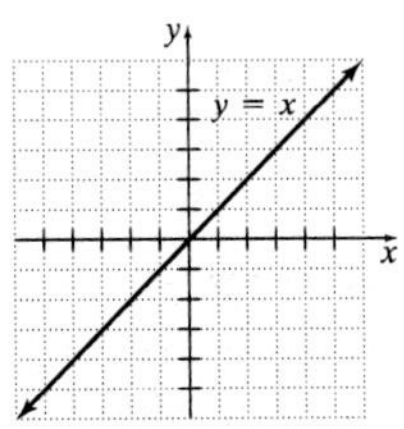


**45.**

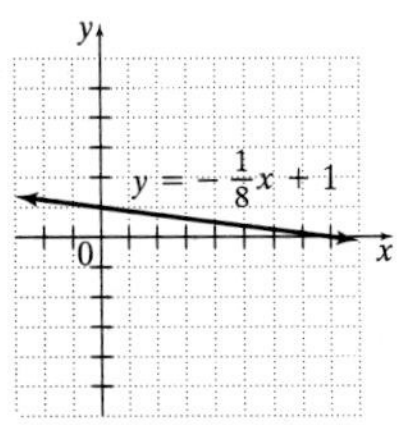


**47.**

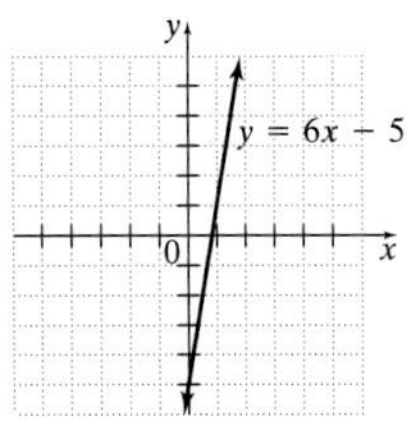


**49.**

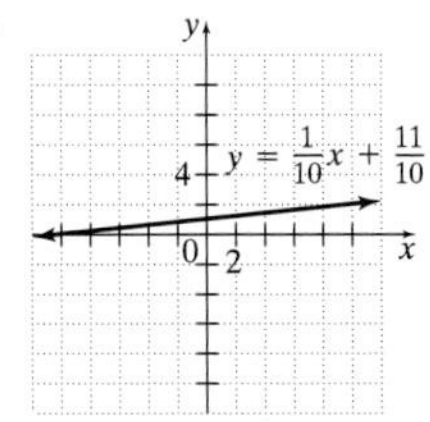


**51.**

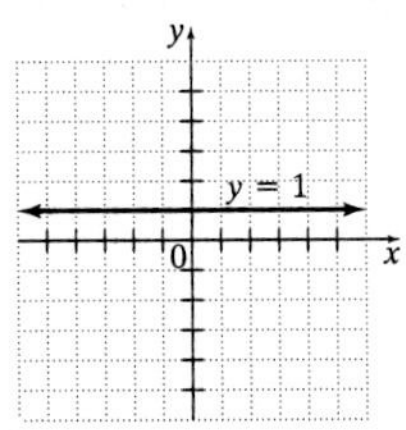


**53.**

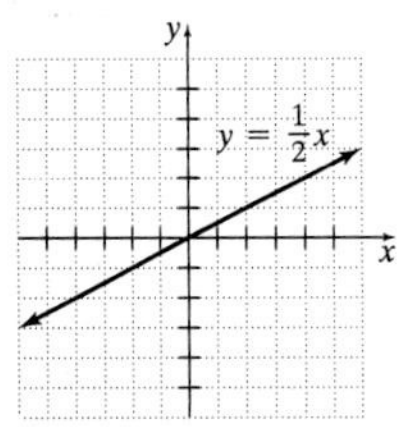


**55.**

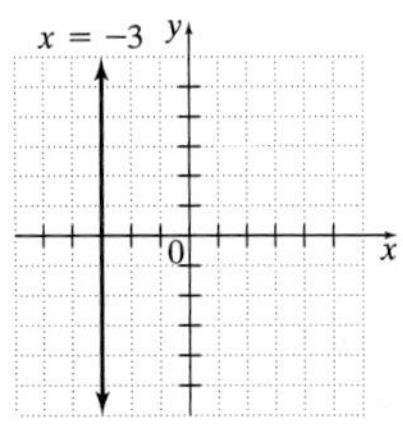


**57.**

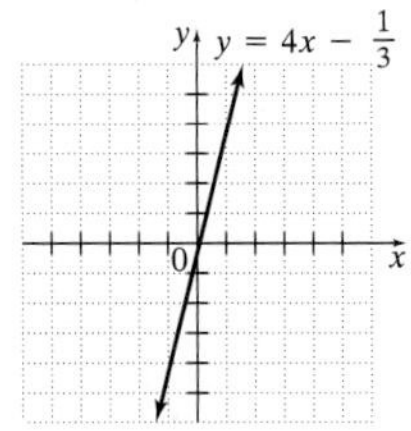


**59.**

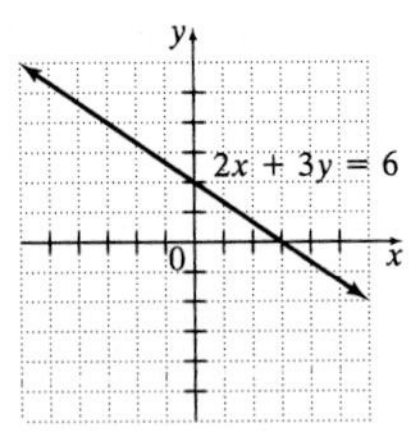


**61. a.** $(0, 500)$; if no tables are produced, 500 chairs can be produced
**b.** $(750, 0)$; if no chairs are produced, 750 tables can be produced
**c.** 466 chairs

**63. a.** \$64
**b.**

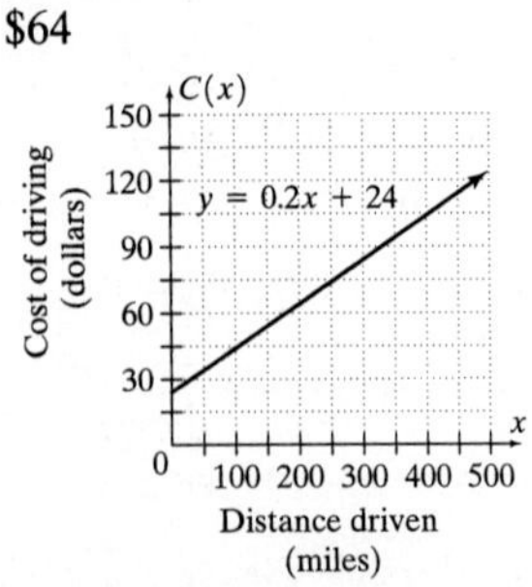


**c.** The line moves upward from left to right.

**65. a.** \$2243.20
**b.** 2007
**c.** Answers may vary.

**67.**

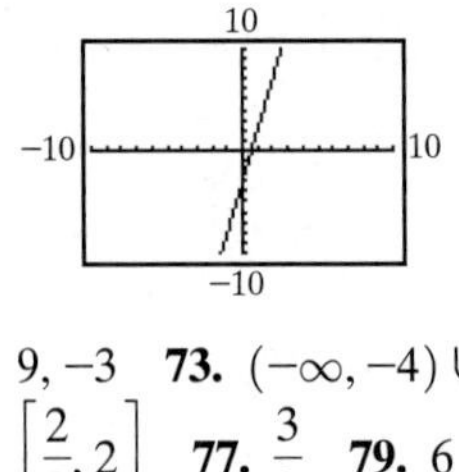


**69.**

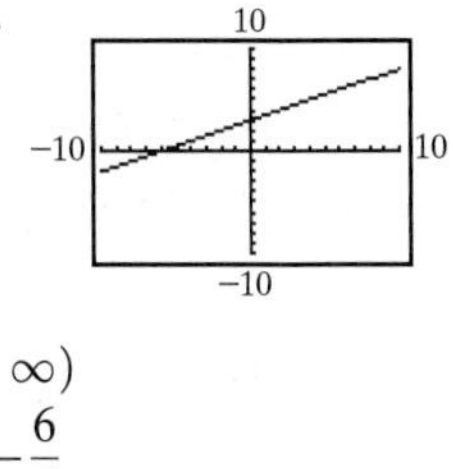


**71.** 9, −3 **73.** $(-\infty, -4) \cup (-1, \infty)$
**75.** $\left[\frac{2}{3}, 2\right]$ **77.** $\frac{3}{2}$ **79.** 6 **81.** $-\frac{6}{5}$

***Graphing Calculator Explorations*** **1.** 18.4 **3.** $-1.5$ **5.** 14.0; 4.2, $-9.4$

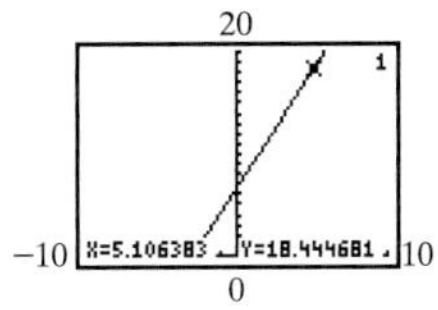


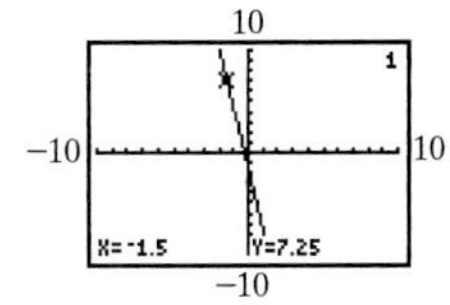


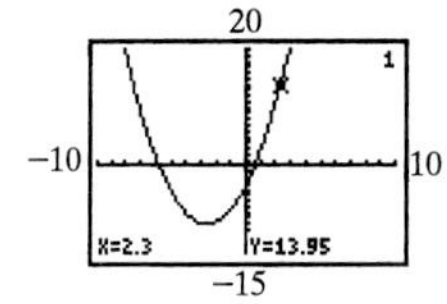


***Mental Math*** **1.** upward **3.** horizontally

***Exercise Set 3.4*** **1.** $\frac{9}{5}$ **3.** $-\frac{7}{2}$ **5.** $-\frac{5}{6}$ **7.** $\frac{1}{3}$ **9.** $-\frac{4}{3}$ **11.** 0 **13.** undefined **15.** 2 **17.** $-1$ **19.** $l_2$ **21.** $l_2$ **23.** $l_2$ **25. a.** $l_1$: $-2$, $l_2$: $-1$, $l_3$: $-\frac{2}{3}$ **b.** lesser **27.** $m = -2, b = 6$ **29.** $m = 5, b = 10$ **31.** $m = -\frac{3}{4}, b = -\frac{3}{2}$ **33.** $m = -\frac{1}{4}, b = 0$ **35.** D **37.** C **39.** 0 **41.** undefined **43.** 0 **45.** Answers may vary. **47.** $m = 1, b = 2$ **49.** $m = \frac{4}{7}, b = -4$ **51.** $m = \frac{1}{2}, b = \frac{7}{2}$ **53.** slope is undefined, no $y$-intercept **55.** $m = \frac{1}{7}, b = 0$ **57.** slope is undefined, no $y$-intercept **59.** $m = 0, b = -\frac{11}{2}$ **61.** parallel **63.** perpendicular **65.** neither **67.** Answers may vary. **69.** $-3$ **71.** 1 **73.** $\frac{3}{25}$ **75.** $\frac{3}{20}$

**77. a.** \$28,559.40 **b.** $m = 1054.7$; The annual average income increases \$1054.70 every year **c.** $b = 23{,}285.9$; At year $x = 0$ or 1991, the annual average income was \$23,285.90. **79. a.** $m = 7.6, b = 113$ **b.** The number of people employed as paralegals increases 7.6 thousand for every 1 year. **c.** There were 113 thousand paralegals employed in 1996. **81. a.** The yearly cost of tuition increases \$72.90 every 1 year. **b.** The yearly cost of tuition in 1990 was \$785.20. **83.** 1 **85.** $-1$ **87.** $\frac{3}{4}$ **89. a.** $(6, 20)$ **b.** $(10, 13)$ **c.** $-\frac{7}{4}$ or $-1.75$ yd per sec **d.** $\frac{3}{2}$ or 1.5 yd per sec

**91.**

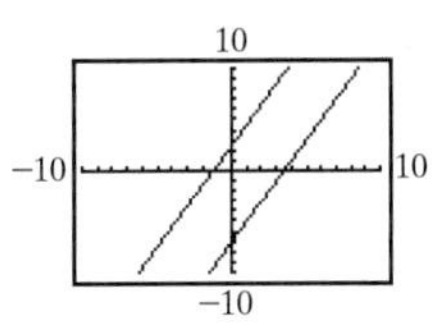


**93. a.**

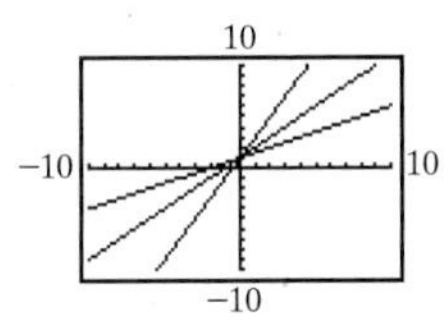


**b.**

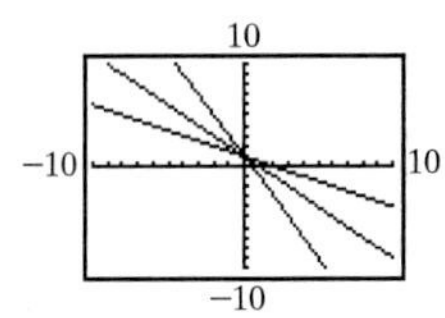


**c.** true **95.** $\frac{2}{11}$ **97.** $\frac{3}{11}$ **99.** $\frac{4}{11}$ **101.** $y = -3x - 30$ **103.** $y = -8x - 23$

***Mental Math*** **1.** $m = -4, b = 12$ **3.** $m = 5, b = 0$ **5.** $m = \frac{1}{2}, b = 6$ **7.** parallel **9.** neither

***Exercise Set 3.5*** **1.** $y = -x + 1$ **3.** $y = 2x + \frac{3}{4}$ **5.** $y = \frac{2}{7}x$

**7.** $y = 5x$

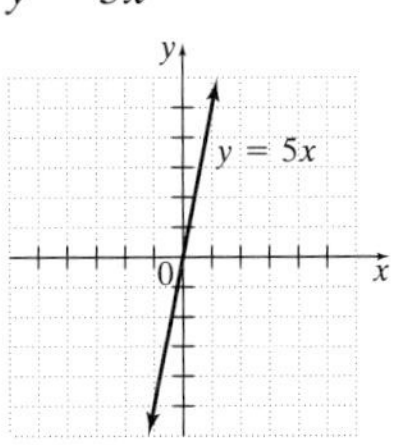


**9.** $x + y = 7$

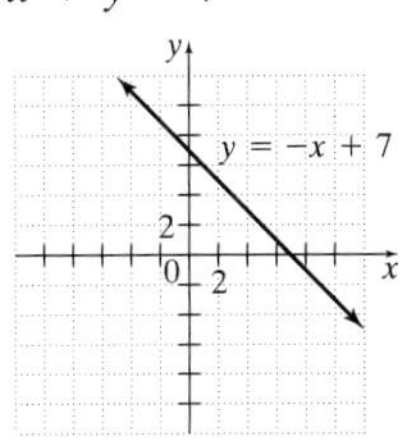


**11.** $-3x + 2y = 3$

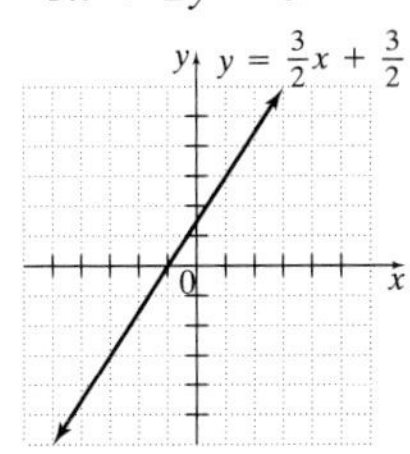


**13.** $y = 3x - 1$ **15.** $y = -2x - 1$ **17.** $y = \frac{1}{2}x + 5$ **19.** $y = -\frac{9}{10}x - \frac{27}{10}$ **21.** $2x + y = 3$ **23.** $2x - 3y = -7$ **25.** $f(x) = 3x - 6$ **27.** $f(x) = -2x + 1$ **29.** $f(x) = -\frac{1}{2}x - 5$ **31.** $f(x) = \frac{1}{3}x - 7$ **33.** Answers may vary. **35.** $-2$ **37.** 2 **39.** $-2$ **41.** $y = -4$ **43.** $x = 4$ **45.** $y = 5$ **47.** $f(x) = 4x - 4$ **49.** $f(x) = -3x + 1$ **51.** $f(x) = -\frac{3}{2}x - 6$ **53.** $2x - y = -7$ **55.** $f(x) = -x + 7$ **57.** $x + 2y = 22$ **59.** $2x + 7y = -42$ **61.** $4x + 3y = -20$ **63.** $x = -2$

**65.** $x + 2y = 2$ **67.** $y = 12$ **69.** $8x - y = 47$ **71.** $x = 5$ **73.** $f(x) = -\frac{3}{8}x - \frac{29}{4}$

**75. a.** $P(x) = 12{,}000x + 18{,}000$ **b.** $102,000 **c.** end of the ninth yr **77. a.** $y = -1000x + 13{,}000$ **b.** 9500 Fun Noodles **79. a.** $4625x + 109{,}900$ **b.** $174,650.67 **81. a.** $y = 16.6x + 225$ **b.** 357.8 thousand people

**83.**

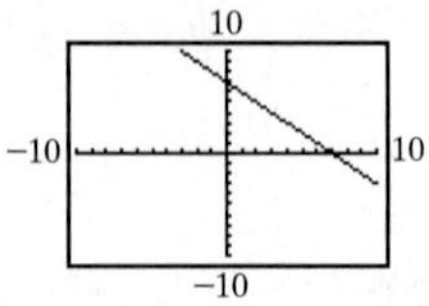


**85.**

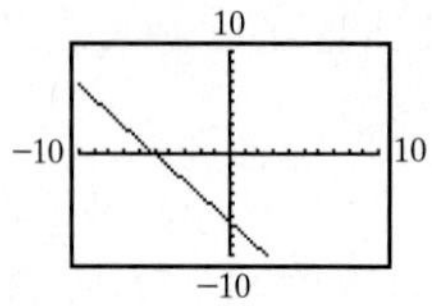


**87.**

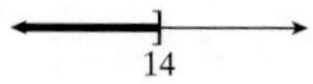


, $(-\infty, 14]$ **89.**

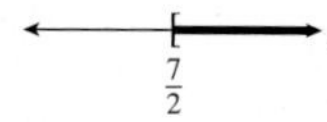


, $\left[\frac{7}{2}, \infty\right)$

**91.**

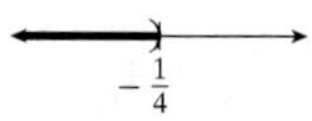


, $\left(-\infty, -\frac{1}{4}\right)$ **93.** $-4x + y = 4$

**95.** $2x + y = -23$ **97.** $3x - 2y = -13$

***Exercise Set 3.6*** **1.**

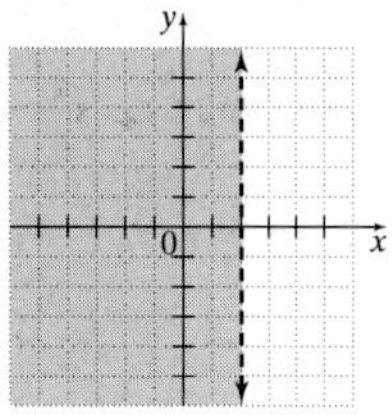

**3.**

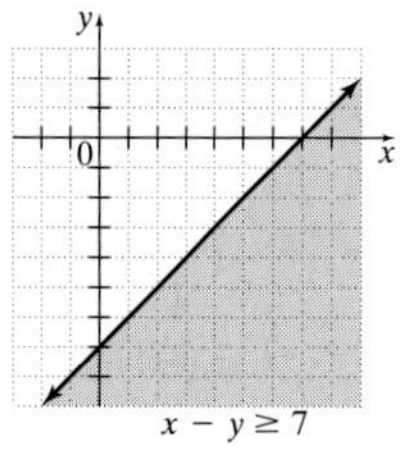


**5.**

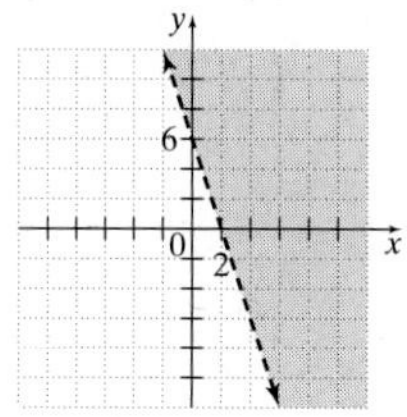


**7.**

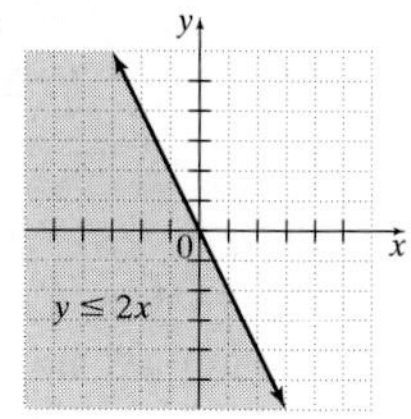


**9.**

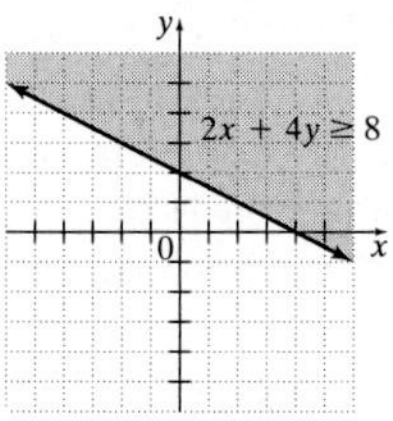


**11.**

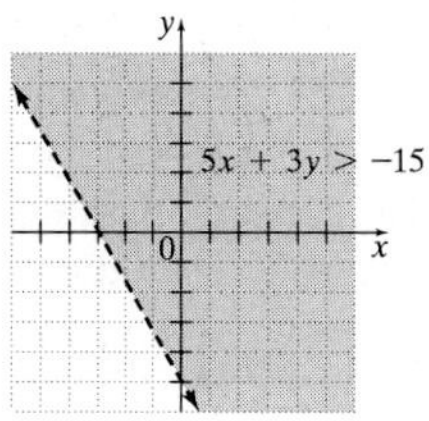


**13.** Answers may vary. **15.**

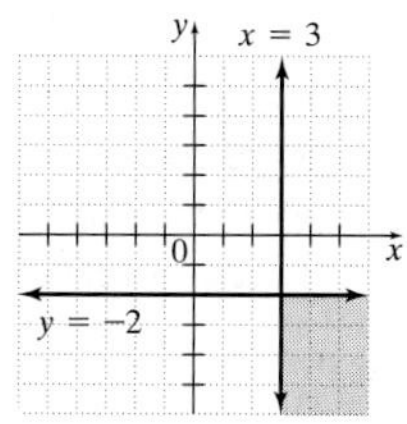


**17.**

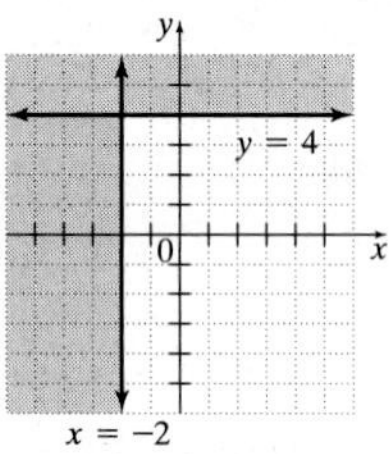


**19.**

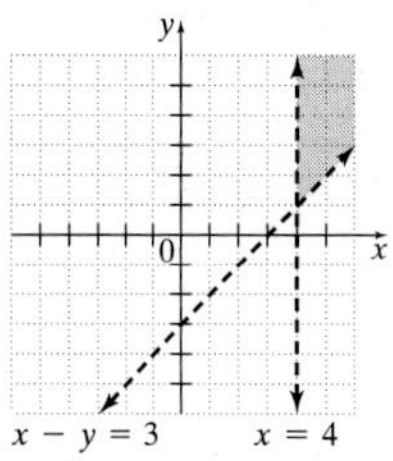


**21.**

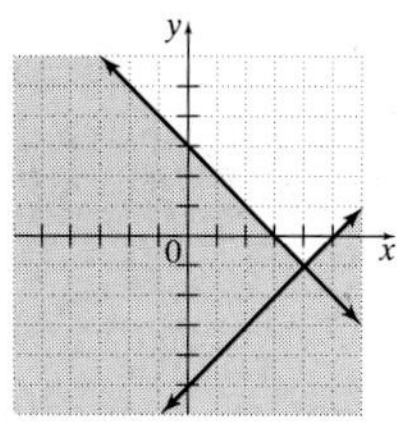

**23.**

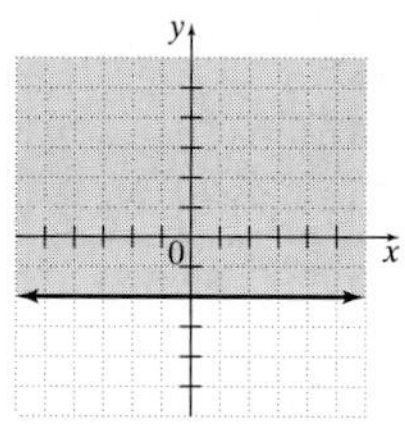

**25.**

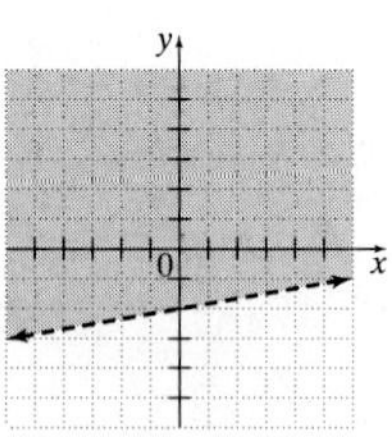

**27.**

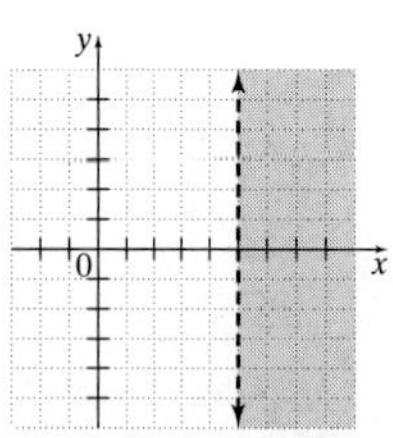

**29.**

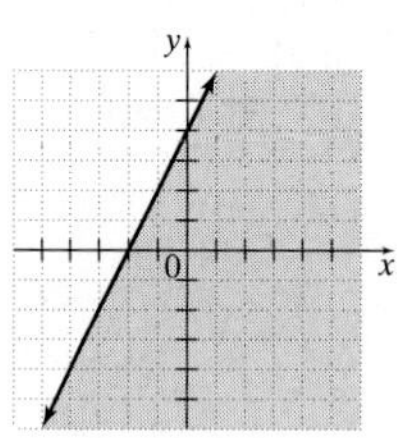

**31.**

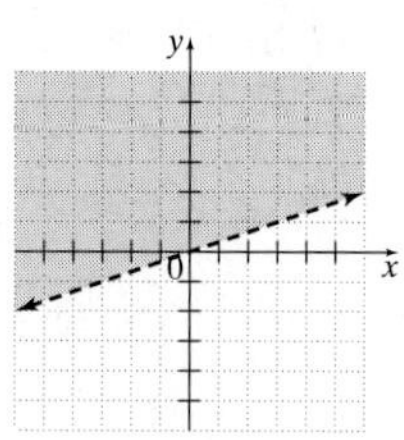

**33.**

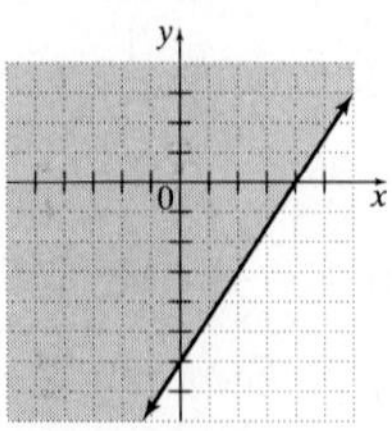

**35.**

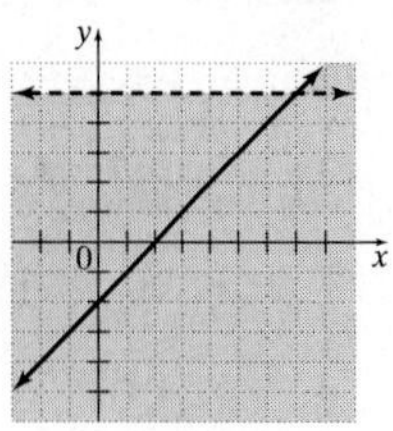

**37.**

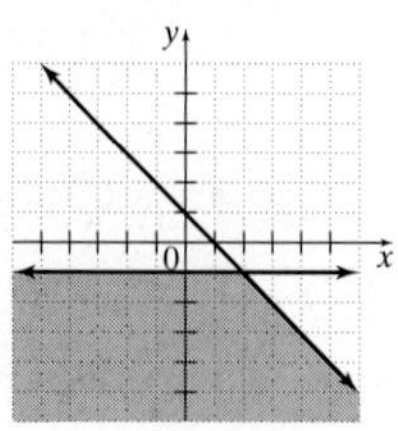

**39.**

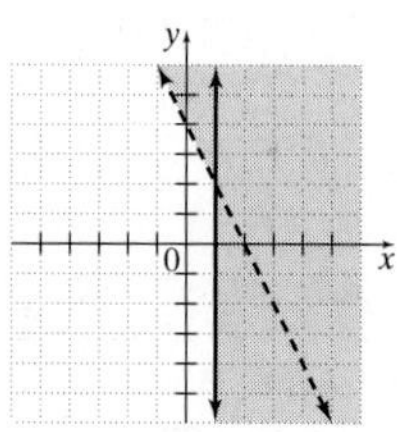

**41.** 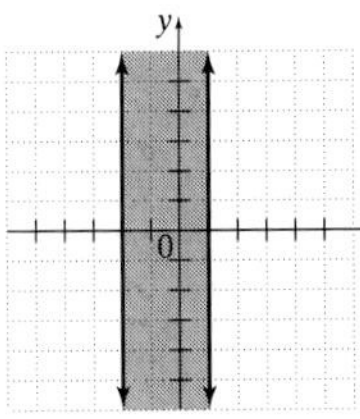

**43.** 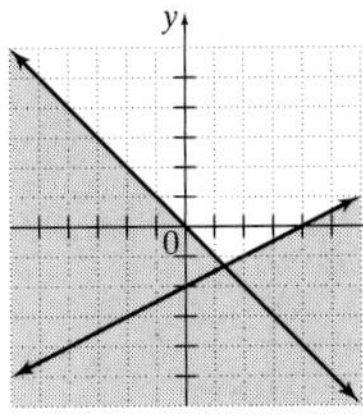

**45.** 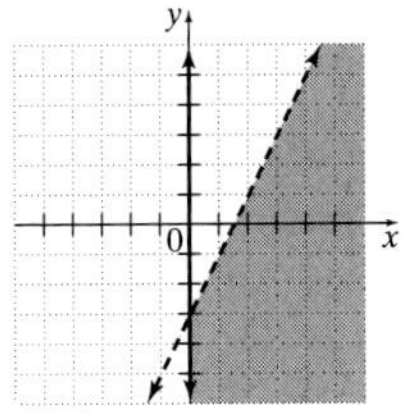

**47.** D **49.** A
**51.** $x \geq 2$ **53.** $y \leq -3$
**55.** $y > 4$ **57.** $x < 1$

**59.** $x \leq 20$ and $y \geq 10$. 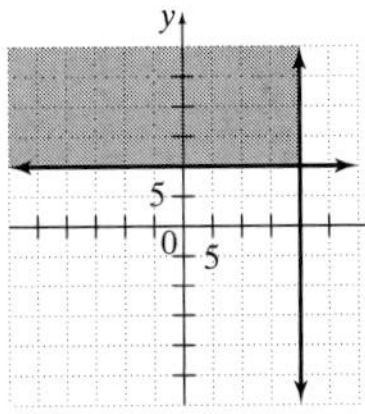

**61.** 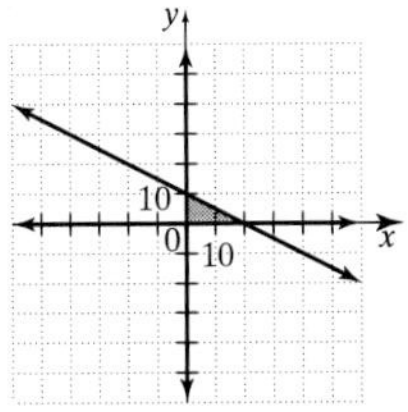

**63.** 9 **65.** 25 **67.** −16 **69.** $\frac{4}{49}$

**71.** domain: $(-\infty, -2] \cup [2, \infty)$; range: $(-\infty, \infty)$; no

***Chapter 3 Review*** **1.** 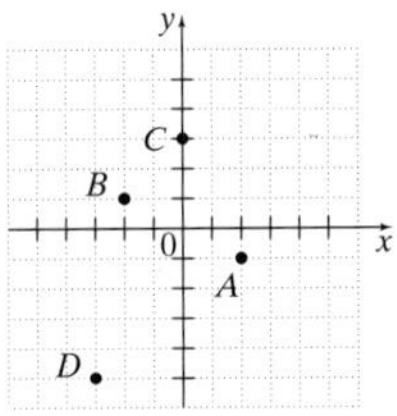


**3.** no, yes **5.** yes, yes **7.** linear

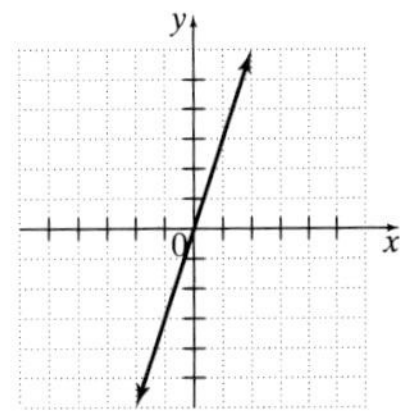

**9.** linear

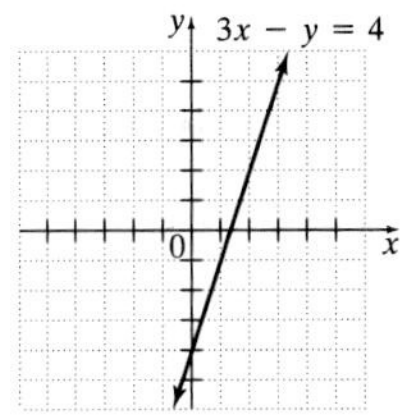


**11.** nonlinear

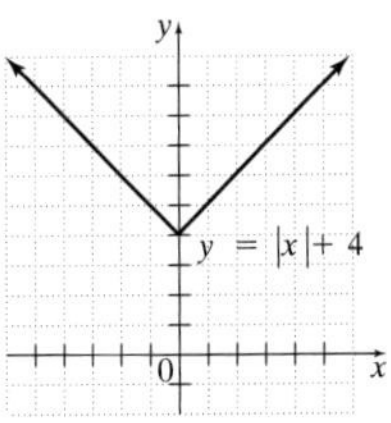


**13.** linear

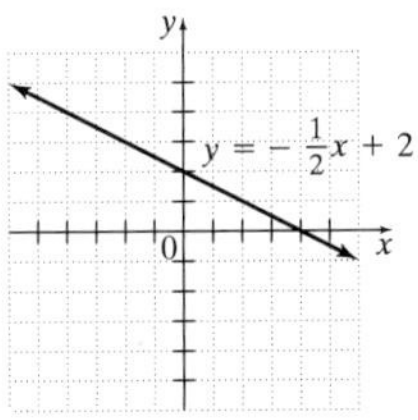


**15.** linear

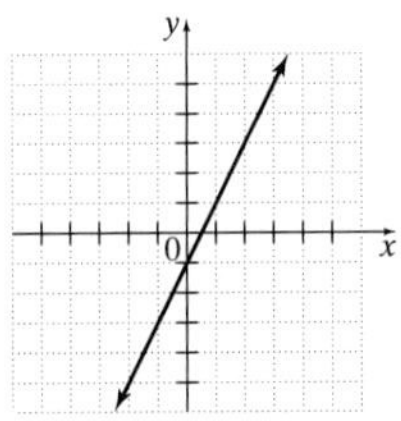

**17.** linear

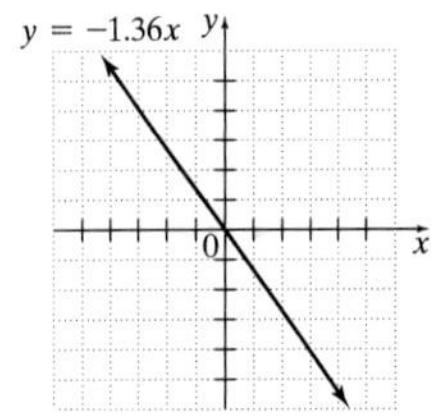


**19.** domain: $\left\{-\frac{1}{2}, 6, 0, 25\right\}$; range: $\left\{\frac{3}{4} \text{ or } 0.75, -12, 25\right\}$; function **21.** domain: $\{2, 4, 6, 8\}$; range: $\{2, 4, 5, 6\}$; not a function
**23.** domain: $(-\infty, \infty)$; range: $(-\infty, -1] \cup [1, \infty)$; not a function **25.** domain: $(-\infty, \infty)$; range: $\{4\}$; function **27.** −3
**29.** 18 **31.** −3 **33.** 381 lb **35.** 0 **37.** −2, 4

**39.** 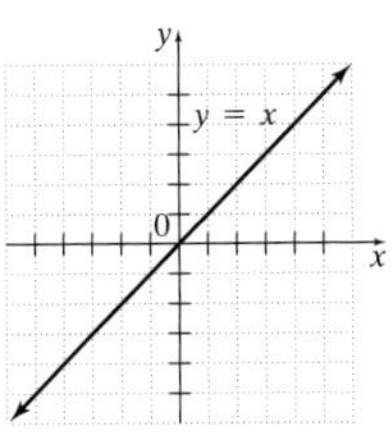


**41.** 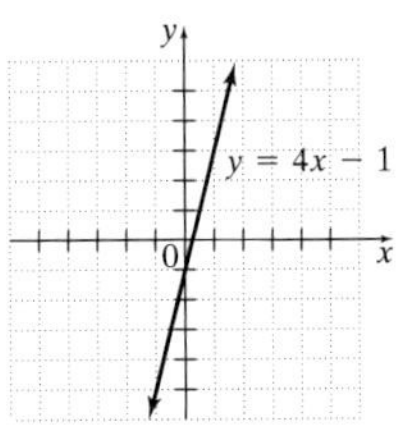


**43.** A **45.** D **47.** 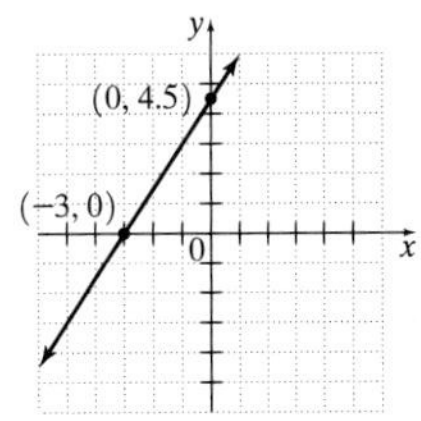


**49.** 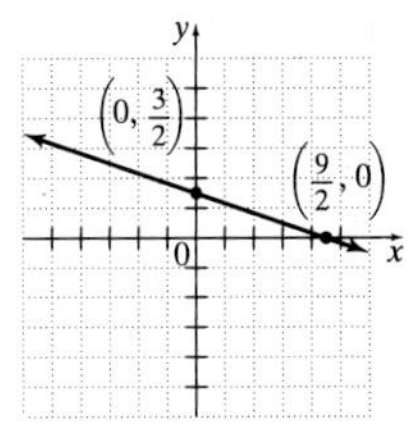


**51.** 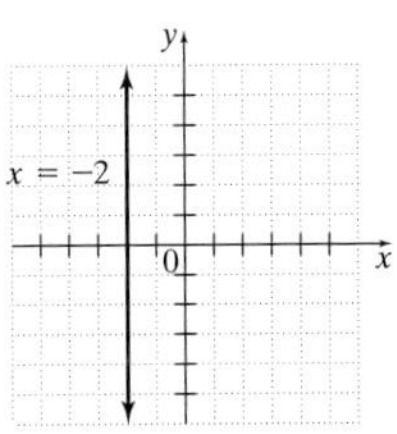


**53.** 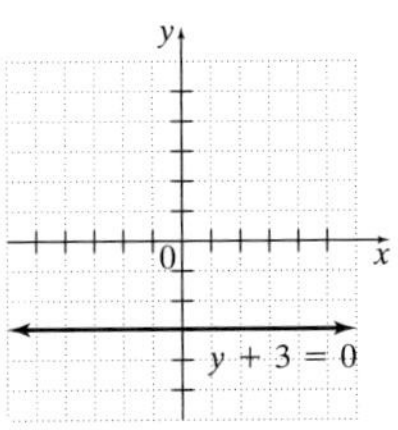


**55.** −3 **57.** $\frac{5}{2}$ **59.** $m = \frac{2}{5}, b = -\frac{4}{3}$
**61.** 0 **63.** $l_2$ **65.** $l_2$

**67. a.** $m = 0.3$; The cost increases by \$0.30 for each additional mile driven. **b.** $b = 42$; The cost for 0 miles driven is \$42.
**69.** parallel **71.**

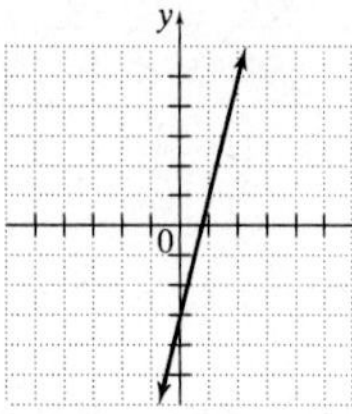

**73.**

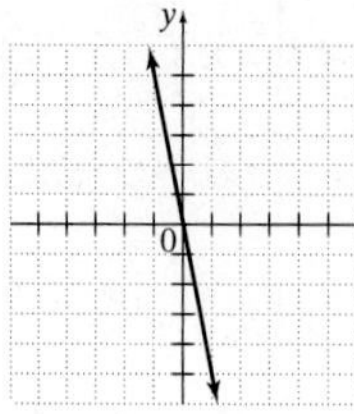

**75.** $x = -2$ **77.** $y = 5$ **79.** $2x - y = 12$
**81.** $11x + y = -52$ **83.** $y = -5$ **85.** $f(x) = -x - 2$
**87.** $f(x) = -\frac{3}{2}x - 8$ **89.** $f(x) = -\frac{3}{2}x - 1$
**91. a.** $y = \frac{17}{22}x + 43$ **b.** 52 million

**93.**

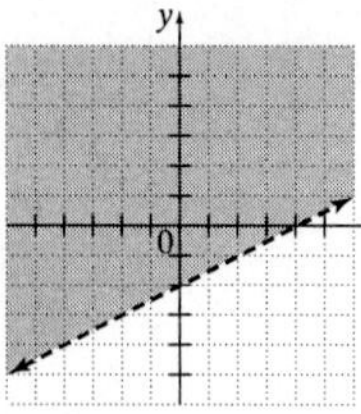

**95.**

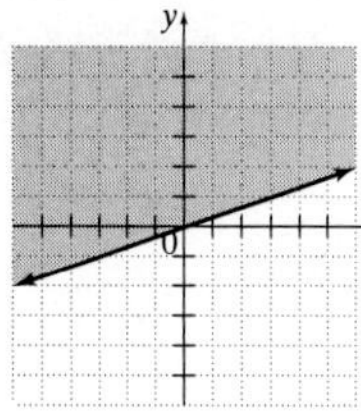

**97.**

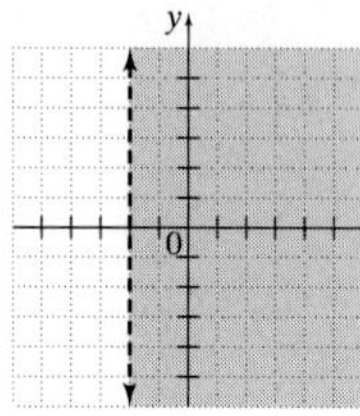

**99.**

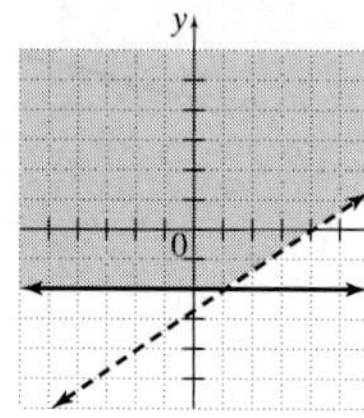

***Chapter 3 Test*** **1.**

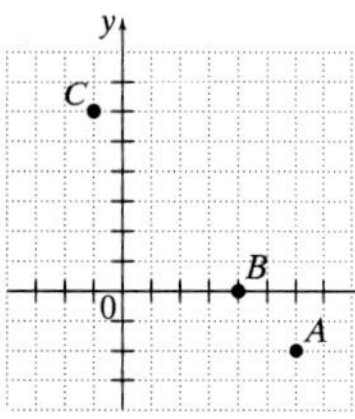


**2.** $(-6, -3)$ **3.**

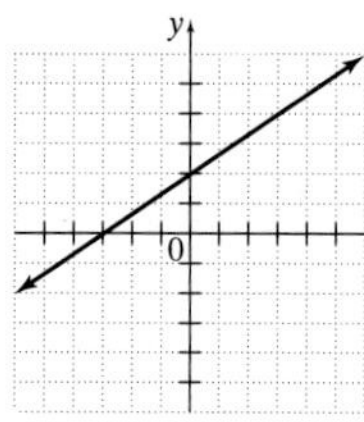

**4.**

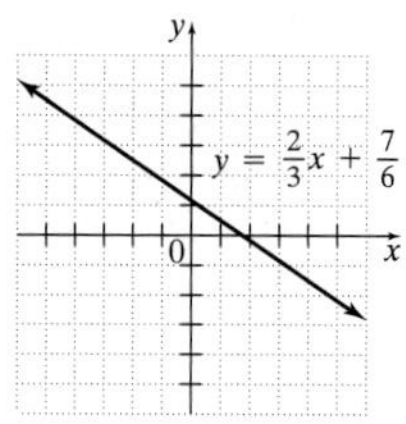


**5.**

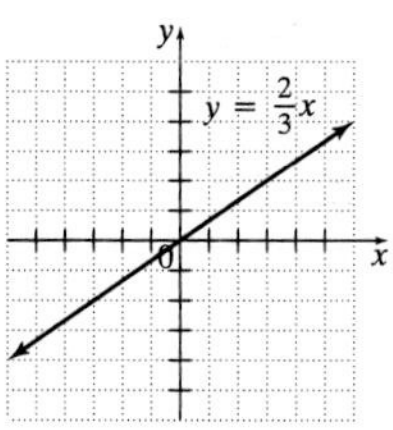


**6.**

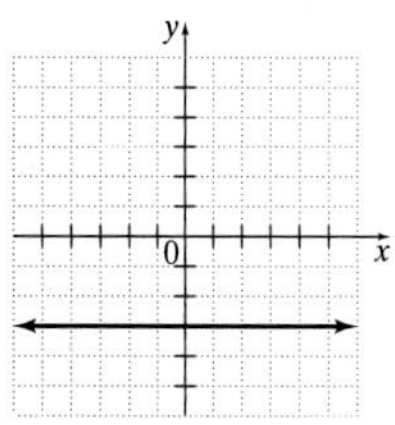

**7.** $-\frac{3}{2}$ **8.** $m = -\frac{1}{4}, b = \frac{2}{3}$ **9.**

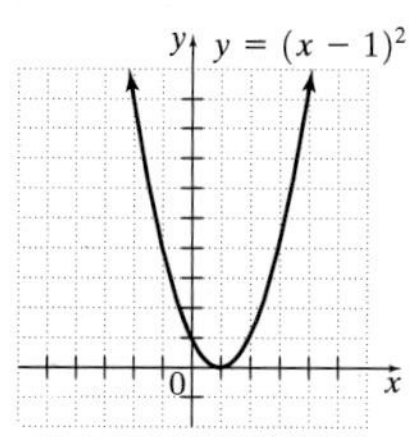


**10.**

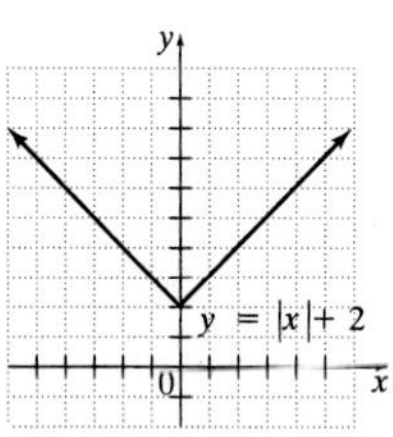


**11.** $y = -8$ **12.** $x = -4$ **13.** $y = -2$ **14.** $3x + y = 11$ **15.** $5x - y = 2$
**16.** $f(x) = -\frac{1}{2}x$ **17.** $f(x) = -\frac{1}{3}x + \frac{5}{3}$ **18.** $f(x) = -\frac{1}{2}x - \frac{1}{2}$ **19.** neither
**20.**

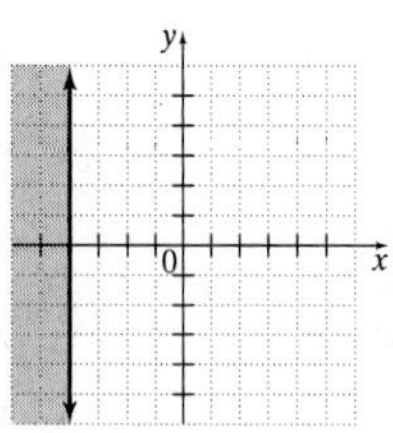

**21.**

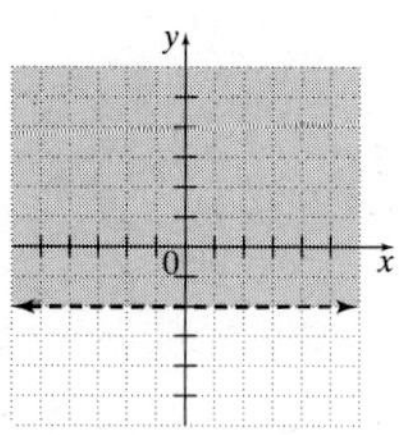

**22.**

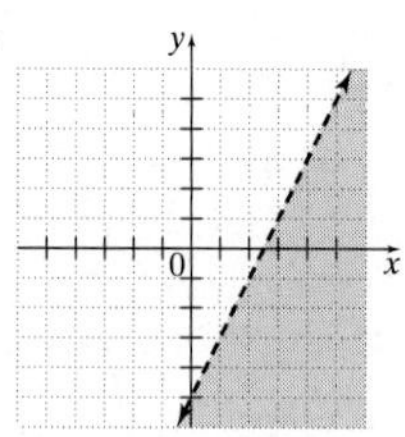

**23.**

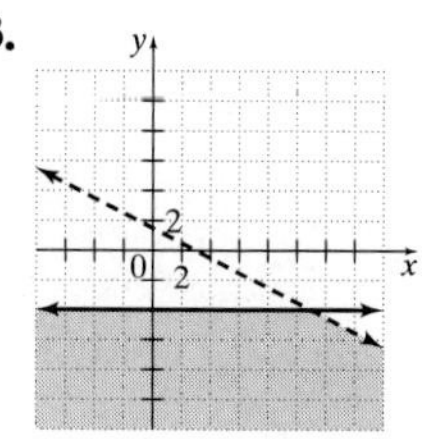

**24.** domain: $(-\infty, \infty)$; range: $\{5\}$; function **25.** domain: $\{-2\}$; range: $(-\infty, \infty)$; not a function
**26.** domain: $(-\infty, \infty)$; range: $[0, \infty)$; function **27.** domain: $(-\infty, \infty)$; range: $(-\infty, \infty)$; function
**28. a.** \$22,892 **b.** \$28, 016 **c.** 2008 **d.** The average yearly earnings for high school graduates increases \$732 per year. **e.** The average yearly earnings for a high school graduate in 1996 was \$21,428.

***Chapter 3 Cumulative Review*** **1.** 41; Sec. 1.2, Ex. 2 **2. a.** true **b.** false **c.** false **d.** false; Sec. 1.2, Ex. 5
**3. a.** $-6$ **b.** $-7$ **c.** $-16$ **d.** 20.5 **e.** $\frac{1}{6}$ **f.** 0.94 **g.** $-3$; Sec. 1.3, Ex. 2 **4. a.** 9 **b.** $\frac{1}{16}$ **c.** $-25$ **d.** 25 **e.** $-125$
**f.** $-125$; Sec. 1.3, Ex. 6 **5. a.** $>$ **b.** $=$ **c.** $<$ **d.** $<$; Sec. 1.4, Ex. 2 **6. a.** $\frac{1}{11}$ **b.** $-\frac{1}{9}$ **c.** $\frac{4}{7}$; Sec. 1.4, Ex. 5

**7.** 0.4; Sec. 2.1, Ex. 2 **8.** { } or ∅; Sec. 2.1, Ex. 8 **9.** 4; Sec. 2.2, Ex. 4 **10.** 86 and 88; Sec. 2.2, Ex. 7 **11.** $\frac{V}{lw} = h$; Sec. 2.3, Ex. 1 **12.** $\{x|x < 7\}$; 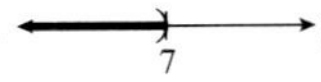; Sec. 2.4, Ex. 2 **13.** $\left(-\infty, -\frac{7}{3}\right]$; Sec. 2.4, Ex. 6 **14.** ∅; Sec. 2.5, Ex. 3 **15.** $\left(-\infty, \frac{13}{5}\right] \cup [4, \infty)$; Sec. 2.5, Ex. 7 **16.** $-2, \frac{4}{5}$; Sec. 2.6, Ex. 2 **17.** $\frac{3}{4}, 5$; Sec. 2.6, Ex. 8 **18.** $\left[-2, \frac{8}{5}\right]$; Sec. 2.7, Ex. 3 **19.** $(-\infty, -4) \cup (10, \infty)$; Sec. 2.7, Ex. 5 **20.** solutions: $(2, -6), (0, -12)$; not a solution $(1, 9)$; Sec. 3.1, Ex. 2 **21.** yes; Sec. 3.2, Ex. 3 **22. a.** $\left(0, \frac{3}{7}\right)$ **b.** $(0, -3.2)$; Sec. 3.3, Ex. 3 **23.** $\frac{2}{3}$; Sec. 3.4, Ex.3 **24.** $y = \frac{1}{4}x - 3$; Sec. 3.5 Ex. 1 **25.**

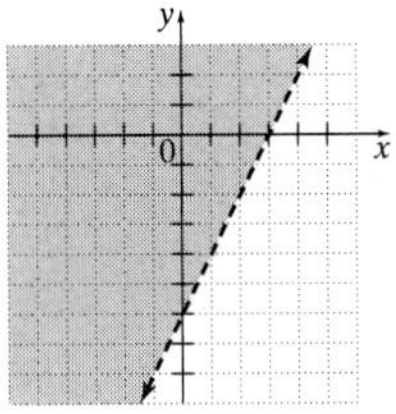

## ■ CHAPTER 4 SYSTEMS OF EQUATIONS

***Graphing Calculator Explorations*** **1.** $(2.11, 0.17)$ **3.** $(0.57, -1.97)$

***Mental Math*** **1.** B **3.** A

***Exercise Set 4.1*** **1.** yes **3.** no **5.** yes

**7.** $(2, -1)$

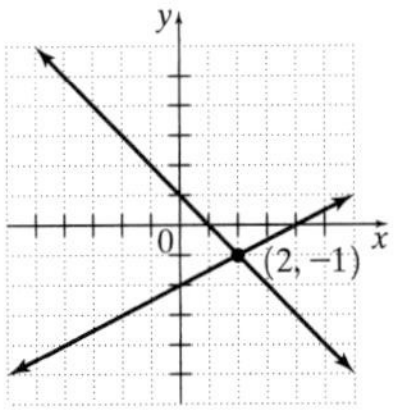


**9.** $(1, 2)$

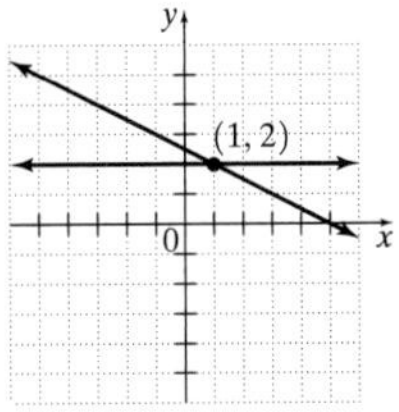


**11.** ∅

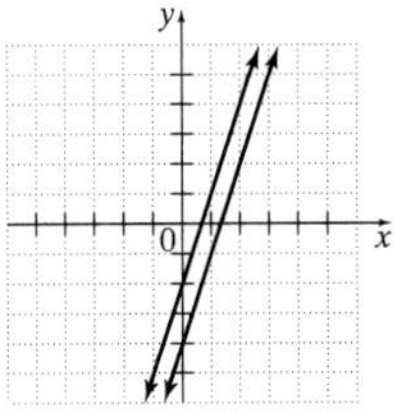

**13.** No **15.** $(2, 8)$ **17.** $(0, -9)$ **19.** $(1, -1)$ **21.** $(-5, 3)$ **23.** $\left(\frac{5}{2}, \frac{5}{4}\right)$ **25.** $(1, -2)$ **27.** $(9, 9)$ **29.** $(7, 2)$ **31.** ∅ **33.** $\{(x, y)|3x + y = 1\}$

**35.** Answers may vary; One possibility: $\begin{cases} -2x + y = 1 \\ x - 2y = -8 \end{cases}$ **37.** $\left(\frac{3}{2}, 1\right)$ **39.** $(2, -1)$ **41.** $(-5, 3)$ **43.** $\{(x, y)|3x + 9y = 12\}$ **45.** ∅ **47.** $\left(\frac{1}{2}, \frac{1}{5}\right)$ **49.** $(8, 2)$ **51.** $\{(x, y)|x = 3y + 2\}$ **53.** $\left(-\frac{1}{4}, \frac{1}{2}\right)$ **55.** $(3, 2)$ **57.** $(7, -3)$ **59.** ∅ **61.** $(3, 4)$ **63.** $(-2, 1)$ **65.** $(1.2, -3.6)$ **67.** 5000 ties; $21 **69.** Supply is greater than demand. **71.** (1875; 4687.5) **73.** makes money **75.** for $x$-values greater than 1875 **77. a.** Consumption of red meat is decreasing while consumption of poultry is increasing. **b.** (14, 113) **c.** In the year 2009, red meat and poultry consumption will each be about 113 pounds per person. **79.** 1993 **81.** false **83.** true **85.** $3x - 7z = 3$ **87.** $-4y - 2z = 43$ **89.** $\left(-3, \frac{1}{5}\right)$ **91.** $(1, 1)$ **93.** $\left(-\frac{1}{4}, \frac{1}{3}\right)$ **95.** ∅

***Exercise Set 4.2*** **1.** $(-2, 5, 1)$ **3.** $(-2, 3, -1)$ **5.** $\{(x, y, z)|x - 2y + z = -5\}$ **7.** ∅ **9.** Answers may vary; One possibility is: $\begin{cases} 3x = -3 \\ 2x + 4y = 6 \\ x - 3y + z = -11 \end{cases}$ **11.** $(0, 0, 0)$ **13.** $(-3, -35, -7)$ **15.** $(6, 22, -20)$ **17.** ∅ **19.** $(3, 2, 2)$ **21.** $\{(x, y, z)|x + 2y - 3z = 4\}$ **23.** $(-3, -4, -5)$ **25.** $(12, 6, 4)$ **27.** $(1, 1, -1)$ **29.** 15 and 30 **31.** 5 **33.** $-\frac{5}{3}$ **35.** $(1, 1, 0, 2)$ **37.** $(1, -1, 2, 3)$

***Exercise Set 4.3*** **1.** 10 and 8 **3.** plane, 520 mph; wind, 40 mph **5.** 20 quarts of 4%; 40 quarts of 1% **7.** 9 large frames; 13 small frames **9.** −10 and −8 **11.** tablets, $0.80; pens, $0.20 **13.** plane, 630 mph; wind, 90 mph **15.** 5 in., 7 in., 7 in., and 10 in. **17.** 18, 13, and 9 **19.** $2000 in sales **21.** $1.90 for a template; $0.75 for a pencil; $2.25 for a pad of paper **23.** 750 units **25.** 750 units **27.** 500 units **29. a.** $R(x) = 31x$ **b.** $C(x) = 15x + 500$ **c.** 31.25, or 32 baskets **31.** $x = 40$; $y = 70$ **33.** 40 oz of the 20% solution and 20 oz of the 50% solution **35.** 120 liters of 25%, 60 liters of 40%, 20 liters of 50% **37.** 4 free throws; 8 two-point field goals; 2 three-point field goals **39.** Answers may vary. **41.** $a = 1, b = -2, c = 3$ **43.** $x = 95$; $y = 123$; $z = 70$ **45.** $a = 0.28, b = -3.71, c = 12.83$; 2.12 inches in September **47.** $3y + 8z = 18$ **49.** $-5x - 5z = -16$ **51.** $\frac{3}{8}$ **53.** $\frac{5}{8}$

***Exercise Set 4.4*** **1.** (2, −1) **3.** (−4, 2) **5.** ∅ **7.** $\{(x, y)|x - y = 3\}$ **9.** (−2, 5, −2) **11.** (1, −2, 3) **13.** (4, −3) **15.** (2, 1, −1) **17.** (9, 9) **19.** ∅ **21.** ∅ **23.** (1, −4, 3) **25.** Answers may vary. **27.** function **29.** not a function **31.** −13 **33.** −36 **35.** 0

***Exercise Set 4.5*** **1.** 26 **3.** −19 **5.** 0 **7.** (1, 2) **9.** $\{(x, y)|3x + y = 1\}$ **11.** (9, 9) **13.** 8 **15.** 0 **17.** 54 **19.** (−2, 0, 5) **21.** (6, −2, 4) **23.** 16 **25.** 15 **27.** $\frac{13}{6}$ **29.** 0 **31.** 56 **33.** 0 **35.** 5 **37.** (−3, −2) **39.** ∅ **41.** (−2, 3, −1) **43.** (3, 4) **45.** (−2, 1) **47.** $\{(x, y, z)|x - 2y + z = -3\}$ **49.** (0, 2, −1)

**51.** $\begin{matrix} + & - & + & - \\ - & + & - & + \\ + & - & + & - \\ - & + & - & + \end{matrix}$ **53.** $6x - 18$ **55.** $9x - 15$ **57.**

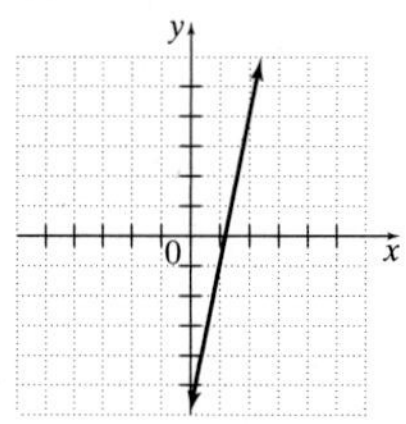

**59.**

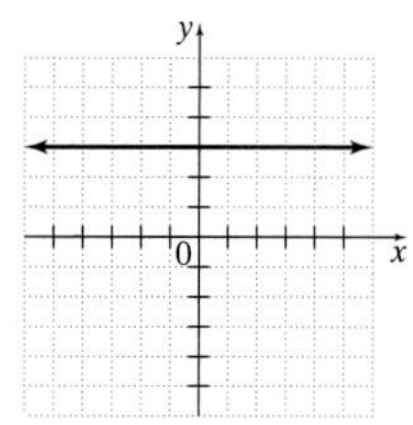

**61.** −125 **63.** 24

***Chapter 4 Review*** **1.** (−3, 1)

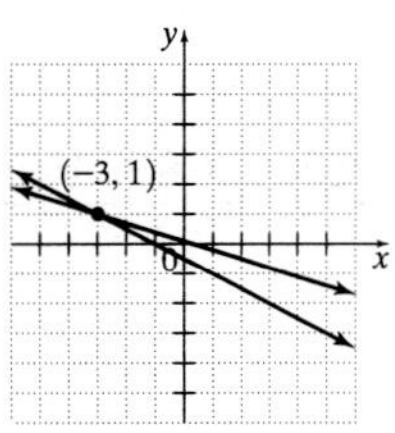


**3.** ∅

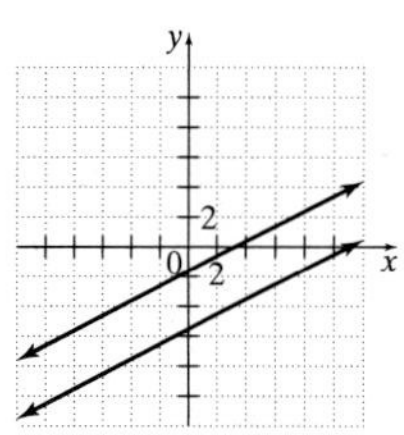

**5.** $\left(3, \frac{8}{3}\right)$

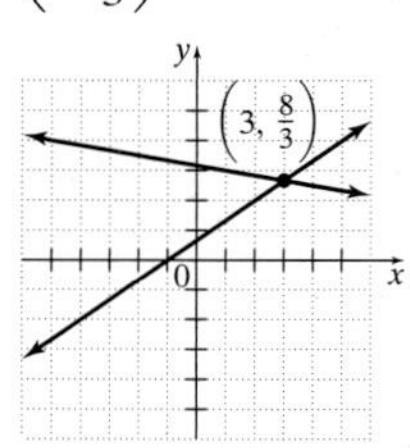


**7.** (2, 0, 2) **9.** $\left(-\frac{1}{2}, \frac{3}{4}, 1\right)$ **11.** ∅ **13.** (1, 1, −2) **15.** 10, 40, and 48 **17.** 58 mph, 65 mph **19.** 20 liters of 10% solution, 30 liters of 60% solution **21.** 17 pennies, 20 nickels, and 16 dimes **23.** Two sides are 22 cm each; third side is 29 cm. **25.** (−3, 1) **27.** $\left(-\frac{2}{3}, 3\right)$ **29.** $\left(\frac{5}{4}, \frac{5}{8}\right)$ **31.** (1, 3) **33.** (1, 2, 3) **35.** (3, −2, 5) **37.** (1, 1, −2) **39.** −17 **41.** 34 **43.** $\left(-\frac{2}{3}, 3\right)$ **45.** (−3, 1) **47.** ∅ **49.** (1, 2, 3) **51.** (2, 1, 0) **53.** ∅

***Chapter 4 Test*** **1.** 34 **2.** −6 **3.** (1, 3)

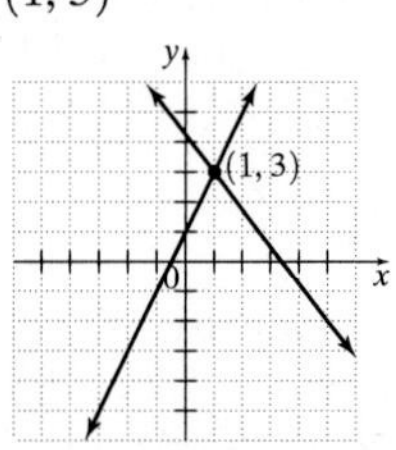


**4.** ∅

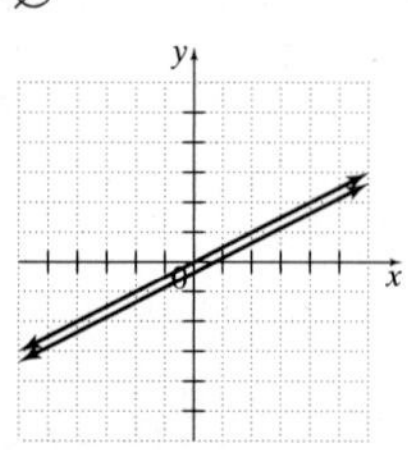

**5.** (2, −3) **6.** $\{(x, y)|10x + 4y = 10\}$ **7.** (−1, −2, 4) **8.** ∅ **9.** $\left(\frac{7}{2}, -10\right)$ **10.** (2, −1) **11.** (3, 6) **12.** (3, −1, 2) **13.** (5, 0, −4) **14.** $\{(x, y)|x - y = -2\}$ **15.** (5, −3) **16.** (−1, −1, 0) **17.** ∅ **18.** 275 frames **19.** 53 double rooms and 27 single rooms **20.** 5 gal of 10%, 15 gal of 20% **21.** 800 packages

***Chapter 4 Cumulative Review*** **1. a.** true **b.** true; Sec. 1.2, Ex. 4 **2. a.** 6 **b.** −7; Sec. 1.3, Ex. 3 **3. a.** −8 **b.** $-\frac{1}{5}$ **c.** 9.6; Sec. 1.4, Ex. 4 **4. a.** $6x + 3y$ **b.** $-3x + 1$; Sec. 1.4, Ex. 8 **5. a.** $-2x + 4$ **b.** $8yz$ **c.** $4z + 6.1$; Sec. 1.4, Ex. 11 **6.** −4; Sec. 2.1, Ex. 3 **7.** −4; Sec. 2.1, Ex. 7 **8.** 25 cm, 62 cm, 62 cm; Sec. 2.2, Ex. 6 **9.** $\{x|x \geq -10\}$

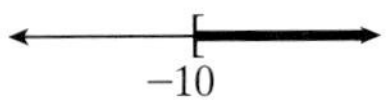


; Sec. 2.4, Ex. 3 **10.** (−3, 2); Sec 2.5, Ex. 4 **11.** 1, −1; Sec. 2.6, Ex. 4 **12.** (4, 8); Sec. 2.7, Ex. 2 **13. a.** IV **b.** $y$-axis **c.** II **d.** $x$-axis **e.** III **f.** I;

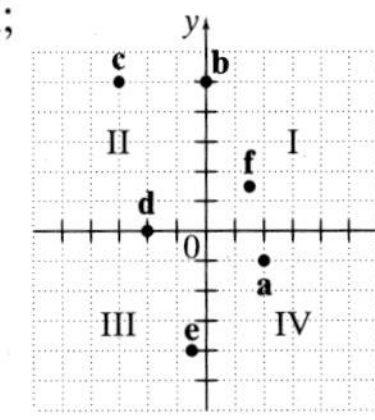


; Sec. 3.1, Ex. 1

**14. a.** $1580 **b.** greater than $1000; Sec. 3.1, Ex. 3 **15. a.** 5 **b.** 1 **c.** 35 **d.** −2; Sec 3.2, Ex. 7

**16.**

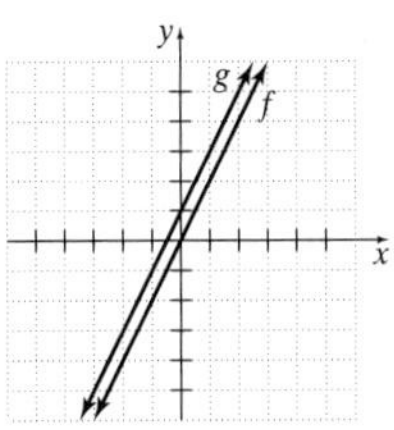


; Sec. 3.3, Ex. 1 **17.** slope: $\frac{3}{4}$; $y$-intercept: (0, −1); Sec 3.4, Ex. 4 **18. a.** parallel **b.** neither; Sec. 3.4, Ex. 8 **19.** $f(x) = \frac{5}{8}x - \frac{5}{2}$; Sec. 3.5, Ex. 5

**20.**

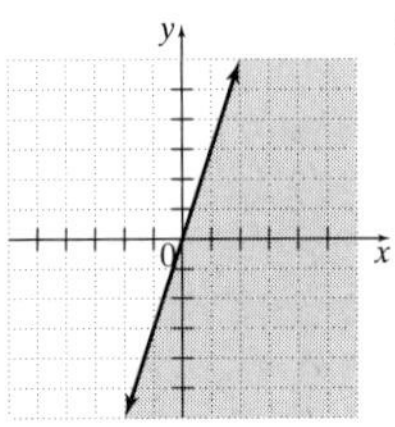


; Sec. 3.6, Ex. 2 **21. a.** yes **b.** no; Sec 4.1, Ex. 1 **22.** (−4, 2, 1); Sec. 4.2, Ex. 1 **23.** (−1, 2); Sec. 4.4, Ex. 1 **24. a.** −2 **b.** −10; Sec. 4.5, Ex. 1

## ■ CHAPTER 5 EXPONENTS, POLYNOMIALS, AND POLYNOMIAL FUNCTIONS

***Graphing Calculator Explorations*** **1.** $6 \times 10^{43}$ **3.** $3.796 \times 10^{28}$

***Mental Math*** **1.** $\frac{5}{xy^2}$ **3.** $\frac{a^2}{bc^5}$ **5.** $\frac{x^4}{y^2}$

***Exercise Set 5.1*** **1.** $4^5$ **3.** $x^8$ **5.** $-140x^{12}$ **7.** $-20x^2y$ **9.** $-16x^6y^3p^2$ **11.** −1 **13.** 1 **15.** 6 **17.** Answers may vary. **19.** $a^3$ **21.** $x$ **23.** $-13z^4$ **25.** $-6a^4b^4c^6$ **27.** $\frac{1}{16}$ **29.** $\frac{1}{x^8}$ **31.** $\frac{5}{a^4}$ **33.** $\frac{1}{x^7}$ **35.** $4r^8$ **37.** 1 **39.** $\frac{13}{36}$ **41.** 9 **43.** $x^{15}$ **45.** $10x^{10}$ **47.** $\frac{1}{z^3}$ **49.** $y^4$ **51.** $\frac{3}{x}$ **53.** −2 **55.** $r^8$ **57.** $\frac{1}{x^9y^4}$ **59.** $\frac{b^7}{9a^7}$ **61.** $\frac{6x^{16}}{5}$ **63.** $3.125 \times 10^7$ **65.** $1.6 \times 10^{-2}$ **67.** $6.7413 \times 10^4$ **69.** $1.25 \times 10^{-2}$ **71.** $5.3 \times 10^{-5}$ **73.** 0.0000000036 **75.** 93,000,000 **77.** 1,278,000 **79.** 7,350,000,000,000 **81.** 0.000000403 **83.** Answers may vary. **85.** **a, c, d** **87.** $7.783 \times 10^8$ **89.** $4.3141 \times 10^7$ **91.** $1.13 \times 10^9$ **93.** $1.0 \times 10^{-3}$ **95.** $x^{7a+5}$ **97.** $x^{2t-1}$ **99.** $x^{4a+7}$ **101.** $z^{6x-7}$ **103.** $x^{6t-1}$ **105.** $x^{3a+9}$ **107.** $7^{13}$ **109.** $7^{-11}$ **111.** 100 **113.** $\frac{27}{64}$ **115.** 64 **117.** $\frac{1}{16}$

***Mental Math*** **1.** $x^{20}$ **3.** $x^9$ **5.** $y^{42}$ **7.** $z^{20}$ **9.** $z^{18}$

***Exercise Set 5.2*** **1.** $\frac{1}{9}$ **3.** $\frac{1}{x^{36}}$ **5.** $\frac{1}{y^5}$ **7.** $9x^4y^6$ **9.** $16x^{20}y^{12}$ **11.** $\frac{c^{18}}{a^{12}b^6}$ **13.** $\frac{y^{15}}{x^{35}z^{20}}$ **15.** $\frac{1}{a^2}$ **17.** $4a^8b^4$ **19.** $\frac{x^4}{4z^2}$ **21.** yes; $a = \pm 1$ **23.** $\frac{1}{125}$ **25.** $\frac{1}{x^{63}}$ **27.** $\frac{343}{512}$ **29.** $16x^4$ **31.** $-\frac{y^3}{64}$ **33.** $4^8x^2y^6$ **35.** $\frac{36}{p^{12}}$ **37.** $-\frac{a^6}{512x^3y^9}$ **39.** $\frac{x^{14}y^{14}}{a^{21}}$

**41.** $\frac{x^4}{16}$ **43.** 64 **45.** $\frac{1}{y^{15}}$ **47.** $\frac{2}{p^2}$ **49.** $\frac{3}{8x^8y^7}$ **51.** $\frac{1}{x^{30}b^6c^6}$ **53.** $\frac{25}{8x^5y^4}$ **55.** $\frac{2}{x^4y^{10}}$ **57.** $1.45 \times 10^9$ **59.** $8 \times 10^{15}$ **61.** $4 \times 10^{-7}$ **63.** $3 \times 10^{-1}$ **65.** $2 \times 10^1$ **67.** $1 \times 10^1$ **69.** $8 \times 10^{-5}$ **71.** $1.1 \times 10^7$ **73.** $1.5 \times 10^{22}$ **75.** $0.002 = 2\times10^{-3}$ sec **77.** $1.331928\times10^{13}$ tons **79.** $\frac{8}{x^6y^3}$ cu. m **81.** $2.5808 \times 10^{-5}$ sq. m **83.** Answers may vary. **85.** 7 times **87.** $x^{4b+14}$ **89.** $x^{-3y+1}$ **91.** $c^{6a+9}$ **93.** $y^{26a+1}$ **95.** $9y^{12a-2}$ **97.** $y^{3b-a}$ **99.** $x^{-3a-3b}y^{b-a}$ **101.** $-3m - 15$ **103.** $-3y - 5$ **105.** $-3x + 5$

***Graphing Calculator Explorations*** **1.** $x^3 - 4x^2 + 7x - 8$ **3.** $-2.1x^2 - 3.2x - 1.7$ **5.** $7.69x^2 - 1.26x + 5.3$

***Exercise Set 5.3*** **1.** 0 **3.** 2 **5.** 3 **7.** degree 1; binomial **9.** degree 2; trinomial **11.** degree 3; monomial **13.** degree 3; none of these **15.** Answers may vary. **17.** 57 **19.** 499 **21.** 1 **23.** 989 ft **25.** 477 ft **27.** $6y$ **29.** $11x - 3$ **31.** $xy + 2x - 1$ **33.** $18y^2 - 17$ **35.** $3x^2 - 3xy + 6y^2$ **37.** $x^2 - 4x + 8$ **39.** $y^2 + 3$ **41.** $-2x^2 + 5x$ **43.** $-2x^2 - 4x + 15$ **45.** $4x - 13$ **47.** $x^2 + 2$ **49.** $12x^3 + 8x + 8$ **51.** $7x^3 + 4x^2 + 8x - 10$ **53.** $-18y^2 + 11yx + 14$ **55.** $-x^3 + 8a - 12$ **57.** $5x^2 - 9x - 3$ **59.** $-3x^2 + 3$ **61.** $8xy^2 + 2x^3 + 3x^2 - 3$ **63.** $7y^2 - 3$ **65.** $5x^2 + 22x + 16$ **67.** $-q^4 + q^2 - 3q + 5$ **69.** $15x^2 + 8x - 6$ **71.** $x^4 - 7x^2 + 5$ **73.** $4x^{2y} + 2x^y - 11$ **75.** 404 ft per sec **77. a.** \$1783.05 **b.** \$3515.05 **c.** \$6274.30 **d.** No, $f(x)$ is not linear. **79.** \$26,000 **81. a.** 284 ft **b.** 536 ft **c.** 756 ft **d.** 944 ft **83.** 19 sec **85. a.** 578 HMO's **b.** 545 HMO's **c.** 1732 HMO's **87.** $4x^2 - 3x + 6$ **89.** $-x^2 - 6x + 10$ **91.** $3x^2 - 12x + 13$ **93.** $15x^2 + 12x - 9$ **95.** $43x^2 - 12x - 43$ **97.** B **99.** C **101.** $(z^3 + 2z^2 - 2z + 3)$ units **103.** $-14z + 42y$ **105.** $-15y^2 - 10y + 35$ **107. a.** $8a + 3$ **b.** $-8x + 3$ **c.** $8x + 8h + 3$ **109. a.** $-4a$ **b.** $4x$ **c.** $-4x - 4h$ **111. a.** $3a - 2$ **b.** $-3x - 2$ **c.** $3x + 3h - 2$

***Graphing Calculator Explorations*** **1.** $x^2 - 16$ **3.** $9x^2 - 42x + 49$ **5.** $5x^3 - 14x^2 - 13x - 2$

***Exercise Set 5.4*** **1.** $-12x^5$ **3.** $12x^2 + 21x$ **5.** $-24x^2y - 6xy^2$ **7.** $-4a^3bx - 4a^3by + 12ab$ **9.** $2x^2 - 2x - 12$ **11.** $2x^4 + 3x^3 - 2x^2 + x + 6$ **13.** $15x^2 - 7x - 2$ **15.** $15m^3 + 16m^2 - m - 2$ **17.** Answers may vary. **19.** $x^2 + x - 12$ **21.** $10x^2 + 11xy - 8y^2$ **23.** $3x^2 + 8x - 3$ **25.** $9x^2 - \frac{1}{4}$ **27.** $x^2 + 8x + 16$ **29.** $36y^2 - 1$ **31.** $9x^2 - 6xy + y^2$ **33.** $9b^2 - 36y^2$ **35.** $16b^2 + 32b + 16$ **37.** $4s^2 - 12s + 8$ **39.** $x^2y^2 - 4xy + 4$ **41.** Answers may vary. **43.** $9x^2 + 18x + 5$ **45.** $10x^5 + 8x^4 + 2x^3 + 25x^2 + 20x + 5$ **47.** $49x^2 - 9$ **49.** $9x^3 + 30x^2 + 12x - 24$ **51.** $16x^2 - \frac{2}{3}x - \frac{1}{6}$ **53.** $36x^2 + 12x + 1$ **55.** $x^4 - 4y^2$ **57.** $-30a^4b^4 + 36a^3b^2 + 36a^2b^3$ **59.** $2a^2 - 12a + 16$ **61.** $49a^2b^2 - 9c^2$ **63.** $m^2 - 8m + 16$ **65.** $9x^2 + 6x + 1$ **67.** $y^2 - 7y + 12$ **69.** $2x^3 + 2x^2y + x^2 + xy - x - y$ **71.** $9x^4 + 12x^3 - 2x^2 - 4x + 1$ **73.** $12x^3 - 2x^2 + 13x + 5$ **75.** $5x^2 + 25x$ **77.** $x^4 - 4x^2 + 4$ **79.** $x^3 + 5x^2 - 2x - 10$ **81. a.** $6x + 12$ **b.** $9x^2 + 36x + 35$ **83.** $\pi(25x^2 - 20x + 4)$sq. km **85.** $a^2 - 3a$ **87.** $a^2 + 2ah + h^2 - 3a - 3h$ **89.** $b^2 - 7b + 10$ **91. a.** $a^2 + 2ah + h^2 + 3a + 3h + 2$ **b.** $a^2 + 3a + 2$ **c.** $2ah + h^2 + 3h$ **93.** $30x^2y^{2n+1} - 10x^2y^n$ **95.** $x^{3a} + 5x^{2a} - 3x^a - 15$ **97.** $-2$ **99.** $\frac{3}{5}$ **101.** function

***Mental Math*** **1.** 6 **3.** 5 **5.** $x$ **7.** $7x$

***Exercise Set 5.5*** **1.** $a^3$ **3.** $y^2z^2$ **5.** $3x^2y$ **7.** $5xz^3$ **9.** $6(3x - 2)$ **11.** $4y^2(1 - 4xy)$ **13.** $2x^3(3x^2 - 4x + 1)$ **15.** $4ab(2a^2b^2 - ab + 1 + 4b)$ **17.** $(x + 3)(6 + 5a)$ **19.** $(z + 7)(2x + 1)$ **21.** $(x^2 + 5)(3x - 2)$ **23.** Answers may vary. **25.** $(a + 2)(b + 3)$ **27.** $(a - 2)(c + 4)$ **29.** $(x - 2)(2y - 3)$ **31.** $(4x - 1)(3y - 2)$ **33.** $3(2x^3 + 3)$ **35.** $x^2(x + 3)$ **37.** $4a(2a^2 - 1)$ **39.** $-4xy(5x - 4y^2)$ **41.** $5ab^2(2ab + 1 - 3b)$ **43.** $3b(3ac^2 + 2a^2c - 2a + c)$ **45.** $(y - 2)(4x - 3)$ **47.** $(2x + 3)(3y + 5)$ **49.** $(x + 3)(y - 5)$ **51.** $(2a - 3)(3b - 1)$ **53.** $(6x + 1)(2y + 3)$ **55.** $(n - 8)(2m - 1)$ **57.** $3x^2y^2(5x - 6)$ **59.** $(2x + 3y)(x + 2)$ **61.** $(5x - 3)(x + y)$ **63.** $(x^2 + 4)(x + 3)$ **65.** $(x^2 - 2)(x - 1)$ **67.** $2\pi r(r + h)$ **69.** $A = P(1 + RT)$ **71. a.** $h(t) = -16t(t - 4)$ **b.** 48 ft **c.** Answers may vary. **73.** none **75.** a **77.** $-14y^4$ **79.** $16y^{12}$ **81.** $x^2 - 8x + 7$ **83.** $x^2 - 2x - 8$ **85.** $s^2 + 18s + 80$ **87.** $y^n(3 + 3y^n + 5y^{7n})$ **89.** $3x^{2a}(x^{3a} - 2x^a + 3)$

***Mental Math*** **1.** 5 and 2 **3.** 8 and 3

***Exercise Set 5.6*** **1.** $(x + 3)(x + 6)$ **3.** $(x - 8)(x - 4)$ **5.** $(x + 12)(x - 2)$ **7.** $(x - 6)(x + 4)$ **9.** $3(x - 2)(x - 4)$ **11.** $4z(x + 2)(x + 5)$ **13.** $2(x + 18)(x - 3)$ **15.** $\pm 5, \pm 7$ **17.** $(5x + 1)(x + 3)$ **19.** $(2x - 3)(x - 4)$ **21.** prime polynomial **23.** $(2x - 3)^2$ **25.** $2(3x - 5)(2x + 5)$ **27.** $y^2(3y + 5)(y - 2)$ **29.** $2x(3x^2 + 4x + 12)$ **31.** $(x + 7z)(x + z)$ **33.** $(2x + y)(x - 3y)$ **35.** $(x - 4)(x + 3)$ **37.** $2(7y + 2)(2y + 1)$ **39.** $(2x - 3)(x + 9)$ **41.** $\pm 8, \pm 16$ **43.** $(x^2 + 3)(x^2 - 2)$ **45.** $(5x + 8)(5x + 2)$ **47.** $(x^3 - 4)(x^3 - 3)$ **49.** $(a - 3)(a + 8)$ **51.** $x(3x + 4)(x - 2)$ **53.** $(x - 27)(x + 3)$ **55.** $(x - 18)(x + 3)$ **57.** $3(x - 1)^2$ **59.** $(3x + 1)(x - 2)$ **61.** $(4x - 3)(2x - 5)$ **63.** $3x^2(2x + 1)(3x + 2)$ **65.** $3(a + 2b)^2$ **67.** prime polynomial **69.** $(2x + 13)(x + 3)$ **71.** $(3x - 2)(2x - 15)$ **73.** $(x^2 - 6)(x^2 + 1)$ **75.** $x(3x + 1)(2x - 1)$ **77.** $(4a - 3b)(3a - 5b)$ **79.** $(3x + 5)^2$ **81.** $y(3x - 8)(x - 1)$ **83.** $2(x + 3)(x - 2)$ **85.** $(x + 2)(x - 7)$ **87.** $(2x^3 - 3)(x^3 + 3)$ **89.** $2x(6y^2 - z)^2$ **91.** $x^2(x + 5)(x + 1)$ **93.** $3x(5x - 1)(2x + 1)$ **95.** $x^3 - 8$ **97.** $-9$ **99.** $-8$ **101.** $(x^n + 8)(x^n + 2)$ **103.** $(x^n - 6)(x^n + 3)$ **105.** $(2x^n + 1)(x^n + 5)$ **107.** $(2x^n - 3)^2$

***Exercise Set 5.7*** **1.** $(x + 3)^2$ **3.** $(2x - 3)^2$ **5.** $3(x - 4)^2$ **7.** $x^2(3y + 2)^2$ **9.** $(x + 5)(x - 5)$ **11.** $(3 + 2z)(3 - 2z)$ **13.** $(y + 9)(y - 5)$ **15.** $4(4x + 5)(4x - 5)$ **17.** $(x + 3)(x^2 - 3x + 9)$ **19.** $(z - 1)(z^2 + z + 1)$ **21.** $(m + n)(m^2 - mn + n^2)$ **23.** $y^2(x - 3)(x^2 + 3x + 9)$ **25.** $b(a + 2b)(a^2 - 2ab + 4b^2)$ **27.** $(5y - 2x)(25y^2 + 10yx + 4x^2)$ **29.** $(x + 3 + y)(x + 3 - y)$ **31.** $(x - 5 + y)(x - 5 - y)$ **33.** $(2x + 1 + z)(2x + 1 - z)$ **35.** $(3x + 7)(3x - 7)$ **37.** $(x - 6)^2$ **39.** $(x^2 + 9)(x + 3)(x - 3)$ **41.** $(x + 4 + 2y)(x + 4 - 2y)$ **43.** $(x + 2y + 3)(x + 2y - 3)$ **45.** $(x - 6)(x^2 + 6x + 36)$ **47.** $(x + 5)(x^2 - 5x + 25)$ **49.** prime polynomial **51.** $(2a + 3)^2$ **53.** $2y(3x + 1)(3x - 1)$ **55.** $(2x + y)(4x^2 - 2xy + y^2)$ **57.** $(x^2 - y)(x^4 + x^2y + y^2)$ **59.** $(x + 8 + x^2)(x + 8 - x^2)$ **61.** $3y^2(x^2 + 3)(x^4 - 3x^2 + 9)$ **63.** $(x + y + 5)(x^2 + 2xy + y^2 - 5x - 5y + 25)$ **65.** $(2x - 1)(4x^2 + 20x + 37)$ **67.** $\pi R^2 - \pi r^2 = \pi(R + r)(R - r)$ **69.** $V = \frac{4}{3}\pi R^3 - \frac{4}{3}\pi 6^3 = \frac{4}{3}\pi(R - 6)(R^2 + 6R + 36)$ **71.** $(1 - y)(1 + y + y^2)$ **73.** $(3x + 1)^2$ **75.** $(x - 4 + y)(x - 4 - y)$ **77.** $x(x - 1)(x^2 + x + 1)$ **79.** $2xy(7x - 1)$ **81.** $4(x + 2)(x - 2)$ **83.** $2(4a - b)(16a^2 + 4ab + b^2)$ **85.** $(3x - 11)(x + 1)$ **87.** $4(x + 3)(x - 1)$ **89.** $(2x + 9)^2$ **91.** $(2x + 3y)(4x^2 - 6xy + 9y^2)$ **93.** $8x^2(2y - 1)(4y^2 + 2y + 1)$ **95.** $(x + 5 + y)(x^2 + 10x + 25 - xy - 5y + y^2)$ **97.** $(5a - 6)^2$ **99.** $c = 9$ **101.** $c = 49$ **103.** $c = \pm 8$ **105. a.** $(x + 1)(x^2 - x + 1)(x - 1)(x^2 + x + 1)$ **b.** $(x + 1)(x - 1)(x^4 + x^2 + 1)$ **107.** $-7$ **109.** 3 **111.** 0 **113.** $-4$ **115.** $(x^n + 6)(x^n - 6)$ **117.** $(5x^n + 9)(5x^n - 9)$ **119.** $(x^{2n} + 25)(x^n + 5)(x^n - 5)$

***Graphing Calculator Explorations*** **1.** $-3.562, 0.562$ **3.** $-0.874, 2.787$ **5.** $-0.465, 1.910$

***Mental Math*** **1.** $3, -5$ **3.** $3, -7$ **5.** $0, 9$

***Exercise Set 5.8*** **1.** $-3, \frac{4}{3}$ **3.** $\frac{5}{2}, -\frac{3}{4}$ **5.** $-3, -8$ **7.** $\frac{1}{4}, -\frac{2}{3}$ **9.** $1, 9$ **11.** $\frac{3}{5}, -1$ **13.** 0 **15.** $6, -3$ **17.** $\frac{2}{5}, -\frac{1}{2}$ **19.** $\frac{3}{4}, -\frac{1}{2}$ **21.** $-2, 7, \frac{8}{3}$ **23.** $0, 3, -3$ **25.** $2, 1, -1$ **27.** Answers may vary. **29.** $-\frac{7}{2}, 10$ **31.** $0, 5$ **33.** $-3, 5$ **35.** $-\frac{1}{2}, \frac{1}{3}$ **37.** $-4, 9$ **39.** $\frac{4}{5}$ **41.** $-5, 0, 2$ **43.** $-3, 0, \frac{4}{5}$ **45.** $\varnothing$ **47.** $-7, 4$ **49.** $4, 6$ **51.** $-\frac{1}{2}$ **53.** $-4, -3, 3$ **55.** $-5, 0, 5$ **57.** $-6, 5$ **59.** $-\frac{1}{3}, 0, 1$ **61.** $-\frac{1}{3}, 0$ **63.** $-\frac{7}{8}$ **65.** $\frac{31}{4}$ **67.** 1 **69. a.** incorrect **b.** correct **c.** correct **d.** incorrect **71.** $-11$ and $-6$ or 6 and 11 **73.** 75 ft **75.** 105 units **77.** 12 cm and 9 cm **79.** 2 in. **81.** 10 sec **83.** width: 7 ft; length: 13 ft **85.** 10 in. sq. tier **87.** E **89.** F **91.** B **93.** Answers may vary. Ex.: $f(x) = x^2 - 8x + 15$ **95.** Answers may vary. Ex.: $f(x) = x^2 - x - 2$ **97.** $(-3, 0), (0, 2)$; function **99.** $(-4, 0), (0, 2), (4, 0), (0, -2)$; not a function **101.** Answers may vary.

***Mental Math*** **1.** upward **3.** downward

***Exercise Set 5.9*** **1. a.** domain: $(-\infty, \infty)$; range: $(-\infty, 5]$ **b.** $x$-intercepts: $(-2, 0), (6, 0)$; $y$-intercept: $(0, 5)$ **c.** $(0, 5)$ **d.** There is no such point. **e.** $-2, 6$ **f.** between $x = -2$ and $x = 6$ **g.** $-2, 6$ **3. a.** domain: $(-\infty, \infty)$; range: $[-4, \infty)$ **b.** $x$-intercepts: $(-3, 0), (1, 0)$; $y$-intercept: $(0, -3)$ **c.** There is no such point. **d.** $(-1, -4)$ **e.** $-3, 1$ **f.** $x < -3$ or $x > 1$ **g.** $-3, 1$ **5. a.** domain: $(-\infty, \infty)$; range: $(-\infty, \infty)$ **b.** $x$-intercepts: $(-2, 0), (0, 0), (2, 0)$; $y$-intercept: $(0, 0)$ **c.** There is no such point. **d.** There is no such point. **e.** $-2, 0, 2$ **f.** between $x = -2$ and 0; $x > 2$ **g.** $-2, 0, 2$

**7.**
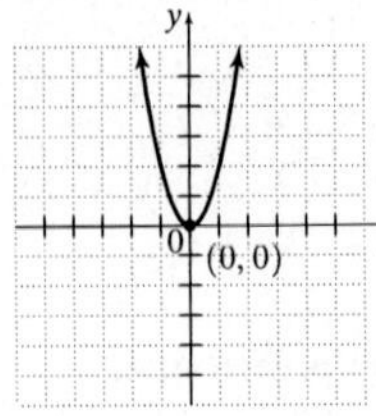


**9.**
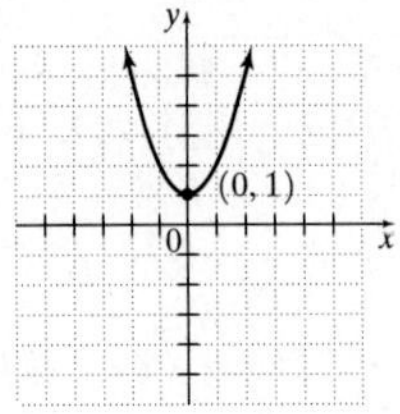


**11.**
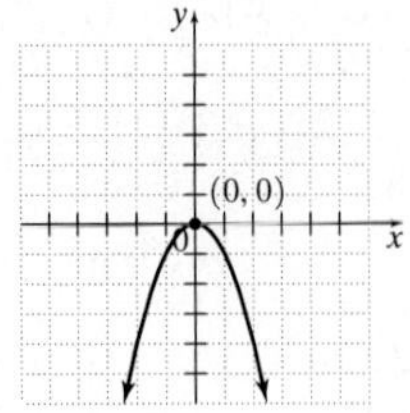


**13.** $(-4, -9)$
**15.** $(-1, 1)$
**17.** $(5, 30)$
**19.** 2
**21.** 0
**23.** 1 $x$-intercept; 1 $y$-intercept

**25.**
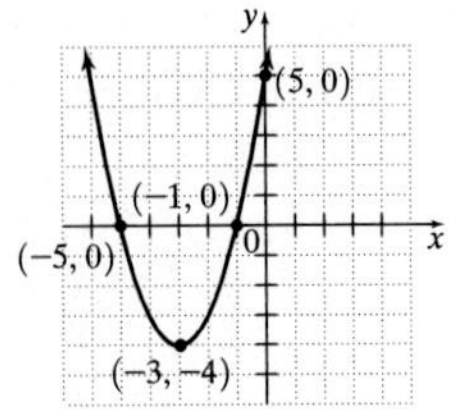


**27.**
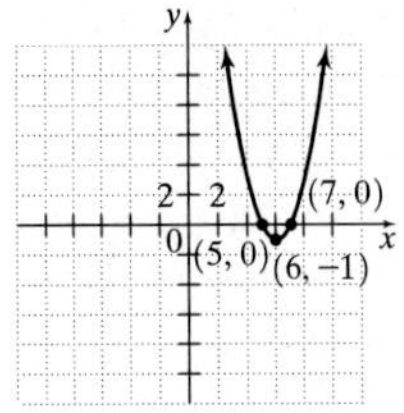


**29.**
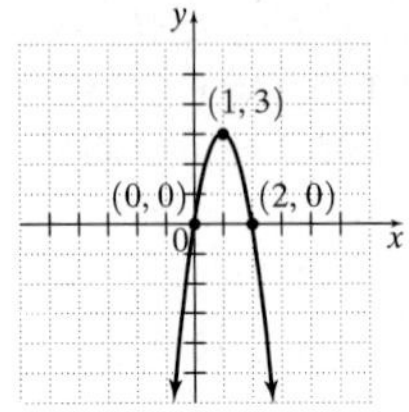


**31.**
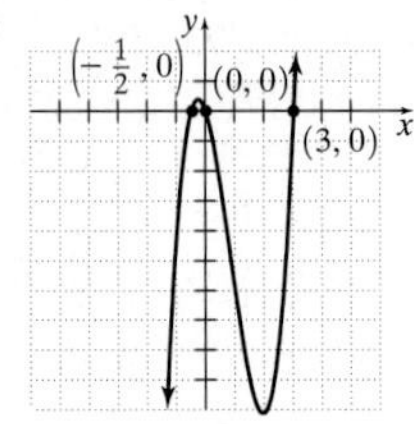


**33.**
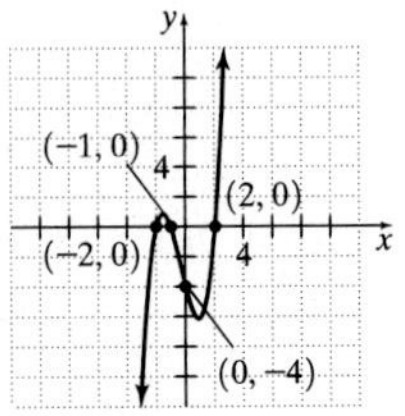


**35.** no **37.**
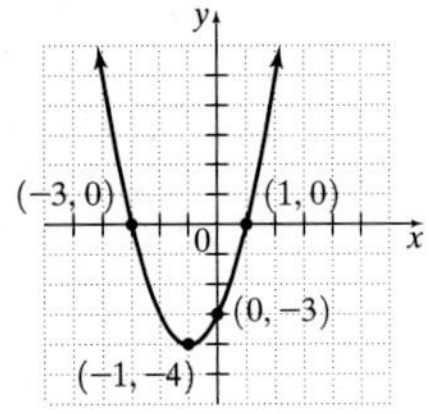


**39.**
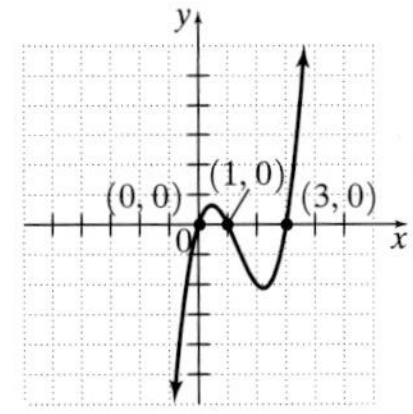


**41.**
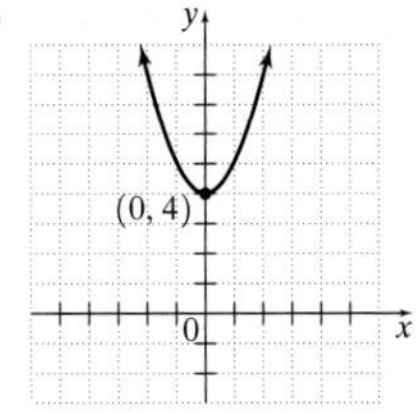


**43.**
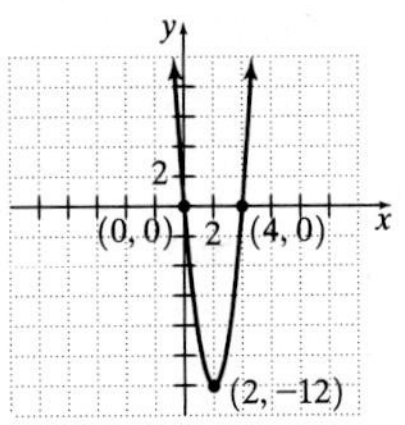


**45.**
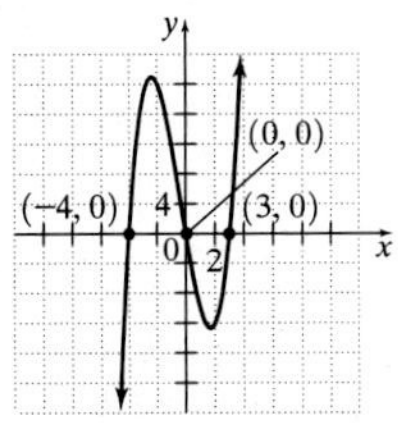


**47.**
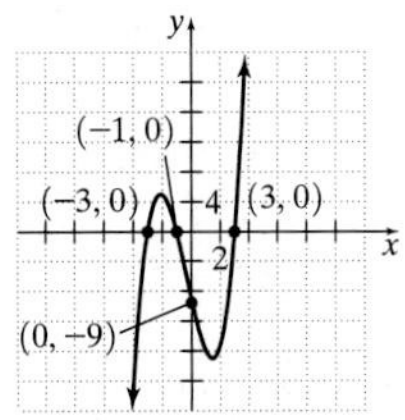


**49.**
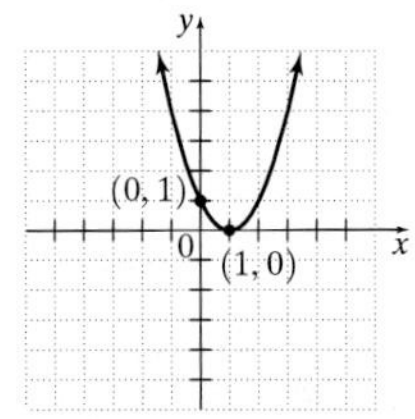


**51.**
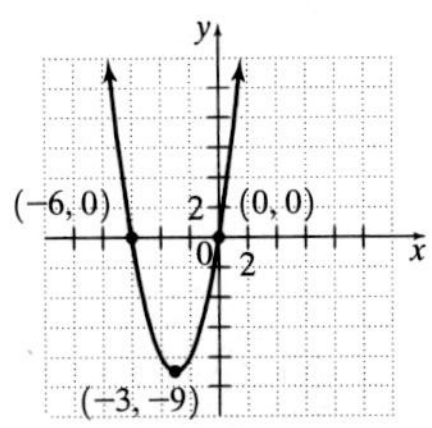


**53.**
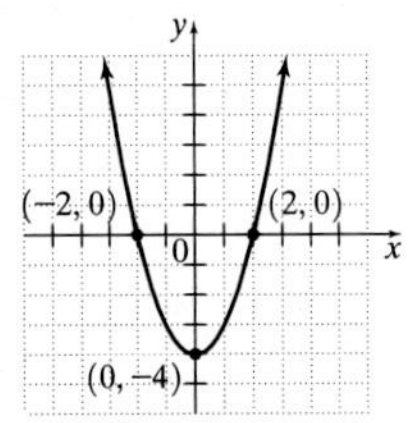


**55.**
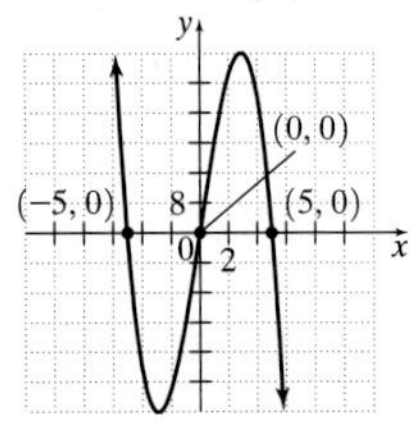


**57.**
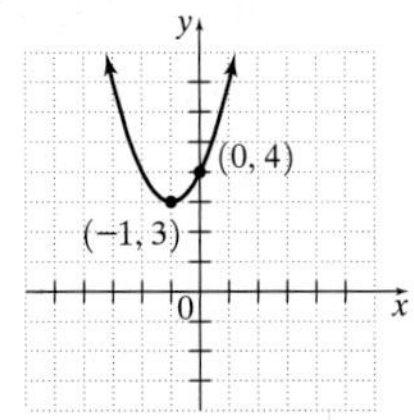


**59.**
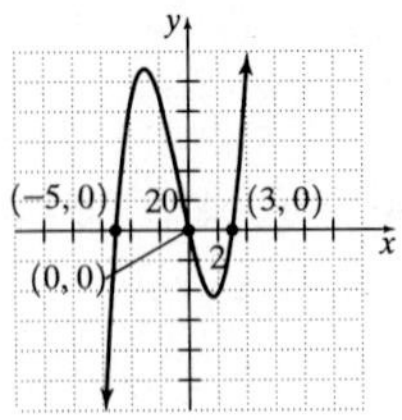


**61.**
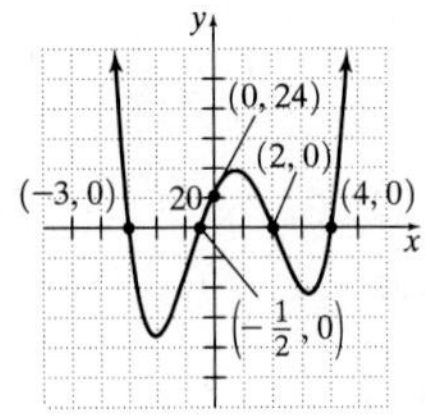


**63.**
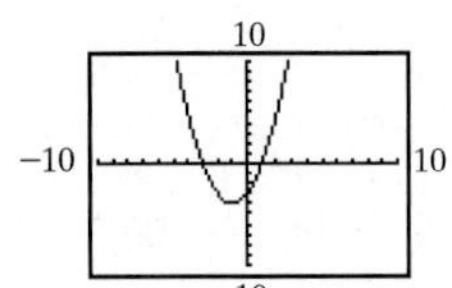


**65.**
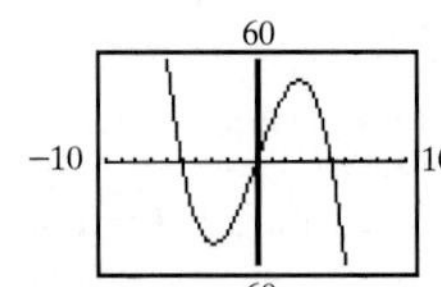


**67.**
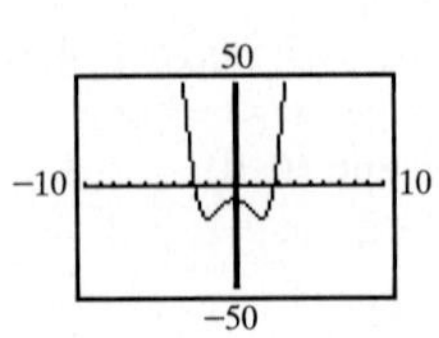


; $x = -2.7x, x - 2.7$ **69.** $-\dfrac{9}{20}$ **71.** $\dfrac{a^{13}}{b^2}$ **73.** $\dfrac{4y^7}{5x^4}$

***Chapter 5 Review*** **1.** 4 **3.** −4 **5.** 1 **7.** $-\frac{1}{16}$ **9.** $-x^2y^7z$ **11.** $\frac{1}{a^9}$ **13.** $\frac{1}{x^{11}}$ **15.** $\frac{1}{y^5}$ **17.** $-3.62 \times 10^{-4}$ **19.** 410,000 **21.** $\frac{a^2}{16}$ **23.** $\frac{1}{16x^2}$ **25.** $\frac{1}{8^{18}}$ **27.** $-\frac{1}{8x^9}$ **29.** $\frac{-27y^6}{x^6}$ **31.** $\frac{xz}{4}$ **33.** $\frac{2}{27z^3}$ **35.** $2y^{x-7}$ **37.** $-2.21 \times 10^{-11}$ **39.** $\frac{x^3y^{10}}{3z^{12}}$ **41.** 5 **43.** $12x - 6x^2 - 6x^2y$ **45.** $4x^2 + 8y + 6$ **47.** $8x^2 + 2b - 22$ **49.** $12x^2y - 7xy + 3$ **51.** $x^3 + x - 2xy^2 - y - 7$ **53.** 58 **55.** $x^2 + 4x - 6$ **57.** $(6x^2y - 12x + 12)$ cm **59.** $-12a^2b^5 - 28a^2b^3 - 4ab^2$ **61.** $9x^2a^2 - 24xab + 16b^2$ **63.** $15x^2 + 18xy - 81y^2$ **65.** $x^4 + 18x^3 + 83x^2 + 18x + 1$ **67.** $16x^2 + 72x + 81$ **69.** $16 - 9a^2 + 6ab - b^2$ **71.** $(9y^2 - 49z^2)$ sq. units **73.** $16x^2y^{2z} - 8xy^zb + b^2$ **75.** $8x^2(2x - 3)$ **77.** $2ab(3b + 4 - 2ab)$ **79.** $(a + 3b)(6a - 5)$ **81.** $(x - 6)(y + 3)$ **83.** $(p - 5)(q - 3)$ **85.** $x(2y - x)$ **87.** $(x - 4)(x + 20)$ **89.** $3(x + 2)(x + 9)$ **91.** $(3x + 8)(x - 2)$ **93.** $(15x - 1)(x - 6)$ **95.** $3(x - 2)(3x + 2)$ **97.** $(x + 7)(x + 9)$ **99.** $(x^2 - 2)(x^2 + 10)$ **101.** $(x + 9)(x - 9)$ **103.** $6(x + 3)(x - 3)$ **105.** $(4 + y^2)(2 + y)(2 - y)$ **107.** $(x - 7)(x + 1)$ **109.** $(y + 8)(y^2 - 8y + 64)$ **111.** $(1 - 4y)(1 + 4y + 16y^2)$ **113.** $2x^2(x + 2y)(x^2 - 2xy + 4y^2)$ **115.** $(x - 3 - 2y)(x - 3 + 2y)$ **117.** $(4a - 5b)^2$ **119.** $\frac{1}{3}, -7$ **121.** $0, 4, \frac{9}{2}$ **123.** 0, 6 **125.** $-\frac{1}{3}, 2$ **127.** −4, 1 **129.** 0, 6, −3 **131.** 0, −2, 1 **133.** $-\frac{15}{2}, 7$ **135.** 5 sec **137.** $x$-intercepts: $(-4, 0)$, $(2, 0)$; $y$-intercept: $(0, 3)$ **139.** between $x = -4$ and $x = 2$

**141.**

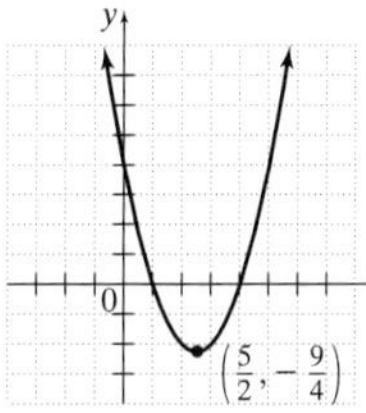


**143.**

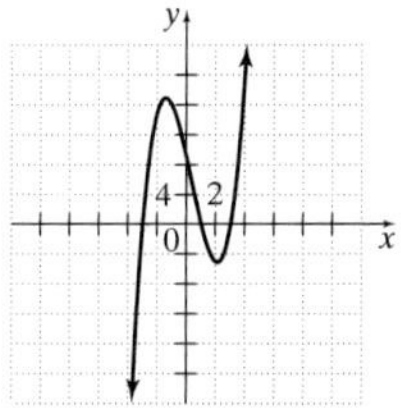


**145.**

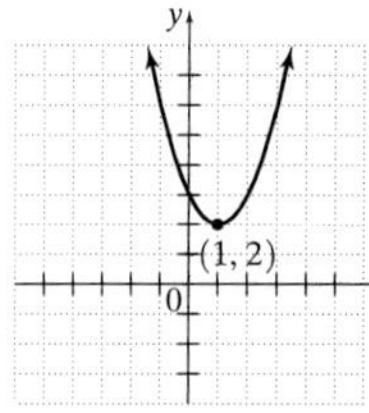


**147.**

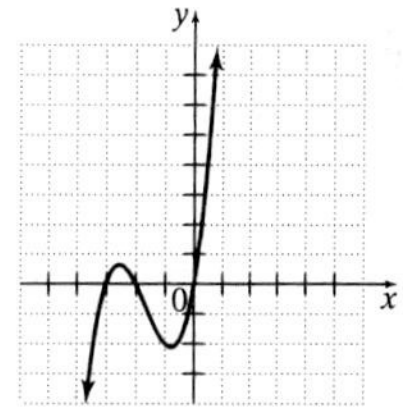


***Chapter 5 Test*** **1.** $\frac{1}{81x^2}$ **2.** $-12x^2z$ **3.** $\frac{3a^7}{2b^5}$ **4.** $-\frac{y^{40}}{z^5}$ **5.** $6.3 \times 10^8$ **6.** $1.2 \times 10^{-2}$ **7.** 0.000005 **8.** 0.0009 **9.** $-5x^3 - 11x - 9$ **10.** $-12x^2y - 3xy^2$ **11.** $12x^2 - 5x - 28$ **12.** $25a^2 - 4b^2$ **13.** $36m^2 + 12mn + n^2$ **14.** $2x^3 - 13x^2 + 14x - 4$ **15.** $4x^2y(4x - 3y^3)$ **16.** $(x - 15)(x + 2)$ **17.** $(2y + 5)^2$ **18.** $3(2x + 1)(x - 3)$ **19.** $(2x + 5)(2x - 5)$ **20.** $(x + 4)(x^2 - 4x + 16)$ **21.** $3y(x + 3y)(x - 3y)$ **22.** $6(x^2 + 4)$ **23.** $2(2y - 1)(4y^2 + 2y + 1)$ **24.** $(x + 3)(x - 3)(y - 3)$ **25.** $4, -\frac{8}{7}$ **26.** −3, 8 **27.** $-\frac{5}{2}, -2, 2$ **28.** $(x + 2y)(x - 2y)$ **29. a.** 960 ft **b.** 953.44 ft **c.** 11 sec

**30.**

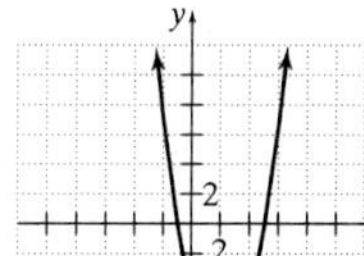


**31.**

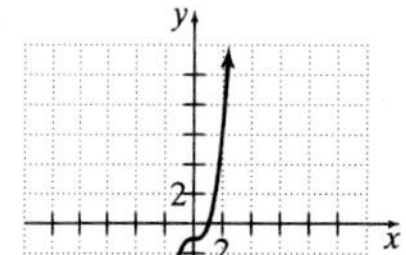


***Chapter 5 Cumulative Review*** **1. a.** 3 **b.** 1 **c.** 2; Sec. 1.3, Ex. 8 **2.** 1; Sec. 2.1, Ex. 4 **3.** $11,607.55; Sec. 2.3, Ex. 4

**4. a.** $\left\{x \,\middle|\, x \le \frac{3}{2}\right\}$

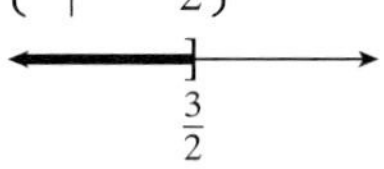


**b.** $\{x|x > -3\}$; Sec. 2.4, Ex. 8

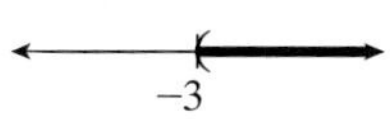


**5.** $\left[-9, -\frac{9}{2}\right]$; Sec. 2.5, Ex. 5 **6.** 0; Sec. 2.6, Ex. 5

**7.** ∅; Sec. 2.7, Ex. 4

**8.**

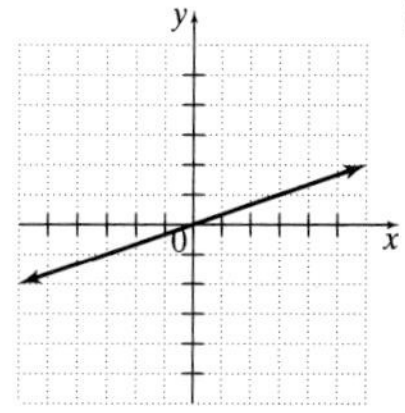


; Sec. 3.1, Ex. 5

**9.** no; Sec. 3.2, Ex. 4 **10.** 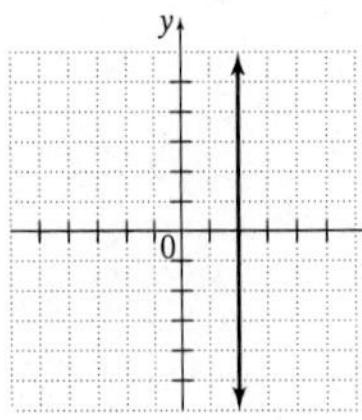

; Sec. 3.3, Ex. 6 **11.** 0; Sec. 3.4, Ex. 7 **12.** $y = 3$; Sec. 3.5, Ex. 7

**13.** 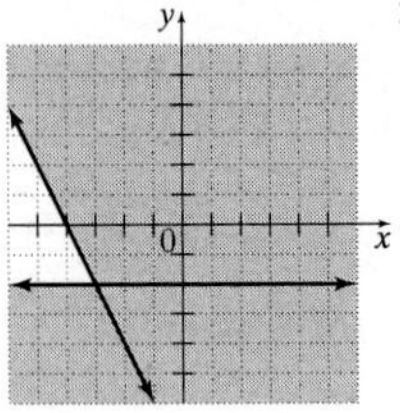

; Sec. 3.6, Ex. 4 **14.** $\left(-4, \frac{1}{2}\right)$; Sec. 4.1, Ex. 3 **15.** $\left(\frac{1}{2}, 0, \frac{3}{4}\right)$; Sec. 4.2, Ex. 3
**16.** 7, 11; Sec. 4.3, Ex. 1 **17.** $\varnothing$; Sec. 4.4, Ex. 2 **18.** $(1, -2, -1)$; Sec. 4.5, Ex. 4
**19. a.** $7.3 \times 10^5$ **b.** $1.04 \times 10^{-6}$; Sec. 5.1, Ex. 8
**20. a.** $\frac{y^6}{4}$ **b.** $x^9$ **c.** $\frac{49}{4}$ **d.** $\frac{y^{16}}{25x^5}$; Sec. 5.2, Ex. 3 **21.** 4; Sec. 5.3, Ex. 3
**22. a.** $10x^9$ **b.** $-7xy^{15}z^9$; Sec. 5.4, Ex. 1 **23.** $17x^3y^2(1 - 2x)$; Sec. 5.5, Ex. 3
**24.** $(x + 2)(x + 8)$; Sec. 5.6, Ex. 1 **25.** $-5, \frac{1}{2}$; Sec. 5.8, Ex. 2

## ■ CHAPTER 6 RATIONAL EXPRESSIONS

***Graphing Calculator Explorations*** **1.** $\{x|x$ is a real number and $x \neq -2, x \neq 2\}$

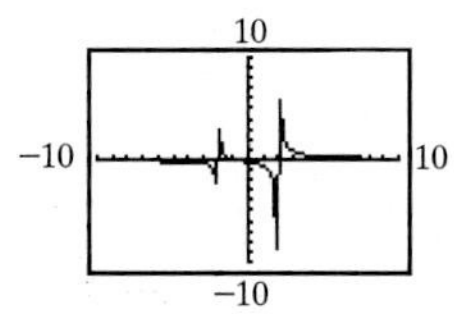


**3.** $\left\{x \,\middle|\, x \text{ is a real number and } x \neq -4, x \neq \frac{1}{2}\right\}$

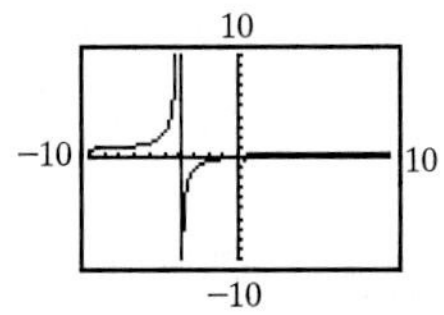


***Exercise Set 6.1*** **1.** $\frac{10}{3}, -8, -\frac{7}{3}$ **3.** $-\frac{17}{48}, \frac{2}{7}, -\frac{3}{8}$ **5.** $\{x|x$ is a real number$\}$ **7.** $\{t|t$ is a real number and $t \neq 0\}$
**9.** $\{x|x$ is a real number and $x \neq 7\}$ **11.** $\{x|x$ is a real number and $x \neq -2, x \neq 0, x \neq 1\}$
**13.** $\{x|x$ is a real number and $x \neq 2, x \neq -2\}$ **15.** Answers may vary. **17.** $\frac{4}{3}$ **19.** $-2$ **21.** $\frac{x + 1}{x - 3}$ **23.** $\frac{2(x + 3)}{x - 3}$
**25.** $\frac{3}{x}$ **27.** $\frac{x + 1}{x^2 + 1}$ **29.** $\frac{1}{2(q - 1)}$ **31.** $x - 4$ **33.** $-x^2 - 5x - 25$ **35.** $\frac{4x^2 + 6x + 9}{2}$ **37.** $-\frac{2}{3x^3y^2}$ **39.** $\frac{4}{ab^6}$
**41.** $\frac{1}{4a(a - b)}$ **43.** $\frac{(x + 2)(x + 3)}{4}$ **45.** $\frac{3}{2(x - 1)}$ **47.** $\frac{4a^2}{a - b}$ **49.** $\frac{2(x + 3)(x - 3)}{5(x^2 - 8x - 15)}$ **51.** $\frac{x + 2}{x + 3}$ **53.** $\frac{3b}{a - b}$
**55.** $\frac{3a}{a - b}$ **57.** $\frac{1}{4}$ **59.** $-1$ **61.** $\frac{8}{3}$ **63.** $\frac{8(a - 2)}{3(a + 2)}$ **65.** $\frac{8}{x^2y}$ **67.** $\frac{(y + 5)(2x - 1)}{(y + 2)(5x + 1)}$ **69.** $\frac{5(3a + 2)}{a}$ **71.** $\frac{5x^2 - 2}{(x - 1)^2}$
**73.** $\frac{5}{x - 2}$ sq. m **75.** Answers may vary. **77.** $\frac{(x + 2)(x - 1)^2}{x^5}$ ft
**79.** $0, \frac{20}{9}, \frac{60}{7}, 20, \frac{140}{3}, 180, 380, 1980$; 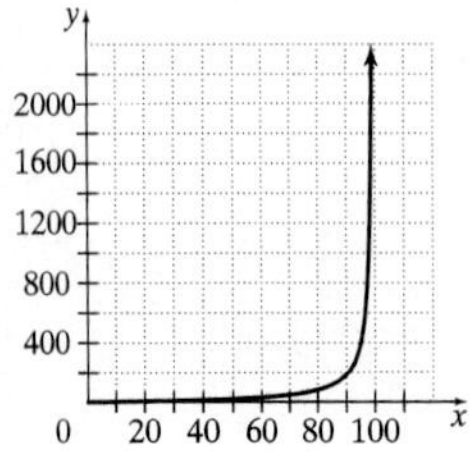


**81. a.** $\{x|0 \leq x < 100\}$
**b.** \$42,857.14
**c.** \$150,000; \$400,000
**d.** \$900,000; \$1,900,000; \$9,900,000; Answers may vary.
**83.** $\frac{7}{5}$ **85.** $\frac{1}{12}$ **87.** $\frac{11}{16}$ **89.** $2x^2(x^n + 2)$
**91.** $\frac{1}{10y(y^n + 3)}$ **93.** $\frac{y^n + 1}{2(y^n - 1)}$

***Exercise Set 6.2*** **1.** $-\frac{3}{x}$ **3.** $\frac{x+2}{x-2}$ **5.** $x-2$ **7.** $\frac{1}{2-x}$ **9.** $\frac{4x}{x+5}$ ft; $\frac{x^2}{x^2+10x+25}$ sq. ft **11.** $35x$ **13.** $x(x+1)$ **15.** $(x+7)(x-7)$ **17.** $6(x+2)(x-2)$ **19.** $(3x-1)(x+2)$ **21.** $(a+b)(a-b)^2$ **23.** $-4x(x+3)(x-3)$ **25.** Answers may vary. **27.** $\frac{17}{6x}$ **29.** $\frac{35-4y}{14y^2}$ **31.** $\frac{-13x+4}{(x+4)(x-4)}$ **33.** $\frac{2x+4}{(x-5)(x+4)}$ **35.** 0 **37.** $-\frac{x}{x-1}$ **39.** $\frac{-x+1}{x-2}$ **41.** $\frac{y^2+2y+10}{(y+4)(y-4)(y-2)}$ **43.** $\frac{5(x^2+x-4)}{(3x+2)(x+3)(2x-5)}$ **45.** $\frac{x^2+5x+21}{(x-2)(x+1)(x+3)}$ **47.** $\frac{-2x+5}{2(x+1)}$ **49.** $\frac{2(x^2+x-21)}{(x+3)^2(x-3)}$ **51.** $\frac{3}{x^2y^3}$ **53.** $-\frac{5}{x}$ **55.** $\frac{25}{6(x+5)}$ **57.** $\frac{-2x-1}{x^2(x-3)}$ **59.** $\frac{2ab-b^2}{(a+b)(a-b)}$ **61.** $\frac{2x+16}{(x+2)^2(x-2)}$ **63.** $\frac{5a+1}{(a+1)^2(a-1)}$ **65.** Answers may vary. **67.** Answers may vary. **69.** $\frac{2x^2+9x-18}{6x^2}$ **71.** $\frac{4}{3}$ **73.** $\frac{4a^2}{9(a-1)}$ **75.** 4 **77.** $\frac{6x}{(x+3)(x-3)^2}$ **79.** $-\frac{4}{x-1}$ **81.** $-\frac{32}{x(x+2)(x-2)}$

**83.**

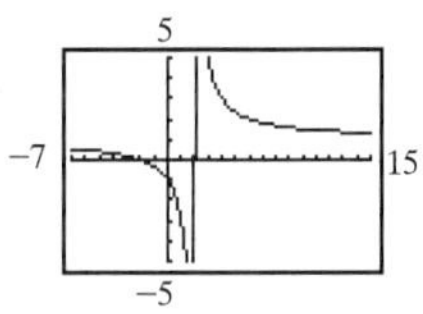


**85.**

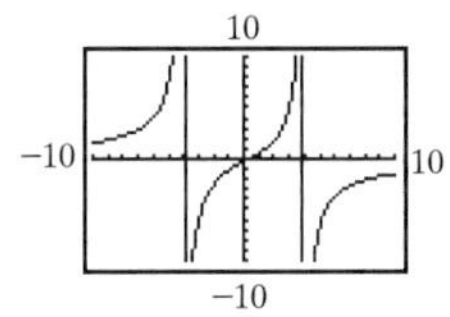


**87.** 10 **89.** $4+x^2$ **91.** 10 **93.** 2 **95.** 3 **97.** 5 m **99.** $\frac{3}{2x}$ **101.** $\frac{4-3x}{x^2}$ **103.** $\frac{1-3x}{x^3}$

***Exercise Set 6.3*** **1.** $\frac{5}{6}$ **3.** $\frac{8}{5}$ **5.** 4 **7.** $\frac{7}{13}$ **9.** $\frac{4}{x}$ **11.** $\frac{9x-18}{9x^2-4}$ **13.** $\frac{1-x}{1+x}$ **15.** $\frac{xy^2}{x^2+y^2}$ **17.** $\frac{2b^2+3a}{b^2-ab}$ **19.** $\frac{x}{x^2-1}$ **21.** $\frac{x+1}{x+2}$ **23.** $\frac{10}{69}$ **25.** $\frac{2(x+1)}{2x-1}$ **27.** $\frac{x(x+1)}{6}$ **29.** $\frac{x}{2-3x}$ **31.** $-\frac{y}{x+y}$ **33.** $-\frac{2x^3}{y(x-y)}$ **35.** $\frac{2x+1}{y}$ **37.** $\frac{x-3}{9}$ **39.** $\frac{1}{x+2}$ **41.** $\frac{x}{5x-10}$ **43.** $\frac{x-2}{2x-1}$ **45.** $-\frac{x^2+4}{4x}$ **47.** $\frac{x-3y}{x+3y}$ **49.** $\frac{1+a}{1-a}$ **51.** $\frac{x^2+6xy}{2y}$ **53.** $\frac{5a}{2a+4}$ **55.** $5xy^2+2x^2y$ **57.** $\frac{xy}{2x+5y}$ **59.** $\frac{xy}{x+y}$ **61.** $x^2+x$ **63.** $\frac{770a}{770-s}$ **65. a.** $\frac{1}{a+h}$ **b.** $\frac{1}{a}$ **c.** $\frac{\frac{1}{a+h}-\frac{1}{a}}{h}$ **d.** $\frac{-1}{a(a+h)}$ **67. a.** $\frac{3}{a+h+1}$ **b.** $\frac{3}{a+1}$ **c.** $\frac{\frac{3}{a+h+1}-\frac{3}{a+1}}{h}$ **d.** $\frac{-3}{(a+h+1)(a+1)}$ **69.** $\frac{x^2y^2}{4}$ **71.** $-9x^3y^4$ **73.** $-4, 14$ **75.** $(-4, 14)$ **77.** $\frac{x-1}{x}$ **79.** $2x$ **81.** $3a^2+4a+4$

***Exercise Set 6.4*** **1.** $2a+4$ **3.** $3ab+4$ **5.** $2y+\frac{3y}{x}-\frac{2y}{x^2}$ **7.** $x^2+2x+1$ **9.** $(x^4+2x^2-6)$ m **11.** $x+1$ **13.** $2x-8$ **15.** $x-\frac{1}{2}$ **17.** $2x^2-\frac{1}{2}x+5$ **19.** $(3x-7)$ in. **21.** $\frac{5b^5}{2a^3}$ **23.** $x^3y^3-1$ **25.** $a+3$ **27.** $2x+5$ **29.** $4y-6y^2$ **31.** $2x+23+\frac{130}{x-5}$ **33.** $10x+3y-6x^2y^2$ **35.** $2x+4$ **37.** $y+5$ **39.** $2x+3$ **41.** $2x^2-8x+38-\frac{156}{x+4}$ **43.** $3x+3-\frac{1}{x-1}$ **45.** $-2x^3+3x^2-x+4$ **47.** $3x^3+5x+4-\frac{2x}{x^2-2}$ **49.** $x-\frac{5}{3x^2}$ **51.** 4 **53.** 372 **55.** Answers may vary. **57.** $3x^2+10x+8+\frac{4}{x-2}$ **59.** $=$ **61.** $=$ **63.** $(-9, -1)$ **65.** $(-\infty, -8] \cup [1, \infty)$ **67.** $x^3+\frac{5}{3}x^2+\frac{5}{3}x+\frac{8}{3}+\frac{8}{3(x-1)}$ **69.** $\frac{3}{2}x^3+\frac{1}{4}x^2+\frac{1}{8}x-\frac{7}{16}+\frac{1}{16(2x-1)}$ **71.** $x^3-\frac{2}{5}x$

***Exercise Set 6.5*** **1.** $x + 8$ **3.** $x - 1$ **5.** $x^2 - 5x - 23 - \frac{41}{x - 2}$ **7.** $4x + 8 + \frac{7}{x - 2}$ **9.** 3 **11.** 73 **13.** −8 **15.** $x^2 + \frac{2}{x - 3}$ **17.** $6x + 7 + \frac{1}{x + 1}$ **19.** $2x^3 - 3x^2 + x - 4$ **21.** $3x - 9 + \frac{12}{x + 3}$ **23.** $3x^2 - \frac{9}{2}x + \frac{7}{4} + \frac{47}{8(x - 1/2)}$ **25.** $3x^2 + 3x - 3$ **27.** $3x^2 + 4x - 8 + \frac{20}{x + 1}$ **29.** $x^2 + x + 1$ **31.** $x - 6$ **33.** 1 **35.** −133 **37.** 3 **39.** $-\frac{187}{81}$ **41.** $\frac{95}{32}$ **43.** Answers may vary. **45.** $(x + 3)(x^2 + 4) = x^3 + 3x^2 + 4x + 12$ **47.** 0 **49.** $x^3 + 2x^2 + 7x + 28$ **51.** $(x - 1)$ m **53.** $\frac{13}{3}$ **55.** −3, 1 **57.** −1 **59.** $(2y + 1)(4y^2 - 2y + 1)$ **61.** $(a - 3)(a^2 + 3a + 9)$ **63.** $(x - 1)(x + y)$ **65.** $2x(x + 4)(x - 4)$

***Exercise Set 6.6*** **1.** 72 **3.** 2 **5.** 6 **7.** 2 **9.** 3 **11.** ∅ **13.** 15 **15.** 4 **17.** ∅ **19.** 1 **21.** −1 **23.** −3 **25.** $\frac{5}{3}$ **27.** 10, 2 **29.** 2 **31.** 3 **33.** ∅ **35.** ∅ **37.** −1 **39.** 9 **41.** 1, 7 **43.** $\frac{1}{10}$ **45.** 800 pencil sharpeners **47.** $\frac{1}{9}, -\frac{1}{4}$ **49.** 3, 2 **51.** 1.39 **53.** −0.08 **55.**

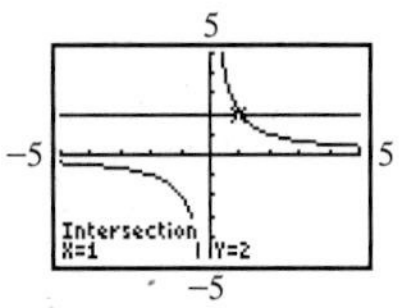


**57.**

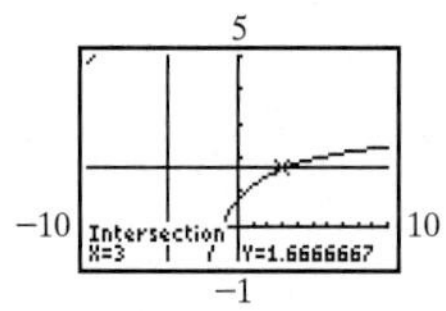


**59.** 73 and 74 **61.** $\frac{1}{2}$ and 2 **63.** 3% **65.** 54% **67.** −1, 0 **69.** −2

***Supplementary Exercises on Expressions and Equations*** **1.** $\frac{1}{2}$ **3.** $\frac{1 + 2x}{8}$ **5.** $\frac{2(x - 4)}{(x + 2)(x - 1)}$ **7.** 4 **9.** −5 **11.** $\frac{2x + 5}{x(x - 3)}$ **13.** −2 **15.** $\frac{(a + 3)(a + 1)}{a + 2}$ **17.** $-\frac{1}{5}$ **19.** $\frac{4a + 1}{(3a + 1)(3a - 1)}$ **21.** $-1, \frac{3}{2}$ **23.** $\frac{3}{x + 1}$ **25.** −1

***Exercise Set 6.7*** **1.** $C = \frac{5}{9}(F - 32)$ **3.** $I = A - QL$ **5.** $R = \frac{R_1R_2}{R_1 + R_2}$ **7.** $n = \frac{2S}{a + L}$ **9.** $b = \frac{2A - ah}{h}$ **11.** $T_2 = \frac{P_2V_2T_1}{P_1V_1}$ **13.** $f_2 = \frac{f_1 f}{f_1 - f}$ **15.** $L = \frac{n\lambda}{2}$ **17.** $c = \frac{2L\omega}{\theta}$ **19.** 1 and 5 **21.** 5 **23.** 6 ohms **25.** $\frac{1}{R} = \frac{1}{R_1} + \frac{1}{R_2} + \frac{1}{R_3}$; $R = \frac{15}{13}$ ohms **27.** 12 hr **29.** $1\frac{1}{3}$ hr **31.** 50 mph **33.** 6 mph **35.** 9 and 11 **37.** $3\frac{1}{3}$ hr **39.** 13 mph **41.** $\frac{6}{10}$ **43.** 10 mph; 8 mph **45.** 3 hr **47.** 22,500 mi **49.** $2\frac{2}{9}$ hr **51.** 2 hr **53.** \$108 **55.** $2\frac{2}{9}$ days **57.** 1 hr **59.** 60 in. or 5 ft **61.** −5 **63.** 2

***Exercise Set 6.8*** **1.** $k = \frac{1}{5}$; $y = \frac{1}{5}x$ **3.** $k = \frac{3}{2}$; $y = \frac{3}{2}x$ **5.** $k = 14$; $y = 14x$ **7.** $k = 0.25$; $y = 0.25x$ **9.** 4.05 lb **11.** $P = 566{,}222$ tons **13.** $k = 30$; $y = \frac{30}{x}$ **15.** $k = 700$; $y = \frac{700}{x}$ **17.** $k = 2$; $y = \frac{2}{x}$ **19.** $k = 0.14$; $y = \frac{0.14}{x}$ **21.** 54 mph **23.** 72 amps **25.** divided by 4 **27.** $x = kyz$ **29.** $r = kst^3$ **31.** 22.5 tons **33.** $15\pi$ cu. in. **35.** 90 hp **37.** 800 millibars **39.** multiplied by 2 **41.** multiplied by 4 **43.**

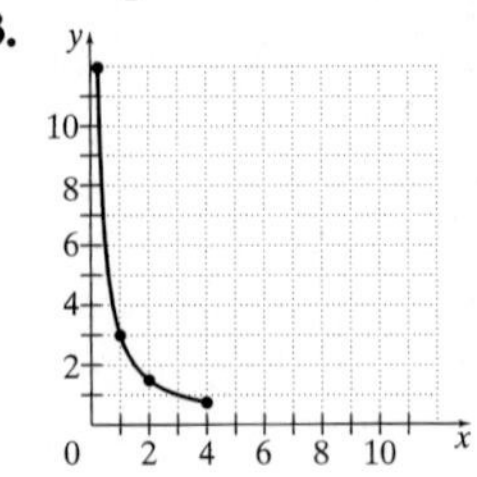


**45.**

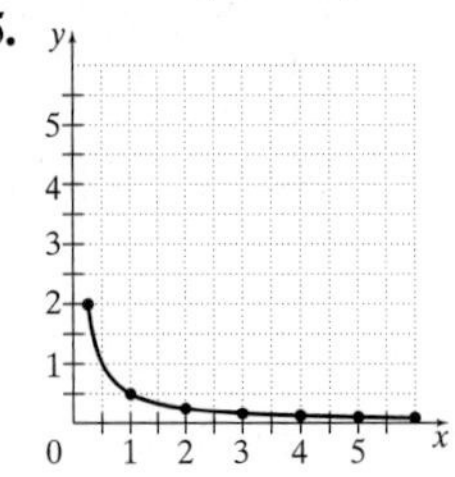


**47.** $C = 12\pi$ cm; $A = 36\pi$ sq. cm **49.** $C = 14\pi$ m; $A = 49\pi$ sq. m **51.** 6 **53.** 2 **55.** $\frac{1}{5}$ **57.** $\frac{5}{11}$

***Chapter 6 Review*** **1.** $\{x|x \text{ is a real number}\}$ **3.** $\{x|x \text{ is a real number and } x \neq 5\}$
**5.** $\{x|x \text{ is a real number and } x \neq 0, x \neq -8\}$ **7.** $\frac{x^2}{3}$ **9.** $\frac{9m^2p}{5}$ **11.** $\frac{1}{5}$ **13.** $\frac{1}{x-1}$ **15.** $\frac{2(x-3)}{x-4}$ **17. a.** \$119 **b.** \$77 **c.** decrease **19.** $\frac{2x^3}{z^3}$ **21.** $\frac{2}{5}$ **23.** $\frac{1}{6}$ **25.** $\frac{3x}{16}$ **27.** $\frac{3c^2}{14a^2b}$ **29.** $\frac{(x+4)(x+5)}{3}$ **31.** $\frac{7(x-4)}{2(x-2)}$ **33.** $-\frac{1}{x}$ **35.** $\frac{8}{9a^2}$ **37.** $\frac{6}{a}$ **39.** $60x^2y^5$ **41.** $5x(x-5)$ **43.** $\frac{2}{5}$ **45.** $\frac{2}{x^2}$ **47.** $\frac{1}{x-2}$ **49.** $\frac{5x^2-3y^2}{15x^4y^3}$ **51.** $\frac{-x+5}{(x+1)(x-1)}$ **53.** $\frac{2x^2-5x-4}{x-3}$ **55.** $\frac{3x^2-7x-4}{(3x-4)(9x^2+12x+16)}$ **57.** $-\frac{12}{x(x+1)(x-3)}$ **59.** $\frac{14x-40}{(x+4)^2(x-4)}$ **61.** $\frac{2}{3}$ **63.** $\frac{2}{15-2x}$ **65.** $\frac{y}{2}$ **67.** $\frac{20x-15}{10x^2-4}$ **69.** $\frac{5xy+x}{3y}$ **71.** $\frac{1+x}{1-x}$ **73.** $\frac{x-1}{3x-1}$ **75.** $-\frac{x^2+9}{6x}$ **77. a.** $\frac{3}{a+h}$ **b.** $\frac{3}{a}$ **c.** $\frac{\frac{3}{a+h}-\frac{3}{a}}{h}$ **d.** $\frac{-3}{a(a+h)}$ **79.** $\frac{9b^2z^3}{4a}$ **81.** $\frac{3}{b}+4b$ **83.** $2x^3-4x^2+7x-9+\frac{6}{x+2}$ **85.** $2x^2-2+\frac{5}{x+3/2}$ **87.** $3x^2+6$ **89.** $3x^2-\frac{5}{2}x-\frac{1}{4}-\frac{5}{8\left(x+\frac{3}{2}\right)}$ **91.** $x^2+3x+9-\frac{54}{x-3}$ **93.** $3x^3-6x^2+10x-20+\frac{50}{x+2}$ **95.** $-9323$ **97.** $\frac{365}{32}$ **99.** 6 **101.** 2 **103.** $\frac{3}{2}$ **105.** $\frac{5}{3}$ **107.** $-\frac{1}{3}, 2$ **109.** $a=\frac{2A-hb}{h}$ **111.** $R=\frac{E-Ir}{I}$ **113.** $A=\frac{HL}{k(T_1-T_2)}$ **115.** 7 **117.** $-10$ and $-8$ **119.** 12 hr **121.** 490 mph **123.** 8 mph **125.** 4 mph **127.** 9 **129.** 3.125 cu. ft

***Chapter 6 Test*** **1.** $\{x \mid x \text{ is a real number and } x \neq 1\}$ **2.** $\{x \mid x \text{ is a real number and } x \neq -3, x \neq -1\}$ **3.** $\frac{5x^3}{3}$ **4.** $-\frac{7}{8}$ **5.** $\frac{x}{x+9}$ **6.** $\frac{x+2}{5}$ **7.** $\frac{5}{3x}$ **8.** $\frac{4a^3b^4}{c^6}$ **9.** $\frac{x+2}{2(x+3)}$ **10.** $-\frac{4(2x+9)}{5}$ **11.** $\frac{3}{x^3}$ **12.** $-1$ **13.** $\frac{5x-2}{(x-3)(x+2)(x-2)}$ **14.** $-\frac{x+30}{6(x-7)}$ **15.** $\frac{3}{2}$ **16.** $\frac{1}{5}$ **17.** $\frac{64}{3}$ **18.** $\frac{(x-3)^2}{x-2}$ **19.** $\frac{4xy}{3z}+\frac{3}{z}+\frac{1}{3x}$ **20.** $x^5+5x^4+8x^3+16x^2+33x+63+\frac{128}{x-2}$ **21.** $4x^3-15x^2+47x-142+\frac{425}{x+3}$ **22.** 91 **23.** 7 **24.** $2, -2$ **25.** 8 **26.** $x=\frac{7a^2+b^2}{4a-b}$ **27.** 5 **28.** $\frac{6}{7}$ hr **29.** 16 **30.** 9 **31.** 256 ft

***Chapter 6 Cumulative Review*** **1. a.** $8x$ **b.** $8x+3$ **c.** $x \div -7$ or $\frac{x}{-7}$ **d.** $2x-1.6$; Sec. 1.2, Ex. 8 **2.** 2; Sec. 2.1, Ex. 5 **3.** after 2013; Sec. 2.4, Ex. 9 **4.** 0; Sec. 2.6, Ex. 7 **5.** $-1$; Sec. 2.7, Ex. 8
**6.**

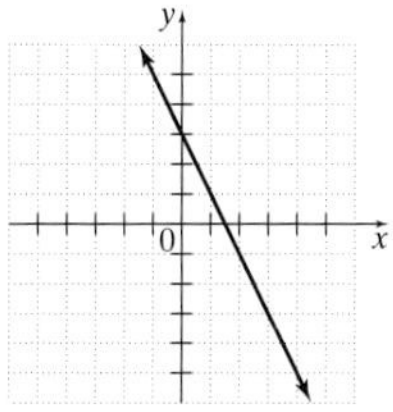


; Sec. 3.1, Ex. 4 **7. a.** function **b.** not a function **c.** function; Sec. 3.2, Ex. 2
**8.**

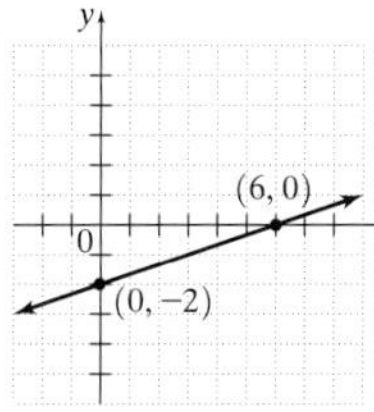


; Sec. 3.3, Ex. 4 **9.** $y=-3x-2$; Sec. 3.5, Ex. 4
**10.**

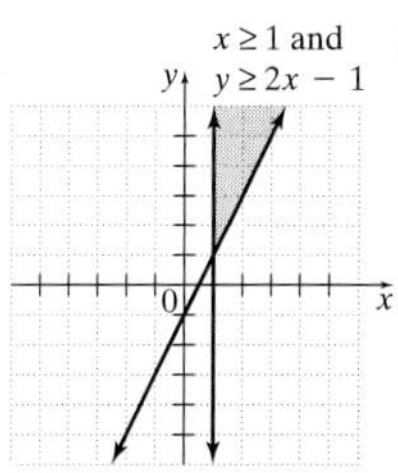


; Sec. 3.6, Ex. 3
**11.** $(0, -5)$; Sec. 4.1, Ex. 6
**12.** $\varnothing$; Sec. 4.2, Ex. 2
**13.** 30°, 110°, 40°; Sec. 4.3, Ex. 5
**14.** $(1, -1, 3)$; Sec. 4.4, Ex. 3
**15. a.** 1 **b.** $-1$ **c.** 1 **d.** 2; Sec. 5.1, Ex. 3 **16. a.** $4x^b$ **b.** $y^{5a+6}$; Sec 5.2, Ex. 5
**17. a.** 2 **b.** 5 **c.** 1 **d.** 6 **e.** 0; Sec. 5.3, Ex. 1 **18.** $9+12a+6b+4a^2+4ab+b^2$; Sec. 5.4, Ex. 9

**19.** $(b - 6)(a + 2)$; Sec 5.5, Ex. 7 **20.** prime polynomial; Sec. 5.6, Ex. 4
**21.** $(x + 2 + y)(x + 2 - y)$; Sec. 5.7, Ex. 5 **22.** $-2, 6$; Sec. 5.8, Ex. 1
**23.**

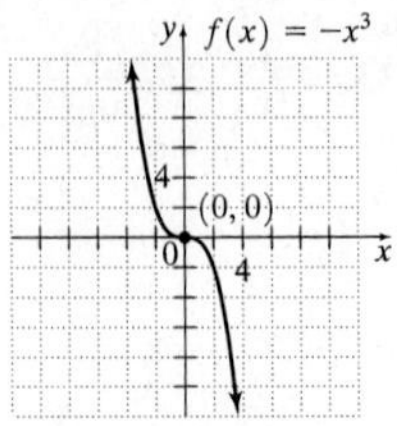


; Sec. 5.9, Ex. 6 **24.** $\dfrac{5k^2 - 7k + 4}{(k + 2)(k - 2)(k - 1)}$; Sec. 6.2, Ex. 4 **25.** $-2$; Sec. 6.6, Ex. 2

## ■ CHAPTER 7 RATIONAL EXPONENTS, RADICALS, AND COMPLEX NUMBERS

***Exercise Set 7.1*** **1.** 10 **3.** $\frac{1}{2}$ **5.** 0.01 **7.** $-6$ **9.** $x^5$ **11.** $4y^3$ **13.** 2.646 **15.** 6.164 **17.** 14.142 **19.** 4 **21.** $\frac{1}{2}$
**23.** $-1$ **25.** $x^4$ **27.** $-3x^3$ **29.** $-2$ **31.** not a real number **33.** $-2$ **35.** $x^4$ **37.** $2x^2$ **39.** $9x^2$ **41.** $4x^2$ **43.** 8
**45.** $-8$ **47.** $2|x|$ **49.** $x$ **51.** $|x - 5|$ **53.** $|x + 2|$ **55.** $-11$ **57.** $2x$ **59.** $y^6$ **61.** $5ab^{10}$ **63.** $-3x^4y^3$ **65.** $a^4b$
**67.** $-2x^2y$ **69.** $\frac{5}{7}$ **71.** $\frac{x}{2y}$ **73.** $-\frac{z^7}{3x}$ **75.** $\frac{x}{2}$ **77.** $\sqrt{3}$ **79.** $-1$ **81.** $-3$ **83.** $\sqrt{7}$

**85.** $[0, \infty)$;

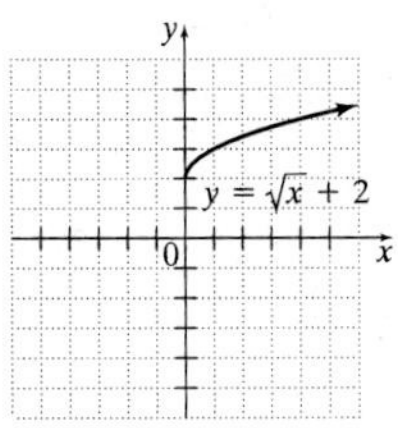


**87.** $[3, \infty)$;

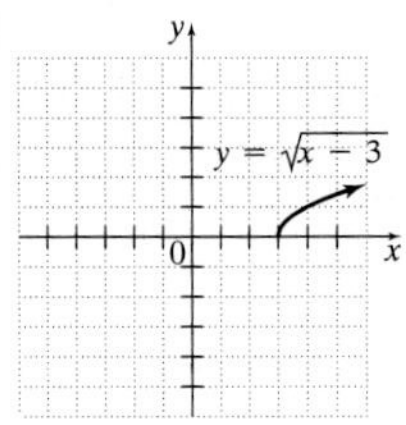


**89.** $(-\infty, \infty)$;

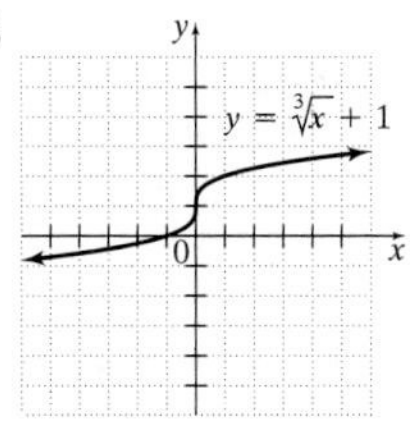


**91.** $(-\infty, \infty)$; 0, 1, $-1$, 2, $-2$

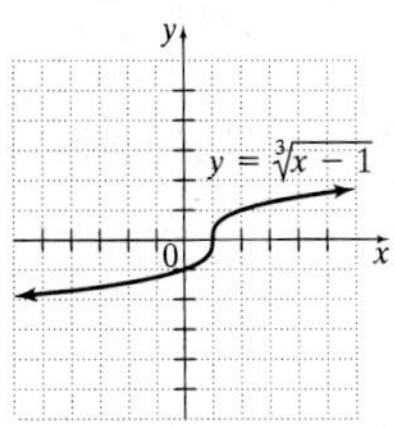


**93.** Answers may vary. **95.**

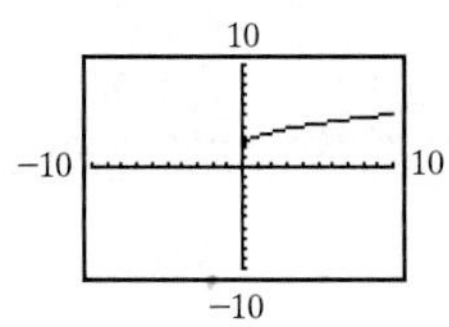


**97.**

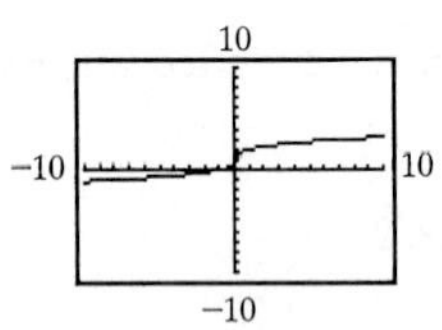


**99.** $-32x^{15}y^{10}$ **101.** $-60x^7y^{10}z^5$ **103.** $\frac{x^9y^5}{2}$

***Exercise Set 7.2*** **1.** 7 **3.** 3 **5.** $\frac{1}{2}$ **7.** 13 **9.** $2\sqrt[3]{m}$ **11.** $3x^2$ **13.** $-3$ **15.** $-2$ **17.** 8 **19.** 16 **21.** not a real number
**23.** $\sqrt[5]{(2x)^3}$ **25.** $\sqrt[3]{(7x + 2)^2}$ **27.** $\frac{64}{27}$ **29.** $\frac{1}{16}$ **31.** $\frac{1}{16}$ **33.** not a real number **35.** $\frac{1}{x^{1/4}}$ **37.** $a^{2/3}$ **39.** $\frac{5x^{3/4}}{7}$
**41.** Answers may vary. **43.** $a^{7/3}$ **45.** $x$ **47.** $3^{5/8}$ **49.** $y^{1/6}$ **51.** $8u^3$ **53.** $-b$ **55.** $27x^{2/3}$ **57.** $y - y^{7/6}$
**59.** $2x^{5/3} - 2x^{2/3}$ **61.** $4x^{2/3} - 9$ **63.** $x^{8/3}(1 + x^{2/3})$ **65.** $x^{1/5}(x^{1/5} - 3)$ **67.** $x^{-1/3}(5 + x)$ **69.** $\sqrt{x}$ **71.** $\sqrt[3]{2}$ **73.** $2\sqrt{x}$
**75.** $\sqrt{xy}$ **77.** $\sqrt[15]{y^{11}}$ **79.** $\sqrt[12]{b^5}$ **81.** $\sqrt{a}$ **83.** $\sqrt[6]{432}$ **85.** $\sqrt[15]{343y^5}$ **87.** $\frac{t^{1/2}}{u^{1/2}}$ **89.** 1316 calories
**91. a.** \$43,340 million **b.** \$63,374 million **93.** $x^{3/8}$ **95.** $y^{1/4}$ **97.** 1.8206 **99.** 22.0515 **101.** $4 \cdot 5$ **103.** $9 \cdot 5$
**105.** $8 \cdot 7$ **107.** $8 \cdot 10$

***Exercise Set 7.3*** **1.** $\sqrt{14}$ **3.** 2 **5.** $\sqrt[3]{36}$ **7.** $\sqrt{6x}$ **9.** $\sqrt{\frac{14}{xy}}$ **11.** $\sqrt[4]{20x^3}$ **13.** $\frac{\sqrt{6}}{7}$ **15.** $\frac{\sqrt{2}}{7}$ **17.** $\frac{\sqrt[4]{x^3}}{2}$ **19.** $\frac{\sqrt[3]{4}}{3}$ **21.** $\frac{\sqrt[4]{8}}{x^2}$ **23.** $\frac{\sqrt[3]{2x}}{3y^4\sqrt[3]{3}}$ **25.** $\frac{x\sqrt{y}}{10}$ **27.** $\frac{\sqrt{5x}}{2y}$ **29.** $-\frac{z^2\sqrt[3]{z}}{3x}$ **31.** $4\sqrt{2}$ **33.** $4\sqrt[3]{3}$ **35.** $25\sqrt{3}$ **37.** $2\sqrt{6}$ **39.** $10x^2\sqrt{x}$ **41.** $2y^2\sqrt[3]{2y}$ **43.** $a^2b\sqrt[4]{b^3}$ **45.** $y^2\sqrt{y}$ **47.** $5ab\sqrt{b}$ **49.** $-2x^2\sqrt[5]{y}$ **51.** $x^4\sqrt[3]{50x^2}$ **53.** $-4a^4b^3\sqrt{2b}$ **55.** $3x^3y^4\sqrt{xy}$ **57.** $5r^3s^4$ **59.** $\sqrt{2}$ **61.** 2 **63.** 10 **65.** $x^2y$ **67.** $24m^2$ **69.** $\frac{15x\sqrt{2x}}{2}$ or $\frac{15x}{2}\sqrt{2x}$ **71.** $2a^2$ **73.** **a.** $20\pi$ sq. cm **b.** 211.57 sq. ft **75.** **a.** 3.8 times **b.** 2.9 times **c.** Answers may vary. **77.** $48x^2$ **79.** $3x - 2$ **81.** $-72y^4$ **83.** $x + 2$ **85.** $4x^2 + 4x + 1$

***Mental Math*** **1.** $6\sqrt{3}$ **3.** $3\sqrt{x}$ **5.** $12\sqrt[3]{x}$

***Exercise Set 7.4*** **1.** $-2\sqrt{2}$ **3.** $10x\sqrt{2x}$ **5.** $17\sqrt{2} - 15\sqrt{5}$ **7.** $-\sqrt[3]{2x}$ **9.** $5b\sqrt{b}$ **11.** $\frac{31\sqrt{2}}{15}$ **13.** $\frac{\sqrt[3]{11}}{3}$ **15.** $\frac{5\sqrt{5x}}{9}$ **17.** $14 + \sqrt{3}$ **19.** $7 - 3y$ **21.** $6\sqrt{3} - 6\sqrt{2}$ **23.** $-23\sqrt[3]{5}$ **25.** $2b\sqrt{b}$ **27.** $20y\sqrt{2y}$ **29.** $2y\sqrt[3]{2x}$ **31.** $6\sqrt[3]{11} - 4\sqrt{11}$ **33.** $4x\sqrt[4]{x^3}$ **35.** $\frac{2\sqrt{3}}{3}$ **37.** $\frac{5x\sqrt[3]{x}}{7}$ **39.** $\frac{5\sqrt{7}}{2x}$ **41.** $\frac{\sqrt[3]{2}}{6}$ **43.** $\frac{14x\sqrt[3]{2x}}{9}$ **45.** $15\sqrt{3}$ in. **47.** $\sqrt{35} + \sqrt{21}$ **49.** $7 - 2\sqrt{10}$ **51.** $3\sqrt{x} - x\sqrt{3}$ **53.** $6x - 13\sqrt{x} - 5$ **55.** $\sqrt[3]{a^2} + \sqrt[3]{a} - 20$ **57.** $6\sqrt{2} - 12$ **59.** $2 + 2x\sqrt{3}$ **61.** $-16 - \sqrt{35}$ **63.** $x - y^2$ **65.** $3 + 2x\sqrt{3} + x^2$ **67.** $5x - 3\sqrt{15x} - 3\sqrt{10x} + 9\sqrt{6}$ **69.** $2\sqrt[3]{2} - \sqrt[3]{4}$ **71.** $-4\sqrt[6]{x^5} + \sqrt[3]{x^2} + 8\sqrt[3]{x} - 4\sqrt{x} + 7$ **73.** **a.** $22\sqrt{5}$ ft **b.** 150 sq. ft **75.** Answers may vary. **77.** $x - 7$ **79.** $\frac{7}{x + y}$ **81.** $2a - 3$ **83.** $\frac{-2 + \sqrt{3}}{3}$

***Mental Math*** **1.** $\sqrt{2} - x$ **3.** $5 + \sqrt{a}$ **5.** $7\sqrt{5} - 8\sqrt{x}$

***Exercise Set 7.5*** **1.** $\frac{\sqrt{14}}{7}$ **3.** $\frac{\sqrt{5}}{5}$ **5.** $\frac{\sqrt[3]{6}}{2}$ **7.** $\frac{4\sqrt[3]{9}}{3}$ **9.** $\frac{3\sqrt{2x}}{4x}$ **11.** $\frac{3\sqrt[3]{2x}}{2x}$ **13.** $\frac{2\sqrt{x}}{x}$ **15.** $\frac{3\sqrt{3a}}{a}$ **17.** $\frac{3\sqrt[3]{4}}{2}$ **19.** $\frac{2\sqrt{21}}{7}$ **21.** $\frac{\sqrt{10xy}}{5y}$ **23.** $\frac{3\sqrt[4]{2}}{2}$ **25.** $\frac{2\sqrt[4]{9x}}{3x^2}$ **27.** $\frac{5a\sqrt[5]{4ab^4}}{2a^2b^3}$ **29.** $\frac{5}{\sqrt{15}}$ **31.** $\frac{6}{\sqrt{10}}$ **33.** $\frac{2x}{7\sqrt{x}}$ **35.** $\frac{5y}{\sqrt[3]{100xy}}$ **37.** $\frac{2}{\sqrt{10}}$ **39.** $\frac{2x}{11\sqrt{2x}}$ **41.** $\frac{7}{2\sqrt[3]{49}}$ **43.** $\frac{3x^2}{10\sqrt[3]{9x}}$ **45.** $\frac{6x^2y^3}{\sqrt{6z}}$ **47.** Answers may vary. **49.** $-2(2 + \sqrt{7})$ **51.** $\frac{7(3 + \sqrt{x})}{9 - x}$ **53.** $-5 + 2\sqrt{6}$ **55.** $\frac{2a + 2\sqrt{a} + \sqrt{ab} + \sqrt{b}}{4a - b}$ **57.** $-\frac{8(1 - \sqrt{10})}{9}$ **59.** $\frac{x - \sqrt{xy}}{x - y}$ **61.** $\frac{5 + 3\sqrt{2}}{7}$ **63.** $\frac{-7}{12 + 6\sqrt{11}}$ **65.** $\frac{3}{10 + 5\sqrt{7}}$ **67.** $\frac{x - 9}{x - 3\sqrt{x}}$ **69.** $\frac{1}{3 + 2\sqrt{2}}$ **71.** $\frac{x - 1}{x - 2\sqrt{x} + 1}$ **73.** $r = \frac{\sqrt{A\pi}}{2\pi}$ **75.** Answers may vary. **77.** 5 **79.** $-\frac{1}{2}, 6$ **81.** 2, 6

***Graphing Calculator Explorations*** **1.** 3.19 **3.** $\varnothing$ **5.** 3.23

***Exercise Set 7.6*** **1.** 8 **3.** 7 **5.** $\varnothing$ **7.** 7 **9.** 6 **11.** $-\frac{9}{2}$ **13.** 29 **15.** 4 **17.** $-4$ **19.** $\varnothing$ **21.** 7 **23.** 9 **25.** 50 **27.** $\varnothing$ **29.** $\frac{15}{4}$ **31.** 13 **33.** 5 **35.** $-12$ **37.** 9 **39.** $-3$ **41.** 1 **43.** 1 **45.** $\frac{1}{2}$ **47.** 0, 4 **49.** $\frac{37}{4}$ **51.** Answers may vary. **53.** $3\sqrt{5}$ ft **55.** $2\sqrt{10}$ m **57.** $2\sqrt{131}$ m $\approx$ 22.9 m **59.** $\sqrt{100.84}$ mm $\approx$ 10.0 mm **61.** 17 ft **63.** 13 ft **65.** 14,657,415 sq. mi **67.** 100 ft **69.** 1 **71.** 2743 deliveries **73.** **a.** Answers may vary. **b.** Answers may vary. **75.** not a function **77.** not a function **79.** not a function **81.** $-\frac{20 + 16y}{3y}$ **83.** $\frac{x + y}{x - y}$ **85.** $-1, 2$ **87.** $-8, -6, 0, 2$

***Mental Math*** **1.** $9i$ **3.** $i\sqrt{7}$ **5.** $-4$ **7.** $8i$

***Exercise Set 7.7*** **1.** $2i\sqrt{6}$ **3.** $-6i$ **5.** $24i\sqrt{7}$ **7.** $-3\sqrt{6}$ **9.** $-\sqrt{14}$ **11.** $-5\sqrt{2}$ **13.** $4i$ **15.** $i\sqrt{3}$ **17.** $2\sqrt{2}$ **19.** $6-4i$ **21.** $-2+6i$ **23.** $-2-4i$ **25.** $18+12i$ **27.** $7$ **29.** $12-16i$ **31.** $-4i$ **33.** $\frac{28}{25}-\frac{21}{25}i$ **35.** $4+i$ **37.** $\frac{17}{13}+\frac{7}{13}i$ **39.** $63$ **41.** $2-i$ **43.** $20$ **45.** $10$ **47.** $2$ **49.** $-5-\frac{16}{3}i$ **51.** $17+144i$ **53.** $\frac{3}{5}-\frac{1}{5}i$ **55.** $5-10i$ **57.** $\frac{1}{5}-\frac{8}{5}i$ **59.** $8-i$ **61.** $1$ **63.** $i$ **65.** $-i$ **67.** $-1$ **69.** $1-i$ **71.** $0$ **73.** $2+3i$ **75.** $2+i\sqrt{2}$ **77.** $\frac{1}{2}-\frac{\sqrt{3}}{2}i$ **79.** Answers may vary. **81.** $6-6i$ **83.** yes **85.** $33°$ **87.** $5x^3-10x^2+17x-34+\frac{70}{x+2}$ **89.** $5$ **91.** $11$ **93.** $16.7\%$

***Chapter 7 Review*** **1.** $9$ **3.** $-2$ **5.** $-\frac{1}{7}$ **7.** $-6$ **9.** $-a^2b^3$ **11.** $2ab^2$ **13.** $\frac{x^6}{6y}$ **15.** $|-x|$ **17.** $-27$ **19.** $-x$ **21.** $5|(x-y)^5|$ **23.** $-x$ **25.** $(-\infty,\infty)$; $-2,-1,0,1,2$

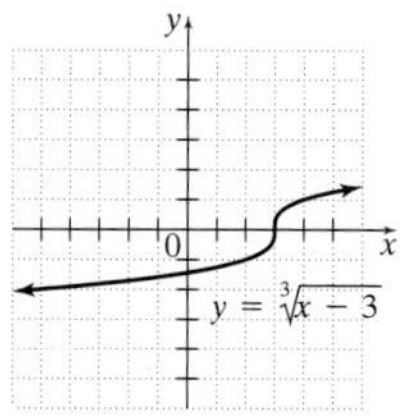


**27.** $-\frac{1}{3}$ **29.** $-\frac{1}{4}$ **31.** $\frac{1}{4}$ **33.** $\frac{343}{125}$ **35.** not a real number **37.** $5^{1/5}x^{2/5}y^{3/5}$ **39.** $5\sqrt[3]{xy^2z^5}$ **41.** $a^{13/6}$ **43.** $\frac{1}{a^{9/2}}$ **45.** $a^4b^6$ **47.** $\frac{b^{5/6}}{49a^{1/4}c^{5/3}}$ **49.** $4.472$ **51.** $5.191$ **53.** $-26.246$ **55.** $\sqrt[6]{1372}$ **57.** $2\sqrt{6}$ **59.** $2x$ **61.** $2\sqrt{15}$ **63.** $3\sqrt[3]{6}$ **65.** $6x^3\sqrt{x}$ **67.** $\frac{p^8\sqrt{p}}{11}$ **69.** $\frac{y\sqrt[4]{xy^2}}{3}$ **71. a.** $\frac{5}{\sqrt{\pi}}$ m or $\frac{5\sqrt{\pi}}{\pi}$ m **b.** 5.75 in. **73.** $xy\sqrt{2y}$ **75.** $3a\sqrt[4]{2a}$ **77.** $\frac{3\sqrt{2}}{4x}$ **79.** $-4ab\sqrt[4]{2b}$ **81.** $x-6\sqrt{x}+9$ **83.** $4x-9y$ **85.** $\sqrt[3]{a^2}+4\sqrt[3]{a}+4$ **87.** $a+64$ **89.** $\frac{\sqrt{3x}}{6}$ **91.** $\frac{2x^2\sqrt{2x}}{y}$ **93.** $-\frac{10+5\sqrt{7}}{3}$ **95.** $-5+2\sqrt{6}$ **97.** $\frac{6}{\sqrt{2y}}$ **99.** $\frac{4x^3}{y\sqrt{2x}}$ **101.** $\frac{x-25}{-3\sqrt{x}+15}$ **103.** $\varnothing$ **105.** $\varnothing$ **107.** $16$ **109.** $\sqrt{241}$ **111.** 4.24 ft **113.** $-i\sqrt{6}$ **115.** $-\sqrt{10}$ **117.** $-13-3i$ **119.** $10+4i$ **121.** $1+5i$ **123.** $87$ **125.** $-\frac{1}{3}+\frac{1}{3}i$

***Chapter 7 Test*** **1.** $6\sqrt{6}$ **2.** $-x^{16}$ **3.** $\frac{1}{5}$ **4.** $5$ **5.** $\frac{4x^2}{9}$ **6.** $-a^6b^3$ **7.** $\frac{8a^{1/3}c^{2/3}}{b^{5/12}}$ **8.** $a^{7/12}-a^{7/3}$ **9.** $|4xy|$ or $4|xy|$ **10.** $-27$ **11.** $\frac{3\sqrt{y}}{y}$ **12.** $\frac{8-6\sqrt{x}+x}{8-2x}$ **13.** $\frac{\sqrt[3]{b^2}}{b}$ **14.** $\frac{6-x^2}{8(\sqrt{6}-x)}$ **15.** $-x\sqrt{5x}$ **16.** $4\sqrt{3}-\sqrt{6}$ **17.** $x+2\sqrt{x}+1$ **18.** $\sqrt{6}-4\sqrt{3}+\sqrt{2}-4$ **19.** $-20$ **20.** $23.685$ **21.** $0.019$ **22.** $2, 3$ **23.** $\varnothing$ **24.** $6$ **25.** $i\sqrt{2}$ **26.** $-2i\sqrt{2}$ **27.** $-3i$ **28.** $40$ **29.** $7+24i$ **30.** $-\frac{3}{2}+\frac{5}{2}i$ **31.** $x=\frac{5\sqrt{2}}{2}$ **32.** $[-2,\infty)$;

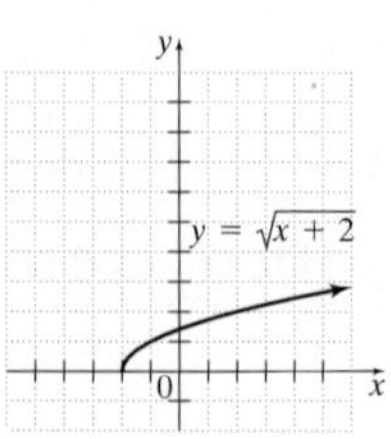


; $0, 1, 2, 3$ **33.** 27 mph **34.** 360 ft

***Chapter 7 Cumulative Review*** **1. a.** $2xy - 2$ **b.** $2x^2 + 23$ **c.** $3.1x - 0.3$; Sec. 1.4, Ex. 12 **2.** $\frac{21}{11}$; Sec. 2.1, Ex. 6 **3.** \$4500 per month; Sec. 2.4, Ex. 8 **4.** $\varnothing$; Sec. 2.6, Ex. 6 **5.** $(-\infty, -3] \cup [9, \infty)$; Sec. 2.7, Ex. 7

**6.** 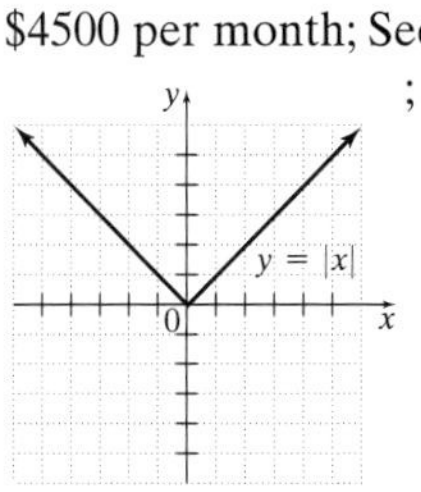

; Sec. 3.1, Ex. 7 **7. a.** domain: $\{2, 0, 3\}$; range: $\{3, 4, -1\}$ **b.** domain: $\{-4, -3, -2, -1, 0, 1, 2, 3\}$; range: $\{1\}$ **c.** domain: {Erie, Escondido, Gary, Miami, Waco}; range: $\{104, 109, 117, 359\}$; Sec. 3.2, Ex. 1

**8.** 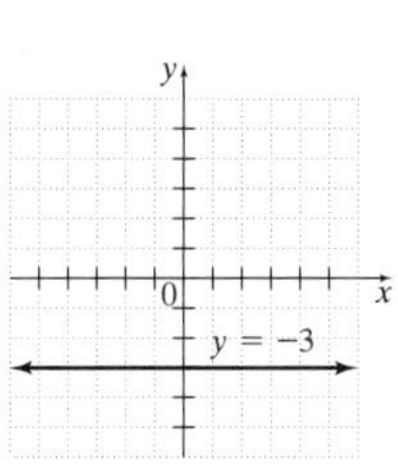

; Sec. 3.3, Ex. 7 **9.** udefined; Sec. 3.4, Ex. 6 **10.** $\left(-\frac{21}{10}, \frac{3}{10}\right)$; Sec. 4.1, Ex. 4 **11. a.** $2^7$ **b.** $x^{10}$ **c.** $y^7$; Sec. 5.1, Ex. 1 **12.** $6 \times 10^{-5}$; Sec. 5.2, Ex. 7 **13. a.** $-4$ **b.** 11; Sec. 5.3, Ex. 4 **14. a.** $2x^2 + 11x + 15$ **b.** $10x^3 - 27x^2 + 32x - 21$; Sec. 5.4, Ex. 3 **15.** $5x^2$; Sec. 5.5, Ex. 1 **16. a.** $x^2 - 2x + 4$ **b.** $\frac{2}{y - 5}$; Sec. 6.1, Ex. 5

**17. a.** $\frac{6x + 5}{3x^3y}$ **b.** $\frac{5x^2 - 2x}{(x + 2)(x - 2)}$ **c.** $\frac{x + 4}{x - 1}$; Sec. 6.2, Ex. 3 **18. a.** $\frac{x(x - 2)}{2(x + 2)}$ **b.** $\frac{x^2}{y^2}$; Sec. 6.3, Ex. 2 **19.** $2x^2 - x + 4$; Sec. 6.4, Ex. 1 **20.** $2x^2 + 5x + 2 + \frac{7}{x - 3}$; Sec. 6.5, Ex. 1 **21.** $\varnothing$; Sec. 6.6, Ex. 3 **22.** $x = \frac{yz}{y - z}$; Sec. 6.7, Ex. 1 **23.** constant of variation: 15; $u = \frac{15}{w}$; Sec. 6.8, Ex. 3 **24. a.** $\frac{1}{8}$ **b.** $\frac{1}{9}$; Sec. 7.2, Ex. 3 **25.** $\frac{x - 4}{5(\sqrt{x} - 2)}$; Sec. 7.5, Ex. 7

## ■ CHAPTER 8 QUADRATIC EQUATIONS AND FUNCTIONS

***Graphing Calculator Explorations***

**1.** $-1.27, 6.27$

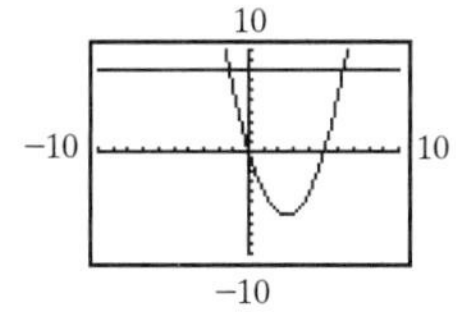


**3.** $-1.10, 0.90$

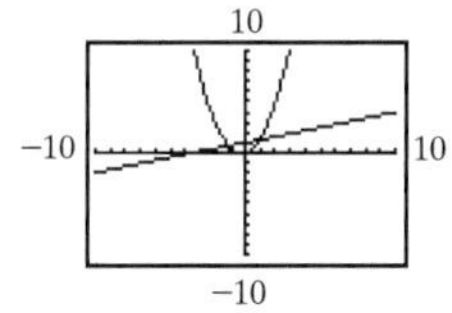


**5.** $\varnothing$

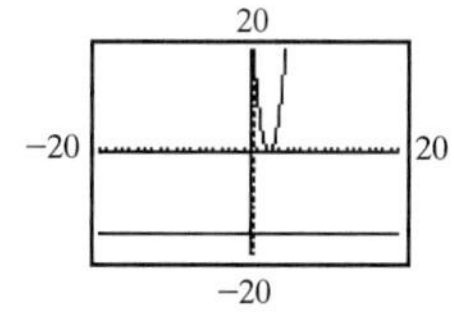


***Exercise Set 8.1*** **1.** $-4, 4$ **3.** $-\sqrt{7}, \sqrt{7}$ **5.** $-3\sqrt{2}, 3\sqrt{2}$ **7.** $-\sqrt{10}, \sqrt{10}$ **9.** $-8, -2$ **11.** $6 - 3\sqrt{2}, 6 + 3\sqrt{2}$ **13.** $\frac{3 - 2\sqrt{2}}{2}, \frac{3 + 2\sqrt{2}}{2}$ **15.** $-3i, 3i$ **17.** $-\sqrt{6}, \sqrt{6}$ **19.** $-2i\sqrt{2}, 2i\sqrt{2}$ **21.** $1 - 4i, 1 + 4i$ **23.** $-7 - \sqrt{5}, -7 + \sqrt{5}$ **25.** $-3 - 2i\sqrt{2}, -3 + 2i\sqrt{2}$ **27.** $x^2 + 16x + 64 = (x + 8)^2$ **29.** $z^2 - 12z + 36 = (z - 6)^2$ **31.** $p^2 + 9p + \frac{81}{4} = \left(p + \frac{9}{2}\right)^2$ **33.** $x^2 + x + \frac{1}{4} = \left(x + \frac{1}{2}\right)^2$ **35.** $-8x, 8x$ **37.** $-5z, 5z$ **39.** $-5, -3$ **41.** $-3 - \sqrt{7}, -3 + \sqrt{7}$ **43.** $\frac{-1 - \sqrt{5}}{2}, \frac{-1 + \sqrt{5}}{2}$ **45.** $-1 - \sqrt{6}, -1 + \sqrt{6}$ **47.** $\frac{6 - \sqrt{30}}{3}, \frac{6 + \sqrt{30}}{3}$ **49.** $\frac{3 - \sqrt{11}}{2}, \frac{3 + \sqrt{11}}{2}$ **51.** $-4, \frac{1}{2}$ **53.** $-1, 5$ **55.** $-4 - \sqrt{15}, -4 + \sqrt{15}$ **57.** $\frac{-3 - \sqrt{21}}{3}, \frac{-3 + \sqrt{21}}{3}$ **59.** $-1, \frac{5}{2}$

**61.** $-1 - i, -1 + i$ **63.** $3 - \sqrt{6}, 3 + \sqrt{6}$ **65.** $-2 - i\sqrt{2}, -2 + i\sqrt{2}$ **67.** $\frac{-15 - 7\sqrt{5}}{10}, \frac{-15 + 7\sqrt{5}}{10}$
**69.** $\frac{1 - i\sqrt{47}}{4}, \frac{1 + i\sqrt{47}}{4}$ **71.** $-5 - i\sqrt{3}, -5 + i\sqrt{3}$ **73.** $-4, 1$ **75.** $\frac{2 - i\sqrt{2}}{2}, \frac{2 + i\sqrt{2}}{2}$ **77.** $\frac{-3 - \sqrt{69}}{6}, \frac{-3 + \sqrt{69}}{6}$
**79.** 20% **81.** 11% **83.** Answers may vary. **85.** simple **87.** 8.11 sec **89.** 6.73 sec **91.** 6 in. **93.** $\frac{27\sqrt{2}}{2}$ in.
**95.** 2.828 thousand units **97.** $\frac{7}{5}$ **99.** $\frac{1}{5}$ **101.** $5 - 10\sqrt{3}$ **103.** $\frac{3 - 2\sqrt{7}}{4}$ **105.** $2\sqrt{7}$ **107.** $\sqrt{13}$

***Exercise Set 8.2*** **1.** $-6, 1$ **3.** $-\frac{3}{5}, 1$ **5.** 3 **7.** $\frac{-7 - \sqrt{33}}{2}, \frac{-7 + \sqrt{33}}{2}$ **9.** $\frac{1 - \sqrt{57}}{8}, \frac{1 + \sqrt{57}}{8}$
**11.** $\frac{7 - \sqrt{85}}{6}, \frac{7 + \sqrt{85}}{6}$ **13.** $1 - \sqrt{3}, 1 + \sqrt{3}$ **15.** $-\frac{3}{2}, 1$ **17.** $\frac{3 - \sqrt{11}}{2}, \frac{3 + \sqrt{11}}{2}$ **19.** Answers may vary.
**21.** $\frac{3 - i\sqrt{87}}{8}, \frac{3 + i\sqrt{87}}{8}$ **23.** $-2 - \sqrt{11}, -2 + \sqrt{11}$ **25.** $\frac{-5 - i\sqrt{5}}{10}, \frac{-5 + i\sqrt{5}}{10}$ **27.** Answers may vary.
**29.** two real solutions **31.** one real solution **33.** two real solutions **35.** two complex but not real solutions
**37.** $\frac{-5 - \sqrt{17}}{2}, \frac{-5 + \sqrt{17}}{2}$ **39.** $\frac{5}{2}, 1$ **41.** $\frac{3 - \sqrt{29}}{2}, \frac{3 + \sqrt{29}}{2}$ **43.** $\frac{-1 - \sqrt{19}}{6}, \frac{-1 + \sqrt{19}}{6}$
**45.** $-3 - 2i, -3 + 2i$ **47.** $\frac{-1 - i\sqrt{23}}{4}, \frac{-1 + i\sqrt{23}}{4}$ **49.** 1 **51.** $\frac{19 - \sqrt{345}}{2}, \frac{19 + \sqrt{345}}{2}$ **53.** 14 ft
**55.** $2 + 2\sqrt{2}$ cm, $2 + 2\sqrt{2}$ cm, $4 + 2\sqrt{2}$ cm **57.** width: $-5 + 5\sqrt{17}$ ft; length: $5 + 5\sqrt{17}$ ft
**59.** **a.** $50\sqrt{2}$ m **b.** 5000 sq. m **61.** $\frac{1 + \sqrt{5}}{2}$ **63.** 8.9 sec **65.** 2.8 sec **67.** Sunday to Monday. **69.** Wednesday
**71.** 32; yes **73.** **a.** \$3056 million **b.** 2005
**75.** 8.9 sec: 1200 ; 2.8 sec: 200

0 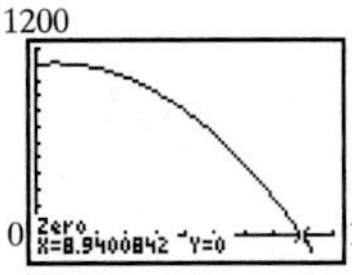 10

−100

0 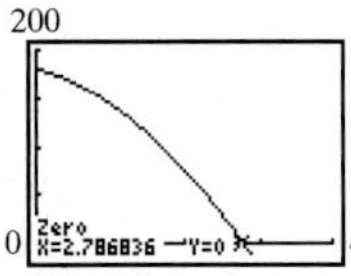 4

−10

**77.** two real solutions **79.** $\frac{11}{5}$ **81.** 15 **83.** $(x^2 + 5)(x + 2)(x - 2)$ **85.** $(z + 3)(z - 3)(z + 2)(z - 2)$
**87.** $\frac{\sqrt{3}}{3}$ **89.** $\frac{-\sqrt{2} - i\sqrt{2}}{2}, \frac{-\sqrt{2} + i\sqrt{2}}{2}$ **91.** $\frac{\sqrt{3} - \sqrt{11}}{4}, \frac{\sqrt{3} + \sqrt{11}}{4}$

***Exercise Set 8.3*** **1.** 2 **3.** 16 **5.** 1, 4 **7.** $3 - \sqrt{7}, 3 + \sqrt{7}$ **9.** $\frac{3 - \sqrt{57}}{4}, \frac{3 + \sqrt{57}}{4}$ **11.** $\frac{1 - \sqrt{29}}{2}, \frac{1 + \sqrt{29}}{2}$
**13.** $-2, 2, -2i, 2i$ **15.** $-\frac{1}{2}, \frac{1}{2}, -i\sqrt{3}, i\sqrt{3}$ **17.** $-3, 3, -2, 2$ **19.** $125, -8$ **21.** $-\frac{4}{5}, 0$ **23.** $-\frac{1}{8}, 27$ **25.** $-\frac{2}{3}, \frac{4}{3}$
**27.** $-\frac{1}{125}, \frac{1}{8}$ **29.** $-\sqrt{2}, \sqrt{2}, -\sqrt{3}, \sqrt{3}$ **31.** $\frac{-9 - \sqrt{201}}{6}, \frac{-9 + \sqrt{201}}{6}$ **33.** 2, 3 **35.** 3 **37.** 27, 125 **39.** $1, -3i, 3i$
**41.** $\frac{1}{8}, -8$ **43.** $-\frac{1}{2}, \frac{1}{3}$ **45.** 4 **47.** $-3$ **49.** $-\sqrt{5}, \sqrt{5}, -2i, 2i$ **51.** $-3, \frac{3 - 3i\sqrt{3}}{2}, \frac{3 + 3i\sqrt{3}}{2}$ **53.** 6, 12
**55.** $-\frac{1}{3}, \frac{1}{3}, -\frac{i\sqrt{6}}{3}, \frac{i\sqrt{6}}{3}$ **57.** 5 mph, then 4 mph **59.** inlet pipe, 15.5 hr; hose, 16.5 hr **61.** 55 mph, 66 mph
**63.** 8.5 hr. **65.** 12 or $-8$ **67.** **a.** $x - 6$ **b.** $300 = (x - 6) \cdot (x - 6) \cdot 3$ **c.** 16 cm by 16 cm
**69.** **a.** 281.0 ft per sec **b.** 284.8 ft per sec **c.** Papis: 191.6 mph; Montoya: 194.2 mph **71.** Answers may vary.
**73.** $(-\infty, 3]$ **75.** $(-5, \infty)$ **77.** domain: $\{x|x$ is a real number$\}$; range: $\{y|y$ is a real number$\}$; function
**79.** domain: $\{x|x$ is a real number$\}$; range: $\{y|y \geq -1\}$; function

***Exercise Set 8.4*** **1.** 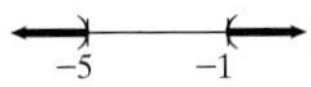

; $(-\infty, -5) \cup (-1, \infty)$ **3.** 

; $[-4, 3]$ **5.** 

; $[2, 5]$

**7.** 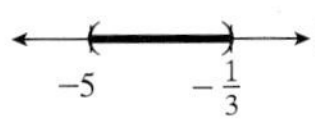

; $\left(-5, -\frac{1}{3}\right)$ **9.** 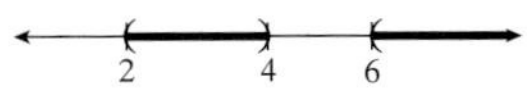

; $(2, 4) \cup (6, \infty)$

**11.** 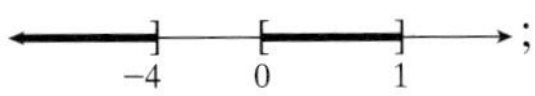

; $(-\infty, -4] \cup [0, 1]$ **13.** 

; $(-\infty, -3) \cup (-2, 2) \cup (3, \infty)$

**15.** 

; $(-7, 2)$ **17.** 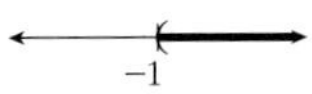

; $(-1, \infty)$ **19.** 

; $(-\infty, -1] \cup (4, \infty)$

**21.** Answers may vary. **23.** 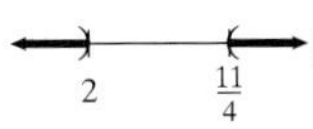

; $(-\infty, 2) \cup \left(\frac{11}{4}, \infty\right)$ **25.** 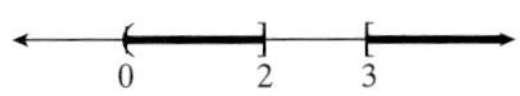

; $(0, 2] \cup [3, \infty)$

**27.** 

; $(-\infty, -7) \cup (8, \infty)$ **29.** 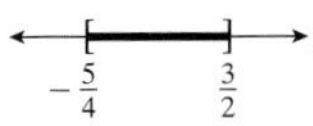

; $\left[-\frac{5}{4}, \frac{3}{2}\right]$ **31.** 

; $(-\infty, 0) \cup (1, \infty)$

**33.** 

; $(-\infty, -4] \cup [4, 6]$ **35.** 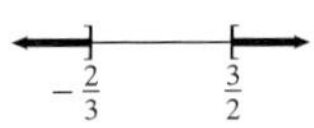

; $\left(-\infty, -\frac{2}{3}\right] \cup \left[\frac{3}{2}, \infty\right)$

**37.** 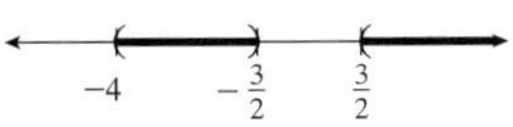

; $\left(-4, -\frac{3}{2}\right) \cup \left(\frac{3}{2}, \infty\right)$ **39.** 

; $(-\infty, -5] \cup [-1, 1] \cup [5, \infty)$

**41.** 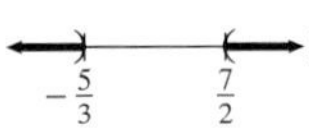

; $\left(-\infty, -\frac{5}{3}\right) \cup \left(\frac{7}{2}, \infty\right)$ **43.** 

; $(0, 10)$ **45.** 

; $(-\infty, -4) \cup [5, \infty)$

**47.** 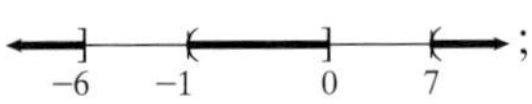

; $(-\infty, -6] \cup (-1, 0] \cup (7, \infty)$ **49.** 

; $(-\infty, 1) \cup (2, \infty)$

**51.** 

; $(-\infty, -8] \cup (-4, \infty)$ **53.** 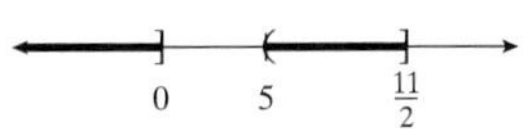

; $(-\infty, 0] \cup \left(5, \frac{11}{2}\right]$

**55.** 

; $(0, \infty)$ **57.** Any number less than −1 or between 0 and 1 **59.** $x$ is between 2 and 11

**61.** 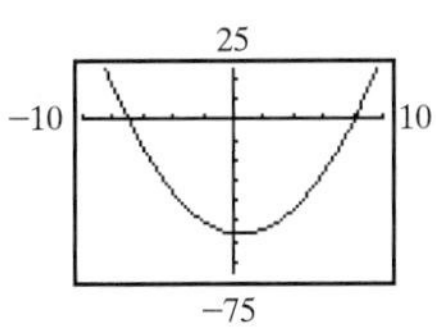


**63.** 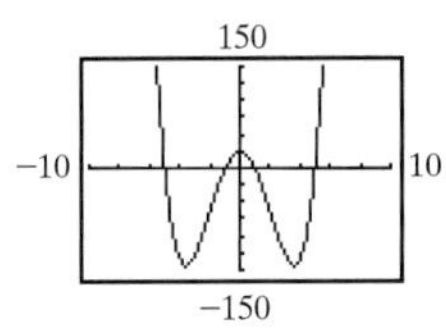


**65.** 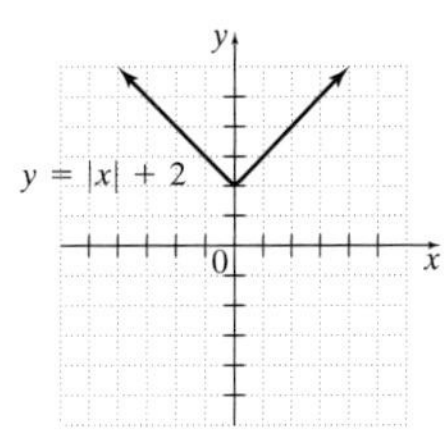


**67.** 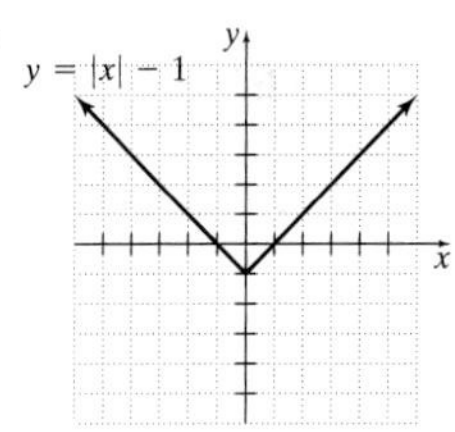


**69.** 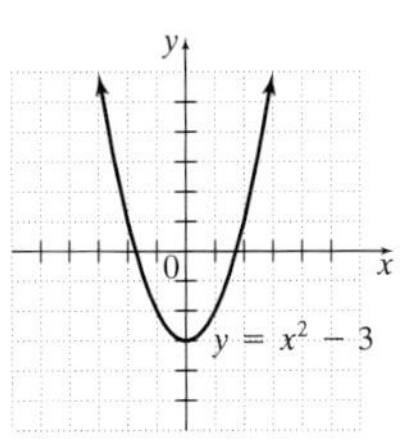


**71.** 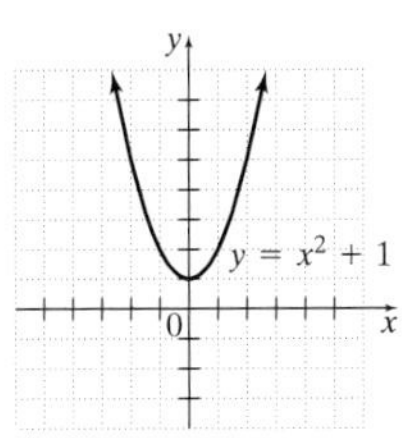


***Graphing Calculator Explorations*** **1.** 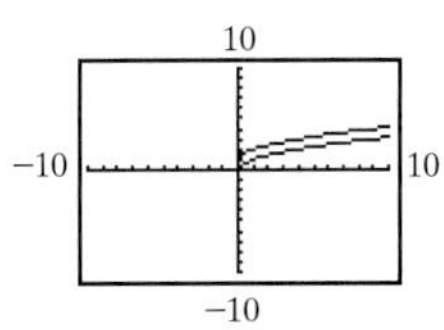


**3.** 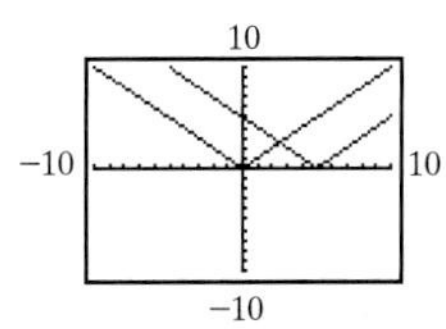


**5.** 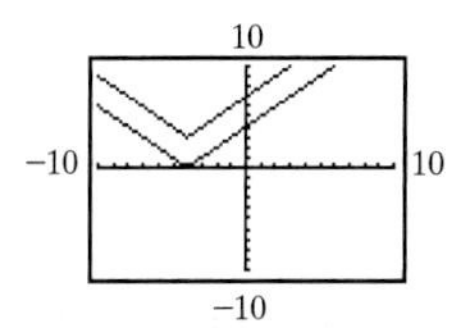


***Mental Math*** **1.** $(0, 0)$ **3.** $(2, 0)$ **5.** $(0, 3)$ **7.** $(-1, 5)$

***Exercise Set 8.5***

**1.** 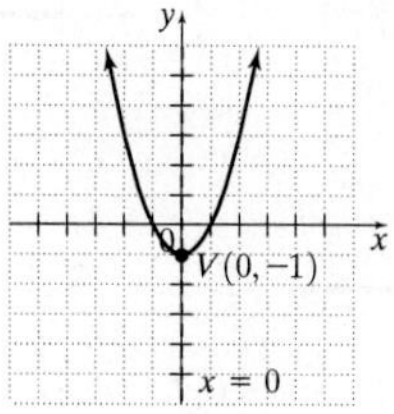


**3.** 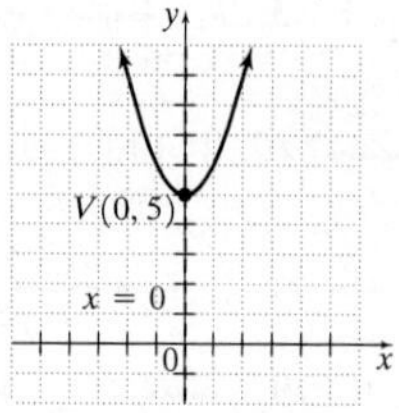


**5.** 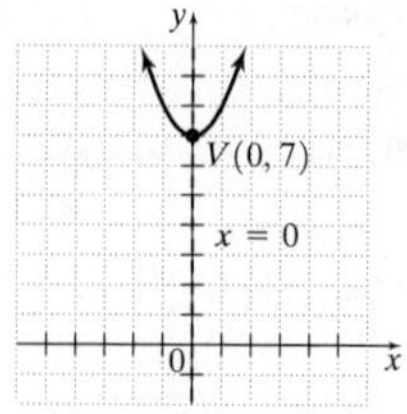


**7.** 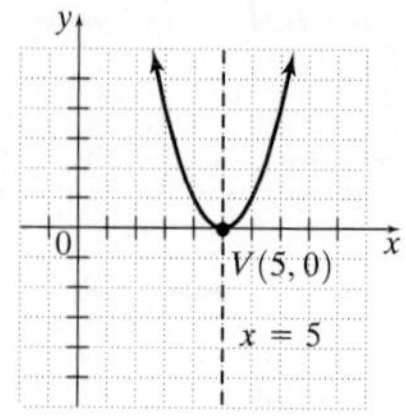


**9.** 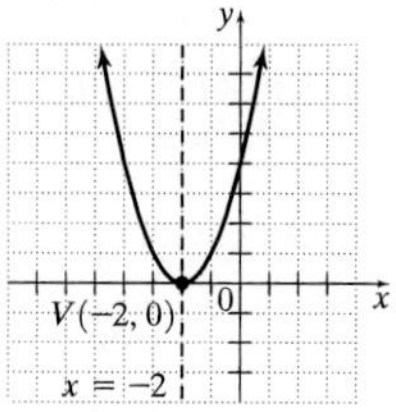


**11.** 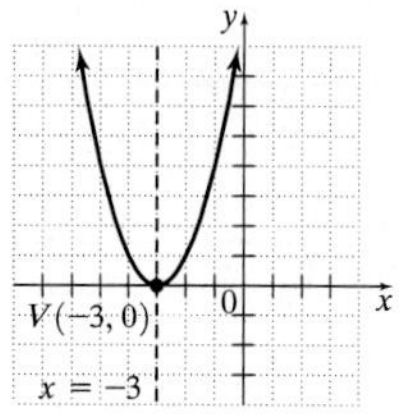


**13.** 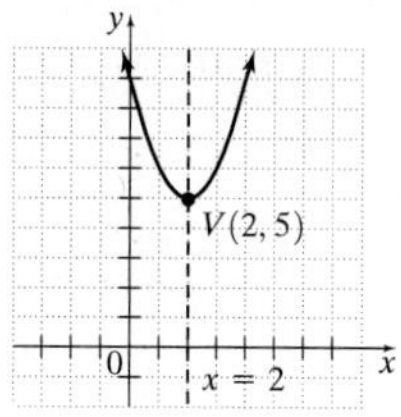


**15.** 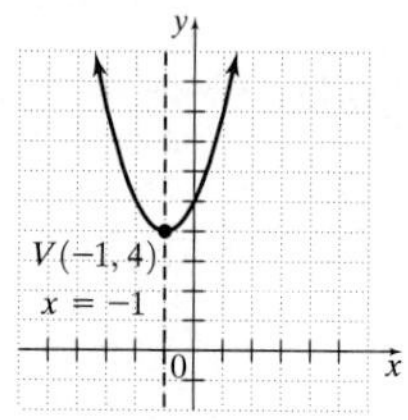


**17.** 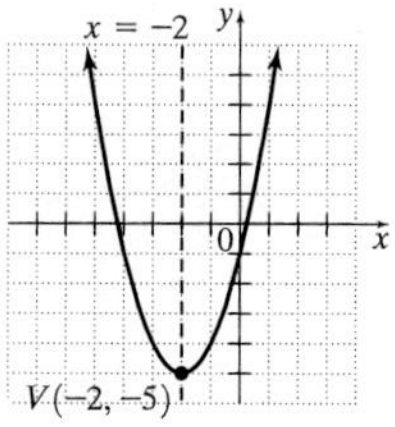


**19.** 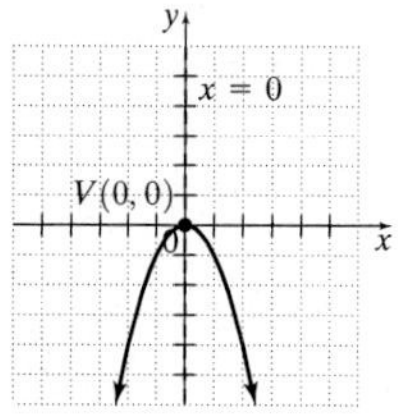


**21.** 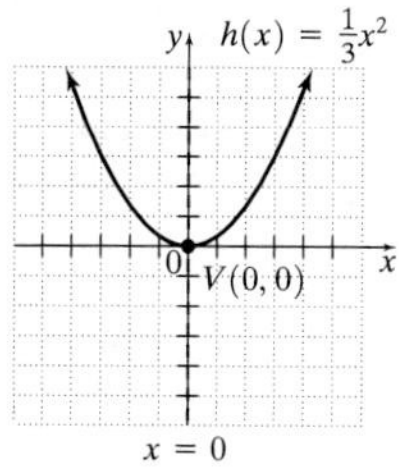


**23.** 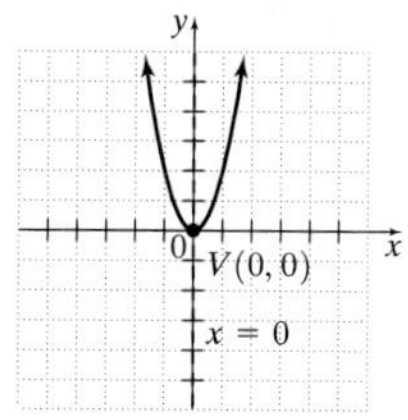


**25.** 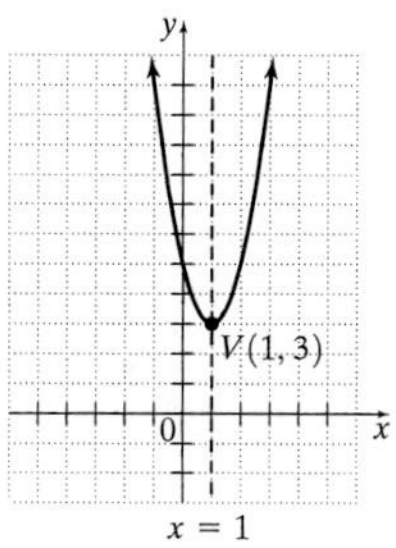


**27.** 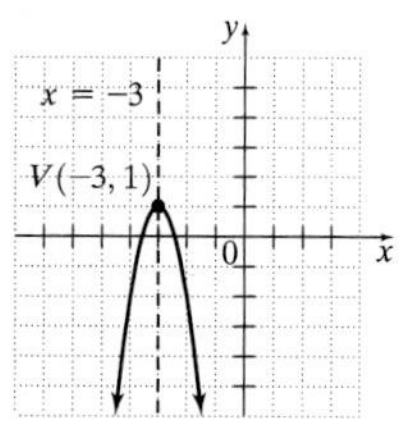


**29.** 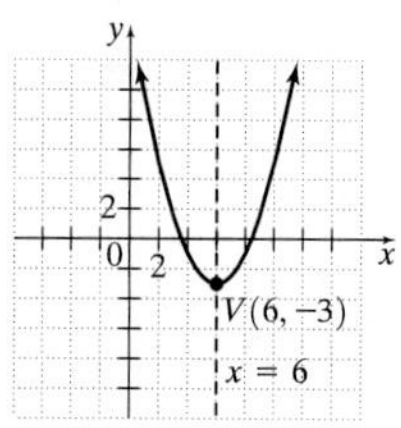


**31.** 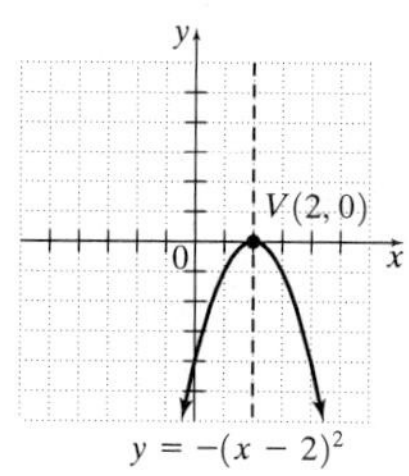


**33.** 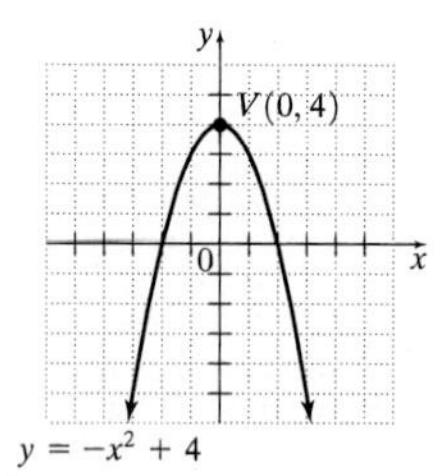


**35.** 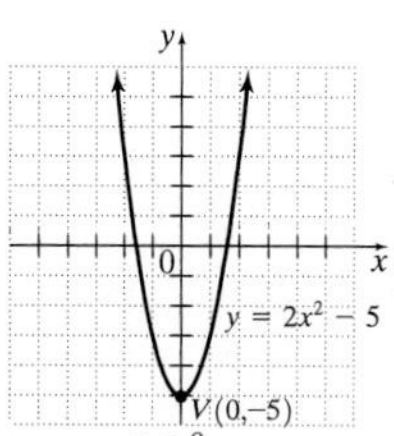


**37.** 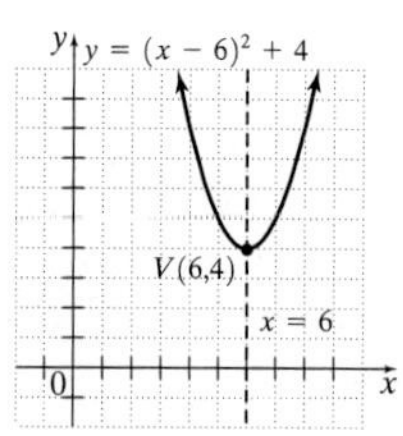


**39.** 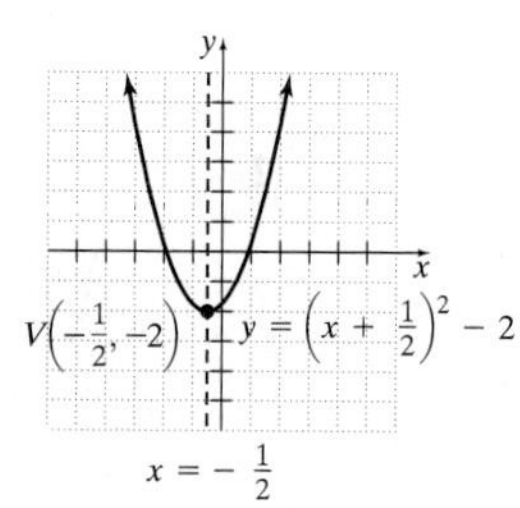


**41.** 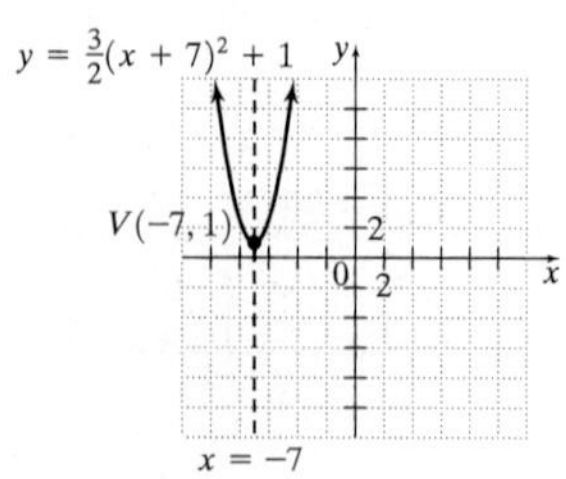


**43.** 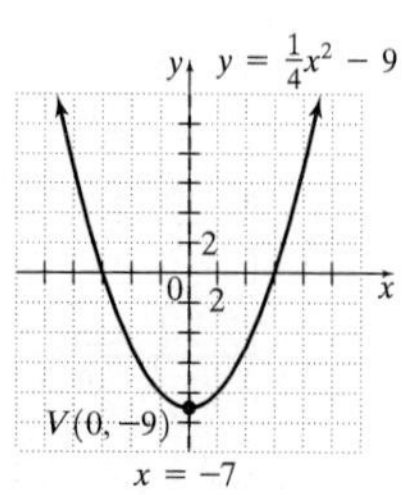


**45.** 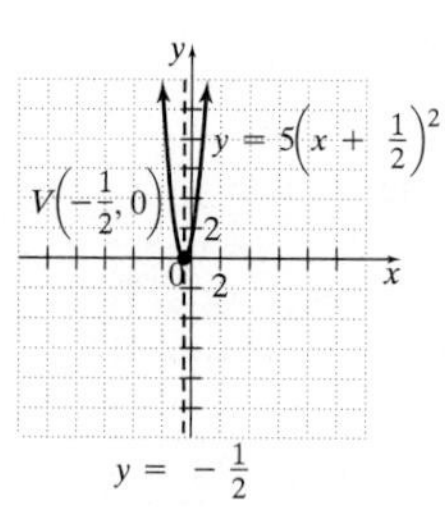


**47.** 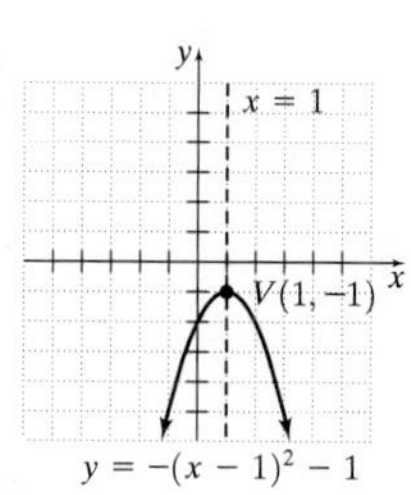

**49.**
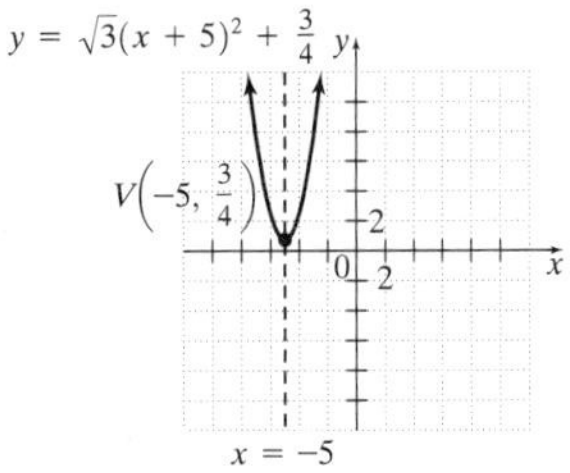


**51.**
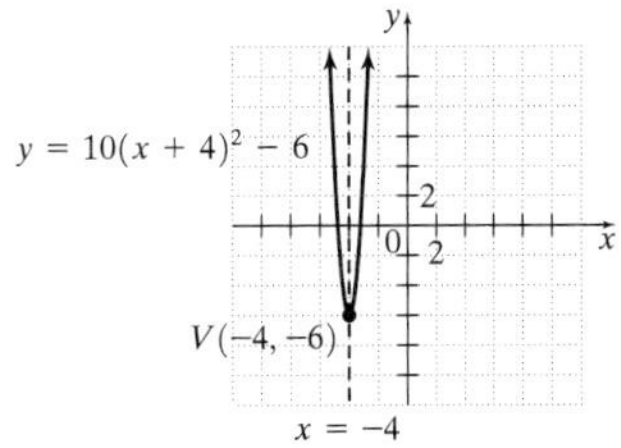


**53.**
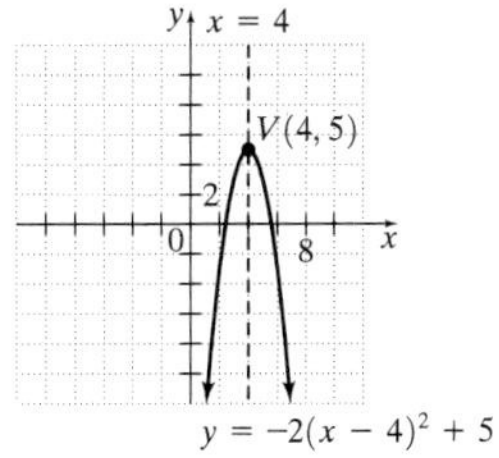


**55.** $f(x) = 5(x - 2)^2 + 3$ **57.** $f(x) = 5(x + 3)^2 + 6$

**59.**
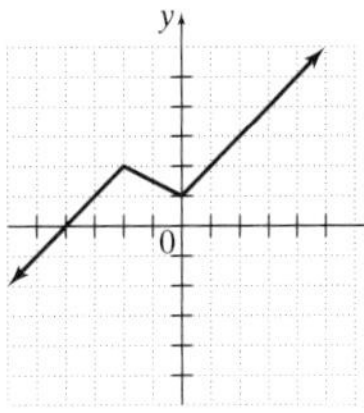

**61.**
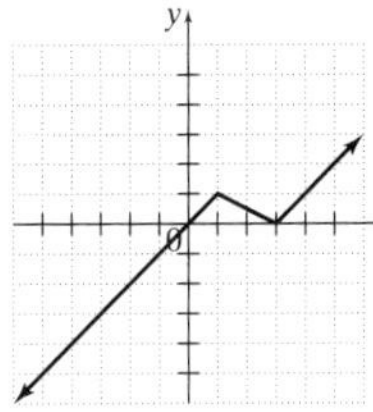

**63.**
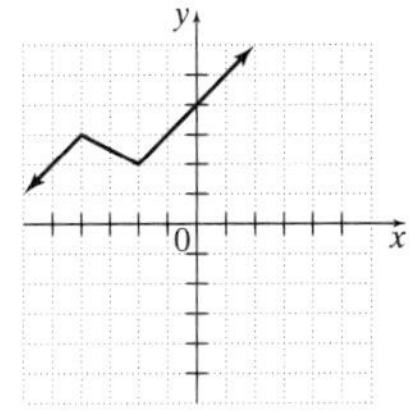

**65.** $x^2 + 8x + 16$
**67.** $z^2 - 16z + 64$
**69.** $y^2 + y + \frac{1}{4}$
**71.** $-6, 2$
**73.** $-5 - \sqrt{26}, -5 + \sqrt{26}$
**75.** $4 - 3\sqrt{2}, 4 + 3\sqrt{2}$

***Exercise Set 8.6*** **1.** $(-4, -9)$ **3.** $(5, 30)$ **5.** $(1, -2)$ **7.** $\left(\frac{1}{2}, \frac{5}{4}\right)$ **9.** D **11.** B

**13.**
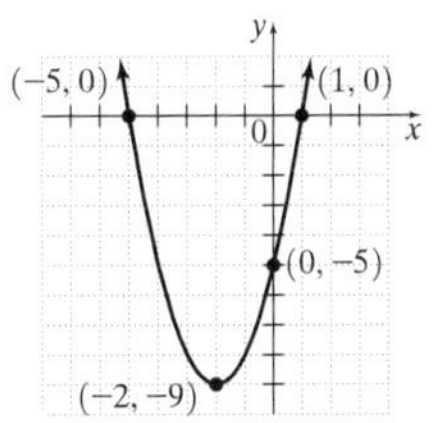


**15.**
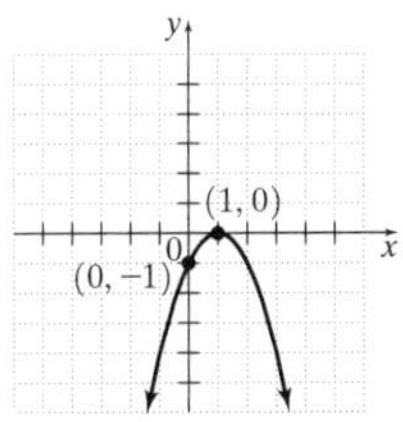


**17.**
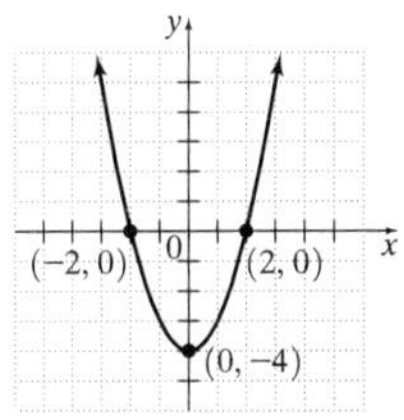


**19.**
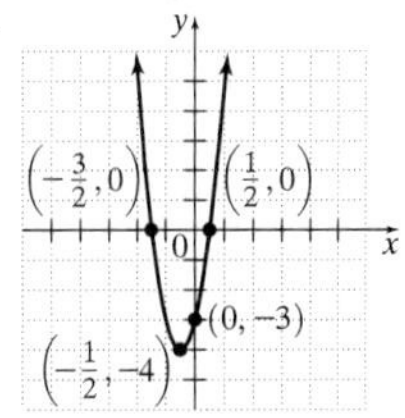


**21.**
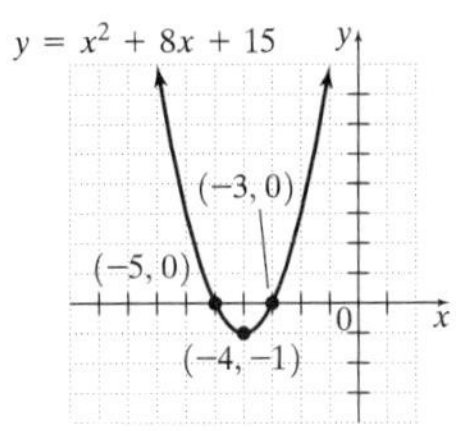


**23.**
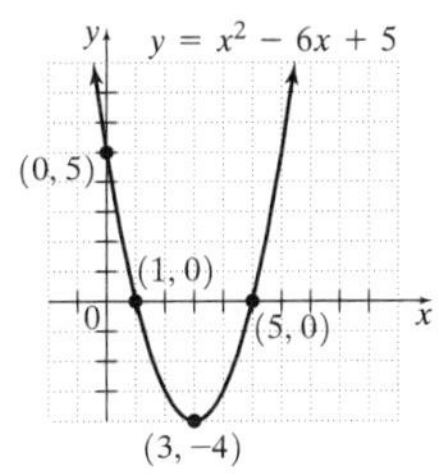


**25.**
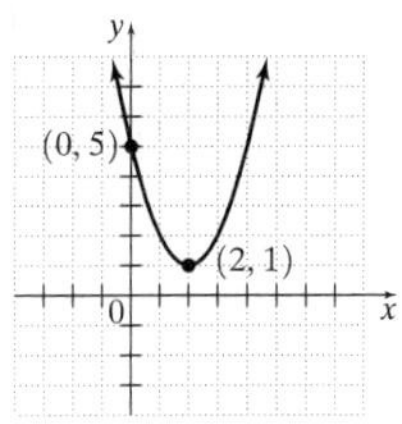


**27.**
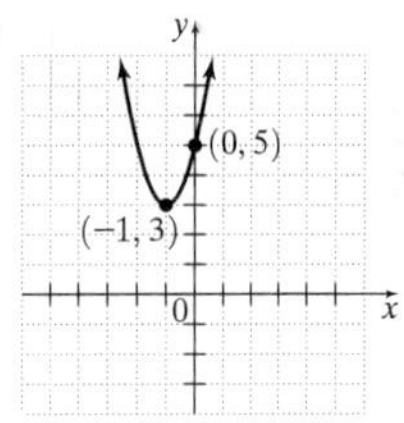


**29.**
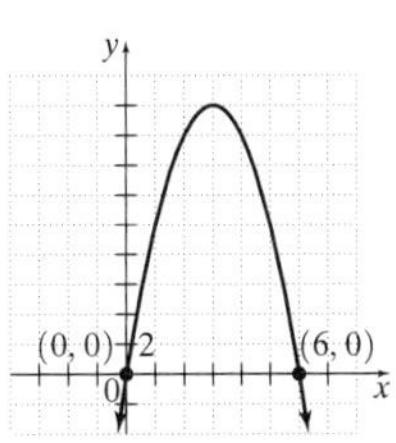


**31.**
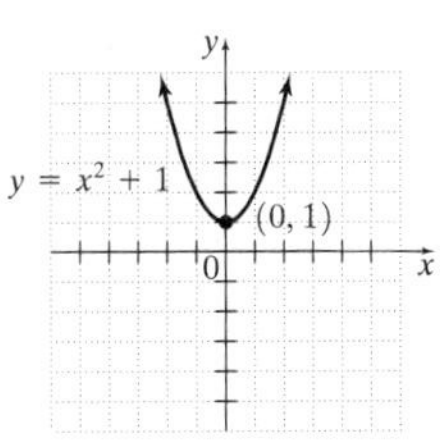


**33.**
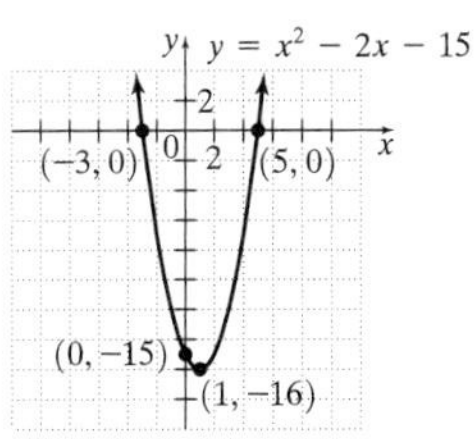


**35.**
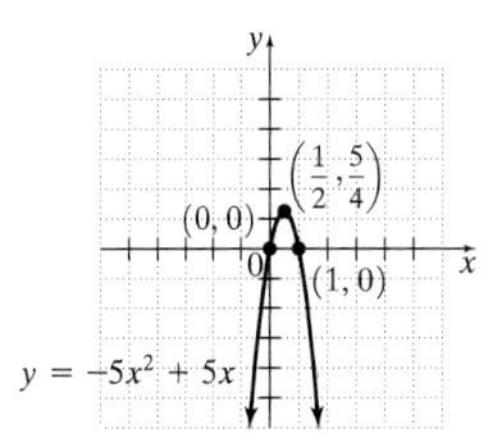


**37.**
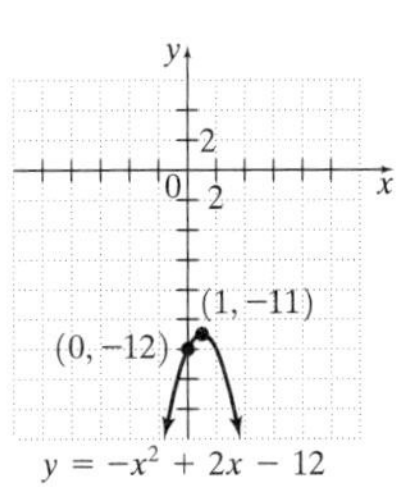


**39.**
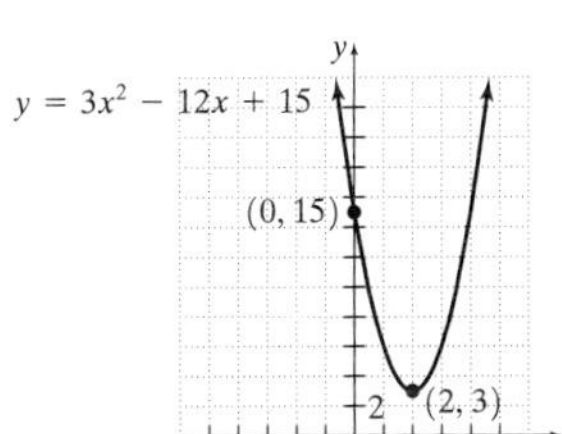


**41.**
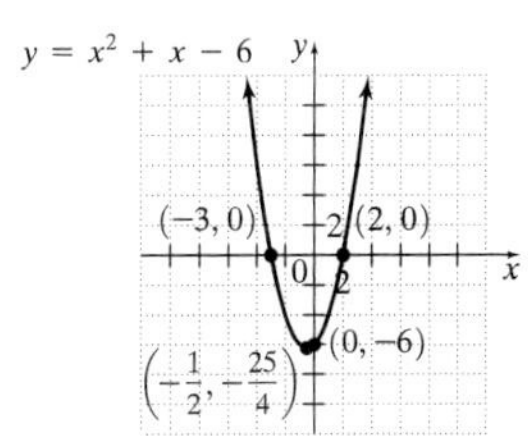


**43.**
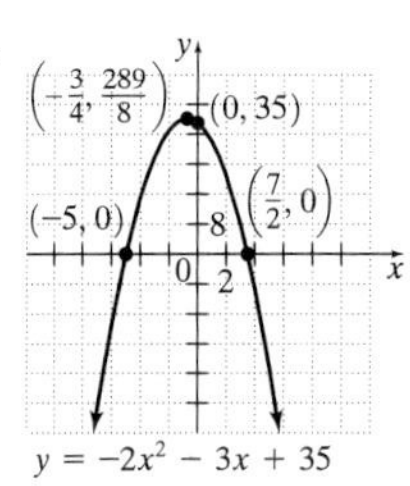


**45. a.** 200 bicycles **b.** \$12,000 **47.** 16 ft **49.** 30 and 30 **51.** 5, −5 **53.** length, 20 units; width, 20 units
**55. a.** 135.27 million metric tons **b.** 1996 **c.** 184.39 million metric tons

**57.** 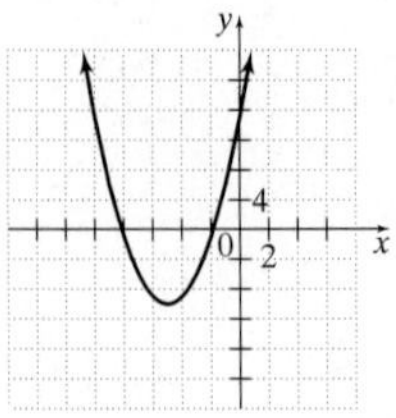


**59.** 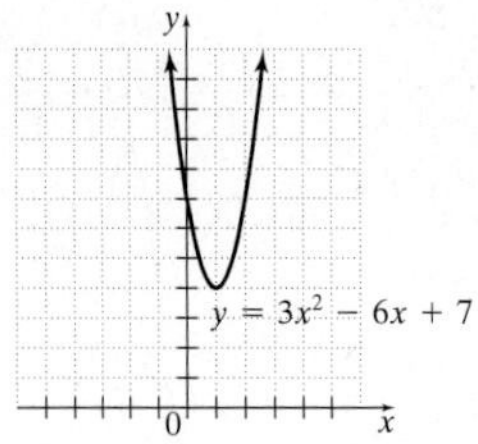


**61.** 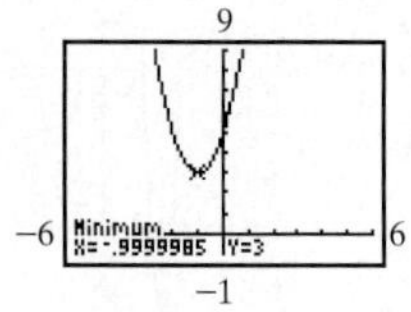


**63.** 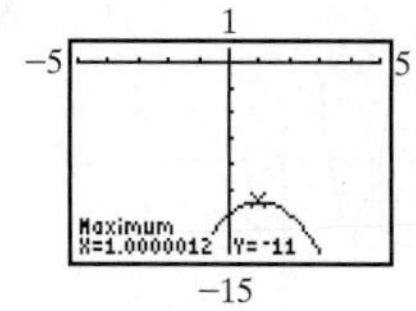


**65.** −0.84 **67.** 1.43 **69.** 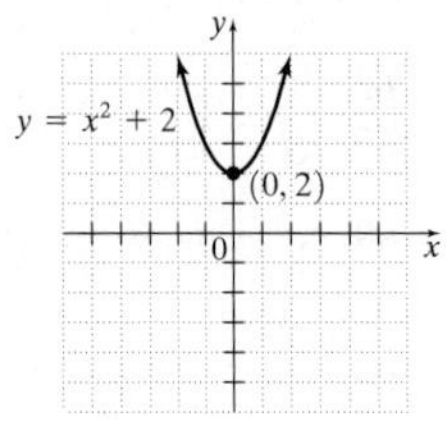


**71.** 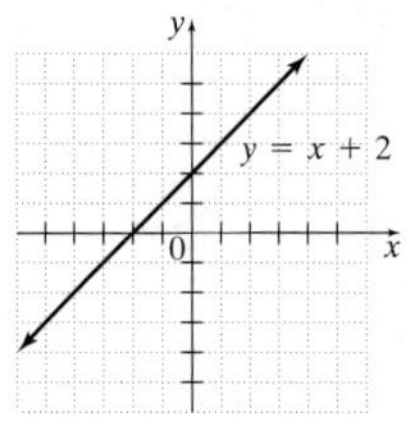


**73.** 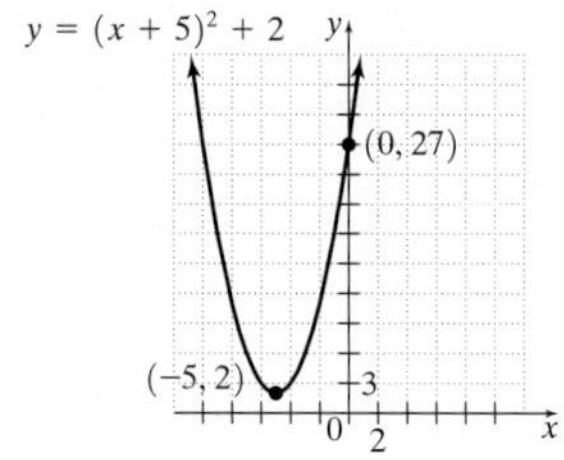


**75.** 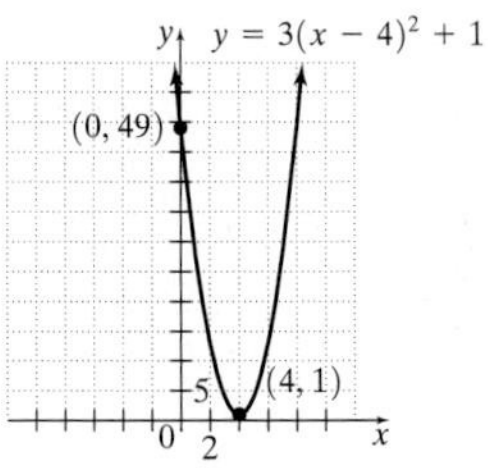


**77.** 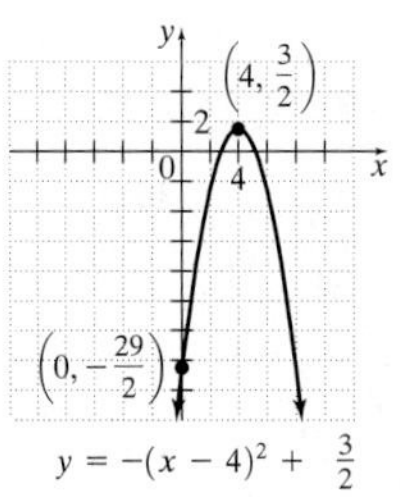


***Chapter 8 Review*** **1.** 14, 1 **3.** $\frac{4}{5}, -\frac{1}{2}$ **5.** −7, 7 **7.** $-\frac{4}{9}, \frac{2}{9}$ **9.** $\frac{-3-\sqrt{5}}{2}, \frac{-3+\sqrt{5}}{2}$ **11.** $\frac{-3-i\sqrt{7}}{8}, \frac{-3+i\sqrt{7}}{8}$ **13.** 4.25% **15.** two complex but not real solutions **17.** two real solutions **19.** 8 **21.** $-i\sqrt{11}, i\sqrt{11}$ **23.** $\frac{5-i\sqrt{143}}{12}, \frac{5+i\sqrt{143}}{12}$ **25.** $\frac{21-\sqrt{41}}{50}, \frac{21+\sqrt{41}}{50}$ **27. a.** 20 ft **b.** $\frac{15+\sqrt{321}}{16}$ sec; 2.1 sec **29.** $3, \frac{-3+3i\sqrt{3}}{2}, \frac{-3-3i\sqrt{3}}{2}$ **31.** $\frac{2}{3}, 5$ **33.** −5, 5, −2*i*, 2*i* **35.** 1, 125 **37.** −1, 1, −*i*, *i* **39.** Jerome: 10.5 hr; Tim: 9.5 hr

**41.** 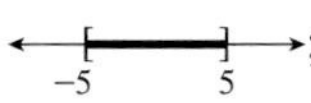

; [−5, 5] **43.** 

; $\left(-\infty, -\frac{5}{4}\right] \cup \left[\frac{3}{2}, \infty\right)$ **45.** 

; (5, 6)

**47.** 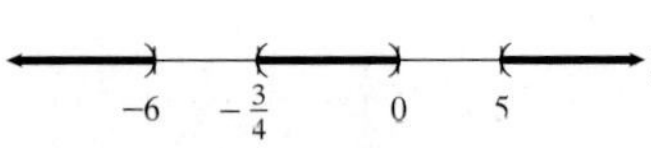

; $(-\infty, -6) \cup \left(-\frac{3}{4}, 0\right) \cup (5, \infty)$ **49.** 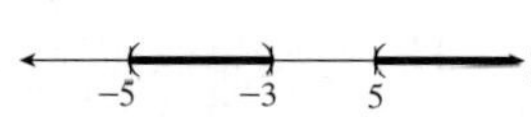

; (−5, −3) ∪ (5, ∞)

**51.** 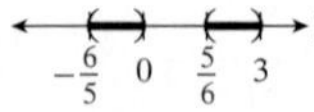

; $\left(-\frac{6}{5}, 0\right) \cup \left(\frac{5}{6}, 3\right)$ **53.** 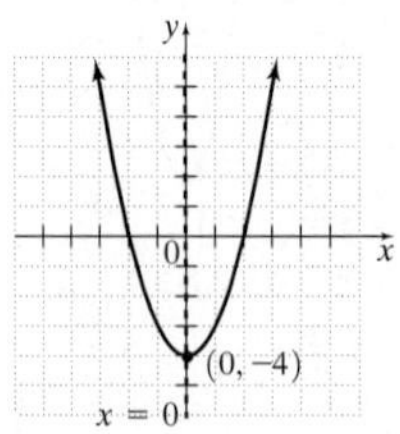


**55.** 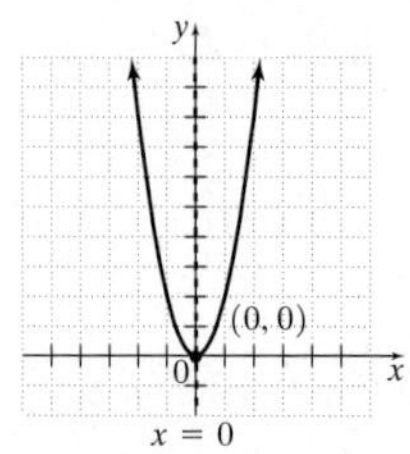


**57.** 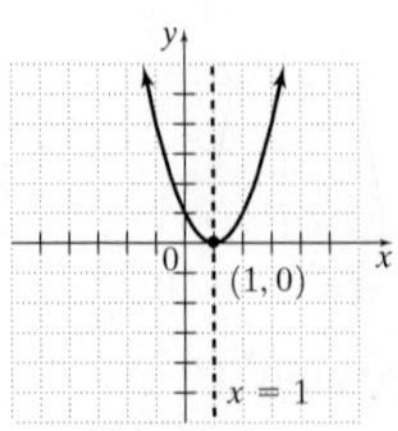


**59.** 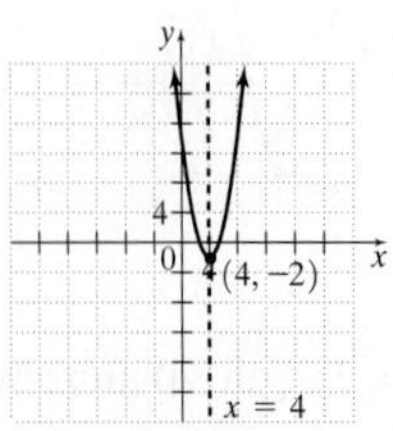


**61.** 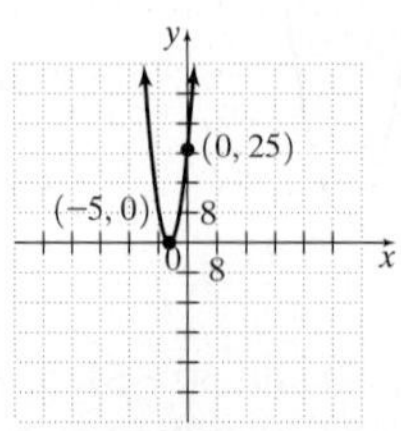


**63.** 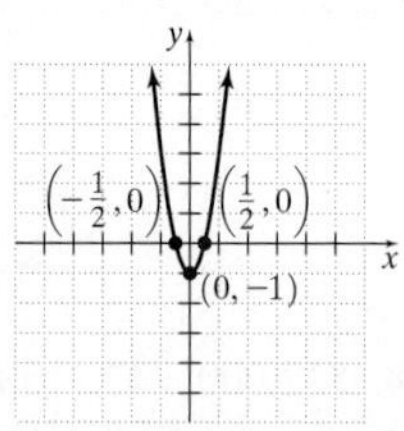

**65.** 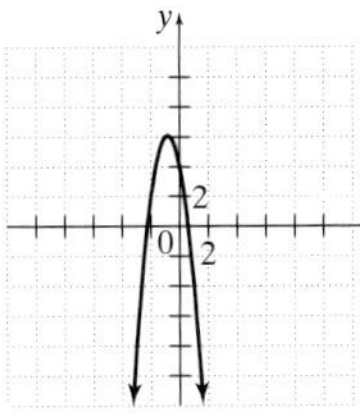

**67.** The numbers are both 210.

***Chapter 8 Test*** **1.** $\frac{7}{5}, -1$ **2.** $-1 - \sqrt{10}, -1 + \sqrt{10}$ **3.** $\frac{1 + i\sqrt{31}}{2}, \frac{1 - i\sqrt{31}}{2}$ **4.** $3 - \sqrt{7}, 3 + \sqrt{7}$ **5.** $-\frac{1}{7}, -1$ **6.** $\frac{3 + \sqrt{29}}{2}, \frac{3 - \sqrt{29}}{2}$ **7.** $-2 - \sqrt{11}, -2 + \sqrt{11}$ **8.** $-3, 3, -i, i$ **9.** $-1, 1, -i, i$ **10.** 6, 7 **11.** $3 - \sqrt{7}, 3 + \sqrt{7}$

**13.** 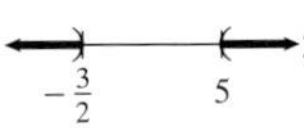 ; $\left(-\infty, -\frac{3}{2}\right) \cup (5, \infty)$ **14.**  ; $(-\infty, -5) \cup (-4, 4) \cup (5, \infty)$

**15.**  ; $(-\infty, -3) \cup (2, \infty)$ **16.** 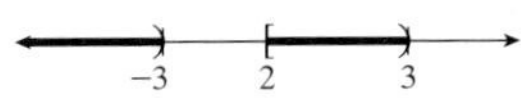 ; $(-\infty, -3) \cup [2, 3)$

**17.** 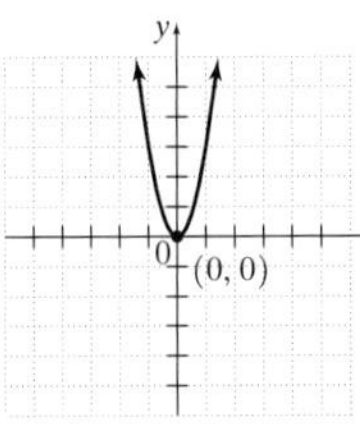

**18.** 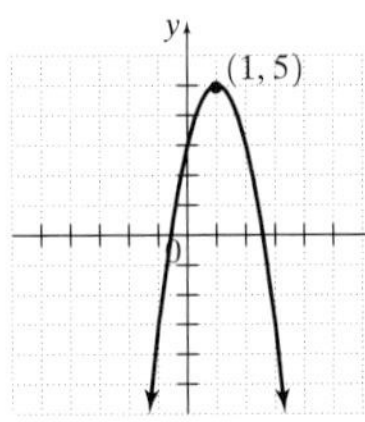

**19.** 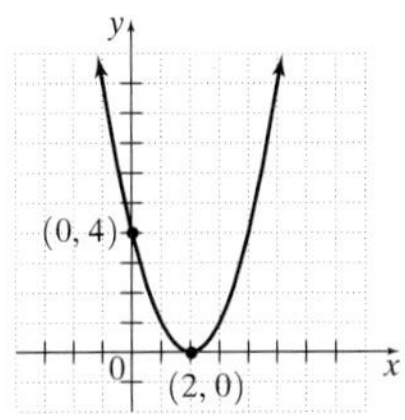

**20.** 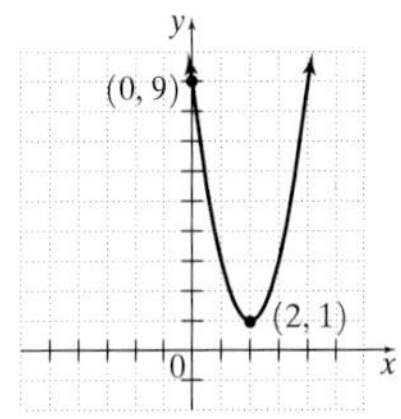

**21.** $(2 + \sqrt{46})$ ft $\approx 8.8$ ft **22.** $(5 + \sqrt{17})$ hr $\approx 9.12$ hr **23. a.** 272 ft **b.** 5.12 sec **24.** 7 ft

***Chapter 8 Cumulative Review*** **1. a.** $5 + y \geq 7$ **b.** $11 \neq z$ **c.** $20 < 5 - 2x$; Sec. 1.4, Ex. 3 **2.** slope: 1; 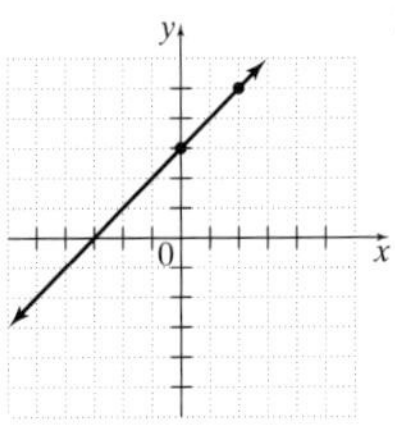 ; Sec. 3.4, Ex. 1 **3.** $(-2, 2)$; Sec. 4.1, Ex. 5 **4. a.** $6x^2 - 29x + 28$ **b.** $15x^2 - xy - 2y^2$; Sec. 5.4, Ex. 6 **5. a.** $4(2x^2 + 1)$ **b.** prime polynomial **c.** $3x^2(2 - x)$; Sec. 5.5, Ex. 2 **6.** $(x - 5)(x - 7)$; Sec. 5.6, Ex. 2 **7.** $3x(a - 2b)^2$; Sec. 5.7, Ex. 2 **8.** $-\frac{2}{3}$; Sec. 5.8, Ex. 4

**9.** 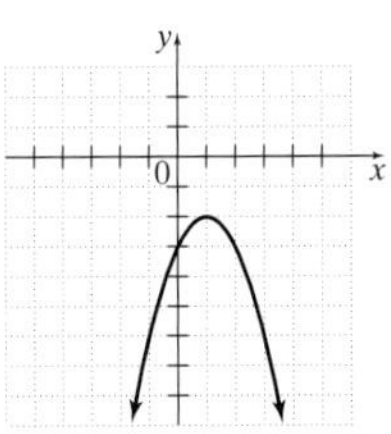 ; Sec. 5.9, Ex. 2 **10.** $\frac{1}{5x - 1}$; Sec. 6.1, Ex. 3 **11.** $\frac{7x^2 - 9x - 13}{(2x + 1)(x - 5)(3x - 2)}$; Sec. 6.2, Ex. 5 **12.** $\frac{xy + 2x^3}{y - 1}$; Sec. 6.3, Ex. 3 **13.** $3x^3y - 15x - 1 - \frac{6}{xy}$; Sec. 6.4, Ex. 2 **14. a.** 5 **b.** 5; Sec. 6.5, Ex. 3 **15.** $-3$; Sec. 6.6, Ex. 1 **16.** 2; Sec. 6.7, Ex. 2 **17.** $\frac{1}{6}$; $y = \frac{1}{6}x$; Sec. 6.8, Ex. 1

**18. a.** 3 **b.** $|x|$ **c.** $|x - 2|$ **d.** $-5$ **e.** $2x - 7$; Sec. 7.1, Ex. 5 **19. a.** $\sqrt[3]{5}$ **b.** $\sqrt{x}$ **c.** $\sqrt{rs^3}$; Sec. 7.2, Ex. 7 **20. a.** $5x\sqrt{x}$ **b.** $3x^2y^2\sqrt[3]{2y^2}$ **c.** $3z^2\sqrt[4]{z^3}$; Sec. 7.3, Ex. 4 **21. a.** $\frac{2\sqrt{5}}{5}$ **b.** $\frac{8\sqrt{x}}{3x}$ **c.** $\frac{\sqrt[3]{4}}{2}$; Sec. 7.5, Ex. 1 **22.** $\frac{2}{9}$; Sec. 7.6, Ex. 5 **23. a.** $\frac{1}{2} + \frac{3}{2}i$ **b.** $-\frac{7}{3}i$; Sec. 7.7, Ex. 5 **24.** $-1 + 2\sqrt{3}, -1 - 2\sqrt{3}$; Sec. 8.1, Ex. 3 **25.** 9; Sec. 8.3, Ex. 1

## ■ CHAPTER 9 CONIC SECTIONS

***Graphing Calculator Explorations*** **1.**

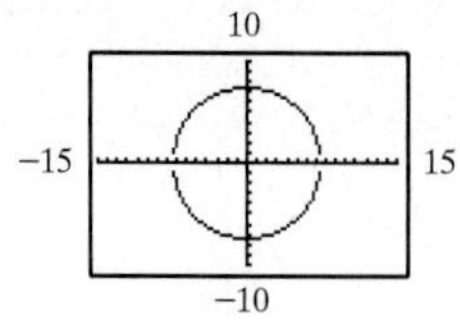


**3.**

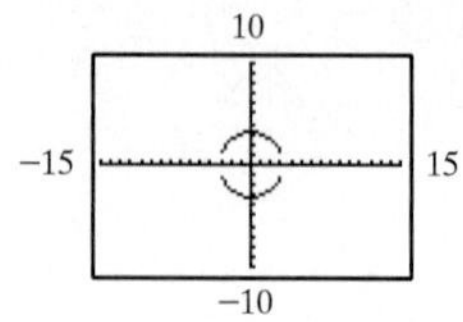


**5.**

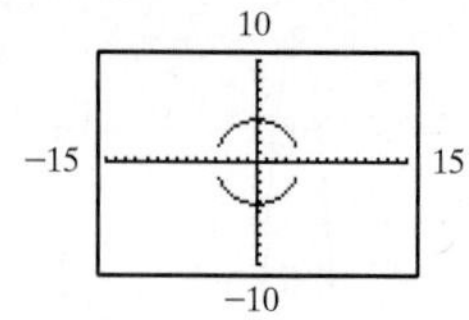


**7.**

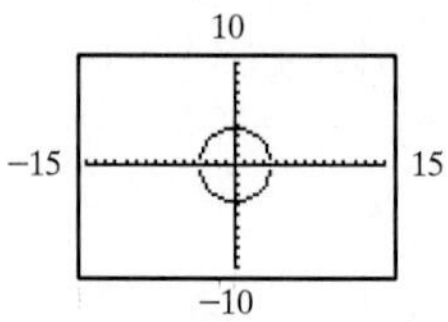


***Mental Math*** **1.** upward **3.** to the left **5.** downward

***Exercise Set 9.1*** **1.**

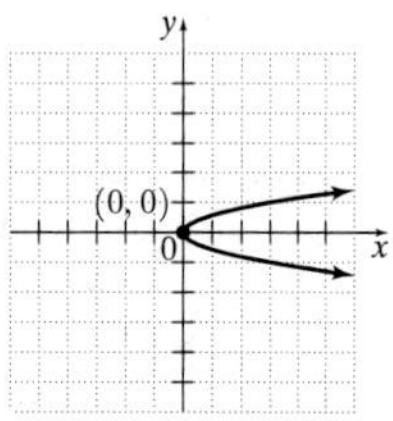


**3.**

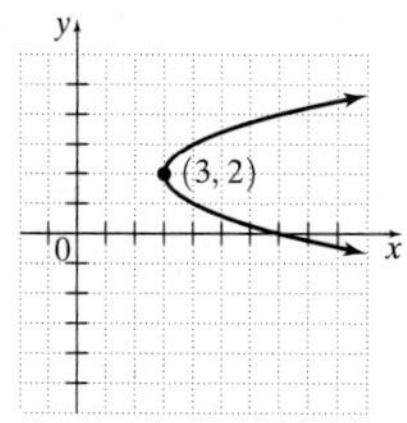


**5.**

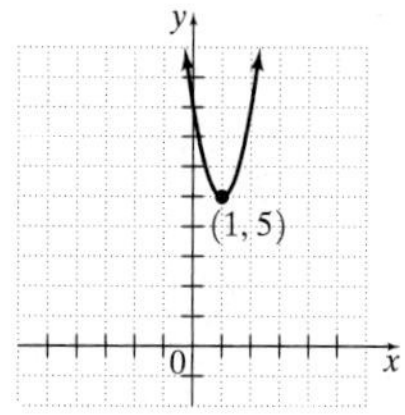


**7.**

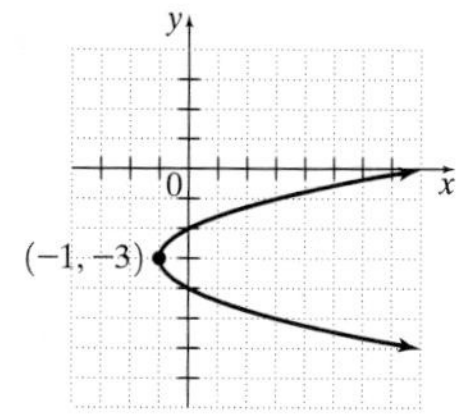


**9.**

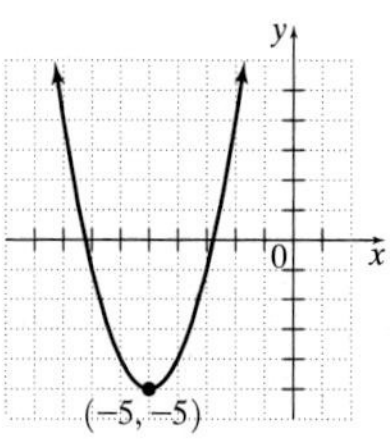


**11.**

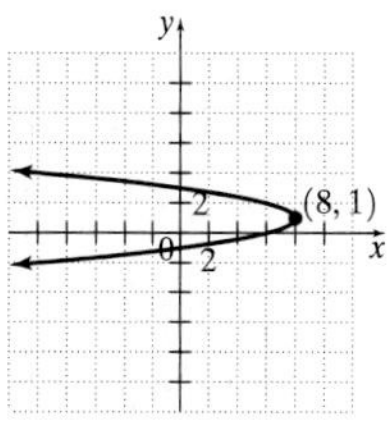


**13.** 5 units **15.** $\sqrt{41}$ units **17.** $\sqrt{10}$ units **19.** $\sqrt{5}$ units **21.** 13.88 units **23.** 9 units **25.** $(4, -2)$ **27.** $\left(-5, \frac{5}{2}\right)$ **29.** $(3, 0)$ **31.** $\left(-\frac{1}{2}, \frac{1}{2}\right)$ **33.** $\left(\sqrt{2}, \frac{\sqrt{5}}{2}\right)$ **35.** $(6.2, -6.65)$

**37.**

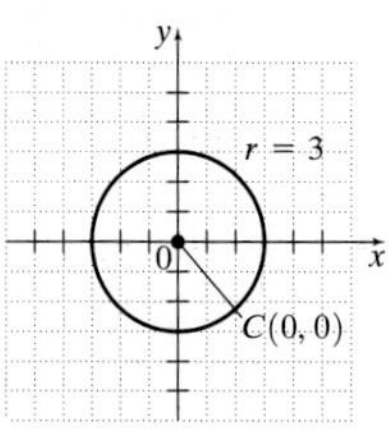


**39.**

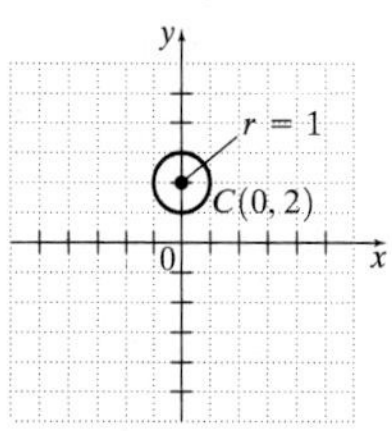


**41.**

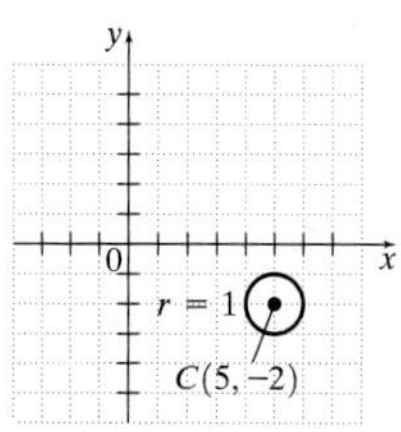


**43.**

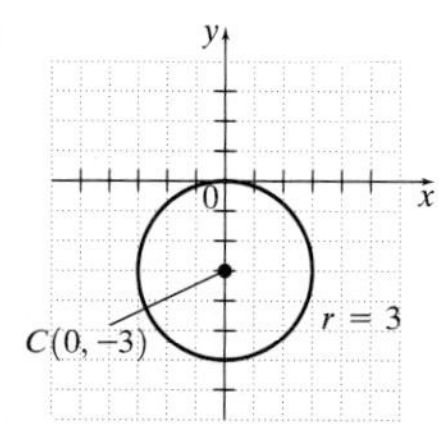


**45.**

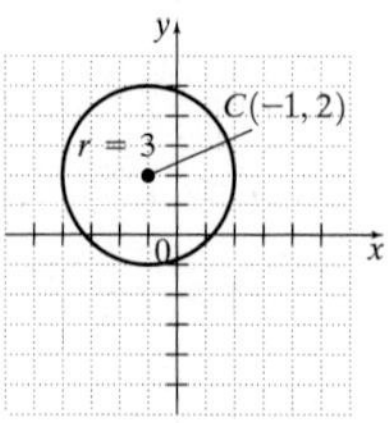


**47.**

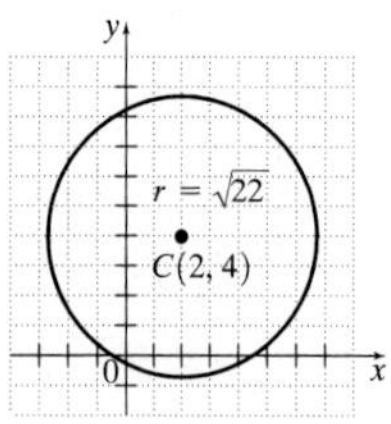


**49.** $(x - 2)^2 + (y - 3)^2 = 36$
**51.** $x^2 + y^2 = 3$
**53.** $(x + 5)^2 + (y - 4)^2 = 45$
**55.** Answers may vary.

**57.**

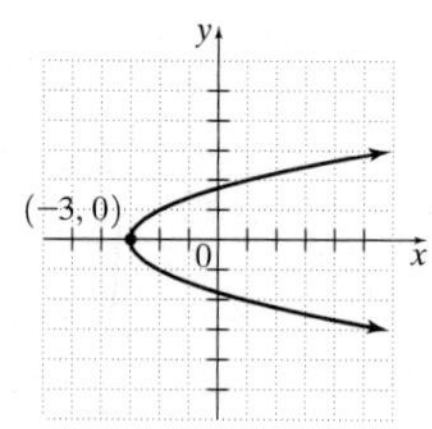


**59.**

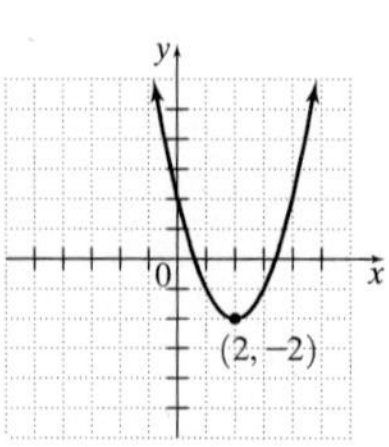


**61.**

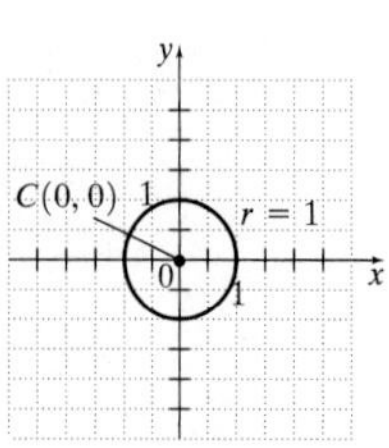


**63.**

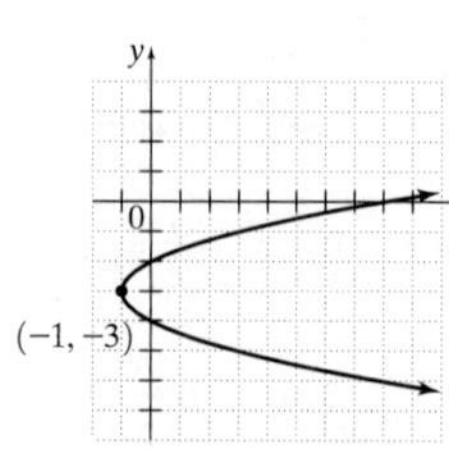


**65.**

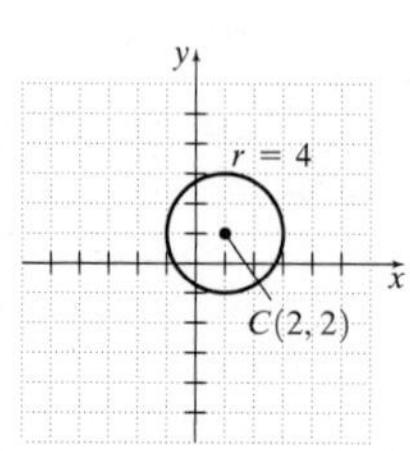

**67.**
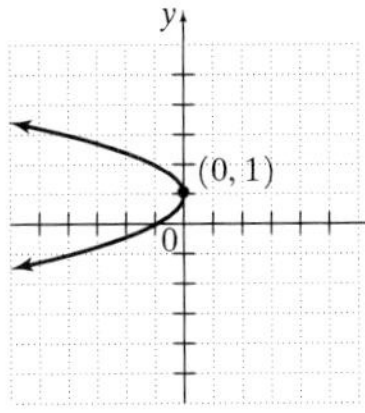


**69.**
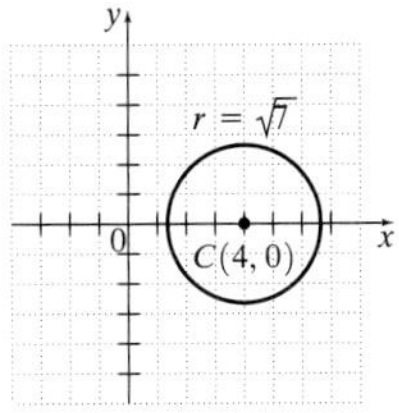


**71.**
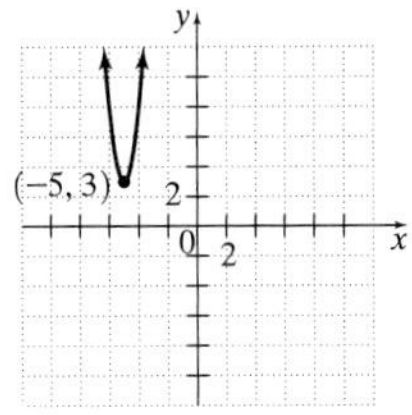


**73.**
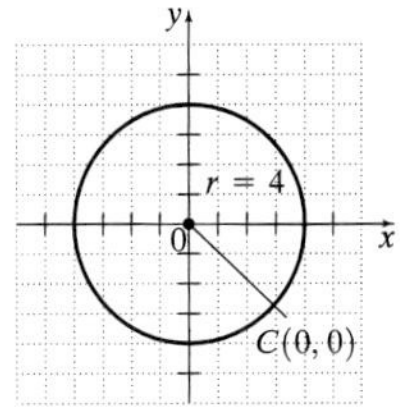


**75.**
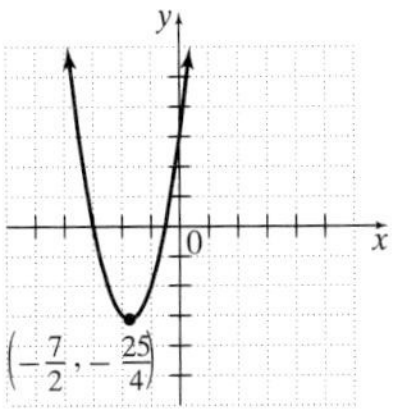


**77.**
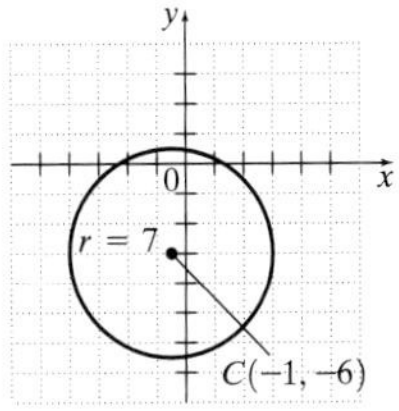


**79.**
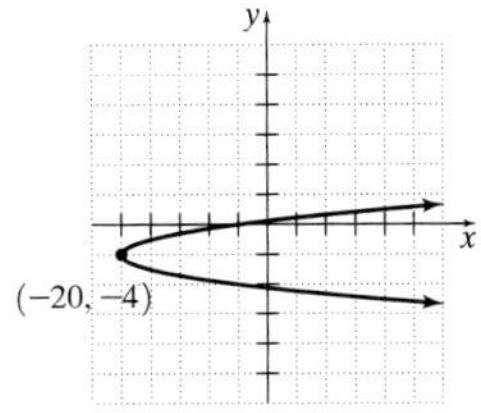


**81.**
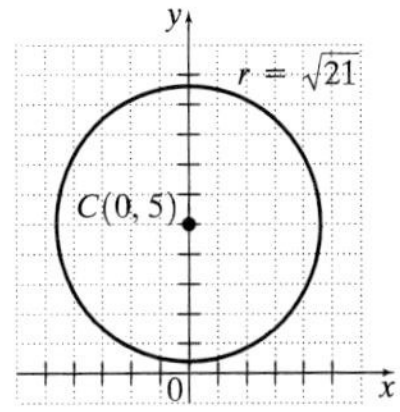


**83.**
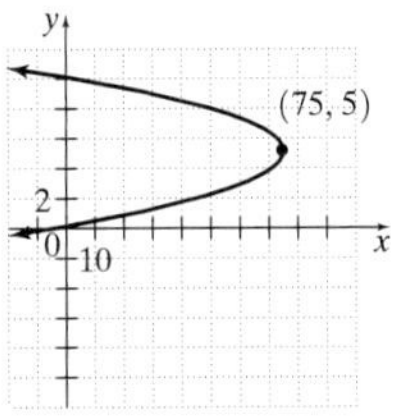


**85.**
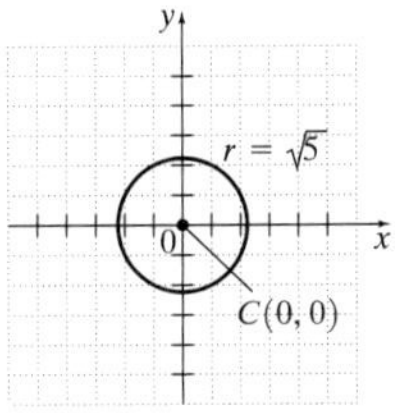


**87.**
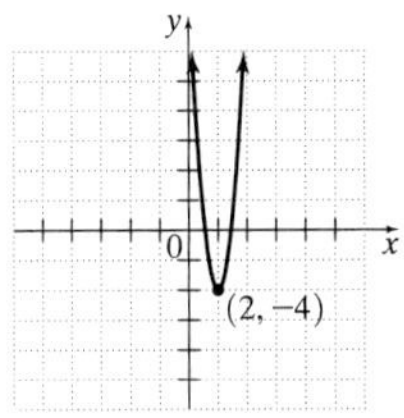


**89.** Yes, it is. **91.** $y = -\dfrac{2}{125}x^2 + 40$

**93.**
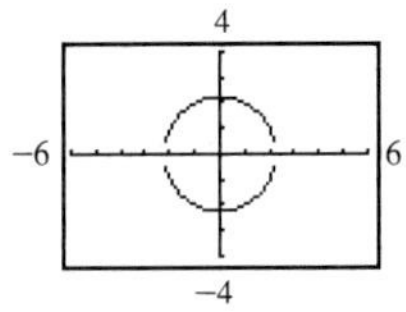


**95.**
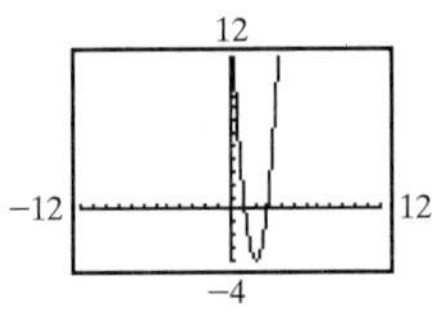


**97.**
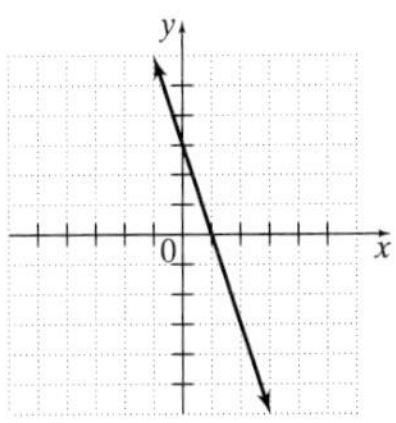


**99.**
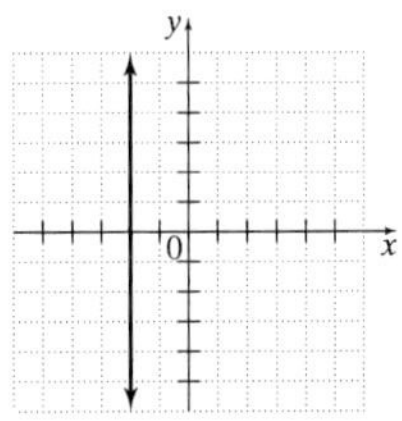


**101.** $\dfrac{\sqrt{10}}{4}$ **103.** $2\sqrt{5}$

***Graphing Calculator Explorations*** **1.**
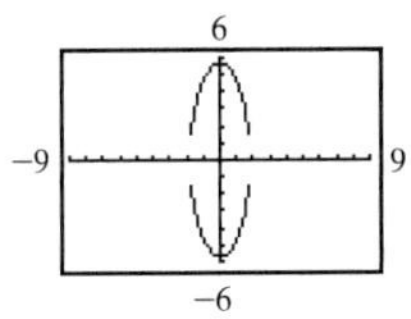


**3.**
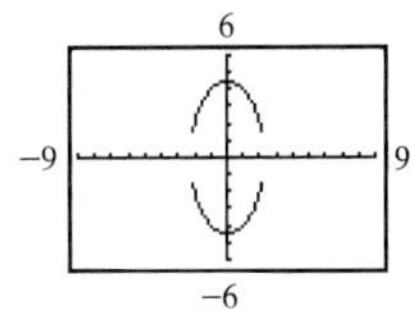


**5.**
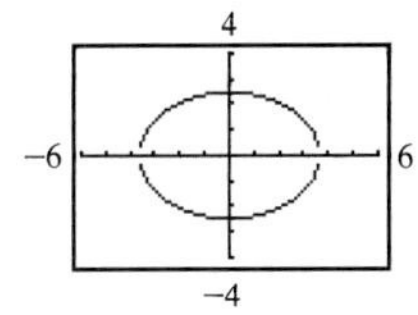


***Exercise Set 9.2*** **1.**
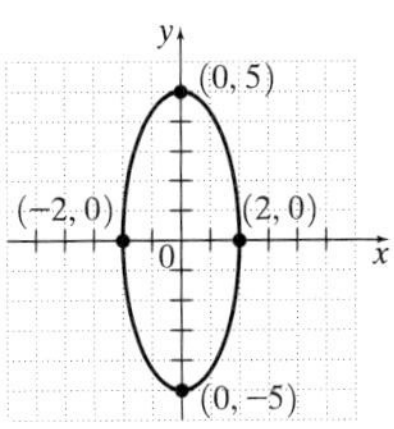


**3.**
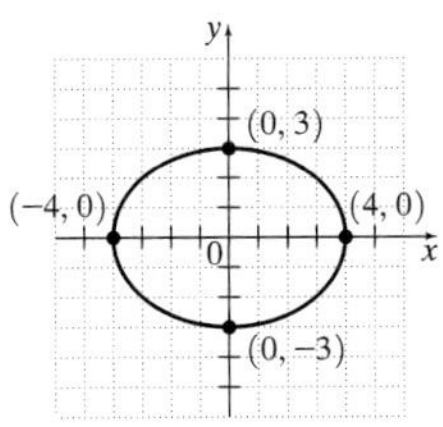


**5.**
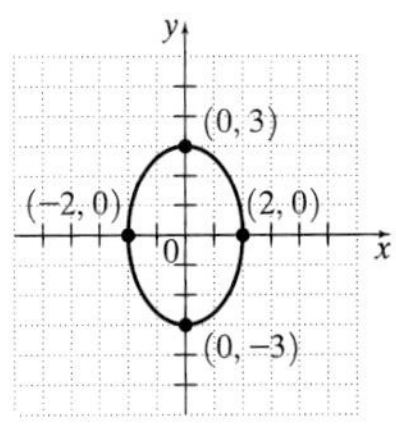


**7.**
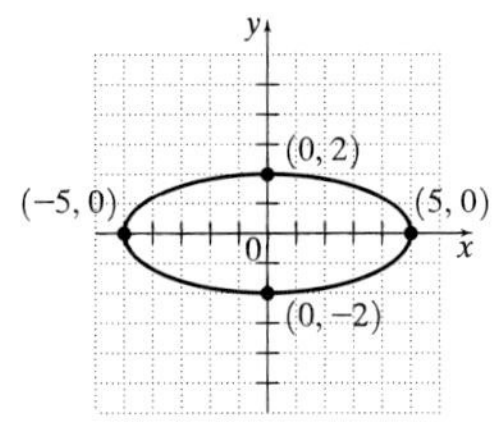


**9.**
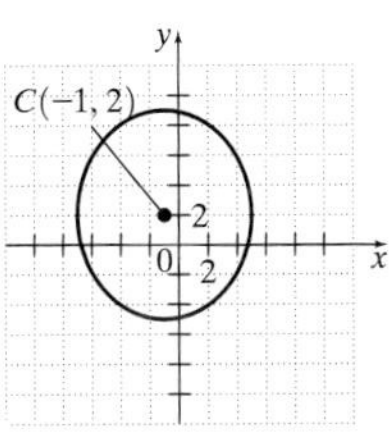


**11.**
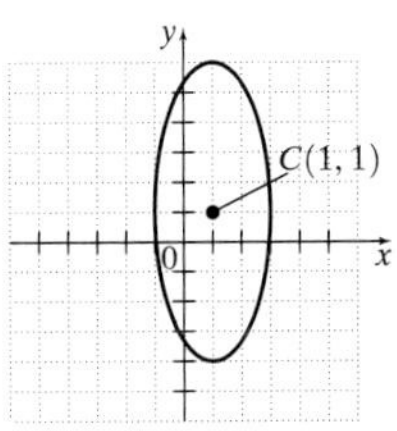


**13.**
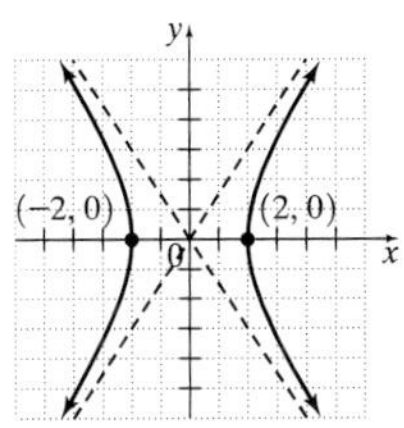


**15.**
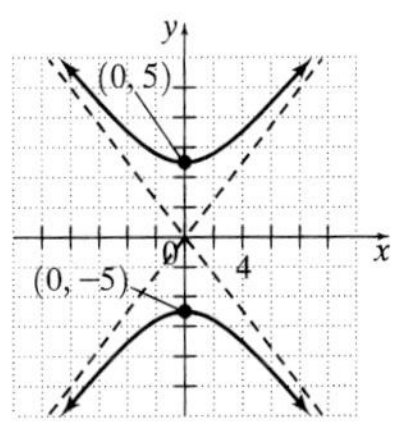

**17.**
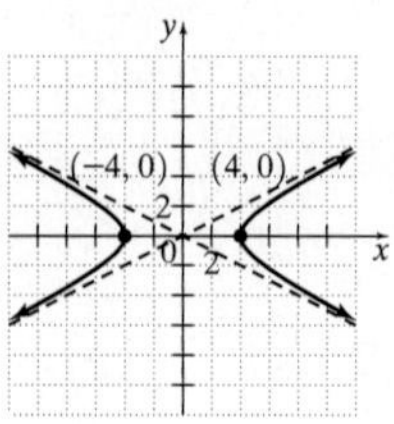


**19.**
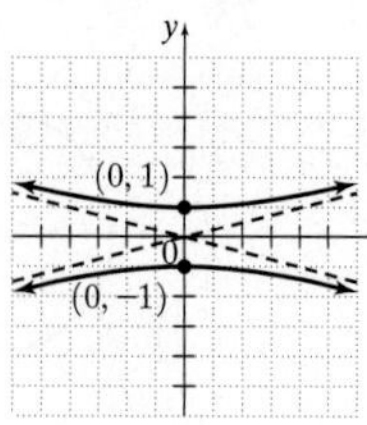


**21.** Answers may vary. **23.** parabola
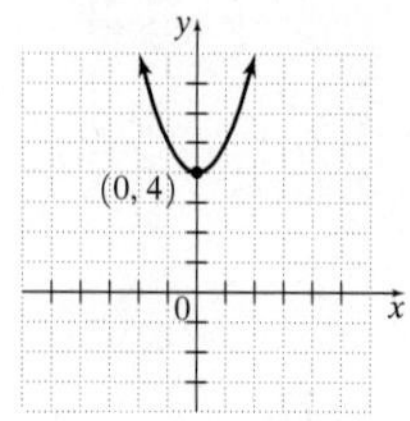


**25.** ellipse
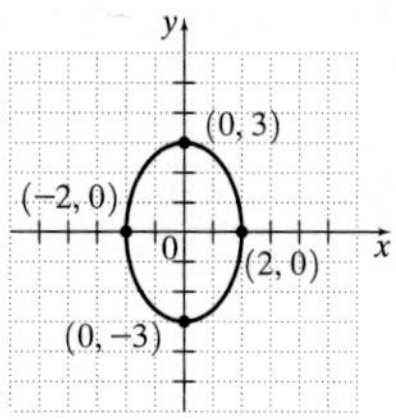


**27.** hyperbola
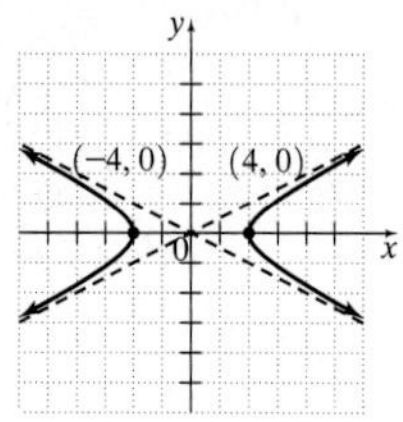


**29.** circle
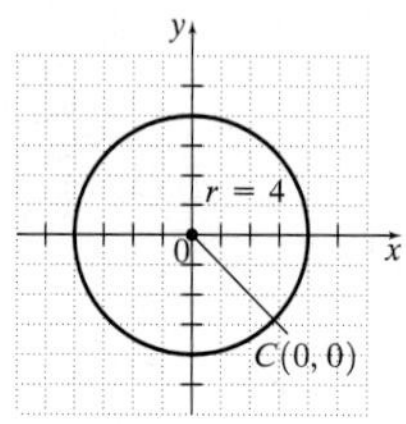


**31.** parabola
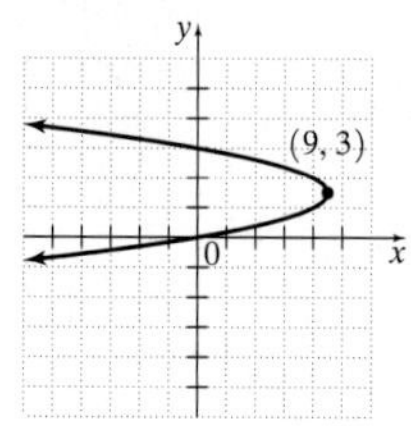


**33.** ellipse
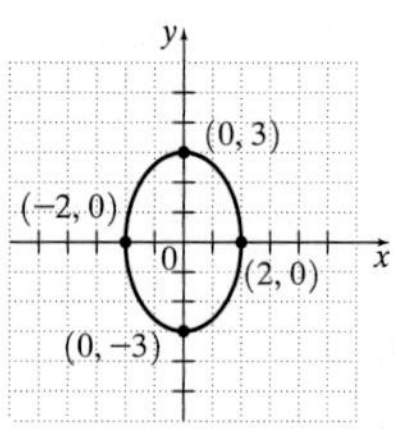


**35.** hyperbola
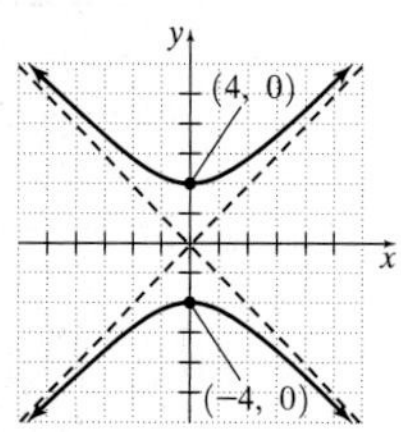


**37.** parabola
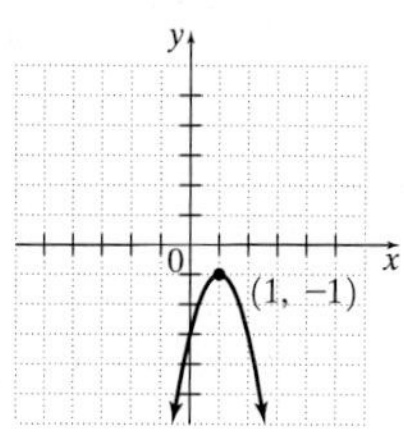


**39.** A: 36, 13; B: 4, 4; C: 25, 16; D: 39, 25; E: 17, 81; F: 36, 36; G: 16, 65; H: 144, 140
**41.** A: 6; B: 2; C: 5; D: 5; E: 9; F: 6; G: 4; H: 12
**43.** greater than zero and less than one
**45.** greater than one
**47.** (1,782,000,000 356,400,000)

**49.**
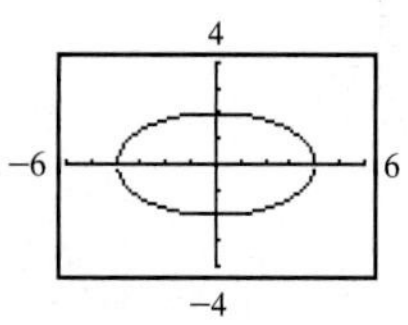


**51.** $(-\infty, 1)$ **53.** $[2, \infty)$ **55.** $-2x^3$ **57.** $-5x^4$

**59.**
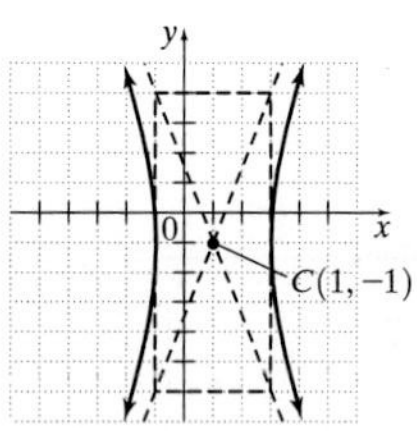


**61.**
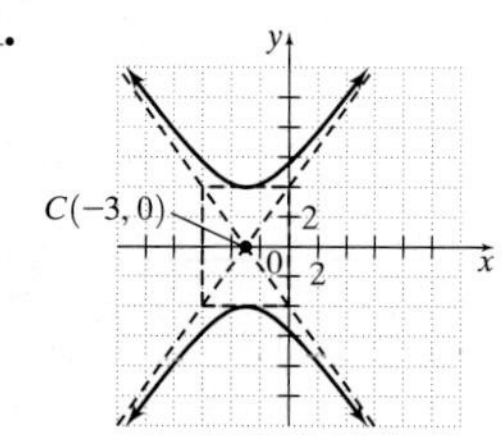


**63.**
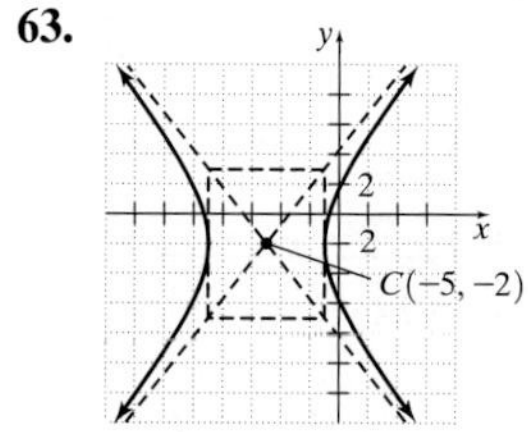


***Exercise Set 9.3*** **1.** $(3, -4), (-3, 4)$ **3.** $(\sqrt{2}, \sqrt{2}), (-\sqrt{2}, -\sqrt{2})$ **5.** $(4, 0), (0, -2)$
**7.** $(-\sqrt{5}, -2), (-\sqrt{5}, 2), (\sqrt{5}, -2), (\sqrt{5}, 2)$ **9.** $\varnothing$ **11.** $(1, -2), (3, 6)$ **13.** $(2, 4), (-5, 25)$ **15.** $\varnothing$
**17.** $(1, -3)$ **19.** $(-1, -2), (-1, 2), (1, -2), (1, 2)$ **21.** $(0, -1)$ **23.** $(-1, 3), (1, 3)$ **25.** $(\sqrt{3}, 0), (-\sqrt{3}, 0)$
**27.** $\varnothing$ **29.** $(-6, 0), (6, 0), (0, -6)$ **31.** 0, 1, 2, 3, or 4 **33.** 9 and 7; 9 and −7; −9 and 7; −9 and −7
**35.** 15 cm by 19 cm **37.** 15 thousand compact discs; price: $3.75

**39.**
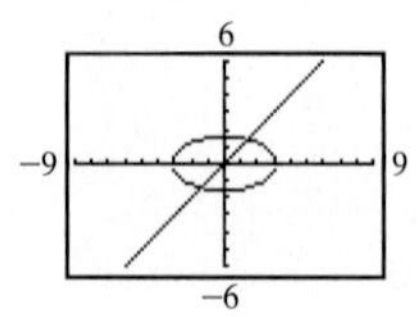


**41.**
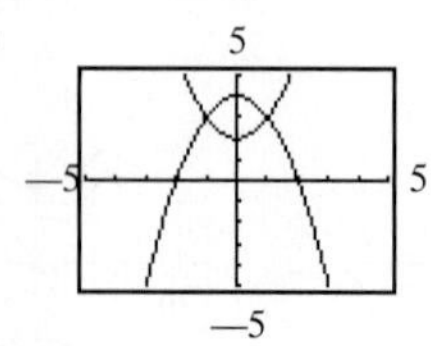


**43.**
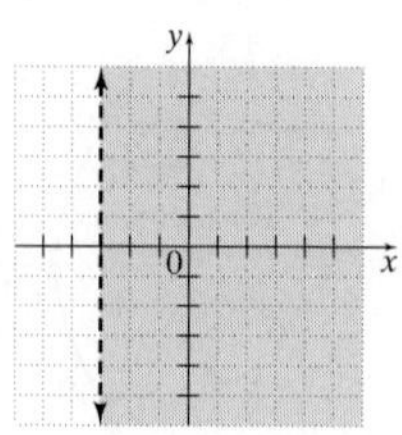


**45.**
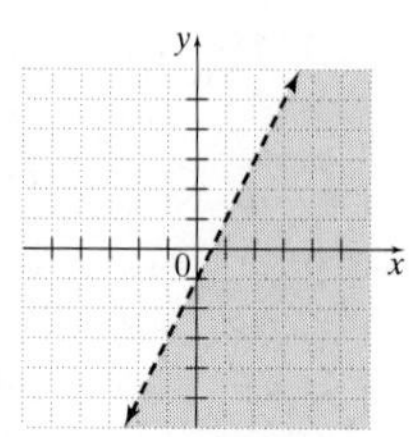


**47.** $(8x - 25)$ in. **49.** $(4x^2 + 6x + 2)$ m

***Exercise Set 9.4***

**1.** 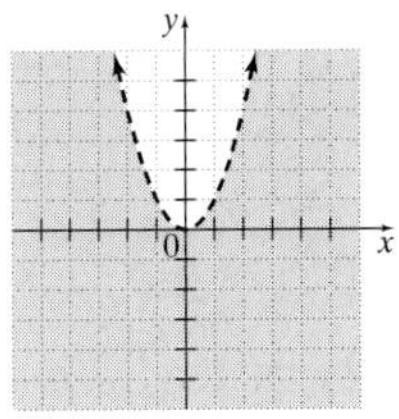

**3.** 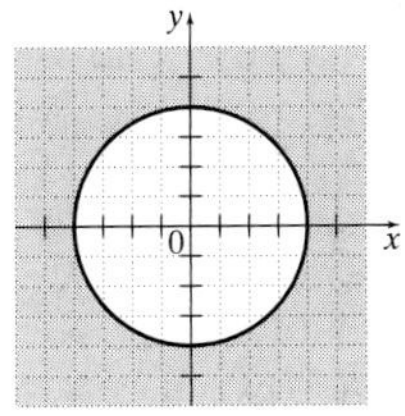

**5.** 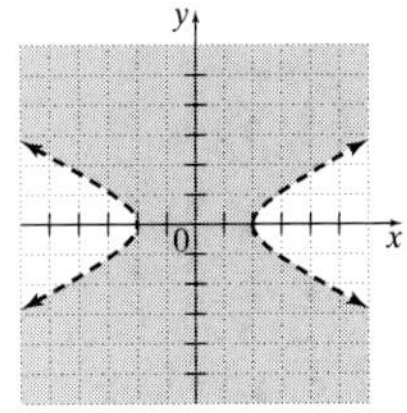

**7.** 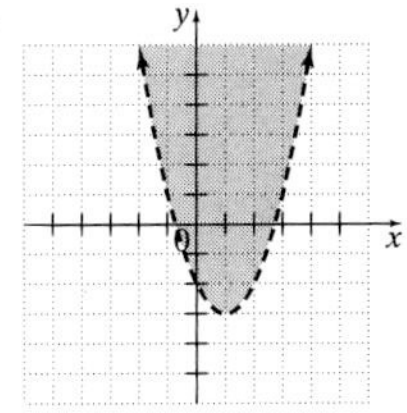

**9.** 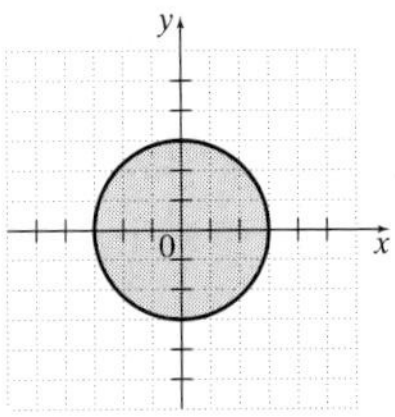

**11.** 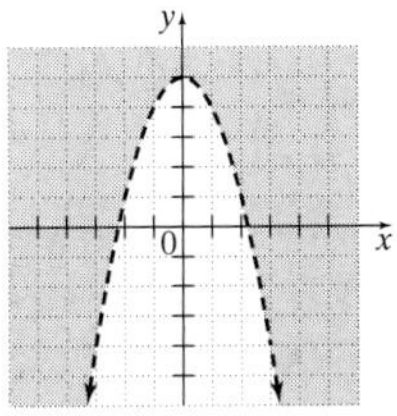

**13.** 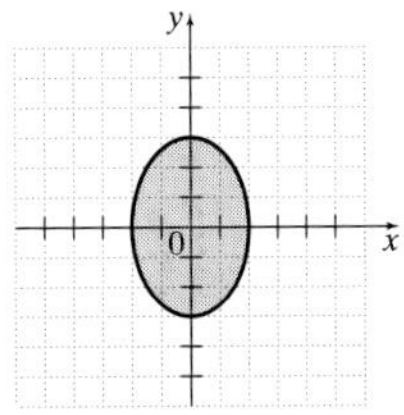

**15.** 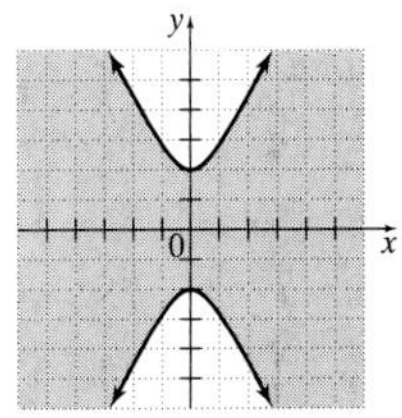

**17.** 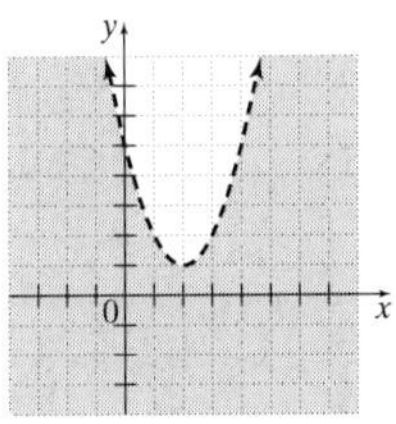

**19.** 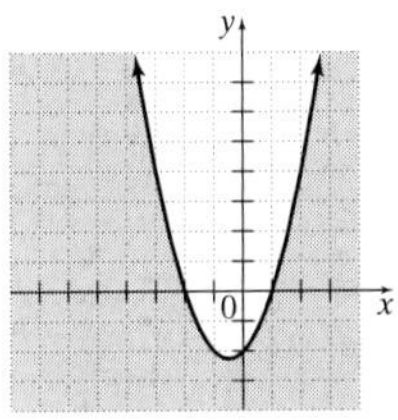

**21.** Answers may vary.

**23.** 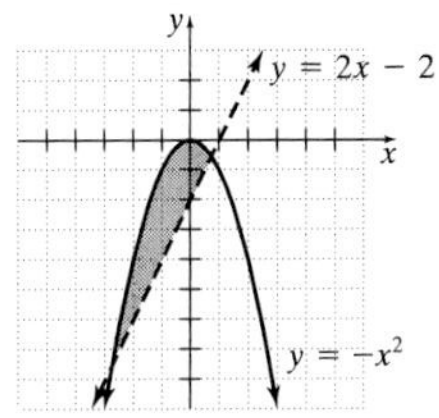


**25.** 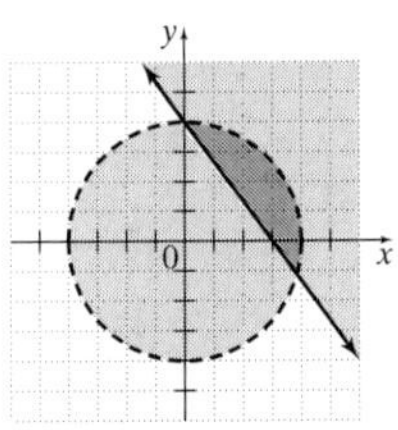

**27.** 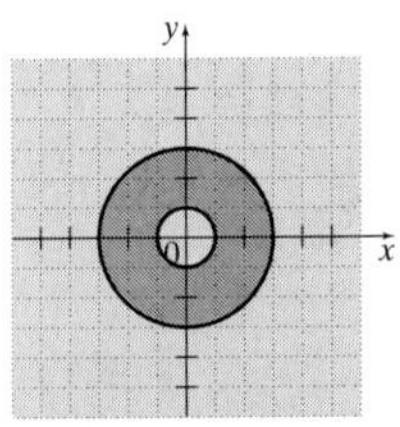

**29.** 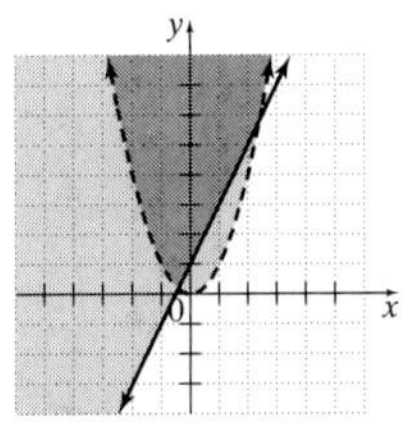

**31.** 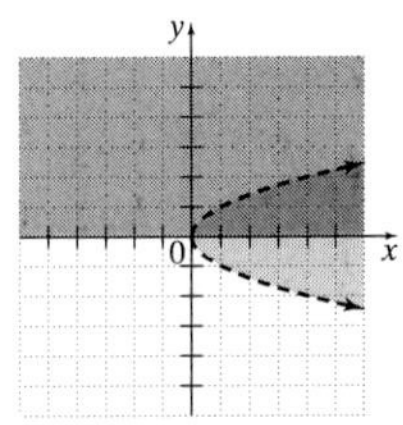

**33.** 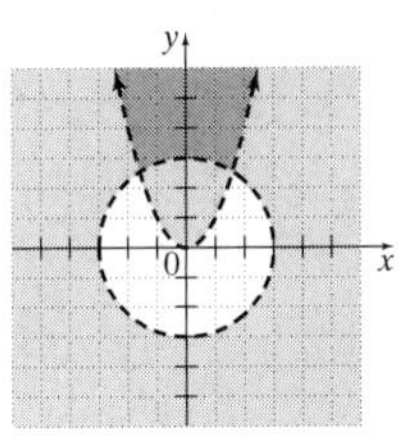

**35.** 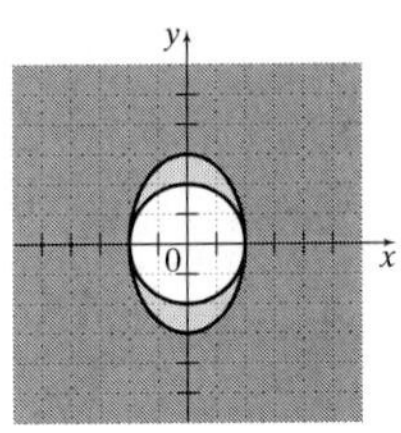

**37.** 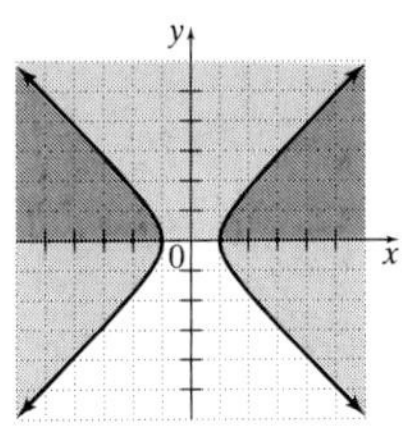

**39.** 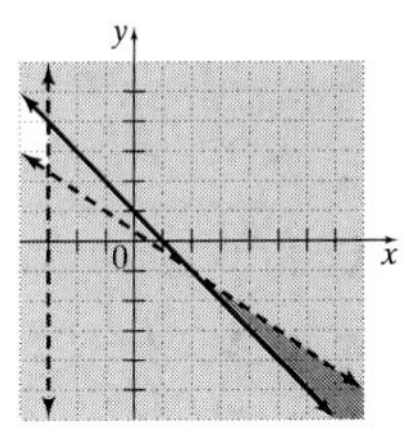

**41.** 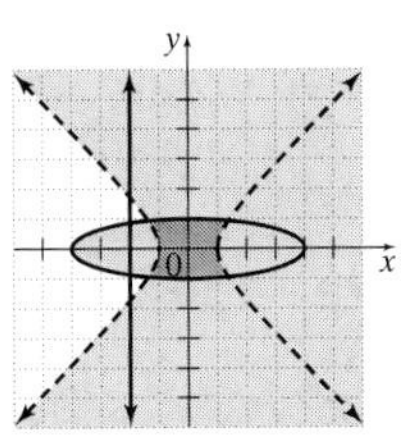

**43.** 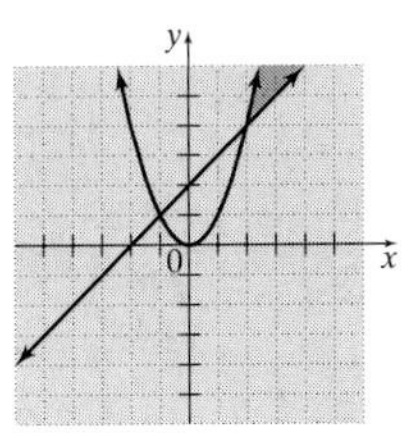

**45.** function **47.** not a function **49.** 25 **51.** $3b^2 - 2$

***Chapter 9 Review*** **1.** $\sqrt{197}$ units **3.** $\sqrt{130}$ units **5.** $7\sqrt{2}$ units **7.** 16.60 units **9.** $(-5, 5)$ **11.** $\left(-\frac{15}{2}, 1\right)$
**13.** $\left(\frac{1}{20}, -\frac{3}{16}\right)$ **15.** $(\sqrt{3}, -3\sqrt{6})$ **17.** $(x + 4)^2 + (y - 4)^2 = 9$ **19.** $(x + 7)^2 + (y + 9)^2 = 11$

**21.**
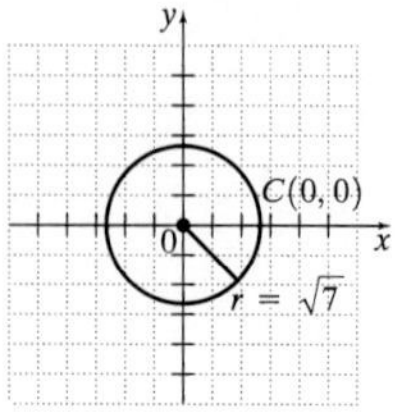


**23.**
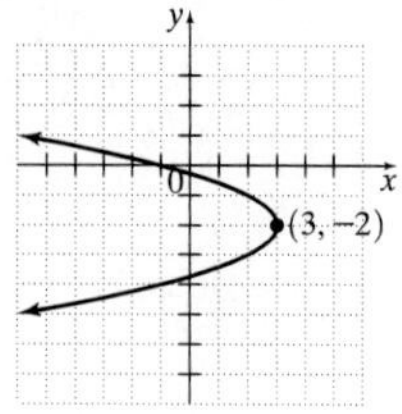


**25.**
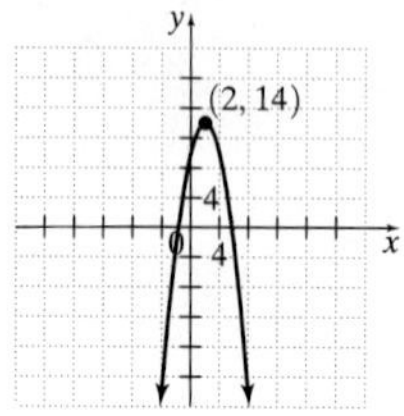


**27.**
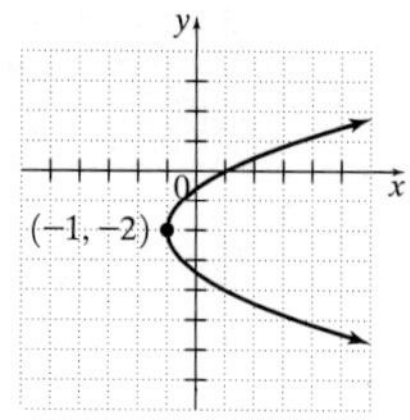


**29.**
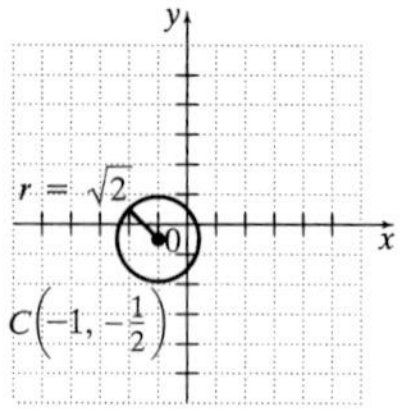


**31.**
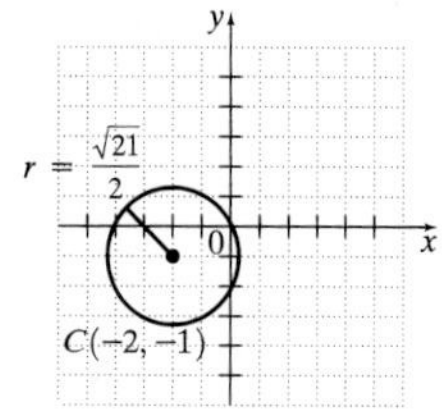


**33.**
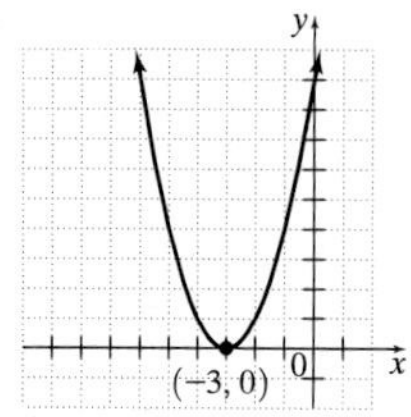


**35.** $(x - 5.6)^2 + (y + 2.4)^2 = 9.61$

**37.**
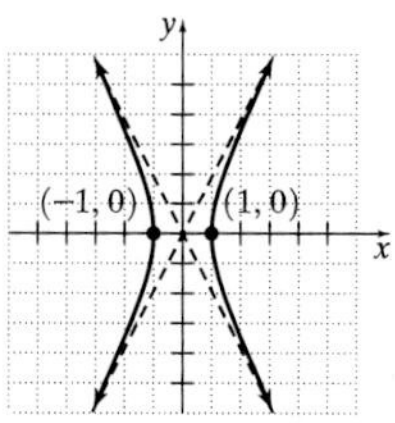


**39.**
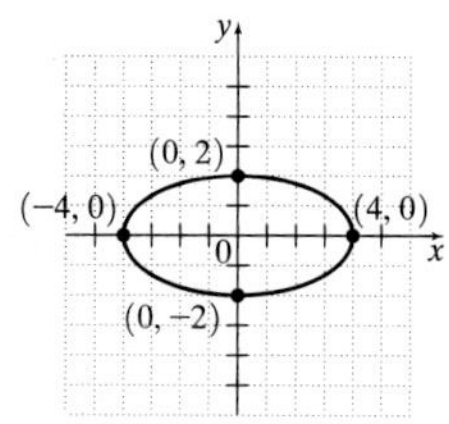


**41.**
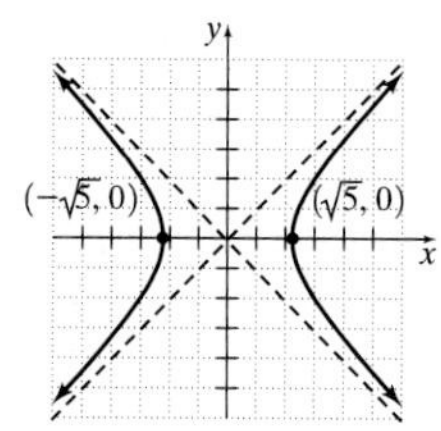


**43.**
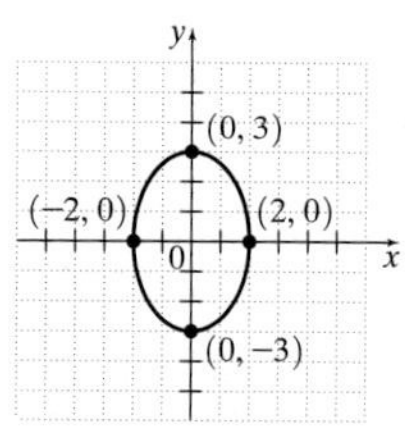


**45.**
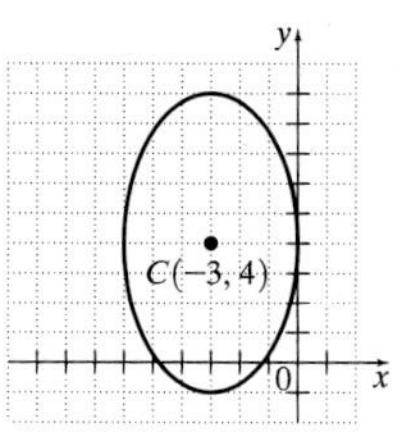


**47.**
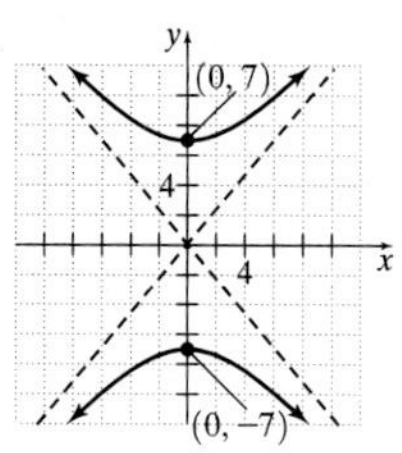


**49.**
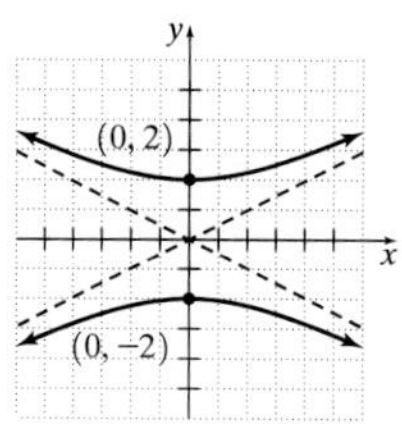


**51.**
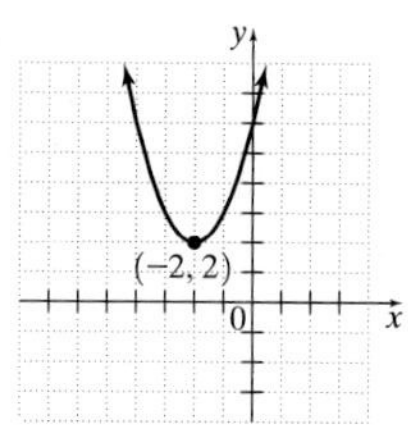


**53.**
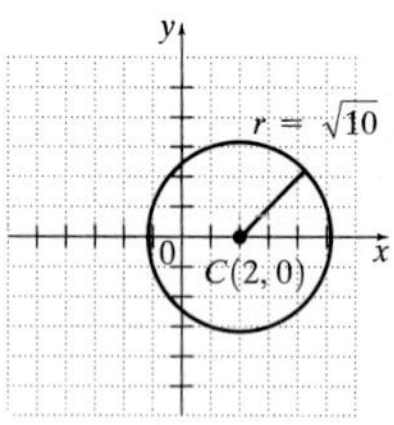


**55.**
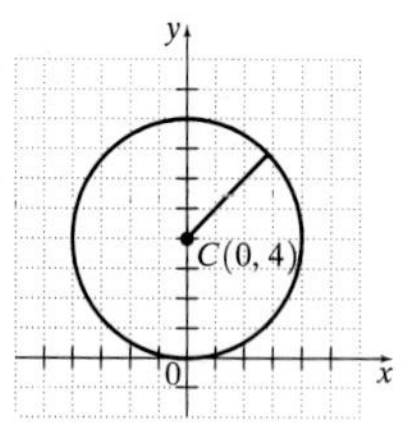


**57.**
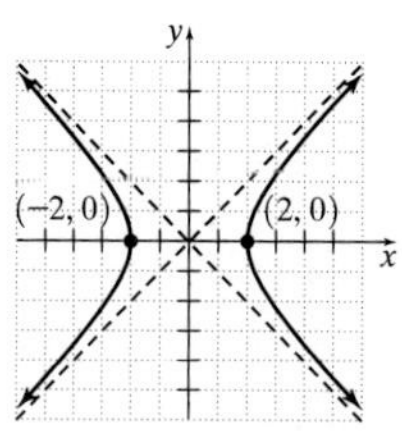


**59.**
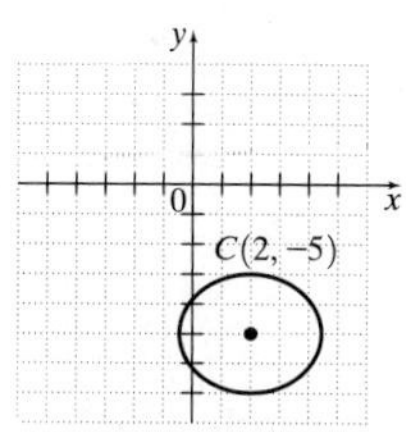


**61.**
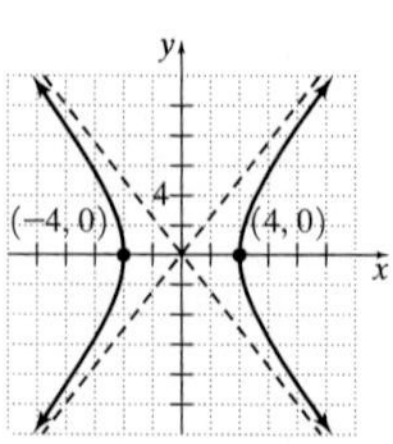


**63.**
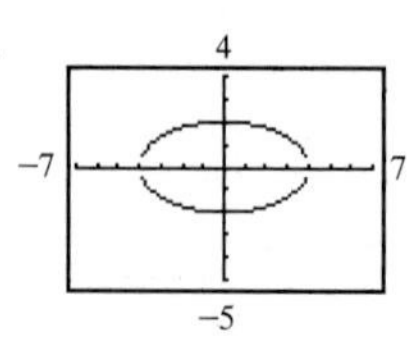


**65.**
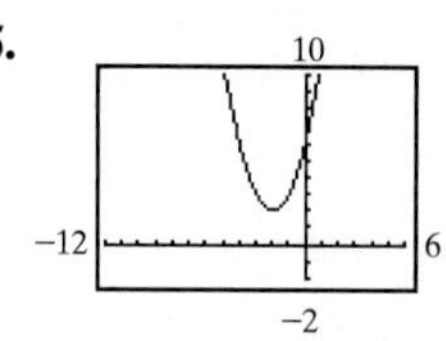


**67.** $(1, -2), (4, 4)$
**69.** $(-1, 1), (2, 4)$
**71.** $(2, 2\sqrt{2}), (2, -2\sqrt{2})$
**73.** $(-1, 3), (-1, -3), (1, 3), (1, -3)$
**75.** $(1, 4)$
**77.** 15 ft by 10 ft

**79.** 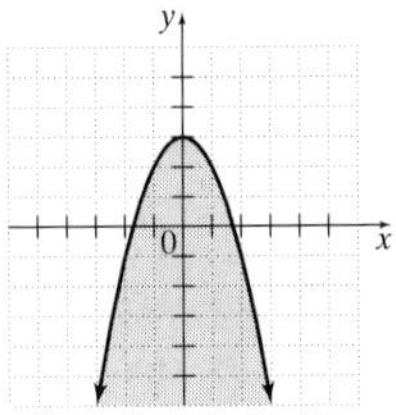
**81.** 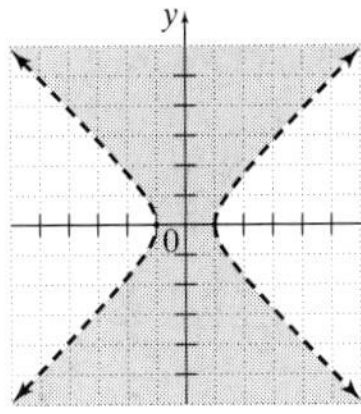
**83.** 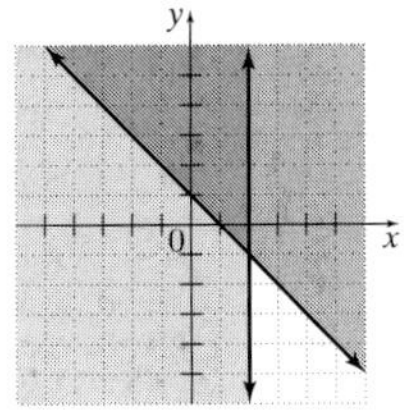
**85.** 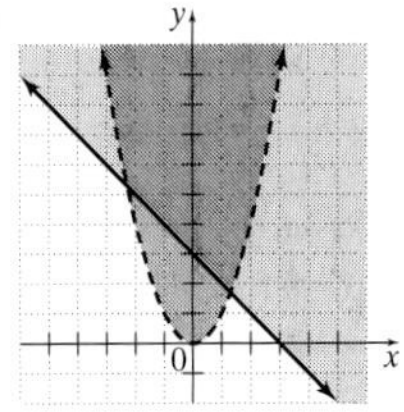

**87.** 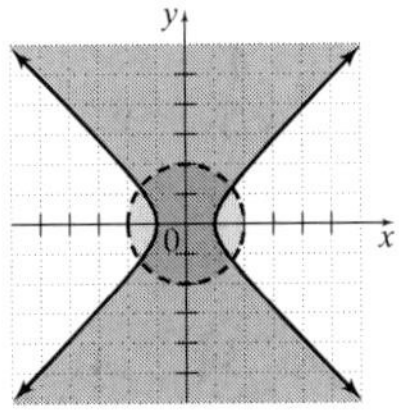

***Chapter 9 Test*** **1.** $2\sqrt{26}$ units **2.** $\sqrt{95}$ units **3.** $\left(-4, \frac{7}{2}\right)$ **4.** $\left(-\frac{1}{2}, \frac{3}{10}\right)$

**5.** 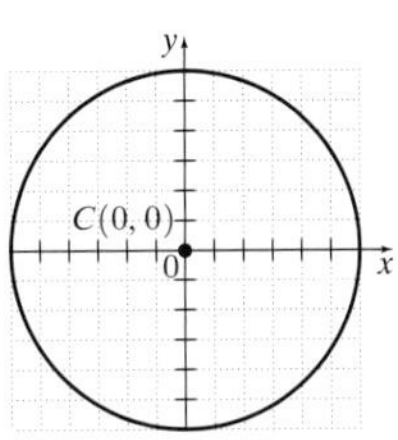

**6.** 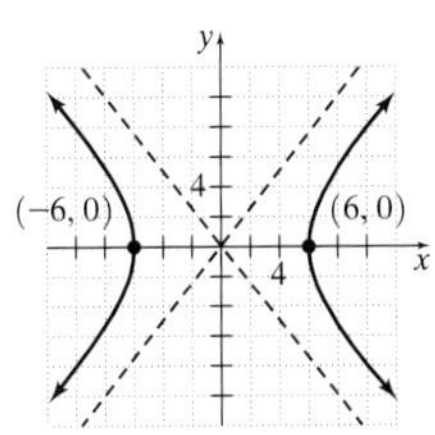

**7.** 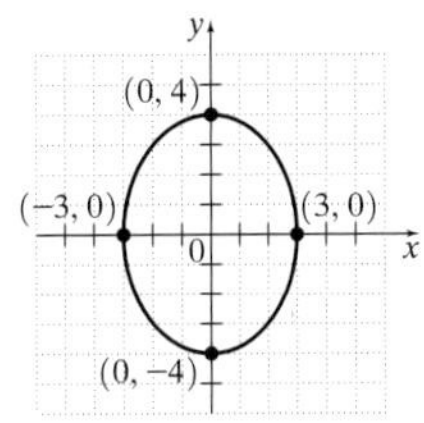

**8.** 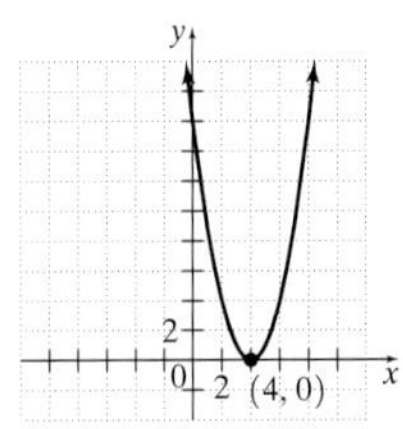


**9.** 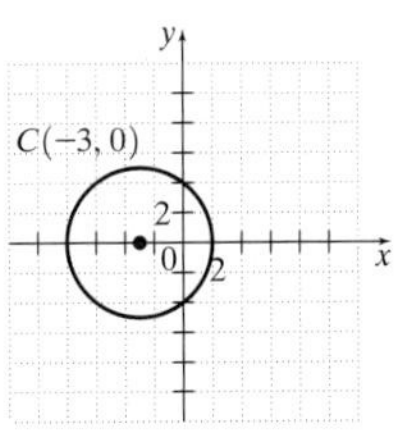

**10.** 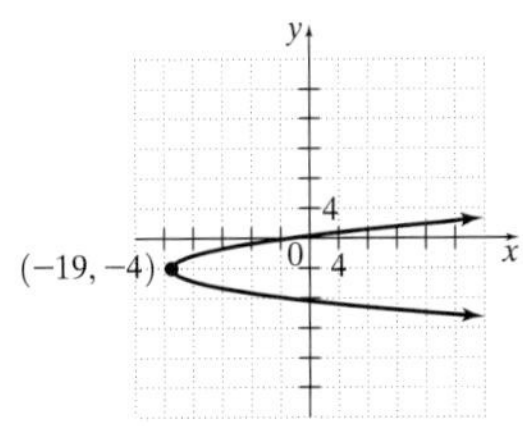

**11.** 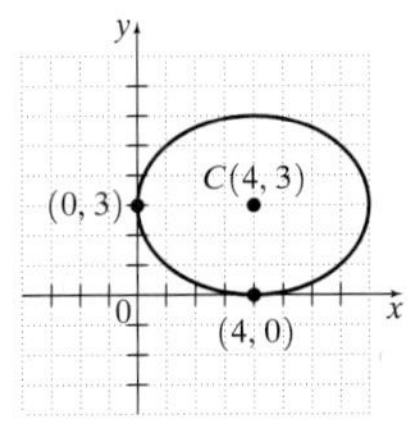

**12.** 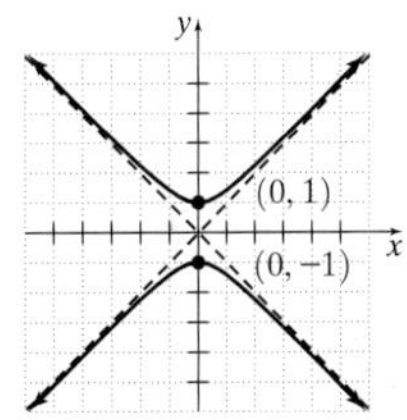


**13.** $(-12, 5), (12, -5)$ **14.** $(-5, -1), (-5, 1), (5, -1), (5, 1)$ **15.** $(6, 12), (1, 2)$ **16.** $(1, 1), (-1, -1)$

**17.** 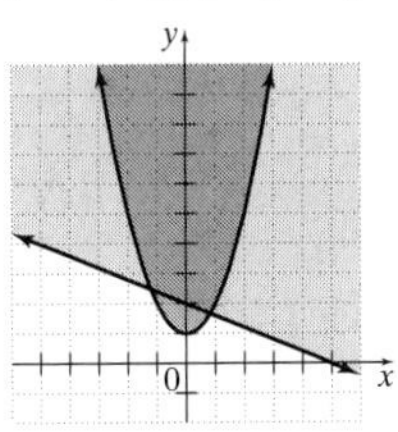
**18.** 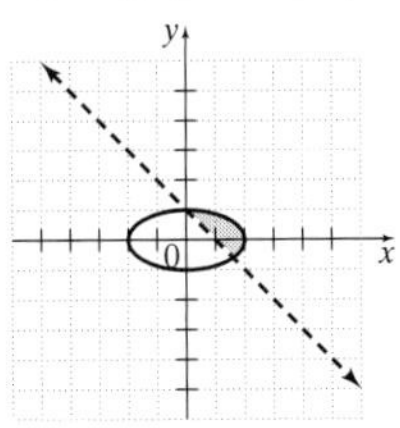
**19.** 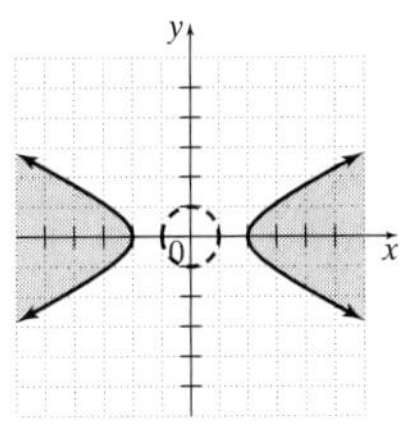
**20.** 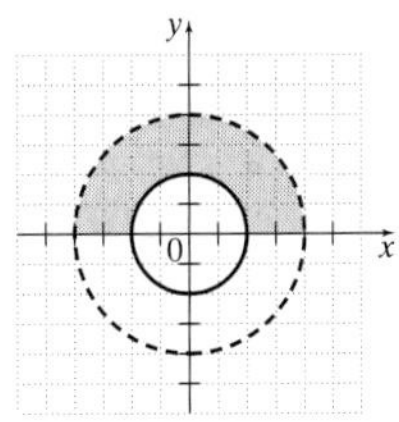

**21.** B **22.** height: 10 ft; width: 30 ft

***Chapter 9 Cumulative Review*** **1.** $(4 \cdot 9)y = 36y$; Sec. 1.4, Ex. 7 **2.** 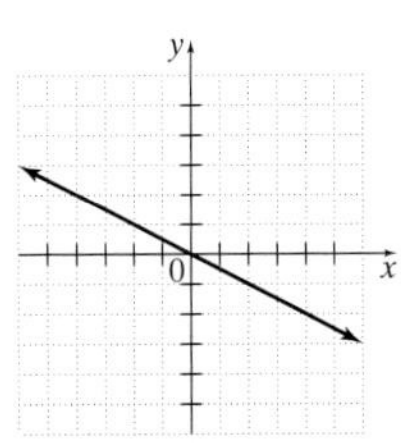 ; Sec. 3.3, Ex. 5

**3.** $\varnothing$; Sec. 4.1, Ex. 7

**4. a.** $125x^6$ **b.** $\frac{8}{27}$ **c.** $\frac{9p^8}{q^{10}}$ **d.** $64y^2$ **e.** $\frac{y^{14}}{x^{35}z^7}$; Sec. 5.2, Ex. 8

**5.** $-\frac{1}{6}, 3$; Sec. 5.8, Ex. 5 **6.** $\frac{12}{x-1}$; Sec. 6.2, Ex. 6

**7. a.** $\frac{1}{9xy^2}$ **b.** $\frac{x(x-2)}{2(x+2)}$ **c.** $\frac{x^2}{y^2}$; Sec. 6.3, Ex. 1 **8.** $2x - 5$; Sec. 6.4, Ex. 3 **9.** 16; Sec. 6.5, Ex. 4

**10.** −6, −1; Sec. 6.6, Ex. 5 **11.** $2\frac{2}{9}$ hr; no; Sec. 6.7, Ex. 4 **12. a.** 1 **b.** −4 **c.** $\frac{2}{5}$ **d.** $x^2$ **e.** $-2x^3$; Sec. 7.1, Ex. 3 **13. a.** $z - z^{17/3}$ **b.** $x^{2/3} - 3x^{1/3} - 10$; Sec. 7.2, Ex. 5 **14. a.** 2 **b.** $\frac{5}{2}\sqrt{x}$ **c.** $14xy^2\sqrt[3]{x}$; Sec. 7.3, Ex. 5 **15. a.** $\frac{5\sqrt{5}}{12}$ **b.** $\frac{5\sqrt[3]{7x}}{2}$; Sec. 7.4, Ex. 2 **16.** $\frac{\sqrt{21xy}}{3y}$; Sec. 7.5, Ex. 2 **17.** 42; Sec. 7.6, Ex. 1 **18. a.** $-i$ **b.** 1 **c.** −1 **d.** 1; Sec. 7.7, Ex. 6 **19.** $-1 + \sqrt{5}, -1 - \sqrt{5}$; Sec. 8.1, Ex. 5 **20.** $2 + \sqrt{2}, 2 - \sqrt{2}$; Sec. 8.2, Ex. 3 **21.** $2, -2, i, -i$; Sec. 8.3, Ex. 3 **22.** $[-2, 3)$; Sec. 8.4, Ex. 4

**23.**

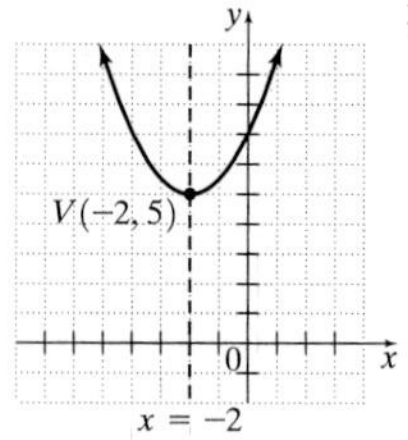


; Sec. 8.5, Ex. 8 **24.** (2, −16); Sec. 8.6, Ex. 4 **25.** $\sqrt{2} \approx 1.414$; Sec. 9.1, Ex. 5

## ■ CHAPTER 10 EXPONENTIAL AND LOGARITHMIC FUNCTIONS

***Exercise Set 10.1*** **1. a.** $3x - 6$ **b.** $-x - 8$ **c.** $2x^2 - 13x - 7$ **d.** $\frac{x - 7}{2x + 1}$, where $x \neq -\frac{1}{2}$ **3. a.** $x^2 + 5x + 1$ **b.** $x^2 - 5x + 1$ **c.** $5x^3 + 5x$ **d.** $\frac{x^2 + 1}{5x}$, where $x \neq 0$ **5. a.** $\sqrt{x} + x + 5$ **b.** $\sqrt{x} - x - 5$ **c.** $x\sqrt{x} + 5\sqrt{x}$ **d.** $\frac{\sqrt{x}}{x + 5}$, where $x \neq -5$ **7. a.** $5x^2 - 3x$ **b.** $-5x^2 - 3x$ **c.** $-15x^3$ **d.** $-\frac{3}{5x}$, where $x \neq 0$ **9.** 42 **11.** −18 **13.** 0 **15.** $(f \circ g)(x) = 25x^2 + 1$; $(g \circ f)(x) = 5x^2 + 5$ **17.** $(f \circ g)(x) = 2x + 11$; $(g \circ f)(x) = 2x + 4$ **19.** $(f \circ g)(x) = -8x^3 - 2x - 2$; $(g \circ f)(x) = -2x^3 - 2x + 4$ **21.** $(f \circ g)(x) = \sqrt{-5x + 2}$; $(g \circ f)(x) = -5\sqrt{x} + 2$ **23.** $H(x) = (g \circ h)(x)$ **25.** $F(x) = (h \circ f)(x)$ **27.** $G(x) = (f \circ g)(x)$ **29.** Answers may vary. **31.** Answers may vary. **33.** Answers may vary. **35.** 6 **37.** 4 **39.** 48 **41.** −1 **43.** $P(x) = R(x) - C(x)$ **45.** $y = x - 2$ **47.** $y = \frac{x}{3}$ **49.** $y = -\frac{x + 7}{2}$

***Exercise Set 10.2*** **1.** one-to-one; $f^{-1} = \{(-1, -1), (1, 1), (2, 0), (0, 2)\}$ **3.** one-to-one; $h^{-1} = \{(10, 10)\}$ **5.** one-to-one; $f^{-1} = \{(12, 11), (3, 4), (4, 3), (6, 6)\}$ **7.** not one-to-one **9.** one-to-one;

| Rank in Population (Input) | 1 | 49 | 12 | 2 | 45 |
|---|---|---|---|---|---|
| State (Output) | CA | VT | VA | TX | SD |

**11. a.** 3 **b.** 1 **13. a.** 1 **b.** −1 **15.** one-to-one **17.** not one-to-one **19.** one-to-one **21.** not one-to-one

**23.** $f^{-1}(x) = \frac{x - 2}{5}$ **25.** $f^{-1}(x) = \frac{x + 3}{2}$ **27.** $f^{-1}(x) = 2x + 2$ **29.** $f^{-1}(x) = \sqrt[3]{x}$

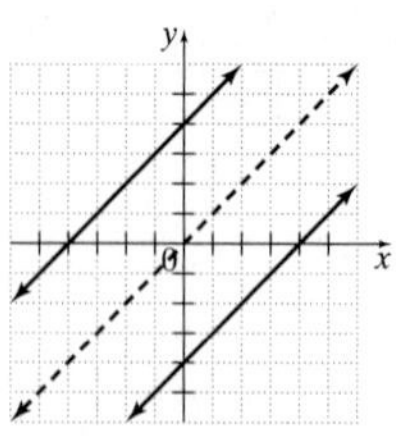

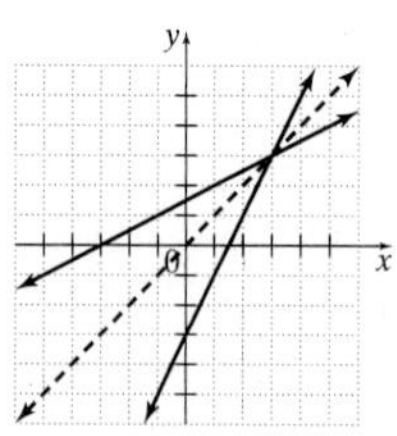

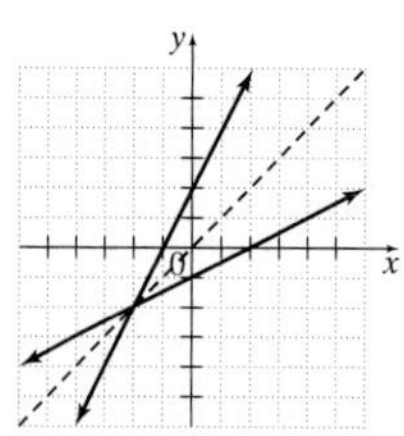

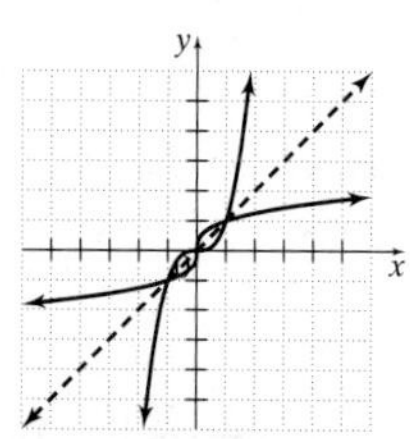

**31.** $f^{-1}(x) = \frac{x - 2}{5}$ **33.** $f^{-1}(x) = 5x + 2$ **35.** $f^{-1}(x) = x^3$ **37.** $f^{-1}(x) = \frac{5 - x}{3x}$ **39.** $f^{-1}(x) = \sqrt[3]{x} - 2$

**41.**
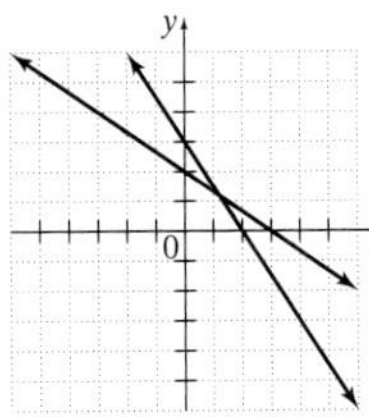

**43.**
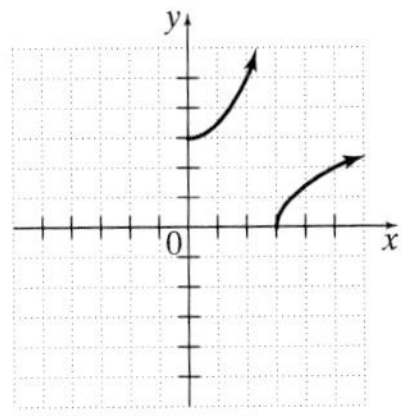

**45.**
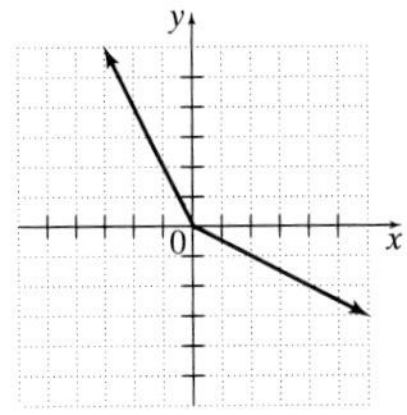

**47.** $(f \circ f^{-1})(x) = x; (f^{-1} \circ f)(x) = x$ **49.** $(f \circ f^{-1})(x) = x; (f^{-1} \circ f)(x) = x$

**51. a.** $\left(-2, \frac{1}{4}\right), \left(-1, \frac{1}{2}\right)$, $(0, 1), (1, 2), (2, 5)$ **b.** $\left(\frac{1}{4}, -2\right), \left(\frac{1}{2}, -1\right)$, $(1, 0), (2, 1), (5, 2)$

**c.**
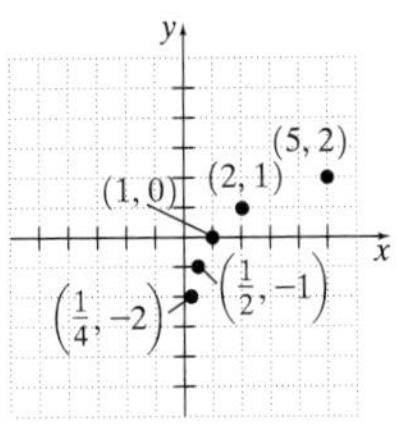


**d.**
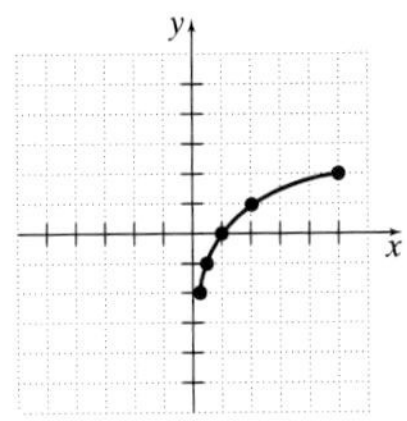

**53.** $f^{-1}(x) = \frac{x - 1}{3}$;
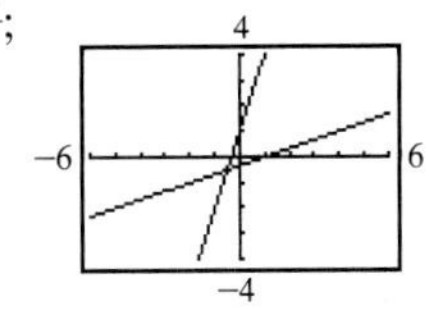


**55.** $f^{-1}(x) = x^3 - 1$;
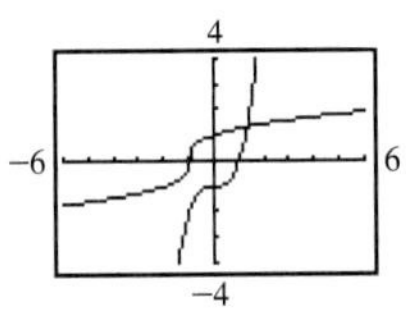


**57.** 5 **59.** 8 **61.** $\frac{1}{27}$ **63.** 9 **65.** $3^{1/2} \approx 1.73$

***Graphing Calculator Explorations:*** **1.** 81.98%;
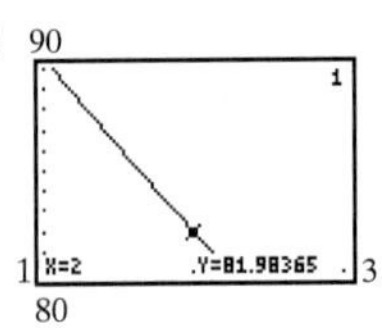


***3.*** 22.54%;
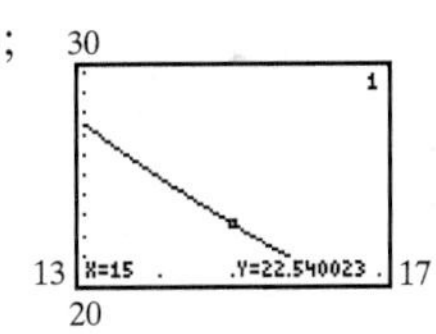


***Exercise Set 10.3*** **1.**
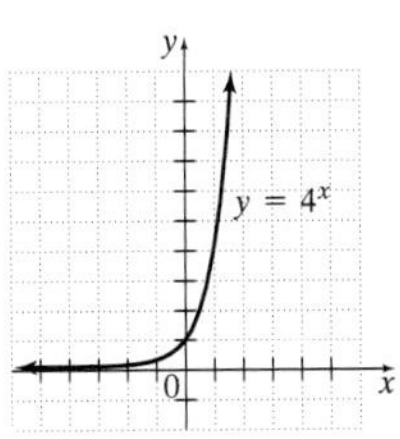


**3.**
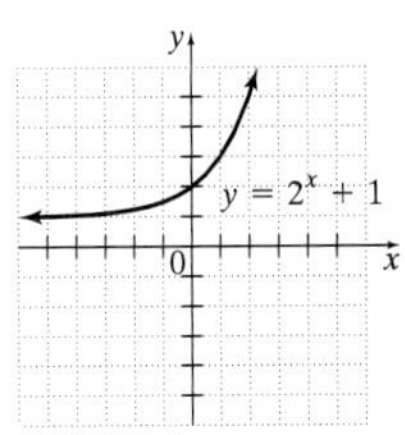


**5.**
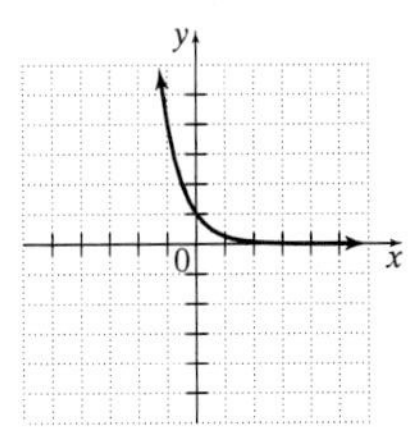

**7.**
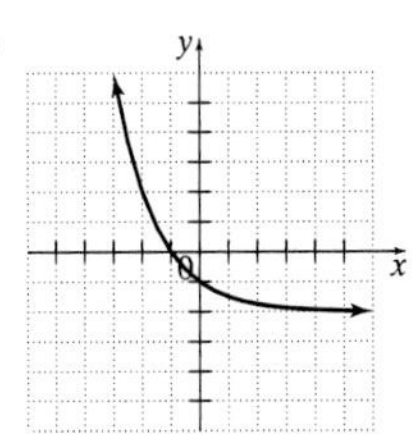

**9.**
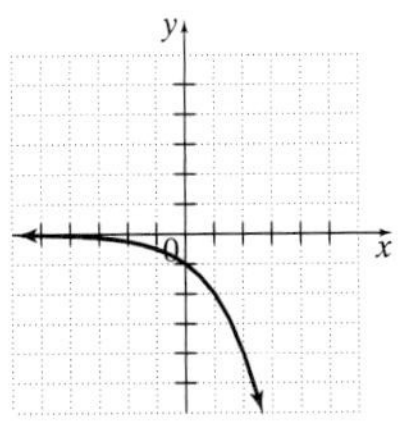

**11.**
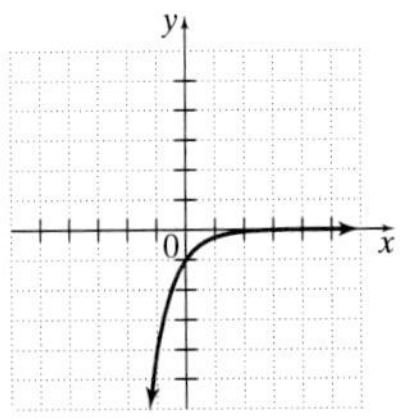

**13.**
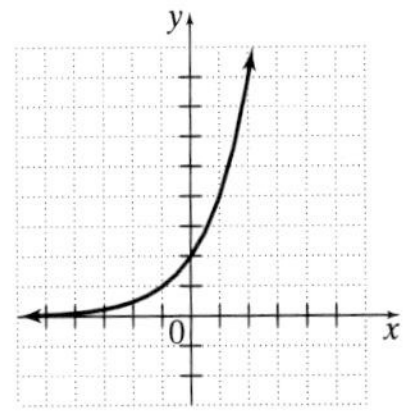

**15.**
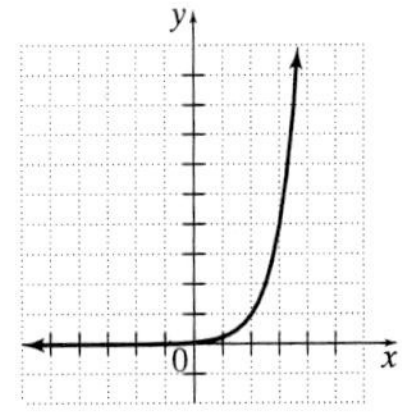

**17.** Answers may vary. **19.** 3 **21.** $\frac{3}{4}$ **23.** $\frac{8}{5}$ **25.** $-\frac{2}{3}$ **27.** 4 **29.** $\frac{3}{2}$ **31.** $-\frac{1}{3}$ **33.** −2 **35.** C **37.** D **39.** 24.6 lb **41.** 519 rats **43.** 1.1 g **45.** 16,190,000 residents **47.** \$7621.42 **49.** \$4065.59 **51.** 24.60 lb;

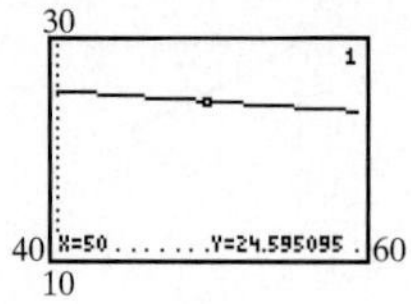


**53.** 18.62 lb;

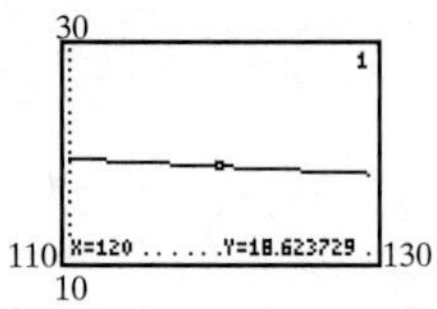


**55.** 50.41 g;

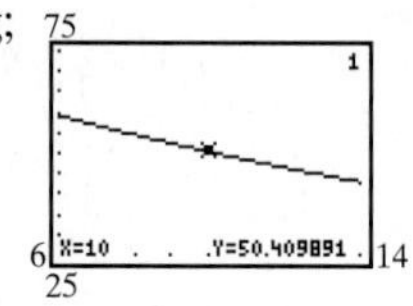


*Exercise Set 10.4* **1.** $6^2 = 36$ **3.** $3^{-3} = \frac{1}{27}$ **5.** $10^3 = 1000$ **7.** $e^4 = x$ **9.** $e^{-2} = \frac{1}{e^2}$ **11.** $7^{1/2} = \sqrt{7}$ **13.** $\log_2 16 = 4$ **15.** $\log_{10} 100 = 2$ **17.** $\log_e x = 3$ **19.** $\log_{10} \frac{1}{10} = -1$ **21.** $\log_4 \frac{1}{16} = -2$ **23.** $\log_5 \sqrt{5} = \frac{1}{2}$ **25.** 3 **27.** −2 **29.** $\frac{1}{2}$ **31.** −1 **33.** 0 **35.** 4 **37.** 2 **39.** 5 **41.** 4 **43.** −3 **45.** Answers may vary. **47.** 2 **49.** 81 **51.** 7 **53.** −3 **55.** −3 **57.** 2 **59.** 2 **61.** $\frac{27}{64}$ **63.** 10 **65.** 3 **67.** 3 **69.** 1

**71.**

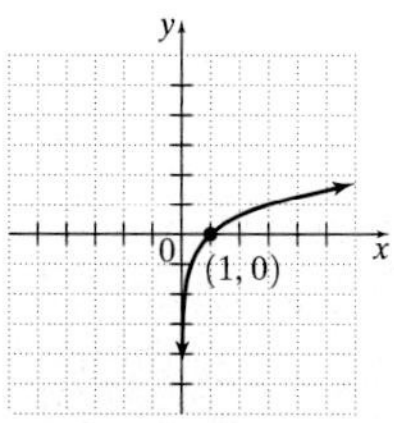


**73.**

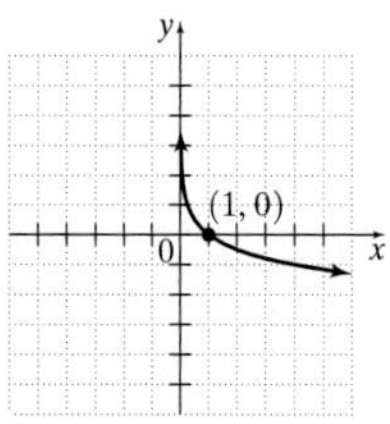


**75.**

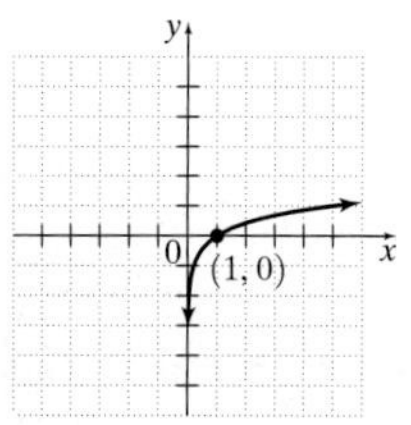


**77.**

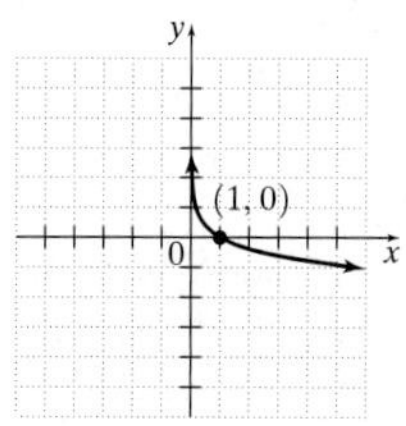


**79.**

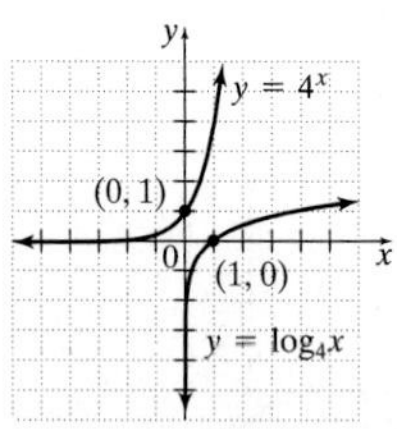


**81.**

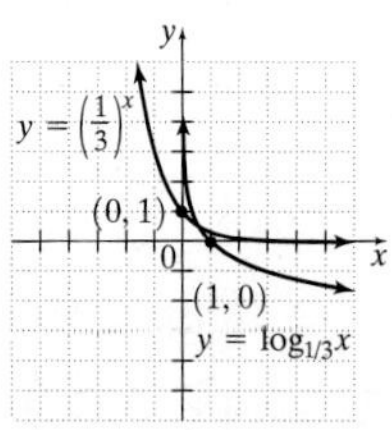


**83.** 0.0827 **85.** 2 and 3 **87.** −1 **89.** $x - 5$ **91.** 3 **93.** $\frac{y - 9}{y^2 - 1}$

*Exercise Set 10.5* **1.** $\log_5 14$ **3.** $\log_4 9x$ **5.** $\log_{10} (10x^2 + 20)$ **7.** $\log_5 3$ **9.** $\log_2 \frac{x}{y}$ **11.** $\log_4 4$, or 1 **13.** $2 \log_3 x$ **15.** $-1 \log_4 5 = -\log_4 5$ **17.** $\frac{1}{2} \log_5 y$ **19.** $\log_2 25$ **21.** $\log_5 x^3z^6$ **23.** $\log_{10} \frac{x^3 - 2x}{x + 1}$ **25.** $\log_4 35$ **27.** $\log_3 4$ **29.** $\log_7 \frac{9}{2}$ **31.** $\log_4 48$ **33.** $\log_2 \frac{x^{7/2}}{(x + 1)^2}$ **35.** $\log_8 x^{16/3}$ **37.** $\log_2 7 + \log_2 11 - \log_2 3$ **39.** $\log_3 4 + \log_3 y - \log_3 5$ **43.** $\frac{1}{2} \log_b 7 + \frac{1}{2} \log_b x$ **45.** $\log_7 5 + \log_7 x - \log_7 4$ **47.** $3 \log_5 x + \log_5(x + 1)$ **49.** $2 \log_6 x - \log_6(x + 3)$ **51.** 0.2 **53.** 1.2 **55.** 0.233 **57.** true **59.** false **61.** false **63.** 1.29 **65.** −0.68 **67.** −0.125

**69.**

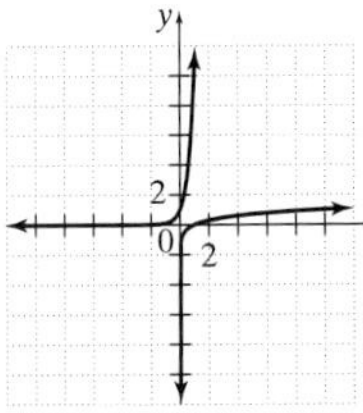


**71.** $-1$ **73.** $\frac{1}{2}$

***Exercise Set 10.6*** **1.** 0.9031 **3.** 0.3636 **5.** 0.6931 **7.** $-2.6367$ **9.** 1.1004 **11.** 1.6094 **13.** 1.6180 **15.** Answers may vary. **17.** 2 **19.** $-3$ **21.** 2 **23.** $\frac{1}{4}$ **25.** 3 **27.** 2 **29.** $-4$ **31.** $\frac{1}{2}$ **33.** Answers may vary. **35.** $10^{1.3} \approx 19.9526$ **37.** $\frac{10^{1.1}}{2} \approx 6.2946$ **39.** $e^{1.4} \approx 4.0552$ **41.** $\frac{4 + e^{2.3}}{3} \approx 4.6581$ **43.** $10^{2.3} \approx 199.5262$ **45.** $e^{-2.3} \approx 0.1003$ **47.** $\frac{10^{-0.5} - 1}{2} \approx -0.3419$ **49.** $\frac{e^{0.18}}{4} \approx 0.2993$ **51.** 1.5850 **53.** $-2.3219$ **55.** 1.5850 **57.** $-1.6309$ **59.** 0.8617 **61.** 4.2 **63.** 5.3 **65.** \$3656.38 **67.** \$2542.50

**69.**

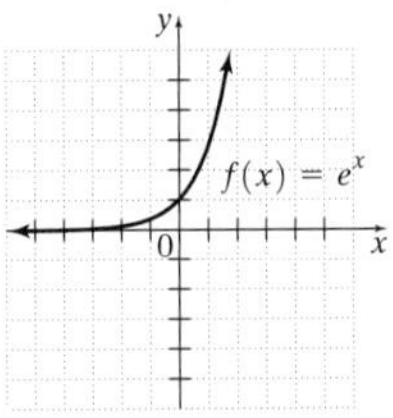


**71.**

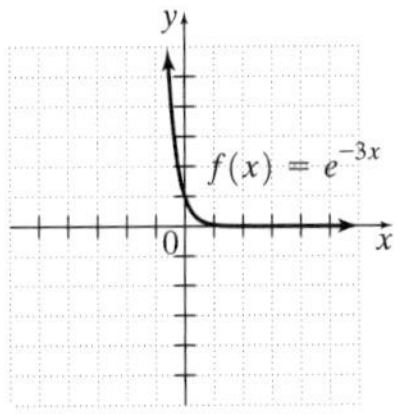


**73.**

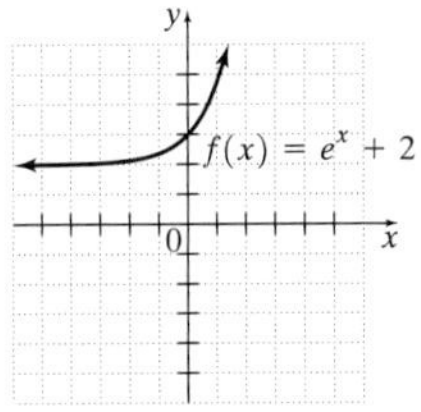


**75.**

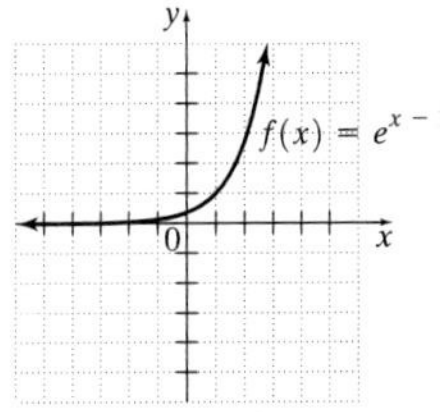


**77.**

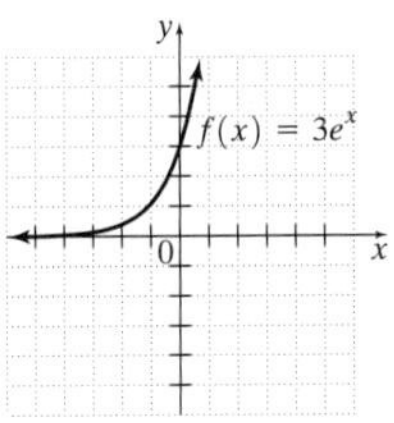


**79.**

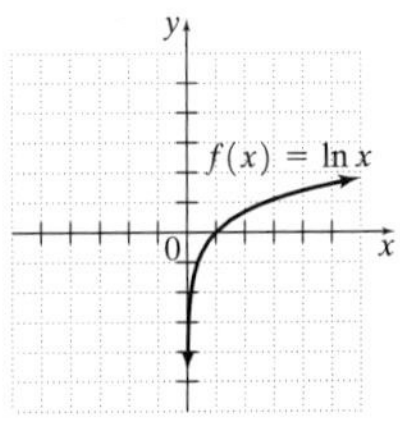


**81.**

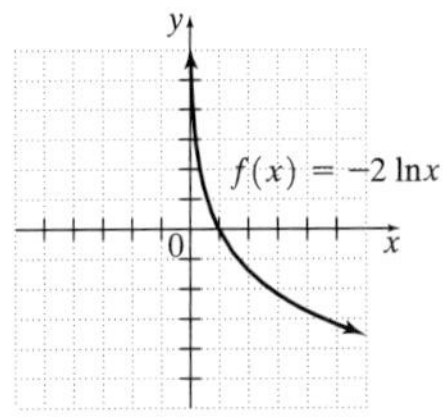


**83.**

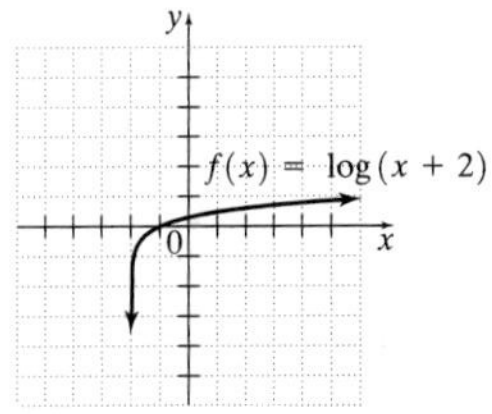


**85.**

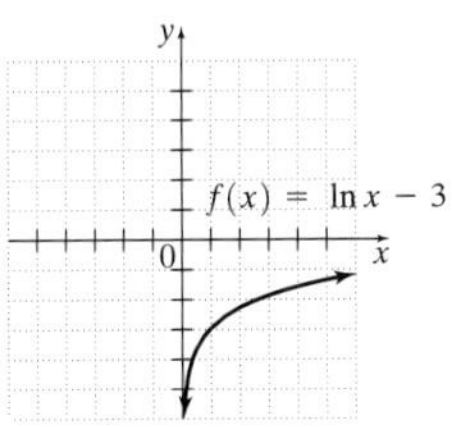


**87.**

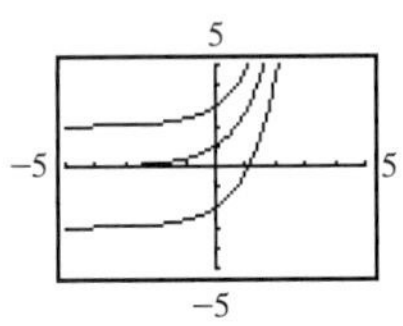


Answers may vary.

**89.** $\frac{4}{7}$ **91.** $x = \frac{3y}{4}$ **93.** $-6, -1$ **95.** $(2, -3)$

***Graphing Calculator Explorations:*** **1.** 3.67 years, or 3 years and 8 months **3.** 23.16 years, or 23 years and 2 months

***Exercise Set 10.7*** **1.** $\frac{\log 6}{\log 3}$; 1.6309 **3.** $\frac{\log 3.8}{2 \log 3}$; 0.6076 **5.** $3 + \frac{\log 5}{\log 2}$; 5.3219 **7.** $\frac{\log 5}{\log 9}$; 0.7325 **9.** $\frac{\log 3}{\log 4} - 7$; $-6.2075$ **11.** $\frac{1}{3}\left(4 + \frac{\log 11}{\log 7}\right)$; 1.7441 **13.** $\frac{\ln 5}{6}$; 0.2682 **15.** 11 **17.** 9, $-9$ **19.** $\frac{1}{2}$ **21.** $\frac{3}{4}$ **23.** 2 **25.** $\frac{1}{8}$ **27.** 11 **29.** 4, $-1$ **31.** $\frac{1}{5}$ **33.** 100 **35.** $\frac{-5 + \sqrt{33}}{2}$ **37.** $\frac{192}{127}$ **39.** $\frac{2}{3}$ **41.** 103 wolves **43.** 11,750,000 inhabitants **45.** 39.8 yr **47.** 10 yr **49.** 1.7 yr **51.** 8.8 yr **55.** 55.7 in. **57.** 11.9 lb/sq. in. **59.** 3.2 mi **61.** 12 weeks **63.** 18 weeks **65.** 6.93 **67.** $-3.68$ **69.** 1.74 **71.** 0.2 **73.** $-\frac{5}{3}$ **75.** $\frac{17}{4}$ **77.** $f^{-1}(x) = \frac{x - 2}{5}$

***Chapter 10 Review*** **1.** $3x - 4$ **3.** $2x^2 - 9x - 5$ **5.** $x^2 + 2x - 1$ **7.** 18 **9.** $-2$
**11.** one-to-one; $h^{-1} = \{(14, -9), (8, 6), (12, -11), (15, 15)\}$
**13.** one-to-one;

| Rank in Automobile Thefts (Input) | 2 | 4 | 1 | 3 |
|---|---|---|---|---|
| US Region (Output) | W | Midwest | S | NE |

**15. a.** 3 **b.** 7 **17.** not one-to-one **19.** not one-to-one

**21.** $f^{-1}(x) = x + 9$ **23.** $f^{-1}(x) = \frac{x - 11}{6}$ **25.** $f^{-1}(x) = \sqrt[3]{x + 5}$ **27.** $g^{-1}(x) = \frac{6x + 7}{12}$

**29.**

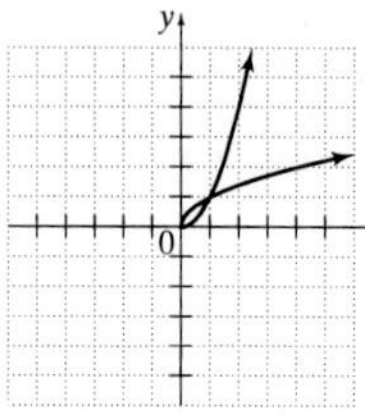


**31.** $f^{-1}(x) = \frac{x + 3}{2}$;

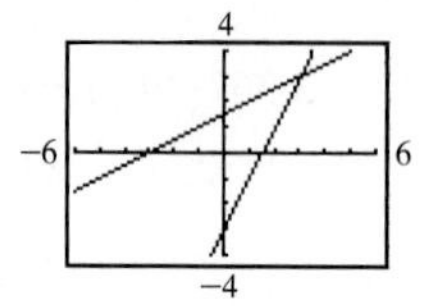


**33.** $-2$ **35.** $\frac{3}{2}$ **37.** $\frac{8}{9}$

**39.**

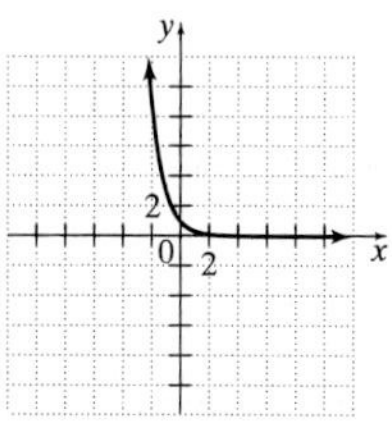


**41.**

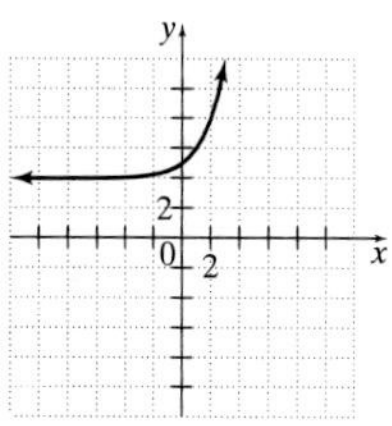


**43.** \$1131.82 **45.** $\log_7 49 = 2$ **47.** $\left(\frac{1}{2}\right)^{-4} = 16$ **49.** $\frac{1}{64}$
**51.** 0 **53.** 8 **55.** 5 **57.** 4 **59.** $\frac{17}{3}$ **61.** $-1, 4$

**63.**

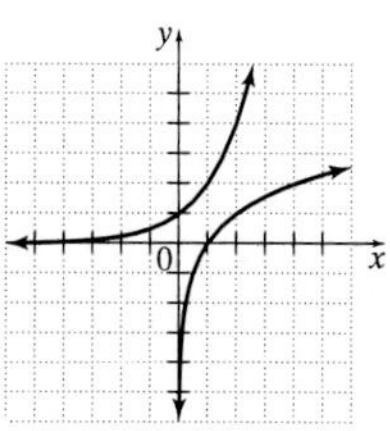


**65.** $\log_3 32$ **67.** $\log_7 \frac{3}{4}$ **69.** $\log_{11} 4$ **71.** $\log_5 \frac{x^3}{(x + 1)^2}$ **73.** $3 \log_3 x - \log_3(x + 2)$
**75.** $\log_2 3 + 2 \log_2 x + \log_2 y - \log_2 z$ **77.** 2.02 **79.** 0.5563 **81.** 0.2231 **83.** 3
**85.** $-1$ **87.** $\frac{e^2}{2}$ **89.** $\frac{e^{-1} + 3}{2}$ **91.** 1.67 mm **93.** 0.2920 **95.** \$1957.30
**97.** $\frac{\log 7}{2 \log 3}$; 0.8856 **99.** $\frac{1}{2}\left(\frac{\log 6}{\log 3} - 1\right)$; 0.3155 **101.** $\frac{1}{3}\left(\frac{\log 4}{\log 5} + 5\right)$; 1.9538

**103.** $-\frac{\log 2}{\log 5} + 1$; 0.5693 **105.** $\frac{25}{2}$ **107.** $\varnothing$ **109.** $2\sqrt{2}$ **111.** 197,044 ducks **113.** 53 yr **115.** 36 yr

**117.** 8.5 yr **119.** 2.82

***Chapter 10 Test*** **1.** 5 **2.** $x - 7$ **3.** $x^2 - 6x - 2$

**4.**

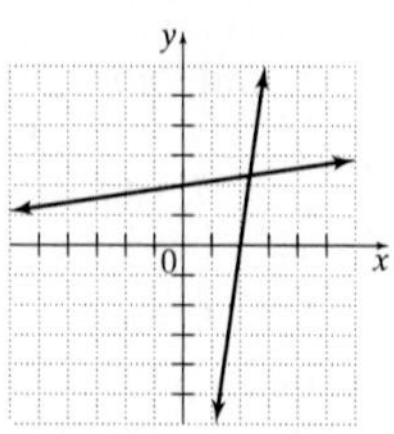


**5.** one-to-one **6.** not one-to-one **7.** one-to-one; $f^{-1}(x) = \frac{-x + 6}{2}$

**8.** one-to-one; $f^{-1} = \{(0, 0), (3, 2), (5, -1)\}$ **9.** not one-to-one **10.** $\log_3 24$

**11.** $\log_5 \frac{x^4}{x + 1}$ **12.** $\log_6 2 + \log_6 x - 3 \log_6 y$ **13.** $-1.53$ **14.** 1.0686 **15.** $-1$

**16.** $\frac{1}{2}\left(\frac{\log 4}{\log 3} - 5\right)$; $-1.8691$ **17.** $\frac{1}{9}$ **18.** $\frac{1}{2}$ **19.** 22 **20.** $\frac{25}{3}$ **21.** $\frac{43}{21}$ **22.** $-1.0979$

**23.** 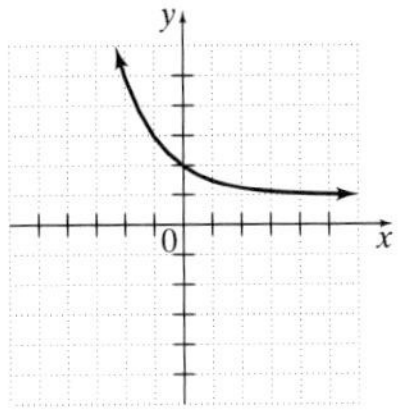

**24.** 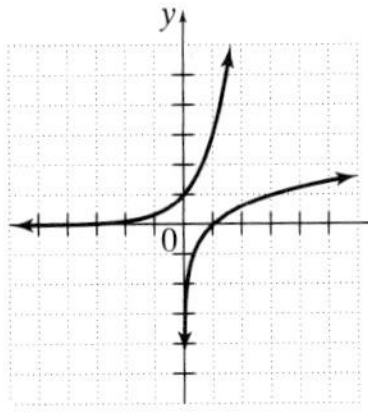

**25.** \$5234.58 **26.** 6 yr **27.** 64,913 prairie dogs **28.** 15 yr **29.** 1.2% **30.** 3.95;

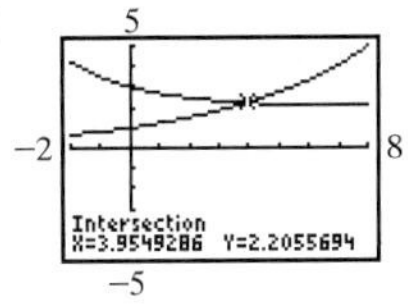


***Chapter 10 Cumulative Review*** **1. a.** 8 **b.** $-\frac{1}{3}$ **c.** $-9$ **d.** 0 **e.** $-\frac{2}{11}$ **f.** 42 **g.** 0; Sec. 1.3, Ex. 4

**2.** 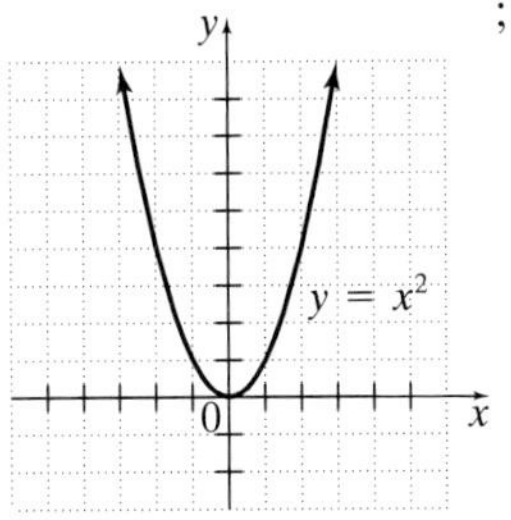


; Section. 3.1, Ex. 6 **3.** $\{(x, y, z) \mid x - 5y - 2z = 6\}$; Sec. 4.2, Ex. 4 **4. a.** $x^3$ **b.** $5^6$ **c.** $5x$ **d.** $\frac{6y^2}{7}$; Sec. 5.1, Ex. 4 **5. a.** \$102.60 **b.** \$12.60; Sec. 6.1, Ex. 1 **6. a.** $\frac{3x}{2}$ **b.** $x - 7$ **c.** $-\frac{1}{3y^2}$; Sec. 6.2, Ex. 1 **7.** $3x^2 + 2x + 3 + \frac{-6x + 9}{x^2 - 1}$; Sec. 6.4, Ex. 5 **8.** $-1$; Sec. 6.6, Ex. 4 **9.** 6 mph; Sec. 6.7, Ex. 5 **10. a.** 3 **b.** $-3$ **c.** $-5$ **d.** not a real number **e.** $4x$; Sec. 7.1, Ex. 4 **11. a.** $\sqrt[4]{x^3}$ **b.** $\sqrt[6]{x}$ **c.** $\sqrt[6]{72}$; Sec. 7.2, Ex. 8 **12. a.** $5\sqrt{3} + 3\sqrt{10}$ **b.** $\sqrt{35} + \sqrt{5} - \sqrt{42} - \sqrt{6}$ **c.** $21x - 7\sqrt{5x} + 15\sqrt{x} - 5\sqrt{5}$ **d.** $49 - 8\sqrt{3}$ **e.** $2x - 25$; Sec. 7.4, Ex. 3 **13.** $\frac{\sqrt[4]{xy^3}}{3y^2}$; Sec. 7.5, Ex. 3 **14.** 3; Sec. 7.6, Ex. 4

**15.** $\frac{9 + i\sqrt{15}}{6}, \frac{9 - i\sqrt{15}}{6}$; Sec. 8.1, Ex. 8 **16.** $\frac{-1 + \sqrt{33}}{4}, \frac{-1 - \sqrt{33}}{4}$; Sec. 8.3, Ex. 2 **17.** [0, 4]; Sec. 8.4, Ex. 2

**18.** 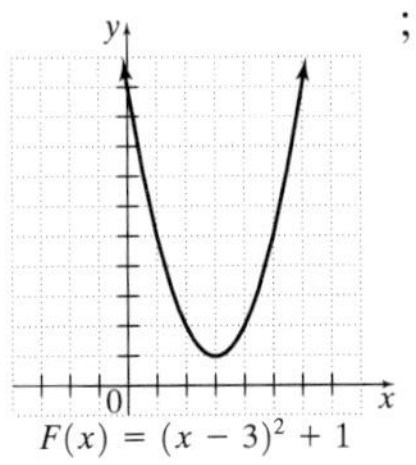


; Sec. 8.5, Ex. 2 **19.** $\left(-1, \frac{3}{2}\right)$; Sec. 9.1, Ex. 6

**20.** 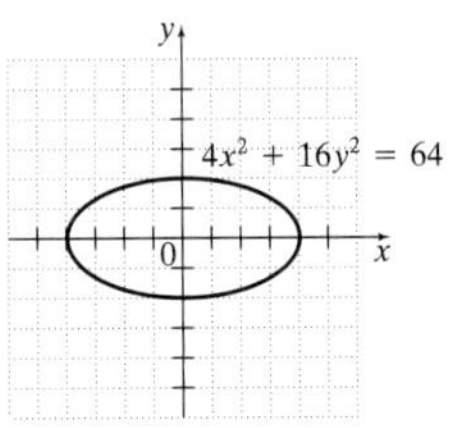


; Sec. 9.2, Ex. 2 **21.** $\varnothing$; Sec. 9.3, Ex. 3 **22.** 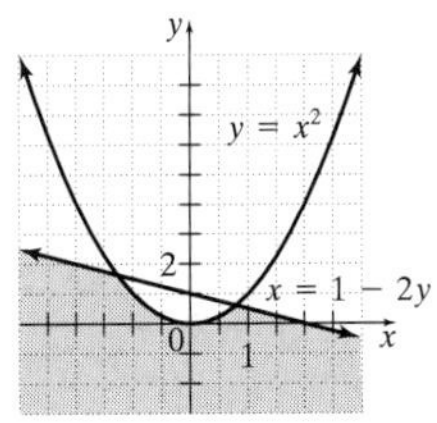


; Sec. 9.4, Ex. 3

**23. a.** $3x - 4$ **b.** $-x + 2$ **c.** $2x^2 - 5x + 3$ **d.** $\frac{x - 1}{2x - 3}$, where $x \neq \frac{3}{2}$; Sec. 10.1, Ex. 1

**24.** $f^{-1}(x) = x - 3$; Sec. 10.2, Ex. 4 **25. a.** 2 **b.** $-1$ **c.** $\frac{1}{2}$; Sec. 10.4, Ex. 3

## ■ CHAPTER 11 SEQUENCES, SERIES, AND THE BINOMIAL THEOREM

***Exercise Set 11.1*** **1.** 5, 6, 7, 8, 9 **3.** −1, 1, −1, 1, −1 **5.** $\frac{1}{4}, \frac{1}{5}, \frac{1}{6}, \frac{1}{7}, \frac{1}{8}$ **7.** 2, 4, 6, 8, 10 **9.** −1, −4, −9, −16, −25 **11.** 2, 4, 8, 16, 32 **13.** 7, 9, 11, 13, 15 **15.** −1, 4, −9, 16, −25 **17.** 75 **19.** 118 **21.** $\frac{6}{5}$ **23.** 729 **25.** $\frac{4}{7}$ **27.** $\frac{1}{8}$ **29.** −95 **31.** $-\frac{1}{25}$ **33.** $a_n = 4n - 1$ **35.** $a_n = -2^n$ **37.** $a_n = \frac{1}{3^n}$ **39.** 48 ft, 80 ft, and 112 ft **41.** $a_n = 0.10(2)^{n-1}$; \$819.20 **43.** 2400 cases; 75 cases **45.** 50 sparrows in 2000: extinct in 2006 or 2007 **47.** 1, 0.7071, 0.5774, 0.5, 0.4472 **49.** 2, 2.25, 2.3704, 2.4414, 2.4883 **51.**

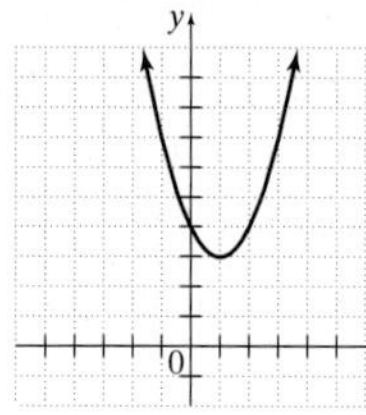


**53.**

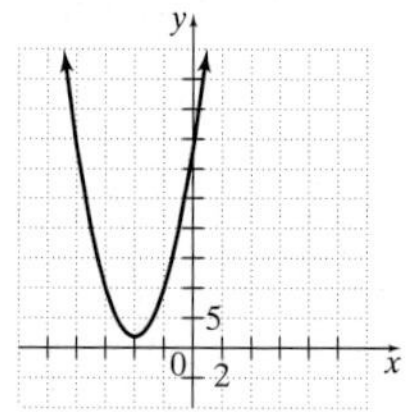


**55.** $\sqrt{13}$ units **57.** $\sqrt{41}$ units

***Exercise Set 11.2*** **1.** 4, 6, 8, 10, 12 **3.** 6, 4, 2, 0, −2 **5.** 1, 3, 9, 27, 81 **7.** 48, 24, 12, 6, 3 **9.** 33 **11.** −875 **13.** −60 **15.** 96 **17.** −28 **19.** 1250 **21.** 31 **23.** 20 **25.** $a_1 = = \frac{2}{3}; r = -2$ **27.** Answers may vary. **29.** $a_1 = 2; d = 2$ **31.** $a_1 = 5; r = 2$ **33.** $a_1 = \frac{1}{2}; r = \frac{1}{5}$ **35.** $a_1 = x; r = 5$ **37.** $a_1 = p; d = 4$ **39.** 19 **41.** $-\frac{8}{9}$ **43.** $\frac{17}{2}$ **45.** $\frac{8}{81}$ **47.** −19 **49.** $a_n = 4n + 50$; 130 seats **51.** $a_n = 6(3)^{n-1}$ **53.** 486, 162, 54, 18, 6; $a_n = \frac{486}{3^{n-1}}$; 6 bounces **55.** $a_n = 4000 + 125(n - 1)$ or $a_n = 3875 + 125n$; \$5375 **57.** 25 grams **59.** \$11,782.40, \$5891.20, \$2945.60, \$1472.80 **61.** 19.652, 19.618, 19.584, 19.55 **63.** Answers may vary. **65.** $\frac{11}{18}$ **67.** 40 **69.** $\frac{907}{495}$

***Exercise Set 11.3*** **1.** −2 **3.** 60 **5.** 20 **7.** $\frac{73}{168}$ **9.** $\frac{11}{36}$ **11.** 60 **13.** 74 **15.** 62 **17.** $\frac{241}{35}$ **19.** $\sum_{i=1}^{5}(2i - 1)$ **21.** $\sum_{i=1}^{4} 4(3)^{i-1}$ **23.** $\sum_{i=1}^{6}(-3i + 15)$ **25.** $\sum_{i=1}^{4}\frac{4}{3^{i-2}}$ **27.** $\sum_{i=1}^{7} i^2$ **29.** −24 **31.** −13 **33.** 82 **35.** −20 **37.** −2 **39.** 1, 2, 3, …, 10; 55 trees **41.** $a_n = 6(2)^{n-1}$; 96 units **43.** $a_n = 50(2)^n$; $n$ represents the number of 12-hour periods; 800 bacteria **45.** 30 opossums; 68 opossums **47.** 6.25 lb; 93.75 lb **49.** 16.4 in.; 134.5 in. **51. a.** $2 + 6 + 12 + 20 + 30 + 42 + 56$ **b.** $1 + 2 + 3 + 4 + 5 + 6 + 7 + 1 + 4 + 9 + 16 + 25 + 36 + 49$ **c.** Answers may vary. **d.** true **53.** 10 **55.** $\frac{10}{27}$ **57.** 45 **59.** 90

***Exercise Set 11.4*** **1.** 36 **3.** 484 **5.** 63 **7.** 2.496 **9.** 55 **11.** 16 **13.** 24 **15.** $\frac{1}{9}$ **17.** −20 **19.** $\frac{16}{9}$ **21.** $\frac{4}{9}$ **23.** 185 **25.** $\frac{381}{64}$ **27.** $-\frac{33}{4}$, or −8.25 **29.** $-\frac{75}{2}$ **31.** $\frac{56}{9}$ **33.** 4000, 3950, 3900, 3850, 3800: 3450 cars; 44,700 cars **35.** Firm *A* (Firm *A*, \$265,000; Firm *B*, \$254,000) **37.** \$39,930; \$139,230 **39.** 20 min; 123 min **41.** 180 ft **43.** Player *A*, 45 points; Player *B*, 75 points **45.** \$3050 **47.** \$10,737,418.23 **49.** $\frac{8}{10} + \frac{8}{100} + \frac{8}{1000} + \cdots; \frac{8}{9}$ **51.** Answers may vary. **53.** 720 **55.** 3 **57.** $x^2 + 10x + 25$ **59.** $8x^3 - 12x^2 + 6x - 1$

***Exercise Set 11.5*** **1.** $m^3 + 3m^2n + 3mn^2 + n^3$ **3.** $c^5 + 5c^4d + 10c^3d^2 + 10c^2d^3 + 5cd^4 + d^5$ **5.** $y^5 - 5y^4x + 10y^3x^2 - 10y^2x^3 + 5yx^4 - x^5$ **7.** Answers may vary. **9.** 8 **11.** 42 **13.** 360 **15.** 56 **17.** $a^7 + 7a^6b + 21a^5b^2 + 35a^4b^3 + 35a^3b^4 + 21a^2b^5 + 7ab^6 + b^7$ **19.** $a^5 + 10a^4b + 40a^3b^2 + 80a^2b^3 + 80ab^4 + 32b^5$ **21.** $q^9 + 9q^8r + 36q^7r^2 + 84q^6r^3 + 126q^5r^4 + 126q^4r^5 + 84q^3r^6 + 36q^2r^7 + 9qr^8 + r^9$ **23.** $1024a^5 + 1280a^4b + 640a^3b^2 + 160a^2b^3 + 20ab^4 + b^5$ **25.** $625a^4 - 1000a^3b + 600a^2b^2 - 160ab^3 + 16b^4$ **27.** $8a^3 + 36a^2b + 54ab^2 + 27b^3$ **29.** $x^5 + 10x^4 + 40x^3 + 80x^2 + 80x + 32$ **31.** $5cd^4$ **33.** $d^7$ **35.** $-40r^2s^3$

**37.** $6x^2y^2$ **39.** $30a^9b$ **41.**

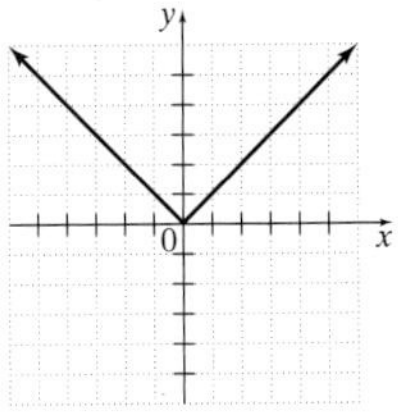


**43.**

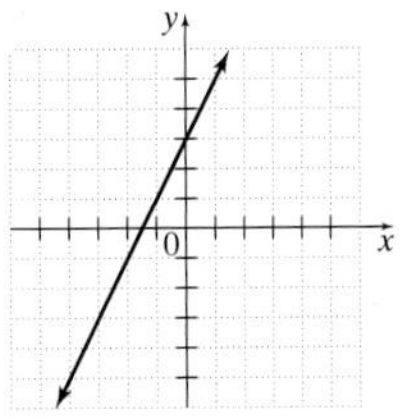


**45.**

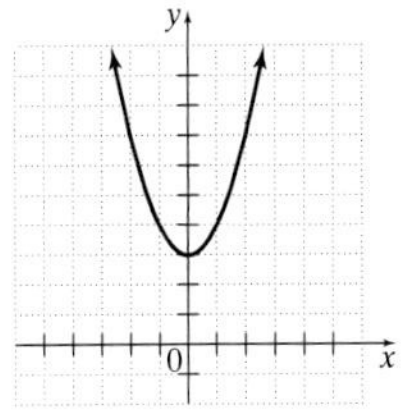


***Chapter 11 Review*** **1.** −3, −12, −27, −48, −75 **3.** $\frac{1}{100}$ **5.** $a_n = \frac{1}{6n}$ **7.** 144 ft, 176 ft, 208 ft **9.** 450, 1350, 4050, 12,150, 36,450; 36,450 infected people in 1999 **11.** $-2, -\frac{4}{3}, -\frac{8}{9}, -\frac{16}{27}, -\frac{32}{81}$ **13.** 111 **15.** −83 **17.** $a_1 = 3; d = 5$ **19.** $a_n = \frac{3}{10^n}$ **21.** $a_1 = \frac{8}{3}, r = \frac{3}{2}$ **23.** $a_1 = 7x, r = -2$ **25.** 8, 6, 4.5, 3.4, 2.5, 1.9; good **27.** $a_n = 2^{n-1}$, \$512, \$536,870,912 **29.** $a_n = 900 + (n - 1)150$ or $a_n = 150n + 750$; \$1650/month **31.** $1 + 3 + 5 + 7 + 9 = 25$ **33.** $\frac{1}{4} - \frac{1}{6} + \frac{1}{8} = \frac{5}{24}$ **35.** −4 **37.** −10 **39.** $\sum_{i=1}^{6} 3^{i-1}$ **41.** $\sum_{i=1}^{4} \frac{1}{4^i}$ **43.** $a_n = 20(2)^n$; $n$ represents the number of 8-hour periods; 1280 yeast **45.** Job *A*, \$48,300; Job *B*, \$46,000 **47.** 150 **49.** 900 **51.** −410 **53.** 936 **55.** 10 **57.** −25 **59.** \$30,418; \$99,868 **61.** \$58; \$553 **63.** 2696 mosquitoes **65.** $\frac{5}{9}$ **67.** $x^5 + 5x^4z + 10x^3z^2 + 10x^2z^3 + 5xz^4 + z^5$ **69.** $16x^4 + 32x^3y + 24x^2y^2 + 8xy^3 + y^4$ **71.** $b^8 + 8b^7c + 28b^6c^2 + 56b^5c^3 + 70b^4c^4 + 56b^3c^5 + 28b^2c^6 + 8bc^7 + c^8$ **73.** $256m^4 - 256m^3n + 96m^2n^2 - 16mn^3 + n^4$ **75.** $35a^4b^3$

***Chapter 11 Test*** **1.** $-\frac{1}{5}, \frac{1}{6}, -\frac{1}{7}, \frac{1}{8}, -\frac{1}{9}$ **2.** −3, 3, −3, 3, −3 **3.** 247 **4.** 39,999 **5.** $a_n = \frac{2}{5}\left(\frac{1}{5}\right)^{n-1}$ **6.** $a_n = (-1)^n 9n$ **7.** 155 **8.** −330 **9.** $\frac{144}{5}$ **10.** 1 **11.** 10 **12.** −60 **13.** $a^6 - 6a^5b + 15a^4b^2 - 20a^3b^3 + 15a^2b^4 - 6ab^5 + b^6$ **14.** $32x^5 + 80x^4y + 80x^3y^2 + 40x^2y^3 + 10xy^4 + y^5$ **15.** $y^8 + 8y^7z + 28y^6z^2 + 56y^5z^3 + 70y^4z^4 + 56y^3z^5 + 28y^2z^6 + 8yz^7 + z^8$ **16.** $128p^7 + 448p^6r + 672p^5r^2 + 560p^4r^3 + 280p^3r^4 + 84p^2r^5 + 14pr^6 + r^7$ **17.** 925 people; 250 people initially **18.** $1 + 3 + 5 + 7 + 9 + 11 + 13 + 15$; 64 shrubs **19.** 33.75 cm, 218.75 cm **20.** 320 cm **21.** 304 ft; 1600 ft **22.** $\frac{14}{33}$

***Chapter 11 Cumulative Review*** **1. a.** −5 **b.** 3 **c.** $-\frac{1}{8}$ **d.** −4 **e.** $\frac{1}{4}$ **f.** undefined; Sec. 1.3, Ex. 5 **2.** \$2350; Sec. 2.2, Ex. 5 **3. a.** (−5, 2) **b.** (1, 0); Sec. 4.5, Ex. 2 **4. a.** $15x^7$ **b.** $-8x^4p^{12}$ **5.** $x^3 - 4x^2 - 3x + 11 + \frac{12}{x + 2}$; Sec. 6.5, Ex. 2 **6. a.** $5\sqrt{2}$ **b.** $2\sqrt[3]{3}$ **c.** $\sqrt{26}$ **d.** $2\sqrt[4]{2}$; Sec. 7.3, Ex. 3 **7.** 10%; Sec. 8.1, Ex. 9 **8.** 2, 7; Sec. 8.3, Ex. 4 **9.** $\left(-\frac{7}{2}, -1\right)$; Sec. 8.4, Ex. 5 **10.** $\frac{25}{4}$ ft; $\frac{5}{8}$ sec; Sec. 8.6, Ex. 5

**11.**

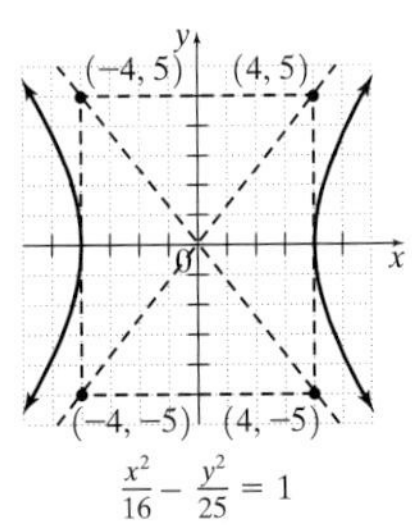


$\frac{x^2}{16} - \frac{y^2}{25} = 1$

; Sec. 9.2, Ex. 4 **12.** $(2, \sqrt{2})$; Sec. 9.3, Ex. 2

**13.** 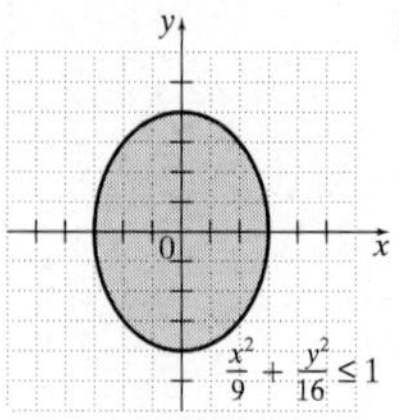

; Sec. 9.4, Ex. 1

**14. a.** 25; 7 **b.** $x^2 + 6x + 9$; $x^2 + 3$; Sec. 10.1, Ex. 2 **15.** $f^{-1} = \{(1, 0), (7, -2), (-6, 3), (4, 4)\}$; Sec. 10.2, Ex. 3 **16. a.** 4 **b.** $\frac{3}{2}$ **c.** 6; Sec. 10.3, Ex. 4 **17. a.** 2 **b.** $-1$ **c.** 3 **d.** 6; Sec. 10.4, Ex. 5 **18. a.** $\log_{11}30$ **b.** $\log_3 6$ **c.** $\log_2(x^2 + 2x)$; Sec. 10.5, Ex. 1 **19.** \$2509.30; Sec. 10.6, Ex. 8 **20.** $\frac{\log 7}{\log 3} \approx 1.7712$; Sec. 10.7, Ex. 1 **21.** 18; Sec. 10.7, Ex. 2 **22.** 0, 3, 8, 15, 24; Sec. 11.1, Ex. 1 **23.** 72; Sec. 11.2, Ex. 3 **24. a.** $\frac{7}{2}$ **b.** 56; Sec. 11.3, Ex. 1 **25.** 465; Sec. 11.4, Ex. 2

## APPENDIX A REVIEW OF ANGLES, LINES, AND SPECIAL TRIANGLES

**1.** 71° **3.** 19.2° **5.** $78\frac{3}{4}°$ **7.** 30° **9.** 149.8° **11.** $100\frac{1}{2}°$ **13.** $m\angle 1 = m\angle 5 = m\angle 7 = 110°$, $m\angle 2 = m\angle 3 = m\angle 4 = m\angle 6 = 70°$ **15.** 90° **17.** 90° **19.** 90° **21.** 45°, 90° **23.** 78°, 90° **25.** $50\frac{1}{4}°$, 90° **27.** $x = 6$ **29.** $x = 4.5$ **31.** 10 **33.** 12

## APPENDIX C REVIEW OF VOLUME AND SURFACE AREA

**1.** $V = 72$ cu. in.; $SA = 108$ sq. in. **3.** $V = 512$ cu. cm; $SA = 384$ sq. cm **5.** $V = 4\pi$ cu. yd $\approx 12\frac{4}{7}$ cu. yd; $SA = (2\sqrt{13}\pi + 4\pi)$ sq yd $\approx 35.20$ sq. yd **7.** $V = \frac{500}{3}\pi$ cu. in. $\approx 523\frac{17}{21}$ cu. in.; $SA = 100\pi$ sq. in. $\approx 314\frac{2}{7}$ sq. in. **9.** $V = 48$ cu. cm; $SA = 96$ sq. cm **11.** $2\frac{10}{27}$ cu. in. **13.** 26 sq. ft **15.** $10\frac{5}{6}$ cu. in. **17.** 960 cu. cm **19.** $196\pi$ sq. in. **21.** $7\frac{1}{2}$ cu. ft **23.** $12\frac{4}{7}$ cu. cm

## APPENDIX D AN INTRODUCTION TO USING A GRAPHING UTILITY

***Viewing Window and Interpreting Window Settings Exercise Set***

**1.** yes **3.** no **5.** Answers may vary. **7.** Answers may vary. **9.** Answers may vary.

**11.** Xmin = −12 Ymin = − 12
Xmax = 12 Ymax = 12
Xscl = 3 Yscl = 3

**13.** Xmin = −9 Ymin = −12
Xmax = 9 Ymax = 12
Xscl = 1 Yscl = 2

**15.** Xmin = −10 Ymin = −25
Xmax = 10 Ymax = 25
Xscl = 2 Yscl = 5

**17.** Xmin = −10 Ymin = −30
Xmax = 10 Ymax = 30
Xscl = 1 Yscl = 3

**19.** Xmin = −20 Ymin = −30
Xmax = 30 Ymax = 50
Xscl = 5 Yscl = 10

### *Graphing Equations and Square Viewing Window Exercise Set*

**1.** Setting B **3.** Setting B **5.** Setting B

**7.**
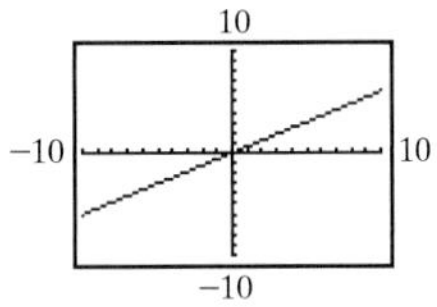


**9.**
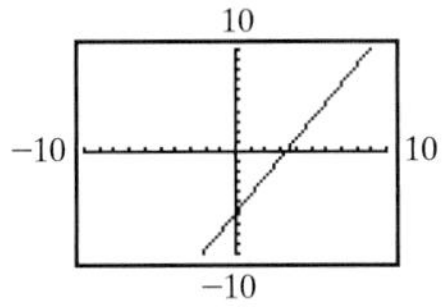


**11.**
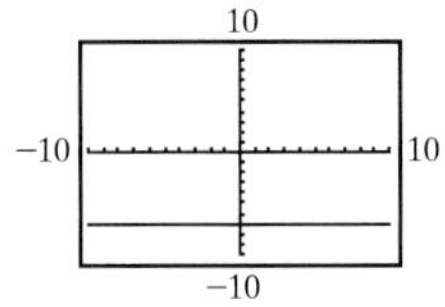


**13.**
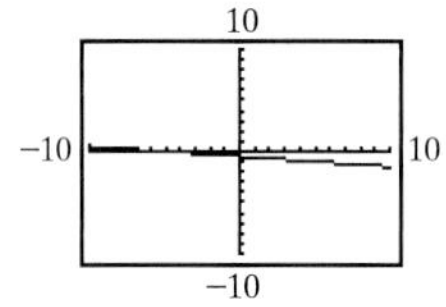


**15.**
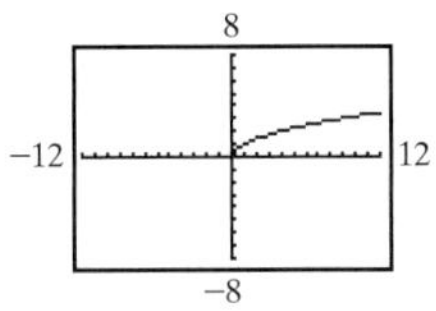


**17.**
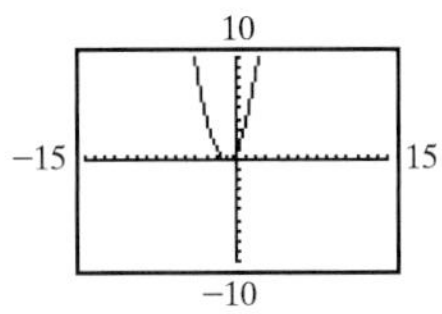


**19.**
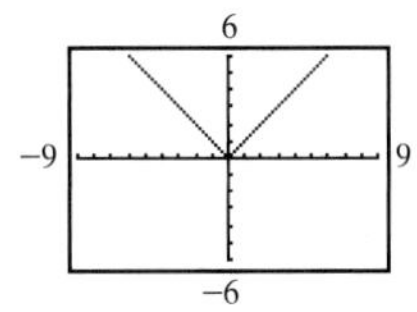


**21.**
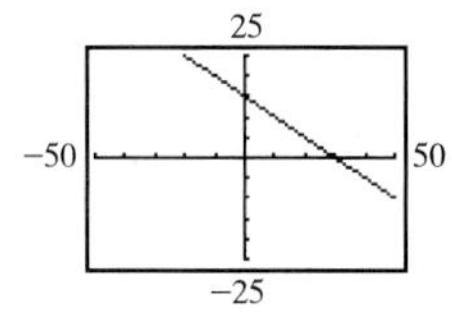

# SUBJECT INDEX


**A**

Abscissa, 118
Absolute value, 13–14, 41
Absolute value bars, 409
Absolute value equations, 110
  solving, 93–97
Absolute value inequalities, 98–102, 110
Acute angle, 665
Addition
  associative property of, 33, 42
  commutative property of, 33, 42
  of complex numbers, 456
  of exponential expressions, 244
  of functions, 566
  of polynomials, 262, 319
  of radical expressions, 426–27, 455
  of rational expressions, 341–42, 393
  of rational expressions with unlike denominators, 343
  of real numbers, 19, 41
  of two numbers with different signs, 19, 41
  of two numbers with same sign, 19, 41
Addition method (or elimination method), for solving nonlinear systems of equations, 547–48, 557
Addition property of equality, 49, 50, 52, 106
Addition property of inequality, 77, 108
Additive identity, 32, 42
Additive inverse (or opposite), 32, 262
Adjacent angles, 667
Algebra, 3
Algebraic expressions, 3, 41
  identifying/evaluating, 8–9
  phrases translated to, 15
  and sets of numbers, 16
  simplifying, 56
  written from word phrases, 34–35
Algebra of functions, 566, 617
Alternate interior angles, 667
American Dietetic Association Web site, 462
American Farmland Trust, 190
American Geological Institute Web site, 2
American Institute of Architects Web site, 524
American Planning Association Web site, 190
American Psychological Association Web site, 404
American Society for Microbiology Web site, 242
American Society of Newspaper Editors Web site, 46
Angles
  finding measures of, 215–16
  types of, 665
Application problems, 81–83
Architect Registration Examination, 525
Architects, 525
Architecture careers, 524
Arithmetic, 3
Arithmetic progression, 634
Arithmetic sequences, 659
  and common differences, 634–36, 659
  partial sum of, 646–47, 660
Arithmetic sequences, 659
Array of signs, 228
Associative property, 36
  of addition, 33, 42
  and combining like terms, 261
  of multiplication, 33, 42
Asymptotes of the hyperbola, 539
Axis of symmetry, 311, 500, 504, 505

**B**

Baby Boomers, and financial management, 629
Base, 23
Base 10, logarithms to, 603
Binomial expansion, finding *n*th term in, raised to positive power, 656
Binomial formula, 655
Binomials, 259
  expanding, 654
  multiplying by another binomial, 270
  Pascal's triangle for expanding, 654
  square of, 271–72
Binomial theorem, 661
  for expanding binomials, 655–56
Boundary equation, 172
Boundary line, 172, 173, 174
Boyle's law, and inverse variation, 385
Break-even point, 192
  finding, 214–15

**C**

Calculator. (*See also* Graphing calculator explorations; Graphing utility)
  approximating common logarithms by, 603
  approximating natural logarithms by, 605
Cartesian coordinate system, 118, 179, 529
Center
  of circle, 531, 532, 674
  in ellipse, 535, 536
  in hyperbola, 538
Certified Financial Planner Board of Standards Web site, 628
Certified financial planners, 629
Change of base
  and common logarithms, 603

Change of base (*cont.*)
formula for, 607
and natural logarithms, 603
Chapter projects
children living with both parents, 177–78
college tuition modeling, 658
Earth's water, 317–18
length and period of pendulum, 452–53
locating lightning strikes, 232–33
modeling conic sections, 554–55
modeling electricity production, 391–92
municipal budgets analysis, 104–105
newspaper circulation analysis, 39–40
poultry consumption, 515
temperature modeling, 615–16
Chernobyl nuclear accident, estimating percent of radioactive material, 586–87
Circle, 67, 524, 531–32, 540, 556, 674
Circular cylinder, volume and surface area formula for, 678
Coefficients, 258, 264, 285, 538
College Board, 658
College tuition, 658
Columns, of matrices, 220
Combined variation, 386
Combining like terms, 35, 36, 261, 320
simplifying linear equations by, 50–51
Common denominators. (*See* Least common denominator)
Common difference, and arithmetic sequences, 634–36, 659
Common logarithms, 603, 620
Common ratio, and geometric sequence, 637–38, 660
Commutative property, 36
of addition, 33, 42
and combining like terms, 261
of multiplication, 33, 42
Complementary angles, 665, 666
Complements, of angles, 665, 666
Completing the square, 466–69, 474, 475
quadratic equations solved by, 516
Complex conjugates, 448, 449, 456
Complex fractions, 348
methods, 349–51
simplifying, 348–52, 394
Complex numbers, 405, 445–50, 456
addition of, 445–47, 456
defined, 446
division of, 449, 456
multiplication of, 448, 456
raising to *i* powers, 450
subsets of, 447
subtraction of, 455–57, 456
system of, 445
Complex rational expressions, 348
Composite functions, 568, 617
Compound inequalities, 86–90, 109
Compound interest, 470
formula for, 69–70, 586
Computer-aided design and drafting (CADD), 525
Conditional equations, 53
Cone, volume and surface area formula, 678
Congruent triangles, 668, 669
Conic sections, 524, 525–58
circles, 531–32, 540, 556
ellipse, 535–37, 540, 557
hyperbolas, 538–40, 541, 557
modeling, 554–55
parabolas, 526–29, 540, 556
Conjugates, 432–34, 455
complex, 448, 449, 456
of denominator, 433
of numerator, 434
Consecutive integers, 56
Consistent systems, 194, 234
in three variables, 204
in two variables, 194, 195
Constant of proportionality, 382, 384
Constant of variation, 382, 384
Constant term (constant), 258
Continuously compounded interest formula, 620
Contradictions, 53
Convex solid, 677
Coplanar lines, 666
Corresponding angles, 667, 669
Corresponding matrix of the system, 236
Cost and revenue functions, problem solving with, 213–15
Cramer's rule, 225
for three equations in three variables, 229–30, 238
for two linear equations in two variables, 227–28, 238
Cube, 675
volume and surface area formula for, 678
volume of, 252
Cube root, 24, 407–8, 454
Cube root functions, graphing, 411
Cubic centimeter, 677
Cubic functions
degree of, 309
graphing, 314–15
Cubic inch, 677
Cubic units, 677

**D**

Decimal approximations, 407
Decimals
clearing equation of, 53
Degree
of polynomial function, 309
of polynomial, 259, 260, 264–65
of term, 259, 260
Denominators
conjugate of, 433
rationalizing, 430–31, 455
rationalizing for radical expressions, 430–31
rationalizing the numerator, 432, 455
Dependent equations, 194, 195, 199, 205, 234
Dependent variables, 133
Descartes, René, 118
Descending order, 259
Determinants
defined, 225
solving systems of equations by, 225–30, 237–38
of $2 \times 2$ matrix, 225–28, 237
of $3 \times 3$ matrix, 228–30, 238
Dieticians, 463
Dietetics, careers in, 462
Difference
of complex numbers, 447
of functions, 566
of two cubes, 292–93
of two squares, 290–92
Direct proportionality, 383
Direct variation, 396
and Hooke's law, 383–84
problem solving involving, 382–84
Discriminant, 478
Distance formula, 67, 70, 529, 556
Distance saved, calculating, 479–80
Distributive property, 33–34
for adding like radicals, 455
for combining like terms, 261, 320

and factoring polynomials, 277
for multiplying two polynomials, 269
for removing parentheses, 51, 52, 54
for simplifying expressions, 36
Division
of complex numbers, 456
of functions, 566
of polynomial by monomial, 354–55
of polynomials, 355–58, 394
of rational expressions, 336, 393
of real numbers, 21–23, 41
synthetic, 360–62
Domain, 410, 411
of function, 583
of logarithmic function, 593, 594, 619
of rational expression, 331
of relation, 129, 132, 133, 180
of sequence, 630, 659
Double negative property, 15

**E**

Earthquakes, finding magnitude of, 604–5
Eccentricity of comet, 542
Edge, of polyhedron, 677
Electricians, 328
Elements, 10, 41, 220
Elimination method (or addition method), 197–99
solving nonlinear systems of equations by, 547–48
solving system of two linear equations using, 197
solving system of three linear equations by, 206
Ellipses, 10, 41, 524, 535–37, 540, 557
Empty set, 10
Equality, words/phrases for, 30
Equation of line, point-slope form of, 164
Equations, 48, 106. (*See also* Exponential equations; Linear equations; Quadratic equations)
absolute value, 93–97
of circles, 531–32, 532
conditional, 53
dependent, 194, 195, 199, 205, 234
difference between expressions and, 364, 371
of ellipses, 536–37
equivalent, 106
graphing, 118–24, 179–80
graphing with graphing utility, 685–87
of hyperbolas, 538
independent, 195, 234
of lines, 162–68, 183
with no solutions, 95, 96, 97
of parabolas, 526–29
of parallel and perpendicular lines, 166–67
solving by using quadratic methods, 517
solving form $b^x = b^y$, 584–85
solving for specified variable, 68
solving those containing rational expressions, 364–69, 395
of vertical and horizontal lines, 166
Equations containing rational expressions, problem solving by writing, 373–78
Equilateral triangle, 673
Equivalent compound inequality, 101
Equivalent equations, 49, 106
Escape velocity, 411
Evaluating expressions, 9, 25–26, 41
Expanding by minors, 228
Exponential equations
solving, 584–85, 609
writing with logarithmic notation, 590–92
Exponential expressions, 244
evaluating those raised to $n$th power, 246–48
simplifying, 255
Exponential functions, 564, 582–86, 619
graphing, 582–84
and uniqueness of $b^x$, 619
Exponential notation
solving logarithmic equations by using, 592–93
writing logarithmic equations with, 590–92
Exponents, 23, 42, 454
power rule for, 252–53
product rule for, 244–45
quotient rule for, 245–46
summary of rules for, 416
Expressions, 364. (*See also* Exponential expressions; Rational expressions)
difference between equation containing rational expressions and, 364, 371
evaluating, 9, 25–26, 41
simplifying, 25, 42
Exterior angles, 667
Extraneous solutions, 365, 395, 436, 438

**F**

Face, of polyhedron, 677
Factorials
evaluating, 654–55
of $n$, 654–55, 661
Factoring, 276
by grouping, 278–79, 320
method, 297
polynomial equations solved by, 297–300, 321
polynomials, 293–95
by special products, 320–21
strategies, 321
by substitution, 286–87
trinomials, by grouping, 287
trinomials of form $ax^2 + bx + c$, 283–86, 320
trinomials of form $x^2 + bx + c$, 282–83
Factoring out the greatest common factor, 277
Factors, 23. (*See also* Greatest common factor)
Fibonacci sequence, 633
Financial planning, 628, 629
Finite sequence, 630, 659
Finite series, 642
First-degree equations, 48
Focus (or foci), 535
in ellipse, 535
in hyperbola, 538
FOIL method, 320, 428
FOIL order, 270–71
Formulas
binomial, 655
compound interest, 470, 586
continuously compounded interest, 620
distance, 529, 556
for factoring perfect square trinomials, 289–90
midpoint, 530, 556
and problem solving, 67–71, 108
simple interest, 470
vertex, 511
volume and surface area, 678
Fourth-degree polynomials, 361
Fractions. (*See also* Complex fractions)
clearing, 51
fundamental principle of, 430
simplifying complex, 348–52
Function composition, 567
Function notation, 133, 180
Functions, 117, 129, 130, 180
Fundamental principle of fractions, 430

Fundamental principle of rational expressions, 331–32
$F(x)$ symbol, 134, 180

**G**

GCF. (*See* Greatest common factor)
General term
  of arithmetic sequence, 635–36, 659
  of sequence, 631–32
  terms of sequence given for, 630–31
Geology, 2
Geometric figures, types of, 673–75
Geometric progression, 637
Geometric sequences
  and common ratio, 637–38, 660
  partial sum of, 647–48, 660
Geometry, origin of word, 665
Global warming, 317
Graphing
  cube root functions, 411
  of cubic functions, 314–15
  equations, 118–24, 179–80
  equations with graphing utility, 685–87
  exponential functions, 582–84, 619
  intersection or union of two linear inequalities, 174–75
  inverses of functions, 578
  linear equations in two variables, 120–22
  linear functions, 142–46, 181–82
  linear inequalities, 172–75, 183–84
  linear inequalities in two variables, 173
  of line given slope and $y$-intercept, 162–63
  logarithmic functions, 594–95
  nonlinear equations, 123–24
  nonlinear inequalities, 550–51
  numbers on number line, 10
  parabolas, 526–29
  polynomial functions, 321–22
  quadratic functions, 310–12
  solutions of equations, 368
  solving systems of linear equations in two variables by, 192–95
  square root functions, 410
  systems of inequalities, 558
  systems of nonlinear inequalities, 551–52
  vertical and horizontal lines, 146
Graphing calculator, 683
Graphing calculator explorations
  addition of functions, 570
  addition/subtraction of polynomials in one variable, 265
  approximating solutions of systems of equations, 200
  features of graphing calculators, 124
  graphing circles, 533
  graphing equations, 147
  graphing functions, 136
  graphing ellipses, 541
  graphing quadratic functions, 505
  inverse functions, 579
  multiplication of polynomials in one variable, 273
  operations on rational expressions, 346
  percent of radioactive material, 587
  quadratic equation in standard form, 304–5
  radical equations, 441
  rational functions, 338
  solving quadratic equations, 471–72
  Trace feature, 157
  tripling of investment, 612–13
Graphing utility, 683–87
Graphs
  of functions, 567
  of one-to-one functions, 573–74
  of polynomial functions, 264, 303–4, 309–310
  of quadratic equations, 479
  of quadratic functions, 322, 493, 499–505, 518
Greater than or equal to symbol, 31, 42
Greater than symbol, 30
Greatest common factor (GCF), 320
  factoring out, 277
  of list of monomials, 277
Greenhouse gases, 317

**H**

Half-plane, 172, 173, 174, 184
Hooke's law, and direct variation, 383–84
Horizontal change, ratio of vertical change to, 150
Horizontal lines
  equation of, 166, 168
  graphing, 146
  slope of, 155, 183
Horizontal line test, 573, 574, 583, 618
Hyperbolas, 524, 538–40, 541, 557
  asymptotes of, 539
  graphing, 551
Hypotenuse, 439, 671
  of right triangle, 301

**I**

Identities, 53
Imaginary numbers, 446, 447
Imaginary unit, 445
Inconsistent system, 194, 234
  in two variables, 194, 195
  in three variables, 204
Independent equations, 195, 234
Independent variable, 133
Index, 408
Index of summation, 642
Inequalities. (*See* Linear inequalities; Nonlinear inequalities; Polynomial inequalities; Rational inequalities)
Inequality symbol, 99, 100
  reversing direction of, 78, 79, 108
Infinite geometric sequence, sum of terms of, 649–50, 660
Infinite sequence, 630, 659
Infinite series, 642
  of geometric sequence, 649–50
Inputs, set of, 128
Institute of Transportation Studies, 190
Integers, 10, 11, 12, 41. (*See also* Rational numbers)
Intercept points, graphing linear functions by finding, 143–45. (*See also* $x$-intercept points; $y$-intercept points)
Intercepts. (*See* $x$-intercept; $y$-intercept)
Interest rates, finding, 470–71
Interior angles, 667
Intersect feature, on graphing calculator, 441, 613, 687
Intersecting lines, 666
Intersection, of two sets, 86–89, 109
Interval notation, 76
  for writing solutions of linear inequalities, 77
Inverse, finding for one-to-one function $f(x)$, 576
Inverse functions, 571–78, 618
  of one-to-one function, 572–73
Inverse variation, 384–85, 396
  and Boyle's law, 385
Investments, doubling, 611–12
Irrational numbers, 11, 12, 41, 407, 447
Is not equal to symbol, 30, 42
Isosceles trapezoid, 674
Isosceles triangle, 673

**J**

*Jobs Rated Almanac*, 565

Joint variation, 386, 396
Journalism careers, 46

**K**

Kelvin, Lord, 567
Kelvin scale, 567

**L**

Least common denominator (LCD), 51, 52
  identifying for rational expressions, 342–43
Least squares method, 135, 153
Legs, of right triangle, 301, 439, 671
Less than or equal to symbol, 31, 42
Like radicals, 426, 427, 455
Like terms, 36, 261
Limits, 649
Linear equations, 243
  contradictions, identities and, 53–54
  forms of, 168
  simplifying by combining like terms, 50–51
  solving those containing fractions, 51–53
  solving using properties of equality, 48–50
Linear equations in one variable, 48–54, 106–7
  defined, 49
  solving, 52, 107
Linear equations in three variables
  solving systems of, 204–8, 235
Linear equations in two variables, 68, 117, 118, 179
  graphing, 120–22
  solving systems by substitution method, 195–97
  solving systems of, 192–99, 234
  solving systems of by graphing, 192–95
Linear functions, 309. (*See also* Functions; One-to-one functions)
  defined, 142
  graphing, 142–46, 181–82
Linear inequalities
  graphing, 172–75, 183–84
  multiplication property of inequality for solving, 78–81
  and problem solving, 75–83, 108–9
  solving applications involving, 81–83
Linear inequality in one variable, 75, 108
  solving, 80
Linear inequality in two variables, 117, 183
  graphing, 172, 173
Lines
  boundary, 172, 173, 174
  equations of, 162–68, 183
  and graphs of linear equations, 120, 179
  horizontal. *See* Horizontal lines
  parallel. *See* Parallel lines
  perpendicular. *See* Perpendicular lines
  point-slope form. *See* Point-slope form
  slope of, 149–57, 182–83
  slope-intercept form. *See* Slope-intercept form
  vertical. *See* Vertical lines
Loan payments, final, 606
Logarithmic definition, 590, 619
Logarithmic equations
  solving, 609–611
  solving by using exponential notation, 592–93
  writing with exponential notation, 590–92
Logarithmic form, 619
Logarithmic functions, 564, 590–95, 619
  characteristics of, 595
  graphing, 594–95
Logarithmic notation, 590
Logarithm property of equality, 609, 620
Logarithms, 585. (*See also* Natural logarithms)
  power property of, 599
  product property of, 597–98
  properties of, 593, 599–601, 619, 620
  quotient property of, 598–99
Long division, 355, 394

**M**

Mathematics, tips for success in, 4–7
Matrices, 219, 220, 236
  square, 225, 237
  to solve system of three equations, 222–23
  to solve system of two equations, 220–22
  solving systems of equations by, 236–37
Maximum value, of quadratic function, 512
Meteorologists, 116
Meteorology, 116
Microbes, 242
Microbiologists, 242
Midpoint, of line segment, 530
Midpoint formula, 530
Minimum value, of quadratic function, 512
Minor, 228, 238
Missing powers, 357, 362
Monomials, 259. (*See also* Polynomials)
  greatest common factor of list of, 277
  multiplying by polynomial other than monomial, 269
Multiplication
  associative property of, 33, 42
  of binomial by binomial, 270–71
  commutative property of, 33, 42
  of complex numbers, 448, 456
  of exponential expressions, 244
  of functions, 566
  of monomial by polynomial other than monomial, 269
  of polynomials, 268–73, 320
  of radical expressions, 428, 455
  of rational expressions, 334–35, 393
  of real numbers, 20–21, 41
  of square roots of negative numbers, 446
  of two polynomials, 269
Multiplication identity, 32, 42
Multiplication property of equality, 49, 50, 52, 106
Multiplication property of inequality, 108
  for solving linear inequalities, 78–81
Multiplicative inverses (or reciprocals), 32

**N**

National Electrical Contractors Association Web site, 328
National Weather Association Web site, 116
Natural logarithms, 603, 605, 620
  and change of base, 607
  evaluating powers of *e*, 605–6
  identifying and approximating by calculator, 605–6
Natural numbers, 10, 12, 41
Natural phenomena, and common logarithms, 604–5
Natural Science Foundation, 2
Negative constant, in trinomials, 285
Negative exponents, 246, 254, 351, 416
Negative numbers, 10, 41, 79, 108, 446
Negative product, 21, 41
Negative quotient, 22
Negative radical sign, 406
Negative radicands, 408
Negative slope, 155
Negative square root, 406, 454
Newton's law of cooling, 615
Nonlinear equations, graphing, 123–24
Nonlinear inequalities, 558
Nonlinear inequalities in one variable, 492–97, 517, 518
  solving inequalities containing rational expressions, 496–97

Nonlinear inequalities in one variable, (*cont.*)
solving polynomial inequalities of degree 2 or greater, 492–96
Nonlinear inequalities in two variables, graphing, 550–51
Nonlinear systems of equations
solving by elimination method, 547–48, 557
solving by substitution method, 544–47, 557
Nonlinear systems of equations in two variables, 544
Nonnegative number, principal square root of, 454
Nonnegative square root, 406
Nonvertical parallel lines, slope of, 182
*N*th power, raising to, 23
*N*th root of number, 408
Null set, 10
Number line, 9–10, 13
Numbers
*n*th root of, 408
relationships among sets of, 13
in scientific notation, 249, 319
sets of, 41
square roots of, 24
Numerator
rationalizing, 455
rationalizing for radical expressions, 432
Numerical coefficient, 258
Nutrition, 463

**O**

Obtuse angle, 665
Ocean levels, increase in, 317–18
One-to-one functions, 572, 583, 618
inverse of, 575
Opposite (or additive inverse), 14–15, 32, 262
Ordered pairs
plotting, 118–19
solutions of, 193, 194, 195, 197, 212
Ordered triple, 204, 223, 230, 235, 238
Order of operations, 24–25, 42
Ordinate, 118
Origin, 9, 118, 179
Outputs, set of, 128

**P**

Parabolas, 311, 524, 540, 556
graphing, 500–505, 526–29
left-/right-opening, 526–29
standard form of, 527, 528, 540
vertex of, 507–10
vertex formula for, 510–11
Parallel lines, 666
equations of, 166–67
and slope, 155, 168
Parallelogram, 674
Partial products, 270
Partial sum, 643
of arithmetic sequence, 646–47, 660
of geometric sequence, 647–48, 660
Pascal's triangle, for expanding binomials, 654
Pendulum
distance traveled by, 650
length and period of, 452
Percents, 59
Perfect cubes, 423, 432
Perfect fourth powers, 423
Perfect powers, 423
Perfect square factors, 422, 423
Perfect squares, 407
Perfect square trinomials, 286, 320, 466, 467
factoring, 289–90
Period, of pendulum, 452
Perpendicular lines, 666
equations of, 166–67
and slope, 156, 168, 183
Phrases
translating into algebraic expressions, 15
writing as algebraic expressions, 34–35
Pi ($\pi$), 11
Pixels, 683
Plane, 666
Plane figures, 666
Plotting ordered pairs, 118–19, 179
Points, 530
Point-slope form, 194
of equation of line, 164
of linear equation, 168
Polygon, 668, 673
Polyhedrons, 677
Polynomial equation in one variable, solution of, 297
Polynomial equations
problems modeled by, 300–303
solving by factoring, 297–300, 321
Polynomial function, 243, 260, 319
finding $x$-intercepts of, 303–4
graphing, 309–315, 321–22
roots of zeros of, 304
Polynomial inequalities, solving, 495, 517
Polynomial in one variable, 259
Polynomial in $x$, 259
Polynomials, 243, 258, 319, 329
addition of, 262, 319
dividing by, 355–58, 394
dividing by monomials, 354–55
factoring, 277, 293–95, 321
multiplication of, 268–73, 320
subtraction of, 262–64, 319
and synthetic division, 360–62
Polynomials in one variable, and grapher, 265
Population size, estimating, 611
Positive constant, in trinomials, 285
Positive exponents, 247, 253, 254, 255, 351
Positive numbers, 10, 41, 108
Positive (or principal) square root, 24, 41, 454
Positive product, 21, 41
Positive quotient, 22
Positive slope, 155
Power of a product rule for exponents, 253, 254
Power of a quotient rule for exponents, 253, 254
Power property of logarithms, 599, 601, 620
Power rules, 319
for exponents, 252–53, 254, 416
for products and quotients, 416
for solving radical equations, 436
Powers of 10, evaluating common logarithms of, 603–4
Prime polynomial, 283, 284
Principal square root, 406, 409, 454
Problem solving
and algebraic expressions, 9
formulas and, 67–71, 108
general strategy for, 58, 107–8
introduction to, 56–62
and linear inequalities, 75–83, 108–9
logarithmic and exponential functions and, 611–12
and quadratic equations, 470–71, 487–89
and radical equations, 436–40, 455
and rational equations, 372–78, 395–96
and systems of linear equations, 209–216, 236
and variations, 382–87, 396
Product, of functions, 566
Product property of logarithms, 597–98, 601, 620

Product rule for exponents, 244–45, 254, 268, 416
Product rule for radicals, 420–21
Product rule for radical expressions, 454
Profit function, 264
Psychology careers, 404
Pythagorean theorem, 301, 302, 439, 529, 671

**Q**

Quadrants, 118, 179
Quadratic (or second-degree) equations, 297, 462
  and discriminant, 478
  solving by completing the square, 464–69, 516
  solving by factoring method, 367
  solving by quadratic formula, 474–77, 517
  solving problems leading to, 487–89
Quadratic equations in two variables, 524
Quadratic formula, 546
  quadratic equations solved by, 474–77, 517
Quadratic functions, 322
  defined, 499
  degree of, 309
  and formula for vertex of parabola, 510–11
  in form $y = a(x - h)^2 + k$, 507–510
  graphing, 310–12, 493, 499–505, 518
  minimum/maximum value of, 512
Quadratic inequalities in one variable, solving, 492
Quadratic methods, equations solved by, 483–86, 517
Quadrilateral, 674
Quartic function, degree of, 309
Quotient
  of functions, 566
  negative, 22
  positive, 22
Quotient property of logarithms, 598–99, 601, 620
Quotient rule for exponents, 245–46, 254, 416
Quotient rule for radicals, 421–22
Quotient rules for radical expressions, 454

**R**

Radical equations
  and problem solving, 436–40, 455
  solving, 438, 484
Radical expressions, 406
  addition of, 426–27, 455
  multiplication of, 428, 455
  rationalizing denominator of, 430–31, 455
  rationalizing numerator of, 432, 455
  simplifying, 420–24, 454
  solving equations containing, 436–39
  subtracting, 426–27, 455
Radical functions, 454
Radicals, 454
  product rule for, 420–21
  quotient rule for, 421–22
Radical sign, 406
Radicand, 406
  largest perfect power factors of, 423
Radioactive material, estimating percent of, 586–87
Radius, of circle, 531, 532
Raising to powers, 23
Range
  of functions, 583
  of logarithmic function, 593, 594, 619
  of relation, 129, 132, 133, 180
Rate of speed, finding, 210–12
Rational equations, and problem solving, 372–78, 395–96
Rational exponents, 405, 454
  meaning of *a-m/n*, 415–16
  meaning of *am/n*, 414–15
  meaning of *a1/n*, 413–14
  for simplifying radical expressions, 417–18
Rational expressions, 329
  addition of, 341–42, 393
  addition/subtraction of with unlike denominators, 343
  defined, 330
  division of, 336, 393
  fundamental principle of, 331–32
  identifying least common denominators for, 342–43
  multiplication of, 334–35, 393
  simplifying/writing in lowest terms, 332
  solving equations containing, 364–69, 395
  subtraction of, 341–42, 393
  undefined, 330–31, 365
Rational expressions with variables in denominator, inequalities containing, 496
Rational functions, 329, 330, 393
Rational inequalities, solving, 497, 518
Rationalizing
  denominator, 430–31, 455
  numerator, 432, 455
Rational numbers, 11, 12, 41, 447
Real numbers, 11, 41, 447
  absolute value of, 13–14
  addition of, 19, 41
  division of, 21–23, 41
  multiplication of, 20–21, 41
  properties of, 30–36
  subtraction of, 20, 41
Reciprocals (or multiplicative inverses), 22, 32, 335
Rectangle, 674
  area formula, 67
Rectangular coordinate system, 118, 179
Rectangular solid, 675
  volume and surface area formula, 678
  volume formula, 67
Relations, 117, 128, 129, 130, 180
Remainder theorem, 362
Renewable energy sources, 391
Revenue and cost, 192
Revenue function, 264
Rhombus, 674
Richter scale, 604
Right angle, 665
Right circular cone, 675
Right circular cylinder, 675
Right triangle, 301, 439, 440, 671, 673
Rise, 150
Roots, of polynomial function, 304
Roster form, 10
Row operations, 220, 237
Rows, of matrices, 220
Rules for exponents, for simplifying expressions containing rational exponents, 416–17

**S**

Salaries, finding, 636
Scalene triangle, 673
Scientific calculator explorations, scientific notation, 250
Scientific notation, 319
  converting between standard notation and, 248–49
  operations on numbers in, 255–56
Second-degree equations, 464. (*See also* Quadratic equations)
Sequences, 628, 630–32, 659
  finite, 630, 659
  infinite, 630, 659
Series, 628, 641–44, 660
  defined, 641
Set builder notation, 10, 41

Set notation, 76
Sets
  elements of, 10
  intersection of two, 86
  of numbers, 41
  union of two, 89–90
Sierra Club Foundation, 190
Sigma, 642
Sign patterns, 20–21
  and sums/differences of cubes, 293
  in trinomials, 285
Similar triangles, 669, 670
Simple interest, 470
Simplifying expressions, 25, 35–36, 42
Simplifying radicals/radical expressions, 422–24, 454
Simplifying rational expressions, 331, 393
Slope
  of horizontal lines, 155, 183
  of a line, 149–57, 182–83
  of parallel lines, 155, 166
  of perpendicular lines, 168, 183
  positive, 155
  of vertical lines, 154, 155, 183
Slope-intercept, interpreting in application, 153–54
Slope-intercept form, 152
  of linear equation, 168, 182
  to write equation of line, 162
Slope of a line, 149–57, 182–83
  finding, given two points on line, 149–51
  given equation of line, 151–52
Solution of an inequality, 76
Solution of equation in three variables, 235
Solution of equation in two variables, 179
Solution of system of two equations in two variables, 234
Solution of the equation, 49, 106
Solution of the system, 192
Solution region, 172
Solutions, mixing, 212–13
Solution set
  of compound inequality, 109
  of equation, 49
  of an inequality, 76
Solutions of equations in two variables, 119–20
Solving formula for specified variable, 108
Solving the equation, 48
Special products, 272, 320
  factoring by, 289–93
Sphere, 675
  volume and surface area formula, 678
Spotlight on decision making
  algebraic expressions and sets of numbers, 16
  altimeter reading, 378
  bacterial culture, 639
  bottled water *versus* filtered water, 199
  bridge design, 553
  CD-player reliability, 601
  cellular telephone subscriptions, 418
  color-coding ocean map for lava flow, 37
  *Ebola Zaire* virus, 250
  eccentricity of comet, 542
  escape velocity, 411
  fluoride doses for children, 71
  hunger story for newspaper, 63
  hurricane tracking, 125
  Internet service providers, 274
  IQ test subject screening, 441
  job offers, 632
  long-distance telephone plans, 216
  mumps eradication, 168
  painting project, 388
  parts list resistors, 337
  public transportation planning, 223
  rental increases, 158
  Roth IRA contributions, 651
  support stocking sizing, 175
  temporary classrooms, 83
  ticket prices for theater group, 513
  tour packages, 27
  vitamin A intake for child, 480
  Web site usage, 595
Square, 674
Square-based pyramid, volume and surface area formula, 678
Square matrix, 225, 237
Square of binomial, 271–72
Square root, 11
  finding, 406–7
  of negative number in terms of *i*, 445
  of number, 24
Square root functions, graphing, 410
Square root property, 464–66, 516
Square setting, graphing utility, 686
Standard form, 120, 179, 527, 528
  equations for hyperbolas, 538, 541
  equations for circles, 531, 532, 540
  equations for ellipses, 536, 540
  equations for parabolas, 527, 528, 540
  of linear equation, 168
  polynomial equation in, 297
  quadratic inequalities in one variable in, 492
Standard notation, converting between scientific notation and, 248–49
Standard viewing window (or rectangle), 684, 685–87
Standard window, for graphing, 124
Straight angle, 665
Subscripts, 26
Subsets, 41
Substitution method, 195
  factoring by, 286–87
  solving nonlinear system of equations by, 544–47, 557
  solving system of two equations using, 196
Subtraction
  of complex numbers, 456
  of functions, 566
  of polynomials, 262–64, 319
  of radical expressions, 426–27, 455
  of rational expressions, 341–42, 393
  of rational expressions with unlike denominators, 343
  of real numbers, 20, 41
Sum
  of complex numbers, 447
  of functions, 566
Summation notation, 642, 643, 660
Sum of two cubes, 292–93
Sum of two squares, 291
Supplementary angles, 665, 666
Supplements, of angles, 665
Surface area
  formulas of common solids, 678
  of polyhedron, 677
Symbols
  approximation, 407
  cube root, 407
  for greater than, 30, 42
  greater than or equal to, 31, 42
  inverse function, 575
  is not equal to, 30, 42
  less than or equal to, 31, 42
  null set, 10

polynomial functions, 260
square roots, 24, 42
Synthetic division, 360–62, 394
System of three equations, matrices for solving, 222–23
System of three equations in three variables, 3 x 3 determinant for solving, 228–29
System of three linear equations, solving problems modeled by, 215–16
System of two equations, matrices for solving, 220–22
System of two linear equations, solving problems modeled by, 209–213
Systems of equations, 191, 192
solving by determinants, 225–30, 237–38
solving by elimination method, 197–99
solving by matrices, 219–23, 236–37
Systems of inequalities, graphing, 558
Systems of linear equations, 234
and problem solving, 209–216, 236
Systems of nonlinear inequalities, graphing, 551–52

**T**

Table feature, on graphing calculator, 587, 687
Temperature
function composition and conversions in, 567–69
and one-to-one function, 572
Term, 258
degree of, 259, 260
of expression, 35
product of sum/difference of two, 272
and rationalizing numerator/denominator, 432–34
of sequence, 630–31
Test point
for graphing nonlinear inequalities, 550
values, 493–95, 497, 518
Third-degree polynomials, 361
Three equations in three variables, Cramer's rule for, 238
Time management, 7
Trace feature, on graphing calculators, 157, 441, 472, 587, 613, 687
Transversal, parallel lines cut by, 667
Trapezoid, 674
area formula for, 68
Triangle, 668, 673
Trinomials, 259. (*See also* Polynomials)
factoring, 282–87, 320
sign patterns in, 285
Two linear equations in two variables, Cramer's rule for, 238

**U**

Undefined rational expression, 330–31
Undefined slope, 155, 166, 183. (*See also* Vertical lines)
Union of two sets, 89–90, 109
Unknown numbers, finding, 58–59, 209–210, 373–74
Urban and regional planners, 190
U.S. Department of Energy, 391

**V**

Value, 86, 89
Value of expression, 25, 41
for given replacement value, 9
Variables, 3, 8, 41
equations solved for specific, 68, 372
formula solved for specific, 108
undefined rational expression for value of, 330
Variations, and problem solving, 382–87, 396
Vertex
of parabola, 311, 500, 504, 507–510
of polyhedron, 677
Vertex formula, 312
Vertical angles, 667
Vertical change, ratio of horizontal change to, 150
Vertical lines
equations of, 166, 168
graphing, 146
slope of, 154, 155, 183
Vertical line test, 131, 142, 180, 310, 311, 410, 411, 573, 574, 583, 618
Viewing window (or rectangle), 683
Volume, 677
Volume formulas
of common solids, 678

**W**

Webmasters, 564, 565
Whole numbers, 10, 12, 41
Wind energy, 391
Window, graphing calculator, 124
Window setting, of graphing utility, 683
Work hours calculation, 375–77
World Organization of Webmasters Web site, 564
World Wide Web, 565

**X**

*x*-axis, 118, 179
on graphing utility, 686
*x*-coordinate, 119
*x*-intercept, 122, 181
*x*-intercept points, finding, 510, 519
*x*-intercepts
finding, 143
of parabola, 312, 313
of polynomial functions, 303–4, 310
*x*-value, 121, 122, 123
in one-to-one functions, 583
in polynomial functions, 310
relations between *y*-value and, 128

**Y**

*y*-axis, 118, 179
on graphing utility, 686
*y*-coordinate, 119
*y*-intercept points, finding, 510, 519
*y*-intercepts, 122, 181
finding, 143
of parabola, 312, 313
and slope, 152, 162
*y*-value, 121, 123
in one-to-one functions, 583
in polynomial functions, 310
relations between *x*-value and, 128

**Z**

Zero exponents, 245, 254, 416
Zero-factor property, 297–98
Zero-factor theorem, 464
Zeros, of polynomial function, 304
Zero slope, 155, 183. (*See also* Horizontal lines)
Zoom feature, on graphing calculator, 441, 472, 587, 687

# Photo Credits

Chapter 1 CO Kerry Sieh/California Institute of Technology/Palomar/Hale Observatory, (p.5) John A. Rizzo/PhotoDisc, Inc.

Chapter 2 CO Robert Brenner/PhotoEdit, (p.57) Peter J. Schulz/Liaison Agency, Inc., (p.60) Michael Newman/PhotoEdit, (p.62) Jeremy Woodhouse/PhotoDisc, Inc., (p.73) Jeffrey Stevenson Studio, (p.74) Photo Researchers, Inc., (p.114) © Mark Downey/Lucid Images/PictureQuest

Chapter 3 CO © Bernard Boutril/Woodfin Camp/PictureQuest, (p.118) Porter Gifford/Liaison Agency, Inc., (p.165) Bill Bachmann/Photo Researchers, Inc., (p.186) Photo Courtesy of Motorola, Inc.

Chapter 4 CO Frank Fisher/Liaison Agency, Inc., (p.218) Doug Densinger/Allsport Photography (USA), Inc., (p.218) Tony Gutierrez/AP/Wide World Photos, (p.233) T.A. Wiewandt/DRK Photo

Chapter 5 CO © Jagdish Agarwal/Stock Connection/PictureQuest, (p.250) Corbis Digital Stock, (p.252) Chris Butler/Science Photo Library/Photo Researchers, Inc., (p.252) Alan Schen/The Stock Market, (p.307) PhotoDisc, Inc.

Chapter 6 CO Bryan F. Peterson/The Stock Market, (p.353) Gary Benson/Gary J. Benson Photography, (p.359) Courtesy Gateway 2000, Inc., (p.369) Spike Matford/PhotoDisc, Inc., (p.376) Amy C. Etra/PhotoEdit, (p.377) John Serafin/Pearson Education Corporate Digital Archive, (p.380) Michael Gadomski/Photo Researchers, Inc., (p.380) Tardos Camesi/The Stock Market, (p.381) Ed Lallo/Liaison Agency, Inc., (p.381) Treat Davidson/National Audubon Society/Photo Researchers, Inc., (p.385) Richard A. Cooke III/Stone, (p.389) Mary Teresa Giancoli

Chapter 7 CO David Young-Wolff/PhotoEdit, (p.418) John Henley/The Stock Market, (p.441) Steve Gottlieb/FPG International LLC

Chapter 8 CO Henley & Savage/The Stock Market, (p.471) Tony Freeman/PhotoEdit, (p.491) Tim Flach/Stone, (p.491) Arthur S. Aubry Photography/PhotoDisc, Inc., (p.514) Jody Dole/The Image Bank, (p.514) Simon Fraser/Northumbrian Environmental Management, LTD/SPL/Photo Researchers, Inc., (p.520) AP/Wide World Photos

Chapter 9 CO Telegraph Colour Library/FPG International LLC

Chapter 10 CO Gary Landsman/The Stock Market, (p.589) PhotoDisc, Inc.

Chapter 11 CO Steve Smith/FPG International LLC

# Common Graphs and Models

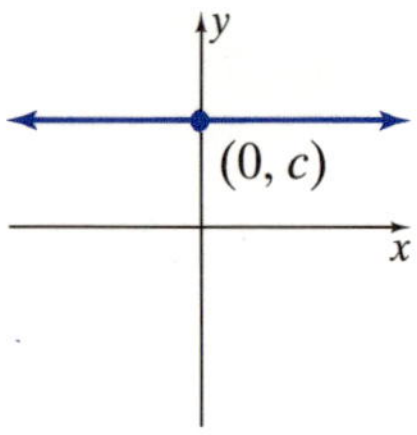


Horizontal Line;
Zero Slope
$y = c$

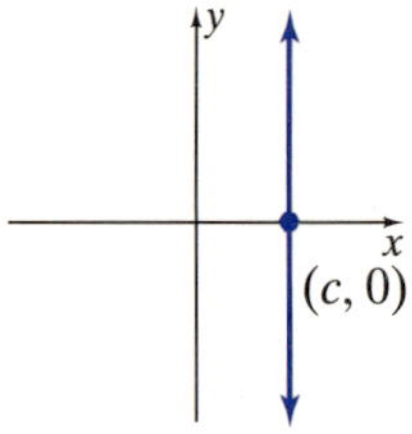


Vertical Line;
Undefined Slope
$x = c$

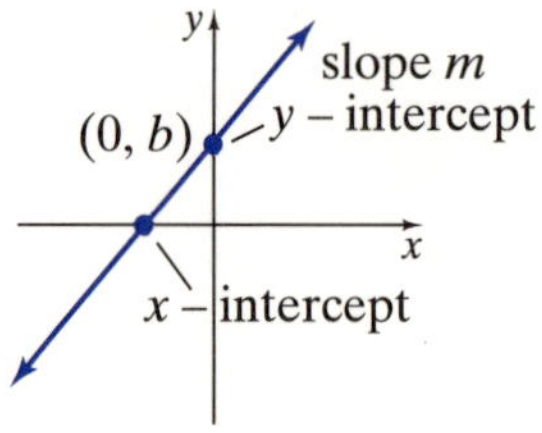


Linear Function;
Positive Slope
$y = mx + b; m > 0$

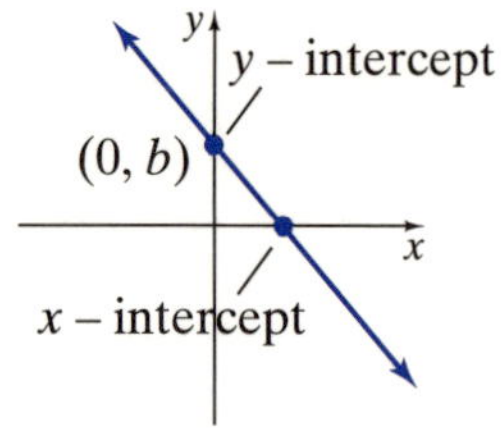


Linear Function;
Negative Slope
$y = mx + b; m < 0$

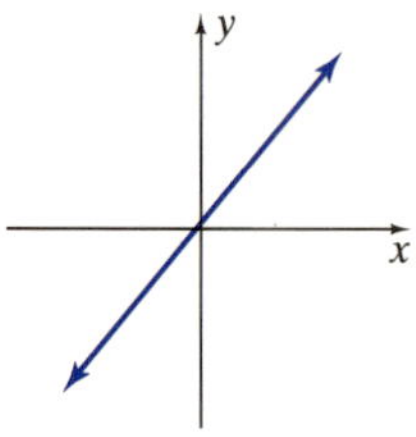


$y = x$

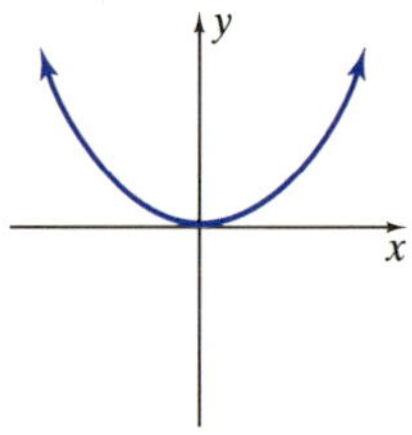


$y = x^2$

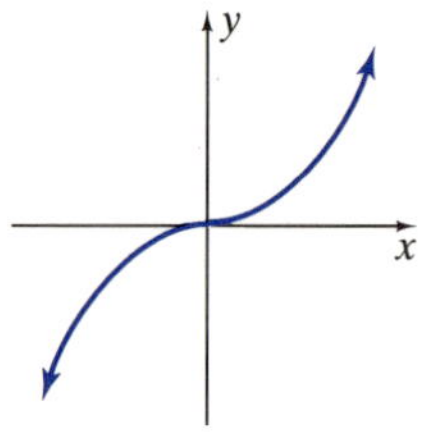


$y = x^3$

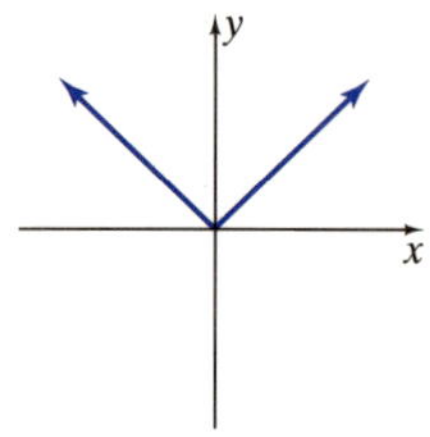


$y = |x|$

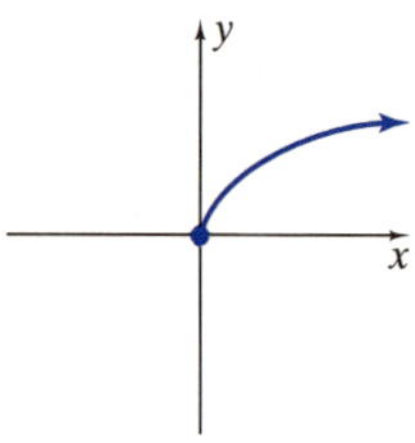


$y = \sqrt{x}; x \geq 0$

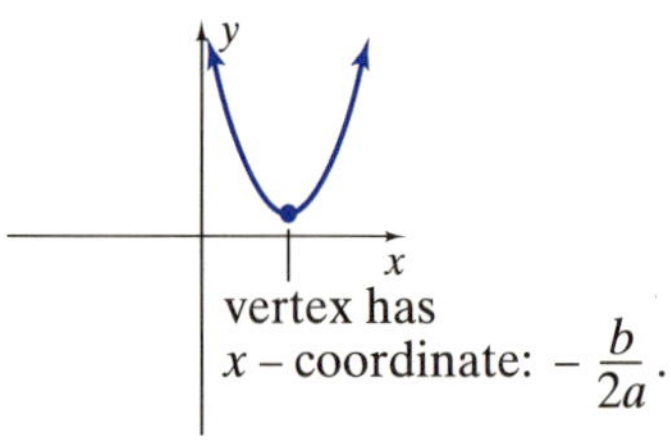


Quadratic Function
$y = ax^2 + bx + c; a \neq 0$
Parabola opens upward if $a > 0$
Parabola opens downward if $a < 0$

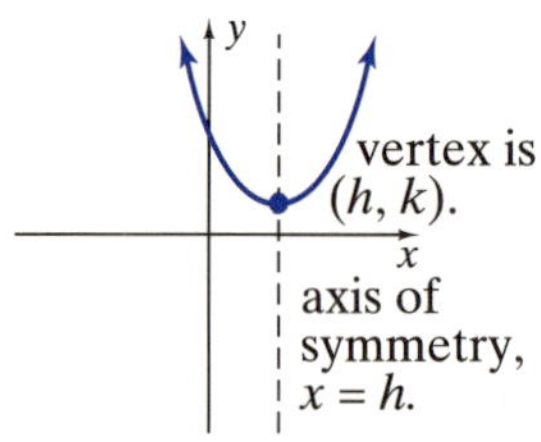


Quadratic Function
$y = a(x - h)^2 + k; a \neq 0$
Parabola opens upward if $a > 0$
Parabola opens downward if $a < 0$

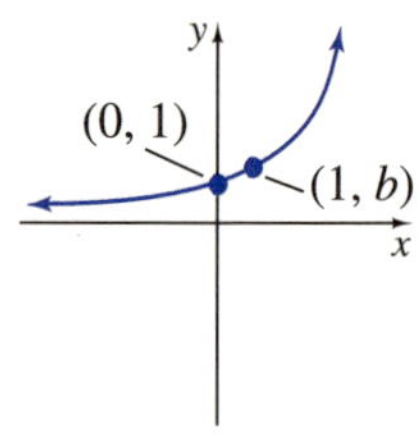


Exponential
Function
$y = b^x$ for $b > 1$

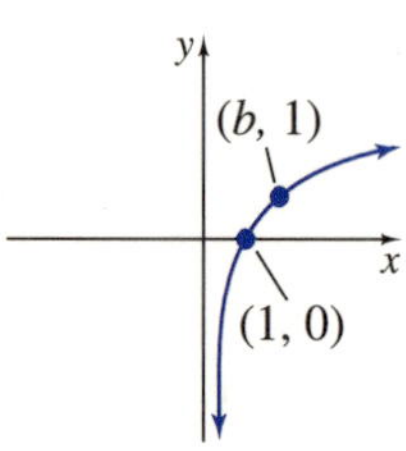


Logarithmic Function
$y = \log_b x$ for $b > 1$

## Systems of Linear Equations

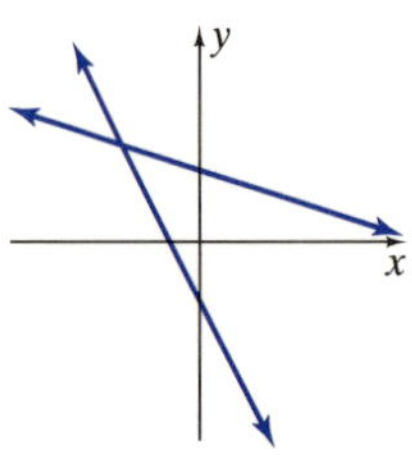


Independent and
consistent; one solution

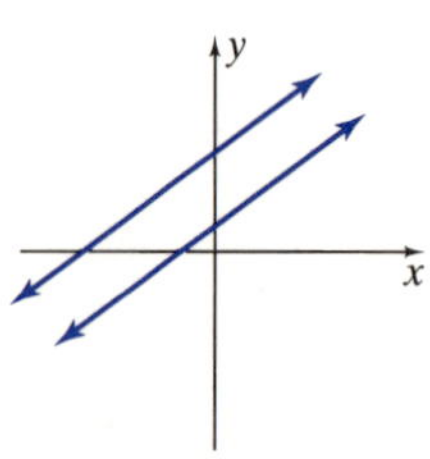


Independent and
inconsistent, no solution

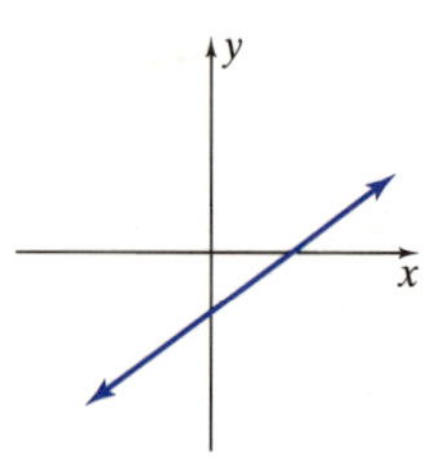


Dependent and consistent;
infinitely many solutions

# ALGEBRAIC FORMULAS

**Slope of a Line Containing Points $(x_1, y_1)$ and $(x_2, y_2)$**

$$m = \frac{y_2 - y_1}{x_2 - x_1}, \text{ if } x_1 \neq x_2$$

The slope is $m$.

**Slope-Intercept Form of the Equation of a Line**

$$y = mx + b$$

The slope is $m$, and the $y$-intercept is $b$.

**Point-Slope Form of the Equation of a Line**

$$y - y_1 = m(x - x_1)$$

The slope is $m$, and point $(x_1, y_1)$ is on the line.

**Quadratic Formula**

$$x = \frac{-b \pm \sqrt{b^2 - 4ac}}{2a}$$

Yields solutions of an equation of the form $ax^2 + bx + c = 0$, where $a \neq 0$.

# REFERENCE TO SELECTED KEYS ON SCIENTIFIC AND GRAPHING CALCULATORS

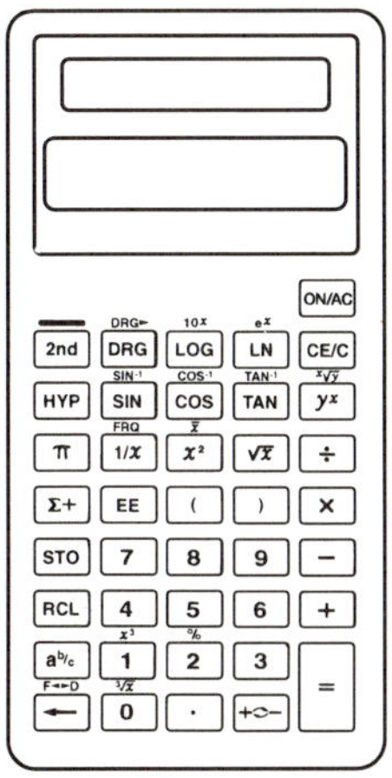


Scientific Calculator

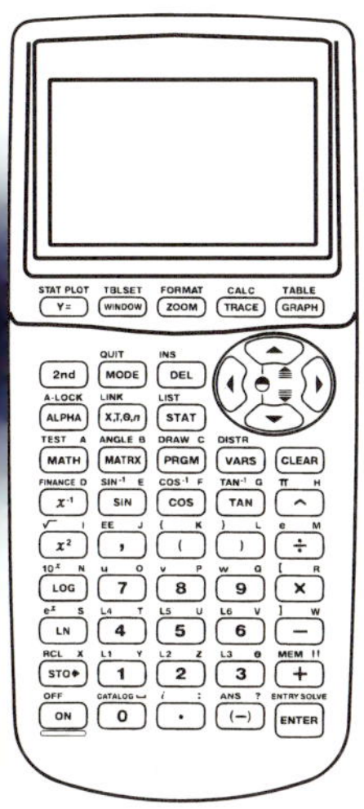

Graphing Calculator (Optional)

| Key(s) | Description |
| --- | --- |
| [=] or [ENTER] | Gives the result of the current operation. |
| [$y^x$] or [$a^b$] or [^] | Raises any base to a power. For example, $4^5$ as 1024. |
| [EE] or [EXP] | Use to enter a number in scientific notation. For example, $6.1 \times 10^8$. |
| [LOG] | Takes the common logarithm of the number in the display. |
| [LN] | Takes the natural logarithm of the number in the display. |
| [+/−] or [+ −] or [(−)] | Changes the sign of the displayed number. Also, use to enter a negative value. (This is not the subtraction key.) |
| [1/$x$] or [$x^{-1}$] | Finds the reciprocal of a displayed number. |
| [STO] or [RCL] or RCL [STO→] | Stores and recalls a number stored in memory. |
| [$x^2$] | Squares the displayed number. |
| [DEL] | On a graphing calculator, deletes previously entered characters. |
| [$X, T$] or [$X, T, \theta$] or [$X, T, \theta, n$] | On a graphing calculator, use to write the variable $x$. |
| [$y =$] | On a graphing calculator, use to enter equation(s) to be graphed. |
| [WINDOW] or [RANGE] | On a graphing calculator, use to set the viewing window to choose the part of the coordinate plane to be displayed. |
| [ZOOM] | On a graphing calculator, use to "zoom-in" or "zoom-out" to magnify or reduce a part of the curve. Can also be used to obtain a "square" setting, providing equal spacing between tick marks on the $x$- and $y$-axes to obtain a true geometric perspective. |
| [TRACE] | On a graphing calculator, use to find the $x$-coordinates and $y$-coordinates of points on a curve. |
| [GRAPH] | On a graphing calculator, use to graph a previously entered equation or equations. |
| TABLE [GRAPH] or [2nd] [GRAPH] | On a graphing calculator, use for a table of values relating to an entered equation. The table can be used to estimate an answer to a problem. |
| [OFF] or [2nd] [OFF] | Turns calculator off. |